Pro Tableau

A Step-by-Step Guide

Seema Acharya
Subhashini Chellappan

Apress®

Pro Tableau: A Step-by-Step Guide

Seema Acharya
Pune, Maharashtra, India

Subhashini Chellappan
Bangalore, Karnataka, India

ISBN-13 (pbk): 978-1-4842-2351-2
ISBN-13 (electronic): 978-1-4842-2352-9
DOI 10.1007/978-1-4842-2352-9

Library of Congress Control Number: 2016961342

Managing Director: Welmoed Spahr
Lead Editor: Celestin Suresh John
Technical Reviewer: Rajeev
Editorial Board: Steve Anglin, Pramila Balan, Laura Berendson, Aaron Black, Louise Corrigan, Jonathan Gennick, Robert Hutchinson, Celestin Suresh John, Nikhil Karkal, James Markham, Susan McDermott, Matthew Moodie, Natalie Pao, Gwenan Spearing
Coordinating Editor: Sanchita Mandal
Copy Editor: Alexander Krider
Compositor: SPi Global
Indexer: SPi Global
Artist: SPi Global

Distributed to the book trade worldwide by Springer Science+Business Media New York, 233 Spring Street, 6th Floor, New York, NY 10013. Phone 1-800-SPRINGER, fax (201) 348-4505, e-mail `orders-ny@springer-sbm.com`, or visit `www.springeronline.com`. Apress Media, LLC is a California LLC and the sole member (owner) is Springer Science + Business Media Finance, Inc. (SSBM Finance, Inc.). SSBM Finance, Inc. is a **Delaware** corporation.

For information on translations, please e-mail `rights@apress.com`, or visit `www.apress.com`.

Apress and friends of ED books may be purchased in bulk for academic, corporate, or promotional use. eBook versions and licenses are also available for most titles. For more information, reference our Special Bulk Sales–eBook Licensing web page at `www.apress.com/bulk-sales`.

Any source code or other supplementary materials referenced by the author in this text are available to readers at `www.apress.com`. For detailed information about how to locate your book's source code, go to `www.apress.com/source-code/`. Readers can also access source code at SpringerLink in the supplementary material section for each chapter.

Printed on acid-free paper

Contents at a Glance

About the Authors....xv

About the Technical Reviewer....xvii

Acknowledgments....xix

Introduction....xxi

■ Chapter 1: Introducing Visualization and Tableau....1

■ Chapter 2: Working with Single and Multiple Data Sources....49

■ Chapter 3: Simplifying and Sorting Your Data....121

■ Chapter 4: Measure Names and Measure Values....237

■ Chapter 5: Table Calculations....321

■ Chapter 6: Customizing Data....433

■ Chapter 7: Statistics....495

■ Chapter 8: Chart Forms....547

■ Chapter 9: Advanced Visualization....665

■ Chapter 10: Dashboard and Stories....729

■ Chapter 11: Integration of Tableau with R....795

Index....835

Contents

About the Authors ... xv

About the Technical Reviewer ... xvii

Acknowledgments ... xix

Introduction ... xxi

Chapter 1: Introducing Visualization and Tableau ... 1

1.1 Why data visualization? ... 2

1.2 What can data visualization help with? ... 2

1.3 An introduction to visualization ... 3

1.3.1 Which domain is leveraging the power of data visualization? ... 6

1.3.2 Who is using data visualization? ... 6

1.3.3 Top data visualization tools ... 6

1.3.4 History of data visualization ... 7

1.3.5 What are the expectations for a data visualization tool? ... 7

1.3.6 Let us see how Tableau fulfils the expectations. ... 7

1.3.7 Reasons to make a switch to Tableau ... 8

1.4 Positioning of Tableau ... 8

1.5 Tableau product line ... 9

1.6 File types in Tableau ... 13

1.6.1 Tableau Workbook (twb) ... 13

1.6.2 Tableau Packaged Workbook (twbx) ... 17

1.6.3 Tableau Data Source (tds) file ... 19

1.6.4 Tableau Packaged Data Source (tdsx) file ... 20

1.6.5 Tableau bookmark ... 28
1.6.6 Tableau data extract ... 34
1.7 Points to remember ... 46
1.8 Assignments ... 46
1.9 Next steps ... 48
Chapter 2: Working with Single and Multiple Data Sources ... 49
2.1 Desktop architecture ... 49
2.1.1 Data layer ... 50
2.1.2 Data connectors ... 50
2.1.3 Live connection ... 50
2.1.4 In-memory ... 50
2.2 Tableau environment ... 51
2.2.1 To open ... 51
2.2.2 To close ... 51
2.2.3 Start page ... 52
2.2.4 Data Source Page ... 55
2.2.5 Workspace ... 59
2.2.6 Workbooks and Sheets ... 60
2.2.7 Visual Cues and Icons in Tableau ... 60
2.3 Connect to a File ... 62
2.3.1 Connect to a Text File ... 62
2.3.2 Connect to MS Access ... 64
2.3.3 Connecting to RData files ... 67
2.4 Connect to a Server ... 73
2.4.1 Connecting to MS SQL Server 2014 Management Studio ... 73
2.4.2 Connecting to MySQL ... 77
2.4.3 Connecting to NoSQL Databases ... 80
2.5 Metadata Grid ... 92

2.6 Joins....93
2.6.1 Adding Fields to the Data Pane....93
2.6.2 Exploring different types of Join....95
2.6.3 Union....100
2.7 Custom SQL....103
2.7.1 Demo 1....103
2.8 Data Blending....106
2.8.1 Demo 1....107
2.9 Data Extracts....114
2.9.1 Demo 1....114
2.10 Points to Remember....120
2.11 Next Step....120
Chapter 3: Simplifying and Sorting Your Data....121
3.1 Filtering....121
3.1.1 Why filtering?....121
3.1.2 What is filtering?....122
3.1.3 How to apply "Filter"?....122
3.2 Sorting....159
3.2.1 Why sorting?....159
3.2.2 What is sorting?....159
3.2.3 How to apply sorting?....159
3.2.4 Discrete and Continuous Dates....178
3.2.5 Why and what?....178
3.3 Groups....192
3.3.1 Why groups?....192
3.3.2 What is a group?....192
3.3.3 How to create a group?....192
3.3.4 Editing an existing group....202
3.3.5 Creating Hierarchies....212
3.3.6 Sets....215

3.4 Difference between a set and group 231
3.4.1 Group 231
3.4.2 Set 231
3.4.3 Creating parameters 231
3.5 Points to remember 236
3.6 Next step 236
Chapter 4: Measure Names and Measure Values 237
4.1 Why are measure names and measure values required? 237
4.1.1 What are measure names and measure values? 237
4.1.2 Where do these fields come from? 238
4.1.3 Measures on an independent axis 251
4.1.4 Blended axes 253
4.1.5 Dual axis 291
4.2 Points to Remember 319
4.3 Next steps 319
Chapter 5: Table Calculations 321
5.1 What is a table calculation? 321
5.2 Running Total of Sales 324
5.2.1 Demo 1 324
5.3 Profitability as Percent of Total 333
5.3.1 Demo 1 333
5.4 Moving average 341
5.4.1 Where is it used? 341
5.4.2 Types of moving average 341
5.4.3 Demo 1 342
5.5 Rank 351
5.5.1 Demo 1 351
5.6 LOD (Level of Detail) 360
5.6.1 Demo 1 364
5.6.2 Demo 2 374

5.7 Percentile 400
5.7.1 Demo 1 401
5.8 Year over Year Growth 406
5.8.1 Demo 1 406
5.8.2 Demo 2 417
5.8.3 Demo 3 426
5.9 Points to remember 431
5.10 Next Steps 431
Chapter 6: Customizing Data 433
6.1 Number functions 433
6.1.1 CEILING(number) and FLOOR(number) 434
6.1.2 MAX(number, number), MIN(number, number) 438
6.1.3 ABS(number) 442
6.2 String functions 446
6.2.1 Concatenation 447
6.2.2 Left() and Find() functions 449
6.2.3 Contains() function 451
6.2.4 Len() function 453
6.3 Logical Functions 456
6.3.1 CASE 456
6.3.2 IIF() function 459
6.3.3 IF ELSE 461
6.3.4 IF ELSEIF 463
6.4 Date functions 467
6.4.1 DATEDIFF() 468
6.4.2 DATEADD() function 474
6.4.3 DATENAME 482
6.5 Aggregate functions 484
6.5.1 ATTR(expression) 484

6.6 Table calculation functions 487
6.6.1 First(), Index() 487
6.7 Points to remember 492
6.8 Next steps 493
Chapter 7: Statistics 495
7.1 Why use statistics? 495
7.2 What is statistics? 497
7.3 Descriptive statistics 497
7.4 Inferential statistics 497
7.5 Few terms in statistics 497
7.6 Why do we use inferential statistics? 498
7.7 Why do we use descriptive statistics? 498
7.7.1 What is the measure of central tendency here? 498
7.8 Five magic number summary 501
7.8.1 Mean 501
7.8.2 Median 502
7.8.3 Mode 503
7.8.4 When to use which average? 504
7.9 Spread of data 504
7.9.1 Range 505
7.9.2 Interquartile range 505
7.9.3 Variance and standard deviation 505
7.9.4 Standard deviation 506
7.9.5 Assignment 1 507
7.9.6 Assignment 2 508
7.10 Box plot 515
7.10.1 Plotting box and whiskers plot in Tableau 517

7.11 Statistics tools in Tableau 519
7.11.1 Reference lines 519
7.12 Trend lines 528
7.12.1 Answering questions with trend lines 529
7.13 Forecasting 534
7.13.1 Demo 1 535
7.14 Points to remember 545
7.15 Next steps 545
Chapter 8: Chart Forms 547
8.1 Pie chart 547
8.1.1 What is a pie chart? 547
8.1.2 When to use a pie chart? 547
8.1.3 How to read a pie chart? 548
8.1.4 Pros 548
8.1.5 Cons 548
8.1.6 Five tips for using pie charts 548
8.1.7 A critique's view 549
8.1.8 An alternative for a pie chart 549
8.1.9 What can further add to the woes? 550
8.2 Treemaps 560
8.2.1 Pros 561
8.2.2 References 561
8.3 Heat Map 568
8.3.1 Why use heat maps? 568
8.3.2 How to create a heat map? 568
8.4 Highlight Table 577
8.4.1 Demo 1 578
8.4.2 Demo 2 586

8.5 Line Graph ... 592
8.5.1 Demo 1 ... 592
8.5.2 Demo 2 ... 599
8.6 Stacked Bar Chart ... 601
8.6.1 Demo 1 ... 601
8.6.2 Steps to create a stacked bar chart ... 602
8.7 Gantt chart ... 610
8.7.1 Shortcomings of Gantt charts ... 610
8.7.2 Demo 2 ... 620
8.8 Scatter plot ... 629
8.8.1 Why use a scatter plot? ... 629
8.8.2 What is a scatter plot? ... 629
8.8.3 Correlation coefficient ... 629
8.8.4 How to plot scatter plots in Tableau? ... 631
8.9 Histogram ... 643
8.9.1 What is required to plot a histogram? ... 643
8.9.2 Difference with bar charts ... 644
8.9.3 Pros of histogram ... 644
8.9.4 Plotting a histogram (customized bin size) ... 644
8.10 Word Cloud ... 655
8.10.1 Why should you use a word cloud? ... 655
8.10.2 When should you use a word cloud? ... 656
8.10.3 For what should you use a word cloud? ... 656
8.10.4 Where should you not use a word cloud? ... 656
8.10.5 How to plot a word cloud? ... 659
8.11 Points to remember ... 663
8.12 Next steps ... 663

Chapter 9: Advanced Visualization 665

9.1 Waterfall charts 665

9.1.1 Where can waterfall charts be used? 665

9.2 Bump charts 698

9.2.1 Where to use a bump chart? 698

9.3 Bullet graph 712

9.3.1 Demo 1 713

9.3.2 Demo 2 721

9.4 Points to remember 727

9.5 Next steps 727

Chapter 10: Dashboard and Stories 729

10.1 Why use a dashboard? 729

10.2 What is a dashboard? 729

10.3 Creating a dashboard 730

10.3.1 Opening a dashboard sheet 730

10.3.2 Adding views to the Dashboard 732

10.3.3 Adding interactivity to the dashboard 736

10.3.4 Adding an object to the dashboard 747

10.3.5 Remove a view or an object from the dashboard 749

10.3.6 Organizing a dashboard 752

10.4 Dashboard actions 765

10.4.1 Filter action 765

10.4.2 Highlight Action 779

10.5 Creating a story 785

10.6 What is a story? 785

10.6.1 How to create a story? 786

10.6.2 Description 791

10.7 Points to remember 793

10.8 Next steps 793

Chapter 11: Integration of Tableau with R 795
11.1 Steps to bring about this integration 795
11.1.1 SCRIPT_STR function 797
11.1.2 SCRIPT_BOOL function 802
11.1.3 SCRIPT_REAL function 808
11.1.4 SCRIPT_INT function 818
11.1.5 Market basket analysis 826
11.2 Points to Remember 833

Index 835

About the Authors

Seema Acharya is a lead principal with the education, training and assessment department of Infosys Limited. She is a technology evangelist, a learning strategist, and an author with over 15+ years of information technology industry experience in learning/education services. An educator by choice and vocation, her areas of interest and expertise are centered on business intelligence and big data and analytics technologies, such as data warehousing, data mining, data analytics, text mining and data visualization.

She is the author of the books *Fundamentals of Business Analytics*, ISBN: 978-81-265-3203-2, publisher - Wiley India (2011) and *Big Data and Analytics*, ISBN: 9788126554782, publisher - Wiley India (2015).

She has co-authored a paper on "Collaborative Engineering Competency Development" for ASEE (American Society for Engineering Education).

She holds the patent on "Method and system for automatically generating questions for a programming language".

Subhashini Chellappan is a technology education team lead specialist with the talent division of Accenture. She has rich experience in both academia and the software industry.

She has published couple of papers in various journals and conferences. *She has co-authored the book, Big Data and Analytics*, ISBN: 9788126554782, publisher - Wiley India (2015).

Her areas of interest and expertise are centered on business intelligence, big data and analytics technologies, such as Hadoop, NoSQL databases, Spark and machine learning.

About the Technical Reviewer

I am Rajeev, an author, blogger, a Tableau lover, data evangelist from Hyderabad, India. Working for Deloitte, I am a multidisciplinary designer working in data visualization, interaction design and innovation. With expertise in developing Tableau, Web-focus-based visualization and reporting applications, I love creativity and enjoy experimenting with various technologies. I am a very individualistic person who has a gift for figuring out how people who are different can work together productively. In addition, because I am driven by talent, I can constantly investigate the "hows", and "whys" of a given situation which would be very beneficial to not only myself, but also for other people.

Follow me on

LinkedIn: `https://in.linkedin.com/in/rajvivan`

Twitter : `@rajvivan`

Website : `http://www.tableaulearners.com/`

Acknowledgments

The making of the book was a journey that we are glad we undertook. The journey spanned a few months, but the experience will last a lifetime. We had our families, friends, colleagues, and well-wishers onboard this journey, and we wish to express our deepest gratitude to each one of them. Without their unwavering support and affection, we could not have pulled it off.

We are grateful to the student and teacher community who with their continual bombardment of queries impelled us to learn more, simplify our learnings and findings, and place it neatly in the book. This book is for them.

We wish to thank our friends – the practitioners from the field for their good counsel – for filling us in on the latest in the field of visualization and sharing with us valuable insights on the best practices and methodologies followed therein.

A special thanks to our technical reviewer for his vigilant review and the filling in with his expert opinion.

We have been fortunate to have the support of our teams who sometimes knowingly and at other times unknowingly contributed to the making of the book by lending us their steady support.

We consider ourselves very fortunate for the editorial assistance provided by Apress Media. We are thankful to Celestin Suresh John, senior manager, editorial acquisition, Apress and Springer Science and Business Media Company, for signing us up for this wonderful creation. We wish to acknowledge and appreciate Sanchita Mandal, coordinating editor, Laura Berendson, development editor, and their team of associates who adeptly guided us through the entire process of preparation and publication.

And finally we can never amply thank our families and friends who have been our pillars of strength, our stimulus, and our soundboards all through the process, and tolerated patiently our crazy schedules as we assembled the book.

Introduction

Why this book?

Data visualization is changing the way the world looks at data. This book will help you make sense of data quickly and effectively, make you look at data differently, more imaginatively. It will help you visualize from the end users perspective. It will impel you to dig for more insights. The topics covered along with demonstrations and illustrations are certain to promote creative data exploration. The book has within its scope the following:

- Sourcing data into Tableau from single and multiple data sources (both homogeneous and heterogeneous)
- Statistical analysis in Tableau
- Integration of R Analytics with Tableau
- Concepts behind visualization and industry best practices

Who is this book for?

The audience for this book includes all levels of IT professionals, executives responsible for determining IT strategies, system administrators, data analysts and decision makers responsible for driving strategic initiatives, etc. It will help to chart your journey from a novice to a professional visualization expert.

The book will also make for interesting read for business users / management graduates / business analysts.

How is this book organized?

Our book has 11 chapters. Here is a sneak peek ...

Chapter 1: This chapter explains the meaning of visualization and the role it plays in BI and data science. It covers the various visualization tools available on the market. It highlights Tableau's products lines, such as Tableau Desktop, Tableau Server, Tableau Online, Tableau Public and Tableau Reader. It also throws light on the various file types in Tableau and brings forth the difference between saving the Tableau workbook as .twb or .twbx etc., the Tableau data source as .tds or .tdsx etc.

Chapter 2: The aim of this chapter is to outline the step-by-step process to connect Tableau to varied data sources, elucidate the concept of joins and blends, and enumerate the differences between live connection and working with data extracts.

Chapter 3: This chapter serves to aid in organizing data into groups and sets. It further delves into the difference between groups and sets.

Chapter 4: This chapter will strengthen our comprehension of measure names and measure values and the techniques to have more than one measure depicted on a single view / worksheet.

Chapter 5: This chapter clearly explains the connotation and significance of calculations, such as moving average, year-on-year growth, level of detail (LOD).These quick table calculations have been explained with the help of case scenarios.

Chapter 6: This chapter will help you customize the data using string functions such as concatenation, find, left, etc., number functions which will bring out the difference between aggregated and non-aggregated measures, date functions such as datepart, datediff, dateadd, dateparse, etc.

Chapter 7: This chapter focusses on enunciating the significance of statistics in analysis. It will elaborate the usage of reference line, constant line, trend line, summary card, etc.

Chapter 8: This chapter will provide an easy comprehension and usage of the various chart forms such as pie chart, heat map, treemap, stacked bar chart, line graph, word cloud, etc., the concepts behind each one of them, the pros and cons, the best chart form to use in a particular scenario etc.

Chapter 9: This chapter will build on your knowledge of visualization with advanced chart forms, such as waterfall charts, bump charts and bullet graphs, a chart form that enables visualizing staged progress to your goal.

Chapter 10: This chapter is designed to help learners weave a powerful and insightful story by putting together the various reports (views / worksheets) into an interactive dashboard.

Chapter 11: This chapter talks about the steps involved in integrating R Analytics with Tableau. It introduces data mining and the implementation of data mining algorithms in R and Tableau.

The source code (the .twbx for all the demos and assignments) are also shared. Feel free to download them from Apress site (`www.apress.com`)

How to get the most out of this book?

It is easy to leverage the book to gain the maximum by religiously abiding by the following:

- Read the chapters thoroughly. Get hands-on by following the step-by-step instructions stated in the demonstrations. Do NOT skip any demonstration. If need be, repeat it a second time or till the time the concept is firmly etched.
- Join a Tableau community or discussion forum.
- Read up customer stories provided on Tableau site (`www.Tableau.com`) to learn how customers the world over are enhancing their visualization experience.
- Read the blogs of data visualization experts, such as Stephen Few, Edward Tufte, and Hans Rosling, etc.

Where next?

We have endeavored to unleash the power of Tableau as a data visualization tool and introduce you to several chart forms / visualizations. We recommend that you read the book from cover to cover, but if you are not that kind of person, we have made an attempt to keep the chapters self-contained so that you can go straight to the topics that interest you most.Whichever approach you choose, we wish you well!

A quick word for fellow instructors

We've paid extra attention in to setting the order of the chapters and to the flow of topics within each chapter. This was undertaken in this way to assist our fellow instructors and academicians in carving out a syllabus from the Table of Contents (TOC) of the book. The complete TOC can qualify as the syllabi for a semester or if the college has an existing syllabus on business intelligence or data visualization or analytics and visualization, a few chapters can be added to the syllabi to make it more robust. We leave it to your discretion on how you wish to use these resources for your students.

We have ensured that each tool / component discussed in the book is with adequate hands-on content to enable you to teach better and provide ample hands-on practice to your students.

Happy Learning!!!
Authors:
Seema Acharya
Subhashini Chellappan

CHAPTER 1

Introducing Visualization and Tableau

"Graphical excellence is that which gives to the viewer the greatest number of ideas in the shortest time with the least ink in the smallest space."

— Edward R. Tufte, a pioneer in the field of data visualization

This chapter will introduce data visualization and the importance of visualizing data well. It will acquaint us with one of the market leading data visualization tools, Tableau. We will learn about the Tableau product line, including Tableau Desktop, Tableau Server, Tableau Online, Tableau Reader, etc. The chapter will also detail out the many file types in Tableau.

Imagine that you have been asked to study pages and pages of data, report your findings, and draw your inferences and conclusions. You prepare yourself for the task which looks uninteresting, time-consuming and plain boring. Would life be any easier if this was replaced with a few good visualizations (Data visualization is the pictorial or graphical depiction of data) that allow you to uncover trends, unearth patterns hitherto hidden, quickly and efficiently?

Data visualization is visual communication. It is to make data more comprehensible, much easier to interpret and analyze. In summary, data visualization serves two important purposes:

- To make sense of data (also called data analysis)
- To communicate visually (what you have discovered to others)

Data visualization tools have come a long way from the standard charts and graphs of Excel to the trendier and more sophisticated chart forms such as geographic maps, sparklines, heat maps, tree maps, fever charts, etc.

Electronic supplementary material The online version of this chapter (doi: 10.1007/978-1-4842-2352-9_1) contains supplementary material, which is available to authorized users.

S. Acharya and S. Chellappan, *Pro Tableau*, DOI 10.1007/978-1-4842-2352-9_1

1.1 Why data visualization?

Imagine that your company has decided for the very first time to launch an exquisite jewelry collection. This will mean a huge investment. You want to be very sure about your target customers. You launch an online survey to get to know your customers. Among the various questions are questions regarding their occupation / profession and their preference for the style of jewelry. The survey is launched across several cities and you have quite a large amount of data to analyze. While this can be done manually, it will mean more time. The festive season is nearing and you want to quickly decide on the style so as to have adequate time for getting your collection ready. You decide to plot the findings of the survey graphically. It clearly shows the trend as the preference for contemporary and, at the same time, conservative jewelry. This implies that your jewelry line should be an amalgamation or fusion of contemporary and conservative styles, because most of the prospective customers work for the corporate sector. A graphical depiction of the survey results allowed you to quickly decide on the style of jewelry that your store should market.

This scenario is an example of why data visualization is important. There are many reasons to use data visualization:

- You want to understand the correlation between sales and profit.
- You want to predict sales volume.
- You want to identify areas where your business is booming and areas where it is dwindling.
- You want to predict how the market will react to the launch of your new product.
- You want to understand the factors that influence your customer's behavior.
- You want to understand how your school has performed over the years in the three critical skills (three Rs), i.e. reading, writing and arithmetic.
- You want to clearly and quickly see the year-over-year growth of your business.
- You want to dig for deeper insights.
- You want to tell a story with your data.

1.2 What can data visualization help with?

Data visualization places power in the hands of business users and dramatically shortens the time it takes to transform data to insights. It enables decision makers

- to quickly spot trends, see patterns, and grasp difficult concepts.
- to see things that otherwise would go unnoticed.
- to investigate the cause-effect relationships.
- to ask a question, to get an answer to their question, and then ask follow-up questions as well.
- to ask better questions of their data thereby leading to better data driven decisions.
- to implement infographics, i.e. graphics that are used to convey information.

1.3 An introduction to visualization

Take a moment to look at the data set given below in Table 1-1. It has details on the number of visitors who visited each city in May of 2012 and 2013.

Table 1-1. *"Sample – Visitors" dataset*

	A	B	C	D
1	Country / Territory	City	Visits May 2012	Visits May 2013
2	United Kingdom	London	31,733	81500
3	United States	New York	9451	8090
4	United Kingdom	Manchester	6395	7797
5	India	New Delhi	3879	5430
6	United Kingdom	Southampton	3368	5333
7	United Kingdom	Birmingham	5144	4879
8	Australia	Sydney	3616	4650
9	United States	Chicago	7974	3115
10	United States	San Francisco	2851	3990

The same data is represented with an additional column in Table 1-2. The negative % Change is shown in red.

Table 1-2. *New column, "% Change" added to the "Sample – Visitors" data set (Table 1-1) and conditional formatting applied on it*

	A	B	C	D	E
1	Country / Territory	City	Visits May 2012	Visits May 2013	% Change
2	United Kingdom	London	31,733	81500	157%
3	United States	New York	9451	8090	-14%
4	United Kingdom	Manchester	6395	7797	22%
5	India	New Delhi	3879	5430	40%
6	United Kingdom	Southampton	3368	5333	58%
7	United Kingdom	Birmingham	5144	4879	-5%
8	Australia	Sydney	3616	4650	29%
9	United States	Chicago	7974	3115	-61%
10	United States	San Francisco	2851	3990	40%

Conditional formatting in combination with colors helps one to clearly depict "what the trends are" and facilitates decision-making. It makes it easier for readers to understand the data rather than have them remember numbers which are both good and bad as well as asking them to pick out highlights or concerns from a standard column of data.

Take a look at Fig. 1-1. It is even better at clearly exhibiting the trends. The use of annotations further helps one to draw attention to the negative % change.

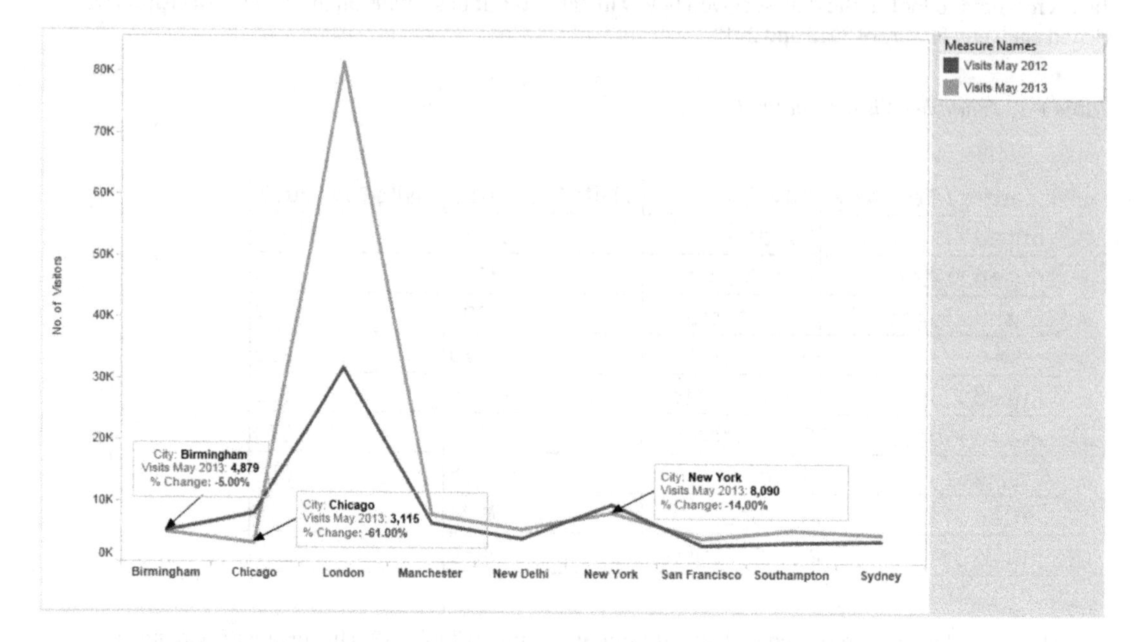

Figure 1-1. *Graphical depiction of the "Sample - Visitors" data set (shown in Table 1-2)*

Now it is useful to watch the TedEx video (17:56 minutes) by David McCandless on data visualization.

at http://www.ted.com/talks/david_mccandless_the_beauty_of_data_visualization to get an easy understanding of the power of visualization. The video highlights the following points:

- Visualization helps one to spot patterns and see the connection between data.
- Readers find it very easy to comprehend the visuals. It calls for much less effort on the part of the readers to make sense.
- A picture is worth a thousand words. The video illustrates how visualization condenses large amount of information into a small space.
- Visualization can get you to the answers much faster and provide clarity very easily.

Let us look at another example to realize the power of visualization.

If we were to ask you to look at around 10,000 records (9,994 to be precise) and tell us which category in each region made the greatest sales, what would be your answer? And more importantly how much time would it take for you to provide the answer?

A subset of the "Sample - Superstore" data set is shown in Figure 1-2:

	A	B	C	D	E	F	G	H	I	J	K	L	M	N	O
1	Row ID	Order ID	Order Date	Ship Date	Ship Mode	Customer ID	Customer Na	Segment	Country	City	State	’ostal Cod	Region	Product ID	Category
2	1	CA-2013-152156	11/9/2013	11/12/2013	Second Class	CG-12520	Claire Gute	Consumer	United States	Henderson	Kentucky	42420	South	FUR-BO-100(	Furniture
3	2	CA-2013-152156	11/9/2013	11/12/2013	Second Class	CG-12520	Claire Gute	Consumer	United States	Henderson	Kentucky	42420	South	FUR-CH-100(	Furniture
4	3	CA-2013-138688	6/13/2013	6/17/2013	Second Class	DV-13045	Darrin Van H	Corporate	United States	Los Angeles	California	90036	West	OFF-LA-1000	Office Supplies
5	4	US-2012-108966	10/11/2012	10/18/2012	Standard Cla	SO-20335	Sean O'Donn	Consumer	United States	Fort Lauderd	Florida	33311	South	FUR-TA-1000	Furniture
6	5	US-2012-108966	10/11/2012	10/18/2012	Standard Cla	SO-20335	Sean O'Donn	Consumer	United States	Fort Lauderd	Florida	33311	South	OFF-ST-1000	Office Supplies
7	6	CA-2011-115812	6/9/2011	6/14/2011	Standard Cla	BH-11710	Brosina Hoffr	Consumer	United States	Los Angeles	California	90032	West	FUR-FU-1000	Furniture
8	7	CA-2011-115812	6/9/2011	6/14/2011	Standard Cla	BH-11710	Brosina Hoffr	Consumer	United States	Los Angeles	California	90032	West	OFF-AR-1000	Office Supplies
9	8	CA-2011-115812	6/9/2011	6/14/2011	Standard Cla	BH-11710	Brosina Hoffr	Consumer	United States	Los Angeles	California	90032	West	TEC-PH-1000	Technology
10	9	CA-2011-115812	6/9/2011	6/14/2011	Standard Cla	BH-11710	Brosina Hoffr	Consumer	United States	Los Angeles	California	90032	West	OFF-BI-1000	Office Supplies
11	10	CA-2011-115812	6/9/2011	6/14/2011	Standard Cla	BH-11710	Brosina Hoffr	Consumer	United States	Los Angeles	California	90032	West	OFF-AP-1000	Office Supplies
12	11	CA-2011-115812	6/9/2011	6/14/2011	Standard Cla	BH-11710	Brosina Hoffr	Consumer	United States	Los Angeles	California	90032	West	FUR-TA-1000	Furniture
13	12	CA-2011-115812	6/9/2011	6/14/2011	Standard Cla	BH-11710	Brosina Hoffr	Consumer	United States	Los Angeles	California	90032	West	TEC-PH-1000	Technology
14	13	CA-2014-114412	4/16/2014	4/21/2014	Standard Cla	AA-10480	Andrew Allen	Consumer	United States	Concord	North Carolin	28027	South	OFF-PA-1000	Office Supplies
15	14	CA-2013-161389	12/6/2013	12/11/2013	Standard Cla	IM-15070	Irene Maddo	Consumer	United States	Seattle	Washington	98103	West	OFF-BI-1000	Office Supplies
16	15	US-2012-118983	11/22/2012	11/26/2012	Standard Cla	HP-14815	Harold Pawla	Home Office	United States	Fort Worth	Texas	76106	Central	OFF-AP-1000	Office Supplies
17	16	US-2012-118983	11/22/2012	11/26/2012	Standard Cla	HP-14815	Harold Pawla	Home Office	United States	Fort Worth	Texas	76106	Central	OFF-BI-1000	Office Supplies
18	17	CA-2011-105893	11/11/2011	11/18/2011	Standard Cla	PK-19075	Pete Kriz	Consumer	United States	Madison	Wisconsin	53711	Central	OFF-ST-1000	Office Supplies
19	18	CA-2011-167164	5/13/2011	5/15/2011	Second Class	AG-10270	Alejandro Gr	Consumer	United States	West Jordan	Utah	84084	West	OFF-ST-1000	Office Supplies

Orders | Returns | People

***Figure 1-2.** Subset of "Sample - Superstore" data subset*

Now, look at the graph in Fig. 1-3 and answer the same question. That is the power of visualization....

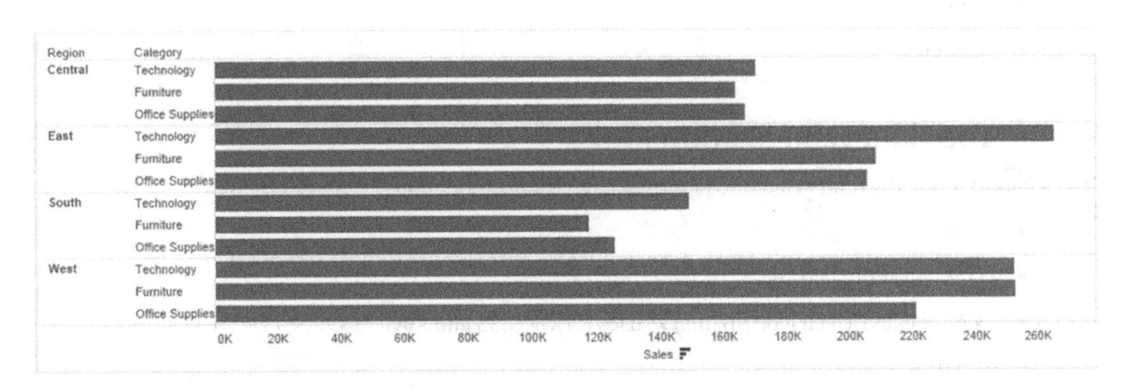

***Figure 1-3.** Visualization that represents which category in each region made for the maximum sales*

1.3.1 Which domain is leveraging the power of data visualization?

Data visualization is used in many areas:

- Education
- Retail
- Banking and Finance
- Healthcare
- Social Media
- Sports
- Entertainment, etc.

1.3.2 Who is using data visualization?

Data visualization is being used by several enterprises. A few are mentioned below:

- Google
- Facebook
- Yahoo
- Cisco Systems Inc.
- Deloitte
- GE, etc.

1.3.3 Top data visualization tools

Take a look at few of the market leading reporting and visualization tools.

- Tableau
- QlikView from QlikTech
- D3.js (Data Driven Documents) ➤ uses HTML, CSS and SVG
- Chart.js
- Tibco SpotFire
- IBM Cognos Visual Analytics
- Roambi Analytics
- Google Charts
- FusionCharts

1.3.4 History of data visualization

Refer Table 1-3 to study the "History of data visualization"

Table 1-3. *History of Data Visualization*

When	Form	For what
2nd Century	Table (textual representation)	To represent astronomical information as a tool for navigation
17th Century (Rene Descartes - French Philosopher and Mathematician)	Graphs	Developed a two-dimensional coordinate system to display values.
18th and 19th Century (William Playfair - Scottish social scientist))	Bar charts, pie charts, etc.	To represent quantitative information
1977, John Tukey of Princeton	Exploratory data analysis	Exploring and making sense of data
1983, Edward Tufte		The visual display of quantitative information

1.3.5 What are the expectations for a data visualization tool?

- Allow goal-oriented visualizations, i.e. plot actuals versus the specific, desired outcomes.
- Make the reports and dashboards available to the right persons at the right time in the right format and on the right device.
- Allow the play with the data such as slicing - dicing, filtering, querying, interacting, etc.
- Able to access real-time or near-real-time data sources to be able to present actionable information and insights in real time or near real time.

1.3.6 Let us see how Tableau fulfils the expectations.

The following salient features make Tableau a market-leading visualization tool:

- Tableau provides a variety of graphs and chart forms to clearly and convincingly present the actuals versus the specific, desired outcomes.
- A number of products from the Tableau product suite allow the reports and dashboards created using Tableau to be made available to decision-makers timely and relatively easily.

The following is a brief summary of the Tableau product line:

- *Tableau Desktop*: a tool that allows one to author/design a report
- *Tableau Server:* a platform that allows the deployment and sharing of reports across an organization
- *Tableau Online*: a secure, cloud-based solution for sharing, distributing, and collaborating on Tableau views and dashboards
- *Tableau Reader:* a free product that anyone can use to view and interact with the Tableau workbooks created by licensed users of Tableau Desktop
- *Tableau Mobile:* A version that enables the accessibility to reports on mobile devices such as iOS devices, etc.

- There are several features available in Tableau to make sense of data, interpret it, run a context sensitive filter, sort as desired, statistically analyze it, etc.
- Tableau can connect to a wide range of underlying data sources from traditional ones such as Excel, text files, etc. to non-traditional ones such as social media data, SalesForce, Microsoft Azure, etc.
- Tableau supports your need for Rapid Fire BI / Agile BI / Self-Service BI and Analytics.

1.3.7 Reasons to make a switch to Tableau

Your current BI visualization tool

- Takes several minutes to refresh the data.
- Runs out of memory.
- Takes several minutes to add a row.
- Forces you to work with a smaller subset of data.
- Cannot accommodate the massive data troves that your business is supposed to handle.
- Does not support the viewing of data on a map.
- Asks you to anticipate all future needs. Has little or no support for data exploration.

1.4 Positioning of Tableau

It is an era of self-service analytics. The dependency on IT is declining. Customers are demanding tools that are easy to use, highly accessible and simple to integrate with existing systems.

As per Gartner's 2016 "Magic Quadrant for Business Intelligence and Analytics Platforms" , Tableau is placed in the "Leaders" quadrant. Refer to Figure 1-4.

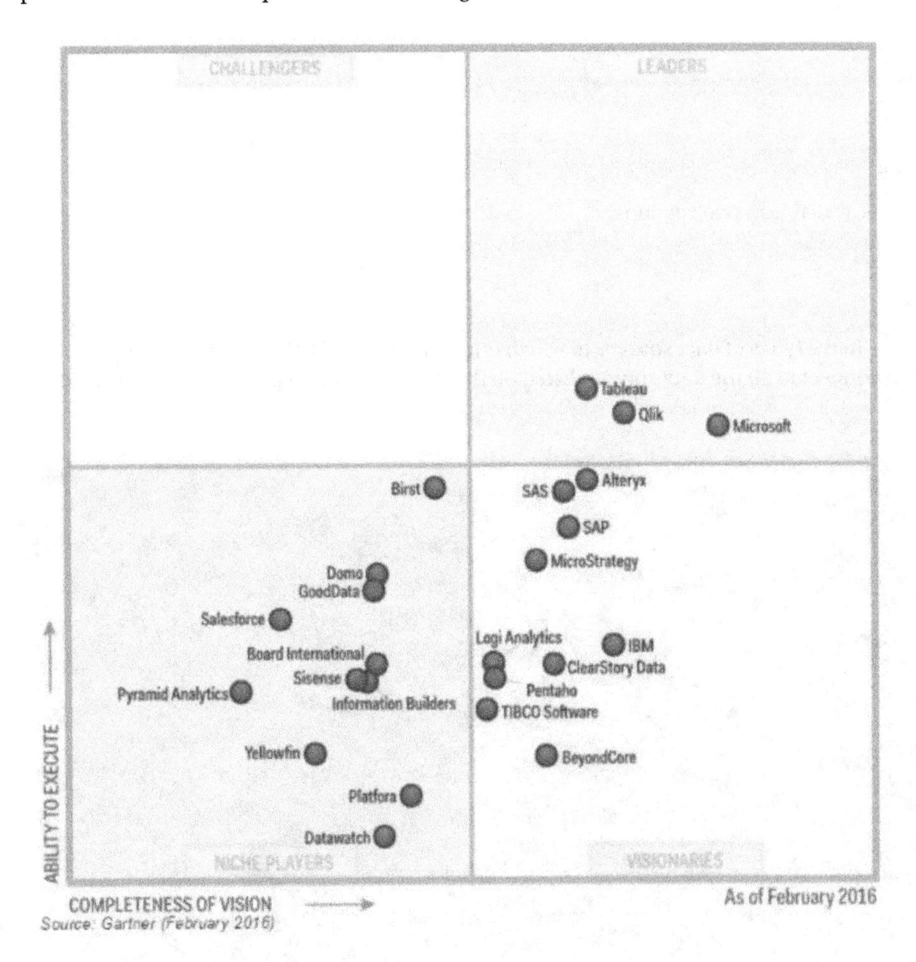

Figure 1-4. *Gartner's Magic Quadrant for Business Intelligence and Analytics Platforms for the year 2016*

Gartner's Magic Quadrant is plotted along "Completeness of Vision" on the X axis and the "Ability to Execute" on the Y axis. The quadrant has four parts, namely, "Visionaries", "Niche Players", "Challengers" and "Leaders".

Tableau placed highest in its ability to execute. The year 2016 is the fourth year in a row that Tableau has placed in the "Leaders" quadrant.

Clearly seen from the Magic Quadrant is a need for leaders to demonstrate excellence in their current execution and the capability and willingness to make progress towards their future mission.

1.5 Tableau product line

Let us broadly classify Tableau tools into the following categories:

- Developer tools (Tableau Desktop and Tableau Public)
- Sharing tools (Tableau Server, Tableau Online, Tableau Reader)

Developer tools will help you create a visualization and/or dashboards.
Sharing tools facilitate the following visualization and/or dashboards tasks:

- Viewing
- Sharing
- Interacting
- Exploring

Tableau Desktop is available in two versions:

- Professional
- Personal

The difference lies in the types of data sources to which one can connect. With Tableau Desktop Professional, one can connect to all the data sources listed on the data connection page (Shown in Fig. 1-5.)

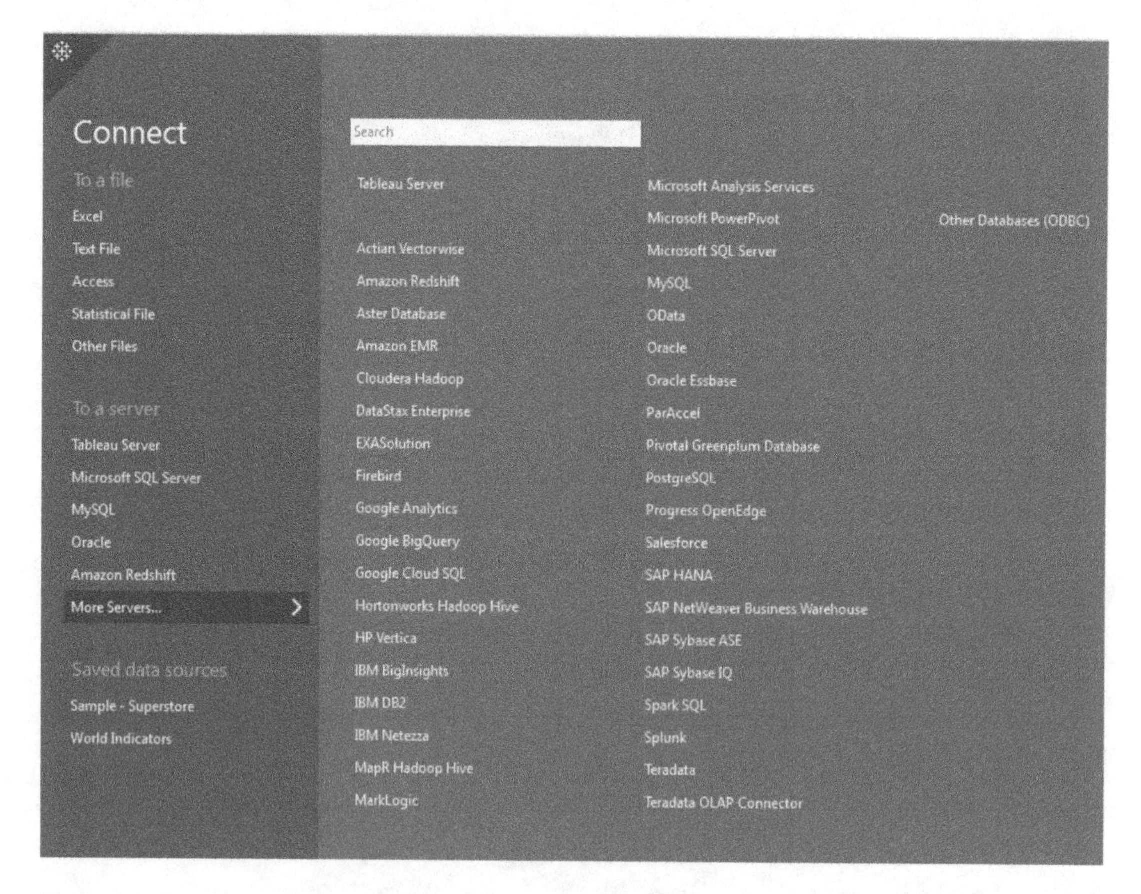

Figure 1-5. *Data connection page of Tableau Desktop Professional*

With Tableau Desktop Personal, one can only connect to OData, Microsoft Windows Azure Marketplace DataMarket, and Tableau Data Extract (.tde) files; however, it is possible to save workbooks locally. It lacks the ability to publish to a Tableau Server (Public or Private).

Tableau Public is a free download from the Tableau website. It is constrained by the data sources to which it can connect. Only the following data sources are supported:

Connect

To a file

- Excel
- Text file
- Access
- Statistical file

To a server

- OData
- More servers
    - Odata
    - Web data connector

With Tableau Public, anyone can find your visualization. It does not support saving workbooks locally. Tableau public can visualize data sets containing up to 1 million rows of data.

Tableau Server is hosted within the organization's premises. It facilitates the sharing of visualizations securely across the organization. However the workbooks that needs to be shared should be published to Tableau Server using Tableau Desktop. Licensed users will then be able to access the visualizations online using a web browser. It can also be used to share the data sources.

Tableau Online has the same functionality as the Tableau Server; however, it is hosted by Tableau in their cloud.

Tableau Reader is a free download from the Tableau Website. It allows one to view or interact with Tableau packaged workbooks (.twbx) ONLY. There is essentially zero security with Tableau Reader. Anyone who has the .twbx file, can use Tableau Reader to open it.

Refer to Fig. 1-6.

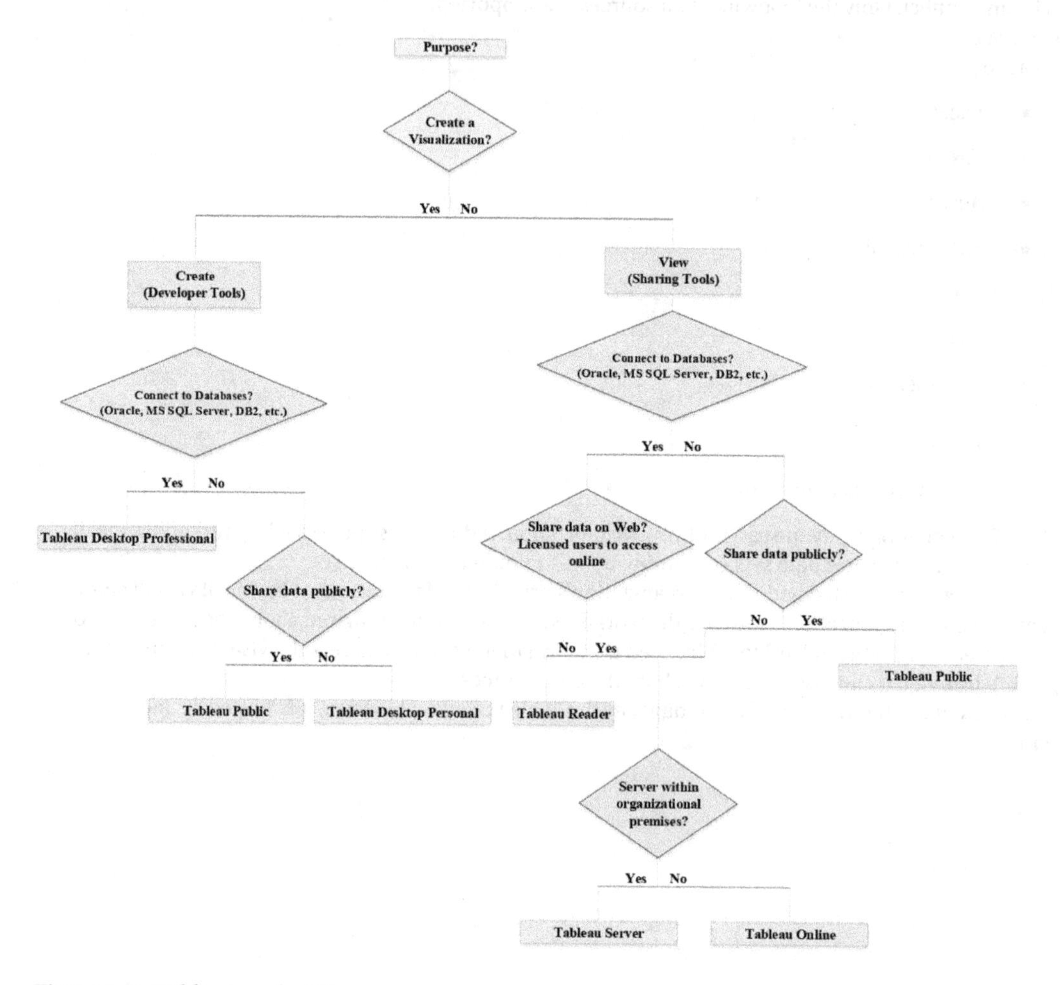

Figure 1-6. *Tableau product line*

The Tableau product line at a glance (Shown in Table 1-4):

Table 1-4. *Tableau products line at a glance*

Tableau Desktop	• can create workbooks comprising of worksheets, dashboards and stories. • is a licensed product. Comes in two versions: Desktop Professional and Desktop Personal. • allows workbooks to be stored locally. • allows workbooks, dashboards and stories to be published to Tableau Server, Tableau Online and Tableau Public.
Tableau Server	• is usually on premises. • allows users to interact directly using browser. • is privately managed.
Tableau Online	• is a hosted version of Tableau Server on the cloud. • has server(s) maintained by Tableau.
Tableau Reader	• allows one to view / interact with Tableau packaged workbooks (.twbx). • does not connect to server. • does not permit modifications to workbooks.
Tableau Public	• is a free product from Tableau. • limits the amount of work with data (number of rows). • can only connect to Excel, Access or text file (no database connectivity) & anything you save in Tableau Public will be saved on the Tableau Public Sever, which anyone can download (no confidentiality).

1.6 File types in Tableau

Let us look at the following file types in Tableau:

- Tableau Workbook (twb)
- Tableau Packaged Workbook (twbx)
- Tableau Data Source (tds)
- Tableau Packaged Data Source (tdsx)
- Tableau Bookmark
- Tableau Data Extract

1.6.1 Tableau Workbook (twb)

When you save your workbook in Tableau, the default file extension is .twb. It is an XML file with instructions to connect and interact with the data source. It has all information to help draw/create the visualization, such as the fields which are displayed on the worksheet or view, the type of aggregations which are used on the measures, the various formatting and styling options, etc. It also has information on any changes made to the worksheet or dashboard such as using a quick filter, etc.

Example: Tableau workbook with the following worksheet/view is saved as a .twb file (Shown in Fig. 1-7). To create this workbook, one performs the following steps:

- Open a Tableau workbook.
- Read in data from "Sample - Superstore.xls" into Tableau. Work with "Orders" sheet within "Sample - Superstore.xlsx".
- Drags the dimension, "Sub-Category" from the dimensions area under the data pane and place it on Rows Shelf.
- Drags the measure "Sales" from the measures area under the data pane and place it on Columns Shelf.
- Save the file as a .twb by going to File ➤ Save As... ➤ .twb

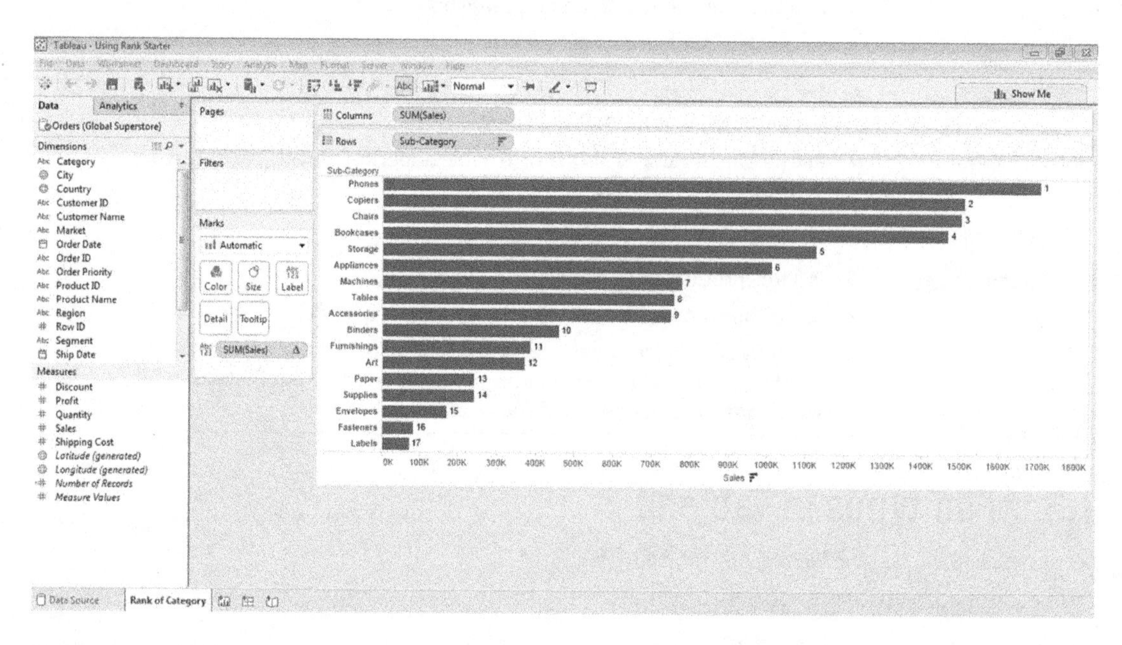

Figure 1-7. *Tableau workbook saved as a .twb file*

A few details about Fig. 1-7 (given in Table 1-5 and Fig. 1-8):

Table 1-5. *Details about a sample Tableau workbook*

Name of the file:	Using Rank Starter
Type of File:	Tableau Workbook (.twb)
Location:	`C:\Users\Seema_Acharya\Desktop`

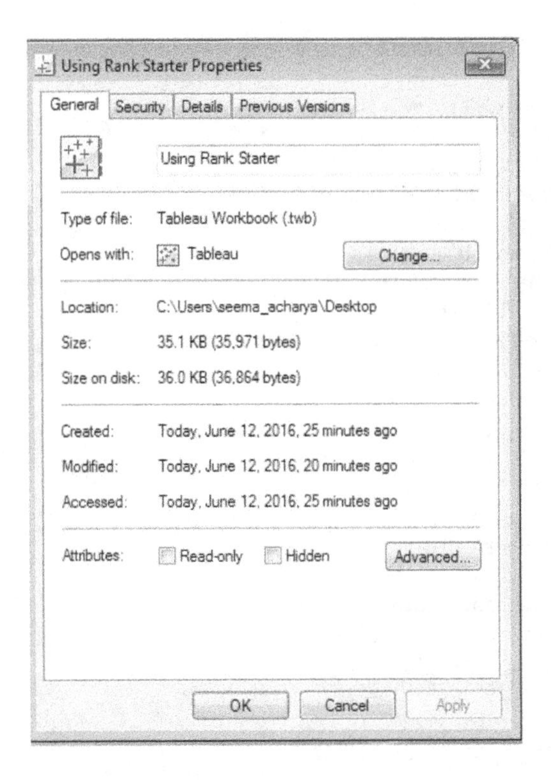

Figure 1-8. *Properties of a sample Tableau workbook (.twb)*

Open the .twb in a text editor such as Notepad / Notepad++ / WordPad. Given below is the content of the saved .twb file (partial content) when opened in WordPad.

```
<?xml version='1.0' encoding='utf-8' ?>

<workbook source-platform='win' version='9.0' xmlns:user='http://www.tableausoftware.com/
xml/user'>
  <!-- build 9000.15.0615.1857                               -->
  <preferences>
    <preference name='ui.encoding.shelf.height' value='24' />
    <preference name='ui.shelf.height' value='26' />
  </preferences>
  <datasources>
    <datasource caption='Orders (Global Superstore)' inline='true' name='excel-
direct.42081.320096539355' version='9.0'>
      <connection class='excel-direct' cleaning='no' compat='no' dataRefreshTime=''
filename='C:\Users\seema_acharya\Desktop\Using Rank Starter.twb Files\Data\Data\Global
Superstore.xls' password='' server='' validate='no'>
```

```
<relation name='Orders$' table='[Orders$]' type='table'>
  <columns header='yes' outcome='6'>
    <column datatype='string' name='Category' ordinal='0' />
    <column datatype='string' name='City' ordinal='1' />
    <column datatype='string' name='Country' ordinal='2' />
    <column datatype='string' name='Customer Name' ordinal='3' />
    <column datatype='string' name='Market' ordinal='4' />
    <column datatype='string' name='Customer ID' ordinal='5' />
    <column datatype='date' name='Order Date' ordinal='6' />
    <column datatype='string' name='Order ID' ordinal='7' />
    <column datatype='string' name='Order Priority' ordinal='8' />
    <column datatype='string' name='Product ID' ordinal='9' />
    <column datatype='string' name='Product Name' ordinal='10' />
    <column datatype='string' name='Region' ordinal='11' />
    <column datatype='integer' name='Row ID' ordinal='12' />
    <column datatype='string' name='Segment' ordinal='13' />
    <column datatype='date' name='Ship Date' ordinal='14' />
    <column datatype='string' name='Ship Mode' ordinal='15' />
    <column datatype='string' name='State' ordinal='16' />
    <column datatype='string' name='Sub-Category' ordinal='17' />
    <column datatype='real' name='Discount' ordinal='18' />
    <column datatype='real' name='Profit' ordinal='19' />
    <column datatype='integer' name='Quantity' ordinal='20' />
    <column datatype='real' name='Sales' ordinal='21' />
    <column datatype='real' name='Shipping Cost' ordinal='22' />
  </columns>
</relation>
```

The connection section provides information on the data source. In the example above, it gives the name and location of the Excel file along with the name of the worksheet within the Excel workbook to which one is connected.

```
<connection class='excel-direct' cleaning='no' compat='no' dataRefreshTime='' filename='C:\
Users\seema_acharya\Desktop\Using Rank Starter.twb Files\Data\Data\Global Superstore.xls'
password='' server='' validate='no'>
        <relation name='Orders$' table='[Orders$]' type='table'>
```

The column section describes the columns (name, data type and its ordinal position) in the 'Orders' worksheet within the "Global Superstore.xls".

The XML file also has details on which columns constitute the dimensions and which ones constitute the measures:

```
<column datatype='string' name='[State]' role='dimension' semantic-role='[State].[Name]'
type='nominal'>

<column datatype='integer' name='[Number of Records]' role='measure' type='quantitative'
user:auto-column='numrec'>
```

The extract below gives information on the type of aggregation used, the sort order and the table calculation applied, etc.

```
<table-calc ordering-type='Columns' rank-options='Competition,Descending' type='Rank' />
            </column-instance>
            <column-instance column='[Sales]' derivation='Sum' name='[sum:Sales:qk]'
pivot='key' type='quantitative' />
          </datasource-dependencies>
          <sort class='computed' column='[excel-direct.42081.320096539355]. [none:Sub-
Category:nk]' direction='DESC' using='[excel-direct.42081.320096539355].[sum:Sales:qk]'>
          </sort>
```

1.6.2 Tableau Packaged Workbook (twbx)

.twbx is a Tableau packaged workbook. It is a package that has the original .twb file grouped together with the data source. It can be considered analogous to a zipped file. It has all the necessary instructions and data to work in Tableau. One can work even without the network / Internet connection to the data as the data is packaged or held within the .twbx itself. The .twbx file can be unpackaged to split it into the .twb file and the data source.

■ **Note** If working with an underlying database or server, it is required to create a Tableau data extract (.tde) file before it can be packaged into a Tableau packaged workbook. Excel file or .csv file can be packaged directly into a Tableau packaged workbook.

The visualization (in Fig. 1-9) is saved as .twbx file.

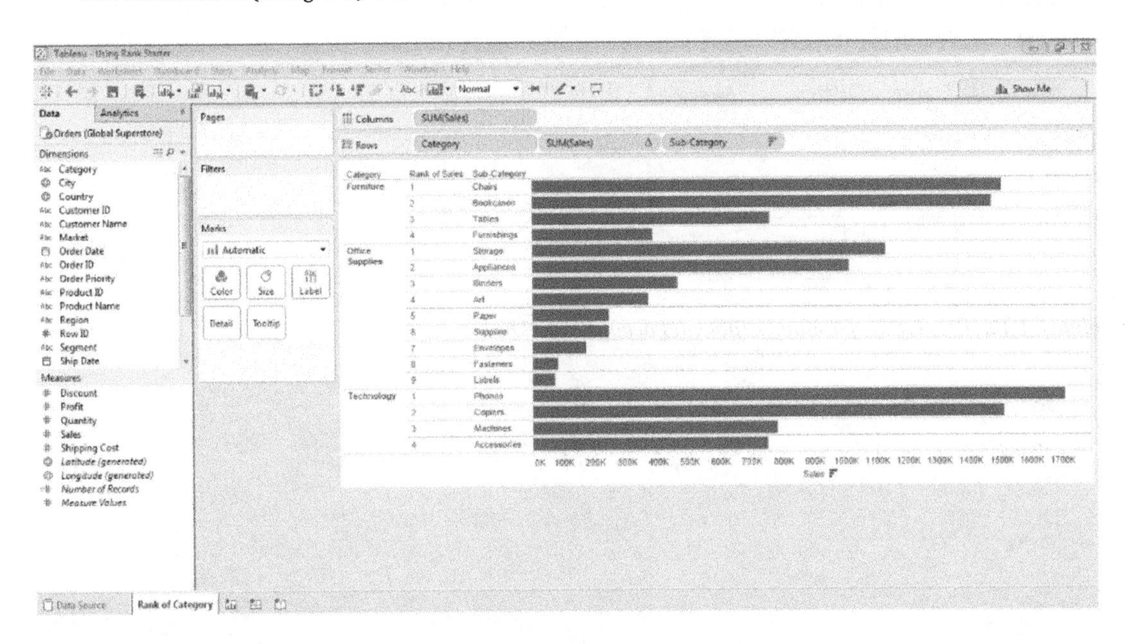

Figure 1-9. Visualization saved as Tableau packaged workbook (.twbx file)

Few details about Fig. 1-9 (refer Table 1-6 and Fig. 1-10):

Table 1-6. *Details about Tableau Packaged Workbook (.twbx)*

Name of the file:	CategoryRankStarter
Type of File:	Tableau Packaged Workbook (.twbx)
Location:	C:\Users\Seema_Acharya\Desktop

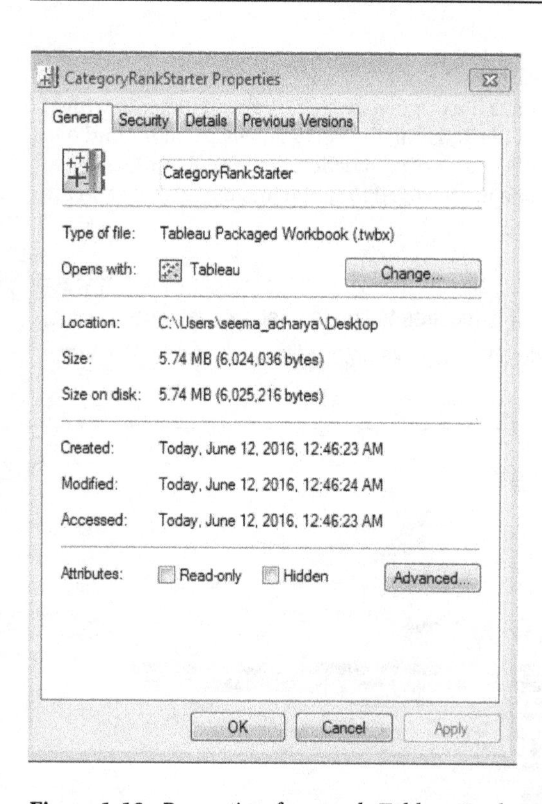

Figure 1-10. *Properties of a sample Tableau Packaged Workbook*

To unpackage the .twbx, perform the following steps:

Step 1

Right click on the file "CategoryRankStarter.twbx" to bring up the menu shown in Fig. 1-11.

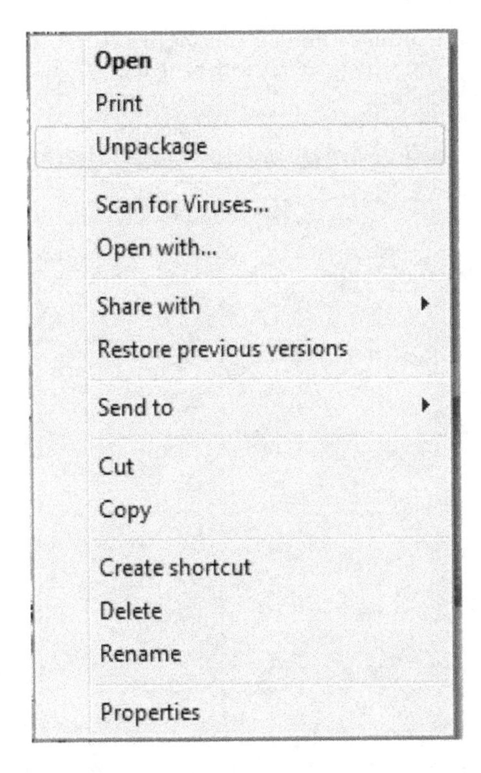

Figure 1-11. *"UnPackage Option" to unpackage the Tableau Packaged Workbook*

Step 2

Click on "Unpackage" to segregate the .twb and the data source.

1.6.3 Tableau Data Source (tds) file

What does it contain?

The Tableau Data Source file includes all of the connection information and metadata about your data source. A TDS file is automatically included as a part of a TWB (Tableau Workbook) file, but it can also be saved as a separate file if you want to share the connection information as well as the metadata for that particular data source. It is an XML file that has the following information:

- Data source type
- Data source connection information specified in the data source page (for example, server, port, location of local files (Excel, text, Extracts, etc.), and tables)
- Groups
- Sets
- Custom Calculated fields
- Bins
- Default field properties (for example, number formats, aggregation, sort order, data types, etc.)

1.6.4 Tableau Packaged Data Source (tdsx) file

It contains all the information in the Data Source (.tds) file as well as any local file data sources (Excel, text, and extracts). This file type is a single zipped file and is good for sharing a data source with people who may not have access to the original data that is stored locally on your computer.

We have a data set stored in "Sample Superstore.xls" (Shown in Table 1-7). It has the following fields:

Table 1-7. *Data set in "Sample Superstore.xls"*

Field	Data type
Row ID	#
Order ID	#
Order Date	Date
Ship Date	Date
Ship Mode	Abc
Customer ID	Abc
Customer Name	Abc
Segment	Abc
Country	Geographic Role
City	Geographic Role
State	Geographic Role
Postal Code	Geographic Role
Region	Abc
Product ID	Abc
Category	Abc
Sub-Category	Abc
Product Name	Abc
Sales	#
Quantity	#
Discount	#
Profit	#

Let us make the following changes:

- The field name , "RowID" to "RowNumber"
- Changed the data type of the fields "RowNumber" from "#" to "Abc"
- Created a group, "StationaryGroup"
- Created a set, "Top10CustomersByProfit"

- Changed the sort order for "Segment" to have the members ordered as follows:
  - Corporate
  - Home Office
  - Consumer
- Created a calculated field, "ProfitRatio"
- Refer Fig. 1-12 and Fig. 1-13.

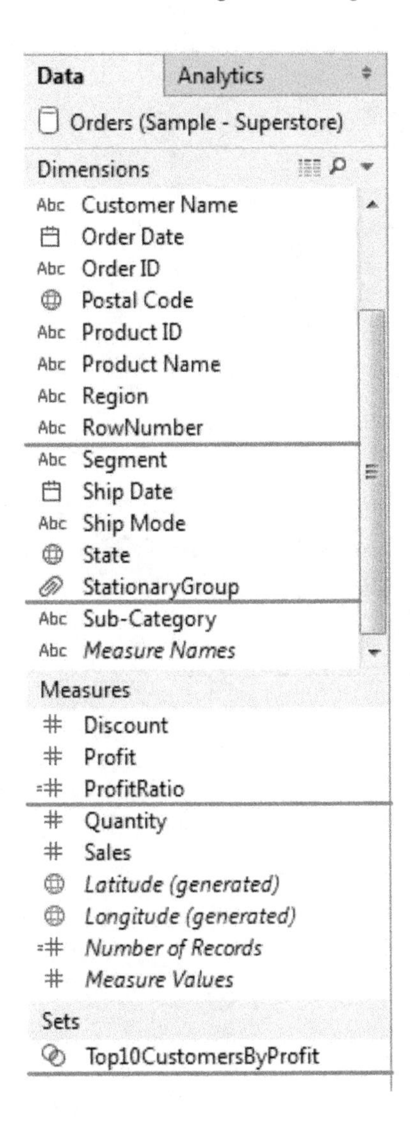

Figure 1-12. *Changes made to the data set in "Sample Superstore.xls"*

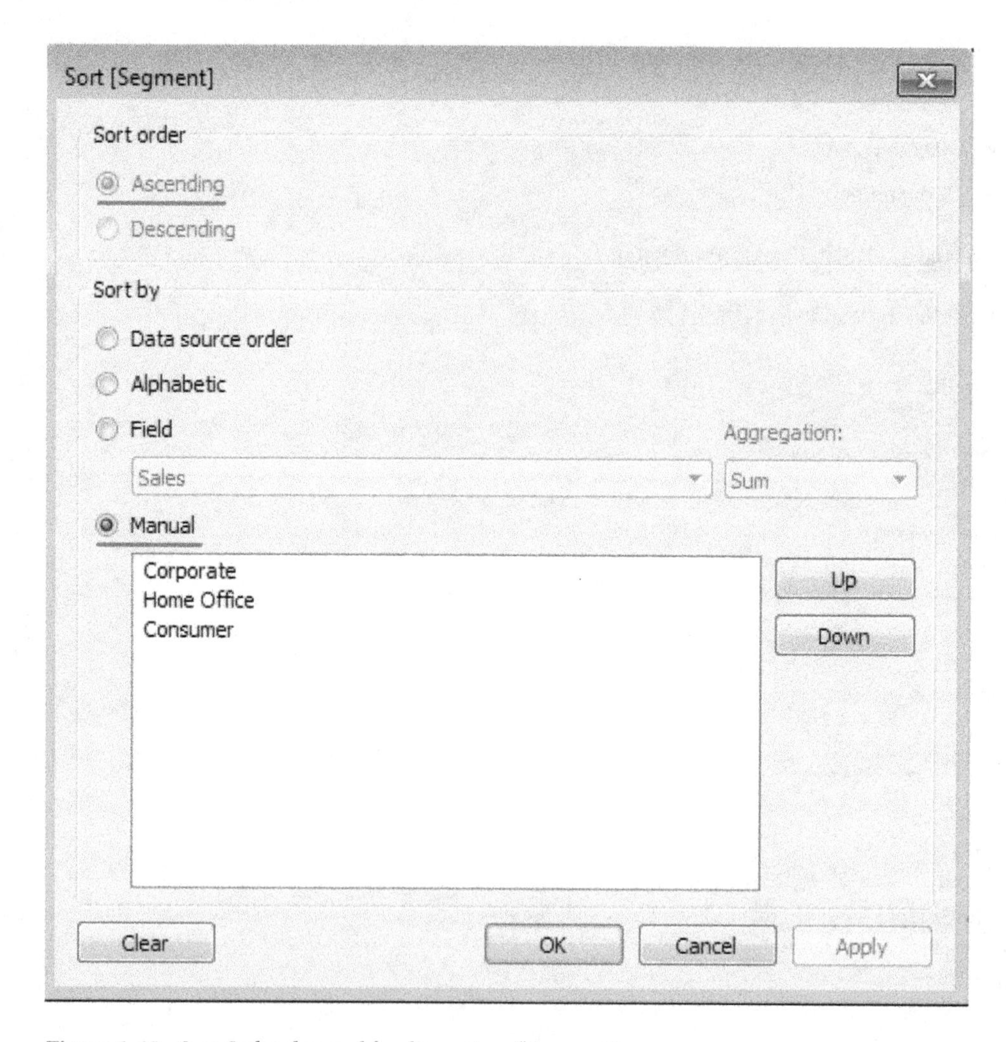

Figure 1-13. *Sort Order changed for dimension, "Segment"*

Save the above changes as .tds

Select Data ➤ Orders(Sample - Superstore) ➤ Add to Saved Data Sources.

Refer to Fig. 1-14 and Fig. 1-15.

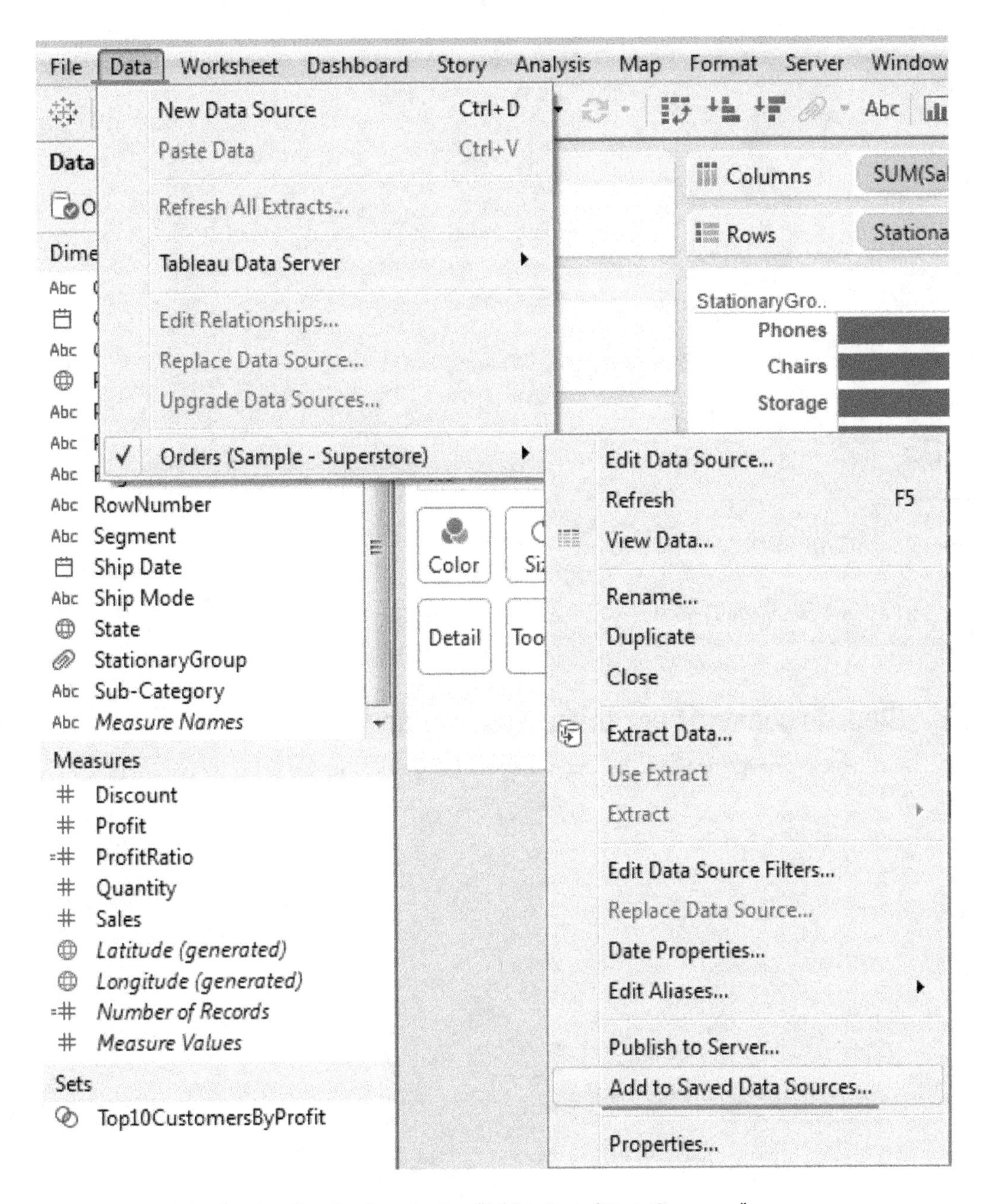

Figure 1-14. *Save the visualization by selecting "Add to Saved Data Sources..."*

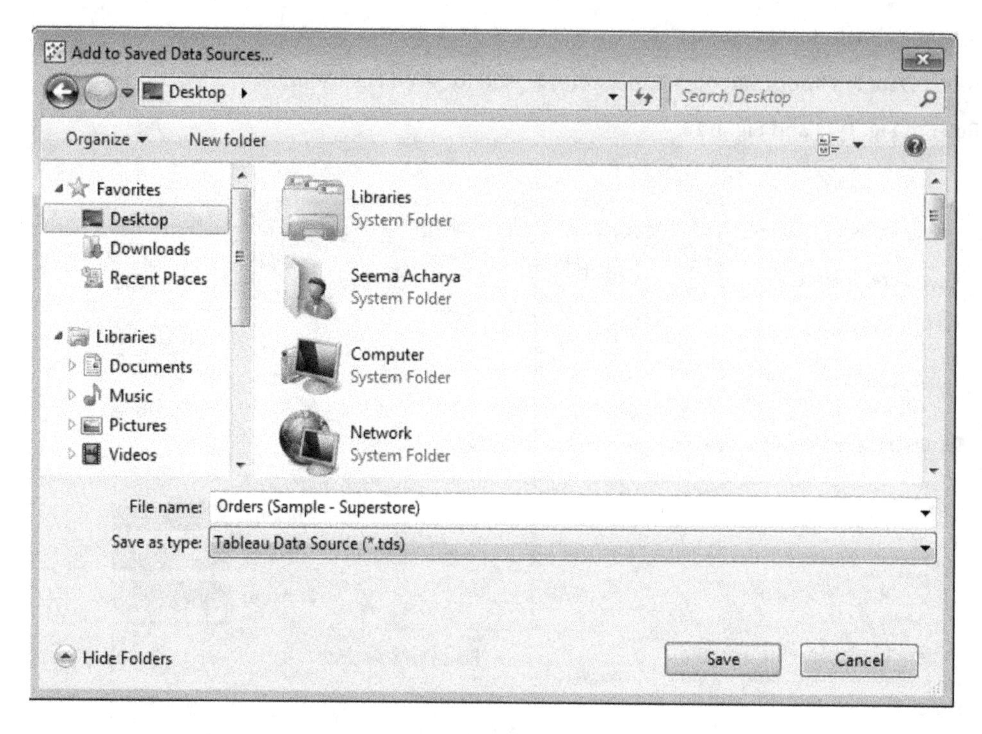

Figure 1-15. *Save the file as "Tableau Data Source (.tds)*

The above connection gets stored by the name, Orders (Sample - Superstore) on the desktop.

1.6.4.1 Steps to connect back to the data source:

- Start Tableau and go to File on the menu bar. Select "Open..." (Shown in Figure 1-16).

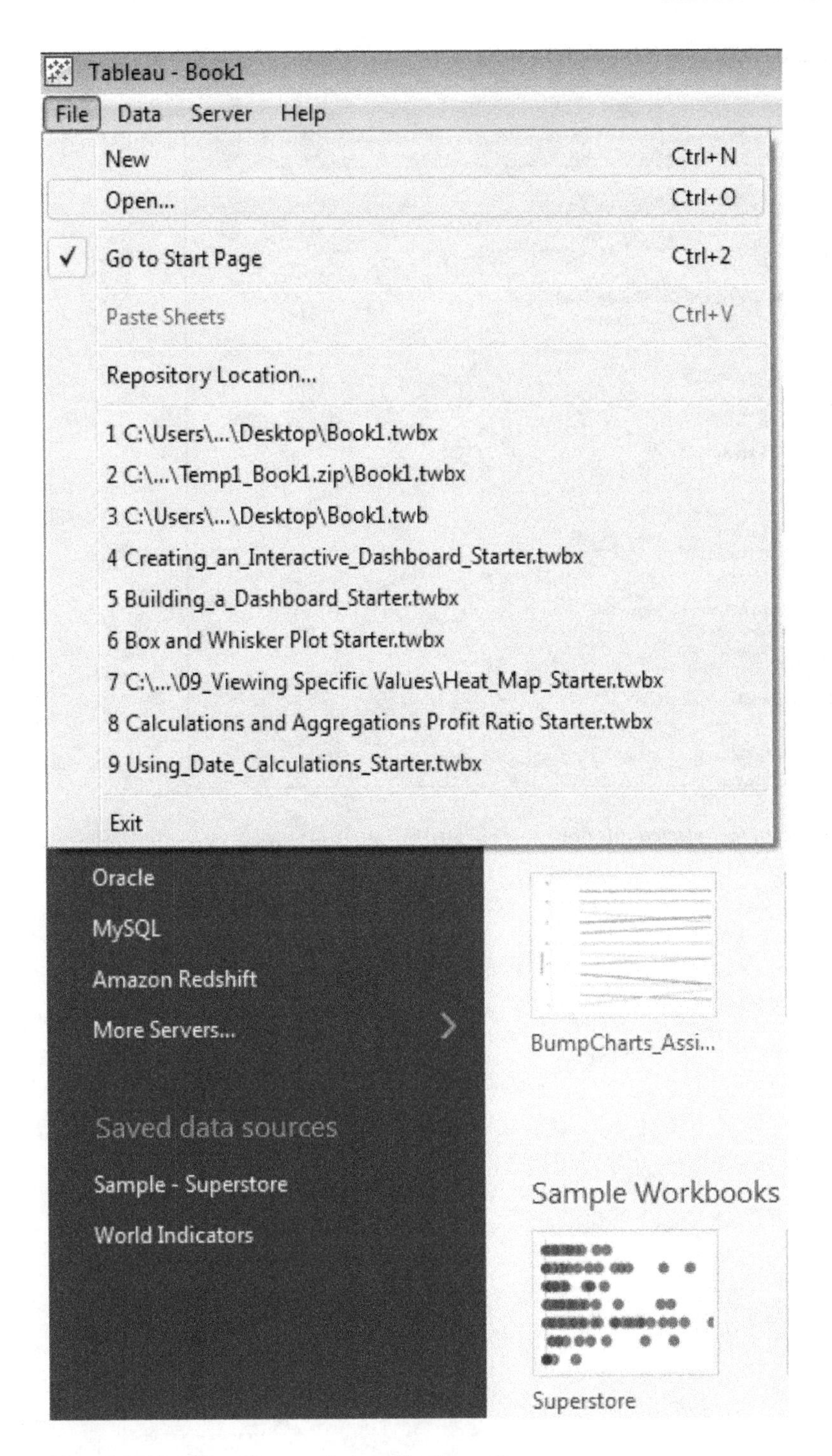

Figure 1-16. *File – Open option on the Menu Bar*

- Select the .tds file and click Open (Shown in Figure 1-17).

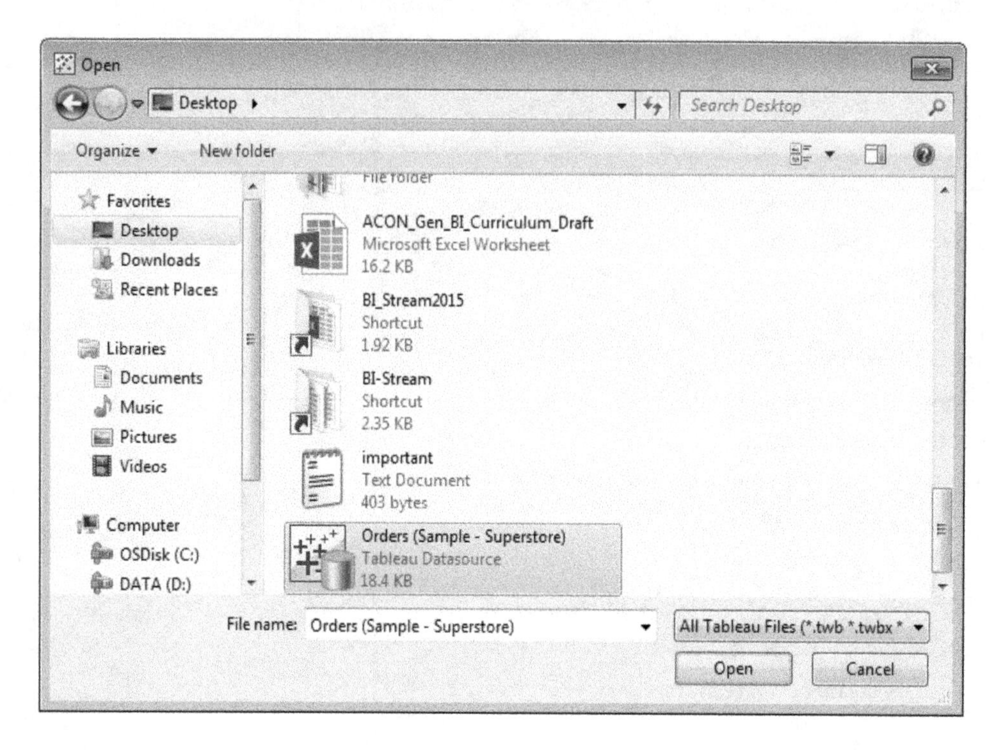

Figure 1-17. *Open dialog box - open the selected .tds file*

- You can see all the changes that you had made to the connection are available as evident from Figure 1-18.

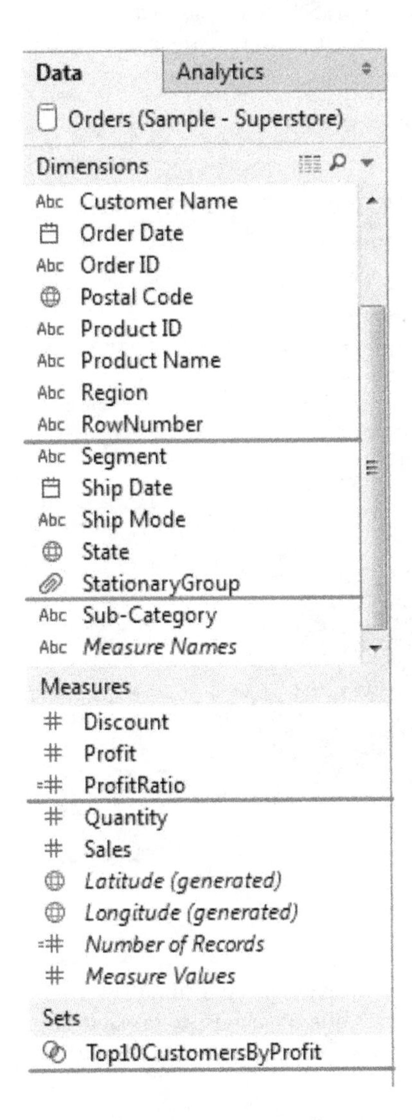

Figure 1-18. *All changes to the metadata of "Sample Superstore" are available*

1.6.5 Tableau bookmark

A single worksheet of a Tableau workbook can be saved as Tableau bookmark. Save a worksheet as a bookmark, if there is a worksheet that you use frequently or if you want to share just a worksheet and not the entire workbook. The bookmark can be accessed from any Tableau bookmark.

To save a Tableau bookmark, go to
Window ➤ Bookmark ➤ Create bookmark (Shown in Figure 1-19).

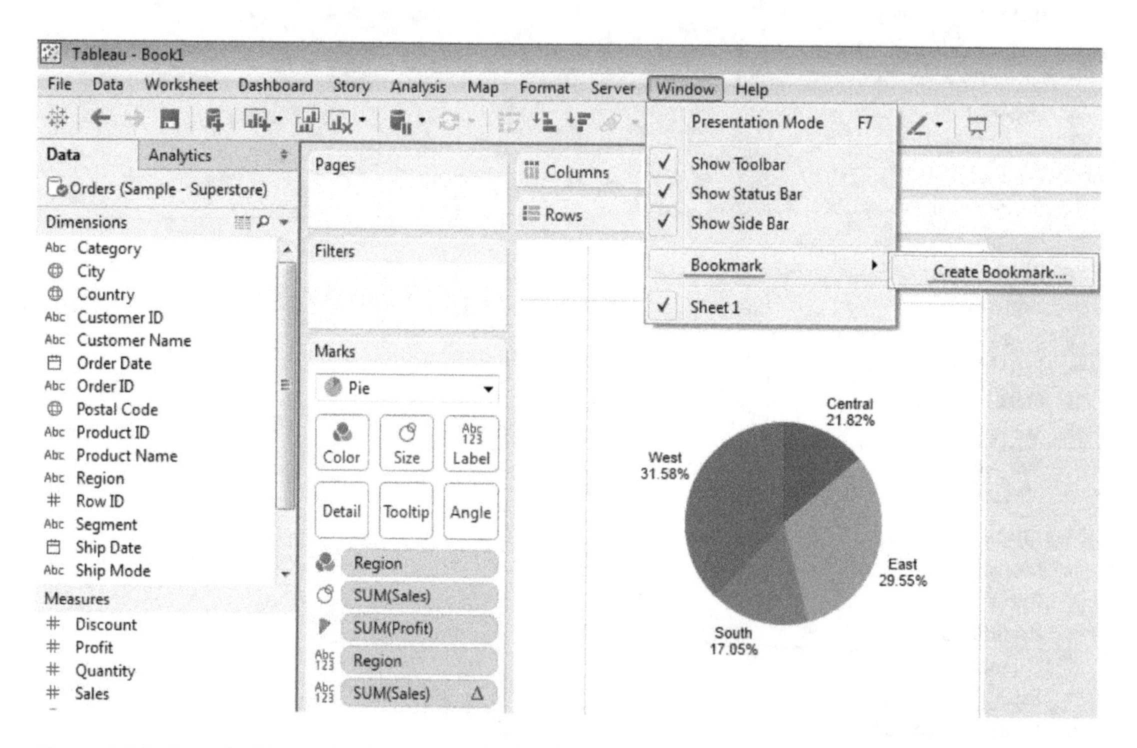

__Figure 1-19.__ Save the file as a bookmark by selecting "Create Bookmark"

Provide the name of the file and the location. The file will be saved with an extension of .tbm. The default location is the bookmarks folder in the Tableau repository. However, one can choose to save it to any location of choice. If it is any location other than the bookmarks folder in the Tableau repository, the file will not show up on the bookmark menu.

Save the bookmark file (.tbm) in the bookmarks folder in the Tableau Repository (Shown in Fig. 1-20).

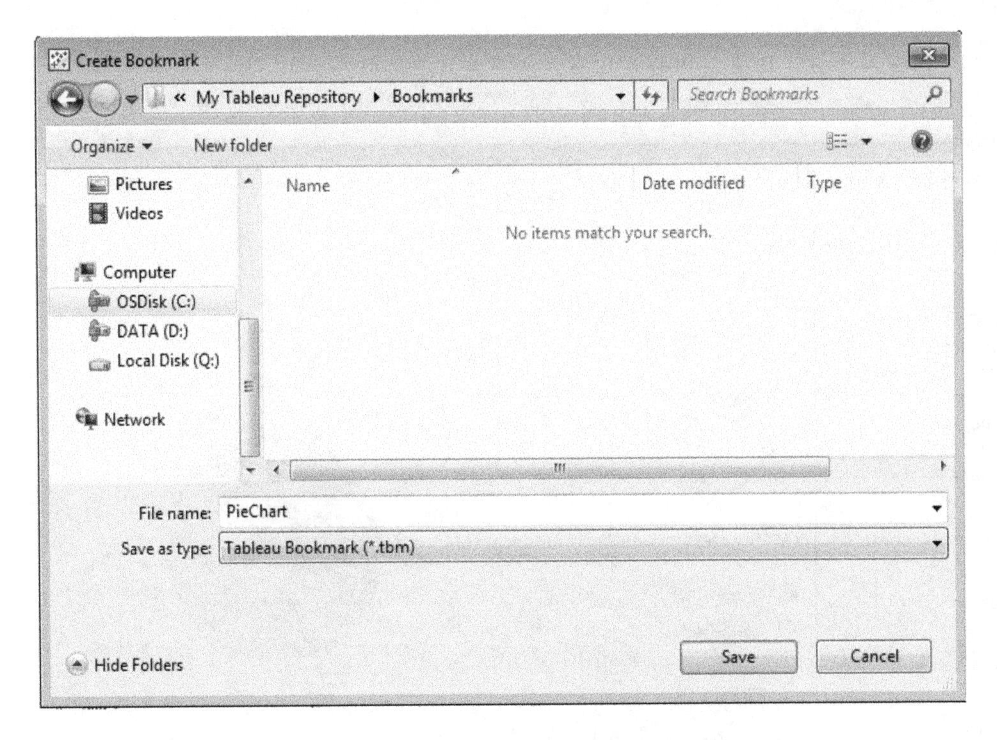

Figure 1-20. *Save the bookmark file in the Tableau Repository*

To insert a bookmark into a new workbook:

(a) Open the workbook.

(b) Go to Window ➤ Bookmark ➤ PieChart (shown in Figure 1-21).

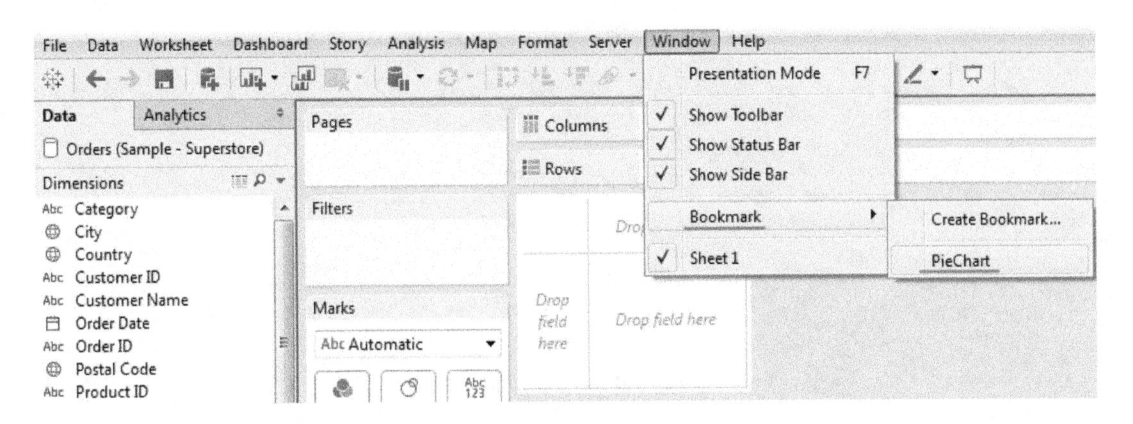

Figure 1-21. *Open the Bookmark in a Tableau workbook*

After insertion of "PieChart.tbm", the worksheet is as shown in Fig. 1-22.

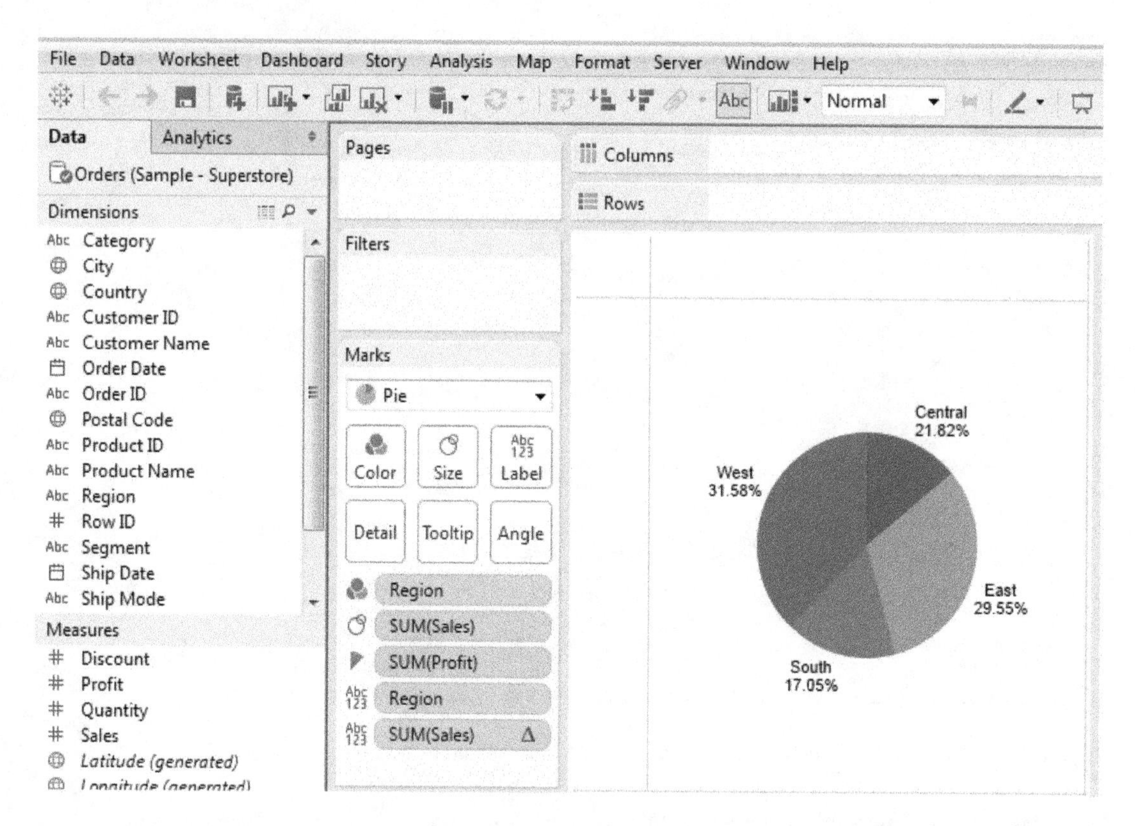

Figure 1-22. Worksheet after inserting "PieChart.tbm" file

1.6.5.1 What is a bookmark file?

It is the snapshot of the current worksheet ONLY. It includes information related to data connection, formatting applied to the current worksheet, calculated fields, groups, etc. However it does not store parameter values, current page settings, etc. It is used as a template to create future workbooks.

You cannot create a bookmark from a dashboard page. E.g.: Pull this worksheet into a dashboard. If we try to bookmark the dashboard page, the option: Window ➤ Bookmark ➤ Create Bookmark appears disabled (Shown in Figure 1-23).

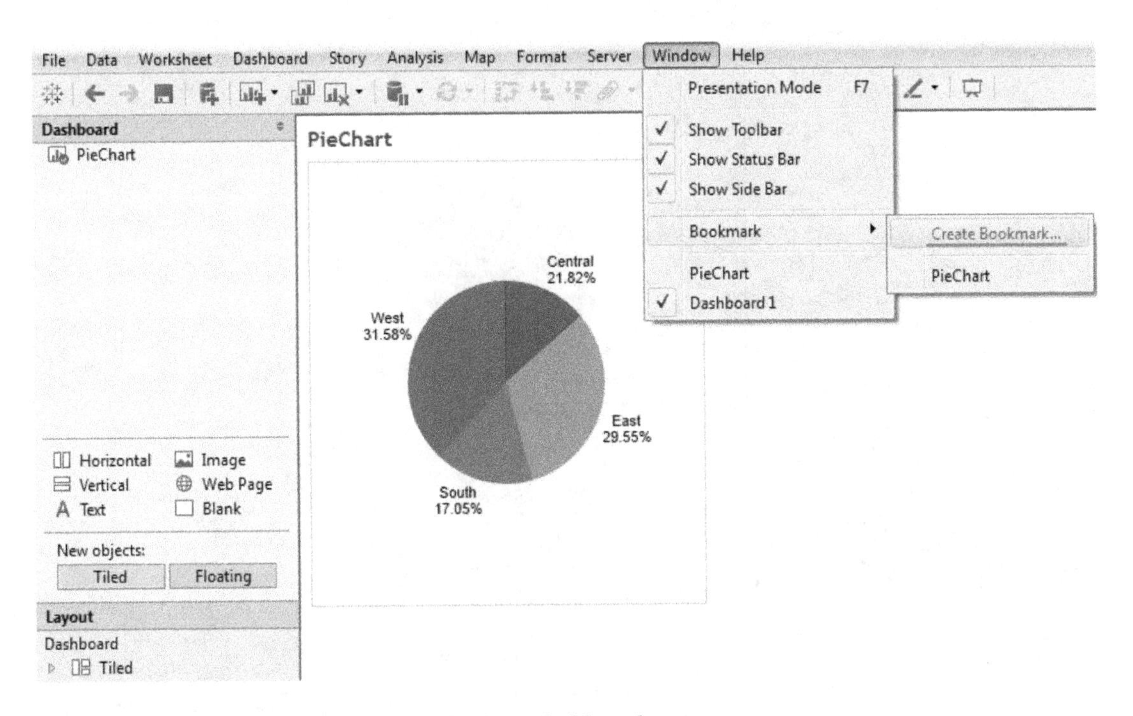

***Figure 1-23.** "Create Bookmark ..." is disabled in a dashboard*

How to then copy a dashboard from one workbook to another workbook?

Open a dashboard sheet in a Tableau workbook. Right click on the dashboard tab at the bottom of the screen. Click on "Copy Sheet" (Shown in Figure 1-24).

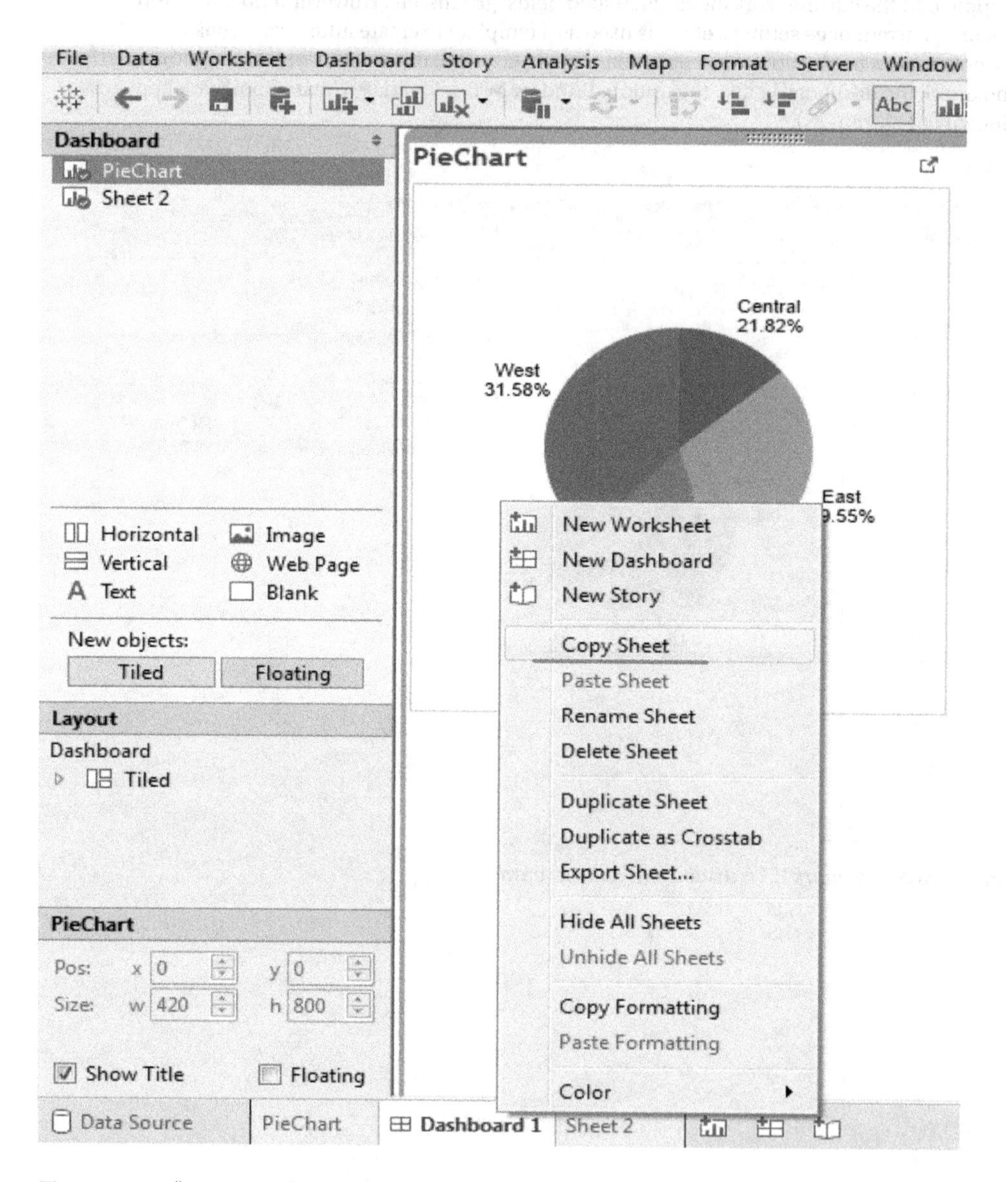

***Figure 1-24.** "Copy Sheet" in Dashboard*

Open the Tableau workbook where you wish to copy the dashboard sheet. Click on File ➤ Paste Sheets (Shown in Figure 1-25).

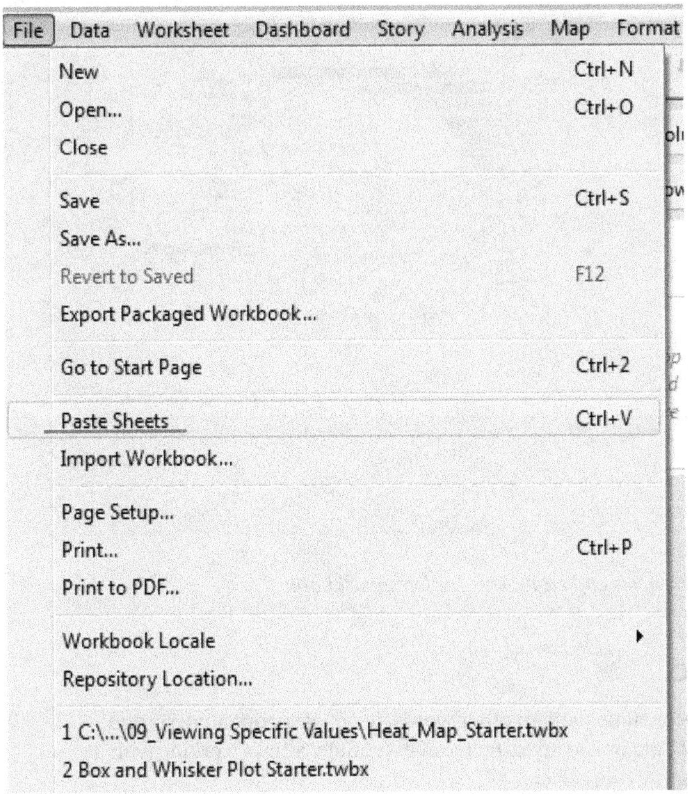

***Figure 1-25.** "Paste Sheets" to copy a dashboard sheet to another Tableau workbook*

All the worksheets along with the dashboard are copied to the new workbook (Shown in Figure 1-26).

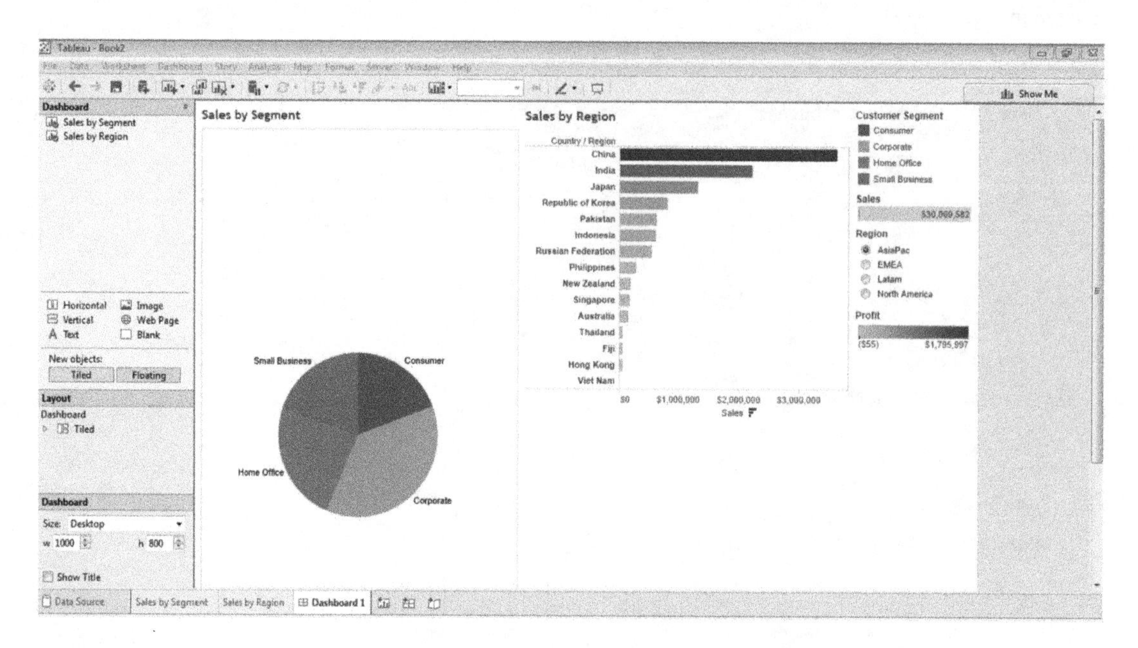

***Figure 1-26.** Dashboard along with worksheets copied to new Tableau workbook*

1.6.6 Tableau data extract

In Tableau, one can either work with live or static data. In other words, it can carry out analysis and visualizations either on the most current data or use an extract that essentially allows working with a snapshot of the results as per the criteria that one selects.

1.6.6.1 When to use live connection and when to work with an extract?

If your analysis is required to show the current state of affairs as it is happening now, you should use a live data connection. However, if your results are required to show data from time frames such as last year or the last quarter, you should work with an extract.

Making a switch from LIVE to EXTRACT means that you will be working with a snapshot of your data.

1.6.6.2 Where is the extract option?

The extract option for connection is available on Tableau's Data Source page. In Fig. 1-27, we have decided to work with an extract; however, we have not set any filters, owing to which all records (9,994) will be selected. However no data analysis or visualizations will be automatically updated when the information in the underlying data source changes.

Figure 1-27. *"Extract" option on Data Source page*

Let us now apply a filter to the extract. Click on the "Edit" next to the "Extract" option. Define the filter condition as shown in Fig. 1-28.

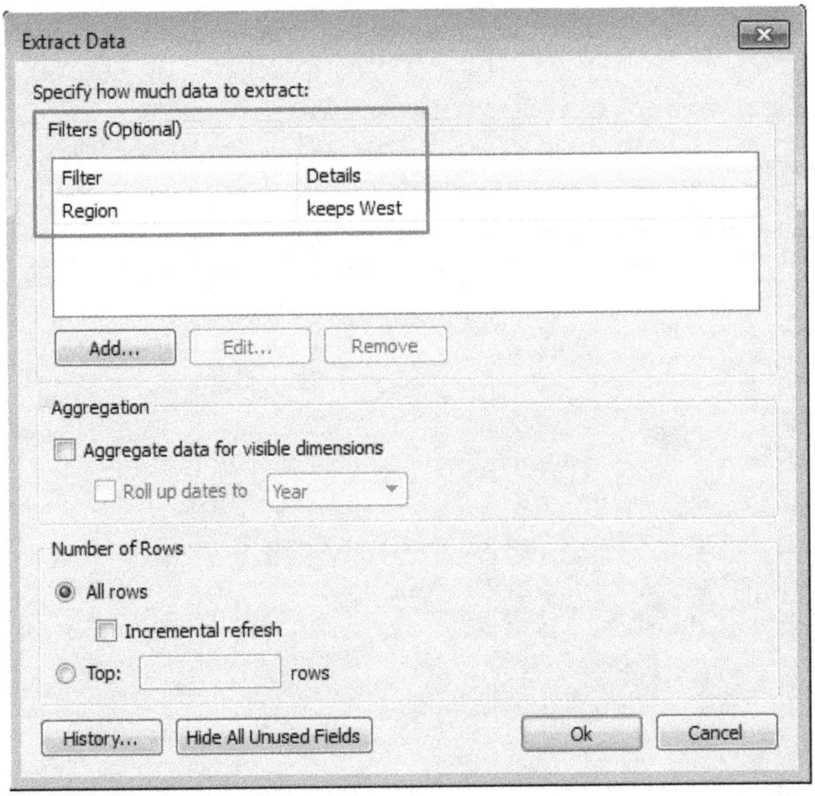

Figure 1-28. *"Filter" condition being set for "Extract" option*

When you proceed to the data sheet after defining the extract, you will be asked to save the extract (Shown in Figure 1-29).

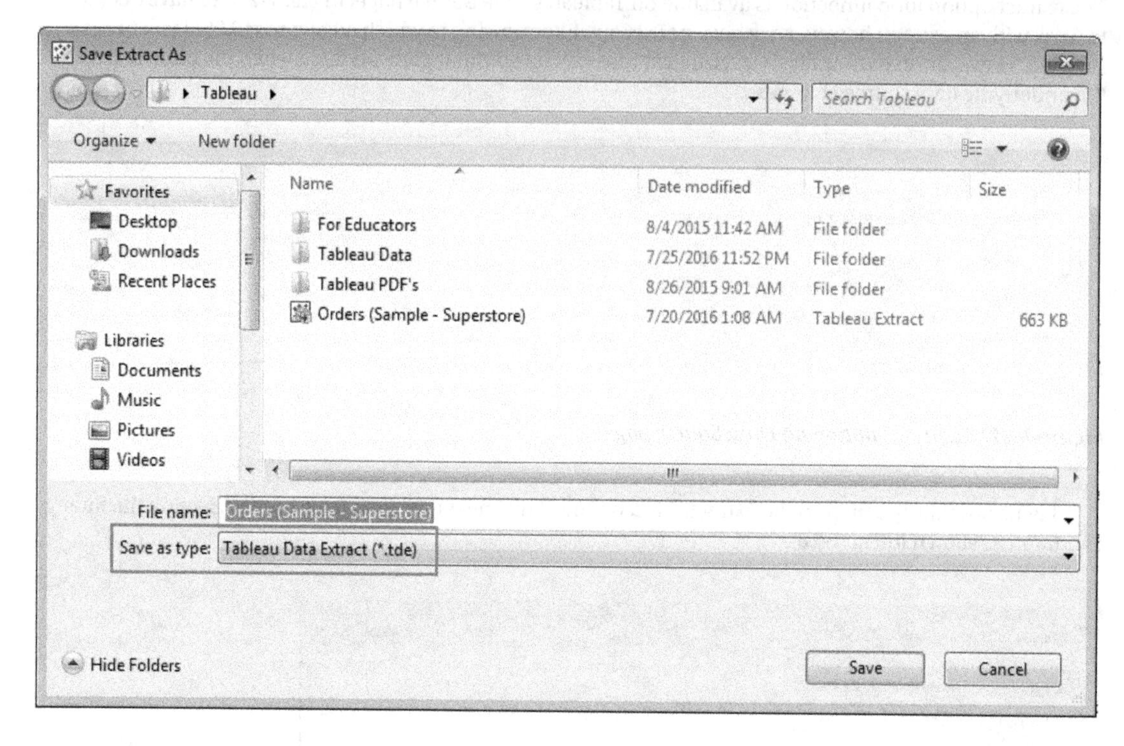

Figure 1-29. *"Save Extract As" dialog box*

Save the extract and continue to the data sheet. As you work with the "Region" dimension on which the filter condition had been defined, you will notice that only the value (s) that qualifies the filter criteria is on display (Shown in Figure 1-30).

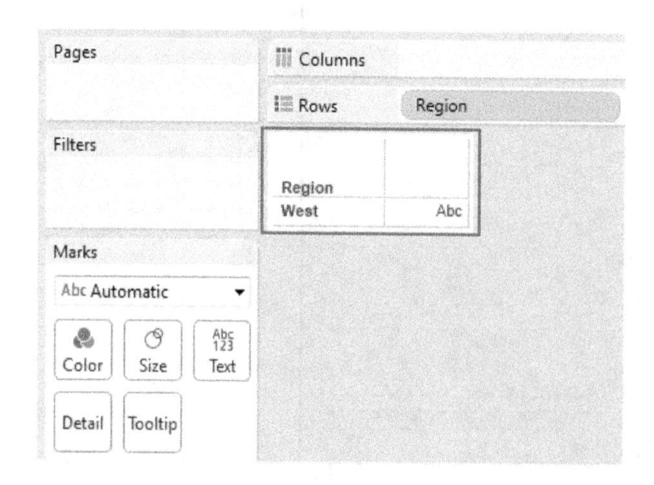

Figure 1-30. *Value(s) that qualify the filter criteria are on display*

In case you further wish to edit the filter condition in the extract or add another filter, right click on the data source in the worksheet/view to bring up the context menu. Select "Extract Data" (Shown in Figure 1-31).

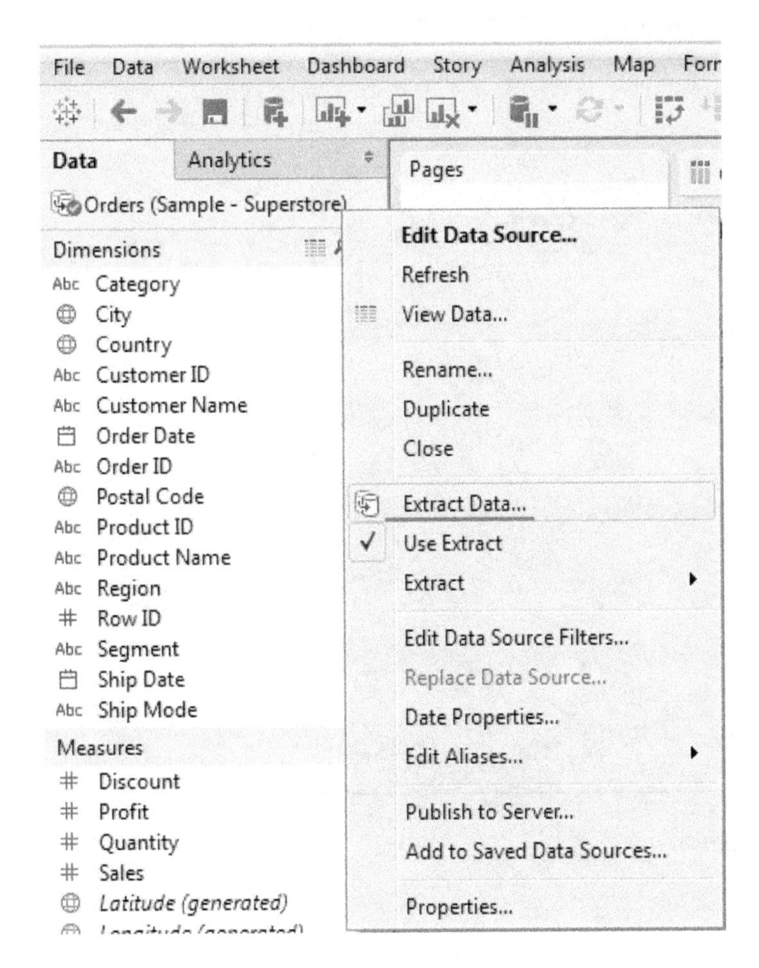

Figure 1-31. *Edit or add another filter to the extract*

Set your filter condition as desired (Shown in Figure 1-32).

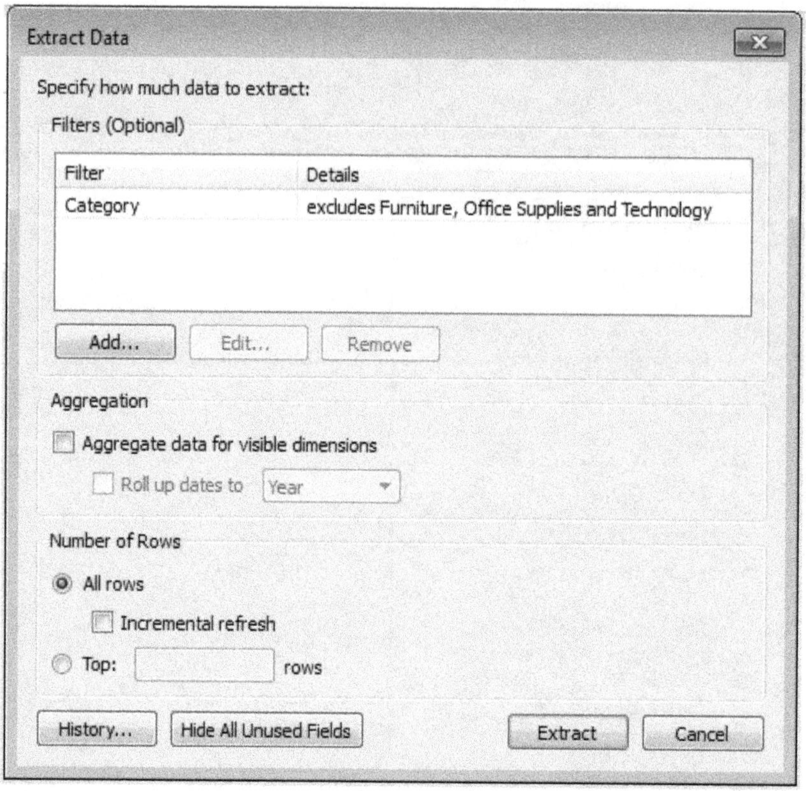

Figure 1-32. *Setting a new "Filter" condition*

To verify and continue working with the extract, perform a quick check as follows:
Right click on the data source in the worksheet / view to bring up the context menu (Shown in Figure 1-33).

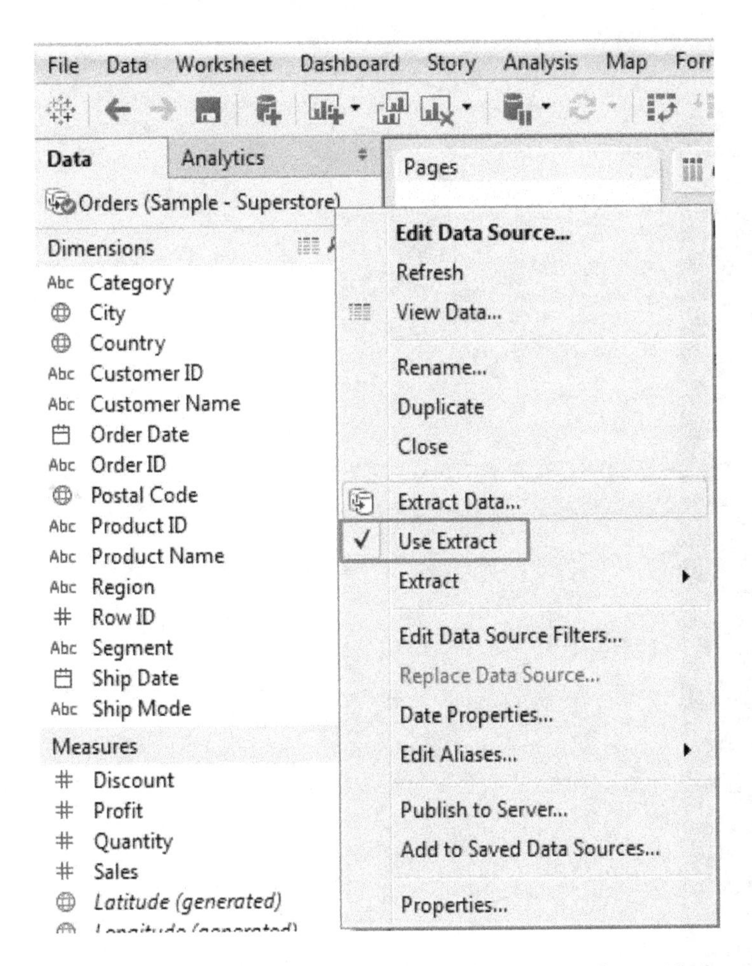

Figure 1-33. *Tick mark next to "Use Extract" implies working with the extract*

Notice the tick mark next to "Use Extract". If you want to stop working with the extract and revert to working with a live data connection, simply click on "Use Extract".

To work with a .tde file, follows the steps below:

Step 1

On the data source connection page, click on "Other Files" (Shown in Figure 1-34).

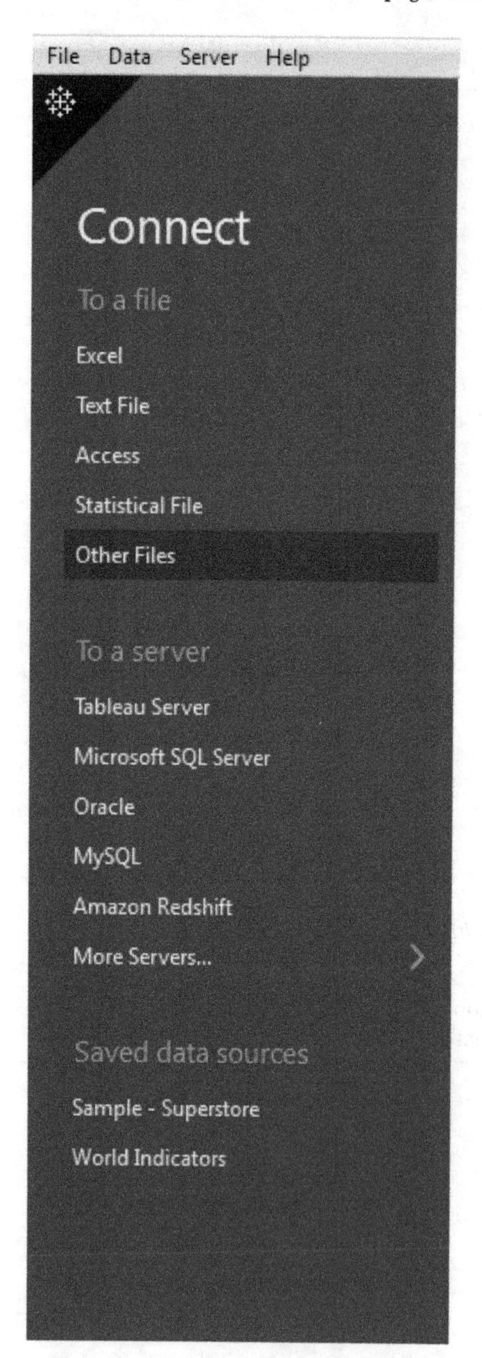

Figure 1-34. *"Other Files" option on Data Source Connection Page*

Step 2

Locate the .tde file (Shown in Figure 1-35).

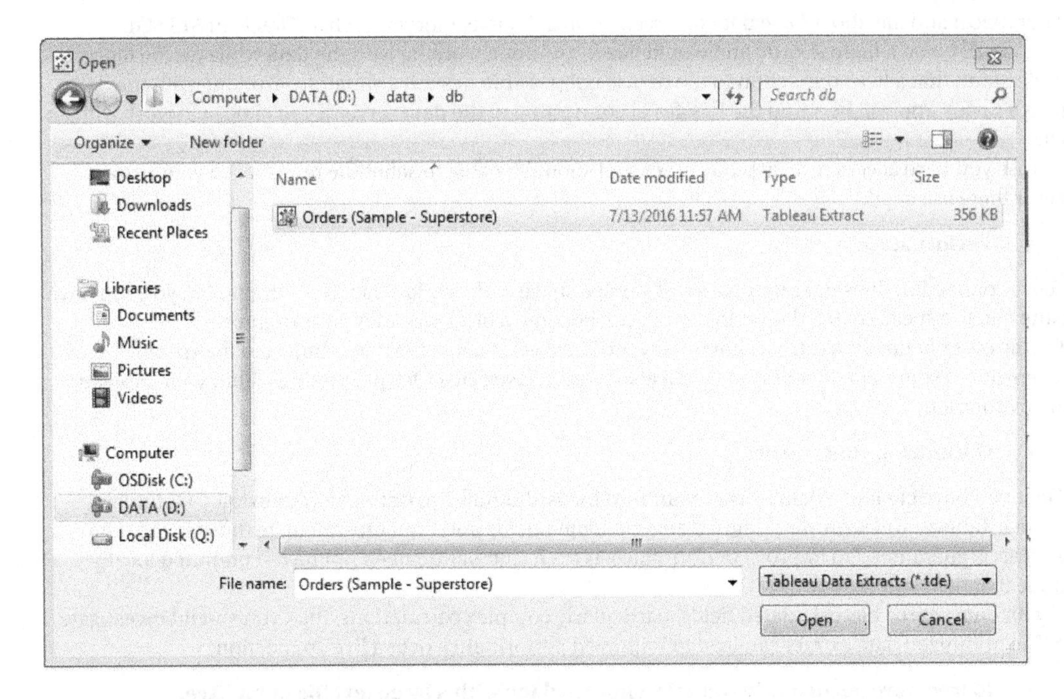

Figure 1-35. *"Open" dialog window to locate and open a .tde file.*

Select the file. Click on Open. The output is as shown in Figure 1-36.

File Data Server Window Help

Orders (Sample - Superstore) Extract

Live Extract

Show aliases Show hidden fields Rows 3,203

	Product ID	Product Name	Profit	Quantity	Region	Row ID	Sales	Segment	Ship Date	Ship Mode	State	Sub-Category
90049	OFF-PA-10002005	Xerox 225	9.33	3	West	5328	19.44	Consumer	1/9/2011	Second Class	California	Paper
94109	FUR-BO-10003034	O'Sullivan Elevations B...	3.93	3	West	4939	334.00	Consumer	1/19/2011	Standard Class	California	Bookcases
94109	OFF-AR-10003514	4009 Highlighters by Sa...	6.57	5	West	4940	19.90	Consumer	1/19/2011	Standard Class	California	Art
94109	OFF-ST-10000078	Tennsco 6- and 18-Co...	238.65	5	West	4938	1,325.85	Consumer	1/19/2011	Standard Class	California	Storage
97477	OFF-AP-10000692	Fellowes Mighty 8 Co...	6.49	4	West	5365	64.86	Corporate	1/22/2011	Second Class	Oregon	Appliances
85254	FUR-BO-10001972	O'Sullivan 4-Shelf Book...	-320.60	5	West	5466	181.47	Consumer	1/21/2011	First Class	Arizona	Bookcases
85254	OFF-BI-10003676	GBC Standard Recycled...	-23.72	10	West	5463	32.34	Consumer	1/21/2011	First Class	Arizona	Binders
85254	OFF-EN-10002504	Tyvek Top-Opening Pe...	36.69	5	West	5465	108.72	Consumer	1/21/2011	First Class	Arizona	Envelopes
85254	OFF-PA-10000659	TOPS Carbonless Recei...	19.62	4	West	5464	56.06	Consumer	1/21/2011	First Class	Arizona	Paper
90049	FUR-FU-10003194	Eldon Expressions Desk...	5.79	2	West	5738	19.30	Consumer	1/26/2011	Standard Class	California	Furnishings
90049	OFF-PA-10002893	Wirebound Service Call...	9.29	2	West	5737	19.36	Consumer	1/26/2011	Standard Class	California	Paper
89115	OFF-PA-10001804	Xerox 195	19.24	6	West	9156	40.08	Consumer	1/29/2011	Standard Class	Nevada	Paper
92037	FUR-TA-10003469	Balt Split Level Comput...	-16.65	3	West	3796	333.00	Consumer	2/3/2011	Standard Class	California	Tables
92037	OFF-AR-10003179	Dixon Ticonderoga Cor...	12.03	4	West	3797	36.44	Consumer	2/3/2011	Standard Class	California	Art

Data Source Sheet1

Figure 1-36. *Data that qualifies the "Filter" condition of "Extract" is displayed*

1.6.6.3 Benefits of working with an extract:

- **Portability**

Your report and dashboard use data that resides on a database server, such as Oracle or MS SQL Server. You will be on a flight shortly and would like to continue working with the data while on the flight. You understand that a live connection to the data is not possible in such a situation. Working with the data extract is a savior. You will be saved the hassles of connecting to the data server if you choose to work with an extract.

In case you have access to a Tableau Server, you should be able to schedule refreshes to your extract whenever there is a need.

- **Performance**

A live connection does not require a lot of storage space within a Tableau deployment, but you will have to ensure that the speed of that data source does not become a bottleneck for performance.

On the other hand, data extracts potentially require a great deal of storage subject to the specific characteristics of your data. In addition, you'll also require faster disks for processing within your Tableau Server environment.

- **Optimizing your extract**

Once you have created a data extract, your next focus should be to optimize your extract. Optimizing implies that Tableau will scan your data for any calculated fields and then pre-calculate those in the data extract. This is guaranteed to improve performance as each calculation need not be re-computed locally each time that field is accessed.

If your data source has calculated fields, particularly complex calculations, then you should investigate to see if an optimized data extract can lead to improved performance over a live connection.

- **Extracts are particularly suitable while working with a large text file or an Excel file**

If your data source is a large text file or an Excel file, you will find immediate benefits from creating a data extract.

Text and Excel data are extracted from the source files and loaded into Tableau high-performance columnar data engine, specifically designed for visual analytics.

- **Filters in the extract help to create a streamlined data subset**

In the **Extract** options under **Data** > **Extract**, you'll be prompted to add **Filters** to your data extract (Shown in Figure 1-37).

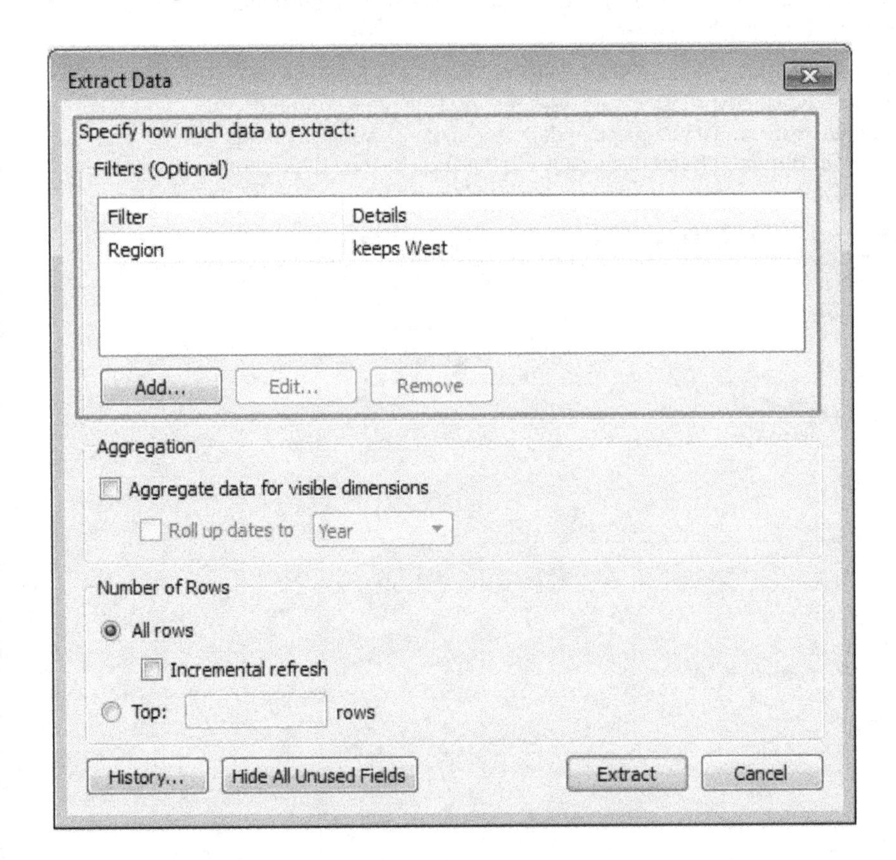

Figure 1-37. *Specify how much data to extract using the "Extract Data" dialog box*

Your data source may be very large; however, you may not need the entire data set to complete the required analysis. Filters help one to zero down to the essential records from a large data source, thus creating a streamlined data sub-set.

Smaller data extracts require less computing power.

- **Incremental extracts**

There are two options to refresh the data in the extract (Shown in Fig. 1-38). They are:

Full extracts: A full extract rewrites the existing data extract in the Tableau data engine with a new file from the data source.

Incremental extracts: Incremental extracts will help to append new records that have been added since the last extract was created. This can be particularly useful if your data extract must be refreshed daily, for instance. You can do this by selecting the **Incremental Refresh** checkbox in the **Data Extract** dialog box.

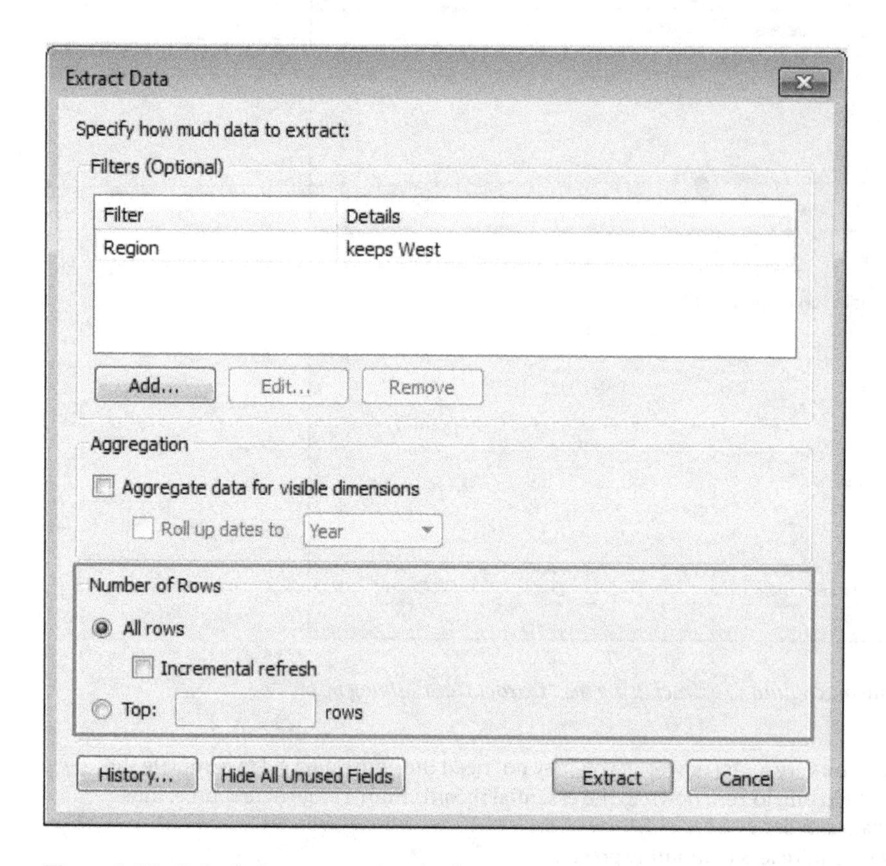

Figure 1-38. *Refresh the extract using the "Extract Data" window*

- ***Creating packaged workbooks***

If you are working with data from SQL Server or Oracle, etc., and need to share the workbook with someone who does not have an installation of Tableau Desktop (Professional or Personal), you will need to package the workbook before sharing it with that person. Data held on remote servers cannot be packaged unless one creates an extract.

- ***Publishing to Tableau Public***

If you wish to publish data to the web, you'll have to use an extract.

- ***Data security***

Imagine you work for a large corporation in the HR division. You are in charge of the employee database. There is some information about employees that you absolutely cannot share. However there are few pieces of information related to employee skills and competencies that can be shared. You have decided to create a packaged workbook using a data extract. The data extract provides you with the feature that "**Hides All Unused Fields**" (Shown in Fig. 1-39). In other words, it removes the dimensions and measures which you have not used in any visualization based on the extract.

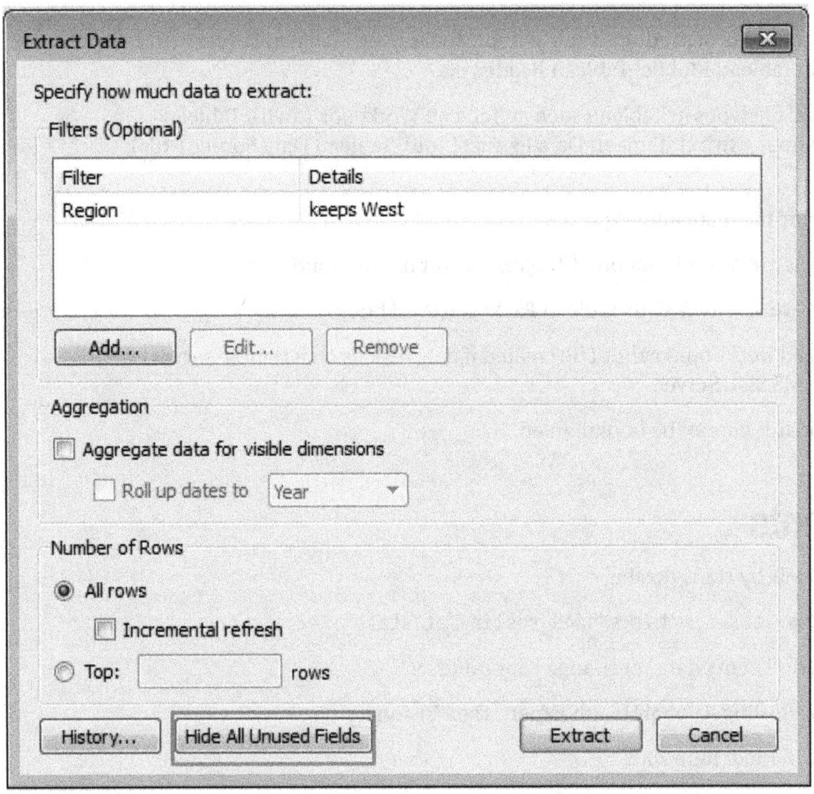

Figure 1-39. *"Hide All Unused Fields" using the "Extract Data" dialog box.*

The extract that you have created can then be safely packaged, reassured that fields that you did not wish to be visible are not even in the extract, and therefore cannot be accessed.

Using this same dialog, one can restrict the ROWS (rather than columns) which are included in your data set by using the filter section.

1.7 Points to remember

- Data visualization is the pictorial or graphical depiction of data.
- Data visualization helps users to ask a question, to get an answer to their question, and then ask follow-up questions as well.
- Data visualization helps to quickly spot trends, see patterns, and grasp difficult concepts.
- Tableau has a rich array of products such as Tableau Desktop, Tableau Server, Tableau Online, Tableau Mobile, Tableau Reader, etc.
- There are several file types in Tableau such as Tableau Workbook (.twb), Tableau Packaged Workbook (.twbx), Tableau Data Extract (.tde), Tableau Data Source (.tds), etc.
- Tableau helps one to create infographics.
- Tableau Online is the hosted version of Tableau Server on the cloud.
- Tableau Reader works ONLY with Tableau Packaged Workbooks.
- Tableau packaged workbooks cannot be created if data resides on a remote server such as Oracle, MS SQL Server.
- Tableau Dashboards cannot be bookmarked.

1.8 Assignments

1. Watch the TED talk by Hans Rosling

   Link: `https://www.ted.com/talks/hans_rosling_at_state`

   Title of the video: "Let my data set change your mindset"

   Presenter: Hans Rosling, Global Health Expert, Data Visionary

   Duration: 19:56, filmed: June 2009

   Why watch this video?

   Watch this video to view an example of amazing use of visualization to represent data in an understandable way.

2. Watch the talk by Hans Rosling

   Link: `http://www.presentationzen.com/presentationzen/2012/01/hans-rosling-the-jedi-master-of-data-visualization-.html`

   Title of the video: "US in a converging world."

   Presenter: Hans Rosling, Global Health Expert, Data Visionary

   Duration: 5:22, a piece on CNN's Fareed Zakaria GPS

Why watch this video?

To learn the focus on the “meaning” rather than on the “data”.

3. Perform the following steps:

   a. Create a folder with the name, “Data” in the “C:” drive. Copy “Sample - Superstore.xlsx” to the “Data” folder in the “C:” drive.

   b. Create a simple visualization based on the data in the “Sample - Superstore.xlsx”.

   c. Store the workbook with the name, “Sample.twb” on the desktop.

   d. Rename the “Sample - Superstore.xlsx” file to “Practice - Superstore.xlsx”.

   e. Try to open the “Sample.twb” in Tableau.

   f. Comment on what happens when you open the “Sample.twb” file and explain the reason behind.

4. Perform the following steps:

   a. Create a folder with the name, “Data” in the “C:” drive. Copy “Sample - Superstore.xlsx” to the “Data” folder in the “C:” drive.

   b. Create a simple visualization based on the data in the “Sample - Superstore.xlsx”.

   c. Store the workbook with the name, “Sample.twbx” on the desktop.

   d. Rename the “Sample - Superstore.xlsx” file to “Practice - Superstore.xlsx”.

   e. Try to open “Sample.twbx” in Tableau.

   f. Comment on what happens when you open the “Sample.twbx” file and explain the reason behind.

5. Perform the following steps:

   a. Start Tableau and Connect to an Oracle database. Select tables with which you wish to work.

   b. Create any simple visualization using data from the selected tables in the Oracle database

   c. Try to create a packaged workbook (.twbx).

   d. Were you able to create a packaged workbook? If yes, why? If no, why not?

6. Read the white paper and write three points of difference between Excel and Tableau.

   Excel: great hammer - lousy screwdriver

   `http://www.tableau.com/learn/whitepapers/excel-great-hammer-lousy-screwdriver`

7. Read up a few customer stories at:

   `https://www.tableau.com/about/customers`

1.9 Next steps

The next chapter will provide detailed steps on sourcing data from varied data sources such as Big Data, NoSQL (MongoDB, Cassandra), R Scripts (.rdata), RDBMSs (MS SQL Server, MySQL, MS Access DB), etc. besides connecting to spreadsheets (Excel), text files(.txt), etc.

CHAPTER 2

Working with Single and Multiple Data Sources

"Most of us need to listen to the music to understand how beautiful it is. But often that's how we present statistics: we just show the notes, we don't play the music."

— Hans Rosling, co-founder and chairman of the Gapminder Foundation, who developed the Trendalyzer software system

Chapter 1 familiarized us with the basic concepts of visualization, the need and significance of visualization, the features of Tableau, the Tableau product line and the various file formats in Tableau. This chapter will help us to understand how to work with single and multiple data sources in Tableau. We will explore the following:

- Desktop architecture
- Tableau environment
- Connect to a file
- Connect to a server
- Metadata grid
- Joins
- Custom SQL
- Data blending
- Data extracts

Let us start with the desktop architecture.

2.1 Desktop architecture

Tableau architecture is based on an n-tier client server architecture (Shown in Fig. 2-1.) Tableau serves as a desktop installed software, web client and mobile client. Tableau Desktop is an authoring and publishing tool. It is used to create shared views on Tableau Server. Tableau offers a scalable solution to create and deliver desktop, web and mobile analytics. Tableau Desktop allows one to explore data and share insights.

S. Acharya and S. Chellappan, *Pro Tableau*, DOI 10.1007/978-1-4842-2352-9_2

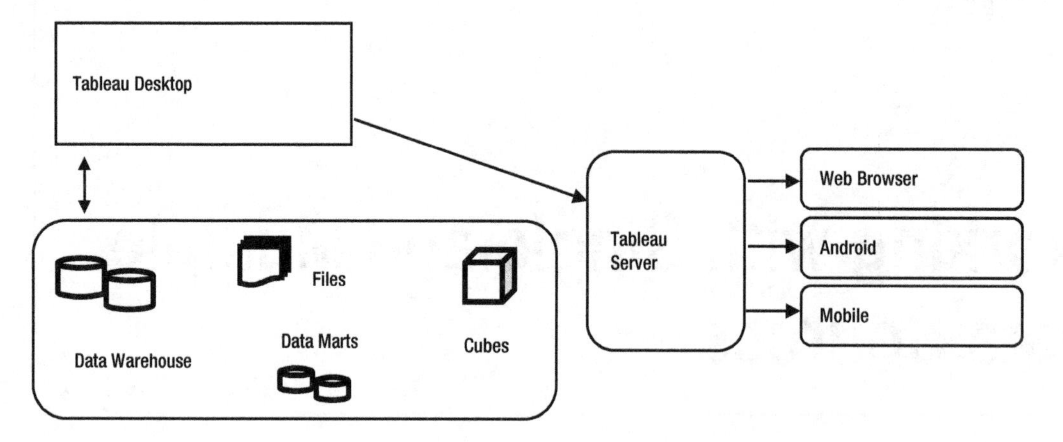

Figure 2-1. *Tableau Desktop Architecture*

Let us discuss the various layers of Tableau desktop architecture in brief.

2.1.1 Data layer

The bedrock of Tableau is its data layer. Tableau allows you to work with a heterogeneous data environment. You can work with databases, servers, data warehouses, cubes and flat files such as Excel, Access, etc. In Tableau, it is not necessary to bring all your data into memory unless it is required. Tableau allows you to leverage your existing environment by applying the database features to answer your questions.

2.1.2 Data connectors

Tableau provides various data connectors to work with databases such as Microsoft SQL Server, Oracle, Teradata, Vertica, Cloudera Hadoop, and many more. In addition to this there are generic ODBC connectors to connect to any system without having a native connector. In Tableau, there are two modes to interact with data: (i) live connection (ii) in-memory. Tableau users can switch between these two with ease.

2.1.3 Live connection

Tableau's data connectors allows you to leverage your existing data infrastructure. This is done by sending dynamic SQL or MDX statements to the source database directly instead of importing all the data. It means, if you have invested in a fast, analytics-optimized database like Vertica, you can get the benefits of those by connecting live to your data. This leaves the detail data in the source system and send the aggregate results of queries to Tableau. Tableau can also utilize unlimited amount of data. Tableau is the front-end analytics client to many of the largest databases in the world. Each connector is optimized to take the unique characteristics of each data source.

2.1.4 In-memory

Tableau has a fast, in-memory data engine for analytics. Tableau allows you to connect your data with one click, extract and bring it in memory. Tableau's data engine fully utilizes the entire system to achieve fast query response on hundreds of millions of rows of data on commodity hardware. Because the data engine can access disk storage as well as RAM and cache memory, it is not limited by the amount of memory on a system. There is no requirement that an entire data set be loaded into memory to achieve its performance goals.

In-memory is ideal when:

- your database is too slow for interactive analytics.
- you need to take load off a transactional database.
- you need to be offline and can't connect to your data live.

But live connections can be preferable when:

- you have a fast database, like Vertica, Teradata, or another analytics-optimized database.
- you need up-to-the minute data.

Refer to link below to learn about in memory and live data: Which is better?

http://www.tableau.com/learn/whitepapers/memory-or-live-data

2.2 Tableau environment

Let us try to understand the Tableau environment.

2.2.1 To open

Double click the Tableau icon on the desktop (Shown in Fig. 2-2).

Figure 2-2. *Tableau shortcut icon*

2.2.2 To close

Click on close button on the right side of the application (Shown in Fig. 2-3).

Figure 2-3. *Close button*

Next, we will learn about the start page.

2.2.3 Start page

The start page is a central location to help connect to data sources, access recent work books and explore tutorials provided by the Tableau community (Shown in Fig. 2-4).

Figure 2-4. *The Tableau start page*

There are three panes in the start page.

1. Connect: Using Connect, you can connect to various data sources such as connect to a file and connect to a server. Also you can open the saved data sources. "Sample - Superstore" is the default saved data source that comes with the Tableau Desktop Edition. Refer to Fig. 2-5.

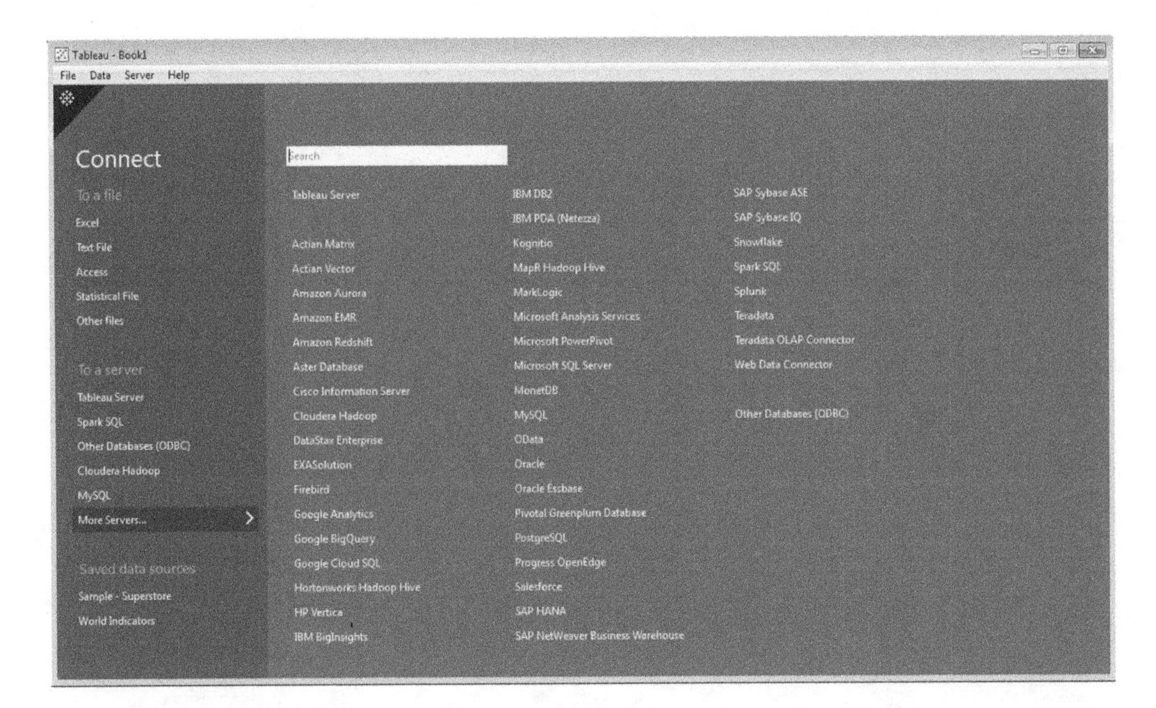

Figure 2-5. *Connect page*

2. Open: When you open Tableau for the first time, this pane will be empty. As you start creating workbooks, you can see the most recently opened workbooks in this pane. You can also open sample workbooks to explore the functionality of Tableau. You can also pin workbooks to the start page by clicking the pin icon that appears in the top-left corner of the workbook thumbnail. Pinned workbooks always appear on the start page, even if they weren't opened recently. To remove a recently opened or pinned workbook, hover over the workbook thumbnail, and then click on the "x" that appears. The workbook thumbnail is removed immediately but will show again with your most recently used workbooks the next time you open Tableau Desktop. Refer to Fig. 2-6.

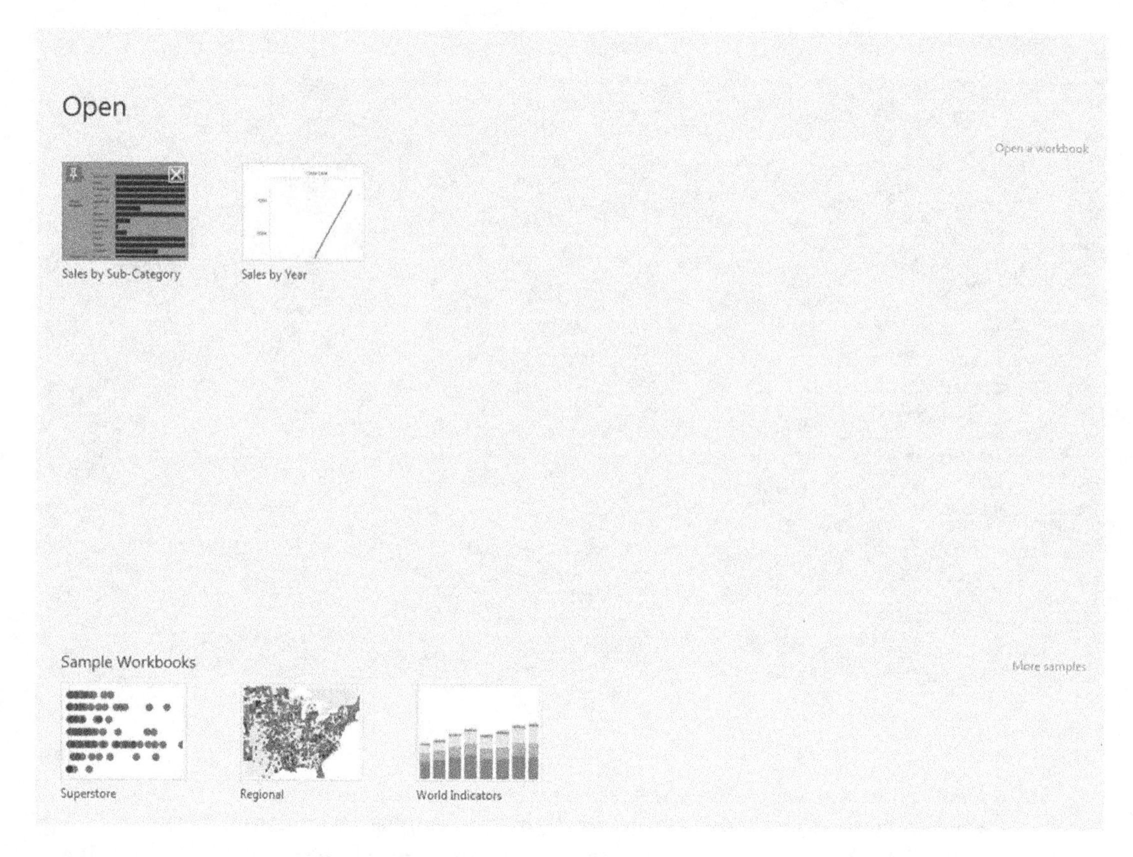

***Figure 2-6.** Open page showing pin option*

3. Discover: You can view details about training provided by Tableau, blogs, conferences and references, etc. Refer to Fig. 2-7.

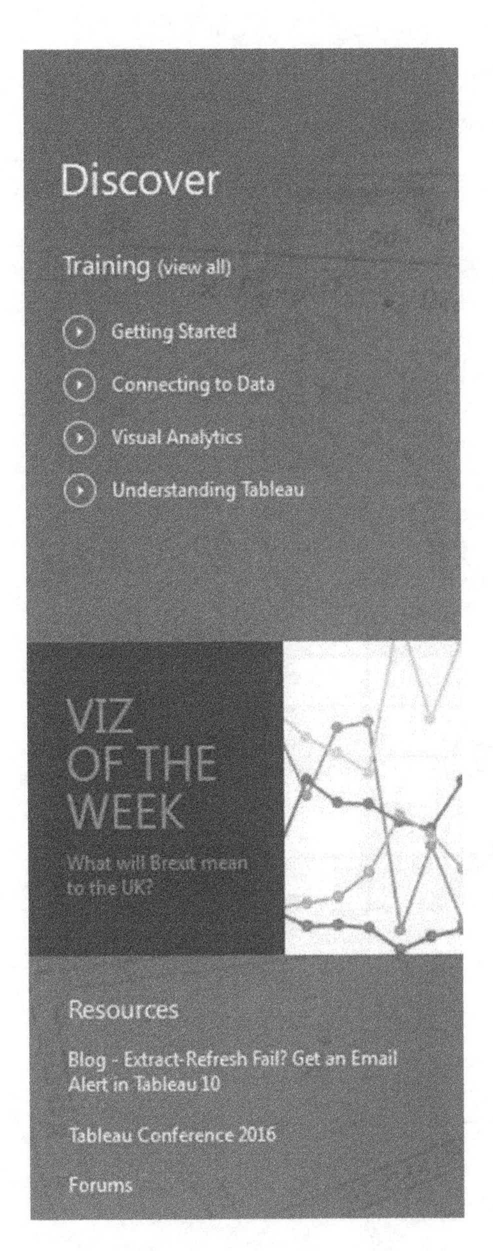

Figure 2-7. "Discover" Page

2.2.4 Data Source Page

You need to establish an initial connection to your data to get the data source page. You can follow the steps below to connect to an Excel file (Sample - Superstore).

Follow these steps:

2.2.4.1 Step 1

On the start page, under "Connect", select To a File ➤ Excel (Shown in Fig. 2-8).

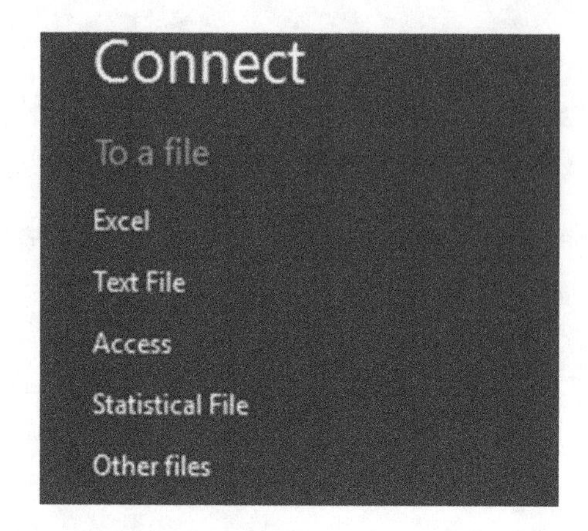

Figure 2-8. *"Connect" Section*

2.2.4.2 Step 2

"Open" dialog box will be opened. Navigate to the folder where Sample-Superstore excel file is present. In Tableau Desktop, the default path is "C:\Users\Username\Documents\My Tableau Repository\Datasources\9.3\en_US-US" (Shown in Fig. 2-9).

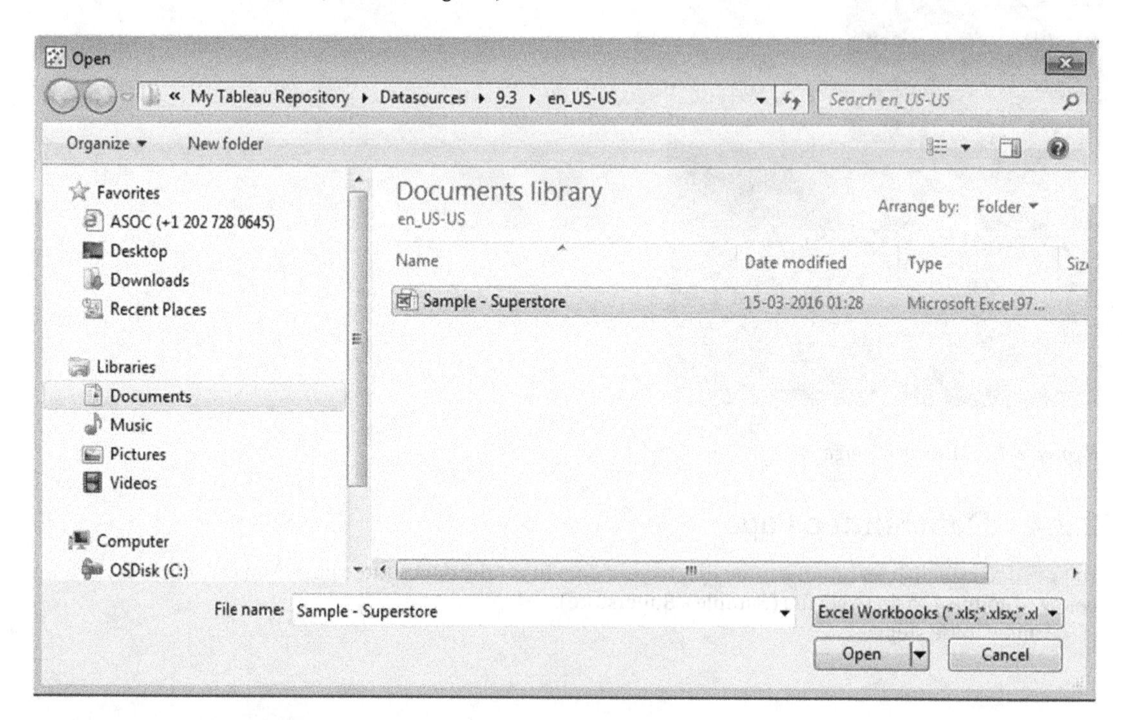

Figure 2-9. *"Open" dialog box to open "Sample - Superstore"*

2.2.4.3 Step 3

When you click on "Open" button, you can see "Processing Request" window as shown in Fig. 2-10.

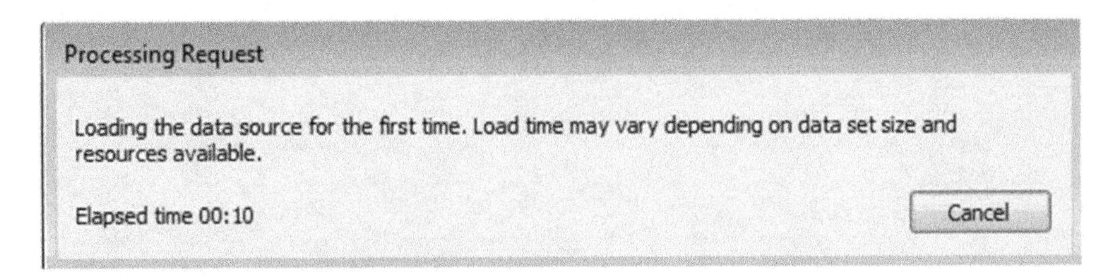

Figure 2-10. *Processing request window*

2.2.4.4 Step 4

Now, you will be able to view the "Data Source Page". (Shown in Fig. 2-11).

Figure 2-11. *Data source page*

2.2.4.5 Step 5

Drag "Orders" sheet from the left pane to the canvas area as shown in Fig. 2-12. You will be able to preview the data as well.

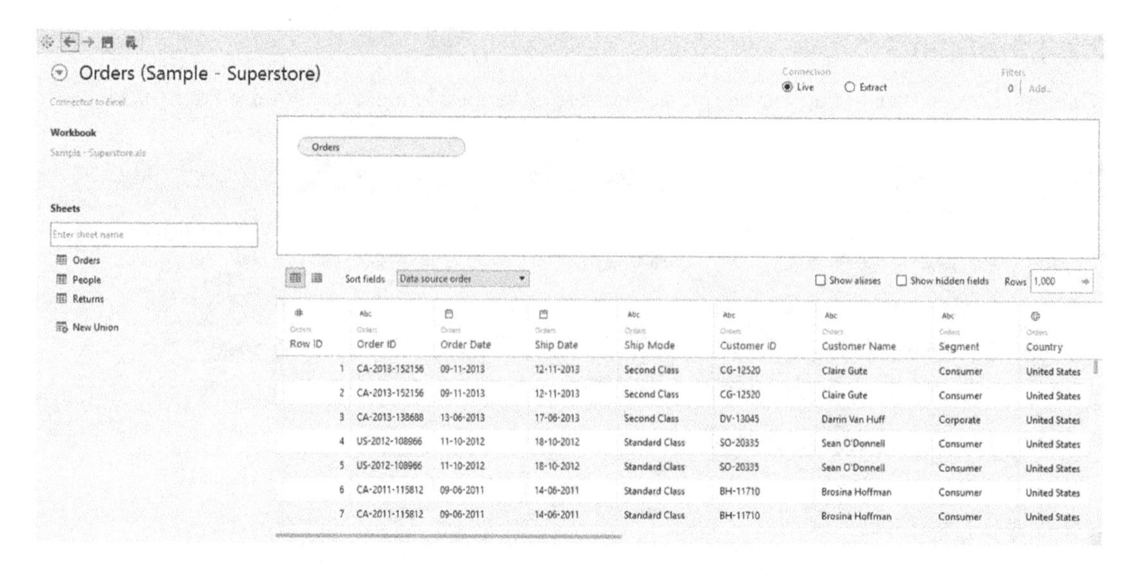

Figure 2-12. *"Orders" sheet placed on the canvas area*

There are four main areas in the Data Source Page. They are (i) left pane, (ii) canvas, (iii) grid, and (iv) metadata grid (Shown in Fig. 2-13).

1. Left pane: displays details about the data to which you are connected. For example, for file-based data, the file name and worksheets will be displayed.
2. Canvas: allows you to drag and drop one or more tables to the canvas area to set up your data source.
3. Grid: allows you to review first 1,000 rows of data that is present in your data source. It also allows you to modify your data source like renaming field names, sorting, creating a field, etc.
4. Metadata grid: allows one to click on the metadata grid to display fields in your data source.

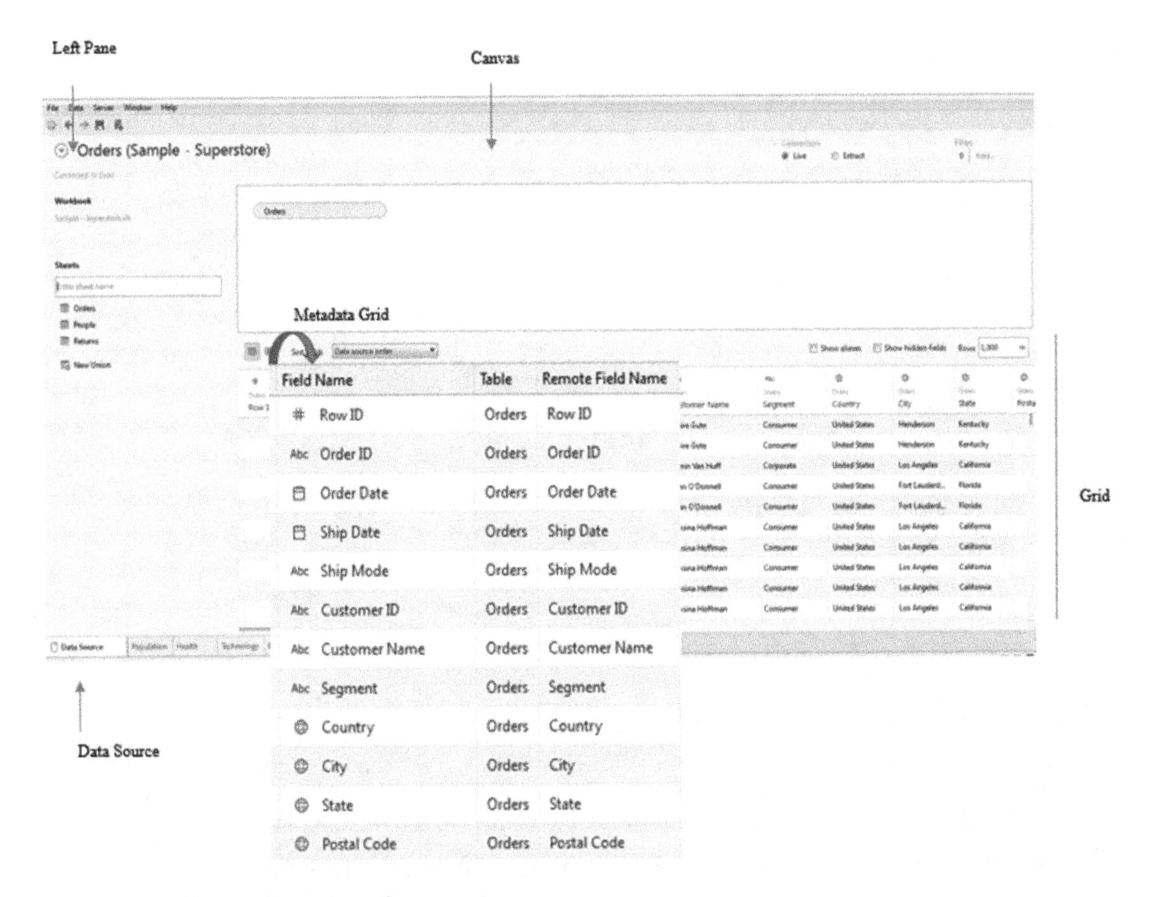

Figure 2-13. *Four main areas on data source page*

Tableau provides two types of connections:

- Live: This allows extracting data in real time.
- Extract: This is about extracting data at regular frequencies.

2.2.5 Workspace

Workspace contains data pane, cards and shelves, and one or more sheets. These sheets can either be worksheets, stories or dashboards. Cards and shelves can be used to build views (Shown in Fig. 2-14).

The workspace for creating a story is quite different from the workspace for creating a dashboard.

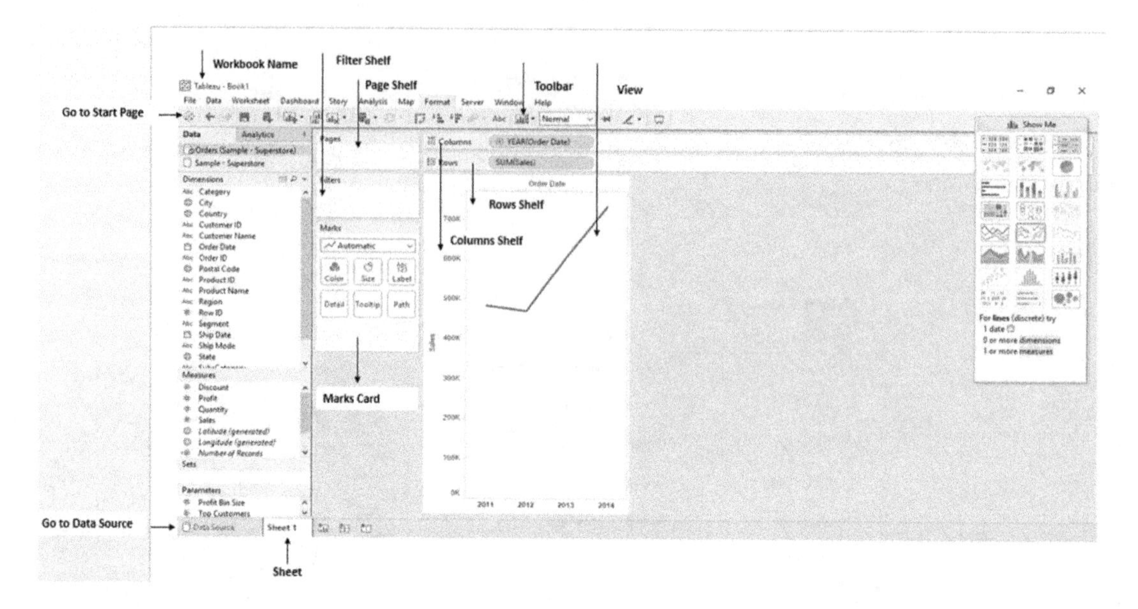

Figure 2-14. *Tableau Workspace*

2.2.6 Workbooks and Sheets

A Tableau workbook is quite similar to a Microsoft Excel workbook. A workbook can contain one or more sheets. These sheet can be a worksheet(s), a story or a dashboard(s). A workbook is a container for all your work. Workbook will help you to organize, perform analysis, and save and share your results.

- Worksheet represents single view with data pane, cards, shelves, and legends.
- Dashboard represents a collection of views from multiple worksheets.
- Story represents a sequence of worksheets or dashboards to convey certain information.

2.2.7 Visual Cues and Icons in Tableau

Let us discuss few visual cues and icons in Tableau.

2.2.7.1 Data Sources in Data Pane

Refer to Table 2-1 for data source icons.

Table 2-1. *Data Source Icons*

Visual Cue	Description
	Primary Data Source.
	Secondary Data Source.

2.2.7.2 Fields in Data Pane

Let us discuss the icons displayed in the data pane.

- Discrete field is indicated by the "blue" color.
- Continuous field is indicated by the "green" color.
- User-defined function is indicated by an equal sign.

Refer to Table 2-2 for icons in data pane.

Table 2-2. *Icons in Data Pane*

Visual Cue	Description
Abc =Abc	Text values.
# =#	Numeric values.
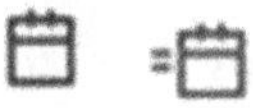	ONLY date values.
	Both date and time values.
	Geographical data.
	User-defined set.
	Numeric bin.
	Group.
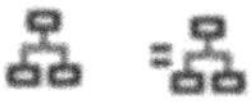	Relational hierarchy.
	The field is blended with a field from another data source.
	The field is not blended with a field from another data source.

2.2.7.3 Sheets in the Dashboard and Worksheet Pane

Let us discuss the sheet(s) which are used in a story.

A blue check mark indicates that sheet is being used in one or more stories. Refer to Table 2-3.

Table 2-3. *Sheets in the Dashboard and Worksheet Pane*

Visual Cue	Description
	Worksheet.
	Dashboard.

Refer to the link below to learn more about the visual cues and icons in Tableau.

`https://onlinehelp.tableau.com/current/pro/desktop/en-us/tips_visualcues.html`

2.3 Connect to a File

Let us explore how to connect to the below-mentioned files.

- Text
- Microsoft Access
- R data file

2.3.1 Connect to a Text File

Follow the steps to connect to a text file.

2.3.1.1 Steps to connect to a text file

2.3.1.1.1 Step 1

On the start page, under Connect, select To a File ➤ Text File (Shown in Fig. 2-15)

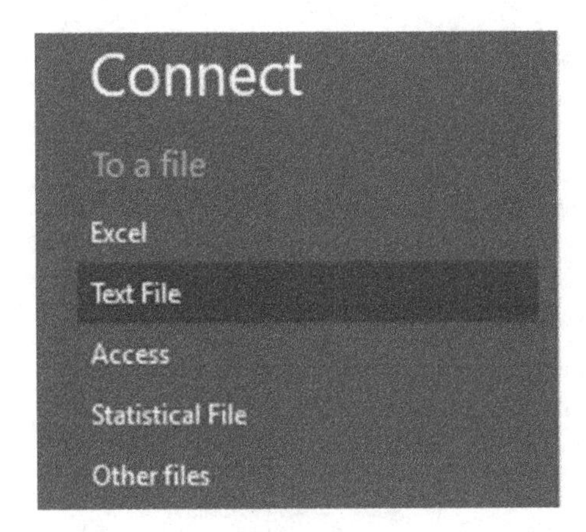

Figure 2-15. *Connect to a text file*

2.3.1.1.2 Step 2

"Open" dialog box will show up. Navigate to the folder where "Sample-Superstore" text file is present. Select the file and click on the "Open" button (Shown in Fig. 2-16).

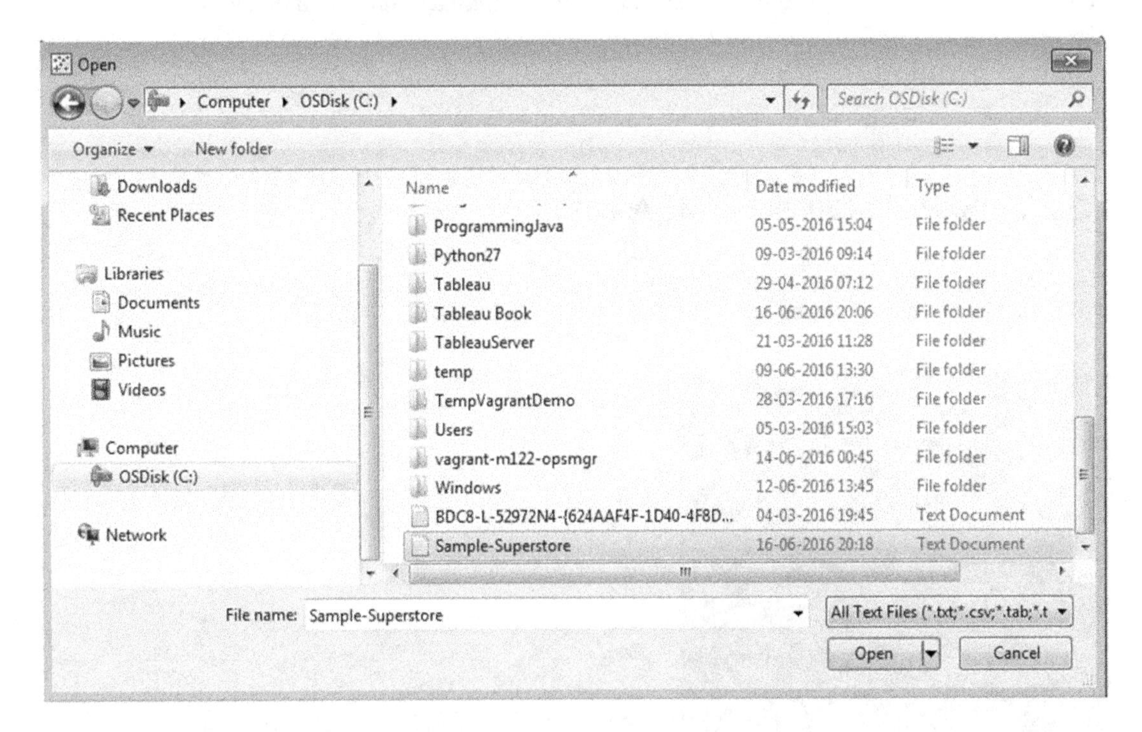

Figure 2-16. *Open dialog box to open the "Sample-Superstore" text file*

2.3.1.1.3 Step 3

Data source page will open as shown in Fig. 2-17.

Figure 2-17. *Data source page showing the connection to "Sample-Superstore.xls" data source*

2.3.2 Connect to MS Access

Follow the steps to connect to MS Access.

2.3.2.1 Steps to connect to MS Access

2.3.2.1.1 Step 1

On the start page, under Connect, select To a File ➤ Access (Shown in Fig. 2-18).

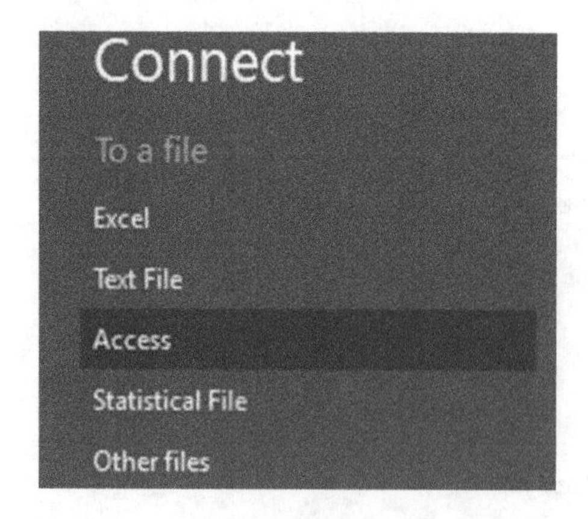

Figure 2-18. *Connect to "Access"*

2.3.2.1.2 Step 2

Access connection wizard will open (Shown in Fig. 2-19). Click on the "Browse" button. "Open dialog box" will show up.

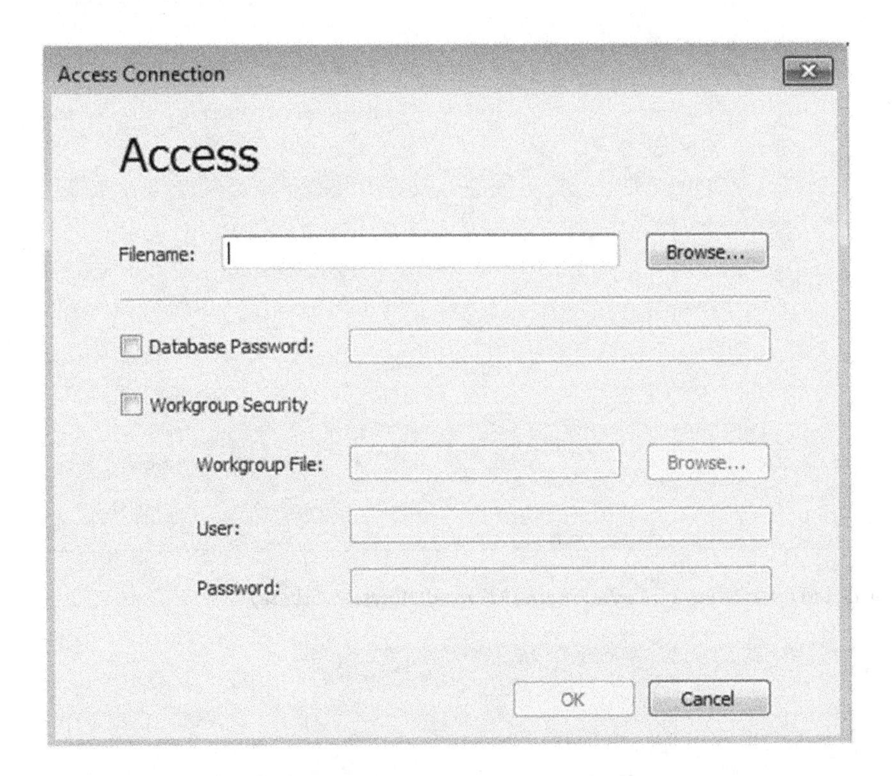

Figure 2-19. *"Access Connection" window*

2.3.2.1.3 Step 3

Navigate to the folder where the required access file is present. Select the file and click on "Open" button (Shown in Fig. 2-20) to open an Access file (Shown in Fig. 2-21).

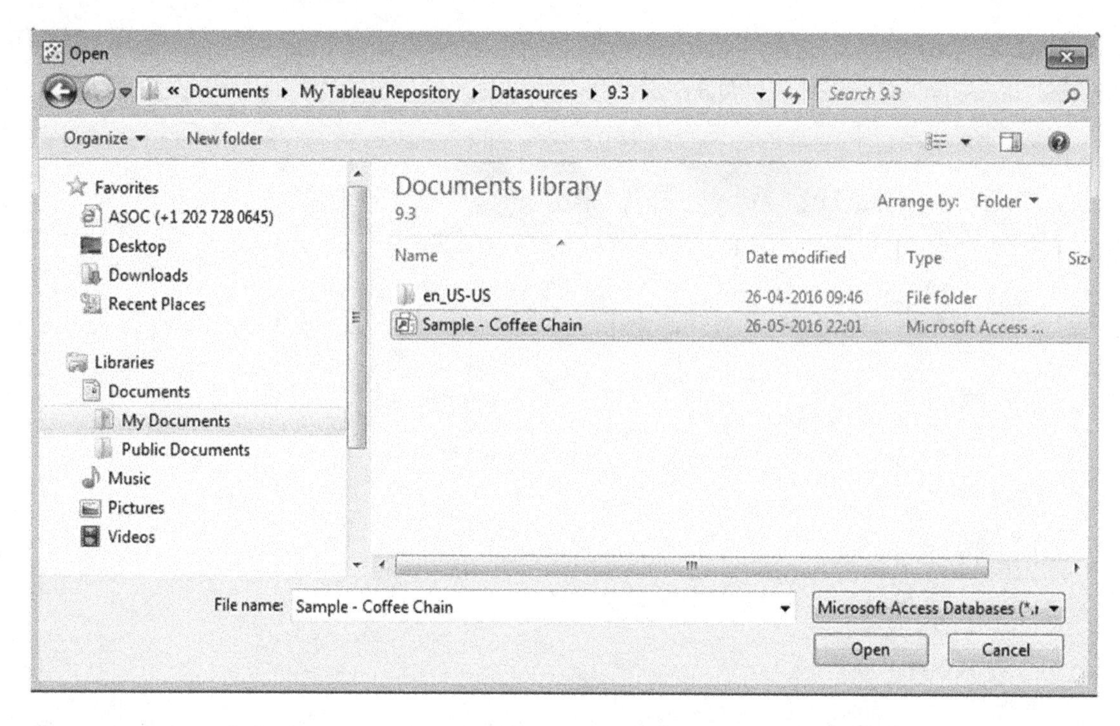

Figure 2-20. Open dialog box to open Sample - Coffee Chain (Microsoft Access database)

Figure 2-21. Access Connection window showing the selected "Sample - Coffee Chain.mdb" file

2.3.2.1.4 Step 4

Click on "OK" to connect to Access file (Shown in Fig. 2-22).

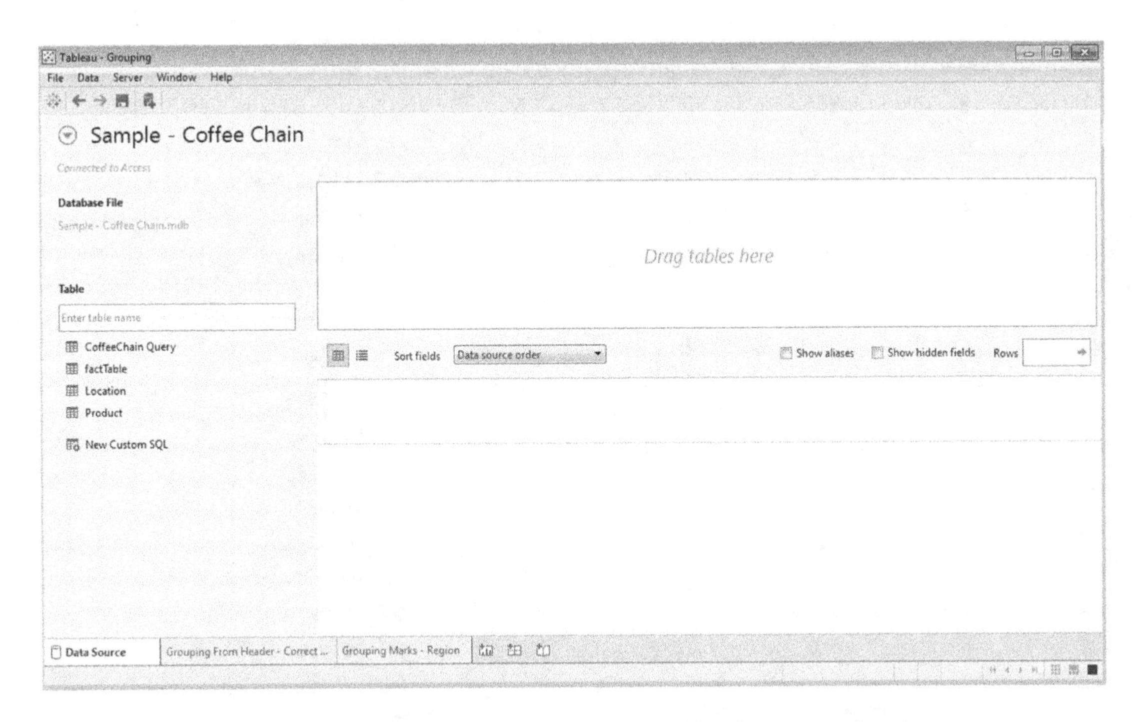

Figure 2-22. *Data source page showing connection to Sample-Coffee Chain Access data source*

2.3.3 Connecting to RData files

Perform the following steps to bring in the data stored in .RData file (created in R Statistical Programming Language) into Tableau.

2.3.3.1 Steps to connect to RData file

We will look at a .CSV file that we will read inside the R interface and save it into a dataset. Then this dataset will be saved to an .RData file. This file will then be read by tableau.

2.3.3.1.1 Step 1

DataSetIris.csv is the comma separated value file, available in the D: drive. Let us look at the data in the .csv file (Shown in Fig. 2-23).

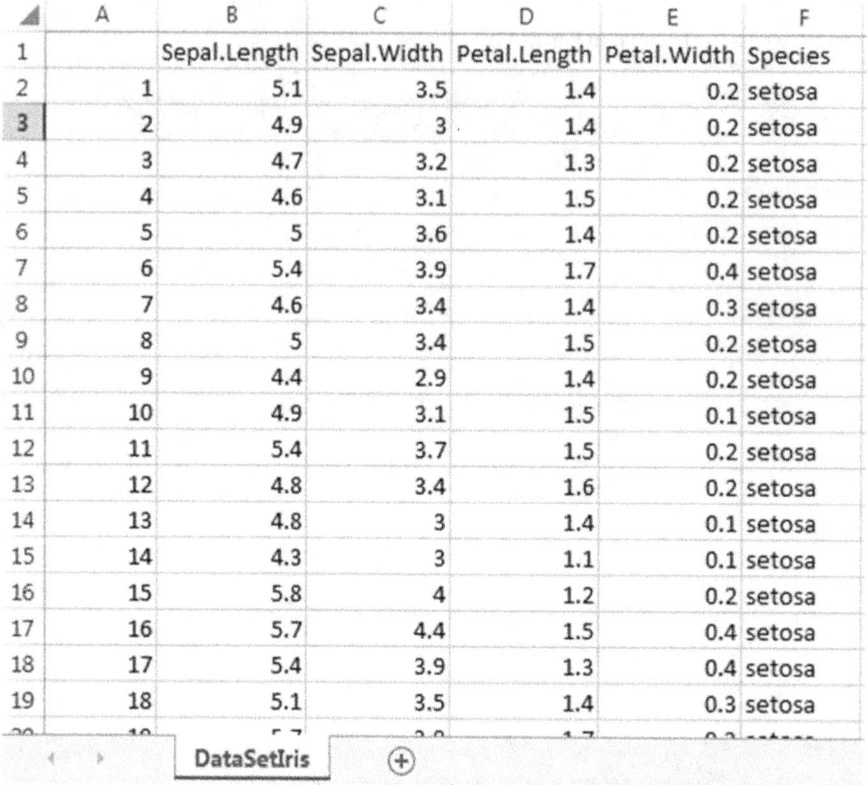

	A	B	C	D	E	F
1		Sepal.Length	Sepal.Width	Petal.Length	Petal.Width	Species
2	1	5.1	3.5	1.4	0.2	setosa
3	2	4.9	3	1.4	0.2	setosa
4	3	4.7	3.2	1.3	0.2	setosa
5	4	4.6	3.1	1.5	0.2	setosa
6	5	5	3.6	1.4	0.2	setosa
7	6	5.4	3.9	1.7	0.4	setosa
8	7	4.6	3.4	1.4	0.3	setosa
9	8	5	3.4	1.5	0.2	setosa
10	9	4.4	2.9	1.4	0.2	setosa
11	10	4.9	3.1	1.5	0.1	setosa
12	11	5.4	3.7	1.5	0.2	setosa
13	12	4.8	3.4	1.6	0.2	setosa
14	13	4.8	3	1.4	0.1	setosa
15	14	4.3	3	1.1	0.1	setosa
16	15	5.8	4	1.2	0.2	setosa
17	16	5.7	4.4	1.5	0.4	setosa
18	17	5.4	3.9	1.3	0.4	setosa
19	18	5.1	3.5	1.4	0.3	setosa

DataSetIris

Figure 2-23. *DataSetIris.csv Data Set*

Let us explore how to create an .RData file.
Steps to read the DataSetIris.csv file into R

2.3.3.1.2 Step 2

Start the R interface. At the R command prompt issue the following command (Shown in Fig. 2-24 and Fig. 2-25).

```
> RDataSet <- read.csv ("D:/DataSetIris.csv")
```

Figure 2-24. *Command to read data from .csv file into a Dataset, "RDataSet"*

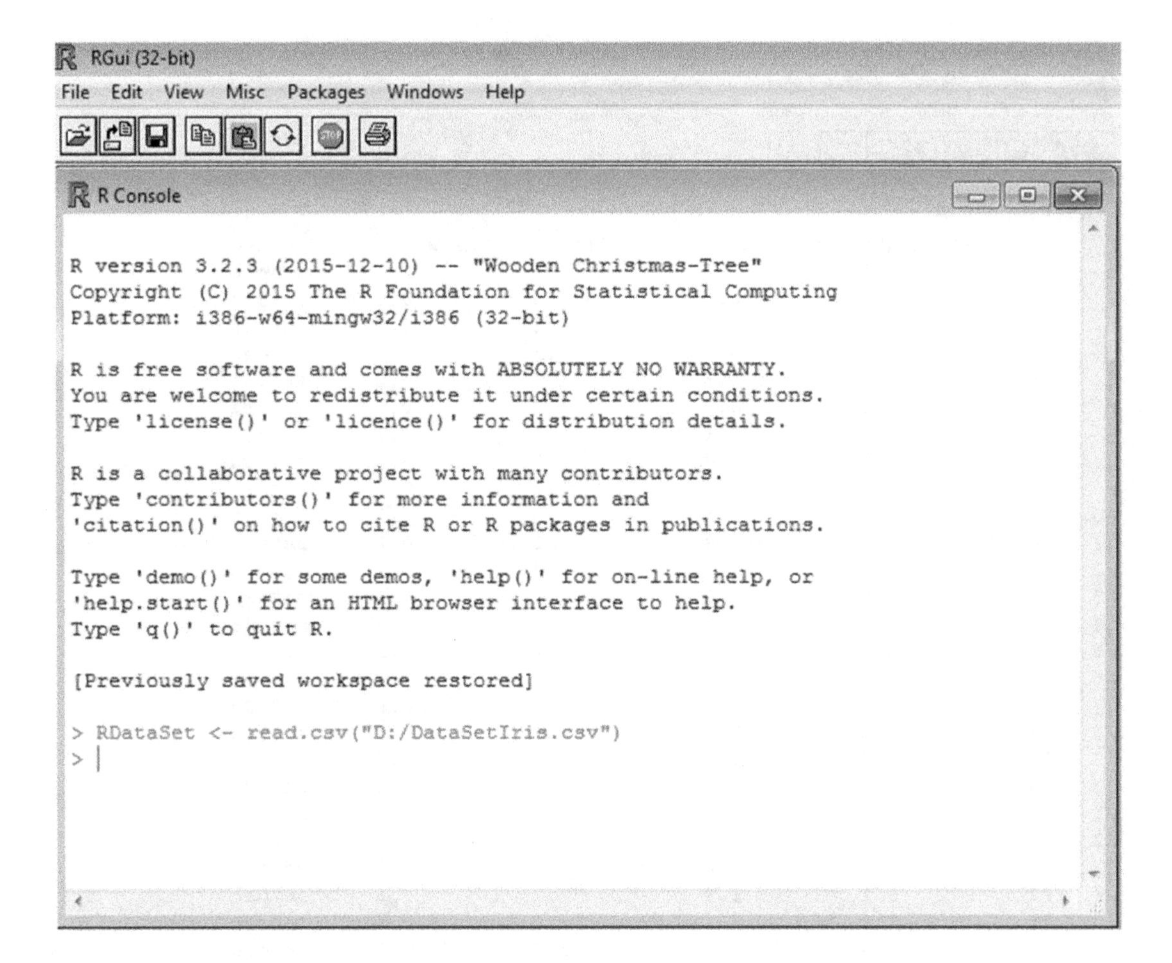

Figure 2-25. *R Interface*

The above command creates a dataset by the name, "RDataSet". And into this data set is read the data values from "DataSetIris.csv" file which is available in the D: drive.

2.3.3.1.3 Step 3

To display the values contained in RDataSet, type the following command (Shown in Fig. 2-26 and Fig. 2-27).

```
> RDataSet
```

Figure 2-26. *Command to display data in "RDataSet"*

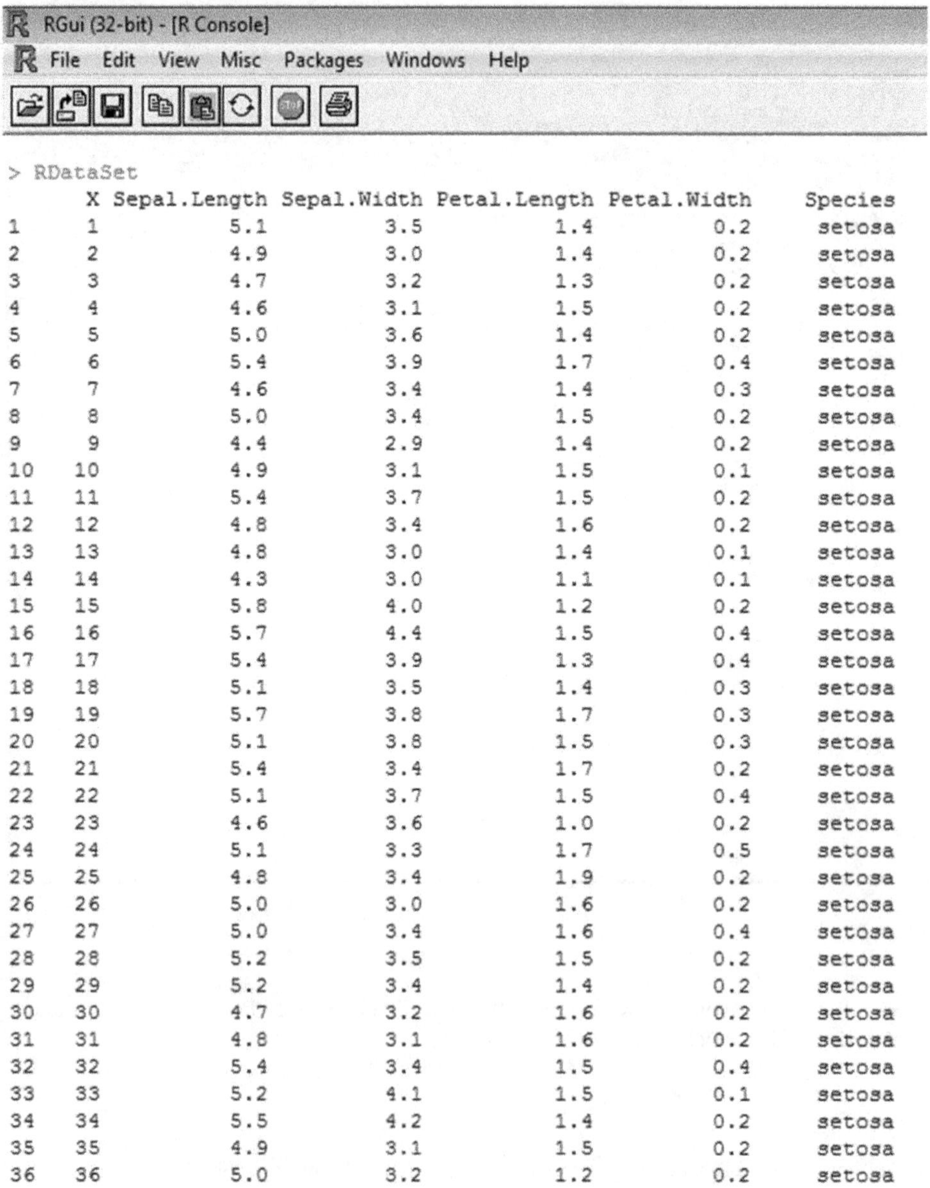

RGui (32-bit) - [R Console]

File Edit View Misc Packages Windows Help

```
> RDataSet
    X Sepal.Length Sepal.Width Petal.Length Petal.Width    Species
1   1          5.1         3.5          1.4         0.2     setosa
2   2          4.9         3.0          1.4         0.2     setosa
3   3          4.7         3.2          1.3         0.2     setosa
4   4          4.6         3.1          1.5         0.2     setosa
5   5          5.0         3.6          1.4         0.2     setosa
6   6          5.4         3.9          1.7         0.4     setosa
7   7          4.6         3.4          1.4         0.3     setosa
8   8          5.0         3.4          1.5         0.2     setosa
9   9          4.4         2.9          1.4         0.2     setosa
10 10          4.9         3.1          1.5         0.1     setosa
11 11          5.4         3.7          1.5         0.2     setosa
12 12          4.8         3.4          1.6         0.2     setosa
13 13          4.8         3.0          1.4         0.1     setosa
14 14          4.3         3.0          1.1         0.1     setosa
15 15          5.8         4.0          1.2         0.2     setosa
16 16          5.7         4.4          1.5         0.4     setosa
17 17          5.4         3.9          1.3         0.4     setosa
18 18          5.1         3.5          1.4         0.3     setosa
19 19          5.7         3.8          1.7         0.3     setosa
20 20          5.1         3.8          1.5         0.3     setosa
21 21          5.4         3.4          1.7         0.2     setosa
22 22          5.1         3.7          1.5         0.4     setosa
23 23          4.6         3.6          1.0         0.2     setosa
24 24          5.1         3.3          1.7         0.5     setosa
25 25          4.8         3.4          1.9         0.2     setosa
26 26          5.0         3.0          1.6         0.2     setosa
27 27          5.0         3.4          1.6         0.4     setosa
28 28          5.2         3.5          1.5         0.2     setosa
29 29          5.2         3.4          1.4         0.2     setosa
30 30          4.7         3.2          1.6         0.2     setosa
31 31          4.8         3.1          1.6         0.2     setosa
32 32          5.4         3.4          1.5         0.4     setosa
33 33          5.2         4.1          1.5         0.1     setosa
34 34          5.5         4.2          1.4         0.2     setosa
35 35          4.9         3.1          1.5         0.2     setosa
36 36          5.0         3.2          1.2         0.2     setosa
```

Figure 2-27. *Data in "RDataset"*

There are 150 such rows in the data set.

2.3.3.1.4 Step 4

You can make an .RData file by issuing the following command at the R prompt.

Save (RDataSet, file="D:/TableauDataSet.RData")

■ **Note** The Data Set, "RDataSet" is saved to the file, "TableauDataSet.RData".

2.3.3.1.5 Step 5

Start Tableau Desktop. Click on "Statistical File" under "To a File" (Shown in Fig. 2-28. Choose the statistical file that you wish to open within R (Shown in Fig. 2-29).

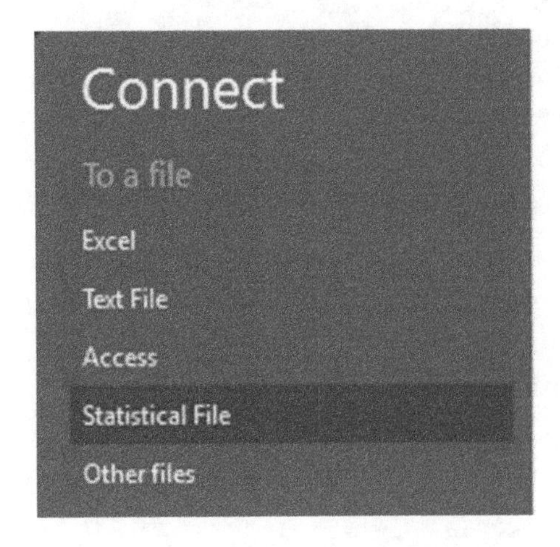

Figure 2-28. *Connecting to a statistical file*

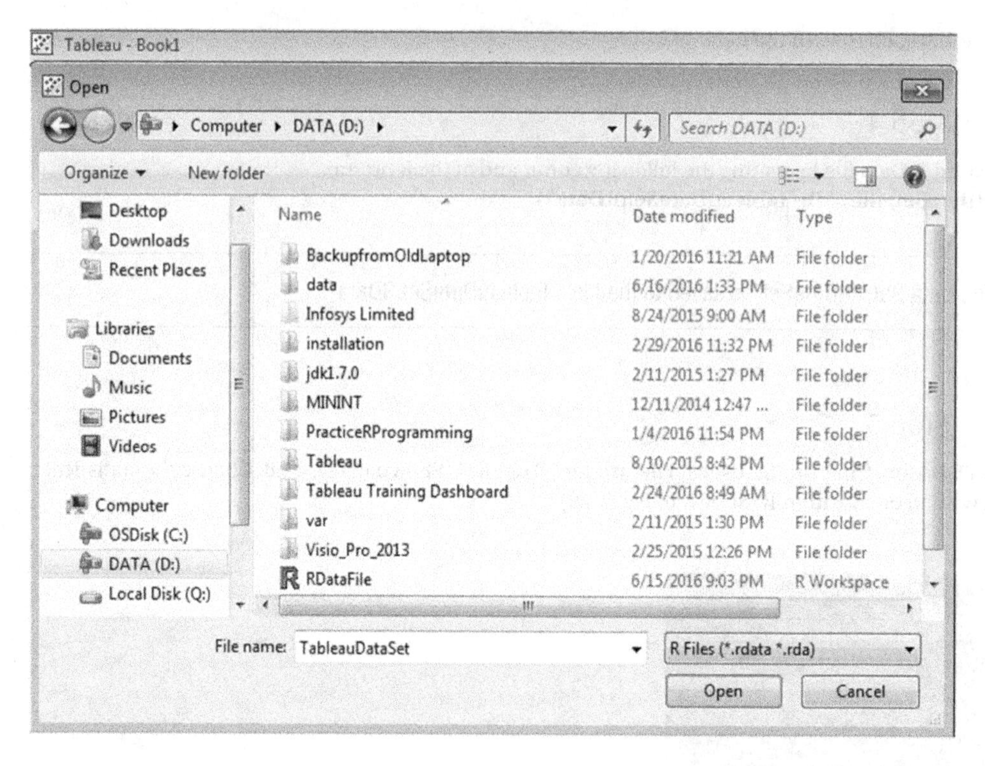

Figure 2-29. Open dialog box to open RDataFile

2.3.3.1.6 Step 5

Now, you can view the RDataset in the data source page (Shown in Fig. 2-30).

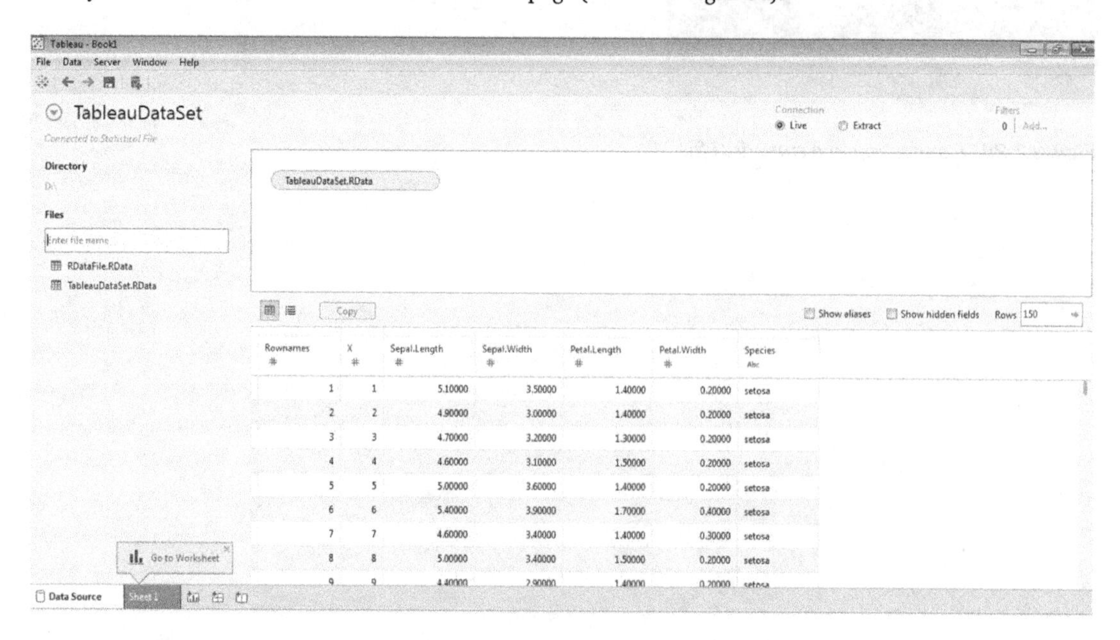

Figure 2-30. Data source page showing the connection to the statistical file

2.4 Connect to a Server

Let us explore how to connect to the below data sources:

- Microsoft SQL Server
- MySQL
- NoSQL Databases
    - MongoDB
    - Cassandra

2.4.1 Connecting to MS SQL Server 2014 Management Studio

We have a table with the name "Employee" in the "Test" Database in MS SQL Server 2014.

We have six records in the table "Employee" as displayed below (Shown in Fig. 2-31).

Results | Messages

	EmpNo	EmpName	Desg
1	101	Seema	SE
2	102	Merrilyn	PM
3	103	Manish	SE
4	104	Vishwas	SSE
5	105	Fedora	PM
6	106	Philips	Consultant

Figure 2-31. *"Employee" Table*

The objective is to read these six records inside tableau.

Follow the steps below.

2.4.1.1 Steps to connect to MS SQL Server

2.4.1.1.1 Step 1

Open a Tableau workbook on the "Connect" page.

Click on Microsoft SQL Server under "To the server" (Shown in Fig. 2-32) to bring up the dialog box displaying the server connection to Microsoft SQL Server (Shown in Fig. 2-33).

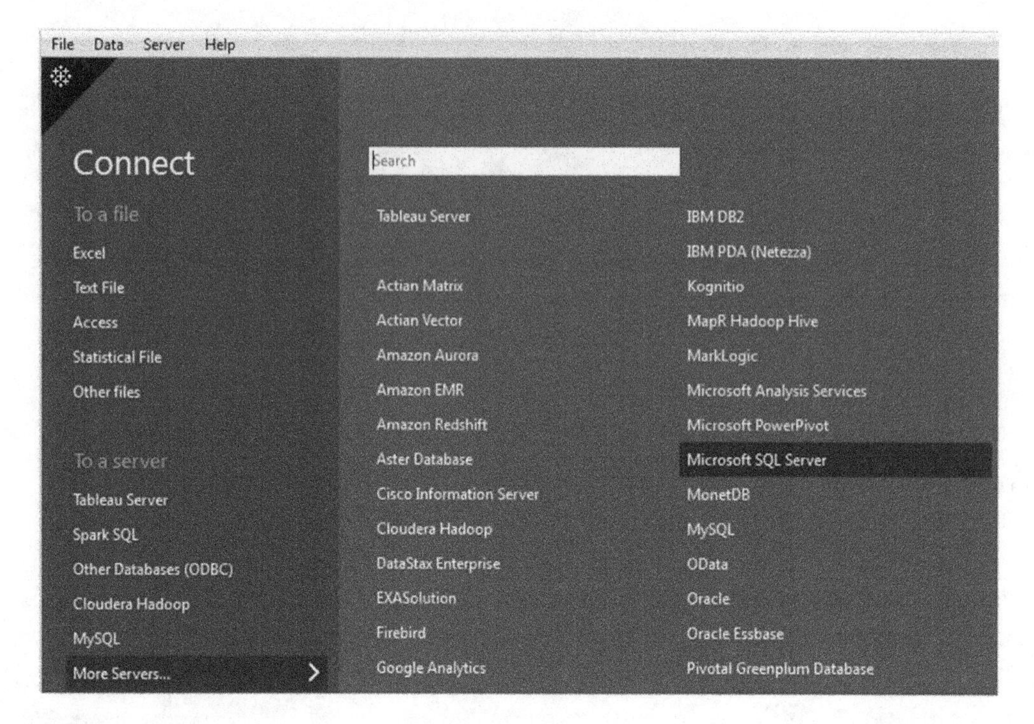

Figure 2-32. *Connect to Microsoft SQL Server*

Figure 2-33. *Microsoft SQL server connection window*

2.4.1.1.2 Step 2

Fill in the details about the database server such as "server name". Select either "Windows Authentication" or provide a specific username and password (Shown in Fig. 2-34).

Figure 2-34. *Microsoft SQL Server "Server Connection" details*

2.4.1.1.3 Step 3

If the connection is successful, it shows the screen below to allow one to select the desired database (Shown in Fig. 2-35).

Figure 2-35. *Data Source Page showing connection to "Microsoft SQL Server"*

2.4.1.1.4 Step 4

The table "Employee" that we wish to work with is in the "Test" database (Shown in Fig. 2-36).

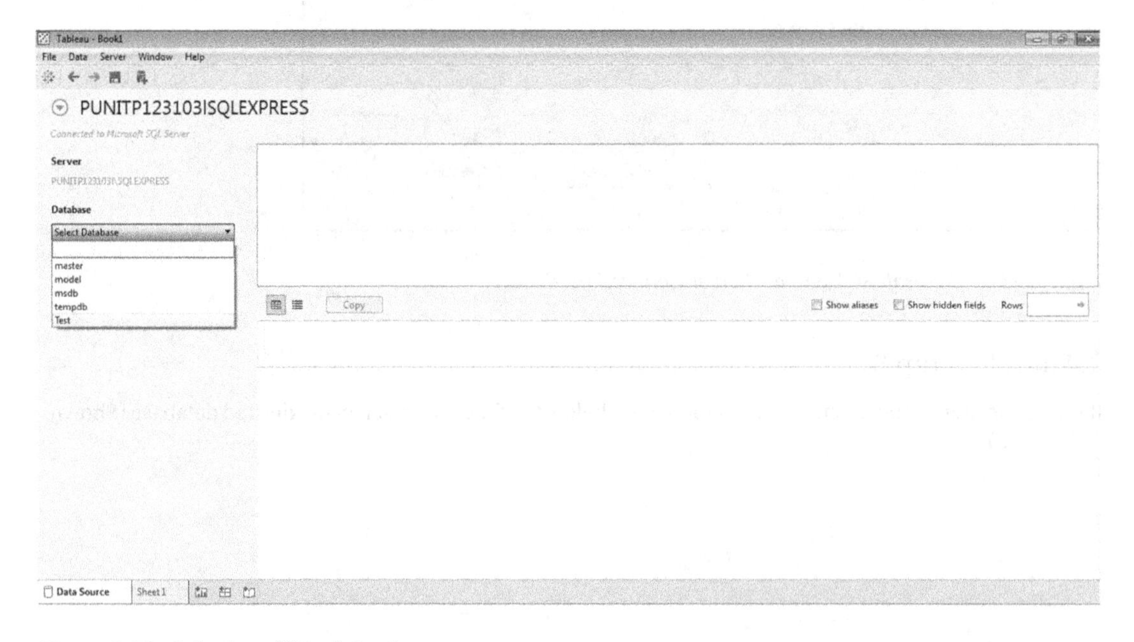

Figure 2-36. *Selection of "Test" database*

2.4.1.1.5 Step 5

Currently, there is only one table, "Employee" in the "Test" database. Drag the table to the canvas area. Select how you wish to have the records updated, either "Update Now" or "Automatically Update" (Shown in Fig. 2-37).

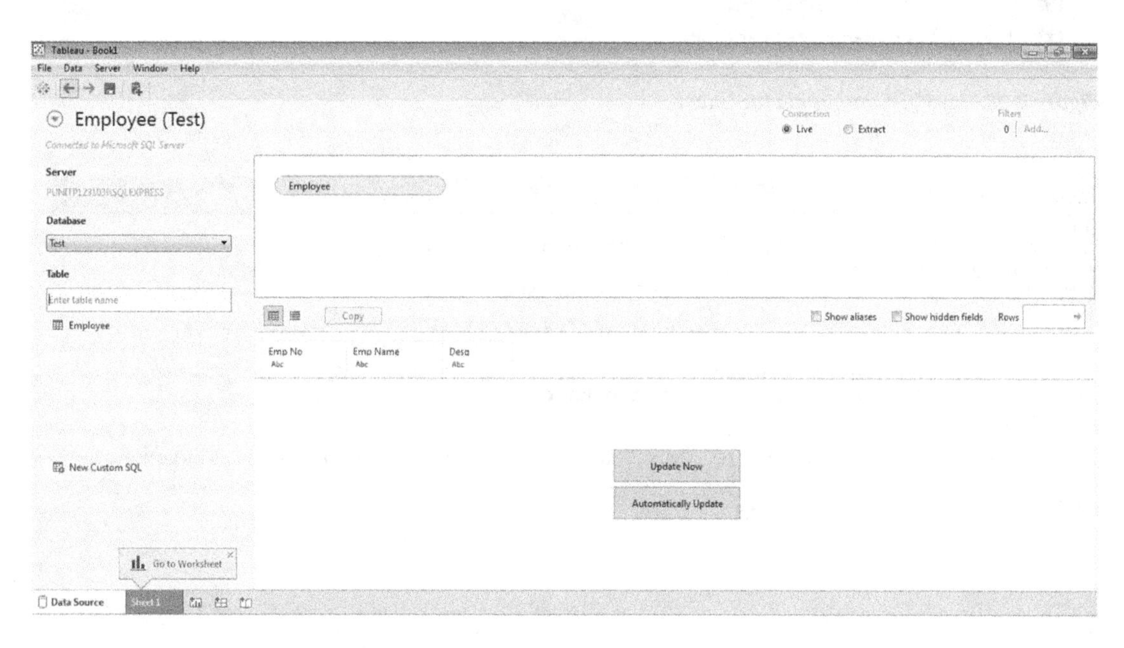

Figure 2-37. *"Employee (Test)" table placed on canvas area*

2.4.2 Connecting to MySQL

Follow the steps to connect to a MySQL Database.

2.4.2.1 Steps to connect to MySQL

2.4.2.1.1 Step 1

Download MySQL installer for windows from below mentioned link and install it.

`https://dev.mysql.com/downloads/installer/`

2.4.2.1.2 Step 2

Go to All Programs ➤ MySQL Server 5.7 ➤ MySQL Command Line Client (Shown in Fig. 2-38). Click on MySQL Command Line Client to start MySQL Server (Shown in Fig. 2-39).

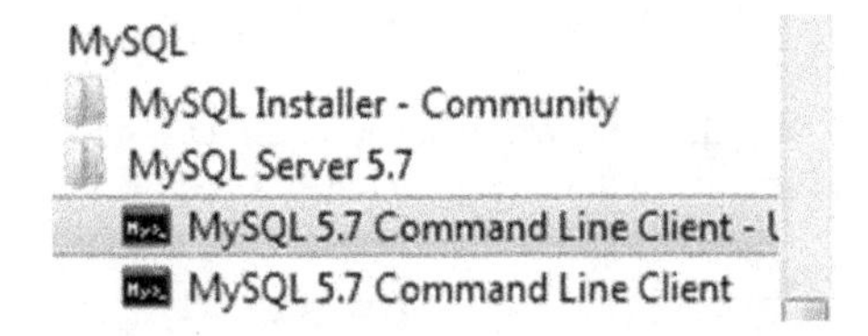

Figure 2-38. *MySQL Command Line Client*

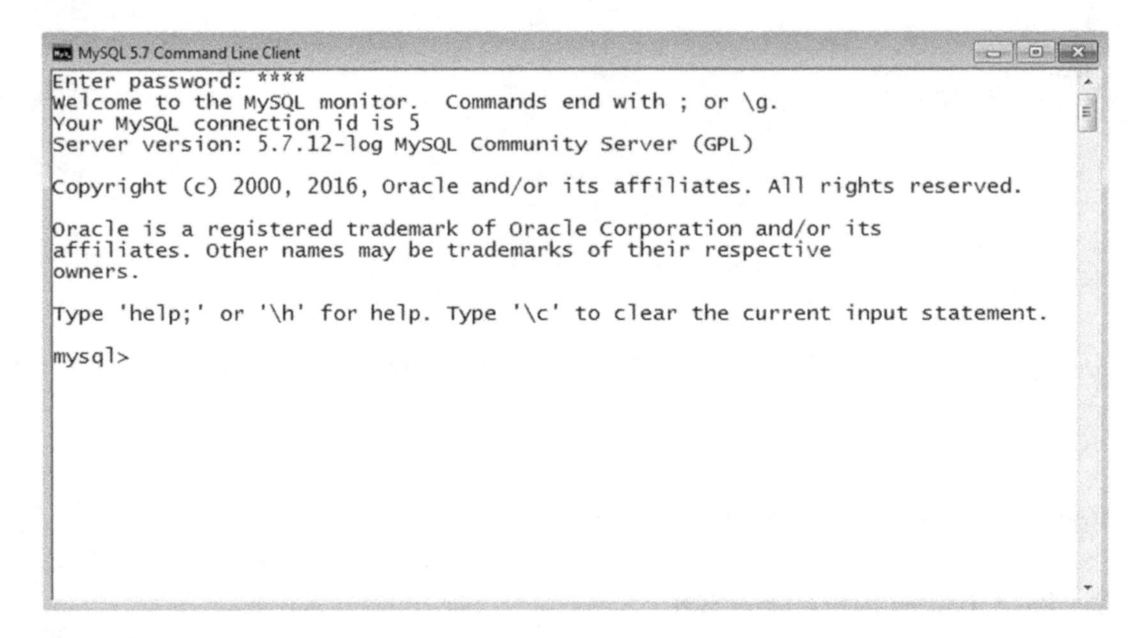

Figure 2-39. *MySQL command prompt*

2.4.2.1.3 Step 3

Download MySQL Driver for Tableau from link below and install it.

`http://www.tableau.com/support/drivers`

2.4.2.1.4 Step 4

Open Tableau Desktop, Under Connect ➤ Select MySQL (Shown in Fig. 2-40).

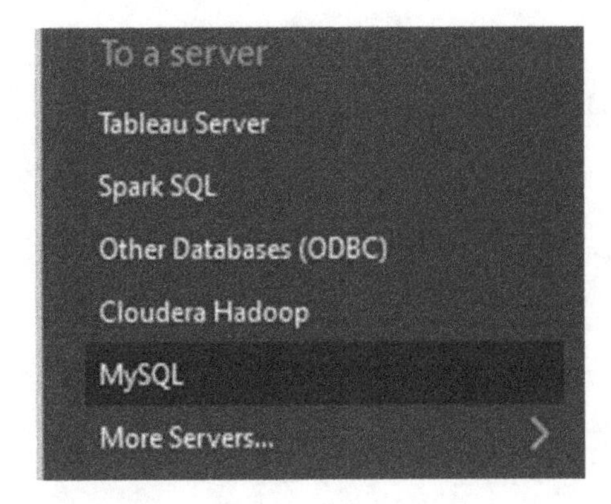

Figure 2-40. *Connect to MySQL server*

2.4.2.1.5 Step 5

MySQL connection wizard will open. Provide inputs for "Server", "Username" and "Password" and click "OK" to connect to MySQL Server (Shown in Fig. 2-41).

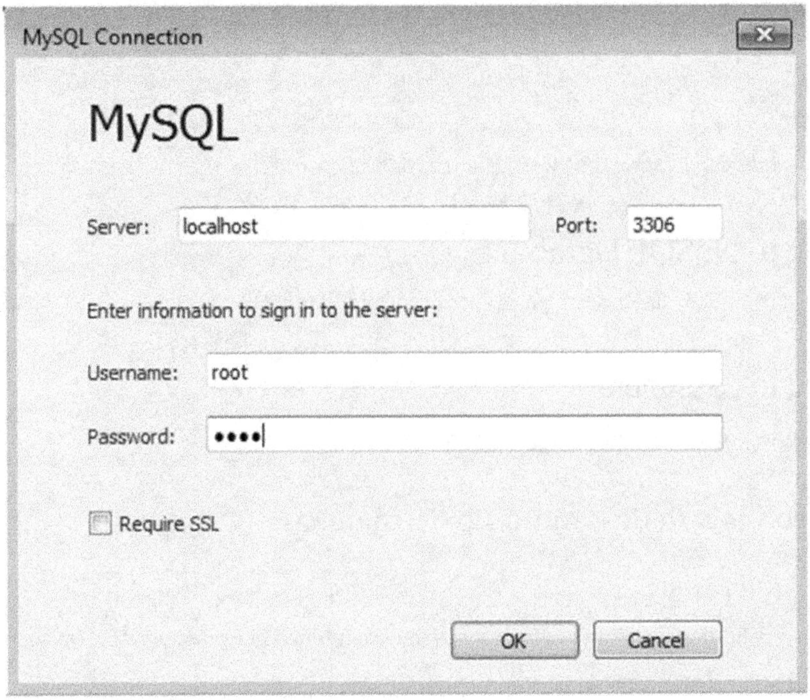

Figure 2-41. *MySQL server connection details*

2.4.2.1.6 Step 6

If the connection is successful, it will show the screen below to allow one to select the desired database (Shown in Fig. 2-42).

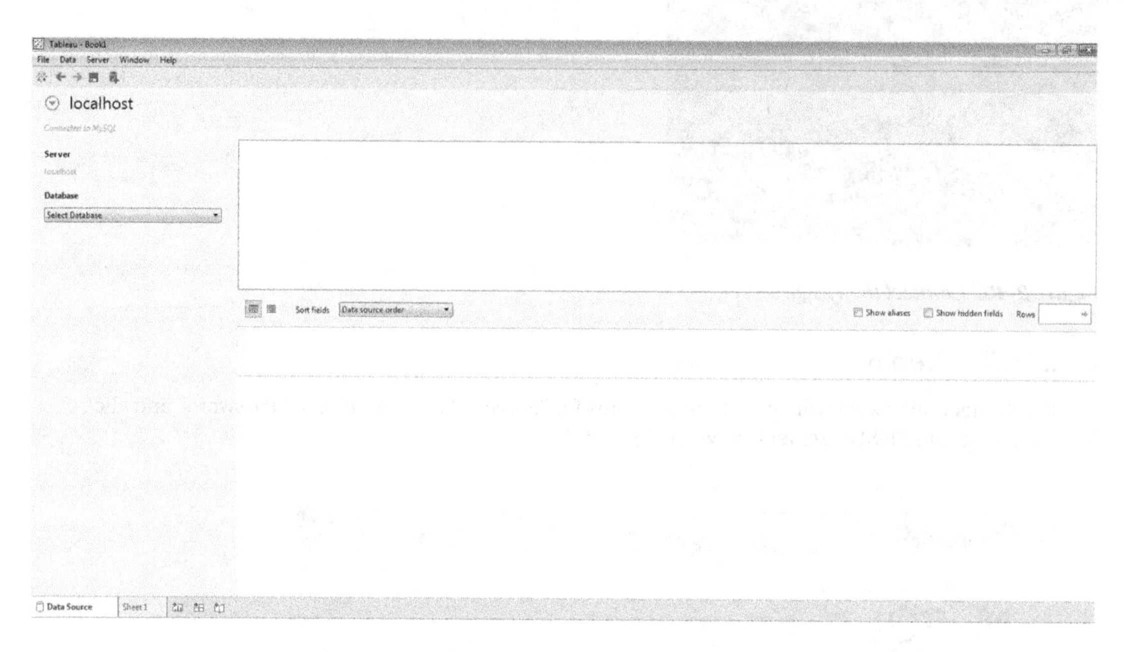

Figure 2-42. *Data source page showing a successful MySQL server connection*

2.4.3 Connecting to NoSQL Databases

Let us discuss how to connect to NoSQL databases such as Cassandra and MongoDB.

2.4.3.1 Connecting to Cassandra

Follow the below steps to connect to a Cassandra NoSQL Database.

2.4.3.1.1 Steps to connect to Cassandra NoSQL database

2.4.3.1.1.1 Step 1

Download DataStax Community Edition for Windows from the below-mentioned link and install it.

```
https://downloads.datastax.com/community/
```

Java 1.8 is required to work with Apache Cassandra.

2.4.3.1.1.2 Step 2

Select All Programs ➤ DataStax Community Edition ➤ Cassandra CQL Shell (Shown in Fig. 2-43) to start Cassandra CQL Shell (Shown in Fig. 2-44).

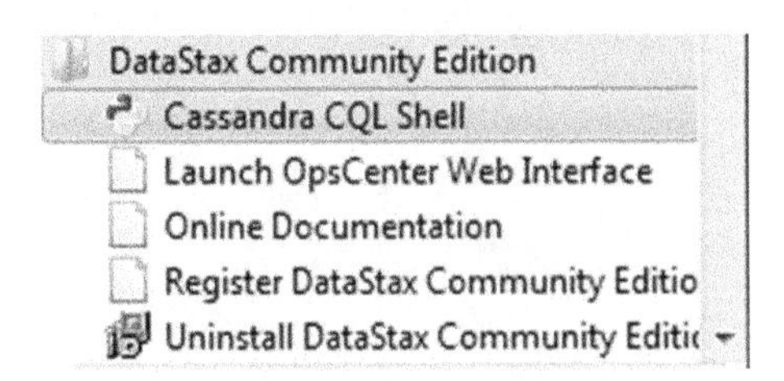

Figure 2-43. *Cassandra CQL Shell option*

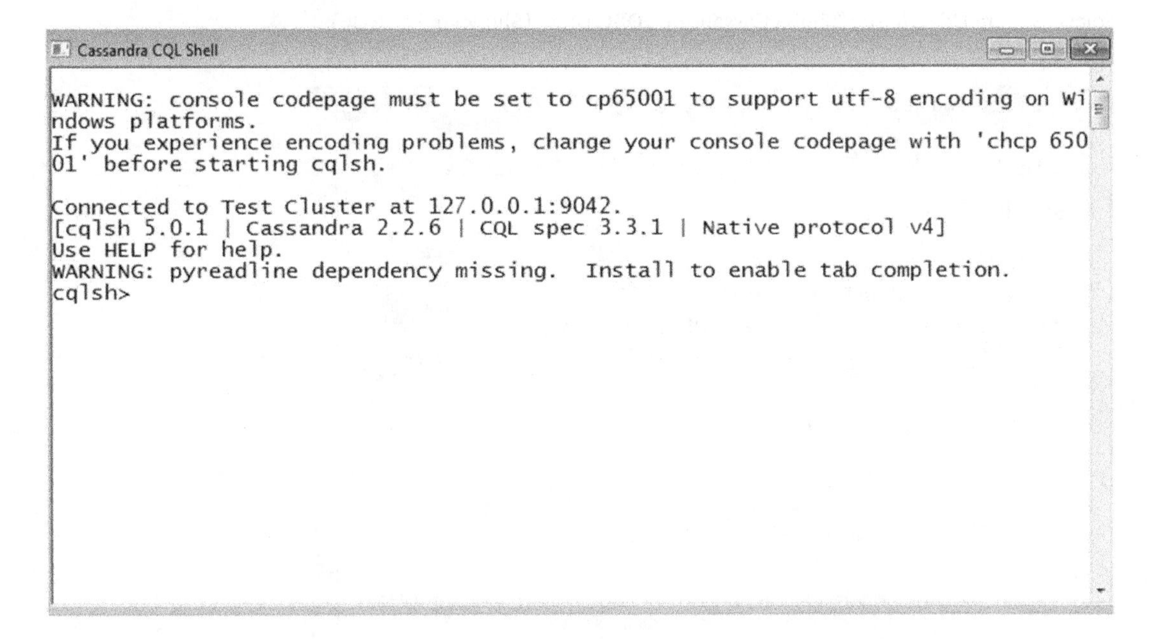

Figure 2-44. *Cassandra CQL shell*

2.4.3.1.1.3 Step 3

Download Cassandra ODBC and JDBC drivers with SQL connector from the below link and install it.

```
http://www.simba.com/drivers/cassandra-odbc-jdbc/
```

You can download 30 days trial version, SimbaApacheCassandraDriver.lic file which will be sent to your registered email.

2.4.3.1.1.4 Step 4

To check Cassandra ODBC and JDBC driver installation, Select All Programs ➤ Simba Cassandra ODBC Driver 2.2 ➤ 64 bit ODBC Driver (Shown in Fig. 2-45).

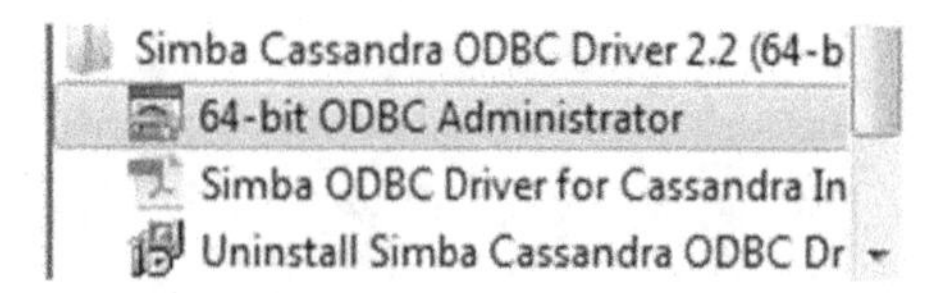

Figure 2-45. *Cassandra ODBC Driver option*

2.4.3.1.1.5 Step 5

Select "System DSN" to see "Simba Cassandra ODBC DSN" (Shown in Fig. 2-46).

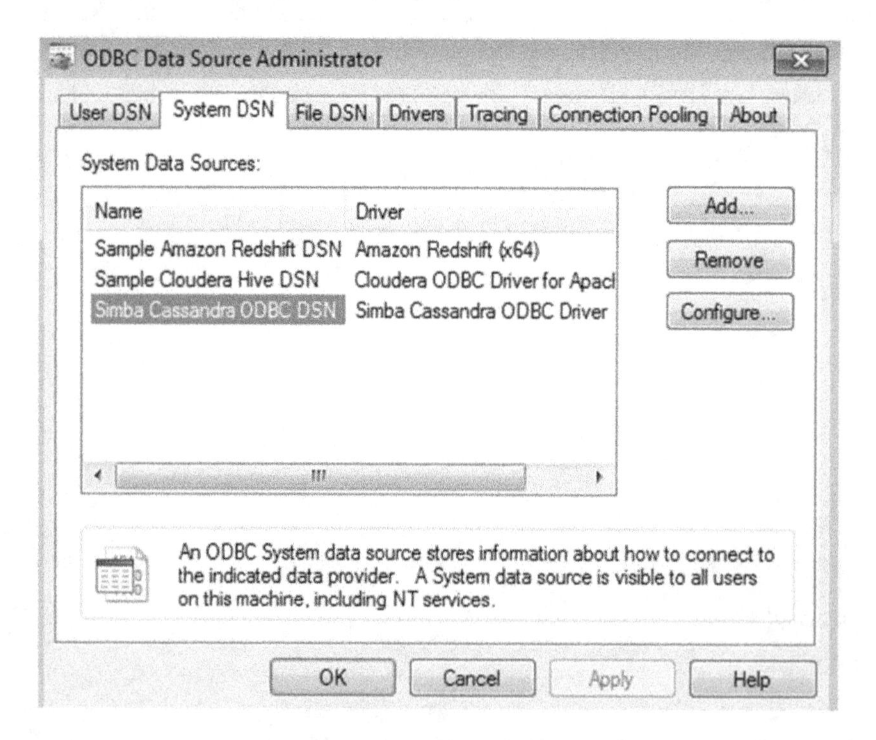

Figure 2-46. *Simba Cassandra OBDC DSN (highlighted)*

Copy SimbaApacheCassandraDriver.lic file to the lib folder of SimbaCassandraODBC Driver. You can find SimbaCassandraODBC Driver folder inside the Program Files.

2.4.3.1.1.6 Step 6

Click on the "Configure" button to configure the Simba Cassandra ODBC DSN (Shown in Fig. 2-47). Specify the host as 127.0.0.1 and click "Test" to check the connectivity (Shown in Fig. 2-48). Once the connection comes through click on the "OK" button.

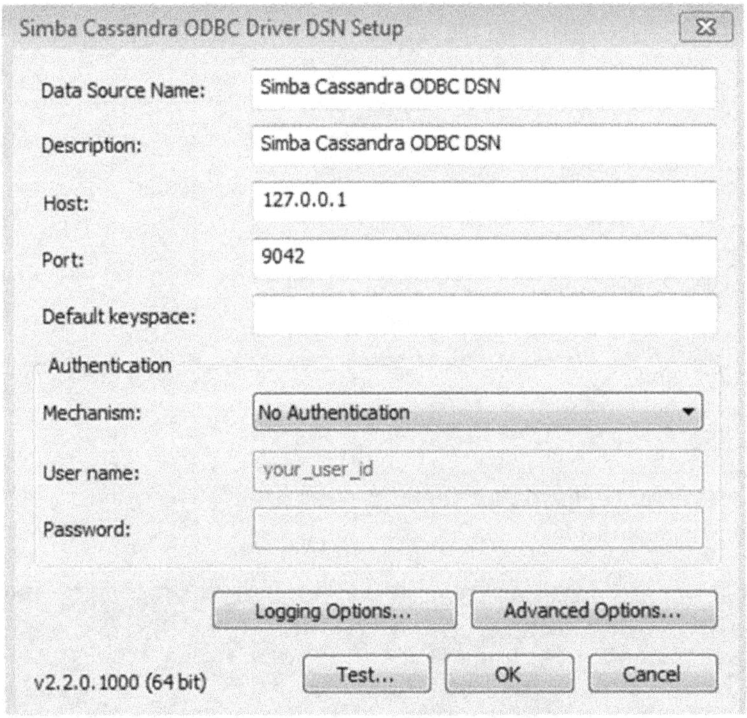

Figure 2-47. *Simba Cassandra ODBC driver DSN setup*

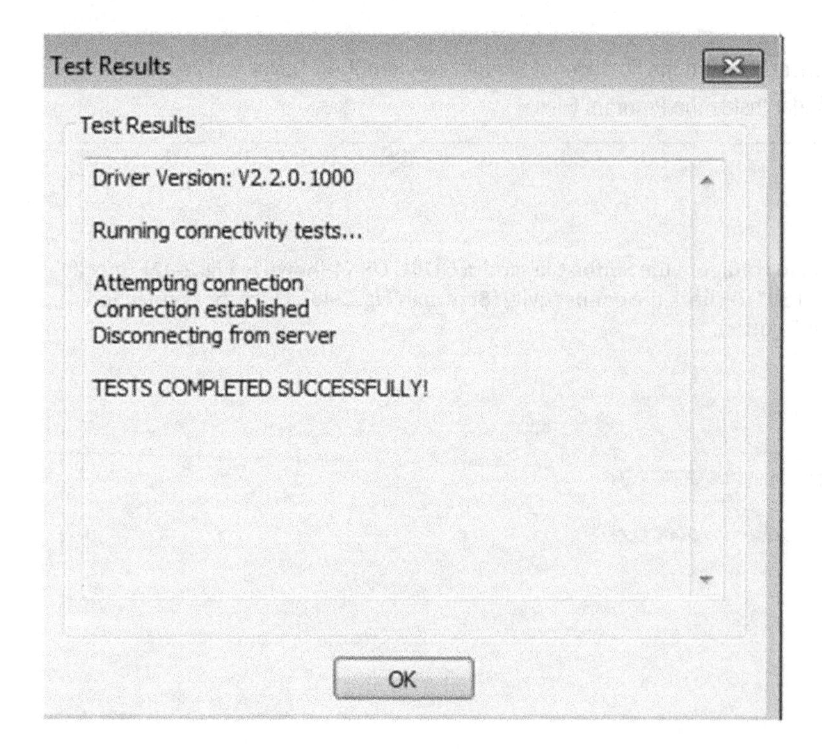

Figure 2-48. *Connection "Test Results" window*

2.4.3.1.1.7 Step 7

Open Tableau Desktop, From Connect, Select More Servers ➤ Other Databases (ODBC) (Shown in Fig. 2-49).

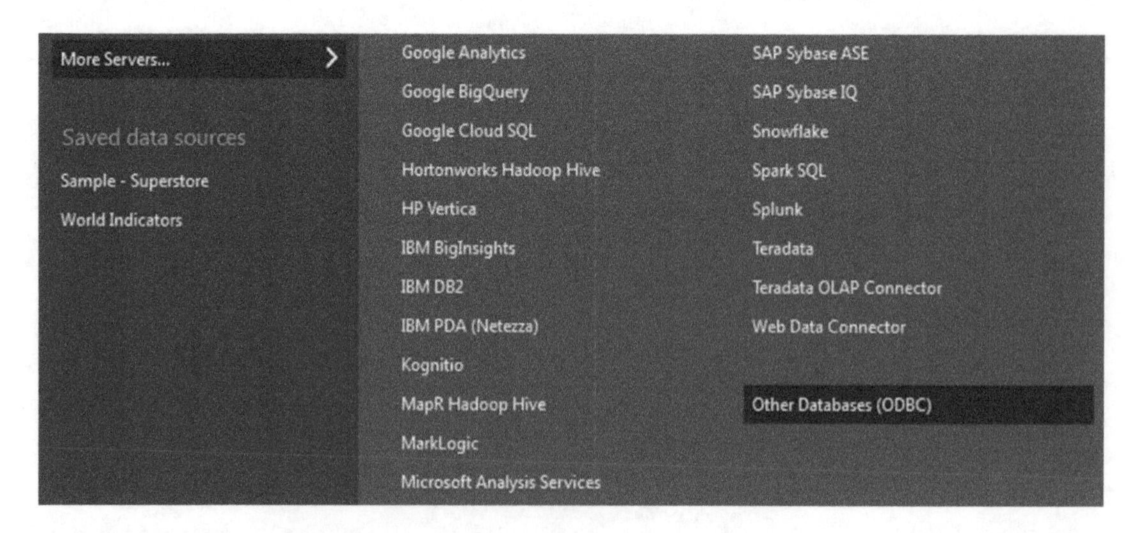

Figure 2-49. *Connect to "Other Databases (ODBC)"*

2.4.3.1.1.8 Step 8

Other Databases(ODBC) Connection Wizard shows up. Select Simba Cassandra ODBC DSN (Shown in Fig. 2-50) and click on “Connect” button. Next, click on “OK” button to connect to the Cassandra database.

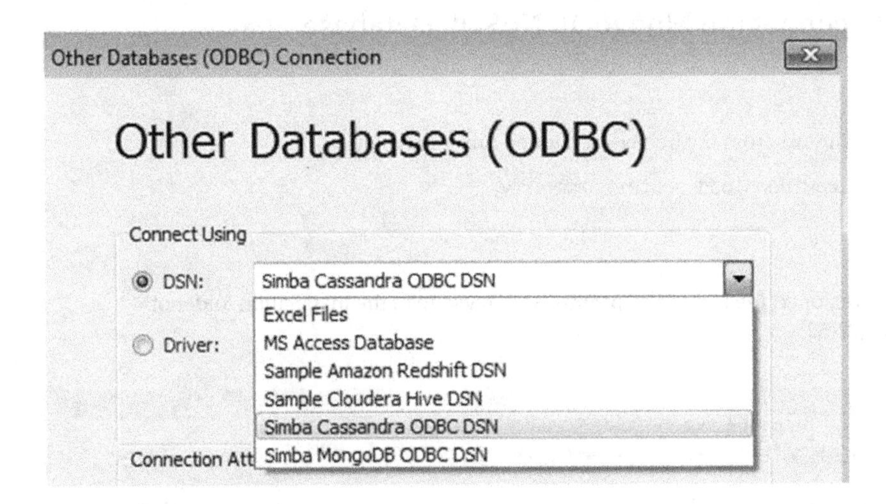

Figure 2-50. *Other Databases (ODBC) connection window*

2.4.3.1.1.9 Step 9

If the connection is successful, it shows the screen below to allow one to select the desired database (Shown in Fig. 2-51).

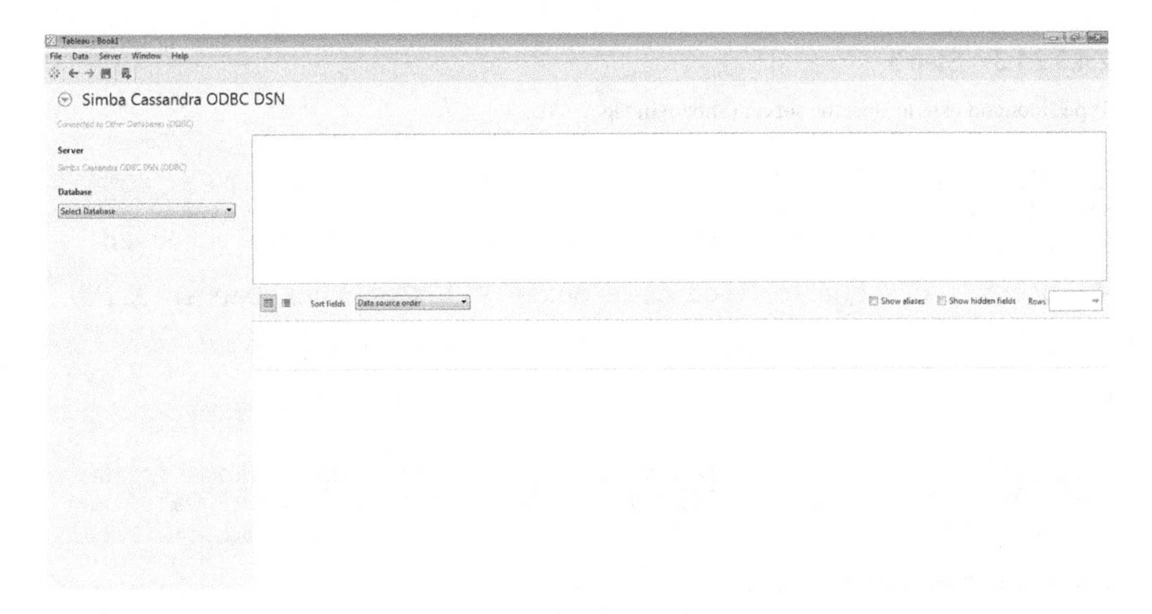

Figure 2-51. *Data source page showing the Cassandra connection*

2.4.3.2 Connecting to MongoDB

Follow the below steps to connect to a MongoDB NoSQL Database.

2.4.3.2.1 Steps to connect to MongoDB NoSQL Database

2.4.3.2.1.1 Step 1

Download MongoDB for Windows from the below-mentioned link and install it.

```
https://www.mongodb.com/download-center#community
```

2.4.3.2.1.2 Step 2

To start the MongoDB Server, open the command prompt and navigate to the installation folder of MongoDB as shown in Fig. 2-52.

```
C:\windows\system32\cmd.exe
Microsoft Windows [Version 6.1.7601]
Copyright (c) 2009 Microsoft Corporation.  All rights reserved.

C:\Users\r.c.subhashini>cd C:\Program Files\MongoDB\Server\3.2\b

C:\Program Files\MongoDB\Server\3.2\bin>
```

Figure 2-52. *MongoDB installation directory path*

2.4.3.2.1.3 Step 3

Type "mongod.exe" to start the server (Shown in Fig. 2-53).

```
C:\windows\system32\cmd.exe
Microsoft Windows [Version 6.1.7601]
Copyright (c) 2009 Microsoft Corporation.  All rights reserved.

C:\Users\r.c.subhashini>cd C:\Program Files\MongoDB\Server\3.2\b

C:\Program Files\MongoDB\Server\3.2\bin>mongod.exe
```

```
iagnostic data capture with directory 'C:/data/db/diagnostic.dat
2016-06-05T00:24:56.875+0530 I NETWORK  [initandlisten] waiting
on port 27017
```

Figure 2-53. *Starting MongoDB Server*

You should get a message stating "waiting on port 27017..."

2.4.3.2.1.4 Step 4

To start the MongoDB Client, open the command prompt and navigate to the installation folder of MongoDB and type mongo.exe as shown in Fig. 2-54.

```
C:\Users\r.c.subhashini>cd C:\Program Files\MongoDB\Server\3.2\b

C:\Program Files\MongoDB\Server\3.2\bin>mongo.exe
2016-06-05T00:28:40.458+0530 I CONTROL  [main] Hotfix KB2731284
is not installed, will zero-out data files
MongoDB shell version: 3.2.3
connecting to: test
>
```

Figure 2-54. *Starting MongoDB Client*

2.4.3.2.1.5 Step 5

Download the MongoDB ODBC and JDBC drivers with SQL connector from the below link and install it.

http://www.simba.com/drivers/mongodb-odbc-jdbc/

You can download 30 days trial version, you will receive the SimbaMongoDBODBDriver.lic file in your registered email.

2.4.3.2.1.6 Step 6

To check MongoDB ODBC and JDBC driver installation, select All Programs ➤ Simba MongoDB ODBC Driver 2.0 ➤ 64 bit ODBC Driver (Shown in Fig. 2-55).

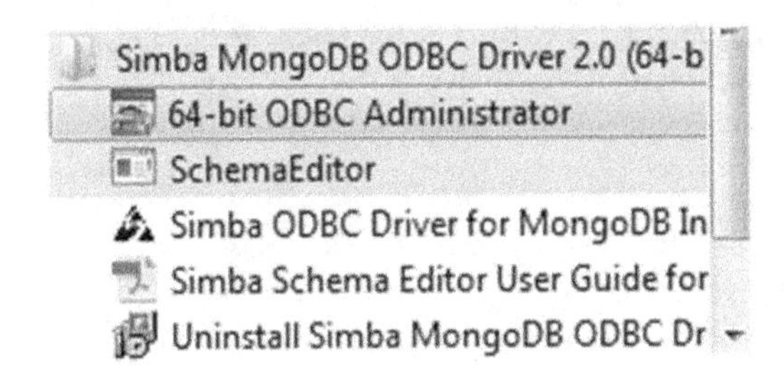

Figure 2-55. *Simba MongoDB ODBC Driver*

2.4.3.2.1.7 Step 7

Select "System DSN" to see "Simba MongoDB ODBC DSN" (Shown in Fig. 2-56).

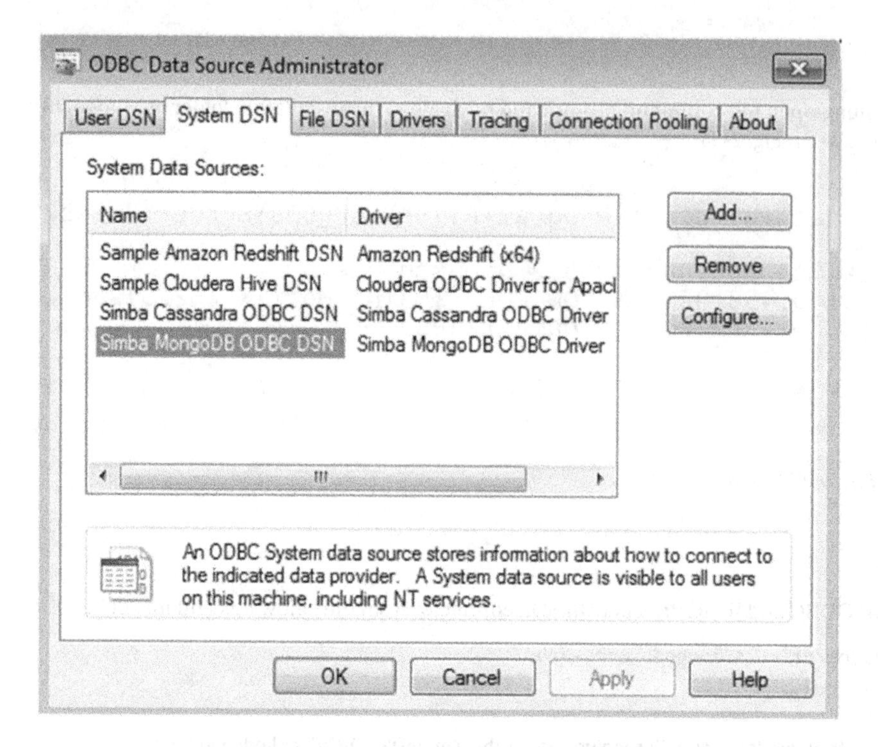

Figure 2-56. *"System DSN" tab showing "Simba MongoDB ODBC DSN"*

Copy SimbaMongoDBODBDriver.lic file to the lib folder of SimbaMongoDBODBC Driver. You can find SimbaMongoDBODBC Driver folder inside the Program Files.

2.4.3.2.1.8 Step 8

Click "Configure..." button to configure the "Simba MongoDB ODBC DSN". Specify the server as "localhost", port as "27017" and the database as "test" (default database of MongoDB) (Shown in Fig. 2-57) and click "Test" to check the connectivity (Shown in Fig. 2-58). Once the connection is successful, click on the "OK" button.

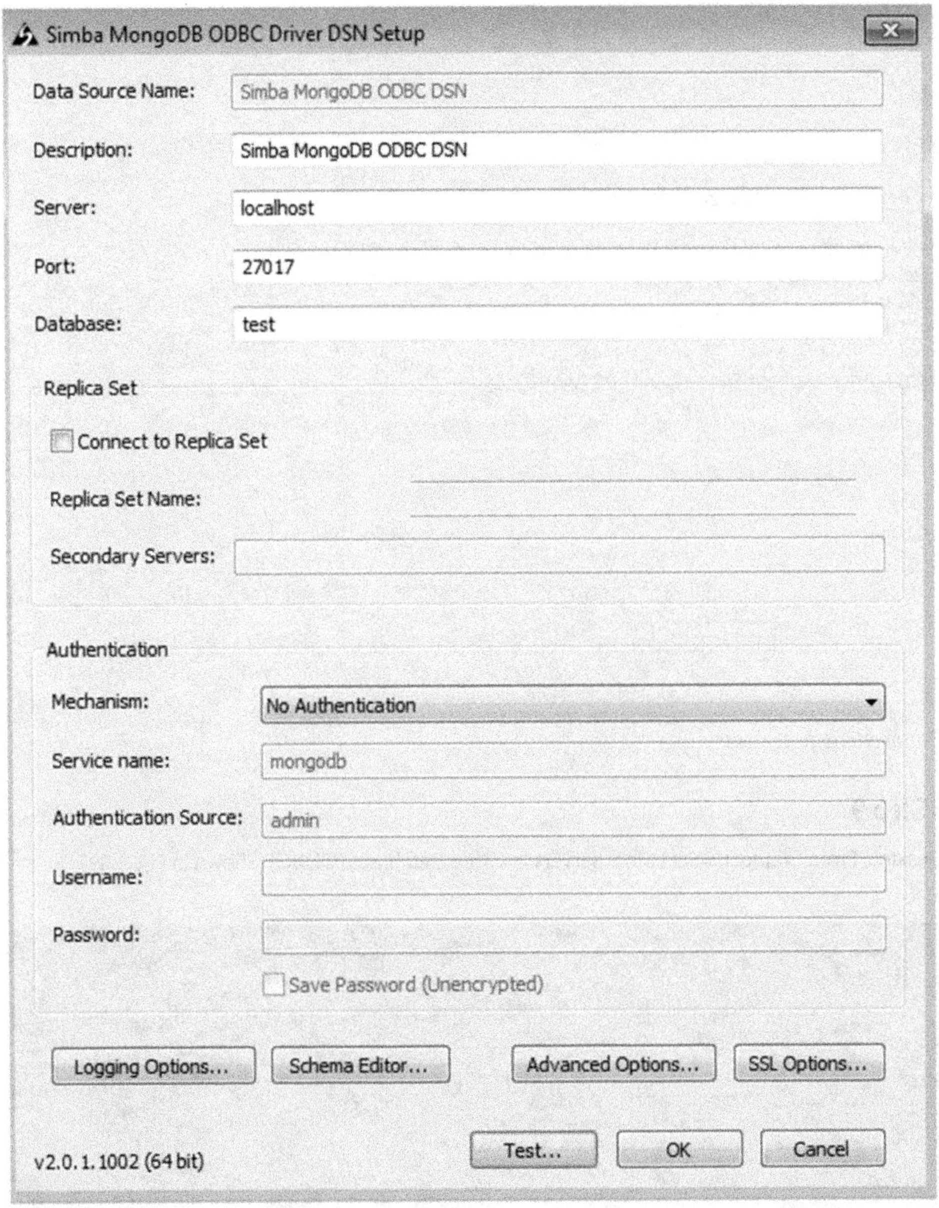

Figure 2-57. *Simba MongoDB ODBC Driver DSN Setup*

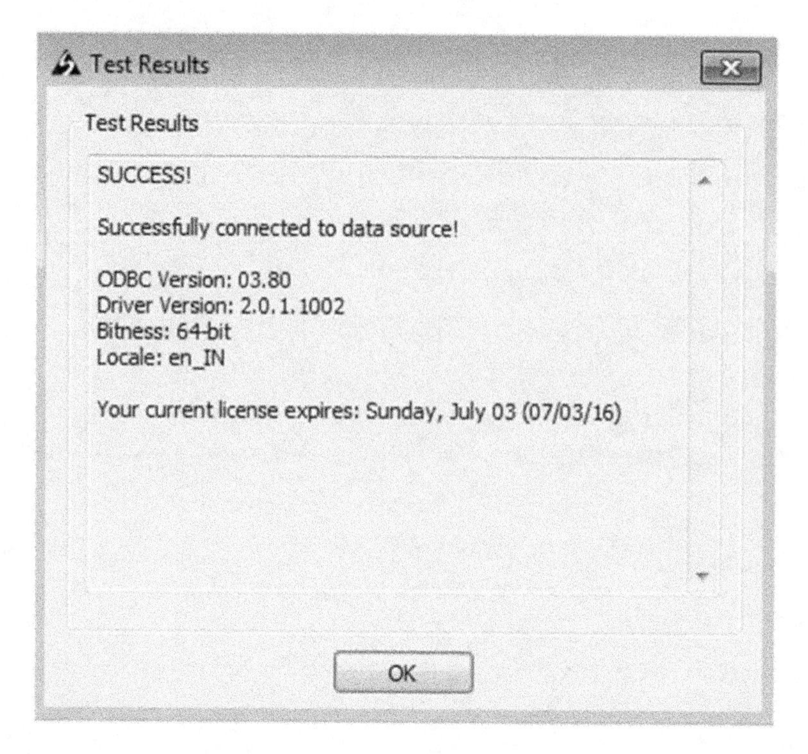

Figure 2-58. *Connectivity test results*

2.4.3.2.1.9 Step 9

Open Tableau Desktop, From Connect, Select More Servers ➤ Other Databases (ODBC) (Shown in Fig. 2-59).

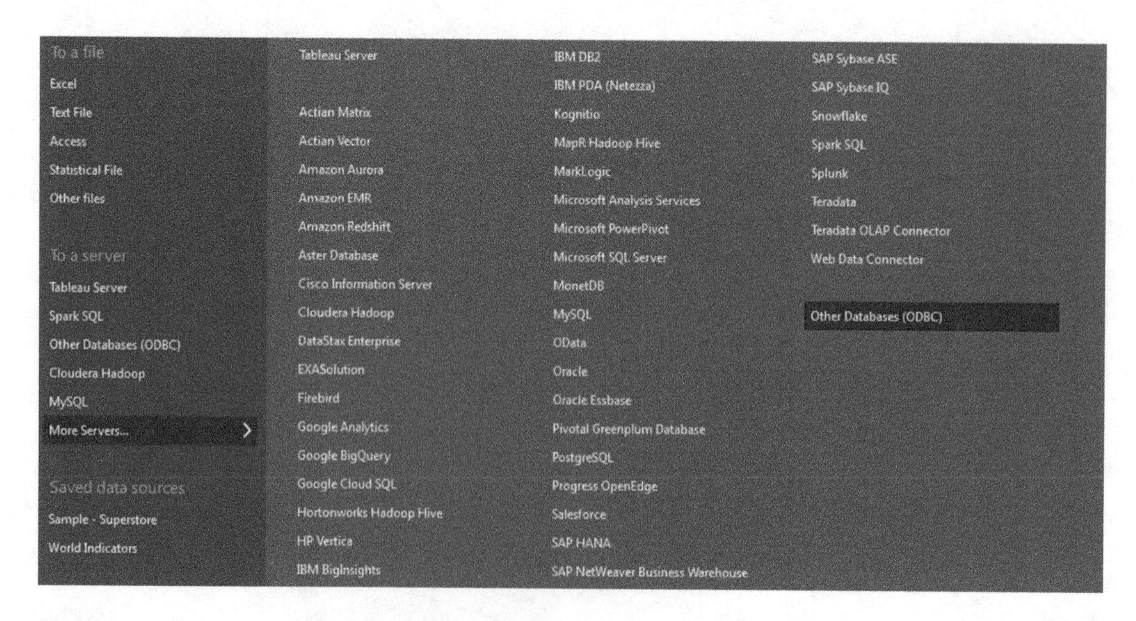

Figure 2-59. *Connection to "Other Databases (ODBC)"*

2.4.3.2.1.10 Step 10

Other Databases (ODBC) Connection Wizard will show up. Select "Simba MongoDB ODBC DSN" and click on the "Connect" button (Shown in Fig. 2-60). Next, click on the "OK" button to connect to MongoDB Database.

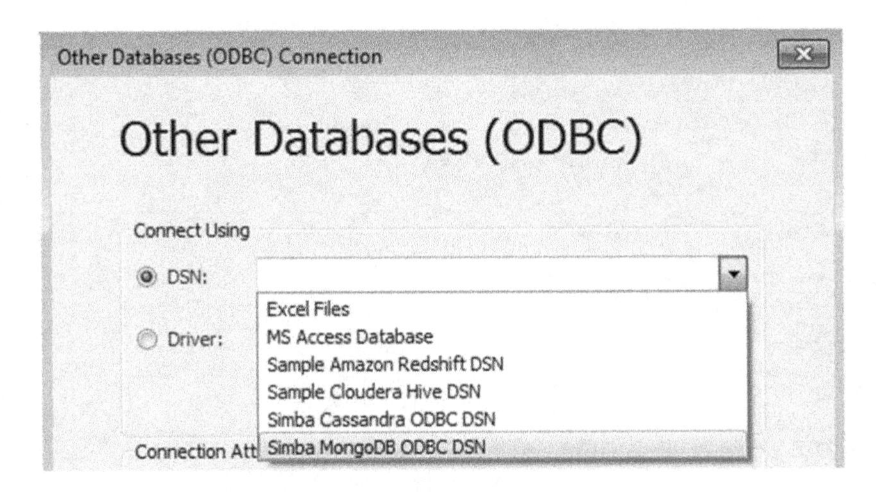

Figure 2-60. *Selection of "Simba MongoDB ODBC DSN"*

2.4.3.2.1.11 Step 11

If the connection is successful, it shows the screen below to allow one to select the desired database (Shown in Fig. 2-61).

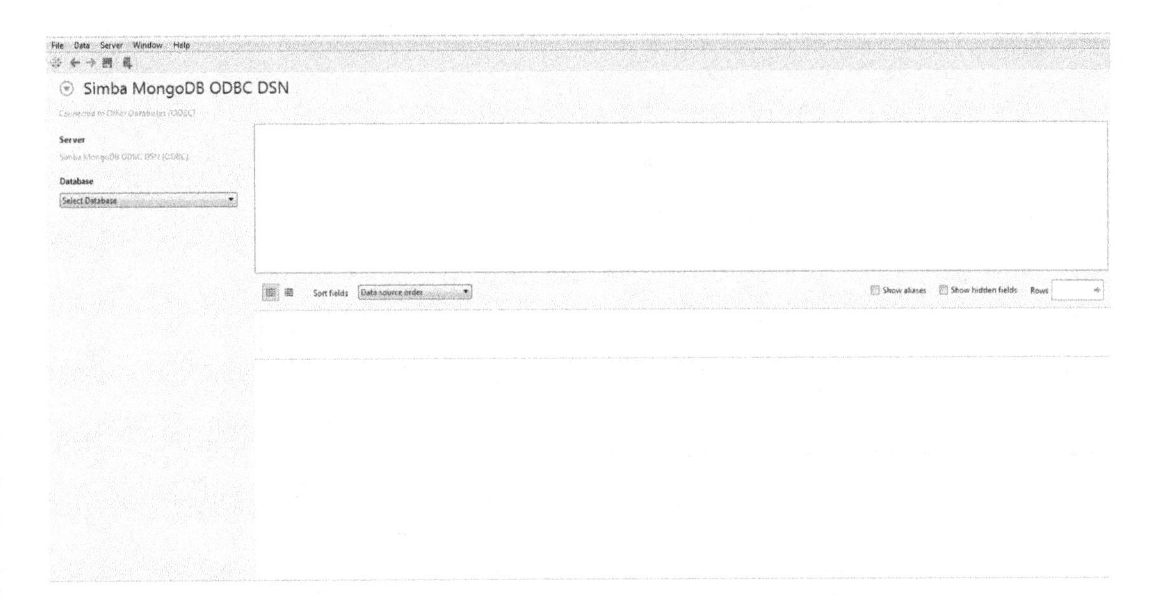

Figure 2-61. *Data source page showing a MongoDB connection*

2.5 Metadata Grid

Click on the metadata grid icon (Shown in Fig. 2-62) to open the metadata grid. Metadata grid displays the fields that are available in the Data Source. Metadata grid helps one to analyse the structure of Tableau data source (Shown in Fig. 2-63), to rename or hide fields at once (Shown in Fig. 2-64), etc.

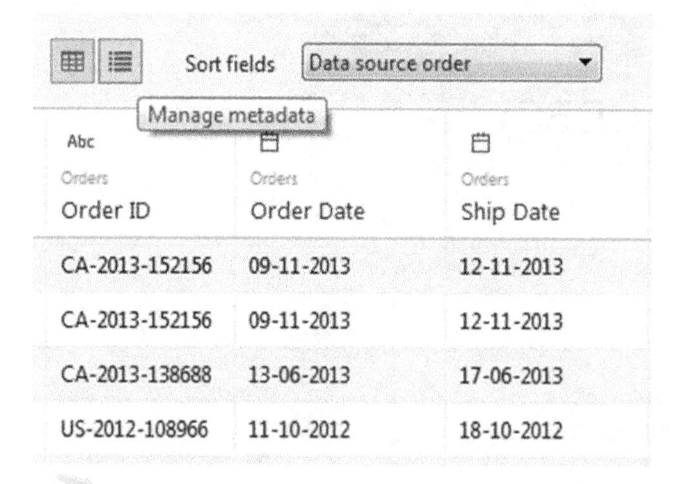

Figure 2-62. *Metadata Grid Icon*

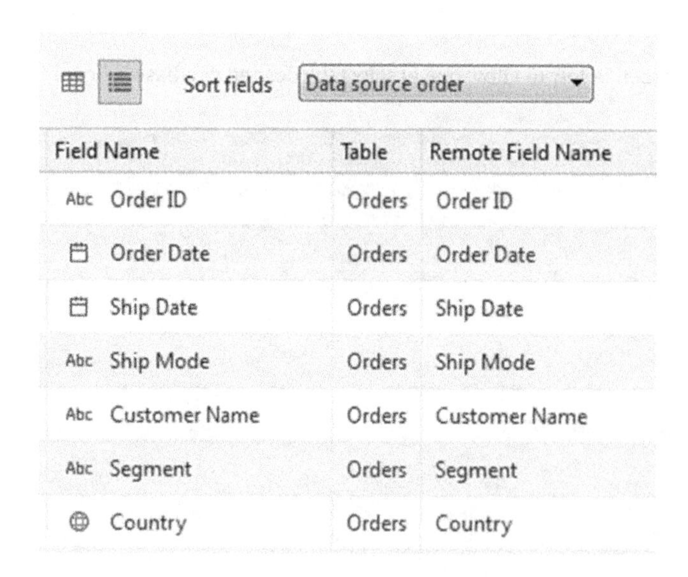

Figure 2-63. *Metadata grid displaying data source fields*

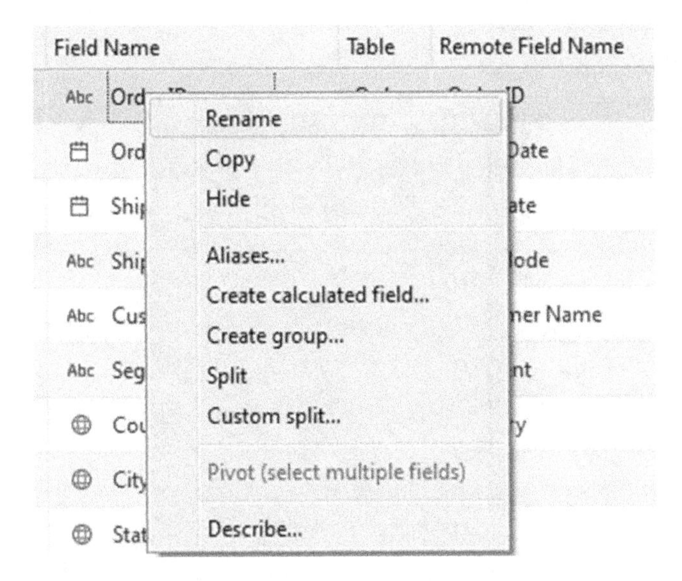

Figure 2-64. *Options to "Rename" or "Hide" field and other options*

2.6 Joins

Relational data source contains collections of tables and tables are related by specific field. For example, let us consider the schema design for a blog website. It contains a table for recording blog entries such as blog id, blog tile, description, URL, likes, and posted by, etc. In addition to this, there could be another table to store details of comments such as blog id, comment id, by user, likes and comments. To analyze and answer questions such as which blog contains the highest likes, you will need to join two tables using a common field such as blog id.

Once you establish a connection, you can use the data source page to connect to multiple tables, and specify joins to perform your analysis.

Tableau Desktop supports different types of joins such as inner, left, right and full outer.

2.6.1 Adding Fields to the Data Pane

You can add or edit a table to add or remove a field, modify join operation from the data pane to specify how your data should look for your analysis.

Follow the steps to add a field to a table.

2.6.1.1 Steps to add a field to a table

2.6.1.1.1 Step 1

Select a data source on the data pane and then right click to get the "Edit data source..." option (Shown in Fig. 2-65).

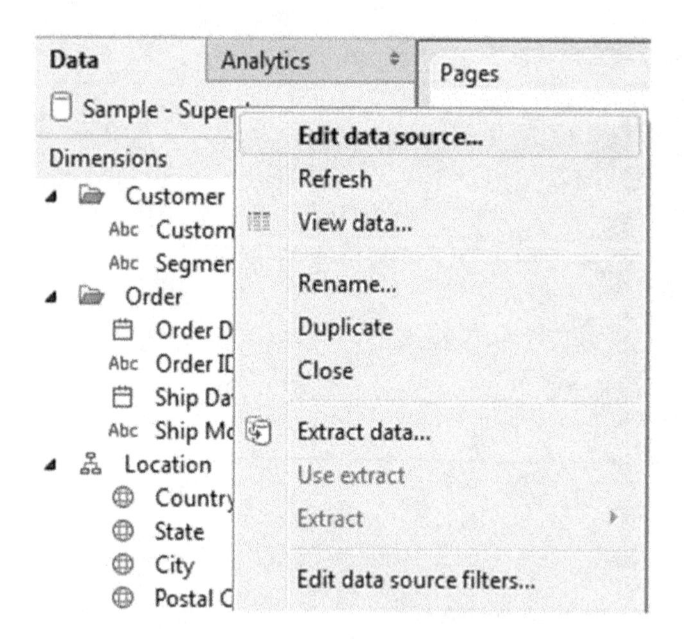

Figure 2-65. *"Edit data source..." option*

2.6.1.1.2 Step 2

You will be able to view the data source page. Drag the required table to the canvas area to perform the join operation (Shown in Fig. 2-66 and Fig. 2-67).

Orders

Sort fields: Data source order | Show aliases | Show hidden fields | Rows 1,000

Order ID	Order Date	Ship Date	Ship Mode	Customer Name	Segment	Country	City	State	Postal Code	Region	Categ
CA-2013-152156	09-11-2013	12-11-2013	Second Class	Claire Gute	Consumer	United States	Henderson	Kentucky	42420	South	Fur
CA-2013-152156	09-11-2013	12-11-2013	Second Class	Claire Gute	Consumer	United States	Henderson	Kentucky	42420	South	Fur
CA-2013-138688	13-06-2013	17-06-2013	Second Class	Darrin Van Huff	Corporate	United States	Los Angeles	California	90036	West	Off
US-2012-108966	11-10-2012	18-10-2012	Standard Class	Sean O'Donnell	Consumer	United States	Fort Lauderd...	Florida	33311	South	Fur
US-2012-108966	11-10-2012	18-10-2012	Standard Class	Sean O'Donnell	Consumer	United States	Fort Lauderd...	Florida	33311	South	Off
CA-2011-115812	09-06-2011	14-06-2011	Standard Class	Brosina Hoffman	Consumer	United States	Los Angeles	California	90032	West	Fur
CA-2011-115812	09-06-2011	14-06-2011	Standard Class	Brosina Hoffman	Consumer	United States	Los Angeles	California	90032	West	Off
CA-2011-115812	09-06-2011	14-06-2011	Standard Class	Brosina Hoffman	Consumer	United States	Los Angeles	California	90032	West	Tec
CA-2011-115812	09-06-2011	14-06-2011	Standard Class	Brosina Hoffman	Consumer	United States	Los Angeles	California	90032	West	Off
CA-2011-115812	09-06-2011	14-06-2011	Standard Class	Brosina Hoffman	Consumer	United States	Los Angeles	California	90032	West	Off

Figure 2-66. *"Orders" sheet placed on canvas area*

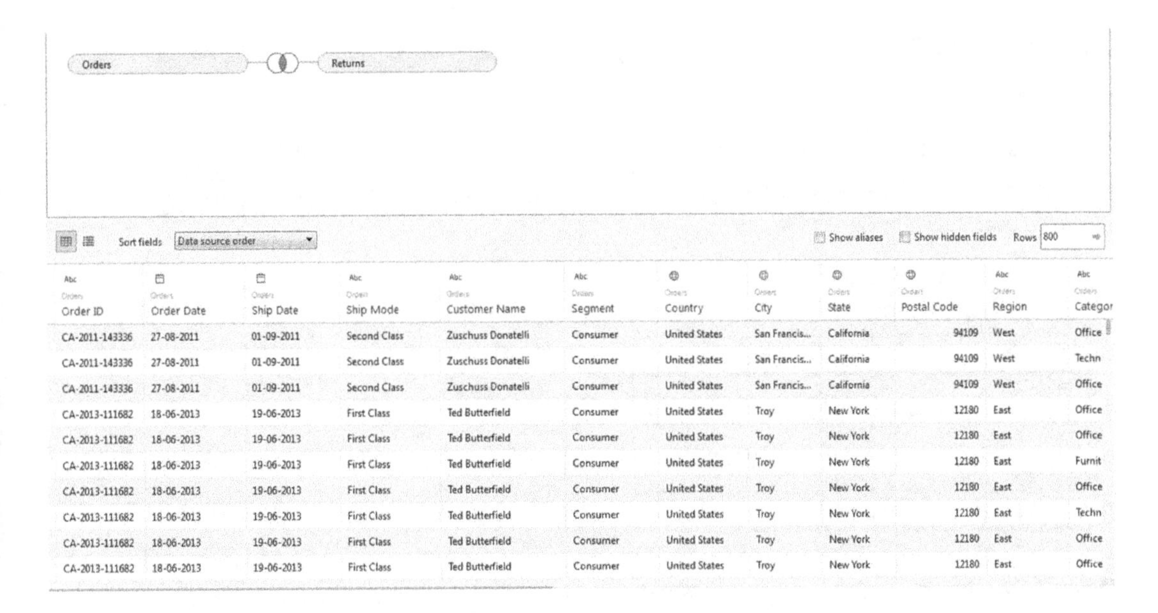

Figure 2-67. *"Returns" sheet placed on canvas area*

2.6.1.1.3 Step 3

Click on join icon to select / edit the type of join operation (Shown in Fig. 2-68).

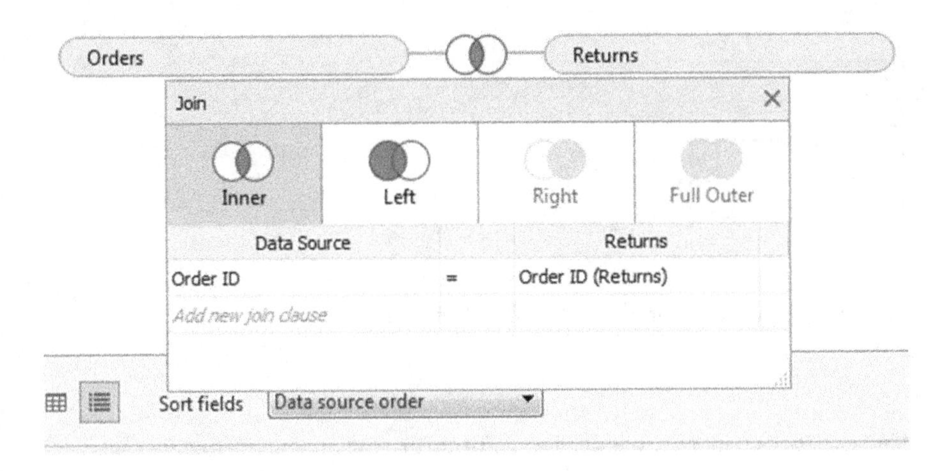

Figure 2-68. *Different types of join operation*

2.6.2 Exploring different types of Join

Let us understand how to perform the Inner Join, Left Join, etc. in Tableau using two tables namely, "Student" and "Grade".

Follow the steps to perform the join operation between two data sources in Tableau.

2.6.2.1 Steps

Follow the below steps.

2.6.2.1.1 Step 1

Create an Excel File and name it as "Sample - Student". Student sheet contains details about students as shown below (Shown in Fig. 2-69).

	A	B
1	StudNo	Student Name
2	1001	John
3	1002	Jack
4	1003	Smith
5	1004	Joshi

Figure 2-69. *Student sheet showing details about students*

Grade Sheet contains the below details (Shown in Fig. 2-70).

	A	B
1	StudNo	Grade
2	1001	A
3	1002	B
4	1003	A
5	1006	C
6	1007	A
7	1008	B

Figure 2-70. *Grade sheet showing details about the grades scored by students*

2.6.2.1.2 Step 2

Connect to "Sample-Student". In the Data Source Page, drag and drop the "Student" and "Grade" Sheets (Shown in Fig. 2-71).

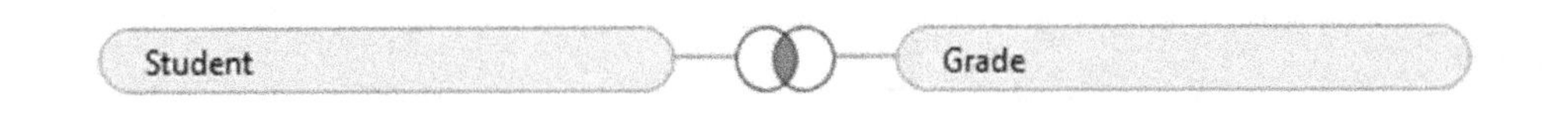

Sort fields: Data source order

Grade StudNo (Grade)	Grade Grade	Student Stud No	Student Student Name
1,001	A	1,001	John
1,002	B	1,002	Jack
1,003	A	1,003	Smith

Figure 2-71. *"Student" and "Grade" tables placed on the canvas*

2.6.2.2 Inner Join

Inner Join fetches all records from one table having a matching entry in the second table based on a common field. By default in Tableau, tables are joined using Inner Join (Shown in Fig. 2-72).

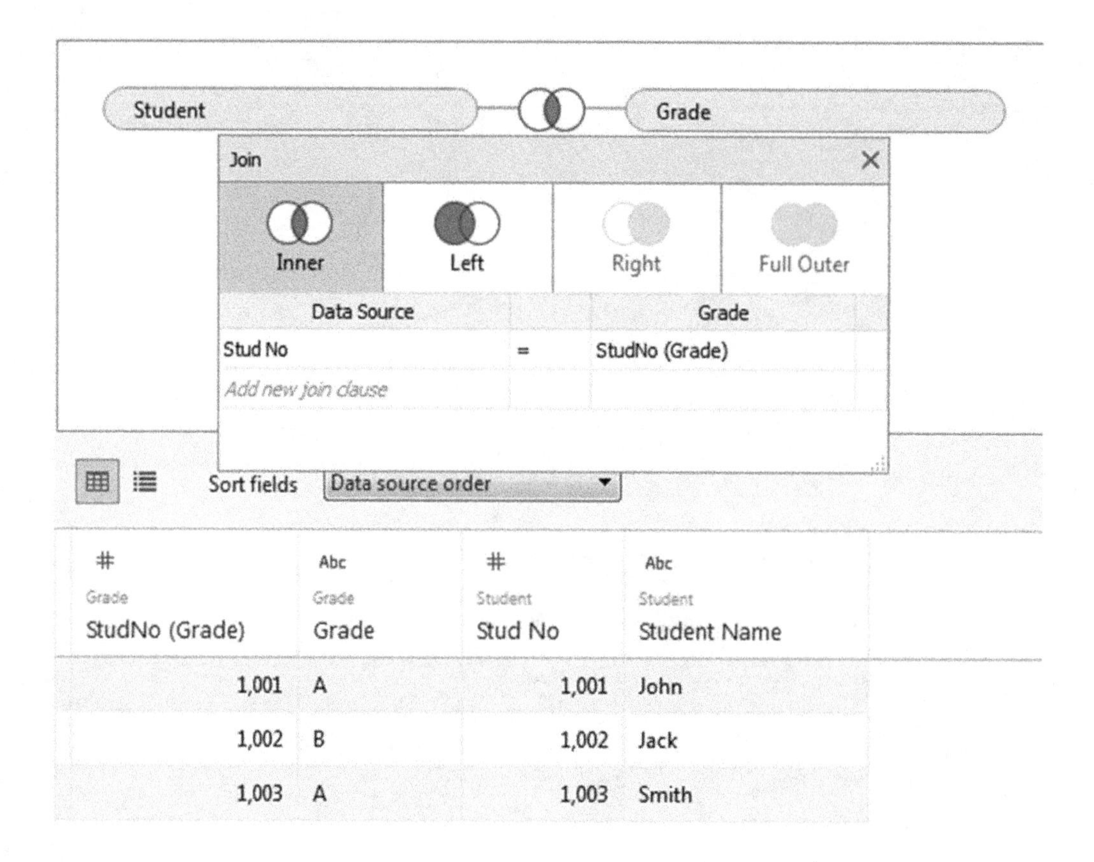

Figure 2-72. *"Inner Join" Operation*

2.6.2.3 Left Join

Left join fetches records from the left table having a matching record(s) in the right table. Null values will be displayed for records where there is no match in the right side table (Shown in Fig. 2-73).

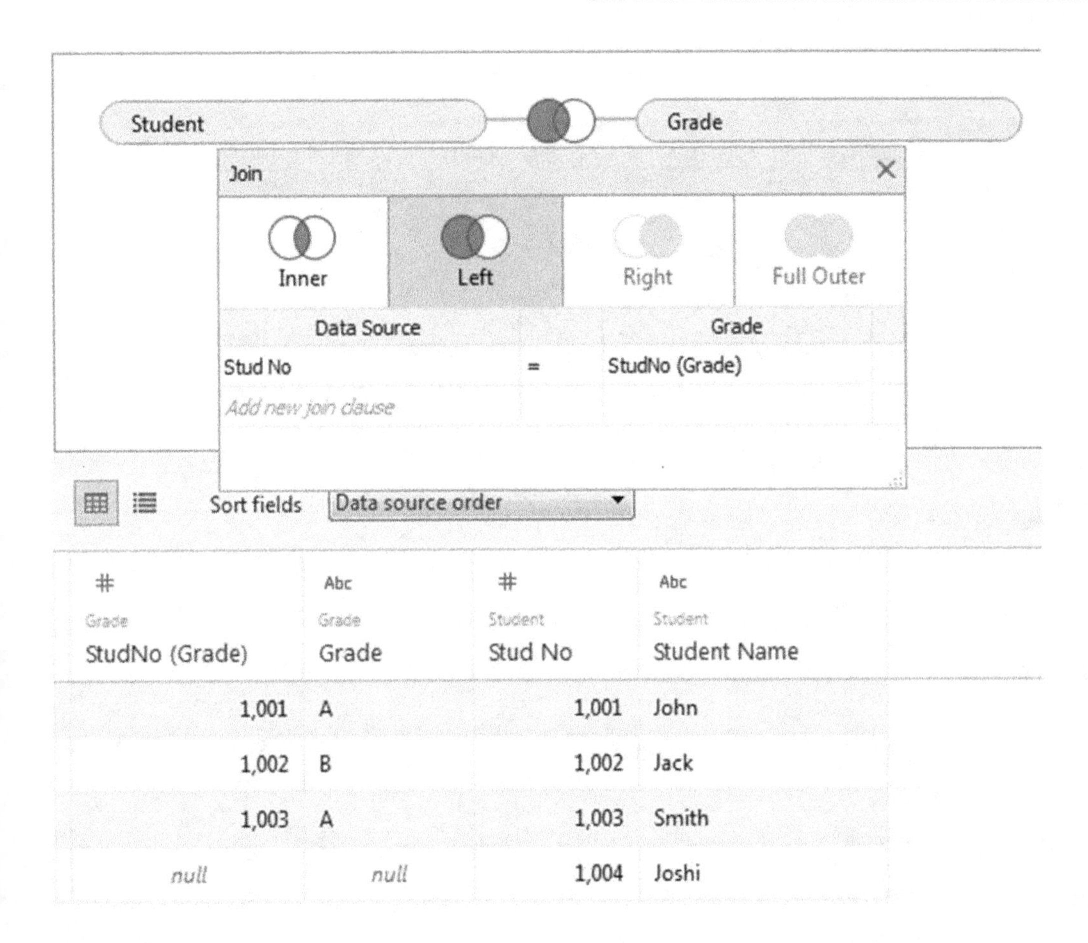

Figure 2-73. *"Left Join" Operation*

2.6.2.4 Right Join

MySQL data source supports Right Join. Refer to connect to server section to connect to MySQL data source. Create two tables namely "Student" and "Grade" with the same structure as shown in the example on "Inner Join".

Right join fetches records from the right table having a matching record(s) in the left table. Null values will be displayed for records where there is no match in the left side table (Shown in Fig. 2-74).

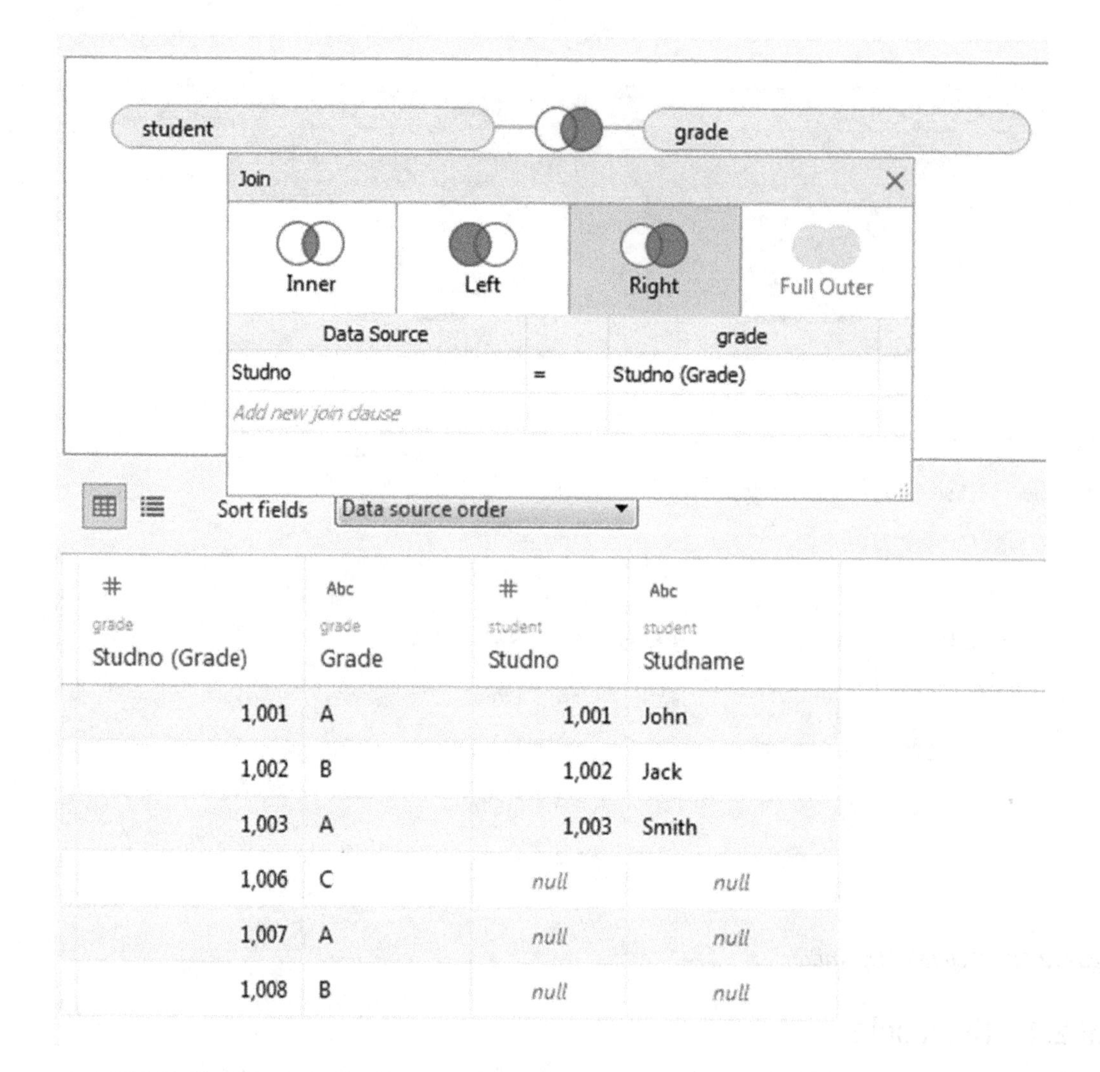

Figure 2-74. "Right Join" Operation

2.6.3 Union

In Tableau, "union" operation supports combining data from different files. Consider the "Sample - Student" Excel file used in Inner Join.

Follow the below steps to perform the "union" operation.

2.6.3.1 Steps

2.6.3.1.1 Step 1

Click on "New Union" option as shown below (Shown in Fig. 2-75).

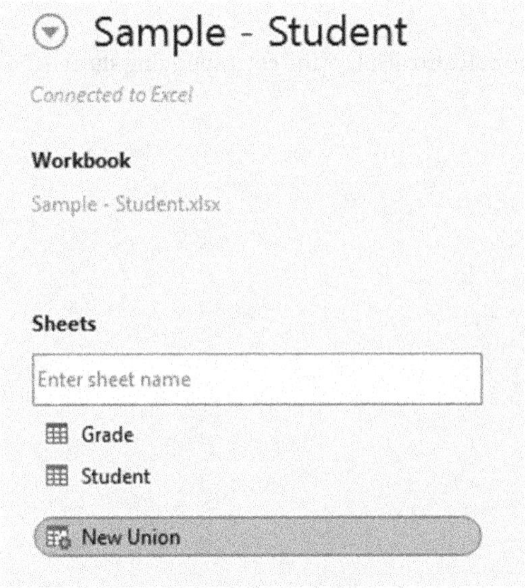

Figure 2-75. "New Union" option

2.6.3.1.2 Step 2

The "Union Window" will open. Specify the name as "Student_Grade", drag and drop the "Student" and "Grade" sheets to the window. Click on "OK" button to combine the data from the two sheets (Shown in Fig. 2-76).

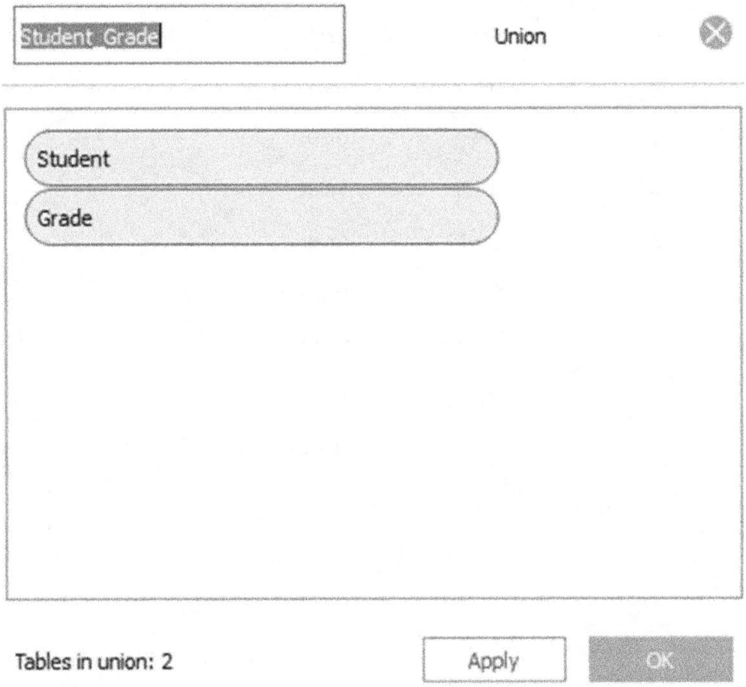

Figure 2-76. Union window

2.6.3.1.3 Step 3

You will be able to view the combined data in the grid window. It also displays the corresponding sheet name for each field (Shown in Fig. 2-77).

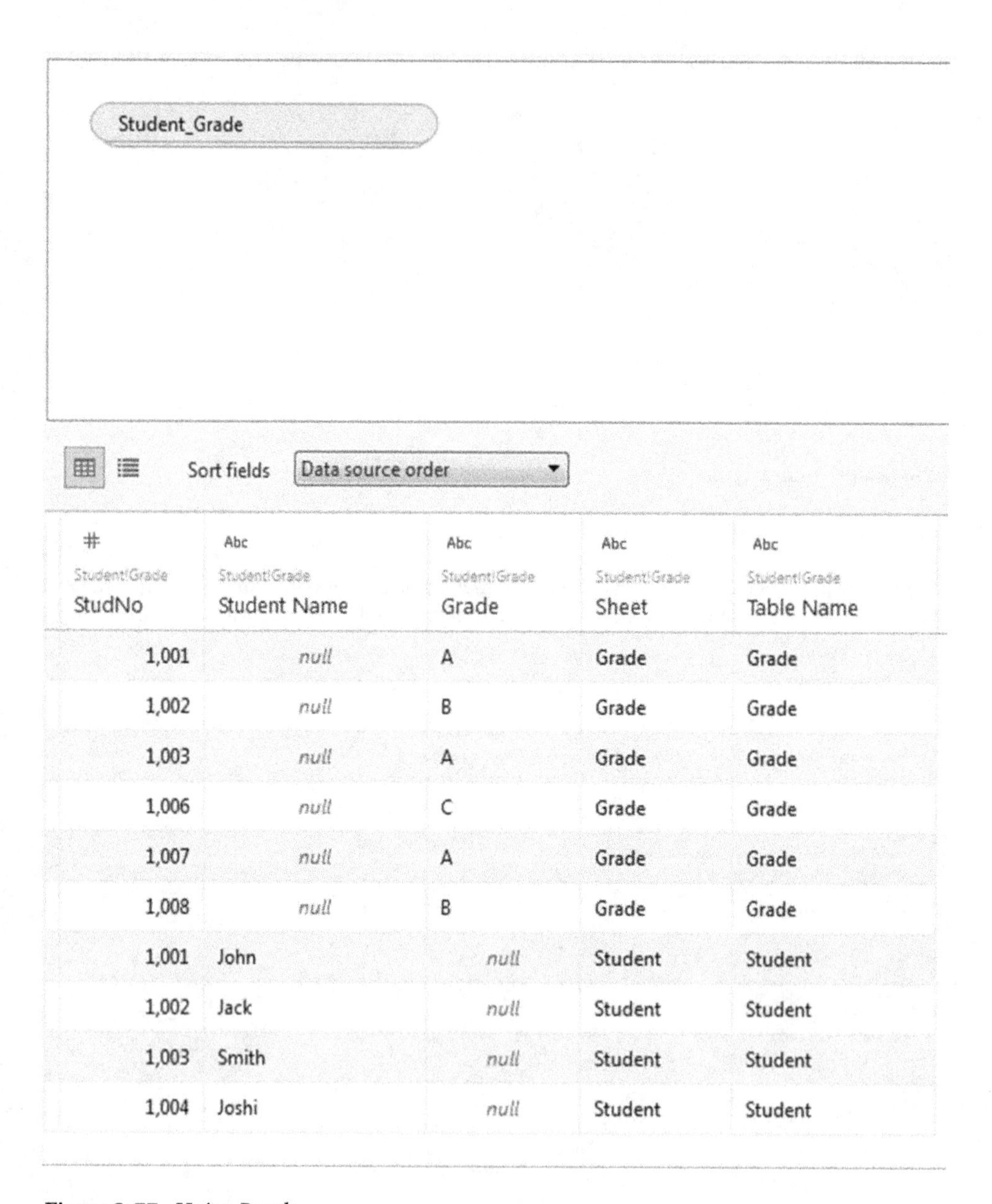

# Student!Grade StudNo	Abc Student!Grade Student Name	Abc Student!Grade Grade	Abc Student!Grade Sheet	Abc Student!Grade Table Name
1,001	null	A	Grade	Grade
1,002	null	B	Grade	Grade
1,003	null	A	Grade	Grade
1,006	null	C	Grade	Grade
1,007	null	A	Grade	Grade
1,008	null	B	Grade	Grade
1,001	John	null	Student	Student
1,002	Jack	null	Student	Student
1,003	Smith	null	Student	Student
1,004	Joshi	null	Student	Student

Figure 2-77. *Union Result*

2.7 Custom SQL

Let us learn to write SQL statements to retrieve data suitable for analysis. Assume, you have connected to a MySQL data source, which has the following structure (Refer to Table 2-4, Table 2-5).

Table 2-4. *Student Table*

StudNo	Student Name
1001	John
1002	Jack
1003	Smith
1004	Joshi

Table 2-5. *Grade Table*

StudNo	Grade
1001	A
1002	B
1003	A
1006	C
1007	A
1008	B

Database Name: student

2.7.1 Demo 1

Follow the below steps to write a Custom SQL.

2.7.1.1 Steps to write custom SQL

2.7.1.1.1 Step 1

Select the "student" database (Shown in Fig. 2-78).

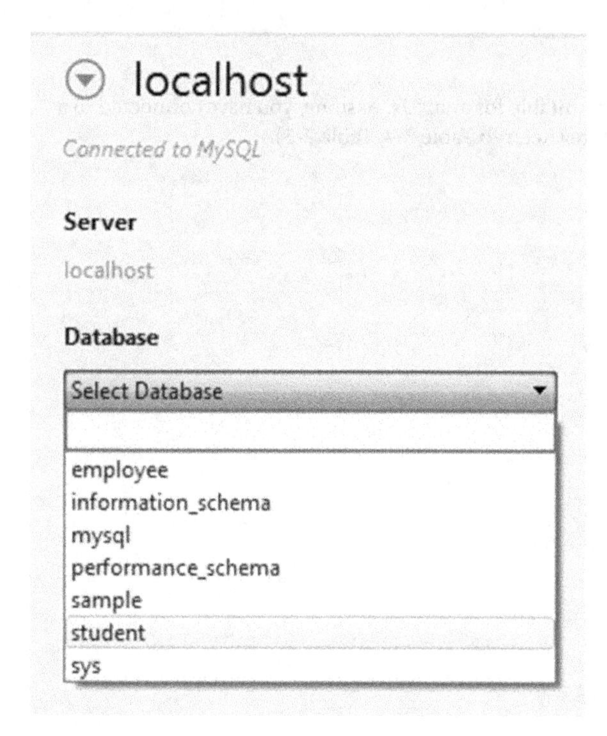

Figure 2-78. *Selection of "student" database*

2.7.1.1.2 Step 2

Double click on "New Custom SQL" (Shown in Fig. 2-79).

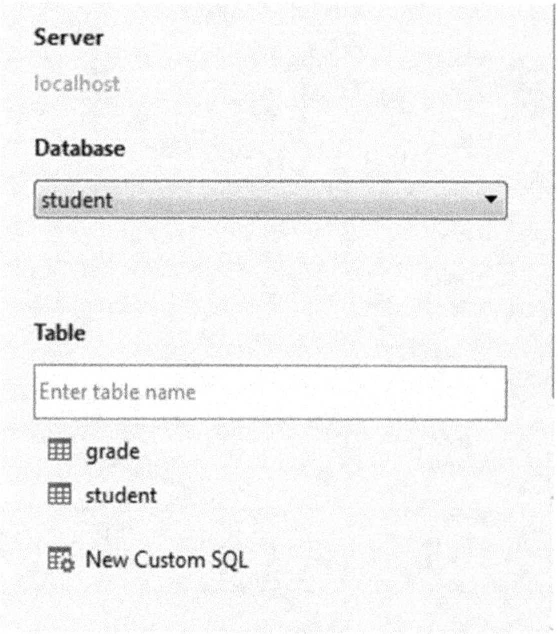

Figure 2-79. *"New Custom SQL" option*

2.7.1.1.3 Step 3

"Edit Custom SQL" dialog box shows up. Type in the required SQL statement (Shown in Fig. 2-80).

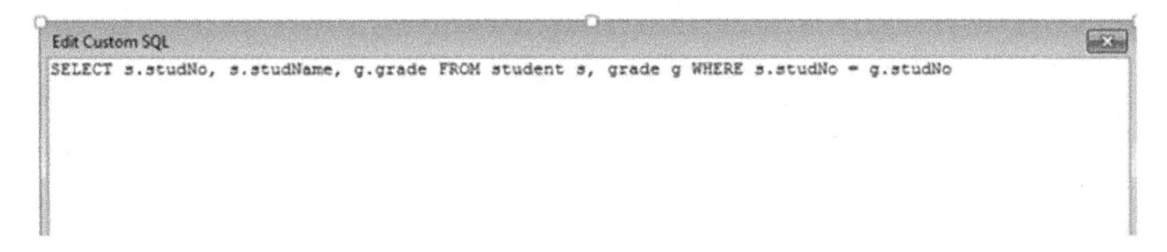

Figure 2-80. *"Edit Custom SQL" Statement*

2.7.1.1.4 Step 4

Click on "Preview Results" to preview the output and click on "OK" to create "Custom SQL" (Shown in Fig. 2-81).

View Data

3 rows

grade	studName	studNo
A	John	1,001
B	Jack	1,002
A	Smith	1,003

Figure 2-81. *"View Data" window showing preview of the data*

2.7.1.1.5 Step 5

Custom SQL Query is displayed in the canvas area (Shown in Fig. 2-82).

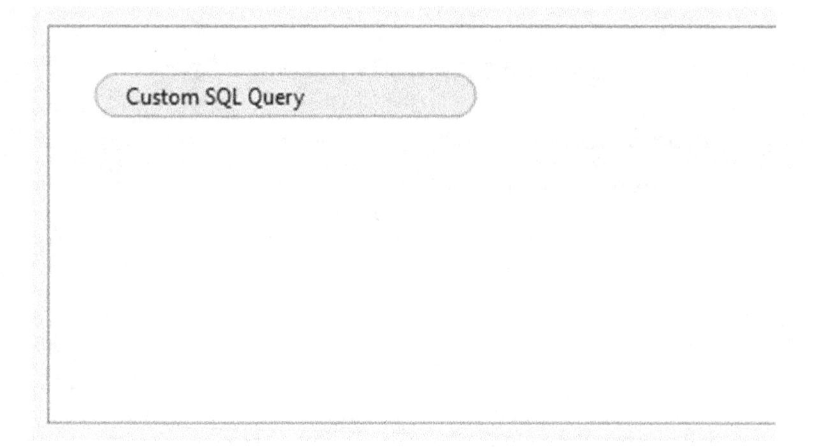

Figure 2-82. *"Custom SQL Query" placed in the canvas area*

2.7.1.1.6 Step 6

Right click on Custom SQL Query, to "Edit Custom SQL Query" (Shown in Fig. 2-83).

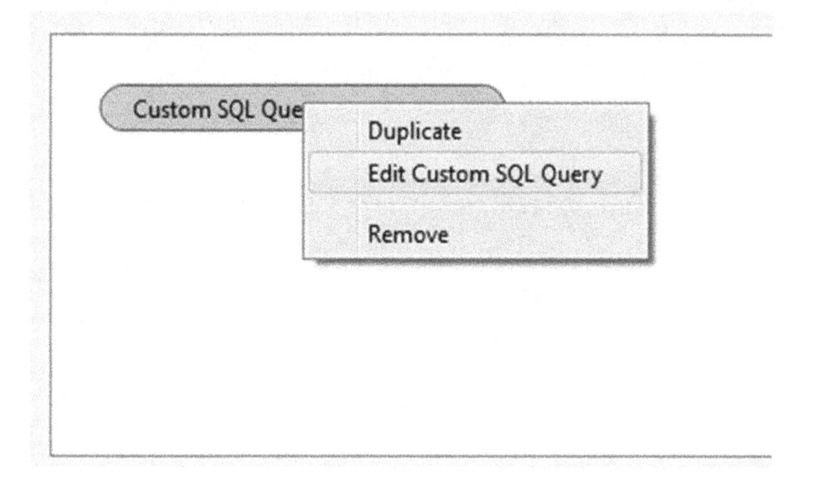

Figure 2-83. *"Edit Custom SQL Query"*

2.8 Data Blending

Data blending is the best choice when you want to use data from more than one data source for your analysis.

To perform data blending, a common field should be present in both the data sources. Here, we will consider "Sample-Superstore Excel data source" and "Sample - CoffeeChain Access data source". They have a common field namely, Market (CoffeeChain) and Region (Superstore).

2.8.1 Demo 1

Follow the below steps to perform Data Blending.

2.8.1.1 Steps

2.8.1.1.1 Step 1

Connect to Sample-Superstore data source.

2.8.1.1.2 Step 2

Drag and drop "Orders" sheet into the Canvas area.

2.8.1.1.3 Step 3

Connect to "Sample-Coffee chain Access" data source.

2.8.1.1.4 Step 4

Drag and drop "CoffeeChain Query" to the Canvas area.

2.8.1.1.5 Step 5

Go to worksheet. To create a relationship, select the data menu, and then select "Edit relationship..." (Shown in Fig. 2-84).

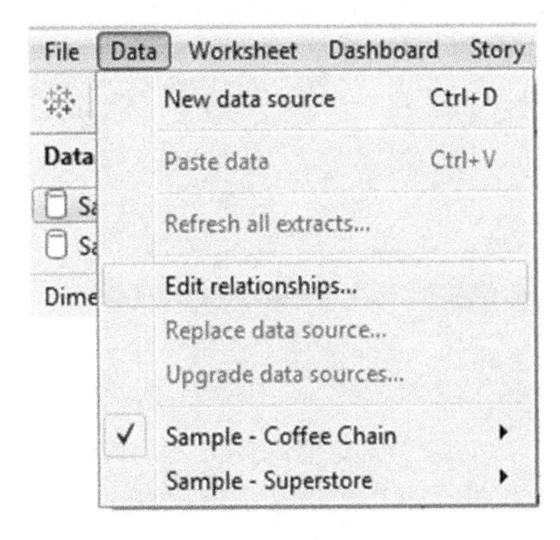

Figure 2-84. *"Edit relationships..." option*

2.8.1.1.6 Step 6

Relationships dialog box will open as shown below (Shown in Fig. 2-85).

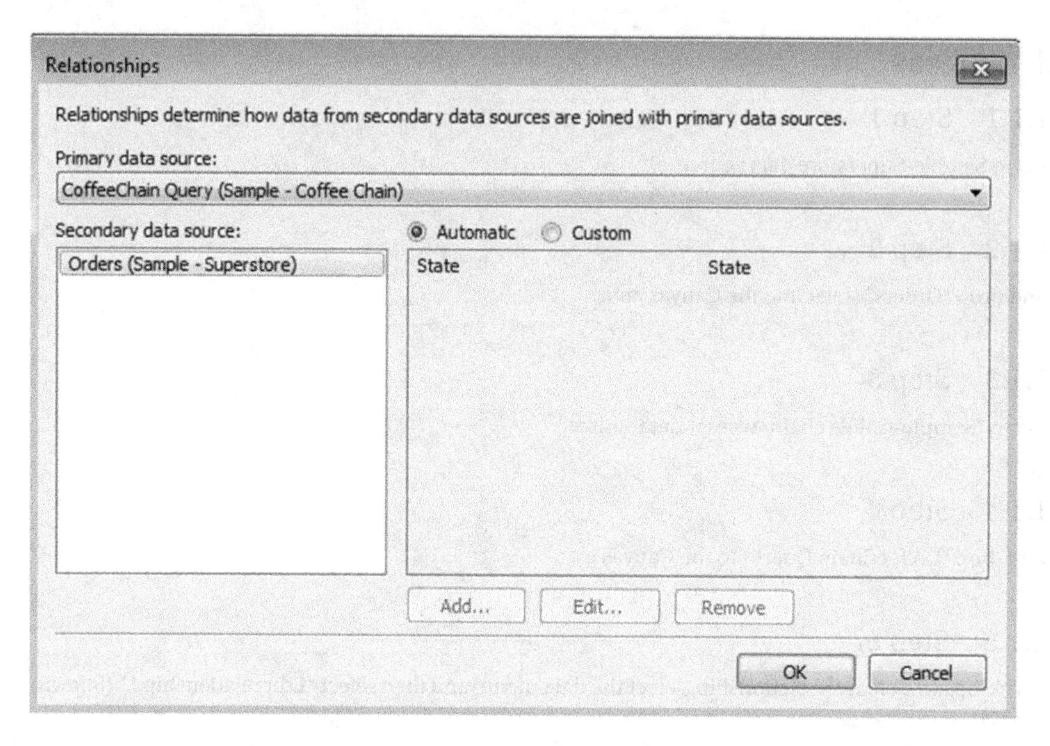

Figure 2-85. *"Relationships" dialog box showing Primary data source and Secondary data source*

2.8.1.1.7 Step 7

Click on the "Custom" radio button to create a custom relationship and then click on the "Add" button to add field mapping (Shown in Fig. 2-86).

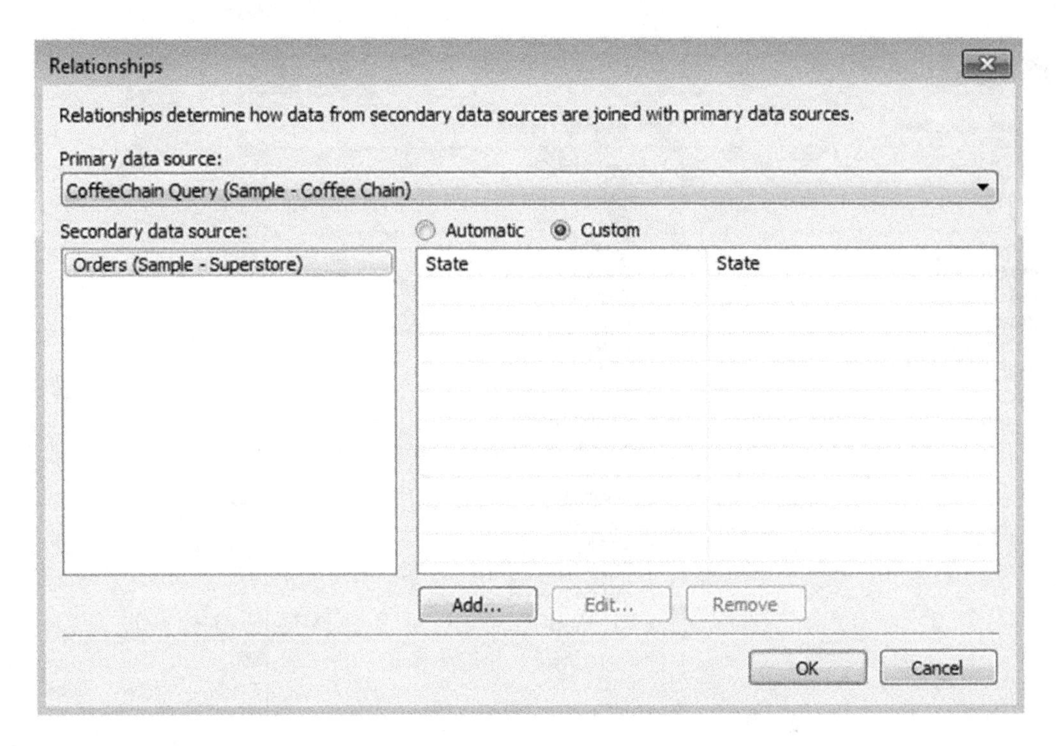

Figure 2-86. *Selected "Custom" option*

2.8.1.1.8 Step 8

From the Add/Edit field mapping dialog box, select "Market" from the primary data source (Coffeechain) and "Region" from the secondary data source (Superstore) and click on "OK" to create the field mapping (Shown in Fig. 2-87).

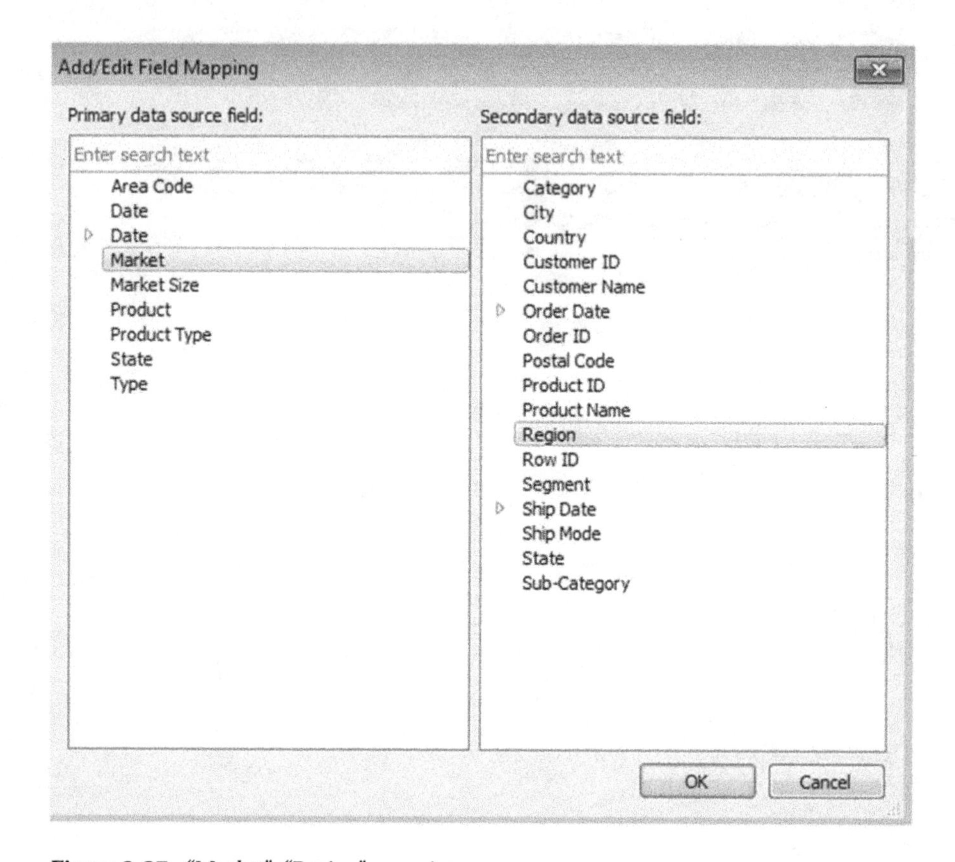

Figure 2-87. *"Market", "Region" mapping*

2.8.1.1.9 Step 9

Observe the mapping field in the "Relationships" dialog box. (Shown in Fig. 2-88).

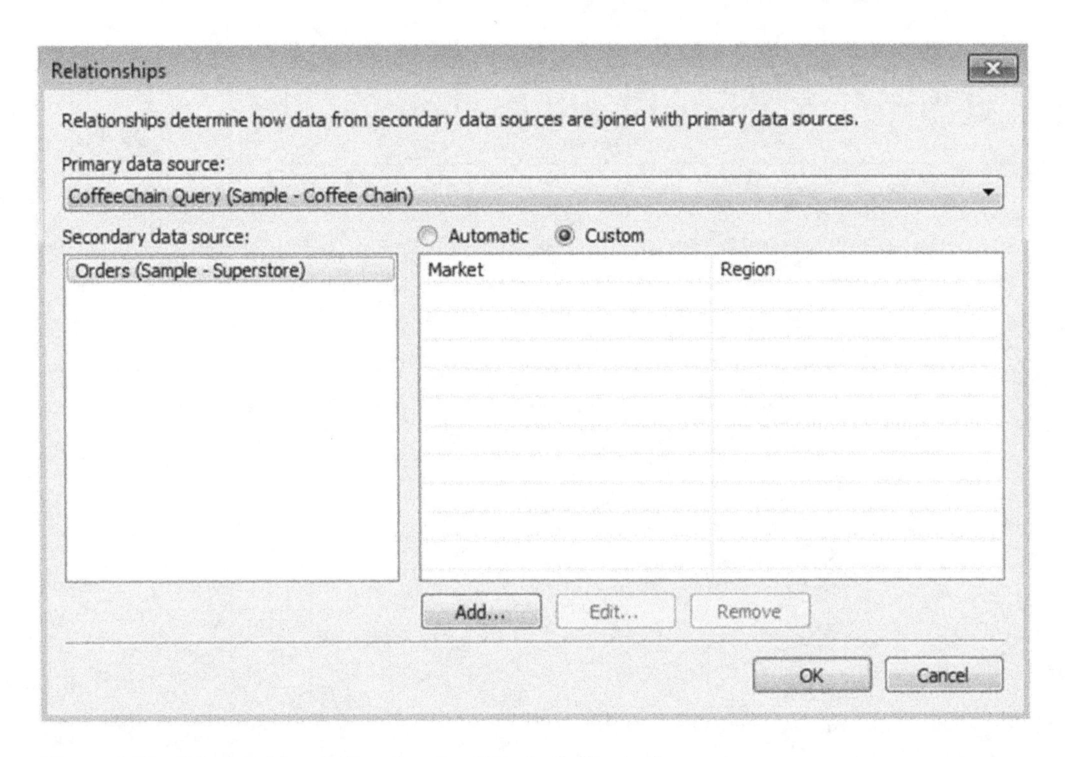

Figure 2-88. *Relationship window showing "Market", "Region" mapping*

2.8.1.1.10 Step 10

From "CoffeeChain Query", select the dimension "Market" and place it on the rows shelf (Shown in Fig. 2-89).

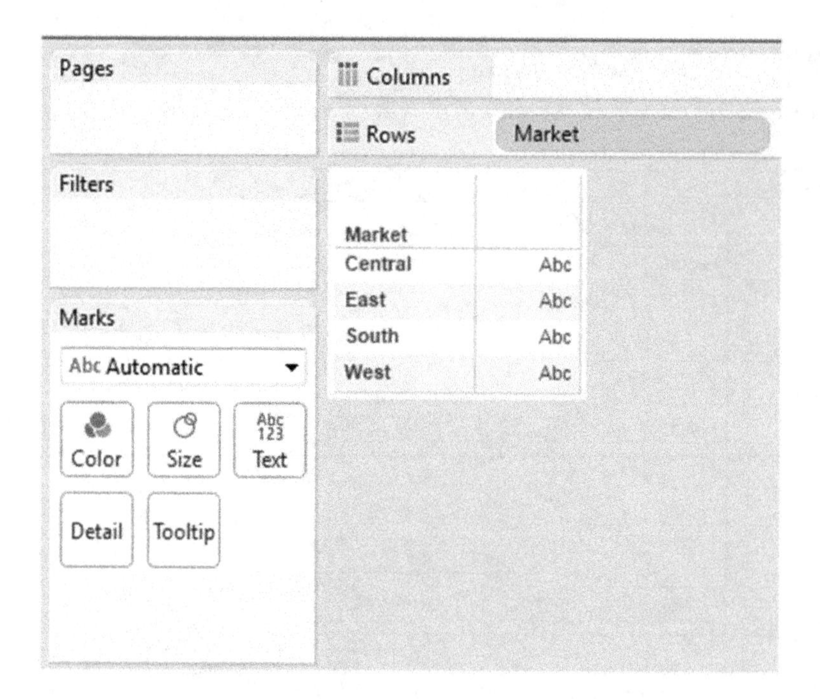

Figure 2-89. *Dimension, "Market" placed on the rows shelf*

2.8.1.1.11 Step 11

Now, click on "Orders" data source. You can see the relationship symbol (paper clip symbol) next to the dimension, "Region" (Fig. 2-90).

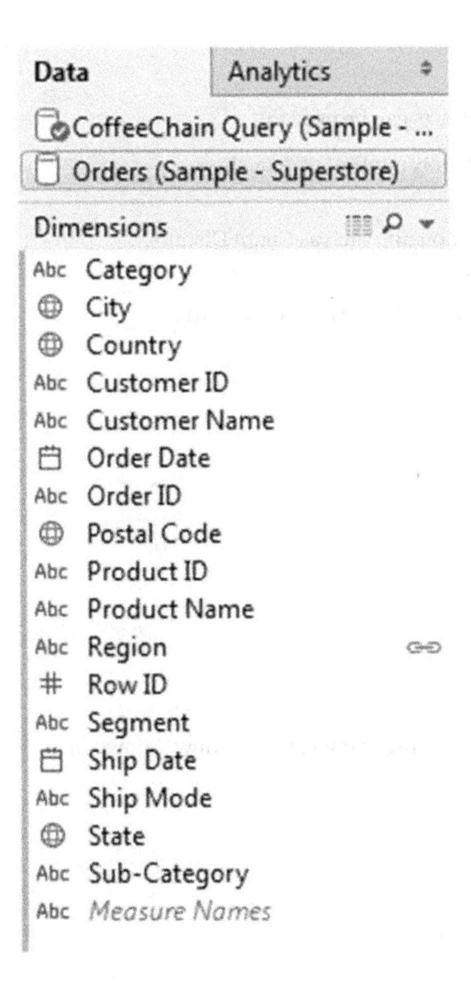

***Figure 2-90.** Relationship symbol for the dimension "Region"*

2.8.1.1.12 Step 12

From Orders, select the measure "Sales" and place it on the columns shelf to construct a view. The view below represents the "Sales" by "Market" (Fig. 2-91).

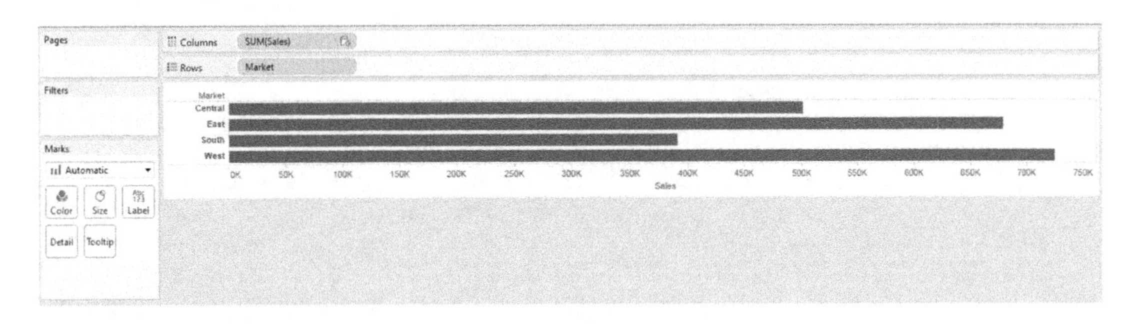

***Figure 2-91.** "Sales" by "Market" view*

2.9 Data Extracts

The saved subsets of data source are known as extracts. Data extracts can be used to:

- improve performance: When you want to work with only a subset of data, filter extract helps you to limit the load on the server.
- use Tableau functionality: You can use Tableau functionality such as Count Distinct, which is not available with the original data source.
- to provide an offline access to data: You can extract the data to a local data source when you don't have access to server.

2.9.1 Demo 1

Let us learn how to create an extract.

2.9.1.1 Steps to create an extract

Follow these steps.

2.9.1.1.1 Step 1

On the data source page, select "Extract" and click on "Edit" to open the Extract Data window (Shown in Figure 2-92).

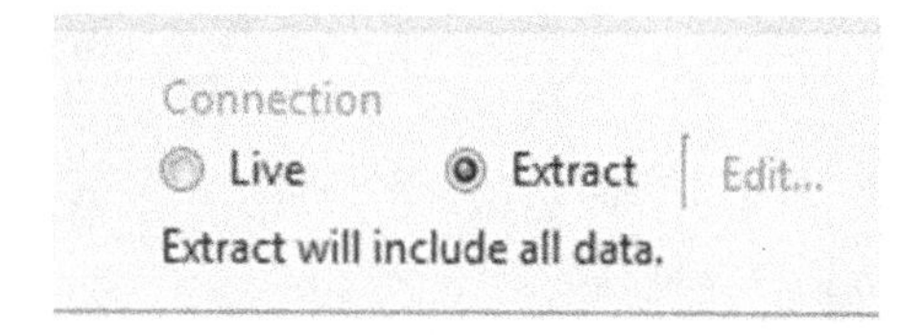

Figure 2-92. *Extract option*

2.9.1.1.2 Step 2

Extract Data dialog box is displayed (Shown in Fig. 2-93).

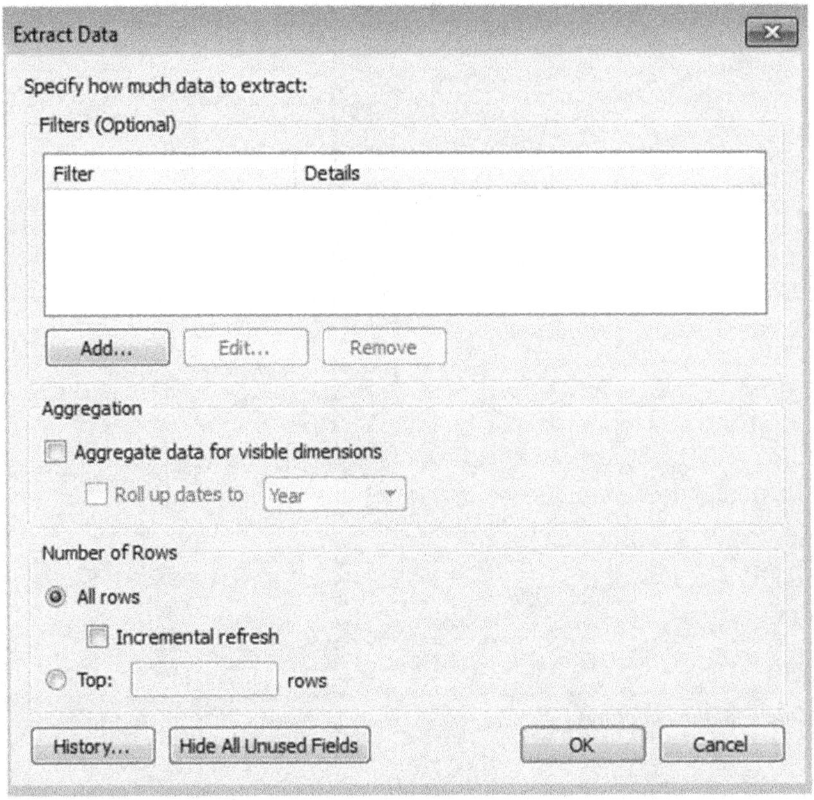

Extract Data
Specify how much data to extract:
Filters (Optional)
Filter
Details
Add...
Edit...
Remove
Aggregation
Aggregate data for visible dimensions
Roll up dates to
Year
Number of Rows
All rows
Incremental refresh
Top:
rows
History...
Hide All Unused Fields
OK
Cancel

Figure 2-93. *"Extract Data" dialog box*

2.9.1.1.3 Step 3

You can limit data for an extract by adding a filter. Click on "Add" button to add a filter. Select "State" field (Shown in Fig. 2-94) as filter criteria and click OK.

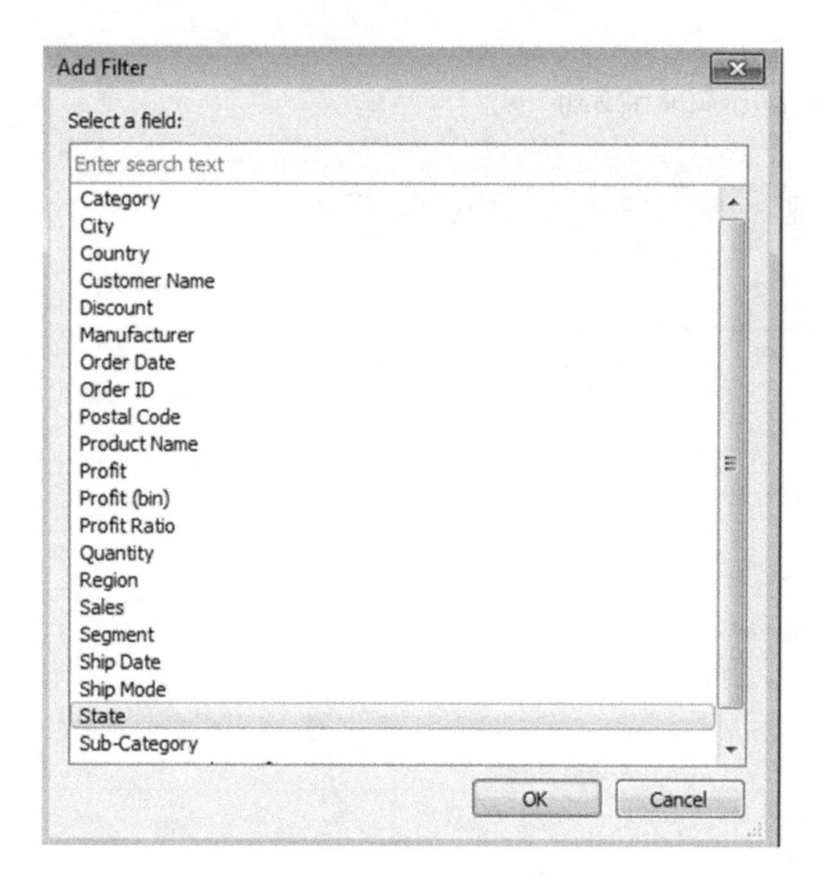

Figure 2-94. *Selected "State" field for filter condition*

2.9.1.1.4 Step 4

Filter [State] dialog box shows up. From the list of states, select only "California" (Shown in Figure 2-95) and click "OK".

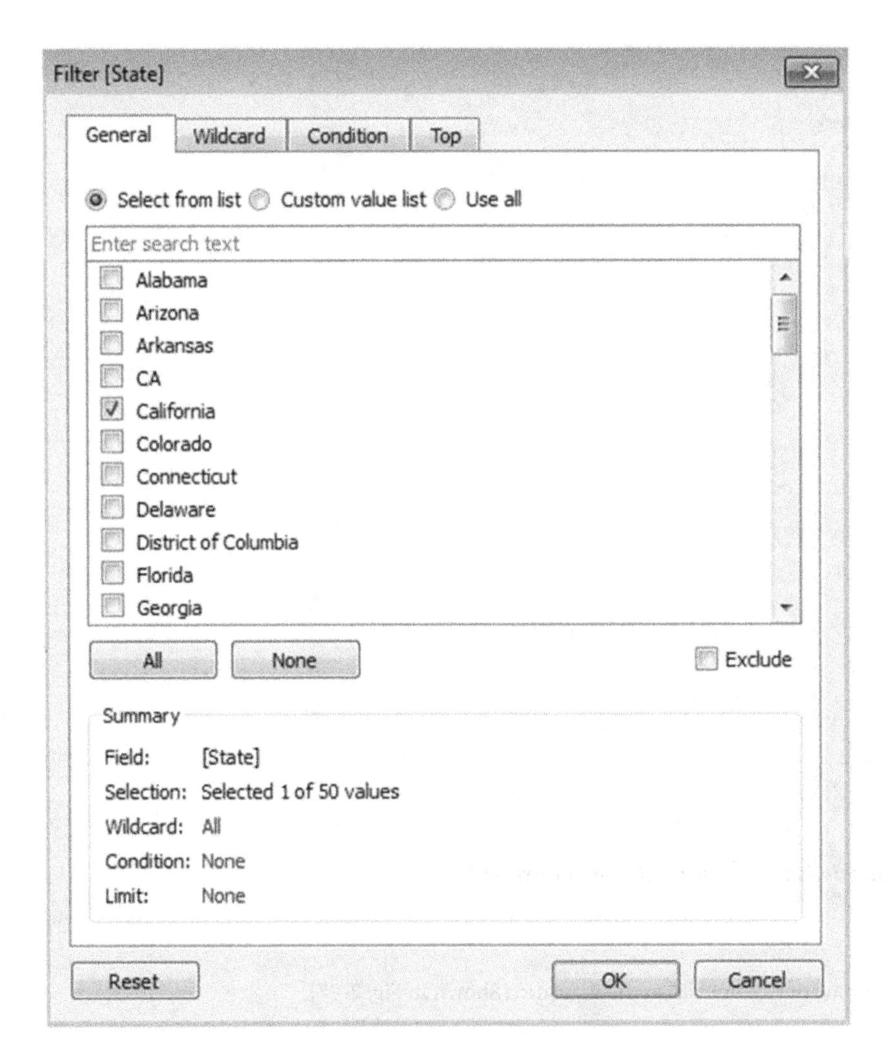

Figure 2-95. *Filter[State] dialog box, California checked*

2.9.1.1.5 Step 5

Check "Aggregate data for visible dimensions" to aggregate measures by their default aggregation. This helps you to minimize the extract file size and to increase performance. Also choose "Roll up dates to" to specify a date level such as Year, Month and select the "Number of Rows" to display a certain number of rows (Shown in Fig. 2-96).

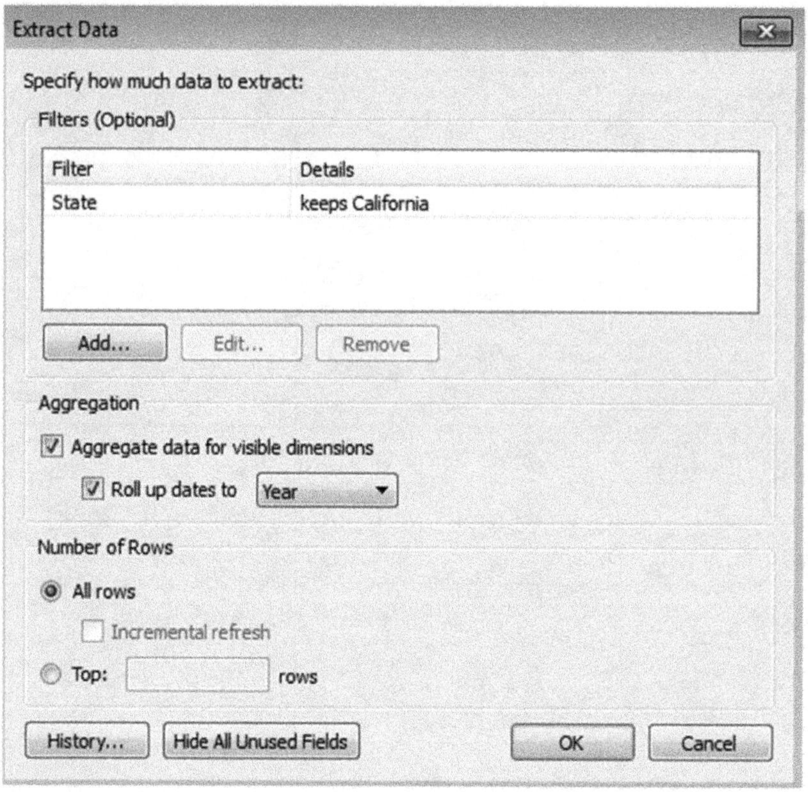

Figure 2-96. Extract data window showing conditions for Extraction

2.9.1.1.6 Step 6

Observe details about the state of "California" in the data grid (Shown in Fig. 2-97).

Orders Country	Orders City	Orders State
United States	Mission Viejo	California
United States	Los Angeles	California
United States	Los Angeles	California
United States	Pasadena	California
United States	Los Angeles	California
United States	San Diego	California
United States	Fresno	California
United States	San Francis...	California
United States	Los Angeles	California
United States	Sacramento	California

Figure 2-97. *Data grid showing only state of California details*

2.9.1.1.7 Step 7

When you go to Sheet 1, you will get a "Save Extract As" dialog box to save your extract (Shown in Fig. 2-98).

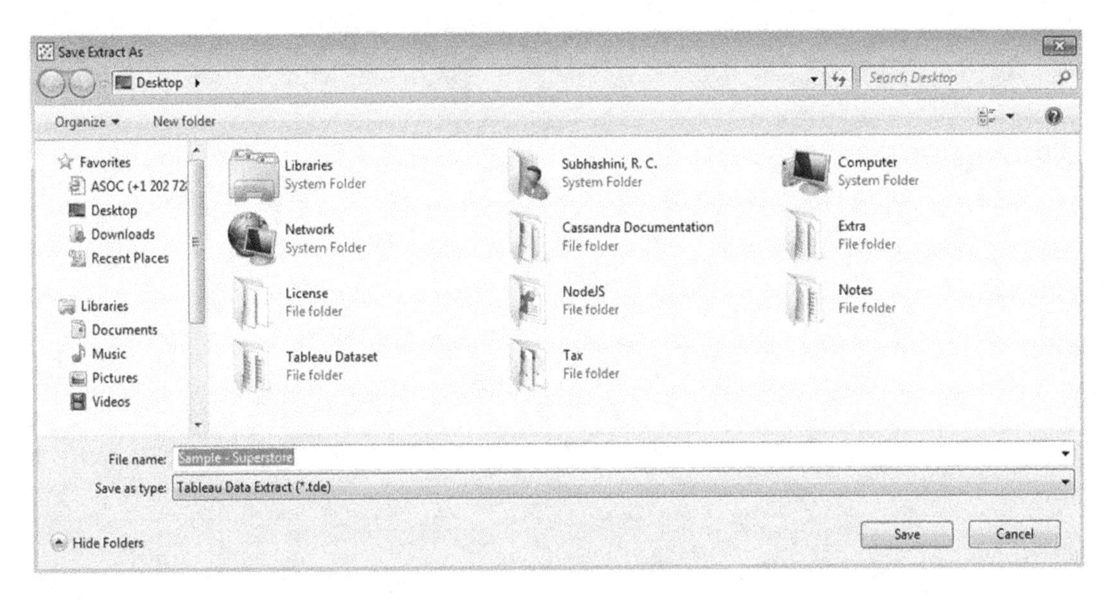

Figure 2-98. *"Save Extract As" dialog box to save extract*

2.10 Points to Remember

- Tableau has a fast, in-memory data engine for analytics.
- Tableau provides various data connectors for databases and a generic ODBC connector to connect to any system not having a native connector.
- All joins are not supported by all databases.
- Tableau extracts helps to improve the performance and provides offline access to your data.

2.11 Next Step

In the next chapter, we will focus on the following:

- Filter
- Sort
- Group
- Hierarchies
- Sets

CHAPTER 3

Simplifying and Sorting Your Data

Chapter 2 introduced us to Tableau Desktop architecture, Tableau environment, connecting to different data sources, joins, custom SQL, data blending and data extracts. This chapter will help us to understand how to simplify, sort and slice data using:

- Filtering
- Sorting
- Discrete and continuous date
- Groups
- Hierarchies
- Sets
- Difference between groups and sets
- Creating parameters

3.1 Filtering

Let us explore how to use filtering to simplify our data.

3.1.1 Why filtering?

Filtering allows one to display records from the data source that meet certain criteria. By applying a filter, you are able to limit the data in a view without altering the design of the underlying object.

3.1.1.1 Picture this…

You work for "XYZ Co.," a leading corporation. You are in charge of maintaining a dashboard for all employees. The dashboard displays the same information to all employees, such as the number of employees working in the unit, the projects that belongs to the unit, the location that their business unit operates from, etc. However, the dashboard has a report that displays an annual performance rating for the employee. This piece of information is unique to the employee. Since this is confidential information, you have the responsibility to restrict the visibility of the annual performance rating to only the employee to which it belongs. This scenario requires you to use a filter to display relevant data to each employee.

S. Acharya and S. Chellappan, *Pro Tableau*, DOI 10.1007/978-1-4842-2352-9_3

3.1.2 What is filtering?

Filtering allows the exclusion or inclusion of certain values for a field. You can use filters to display specific records in a form, report, query, or datasheet, or to print only certain records from a report, table, or query.

3.1.3 How to apply "Filter"?

Tableau provides the following filtering options:

1. Filtering for dimensions
2. Filtering for measures
3. Quick filter
4. Context filter
5. Cascading filter
6. Calculation filter
7. Data Source filter

3.1.3.1 Filtering for dimensions

Dimensions are categorical values. A filter on this type of field allows you to select the values to include or exclude. A filter for dimensions includes:

- Basic categorical filter
- Wildcard match filter
- Conditions for filtering
- Limits to filtering

Let's go through a few demos that will provide step-by-step instructions showing how to filter for dimensions.

3.1.3.2 Basic categorical filter

You can use the "General" tab to include or exclude value for a field.

3.1.3.2.1 Step 1

Connect to the Sample-Superstore data source, drag the dimension "Order Date" from the dimensions area under the data pane to the columns shelf. Set the hierarchy to "Quarter". Drag the measure, "Sales" from the measures area under the data pane to the rows shelf. By default the aggregation is SUM. Refer to Fig. 3-1.

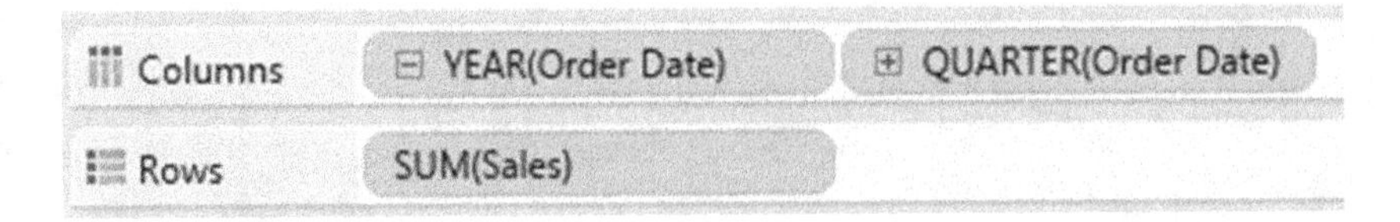

Figure 3-1. *Dimension "Order Date" placed on columns shelf and the measure "Sales" placed on rows shelf*

3.1.3.2.2 Step 2

Drag the dimension "Category" from the dimensions area under the data pane to "Color" on the marks card (Shown in Fig. 3-2).

Figure 3-2. *Dimension, "Category" placed on "Color" on the marks card*

3.1.3.2.3 Step 3

Drag the dimension "Category" from the dimensions area under the data pane to "Filters" Shelf. As you drag and drop, you will be prompted by a filter window. By default, "Select from list" is enabled as shown in Fig. 3-3. From the list select the "Furniture" and "Technology" categories and then click "OK" to include the filter in the view as shown in Fig. 3-4.

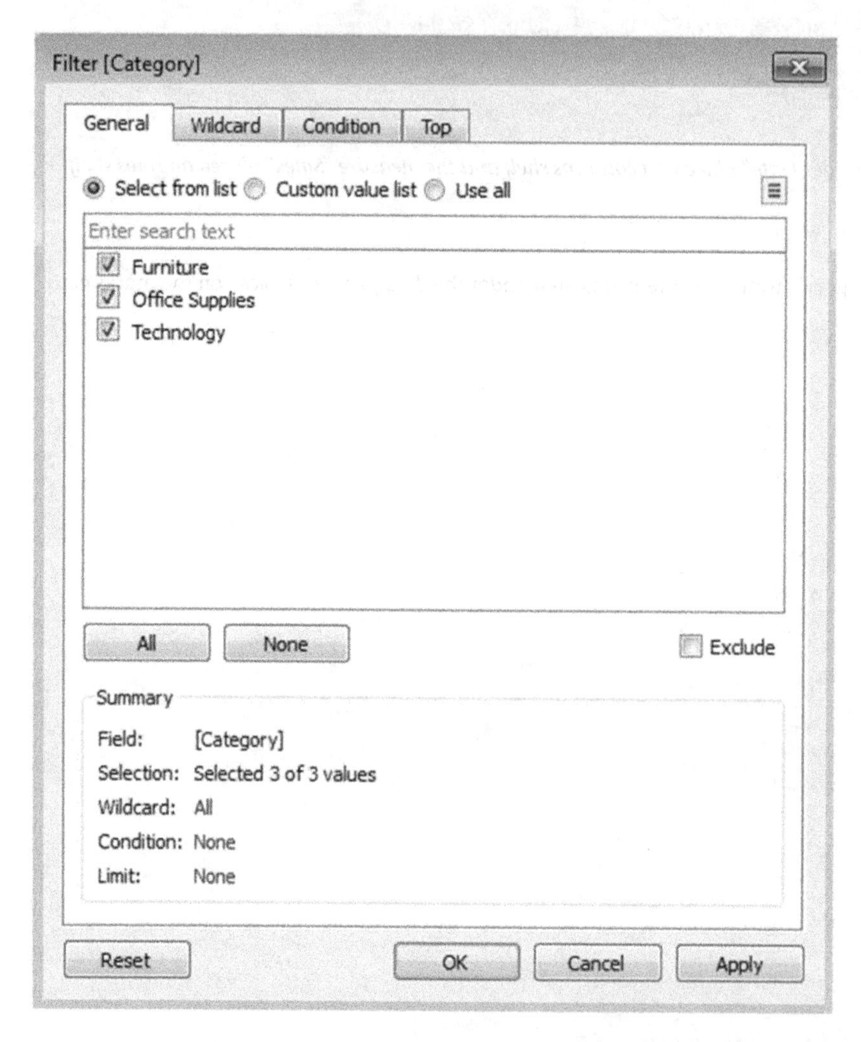

***Figure 3-3.** Filter window showing values for the "Category" dimension*

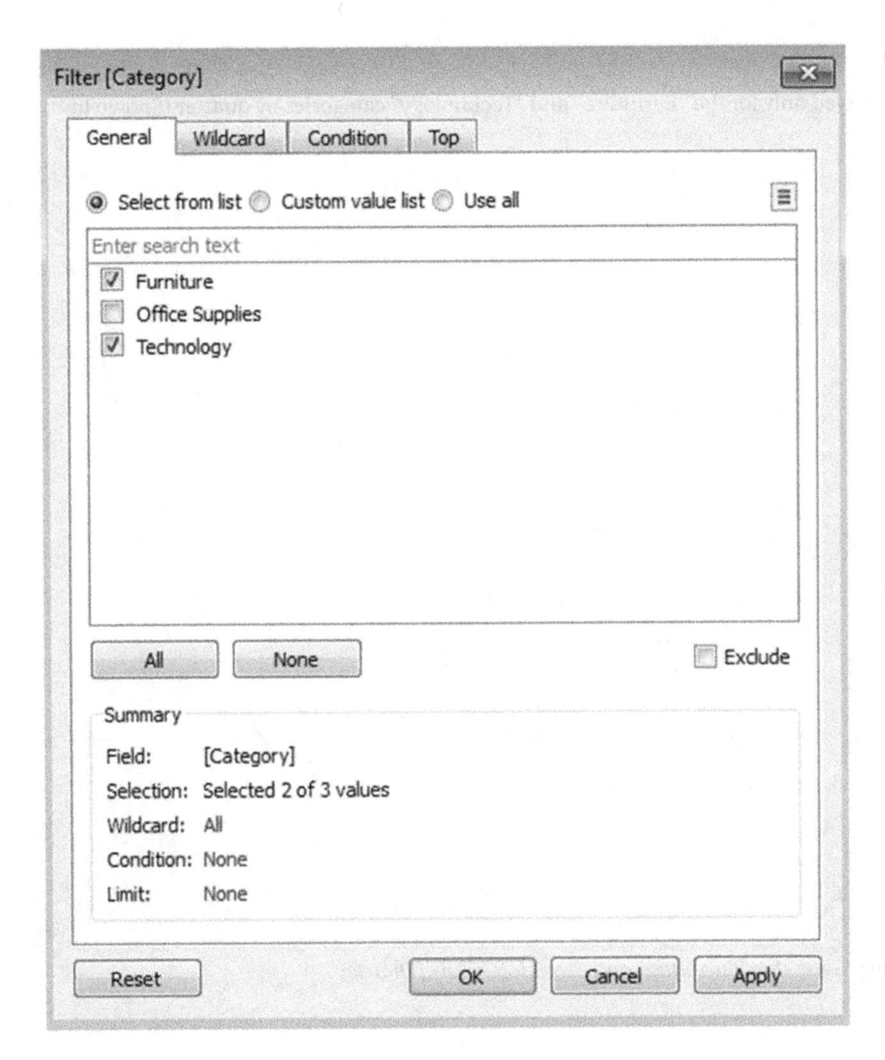

Figure 3-4. *Filter window showing selected categories "Furniture" & "Technology"*

3.1.3.2.4 Step 4

Note that "Sales" are displayed only for the "Furniture" and "Technology" categories by quarter (Shown in Fig. 3-5).

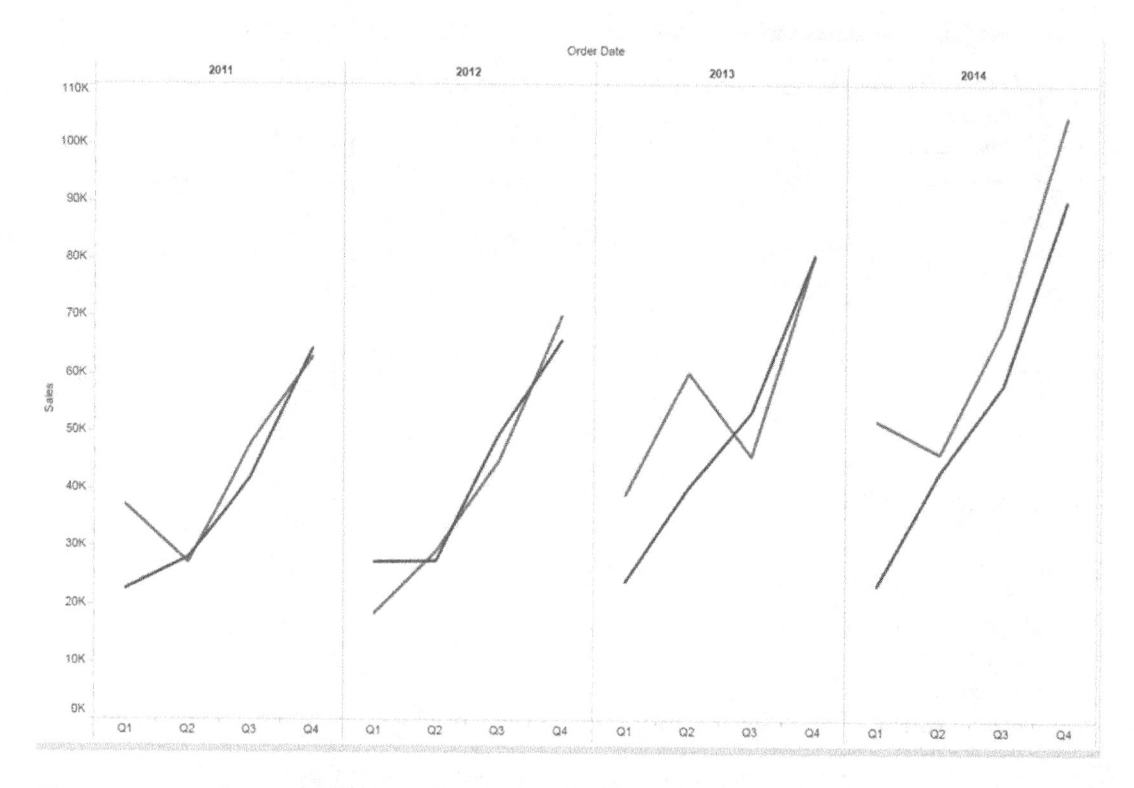

Figure 3-5. *View showing "Sales" for "Furniture" and "Technology" by "Quarter"*

3.1.3.2.5 Step 5

You can also edit the filter to exclude certain values. To edit the filter, right click on the "Category" field and select "Edit filter..." option as shown in Fig. 3-6.

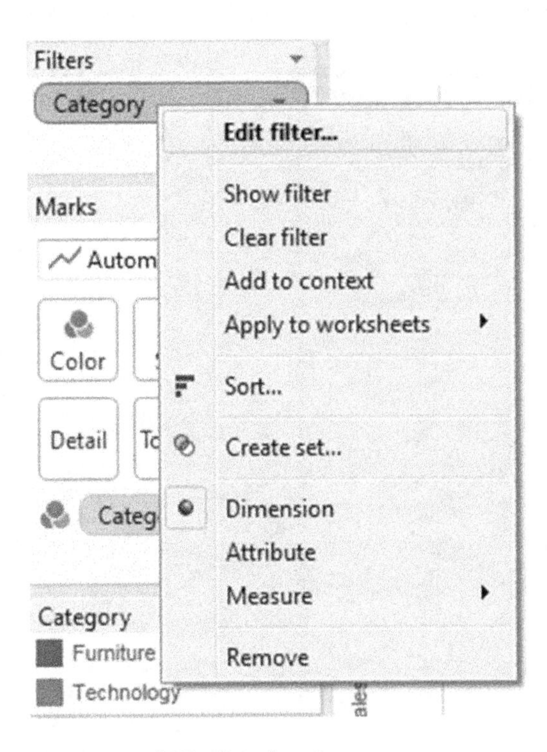

***Figure 3-6.** "Edit filter..." option*

3.1.3.3 Wildcard match to the filter

You can use the "Wildcard" Tab present in the "Filter" dialog box to define pattern for use in the filter.

3.1.3.3.1 Step 1

Drag the dimension "Sub-Category" from the dimensions area under the data pane to the rows shelf, order date to the columns shelf and set the hierarchy to the "Quarter". Drag the measure "Sales" from the measures area under the data pane to the rows shelf (Shown in Fig. 3-7).

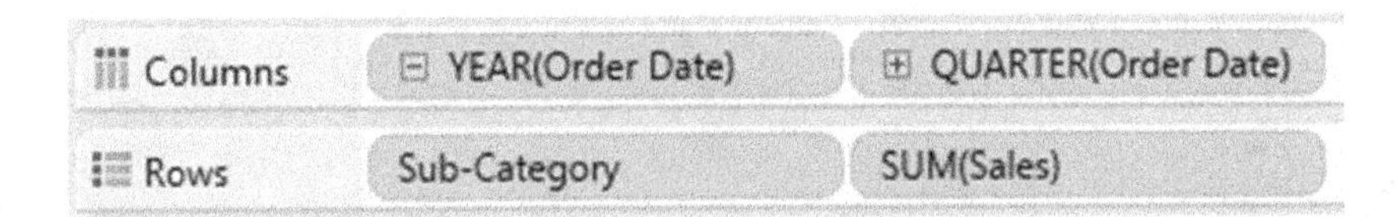

***Figure 3-7.** Dimension "Order Date" placed on the columns shelf, dimension "Sub-Category" and the measure "Sales" placed on the rows shelf*

3.1.3.3.2 Step 2

Drag the dimension "Sub-Category" from the dimensions area under the data pane to the "Filters" Shelf. Select the "wildcard" tab and select "Starts with" option. In the match value dialog box type C to include sub-category value that starts with C (Shown in Fig. 3-8).

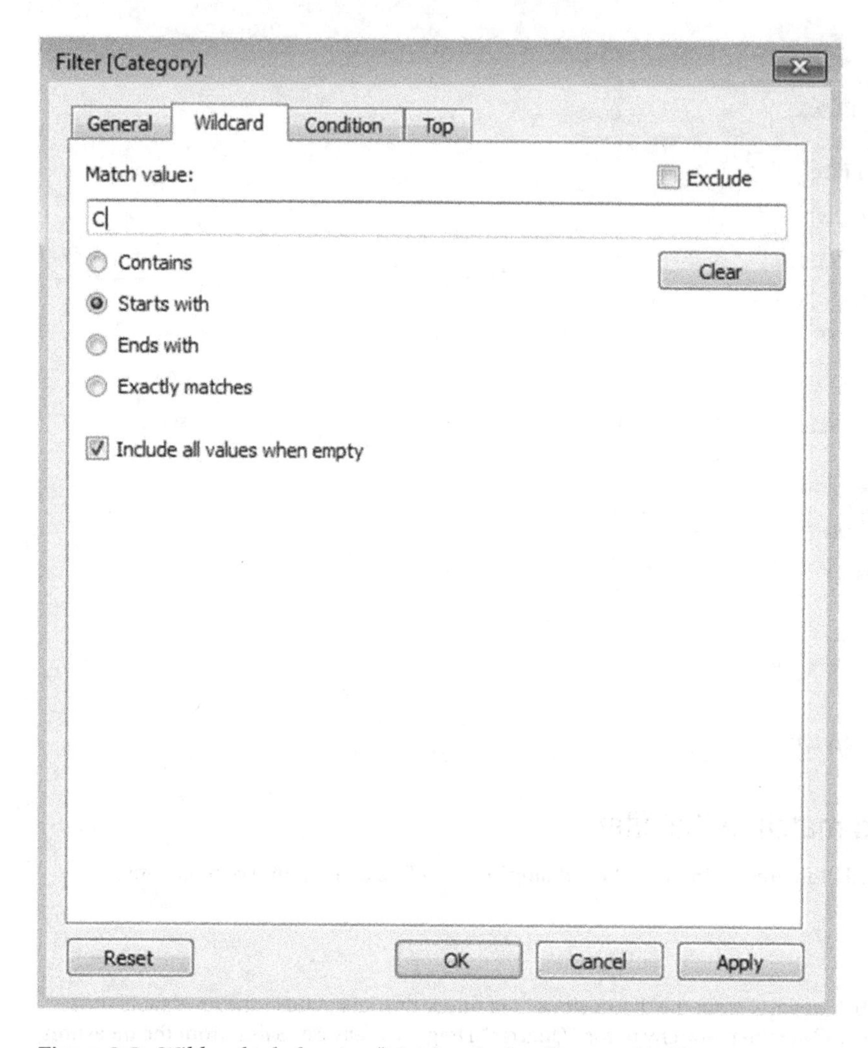

Figure 3-8. *Wildcard tab showing "Match value" as "Starts with C"*

3.1.3.3.3 Step 3

When you click the "OK" button, the view shows only those sub-category values that start with "C".

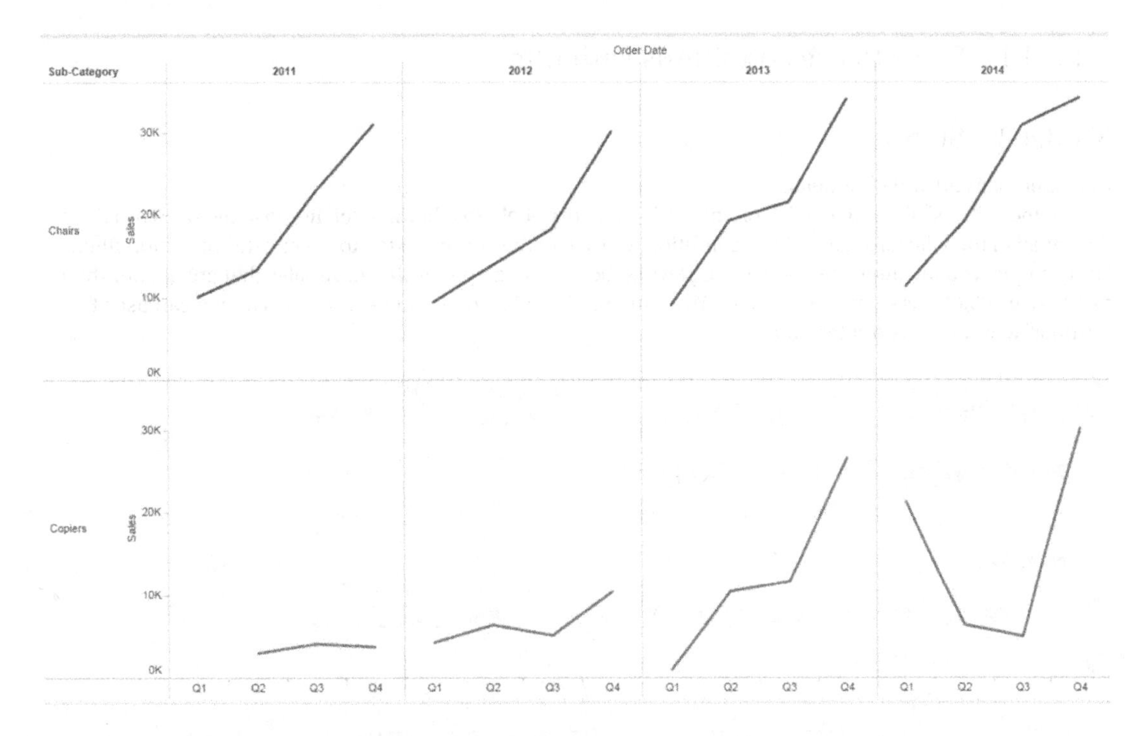

Figure 3-9. *View after applying wildcard option "Starts with C" to the "Sub-Category" field*

3.1.3.4 Conditions to Filter

You can use the condition tab to specify filtering rules.

3.1.3.4.1 Steps to use conditions with filter:

3.1.3.4.2 Step 1

Consider the Wildcard filter demo.

In the "Filters" shelf, right click on the "Sub-Category" field to edit the filter. Remove the wildcard filter described in the wildcard match filter condition and select the condition tab to specify the condition filter. For example, you are interested only in displaying those sub-categories that have sales that are greater than or equal to 200,000 as shown in Fig. 3-10. You can use "By field" to use built-in controls or you can use "By formula" to write a custom formula.

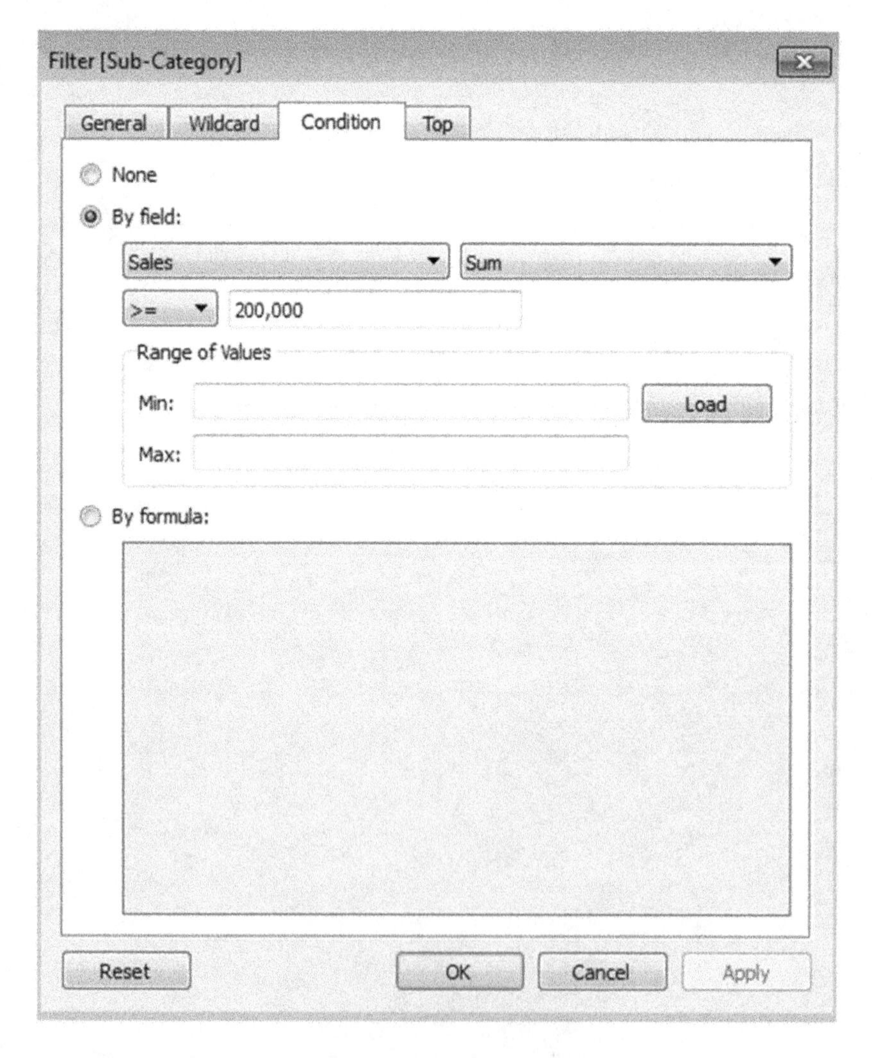

Figure 3-10. *Filter window showing the condition criteria for the "Sales" field*

3.1.3.4.3 Step 2

When you click on “OK,” you can see the updated view as shown in Fig. 3-11.

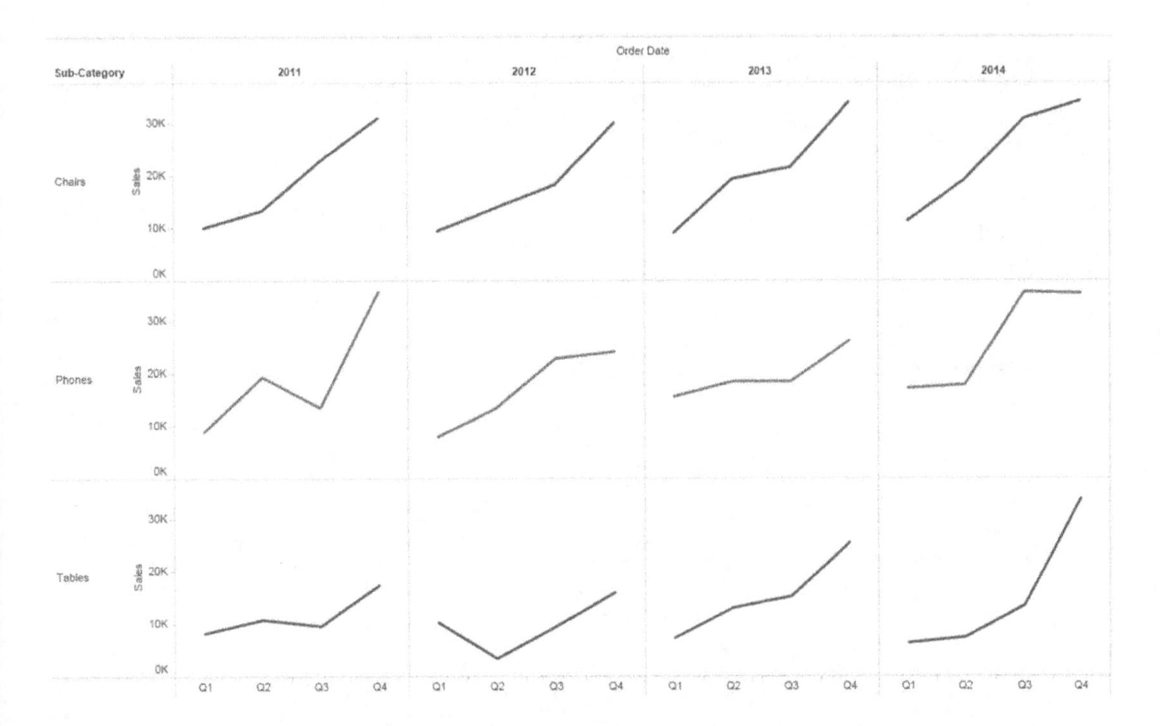

Figure 3-11. *Updated view after applying condition criteria to “Sales” field*

3.1.3.4.4 Step 3

You can also use a custom formula to specify a filter condition. Click on "By formula" and mention the custom formula as shown in Fig. 3-12.

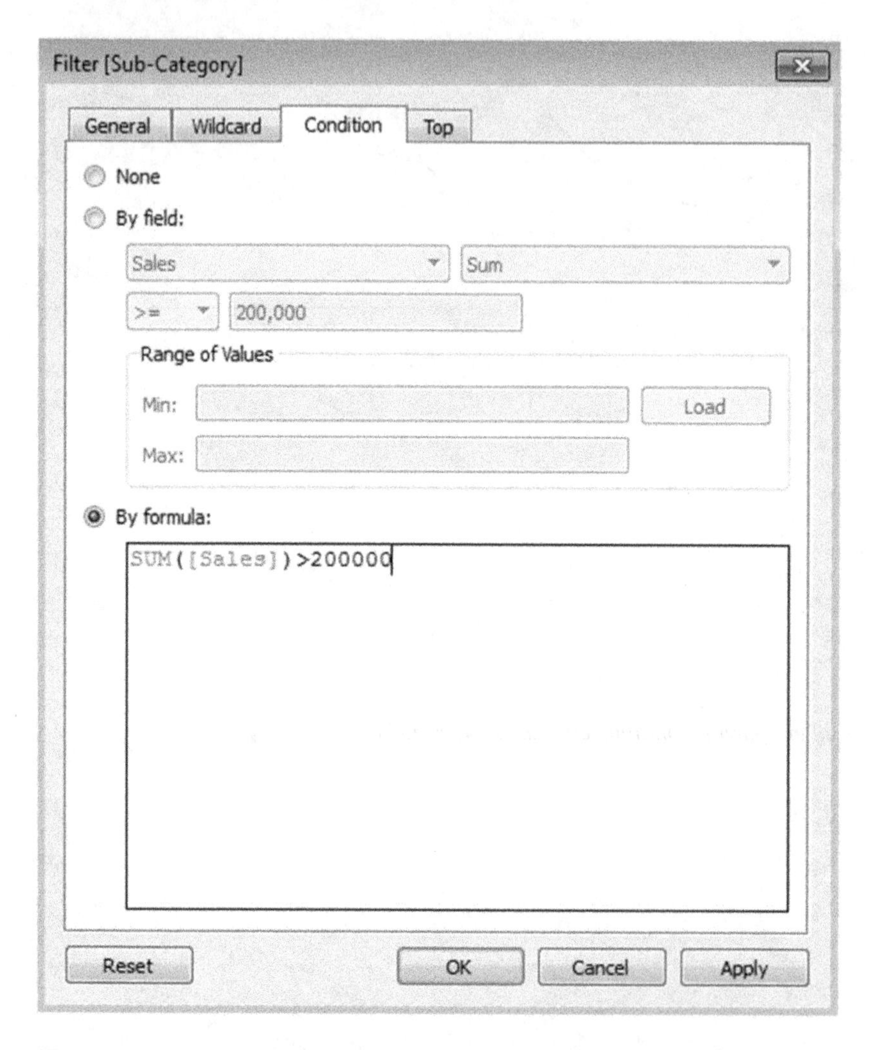

***Figure 3-12.** Filter window showing condition criteria for "Sales" field*

3.1.3.5 Limits to filter

You can use "Top" tab to define a formula which computes the data in the view.

3.1.3.5.1 Steps to use limits with filter

3.1.3.5.2 Step 1

Remove the condition filter described in the "Condition" filter section and select "Top" tab in the filter dialog box to define the formula as shown in Fig. 3-13. For example, you want to show Top 3 sub-category by sales.

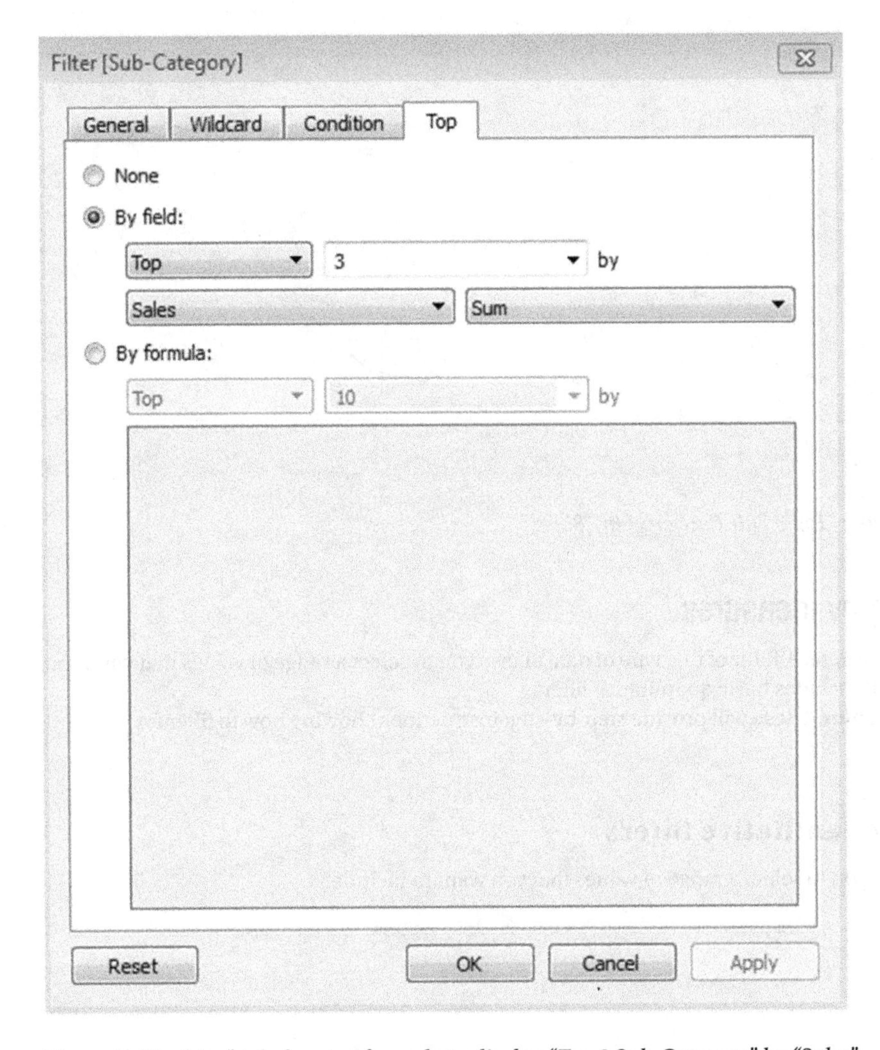

Figure 3-13. *"Top" tab showing formula to display "Top 3 Sub-Category" by "Sales"*

3.1.3.5.3 Step 2

When you click the "OK" button, the view is updated to show top three "Sub-Category" by "Sales" (Shown in Fig. 3-14).

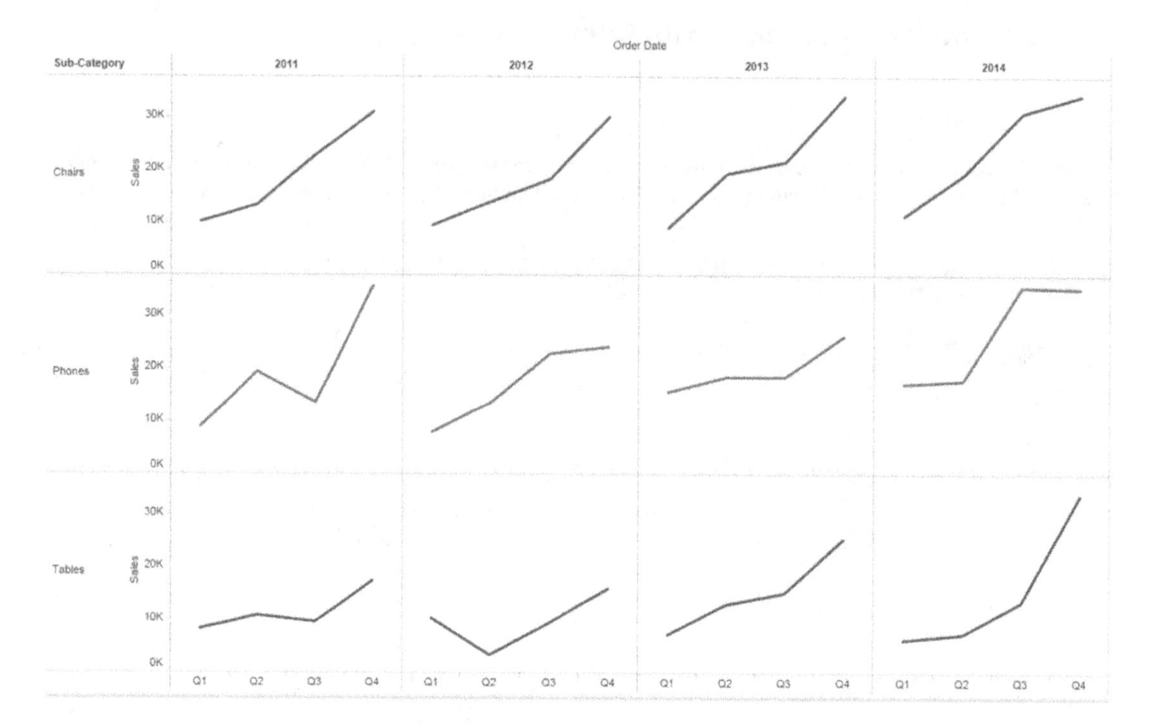

***Figure 3-14.** View showing "Top 3 Sub-Category" by "Sales"*

3.1.3.6 Filtering by measures

Measures are quantitative data. A filter of this type of data allows you to select a range of values that you want to include in your view. It includes basic quantitative filters.

Let's look at a few demos. These will provide step-by-step instructions showing how to filter for measures.

3.1.3.6.1 Basic quantitative filters

Quantitative filters allow you to select a range of values that you want to include.

3.1.3.6.2 Step 1

Connect to the Sample-Superstore data source. Drag the dimension "Sub-Category" from the dimensions area under the data pane to columns shelf and drag the measure "Profit" from the measures area under the data pane to the rows shelf (Shown in Fig. 3-15). The default aggregation applied to the "Profit" field is "SUM".

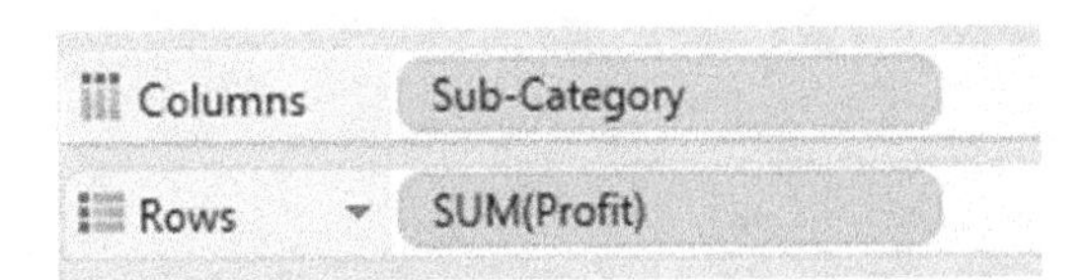

Figure 3-15. *Dimension "Sub-Category" placed on columns shelf and measure "Profit" placed on the rows shelf*

3.1.3.6.3 Step 2

Drag the dimension "Sub-Category" from the dimensions area under the data pane to "Color" on the marks card as shown in Fig. 3-16.

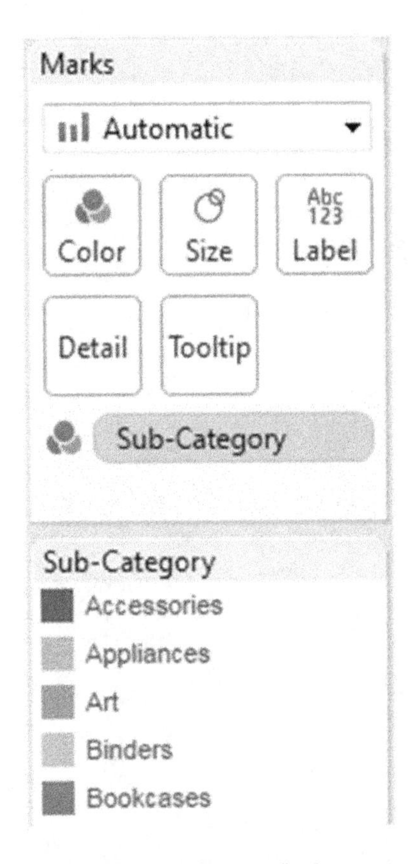

Figure 3-16. *Dimension, "Sub-Category" placed on "Color" on the marks card*

3.1.3.6.4 Step 3

Drag the measure "Profit" from the measures area under the data pane to "Filters" shelf. A "Filter field" dialog box opens, select the required aggregation for your filter condition as shown in Fig. 3-17 and click on the "Next" button.

Figure 3-17. *"Filter Field[Profit]" dialog box showing a list of aggregations*

3.1.3.6.5 Step 4

For example, you wish to display only those "Sub-Category" whose "Profit" value ranges from 1,000 to 20,000. You can use "Range of values" for this kind of view. Use the range slider to specify the values as shown in Fig. 3-18. There are four types of quantitative filters.

- Range of values: Includes all values that are within the minimum and maximum values of the range.
- At least: Includes all values that are greater than or equal to a specified minimum value.
- At most: Includes all values that are less than or equal to a specified maximum value.
- Special: Helps you to filter on null values. Include only null values, non-null values or all values.

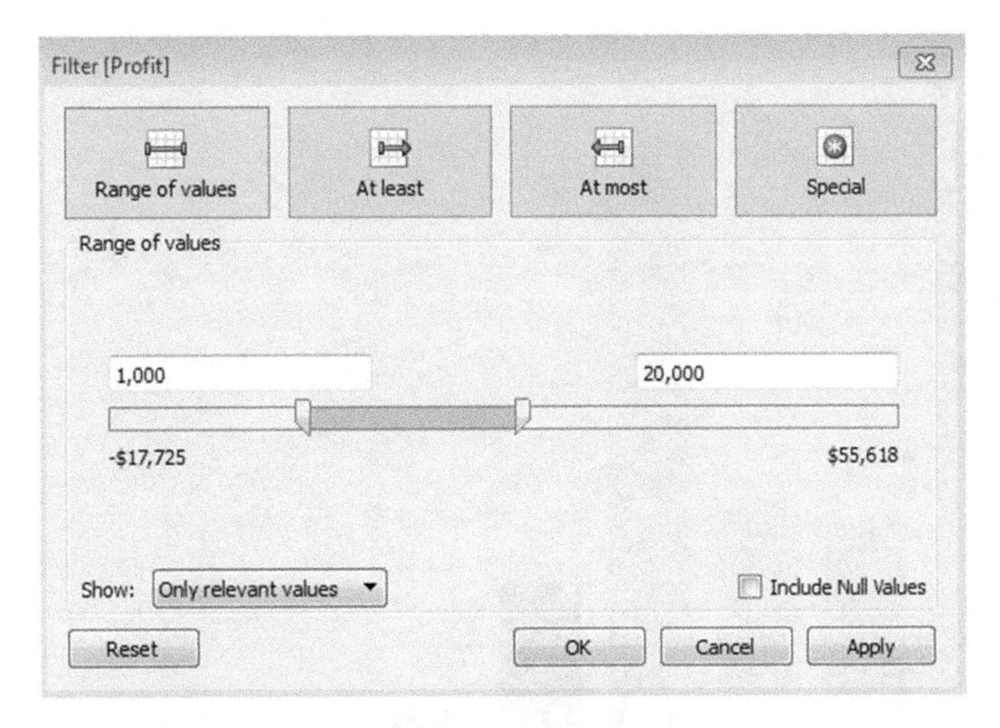

Figure 3-18. *"Filter[Profit]" dialog box showing "Range of values"*

3.1.3.6.6 Step 5

Observe the updated view that shows only the "Sub-Category" whose "Profit" range is from 1,000 to 20,000 (Shown in Fig. 3-19).

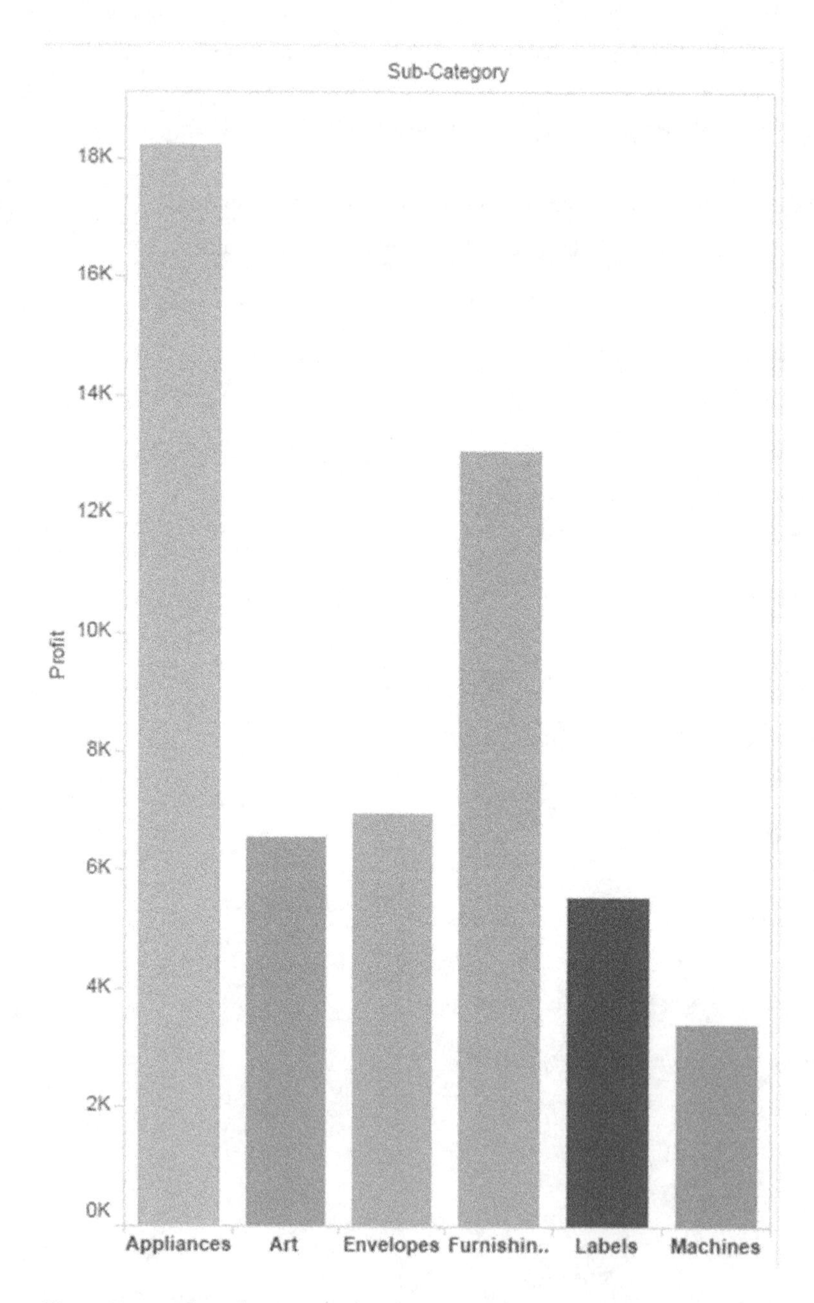

Figure 3-19. *View showing results after applying the range filter(Only "Sub-Category" whose "Profit" range is from 1,000 to 20,000)*

3.1.3.6.7 Step 6

You can use the "Show" option present in left bottom corner, to switch between "Only relevant values" and "All values in database" as shown in Fig. 3-20.

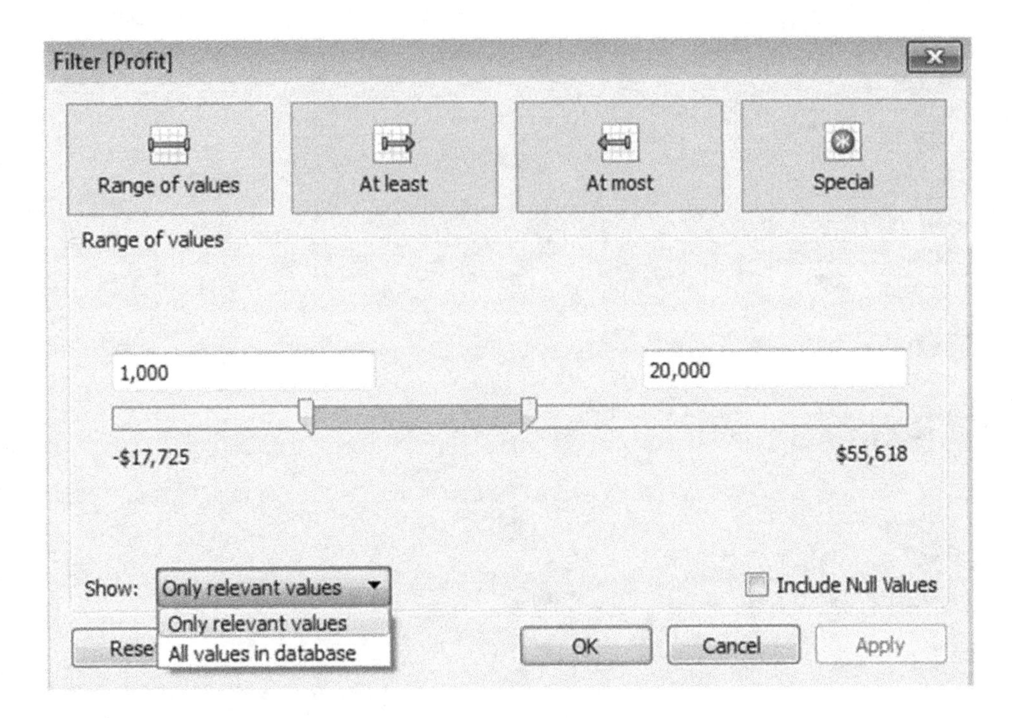

Figure 3-20. "Show" option to select "Only relevant values" and "All values in database"

3.1.3.7 Quick filters

Quick filters help you to modify the filter selection directly from the view.

3.1.3.7.1 Steps

3.1.3.7.2 Step 1

Connect to the Sample-Superstore data source. Drag the dimension "Category," "Sub-Category" from the dimensions area under data pane to the rows shelf. Drag the measure "Sales" from the measures area under the data pane to the columns shelf (Shown in Fig. 3-21).

Figure 3-21. Dimension "Category," "Sub-Category" placed on the rows shelf and measure "Sales" placed on the columns shelf

3.1.3.7.3 Step 2

Drag the dimension “Sub-Category” from the dimensions area under the data pane to “Color” on the marks card (Shown in Fig. 3-22).

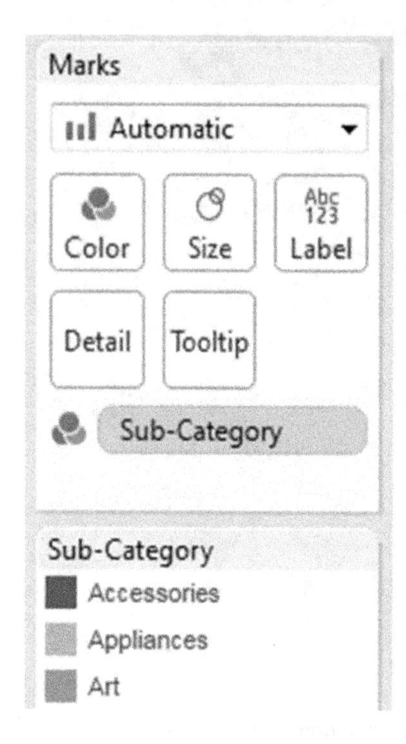

Figure 3-22. *Dimension, "Sub-Category" placed on “Color” on the Marks Card*

3.1.3.7.4 Step 3

Right click on sub-category field anywhere in the view and select the show filter option to display the quick filter as shown in Fig. 3-23.

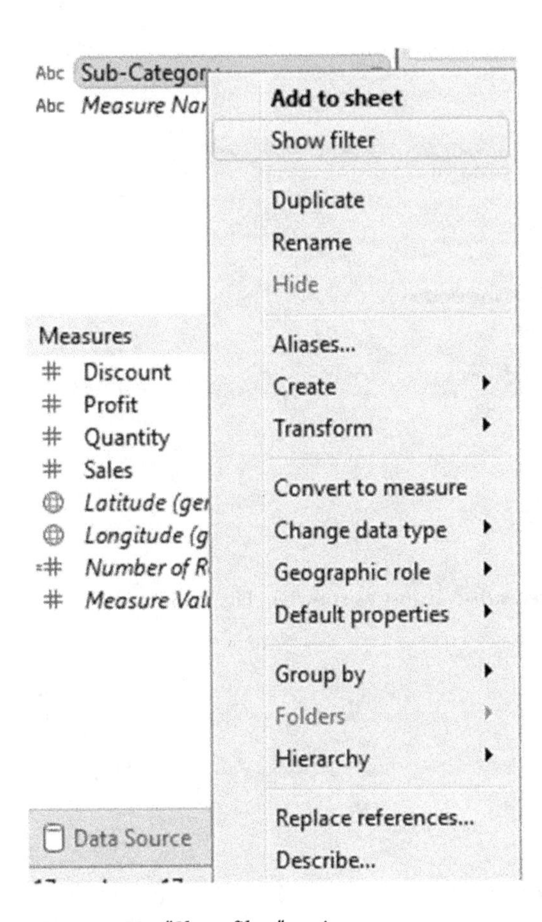

Figure 3-23. *"Show filter" option*

3.1.3.7.5 Step 4

Observe the quick filter for "Sub-Category" on the right-hand side of the sheet and the filter gets automatically added to the filters shelf as shown in Fig. 3-24.

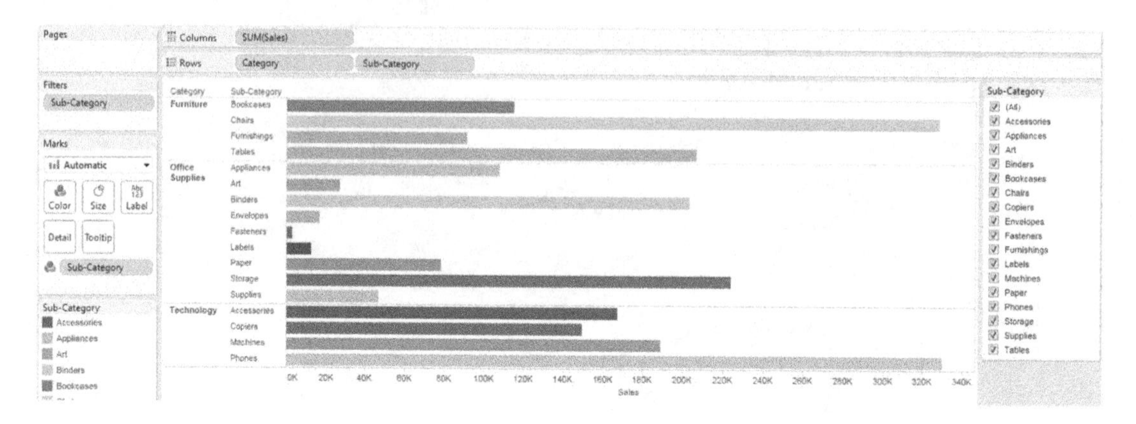

Figure 3-24. *View showing "Quick Filter"*

3.1.3.7.6 Step 5

You can modify the filter selection by selecting the required sub-category as shown in Fig. 3-25.

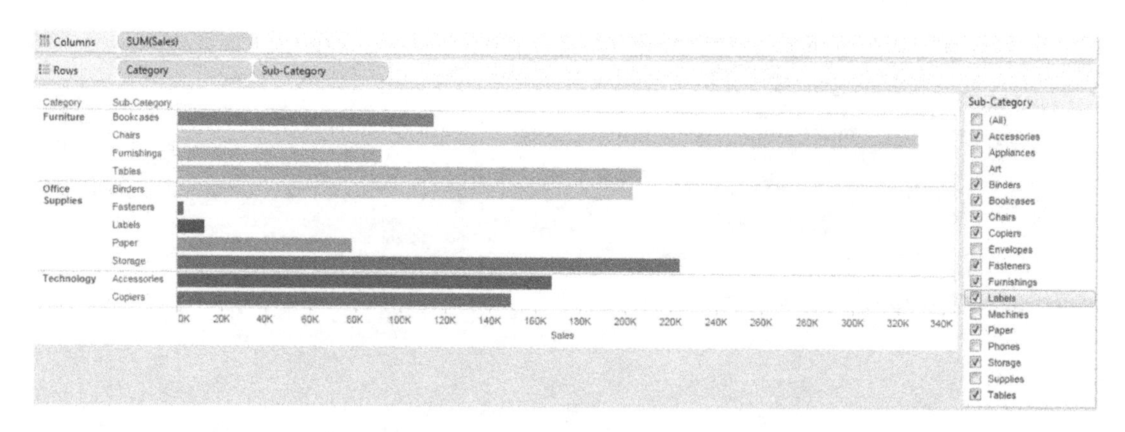

Figure 3-25. *Updated view showing SUM(Sales) for selected "Sub-Category"*

3.1.3.7.7 Step 6

You can also edit the appearance (layout modes) and functions of quick filter by clicking on the caret as shown in Fig. 3-26.

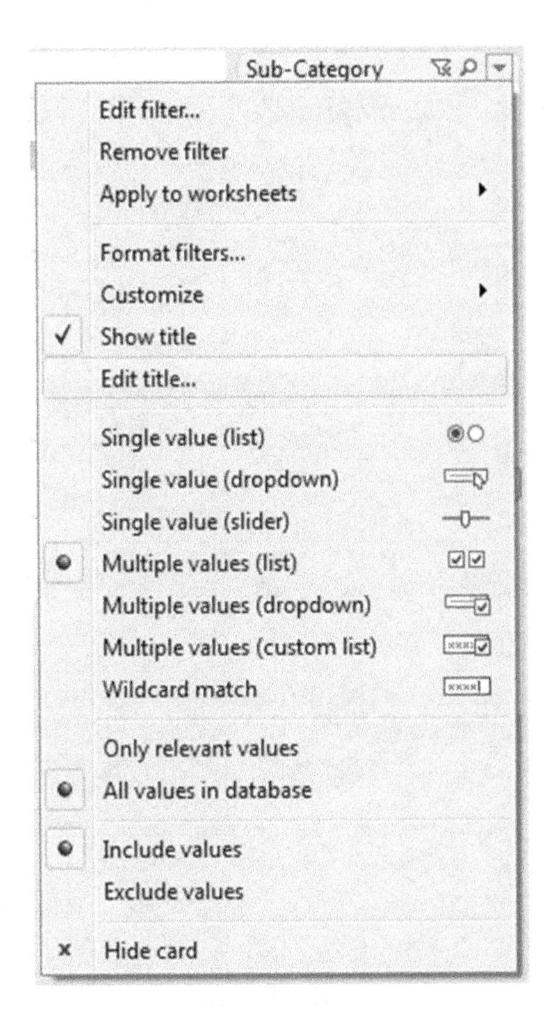

***Figure 3-26.** Options to edit "Quick Filter"*

Layout Modes are single value (list), single value (dropdown), etc.

3.1.3.7.8 Step 7

You can edit the title for the sub-category quick filter by clicking on the caret and selecting edit title as shown in Fig. 3-27.

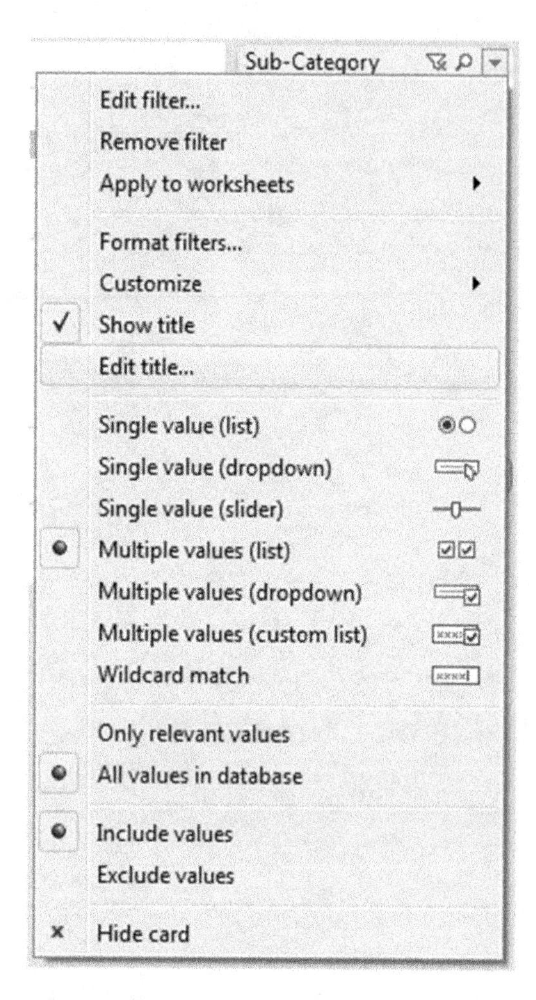

Figure 3-27. *Drop down menu showing "Edit Title" option*

3.1.3.7.9 Step 8

Type the title as shown in Fig. 3-28 and click the "OK" button.

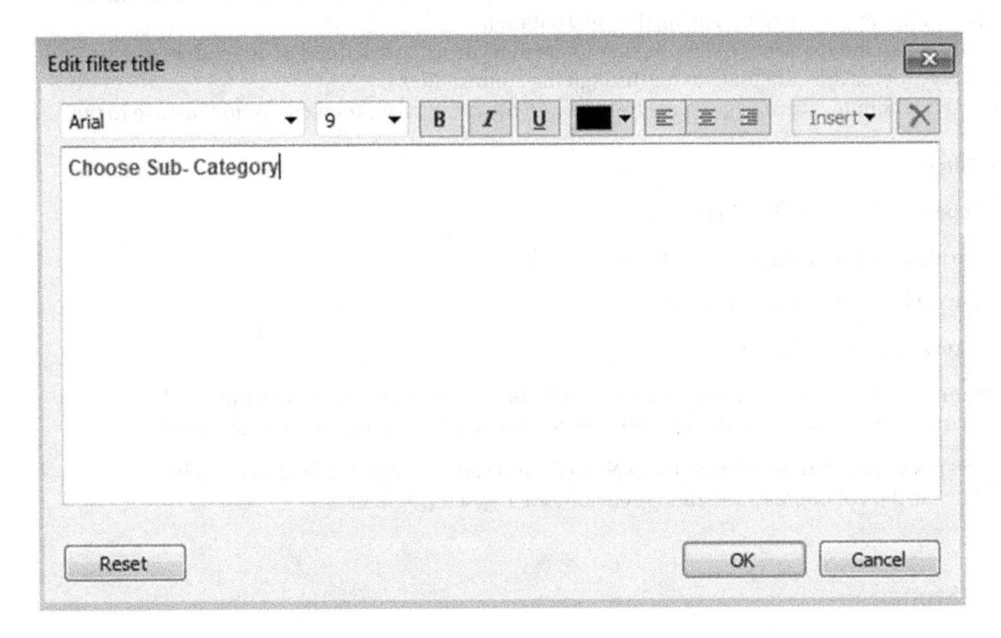

Figure 3-28. *"Edit filter title" dialog box to edit title for the quick filter "Sub-Category"*

3.1.3.7.10 Step 9

Observe the new title for the sub-category quick filter as shown in Fig. 3-29.

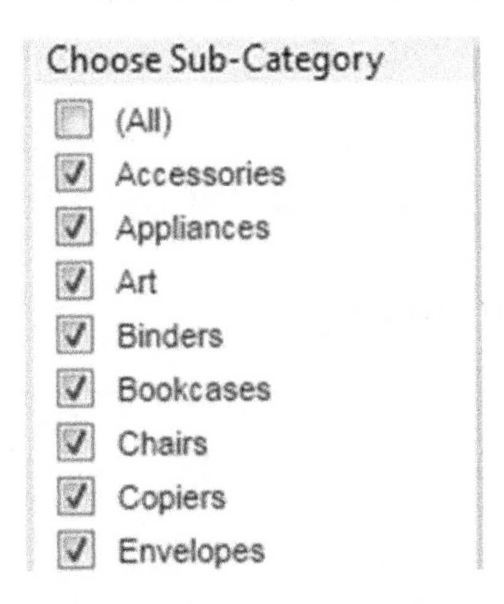

Figure 3-29. *View showing quick filter title as "Choose Sub-Category"*

3.1.3.8 Context filter

The filters that you add to your visualization are each independently calculated, regardless of what the other filters are doing. You can add "context" to your filters by adding a context filter. Once you create a context filter, then all other filters are calculated using this new data set.

Context Filter is an independent filter; all other filters that you set are defined as dependent filters because they only process the data that passes through the context filter.

The context is computed once to generate the view. All other filters are then computed relative to the context.

Context filters:

- Appear at the top of the filters shelf.
- Are identified by a gray color on the filters shelf.
- Cannot be rearranged on the shelf.

You may create a context filter to:

- improve performance - If you set a lot of filters or have a large data source, the queries can be slow. You can set one or more context filters to improve performance.
- create a dependent numerical or top N filter - You can set a context filter to include only the data of interest, and then set a numerical or a top N filter.

3.1.3.8.1 Steps

Objective: To display top 10 customers by their sales for each segment.

3.1.3.8.2 Step 1

Connect to the Sample-Superstore data source. Drag the dimension "Customer Name" from the dimensions area under data pane to the rows shelf. Drag the measure "Sales" from the measures area under the data pane to the columns shelf (Shown in Fig. 3-30).

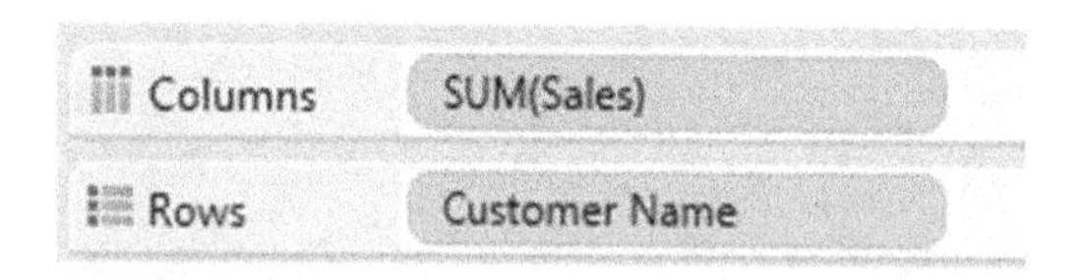

Figure 3-30. *Dimension "Customer Name" placed on the rows shelf and the measure "Sales" placed on the columns shelf*

3.1.3.8.3 Step 2

Click on the sales axis to sort the customer name by the sum of their sales (Shown in Fig. 3-31).

Figure 3-31. *Highlighted sales axis to "Sort"*

3.1.3.8.4 Step 3

Drag the dimension "Customer Name" from the dimensions area under the data pane to the "Filters" shelf to create the Top N filter. Fill in the details as shown in Fig. 3-32.

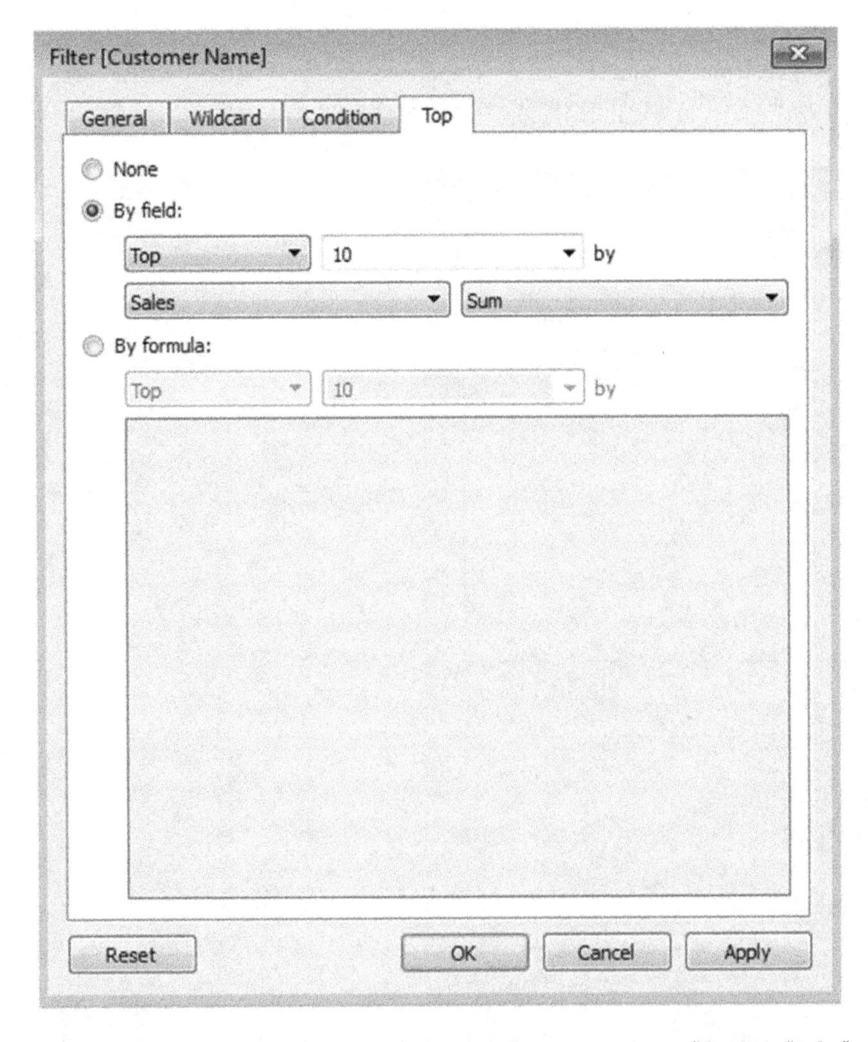

Figure 3-32. *"Filter[Customer Name]" to create "Top 10 Customers" by their "Sales"*

3.1.3.8.5 Step 4

Observe that top 10 customer names by their sales are shown in Fig. 3-33.

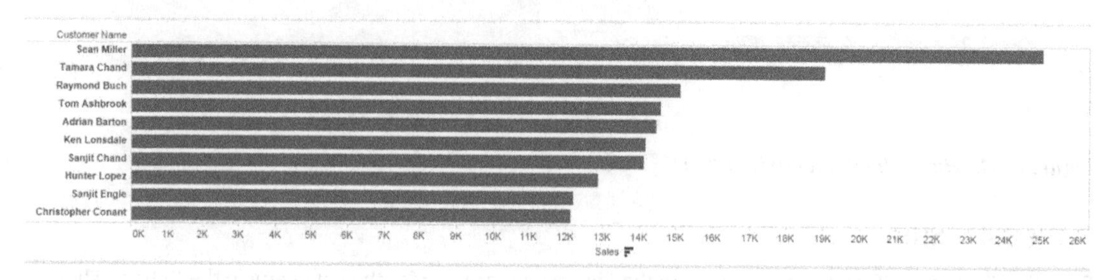

Figure 3-33. *View showing "Top 10 Customers" by their "Sales"*

3.1.3.8.6 Step 5

Create one more filter to display customer by their sales for the corporate segment. Drag the dimension "Segment" from the dimensions area under the data pane to the "Filters" shelf, select corporate as a segment value as shown in Fig. 3-34.

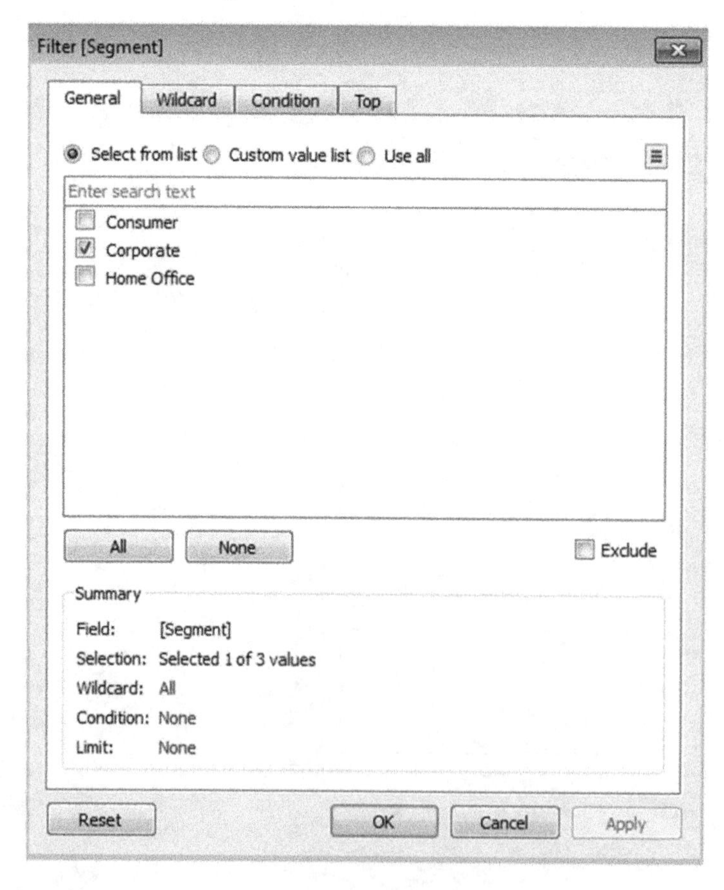

Figure 3-34. *"Filter[Segment]" dialog box to include only the "Corporate" segment*

3.1.3.8.7 Step 6

The filtered view displays only one customer instead of the top 10 customers. (Shown in Fig. 3-35). This is not the result that we want. This is because all filters are evaluated separately and the view is an intersection of results.

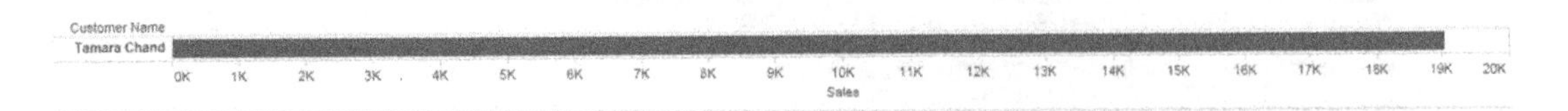

Figure 3-35. *View shows only one customer for the selected "Corporate" segment*

3.1.3.8.8 Step 7

To display the Top 10 Customers by their "Sales" for each "Segment," right click on "Segment" and select "Add to Context" to make "Segment" as Context Filter (Shown in Fig. 3-36).

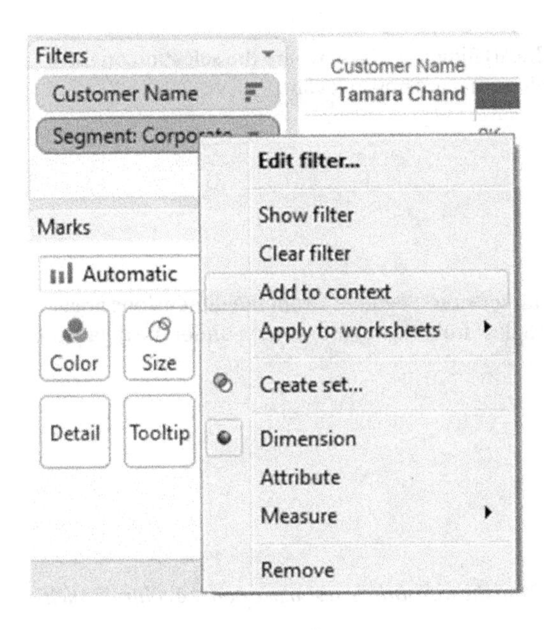

Figure 3-36. *"Add to context" option*

3.1.3.8.9 Step 8

The updated view shows the "Top 10 customers" by their "Sales" for "Corporate Segment" (Shown in Fig. 3-37).

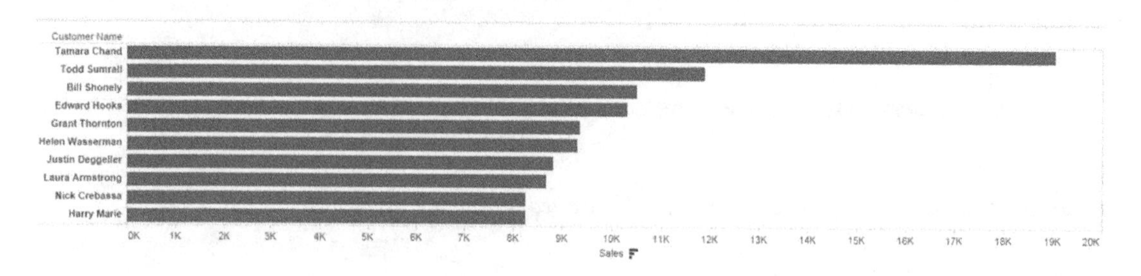

Figure 3-37. *Updated view after applying context filter*

3.1.3.9 Cascading filter

Cascading filters are a set of filters. Cascading (or hierarchical) filters are those where the selection on the first filter causes the second to be limited to only those values that are now relevant.

3.1.3.9.1 Steps

3.1.3.9.2 Step 1

Connect to the Sample-Superstore data source. Drag the dimension "Segment" from the dimensions area under the data pane to the rows shelf. Drag the measure "Sales" from the measures area under the data pane to the columns shelf (Shown in Fig. 3-38).

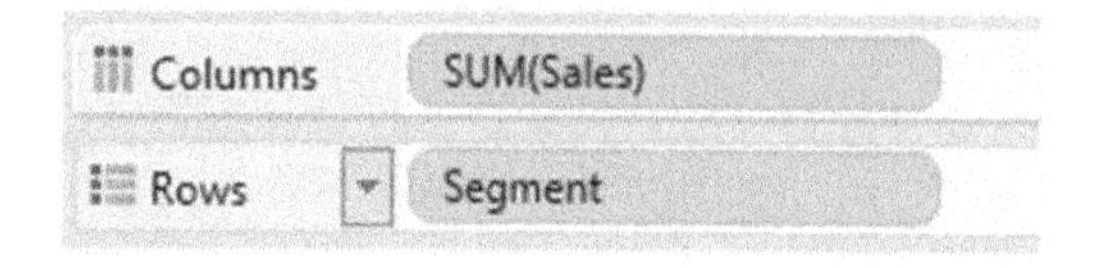

Figure 3-38. *Dimension "Segment" placed on the rows shelf and the measure "Sales" placed on the columns shelf*

3.1.3.9.3 Step 2

Create Quick Filter for category and sub-category (Shown in Fig. 3-39 and Fig. 3-40).

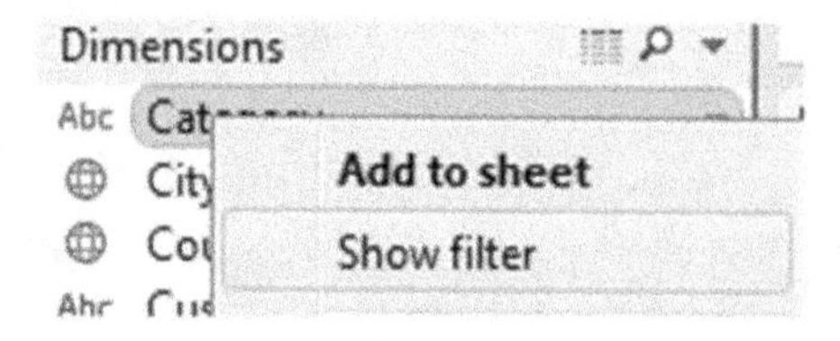

Figure 3-39. *Adding "Quick Filter" for "Category"*

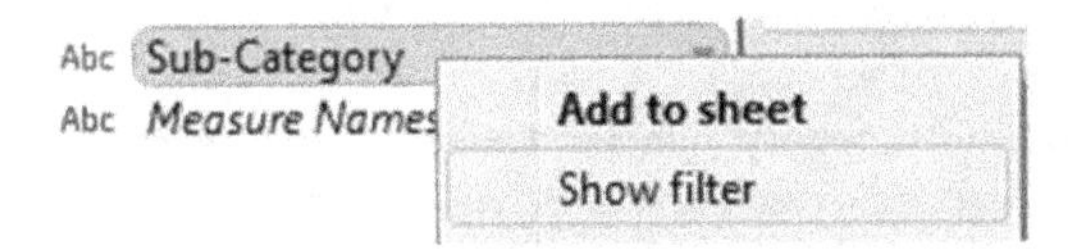

Figure 3-40. *Adding "Quick Filter" for "Sub-Category".*

3.1.3.9.4 Step 3

Quick filter for category and sub-category are displayed (Shown in Fig. 3-41).

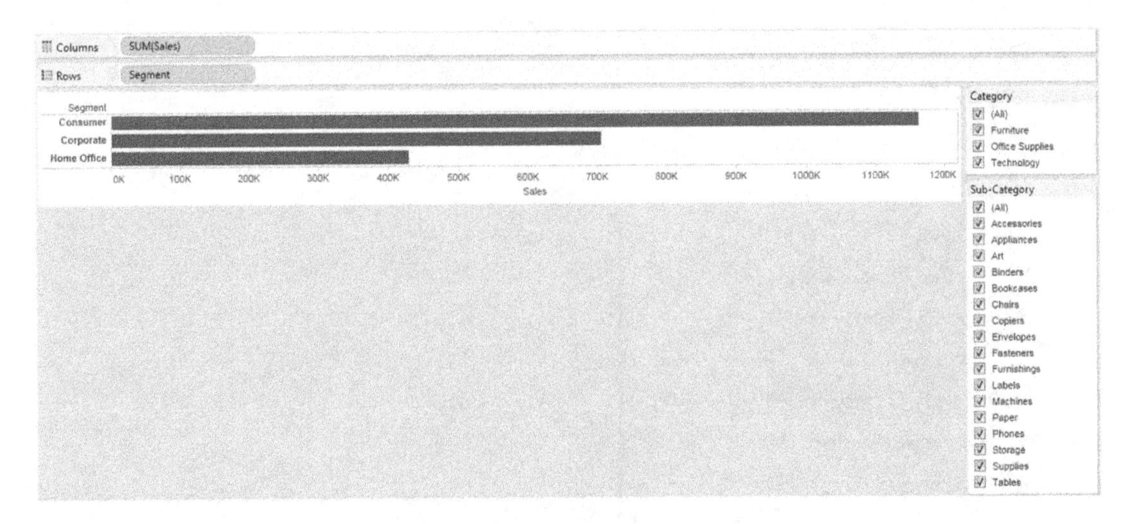

Figure 3-41. *View showing "Quick Filter" for "Category" and "Sub-Category"*

3.1.3.9.5 Step 4

From the "Category" filter select only "Furniture". Click on "Sub-Category" caret and check "Only relevant values" option (Shown in Fig. 3-42).

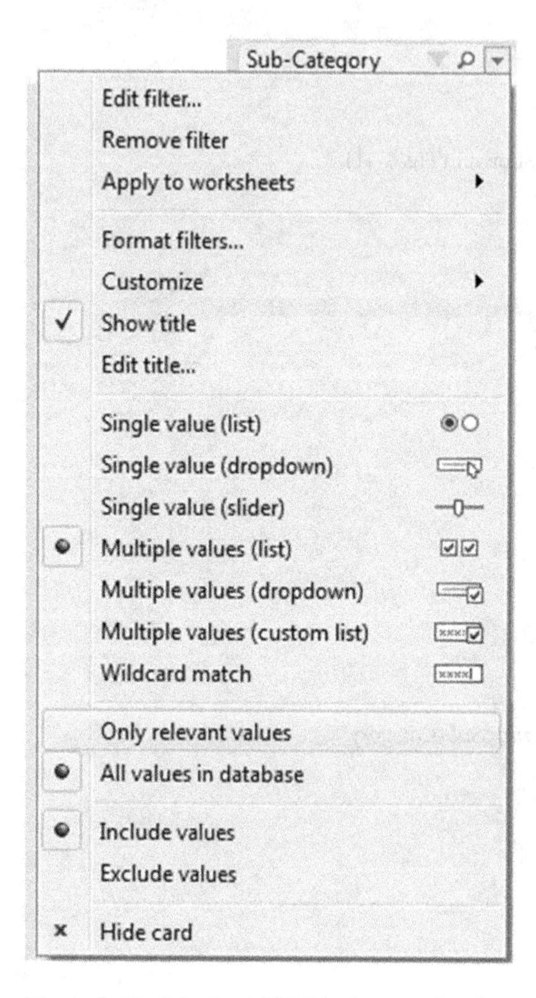

Figure 3-42. *Selection of "Only relevant values"*

3.1.3.9.6 Step 5

"Sub-Category" filter shows all products that are relevant to "Furniture" Category (Shown in Fig. 3-43).

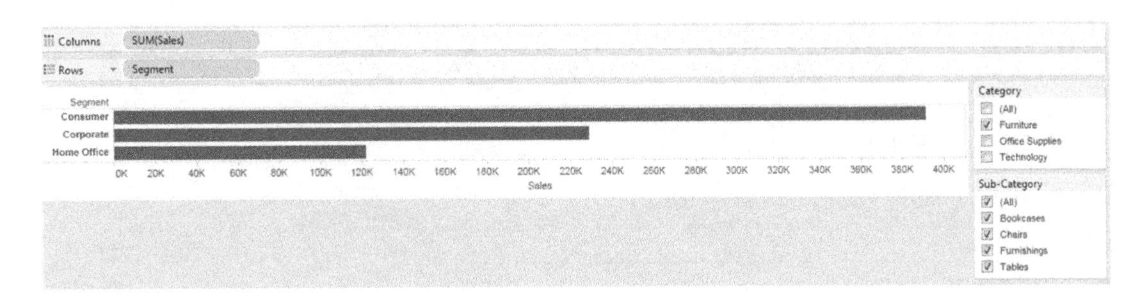

Figure 3-43. *"Sub-Category" shows all products that are relevant to the "Furniture" Category*

3.1.3.10 Calculation filter

Calculation filter allows you to perform calculations on the selected dimension members. The calculation filter is also known as the "Slicing" filter.

3.1.3.10.1 Steps

3.1.3.10.2 Step 1

Connect to the Sample-Superstore data source. Drag the dimension "Segment" from the dimensions area under the data pane to the rows shelf. Drag the measure "Sales" and "Profit" from the measures area under data pane to the columns shelf (Shown in Fig. 3-44). The view shows various segments by sum of sales and sum of profit (Shown in Fig. 3-45).

Figure 3-44. *Dimension "Segment" placed on the rows shelf and measure "Sales," "Profit" placed on the columns shelf*

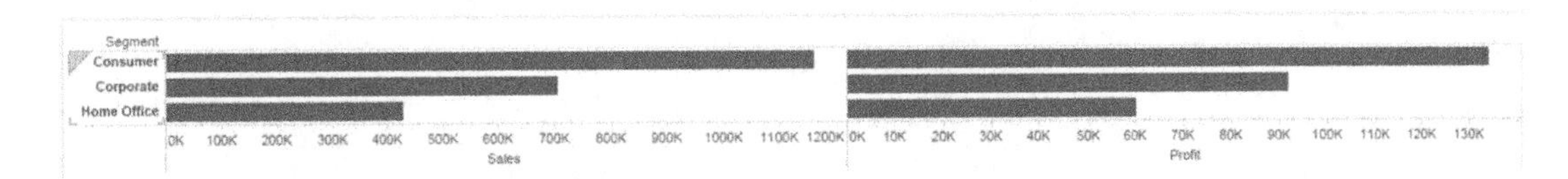

Figure 3-45. *View shows "Segment" by "Sales" and "Profit"*

3.1.3.10.3 Step 2

Drag the dimension "Category" from the dimensions area under data pane to "Filters" Shelf, select the categories, "Furniture" and "Technology" as shown in Fig. 3-46.

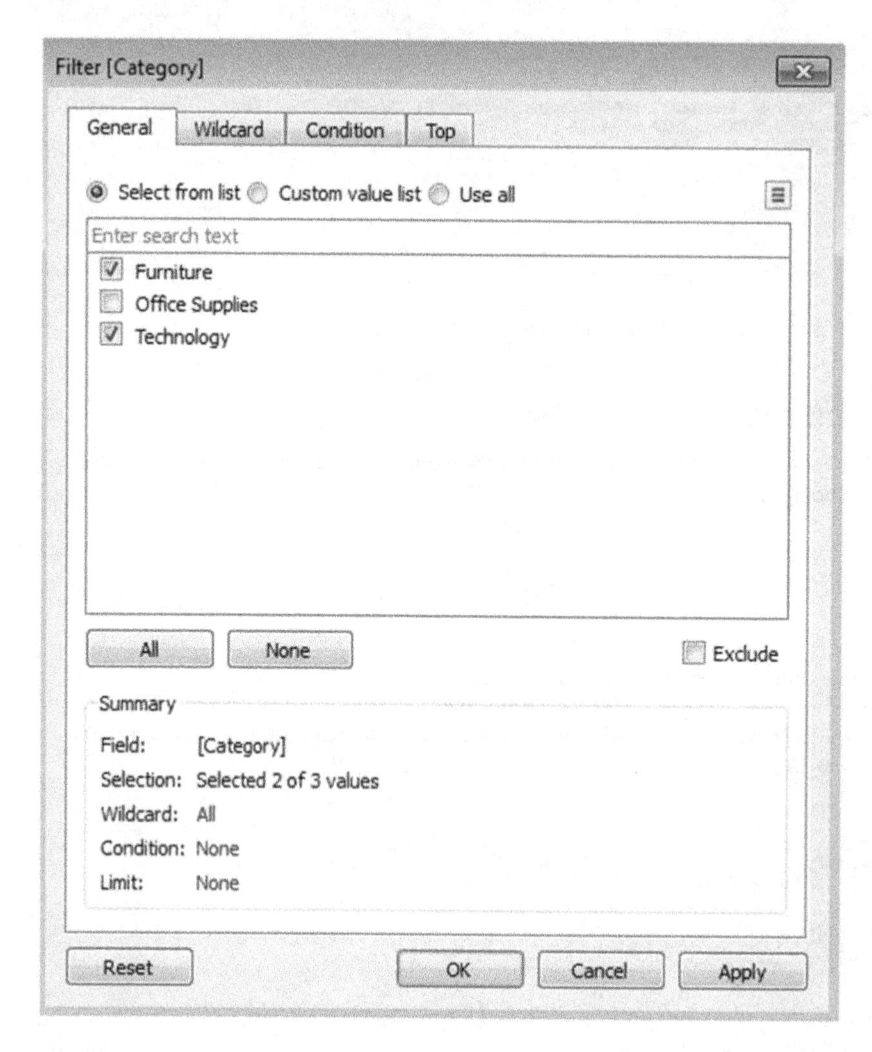

Figure 3-46. *"Filter[Category]" to include "Furniture" and "Technology" category*

3.1.3.10.4 Step 3

When you click on "OK" button, Tableau automatically applies the appropriate calculation to the members of the filter based on the aggregation of each measure shown in Fig. 3-47.

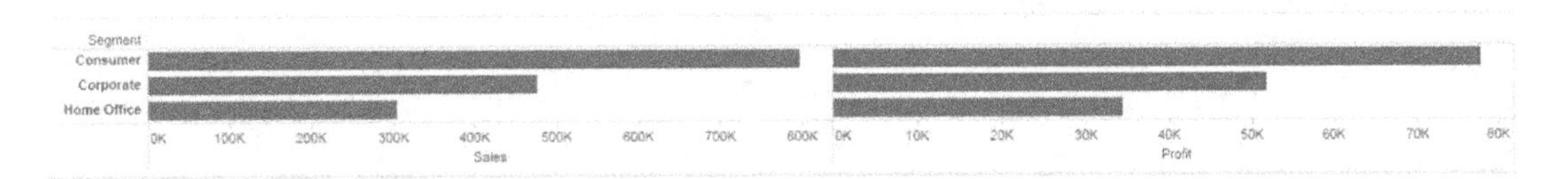

***Figure 3-47.** View after applying the calculation filter*

3.1.3.11 Data source filter

The data source filter helps you to reduce the amount of data in the data source.

3.1.3.11.1 Steps

3.1.3.11.2 Step 1

Connect to the Sample-Superstore data source.

3.1.3.11.3 Step 2

To create data source filter click on "Add..." in the upper right corner of the data source page as shown in Fig. 3-48.

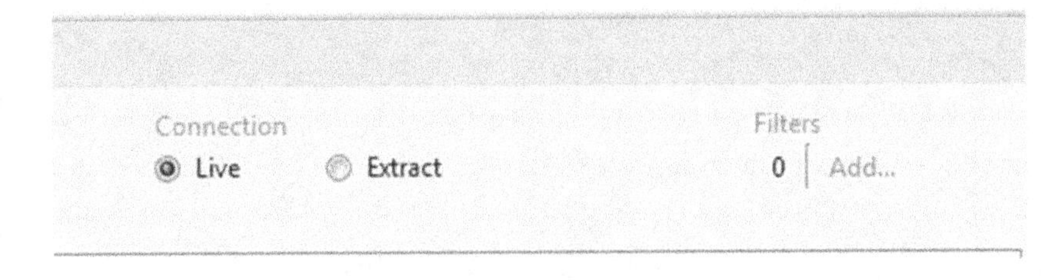

***Figure 3-48.** Data source page showing "Add..." option*

3.1.3.11.4 Step 3

"Edit Data Source Filters" dialog box appears as shown in Fig. 3-49.

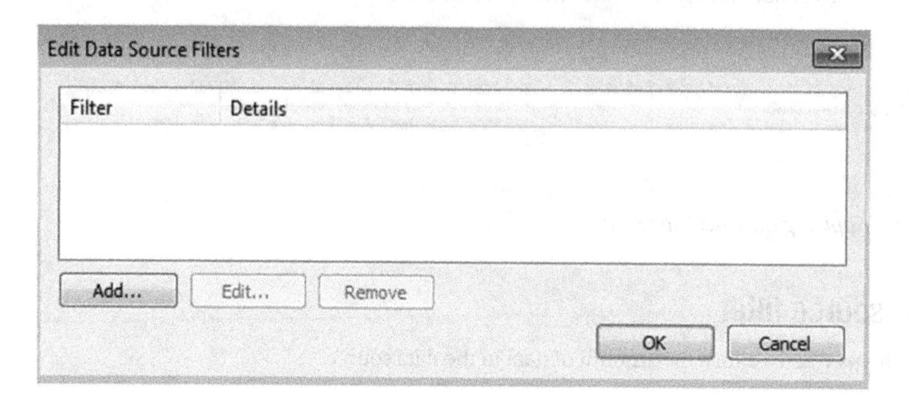

Figure 3-49. *"Edit Data Source Filters" dialog box*

3.1.3.11.5 Step 4

Click on the "Add" button to add the filter and select category field as shown in Fig. 3-50.

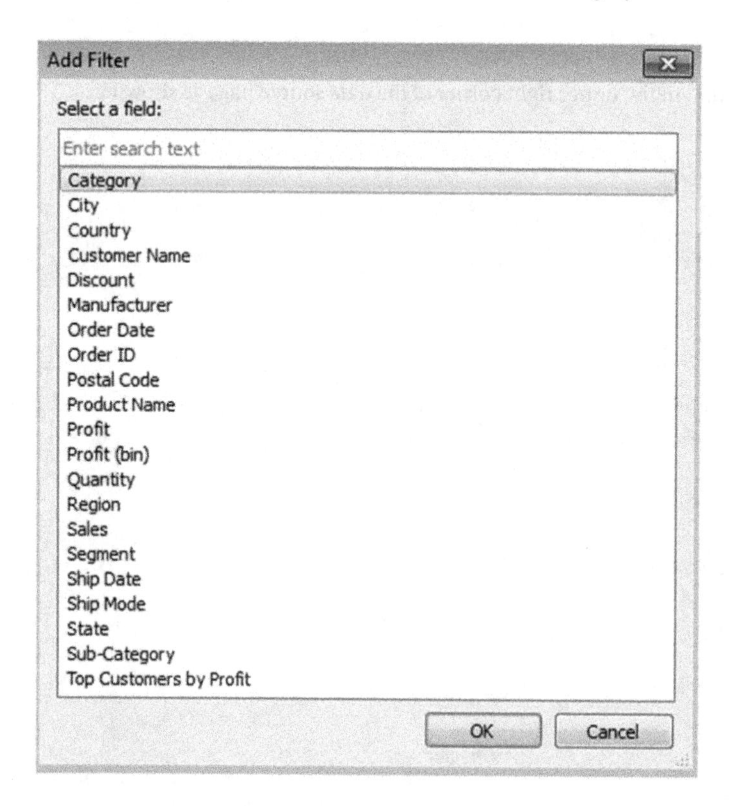

Figure 3-50. *"Add Filter" dialog box showing "Category" field as selected member*

3.1.3.11.6 Step 5

When you click "OK," you can see the filter field dialog box. Select "Furniture" Category as shown in Fig. 3-51 and click "OK".

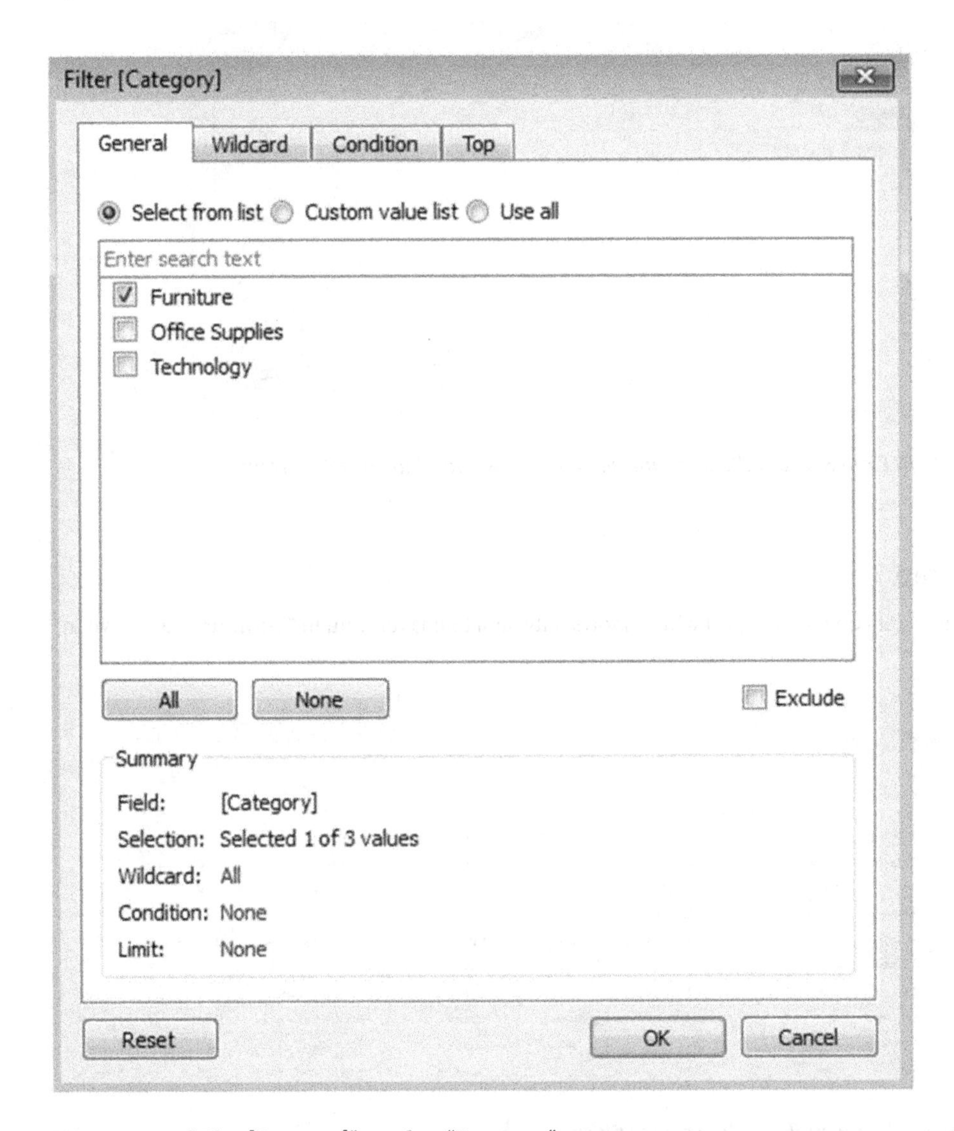

Figure 3-51. *"Filter[Category]" to select "Furniture" category*

3.1.3.11.7 Step 6

Next, you can see the "Category" filter in the "Edit Data Source Filters" as shown in Fig. 3-52.

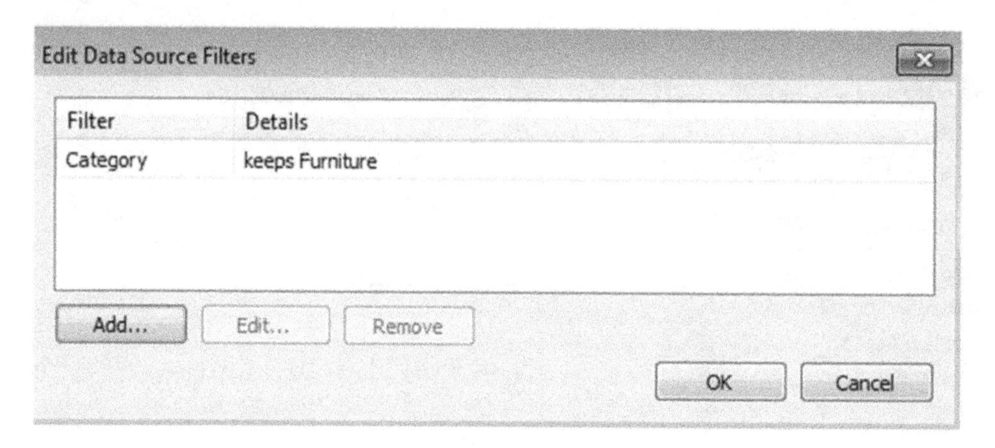

Figure 3-52. *The "Edit Data Source Filters" dialog box that keeps only "Furniture" Category for filter criteria*

3.1.3.11.8 Step 7

Observe the updated data on the data grid which shows only data that is relevant to "Furniture" as shown in Fig. 3-53.

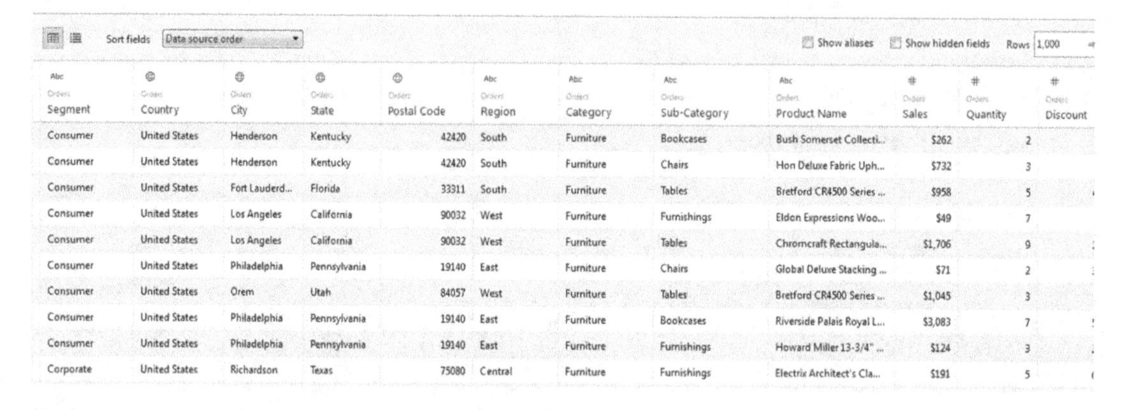

Sort fields Data source order
Show aliases Show hidden fields Rows 1,000

Segment	Country	City	State	Postal Code	Region	Category	Sub-Category	Product Name	Sales	Quantity	Discount
Consumer	United States	Henderson	Kentucky	42420	South	Furniture	Bookcases	Bush Somerset Collecti...	$262	2	
Consumer	United States	Henderson	Kentucky	42420	South	Furniture	Chairs	Hon Deluxe Fabric Uph...	$732	3	
Consumer	United States	Fort Lauderd...	Florida	33311	South	Furniture	Tables	Bretford CR4500 Series ...	$958	5	
Consumer	United States	Los Angeles	California	90032	West	Furniture	Furnishings	Eldon Expressions Woo...	$49	7	
Consumer	United States	Los Angeles	California	90032	West	Furniture	Tables	Chromcraft Rectangula...	$1,706	9	
Consumer	United States	Philadelphia	Pennsylvania	19140	East	Furniture	Chairs	Global Deluxe Stacking ...	$71	2	
Consumer	United States	Orem	Utah	84057	West	Furniture	Tables	Bretford CR4500 Series ...	$1,045	3	
Consumer	United States	Philadelphia	Pennsylvania	19140	East	Furniture	Bookcases	Riverside Palais Royal L...	$3,083	7	
Consumer	United States	Philadelphia	Pennsylvania	19140	East	Furniture	Furnishings	Howard Miller 13-3/4" ...	$124	3	
Corporate	United States	Richardson	Texas	75080	Central	Furniture	Furnishings	Electrix Architect's Cla...	$191	5	

Figure 3-53. *Data Grid after applying data source filter*

Refer to the link below to understand the order of filter execution in Tableau

https://onlinehelp.tableau.com/current/pro/desktop/en-us/order_of_operations.html

---

3.2 Sorting

You can arrange dimension members in a specified order with the help of sorting.

3.2.1 Why sorting?

To display data in an order such as alphabetic order or numeric order.

3.2.1.1 Picture this...

You work for "XYZ" retail store. The company wants to provide some percentage of offers to its TOP 10 customers based on their purchases. The Vice President of "XYZ," asks you to generate a report to show the "TOP 10 customers" by "Sales". In this situation, you can apply sorting to display the names of the TOP 10 customers by their "Sales".

3.2.2 What is sorting?

Sorting allows you to arrange dimensions in a specific order. There are two types of sorting:

- Computed sorting
- Manual sorting

3.2.3 How to apply sorting?

You can apply sorting in different ways:

- Computed sorting
- Manual sorting
- Nested sorting

3.2.3.1 Computed sorting

When you apply some programmatic rules for sorting, it is known as computed sorting. For example: Sorting product names by their alphabetic order. Computed sorting includes sorting on axis and sorting specific fields.

3.2.3.2 Sorting on axis

Use sort buttons on an axis for a quick computed sort.

3.2.3.2.1 Step 1

Connect to the Sample-Superstore data source. Drag the dimension "Category" "Sub-Category" from the dimensions area under the data pane to the rows shelf. Drag the measure "Sales" from the measures area under the data pane to the columns shelf (Shown in Fig. 3-54).

Figure 3-54. *Dimension "Category," "Sub-Category" placed on the rows shelf and the measure "Sales" placed on the columns shelf*

3.2.3.2.2 Step 2

Go to the sub-category field on the view and hover the mouse cursor over the axis. A sort icon is displayed as shown in Fig. 3-55.

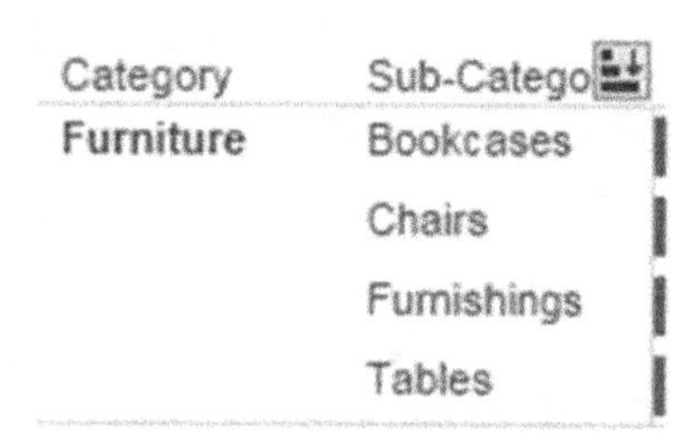

Figure 3-55. *Sort axis for "Sub-Category" field*

3.2.3.2.3 Step 3

Click it once to sort in ascending order, click on the axis again to sort the sub-category field in descending order as shown in Fig. 3-56.

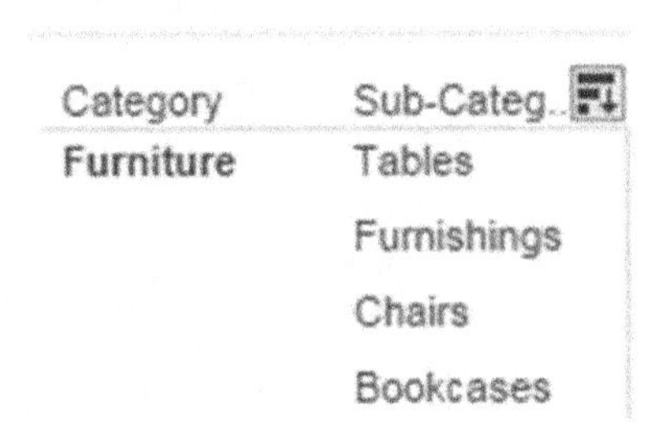

Figure 3-56. *"Sub-Category" field sorted in descending order*

Click a third time on the "Sub-Category" axis to clear the sort.

3.2.3.3 Sorting specific fields

Let's discuss steps to sort specific fields.

3.2.3.3.1 Step 1

Connect to the Sample-Superstore data source. Drag the dimension "Category" "Sub-Category" from the dimensions area under the data pane to the rows shelf. Drag the measure "Sales" from measures area under the data pane to the columns shelf (Shown in Fig. 3-57).

Figure 3-57. *Dimension "Category," "Sub-Category" placed on the rows shelf and the measure "Sales" placed on the columns shelf*

3.2.3.3.2 Step 2

Right click on "Sub-Category" field and select "Sort" option as shown in Fig. 3-58.

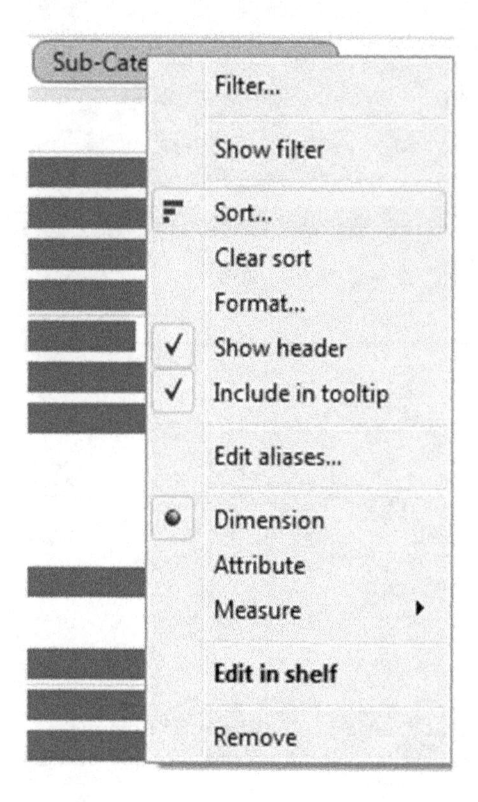

Figure 3-58. *"Sub-Category" field with "Sort" option*

3.2.3.3.3 Step 3

The sort dialog box opens. Specify the sort order as "Descending" and "Sort by" as the field, as shown in Fig. 3-59. You can select "Data source order," which orders the data by data source order. The default data source order is alphabetic order. "Field" orders the data based on the associated values of another field.

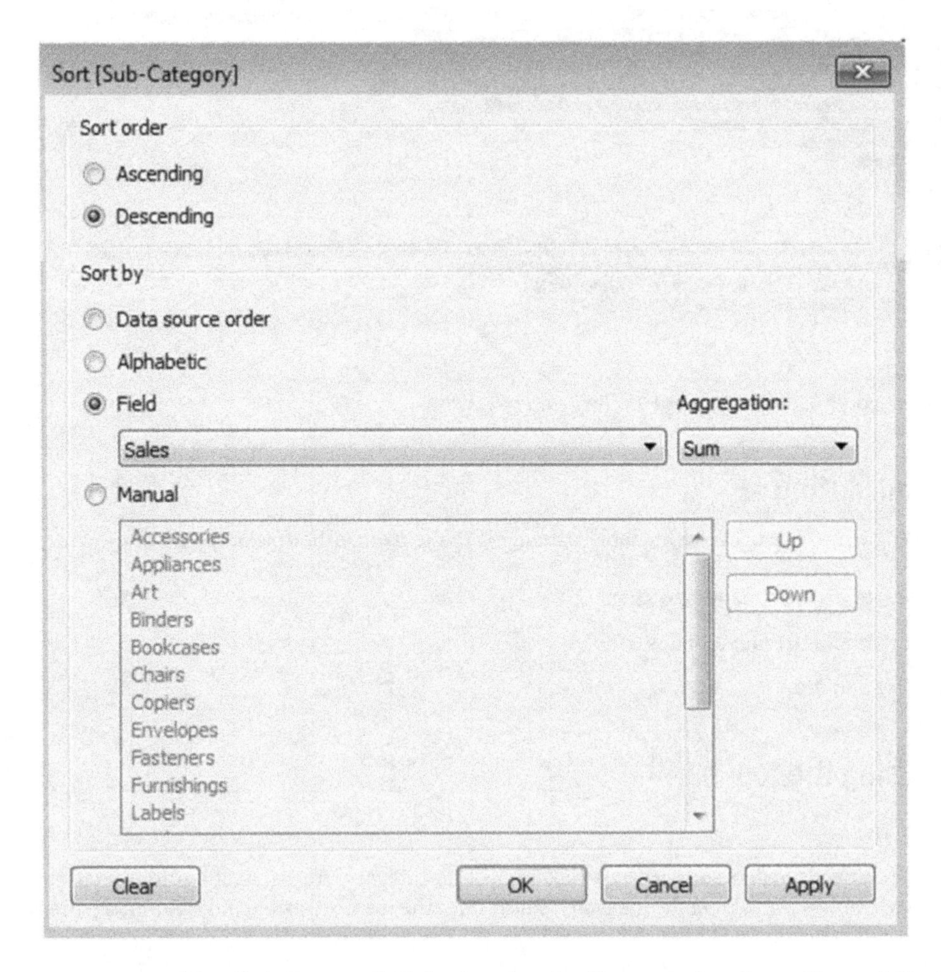

Figure 3-59. *"Sort [Sub-Category]" dialog box showing "Sort order" as descending and "Sort by" as a field*

3.2.3.3.4 Step 4

When you click on "OK," "Sub-Category" field is sorted based on their sum of "Sales" as shown in Fig. 3-60.

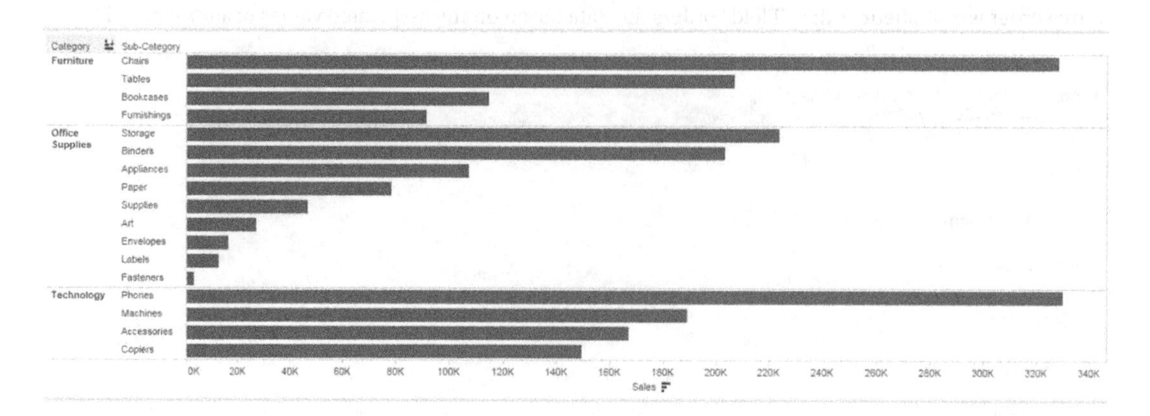

Figure 3-60. *"Sub-Category" field sorted based on their sum of "Sales"*

3.2.3.4 Manual sorting

Another way to rearrange the dimensions in the table is dragging them in an ad-hoc fashion. This is known as manual sorting.

There are two ways to perform manual sorting:

- Sort using the tool bar and tool tips
- Sort by drag and drop

3.2.3.5 Sort using the tool bar

3.2.3.5.1 Step 1

Connect to the Sample-Superstore data source. Drag the dimension "Category" from the dimensions area under the data pane to the rows shelf. Drag the measure "Sales" from the measures area under the data pane to the columns shelf (Shown in Fig. 3-61).

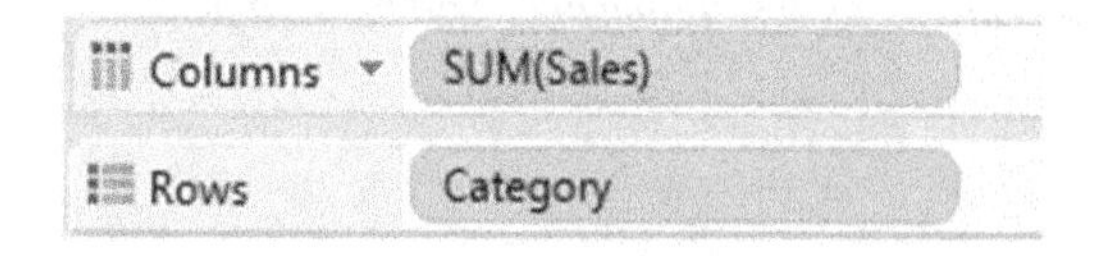

Figure 3-61. *Dimension "Category" placed on the rows shelf and the measure "Sales" placed on the columns shelf*

3.2.3.5.2 Step 2

Use the sort button on the tool bar to sort a field either in ascending or descending order as shown in Fig. 3-62.

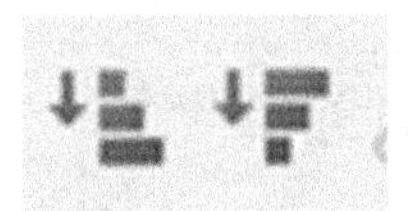

Figure 3-62. *Sort button on the tool bar*

3.2.3.6 Sort by drag and drop

Let's discuss the steps for performing "sort by drag and drop".

3.2.3.6.1 Step 1

Connect to the Sample-Superstore data source. Drag the dimension "Category" from the dimensions area under the data pane to the rows shelf. Drag the measure "Sales" from the measures area under the data pane to the columns shelf (Shown in Fig. 3-63.).

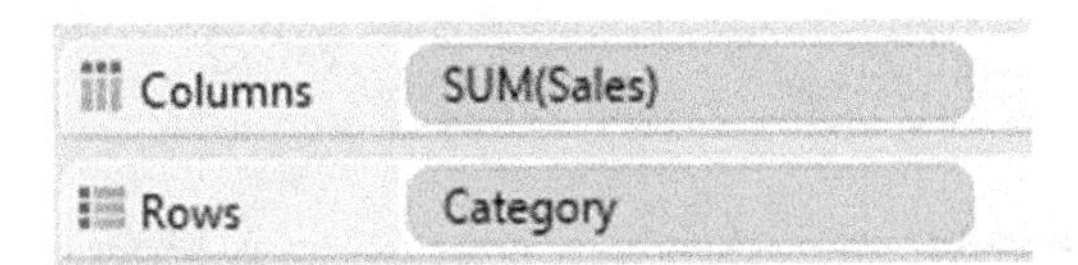

Figure 3-63. *Dimension "Category" placed on the rows shelf and the measure "Sales" placed on the columns shelf*

3.2.3.6.2 Step 2

Select the dimension member that you want to move, for example the technology category as shown in Fig. 3-64.

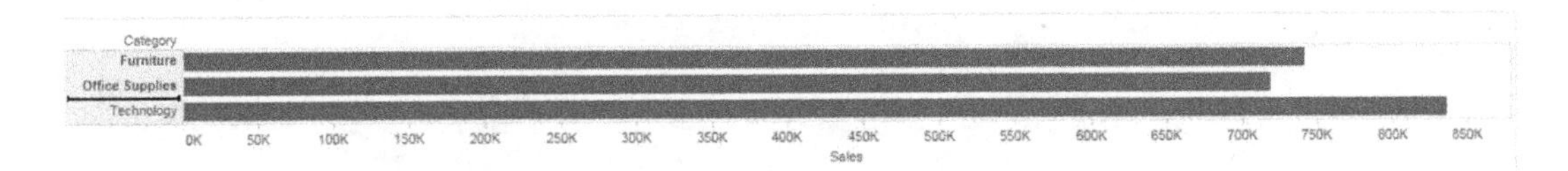

Figure 3-64. *View showing selected category member, "Technology"*

3.2.3.6.3 Step 3

Drag the dimension member "Technology," and drop it in the desired location as shown in Fig. 3-65.

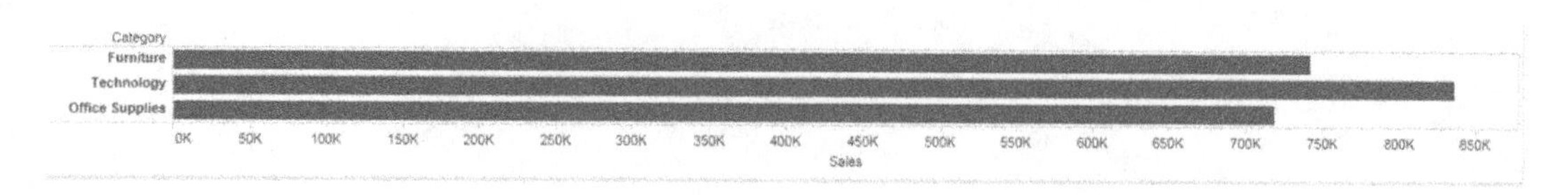

Figure 3-65. *View after dropping the "Technology" field to the desired location*

3.2.3.7 Nested sorting

Let's discuss steps for performing "Nested Sort".

3.2.3.7.1 Step 1

Connect to the Sample-Superstore data source. Drag the dimension "Region," "Sub-Category" from the dimensions area under the data pane to the rows shelf. Drag the measure "Sales" from the measures area under the data pane to the columns shelf (Shown in Fig. 3-66).

Figure 3-66. *Dimension "Region," "Sub-Category" placed on the rows shelf and the measure "Sales" placed on the columns shelf*

3.2.3.7.2 Step 2

Sort the sub-category in descending order by clicking on the sort icon on the tool bar, as shown in Fig. 3-67.

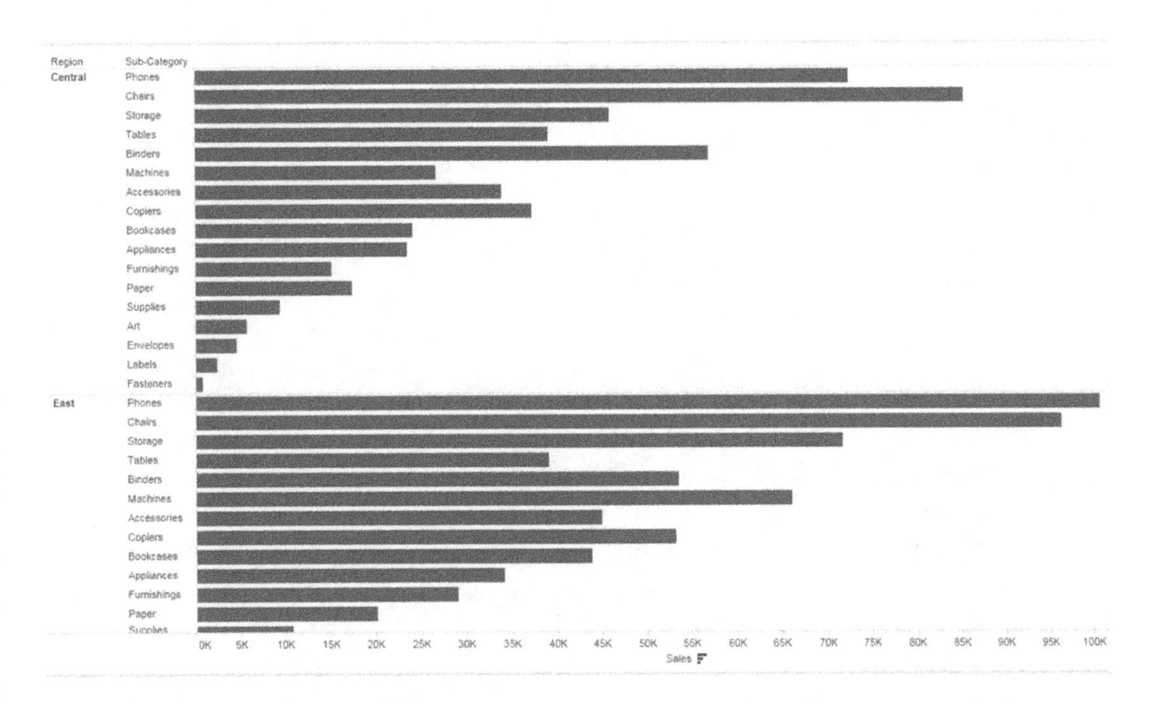

Figure 3-67. *"Sub-Category" sorted in descending order*

3.2.3.7.3 Step 3

The need was to sort each "Sub-Category" within each "Region". However that did not happen. The sort occurred at the "Sub-Category" level.

You can achieve this with the help of nested sorting.

3.2.3.7.4 Step 4a

In the data pane under dimensions, press ctrl key, select the region and the sub-category as shown in Fig. 3-68.

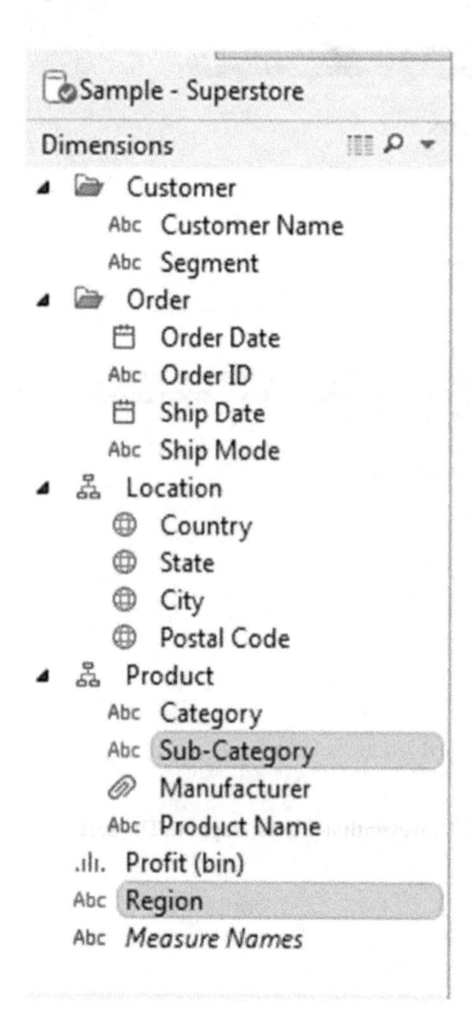

Figure 3-68. *Selection of "Sub-Category" and "Region"*

3.2.3.7.5 Step 4b

Right click on selected field, select create and then select combined field as shown in Fig. 3-69.

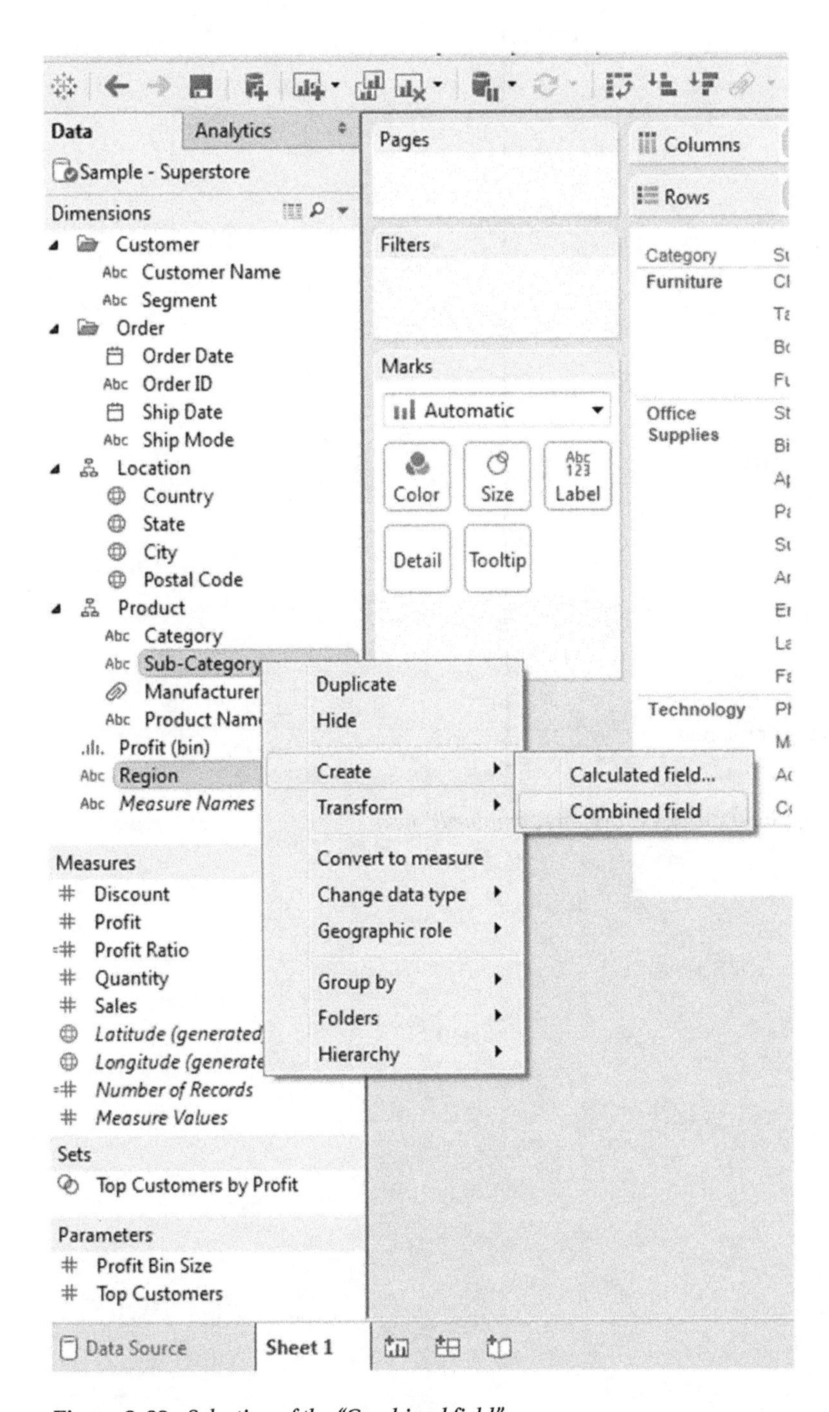

Figure 3-69. *Selection of the "Combined field"*

3.2.3.7.6 Step 5

You can see the "Combined field" on the data pane as shown in Fig. 3-70.

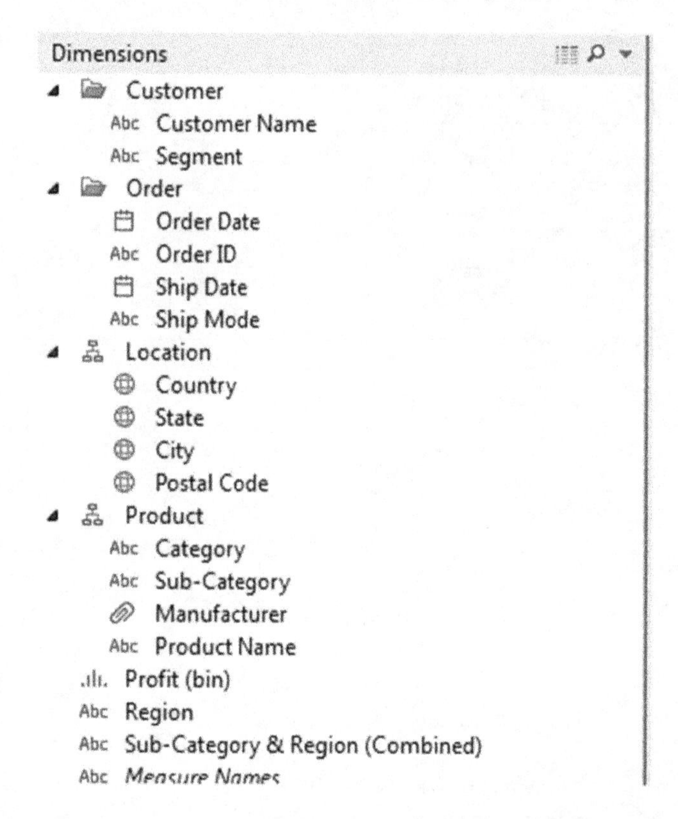

Figure 3-70. *Data pane showing "Sub-Category & Region (Combined)" field*

3.2.3.7.7 Step 6

Right click on "Sub-Category" and select "Clear sort" to clear the sort as shown in Fig. 3-71. You can see the updated view as shown in Fig. 3-72.

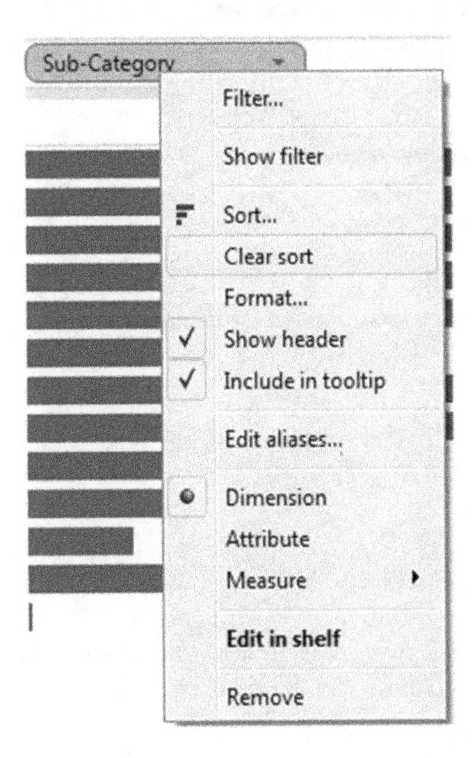

Figure 3-71. *Showing "Clear sort" option for "Sub-Category" field*

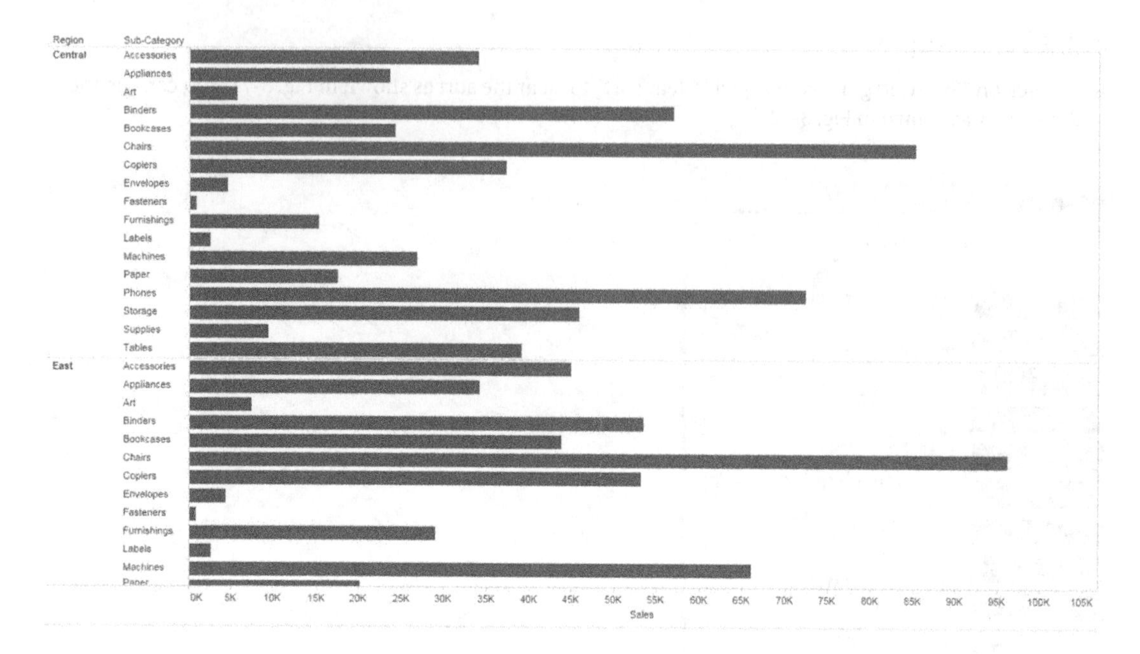

Figure 3-72. *View after applying "Clear sort" option to the "Sub-Category" field*

3.2.3.7.8 Step 7

Drag the "Sub-Category & Region (Combined)" field from the dimensions area under data pane to the rows shelf (Shown in Fig. 3-73).

Figure 3-73. *"Sub-Category & Region (combined field)" placed on the rows shelf*

3.2.3.7.9 Step 8

Right click on combined field, select sort option as shown in Fig. 3-74.

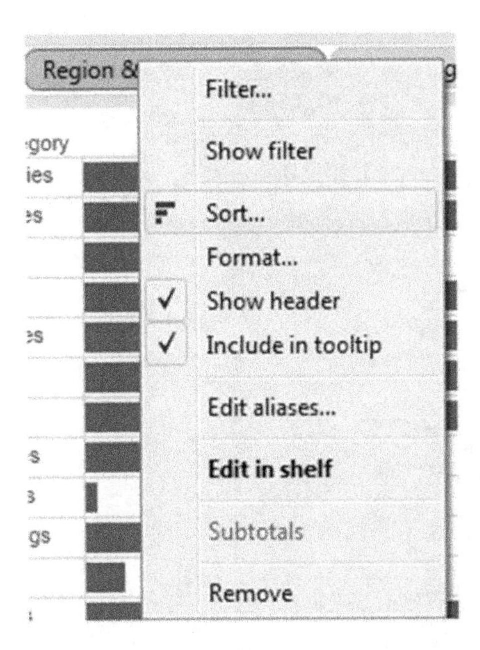

Figure 3-74. *Drop down menu that shows option to sort the "Sub-Category & Region (Combined)" field*

3.2.3.7.10 Step 9

The sort field dialog box opens. Select "Sort order" as "Descending," "Sort by" as "Field," specify field as "Sales" and aggregation as "Sum" as shown in Fig. 3-75.

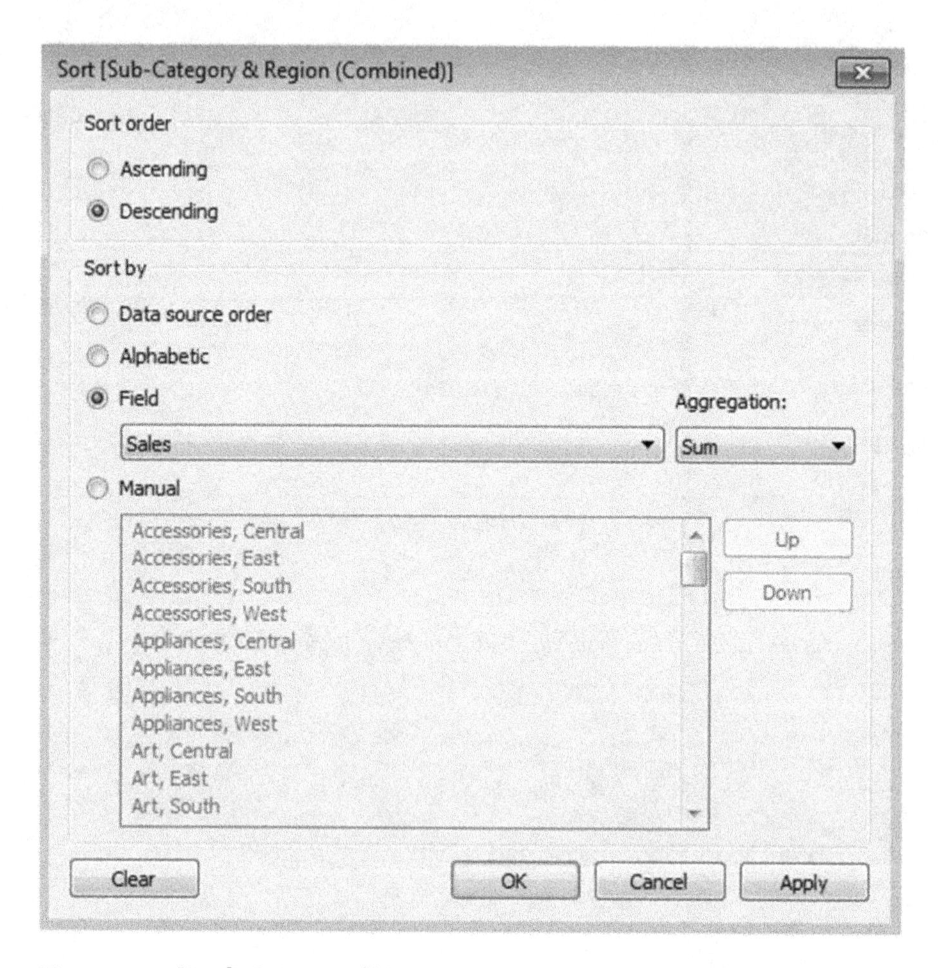

Figure 3-75. *"Sort[Sub-Category]" field dialog box with specfied "Sort order" and "Sort by" option*

3.2.3.7.11 Step 10

Observe the “nested sort” that is “Sub-Category” is sorted within each region by their “Sales” as shown in Fig. 3-76.

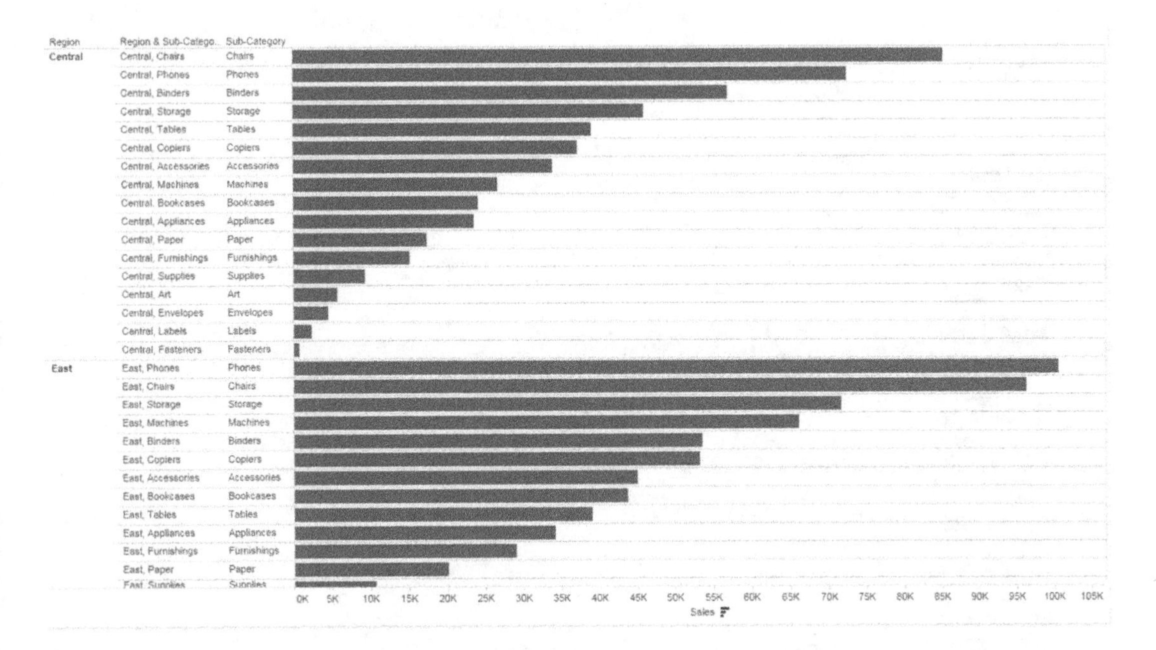

Figure 3-76. *View showing sorting of “Sub-Category” within each “Region” by their “Sales”*

3.2.3.7.12 Step 11

Right click on "Sub-Category & Region (Combined)" field and uncheck "Show header" option as shown in Fig. 3-77.

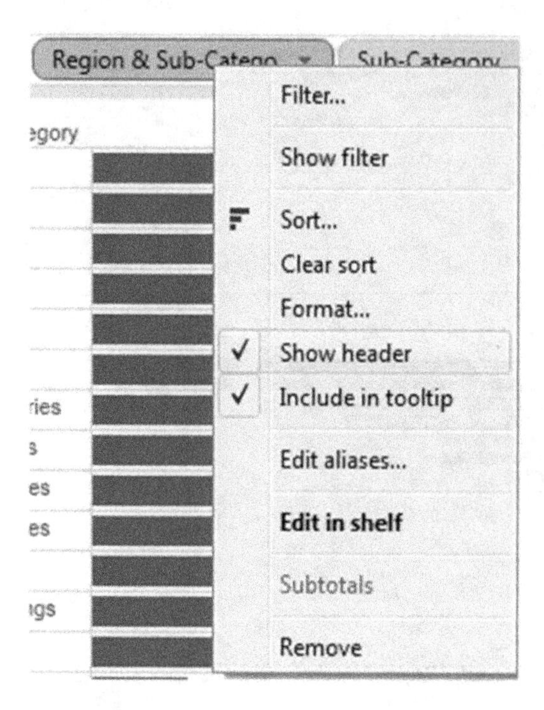

Figure 3-77. *Option to uncheck "Show header"*

3.2.3.7.13 Step 12

The view is improved by showing each "Sub-Category" sorted within each "Region" (Shown in Fig. 3-78).

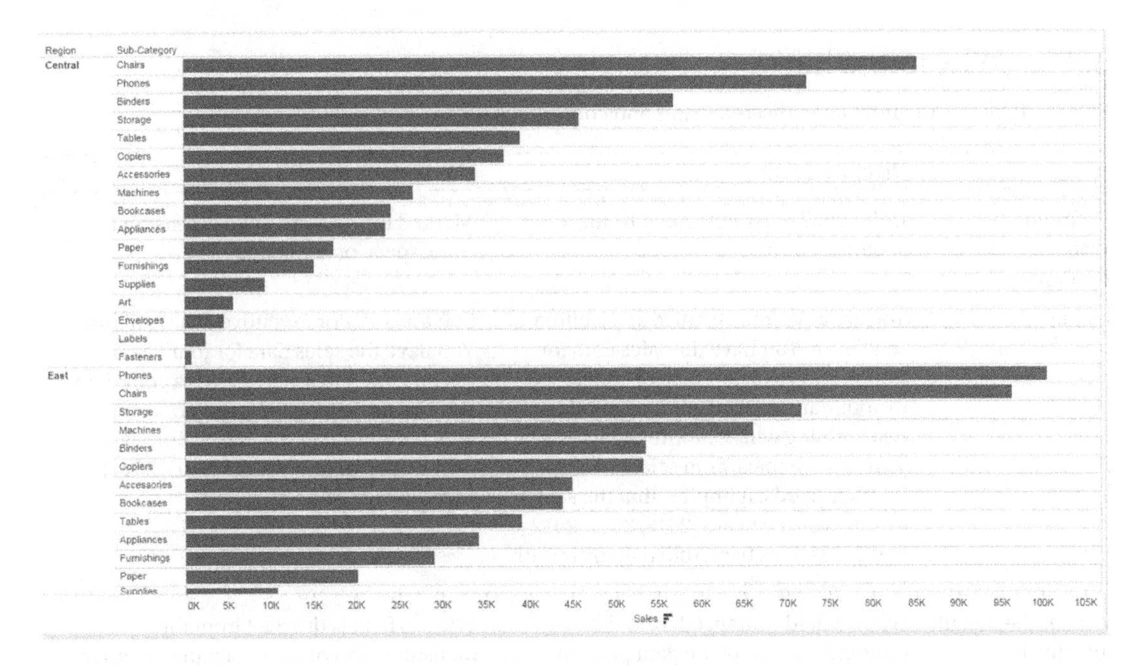

Figure 3-78. *View showing "Sub-Category" sorted within each "Region" by their "Sales"*

3.2.4 Discrete and Continuous Dates

Let's discuss discrete and continuous dates in Tableau.

3.2.5 Why and what?

Refer to Table 3-1 to understand discrete dates and continuous dates.

	Discrete dates	Continuous dates
Definition: Webster's Dictionary	Individually distinct; constituting a separate entity	Marked by uninterrupted extension in time, space or sequence
Example	You are a senior executive at a leading retail firm. You have the sales data for four years (2010, 2011, 2012, and 2013) in your data set. You are interested in determining which month regardless of year has the maximum sales. In other words, it implies that the sales data across the years (2010, 2011, 2012 and 2013) have been rolled up by month. Refer to Fig. 3-79.	You are a senior executive at a retail firm. You have the sales data for four years (2010, 2011, 2012, and 2013) in your data set. You are interested in determining which month had the maximum sales over a span of four years (2010 - 2014) Refer to Fig. 3-80.
What is the default for a field?	When a field is dragged from the dimensions area of the data pane to either the rows shelf or columns shelf, it is "Discrete" by default. Example: Customer ID, Customer Name	When a field is dragged from the measures area of the data pane to either the rows shelf or columns shelf, it is "Continuous" by default. Example: Unit Price, Order Quantity
Tableau creates	Axis headers	Axis
Visual Cue	Blue pill	Green pill
History	Discrete variables can take on only a finite set of values. Example: If you count from 0 to 10, there are 11 distinct values. When dealing with discrete values, you will not consider 2.6 or 9.3, etc.	Continuous variables can take on an infinite set of values. Example: if you count from 0 to 10, there are an infinite number of values between 0 and 10.
Sort Order	E.g. Discrete dates can be sorted in ascending or descending order by sales. Refer to Figure 3-81.	Since the continuous dates form a continuous axis, they are arranged in chronological order by default with the oldest date at the leftmost end and the most recent date at the rightmost end. One is not allowed to change the sequence.
Preferred chart form	Bar Chart	Line Graph

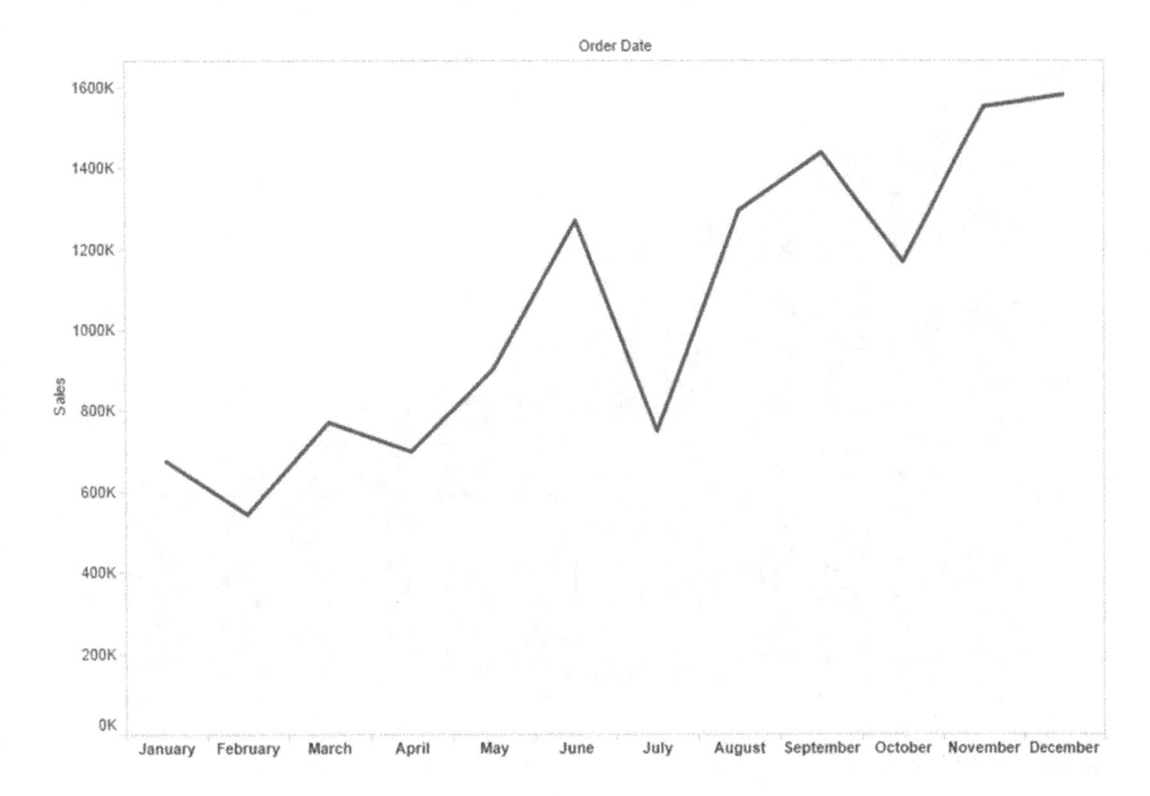

Figure 3-79. *Discrete dates*

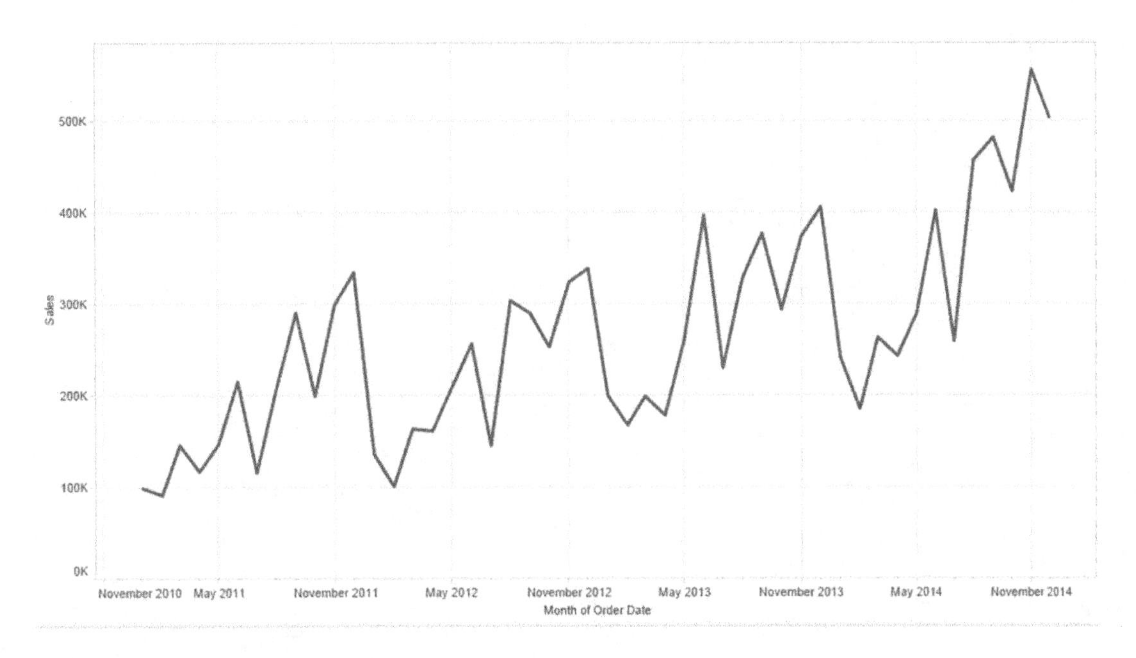

Figure 3-80. *Continuous dates*

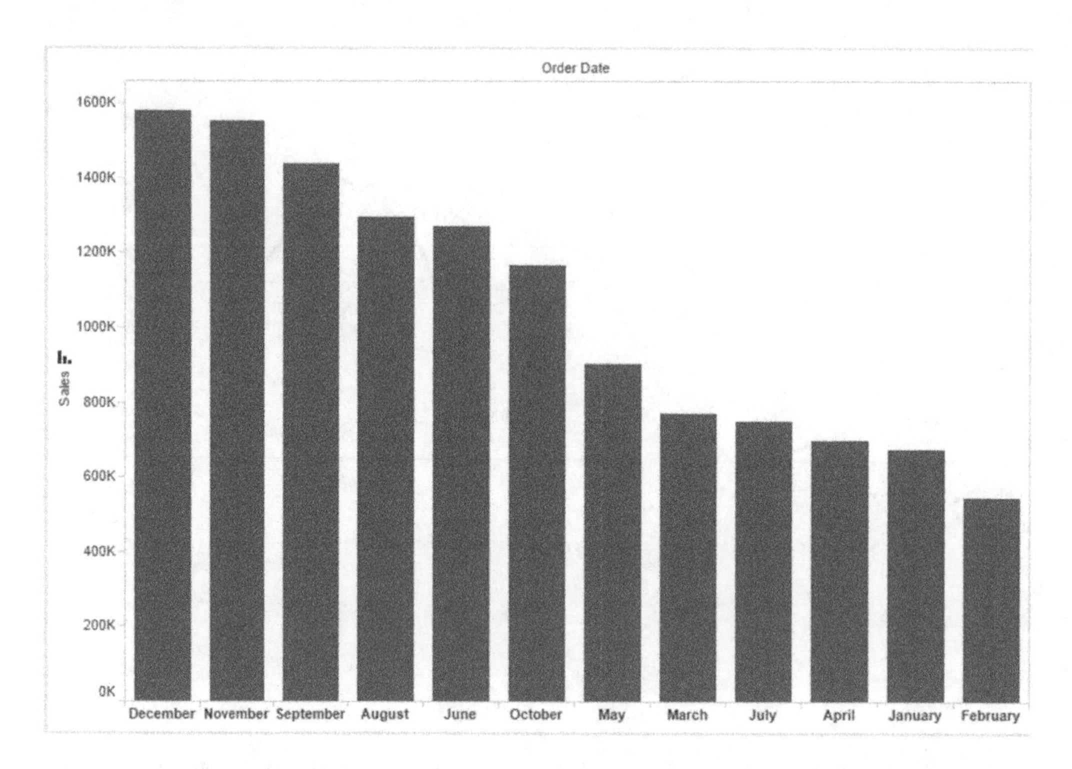

Figure 3-81. *Discrete fields can be sorted: Here December had the highest sales and February the least*

3.2.5.1 How to use a discrete date in Tableau?

Let's discuss steps to create discrete dates.

3.2.5.1.1 Steps

3.2.5.1.2 Step 1

Drag "Order Date" from the dimensions area of the data pane to the Columns Shelf as shown in Fig. 3-82. Tableau by default creates a hierarchy on the date type field. The fields dragged from the dimensions area of the data pane are "Discrete" by default.

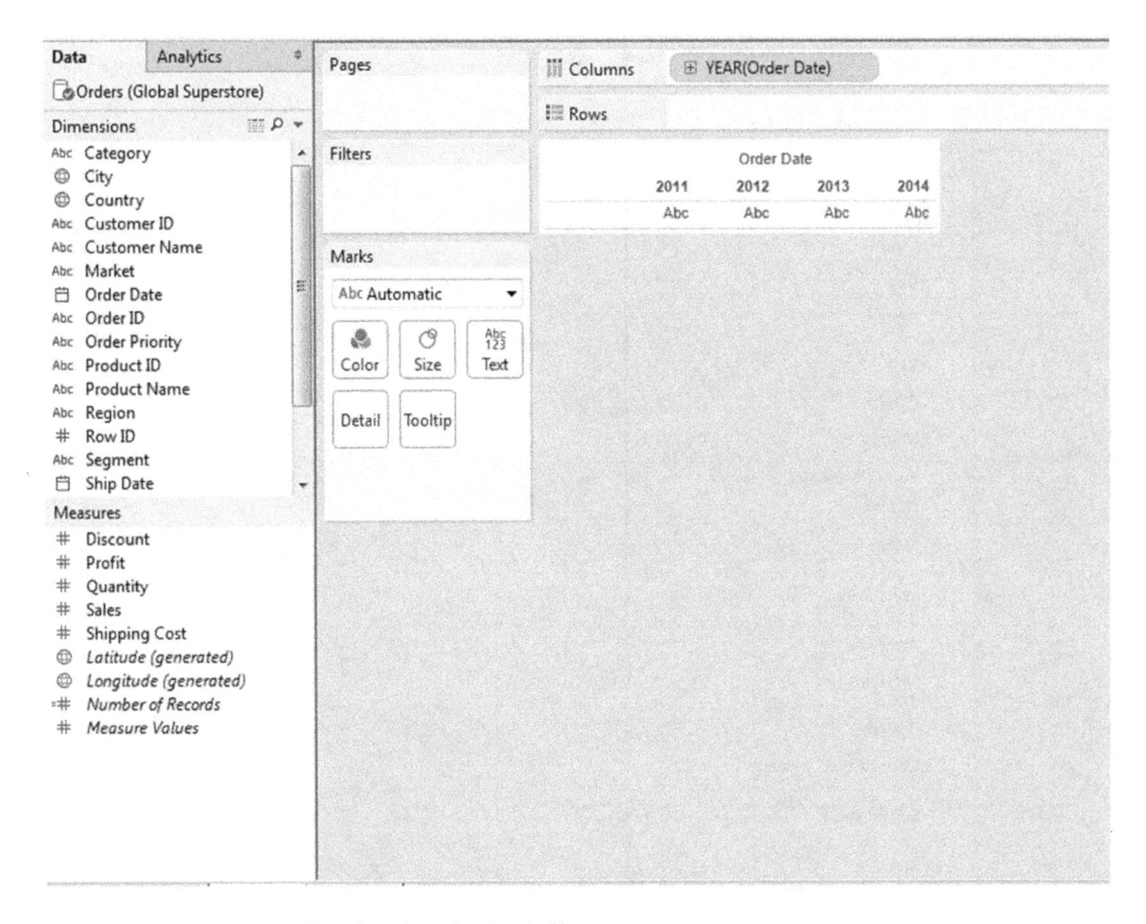

Figure 3-82. *Order Date placed on the columns shelf*

3.2.5.1.3 Step 2

Click on the drop down to select the appropriate date part (Year / Quarter / Month / Day) as shown in Fig. 3-83. The first set of values constitutes the discrete bucket. The next set of values constitutes the continuous bucket.

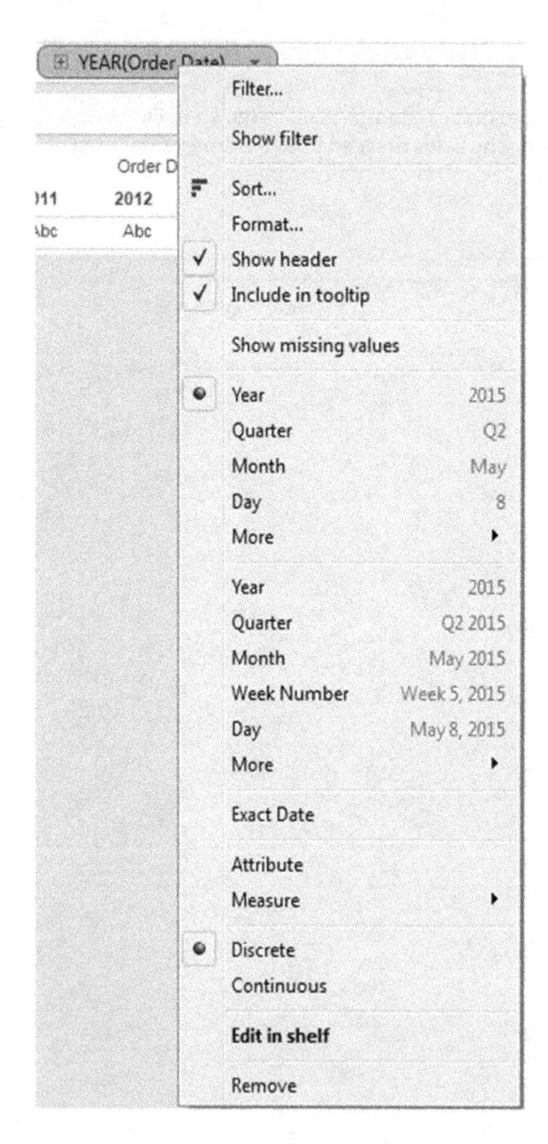

Figure 3-83. *Drop down menu to select appropriate date part*

The discrete bucket is shown in Fig. 3-84.

Year 2015
Quarter Q2
Month May
Day 8
More

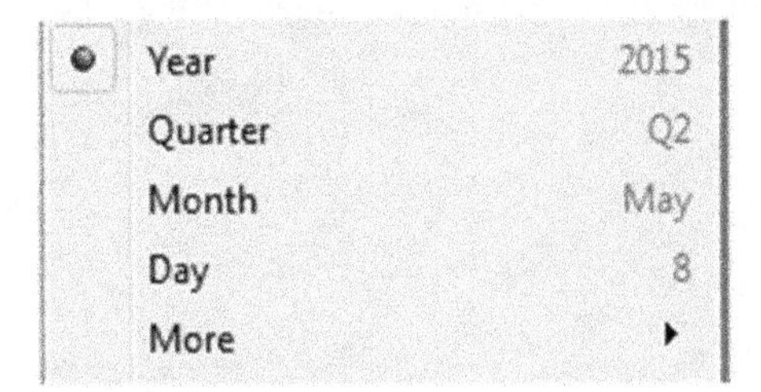

Figure 3-84. *Discrete bucket*

The continuous bucket is shown in Fig. 3-85.

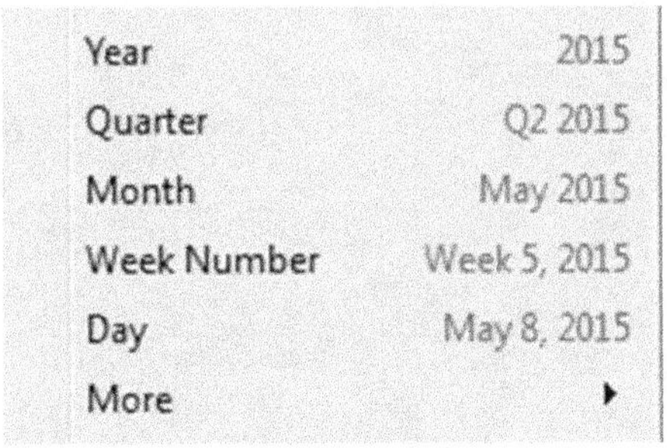

Year 2015
Quarter Q2 2015
Month May 2015
Week Number Week 5, 2015
Day May 8, 2015
More

Figure 3-85. *Continuous bucket*

3.2.5.1.4 Step 3

Let us plot "Sales" by discrete "Month". Select discrete option from the drop down menu as shown in Fig. 3-86.

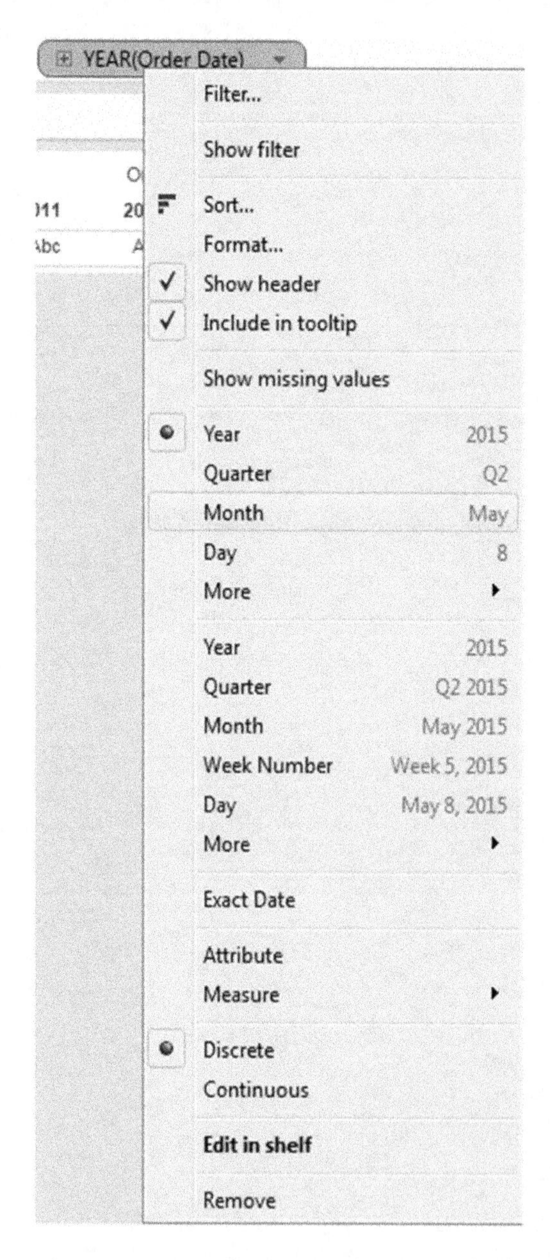

Figure 3-86. *Drop down menu showing "Discrete" option*

3.2.5.1.5 Step 4

Select "Bar" on the marks card, as shown in Fig. 3-87.

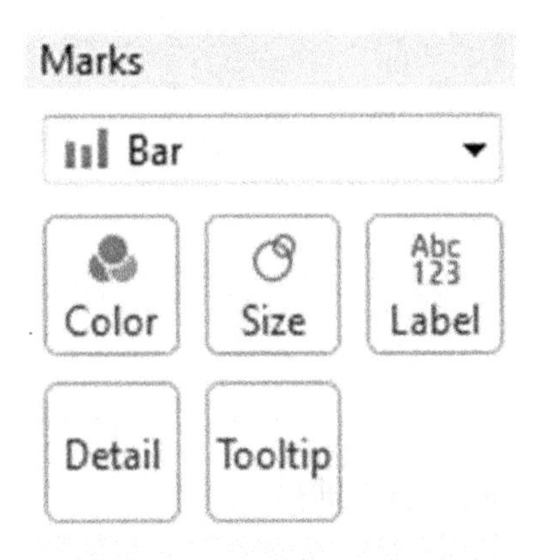

Figure 3-87. *Marks card with "Bar" option*

3.2.5.1.6 Step 5

Drag the measure, "Sales" from the measures area of the data pane on the rows shelf as shown in Fig. 3-88.

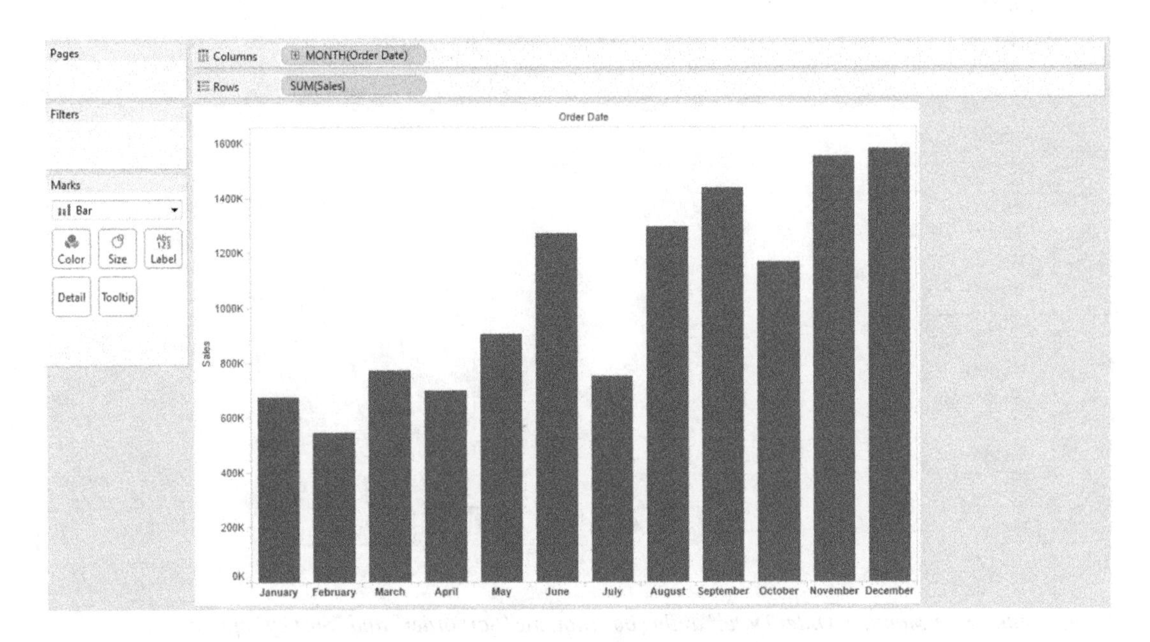

Figure 3-88. *Measure "Sales" on the rows shelf*

3.2.5.1.7 Step 6

Sort the bars representing the "Sales" data in descending order as shown in Fig. 3-89 and Fig. 3-90.

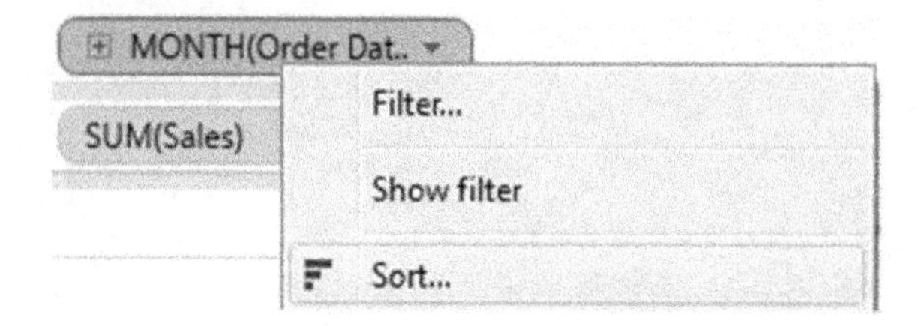

Figure 3-89. *Drop down showing "Sort" option*

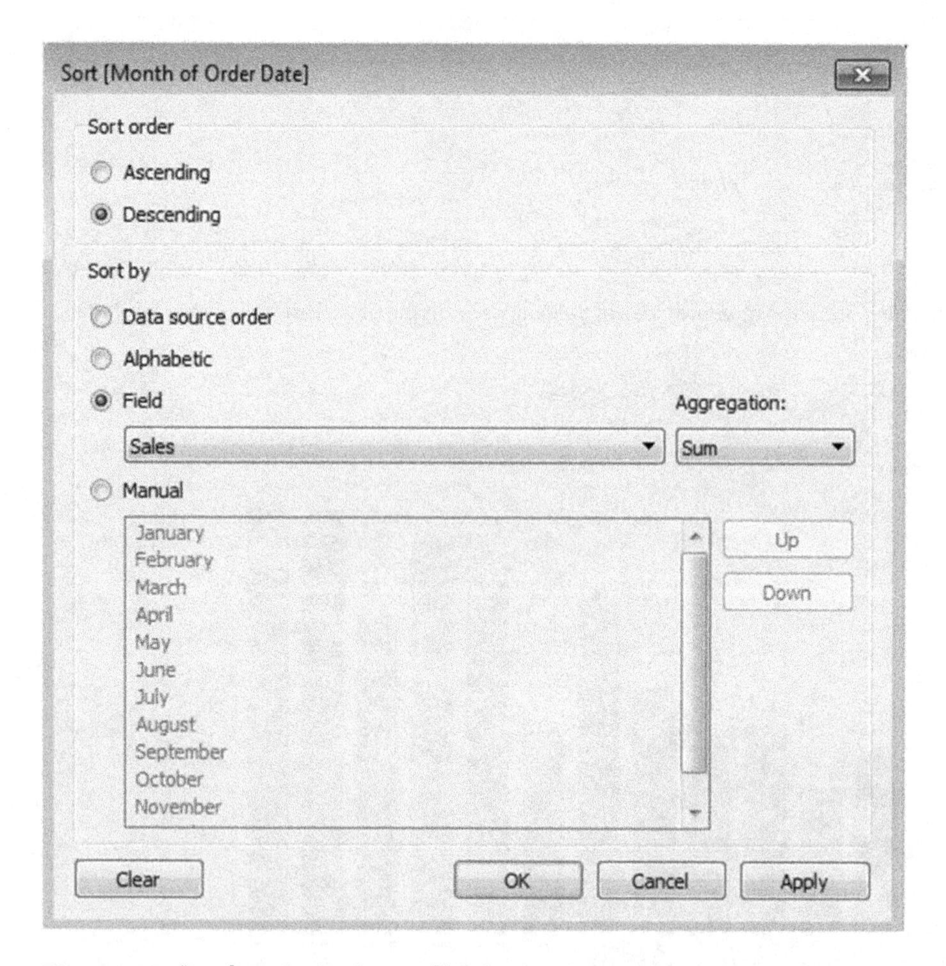

Figure 3-90. *"Sort[Month of Order Date]" dialog box showing "Sort order" and "Sort by" option*

3.2.5.1.8 Step 7

The final output is shown in Fig. 3-91.

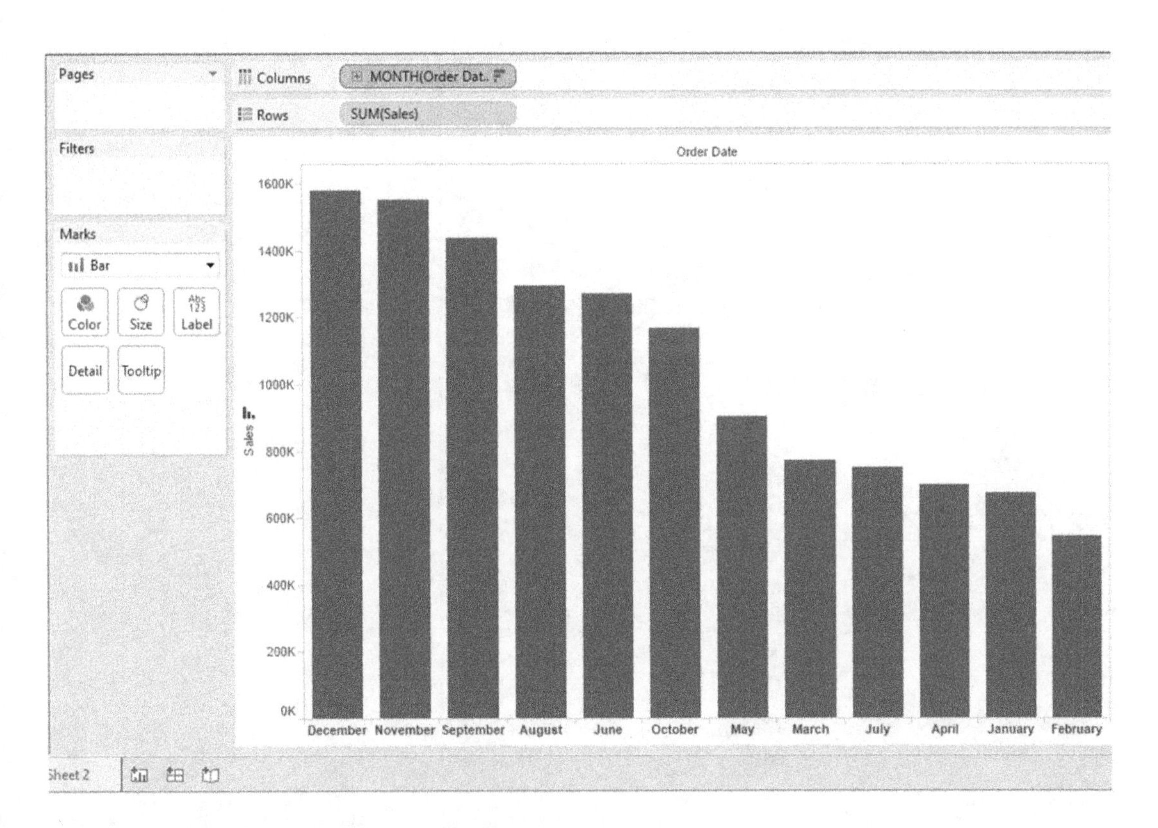

Figure 3-91. *"Sales" sorted in descending order*

3.2.5.1.9 How to use continuous dates in Tableau?

3.2.5.1.10 Step 1

Drag "Order Date" from the dimensions area of the data pane to the columns shelf as shown in Fig. 3-92. Tableau by default creates a hierarchy on the date type field. The fields dragged from the dimensions area of the data pane are "Discrete" by default.

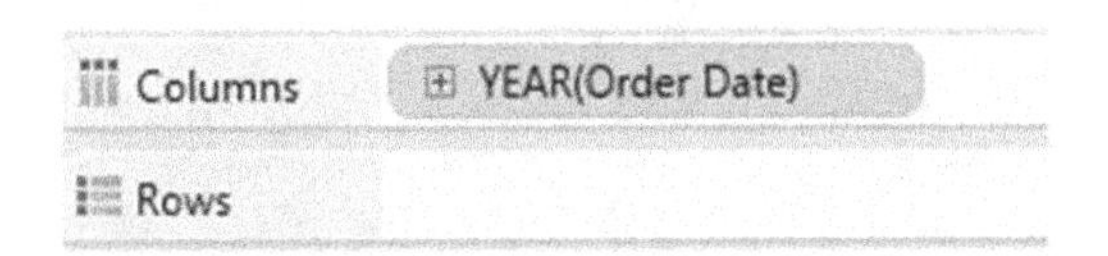

Figure 3-92. *Dimension "Order Date" placed on the columns shelf*

3.2.5.1.11 Step 2

Let us convert the "Discrete" date to "Continuous" date as shown in Fig. 3-93.

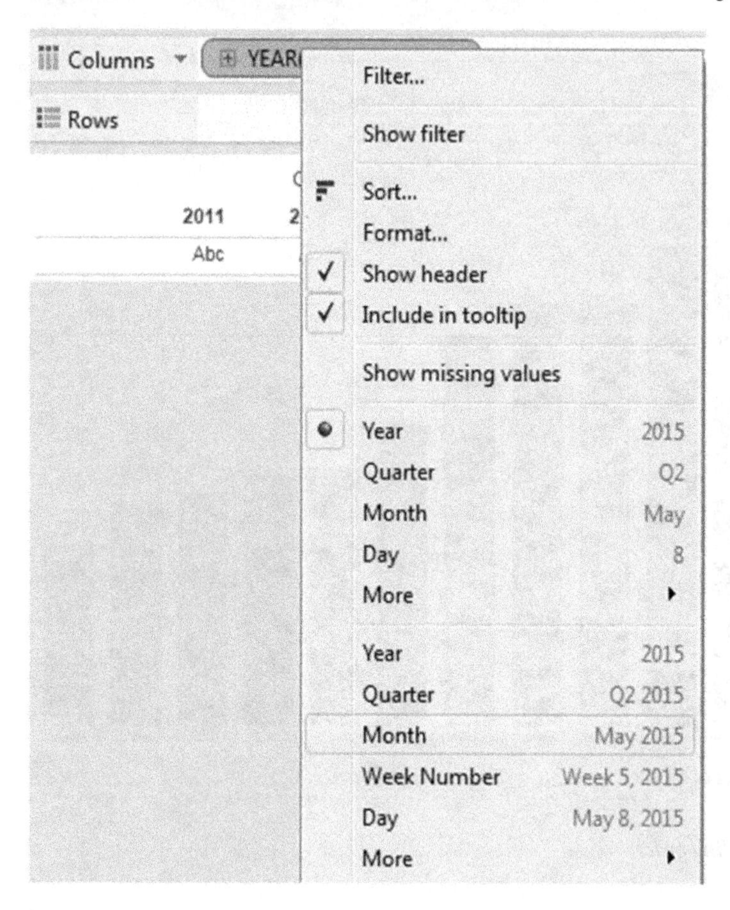

Figure 3-93. *Drop down menu showing "Month" from the "Continuous" bucket*

Select "Line" on the marks card as shown in Fig. 3-94.

Figure 3-94. *Marks card with "Line" option*

3.2.5.1.12 Step 3

Drag the measure, "Sales" from the measures area and drop it on the rows shelf as shown in Fig. 3-95.

Figure 3-95. *View showing continuous month*

3.2.5.2 How to create custom dates in Tableau?

Let's discuss steps to create custom dates.

3.2.5.2.1 Step 1

Select the dimension, "Order Date" from the dimensions area of the data pane. Click on the drop down menu as shown in Fig. 3-96. Select Transform ➤ Create Custom Date.

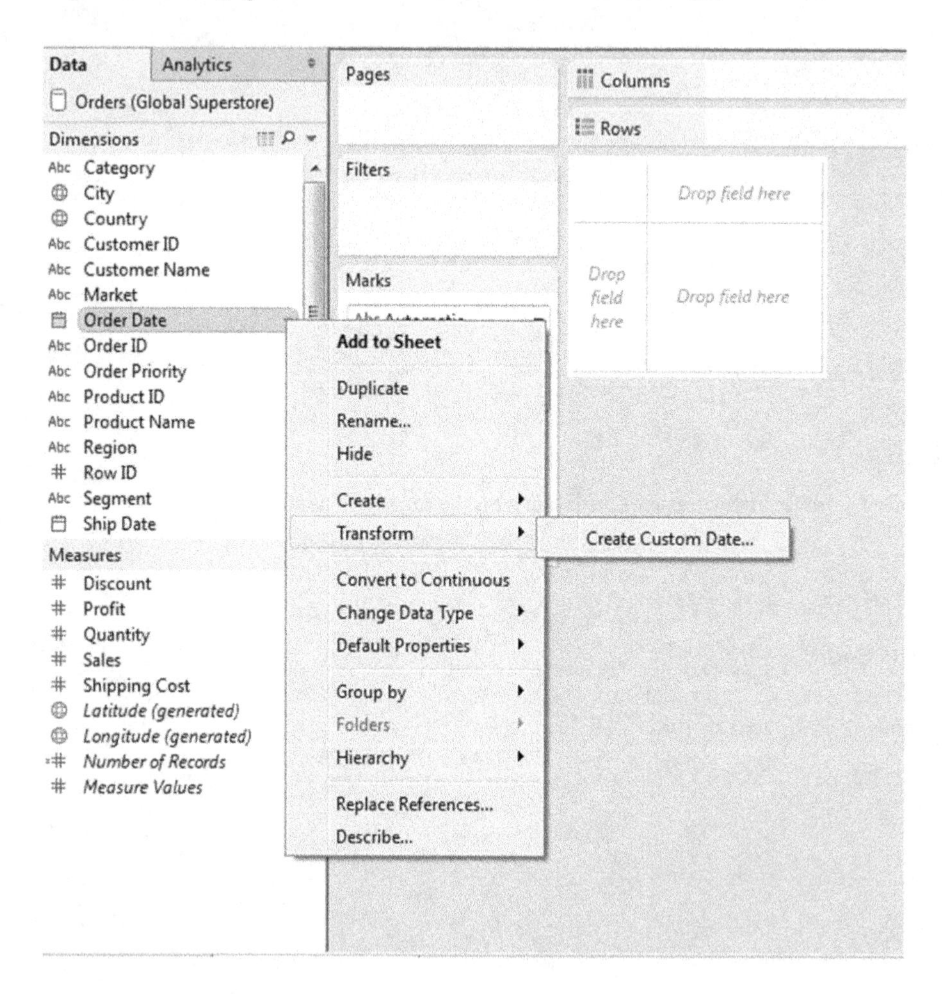

Figure 3-96. *"Create Custom Date" option*

3.2.5.2.2 Step 2

Make the selection as shown in Fig. 3-97 in the "Create Custom Date [Order Date]" dialog box:

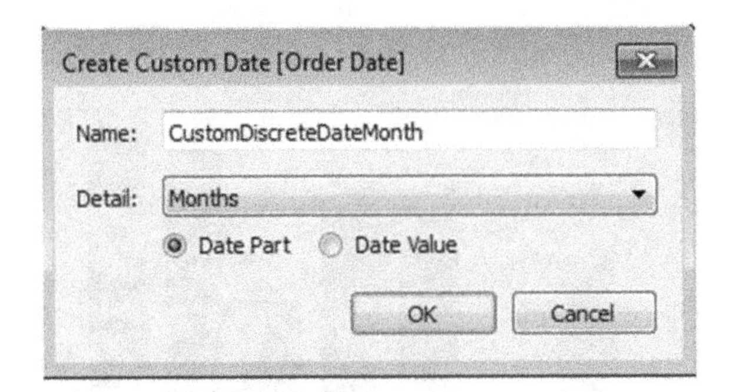

Figure 3-97. *Custom Date "CustomDiscreteDateMonth"*

Select "Date Part" to create discrete date and "Date Value" to create continuous date.

The new custom date gets added as a dimension in the dimensions area of the data pane as shown in Fig. 3-98.

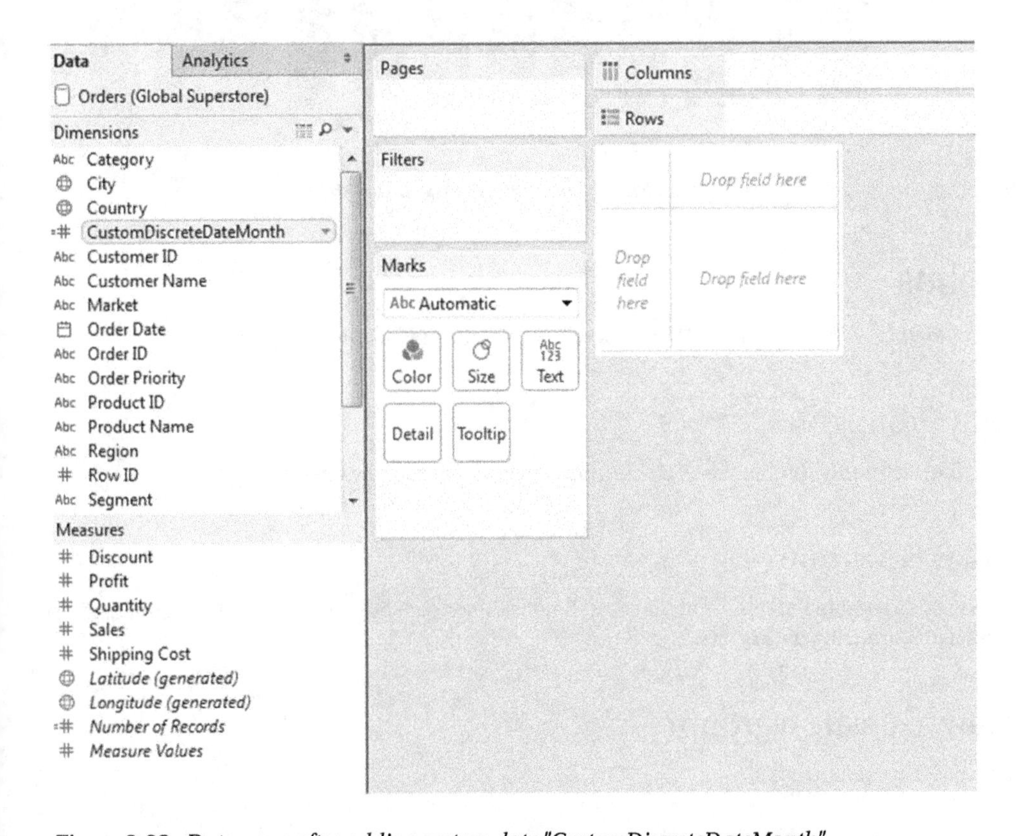

Figure 3-98. *Data pane after adding custom date "CustomDiscreteDateMonth"*

This new dimension can be used as a regular dimension on the rows or columns shelf as shown in Fig. 3-99.

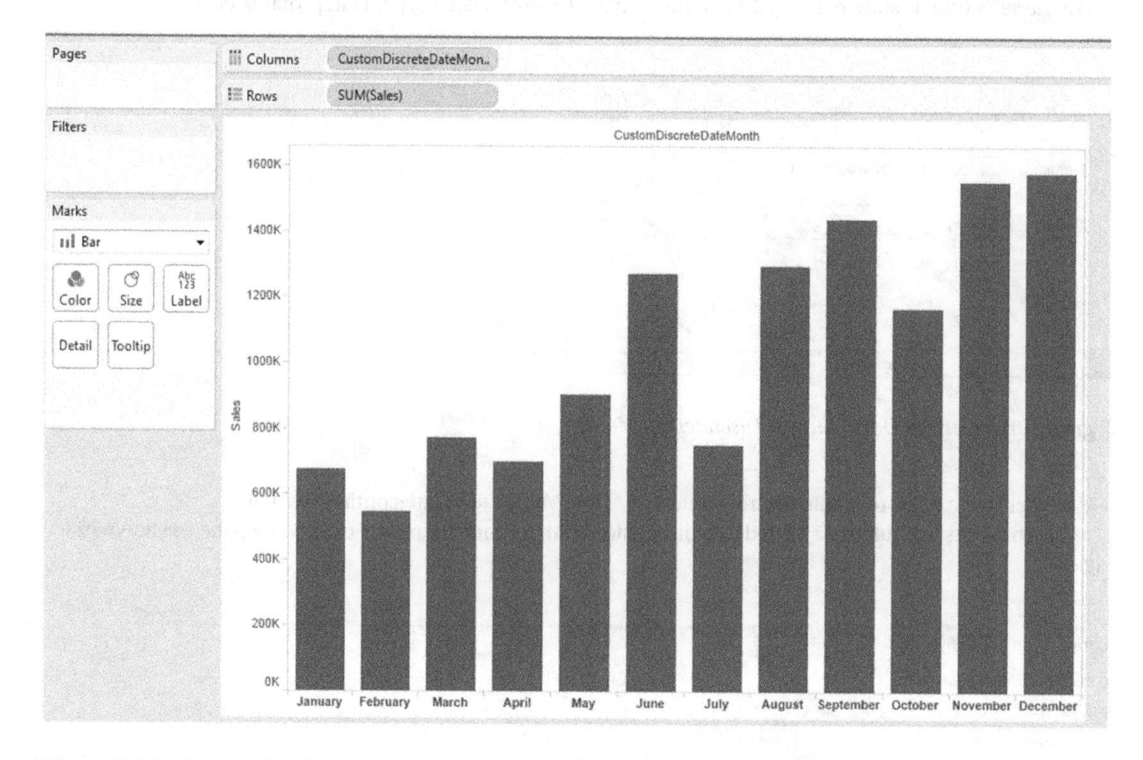

Figure 3-99. *Custom date "CustomDiscreteDateMonth" placed on the columns shelf*

3.3 Groups

A group is a combination of dimension members that will constitute higher level categories.

3.3.1 Why groups?

Use groups to refine views and identify the exact information you want to draw attention to.

3.3.2 What is a group?

A "group" allows you to combine members of different dimensions to constitute higher level categories. Groups are useful for correcting data errors.

3.3.3 How to create a group?

Let's discuss steps for creating groups.

3.3.3.1 Creating groups

Create “Groups” to correct data errors.

3.3.3.1.1 Step 1

Connect to the Sample-Superstore data source.

3.3.3.1.2 Step 2

Consider the view shown in Fig. 3-100.

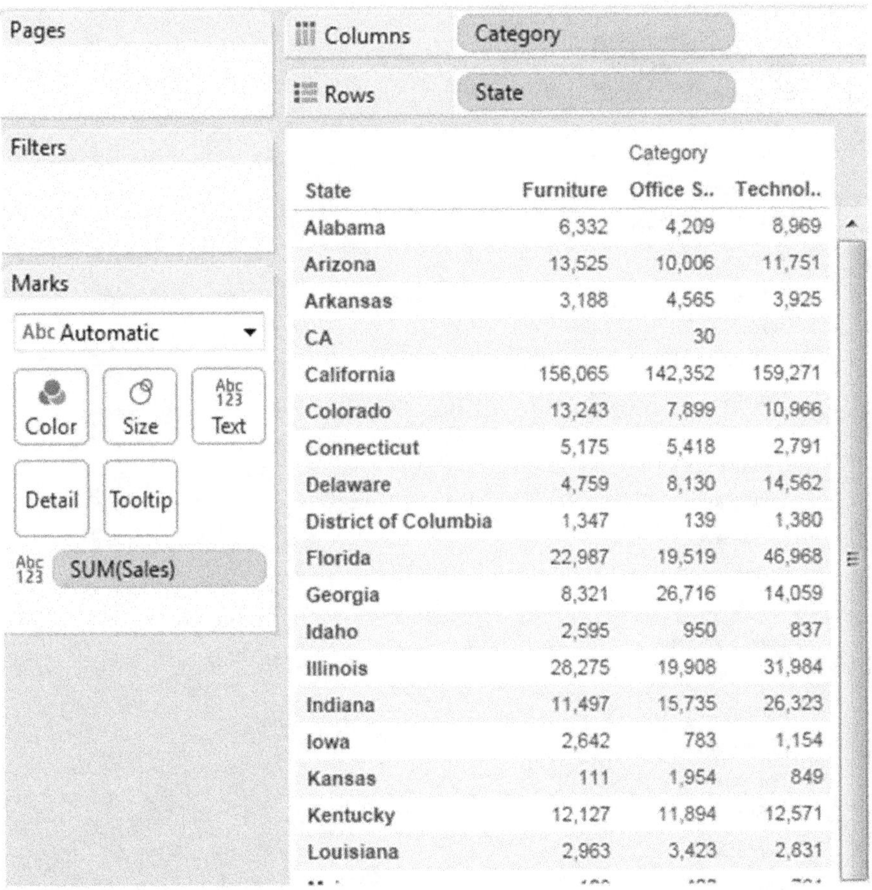

Pages

Columns: Category

Rows: State

Filters

Marks

Abc Automatic

Color | Size | Text

Detail | Tooltip

SUM(Sales)

	Category		
State	Furniture	Office S..	Technol..
Alabama	6,332	4,209	8,969
Arizona	13,525	10,006	11,751
Arkansas	3,188	4,565	3,925
CA		30	
California	156,065	142,352	159,271
Colorado	13,243	7,899	10,966
Connecticut	5,175	5,418	2,791
Delaware	4,759	8,130	14,562
District of Columbia	1,347	139	1,380
Florida	22,987	19,519	46,968
Georgia	8,321	26,716	14,059
Idaho	2,595	950	837
Illinois	28,275	19,908	31,984
Indiana	11,497	15,735	26,323
Iowa	2,642	783	1,154
Kansas	111	1,954	849
Kentucky	12,127	11,894	12,571
Louisiana	2,963	3,423	2,831

Figure 3-100. *View showing how “Category” is performing in various states*

In this view, CA denotes California. But it appears as separate entry. You can correct this by grouping CA and California.

3.3.3.1.3 Step 3

Press and hold the CTRL key, select CA and California as shown in Fig. 3-101.

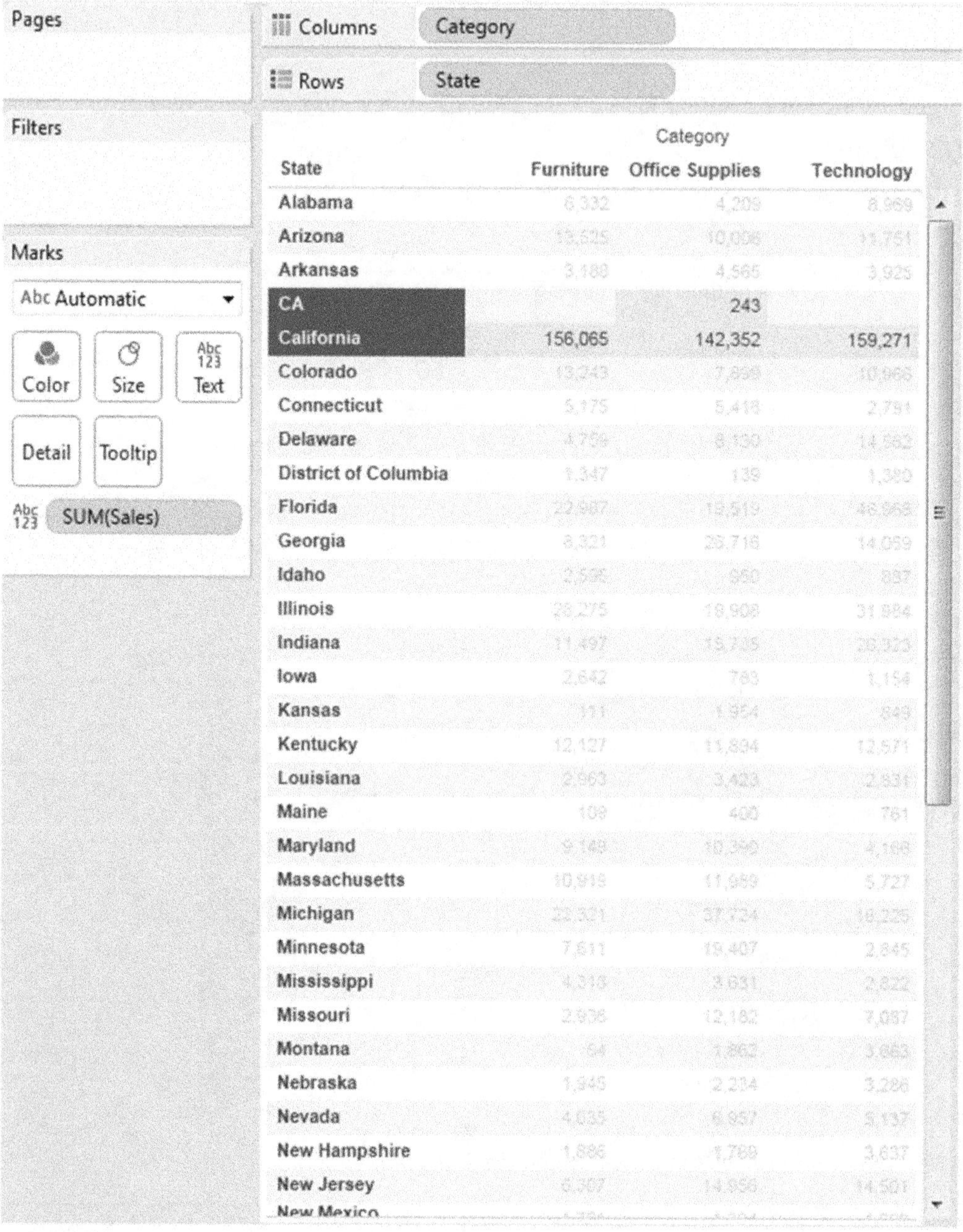

Pages

Columns: Category

Rows: State

Filters

Marks

Abc Automatic

Color | Size | Text

Detail | Tooltip

SUM(Sales)

State	Furniture	Office Supplies	Technology
	Category		
Alabama	6,332	4,209	8,969
Arizona	13,525	10,098	11,751
Arkansas	3,188	4,565	3,925
CA		243	
California	156,065	142,352	159,271
Colorado	13,243	7,898	10,966
Connecticut	5,175	5,418	2,791
Delaware	4,759	8,130	14,582
District of Columbia	1,347	139	1,380
Florida	22,987	19,519	46,968
Georgia	8,321	25,716	14,059
Idaho	2,595	960	887
Illinois	28,275	18,908	31,984
Indiana	11,497	15,735	26,323
Iowa	2,642	783	1,154
Kansas	111	1,954	849
Kentucky	12,127	11,894	12,571
Louisiana	2,963	3,423	2,831
Maine	109	400	761
Maryland	9,149	10,390	4,168
Massachusetts	10,919	11,959	5,727
Michigan	22,321	37,724	18,225
Minnesota	7,611	19,407	2,845
Mississippi	4,318	3,631	2,822
Missouri	2,936	12,182	7,087
Montana	64	1,862	3,663
Nebraska	1,945	2,234	3,286
Nevada	4,035	6,957	5,137
New Hampshire	1,886	1,789	3,637
New Jersey	6,307	14,956	14,501
New Mexico	[illegible]	[illegible]	[illegible]

Figure 3-101. *Selection of states, "CA" and "California"*

3.3.3.1.4 Step 4

One way to group the dimension members is to move the mouse over the selected area to get the pop-up menu and select the paper clip icon to group the selected state. (Shown in Fig. 3-102).

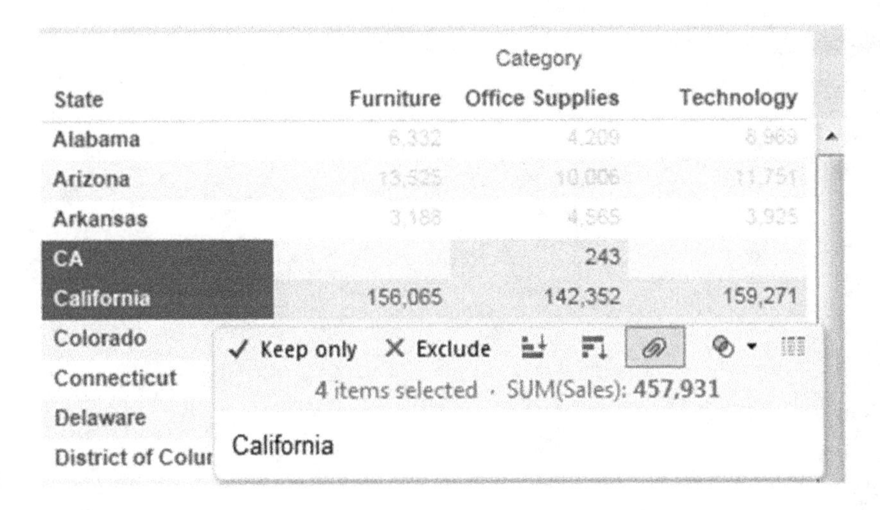

Figure 3-102. *Pop-up menu showing the group members icon*

3.3.3.1.5 Step 5

Another way to group the dimension members is to right click on the dimension members, select "Group" as shown in Fig. 3-103.

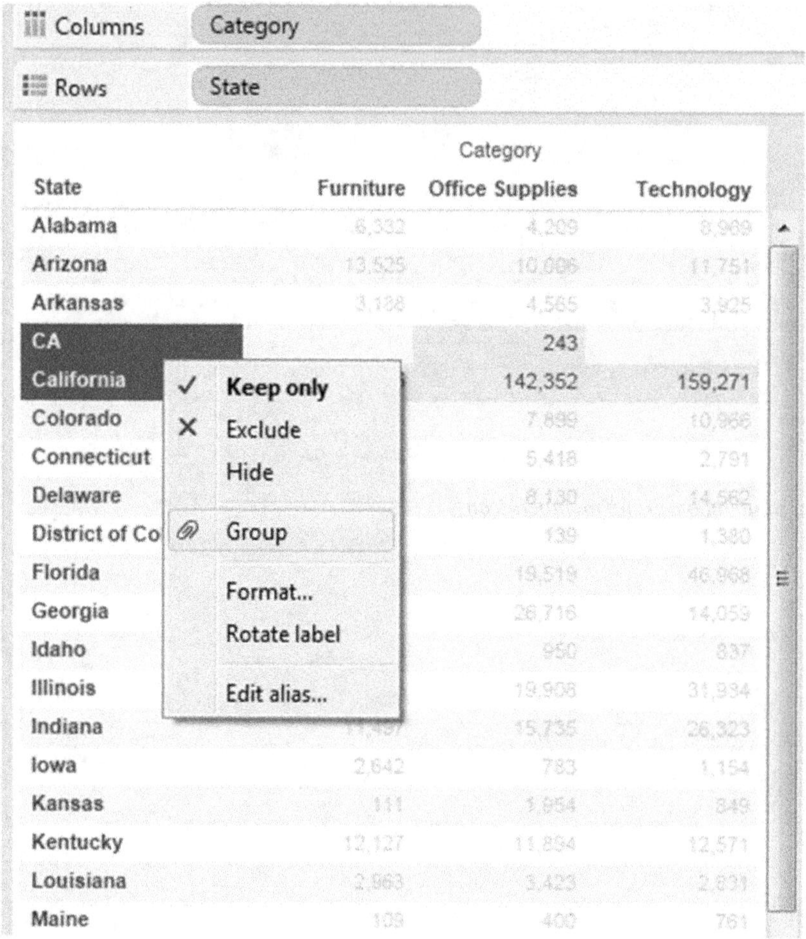

Figure 3-103. *"Group" option*

3.3.3.1.6 Step 6

Selected members are grouped together as shown in Fig. 3-104. This way you can correct the data errors.

Columns: Category

Rows: State (group)

	Category		
State (group)	Furniture	Office Supplies	Technology
Alabama	6,332	4,209	8,969
Arizona	13,525	10,006	11,751
Arkansas	3,188	4,565	3,925
CA & California	156,065	142,595	159,271
Colorado	13,243	7,899	10,966
Connecticut	5,175	5,418	2,791

Figure 3-104. *Group members "CA & California"*

3.3.3.1.7 Step 7

You can see the newly created group in the dimensions area under the data pane (Shown in Fig. 3-105). The rows shelf is replaced with the newly created group field as shown in Fig. 3-106.

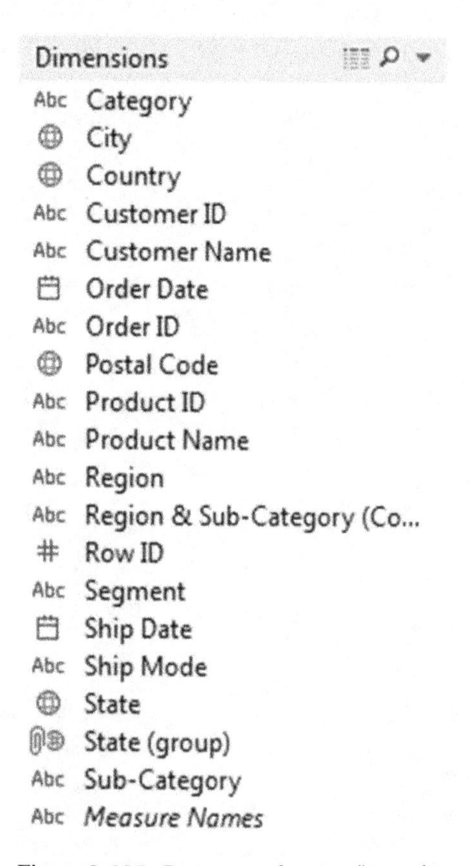

Figure 3-105. *Data pane showing "State (group)"*

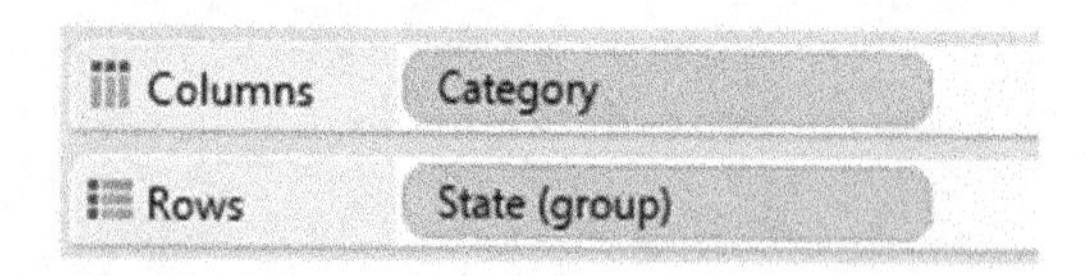

Figure 3-106. *"State(group)" on the rows shelf*

3.3.3.2 Create groups from dimensions in the data pane

Let's discuss how to create groups from dimensions in the data pane.

3.3.3.2.1 Step 1

Connect to the Sample-Superstore data source and go to sheet.

3.3.3.2.2 Step 2

Select "Sub-Category" dimension from the data pane, right click on it and select Create ➤ Group (Shown in Fig. 3-107).

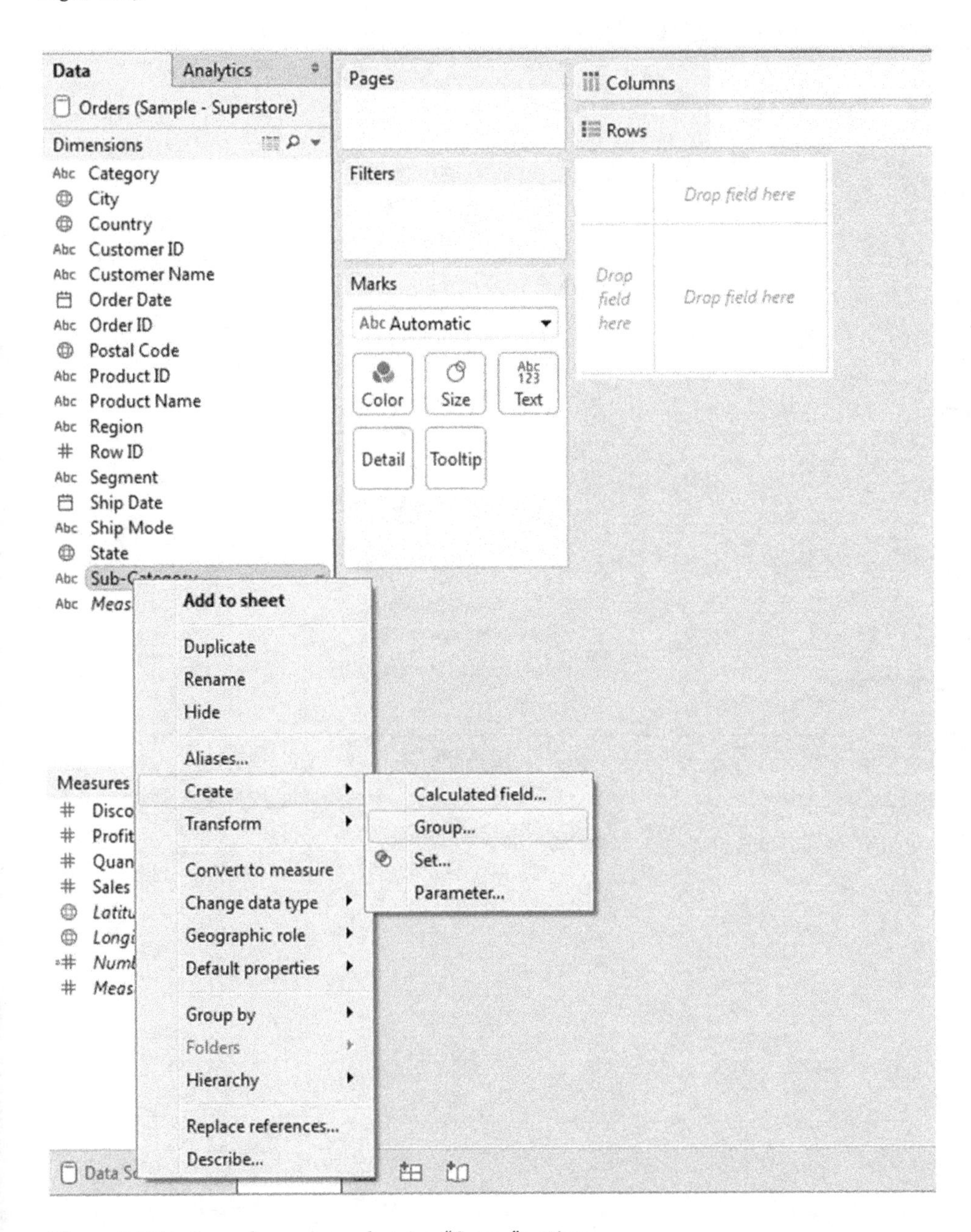

Figure 3-107. *Drop down menu showing "Group" option*

3.3.3.2.3 Step 3

"Create Group [Sub-Category]" dialog box appears (Shown in Fig. 3-108).

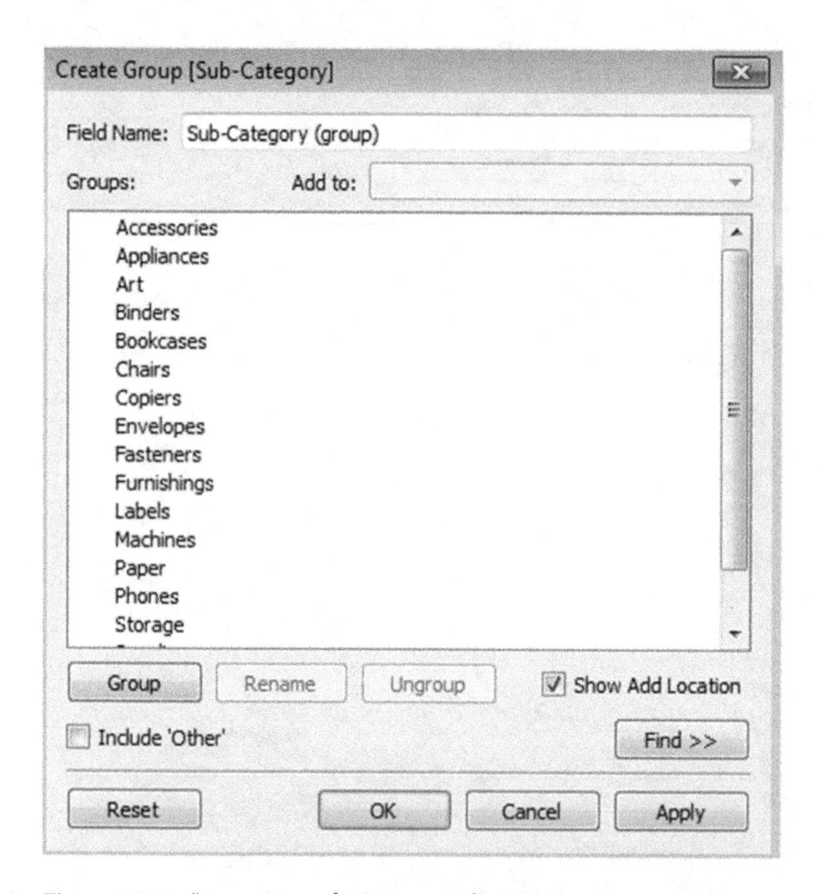

Figure 3-108. *"Create Group [Sub-Category]" dialog box*

3.3.3.2.4 Step 4

Select the dimension members and click on the "Group" button (Shown in Fig. 3-109).

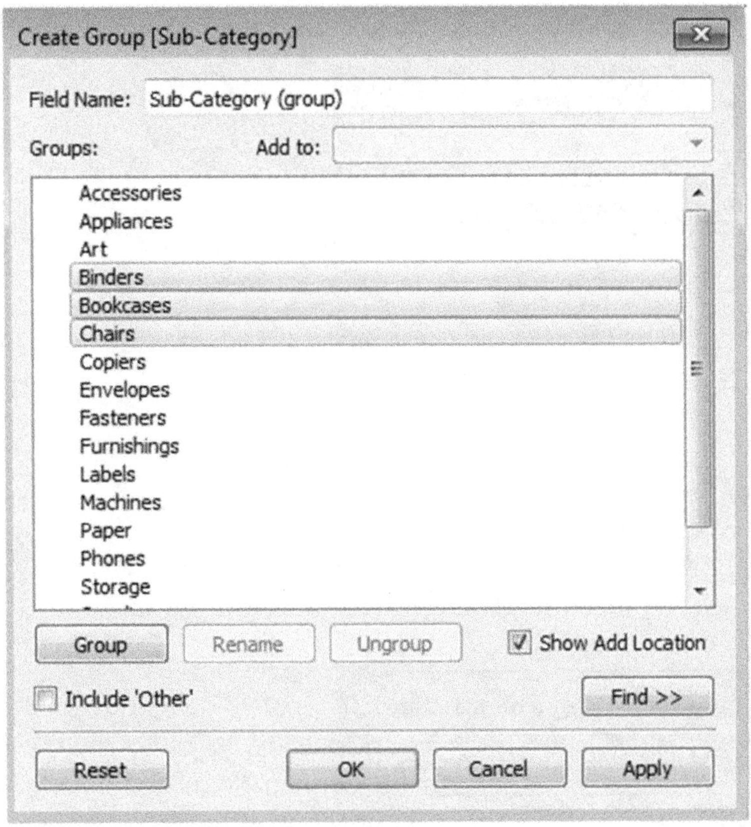

Figure 3-109. *Members selected for "Sub-Category (group)"*

3.3.3.2.5 Step 5

Selected members are grouped together to constitute a single member. The default name for the group is defined automatically by combining all of the members names (Shown in Fig. 3-110). You can rename it by clicking on the "Rename" button.

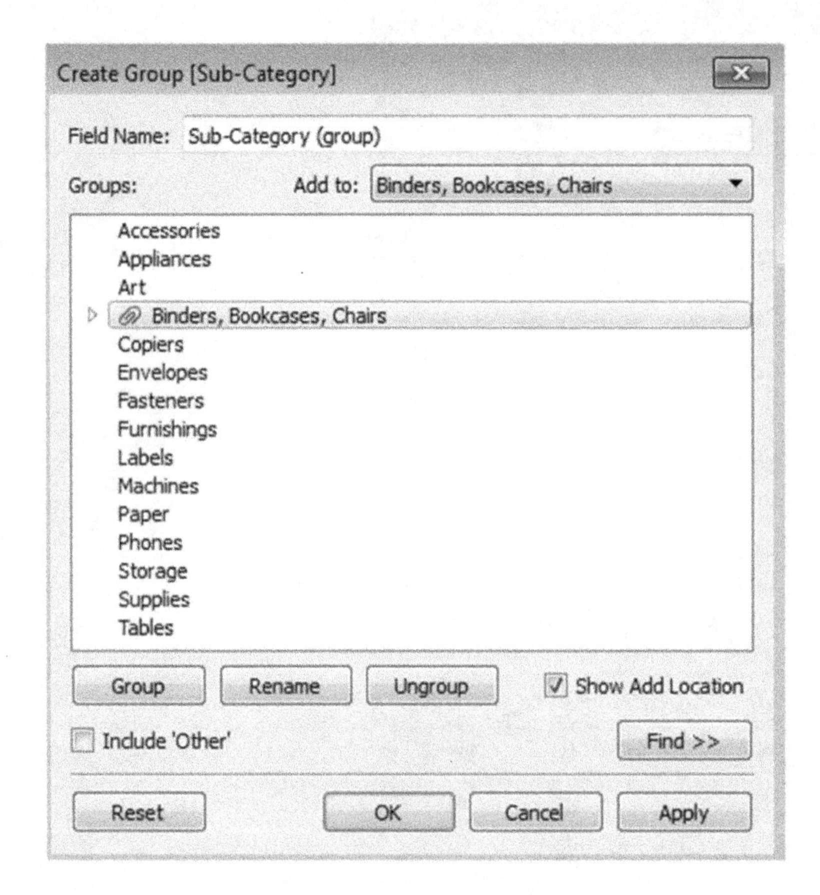

Figure 3-110. *Group "Binders, Bookcases, Chairs"*

3.3.4 Editing an existing group

Let's discuss steps to edit an existing group.

3.3.4.1 Steps

3.3.4.1.1 Step 1

Select “Sub-Category (group)” on the data pane, right click on it and select “Edit group...” (Shown in Fig. 3-111).

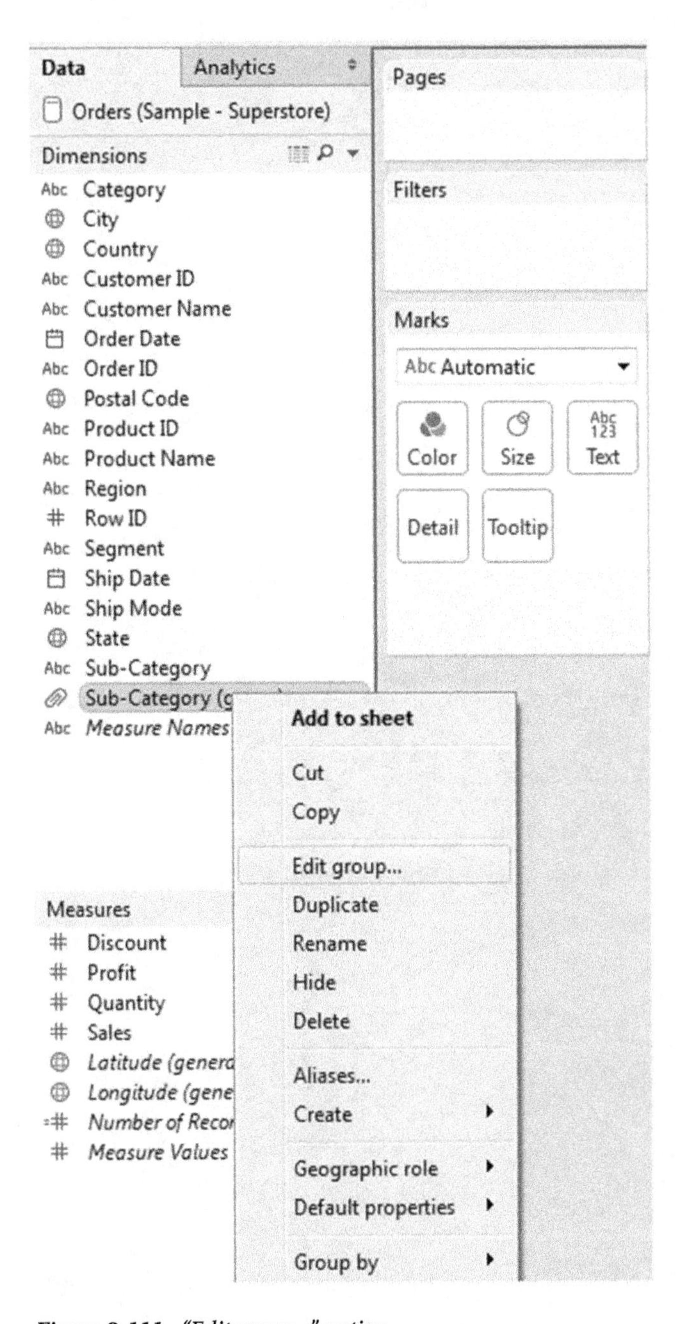

Figure 3-111. “Edit group...” option

3.3.4.1.2 Step 2

The "Edit Group [Sub-Category (group)]" field dialog box appears (Shown in Fig. 3-112).

Figure 3-112. *"Edit Group [Sub-Category(group)]" dialog box*

3.3.4.1.3 Step 3

Select the required dimension members and drag and drop them to the existing group (Shown in Fig. 3-113).

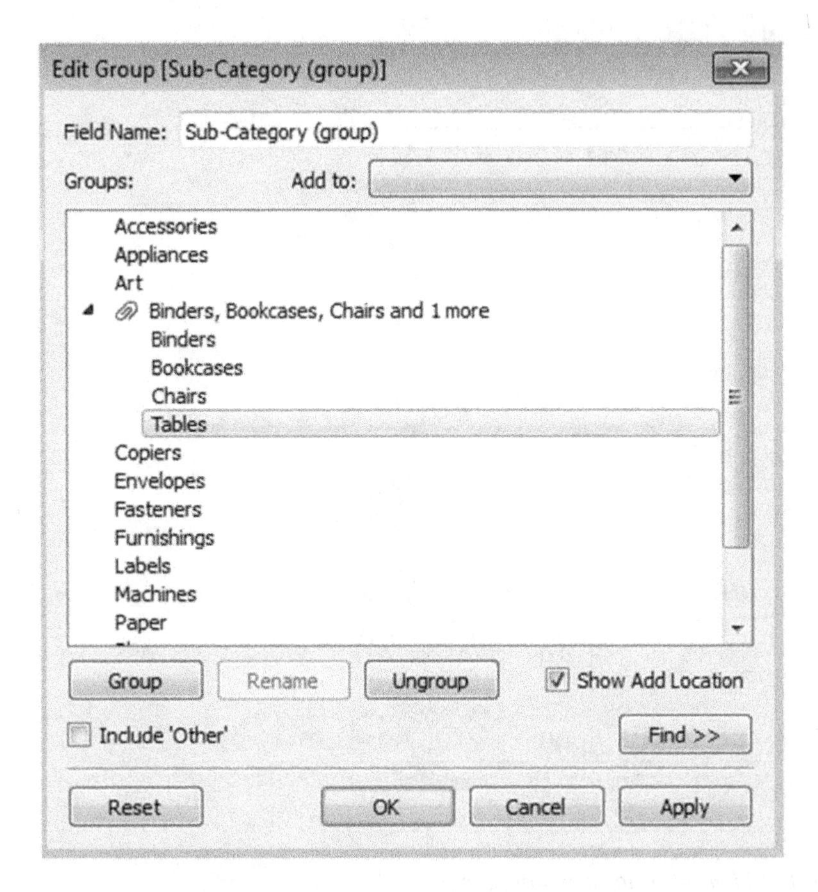

Figure 3-113. *"Edit Group [Sub-Category (group)]" dialog box after adding "Tables" to group "Binders, Bookcases, Chairs and 1 more"*

Or
Select the required dimension, right click on it, select "Add to..." option (Shown in Fig. 3-114).

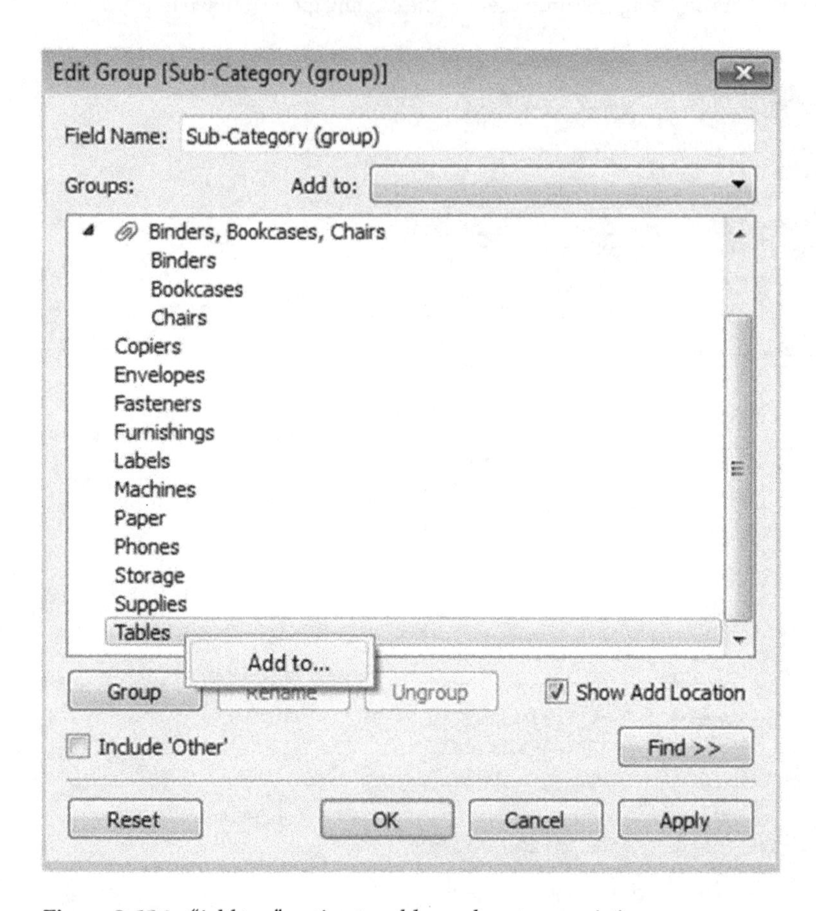

Figure 3-114. *"Add to..." option to add members to an existing group*

3.3.4.1.4 Step 4

"Add to Group" dialog box appears. Select the group (Shown in Fig. 3-115), click "OK" button to add members to the group (Shown in Fig. 3-116).

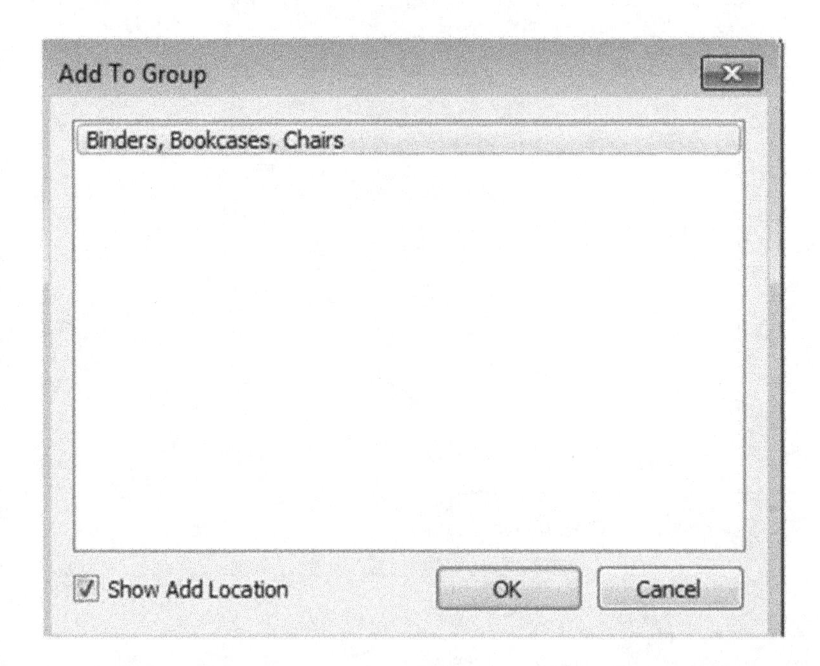

Figure 3-115. *"Add To Group" dialog box showing group "Binders, Bookcases, Chairs"*

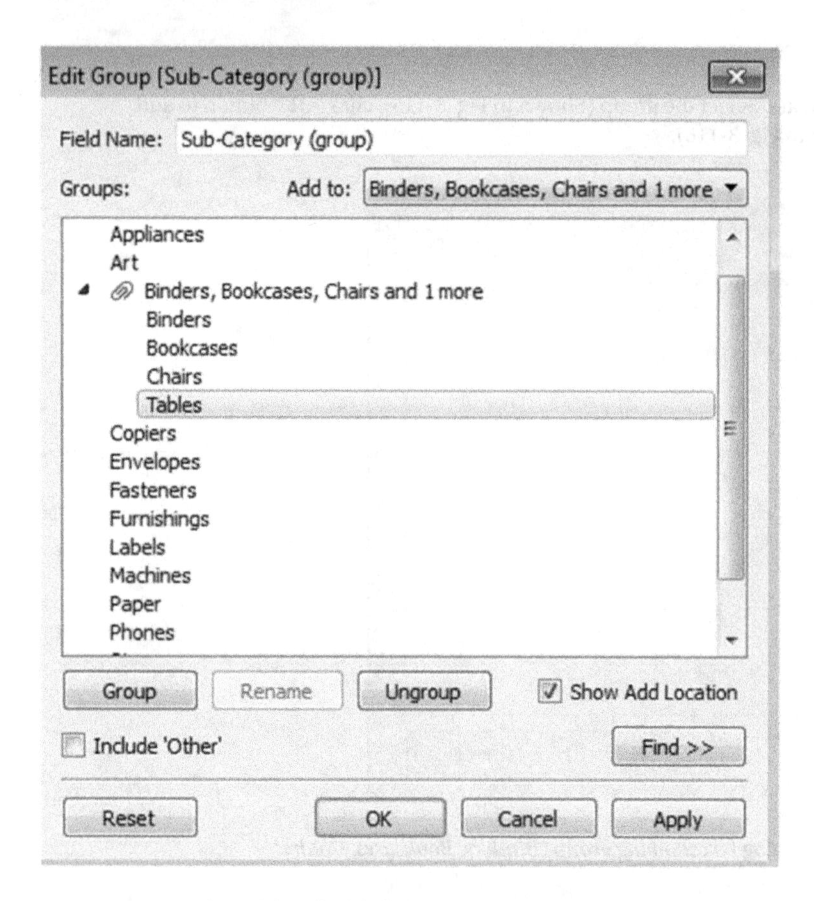

Figure 3-116. *After adding "Tables" to existing group "Binders, Bookcases, Chairs and 1 more"*

3.3.4.2 Removing a member from the group

Let's discuss the steps to remove a group.

3.3.4.2.1 Step 1

To remove a member from the group go to "Edit Group" dialog box, select the member, right click on it and select "Remove" (Shown in Fig. 3-117).

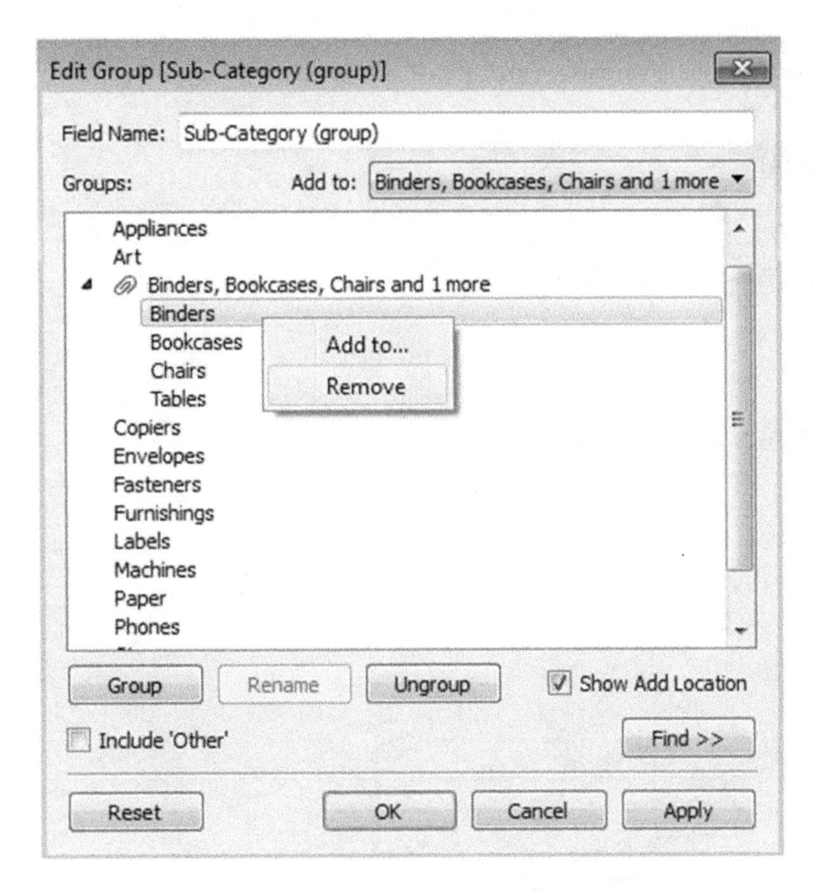

Figure 3-117. *"Remove" option to remove "Binders" from the group "Binders, Bookcases, Chairs and 1 more"*

3.3.4.3 Renaming a group

Let's discuss how to rename a group.

3.3.4.3.1 Step 1

Open "Edit Group" field dialog box, select the group name, and click on "Rename" button (Shown in Fig. 3-118).

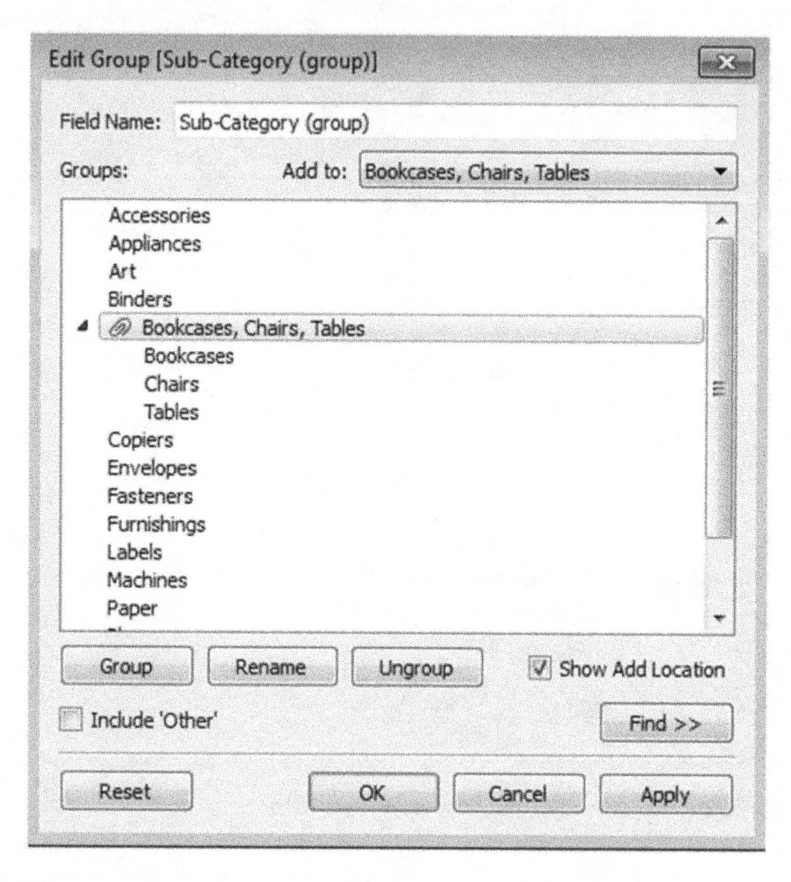

Figure 3-118. *After selecting "Bookcases, Chairs, Tables" group*

3.3.4.3.2 Step 2

Edit dialog box appears for the group name as shown in Fig. 3-119. Type the new name for the group to rename the group (Shown in Fig. 3-120).

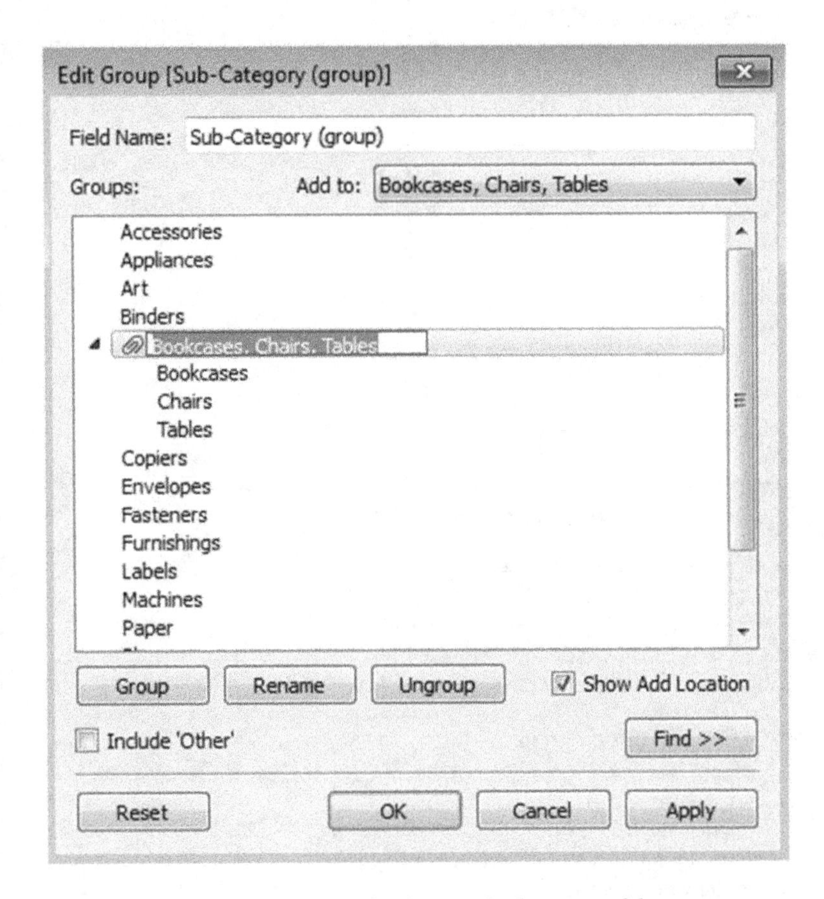

***Figure 3-119.** Edit the group dialog box to edit the name of the group*

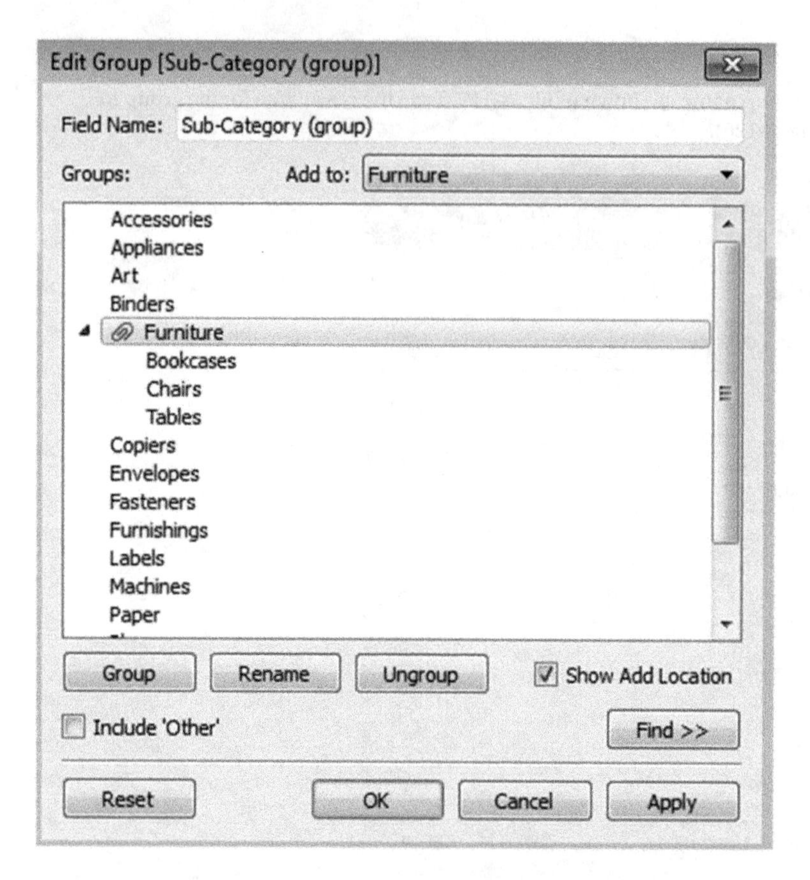

Figure 3-120. *The group "Furniture" after renaming the group*

3.3.5 Creating Hierarchies

Hierarchies in Tableau provide drill down capabilities to the Tableau report.

Tableau allows you to create a hierarchy quite easily.

3.3.5.1 Demo 1

Let's discuss how to create hierarchies.

3.3.5.1.1 Step 1

Connect to the Sample-Superstore Excel data source.

3.3.5.1.2 Step 2

To create hierarchy for "Products," press and hold the CTRL key, select "Category" and "Sub-Category" field as shown in Fig. 3-121.

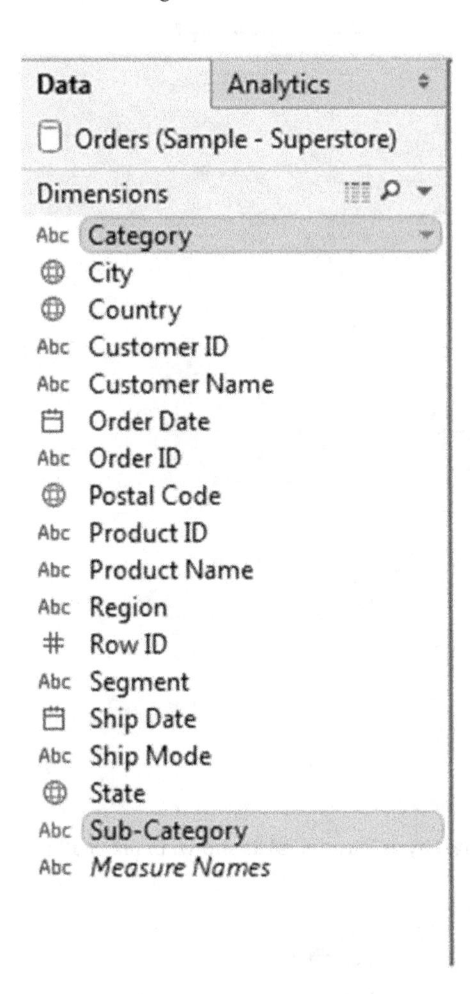

Figure 3-121. *Selection of "Category, Sub-Category"*

3.3.5.1.3 Step 3

Right click on the "Sub-Category," select Hierarchy ➤ Create Hierarchy (Shown in Fig. 3-122).

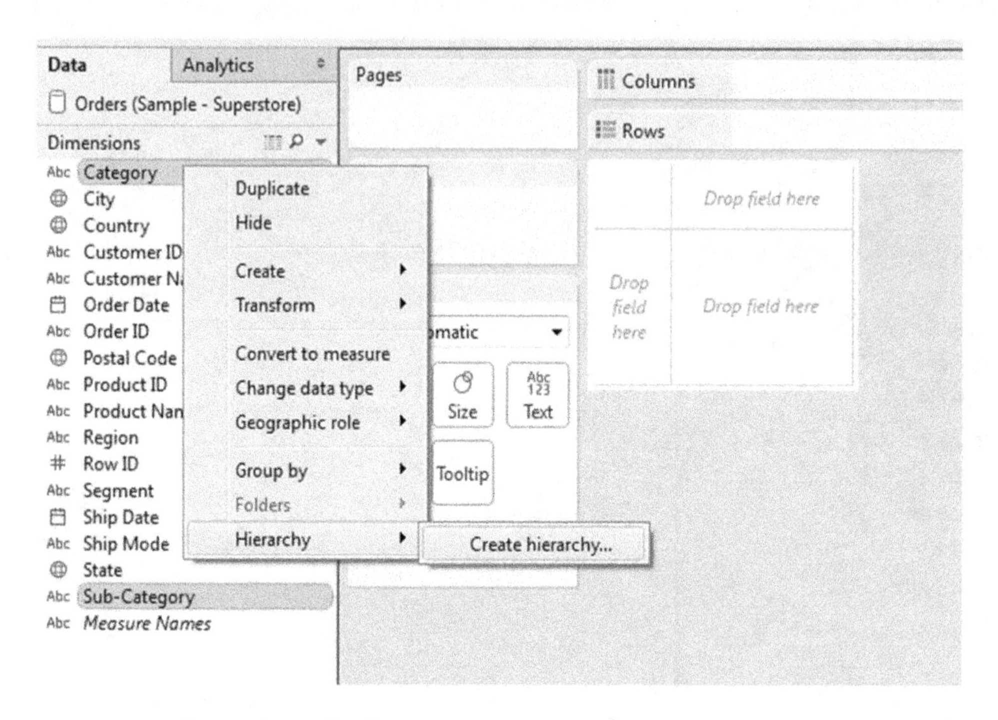

Figure 3-122. "Create hierarchy..." option

3.3.5.1.4 Step 4

"Create Hierarchy" dialog box appears. Specify the name for hierarchy as "Products" as shown in Fig. 3-123.

Figure 3-123. "Create Hierarchy" dialog box, specify "Name" as products

3.3.5.1.5 Step 5

Observe the “Products” hierarchy in the dimensions area under the data pane (Shown in Fig. 3-124).

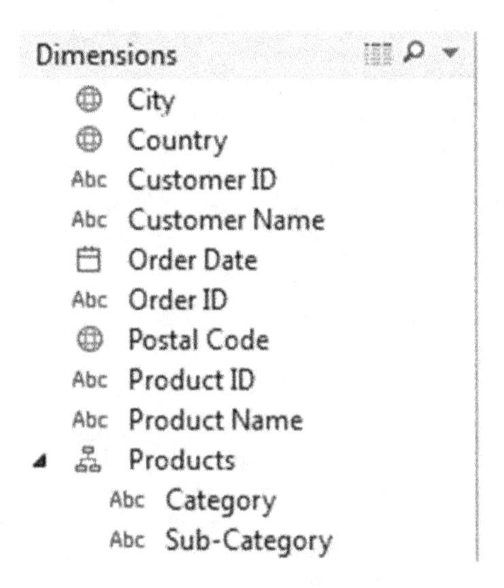

Figure 3-124. *"Products" hierarchy*

3.3.6 Sets

A “Set” is a subset of your data that meets certain conditions based on the existing dimensions. A set allows you to create a subset of data based on some conditions. A set can be a computed set or a constant set.

3.3.6.1 Constant set

In a constant set, members are fixed and they do not change.

Let's discuss the steps.

3.3.6.1.1 Step 1

Connect to the Sample-Superstore Excel data source.

3.3.6.1.2 Step 2

Create the view as shown in Fig. 3-125.

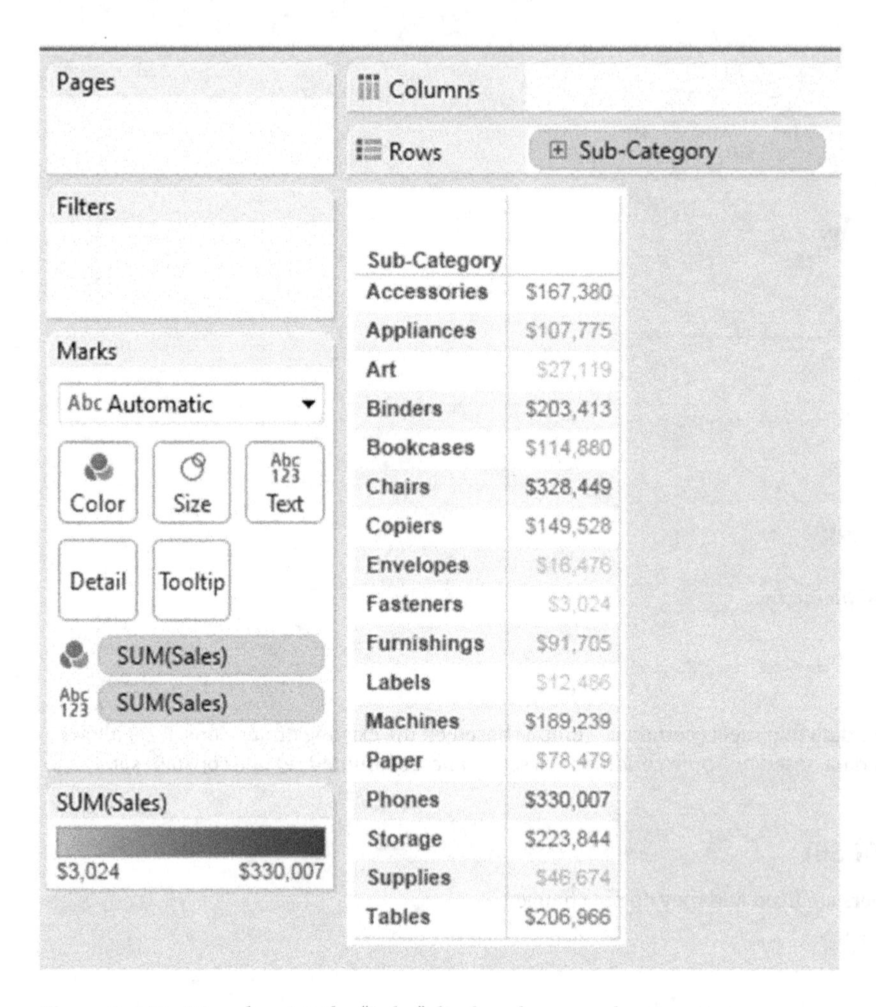

***Figure 3-125.** View showing the "Sales" displayed as per sub-categories*

3.3.6.1.3 Step 3

Press and hold the CTRL key, select six random sub-categories. Move the mouse over the selected area to get the tool tip. Select create set from the tool tip (Shown in Fig. 3-126).

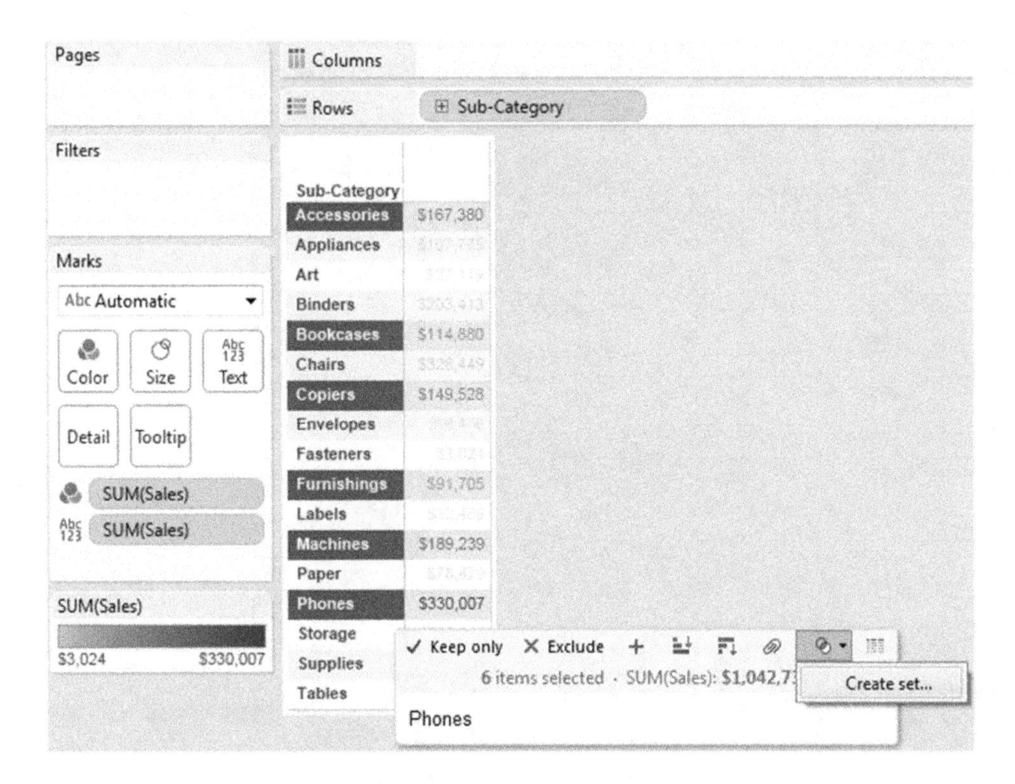

Figure 3-126. "Create set..." option

3.3.6.1.4 Step 4

"Create set" dialog box appears. Specify set name as "Random Sub-Category" and check the "Add to filters shelf" option as shown in Fig. 3-127. Then click "OK".

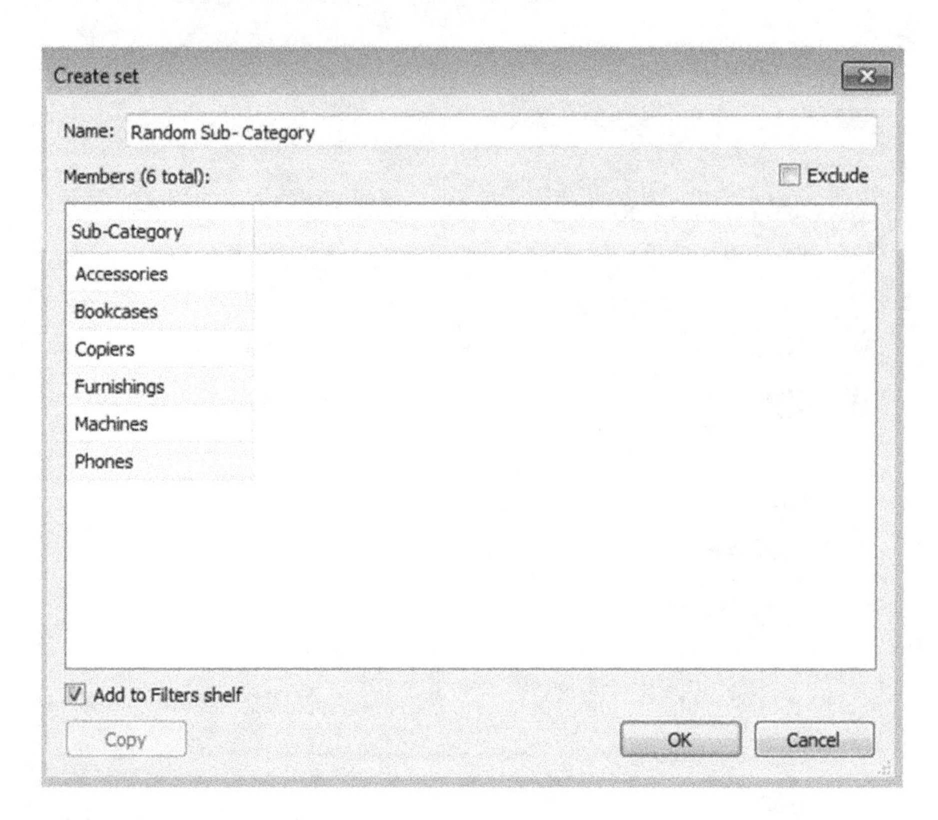

Figure 3-127. *Members of the "Random Sub-Category" set*

3.3.6.1.5 Step 5

Observe the newly created set, "Random Sub-Category" under the data pane as shown in Fig. 3-128.

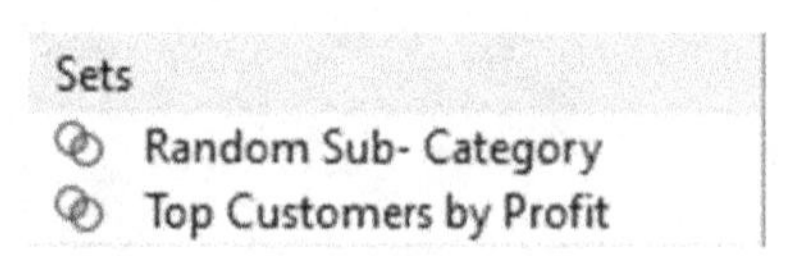

Figure 3-128. *"Random Sub-Category" set*

3.3.6.1.6 Step 6

Remove the dimension, "Sub-Category" from the rows shelf, drag and drop the set, "Random Sub-Category" to the rows shelf as shown in Fig. 3-129.

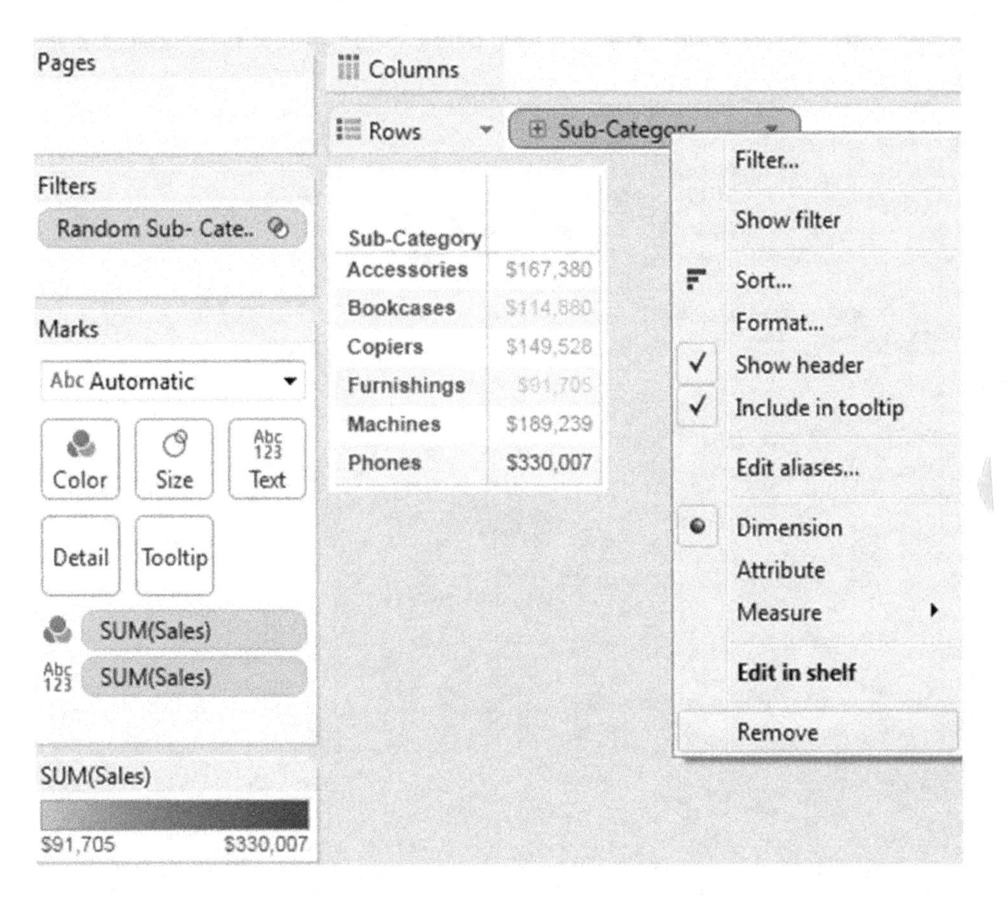

Figure 3-129. *"Remove" option to remove the dimension, "Sub-Category" from the rows shelf*

3.3.6.1.7 Step 7

When you drag and drop the set, "Random Sub-Category" on the rows shelf, you can observe the "IN/OUT (Random Sub-Category)" placed on the rows shelf (Shown in Fig. 3-130). This is because, by default, Tableau creates the IN/OUT mode for sets.

IN is to display the members that are in the set. OUT includes members that are NOT in the set. Here it shows total "Sales" for all the selected members.

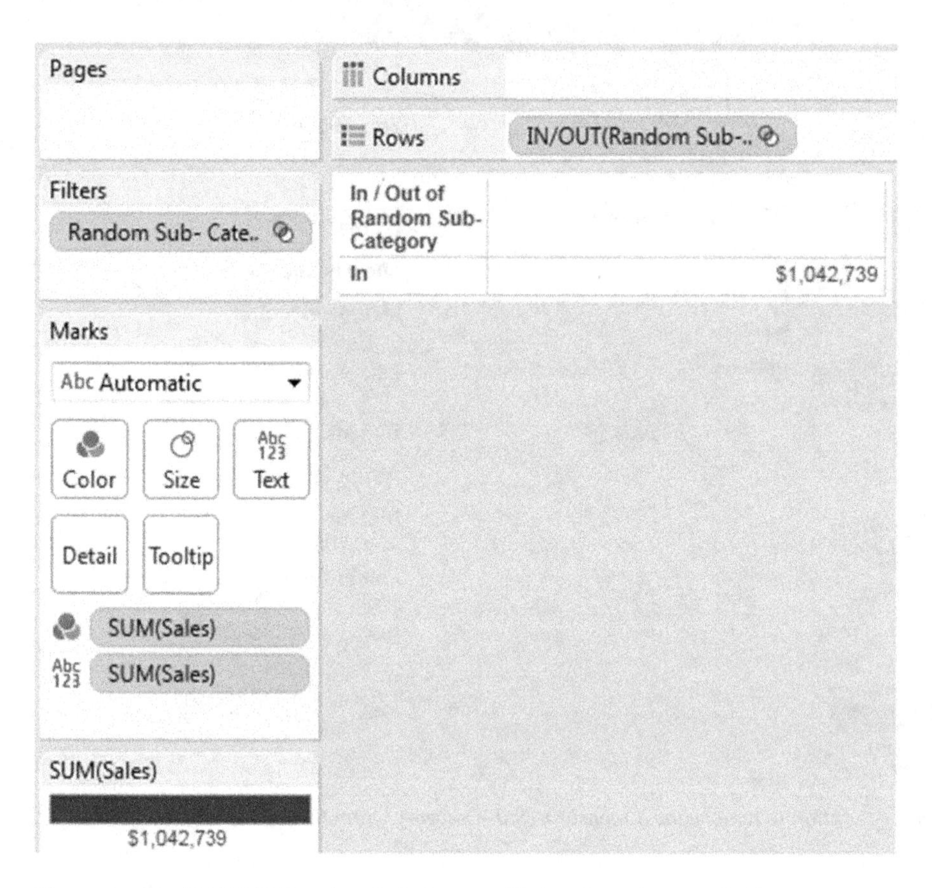

Figure 3-130. *"IN/OUT (Random Sub-Category)" placed on the rows shelf*

3.3.6.1.8 Step 8

To display the "OUT" members, place "Random Sub-Category" on the filters shelf, right click on it and select "Show IN/OUT of set" as shown in Fig. 3-131.

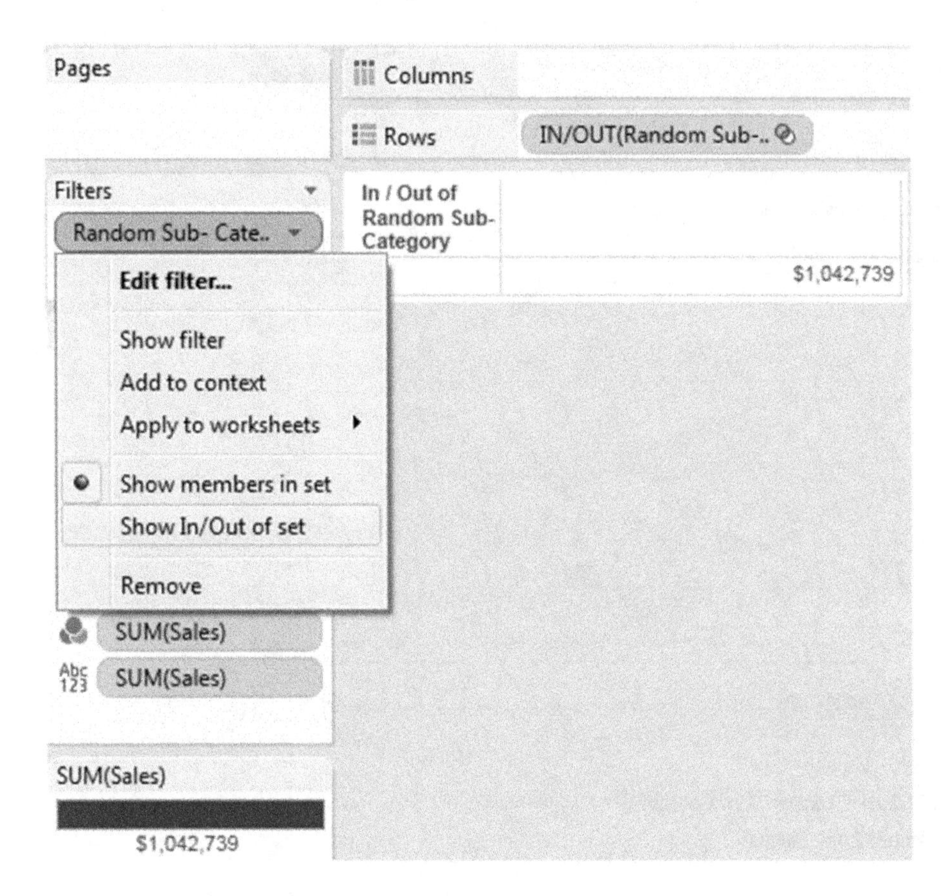

Figure 3-131. *"Show In/Out of set" option*

3.3.6.1.9 Step 9

The "Filter [In / Out of Random Sub-Category]" dialog box appears as shown in Fig. 3-132. Select "IN/OUT" option and click "OK".

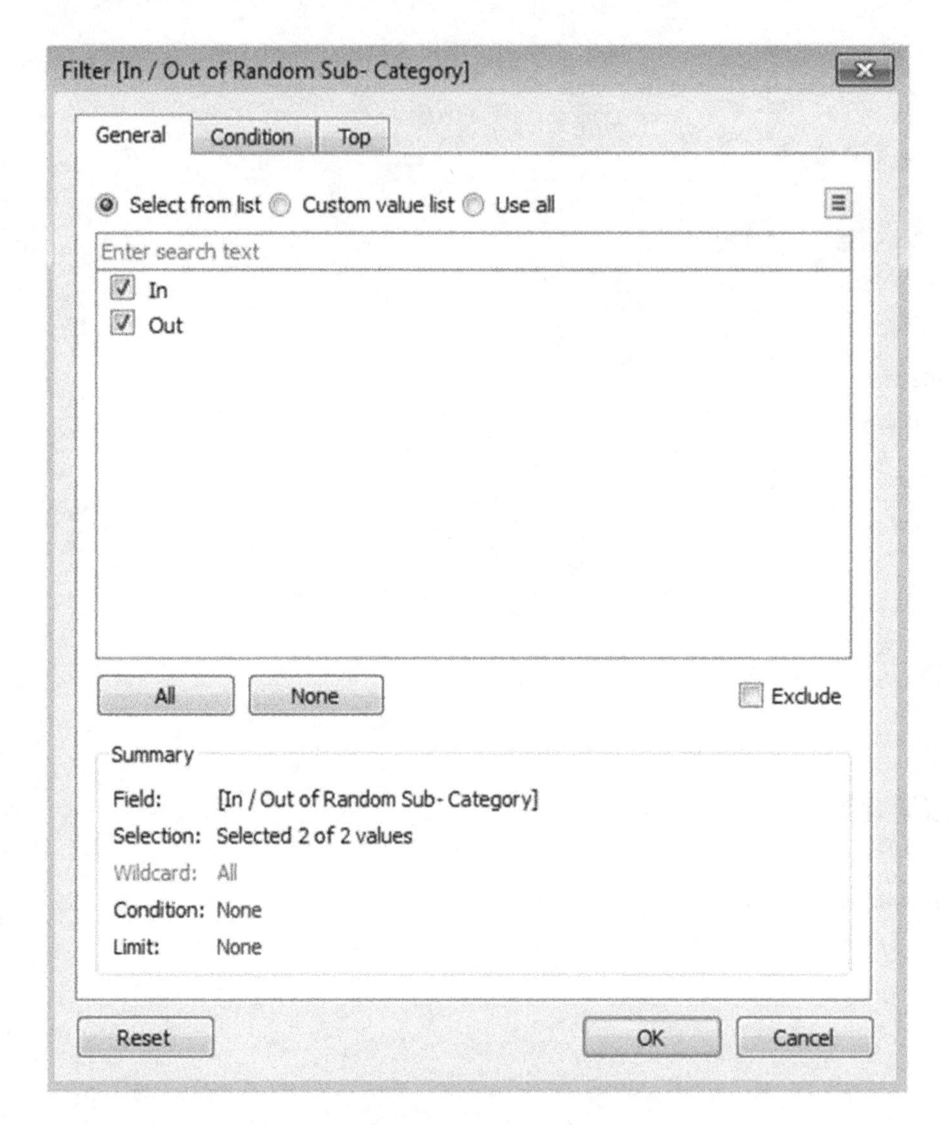

Figure 3-132. *"Filter [In / Out of Random Sub-Category]" dialog box with "In/Out" option*

3.3.6.1.10 Step 10

Observe the total sales of "IN/OUT" members as shown in Fig. 3-133.

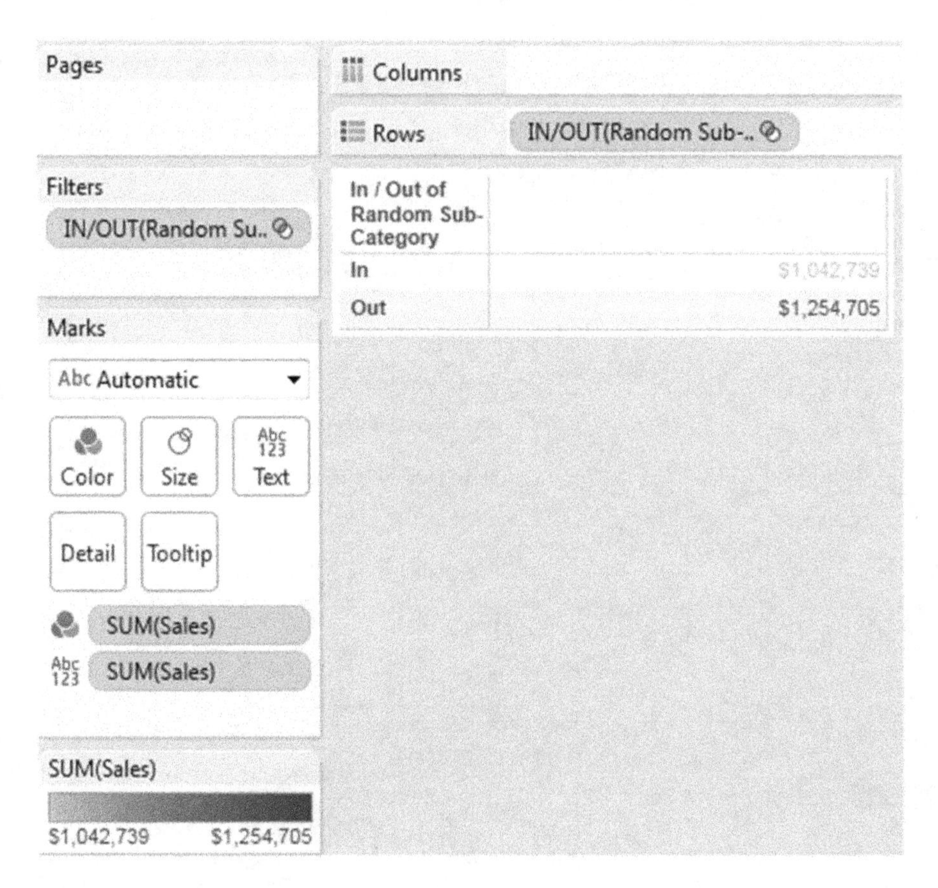

Figure 3-133. *View showing total "Sales" of "IN/OUT" members*

3.3.6.1.11 Step 11

To display members of "IN," right click on "IN/OUT" on the rows shelf and select "Show members in set" as shown in Fig. 3-134 to display members of set (Shown in Fig. 3-135).

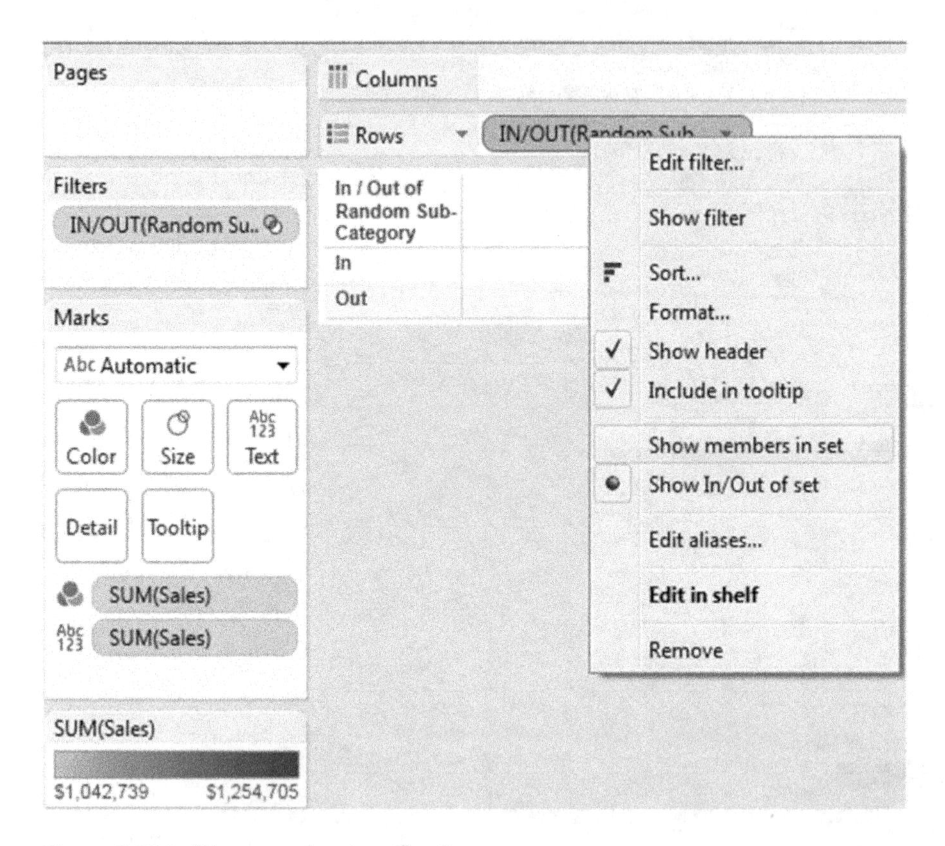

Figure 3-134. *"Show members in set" option*

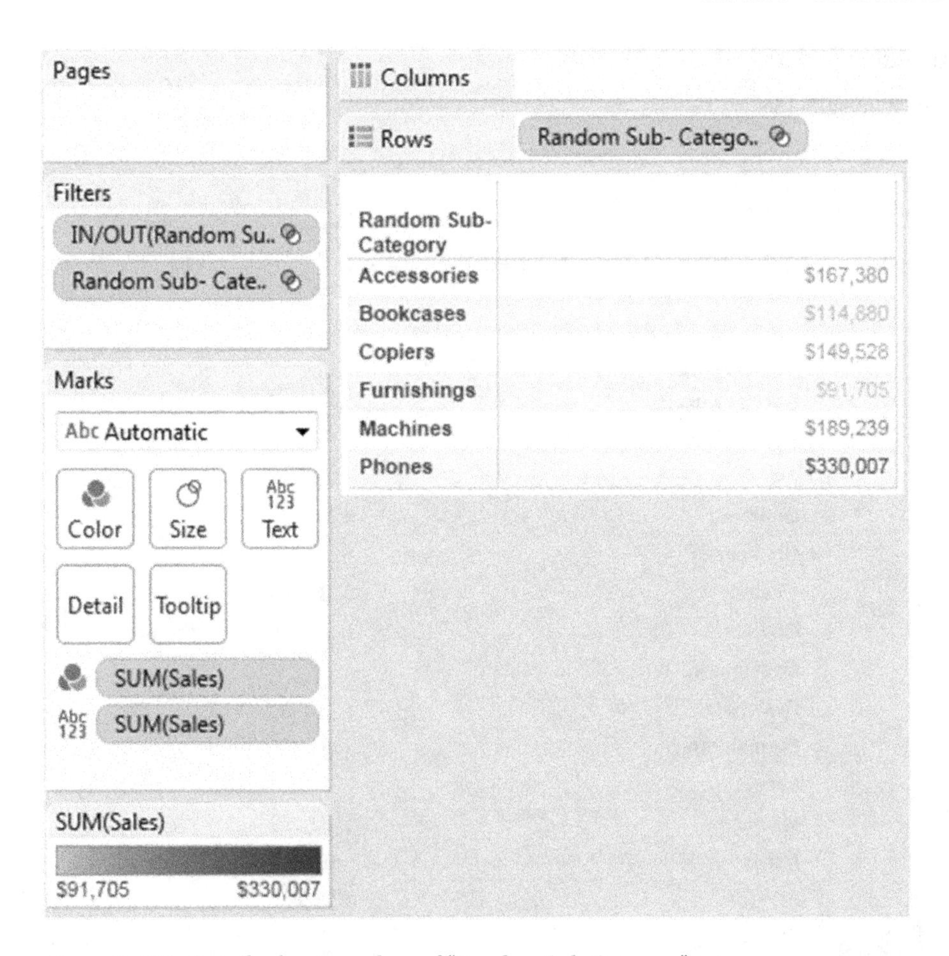

Figure 3-135. Details about members of "Random Sub-Category" set

3.3.6.2 Computed sets

In a "Computed Set," members are dynamic and they change when underlying data is changed.

3.3.6.2.1 Steps

3.3.6.2.2 Step 1

Connect to the Sample-Superstore Excel data source.

3.3.6.2.3 Step 2

Create a view as shown in Fig. 3-136.

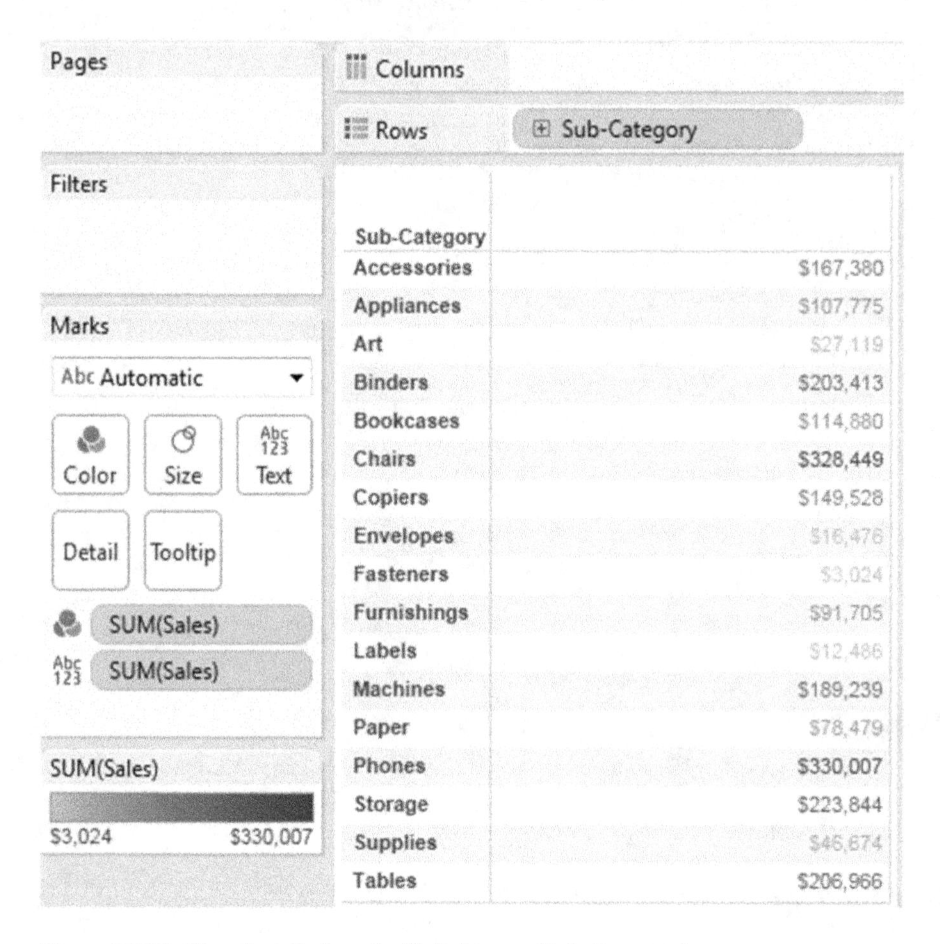

Figure 3-136. *View that displays the "Sales" as per "Sub-Category"*

3.3.6.2.4 Step 3

In the dimensions area under the data pane, select the dimension, "Sub-Category," right click on it and select Create ➤ Set (Shown in Fig. 3-137).

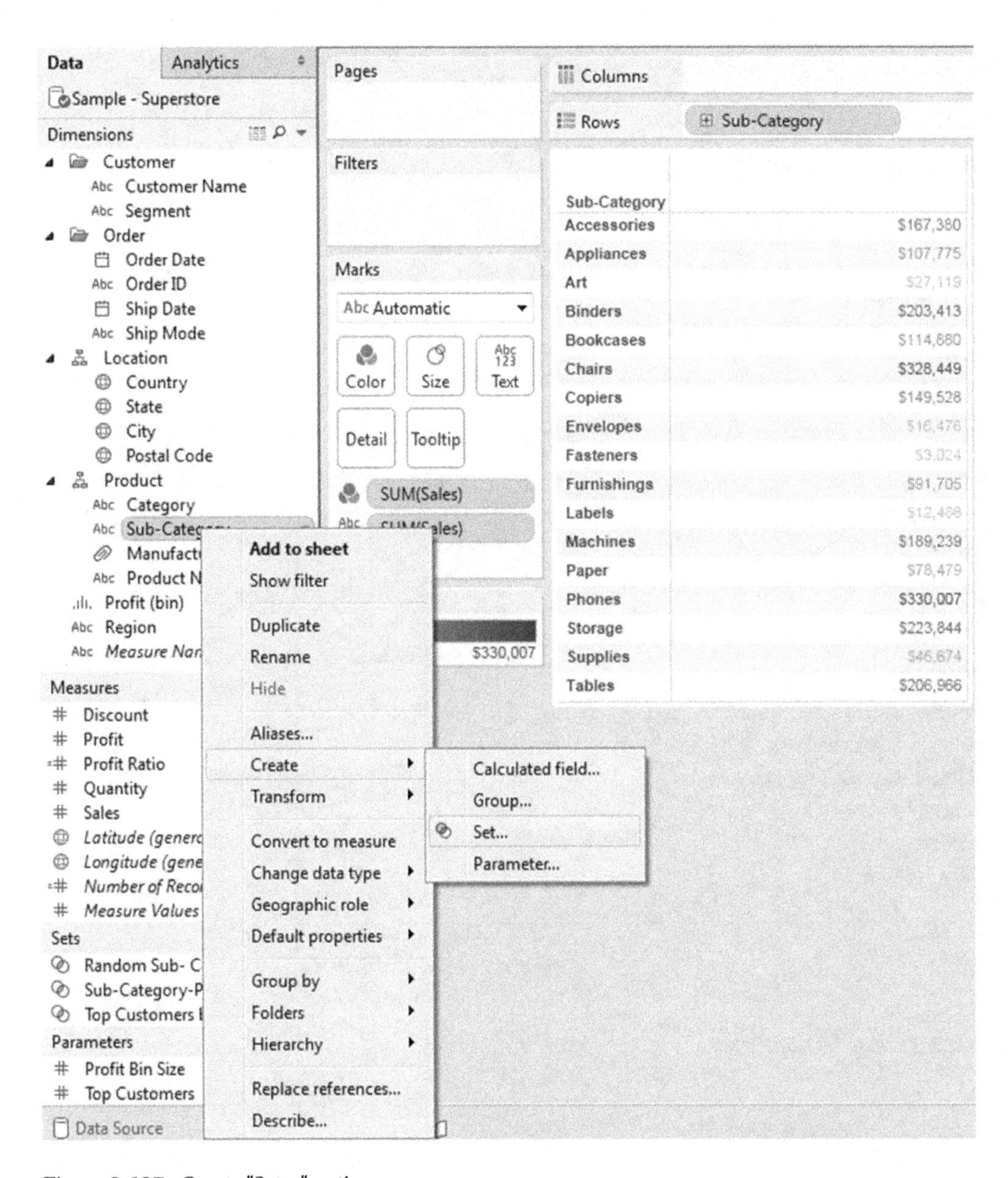

Figure 3-137. *Create "Set..." option*

3.3.6.2.5 Step 4

"Create Set" dialog box appears. Select all members from the list and specify the set name as "Products with sales greater than 30,000" as shown in Fig. 3-138.

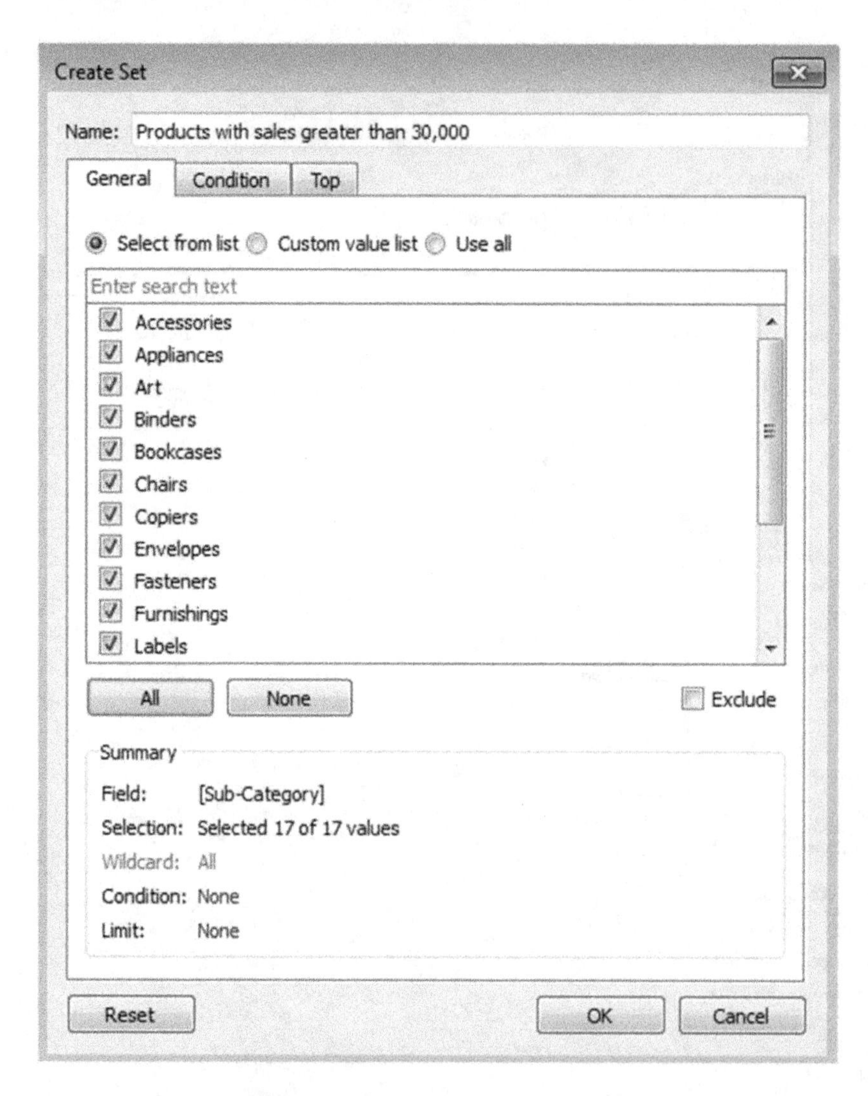

Figure 3-138. *"Create Set" dialog box*

3.3.6.2.6 Step 5

Add a condition to select members based on the outcome of the computation. Go to the Condition tab and specify the condition as shown in Fig. 3-139 and click "OK".

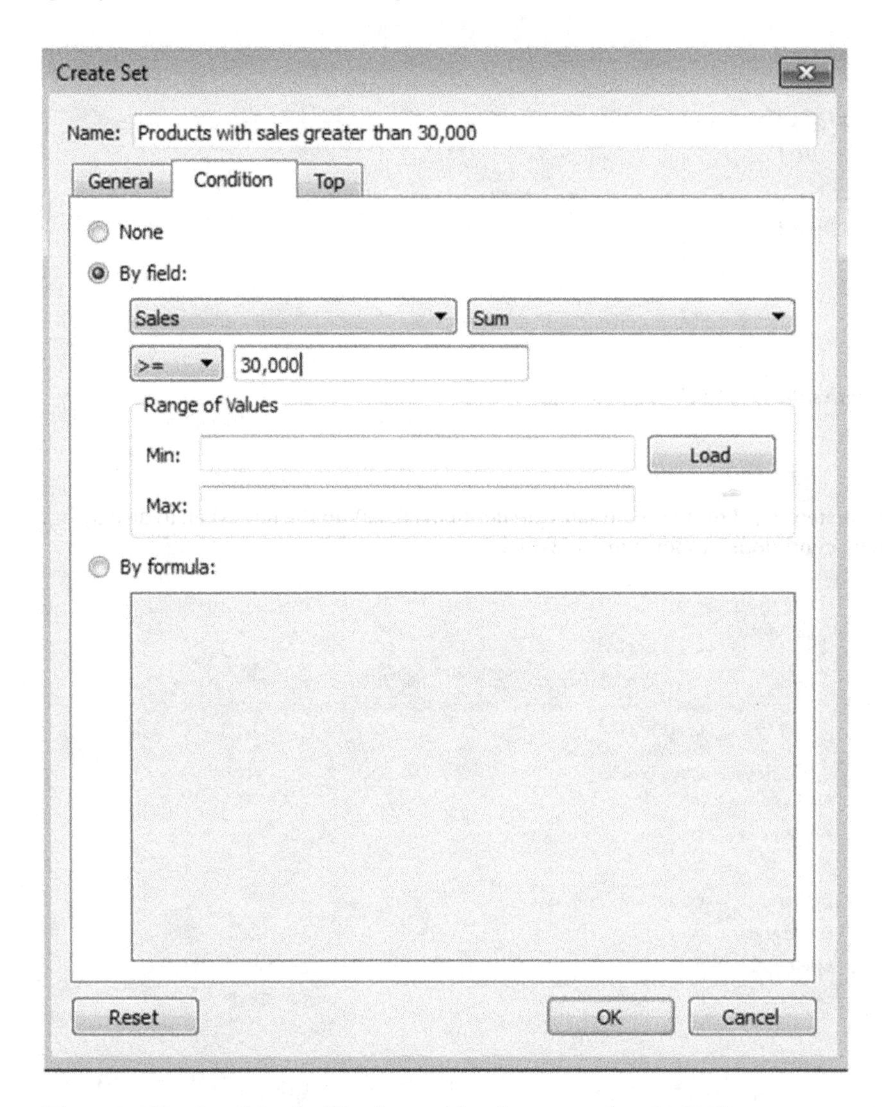

Figure 3-139. *Condition for "Products with sales greater than 30,000" set*

3.3.6.2.7 Step 6

The newly created set, "Products with sales greater than 30,000" is available under the data pane as shown in Fig. 3-140.

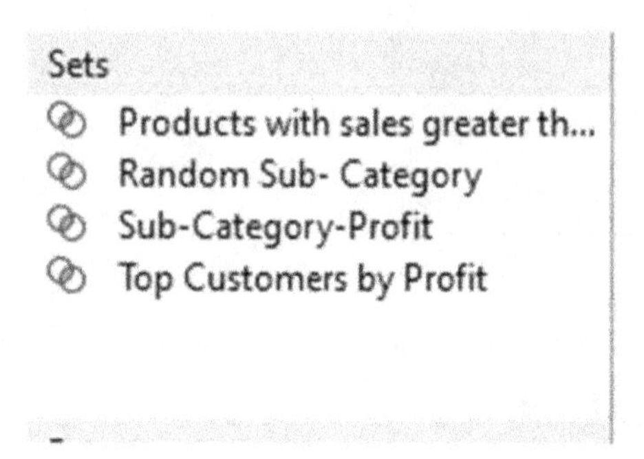

Figure 3-140. *"Products with sales greater than 30,000" set*

3.3.6.2.8 Step 7

Drag and drop the newly created set, "Products with sales greater than 30,000" to the rows shelf to display IN/OUT members based on conditions as shown in Fig. 3-141.

Pages | Columns
Rows: Sub-Category | IN/OUT(Products with..
Filters
Marks: Automatic | Color | Size | Text | Detail | Tooltip | SUM(Sales) | SUM(Sales)
SUM(Sales): $3,024 – $330,007

Sub-Category	In / Out of Products with sales greater..	
Accessories	In	$167,380
Appliances	In	$107,775
Art	Out	$27,119
Binders	In	$203,413
Bookcases	In	$114,880
Chairs	In	$328,449
Copiers	In	$149,528
Envelopes	Out	$16,476
Fasteners	Out	$3,024
Furnishings	In	$91,705
Labels	Out	$12,486
Machines	In	$189,239
Paper	In	$78,479
Phones	In	$330,007
Storage	In	$223,844
Supplies	In	$46,674
Tables	In	$206,966

Figure 3-141. *View showing IN/OUT members of "Products with sales greater than 30000" set*

3.4 Difference between a set and group

3.4.1 Group

- can be created manually
- cannot be used in calculated fields

3.4.2 Set

- can be created either manually or using calculated field
- can be used in calculated field

3.4.3 Creating parameters

Parameters are dynamic values that can be used to replace the constant values in filters, calculations, etc.

3.4.3.1 Demo 1

Let's discuss how to create a parameter.

Objective: To display TOP N Sub-Category based on their "Sales".

3.4.3.1.1 Step 1

Connect to the Sample-Superstore data source.

3.4.3.1.2 Step 2

Drag the dimension "Sub-Category" from the dimensions area under the data pane to the rows shelf. Drag the measure "Sales" from the measures area under the data pane to the columns shelf (Shown in Fig. 3-142).

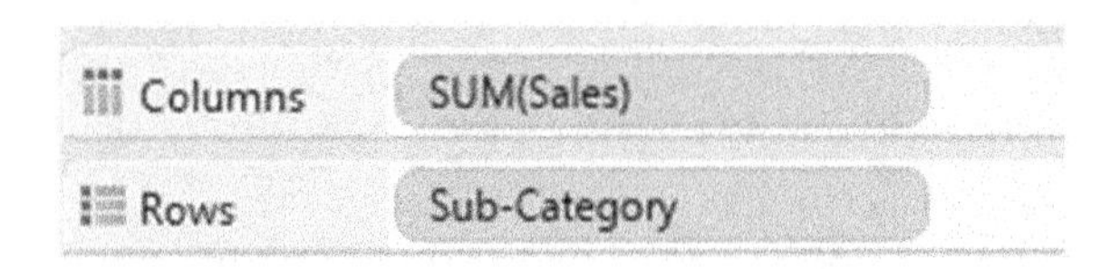

Figure 3-142. *Dimension "Sub-Category" placed on the rows shelf and the measure "Sales" placed on the columns shelf*

3.4.3.1.3 Step 3

In the data pane, Right click on and select "Create parameter..." (Shown in Fig. 3-143).

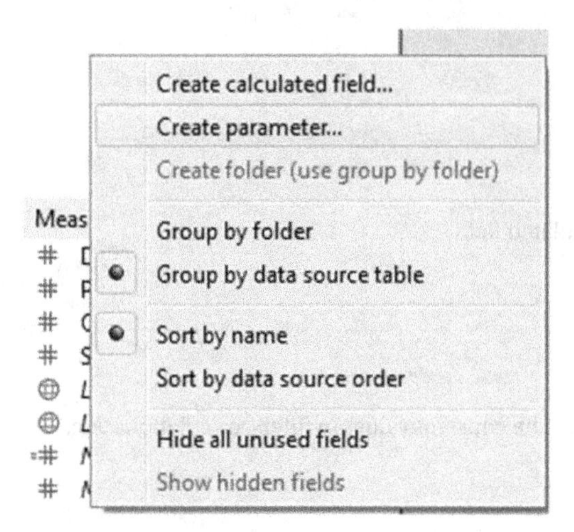

Figure 3-143. *"Create Parameter..." option*

3.4.3.1.4 Step 4

"Create Parameter" dialog box opens. Fill in the details as shown in Fig. 3-144.

Create Parameter

Name: TOP N

Comment >>

Properties

Data type: Integer

Current value: 1

Display format: Automatic

Allowable values: All List Range

Range of values

Minimum: 1

Maximum: 10

Step size: 1

Set from Parameter

Set from Field

OK

Cancel

Figure 3-144. *"Create Parameter" dialog box*

3.4.3.1.5 Step 5

You can see the TOP N parameter in the Parameters area under the data pane (Shown in Fig. 3-145).

Figure 3-145. *"TOP N" parameter in Data Pane*

3.4.3.1.6 Step 6

Right click on "TOP N," select "Show parameter control" to display TOP N parameter in the view (Shown in Fig. 3-146 and Fig. 3-147).

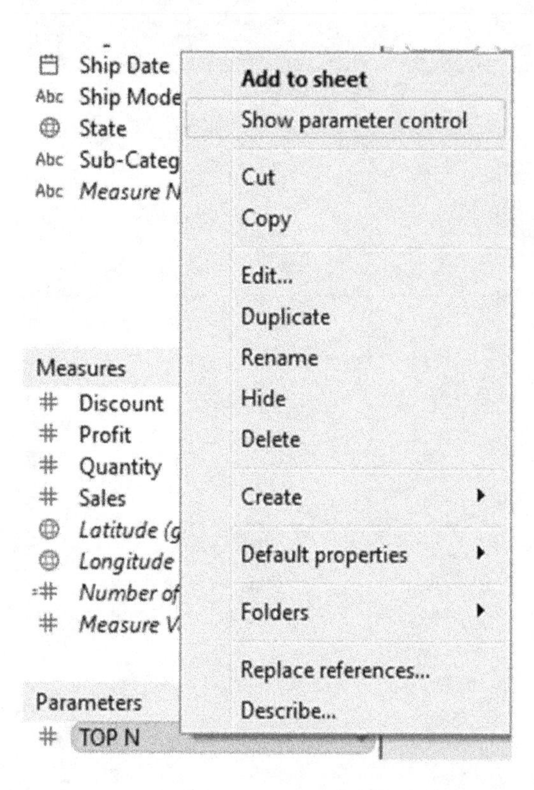

Figure 3-146. *"Show parameter control" option*

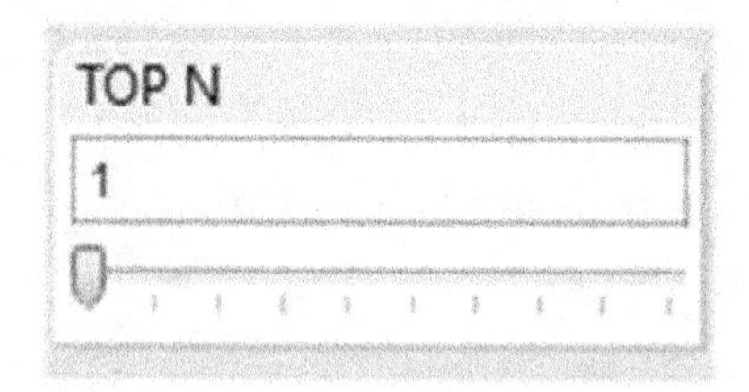

Figure 3-147. *Parameter Control*

3.4.3.1.7 Step 7

Let's see how to use the parameter in the filter. Drag the dimension "Sub-Category" from the dimensions area under the data pane to the filters shelf, Select "TOP" tab and select TOP value as "TOP N" parameter (Shown in Fig. 3-148).

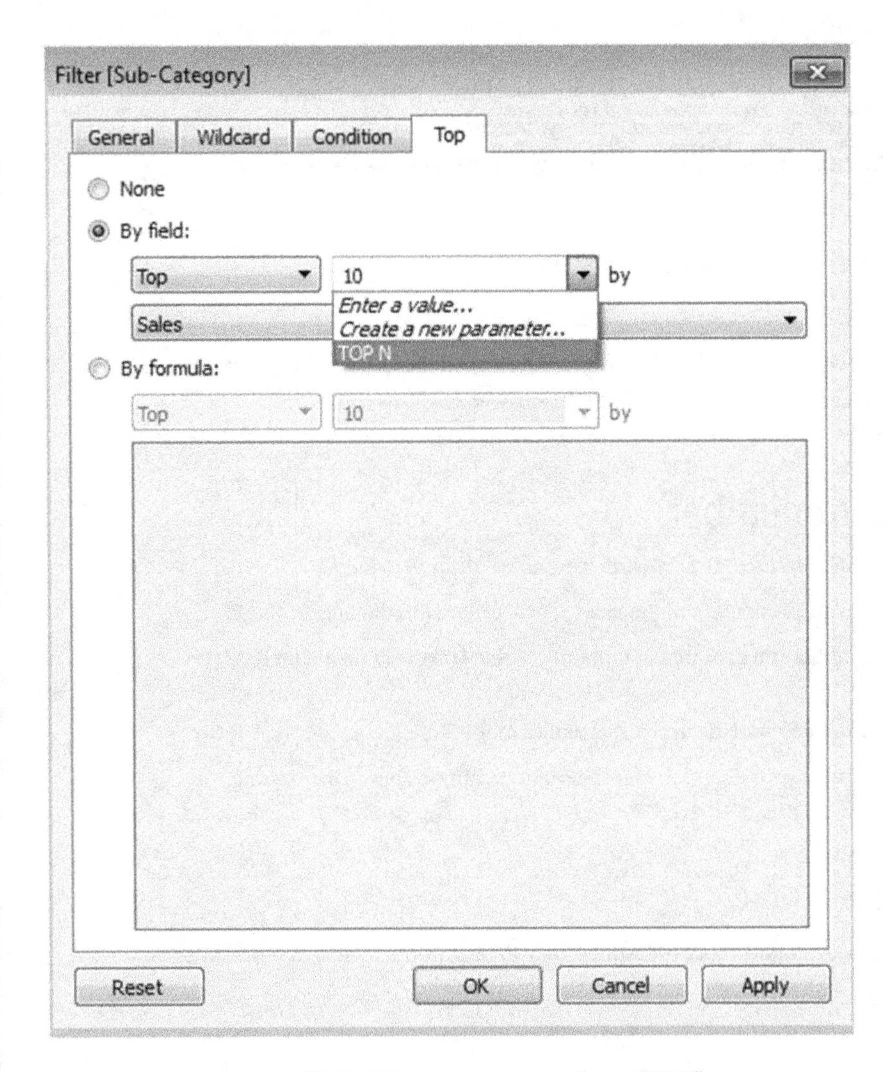

Figure 3-148. *Selection of "TOP N" parameter as a value to "TOP"*

3.4.3.1.8 Step 8

Based on your selection, "TOP N Sub-Category" will be displayed based on their "Sales" (Shown in Fig. 3-149). Click on the sales axis to sort in ascending order.

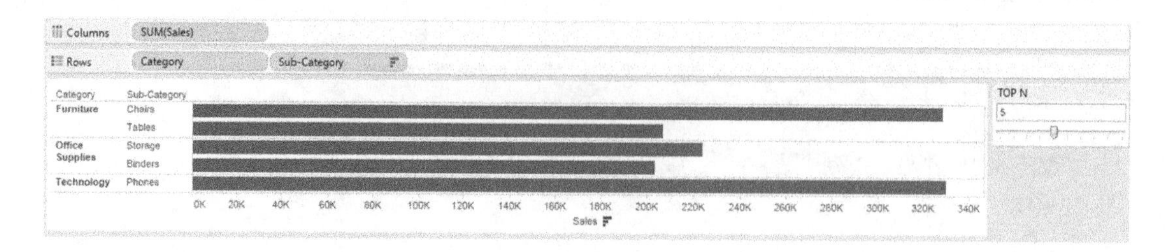

Figure 3-149. *"TOP 5 Sub-Category" based on their "Sales"*

Refer below link to learn how to create a calculated field: `https://onlinehelp.tableau.com/current/pro/desktop/en-us/calculations_calculatedfields_ex1create.html`

3.5 Points to remember

- "Filter" is an option to exclude or include certain values from a field.
- You can arrange dimension members in a specific order with the help of "Sort".
- A group is a combination of dimension members that constitutes a higher level category.
- Hierarchies in Tableau provide drill down action to the Tableau report.
- A "Set" is a subset of your data that meets certain conditions based on existing dimensions.

3.6 Next step

In the next chapter, we will learn more about measures. We will be introduced to two new fields:

- Measure names
- Measure values

CHAPTER 4

Measure Names and Measure Values

Chapter 3 introduced us to working with worksheets and views by using the understanding of dimensions and measures. We learnt to group dimensions, build our own hierarchies of dimensions, and create sets to dynamically select the data to display in a view. This chapter will delve deeper into the understanding of dimensions and measures. It will introduce two new fields, namely, measure names and measure values. We will learn to blend multiple measures on a single axis and also to use dual axis to enhance our presentation of data. In the course of explanation, we will introduce some new chart forms, such as slope graphs and combination charts, such as bar and line together in a view and lollipop charts, etc.

4.1 Why are measure names and measure values required?

These fields are created automatically by Tableau so that a view can quickly be created with multiple measures on it. Few examples are stated below:

- As a senior executive manager of the firm, you would like to compare the measures, "Sales" and "Profits" across "Customer Segments" over a period of time.
- As the head of the firm, you would like to evaluate the performance of the practice units this year against last year's performance. You would like a visualization that clearly and quickly shows the trends, whether the performance of the practice unit has increased, decreased or remained unchanged.
- As a senior sales executive, you would like to conclusively infer, the products that account for top 50% of your sales.

4.1.1 What are measure names and measure values?

These are built-in Tableau fields. "Measure Names" appears as a dimension at the bottom in the dimensions area under the data pane. "Measure Values" appears as a measure at the bottom in the measures area under the data pane.

S. Acharya and S. Chellappan, *Pro Tableau*, DOI 10.1007/978-1-4842-2352-9_4

4.1.2 Where do these fields come from?

When you connect to a data source, Tableau automatically creates these fields to contain all of the measure names and values.

Example:

Consider the table below that shows data in a data source (see Table 4-1.):

Table 4-1. *A sample data set*

Region	Sales	Profit
East	100,000	20,000
West	120,000	12,000
North	150,000	45,000
South	110,000	11,000
Central	90,000	9,000

The "Measure Names" container will contain:

- Sales
- Profit

The "Measure Values" container will contain: (See Table 4-2.)

Table 4-2. *Sample "Measure Values" container*

Sales	Profit
100,000	20,000
120,000	12,000
150,000	45,000
110,000	11,000
90,000	9,000

4.1.2.1 Demo 1

Objective: To plot the dimension "Measure Names" and the measure "Measure Values" in a table in Tableau.

Input: "Sample - Superstore.xls".

4.1.2.1.1 Steps to plot "Measure Names" and "Measure Values" in a table.

4.1.2.1.2 Step 1

Read in the data from "Sample - Superstore.xls" into Tableau (Shown in Fig. 4-1).

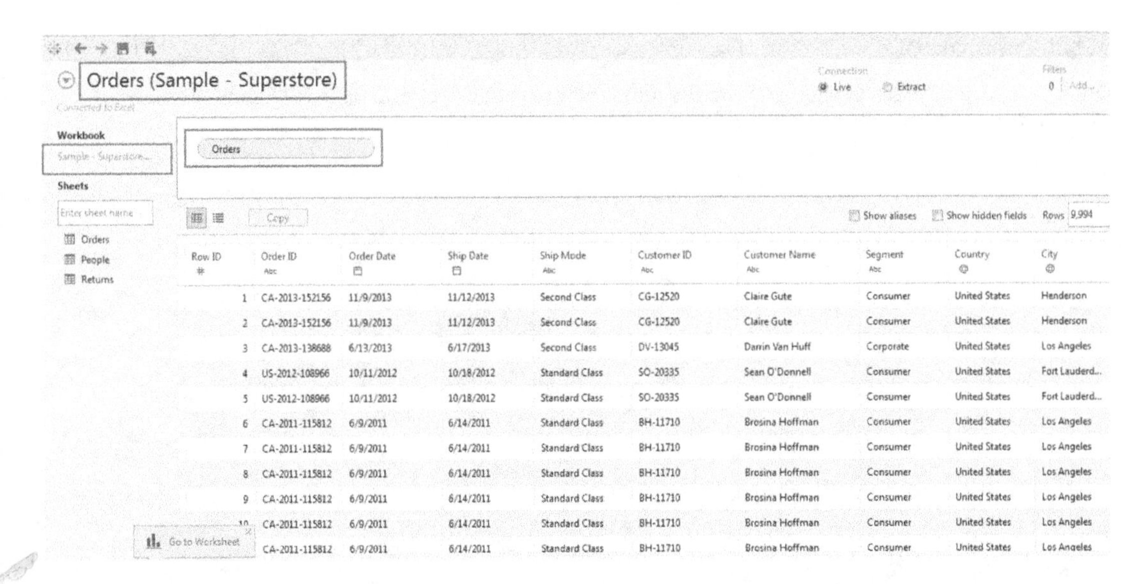

Figure 4-1. *Data from "Sample - Superstore.xls" read into Tableau*

4.1.2.1.3 Step 2

Drag the dimension "Measure Names" from the dimensions area under the data pane to the rows shelf (Shown in Fig. 4-2).

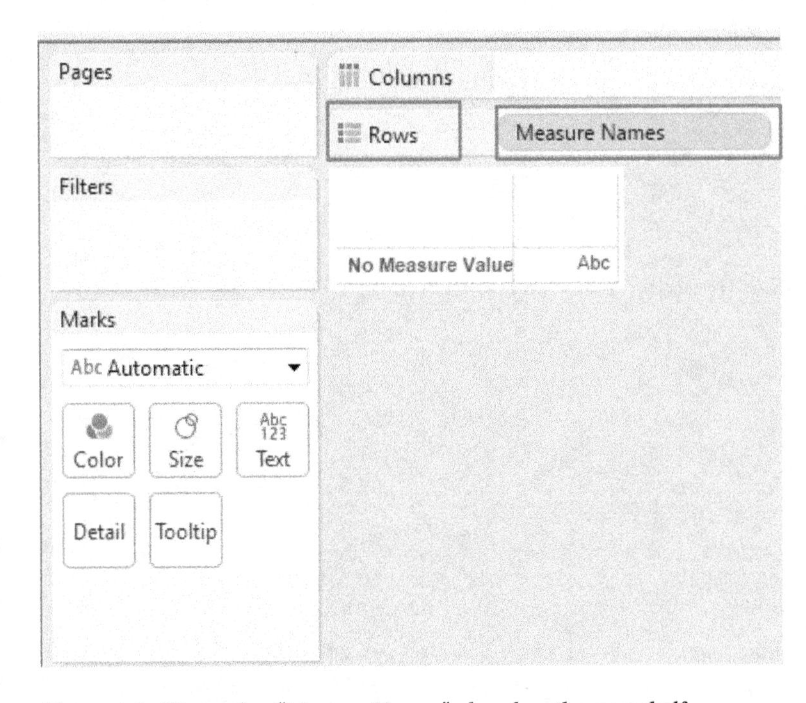

Figure 4-2. *Dimension "Measure Names" placed on the rows shelf*

4.1.2.1.4 Step 3

Drag the measure "Measure Values" from the measures area under the data pane to "Label" on the marks card (Shown in Fig. 4-3).

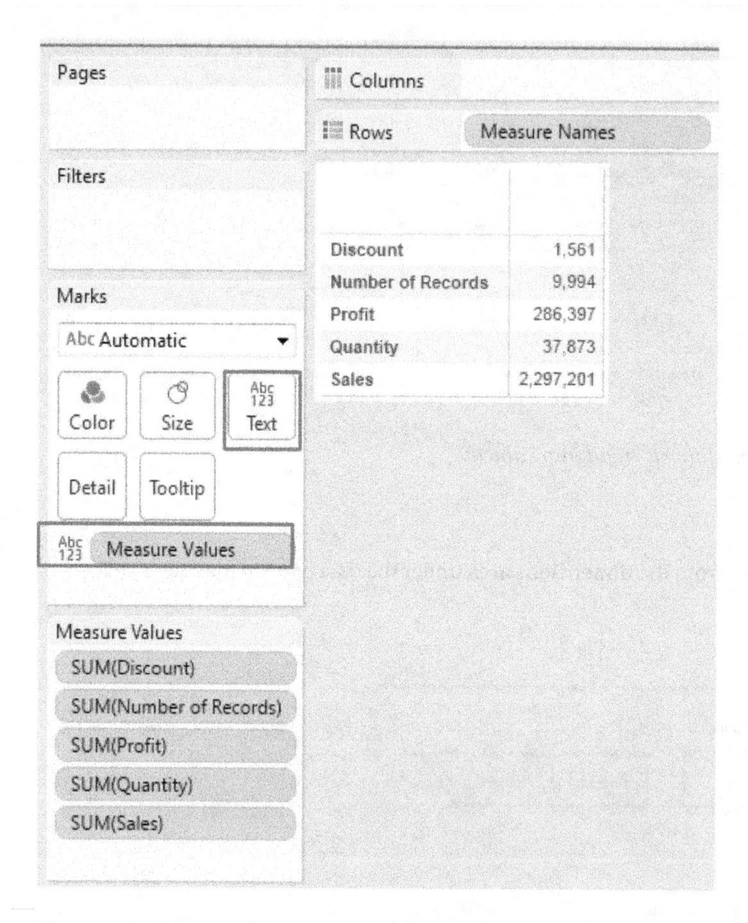

***Figure 4-3.** Measure "Measure Values" placed on "Label" on the marks card*

If one wishes to see the measure values as per the dimension "Region", simply drag the dimension "Region" from the dimensions area under the data pane and place it on the columns shelf (Shown in Fig. 4-4).

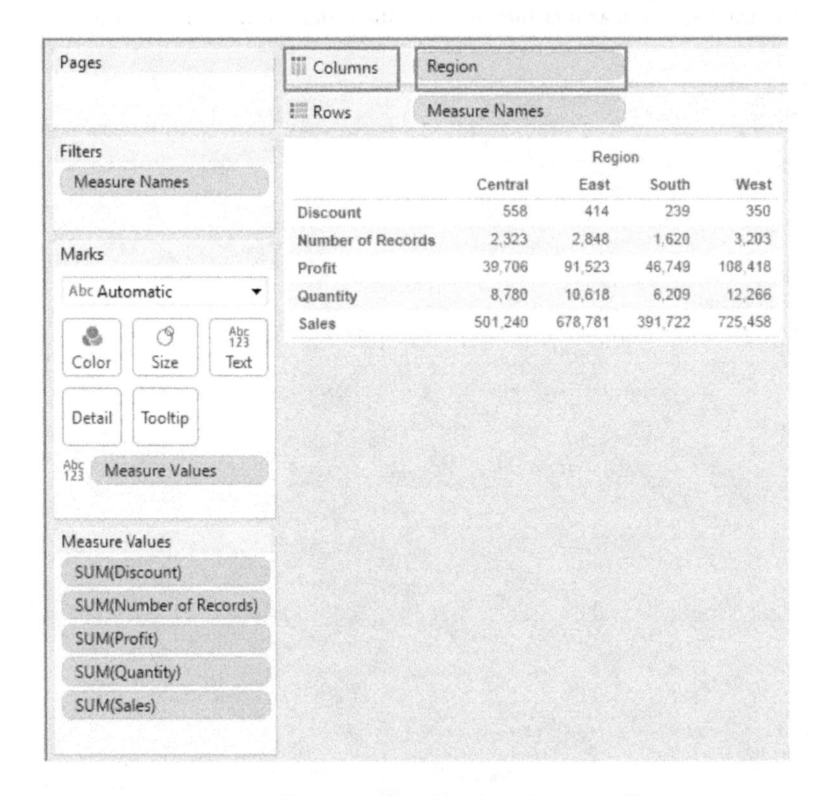

Figure 4-4. *Dimension "Region" placed on the columns shelf*

4.1.2.1.5 Step 4

Notice that "Measure Names" has automatically been placed on the "Filters Shelf," and there is a "Measure Values Shelf" just below the marks card. Let us add a quick filter to "Measure Names" (Shown in Fig. 4-5).

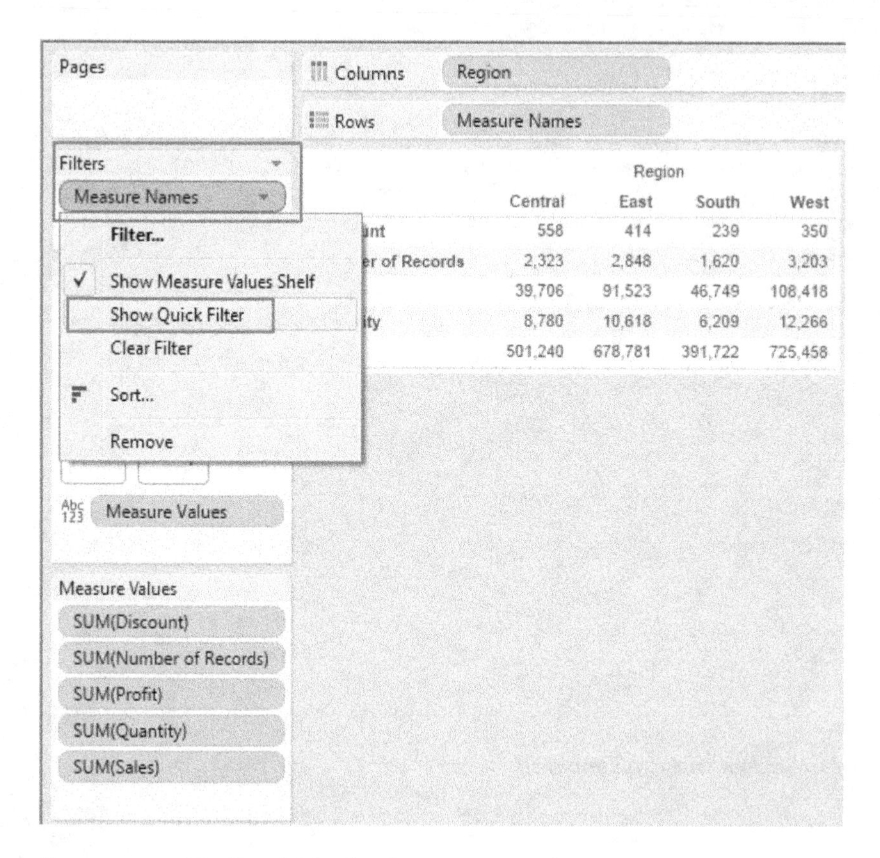

Figure 4-5. *Add a "Quick Filter" to "Measure Names"*

The "Quick Filter" will allow one to select measures to display in the view. Below is the sheet after adding a "Quick Filter" to "Measure Names" (Shown in Fig. 4-6).

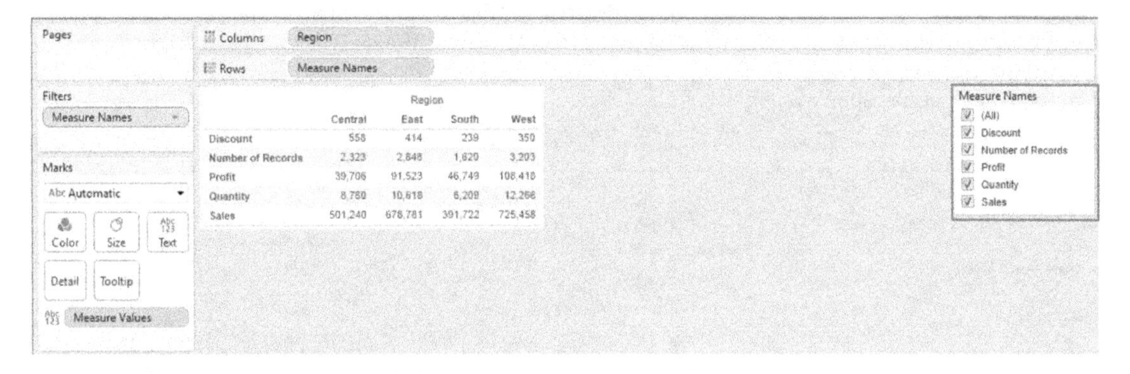

Figure 4-6. *"Quick Filter" on Measure Names*

The output on selecting measures, "Sales" and "Profit" ONLY (See Fig. 4-7).

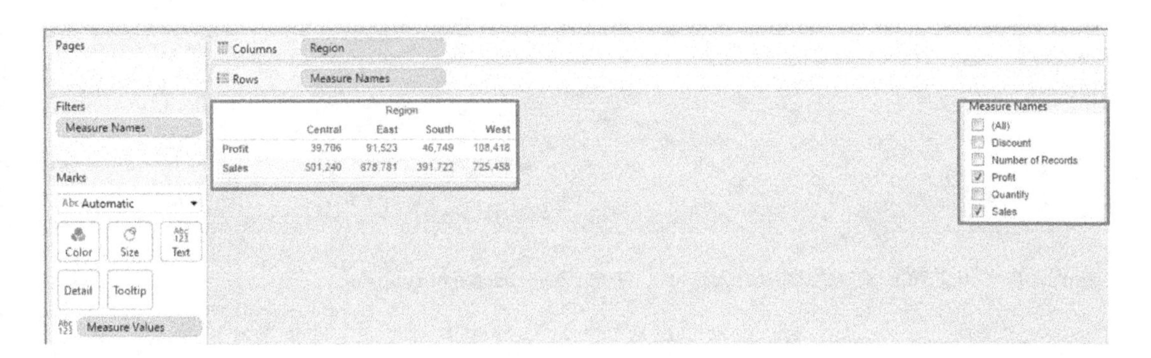

Figure 4-7. *Measures, "Profit" and "Sales" selected in the "Quick Filter"*

4.1.2.2 Demo 2

Objective: Let us create a worksheet / view that allows the user to dynamically select measures to be displayed on the view. Example, the user can choose to have "Profit" displayed over time (2011, 2012, 2013 and 2014) or can choose to have "Discount" or "Sales" displayed over time.

Input: "Sample - Superstore.xls"

Expected output: Shown in Fig. 4-8.

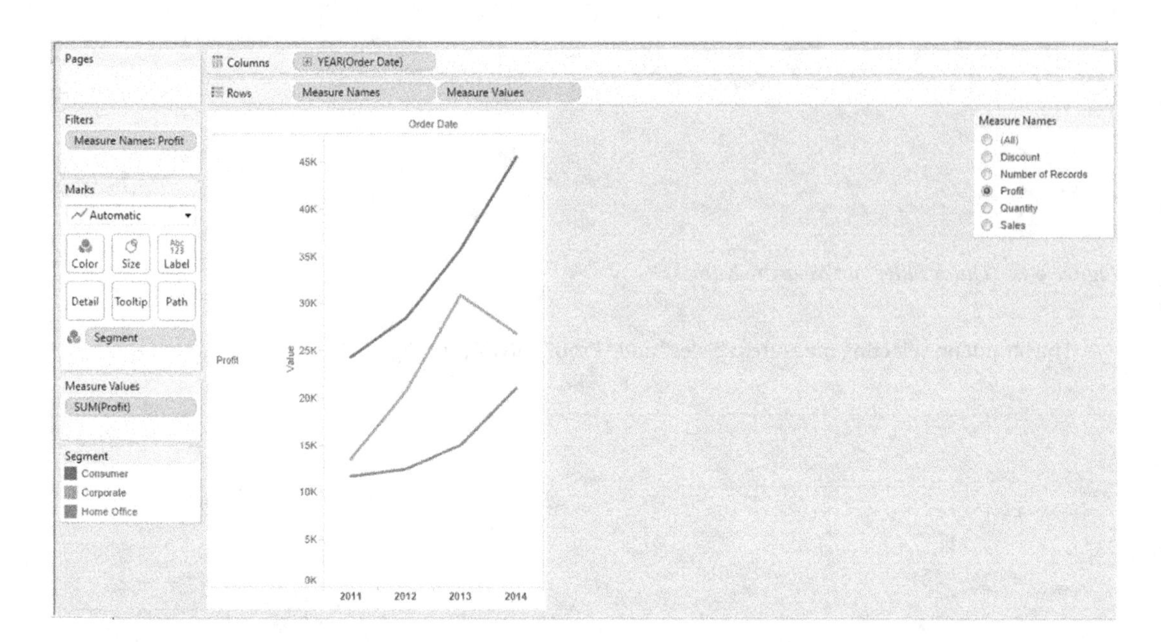

***Figure 4-8.** Quick Filter with "Measure Names" - Demo 2 – expected output*

4.1.2.2.1 Steps to use quick filter with "Measure Names"

4.1.2.2.2 Step 1

Read in data from "Sample - Superstore.xls" into Tableau (Shown in Fig. 4-9).

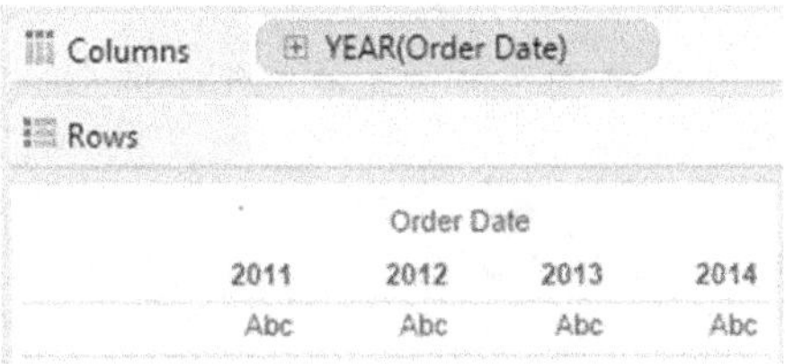

Figure 4-9. *Data from "Sample - Superstore.xls" read into Tableau*

4.1.2.2.3 Step 2

Drag the dimension "Order Date" from the dimensions area under the data pane and place it on the columns shelf (Shown in Fig. 4-10).

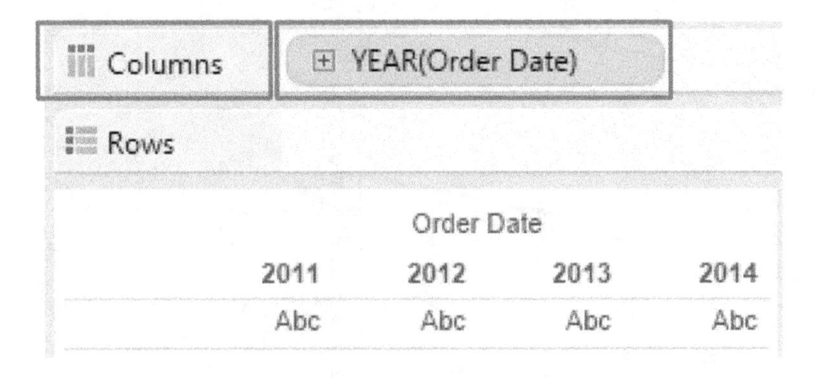

Figure 4-10. *Dimension "Order Date" placed on the columns shelf*

4.1.2.2.4 Step 3

Drag the dimension "Measure Names" from the dimensions area under the data pane and place it on the rows shelf. Drag the measure "Measure Values" from the measures area under the data pane and place it on the rows shelf to the right of "Measure Names" (Shown in Fig. 4-11).

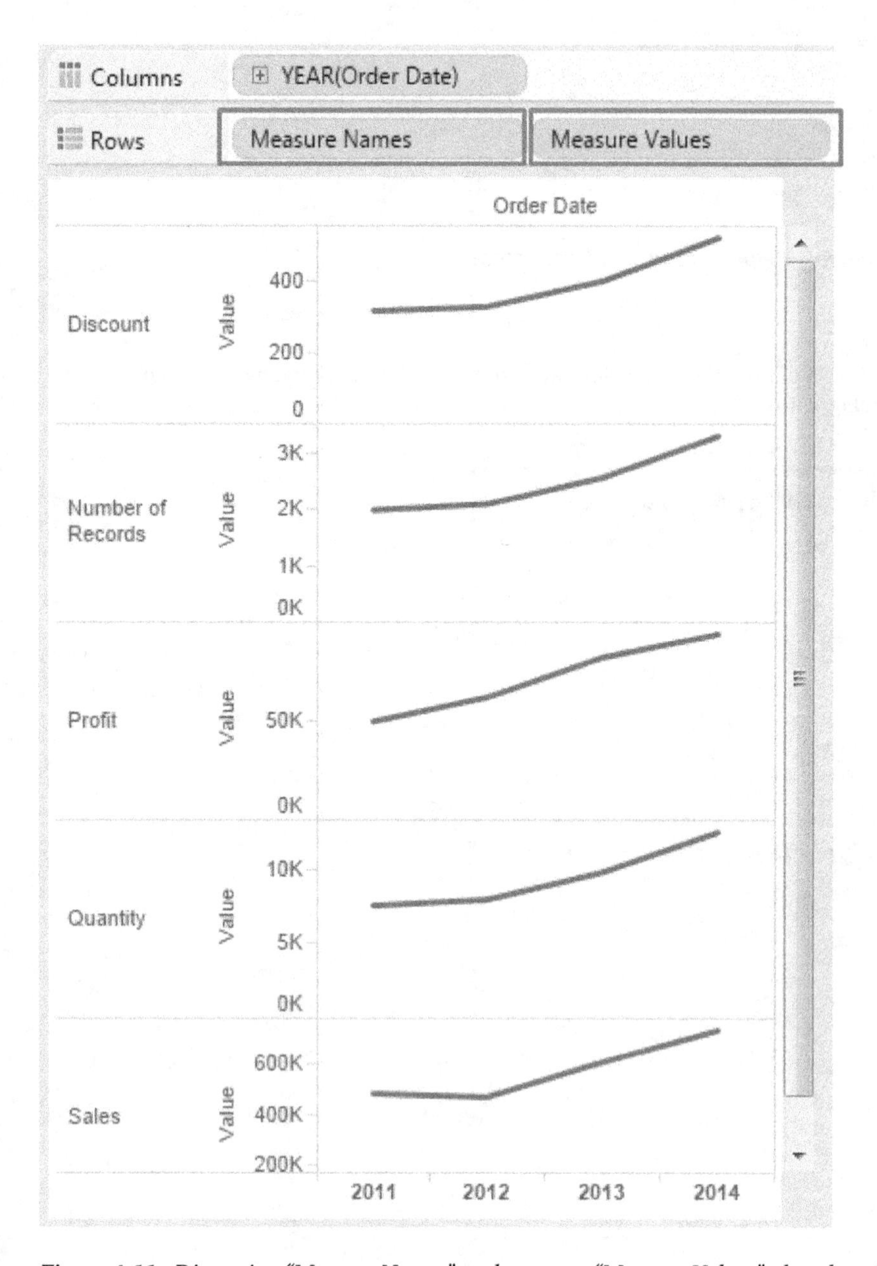

Figure 4-11. *Dimension "Measure Names" and measure "Measure Values" placed on the rows shelf*

4.1.2.2.5 Step 4

Drag the dimension "Segment" from the dimensions area under the data pane and place it on "Color" on the marks card (Shown in Fig. 4-12).

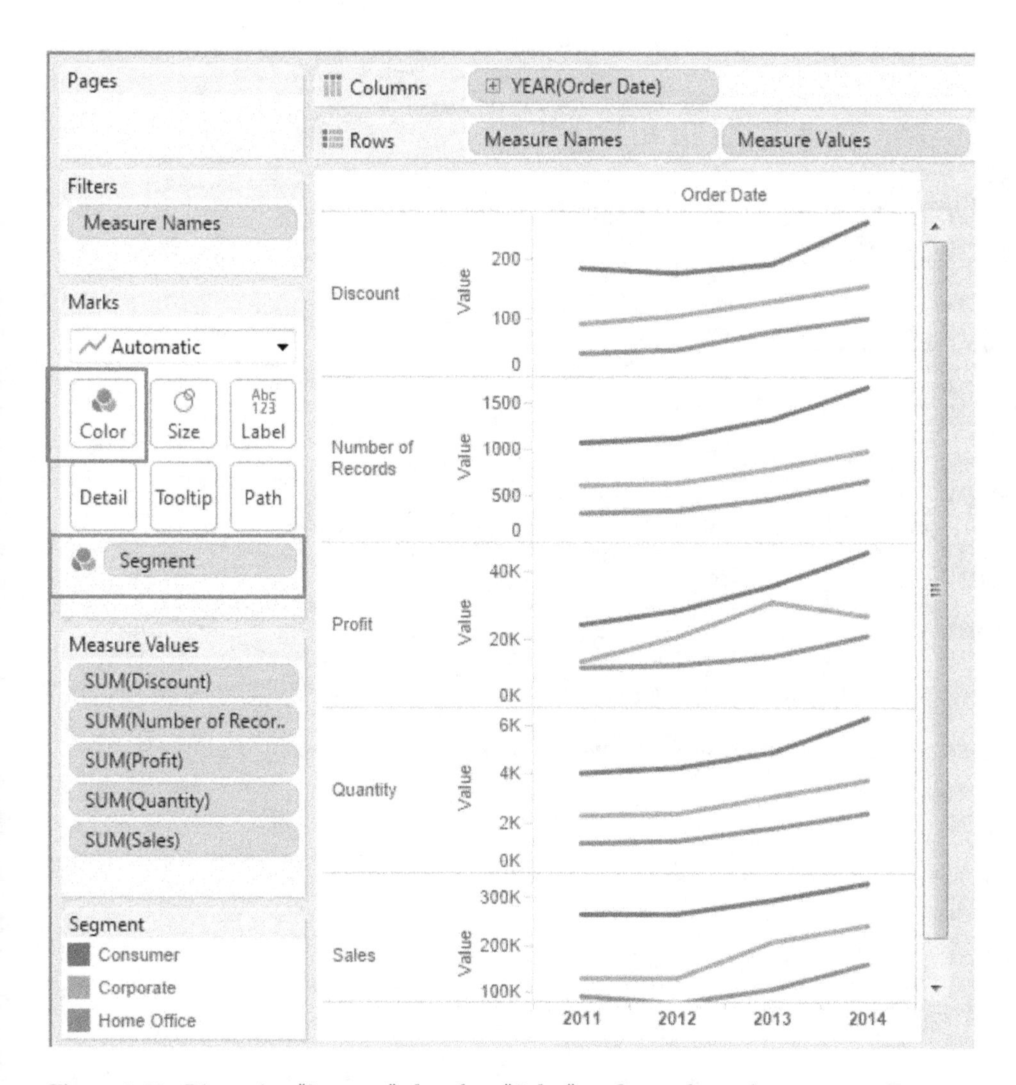

***Figure 4-12.** Dimension "Segment" placed on "Color" on the marks card*

4.1.2.2.6 Step 5

Observe that "Measure Names" has automatically been placed by Tableau on the "Filters Shelf". Let us add a "Quick Filter" to "Measure Names" (Shown in Fig. 4-13).

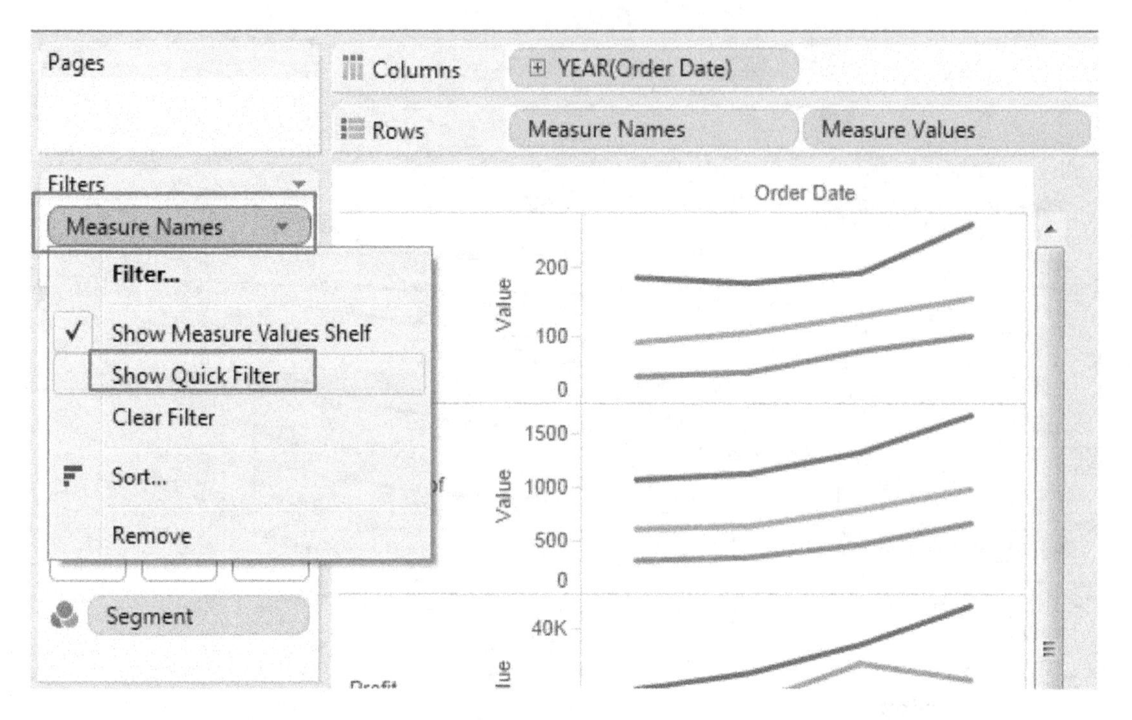

Figure 4-13. Adding a "Quick Filter" to "Measure Names"

4.1.2.2.7 Step 6

Change the "Quick Filter" settings to display a single values list (Shown in Fig. 4-14).

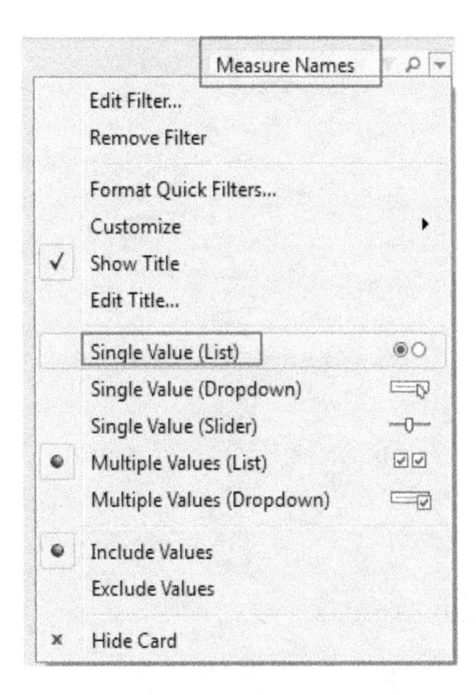

Figure 4-14. *Changing the "Quick Filter" to a "Single Value (List)"*

Select the measure "Profit" from the Single Value (List). The output below shows "Profit" over time by "Segment" (Shown in Fig. 4-15).

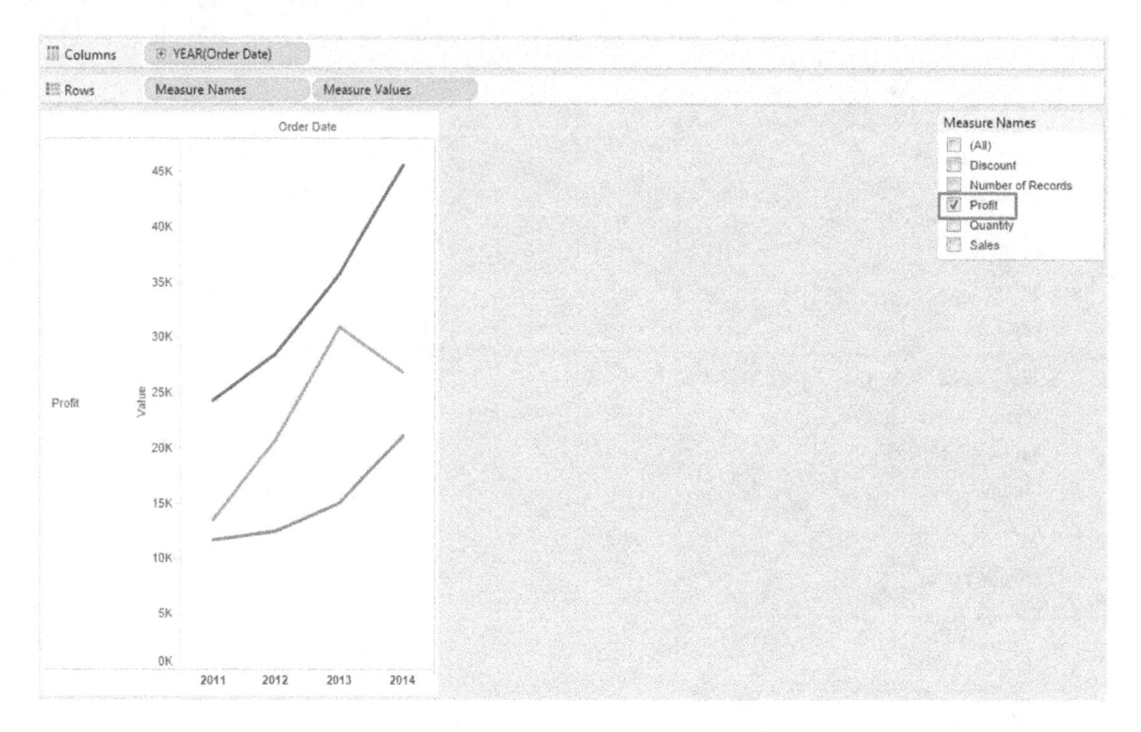

Figure 4-15. *Output shows "Profit" over time by "Segment"*

With this understanding of multiple measures by way of the demos in the previous section, let us further explore how multiple measures can be brought into a worksheet / view. We will begin with plotting each measure on a separate axis, proceed to blend the measures and plot it on a single axis, and then experiment with using dual axis. Measures can be placed on the following:

- Individual axis
- Blend measures and place on single axis
- Dual axis

4.1.3 Measures on an independent axis

One can create individual axis for each measure (Shown in Fig. 4-16). Refer to Table 4-3 for the data used in the Figure 4-16.

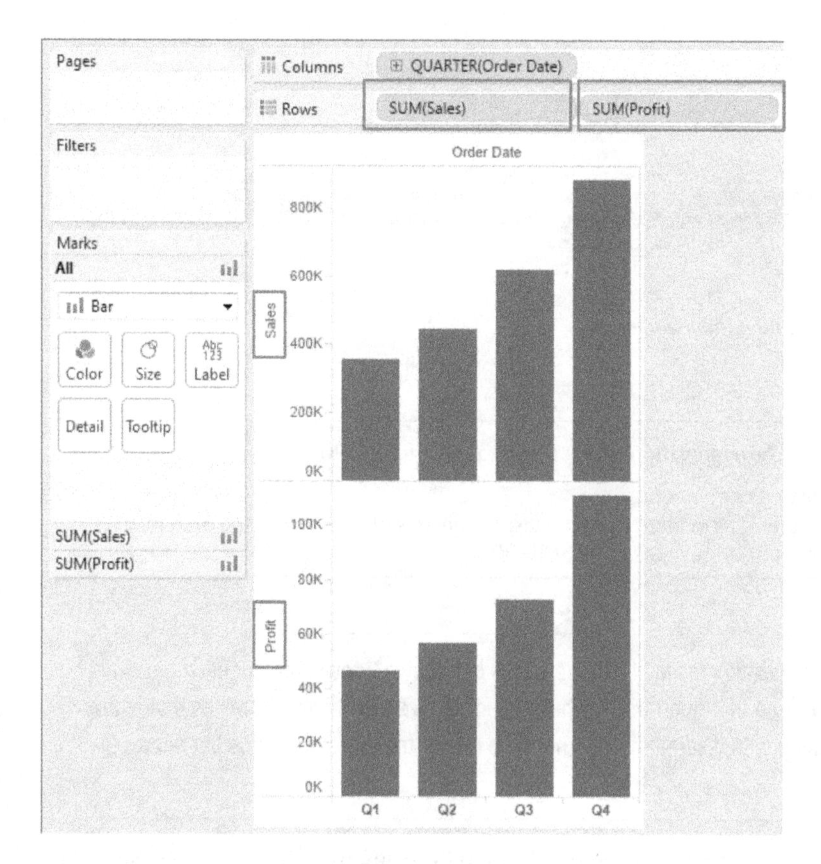

***Figure 4-16.** Independent axis for each measure*

***Table 4-3.** Data used in Fig. 4-16*

Columns Shelf	Quarter(Order Date) Date is "Discrete" as evident from the visual cue. It is blue in color. The preferred chart form to depict discrete dates is "Bar chart".
Rows Shelf	Sales, Profit. The aggregation used on both measures is "SUM".

Each measure on the rows shelf adds an additional axis to the rows of the table. As can be seen from Fig. 4-16 and Fig. 4-17, there are two measures ("Sales" and "Profit"), and they have added two additional axes to the rows of the table. The "Sales" and "Profit" axes are individual rows in the table and have independent scales.

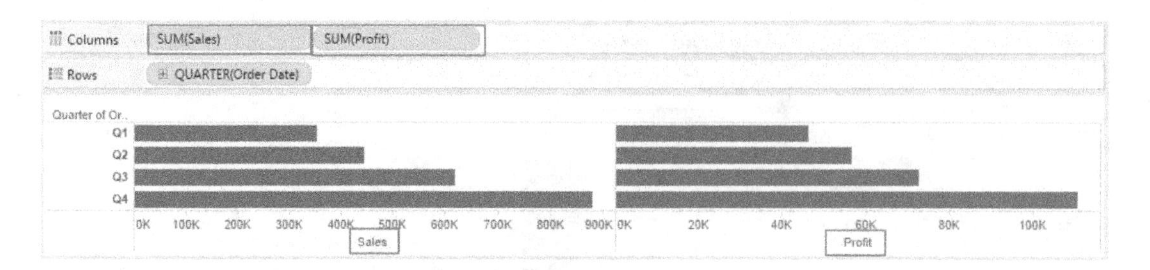

Figure 4-17. *Each measure on its own independent axis*

Table 4-4. *Data used in Fig. 4-17*

Columns shelf	Sales, Profit. The aggregation used on both measures is SUM.
Rows shelf	Quarter(Order Date) Date is "Discrete" as evident from the visual cue. It is blue in color. The preferred chart form to depict discrete dates is "Bar Chart".

■ **Note** Notice that for the visualization in Fig. 4-16, the status bar shows 2 rows (one for each measure) by 4 columns (a year has a maximum of 4 quarters). We have used bar to show the measures as that is the preferred chart form when working with discrete dates. There are 8 bars therefore the status bar shows 8 marks.

Each measure on the columns shelf adds an additional axis to the columns of the table. As can be seen from Fig. 4-17, there are two measures (sales and profit) and they have added two additional axes to the columns of the table. The Sales and Profit axes are individual columns in the table and have independent scales.

■ **Note** Notice that for the visualization in Fig. 4-17, the status bar shows four rows (one for each quarter) by two columns (one for each measure). We have used bar to show the measures as that is the preferred chart form with discrete dates. There are eight bars; therefore, the status bar shows eight marks.

4.1.4 Blended axes

Blend the measures and have them share a common axis. When should one use it? It should be used when one wants to compare measures that have similar scale and units.

4.1.4.1 Demo 1

Objective: Let us create a worksheet / view that displays two measures "Sales" and "Profit" for each year (2011 to 2014) side-by-side using "Side by Side Bars".

Input: "Sample - Superstore.xls".

Expected Output: See Fig. 4-18.

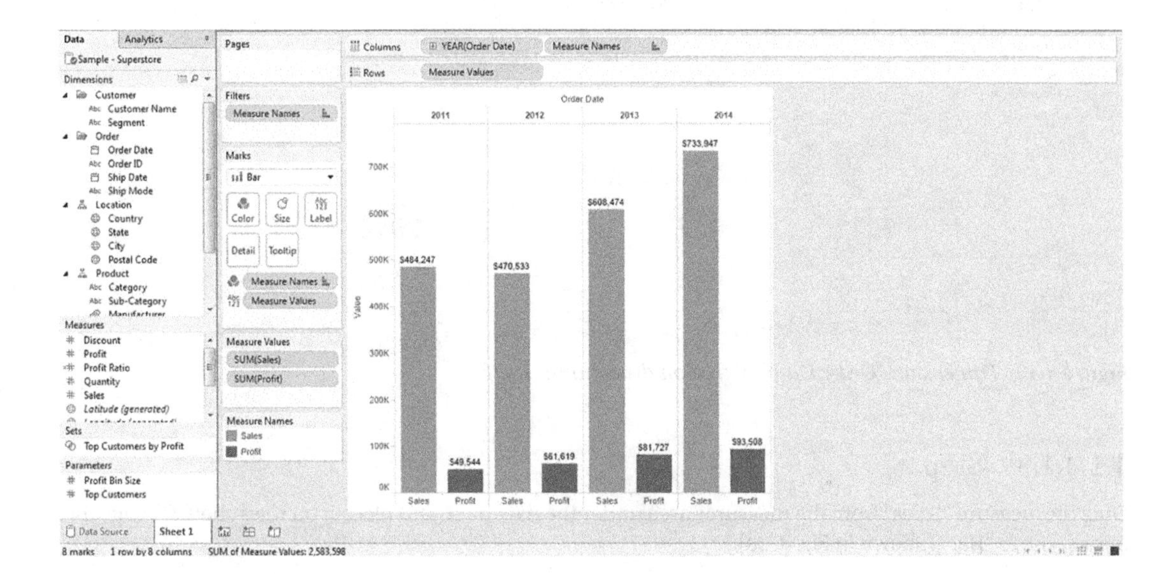

Figure 4-18. Blended Measures - Demo 1 - Expected output

4.1.4.1.1 Steps to displays two measures "Sales" and "Profit" side by side using "Side by Side Bars"

4.1.4.1.2 Step 1

Drag "Order Date" from dimensions area under the data pane and place it on the columns shelf. Dates are always displayed as a hierarchy. By default, when we drag "Order Date" and place it either on the rows or columns shelf, it is "Discrete" (this is evident from the visual cue ("Order Date" appears in blue color) (Shown in Fig. 4-19).

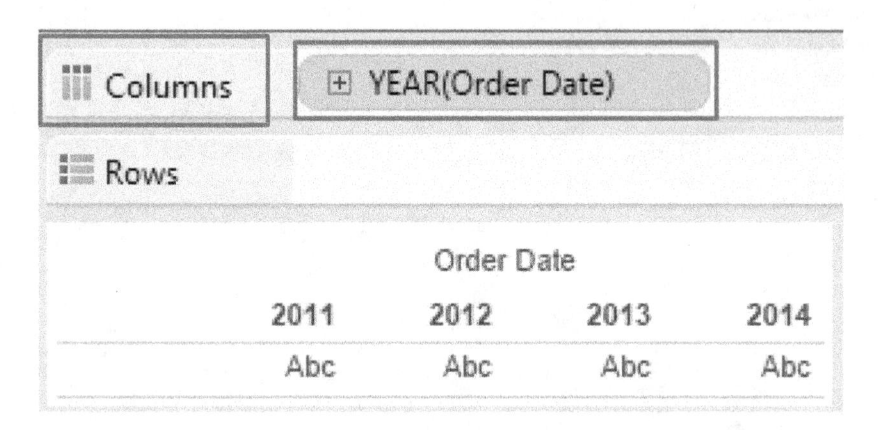

Figure 4-19. *Dimension "Order Date" placed on the columns shelf*

4.1.4.1.3 Step 2

Drag the measure "Sales" from the measures area under the data pane and place it on rows shelf. Change the marks type to "Bar" (Shown in Fig. 4-20).

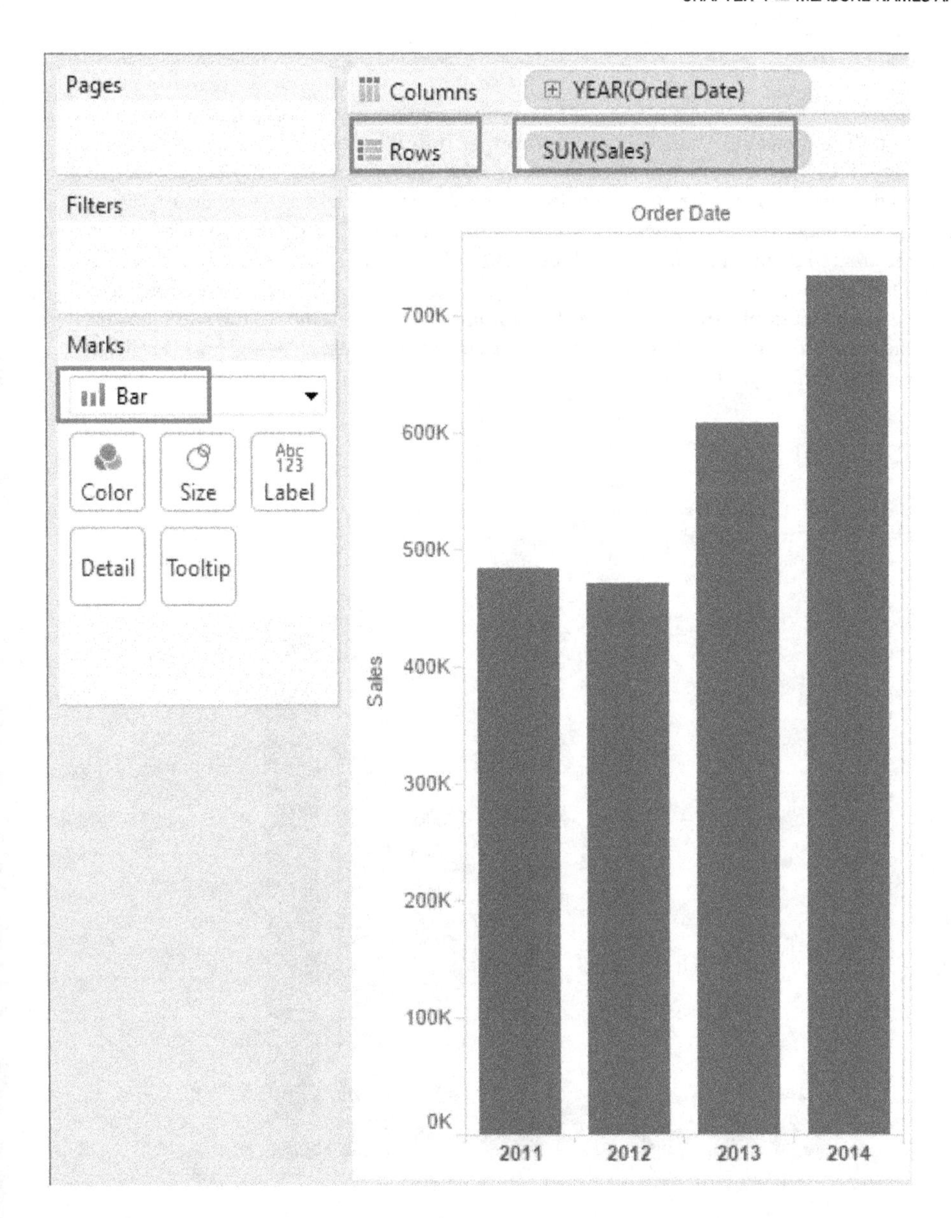

Figure 4-20. *Marks Type changed to "Bar"*

4.1.4.1.4 Step 3

Drag the measure "Profit" from the measures area under the data pane and place it on the same axis as the measure "Sales". Note: The measure "Profit" should be dropped only when you see a ruler or scale image otherwise it will replace the measure "Sales" on the rows shelf. As soon as more than one measure is dropped on the same axis, you will notice a new dimension "Measure Names" and a new measure "Measure Values". In our example, "Measure Names" appears on the columns shelf and "Measure Values" on the rows shelf. The shared axis is created using the "Measure Values" field. What are measure names and measure values? Measure names are a container that contains the names of the measures that has been dragged on the worksheet/view. In our example, the measure names container has the names of the two measures "Sales" and "Profit". The measure values container has the values for the measures, "Sales" and "Profit", i.e. Sum(Sales) and Sum(Profit). (Shown in Fig. 4-21).

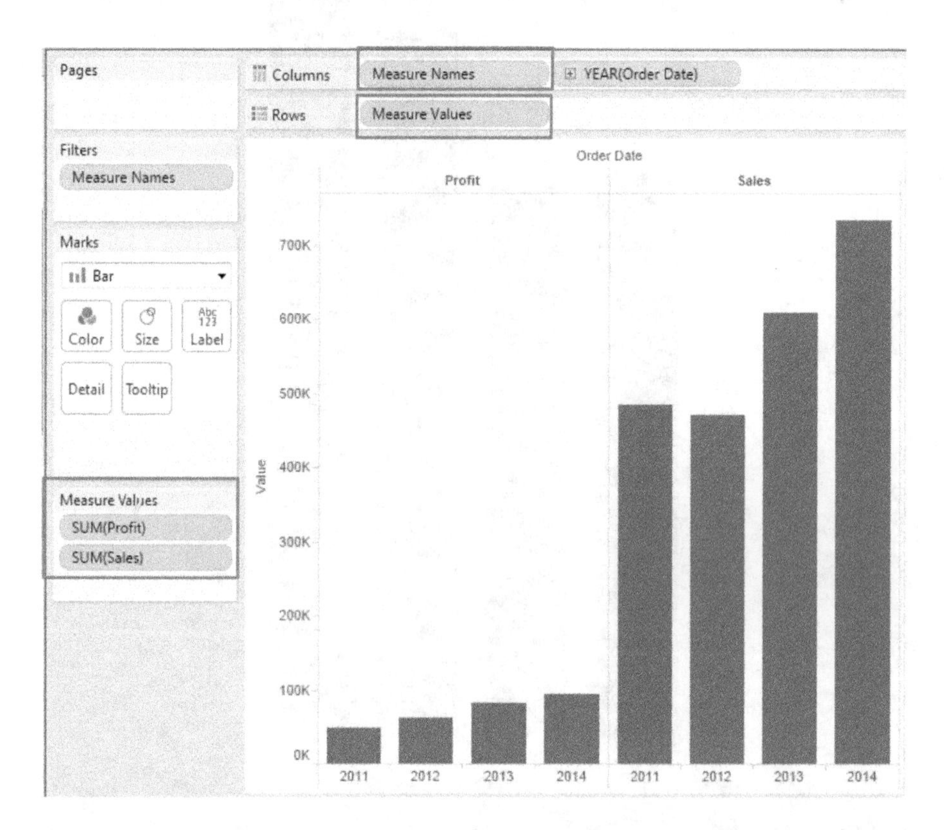

Figure 4-21. *Dimension "Measure Names" and measure "Measure Values" placed on the columns and rows shelf, respectively*

We would like to view the "Sales" and "Profits" bars side by side. Notice in Fig. 4-21, "Measure Names" is displayed first on the columns shelf, followed by "Order Date". Move "Measure Names" to the right of "Order Date". The output is as shown in Fig. 4-22.

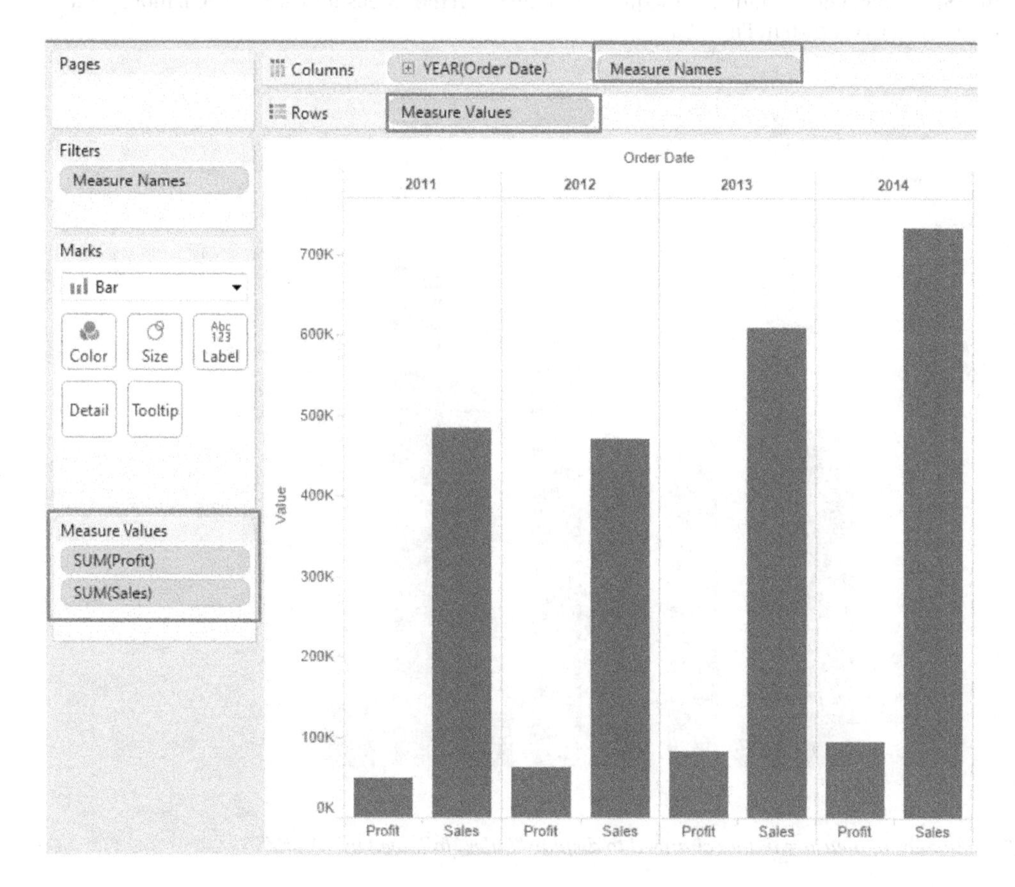

***Figure 4-22.** Measure Values, "Sales" and "Profit" displayed for Years (2011-2014)*

4.1.4.1.5 Step 4

Let us change the sequence in which the bars are displayed. First, the "Profit" bar is displayed for each year followed by the "Sales" bar. Let us change the sequence. To do so, in the "Measure Values" shelf, move sum (Sales) above sum (Profit) (Shown in Fig. 4-23).

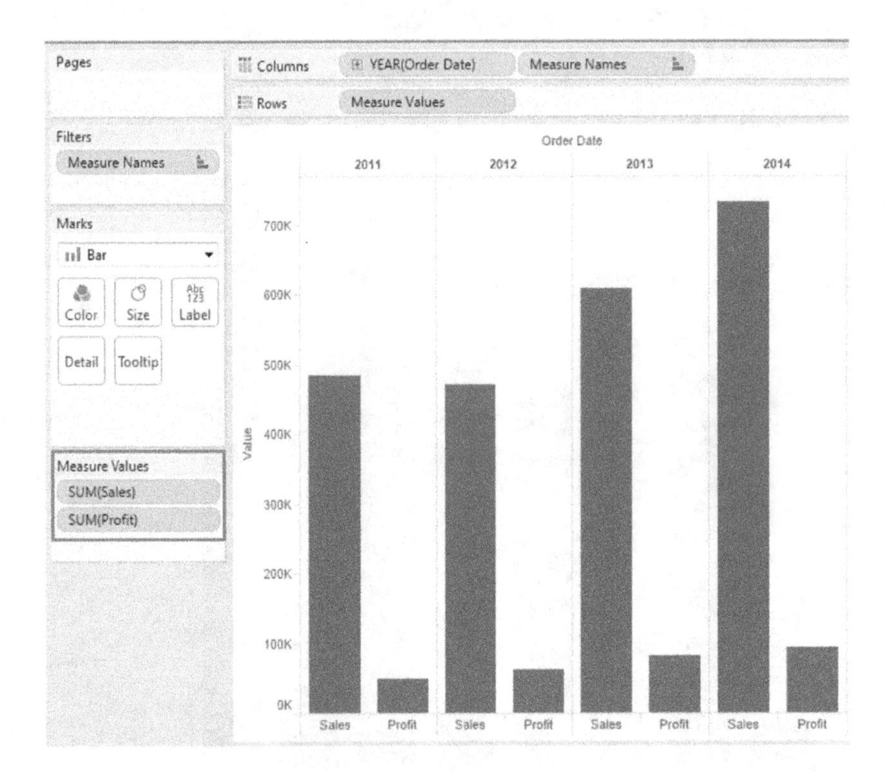

***Figure 4-23.** Sequence of measure values changed to display "Sales" followed by "Profit"*

4.1.4.1.6 Step 5

Keep the CTRL key pressed as you drag "Measure Names" from the columns shelf to "Color" on the "Marks" card (Shown in Fig. 4-24).

■ **Note** If you do NOT keep the CTRL key pressed as you drag "Measure Names" to "Color", "Measure Names" will disappear from the columns shelf and you will get stacked bars in the view.

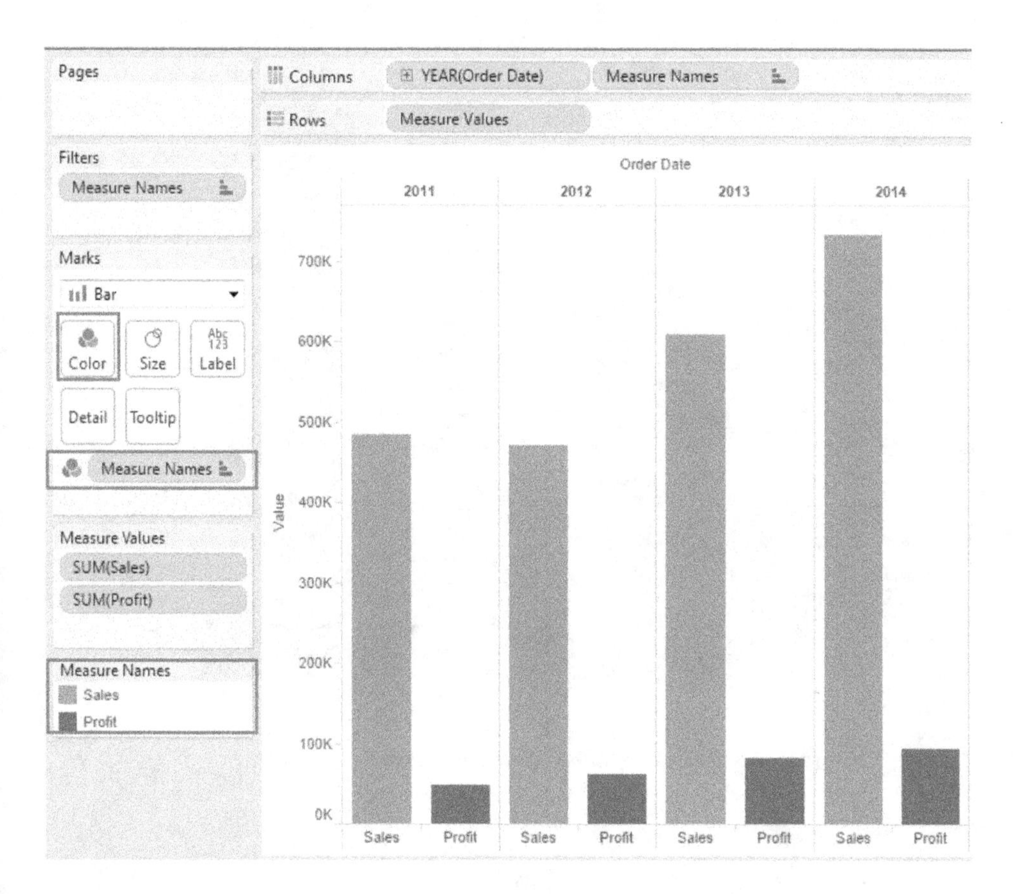

Figure 4-24. *Dimension "Measure Names" placed on "Color" on the marks card*

4.1.4.1.7 Step 6

Keep the CTRL key pressed as you drag “Measure Values” from the rows shelf and drop it on “Label” on the “Marks” card (Shown in Fig. 4-25).

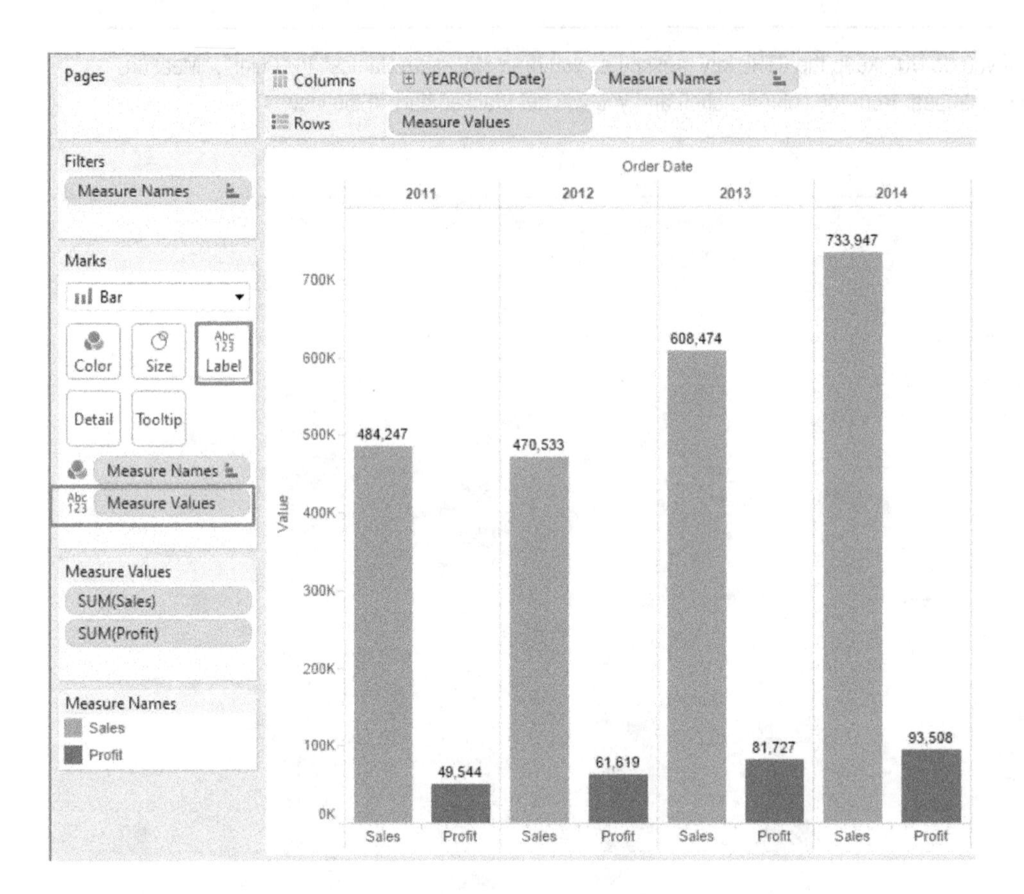

Figure 4-25. *Measure “Measure Values” placed on “Label” on the marks card*

4.1.4.2 Demo 2

Objective: To plot multiple measures (such as "Sales", "Profit" and "Discount") on a single axis.

Input: "Sample Superstore.xls". The Excel sheet has data for 4 years (2011 to 2014).

Expected output: Shown in Fig. 4-26.

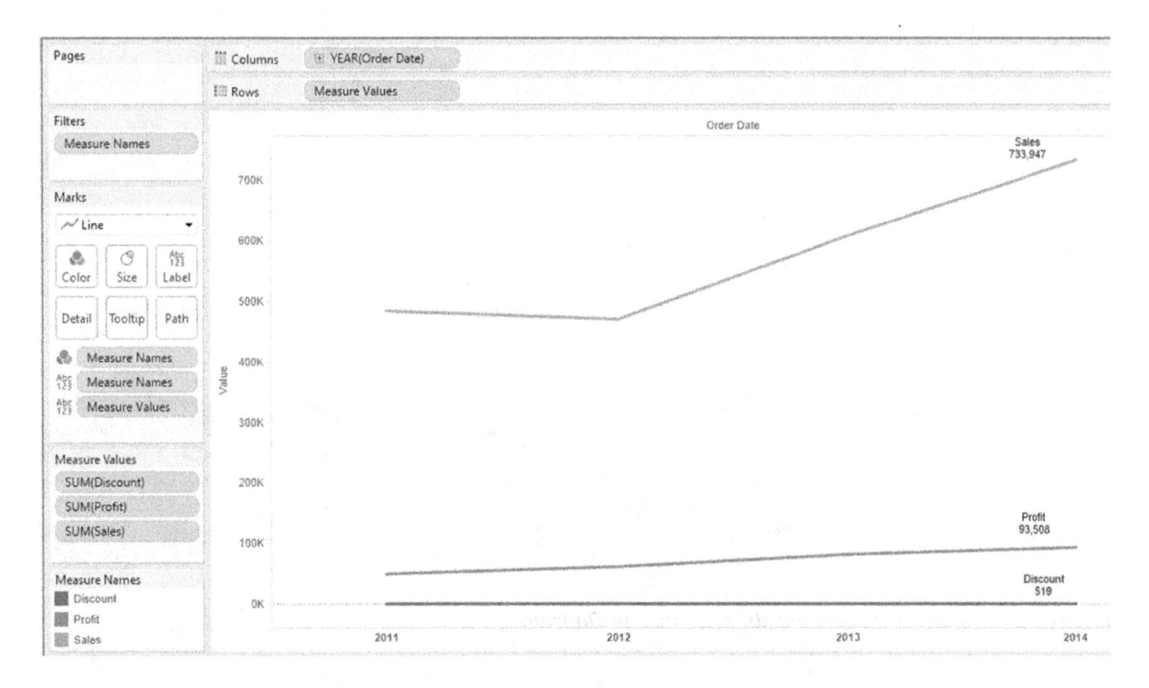

Figure 4-26. Blended measures - Demo 2 – expected output

4.1.4.2.1 Steps to plot multiple measures on the same axis

4.1.4.2.2 Step 1

Read in data from "Sample - Superstore.xls" into Tableau (Shown in Fig. 4-27).

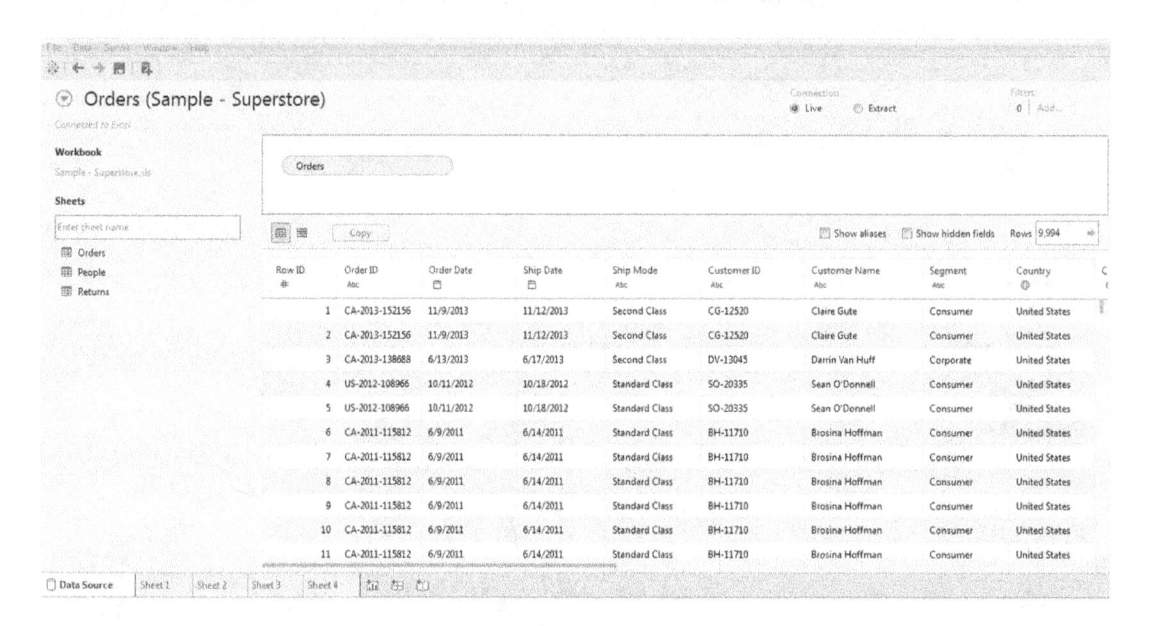

Row ID	Order ID	Order Date	Ship Date	Ship Mode	Customer ID	Customer Name	Segment	Country
1	CA-2013-152156	11/9/2013	11/12/2013	Second Class	CG-12520	Claire Gute	Consumer	United States
2	CA-2013-152156	11/9/2013	11/12/2013	Second Class	CG-12520	Claire Gute	Consumer	United States
3	CA-2013-138688	6/13/2013	6/17/2013	Second Class	DV-13045	Darrin Van Huff	Corporate	United States
4	US-2012-108966	10/11/2012	10/18/2012	Standard Class	SO-20335	Sean O'Donnell	Consumer	United States
5	US-2012-108966	10/11/2012	10/18/2012	Standard Class	SO-20335	Sean O'Donnell	Consumer	United States
6	CA-2011-115812	6/9/2011	6/14/2011	Standard Class	BH-11710	Brosina Hoffman	Consumer	United States
7	CA-2011-115812	6/9/2011	6/14/2011	Standard Class	BH-11710	Brosina Hoffman	Consumer	United States
8	CA-2011-115812	6/9/2011	6/14/2011	Standard Class	BH-11710	Brosina Hoffman	Consumer	United States
9	CA-2011-115812	6/9/2011	6/14/2011	Standard Class	BH-11710	Brosina Hoffman	Consumer	United States
10	CA-2011-115812	6/9/2011	6/14/2011	Standard Class	BH-11710	Brosina Hoffman	Consumer	United States
11	CA-2011-115812	6/9/2011	6/14/2011	Standard Class	BH-11710	Brosina Hoffman	Consumer	United States

Figure 4-27. *"Sample - Superstore.xls" connected to Tableau*

4.1.4.2.3 Step 2

Drag "Measure Names" from the dimensions area under the data pane to "Filters Shelf".

The 'Filter [Measure Names] dialog box shows up. Select the measures "Discount", "Profit" and "Sales" (Shown in Fig. 4-28).

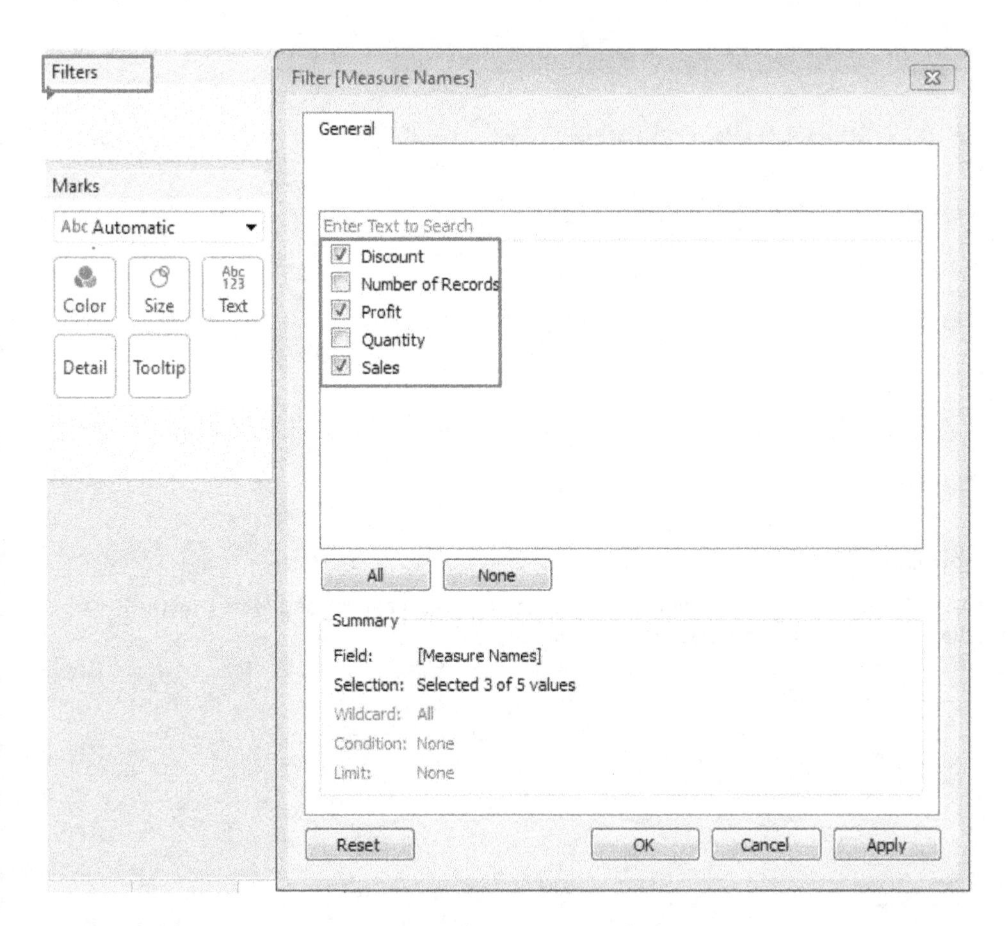

Figure 4-28. *Select measures in the "Filter [Measure Names]" dialog box*

Click on "Apply" and then "OK".

The output after applying the filter is shown in Fig. 4-29.

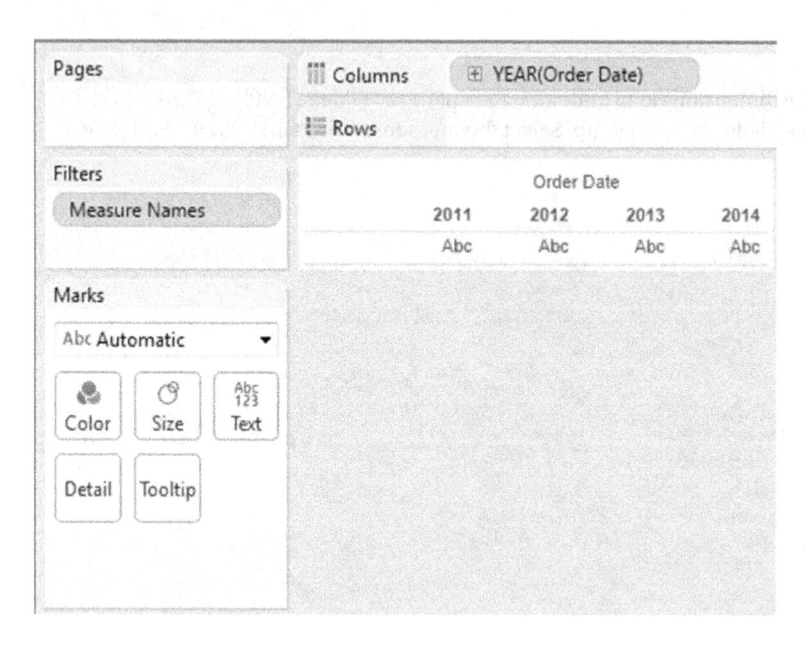

Figure 4-29. *"Measure Names" placed in the "Filters Shelf"*

4.1.4.2.4 Step 3

Drag the dimension "Order Date" from the dimensions area under the data pane and place it on the columns shelf. Retain the default granularity at "Year" (Figure 4-30).

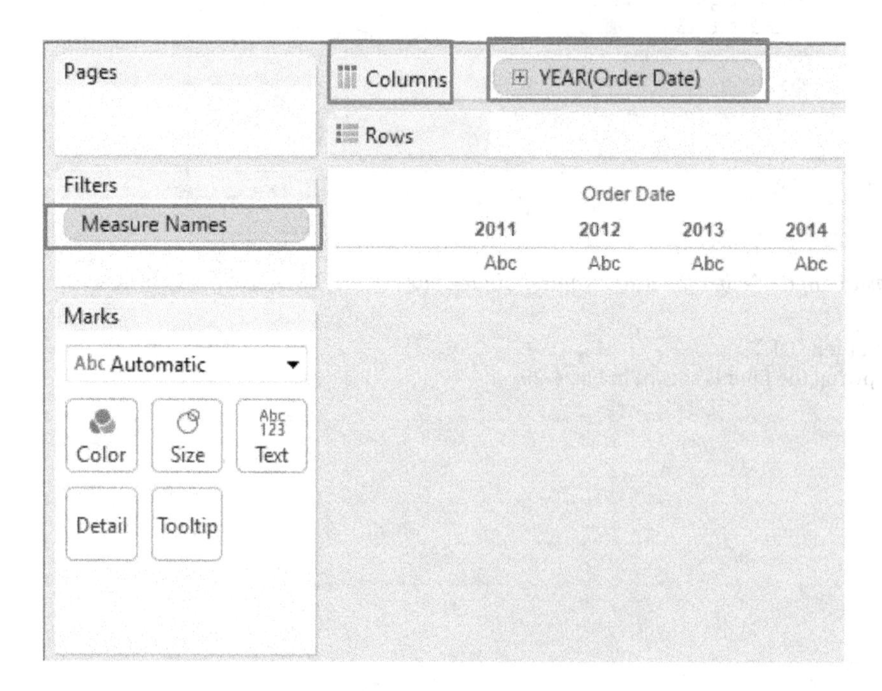

Figure 4-30. *Dimension "Order Date" placed on the columns shelf*

4.1.4.2.5 Step 4

Drag "Measure Values" from the measures area under the data pane and place it on the rows shelf (Shown in Fig. 4-31).

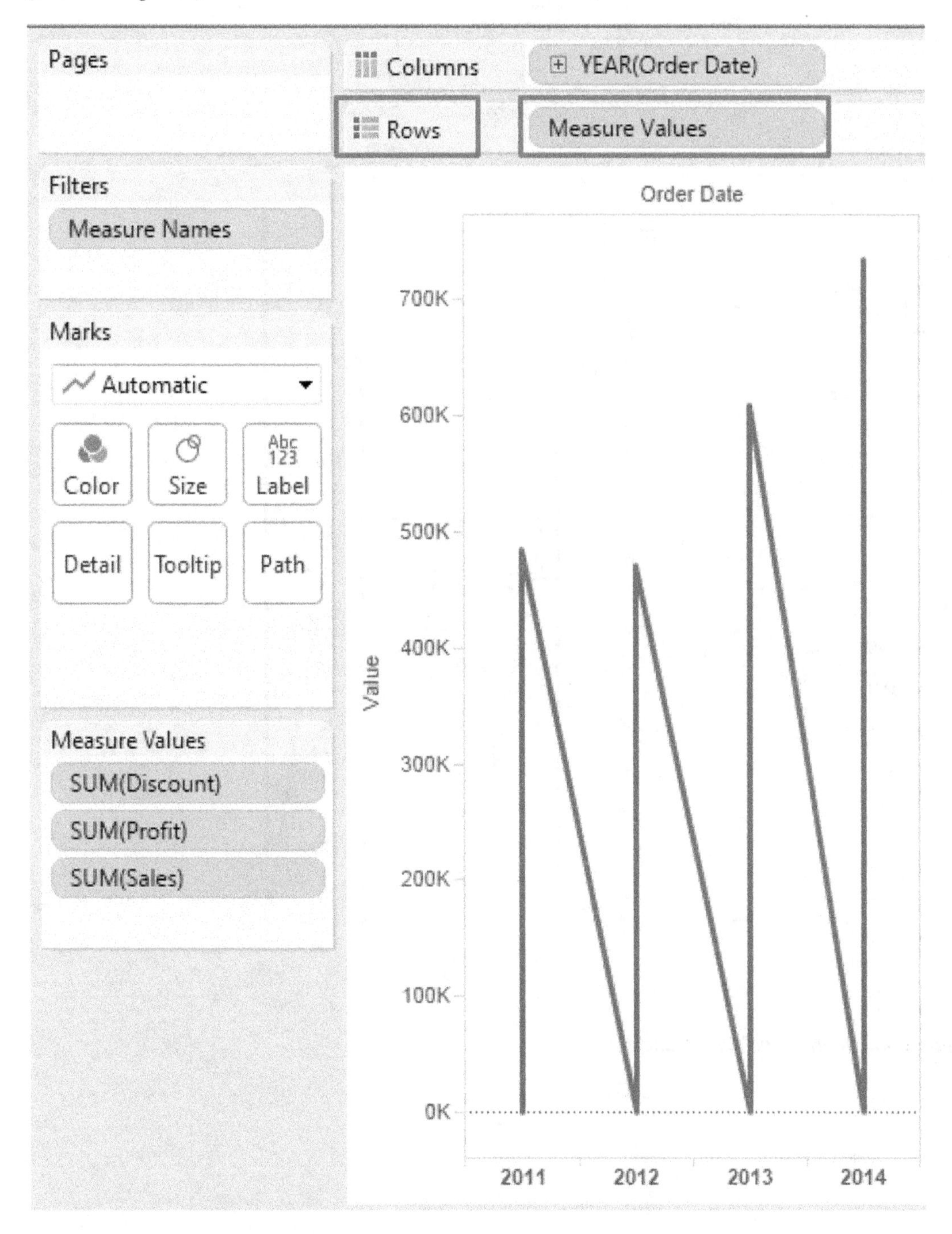

Figure 4-31. *Measure "Measure Values" placed on the rows shelf*

4.1.4.2.6 Step 5

Change the "Marks Type" to "Line" (Shown in Fig. 4-32).

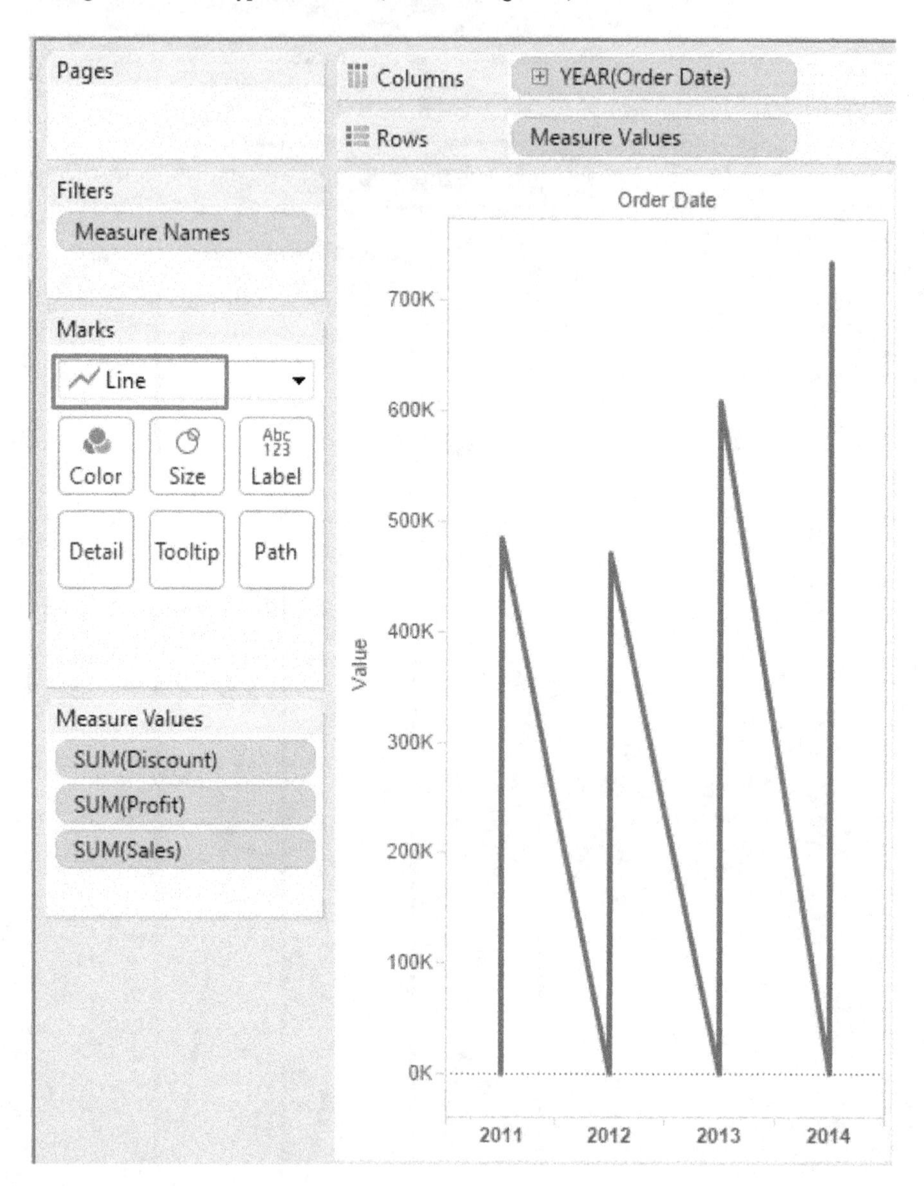

Figure 4-32. *"Marks Type" set to "Line"*

Drag "Measure Names" from the dimensions area under the data pane and place it on "Color" on the marks card (Shown in Fig. 4-33).

Figure 4-33. *"Measure Names" placed on "Color" on the marks card*

4.1.4.2.7 Step 6

Now let us apply some formatting.

Drag and drop "Measure Names" on "Label" on the marks card (Shown in Fig. 4-34).

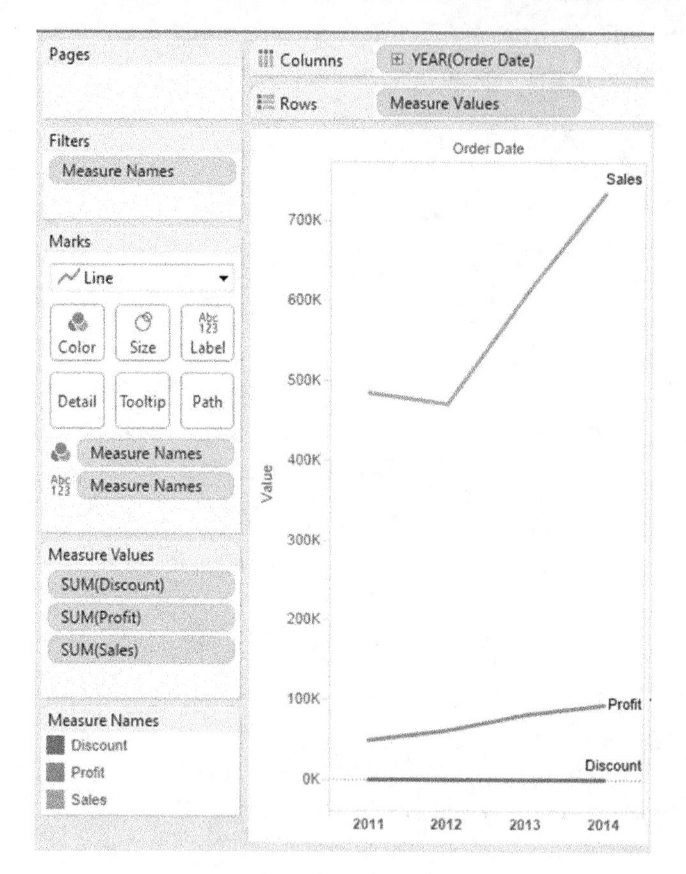

Figure 4-34. *"Measure Names" placed on "Label" on the marks card*

Drag "Measure Values" and place it on "Label" on the marks card (Shown in Fig. 4-35).

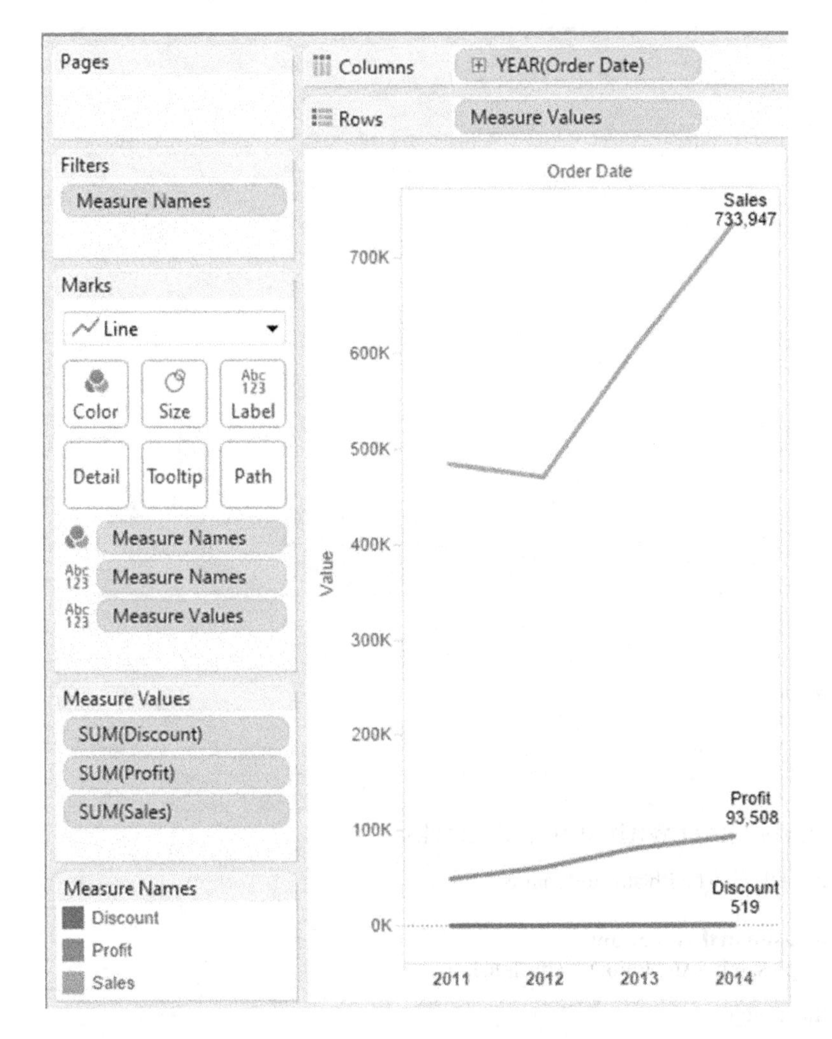

Figure 4-35. "Measure Values" placed on "Label" on the marks card

Change the "Fit" to "Entire View" (Shown in Fig. 4-36).

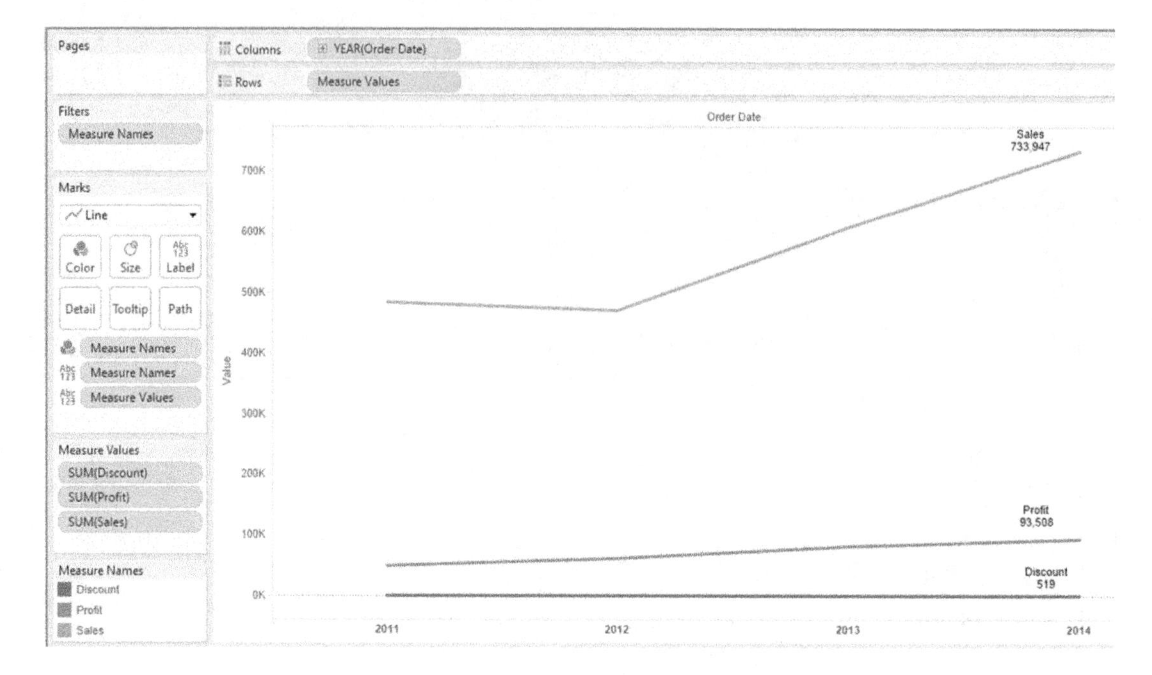

Figure 4-36. *"Fit" set to "Entire View"*

4.1.4.3 Demo 3

4.1.4.3.1 Combined axis chart with stacked marks

Objective: To create a combined axis chart with stacked marks.

Input:

Description of the data set used in this section:

The data set is of 2014 Olympics. It has the following dimensions:

- Athlete: Name of the athlete
- Country: Name of the participating athlete's country
- Sex: Gender of the athlete
- Sport: Name of the sport in which the athlete participated

The measures are as follows:

- Age: Age of the athlete.
- Bronze: Number of bronze medals won by athlete in the sport in which he participated.
- Gold: Number of gold medals won by athlete in the sport in which he participated.
- Silver: Number of silver medals won by athlete in the sport in which he participated.
- Total: Total number of medals won by the athlete in the sport in which he participated.

A subset of the data (Shown in Fig. 4-37).

	A	B	C	D	E	F	G	H	I
1	**Country**	**Athlete**	**Sex**	**Age**	**Sport**	**Gold**	**Silver**	**Bronze**	**Total**
2	Australia	Torah Bright	Female	27	Snowboarding		1		1
3	Australia	David Morris	Male	29	Freestyle Skiing		1		1
4	Australia	Lydia Ierodiaconou-Lassila	Female	32	Freestyle Skiing			1	1
5	Austria	Anna Fenninger	Female	24	Alpine Skiing	1	1		2
6	Austria	Nicole Hosp	Female	30	Alpine Skiing		1	1	2
7	Austria	Dominik Landertinger	Male	26	Biathlon		1	1	2
8	Austria	Julia Dujmovits	Female	26	Snowboarding	1			1
9	Austria	Mario Matt	Male	34	Alpine Skiing	1			1
10	Austria	Matthias Mayer	Male	23	Alpine Skiing	1			1
11	Austria	Thomas Diethart	Male	21	Ski Jumping		1		1
12	Austria	Michael Hayböck	Male	22	Ski Jumping		1		1
13	Austria	Marcel Hirscher	Male	24	Alpine Skiing		1		1
14	Austria	Daniela Iraschko-Stolz	Female	30	Ski Jumping		1		1
15	Austria	Andreas Linger	Male	32	Luge		1		1
16	Austria	Wolfgang Linger	Male	31	Luge		1		1
17	Austria	Thomas Morgenstern	Male	27	Ski Jumping		1		1
18	Austria	Marlies Schild	Female	32	Alpine Skiing		1		1
19	Austria	Gregor Schlierenzauer	Male	24	Ski Jumping		1		1
20	Austria	Christoph Bieler	Male	36	Nordic Combined			1	1
21	Austria	Simon Eder	Male	30	Biathlon			1	1

Figure 4-37. *A subset of the data for Demo 3: Combined Axis Chart with Stacked Marks*

Table 4-5. *Activities to perform*

Columns Shelf	Measure values (Sum(Bronze), Sum(Silver), Sum(Gold))
Rows shelf	Country
Marks card:	
Color	Measure names (Bronze, Silver, Gold)
Label	Measure values

Expected Output: Shown in Fig. 4-38.

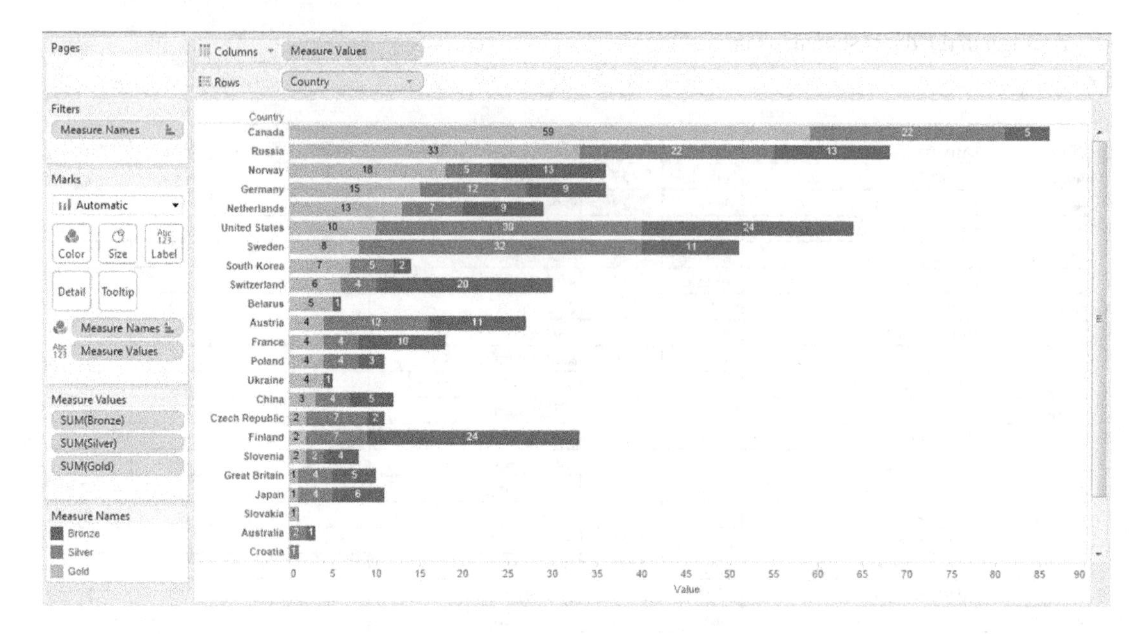

Figure 4-38. *Combined axis with stacked marks - Demo 3 - Expected output*

4.1.4.3.2 Steps to create combined axis chart with stacked marks

4.1.4.3.3 Step 1

Drag the dimension "Country" from the dimensions area under the data pane and place it on the rows shelf. The status bar shows 26 rows by 1 column (the dataset has details about 26 Countries) (Shown in Fig. 4-39).

Figure 4-39. *Dimension "Country" placed on the rows shelf*

4.1.4.3.4 Step 2

Drag the measure "Bronze" from the measures area under the data pane and place it on the columns shelf (Shown in Fig. 4-40).

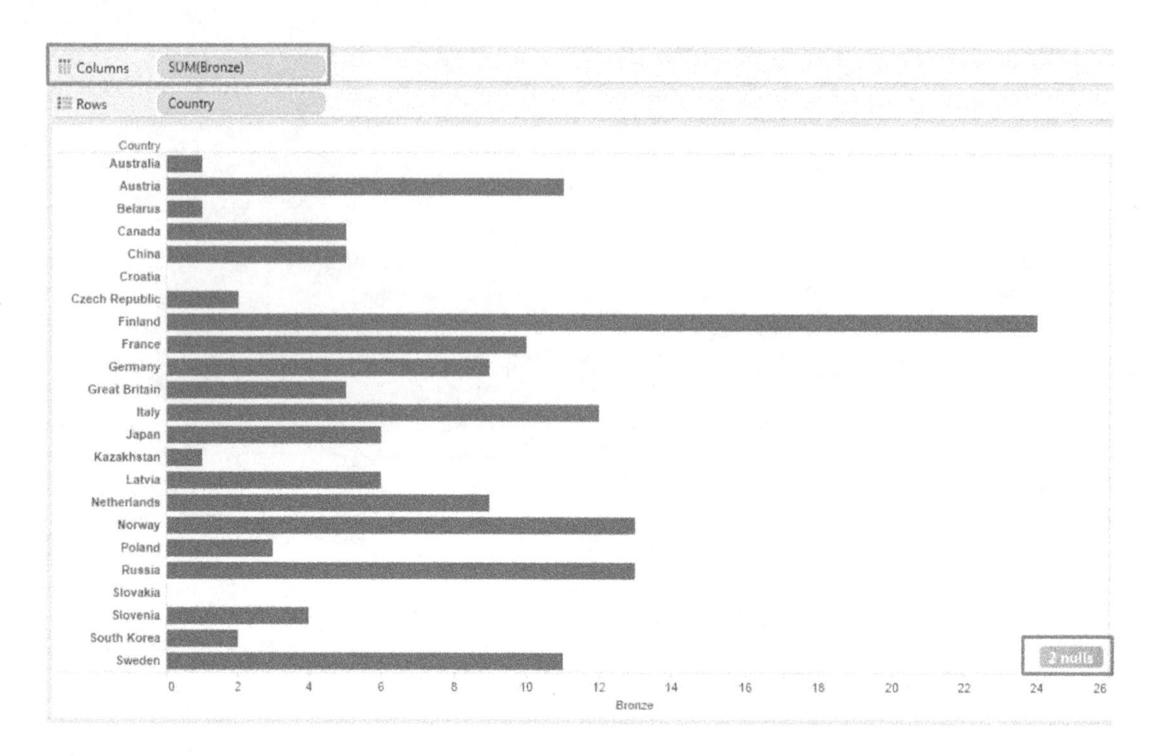

Figure 4-40. *Measure "Bronze" placed on the columns shelf*

Notice the message, "2 nulls" at the bottom right of the screen.

Click on the message "2 nulls". It brings up the "Special Values for [Bronze]" window shown in Fig. 4-41.

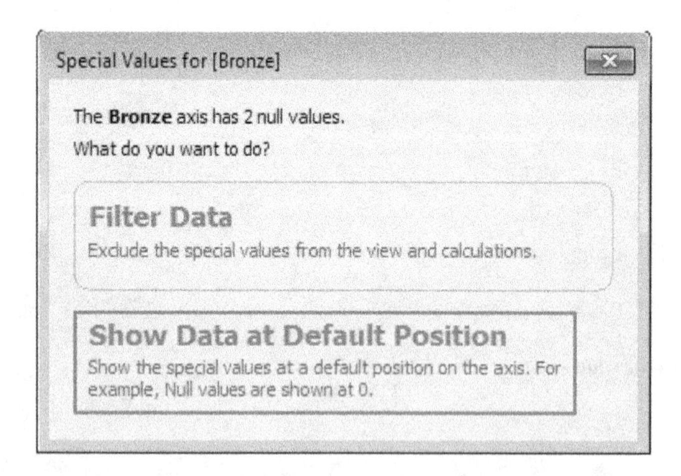

Figure 4-41. *Fixing up "Special Values for [Bronze]"*

"Filter Data" excludes the special values from the view and calculations. "Show Data at Default Position" shows the special values at a default position on the axis. For example, null values are shown at 0.

Select "Show Data at Default Position". The message disappears. The output is as shown in Fig. 4-42.

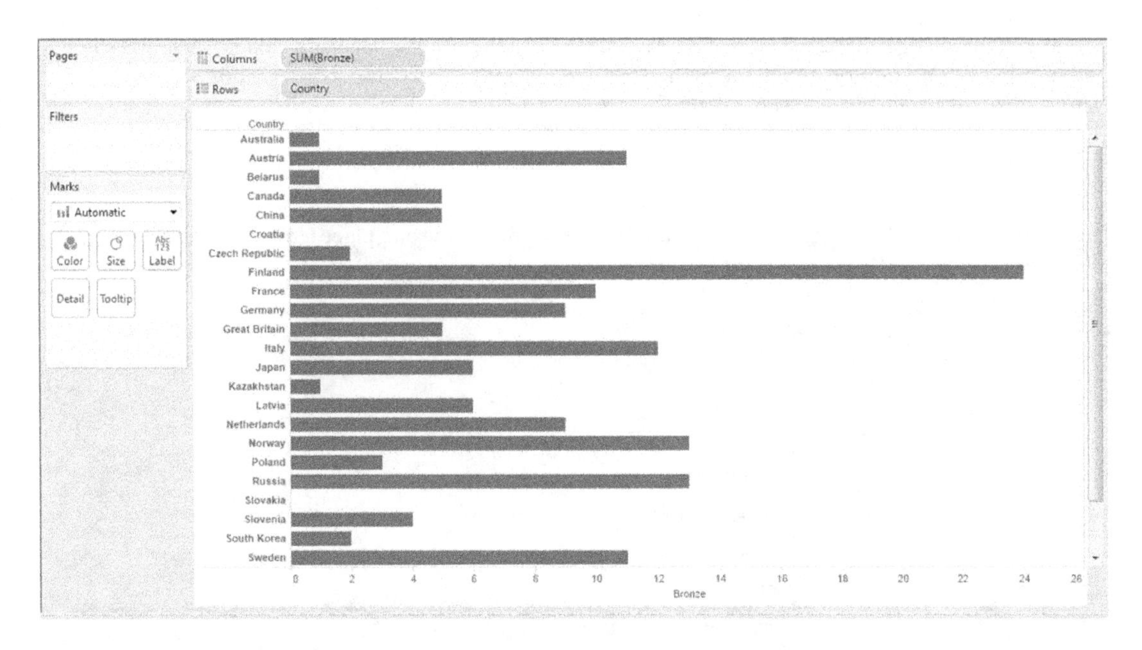

Figure 4-42. *Output after considering the "Special Values" for the measure bronze*

4.1.4.3.5 Step 3

Drag the measure "Silver" from the measures area under the data pane and place it on the same axis as the measure "Bronze" (Shown in Fig. 4-43).

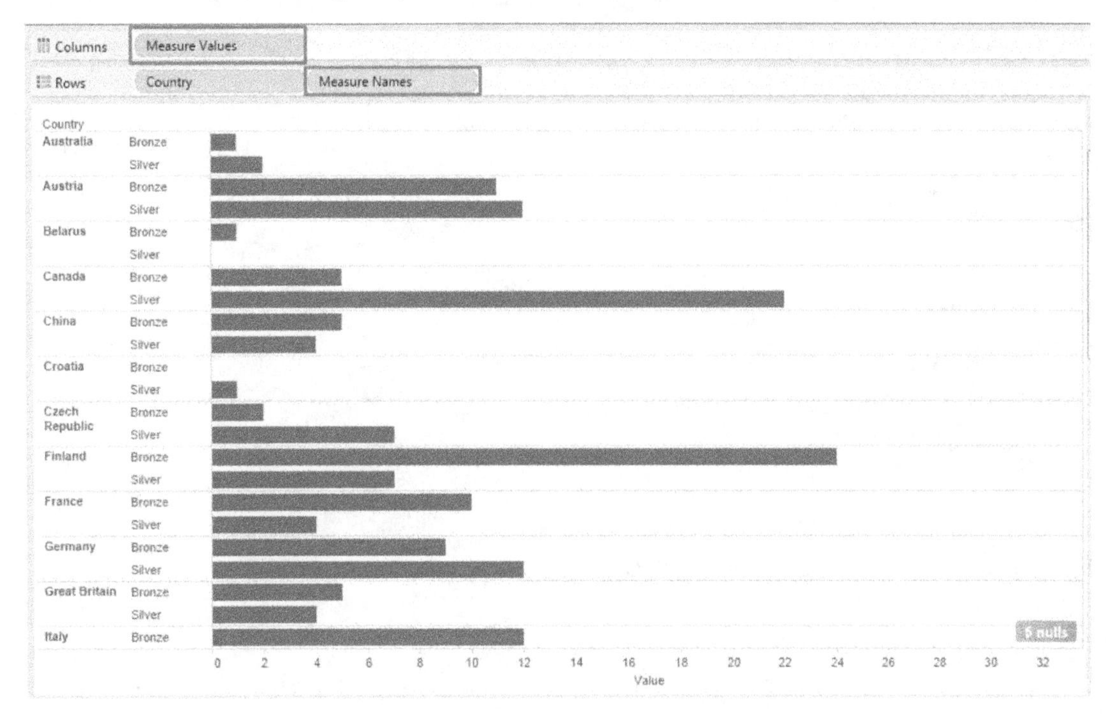

Figure 4-43. *Measure "Bronze" and "Silver" placed on the same axis*

Notice the change on the columns shelf. The columns shelf now has "Measure Values". The rows shelf has the dimension "Country" and to the right of the dimension "Country" is the dimension "Measure Names". The axis has changed to show "Value".

4.1.4.3.6 Step 4

Drag the measure "Gold" from the measures area under the data pane and place it on the same axis as the measures "Bronze" and "Silver" (Shown in Fig. 4-44).

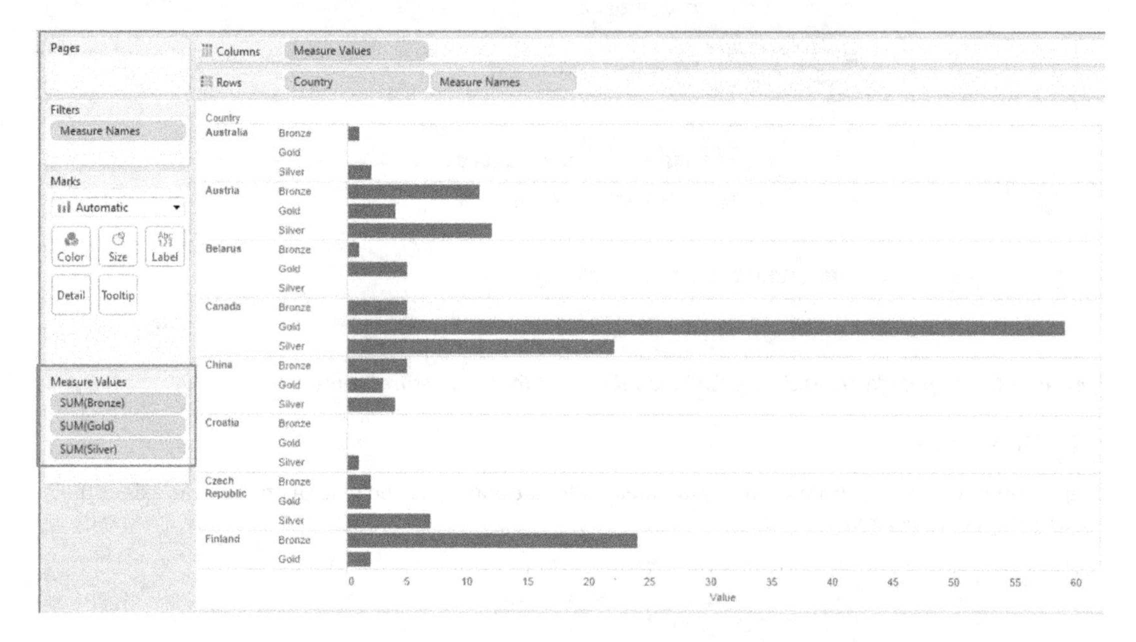

Figure 4-44. *Measures "Bronze", "Silver" and "Gold" placed on the same axis*

Observe the "Measure Values" just below the marks card. The order of the measures is as follows (See Fig. 4-45):

Measure Values
SUM(Bronze)
SUM(Gold)
SUM(Silver)

Figure 4-45. *Measure values below the marks card*

Let us change the order to SUM(Bronze), SUM(Silver) and then SUM(Gold). Drag SUM(Silver) and drop it above SUM(Gold) (See Fig. 4-46).

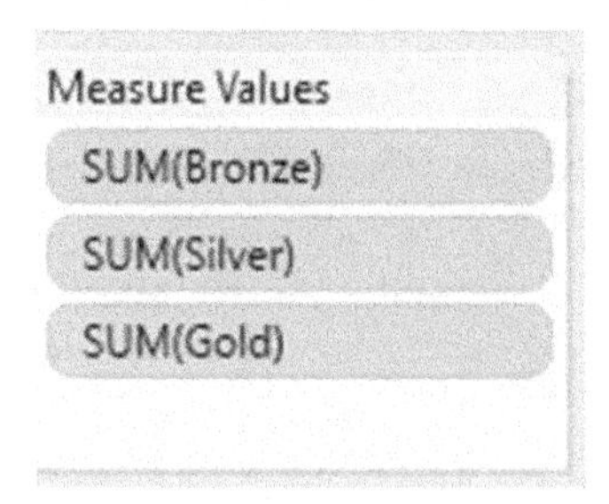

Figure 4-46. *Changed sequence of measure values*

The change reflected on the worksheet / view is as shown in Fig. 4-47.

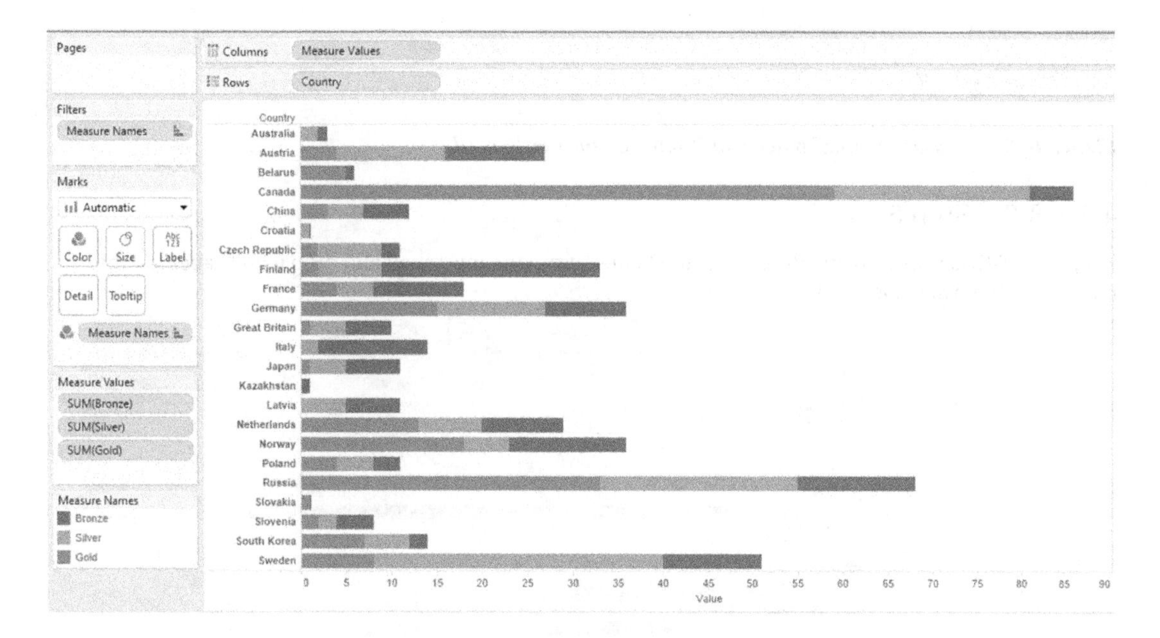

Figure 4-47. *The output after re-sequencing the measure values*

4.1.4.3.7 Step 5

Drag "Measure Names" from the columns shelf and drop it on "Color" on the marks card (Shown in Fig. 4-48).

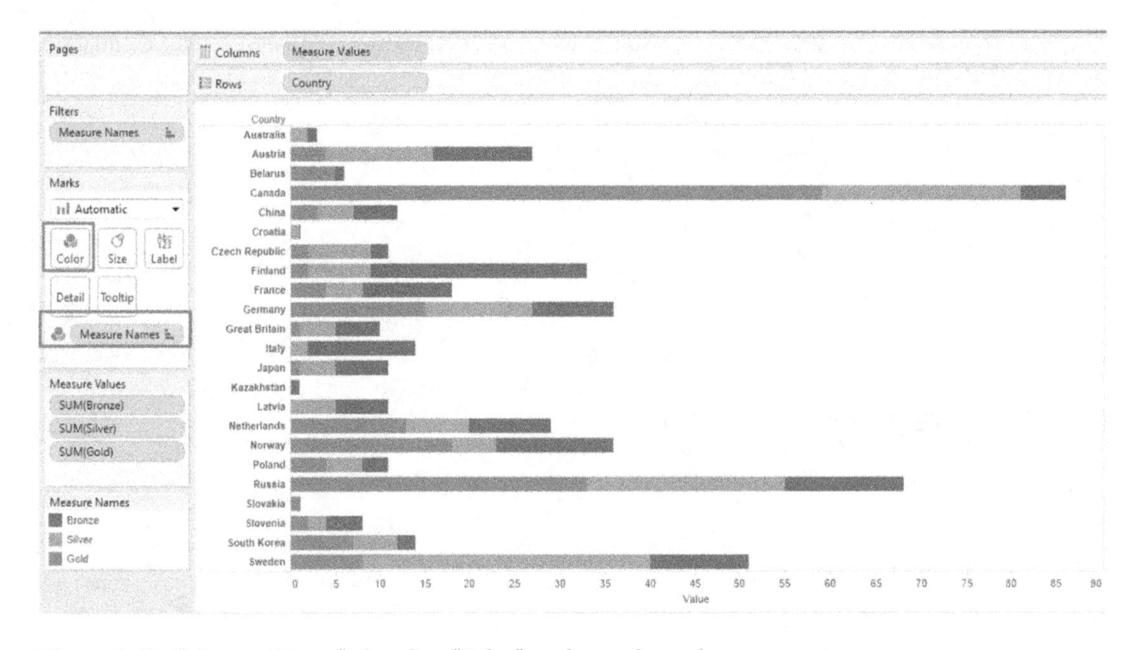

Figure 4-48. *"Measure Names" placed on "Color" on the marks card*

4.1.4.3.8 Step 6

Keep the CTRL key pressed and drag "Measure Values" from the rows shelf and drop it on "Label" on the marks card (Shown in Fig. 4-49).

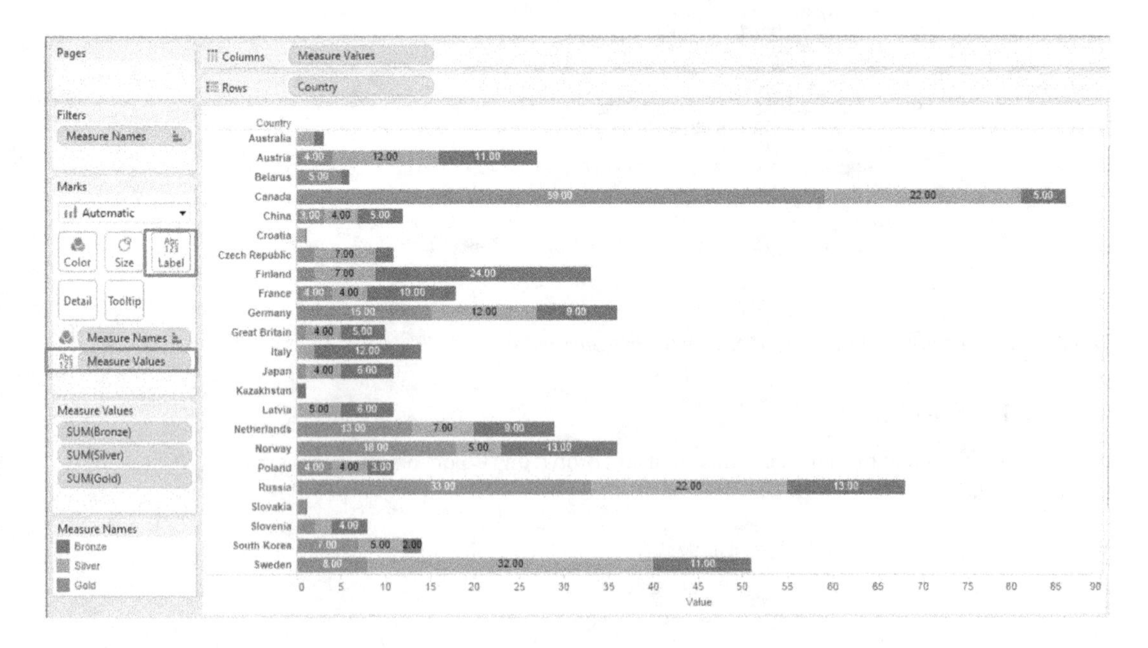

Figure 4-49. *"Measure Values" placed on "Label" on the marks card*

4.1.4.3.9 Step 7

Sort the dimension "Country" by the measure "Gold" in descending order. Right click on the dimension "Country". It brings up the drop down menu. Select "Sort" (Shown in Fig. 4-50).

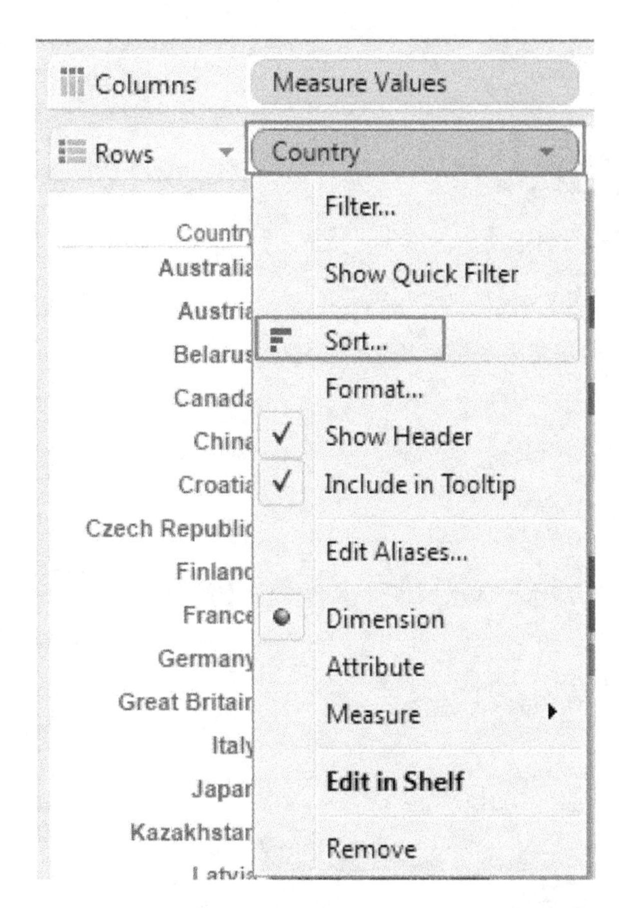

Figure 4-50. *Perform "Sort" on the dimension "Country"*

On selecting "Sort" the screen for "Sort [Country]" shows up. Provide the values as shown in Fig. 4-51.

Figure 4-51. *Perform "Sort" on the dimension "Country" in descending order of measure "Gold"*

Click "Apply" and then "OK".
The output of sort is as follows (Shown in Fig. 4-52).

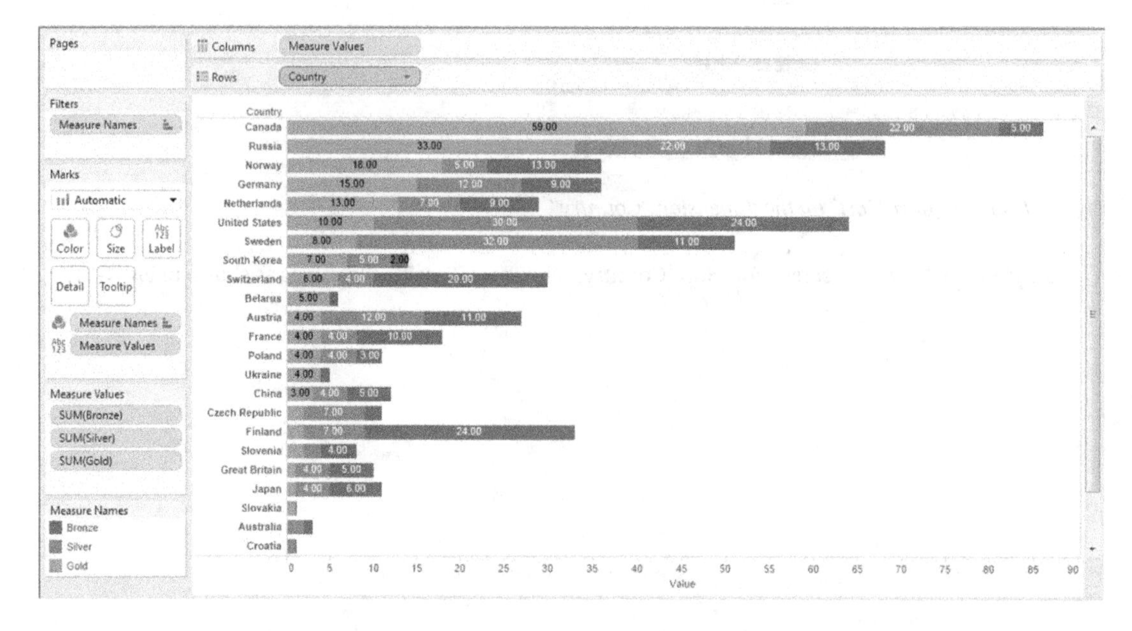

Figure 4-52. *Output after sorting the dimension "Country" in descending order of measure "Gold"*

4.1.4.3.10 Step 8

Now for some formatting.

Change the color of "Bronze", "Silver" and "Gold" bars. Change the number format for all the measure values (Shown in Fig. 4-53 and Fig. 4-54).

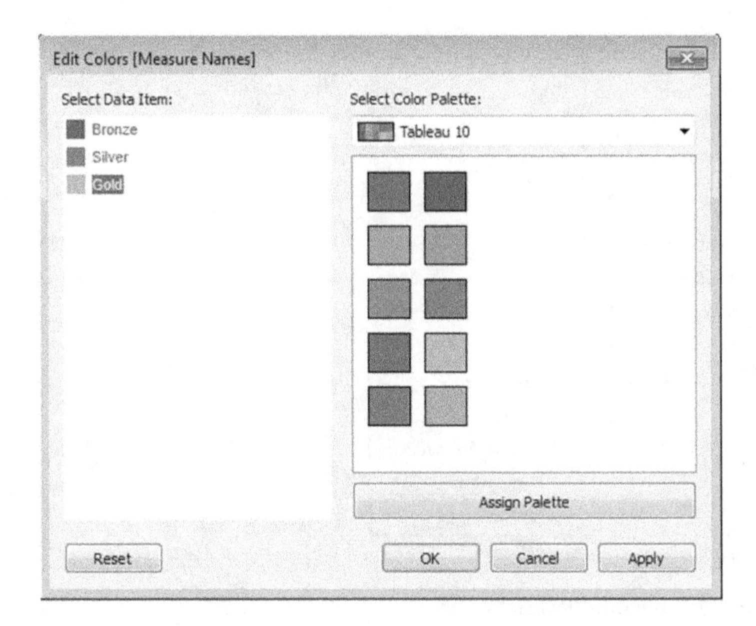

Figure 4-53. *Edit the colors for the measures*

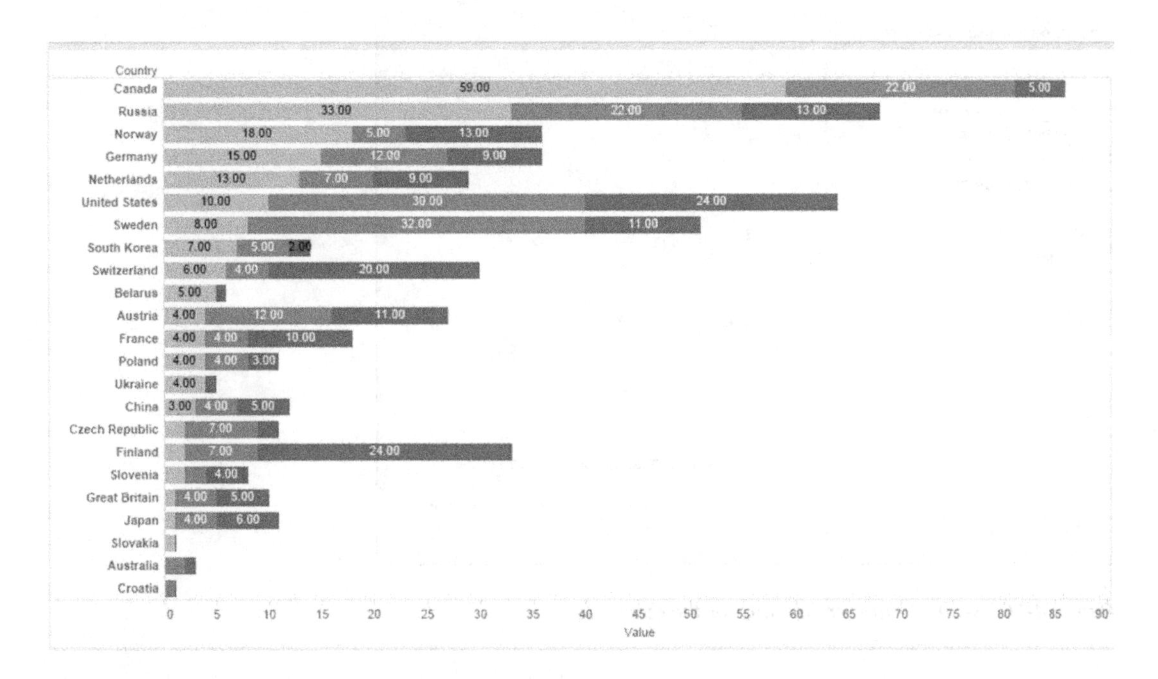

Figure 4-54. *Output after applying the chosen colors to the measures*

Select Number Format for the measure "Bronze". (Shown in Fig. 4-55). Select "Number (Custom)" and select "0" for "Decimal places" (Shown in Fig. 4-56).

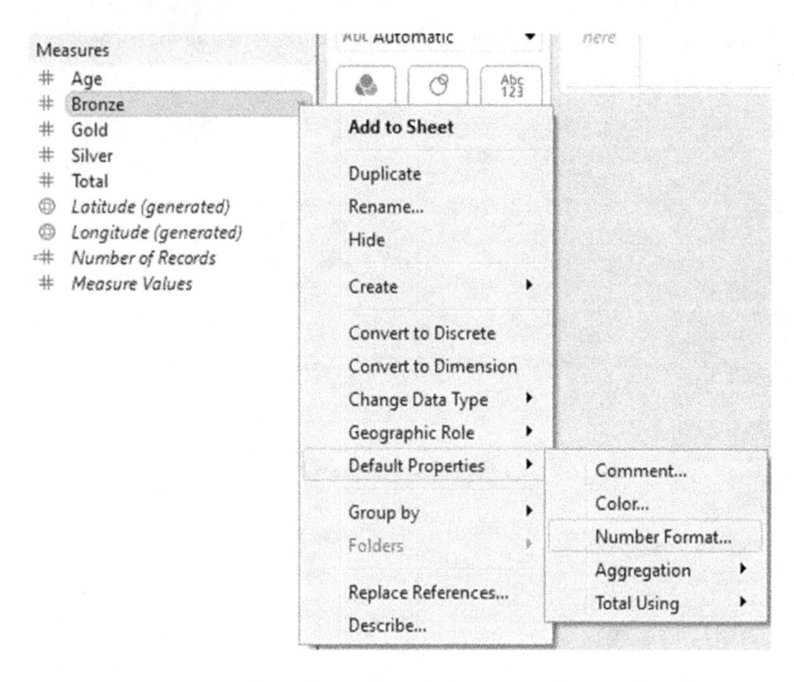

Figure 4-55. *Selecting "Number Format" for measure "Bronze"*

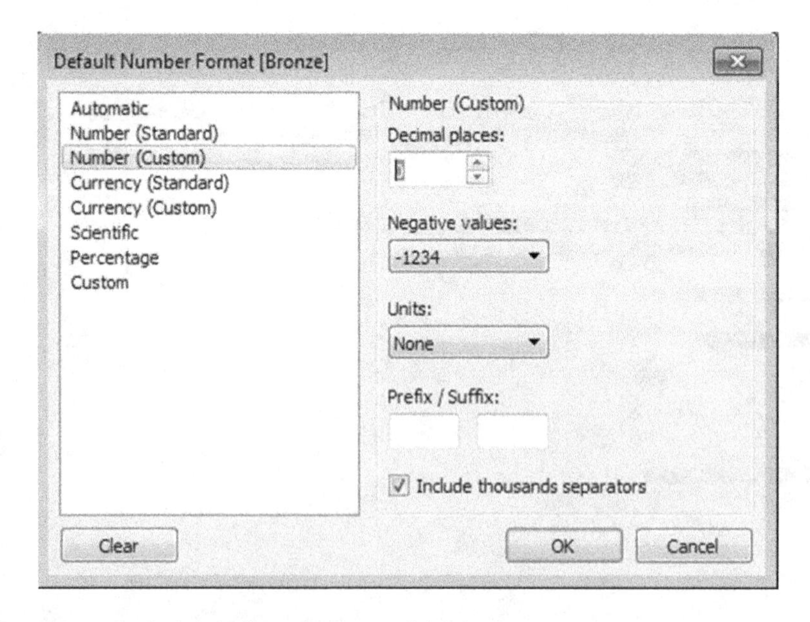

Figure 4-56. *Set the "Number Format" for the measures*

Likewise, change the number format for "Silver" and "Gold" measures as well. The final output is as shown in Fig. 4-57.

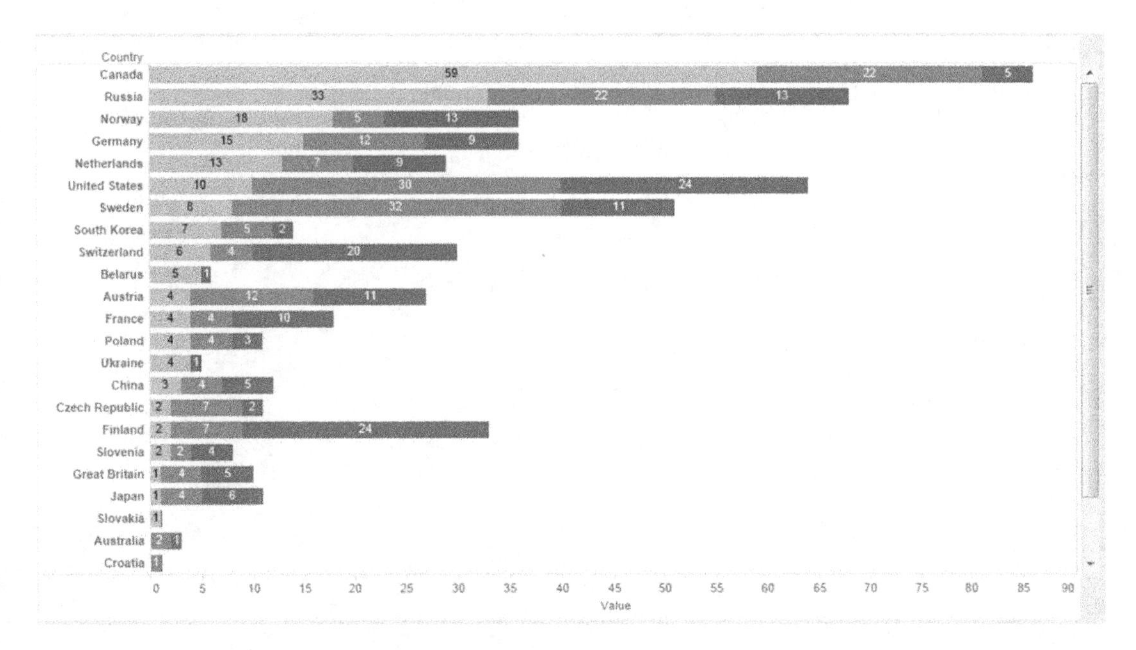

Figure 4-57. *Blended Measures - Demo 3 - Final Output*

From the output, it can be seen that "Canada", "Russia" and "Norway" are placed Nos. 1, 2 and 3 in their gold medal tallies, respectively.

4.1.4.4 Demo 4 (Slope Graph)

Objective: "XYZ" is an enterprise that has six units. Targets are set at the beginning of each year for each unit. At the end of the year the performance of each unit is evaluated. Given below is a data set showing the performance of the units in the year 2014 and 2015 (Table 4-6). Plot a graph to depict the performance of the units showcasing whether the performance has increased, decreased or remained constant/steady.

Input:

Table 4-6. *Blended measures - Demo 3 – data set*

	A	B	C
1	**Units**	**Dec-14**	**Dec-15**
2	Unit 1	78	69
3	Unit 2	82	84
4	Unit 3	65	71
5	Unit 4	70	70
6	Unit 5	73	71
7	Unit 6	65	65

4.1.4.4.1 Steps to create a slope graph

4.1.4.4.2 Step 1

Read in data from "Slope Graph.xls" into Tableau (Shown in Fig. 4-58).

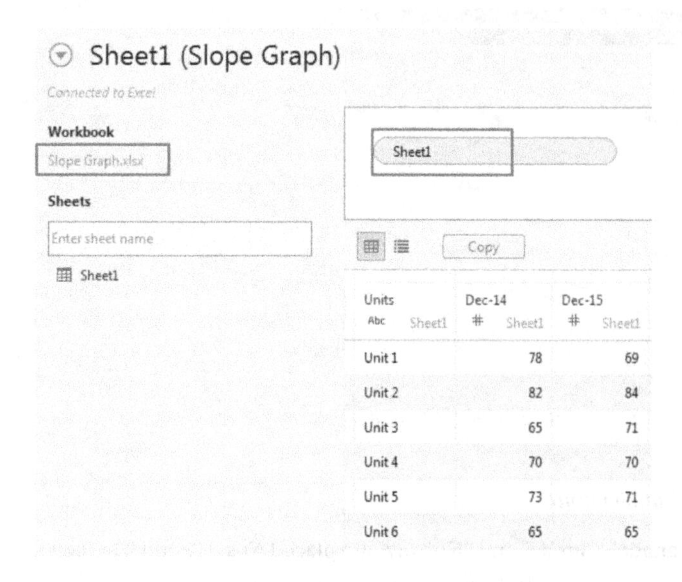

Figure 4-58. *Data from "Slope Graph.xls" read into Tableau*

4.1.4.4.3 Step 2

Drag the dimension "Measure Names" from the dimensions area under the data pane and place it on the columns shelf.

Drag the measure "Measure Values" from the measures area under the data pane and place it on the rows shelf (Shown in Fig. 4-59).

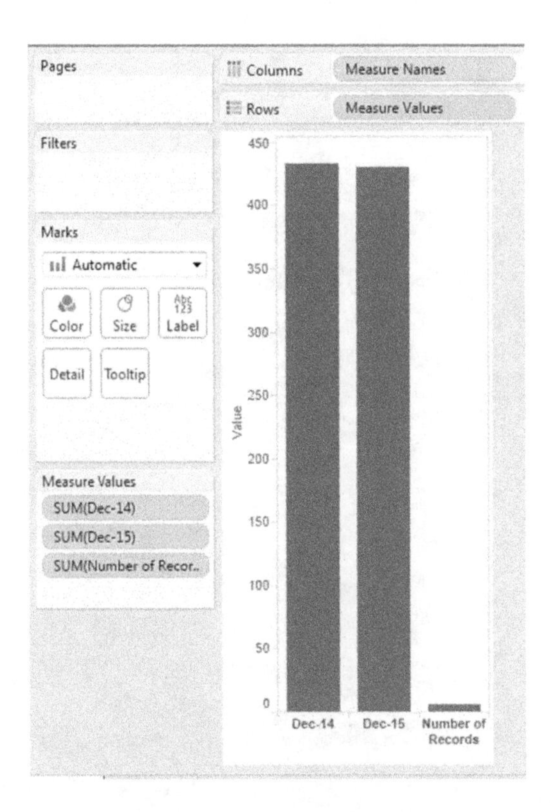

Figure 4-59. *"Measure Names" and "Measure Values" placed on the columns and rows shelf*

4.1.4.4.4 Step 3

Remove the measure "Number of Records" from the "Measure Values" shelf (Shown in Fig. 4-60).

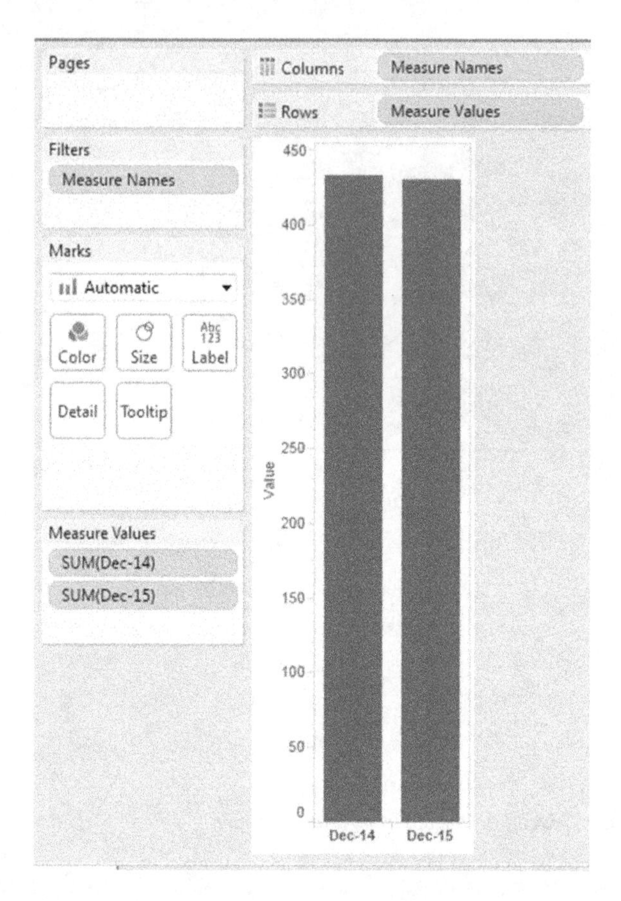

Figure 4-60. *Output after removing the measure "Number of Records" from the "Measure Values" shelf*

4.1.4.4.5 Step 4

Change the "Marks Type" to "Line" (Fig. 4-61).

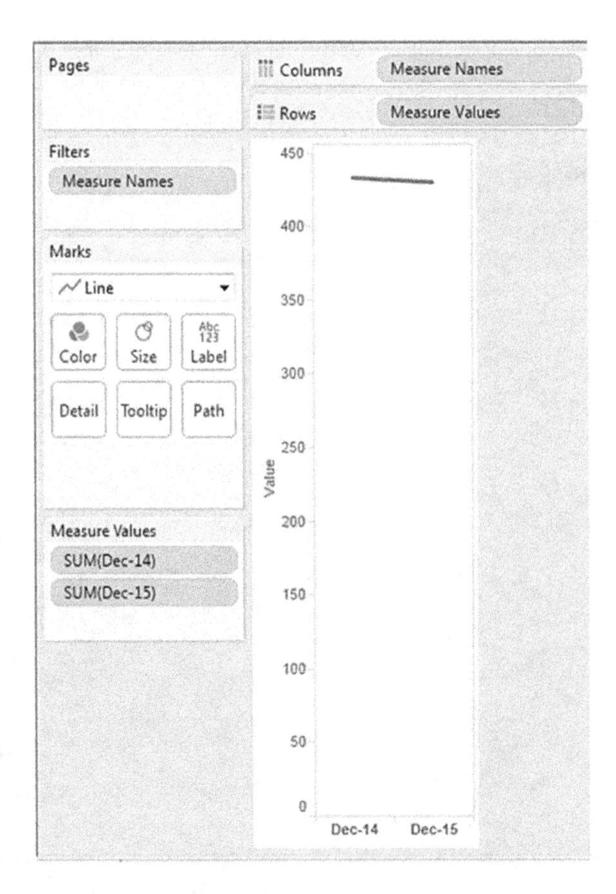

Figure 4-61. *"Marks Type" changed to "Line"*

4.1.4.4.6 Step 5

Create a calculated field "Performance" (Shown in Fig. 4-62).

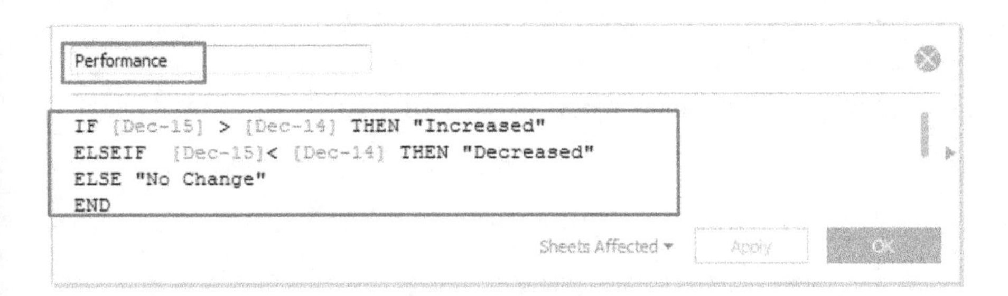

Figure 4-62. *Calculated field "Performance" created*

Drag the calculated field "Performance" from the dimensions area under the data pane to "Color" on the marks card (Shown in Fig. 4-63).

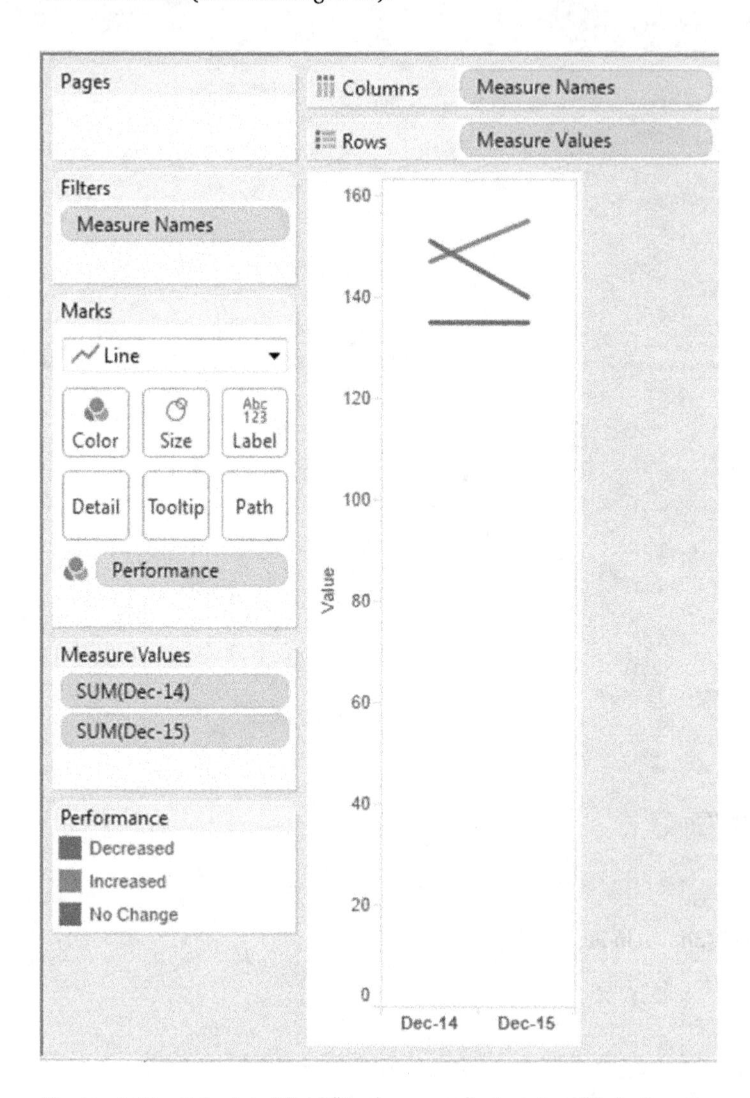

Figure 4-63. *Calculated field "Performance" placed on "Color" on the marks card*

4.1.4.4.7 Step 6

Drag the dimension "Units" from the dimensions area under the data pane and place it on "Detail" on the marks card (Shown in Fig. 4-64).

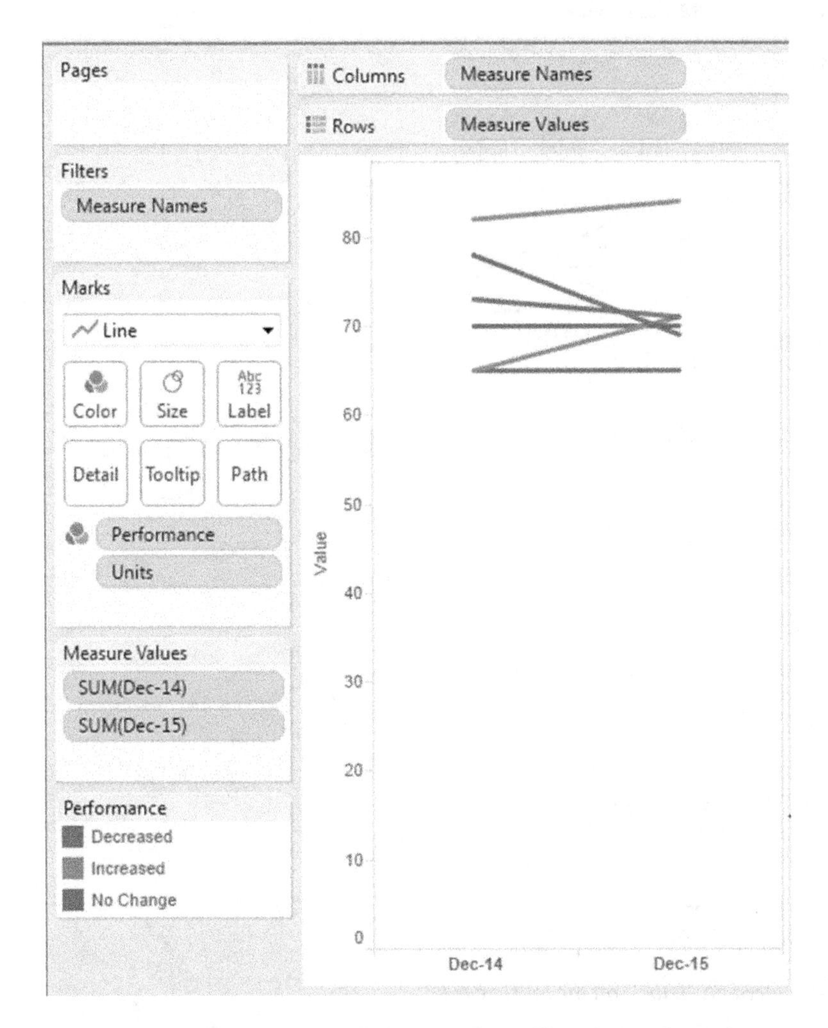

***Figure 4-64.** Dimension "Units" placed on "Detail" on the marks card*

Drag the dimension "Units" from the dimensions area under the data pane and place it on "Label" on the marks card (Shown in Fig. 4-65).

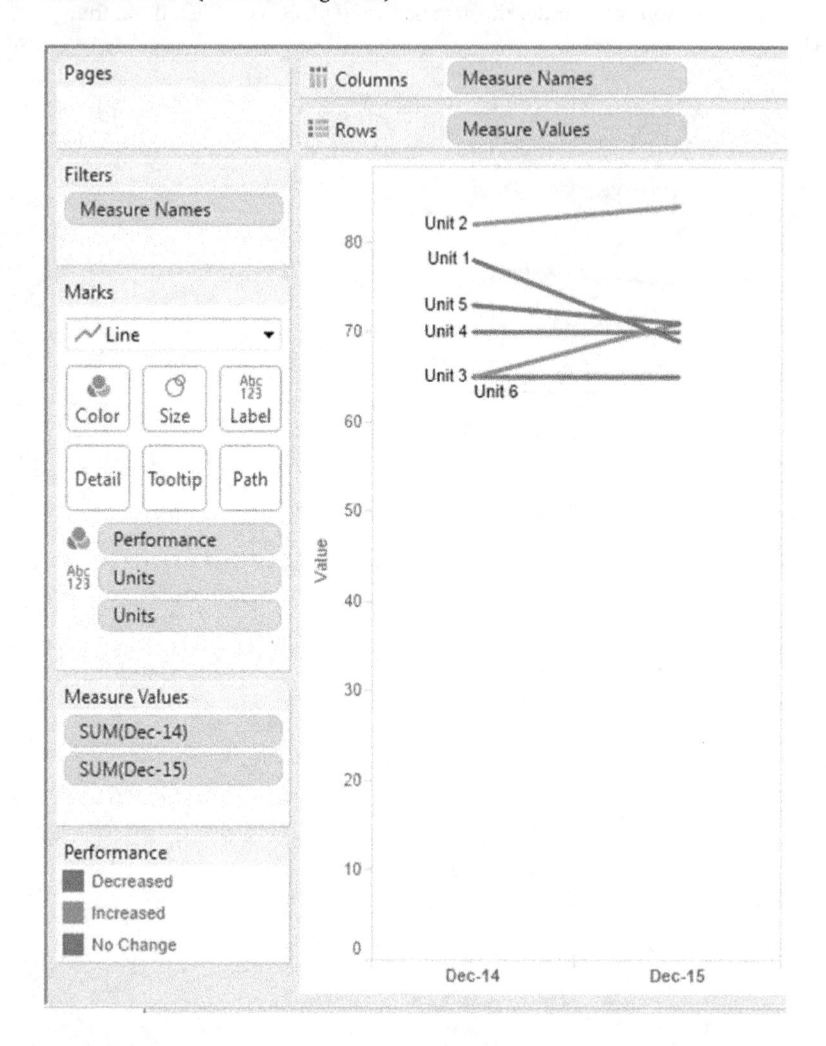

Figure 4-65. *Dimension "Units" placed on "Label" on the marks card*

We can conclude the following from the above figure:

- Performance has shown an increase for Unit 2 and Unit 3
- Performance has shown a decrease for Unit 1 and Unit 5
- Performance has remained steady for Unit 4 and Unit 6

4.1.4.4.8 Assignment 1

You are a student at a post-graduate college. In order to know clearly the subject that you should improve upon, you plot a graph to decipher your performance in the various subjects over the four years spent in a graduate school. Plot a slope graph highlighting the performance of the first and fourth year in graduate school.

Input: "Slope Graph - Assignment.xls"
Expected Output: Shown in Fig. 4-66.

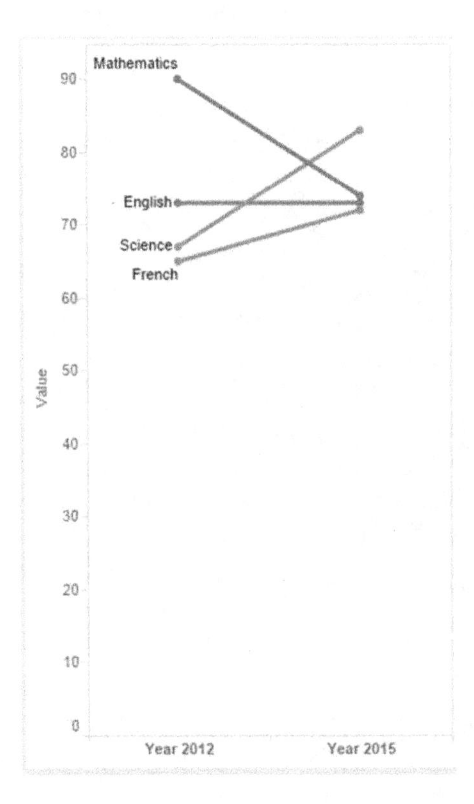

Figure 4-66. *Slope graph - Assignment 1 – expected output*

4.1.5 Dual axis

The previous section familiarized you with blending the measures and placing it on a common axis. However, what if you are required to have a secondary axis in our view.

4.1.5.1 Demo 1

Objective: As a senior executive in a firm, you would like to strengthen the firm's marketing strategies for its products and services. An understanding of how profit fares as the sales are made over the year will hold you in good stead as you pull up innovative marketing strategies. Plot "Sales" and "Profits" in such a way that it helps your understanding.

Input: "Sample - Superstore.xls".

Columns shelf	Month(Order Date) Date is "Discrete" as evident from the visual cue. It is blue in color. The preferred chart form to depict discrete dates is bar chart.
Rows shelf	Sales, Profit. The aggregation used on both measures is SUM.

Expected output: Shown in Fig. 4-67.

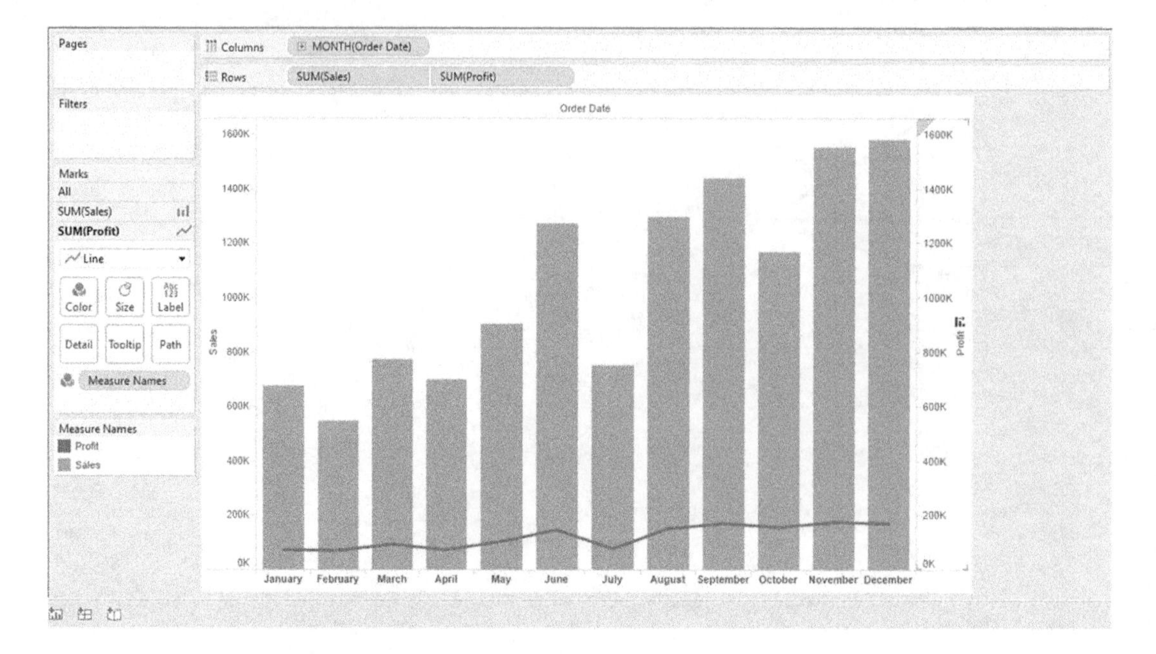

Figure 4-67. Dual Axis - Demo 1 - expected output

4.1.5.1.1 Steps to demonstrate dual axis chart

4.1.5.1.2 Step 1

Drag “Order Date” from dimensions area under the data pane and place it on the columns shelf. Dates are always displayed as hierarchy. By default, when we drag “Order Date” and place it either on the rows shelf or columns shelf, it is “Discrete” (this is evident from the visual cue) (Shown in Fig. 4-68).

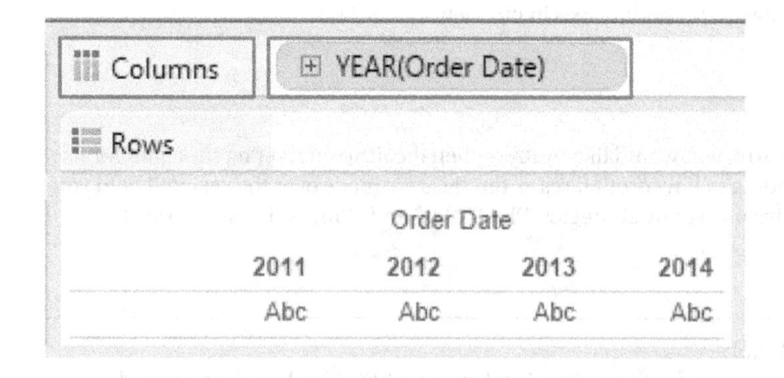

Figure 4-68. Dimension “Order Date” placed on the columns shelf

Right click on Year(Order Date). A drop down in displayed. Select "Month" (Shown in Fig. 4-69).

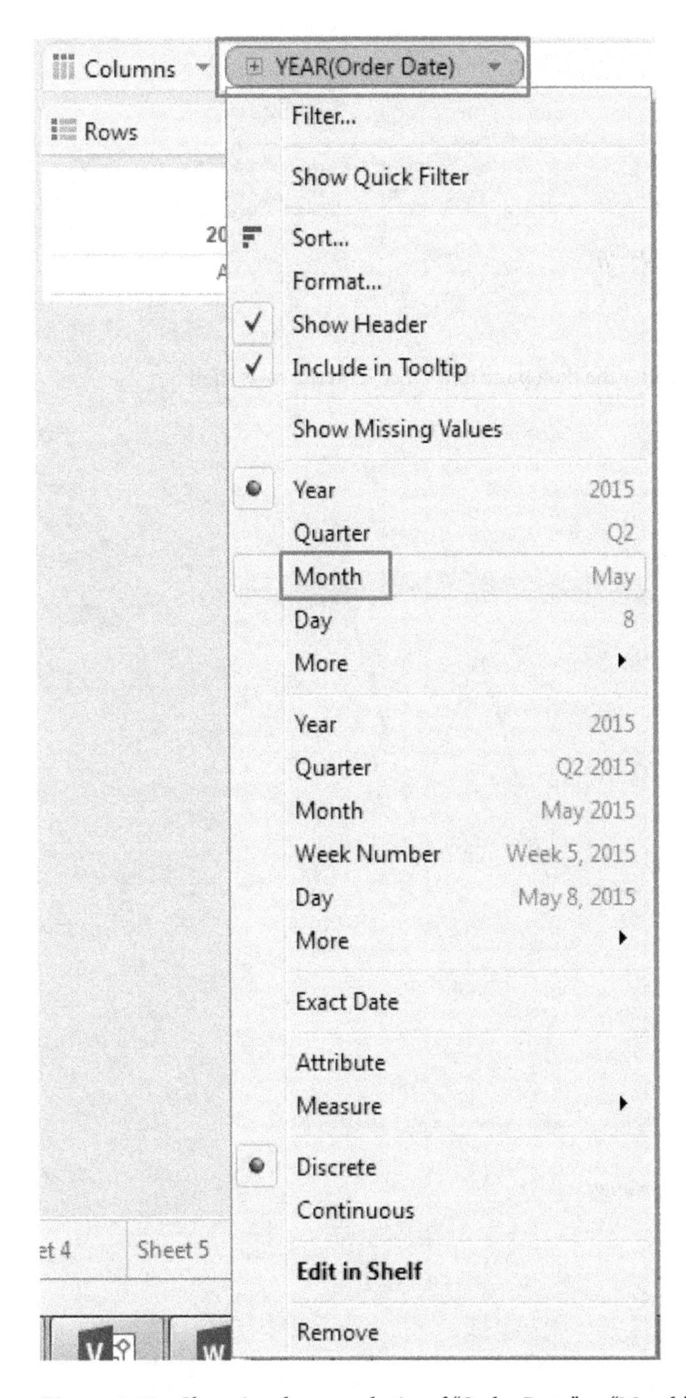

Figure 4-69. *Changing the granularity of "Order Date" to "Month"*

Selecting "Month"(Order Date) will change the display as shown in Fig. 4-70.

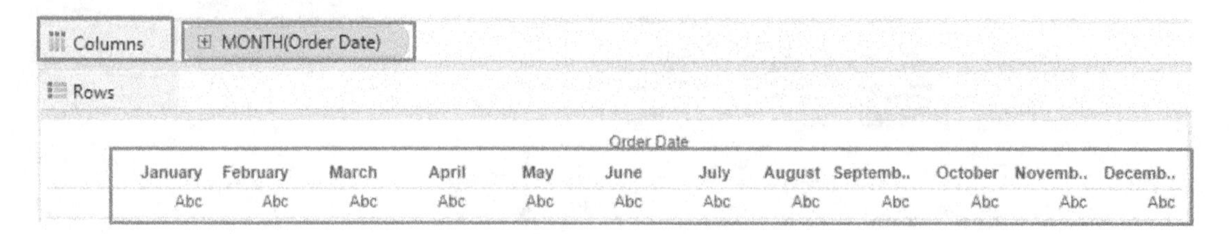

***Figure 4-70.** "Order Date" granularity set to "Month"*

4.1.5.1.3 Step 2

Drag "Sales" from under the measures area under the data pane and place it on the rows shelf (Shown in Fig. 4-71).

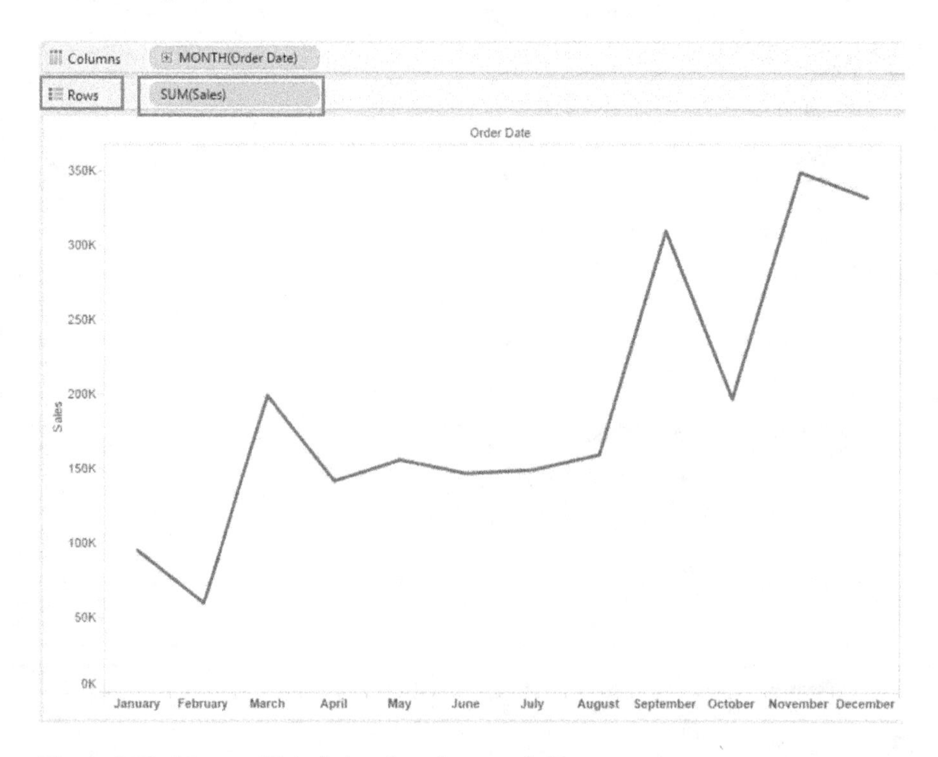

***Figure 4-71.** Measure "Sales" placed on the rows shelf*

Change the chart form to "Bar" in the marks card (Shown in Fig. 4-72).

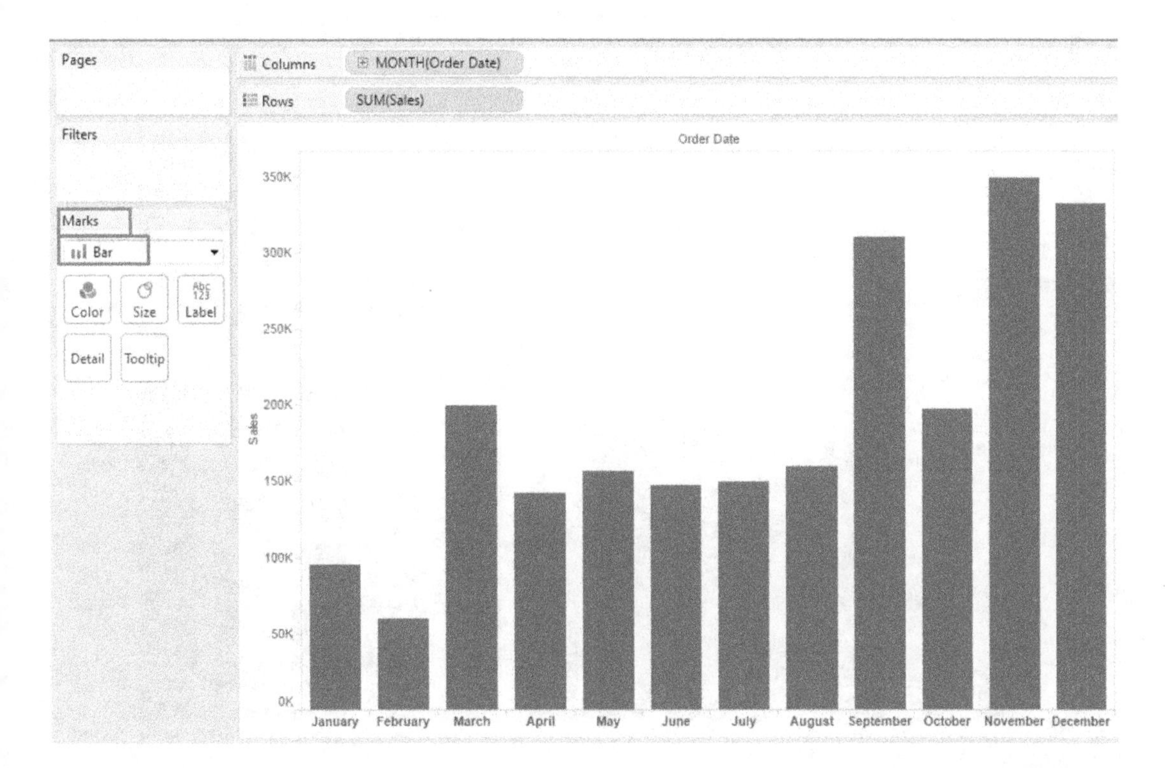

Figure 4-72. *"Marks Type" changed to "Bar"*

4.1.5.1.4 Step 3

Drag "Profit" from under the measures area under the data pane and place it on the opposite axis (the axis opposite to the one on which the "Sales" measure is placed) (Shown in Fig. 4-73).

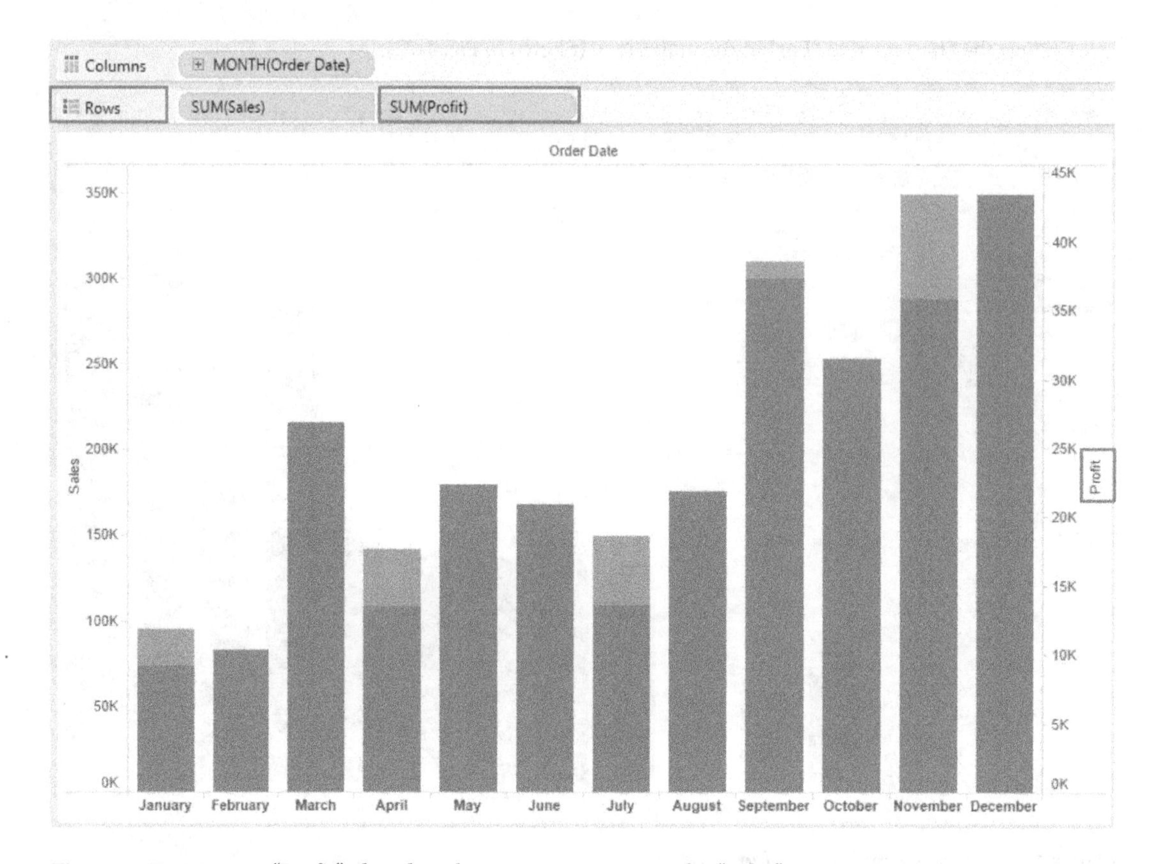

Figure 4-73. *Measure "Profit" placed on the axis opposite to axis for "Sales"*

4.1.5.1.5 Step 4

Synchronize the secondary axis (the axis on which "Profit" measure is placed) (Shown in Fig. 4-74).

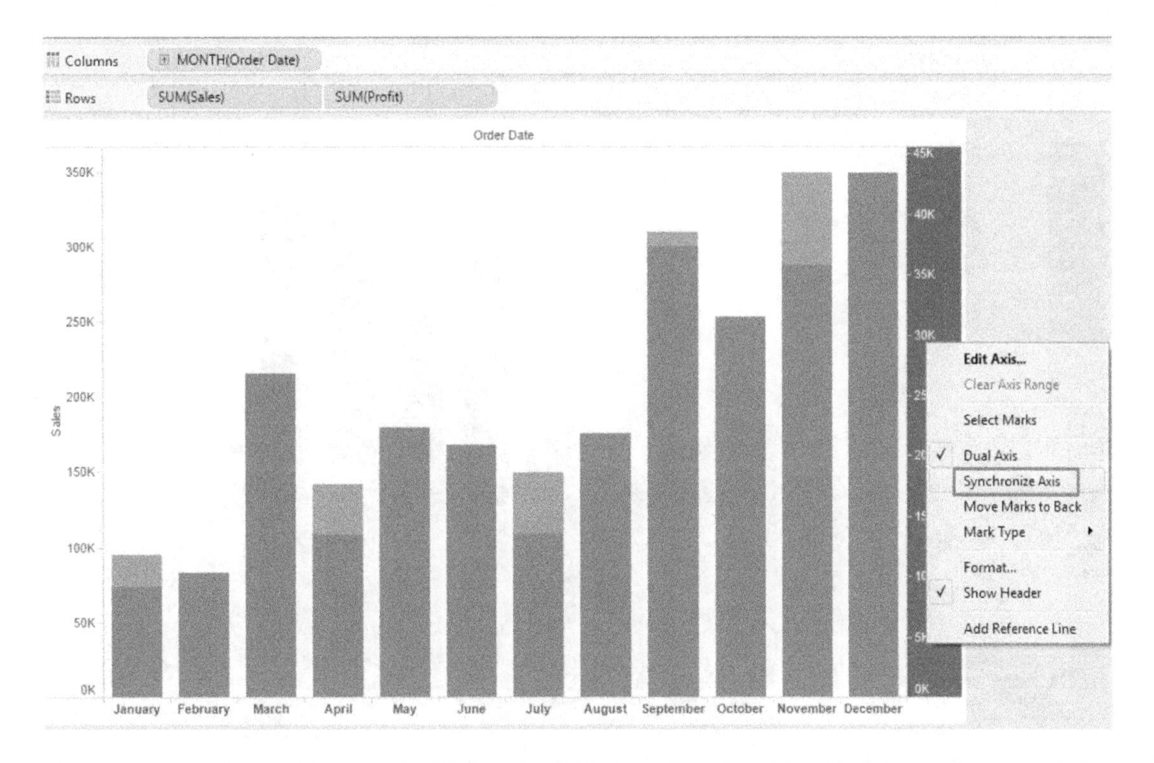

Figure 4-74. *Synchronize the secondary axis to the primary axis*

Note It is always the "Secondary Axis" that can be synchronized with the "Primary Axis". If you select the axis on which the "Sales" measure is placed and right click to show the context menu, you will notice that the "Synchronize Axis" feature is disabled (Shown in Fig. 4-75).

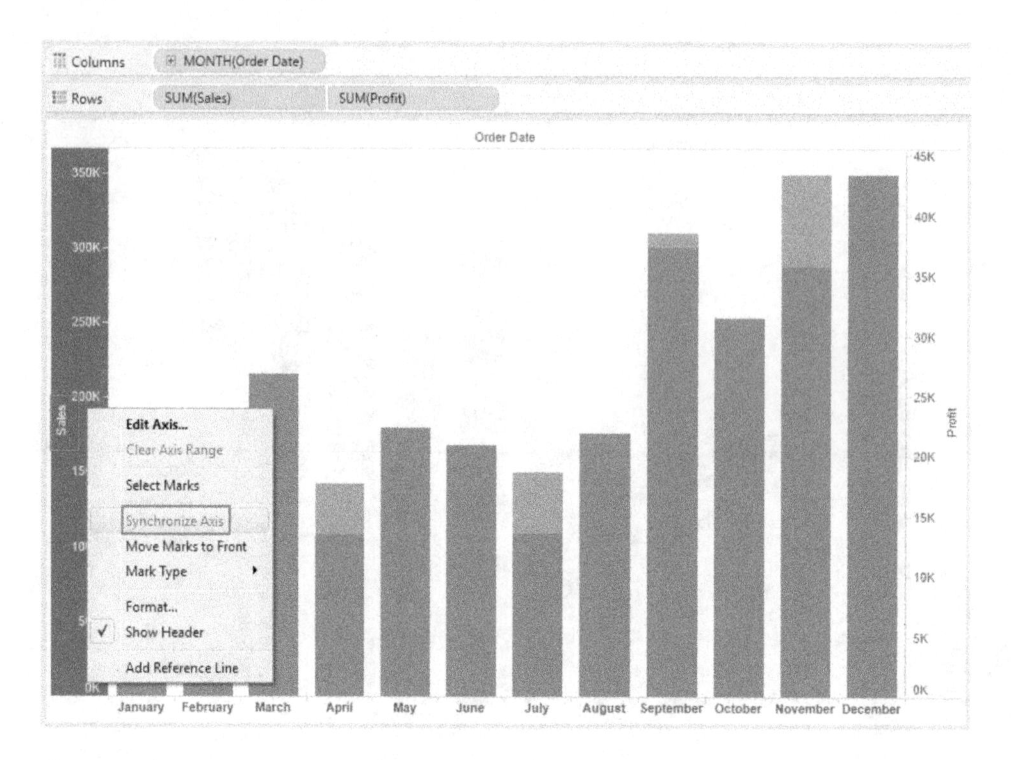

Figure 4-75. *Primary axis cannot be synchronized with the secondary axis*

Let us look at the output of synchronizing the "Profit Axis" with the "Sales Axis" (See Fig. 4-76).

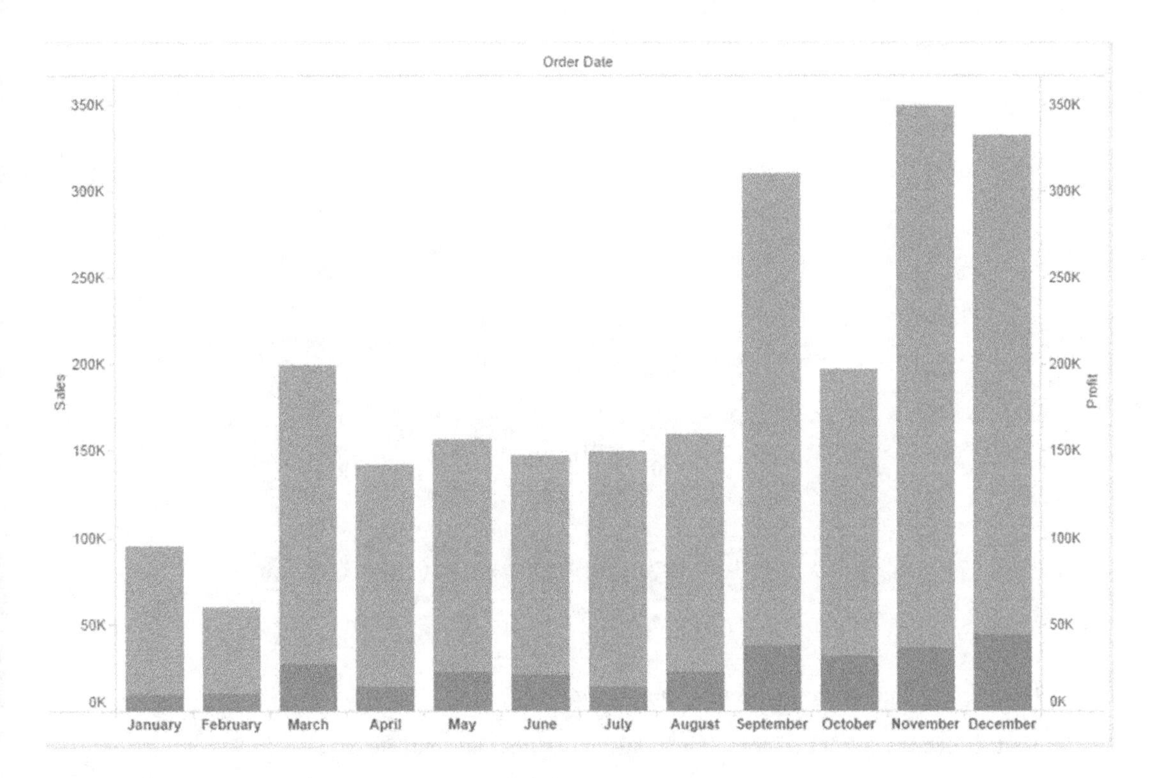

Figure 4-76. *"Profit Axis" synchronised with the "Sales Axis"*

Change the chart form for "Profit" to "Line" (Shown in Fig. 4-77).

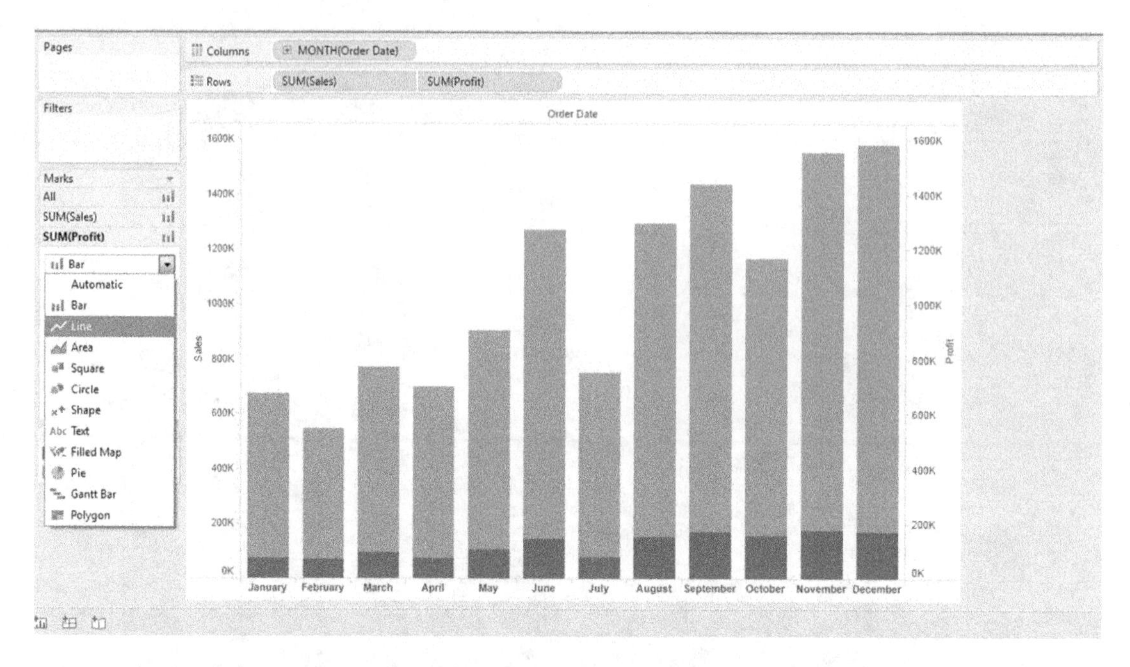

Figure 4-77. *Select "Line" as the marks type for the measure "Profit"*

The output of choosing "Line" graph for "Profit" measure (Shown in Fig. 4-78).

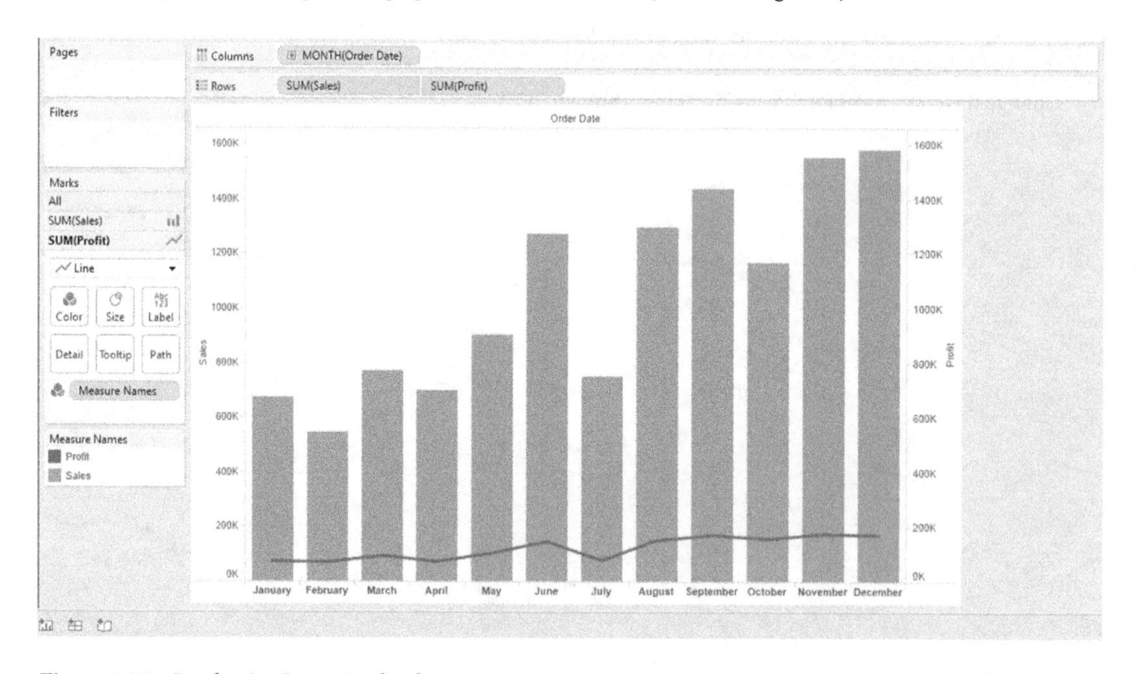

Figure 4-78. *Dual axis - Demo 1 – final output*

Notice a dip in the profit for the month of July.

4.1.5.2 Demo 2

Objective: "ABC" is a retail store that sells many sub categories of products, such as "Phones", "Appliances", "Furnishings", "Paper", and "Art", etc. As a senior analyst, you would like to know which subcategories account for the top 50% of the sales amount. Create a visualization such that it helps your understanding.

Input: "Sample Superstore.xls"

Expected Output: Shown in Fig. 4-79.

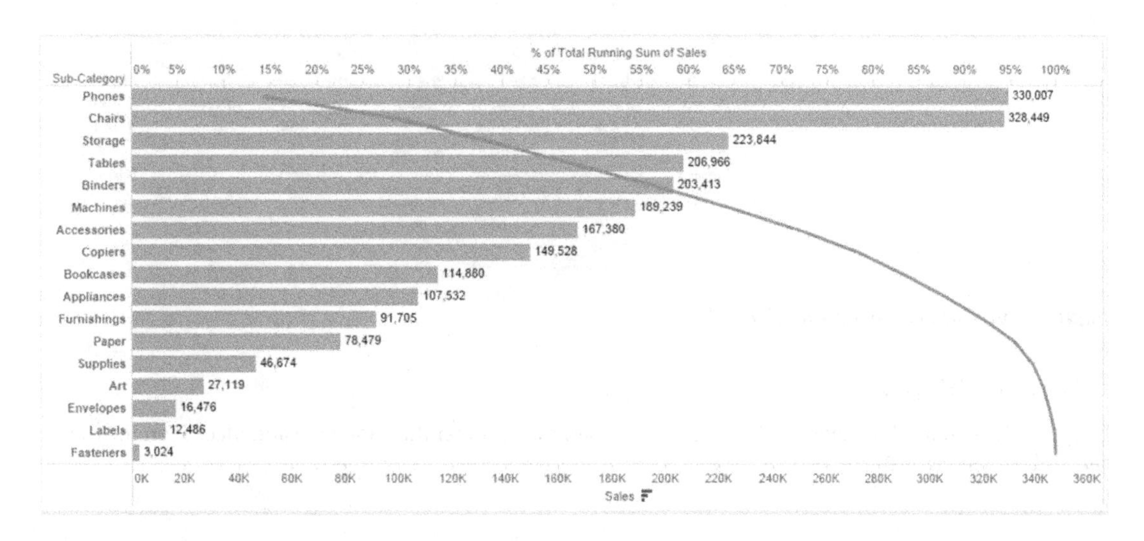

Figure 4-79. *Dual axis - Demo 2 - expected output*

4.1.5.2.1 Steps to create dual axis chart

4.1.5.2.2 Step 1

Read in the data from "Sample - Superstore.xls" into Tableau (Shown in Fig. 4-80).

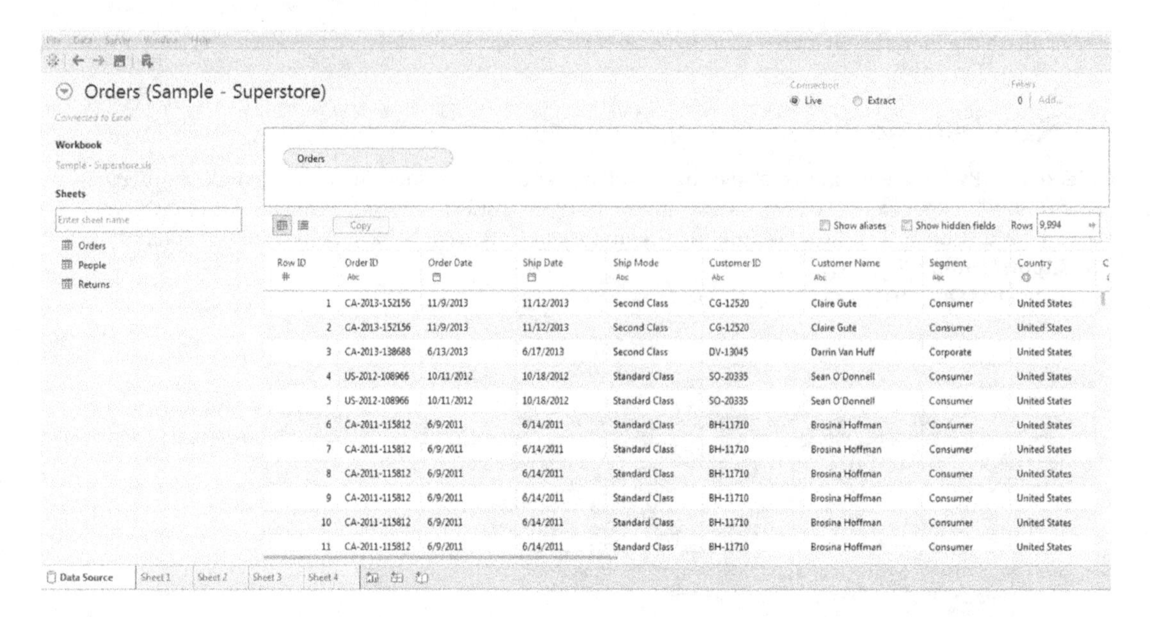

Figure 4-80. *Data for Demo 2 read into Tableau*

4.1.5.2.3 Step 2

Drag the dimension "Sub-Category" from the dimensions area under the data pane and place it on the rows shelf (Shown in Fig. 4-81).

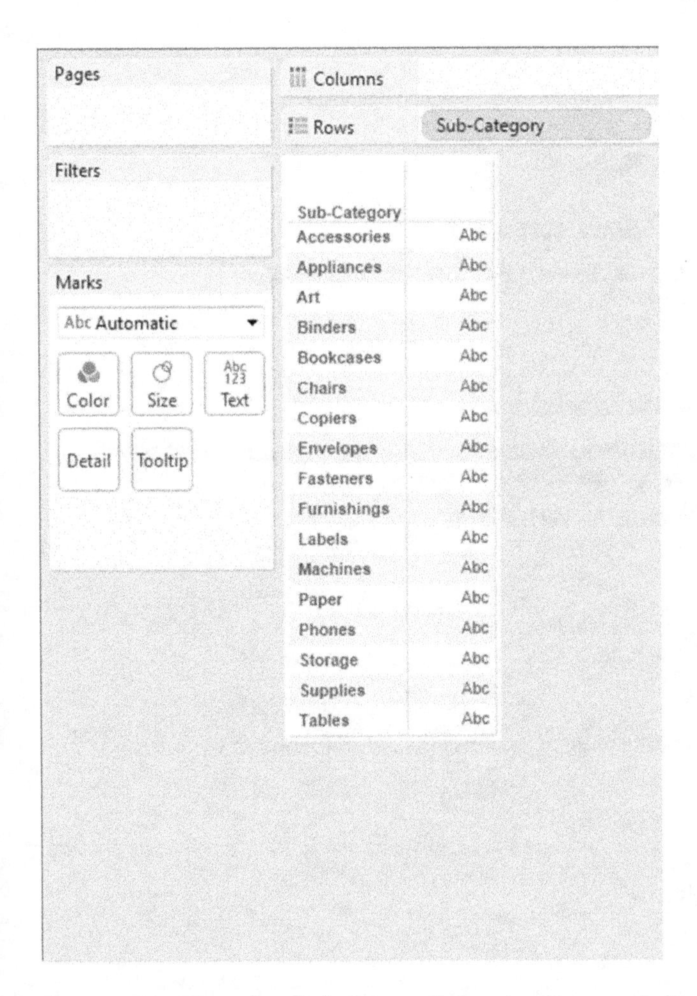

Figure 4-81. *Dimension "Sub-Category" placed on the rows shelf*

4.1.5.2.4 Step 3

Drag the measure "Sales" from the measures area under the data pane and place it on the columns shelf (Shown in Fig. 4-82).

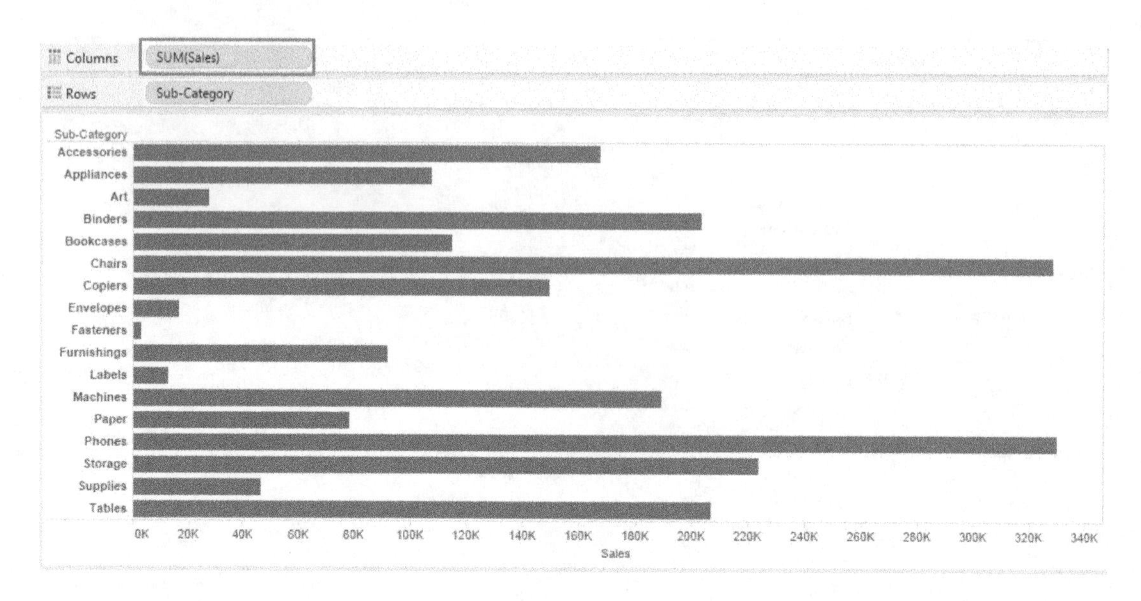

Figure 4-82. *Measure "Sales" placed on the columns shelf*

4.1.5.2.5 Step 4

Sort the dimension "Sub-Category" in descending order of the measure "Sales" (Shown in Fig. 4-83).

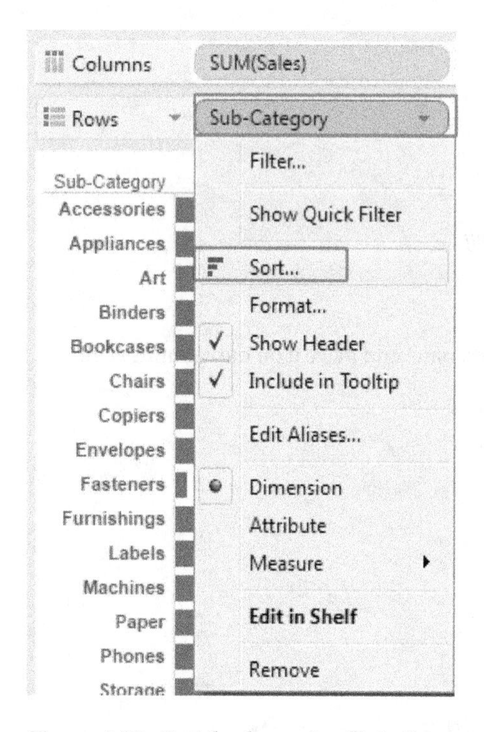

Figure 4-83. *Sort the dimension "Sub-Category"*

Fill in the values in the "Sort Dialog box" as shown in Fig. 4-84.

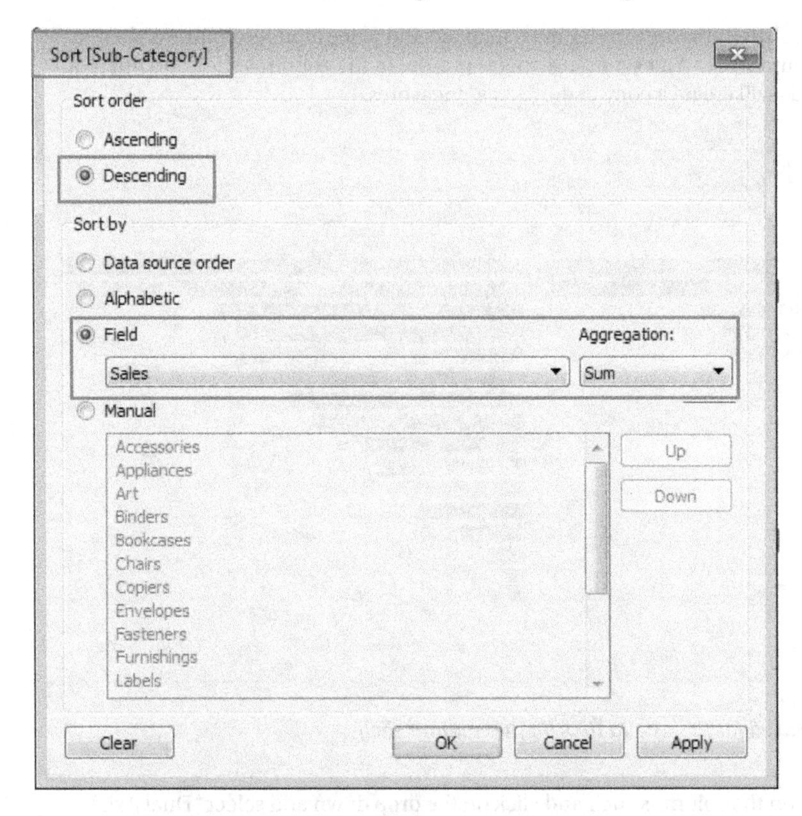

***Figure 4-84.** Fill in the "Sort" dialog box for the dimension "Sub-Category"*

The sorted output is as shown in Fig. 4-85.

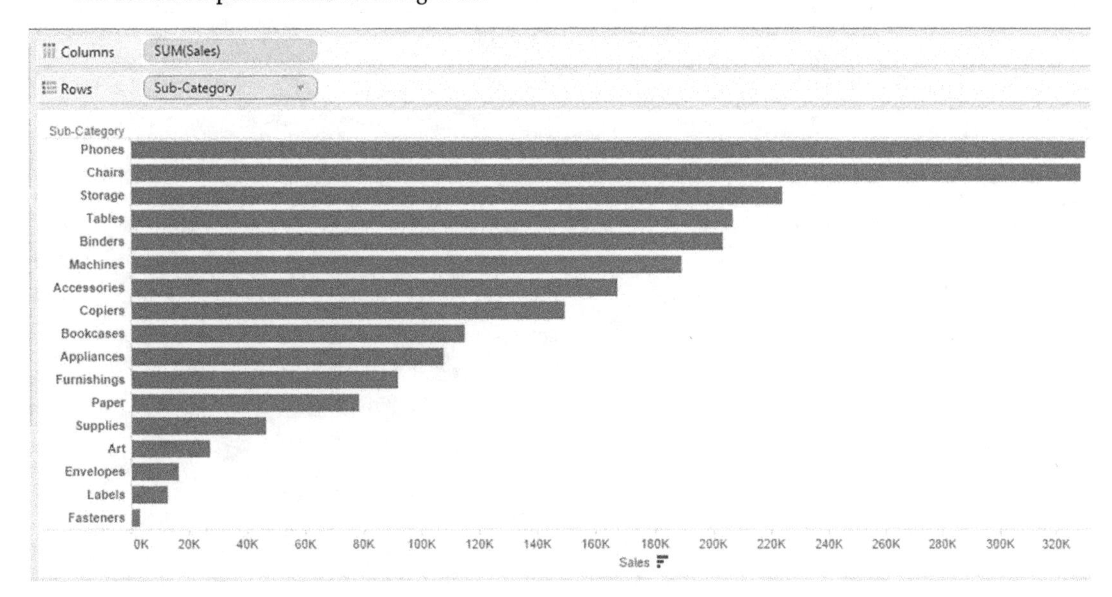

***Figure 4-85.** Dimension "Sub-Category" sorted in descending order of measure "Sales"*

4.1.5.2.6 Step 5

Drag "Sales" once again from the measures area under the data pane and place it on the columns shelf (Shown in Fig. 4-86) or press Control key + Sales measure that is present in the columns shelf and put it on the right side of SUM(Sales). This will create a copy of the "Sales" measures.

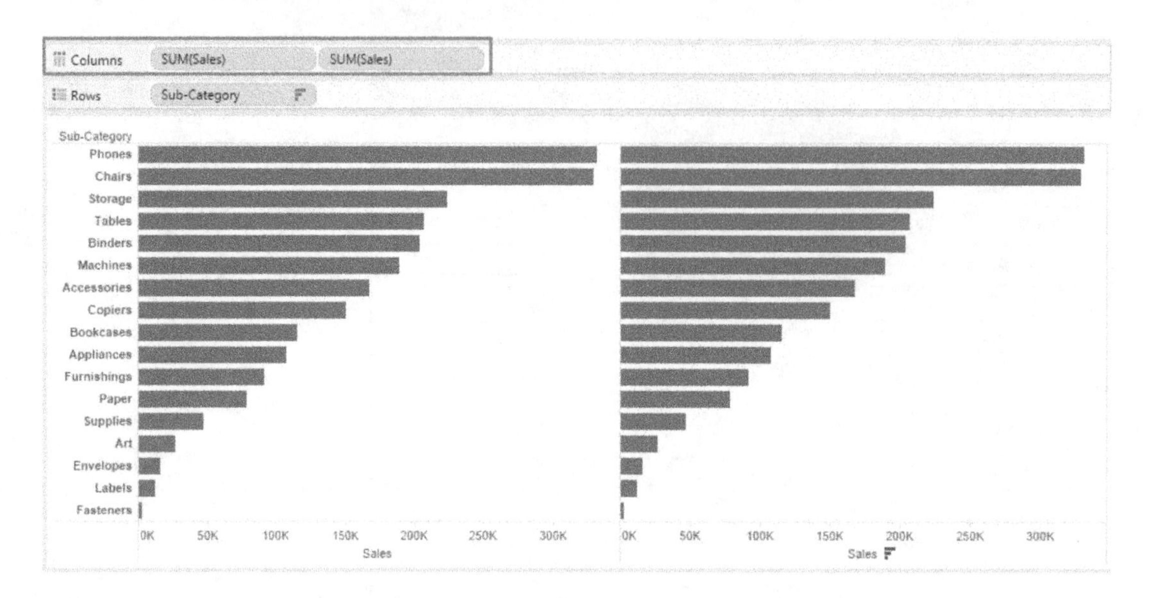

Figure 4-86. *Measure "Sales" placed for the second time on the columns shelf*

Select the second measure on the columns shelf and click on the drop down and select "Dual Axis" (Shown in Fig. 4-87 and Fig. 4-88).

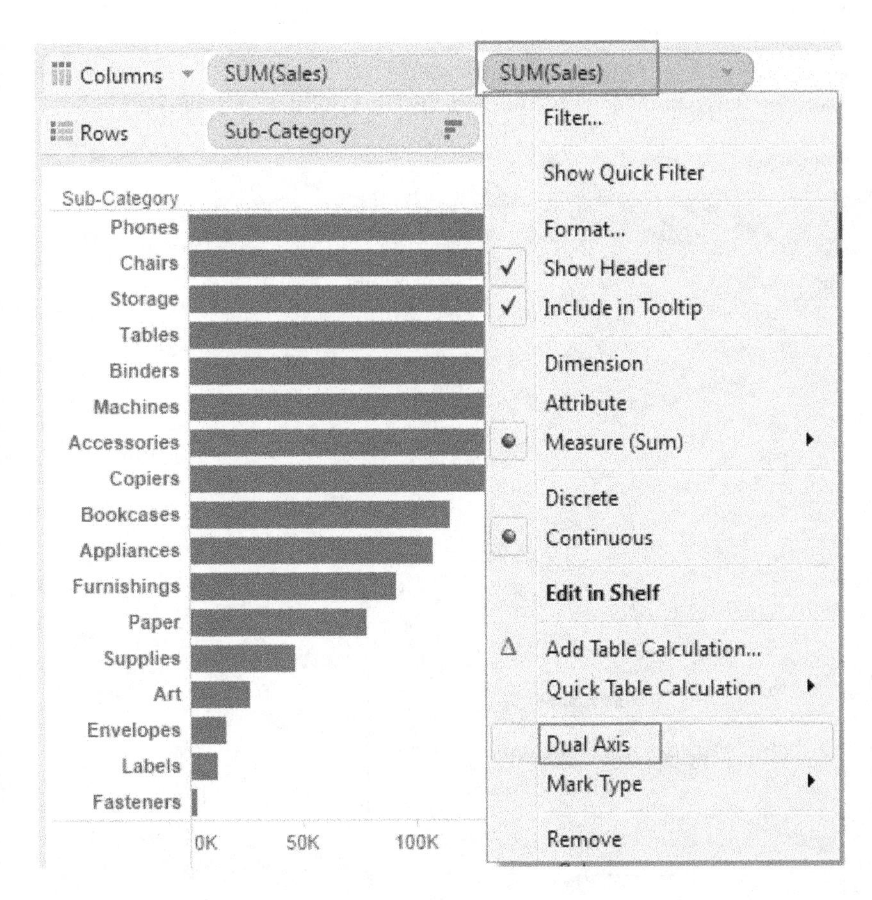

Figure 4-87. Set the second measure "Sales" to "Dual Axis"

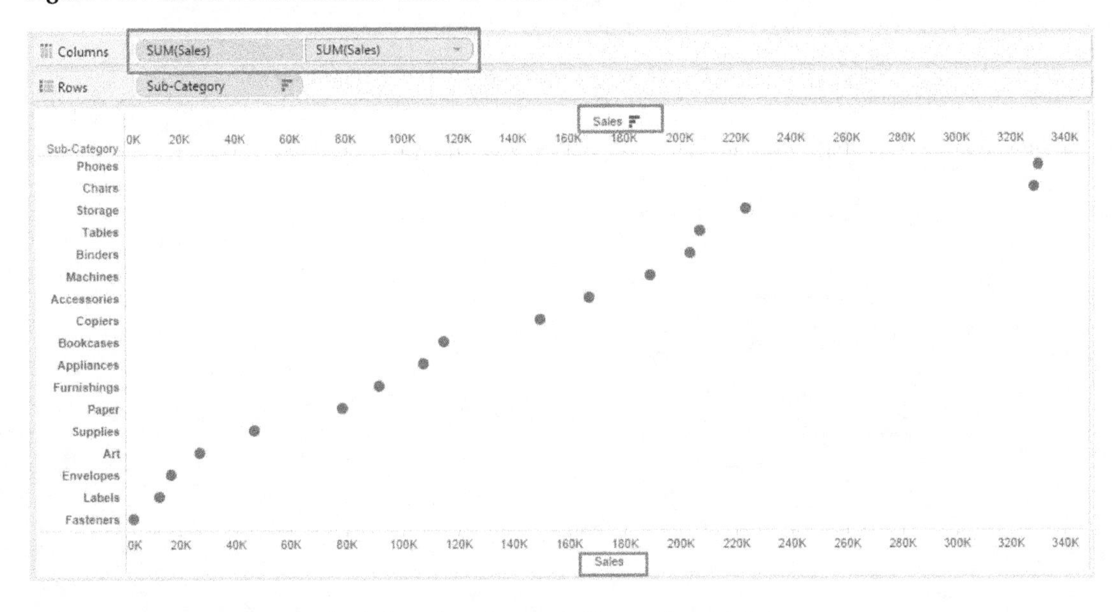

Figure 4-88. Second measure "Sales" on the columns shelf set to "Dual Axis"

4.1.5.2.7 Step 6

Again select the second measure on the columns shelf and perform the table calculation "Running Total" (Shown in Fig. 4-89).

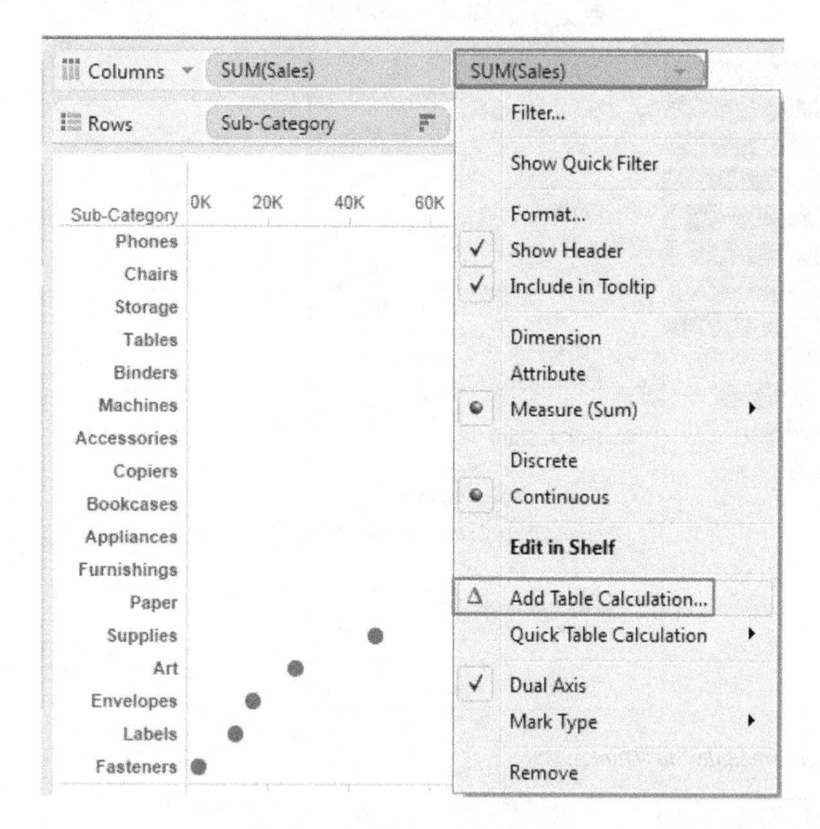

Figure 4-89. *Add a table calculation to the second measure "Sales" on the columns shelf*

Fill in the values in "Table Calculation [Sales]" dialog box as shown in Fig. 4-90.

Table Calculation [Sales]

Calculation Type: Running Total

Calculation Definition

Summarize values using: Sum

Running along: Table (Down)

Restarting every:

Perform a secondary calculation on the result

Secondary Type: Percent of Total

Secondary Calculation Definition

Summarize the values from: Table (Down)

At the level:

Compute total across all pages

OK Cancel Apply

Figure 4-90. *Values filled in the "Table Calculation" dialog box for the measure Sales*

Click Apply and then OK.

The output after computing a table calculation (Shown in Fig. 4-91).

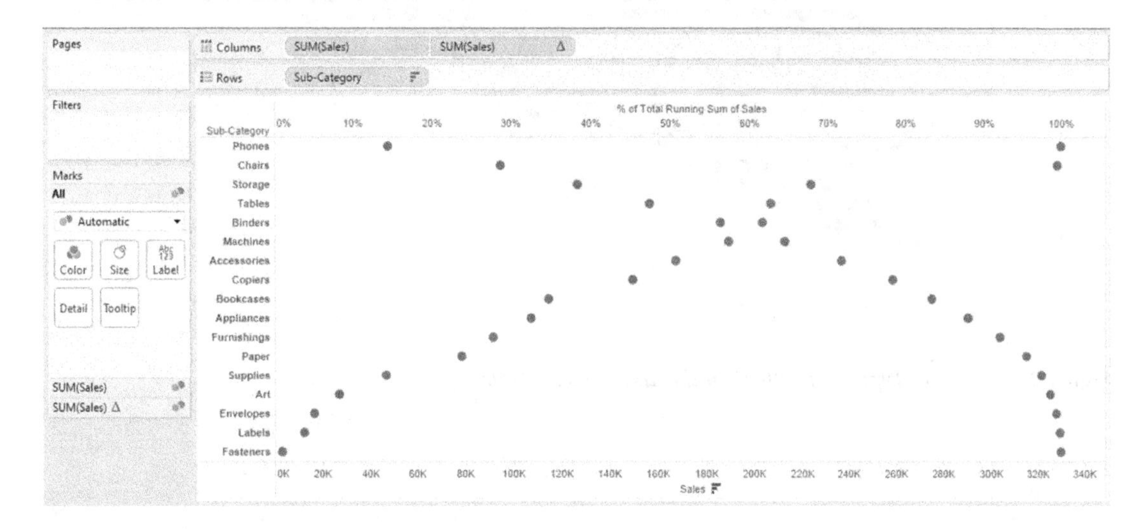

Figure 4-91. *Output after the table calculation is performed on the measure "Sales"*

Select the "Marks Type" for the second measure as "Line" (Shown in Fig. 4-92).

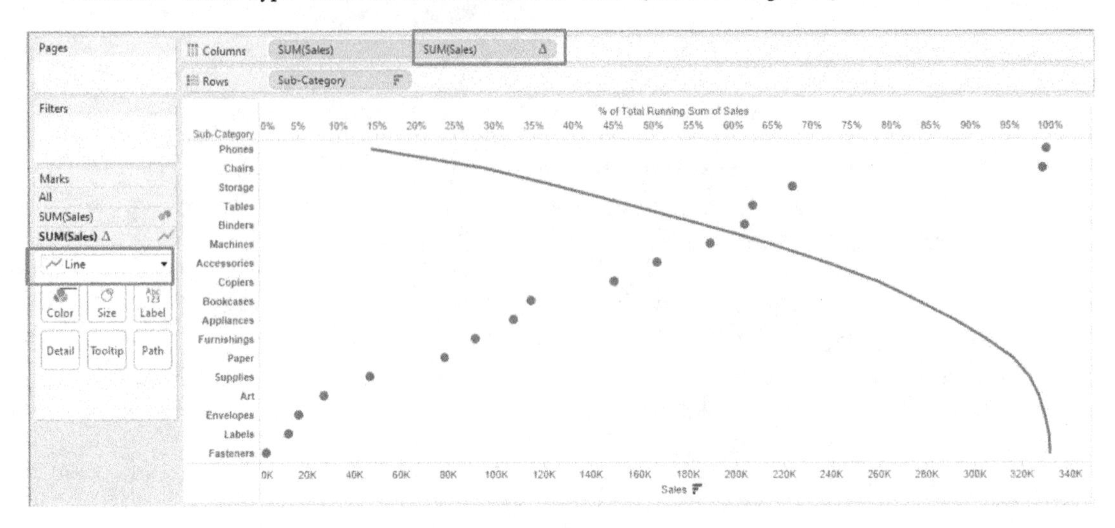

Figure 4-92. *"Marks Type" set to "Line" for the second measure "Sales"*

Select the first measure and select the "Marks Type" as "Bar" (Shown in Fig. 4-93).

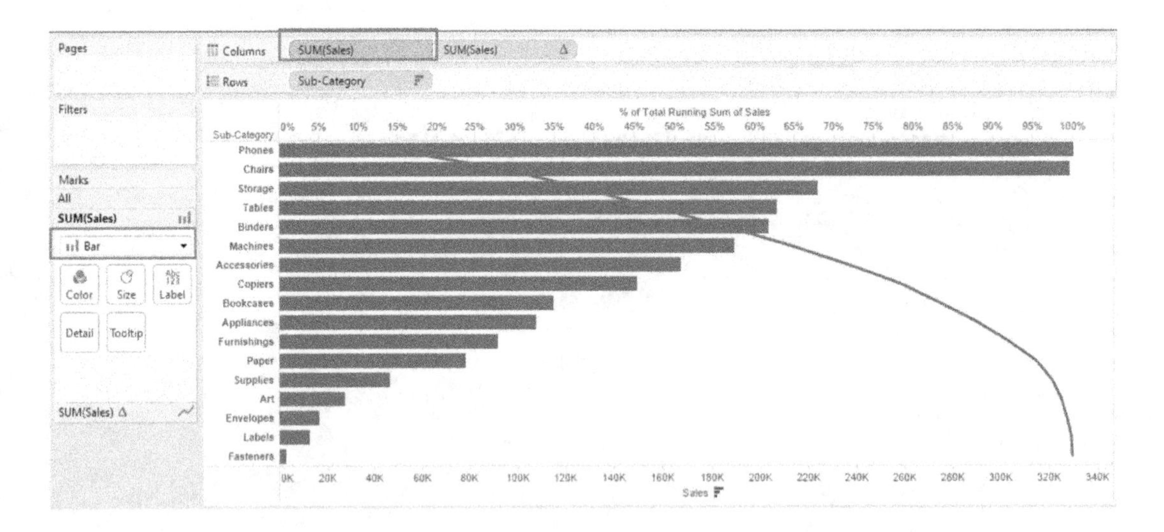

Figure 4-93. *"Marks Type" set to "Bar" for the first measure "Sales"*

4.1.5.2.8 Step 7

Drag "Measure Names" from the dimensions area under the data pane and place it on "Color" on the marks card (Shown in Fig. 4-94).

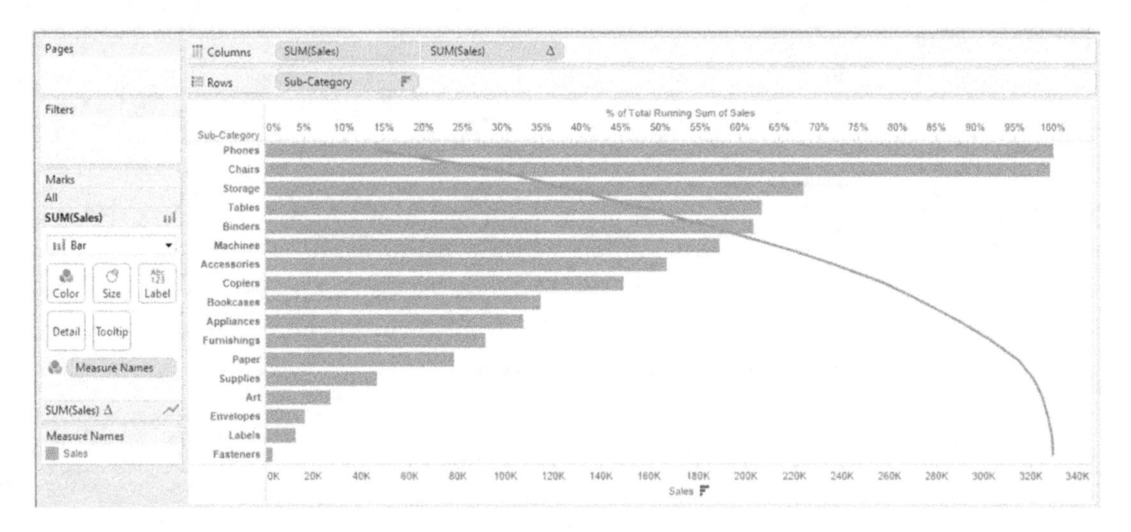

Figure 4-94. *"Measure Names" placed on "Color" on the marks card*

The top five sub-categories contribute to approximately 55% of the sales.

4.1.5.3 Demo 3 (Building a lollipop chart using dual axis)

Objective: You are an analyst employed with "XYZ" corporation. The corporation has office branches in several states. You would like to present the sales amount per segment for each state to the leadership team. You decide to plot a lollipop chart by using the same measure on two axes.

Input: "Sample - Superstore.xls"

Expected output: Shown in Fig. 4-95.

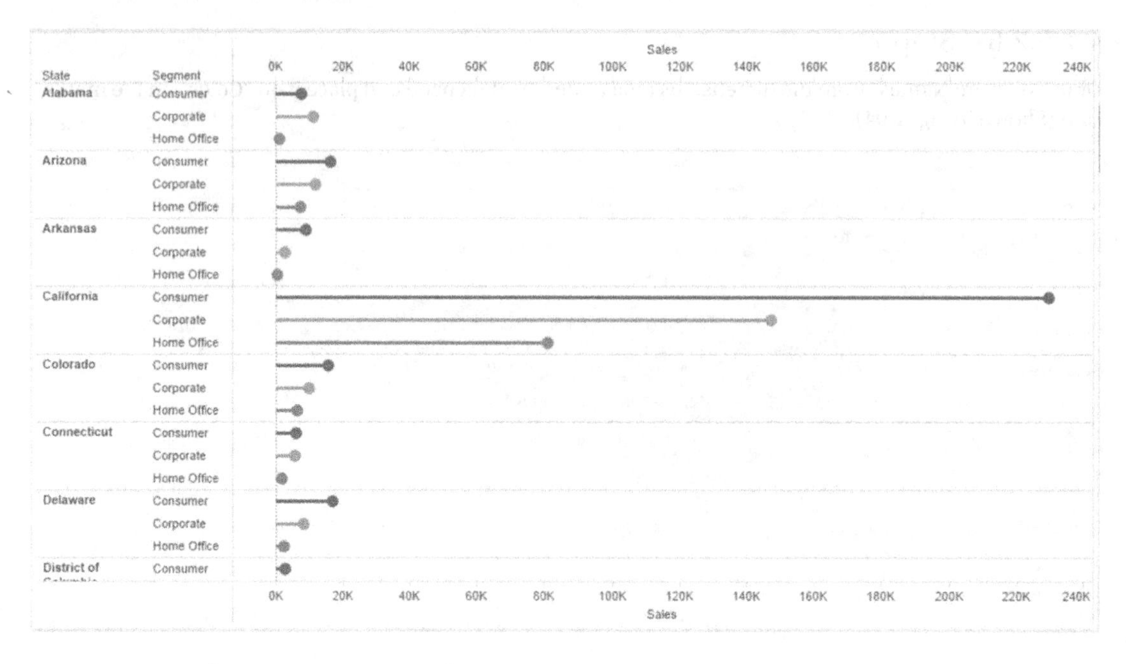

Figure 4-95. Dual axis - Demo 3 - expected output

4.1.5.3.1 Steps to create a lollipop chart

4.1.5.3.2 Step 1

Read in the data from "Sample Superstore.xls" into Tableau (Shown in Fig. 4-96).

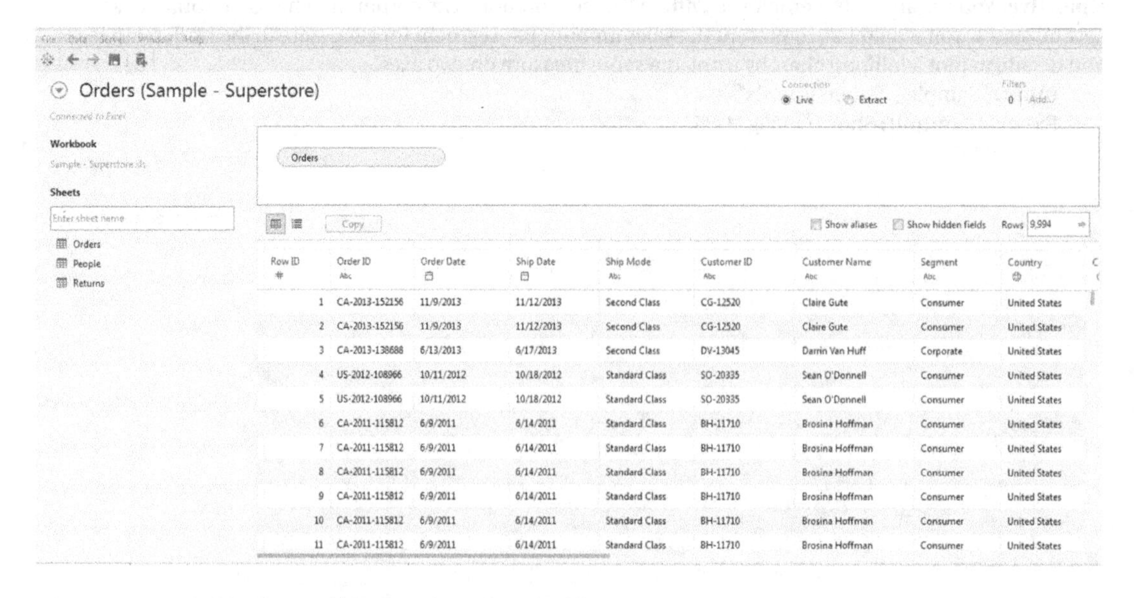

Row ID	Order ID	Order Date	Ship Date	Ship Mode	Customer ID	Customer Name	Segment	Country
1	CA-2013-152156	11/9/2013	11/12/2013	Second Class	CG-12520	Claire Gute	Consumer	United States
2	CA-2013-152156	11/9/2013	11/12/2013	Second Class	CG-12520	Claire Gute	Consumer	United States
3	CA-2013-138688	6/13/2013	6/17/2013	Second Class	DV-13045	Darrin Van Huff	Corporate	United States
4	US-2012-108966	10/11/2012	10/18/2012	Standard Class	SO-20335	Sean O'Donnell	Consumer	United States
5	US-2012-108966	10/11/2012	10/18/2012	Standard Class	SO-20335	Sean O'Donnell	Consumer	United States
6	CA-2011-115812	6/9/2011	6/14/2011	Standard Class	BH-11710	Brosina Hoffman	Consumer	United States
7	CA-2011-115812	6/9/2011	6/14/2011	Standard Class	BH-11710	Brosina Hoffman	Consumer	United States
8	CA-2011-115812	6/9/2011	6/14/2011	Standard Class	BH-11710	Brosina Hoffman	Consumer	United States
9	CA-2011-115812	6/9/2011	6/14/2011	Standard Class	BH-11710	Brosina Hoffman	Consumer	United States
10	CA-2011-115812	6/9/2011	6/14/2011	Standard Class	BH-11710	Brosina Hoffman	Consumer	United States
11	CA-2011-115812	6/9/2011	6/14/2011	Standard Class	BH-11710	Brosina Hoffman	Consumer	United States

Figure 4-96. Data for Demo 3 (dual axis) read into Tableau

4.1.5.3.3 Step 2

Drag the dimension "State" from the dimensions area under the data pane and place it on the rows shelf (Shown in Fig. 4-97).

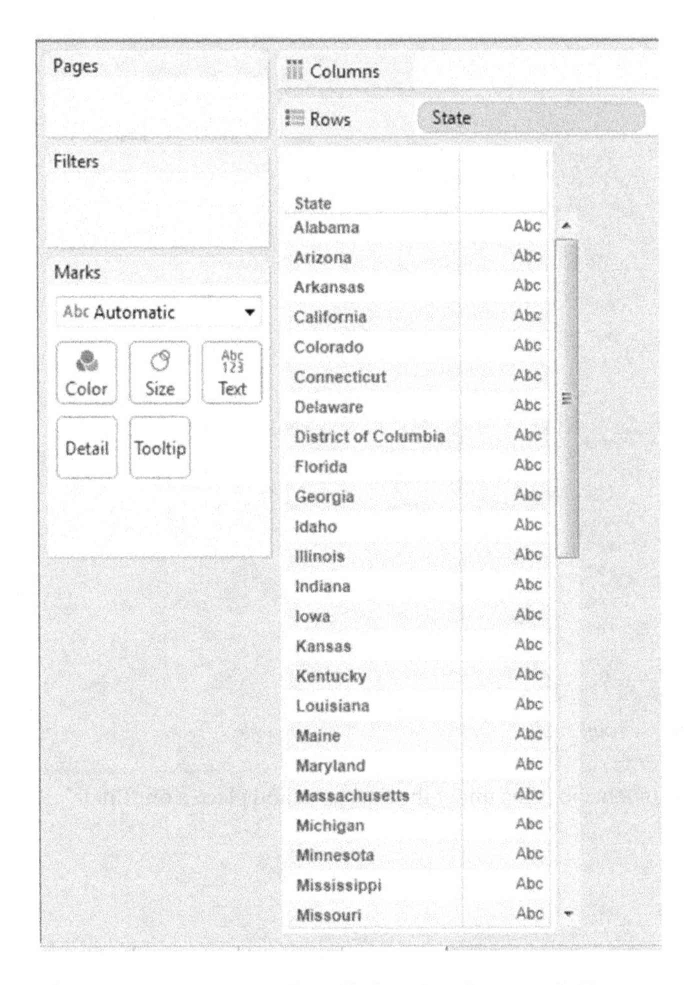

Figure 4-97. *Dimension "State" placed on the rows shelf*

Drag the dimension "Segment" from the dimensions area under the data pane and place it on the rows shelf to the right of the dimension "State" (Shown in Fig. 4-98).

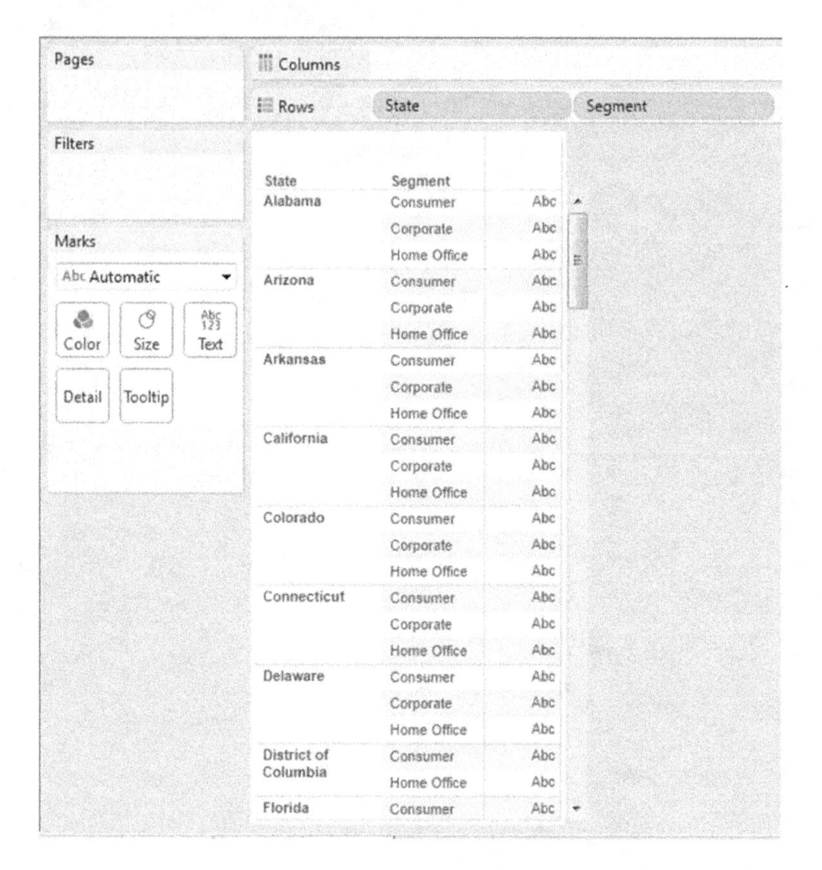

Figure 4-98. *Dimension "Segment" placed on the rows shelf*

Drag the dimension "Segment" from the dimensions area under the data pane and place it on "Color" on the marks card (Shown in Fig. 4-99).

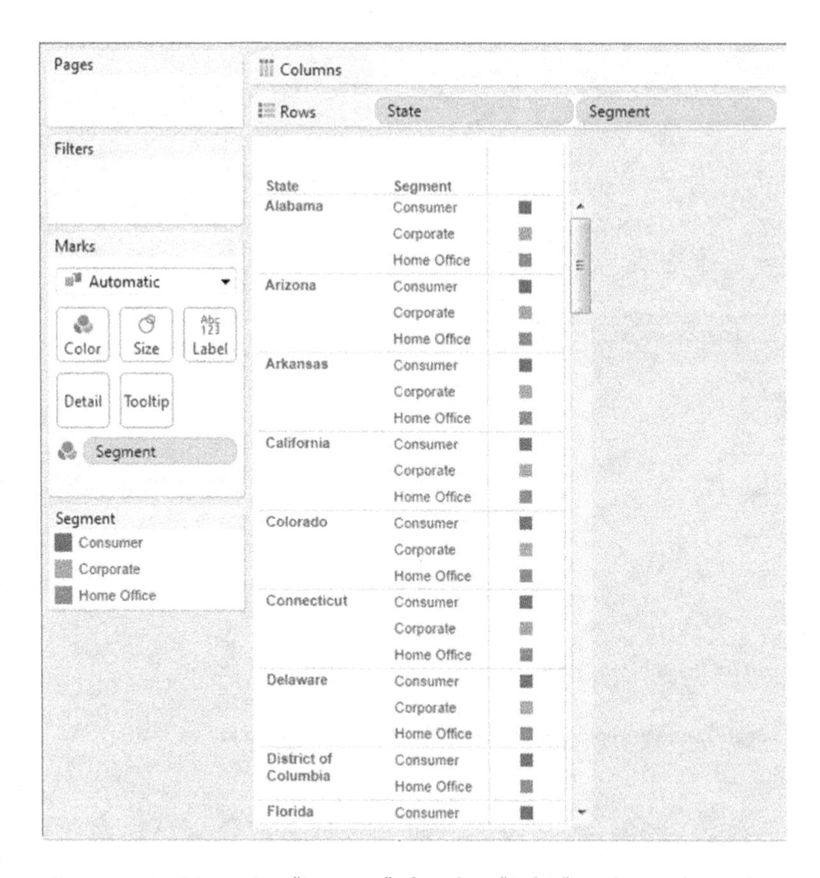

Figure 4-99. *Dimension "Segment" placed on "Color" on the marks card*

4.1.5.3.4 Step 3

Drag the measure "Sales" from the measures area under the data pane and place it on the "Columns Shelf" (Shown in Fig. 4-100).

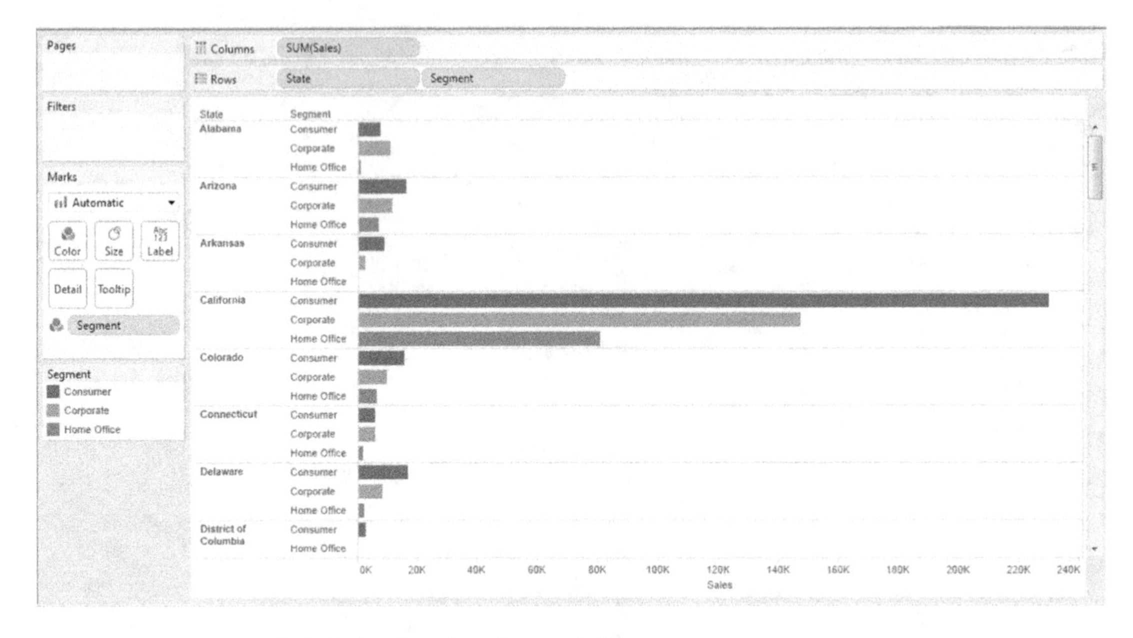

Figure 4-100. *Measure "Sales" placed on the columns shelf*

Drag the measure "Sales" the second time and place it on the columns shelf (Shown in Fig. 4-101).

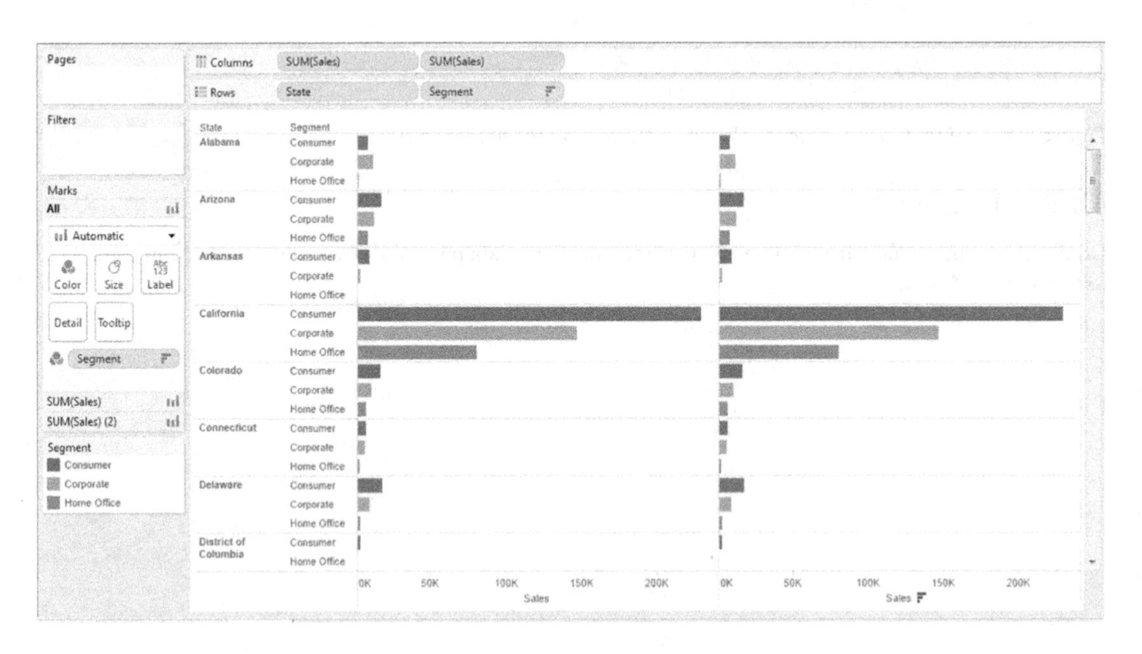

Figure 4-101. *Measure "Sales" placed a second time on the columns shelf*

4.1.5.3.5 Step 4

Select the second measure on the columns shelf (Sum (Sales)), click on the drop down and set it to "Dual Axis" (Shown in Fig. 4-102).

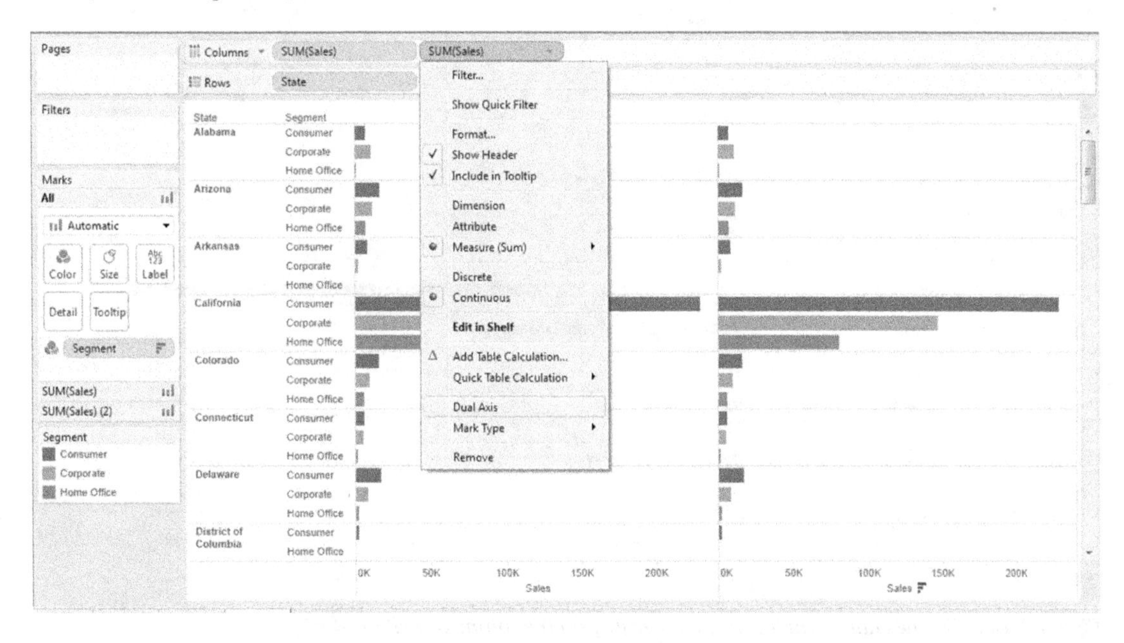

***Figure 4-102.** Second measure "Sales" on the columns shelf set to dual axis*

The output after setting the second measure on the columns shelf, as the dual axis (Shown in Fig. 4-103).

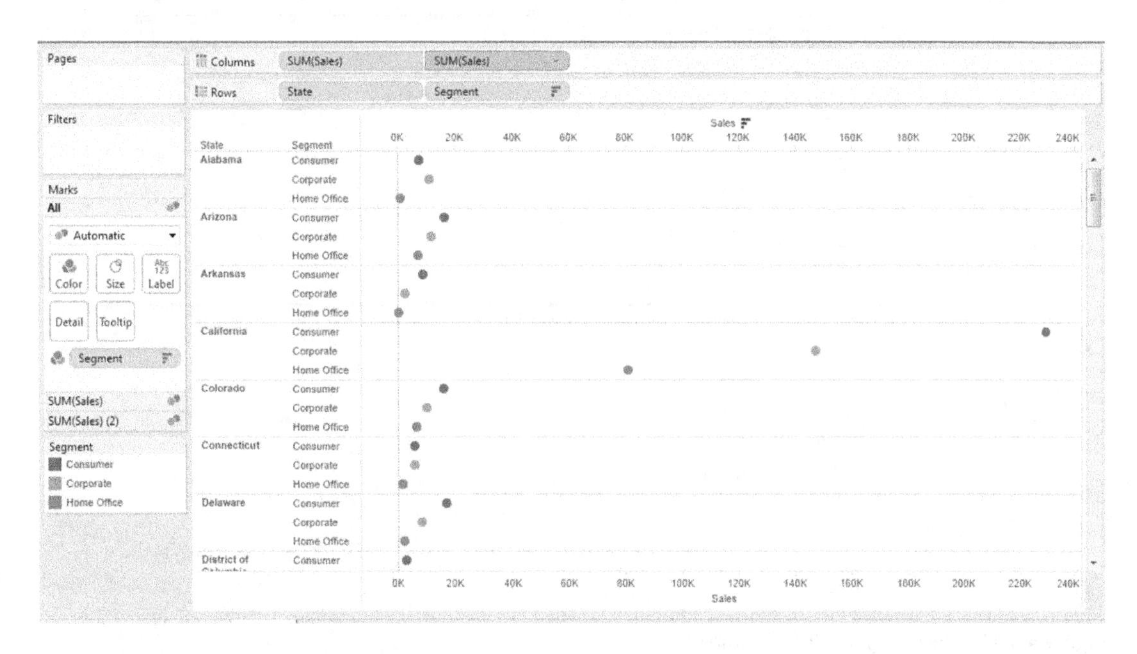

***Figure 4-103.** Output after setting a dual axis*

4.1.5.3.6 Step 5

Select the first measure on the columns shelf and set it to "Bar" chart (Shown in Fig. 4-104).

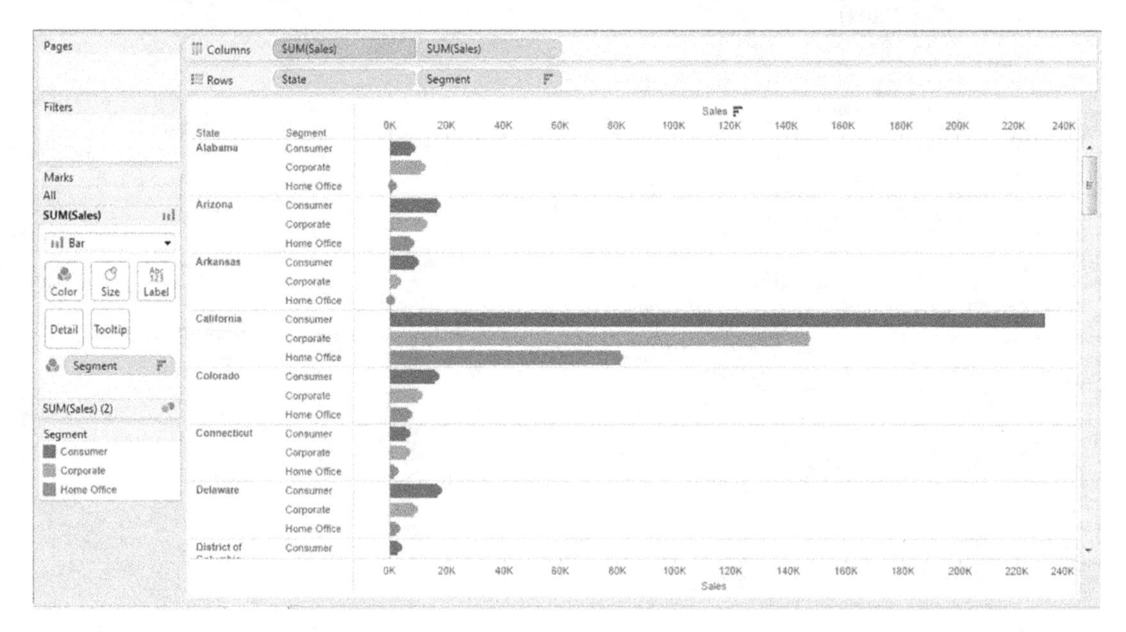

Figure 4-104. *Set the chart form for the first measure on the columns shelf to "Bar"*

Decrease the size of the bars (Shown in Fig. 4-105).

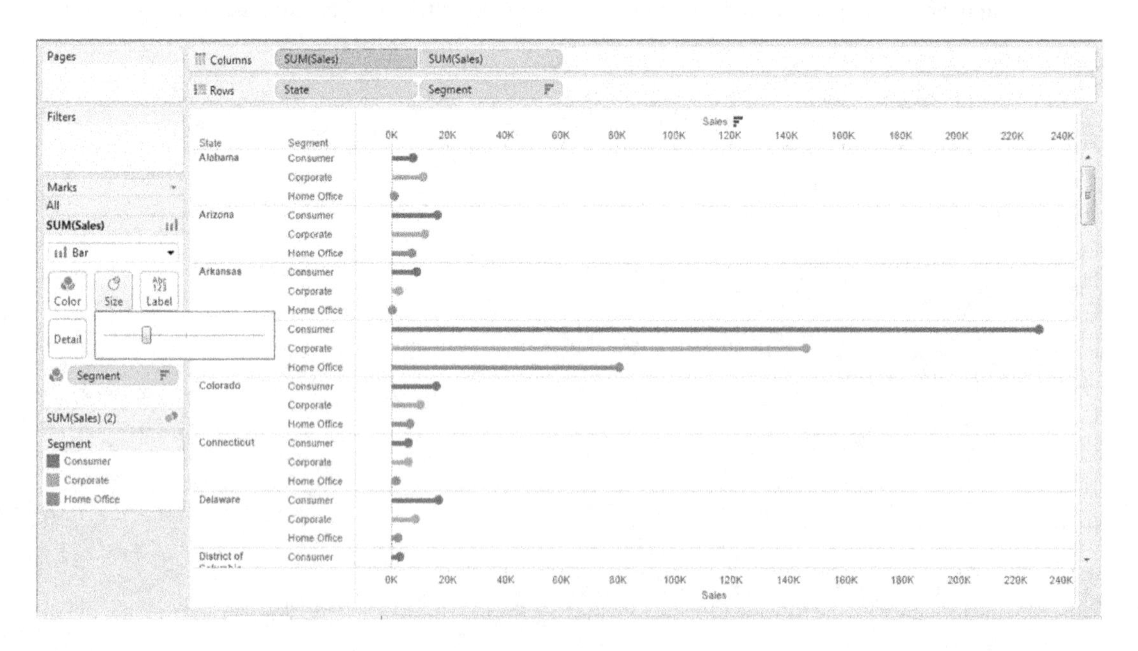

Figure 4-105. *The size of the "Bar" reduced*

4.1.5.3.7 Step 6

Verify that the "Marks Type" for the second instance of "Sales" (on the columns shelf) is set to "Circle".

The final output is as shown in Fig. 4-106.

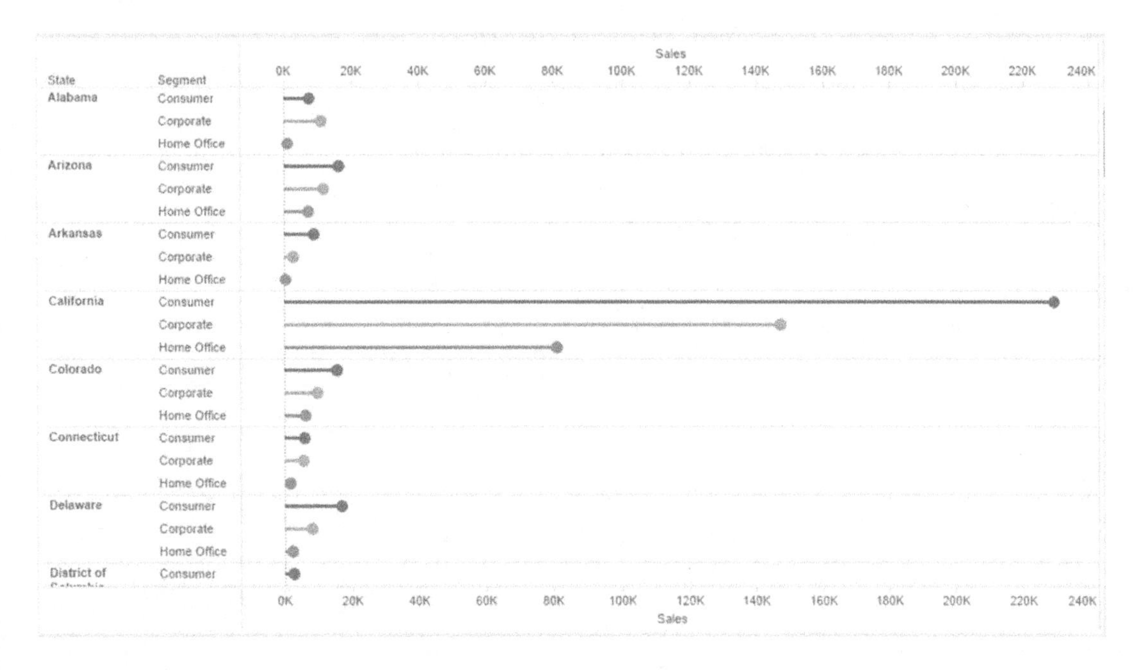

Figure 4-106. *Dual axis - Demo 3 - final output*

Lollipop chart: It makes sense to use the stick of the lollipop only when your data range starts at zero.

4.2 Points to Remember

- "Measure Names" and "Measure Values" are built-in Tableau fields that empower us to work with multiple measures in a worksheet / view.
- Use a slope graph when visualizing a single measure over a period of time.
- Use a combination chart to enhance the visualization by employing the most suitable chart form to present the measure in.

4.3 Next steps

In this chapter we learnt about "Measure Names" and "Measure Values". We were educated on plotting each measure on its own axis, blending measures and plotting it on a common axis, and charting multiple measures in a view by making use of dual axis. The next chapter will familiarize us with running "Table Calculations" on measures.

CHAPTER 5

Table Calculations

The purpose of visualization is insight, not pictures.

—Ben Shneiderman, computer scientist, distinguished university professor in the Department of Computer Science, University of Maryland, College Park, and founding director (1983-2000) of the University of Maryland Human-Computer Interaction Lab

Chapter 4 introduced us to two new fields, namely, measure names and measure values. We learnt to blend multiple measures on a single axis and to use a dual axis to enhance our presentation of data. We were also introduced to some new chart forms, such as slope graphs, combination charts such as bar and line graphs together in a view, lollipop charts, etc. This chapter will help us learn about table calculations that will be performed on measures plotted on the view. In this chapter we will explore the following table calculations:

- Running total of sales
- Percent of total
- Moving average
- Rank
- Level of detail (LOD)
- Percentile
- Year-over-year growth

5.1 What is a table calculation?

Consider a Tableau view (see Fig. 5-1a). For every Tableau view, there is a virtual table determined by the dimensions used in the view. The dimensions can be on the rows shelf, columns shelf, pages, and the marks card (color, size, label, detail and path), in other words, the dimensions within the level of detail.

S. Acharya and S. Chellappan, *Pro Tableau*, DOI 10.1007/978-1-4842-2352-9_5

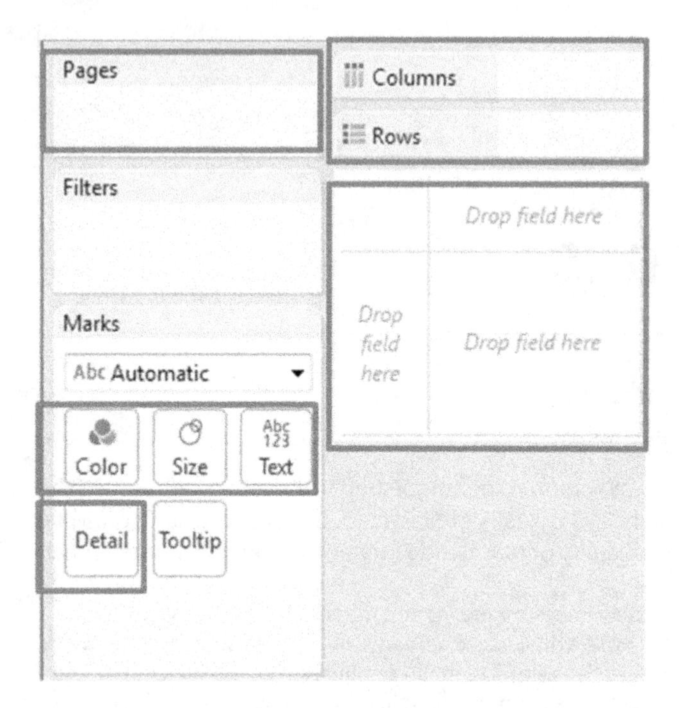

Figure 5-1a. *Dimensions can be placed on the level of details*

A table calculation is a calculation / computation that is applied to all values of a single measure in a view. Table calculations are the computational workhorse of Tableau. They calculate values outside the traditional realm of "Slice by X Dimension". Table calculations allow the user to extend their data. Table calculations are computations that are applied to all values in the entire table and are often dependent on the table structure itself. Example: Table calculations can be used to compute each month's contribution to annual profit.

There are two easy ways to work with table calculations:

- Use quick table calculation. Quick table calculations are a collection of commonly used table calculations (such as running total, difference, percent difference, rank, percentile, etc.).
- Create your own table calculations from scratch using table calculation functions.

Refer to Fig. 5-1b.

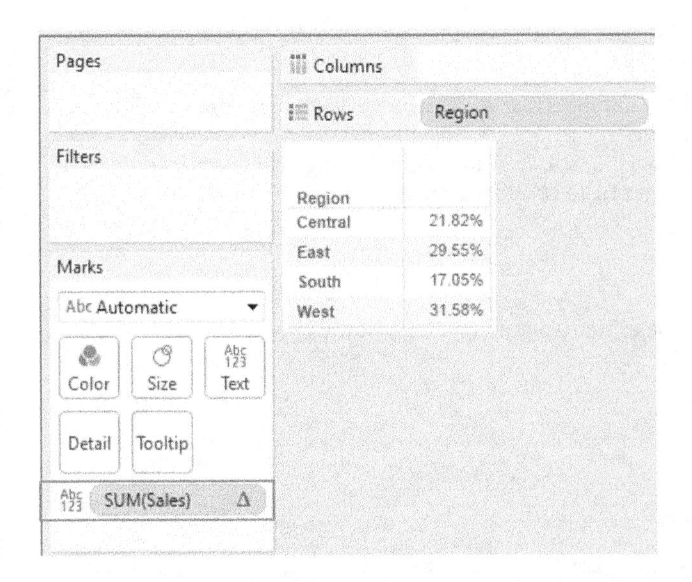

Figure 5-1b. *Table calculation applied on the measure "Sales". Notice the triangular mark next to SUM(Sales)*

In Figure 5-1b, **dimension** "Region" is placed on the rows shelf and m**easure** "Sales" is placed on the "Label" on the marks card. The table calculation, "Percent of Total" is applied to the m**easure** "Sales". The table calculation is "Percent of Total"; therefore, when all the cell values are added up, it aggregates to 100%.

A table calculation makes use of two fields: Partitioning and Addressing fields. In order to understand table calculations, it is important to understand how these fields work. They essentially define "what" a table calculation is and "how" they are performed.

Partitioning field: this field is used to partition the data into buckets. These data buckets are then acted upon by the calculations. In other words, they define the scope or grouping of the calculation. The scope can be the entire table, a pane, a cell, a dimension or it can be customized even further for more advanced calculations.

Addressing field: this field provides the direction in which we want our calculation to proceed. It defines the anchor or the source of each partition. It defines the root of the calculation.

Example: Compute the running total over a period of time (over years) partitioned by a segment. Here "segment" is the partitioning field and "date" is the addressing field.

■ **Note** The sequence in which tableau processes calculated fields, filters and table calculations:

1. Tableau generates a query and sends it for processing to the database.

2. The database processes the query. Tableau takes into consideration all calculated fields, including the level of detail calculations.

3. Lastly, the table calculations are applied.

5.2 Running Total of Sales

A running total is a summation of a sequence of numbers that is updated every time a number is added to the sequence. It is also referred to as "partial sum".

For example: we have a sequence of numbers "5, 2, 4, 7, 8". To get the running total, start by adding 5+2 to get 7. To this 7 add 4 to get 11, to 11 add 7 to get 18, to 18 add 8 to get 26.

5.2.1 Demo 1

Objective: To compute the "Running Total of Sales" (compute using Table Down).

Data set used: "Sample - Superstore.xls"

Expected Output: Shown in Fig. 5-2.

		Order Date							
		2011		2012		2013		2014	
Region	Quarter of O..	Sales	Running Sum	Sales	Running Sum	Sales	Running Sum	Sales	Running Sum
Central	Q1	8,601	8,601	11,768	11,768	20,212	20,212	40,278	40,278
	Q2	17,407	26,008	23,979	35,748	25,709	45,921	26,606	66,884
	Q3	44,171	70,179	24,486	60,233	33,428	79,349	34,042	100,926
	Q4	33,659	103,838	42,641	102,874	68,080	147,429	46,172	147,098
	Total	103,838	103,838	102,874	102,874	147,429	147,429	147,098	147,098
East	Q1	6,579	110,418	17,146	120,020	24,134	171,563	17,341	164,439
	Q2	21,064	131,482	22,703	142,723	52,807	224,371	29,978	194,417
	Q3	33,443	164,925	50,777	193,501	37,528	261,899	67,712	262,129
	Q4	67,594	232,519	65,706	259,206	66,060	327,959	98,209	360,338
	Total	128,680	232,519	156,332	259,206	180,529	327,959	213,239	360,338
South	Q1	44,262	276,781	16,444	275,651	23,934	351,892	9,882	370,219
	Q2	22,524	299,305	16,254	291,905	17,079	368,971	33,137	403,357
	Q3	16,061	315,366	21,460	313,364	22,939	391,910	23,894	427,250
	Q4	20,998	336,364	17,202	330,566	29,588	421,498	56,064	483,314
	Total	103,846	336,364	71,360	330,566	93,539	421,498	122,977	483,314
West	Q1	15,006	351,370	23,493	354,059	24,317	445,815	51,395	534,710
	Q2	25,543	376,913	26,188	380,247	39,774	485,589	44,302	579,011
	Q3	49,957	426,871	33,537	413,784	50,720	536,309	74,786	653,797
	Q4	57,377	484,247	56,748	470,533	72,165	608,474	80,150	733,947
	Total	147,883	484,247	139,966	470,533	186,976	608,474	250,633	733,947

Figure 5-2. *Quarterly sales by region*

5.2.1.1 Steps

Follow the steps as provided.

5.2.1.1.1 Step 1

Read in the data from "Sample - Superstore.xls" into Tableau (Shown in Fig. 5-3).

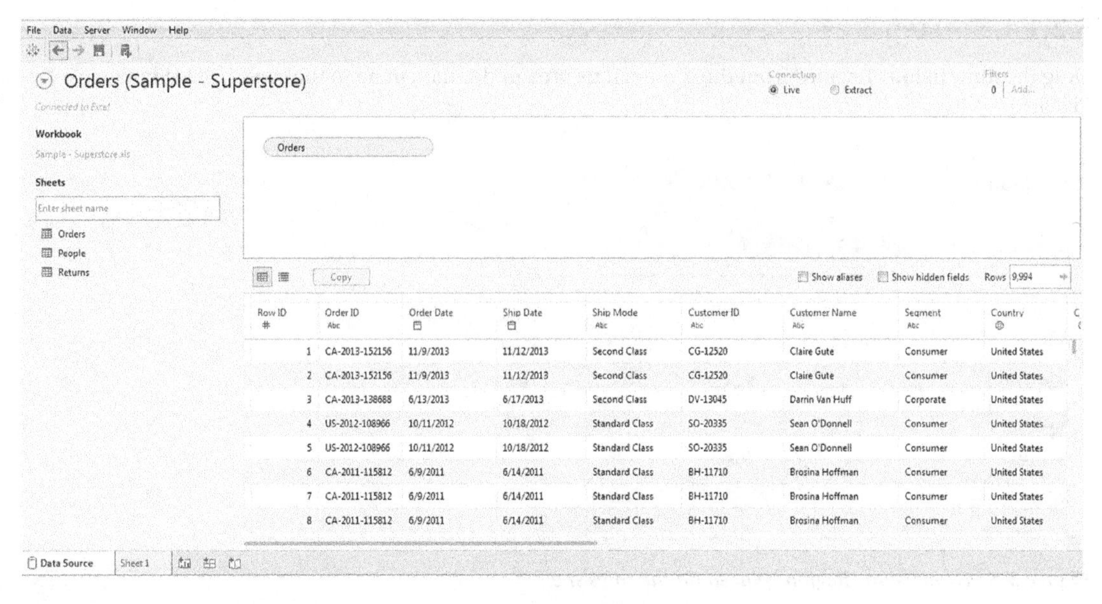

Figure 5-3. *Data read from "Sample - Superstore.xls" into Tableau*

5.2.1.1.2 Step 2

Table 5-1. *Tasks to be performed in the view*

Columns Shelf	Order Date: Set it to "Discrete". The granularity should be "Year". Measure names: Sum(Sales) Sum (Sales): Add a table calculation, "Running Total" and have it compute the running total, "Table Down".
Rows Shelf	Region Order Date: Set it to "Discrete". The granularity should be "Quarter".

Drag the **dimension** "Order Date" from the dimensions area under data pane to the columns shelf. By default it is discrete. This is also evident from the visual cue. It is blue in color. By default dates in tableau have hierarchies defined on it. The hierarchy is set to the highest level, i.e. "Year" (Shown in Fig. 5-4).

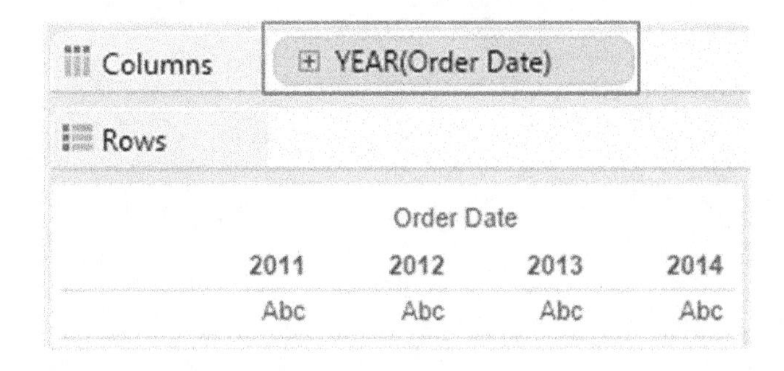

Figure 5-4. *Dimension "Order Date" placed on the columns shelf*

5.2.1.1.3 Step 3

Drag the **dimension** "Region" from the dimensions area under data pane to the rows shelf (Shown in Fig. 5-5).

Columns: YEAR(Order Date)

Rows: Region

	Order Date			
Region	2011	2012	2013	2014
Central	Abc	Abc	Abc	Abc
East	Abc	Abc	Abc	Abc
South	Abc	Abc	Abc	Abc
West	Abc	Abc	Abc	Abc

Figure 5-5. *Dimension "Region" placed on the rows shelf*

5.2.1.1.4 Step 4

Drag the **dimension** "Order Date" from the dimensions area under the data pane to the rows shelf, to the right of "Region". By default it is discrete. This is also evident from the visual cue. It is blue in color. Set the hierarchy to "Quarter" (Shown in Fig. 5-6).

Columns: YEAR(Order Date)

Rows: Region | QUARTER(Order Date)

		Order Date			
Region	Quarter of O..	2011	2012	2013	2014
Central	Q1	Abc	Abc	Abc	Abc
	Q2	Abc	Abc	Abc	Abc
	Q3	Abc	Abc	Abc	Abc
	Q4	Abc	Abc	Abc	Abc
East	Q1	Abc	Abc	Abc	Abc
	Q2	Abc	Abc	Abc	Abc
	Q3	Abc	Abc	Abc	Abc
	Q4	Abc	Abc	Abc	Abc
South	Q1	Abc	Abc	Abc	Abc
	Q2	Abc	Abc	Abc	Abc
	Q3	Abc	Abc	Abc	Abc
	Q4	Abc	Abc	Abc	Abc
West	Q1	Abc	Abc	Abc	Abc
	Q2	Abc	Abc	Abc	Abc
	Q3	Abc	Abc	Abc	Abc
	Q4	Abc	Abc	Abc	Abc

Figure 5-6. *Dimension "Order Date" placed on the rows shelf, to the right of "Region"*

5.2.1.1.5 Step 5

Drag the **measure** "Sales" from the measures area under the data pane to "Label" on the marks card. The default aggregation is "Sum" (Shown in Fig. 5-7).

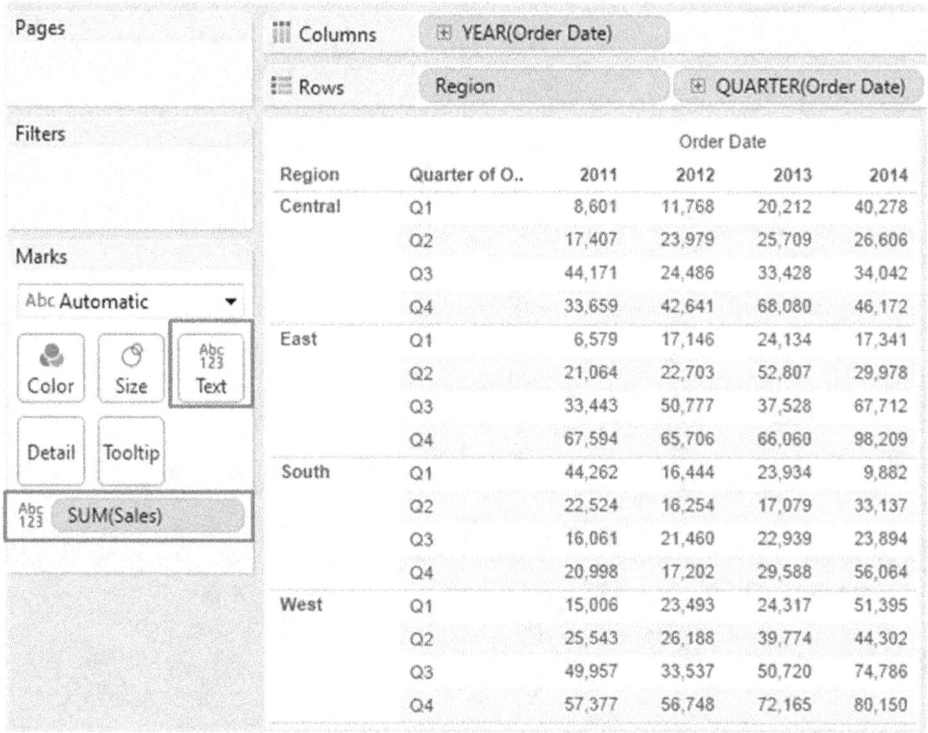

Pages
Columns ⊞ YEAR(Order Date)
Rows Region ⊞ QUARTER(Order Date)
Filters
Marks
Abc Automatic
Color Size Abc 123 Text
Detail Tooltip
Abc 123 SUM(Sales)

Order Date

Region	Quarter of O..	2011	2012	2013	2014
Central	Q1	8,601	11,768	20,212	40,278
	Q2	17,407	23,979	25,709	26,606
	Q3	44,171	24,486	33,428	34,042
	Q4	33,659	42,641	68,080	46,172
East	Q1	6,579	17,146	24,134	17,341
	Q2	21,064	22,703	52,807	29,978
	Q3	33,443	50,777	37,528	67,712
	Q4	67,594	65,706	66,060	98,209
South	Q1	44,262	16,444	23,934	9,882
	Q2	22,524	16,254	17,079	33,137
	Q3	16,061	21,460	22,939	23,894
	Q4	20,998	17,202	29,588	56,064
West	Q1	15,006	23,493	24,317	51,395
	Q2	25,543	26,188	39,774	44,302
	Q3	49,957	33,537	50,720	74,786
	Q4	57,377	56,748	72,165	80,150

Figure 5-7. *Measure "Sales" placed on "Label" on the marks card*

5.2.1.1.6 Step 6

Add a quick table calculation, "Running Total" to the **measure** "Sum (Sales)" and compute the "Running Total" as "Table Down" (Shown in Fig. 5-8).

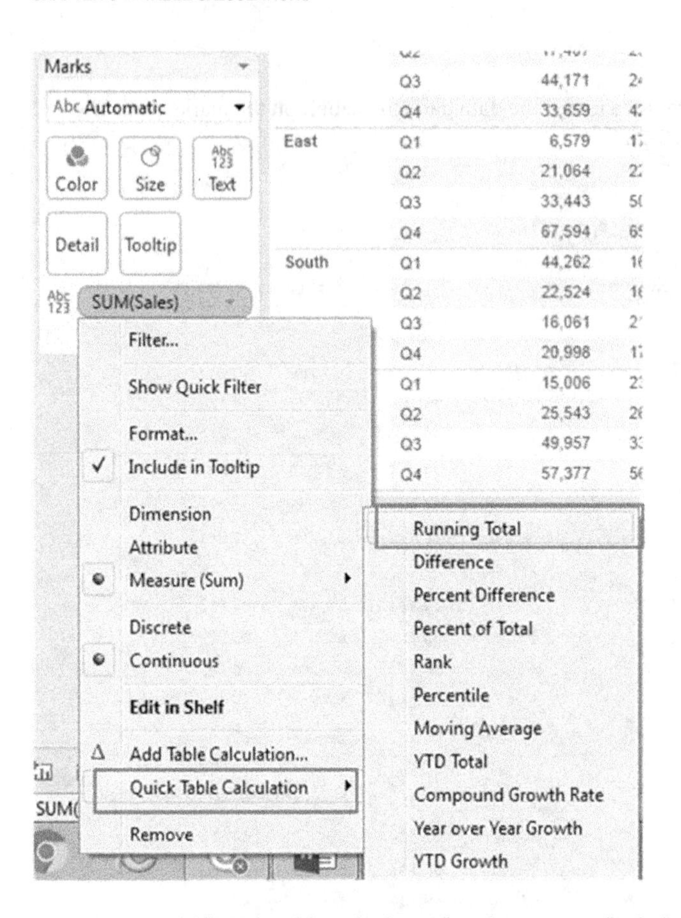

Figure 5-8. *Add "Quick Table Calculation" to the measure "Sales"*

Fill in the values into the "Table Calculation dialog box" as shown in Fig. 5-9.

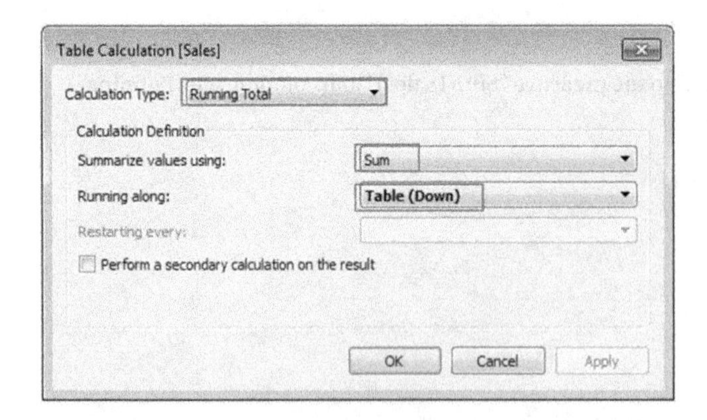

Figure 5-9. *Table Calculation dialog box for measure "Sales"*

Click on "Apply" and then on "OK".

5.2.1.1.7 Step 7

Once again drag the **measure** "Sales" from the measures area under the data pane and drop it into the view area. The default aggregation is "Sum" (Shown in Fig. 5-10).

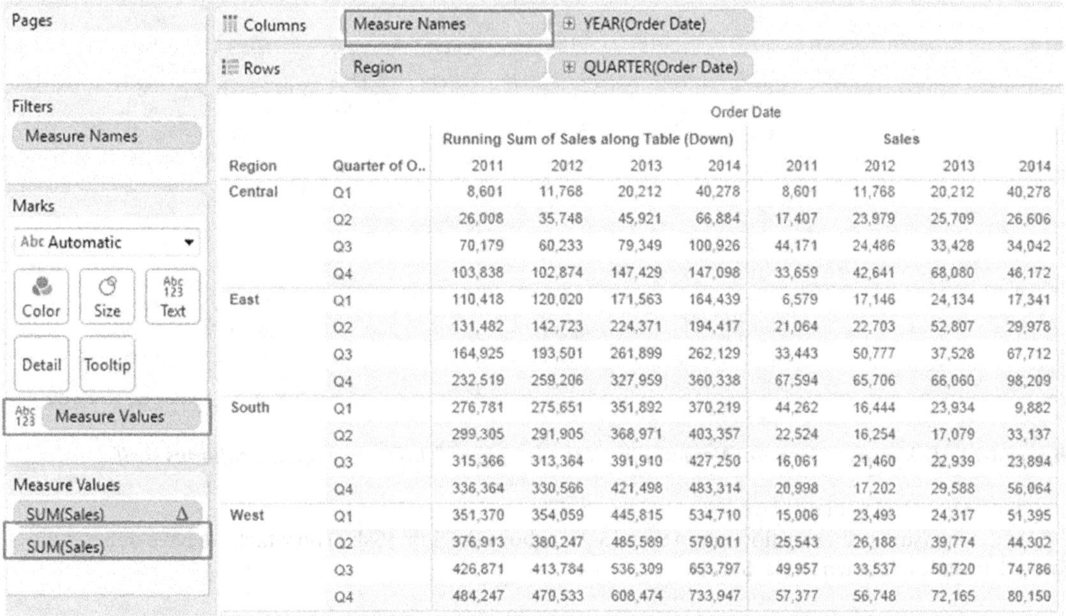
Pages
Columns
Measure Names
YEAR(Order Date)
Rows
Region
QUARTER(Order Date)
Filters
Measure Names
Marks
Abc Automatic
Color
Size
Abc 123 Text
Detail
Tooltip
Measure Values
Measure Values
SUM(Sales)
SUM(Sales)
Order Date
Running Sum of Sales along Table (Down)
Sales
Region Quarter of O.. 2011 2012 2013 2014 2011 2012 2013 2014
Central Q1 8,601 11,768 20,212 40,278 8,601 11,768 20,212 40,278
Q2 26,008 35,748 45,921 66,884 17,407 23,979 25,709 26,606
Q3 70,179 60,233 79,349 100,926 44,171 24,486 33,428 34,042
Q4 103,838 102,874 147,429 147,098 33,659 42,641 68,080 46,172
East Q1 110,418 120,020 171,563 164,439 6,579 17,146 24,134 17,341
Q2 131,482 142,723 224,371 194,417 21,064 22,703 52,807 29,978
Q3 164,925 193,501 261,899 262,129 33,443 50,777 37,528 67,712
Q4 232,519 259,206 327,959 360,338 67,594 65,706 66,060 98,209
South Q1 276,781 275,651 351,892 370,219 44,262 16,444 23,934 9,882
Q2 299,305 291,905 368,971 403,357 22,524 16,254 17,079 33,137
Q3 315,366 313,364 391,910 427,250 16,061 21,460 22,939 23,894
Q4 336,364 330,566 421,498 483,314 20,998 17,202 29,588 56,064
West Q1 351,370 354,059 445,815 534,710 15,006 23,493 24,317 51,395
Q2 376,913 380,247 485,589 579,011 25,543 26,188 39,774 44,302
Q3 426,871 413,784 536,309 653,797 49,957 33,537 50,720 74,786
Q4 484,247 470,533 608,474 733,947 57,377 56,748 72,165 80,150

***Figure 5-10.** Measure "Sales" again placed on the view*

Notice that "Measure Names" appears on the columns shelf. Also, "Measure Values" appears on Label on the marks card.

Drag "Measure Names" and pull it to the right of "Year (Order Date)" on the columns shelf (Shown in Fig. 5-11).

Columns: YEAR(Order Date) | Measure Names

Rows: Region | QUARTER(Order Date)

		Order Date							
		2011		2012		2013		2014	
Region	Quarter of O..	Running Sum	Sales	Running Sum	Sales	Running Sum	Sales	Running Sum	Sales
Central	Q1	8,601	8,601	11,768	11,768	20,212	20,212	40,278	40,278
	Q2	26,008	17,407	35,748	23,979	45,921	25,709	66,884	26,606
	Q3	70,179	44,171	60,233	24,486	79,349	33,428	100,926	34,042
	Q4	103,838	33,659	102,874	42,641	147,429	68,080	147,098	46,172
East	Q1	110,418	6,579	120,020	17,146	171,563	24,134	164,439	17,341
	Q2	131,482	21,064	142,723	22,703	224,371	52,807	194,417	29,978
	Q3	164,925	33,443	193,501	50,777	261,899	37,528	262,129	67,712
	Q4	232,519	67,594	259,206	65,706	327,959	66,060	360,338	98,209
South	Q1	276,781	44,262	275,651	16,444	351,892	23,934	370,219	9,882
	Q2	299,305	22,524	291,905	16,254	368,971	17,079	403,357	33,137
	Q3	315,366	16,061	313,364	21,460	391,910	22,939	427,250	23,894
	Q4	336,364	20,998	330,566	17,202	421,498	29,588	483,314	56,064
West	Q1	351,370	15,006	354,059	23,493	445,815	24,317	534,710	51,395
	Q2	376,913	25,543	380,247	26,188	485,589	39,774	579,011	44,302
	Q3	426,871	49,957	413,784	33,537	536,309	50,720	653,797	74,786
	Q4	484,247	57,377	470,533	56,748	608,474	72,165	733,947	80,150

Figure 5-11. *Dimension "Measure Names" placed to the right of "Order Date" on the columns shelf*

Notice the change in display.

In the measure values section, place Sum (Sales) above the Sum (Sales) on which we have defined the Table Calculation (Shown in Fig. 5-12).

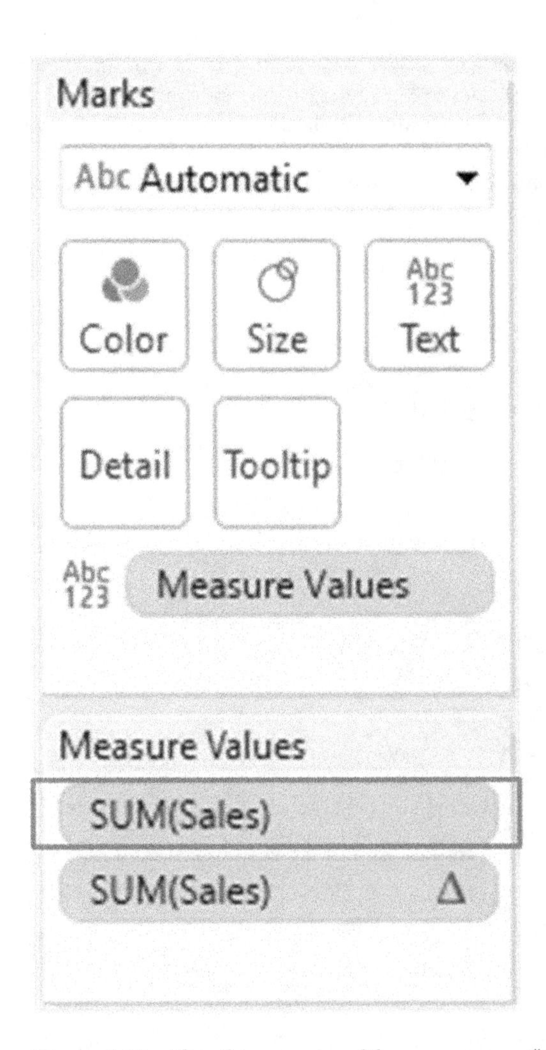

Figure 5-12. *Alter the sequence of the measures on "Label" on the marks card*

As can be seen from the display, first the measure "Sales" is aggregated and then its "Running Total" is displayed.

Let us verify the calculation. For verification, we have picked up the data only for "2011" and for "Central Region".

		2011	
Region	Quarter of O..	Sales	Running Sum
Central	Q1	8,601	8,601
	Q2	17,407	26,008
	Q3	44,171	70,179
	Q4	33,659	103,838

The Sales made in Q1 of 2011 is 8,601. To this we add the sales of Q2, i.e. 17, 407 to give the running sum as 26,008. To this running sum of 26, 008, we add the Q3 sales of 44,171 to give the new running sum of 70,179. To this we add the Sales for Q4, i.e. 33,659 to give the running sum of 103, 838.

Note that the running sum of "Sales" is computed "Table Down".

Let us add Subtotals to the view. For this, select "Analysis" from the menu bar, then click on "Totals" and select "Add All Subtotals" (Shown in Fig. 5-13).

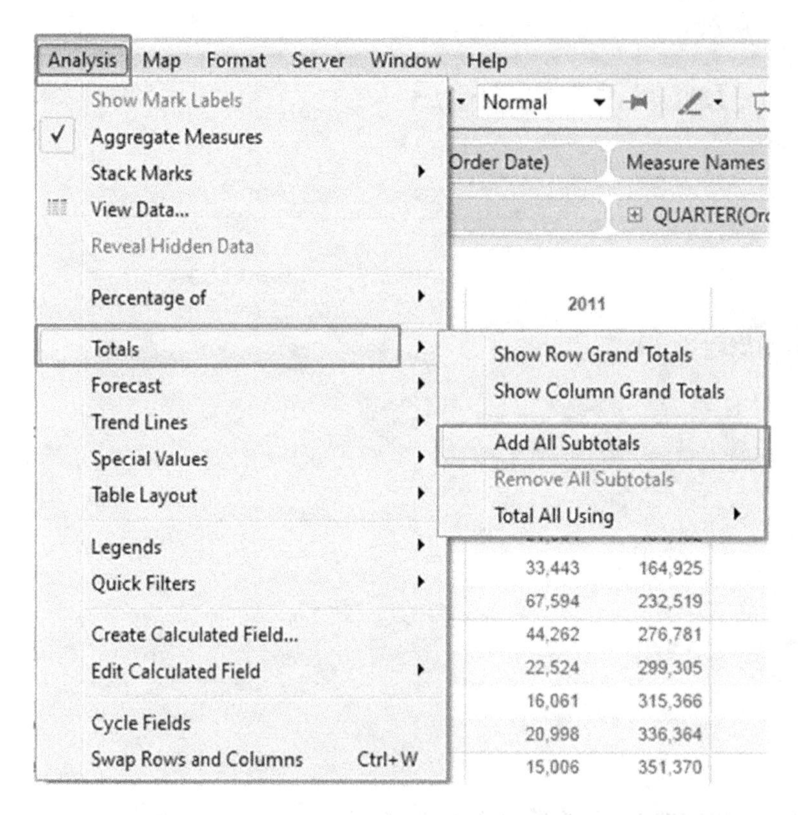

Figure 5-13. *Apply "Add All Subtotals" to measures in the view*

Analysis ➤ Totals ➤ Add All Subtotals

The final output (Shown in Fig. 5-14):

		Order Date							
		2011		2012		2013		2014	
Region	Quarter of O..	Sales	Running Sum	Sales	Running Sum	Sales	Running Sum	Sales	Running Sum
Central	Q1	8,601	8,601	11,768	11,768	20,212	20,212	40,278	40,278
	Q2	17,407	26,008	23,979	35,748	25,709	45,921	26,606	66,884
	Q3	44,171	70,179	24,486	60,233	33,428	79,349	34,042	100,926
	Q4	33,659	103,838	42,641	102,874	68,080	147,429	46,172	147,098
	Total	103,838	103,838	102,874	102,874	147,429	147,429	147,098	147,098
East	Q1	6,579	110,418	17,146	120,020	24,134	171,563	17,341	164,439
	Q2	21,064	131,482	22,703	142,723	52,807	224,371	29,978	194,417
	Q3	33,443	164,925	50,777	193,501	37,528	261,899	67,712	262,129
	Q4	67,594	232,519	65,706	259,206	66,060	327,959	98,209	360,338
	Total	128,680	232,519	156,332	259,206	180,529	327,959	213,239	360,338
South	Q1	44,262	276,781	16,444	275,651	23,934	351,892	9,882	370,219
	Q2	22,524	299,305	16,254	291,905	17,079	368,971	33,137	403,357
	Q3	16,061	315,366	21,460	313,364	22,939	391,910	23,894	427,250
	Q4	20,998	336,364	17,202	330,566	29,588	421,498	56,064	483,314
	Total	103,846	336,364	71,360	330,566	93,539	421,498	122,977	483,314
West	Q1	15,006	351,370	23,493	354,059	24,317	445,815	51,395	534,710
	Q2	25,543	376,913	26,188	380,247	39,774	485,589	44,302	579,011
	Q3	49,957	426,871	33,537	413,784	50,720	536,309	74,786	653,797
	Q4	57,377	484,247	56,748	470,533	72,165	608,474	80,150	733,947
	Total	147,883	484,247	139,966	470,533	186,976	608,474	250,633	733,947

Figure 5-14. *"Running Total" - Demo 1 - final output*

5.3 Profitability as Percent of Total

Percent of total is also called as percent distribution. It is computed using the formula that divides an amount by the total. Example: To find the percent of total for each of the following numbers: 100, 400 and 600, first determine the total by adding up the numbers 100, 400 and 600. The total is (100+400+600) = 1100. Then find what percent of total, 1100 is the number 100. This can be computed as (100 / 1100) *100 = 9.090%.

5.3.1 Demo 1

Objective: To demonstrate "Profitability as Percent of Total" for categories of products per segment per region across several years (2011, 2012, 2013 and 2014).

Data set used: Sample - Superstore.xls

Expected output: (Shown in Fig. 5-15).

Region	Segment	Category	2011	2012	2013	2014
Central	Consumer	Furniture	36.25% 63.75%	46.20% 53.80%	39.47% 60.53%	33.75% 66.25%
		Office Supplies	72.59% 27.41%	64.83% 35.17%	76.63% 23.37%	80.89%
		Technology	49.15% 50.85%	96.84%	90.43%	96.02%
	Corporate	Furniture	90.21%	54.56% 45.44%	66.79% 33.21%	63.24% 36.76%
		Office Supplies	79.14%	92.44%	88.99%	78.07% 21.93%
		Technology	39.36% 60.64%	100.00%	98.46%	87.68%
	Home Office	Furniture	47.54% 52.46%	48.28% 51.72%	41.78% 58.22%	43.65% 56.35%
		Office Supplies	92.60%	95.66%	87.80%	76.02% 23.98%
		Technology	98.65%	99.68%	85.44%	96.67%
East	Consumer	Furniture	43.31% 56.69%	58.75% 41.25%	56.40% 43.60%	59.28% 40.72%
		Office Supplies	85.85%	91.66%	83.38%	73.28% 26.72%
		Technology	86.73%	82.38%	78.08% 21.92%	82.01%
	Corporate	Furniture	47.66% 52.34%	61.86% 38.14%	55.55% 44.45%	64.72% 35.28%
		Office Supplies	80.58%	89.23%	87.70%	86.14%
		Technology	61.49% 38.51%	76.61% 23.39%	86.81%	76.69% 23.31%
	Home Office	Furniture	65.55% 34.45%	67.57% 32.43%	58.35% 41.65%	41.18% 58.82%
		Office Supplies	99.25%	86.98%	89.39%	87.18%
		Technology	100.00%	90.97%	75.15% 24.85%	90.09%
South	Consumer	Furniture	84.23%	57.31% 42.69%	87.56%	60.82% 39.18%
		Office Supplies	81.34%	73.23% 26.77%	93.33%	79.85% 20.15%
		Technology	64.98% 35.02%	96.12%	96.20%	93.04%
	Corporate	Furniture	95.97%	71.84% 28.16%	56.23% 43.77%	44.52% 55.48%

Order Date; axis: 0% 50% 100% — % of Total Sales

Figure 5-15. "Profitability as percent of total" - Demo 1 - expected output

5.3.1.1 Steps

Follow the steps as provided.

5.3.1.1.1 Step 1

Read in the data from "Sample - Superstore.xls" (Shown in Fig. 5-16).

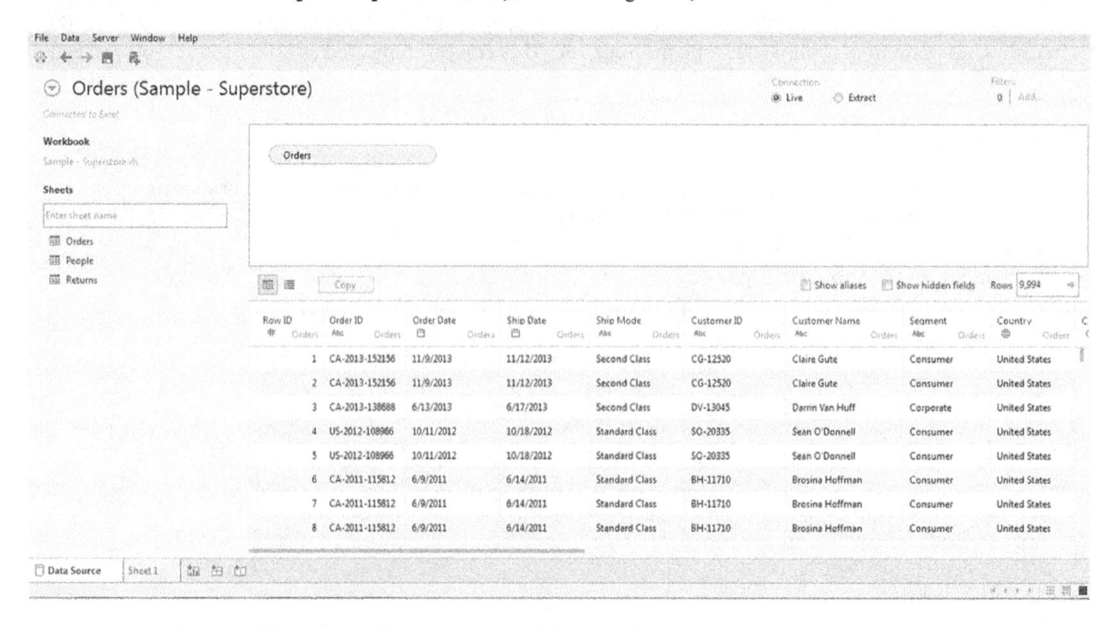

Figure 5-16. Data read from "Sample - Superstore.xls" into Tableau

5.3.1.1.2 Step 2

Table 5-2. Tasks to be performed in the view

columns shelf	Order Date ("Discrete" with the granularity set to "Year") Sales - Aggregation set to "SUM" Add a Table Calculation, "Percent of Total" to the measure "Sales" and summarize the values from - "Cell"
Rows Shelf	Region, Segment and Category

Drag the **dimension** "Order Date" from the dimensions area under the data pane to the columns shelf. Set it to "Discrete". Also set the granularity to "Year" (Shown in Fig. 5-17).

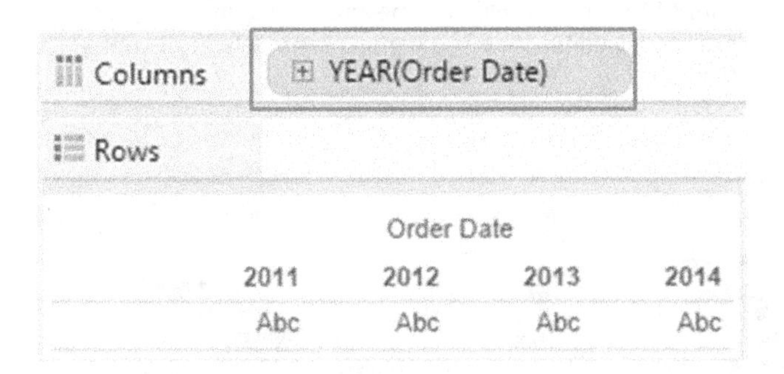

Figure 5-17. *Dimension "Order Date" placed on the columns shelf*

The visual cue ("Order Date" appears blue in color on the columns shelf) indicates that the dimension is discrete. By default, date type fields have a hierarchy defined on it and the default is the highest level in the hierarchy, which in this case is "Year".

Drag the **measure** "Sales" from the measures area under the data pane and drop it to the right of "Order Date". Let it be set to the default aggregation of "SUM" (Shown in Fig. 5-18).

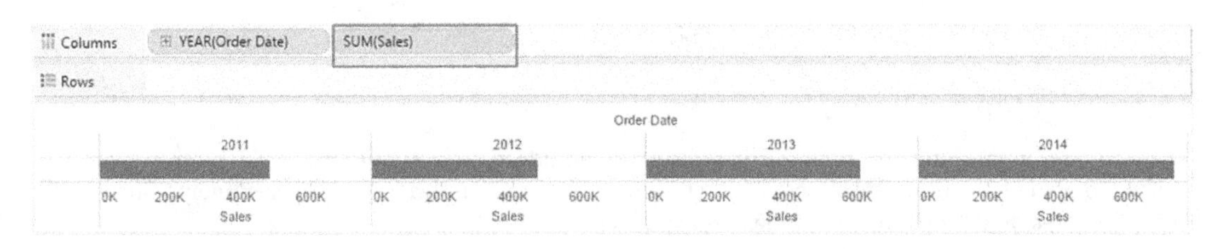

Figure 5-18. *Measure "Sales" placed on the columns shelf*

5.3.1.1.3 Step 3

Drag the **dimension** "Region" from the dimensions area under the data pane and drop it on the rows shelf. The data is available for four regions, namely, "Central", "East", "South" and "West" (Shown in Fig. 5-19).

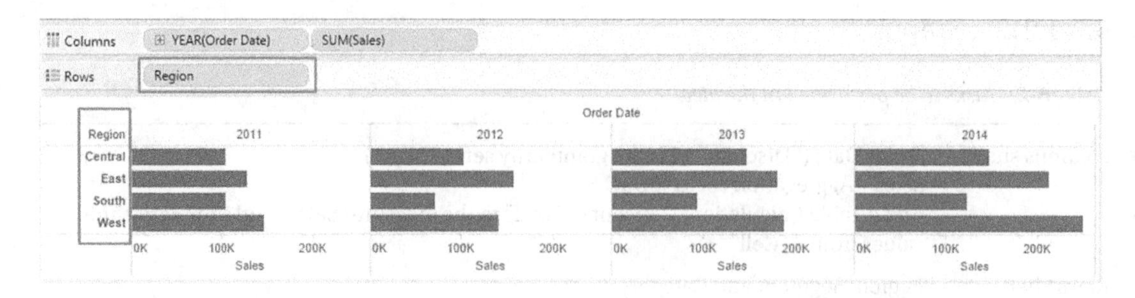

Figure 5-19. Dimension "Region" placed on the rows shelf

5.3.1.1.4 Step 4

Drag the **dimension** "Segment" from the dimensions area under the data pane and drop it on the rows shelf to the right of the **dimension** "Region". The data is available for three Segments, namely, "Consumer", "Corporate", and "Home Office" (Shown in Fig. 5-20).

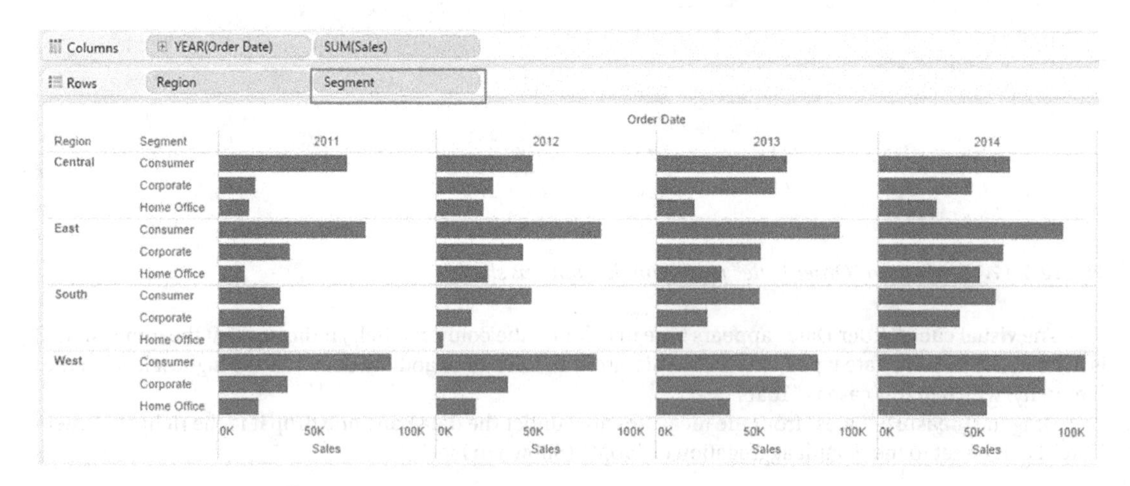

Figure 5-20. Dimension "Segment" placed on the rows shelf

5.3.1.1.5 Step 5

Drag the **dimension** "Category" from the dimensions area under the data pane and drop it on the rows shelf to the right of the **dimension** "Segment". The data is available for three Categories, namely, "Furniture", "Office Supplies", and "Technology" (Shown in Fig. 5-21).

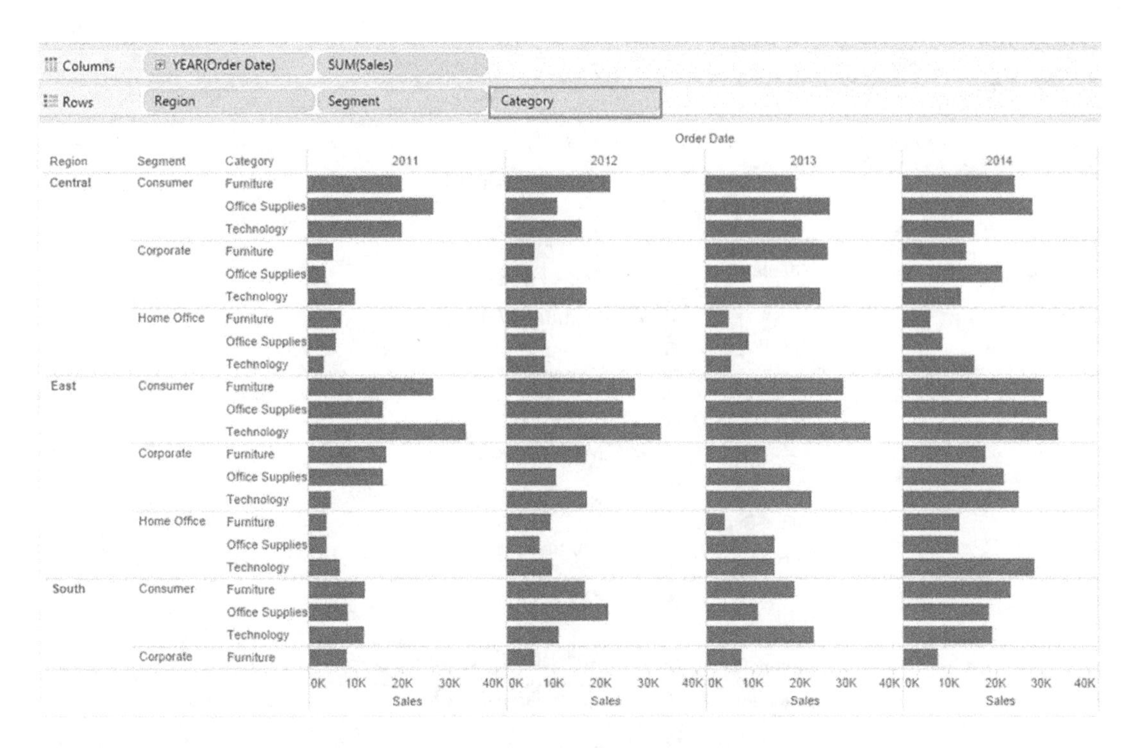

Figure 5-21. *Dimension "Category" placed on the rows shelf*

5.3.1.1.6 Step 6

Add a table calculation, "Percent of Total".

Click on the drop down of the **measure** "Sum (Sales)" and select "Add Table Calculation" (Shown in Fig. 5-22).

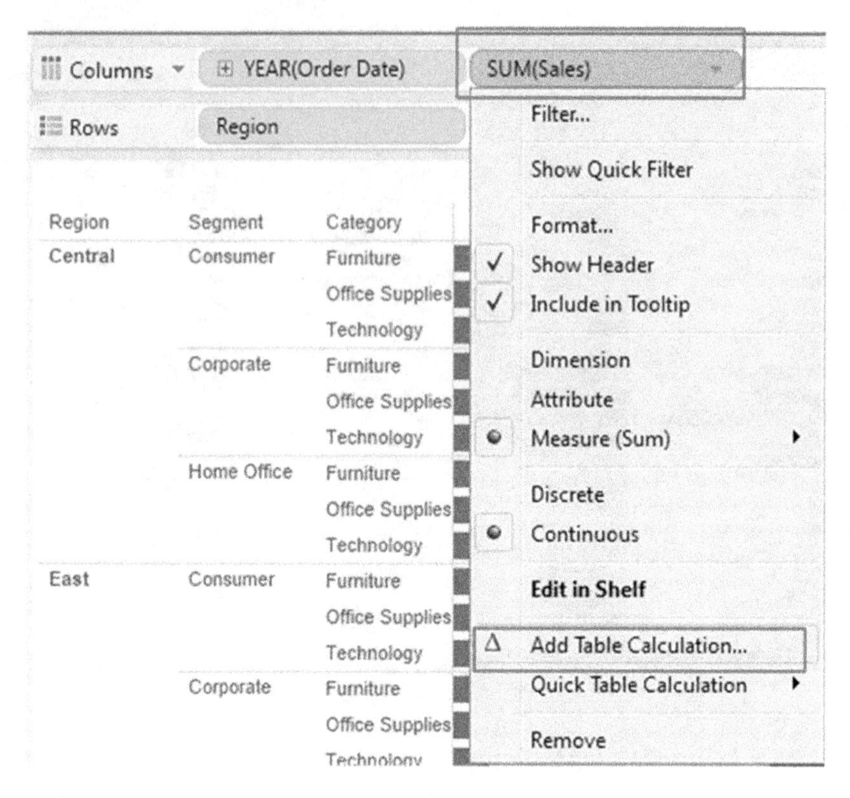

Figure 5-22. *"Add Table Calculation" to the measure "Sales"*

Fill in the values in the "Table Calculation" dialog box as shown in Fig. 5-23.

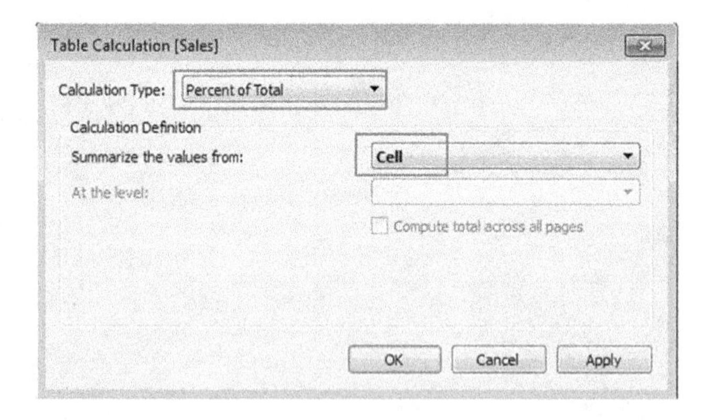

Figure 5-23. *"Table Calculation" dialog box for measure "Sales"*

Click on "Apply" and then click on "OK".
The output will be as shown in Fig. 5-24.

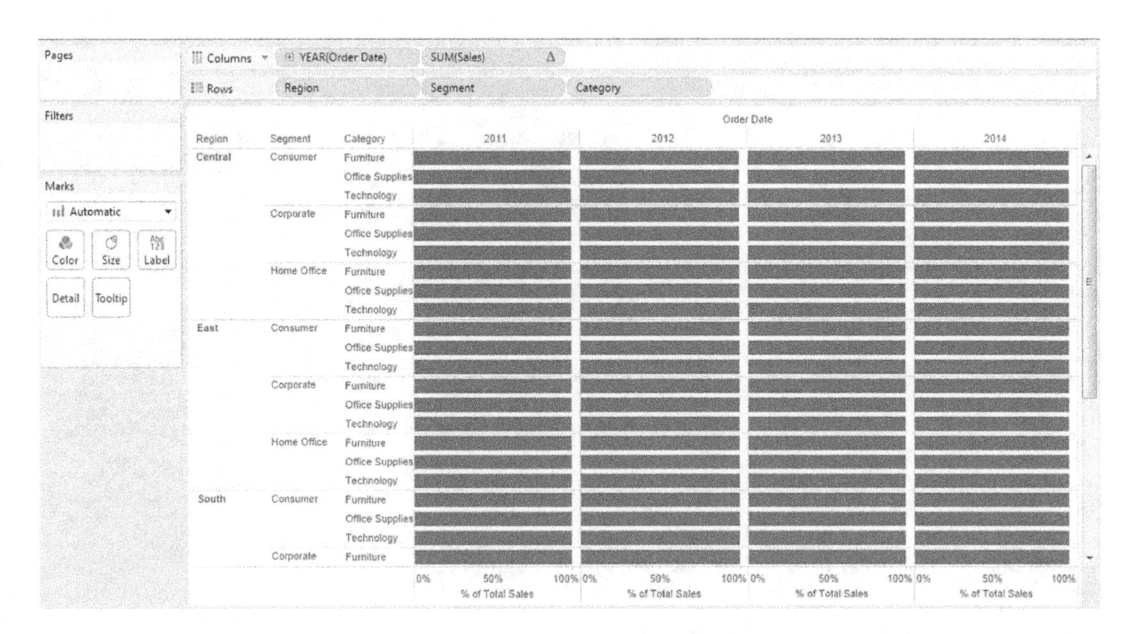

Figure 5-24. *Output after applying the "Table Calculation - Percent of Total" to the measure "Sales"*

5.3.1.1.7 Step 7

Create a calculated field, "Profit or Loss", as shown in Fig. 5-25.

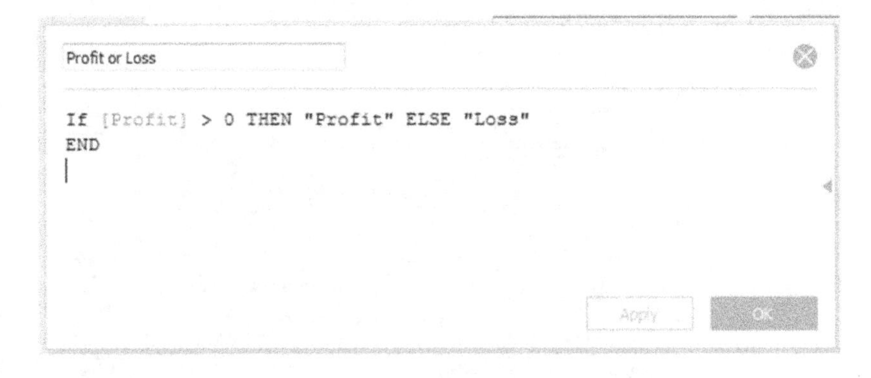

Figure 5-25. *"Calculated Field - Profit or Loss" being created*

Note that a new **dimension** "Profit or Loss" is added to the dimensions area under the data pane.

5.3.1.1.8 Step 8

Drag the newly created d**imension** "Profit or Loss" and drop it on "Color" on the marks card (Shown in Fig. 5-26).

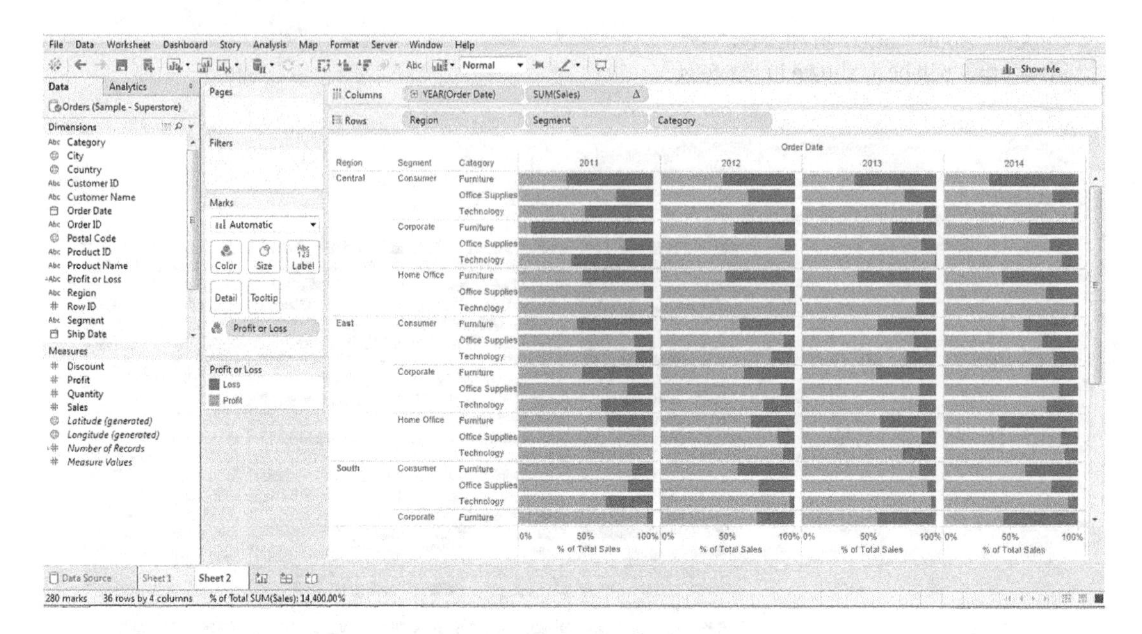

Figure 5-26. *"Calculated Field - Profit or Loss" placed on "Color" on the marks card*

5.3.1.1.9 Step 9

Press control key (CTRL) and drag the measure "Sum (Sales) (where we have added the table calculation)" from the columns shelf and drop it **on the "Label"** on the marks card (Shown in Fig. 5-27).

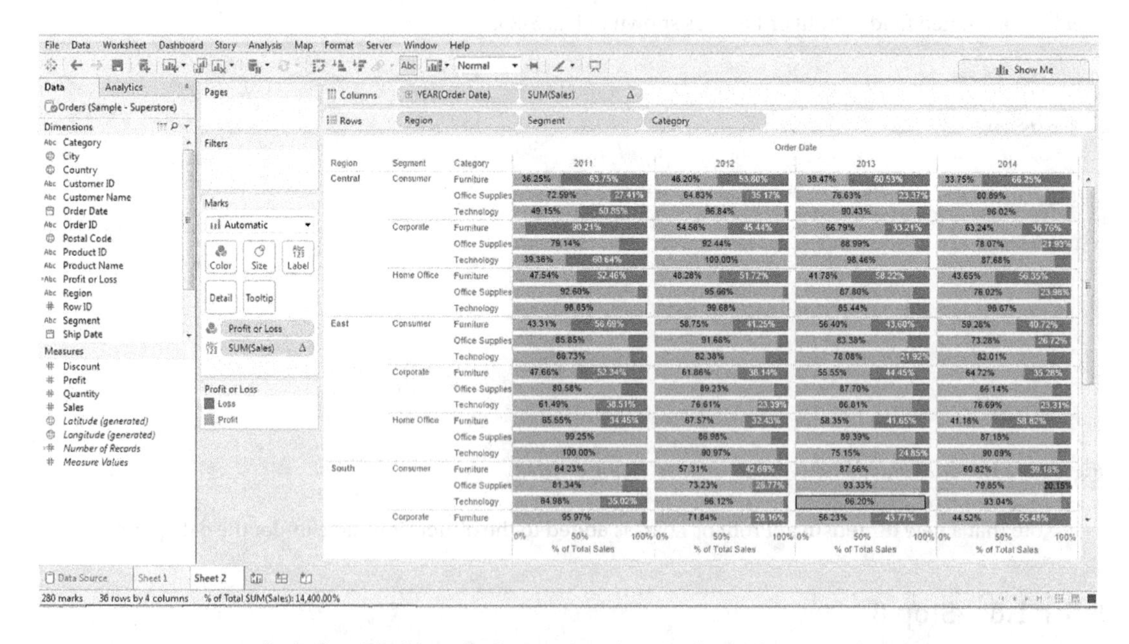

Figure 5-27. *Measure "Sales" with Table Calculation placed on the "Label" on the marks card*

The loss is shown in blue color and the profit in orange color.

5.4 Moving average

It is called by various names, such as rolling average, running average, rolling means, or running means.

5.4.1 Where is it used?

- In the technical analysis of financial data, such as stock prices, returns etc.
- To determine the market conditions. It is used with time-series data to iron out short-term price fluctuations or noises and highlight longer-term trends.
- To identify trends and reversals. Moving averages are lagging indicators. They are never used to predict new trends but confirm trends once they have been established. Example: a stock is termed uptrend when its price is above the moving average and the average slopes upwards. Likewise a stock is considered downtrend when its price is below the moving average and the average is sloping downward.
- To measure the strength of an asset's momentum. This has to do with the time period chosen for computing the moving average.
    - Short-term momentum: <=20 days
    - Medium-term momentum: between 20 to 100 days
    - Long-term momentum: > 100 days
- A valuable tool in planning trading strategy.
- To help with creation of a number of other technical indicators such as moving average convergence divergence (MACD) or Bollinger's bands.
- To help with stochastic measurements.

5.4.2 Types of moving average

Simple Moving Average: This is computed by taking arithmetic mean of a given set of values.

Weighted Moving Average: This is used to ensure that the most recent values have the most impact on the average. It uses values that are linearly weighted. Example: the oldest value is given a weight of 1, the next oldest value a weight of 2, and so on ... all the way up to the most recent value which gets the highest weight.

Exponential Moving Average: This is similar to the simple moving average. The difference lies in the fact that while a simple moving average will remove the older values as the new values become available, the exponential moving average calculates the average of all historical ranges, starting at the points that one specifies.

■ **Points to Note** Moving Averages are Lagging Indicators. They are based on events that have already occurred in the market.

They are not predictive indicators.

5.4.3 Demo 1

Objective: To demonstrate the "Moving Average" of the **measure** "Sales" across several years (2011, 2012, 2013 and 2014).

Input Data Set: "Sample - Superstore.xls"

Expected output: Shown in Fig. 5-28.

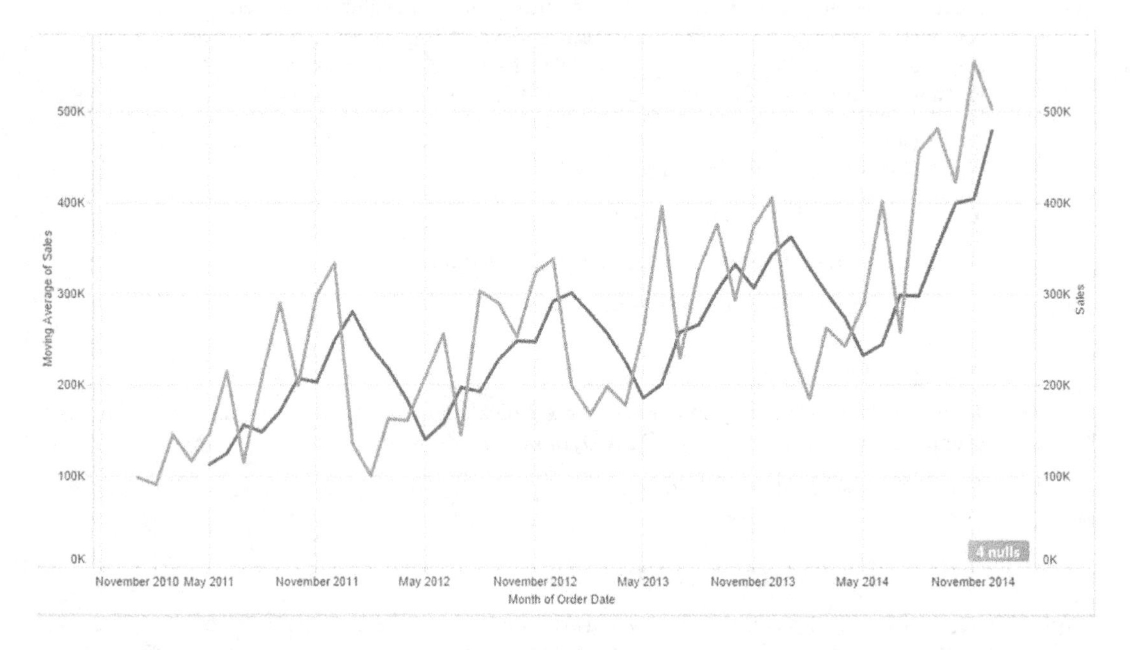

Figure 5-28. *Moving average - Demo 1 - expected output*

5.4.3.1 Steps

Table 5-3. Tasks to perform in the view

Columns shelf	Month(Order Date) : Continuous Date
Rows shelf	Sum(Sales) with calculation type - "Moving Calculation", summarize values using "Average"
Rows shelf	Sum(Sales)

5.4.3.1.1 Step 1

Drag the **dimension** "Order Date" from the dimensions area under data pane and drop it on the columns shelf (Shown in Fig. 5-29).

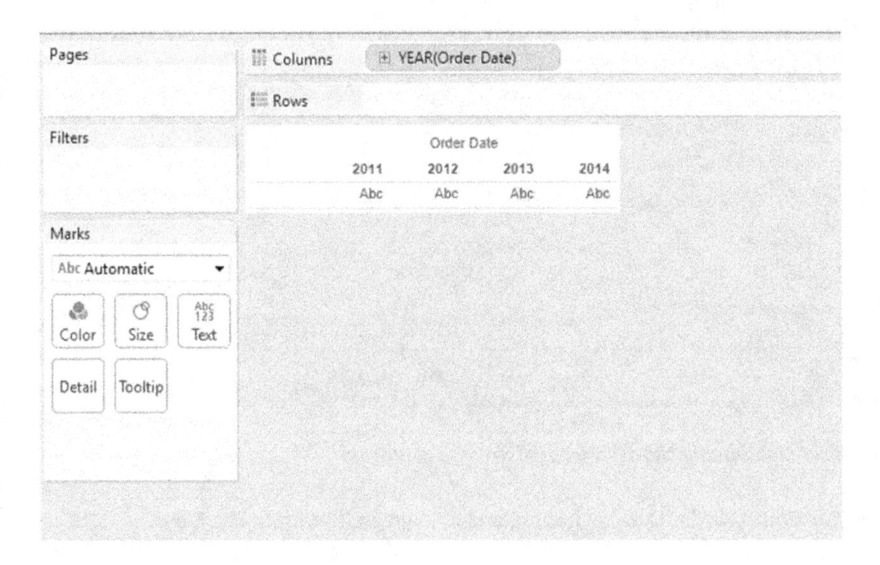

Figure 5-29. Dimension "Order Date" placed on columns shelf

5.4.3.1.2 Step 2

Change the "Order Date" to Continuous Month (Order Date) (Shown in Fig. 5-30).

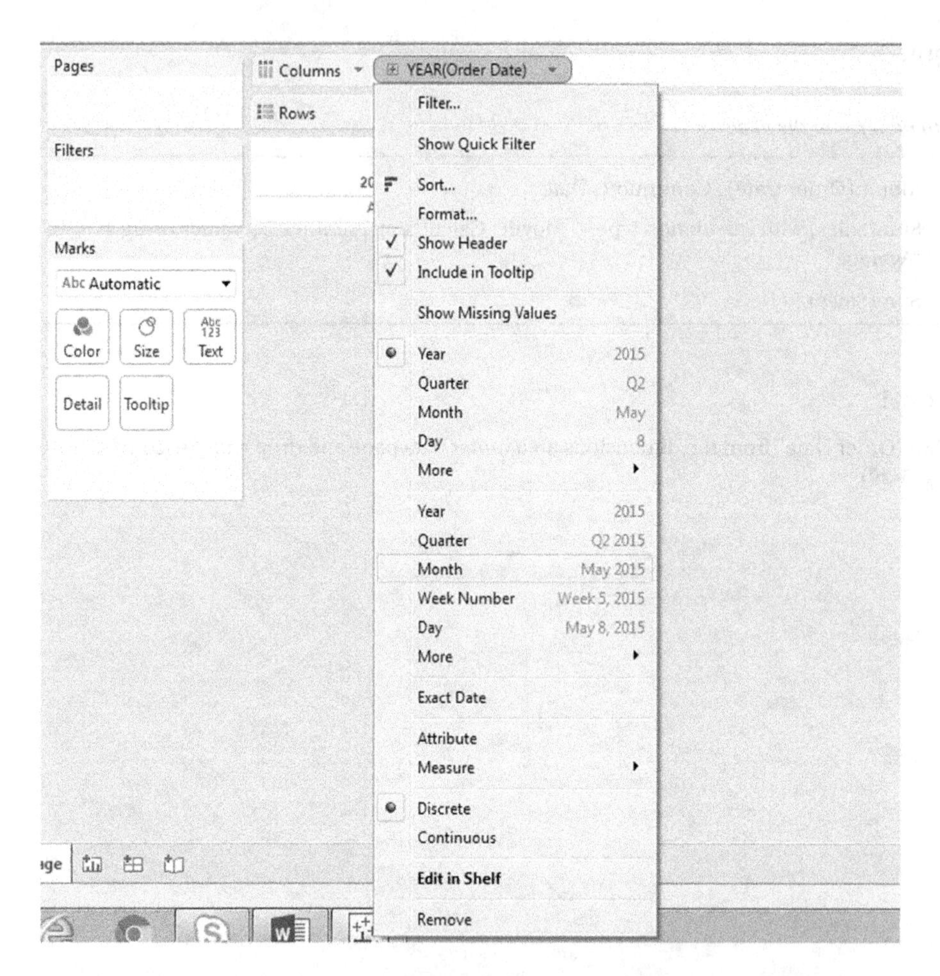

Figure 5-30. *Dimension "Order Date" being changed from discrete to continous*

Notice the visual cue. The Month (Order Date) color is changed to green (Shown in Fig. 5-31).

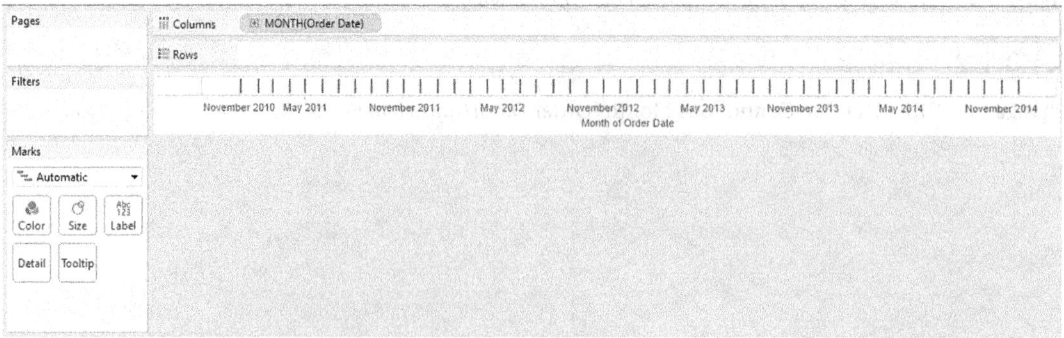

Figure 5-31. *Dimension "Order Date" changed to "Continous - Month"*

5.4.3.1.3 Step 3

Drag the **measure** "Sales" from the measures area under data pane and drop it on the rows shelf (Shown in Fig. 5-32).

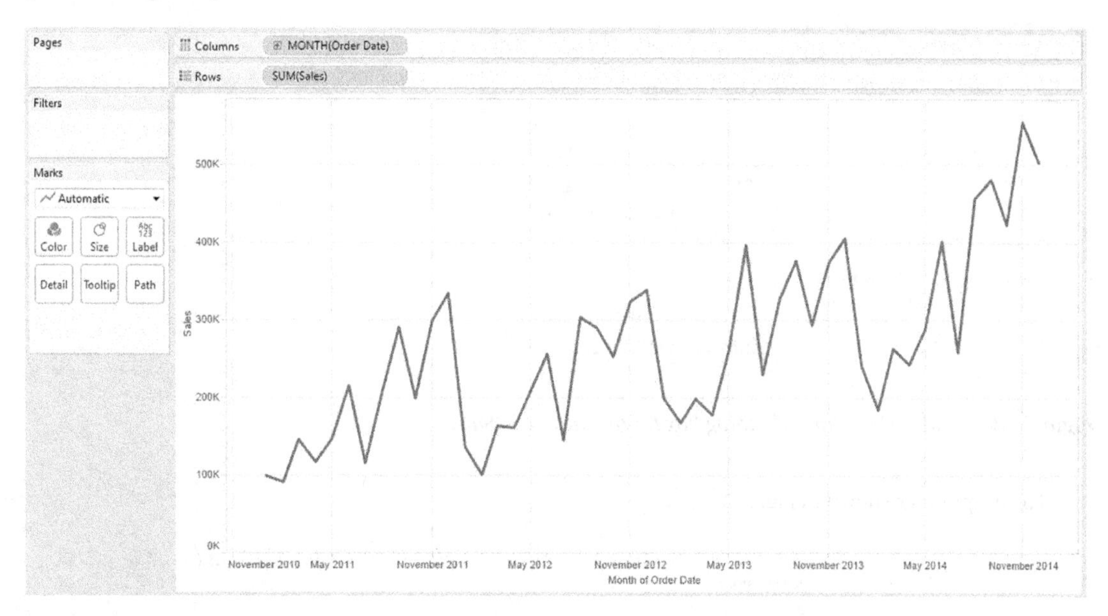

__Figure 5-32.__ Measure "Sales" placed on the rows shelf

5.4.3.1.4 Step 4

Click on the drop down of Sum (Sales) to select "Add Table Calculation" (Shown in Fig. 5-33).

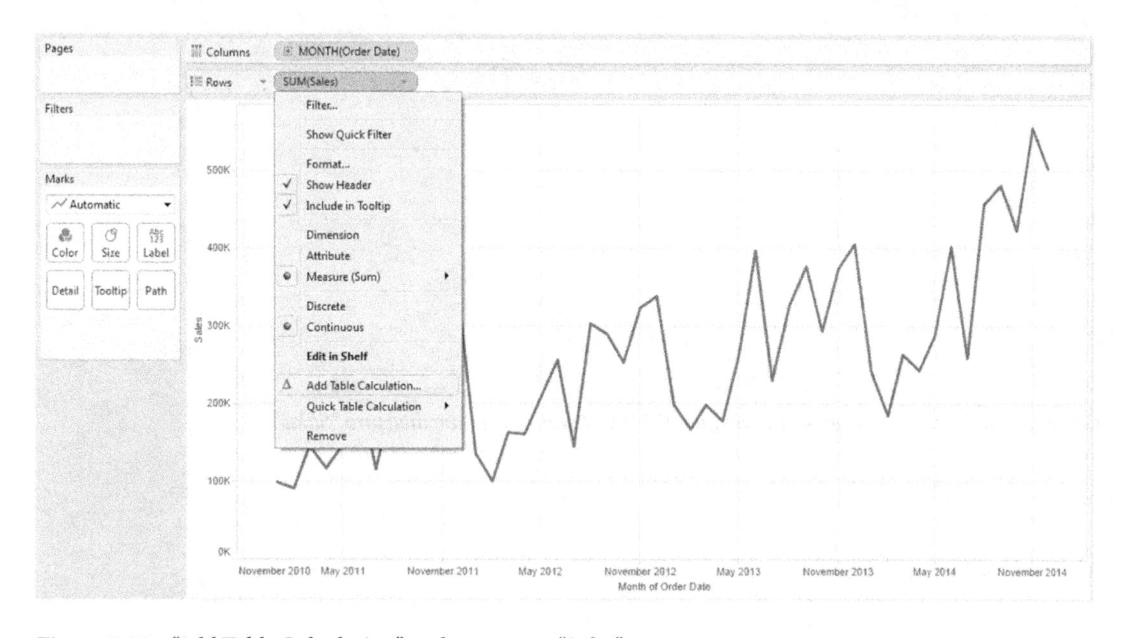

__Figure 5-33.__ "Add Table Calculation" to the measure "Sales"

5.4.3.1.5 Step 5

Fill in the values in the "Table Calculation" dialog box as shown in Fig. 5-34.

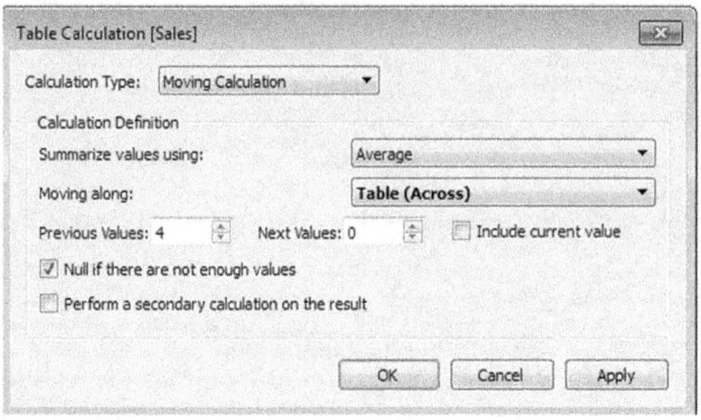

Figure 5-34. *"Table Calculation" dialog box for the measure "Sales"*

The output is as shown in Fig. 5-35.

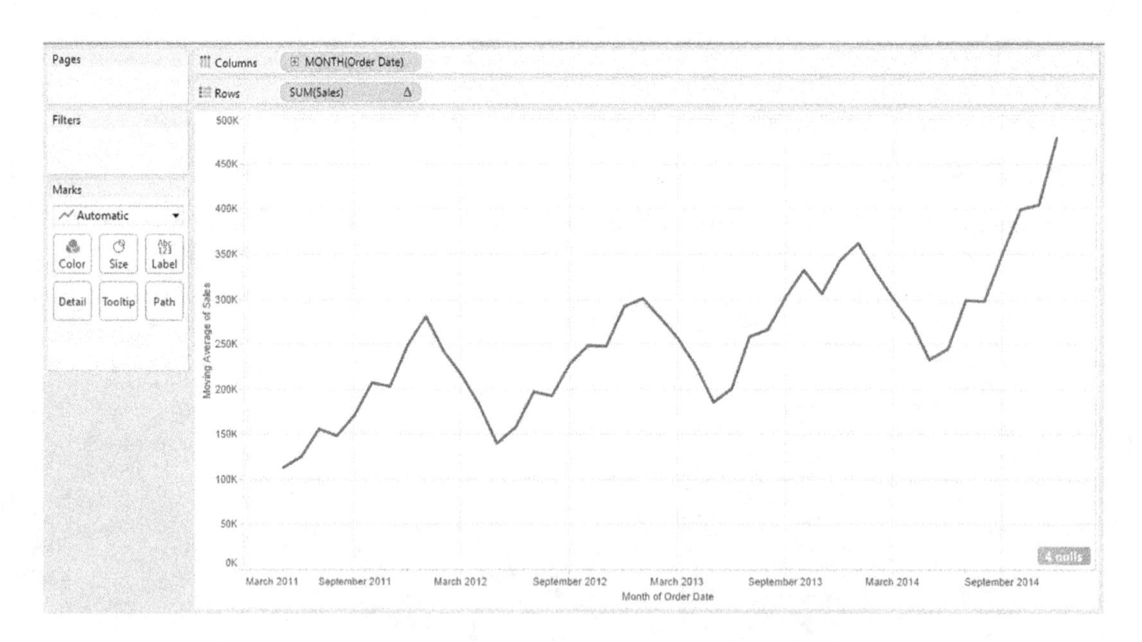

Figure 5-35. *The output after applying the "Table Calculation" to the measure "Sales"*

Click on "4 nulls". It brings up the "Special Values for [Moving Average of Sales]" dialog box (Shown in Fig. 5-36).

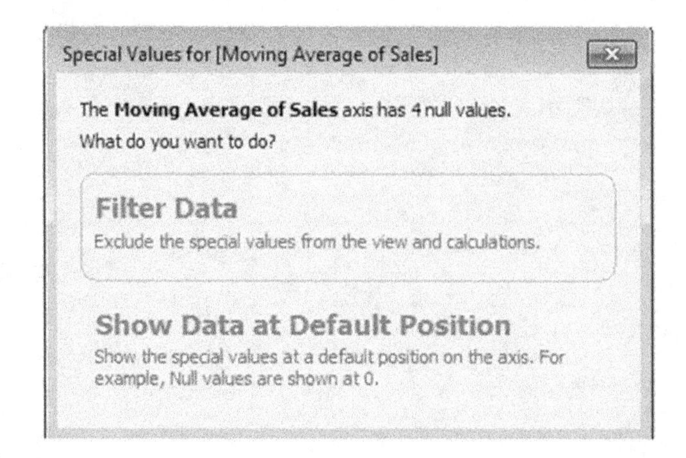

Figure 5-36. *"Special Values for [Moving Average of Sales]" dialog box*

Click on "Show Data at Default Position".
The output changes to the below (Shown in Fig. 5-37).

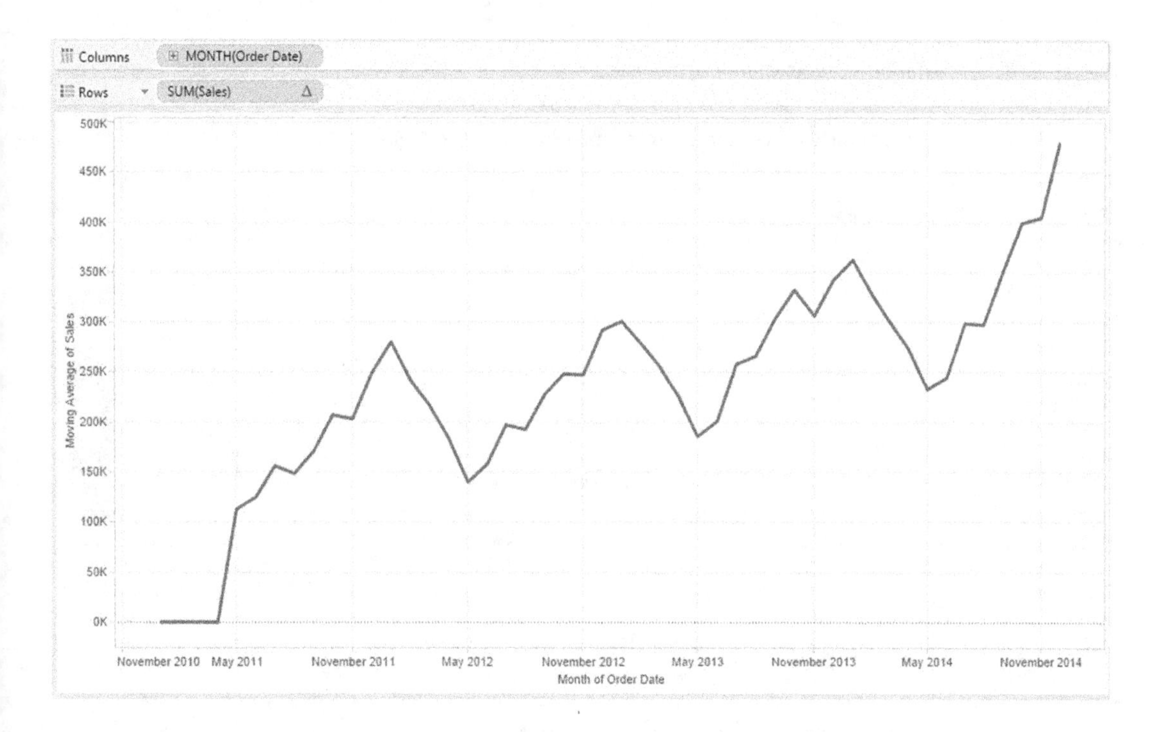

Figure 5-37. *The output after setting the null values*

5.4.3.1.6 Step 6

Drag the **measure** "Sales" from the measures area under the data pane and drop it on the opposite axis (Shown in Fig. 5-38).

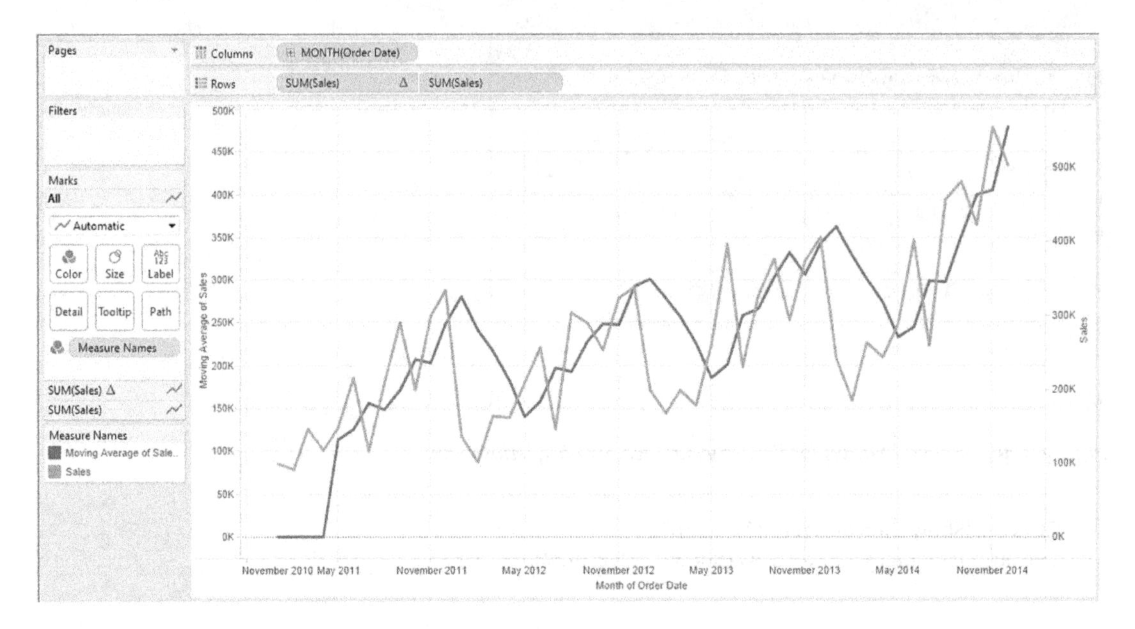

Figure 5-38. Measure "Sales" placed on the rows shelf

Synchronize the secondary axis to the primary one (Shown in Fig. 5-39).

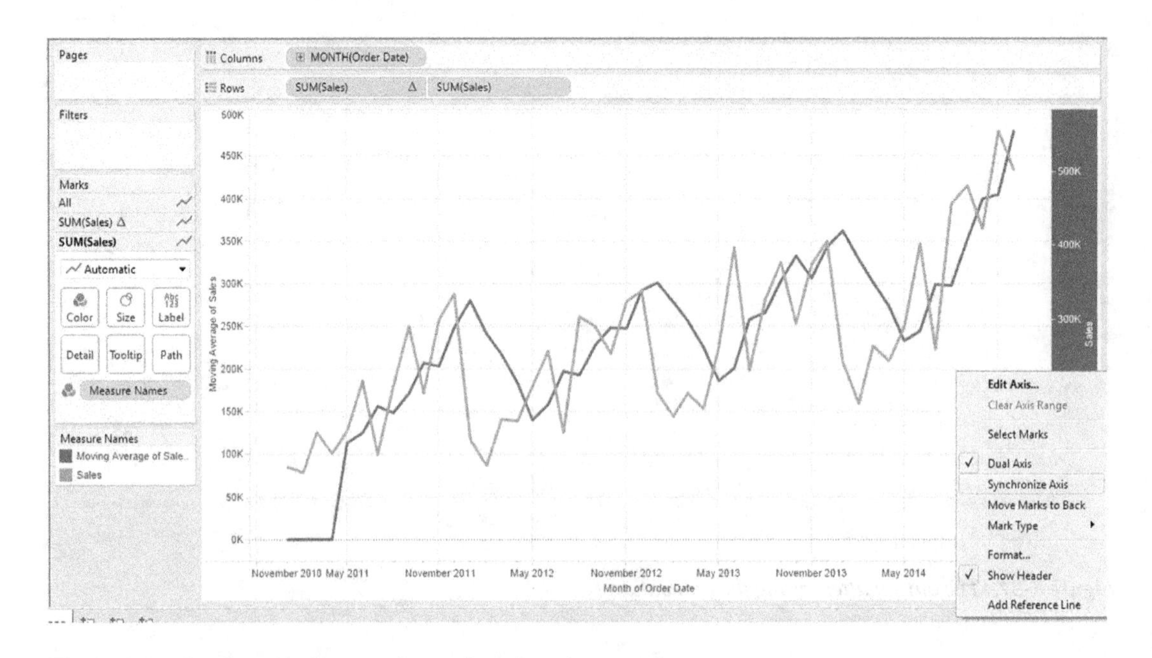

Figure 5-39. Synchronize the secondary axis to the primary axis

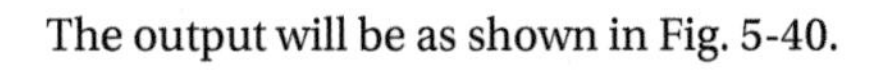

The output will be as shown in Fig. 5-40.

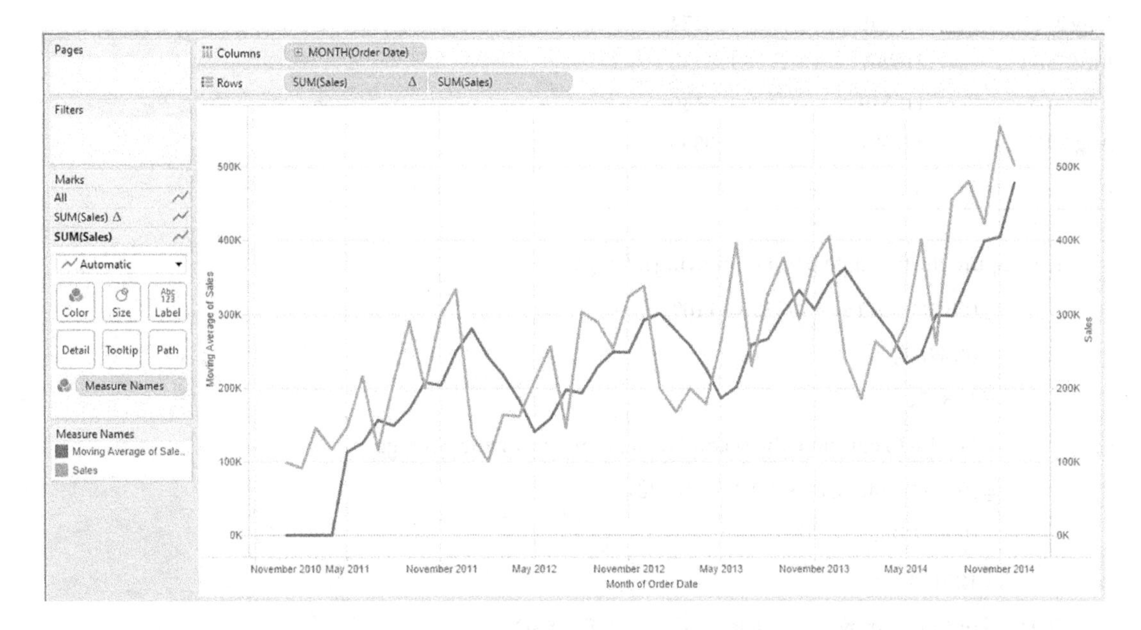

Figure 5-40. *The output after the secondary axis is synchronized with the primary axis*

Let us verify the result:

Month	Sum(Sales)
Jan 2011	98898
Feb 2011	91152
Mar 2011	145729
Apr 2011	116916
May 2011	146748
June 2011	215207
July 2011	115510
....	....

The moving average is calculated moving along, “Table Across” using the previous four values and NOT including the current value. If there are not enough values, it will use Null.

Month	Sum(Sales)	Moving Average
Jan 2011	98898	
Feb 2011	91152	
Mar 2011	145729	
Apr 2011	116916	

Month	Sum(Sales)	Moving Average
May 2011	146748	113174
June 2011	215207	125136
July 2011	115510	156150
Aug 2011	207581	148595
....	...	....

To compute the first data point for moving average:

= (98898 + 91152 + 145729 + 116916) / 4

= (452695)/ 4

= 113174

Likewise to compute the second data point for Moving Average:

= (91152 + 145729 + 116916 + 146748) / 4

= (500545) / 4

= 125136

The final output of moving averages is shown in Fig. 5-41.

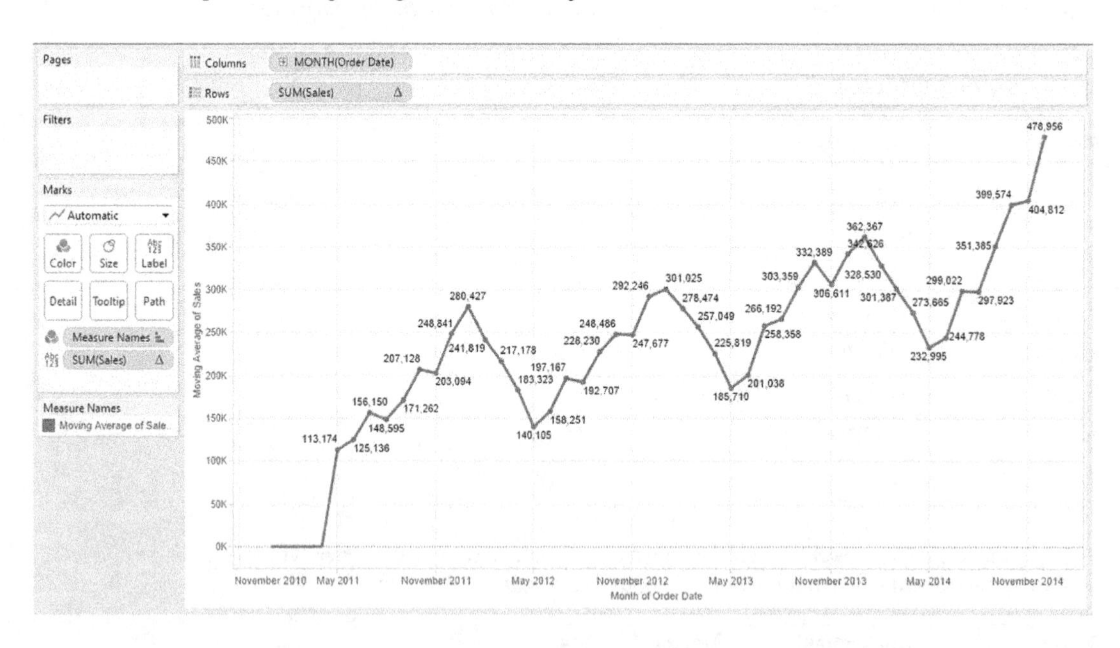

Figure 5-41. *Moving average - Demo 1 - final output*

5.5 Rank

Tableau ignores "nulls" in rank calculations. They appear as blank rows.

The following choices are available on the type of ranking that one can apply:

Table 5-4. *Type of Ranking*

Type of Ranking	Example	
1,2,2,2,5 (Competition)	Below is the input set and the ranks assigned in ascending order	
	Input set	Rank
	50	1
	60	2
	60	2
	65	4
	78	5
	90	6
1,4,4,4,5 (Modified Competition)	Below is the input set and the ranks assigned in ascending order	
	Input Set	Rank
	50	1
	60	3
	60	3
	65	4
	78	5
	90	6
1,2,2,2,3 (Dense)	Below is the input set and the ranks assigned in ascending order	
	Input set	Rank
	50	1
	60	2
	60	2
	65	3
	78	4
	90	5
1,2,3,4,5 (Unique)	Below is the input set and the ranks assigned in ascending order	
	Input set	Rank
	50	1
	60	2
	60	3
	65	4
	78	5
	90	6

5.5.1 Demo 1

Objective: To rank the "Sub-Category" based on the "Sales Amount". The sub-Category with the highest sales amount is ranked one followed by the Sub-Category with the next highest sales amount, which is ranked two and so on...

Input data Set: "Sample - Superstore.xls"

Expected Output: Shown in Fig. 5-42.

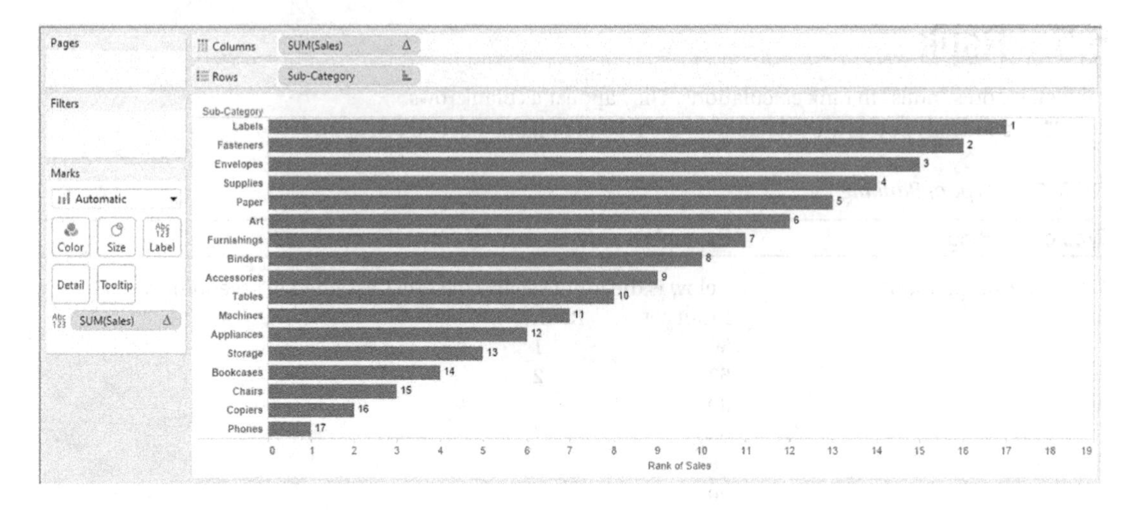

Figure 5-42. *"Sub-Category" ranked in descending order of "Sales" amount*

Table 5-5. *Tasks to perform in the view*

Columns shelf	Sum(Sales)
Row shelf	Sub-Category
Table calculation	Rank Running Along: Table (Down) Sort Order: Descending Rank duplicate Values as: Competition (1,2,2,4)

5.5.1.1 Steps

5.5.1.1.1 Step 1

Drag the **dimension** "Sub-Category' from the dimensions area under data pane and drop it on the rows shelf (Shown in Fig. 5-43).

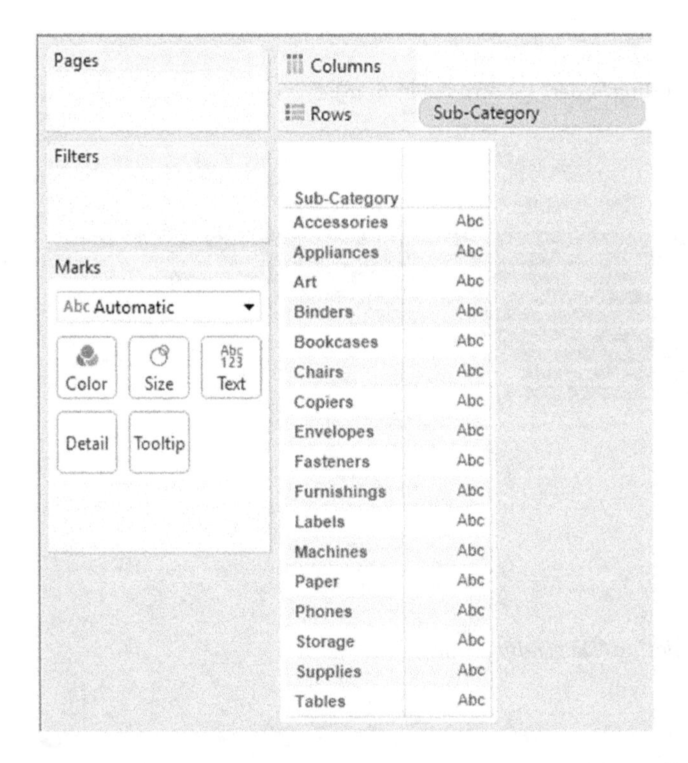

Figure 5-43. *Dimension "Sub-Category" placed on the rows shelf*

5.5.1.1.2 Step 2

Drag the **measure** "Sales" from the measures area under the data pane and drop it on the columns shelf (Shown in 5-44).

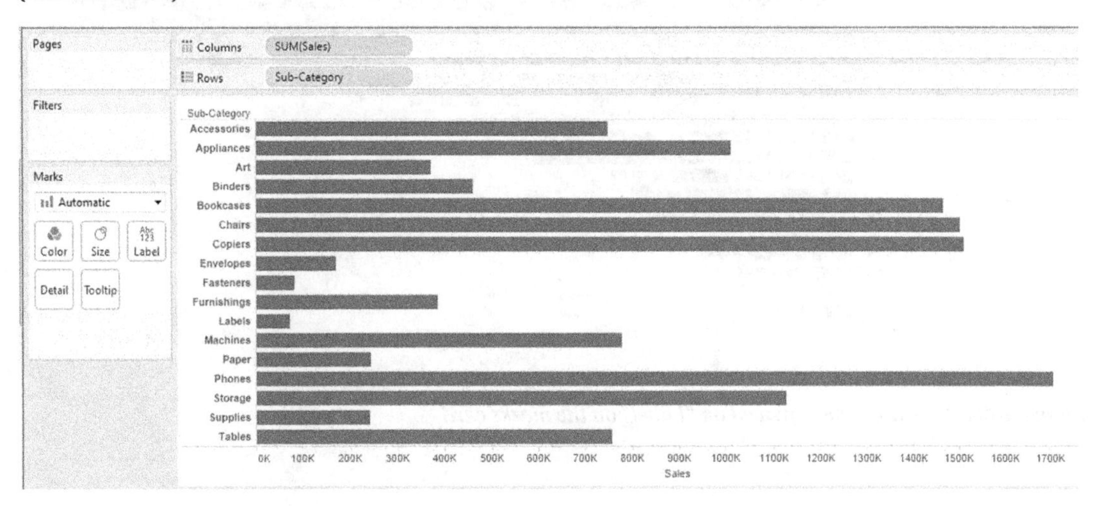

Figure 5-44. *Measure "Sales" placed on the columns shelf*

5.5.1.1.3 Step 3

Sort "Sub-Category" as per "Sales" in "Descending Order" (Shown in Fig. 5-45).

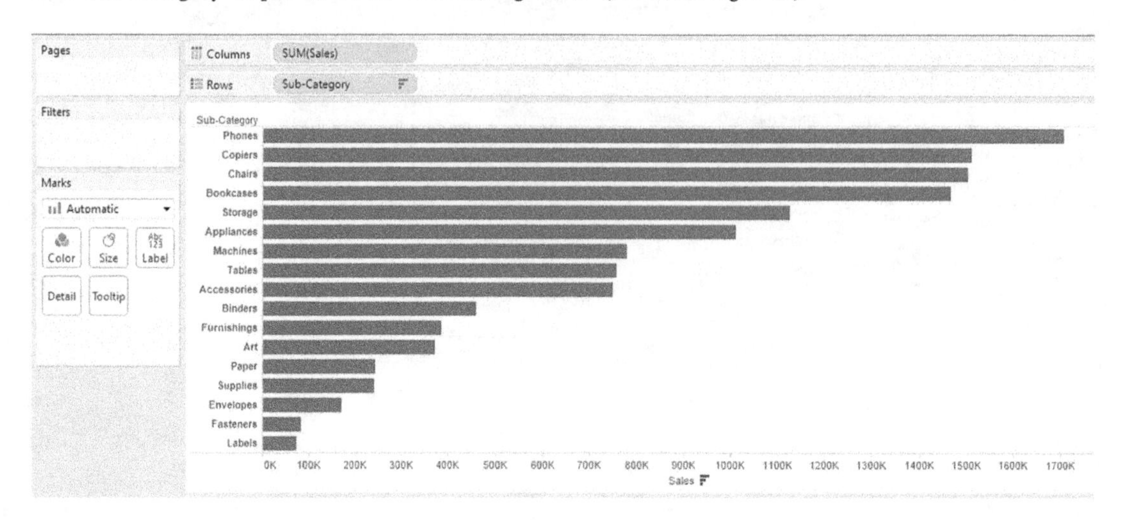

Figure 5-45. *Sort, "Sub-Category" as per "Sales" in "Descending Order"*

5.5.1.1.4 Step 4

Drag "Sales" on the **"Label"** on the marks card (Shown in Fig. 5-46).

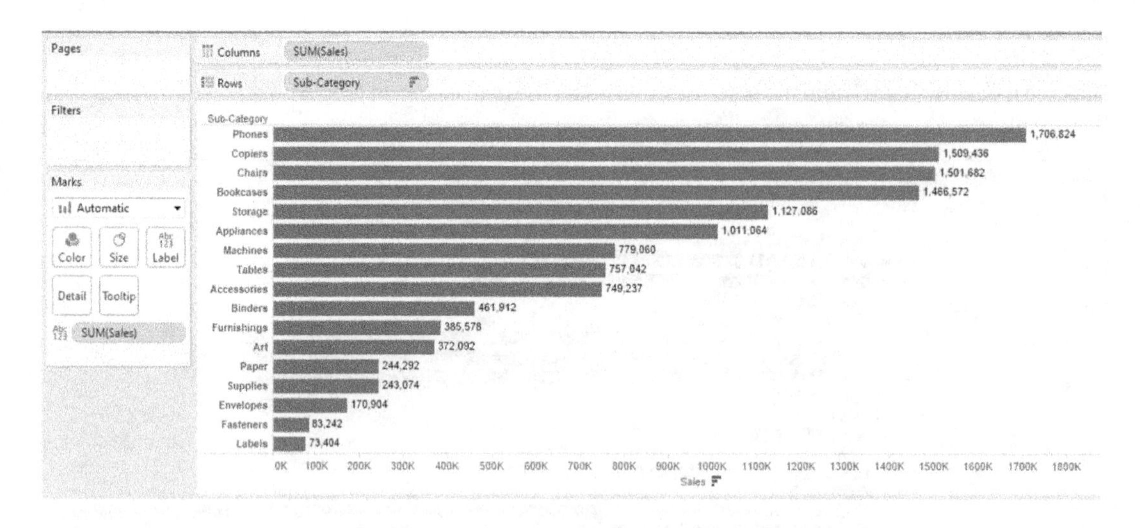

Figure 5-46. *Measure "Sales" placed on "Label" on the marks card*

Click on the drop down button on the Sum (Sales) to bring up the "Add Table Calculation" (Shown in Fig. 5-47).

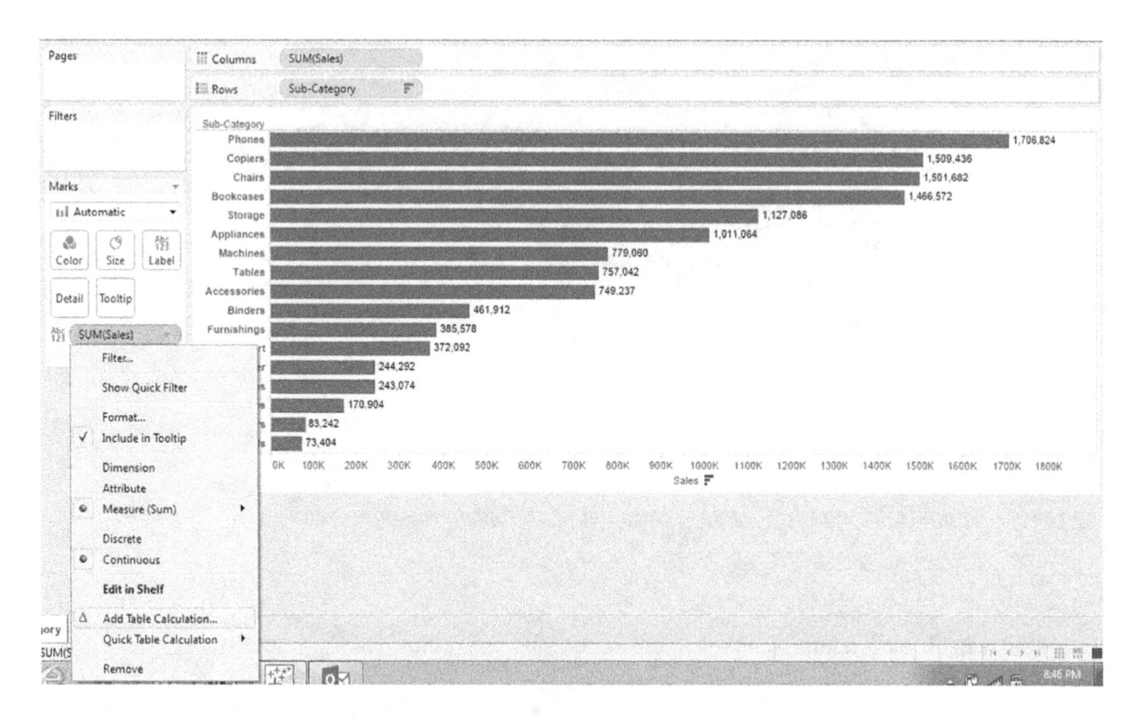

Figure 5-47. *Apply "Add Table Calculation" to the measure "Sales"*

Fill in the values in the "Table Calculation" dialog box as shown in Fig. 5-48.

Table Calculation [Sales]

Calculation Type: Rank

Calculation Definition

Running along: **Table (Down)**

Sort by: SUM(Sales)

Sort order: Descending Ascending

Rank duplicate values as: Competition (1, 2, 2, 4)

OK Cancel Apply

Figure 5-48. *"Table Calculation" dialog box for the measure "Sales"*

The output is as shown in Fig. 5-49.

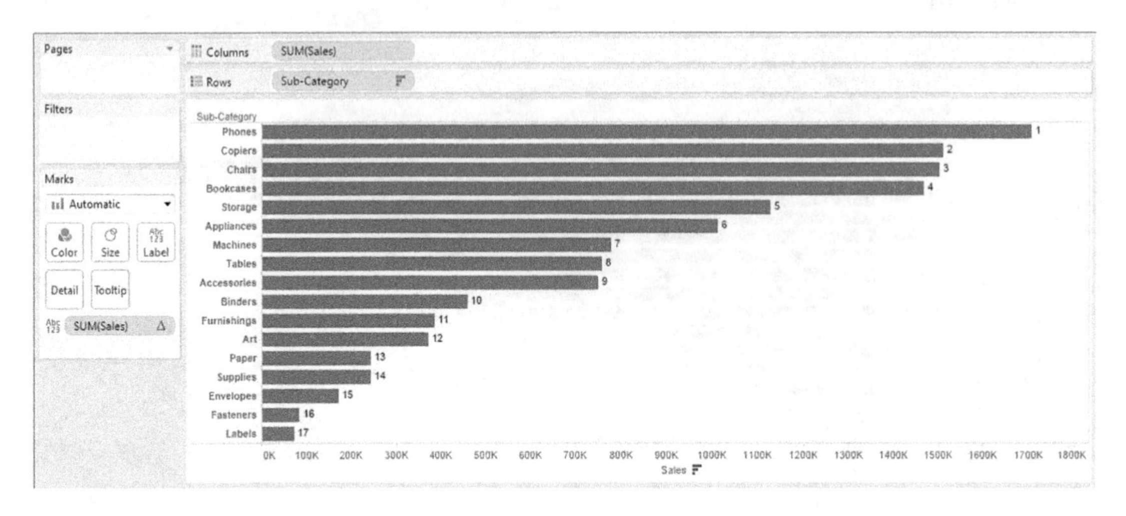

Figure 5-49. *Output after applying "Table Calculation - Rank" to the measure "Sales"*

5.5.1.1.5 Step 5

Let us display Rank as the leftmost column in the worksheet / view (Shown in Fig. 5-50).

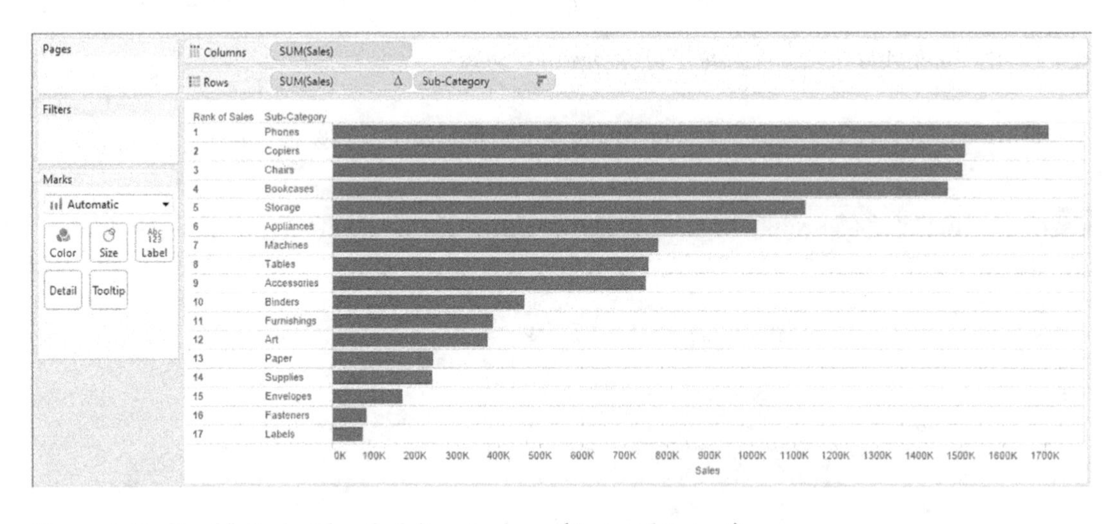

Figure 5-50. *"Rank" displayed as the leftmost column (Expected Output)*

How to achieve the above?

It is required to convert the Sum (Sales) **on the "Label"** on the marks card to "Discrete" (Shown in Fig. 5-51).

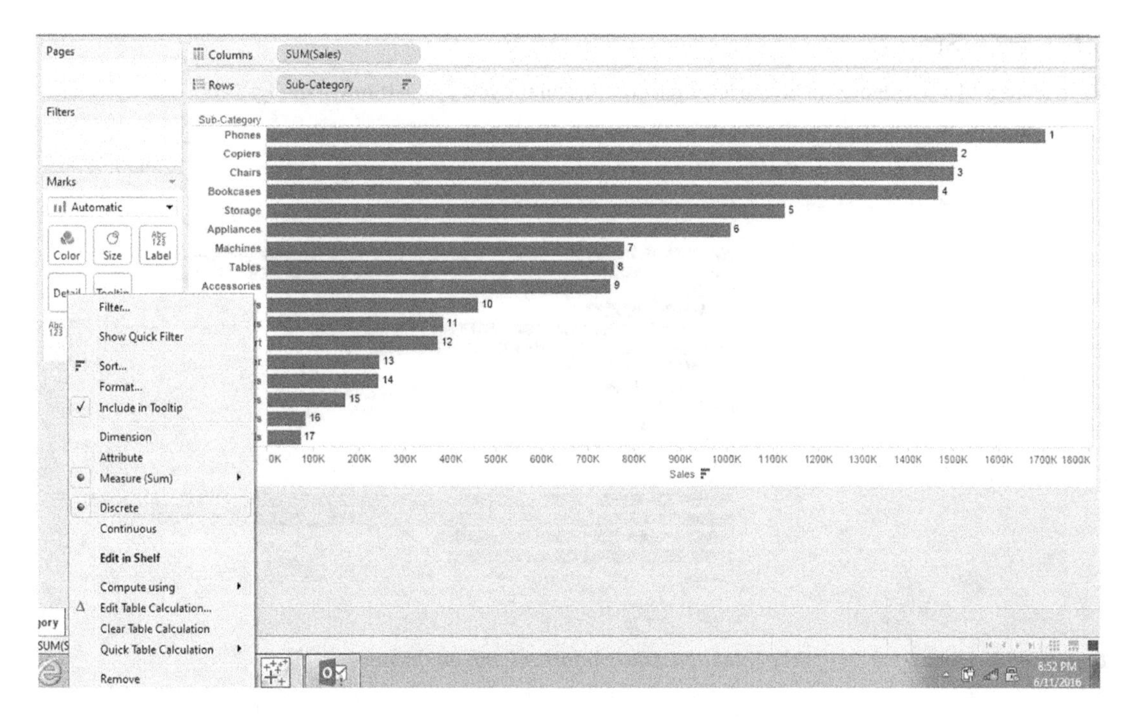

Figure 5-51. *Measure "Sales" converted to "Discrete"*

5.5.1.1.6 Step 6

Now drag the Sum (Sales) which is placed on the **"Label"** on the marks card and place it to the left of "Sub-Category" (Shown in Fig. 5-52).

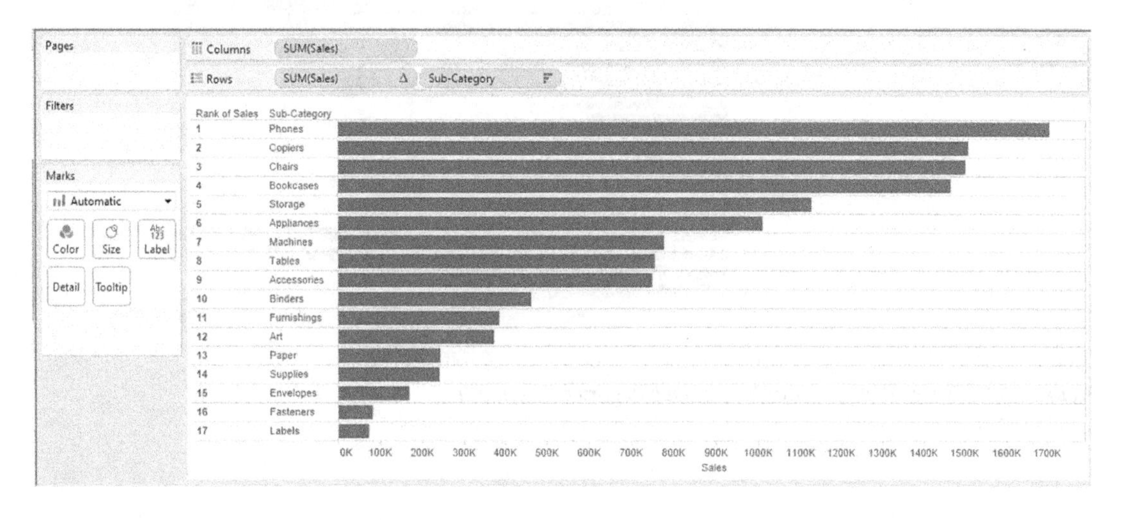

Figure 5-52. *Discrete measure "Sales" placed to the left of dimension "Sub-Category"*

5.5.1.1.7 Step 7

Next, let us add a "Category" to be the leftmost column on the rows shelf (Shown in Fig. 5-53).

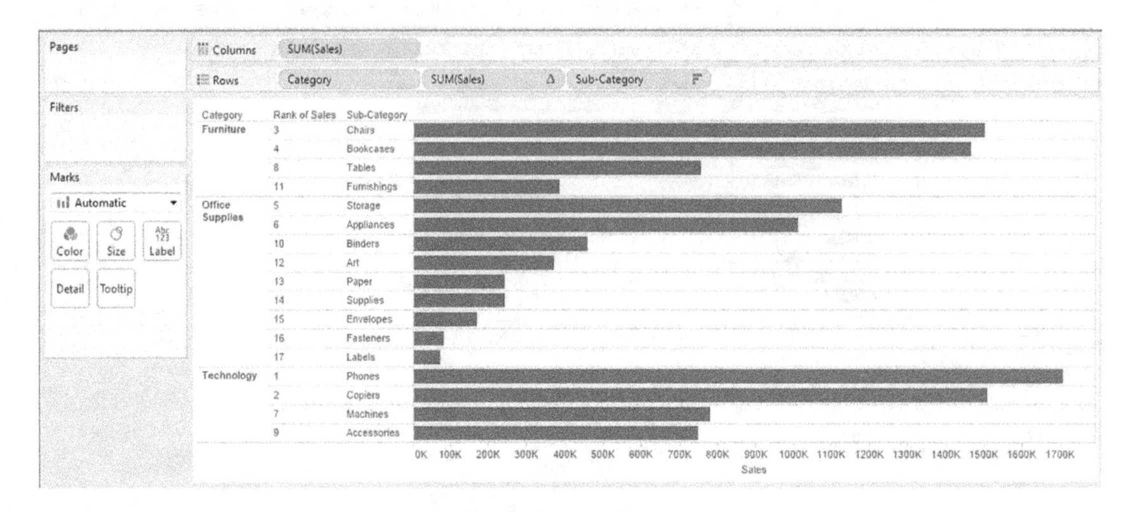

***Figure 5-53.** Dimension "Category" placed as the leftmost column on the rows shelf*

The rank for each category should begin at 1 (Shown in Fig. 5-54).

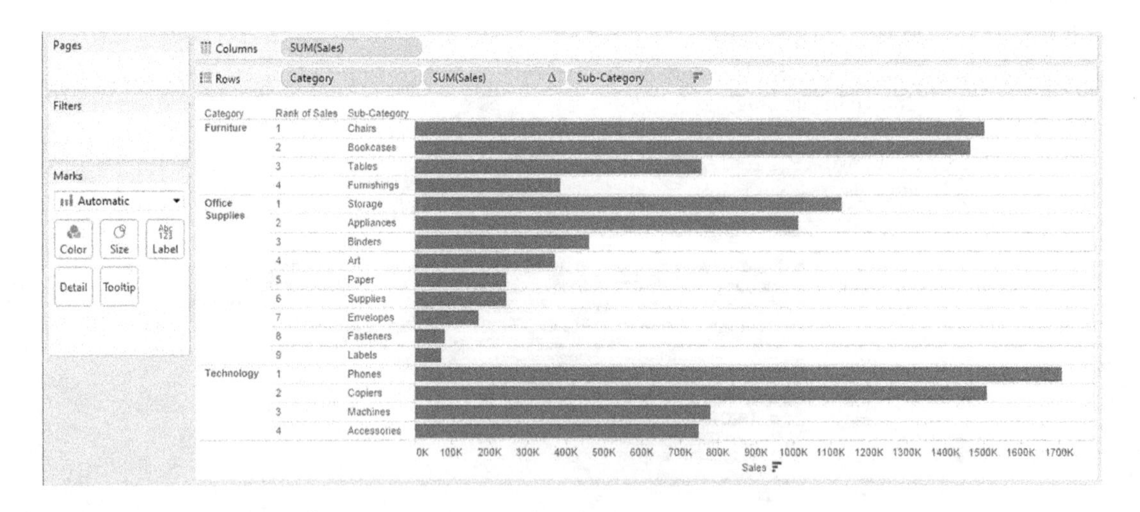

***Figure 5-54.** Expected output: Rank for each category to begin at 1*

How to achieve the above?

5.5.1.1.8 Step 8

Click on the drop down on Sum (Sales) to bring up the "Edit Table Calculation" (Shown in Fig. 5-55).

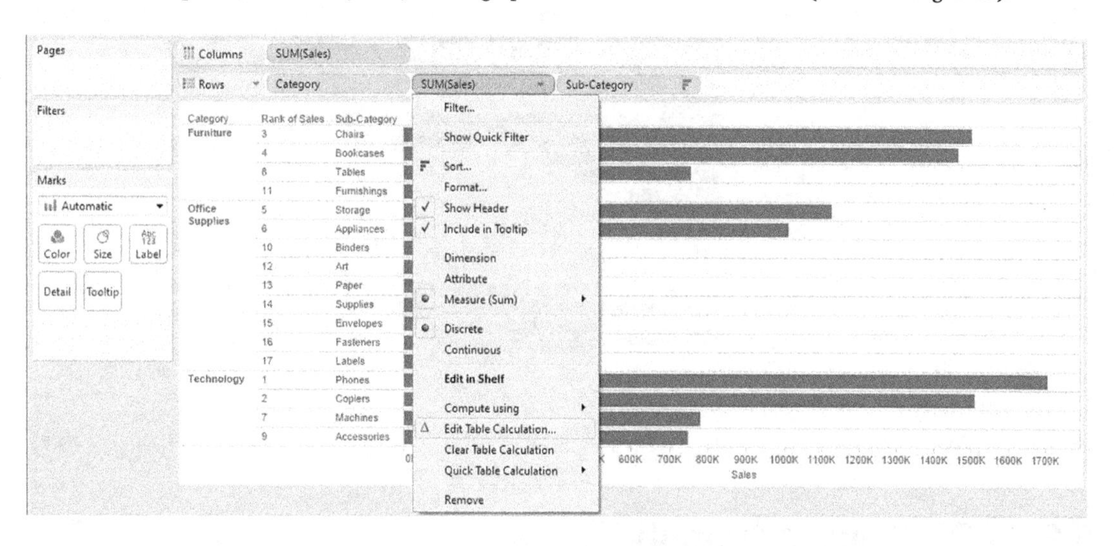

Figure 5-55. *Perform "Edit Table Calculation" to the measure "Sales"*

Change the "Running Along" to "Pane Down" (Figure 5-56).

Table Calculation [Rank of Sales]

Calculation Type: Rank

Calculation Definition

Running along: **Pane (Down)**

Sort by: SUM(Sales)

Sort order: Descending Ascending

Rank duplicate values as: Competition (1, 2, 2, 4)

OK Cancel Apply

Figure 5-56. *"Running Along" for rank of sales changed to pane (Down)*

Click on “Apply” and then “OK”.

The final output is shown in Fig. 5-57.

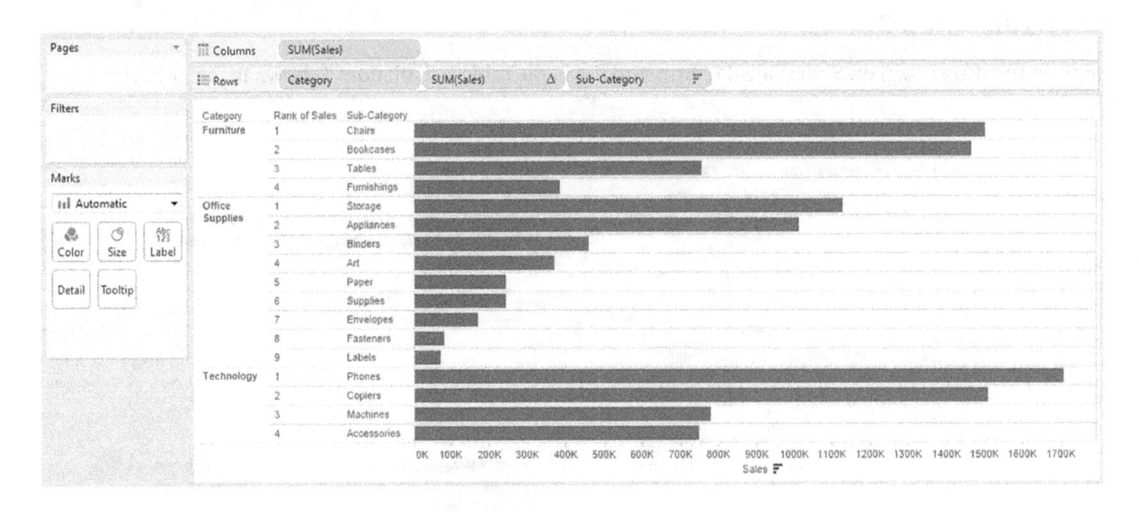

Figure 5-57. *Rank - Demo 1 - final output*

5.6 LOD (Level of Detail)

Level of detail (LOD) is a very important concept. An understanding of the idea of details helps with understanding the level of detail. Let us look at the areas where one can add details to the view / worksheet.

- Columns shelf
- Rows shelf
- Detail on the marks card

Details are defined by the dimensions that are used to segment the measures.

Example:

Drag the **dimension** “Region” from the dimensions area under the data pane and place it on the rows shelf.

Drag the **measure** “Sales” from the measures area under the data pane and place it on the columns shelf (Shown in Fig. 5-58).

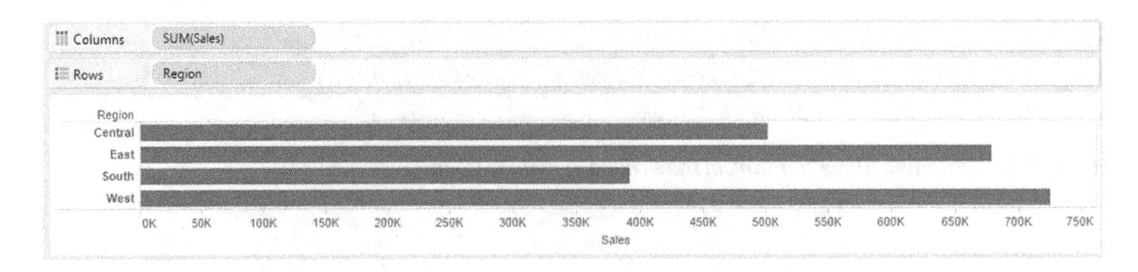

Figure 5-58. *Dimension “Region” & measure “Sales” placed on the rows shelf and the columns shelf, respectively*

The **measure** "Sales" (default aggregation is SUM) is aggregated by the **dimension** "Region".
Let us add another **dimension** "Category" to the rows shelf (Shown in Fig. 5-59).

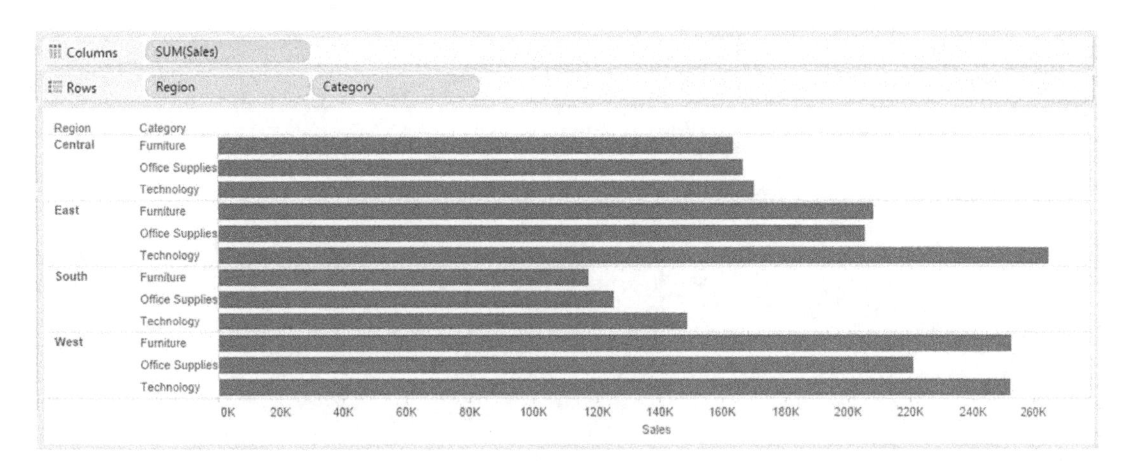

Figure 5-59. *Dimension "Category" placed on the rows shelf*

By placing the **dimension** "Category" on the rows shelf, we have added more granularity and less aggregation to the view / worksheet. It implies that we are adding to our level of detail.

Another area that one can add details is the Details on the marks card. The dimension or dimensions when added to the detail button or the detail shelf affects the visualization in different ways depending on the type of graph in the view or worksheet.

Example:

Drag the **measure** "Sales" from the measures area under the data pane and place it on the columns shelf. Drag the **measure** "Profit" from the measures area under the data pane and place it on the rows shelf (Shown in Fig. 5-60).

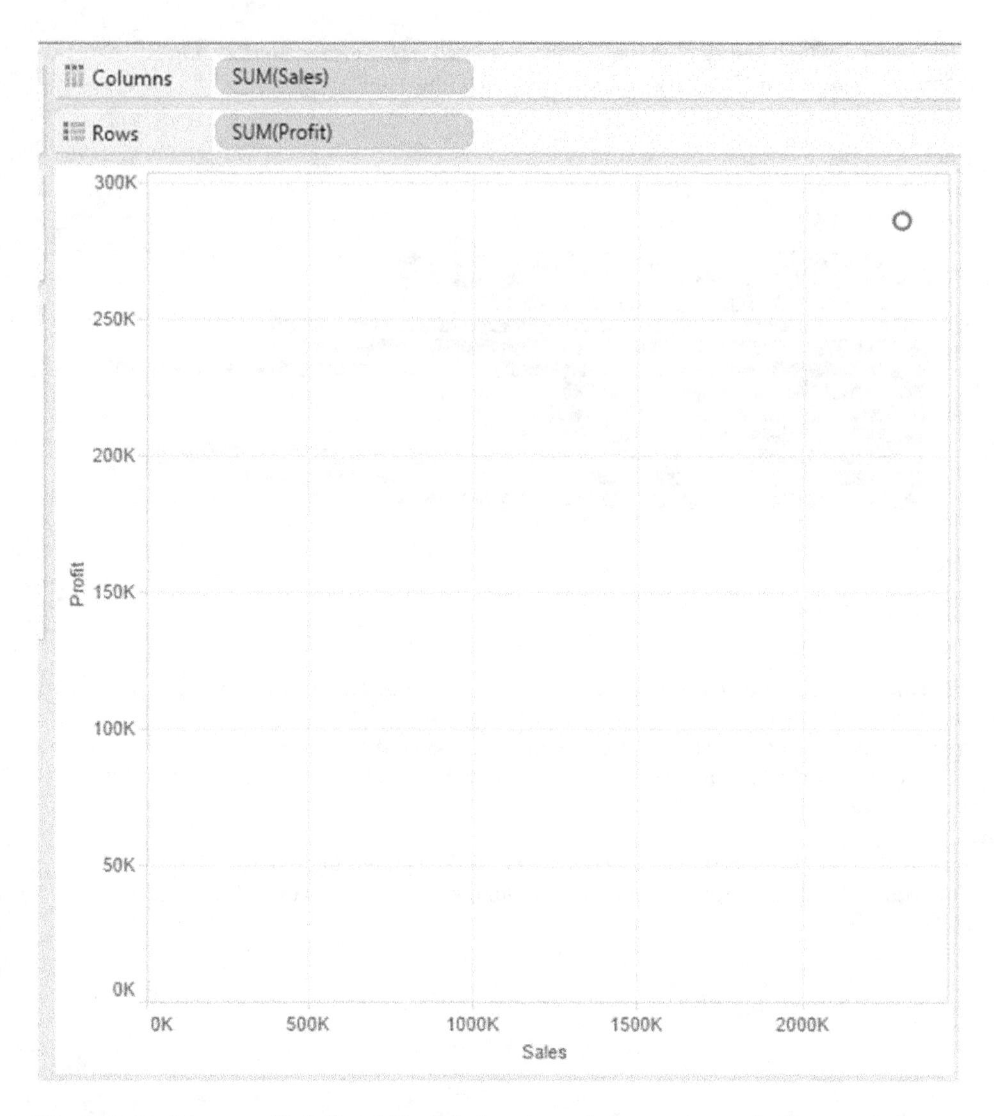

Figure 5-60. *Measures, "Sales" and "Profit" placed on columns and rows shelf, respectively*

The output is as shown in Fig. 5-60. The output is a scatter plot with a single mark on the view. The reason behind the single mark is that the measure is not yet segmented as per any dimension. Drag the **dimension** "Customer Name" from the dimensions area under the data pane and place it on "Details" on the marks card (See Fig. 5-61).

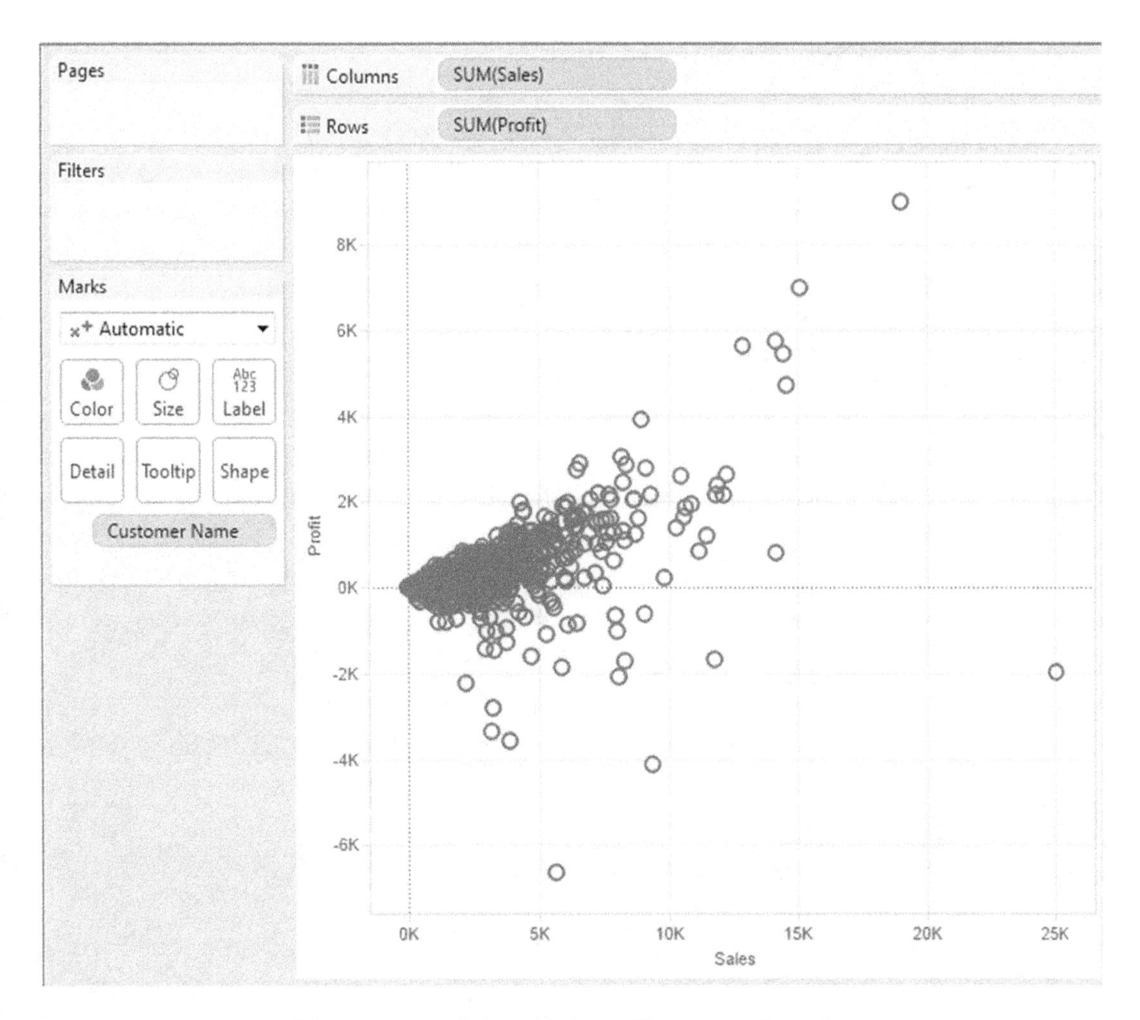

Figure 5-61. *Dimension "Customer Name" placed on "Detail" on the marks card*

The output in Fig. 5-61 shows the measure being segmented by "Customer Name". The view displays mark for every single customer. It makes the view more granular and less aggregated.

Yet another way to add details to the view is by using LOD (Level of Details). LOD expression represents an elegant and powerful way to answer questions involving multiple levels of granularity in a single visualization. Level of detail expressions provide a way to easily compute aggregations that are not at the level of detail of the visualization. You can then integrate those values within visualizations in arbitrary ways. From Tableau v9 onwards, a new concept called level of detail (LOD) expressions has been introduced. LOD expressions can be used to represent the data in different ways at different levels.

When can we use LOD expressions?

Consider using LOD expressions when:

- There is a requirement to show the data at a level different from the dimensions / level present in the view.
- There is a need to obtain some static calculated value that is not affected by any filters that are applied to the view.
- Refer for further reading: `http://www.tableaulearners.com/2016/level-detail-expressions-tableau/`

There are three options available with LOD. They are:

- Include
- Exclude
- Fixed

5.6.1 Demo 1

Objective: To demonstrate the "Level of Detail - Exclude".

Input: "Sample - Superstore.xls"

Expected Output: Shown in Fig. 5-62.

Region	State	City	Region_Sales	State_Sales	City_Sales
Central	Iowa	Dubuque	501,240	4,580	1,687
		Iowa City	501,240	4,580	10
		Marion	501,240	4,580	358
		Urbandale	501,240	4,580	149
		Waterloo	501,240	4,580	30
	Kansas	Garden City	501,240	2,914	312
		Manhattan	501,240	2,914	274
		Olathe	501,240	2,914	896
		Overland Park	501,240	2,914	607
		Wichita	501,240	2,914	825
	Michigan	Ann Arbor	501,240	76,270	889
		Canton	501,240	76,270	818
		Dearborn	501,240	76,270	1,603
		Dearborn Heights	501,240	76,270	1,052
		Detroit	501,240	76,270	42,447
		Grand Rapids	501,240	76,270	526
		Holland	501,240	76,270	138
		Jackson	501,240	76,270	15,420
		Lansing	501,240	76,270	1,610
		Lincoln Park	501,240	76,270	388
		Midland	501,240	76,270	5,292
		Mount Pleasant	501,240	76,270	17
		Oak Park	501,240	76,270	581
		Rochester Hills	501,240	76,270	133
		Roseville	501,240	76,270	638

Figure 5-62. *"LOD Exclude" - Demo 1 - expected output*

Explanation of the output:

The view / worksheet should display the sales by region, by state and by city alongside the dimensions, "Region", "State" and "City".

Let us split the output:

Sales by Region: Shown in Fig. 5-63.

Region	
Central	501,240
East	678,781
South	391,722
West	725,458

Figure 5-63. *Sales by Region*

Sales by State: Shown in Fig. 5-64**.**

Region	State	
Central	Illinois	80,166
	Indiana	53,555
	Iowa	4,580
	Kansas	2,914
	Michigan	76,270
	Minnesota	29,863
	Missouri	22,205
	Nebraska	7,465
	North Dakota	920
	Oklahoma	19,683
	South Dakota	1,316
	Texas	170,188
	Wisconsin	32,115
East	Connecticut	13,384
	Delaware	27,451
	District of Columb..	2,865
	Maine	1,271
	Maryland	23,706
	Massachusetts	28,634
	New Hampshire	7,293
	New Jersey	35,764
	New York	310,876
	Ohio	78,258
	Pennsylvania	116,512

Figure 5-64. *Sales by State*

Sales by City: Shown in Fig. 5-65.

Rows: Region | State | City

Region	State	City	
Central	Illinois	Arlington Heights	14
		Aurora	7,573
		Bloomington	964
		Bolingbrook	218
		Buffalo Grove	831
		Carol Stream	1,306
		Champaign	152
		Chicago	48,540
		Danville	43
		Decatur	3,169
		Des Plaines	1,493
		Elmhurst	892
		Evanston	1,754
		Frankfort	98
		Freeport	216
		Glenview	158
		Highland Park	2,035
		Naperville	1,288
		Normal	367
		Oak Park	10
		Orland Park	340
		Oswego	322
		Palatine	116
		Park Ridge	685

Figure 5-65. *Sales by city*

The challenge is to combine all the three outputs stated above in a single view / worksheet. Let us look at "LOD - Exclude" to accomplish the above output.

5.6.1.1 Steps

5.6.1.1.1 Step 1

Read in the data from "Sample - Superstore.xls" into Tableau (Shown in Fig. 5-66).

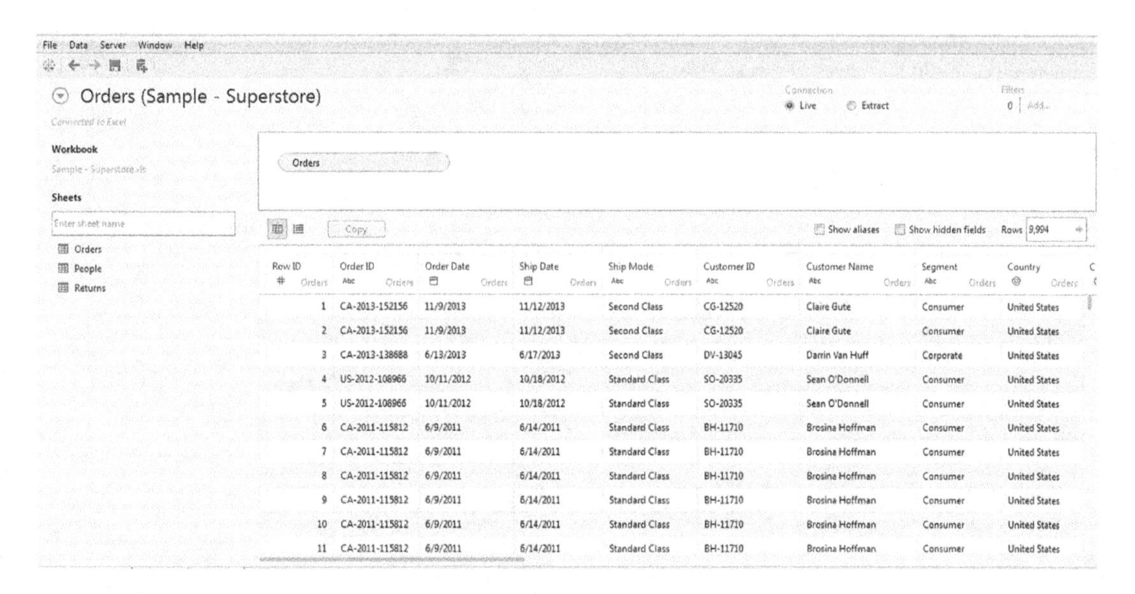

Figure 5-66. *Data from "Sample - Superstore" read into Tableau*

5.6.1.1.2 Step 2

Drag the **dimension** "Region" from the dimensions area under the data pane to the rows shelf (Shown in Fig. 5-67).

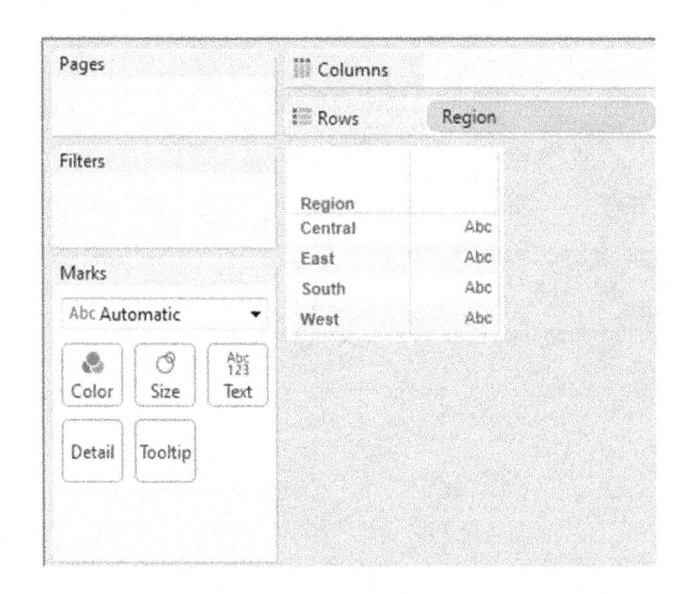

Figure 5-67. *Dimension "Region" placed on the rows shelf*

Drag the **dimension** "State" from the dimensions area under the data pane to the rows shelf (Shown in Fig. 5-68).

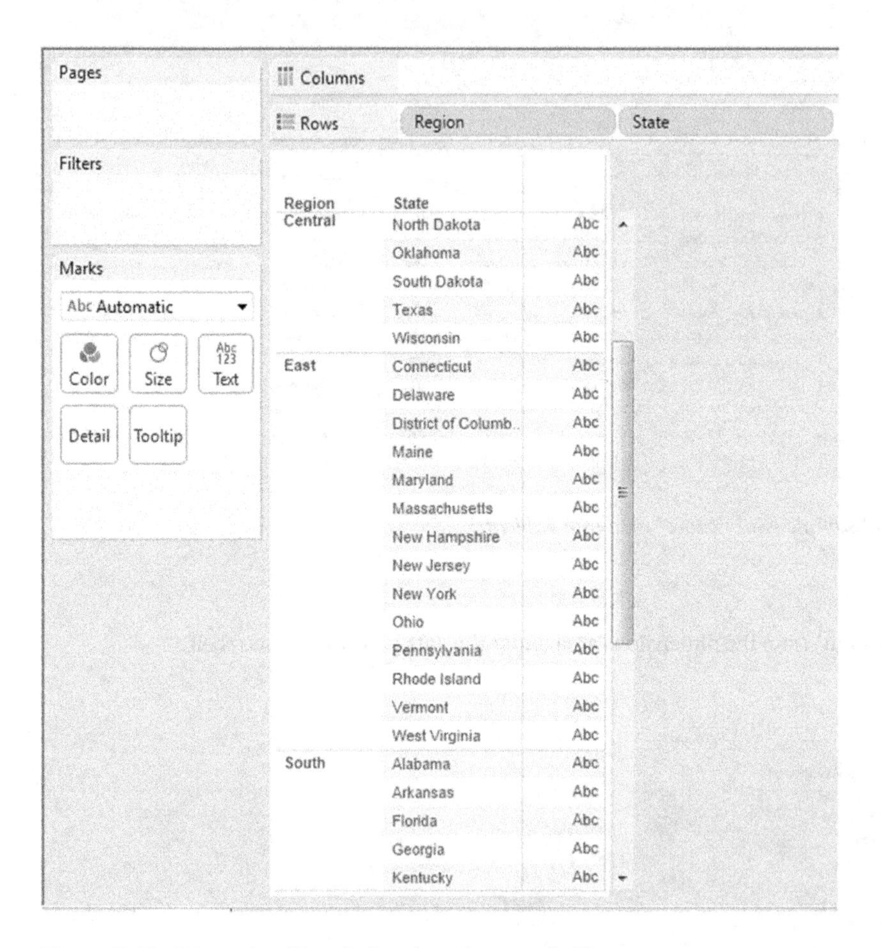

Figure 5-68. *Dimension "State" placed on the rows shelf*

Drag the **dimension** "City" from the dimensions area under the data pane to the rows shelf (Shown in Fig. 5-69).

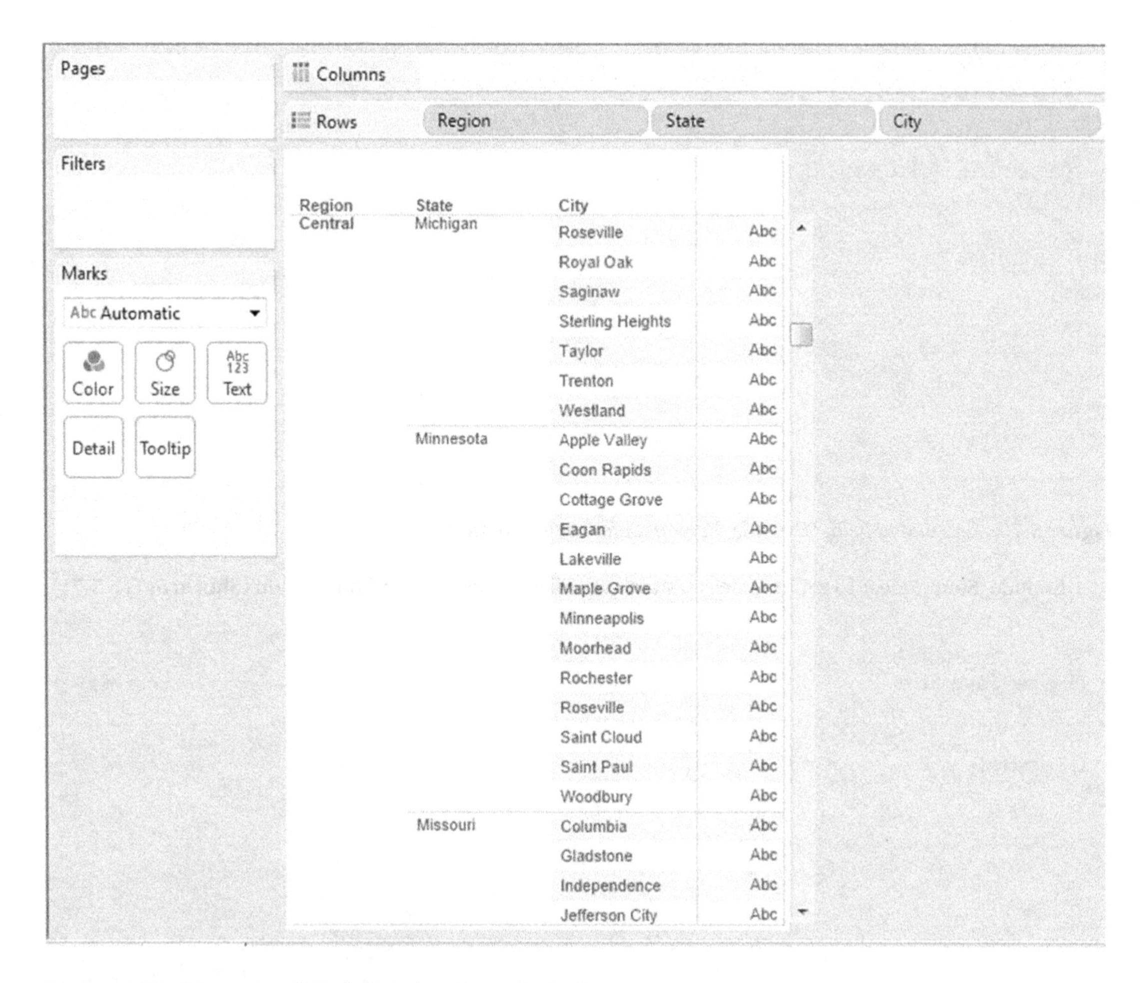

Figure 5-69. *Dimension "City" placed on the rows shelf*

5.6.1.1.3 Step 3

Create two calculated fields, "Exclude_Region_Sales" and "Exclude_State_Sales".

Exclude_Region_Sales: To get the sales by region, we will exclude the "State" and "City" dimensions (Shown in Fig. 5-70).

Exclude_Region_Sales

```
{Exclude [State], [City]: sum([Sales])}
```

Sheets Affected ▾ Apply OK

Figure 5-70. *Calculated field, "Exclude_Region_Sales" being created*

Exclude_State_Sales: To get the sales by state, we will exclude the "City" dimension (Shown in Fig. 5-71).

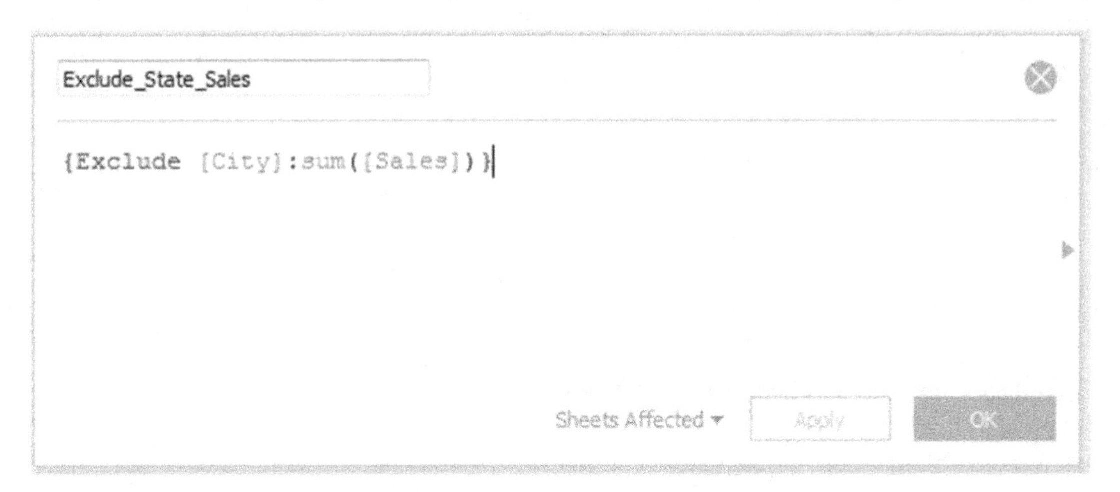

Figure 5-71. *Calculated field field "Exclude_State_Sales" being created*

5.6.1.1.4 Step 4

Drag the **measure** "Exclude_Region_Sales" from the measures area under the data pane and drop it into the view/worksheet (Shown in Fig. 5-72).

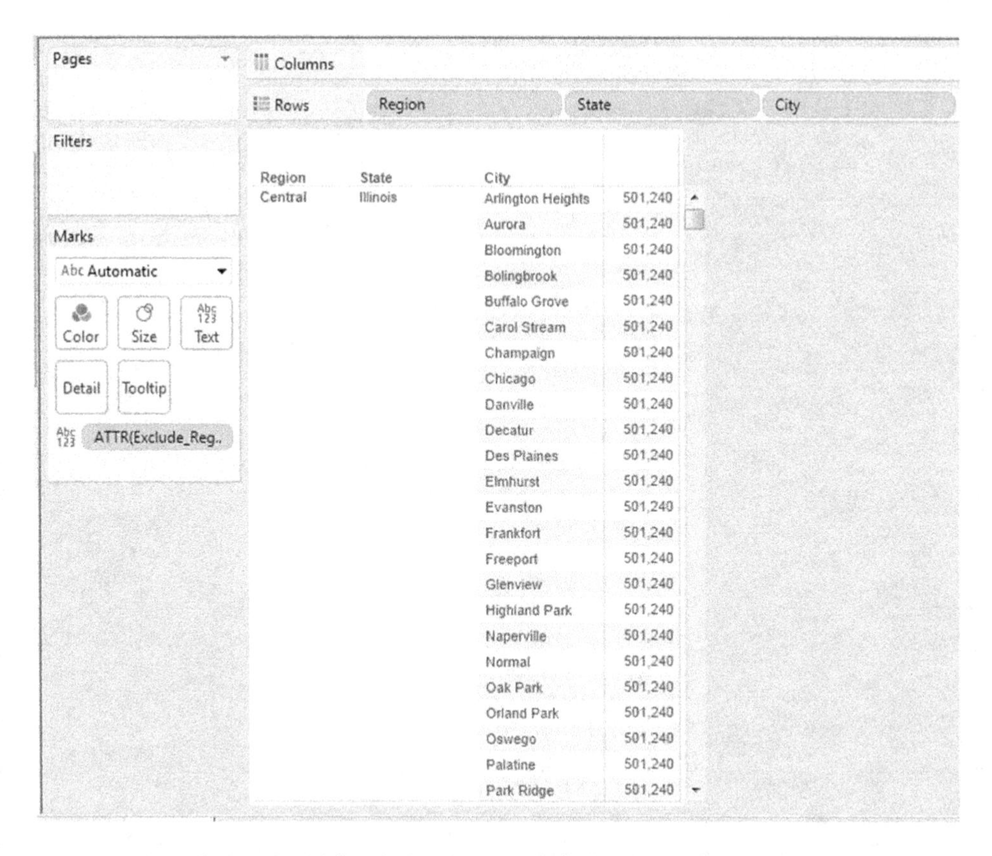

Figure 5-72. *Calculated field, "Exclude_Region_Sales" placed on the view*

Drag the **measure** "Exclude_State_Sales" from the measures area under the data pane and drop it into the view/worksheet (Shown in Fig. 5-73).

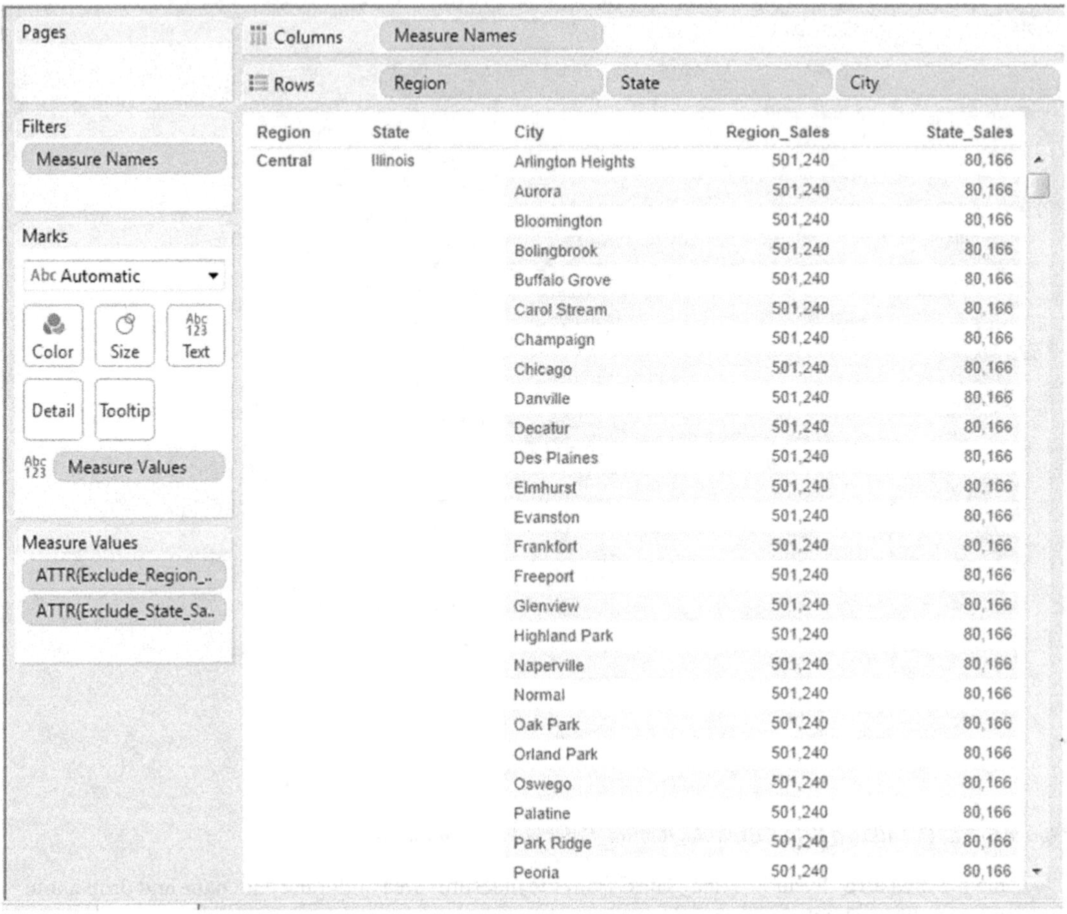

Pages

Columns: Measure Names

Rows: Region | State | City

Filters: Measure Names

Marks: Automatic — Color, Size, Text, Detail, Tooltip — Measure Values

Measure Values: ATTR(Exclude_Region_.. ATTR(Exclude_State_Sa..

Region	State	City	Region_Sales	State_Sales
Central	Illinois	Arlington Heights	501,240	80,166
		Aurora	501,240	80,166
		Bloomington	501,240	80,166
		Bolingbrook	501,240	80,166
		Buffalo Grove	501,240	80,166
		Carol Stream	501,240	80,166
		Champaign	501,240	80,166
		Chicago	501,240	80,166
		Danville	501,240	80,166
		Decatur	501,240	80,166
		Des Plaines	501,240	80,166
		Elmhurst	501,240	80,166
		Evanston	501,240	80,166
		Frankfort	501,240	80,166
		Freeport	501,240	80,166
		Glenview	501,240	80,166
		Highland Park	501,240	80,166
		Naperville	501,240	80,166
		Normal	501,240	80,166
		Oak Park	501,240	80,166
		Orland Park	501,240	80,166
		Oswego	501,240	80,166
		Palatine	501,240	80,166
		Park Ridge	501,240	80,166
		Peoria	501,240	80,166

Figure 5-73. *Calculated field "Exclude_State_Sales" placed on the View*

Drag the **measure** "Sales" from the measures area under the data pane and drop it into the view/ worksheet (Shown in Fig. 5-74).

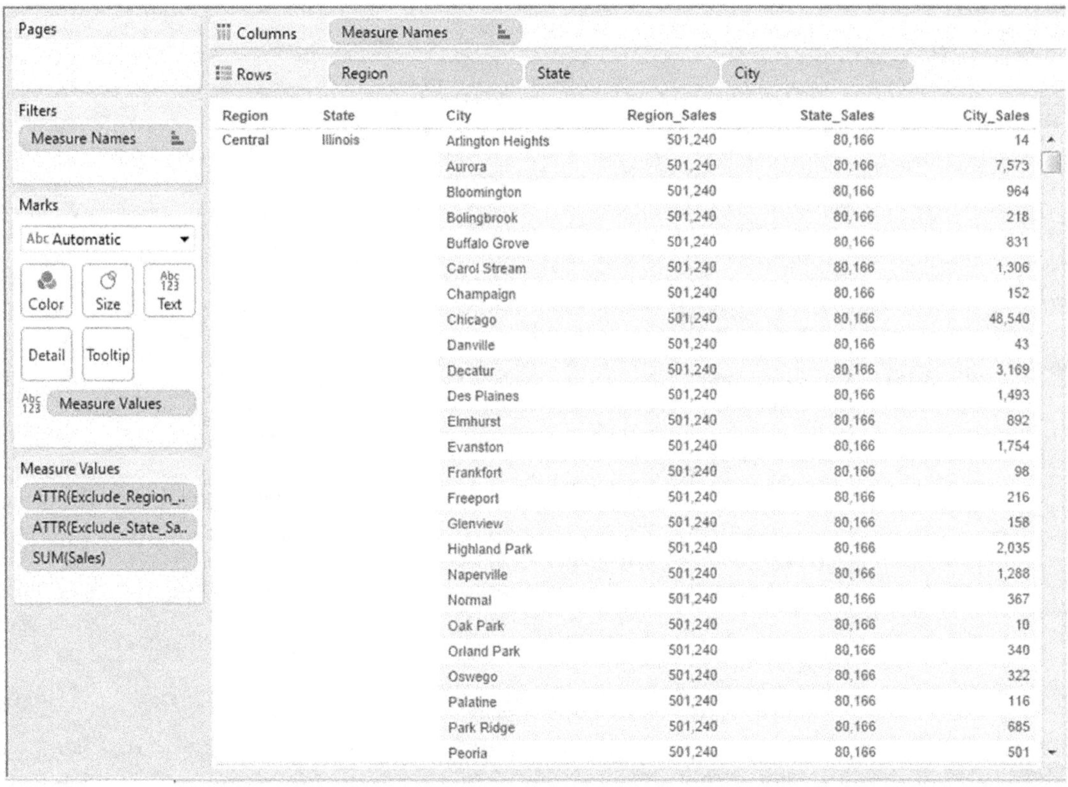

Region	State	City	Region_Sales	State_Sales	City_Sales
Central	Illinois	Arlington Heights	501,240	80,166	14
		Aurora	501,240	80,166	7,573
		Bloomington	501,240	80,166	964
		Bolingbrook	501,240	80,166	218
		Buffalo Grove	501,240	80,166	831
		Carol Stream	501,240	80,166	1,306
		Champaign	501,240	80,166	152
		Chicago	501,240	80,166	48,540
		Danville	501,240	80,166	43
		Decatur	501,240	80,166	3,169
		Des Plaines	501,240	80,166	1,493
		Elmhurst	501,240	80,166	892
		Evanston	501,240	80,166	1,754
		Frankfort	501,240	80,166	98
		Freeport	501,240	80,166	216
		Glenview	501,240	80,166	158
		Highland Park	501,240	80,166	2,035
		Naperville	501,240	80,166	1,288
		Normal	501,240	80,166	367
		Oak Park	501,240	80,166	10
		Orland Park	501,240	80,166	340
		Oswego	501,240	80,166	322
		Palatine	501,240	80,166	116
		Park Ridge	501,240	80,166	685
		Peoria	501,240	80,166	501

Figure 5-74. *The measure "Sales", placed on the view*

The final output (Shown in Fig. 5-75):

Region	State	City	Region_Sales	State_Sales	City_Sales
Central	Iowa	Dubuque	501,240	4,580	1,687
		Iowa City	501,240	4,580	10
		Marion	501,240	4,580	358
		Urbandale	501,240	4,580	149
		Waterloo	501,240	4,580	30
	Kansas	Garden City	501,240	2,914	312
		Manhattan	501,240	2,914	274
		Olathe	501,240	2,914	896
		Overland Park	501,240	2,914	607
		Wichita	501,240	2,914	825
	Michigan	Ann Arbor	501,240	76,270	889
		Canton	501,240	76,270	818
		Dearborn	501,240	76,270	1,603
		Dearborn Heights	501,240	76,270	1,052
		Detroit	501,240	76,270	42,447
		Grand Rapids	501,240	76,270	526
		Holland	501,240	76,270	138
		Jackson	501,240	76,270	15,420
		Lansing	501,240	76,270	1,610
		Lincoln Park	501,240	76,270	388
		Midland	501,240	76,270	5,292
		Mount Pleasant	501,240	76,270	17
		Oak Park	501,240	76,270	581
		Rochester Hills	501,240	76,270	133
		Roseville	501,240	76,270	638

Figure 5-75. *"LOD - Exclude" - Demo 1 - final output*

Can you answer this?

What will happen if a dimension that is not in the view is excluded in the LOD calculation? The answer is nothing will change in the view. The exclude LOD calculation returns results relative to your visualization; this implies that it does matter what dimensions are used in the view.

5.6.2 Demo 2

Objective: To demonstrate level of detail - fixed.

Fixed LOD calculations are not relative to the view. They focus only on the dimension that we use in the "Fixed LOD Calculations", regardless of what is or what is not included in the view.

Input: "Sample - Superstore.xls"

Expected Output: Shown in Fig. 5-76.

Region	State	City	Fixed_Region_S..	Fixed_State_Sales	City_Sales
Central	Illinois	Arlington Heights	501,240	80,166	14
		Aurora	501,240	80,166	7,573
		Bloomington	501,240	80,166	964
		Bolingbrook	501,240	80,166	218
		Buffalo Grove	501,240	80,166	831
		Carol Stream	501,240	80,166	1,306
		Champaign	501,240	80,166	152
		Chicago	501,240	80,166	48,540
		Danville	501,240	80,166	43
		Decatur	501,240	80,166	3,169
		Des Plaines	501,240	80,166	1,493
		Elmhurst	501,240	80,166	892
		Evanston	501,240	80,166	1,754
		Frankfort	501,240	80,166	98
		Freeport	501,240	80,166	216
		Glenview	501,240	80,166	158
		Highland Park	501,240	80,166	2,035
		Naperville	501,240	80,166	1,288
		Normal	501,240	80,166	367
		Oak Park	501,240	80,166	10
		Orland Park	501,240	80,166	340
		Oswego	501,240	80,166	322
		Palatine	501,240	80,166	116
		Park Ridge	501,240	80,166	685
		Peoria	501,240	80,166	501

Figure 5-76. *"LOD - Fixed" - Demo 2 - expected output*

Explanation of the expected output:

We were able to get the above output using LOD - exclude, but we had to be cognizant of the dimensions present in the view. We would like to get the above output without any consideration to the dimensions present in the view. We will accomplish this using LOD - fixed. LOD - fixed provides us with increased flexibility and can be used across worksheets.

5.6.2.1 Steps

Follow the steps as provided.

5.6.2.1.1 Step 1

Read in the data from "Sample - Superstore.xls" into Tableau (Shown in Fig. 5-77).

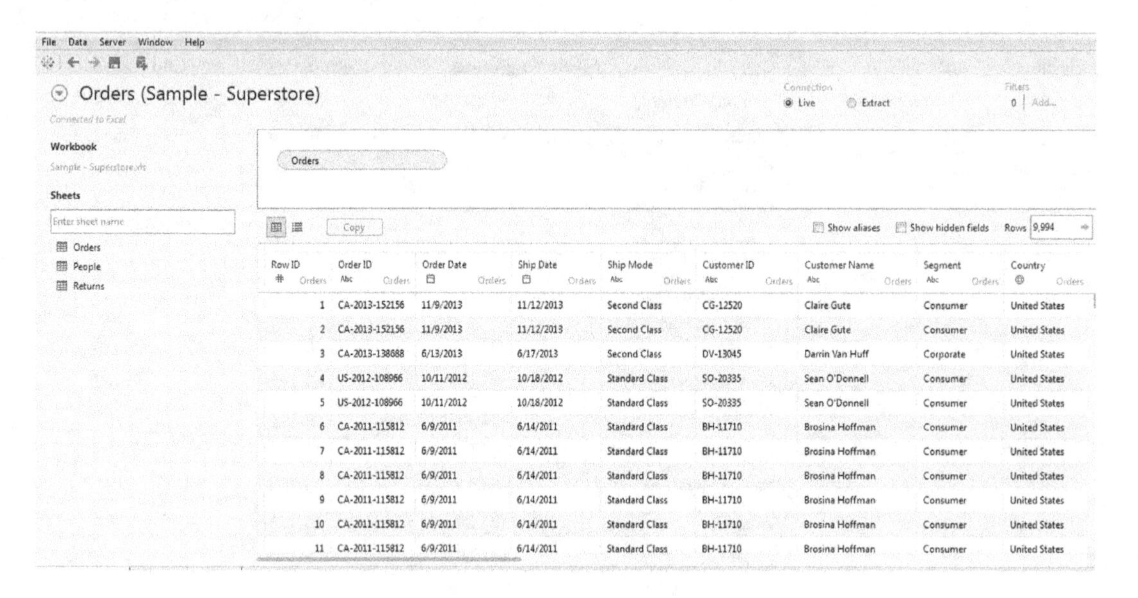

Figure 5-77. *Data from "Sample -Superstore.xls" read into Tableau*

5.6.2.1.2 Step 2

Drag the **dimension** "Region" from the dimensions area under the data pane to the rows shelf (Shown in Fig. 5-78).

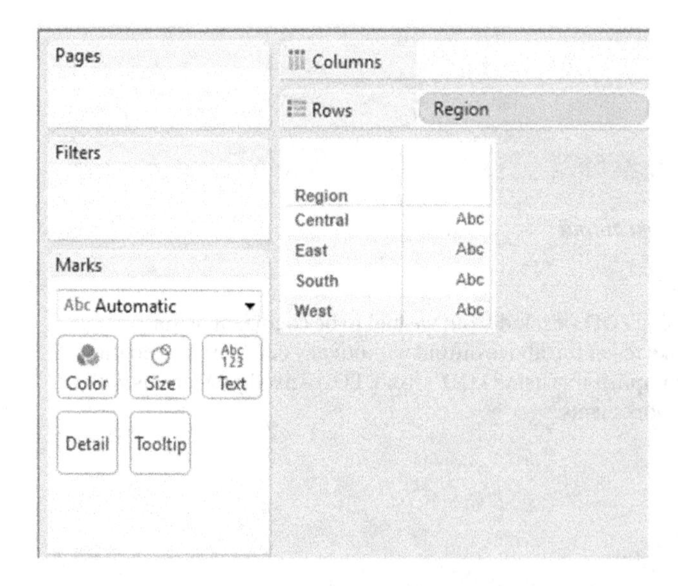

Figure 5-78. *Dimension "Region" placed on the rows shelf*

Drag the **dimension** “State” from the dimensions area under the data pane to the rows shelf (Shown in Fig. 5-79).

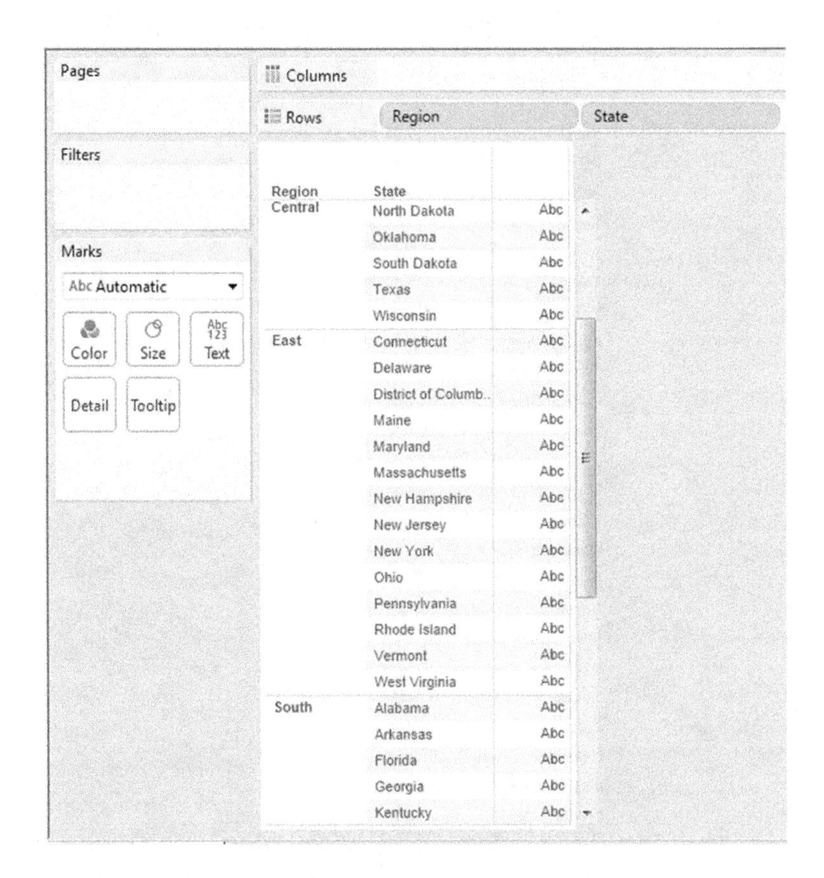

Figure 5-79. *Dimension “State” placed on the rows shelf*

Drag the **dimension** “City” from the dimensions area under the data pane to the rows shelf (Shown in Fig. 5-80).

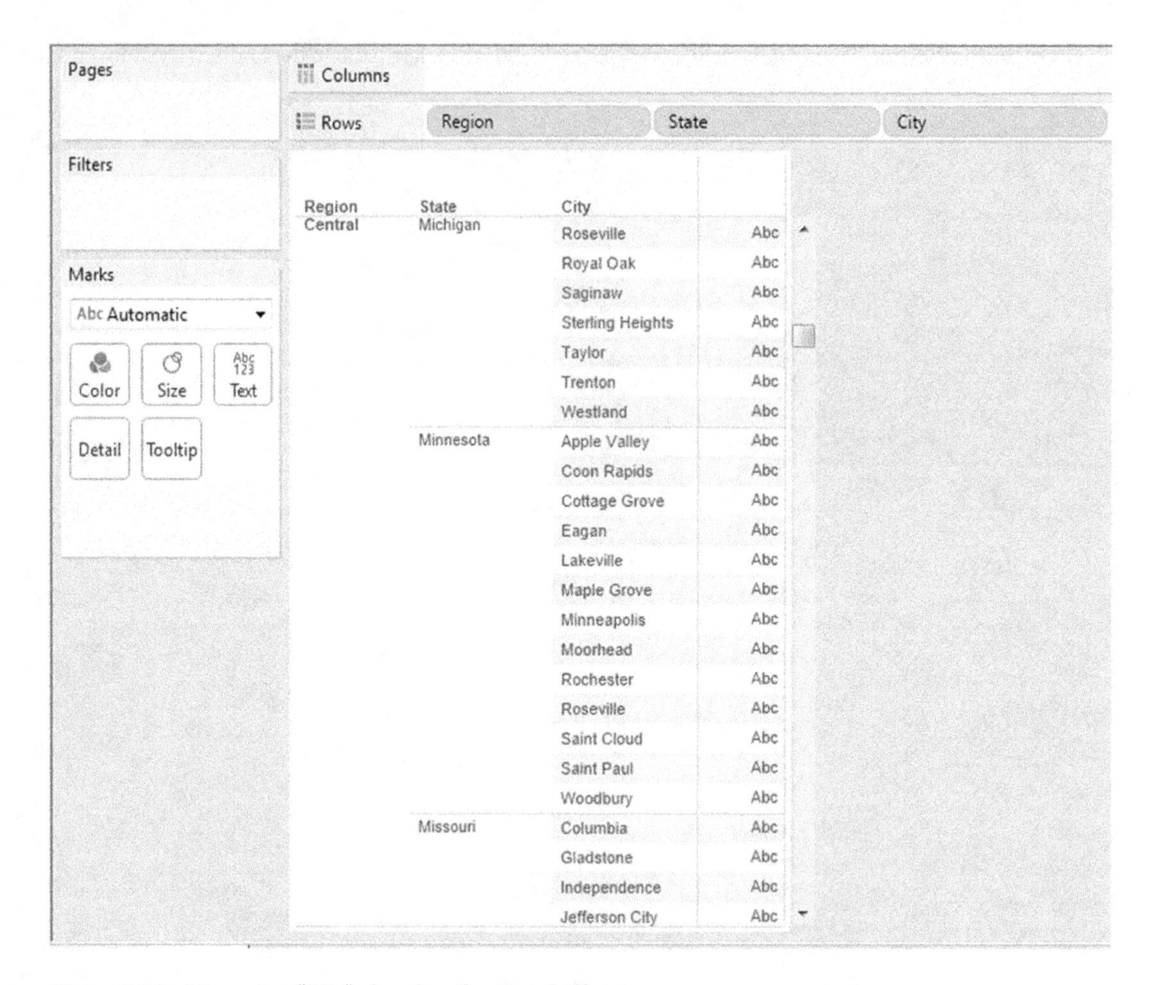

Figure 5-80. *Dimension "City" placed on the rows shelf*

5.6.2.1.3 Step 3

Create two calculated fields, "Fixed_Region_Sales" and "Fixed_State_Sales" (Shown in Fig. 5-81 & Figure 5-82).

Figure 5-81. *Calculated field field "Fixed_Region_Sales" being created*

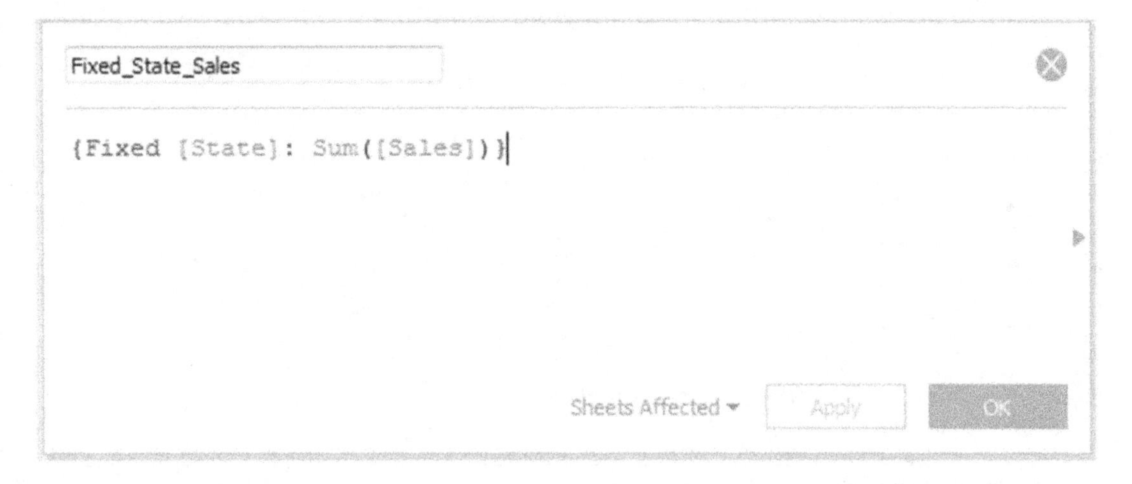

Figure 5-82. *Calculated field field "Fixed_State_Sales" being created*

5.6.2.1.4 Step 4

Drag the **measure** "Fixed_Region_Sales" from the measures area under the data pane and drop it into the view/worksheet (Shown in Fig. 5-83).

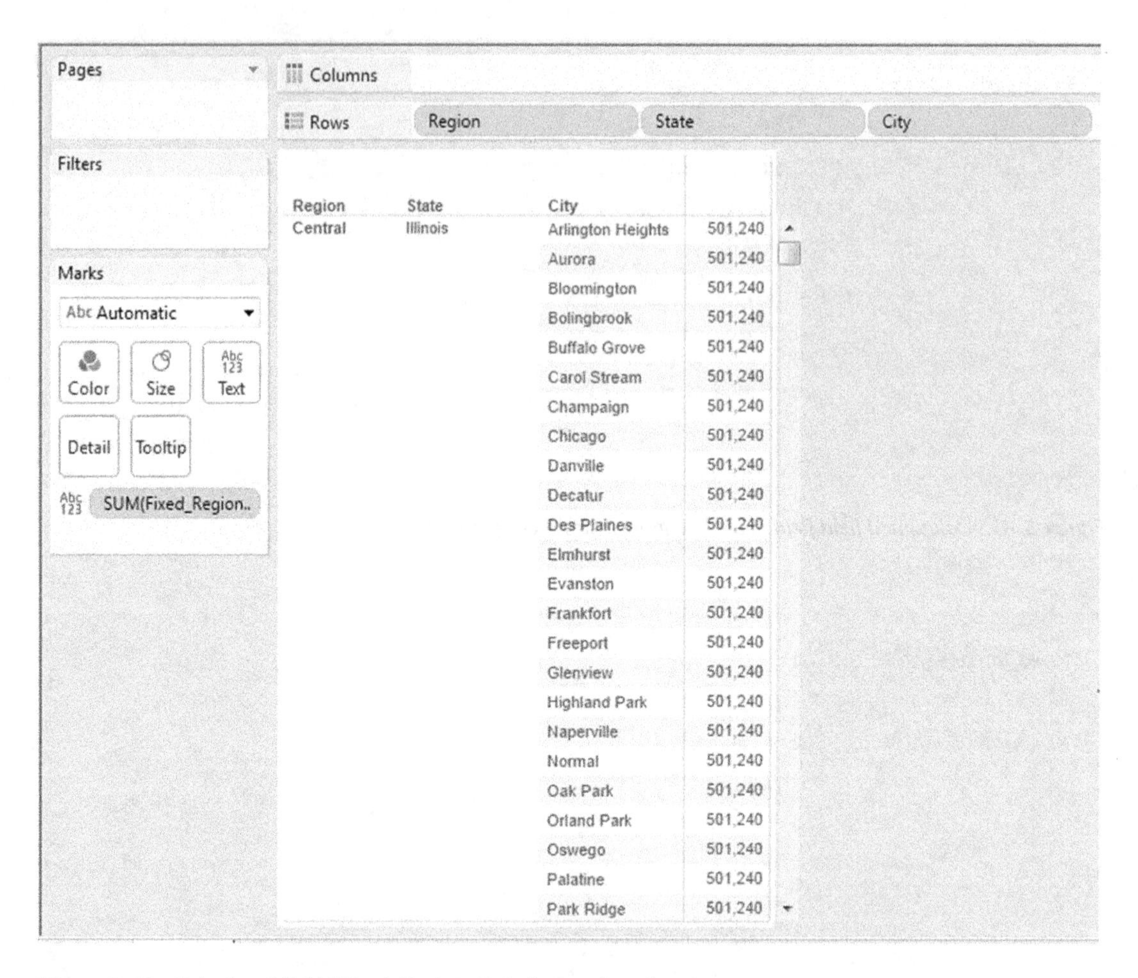

Figure 5-83. *Calculated field "Fixed_Region_Sales" placed on the view*

Drag the **measure** "Fixed_State_Sales" from the measures area under the data pane and drop it into the view/worksheet (Shown in Fig. 5-84).

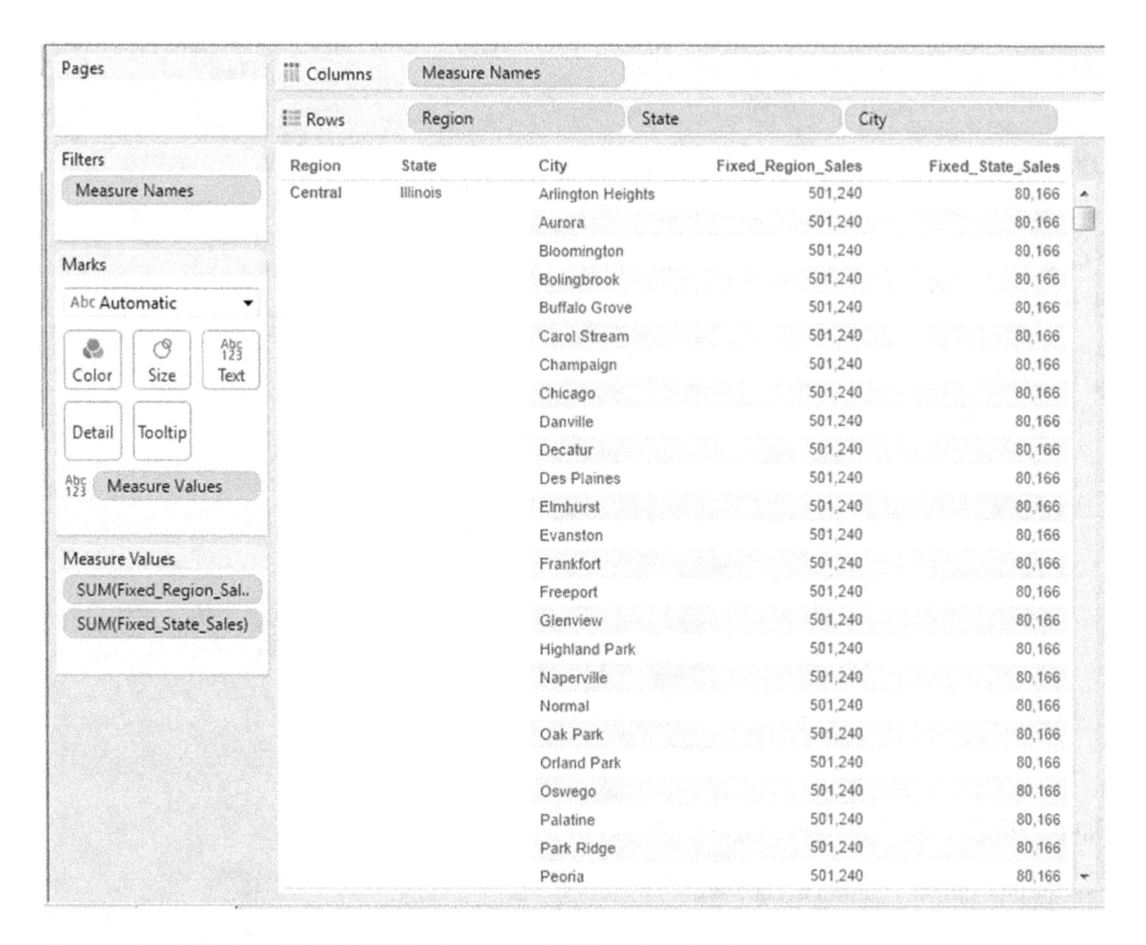

Region	State	City	Fixed_Region_Sales	Fixed_State_Sales
Central	Illinois	Arlington Heights	501,240	80,166
		Aurora	501,240	80,166
		Bloomington	501,240	80,166
		Bolingbrook	501,240	80,166
		Buffalo Grove	501,240	80,166
		Carol Stream	501,240	80,166
		Champaign	501,240	80,166
		Chicago	501,240	80,166
		Danville	501,240	80,166
		Decatur	501,240	80,166
		Des Plaines	501,240	80,166
		Elmhurst	501,240	80,166
		Evanston	501,240	80,166
		Frankfort	501,240	80,166
		Freeport	501,240	80,166
		Glenview	501,240	80,166
		Highland Park	501,240	80,166
		Naperville	501,240	80,166
		Normal	501,240	80,166
		Oak Park	501,240	80,166
		Orland Park	501,240	80,166
		Oswego	501,240	80,166
		Palatine	501,240	80,166
		Park Ridge	501,240	80,166
		Peoria	501,240	80,166

Figure 5-84. *Calculated field field "Fixed_State_Sales" placed on the View*

Drag the **measure** "Sales" from the measures area under the data pane and drop it into the view/worksheet (Shown in Fig. 5-85).

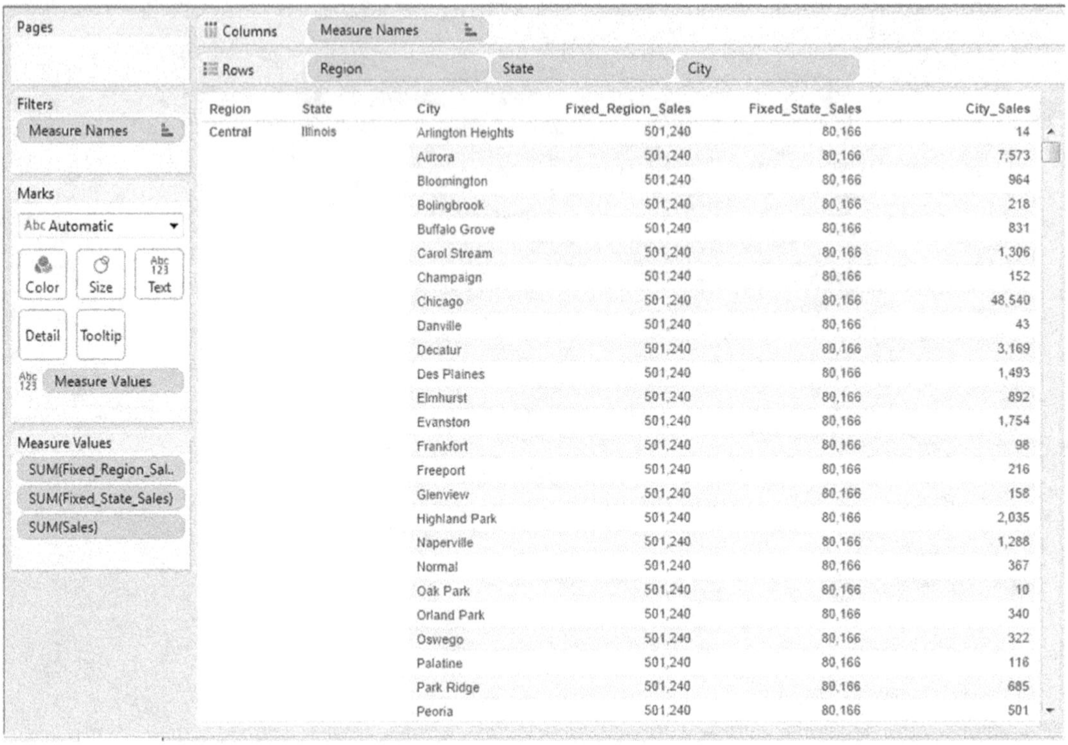

Pages

Columns: Measure Names

Rows: Region, State, City

Filters: Measure Names

Marks: Automatic — Color, Size, Text, Detail, Tooltip — Measure Values

Measure Values: SUM(Fixed_Region_Sal.., SUM(Fixed_State_Sales), SUM(Sales)

Region	State	City	Fixed_Region_Sales	Fixed_State_Sales	City_Sales
Central	Illinois	Arlington Heights	501,240	80,166	14
		Aurora	501,240	80,166	7,573
		Bloomington	501,240	80,166	964
		Bolingbrook	501,240	80,166	218
		Buffalo Grove	501,240	80,166	831
		Carol Stream	501,240	80,166	1,306
		Champaign	501,240	80,166	152
		Chicago	501,240	80,166	48,540
		Danville	501,240	80,166	43
		Decatur	501,240	80,166	3,169
		Des Plaines	501,240	80,166	1,493
		Elmhurst	501,240	80,166	892
		Evanston	501,240	80,166	1,754
		Frankfort	501,240	80,166	98
		Freeport	501,240	80,166	216
		Glenview	501,240	80,166	158
		Highland Park	501,240	80,166	2,035
		Naperville	501,240	80,166	1,288
		Normal	501,240	80,166	367
		Oak Park	501,240	80,166	10
		Orland Park	501,240	80,166	340
		Oswego	501,240	80,166	322
		Palatine	501,240	80,166	116
		Park Ridge	501,240	80,166	685
		Peoria	501,240	80,166	501

Figure 5-85. *Measure "Sales" placed on the view*

The final output as shown in Fig. 5-86.

Region	State	City	Fixed_Region_S..	Fixed_State_Sales	City_Sales
Central	Illinois	Arlington Heights	501,240	80,166	14
		Aurora	501,240	80,166	7,573
		Bloomington	501,240	80,166	964
		Bolingbrook	501,240	80,166	218
		Buffalo Grove	501,240	80,166	831
		Carol Stream	501,240	80,166	1,306
		Champaign	501,240	80,166	152
		Chicago	501,240	80,166	48,540
		Danville	501,240	80,166	43
		Decatur	501,240	80,166	3,169
		Des Plaines	501,240	80,166	1,493
		Elmhurst	501,240	80,166	892
		Evanston	501,240	80,166	1,754
		Frankfort	501,240	80,166	98
		Freeport	501,240	80,166	216
		Glenview	501,240	80,166	158
		Highland Park	501,240	80,166	2,035
		Naperville	501,240	80,166	1,288
		Normal	501,240	80,166	367
		Oak Park	501,240	80,166	10
		Orland Park	501,240	80,166	340
		Oswego	501,240	80,166	322
		Palatine	501,240	80,166	116
		Park Ridge	501,240	80,166	685
		Peoria	501,240	80,166	501

Figure 5-86. *"LOD - Fixed" - Demo 2 - final output*

5.6.2.2 Demo 3

Objective: To demonstrate the "level of detail - include".

Input: "LOD.xls"

The dataset as available in "LOD.xls".

	A	B	C
1	TransactionID	CustomerName	Amount
2	1	Alex Maxwell	1000
3	2	Alex Maxwell	250
4	3	Barbara Mori	1200
5	4	Barbara Mori	300
6	5	Barbara Mori	450
7	6	Ileana D'Souza	350
8	7	Ileana D'Souza	450
9	8	Esha Mathews	600
10	9	Esha Mathews	600
11	10	John Tukey	650
12	11	Kelly M	700
13	12	George T	800

Expected Output: Shown in Fig. 5-87.

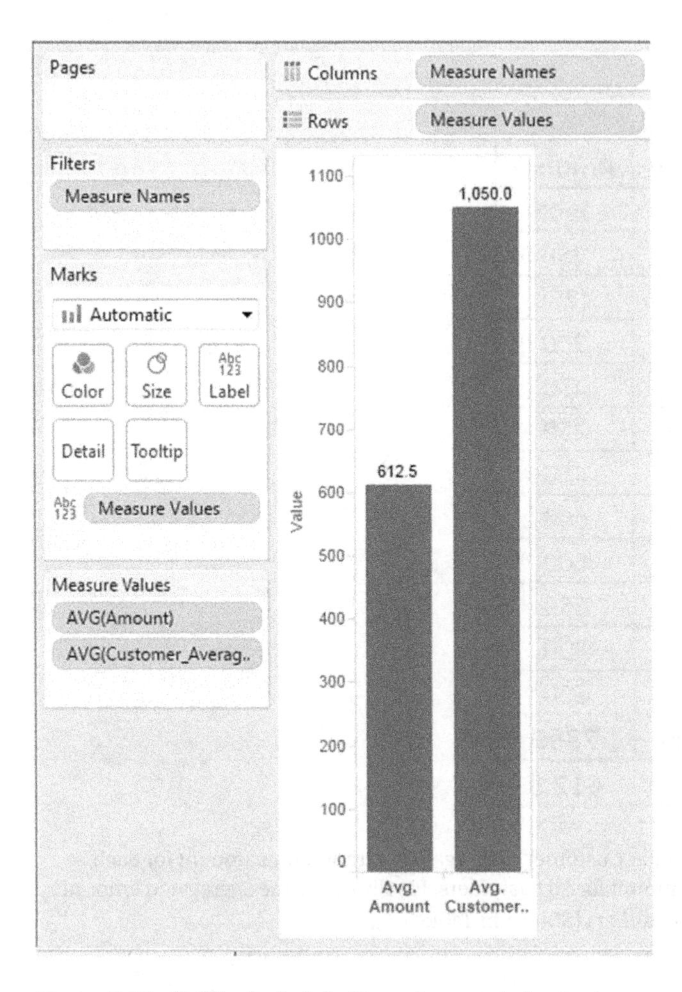

Figure 5-87. "LOD - Include" - Demo 3 - expected output

Explanation of the expected output:

We would like to view the "Average Transaction Amount" of all customers. Alongside this, we would like to view the "Average Transaction Amount per Customer".

To get the "Average Transaction Amount", simply sum up the transaction amounts of all customers and then divide by the number of customers (Shown in Table 5-6).

Table 5-6. *Average Transaction Amount*

	A	B	C
1	**TransactionID**	**CustomerName**	**Amount**
2	1	Alex Maxwell	1000
3	2	Alex Maxwell	250
4	3	Barbara Mori	1200
5	4	Barbara Mori	300
6	5	Barbara Mori	450
7	6	Ileana D'Souza	350
8	7	Ileana D'Souza	450
9	8	Esha Mathews	600
10	9	Esha Mathews	600
11	10	John Tukey	650
12	11	Kelly M	700
13	12	George T	800
14		**Grand Total**	**7350**
15		**Average**	**612.5**

To get the "Average Transaction Amount per Customer", aggregate the transaction amount for each customer and then sum up the aggregated amount for all customers. Finally divide the aggregated amount for all customers by the number of unique customers (Shown in Table 5-7).

Table 5-7. *Average Transaction Amount per Customer*

TransactionID	CustomerName	Amount	Transaction amount per Customer
1	Alex Maxwell	1000	1250
2	Alex Maxwell	250	
3	Barbara Mori	1200	1950
4	Barbara Mori	300	
5	Barbara Mori	450	
6	Ileana D'Souza	350	800
7	Ileana D'Souza	450	
8	Esha Mathews	600	1200
9	Esha Mathews	600	
10	John Tukey	650	650
11	Kelly M	700	700
12	George T	800	800
		Grand Total	**7350**
		Average (Grand Total divided by 7)	**1050**

Let us look at the steps to accomplish the same in Tableau.

5.6.2.3 Steps

Follow the steps as provided.

5.6.2.3.1 Step 1

Read in the data from "LOD.xls" into Tableau (Shown in Fig. 5-88).

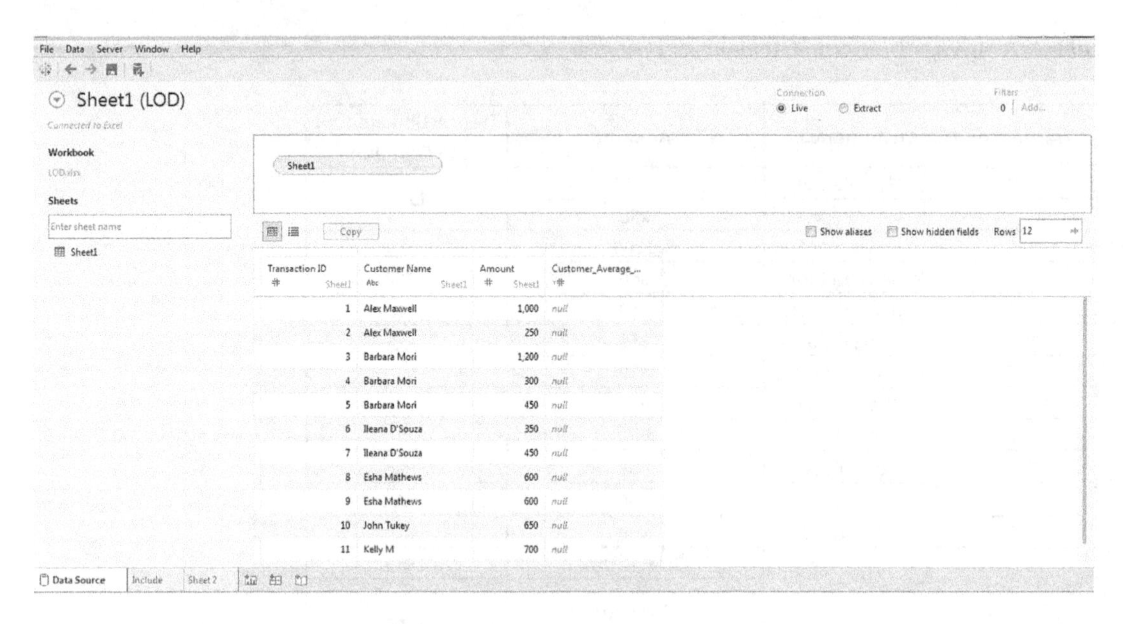

Figure 5-88. *Read in data from "LOD.xls" into Tableau*

5.6.2.3.2 Step 2

Create a calculated field, "Customer_Average_Amount" (Shown in Fig. 5-89).

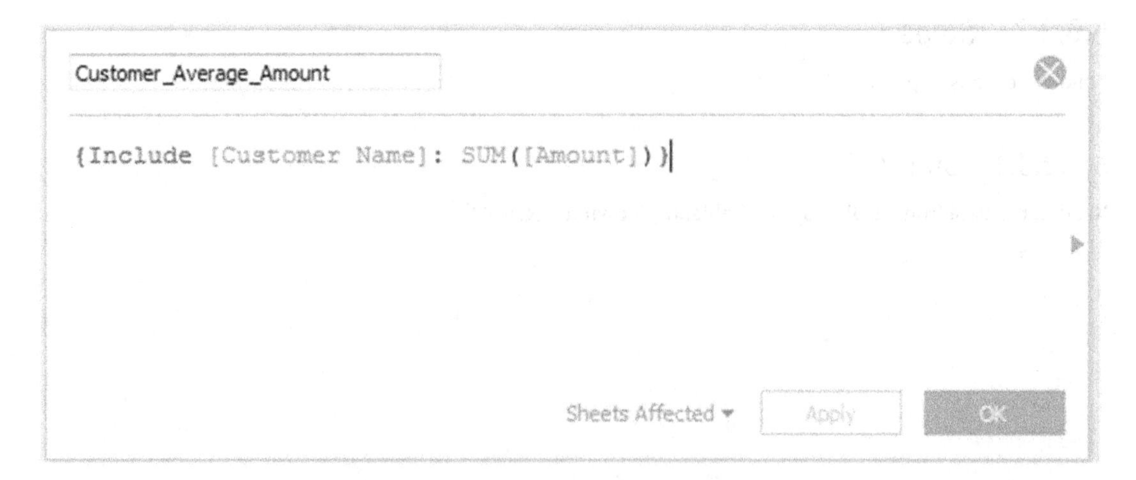

Figure 5-89. *Calculated field "Customer_Average_Amount" being created*

5.6.2.3.3 Step 3

Drag the **dimension** "Measure Names" from the dimensions area under the data pane to the columns shelf (Shown in Fig. 5-90).

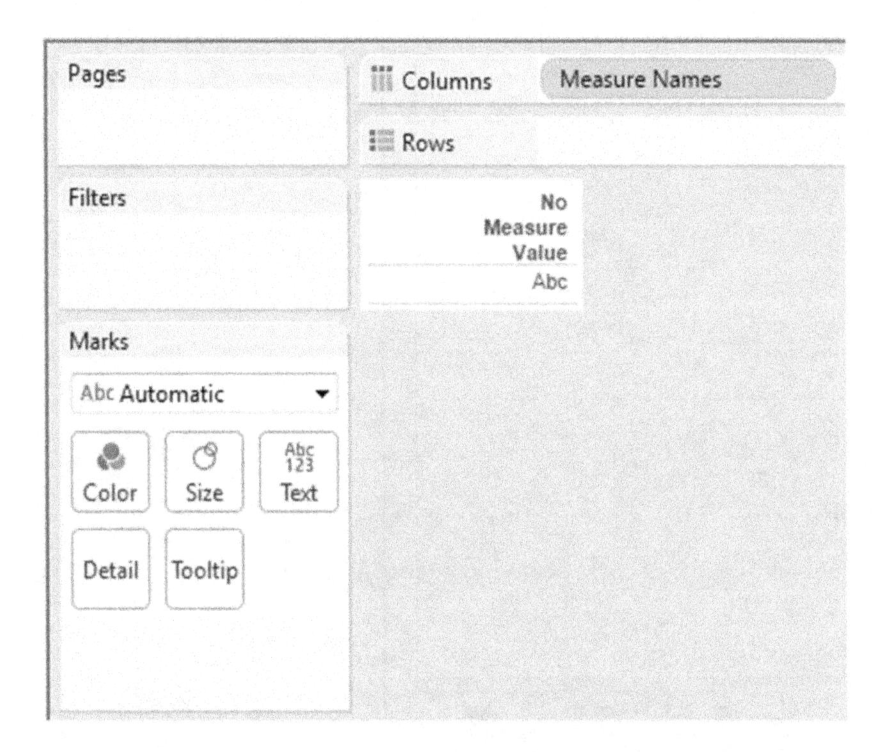

Figure 5-90. *Dimension "Measure Names" placed on the columns shelf*

5.6.2.3.4 Step 4

Drag the **dimension** "Measure Names" from the dimensions area under the data pane to Filters Shelf (Shown in Fig. 5-91).

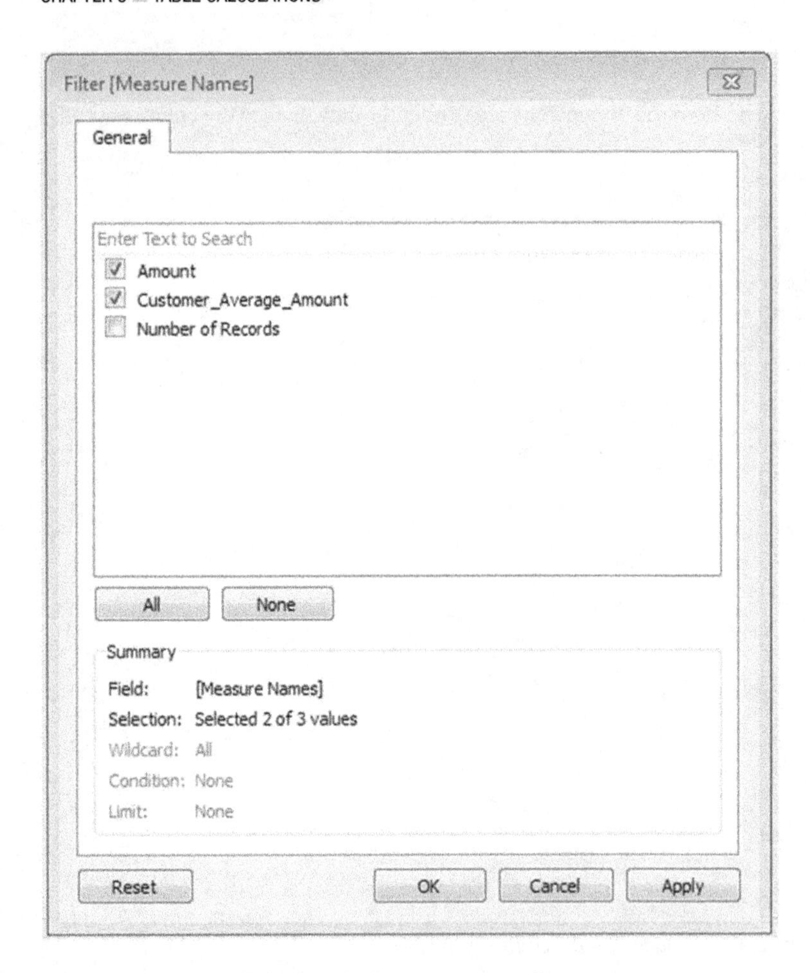

Figure 5-91. *Filter dialog box for "Measure Names"*

5.6.2.3.5 Step 5

Drag the **measure** "Measure Values" from the measures area under the data pane to the rows shelf (Shown in Fig. 5-92).

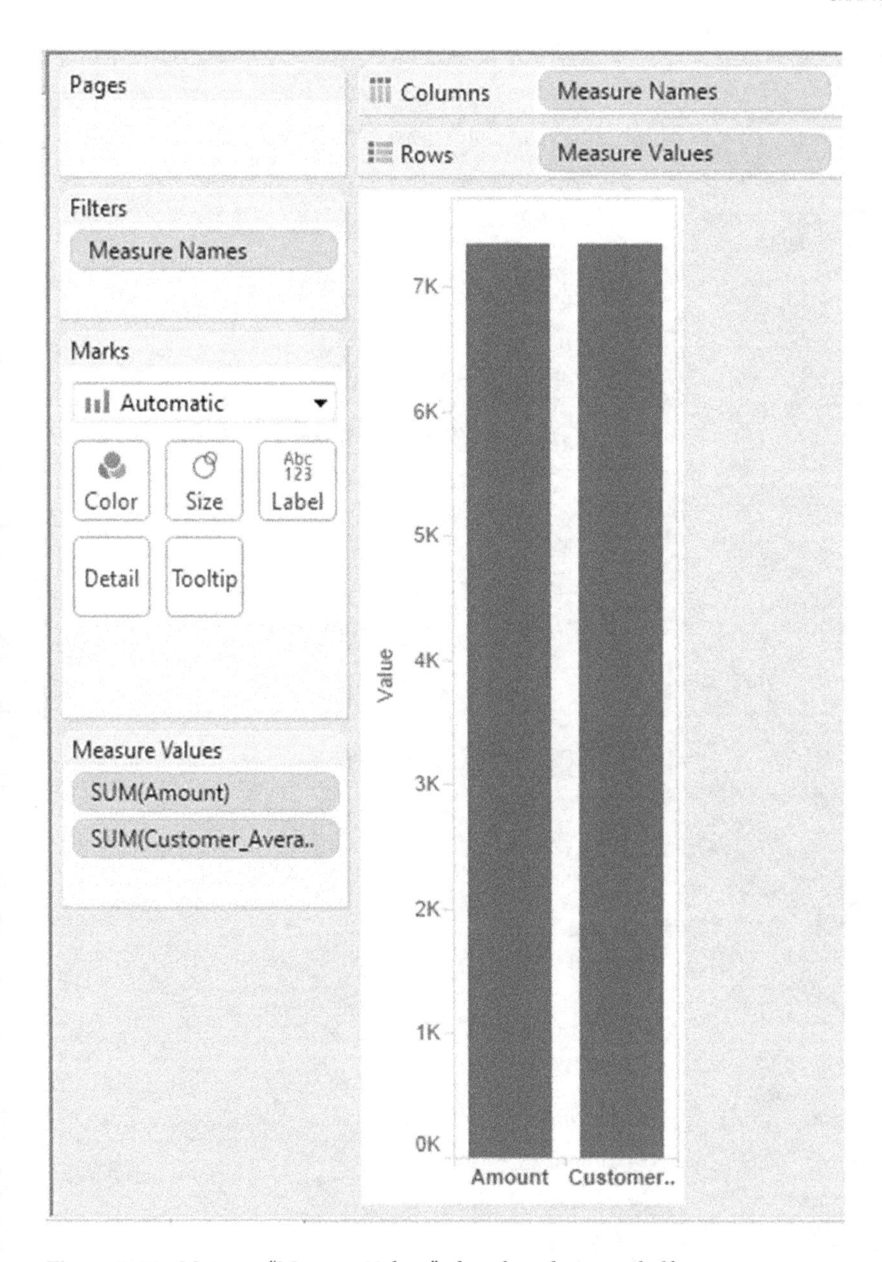

Figure 5-92. *Measure "Measure Values" placed on the rows shelf*

5.6.2.3.6 Step 6

Change the aggregation of both the measures, "Amount" and "Customer_Average_Amount" to "Average" (Shown in Fig. 5-93 & Figure 5-94).

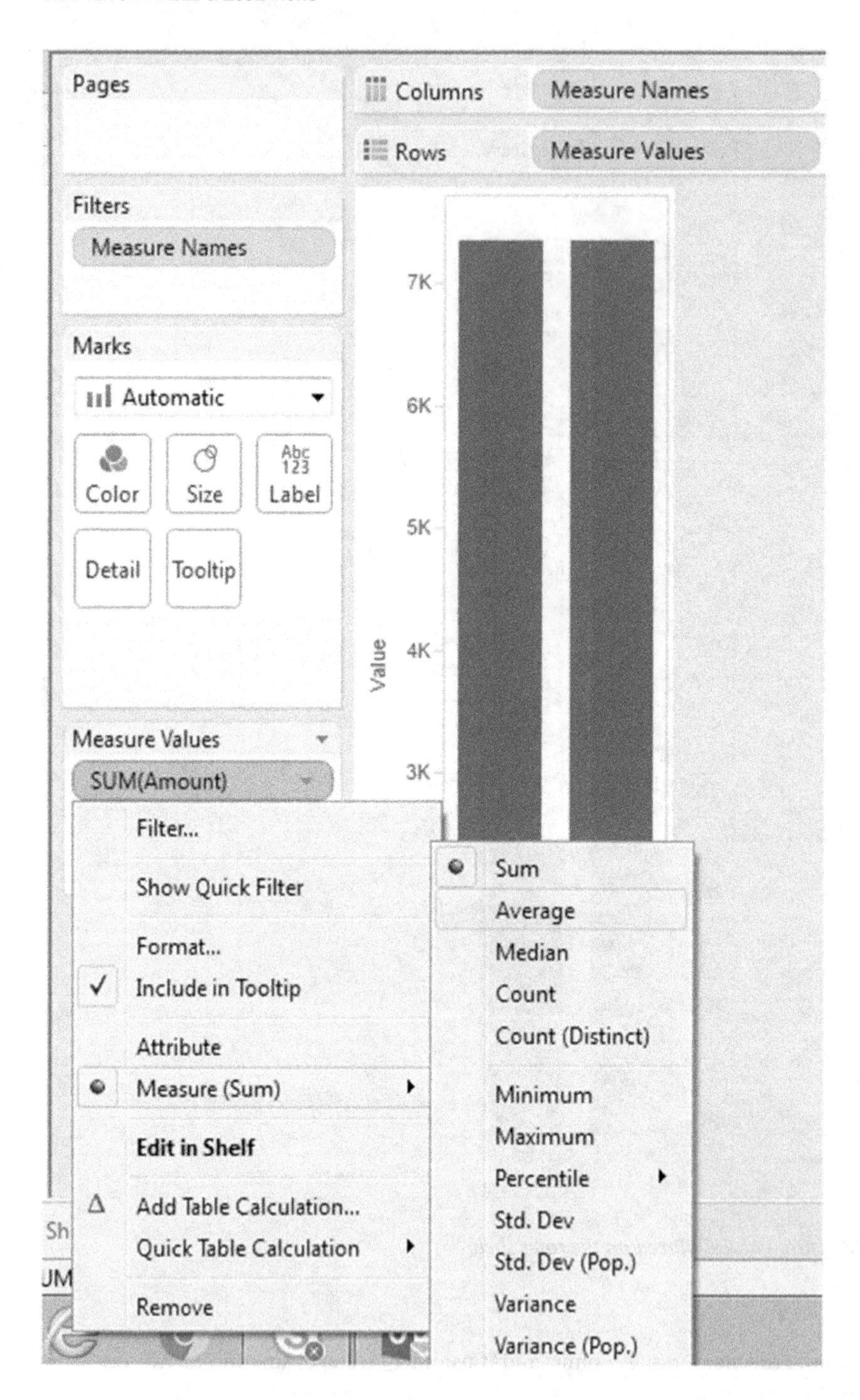

Figure 5-93. *Aggregation for measure "Amount" set to "Average"*

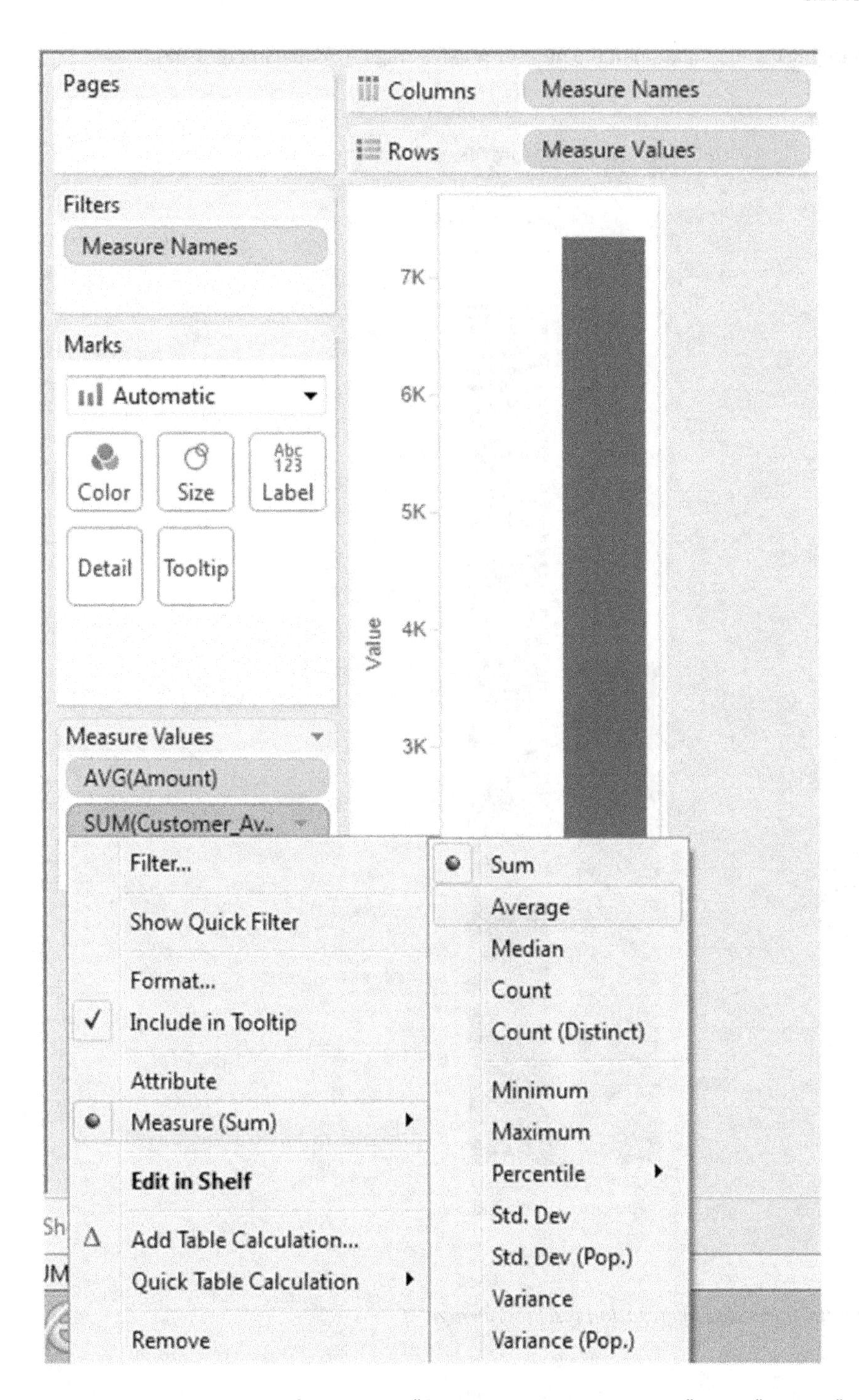

Figure 5-94. *Aggregation for measure "Customer_Average_Amount" set to "Average"*

The output after changing the aggregation of the measures to "Average" (Shown in Fig. 5-95).

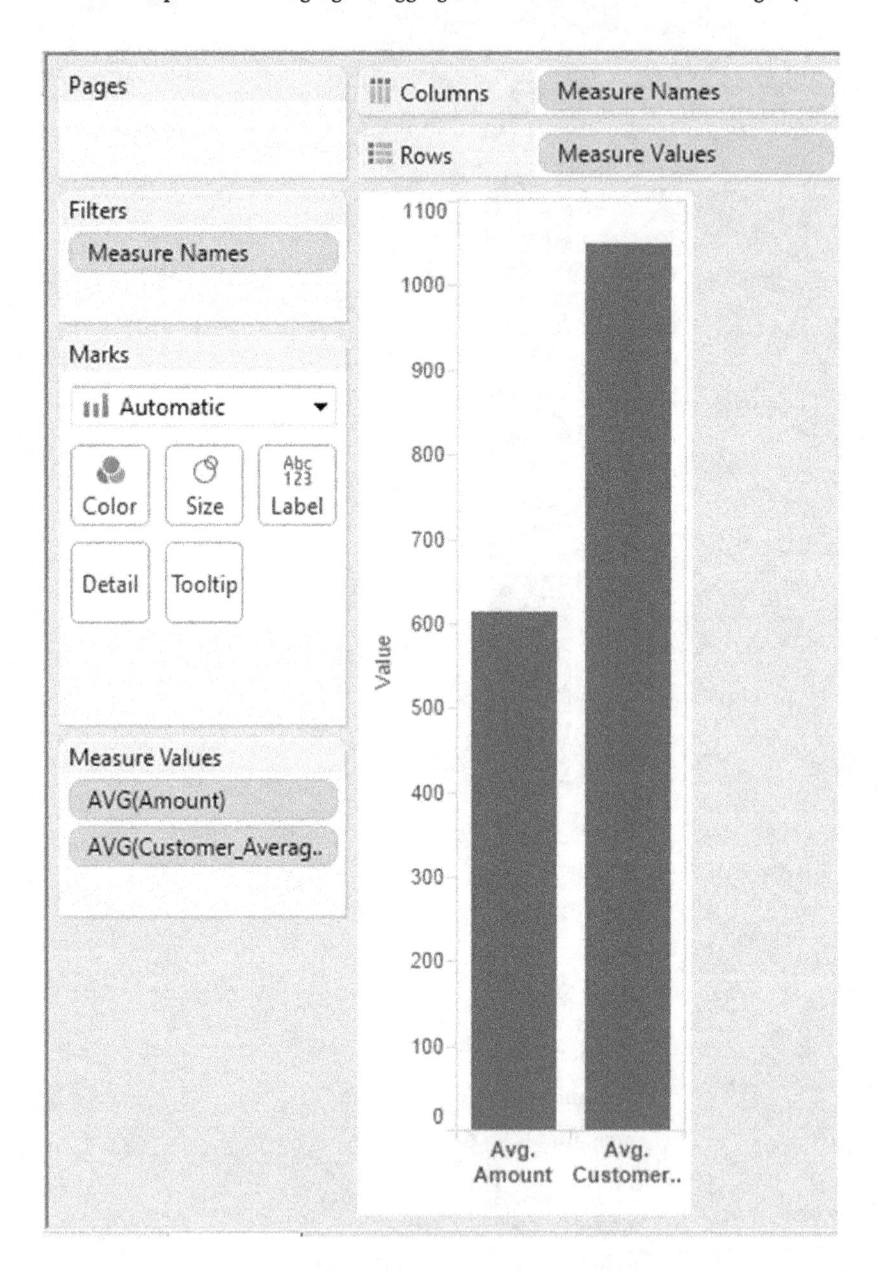

Figure 5-95. *Output after the measure's aggregation is set to "Average"*

5.6.2.3.7 Step 7

Press the CTRL key and drag the "Measure Values" from the rows shelf to "Label" on the marks card (Shown in Fig. 5-96).

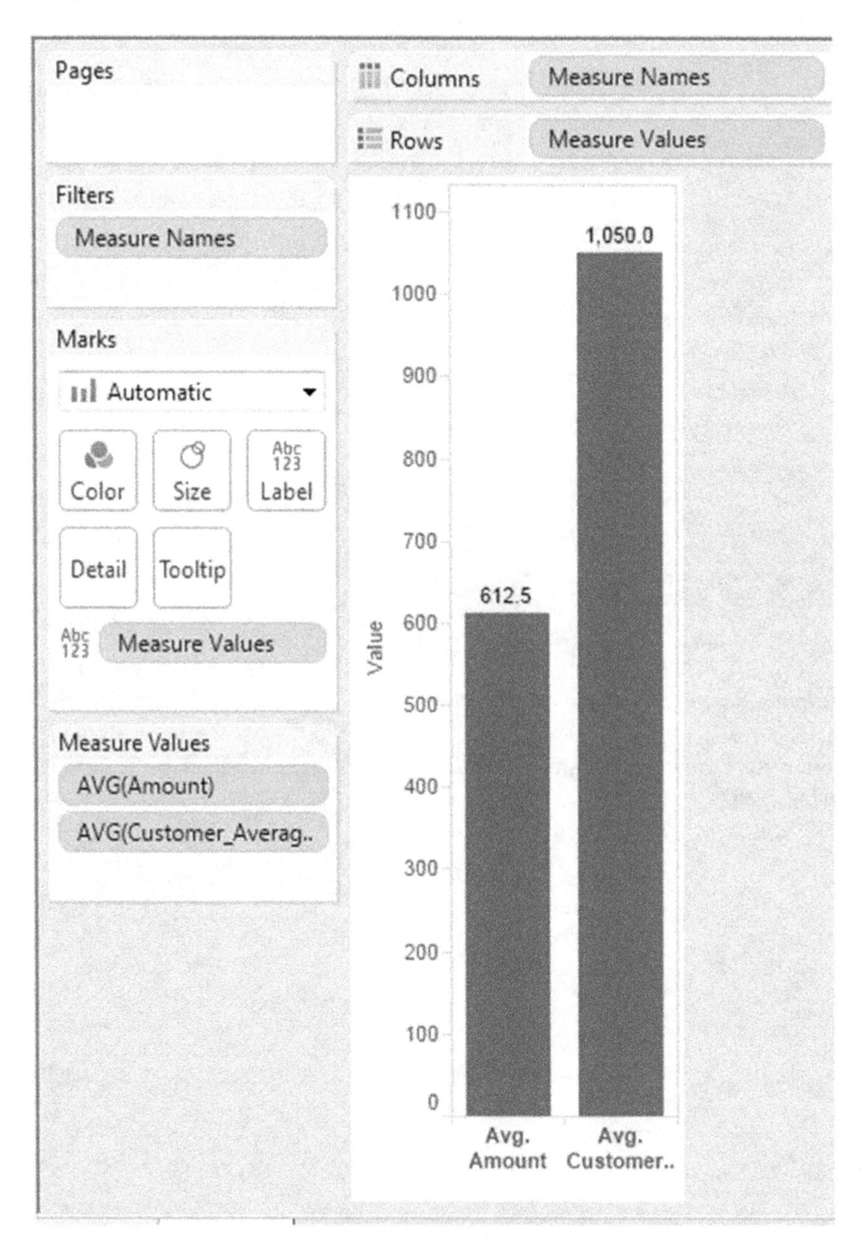

Figure 5-96. "Measure Values" placed on "Label" on the marks card

Just to cross-verify the "Average Sales Amount per Customer", let us perform the below steps:

Drag the **dimension** "Customer Name" from the dimensions area under the data pane and place it on the rows shelf (Shown in Fig. 5-97).

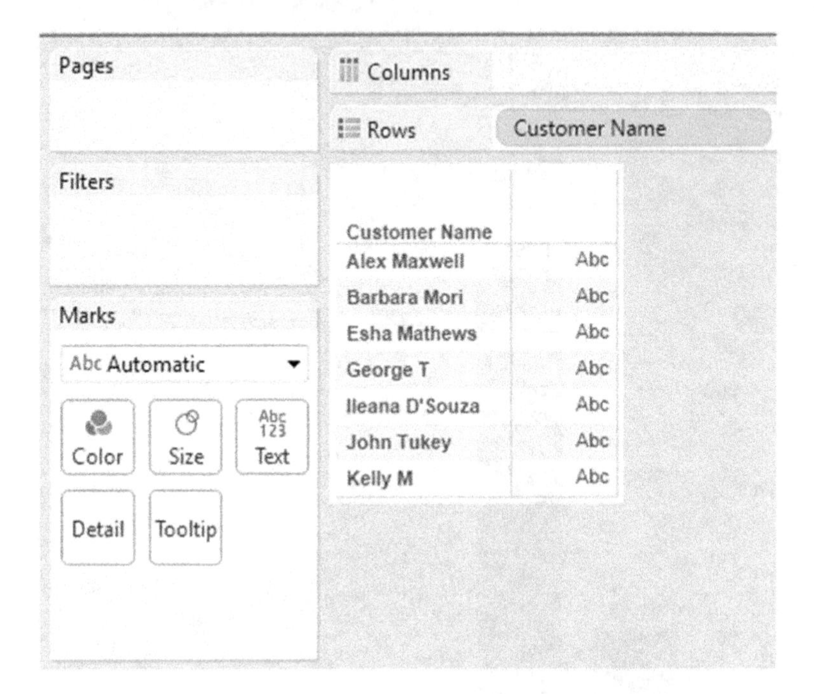

Figure 5-97. *Dimension "Customer Name" placed on the rows shelf*

Drag the **measure** "Amount" from the measures area under the data pane and place it on the **"Label"** on the marks card (Shown in Fig. 5-98).

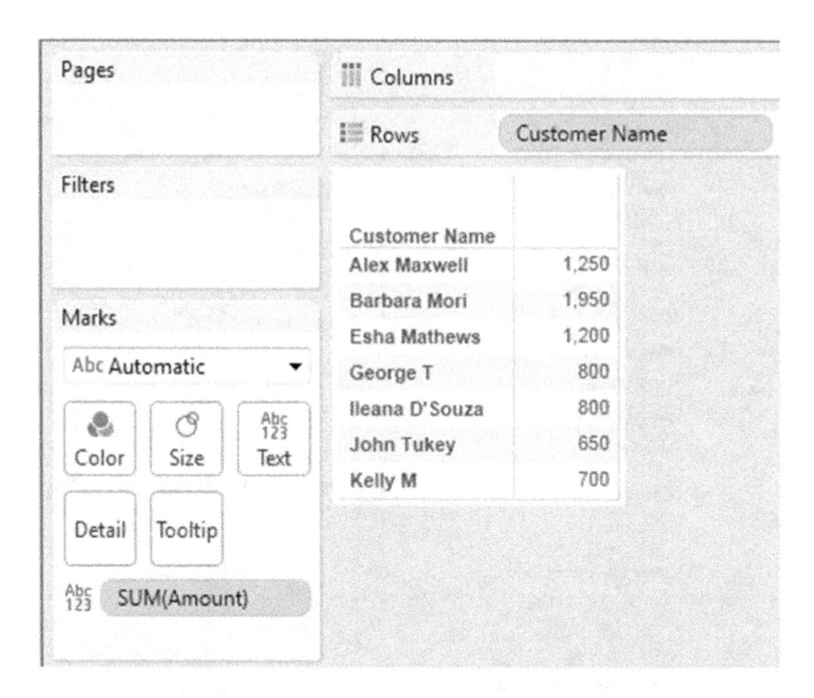

Figure 5-98. *Measure "Amount" placed on "Label" on the marks card*

Select "Analysis" on the menu bar. Select "Totals" and then select "Show Column Grand Totals".
Analysis ➤ Totals ➤ Show Column Grand Totals.
Refer to Fig. 5-99.

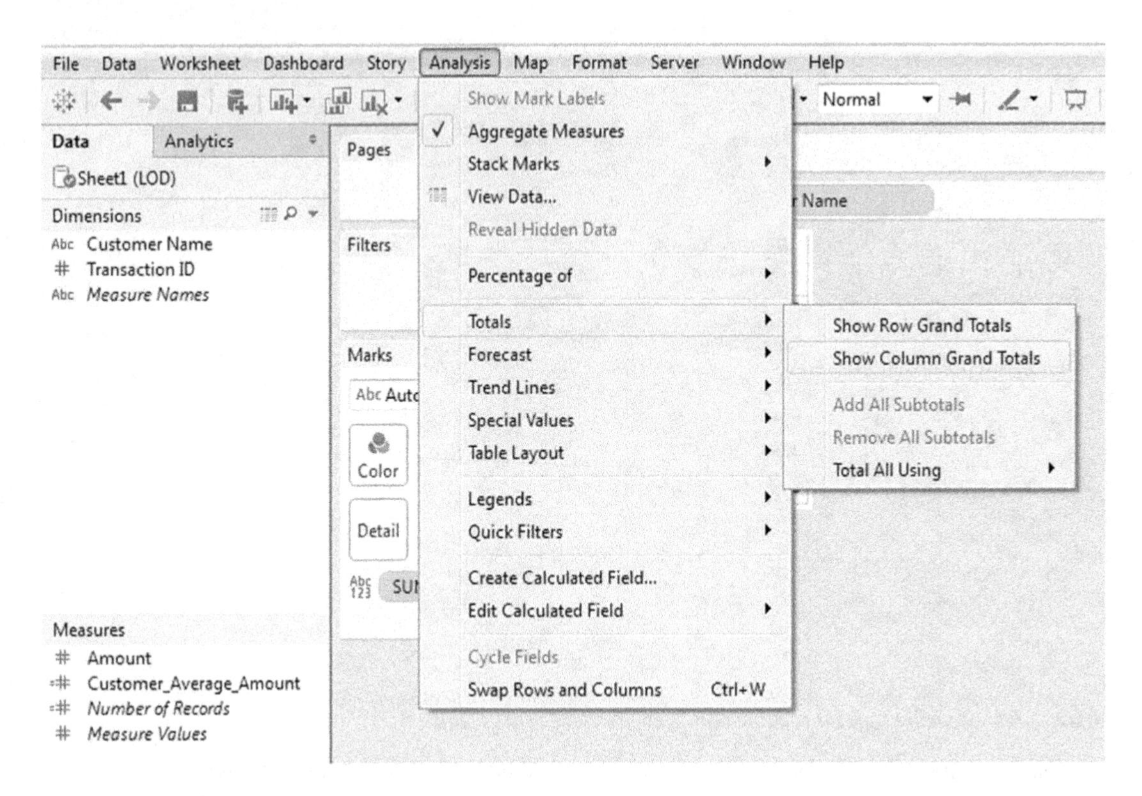

Figure 5-99. *"Show Column Grand Totals" for the measure on the view*

The output is shown in Fig. 5-100.

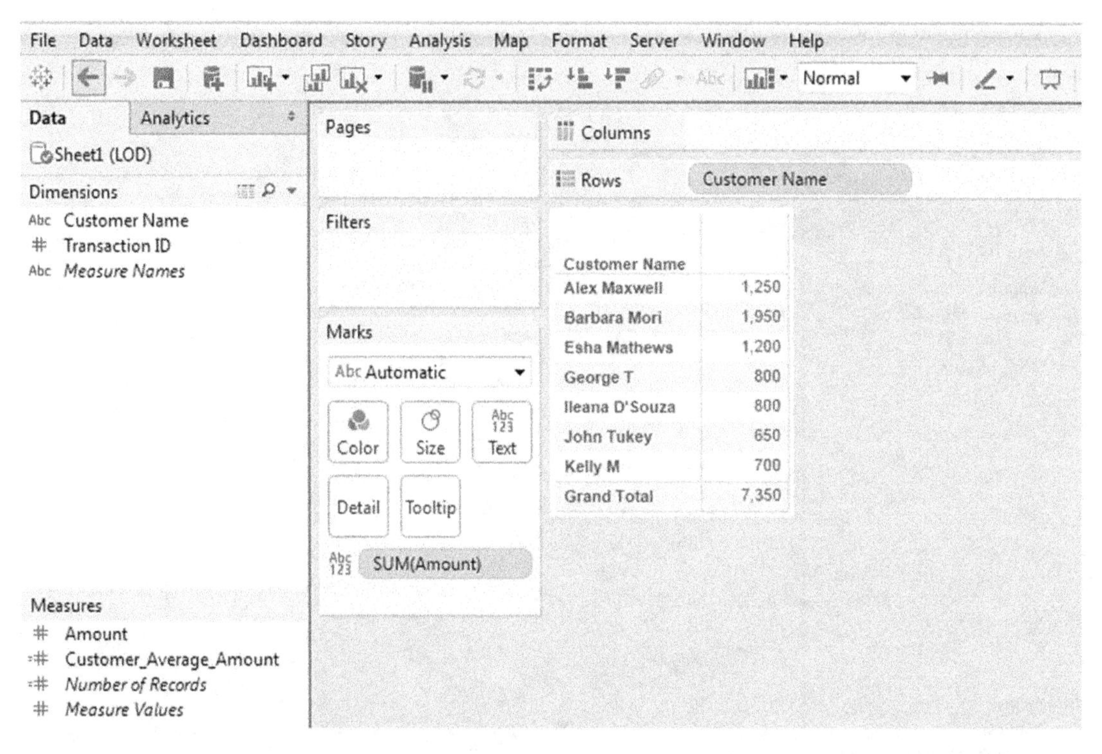

Figure 5-100. *Output with "Column Grand Totals"*

Use "Average" aggregation for grand total (Shown in Fig. 5-101).

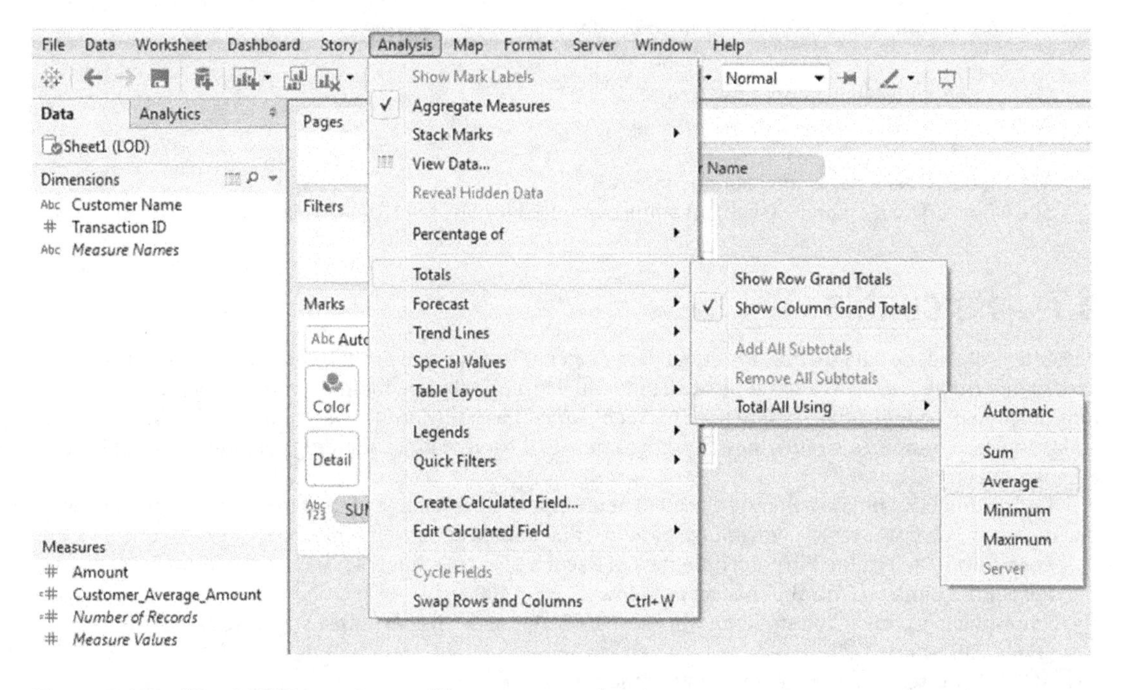

Figure 5-101. *"Total All Using - Average" for measure on the view*

The output after the aggregation for grand total was changed to "Average" (Shown in Fig. 5-102).

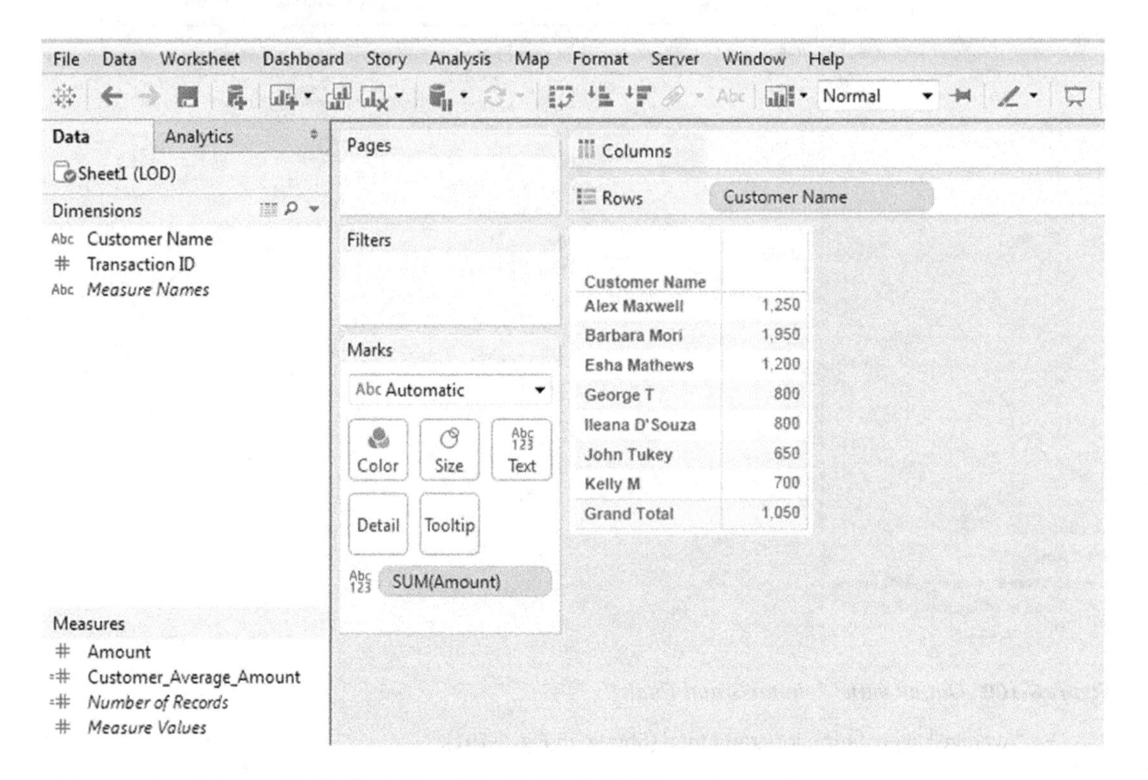

Figure 5-102. *Output with "Column Grand Total - Total All Using - Average"*

This can be achieved in another two ways:
WINDOW_AVG(Sum(Amount))
Or
AVG({Fixed [Customer Name] : Sum(Amount)})
The value of "1,050" matches with our value computed for "Average Sales Amount per Customer".

5.7 Percentile

Percentiles divide the data set into 100 equal parts. Percentiles measure position from the bottom. They are used to determine the relative standing of an individual in a population. In other words, they provide the rank position of an individual. Where have we seen percentiles being used? It is usually used with test scores and graduation standings. Graduation standings refers to the individual's standing at graduation relative to other graduate students.

Definition: Percentile is a measure used to determine the percentage of total frequency scored below that measure. Percentile rank is percentage of scores that fall below a given score.

Formula: To determine the percentile rank of a score, x, out of a total of n scores, the formula is
Percentile Rank = ((Number of scores below x) / n) * 100
Example: In a class of 200 students, Mason scored 25th rank. His percentile standing in the class is:
(175 / 200) * 100 = 87.5%
At 87.5%, his scores are better than 88% of the class.

5.7.1 Demo 1

Objective: To compute the percentile for students of VIII grade.

Input: "Percentile.xlsx"

The sample data set as available in "Percentile.xlsx".

	A	B
1	RollNo	CGPA
2	1	4.6
3	2	4.2
4	3	4.4
5	4	4.3
6	5	3.9
7	6	5
8	7	4.3
9	8	4.4
10	9	4.6
11	10	4.7

Expected output: Shown in Fig. 5-103.

Roll No	CGPA	Percentile
1	4.600	80.00%
2	4.200	20.00%
3	4.400	60.00%
4	4.300	40.00%
5	3.900	10.00%
6	5.000	100.00%
7	4.300	40.00%
8	4.400	60.00%
9	4.600	80.00%
10	4.700	90.00%

Figure 5-103. *Percentile - Demo 1 - expected output*

5.7.1.1 Steps

Follow the steps as provided.

5.7.1.1.1 Step 1

Read in the data from “Percentile.xlsx” into Tableau (Shown in Fig. 5-104).

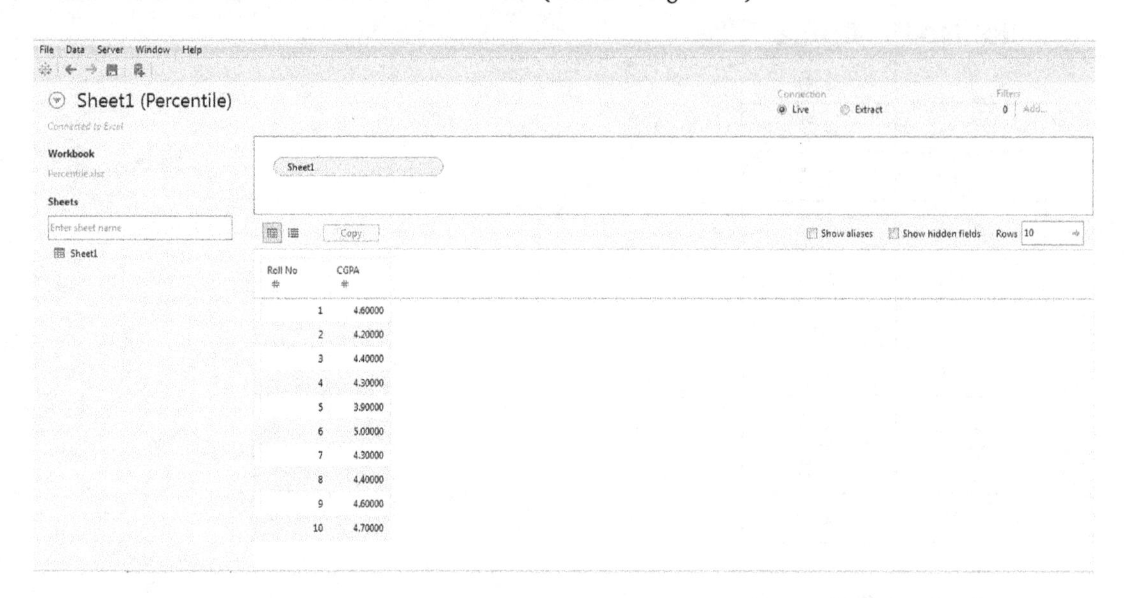

Figure 5-104. *Data from “Percentile.xls” read into Tableau*

5.7.1.1.2 Step 2

Drag the **dimension** “Roll No” from the dimensions area under the data pane and place it on the rows shelf (Shown in Fig. 5-105).

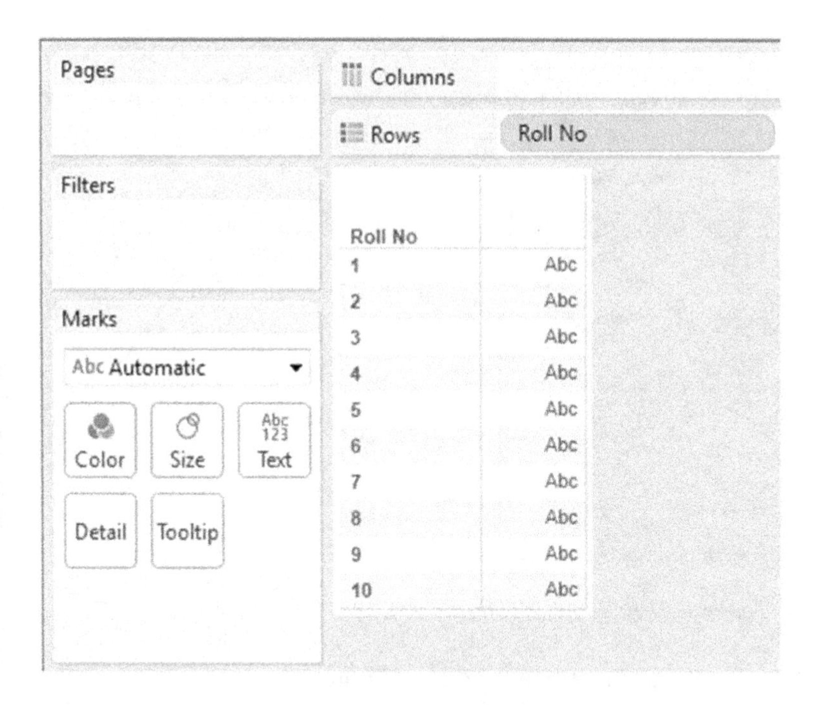

Figure 5-105. *Dimension "Roll No" placed on the rows shelf*

5.7.1.1.3 Step 3

Drag the **measure** "CGPA" from the measures area under the data pane and place it on the **"Label"** on the marks card (Shown in Fig. 5-106).

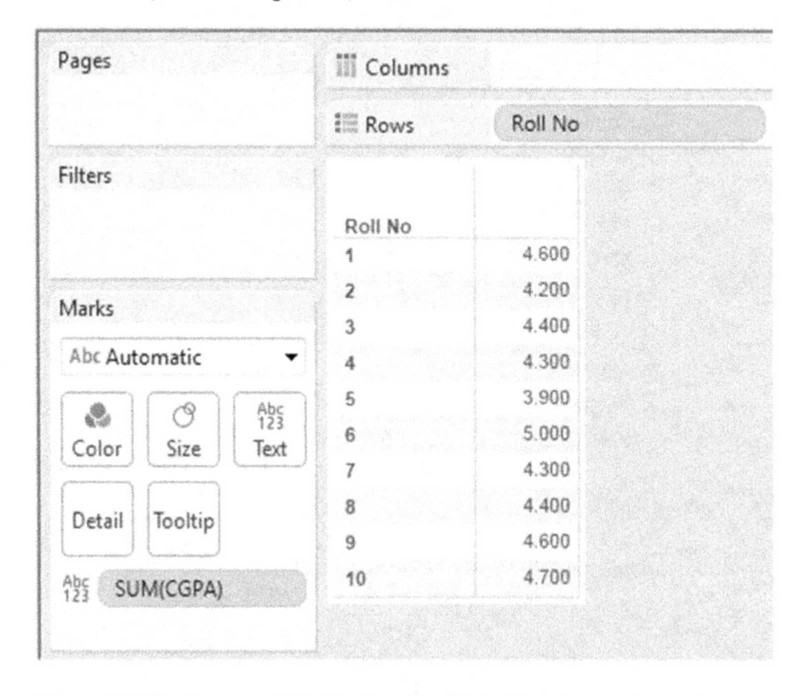

Figure 5-106. *Measure "CGPA" placed on "Label" on the marks card*

5.7.1.1.4 Step 4

Add a table calculation "Percentile" to the measure "CGPA" placed on the "Label" on the marks card (Shown in Fig. 5-107).

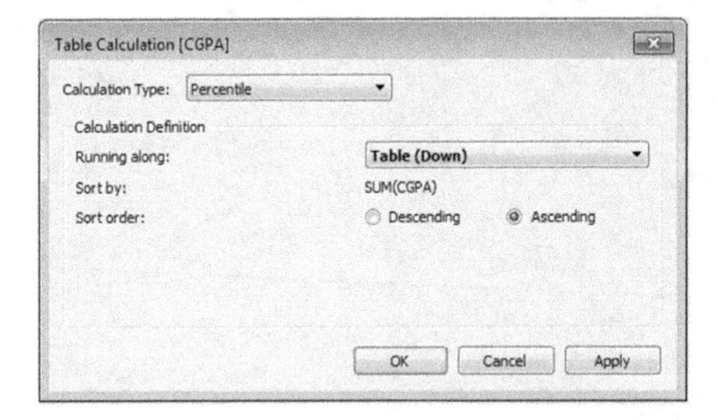

Figure 5-107. *"Table Calculation - Percentile" being applied to measure "CGPA"*

The output after adding the "Table Calculation – Percentile" to the measure "CGPA" (Shown in Fig. 5-108).

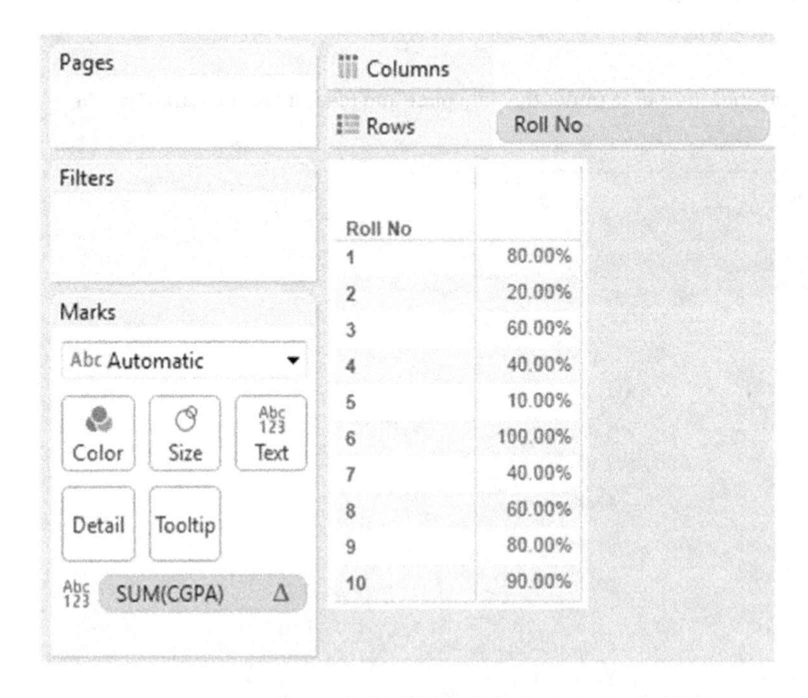

Figure 5-108. *Output after applying "Table Calculation - Percentile" to measure "CGPA"*

5.7.1.1.5 Step 5

Drag the measure "CGPA" from the measures area under the data pane and place it in the text area (Shown in Fig. 5-109).

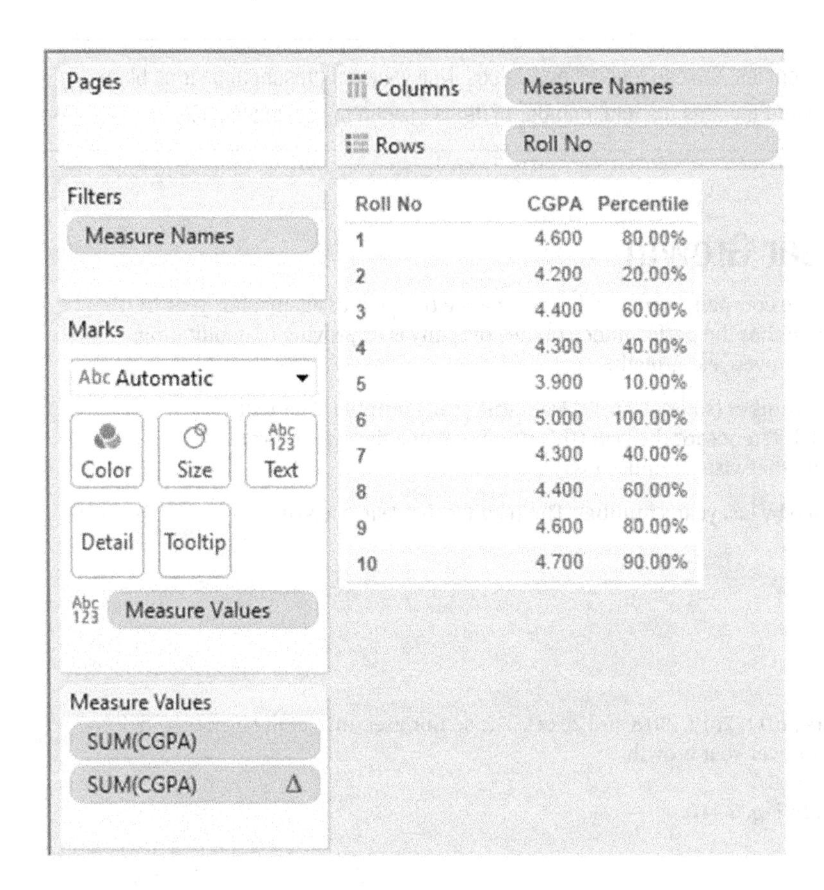

Roll No	CGPA	Percentile
1	4.600	80.00%
2	4.200	20.00%
3	4.400	60.00%
4	4.300	40.00%
5	3.900	10.00%
6	5.000	100.00%
7	4.300	40.00%
8	4.400	60.00%
9	4.600	80.00%
10	4.700	90.00%

Figure 5-109. *Measure "CGPA" placed in the view*

The formula used in the calculation:
(Number of values less than or equal to the value under consideration / Total number of values) * 100
Example: Let us consider the CGPA score for student whose roll number is 1. The Student's CGPA score is 4.6. There are eight CGPA scores that are less than or equal to the CGPA score of 4.6.
(8 / 10) * 100 = 80%

Now, just to reconfirm the formula, let us consider another student's CGPA value. This time let us consider the CGPA value of 3.9.

This is the least CGPA score that a student has attained.

(1/10)*100 = 10%

■ **Note** While computing percentiles, Tableau ignores null values. Null values if present appear as blank rows in a cross-tab and do not count towards the total number of items used in the calculation (%).

5.8 Year over Year Growth

In layman's terms, YOY means the company's financial performance this year as against last year. YOY performance is used to gauge whether the performance of the company is improving or debilitating.

Formula to compute the Year over year growth:

- Subtract last year's number (sales or profit) from this year's number. This will constitute the total difference for the year (this number if positive will indicate a year-over-year gain otherwise it implies loss).
- Divide the difference by last year's number. The result is the year-over-year growth rate.

5.8.1 Demo 1

Objective:

Data is provided for 4 years (2011, 2012, 2013 and 2014). The senior executive at the firm would like a visualization that shows the Year over Year growth.

Input: "Sample - Superstore.xls"

Expected Output: Shown in Fig. 5-110.

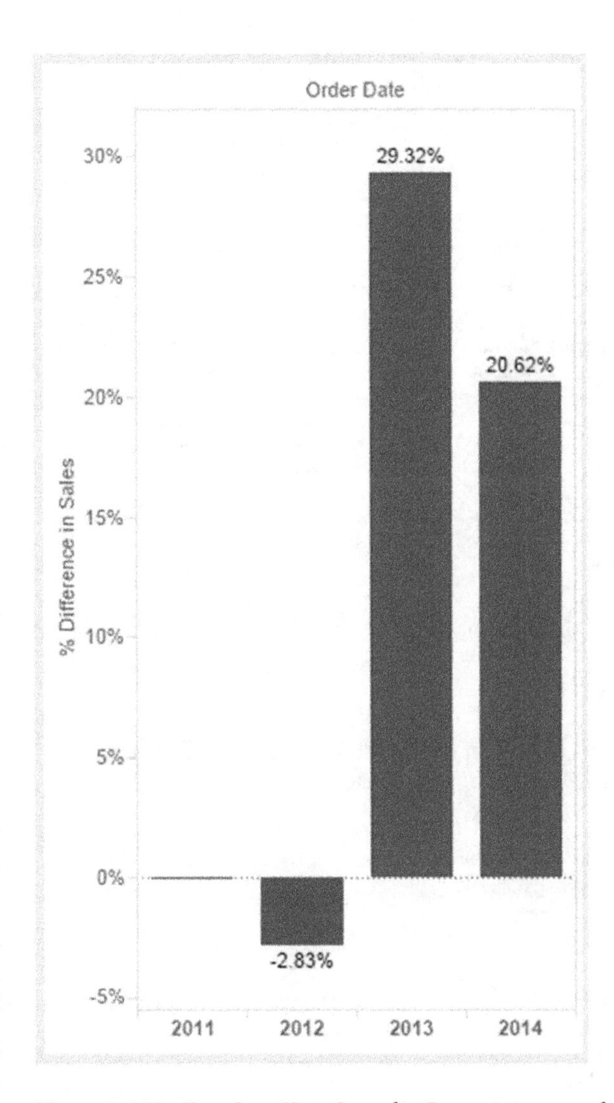

Figure 5-110. *Year Over Year Growth - Demo 1 - expected output*

5.8.1.1 Steps

Follow the steps as provided.

5.8.1.1.1 Step 1

Read in data from "Sample - Superstore.xls" into Tableau (Shown in Fig. 5-111).

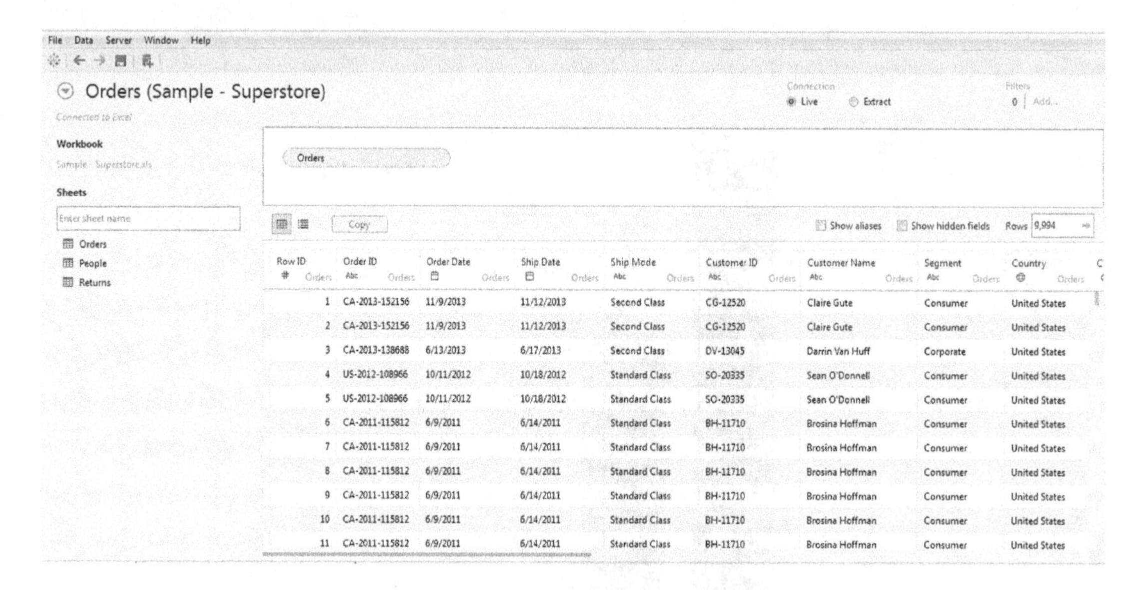

Row ID	Order ID	Order Date	Ship Date	Ship Mode	Customer ID	Customer Name	Segment	Country
1	CA-2013-152156	11/9/2013	11/12/2013	Second Class	CG-12520	Claire Gute	Consumer	United States
2	CA-2013-152156	11/9/2013	11/12/2013	Second Class	CG-12520	Claire Gute	Consumer	United States
3	CA-2013-138688	6/13/2013	6/17/2013	Second Class	DV-13045	Darrin Van Huff	Corporate	United States
4	US-2012-108966	10/11/2012	10/18/2012	Standard Class	SO-20335	Sean O'Donnell	Consumer	United States
5	US-2012-108966	10/11/2012	10/18/2012	Standard Class	SO-20335	Sean O'Donnell	Consumer	United States
6	CA-2011-115812	6/9/2011	6/14/2011	Standard Class	BH-11710	Brosina Hoffman	Consumer	United States
7	CA-2011-115812	6/9/2011	6/14/2011	Standard Class	BH-11710	Brosina Hoffman	Consumer	United States
8	CA-2011-115812	6/9/2011	6/14/2011	Standard Class	BH-11710	Brosina Hoffman	Consumer	United States
9	CA-2011-115812	6/9/2011	6/14/2011	Standard Class	BH-11710	Brosina Hoffman	Consumer	United States
10	CA-2011-115812	6/9/2011	6/14/2011	Standard Class	BH-11710	Brosina Hoffman	Consumer	United States
11	CA-2011-115812	6/9/2011	6/14/2011	Standard Class	BH-11710	Brosina Hoffman	Consumer	United States

Figure 5-111. *Data from "Sample - Superstore.xls" read into Tableau*

5.8.1.1.2 Step 2

Drag the **dimension** "Order Date" from the dimensions area under the data pane to the columns shelf. Retain the date hierarchy at the default, i.e. "Year". Retain the "Order Date" at "Discrete" (Shown in Fig. 5-112).

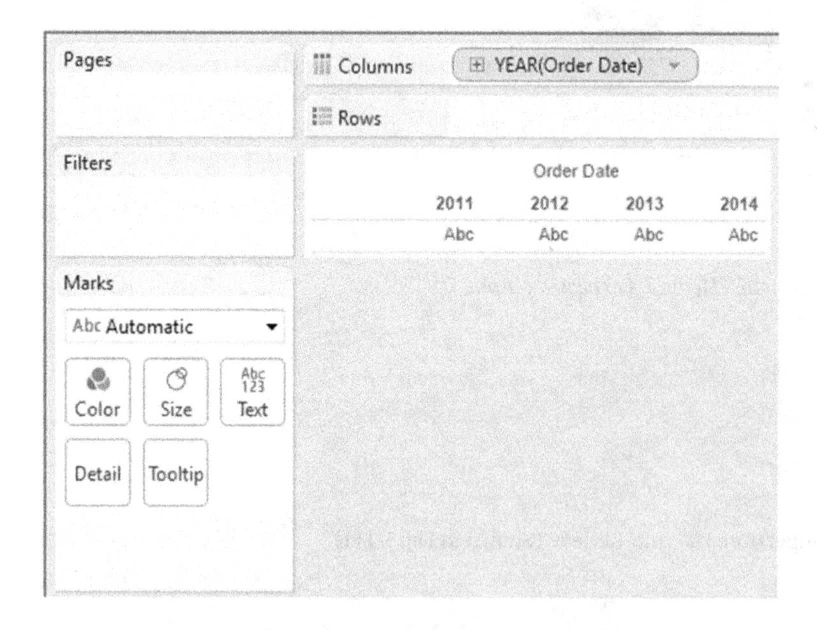

Figure 5-112. *Dimension "Order Date" placed on the columns shelf*

5.8.1.1.3 Step 3

Drag the measure "Sales" from the measures area under the data pane and place it on the rows shelf (Shown in Fig. 5-113).

***Figure 5-113.** Measure "Sales" placed on the rows shelf*

5.8.1.1.4 Step 4

Change the "Mark Type" to "Bar" (Shown in Fig. 5-114).

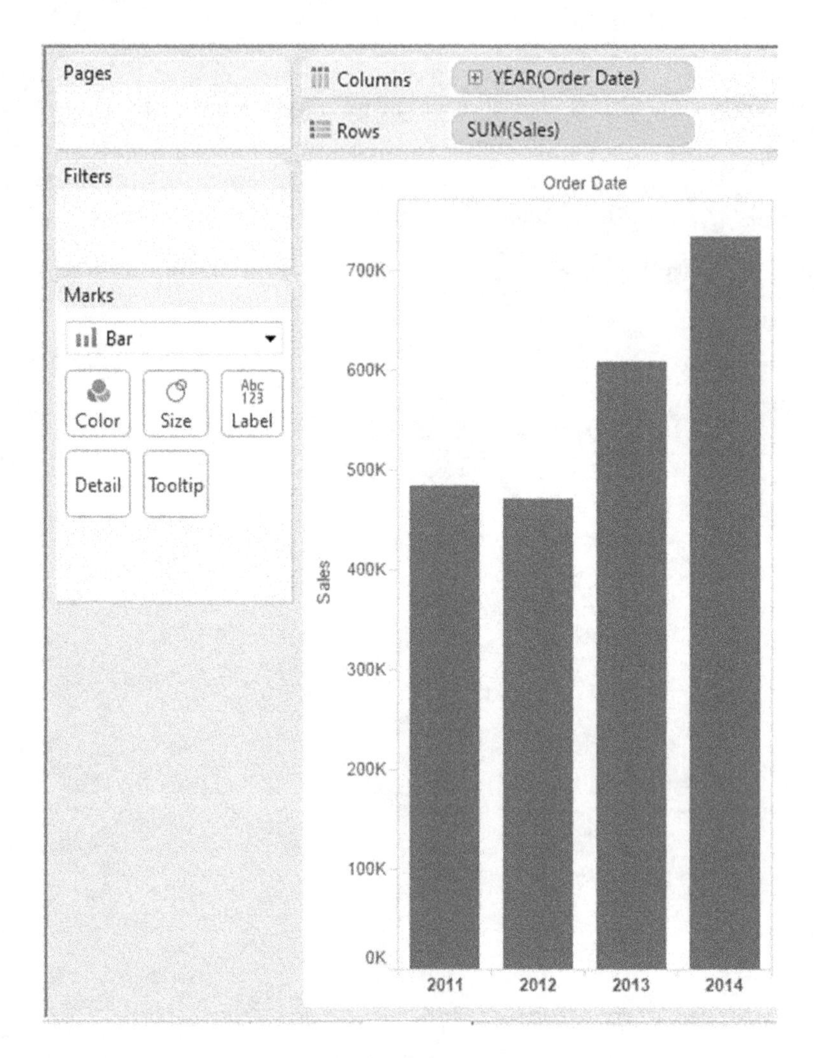

Figure 5-114. *Mark Type changed to "Bar"*

5.8.1.1.5 Step 5

Add a "Quick Table Calculation - Year over Year Growth" to the measure "Sales" on the rows shelf (Shown in Fig. 5-115 & Figure 5-116).

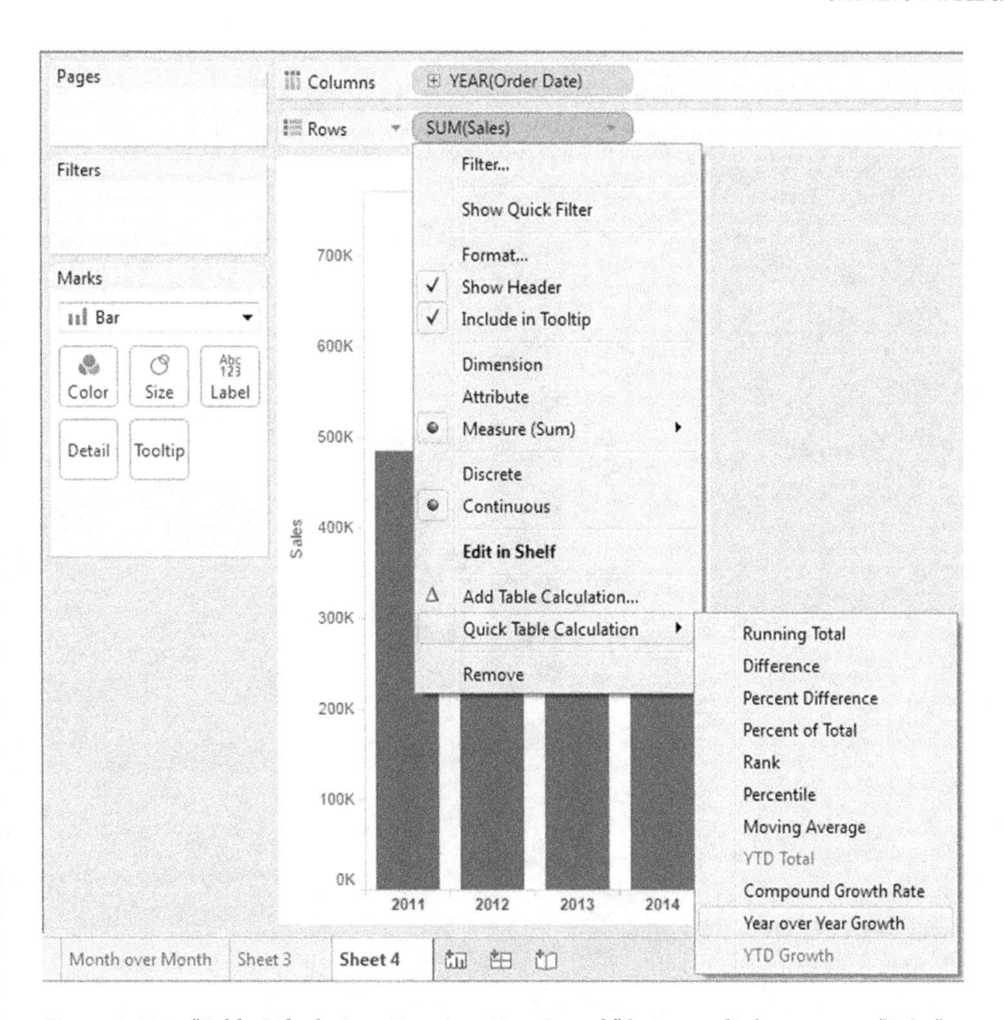

Figure 5-115. *"Table Calculation - Year Over Year Growth" being applied to measure "Sales"*

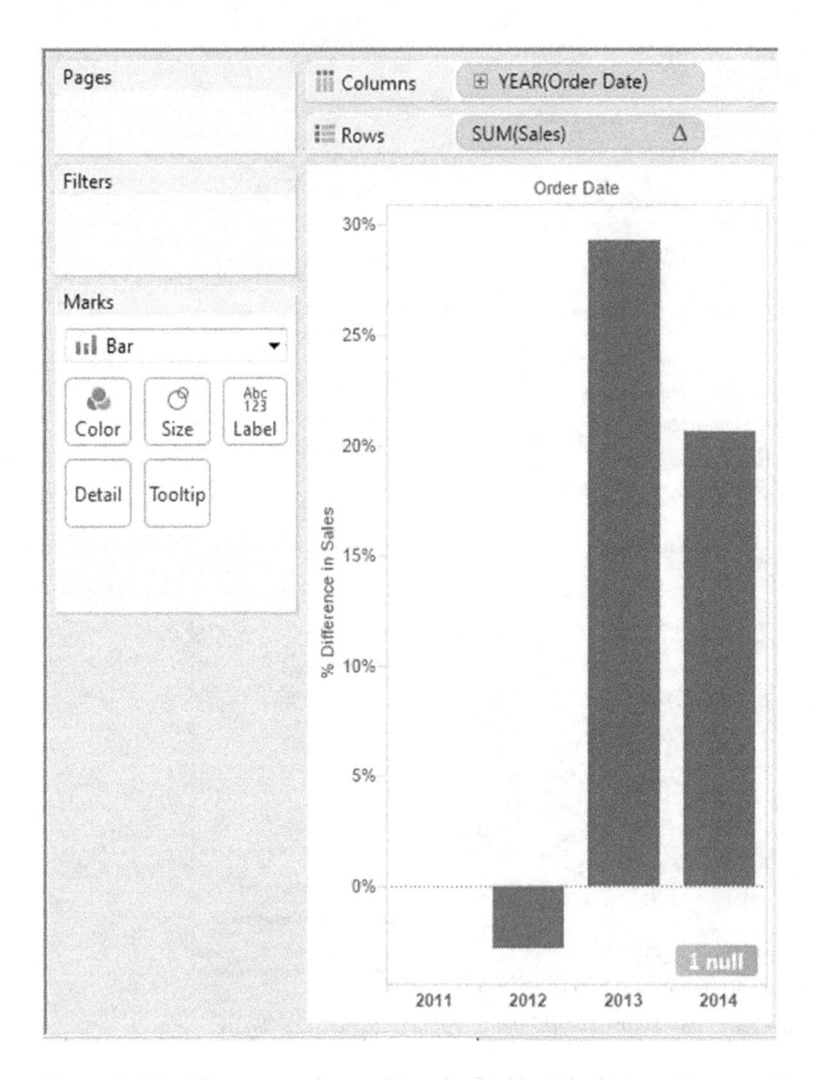

Figure 5-116. *The output after applying the "Table Calculation - Year over Year Growth"*

Click on the message, "1 null" at the bottom of the view/worksheet to bring up the "Special Values for [% Difference in Sales]" dialog box (Shown in Fig. 5-117).

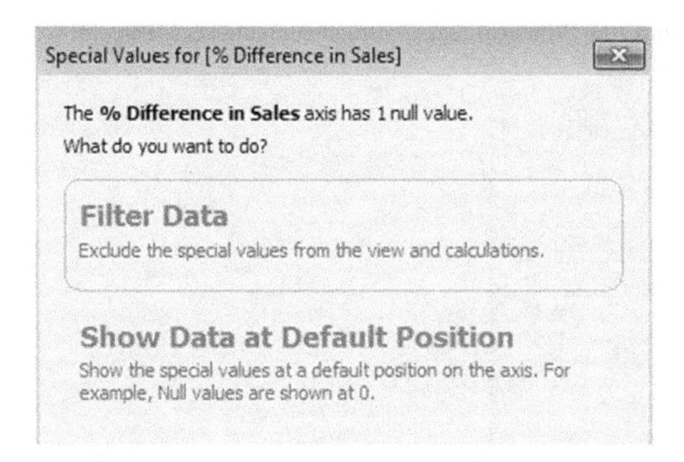

Figure 5-117. *"Special Values for [% Difference in Sales]" dialog box*

Click on "Show Data at Default Position". The output after considering the null value (Shown in Fig. 5-118).

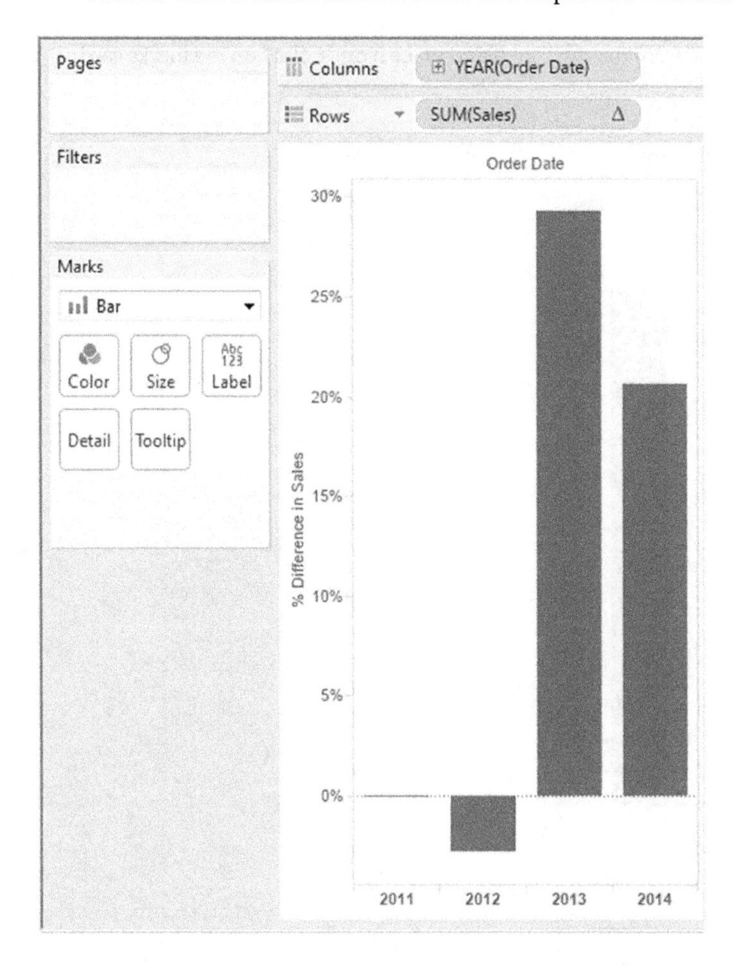

Figure 5-118. *The output after taking the null values into consideration*

A look at the "Table Calculation" dialog box. The "Year over Year Growth" is computed as a "Percent Difference" from "Previous" (Shown in Fig. 5-119).

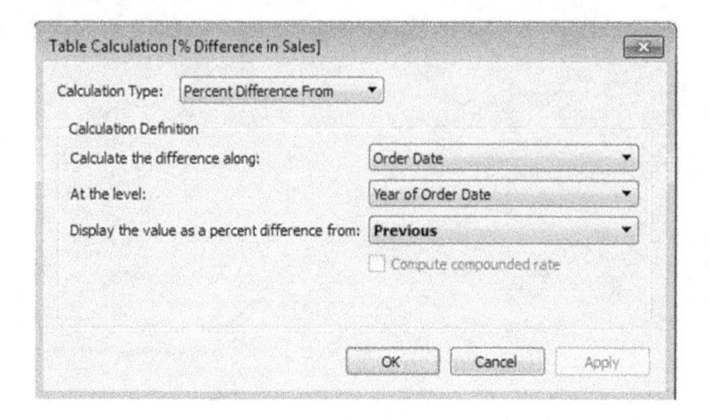

Figure 5-119. *"The "Year over Year Growth" is computed as a "Percent Difference" from "Previous"*

5.8.1.1.6 Step 6

Press "CTRL" and drag the measure "Sales" from the rows shelf and place it on "Color" on the marks card (Shown in Fig. 5-120).

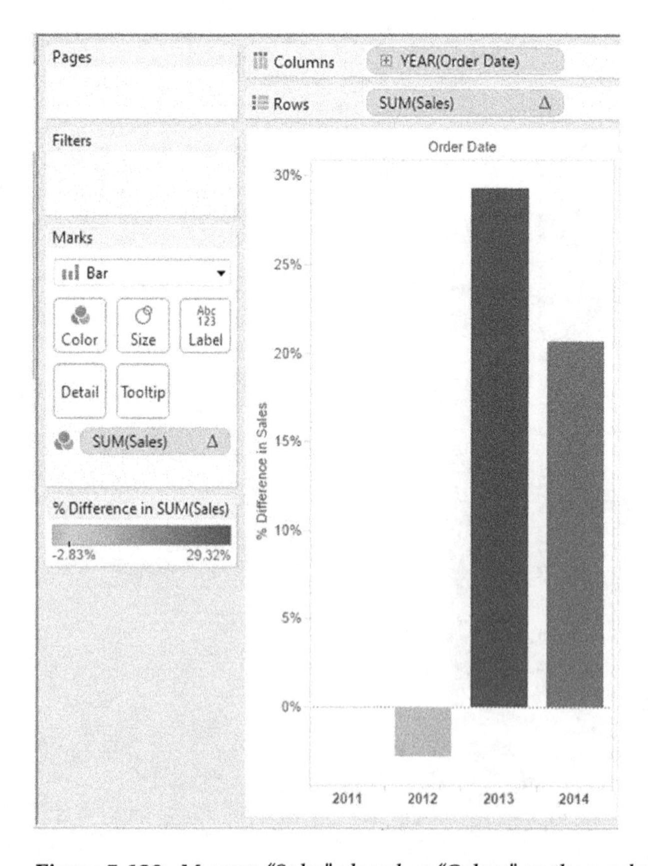

Figure 5-120. *Measure "Sales" placed on "Colors" on the marks card*

Change the stepped color to 2 (Shown in Fig. 5-121).

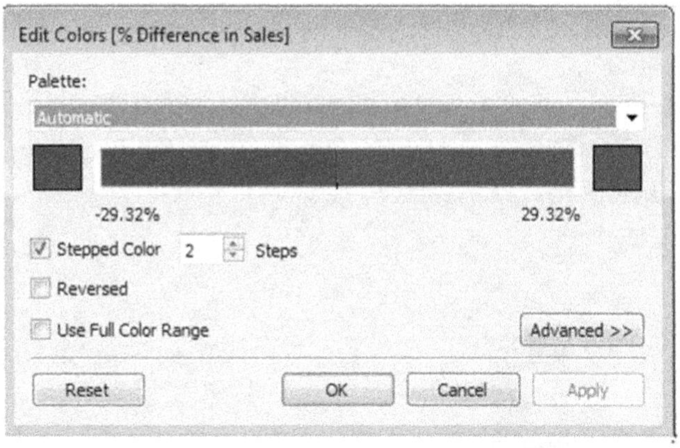

Edit Colors [% Difference in Sales]
Palette:
Automatic
-29.32%
29.32%
Stepped Color 2 Steps
Reversed
Use Full Color Range
Advanced >>
Reset
OK
Cancel
Apply

Figure 5-121. *"Stepped Color" changed to 2*

The output after changing the stepped color to 2 (Shown in Fig. 5-122).

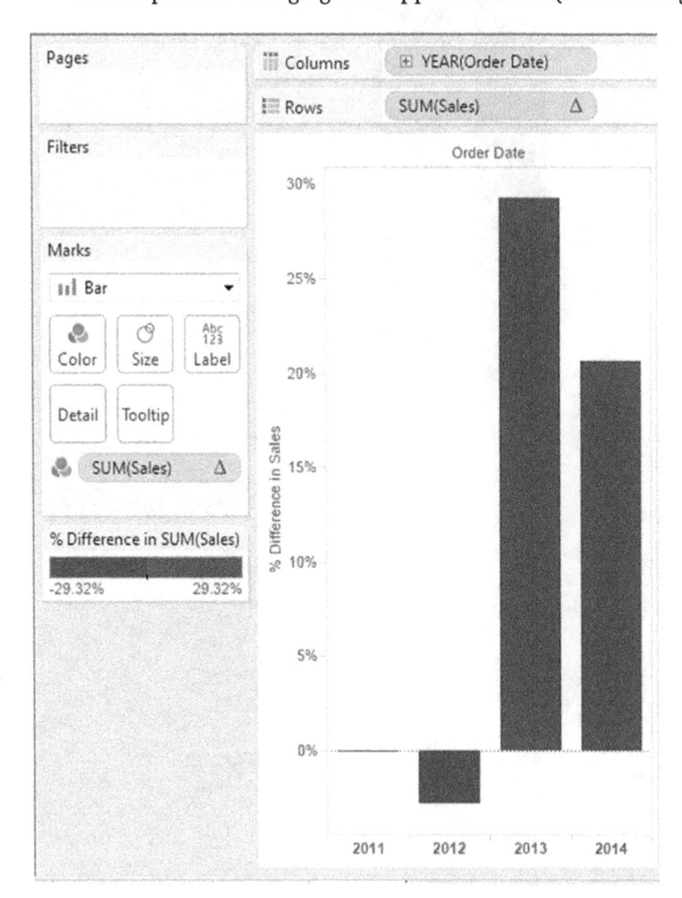

Figure 5-122. *Output after changing the "Stepped Color" to two*

5.8.1.1.7 Step 7

Press "CTRL" and drag the **measure** "Sales" from the rows shelf to "Label" on the marks card (Shown in Fig. 5-123).

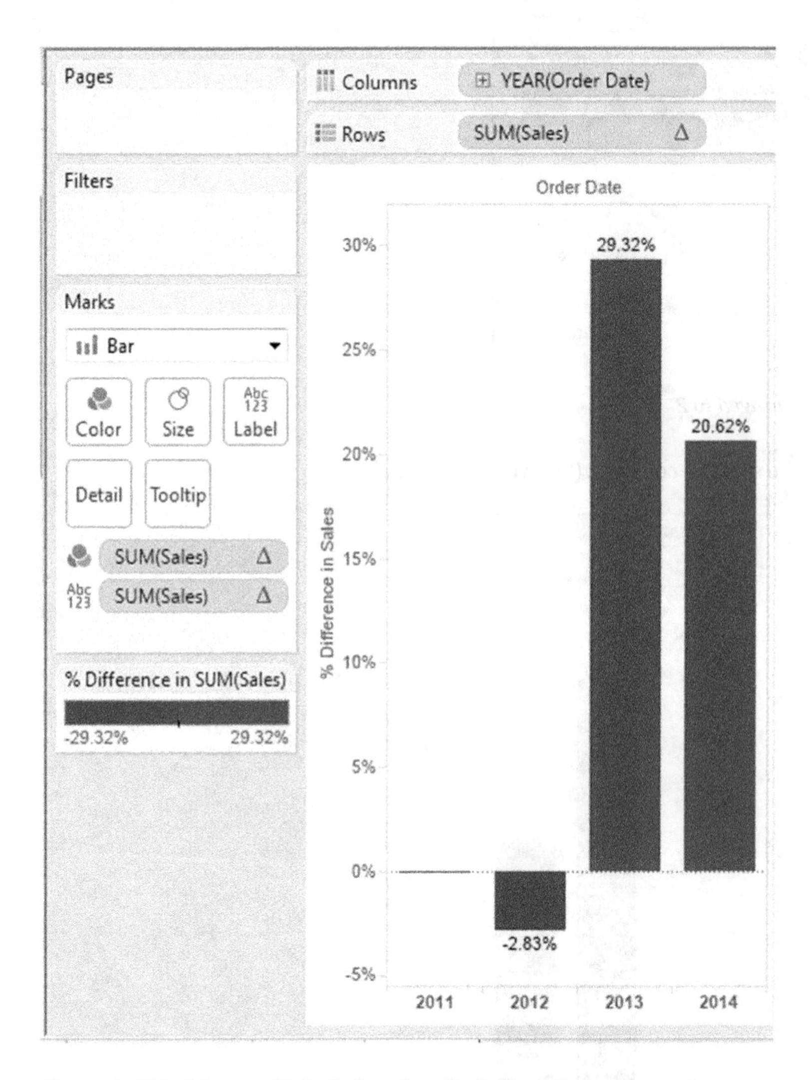

Figure 5-123. *Measure "Sales" placed on "Label" on the marks card*

The final output: Shown in Fig. 5-124.

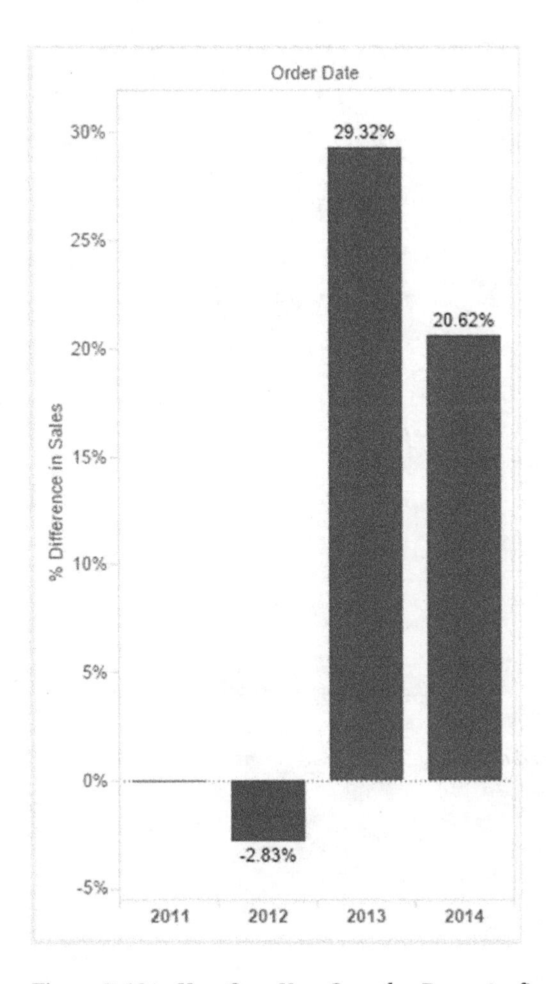

Figure 5-124. *Year Over Year Growth - Demo 1 - final output*

5.8.2 Demo 2

Objective:

The senior sales executive of "XYZ" corporation would like a visualization that presents the "Month over Month Growth" for the years 2011 and 2012.

Input: "Sample - Superstore.xls"

Expected Output: Shown in Fig. 5-125.

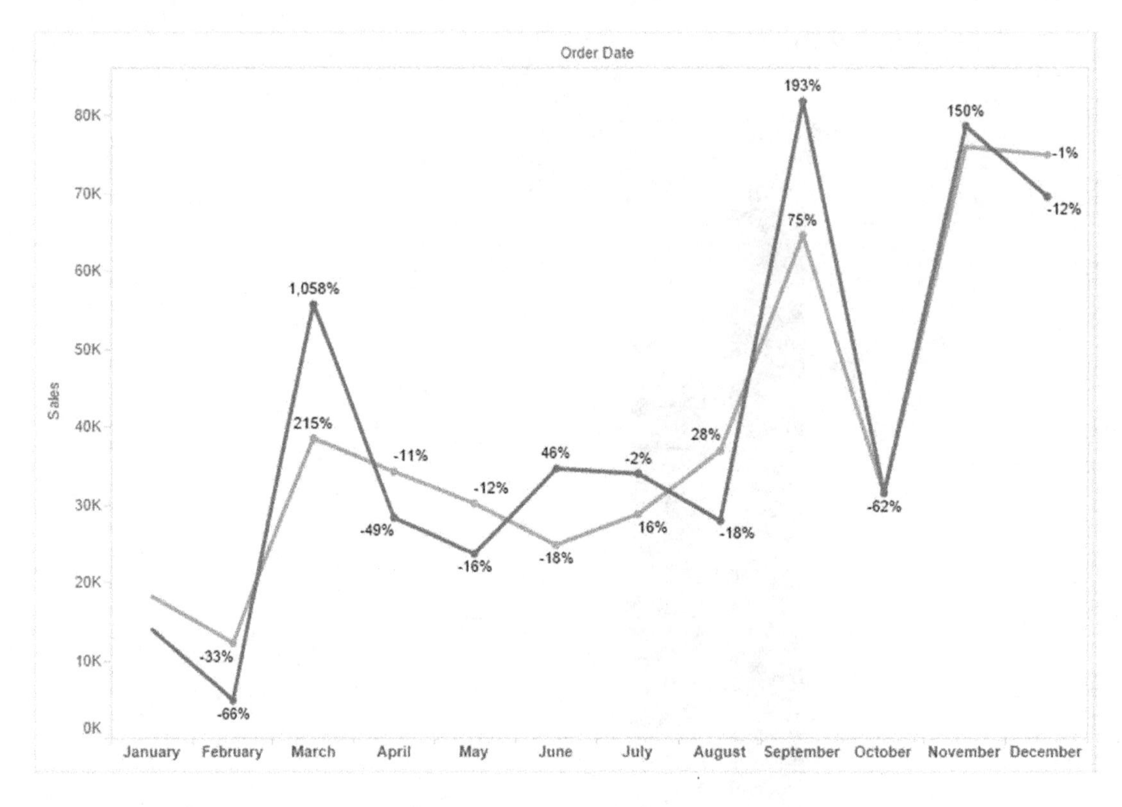

Figure 5-125. "Year Over Year Growth" - Demo 2 – expected output

5.8.2.1 Steps

Follow the steps as provided.

5.8.2.1.1 Step 1

Create a calculated field, "YearToDisplay" (Shown in Fig. 5-126).

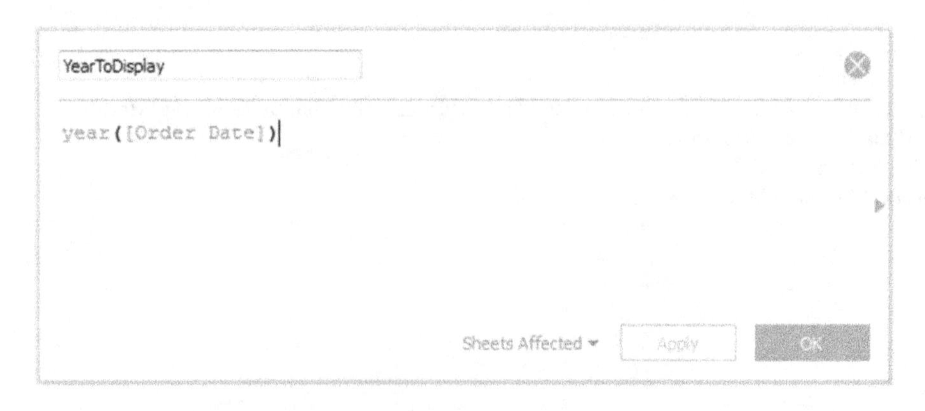

Figure 5-126. Calculated Field - "YearToDisplay" being created

5.8.2.1.2 Step 2

Convert the calculated field, "YearToDisplay" to "Dimension" (Shown in Fig. 5-127).

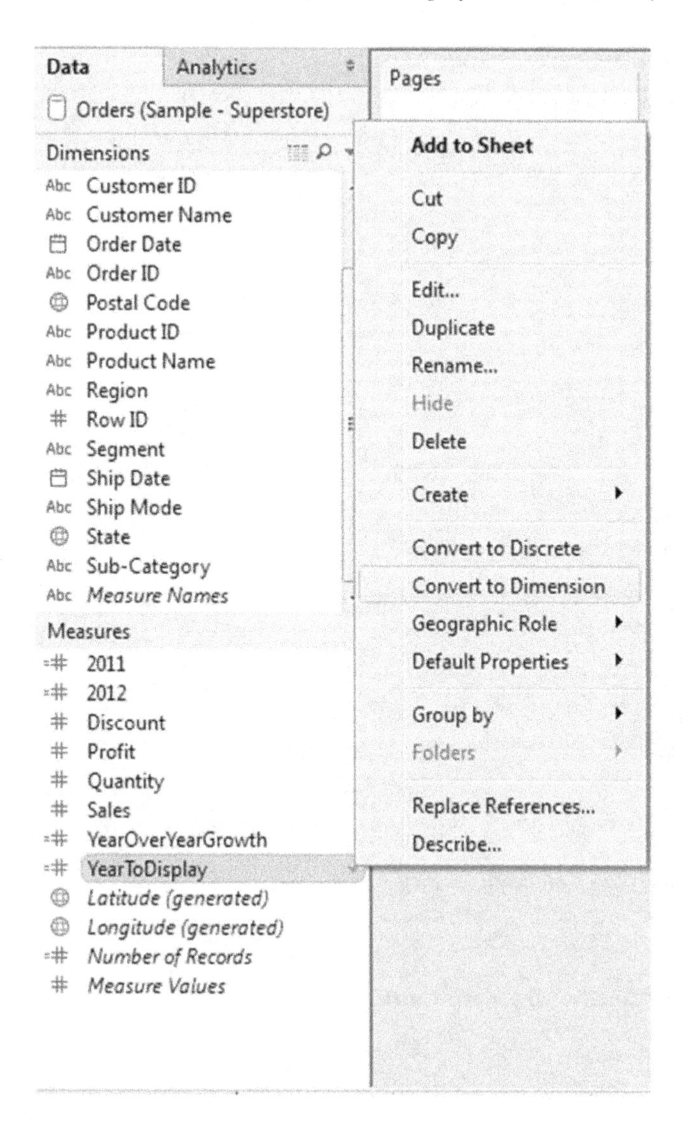

Figure 5-127. *Calculated field field "YearToDisplay" being converted to "Dimension"*

5.8.2.1.3 Step 3

Place the calculated field, "YearToDisplay" on the filters shelf.

Select the years "2011" and "2012" in the filter dialog box (Shown in Fig. 5-128).

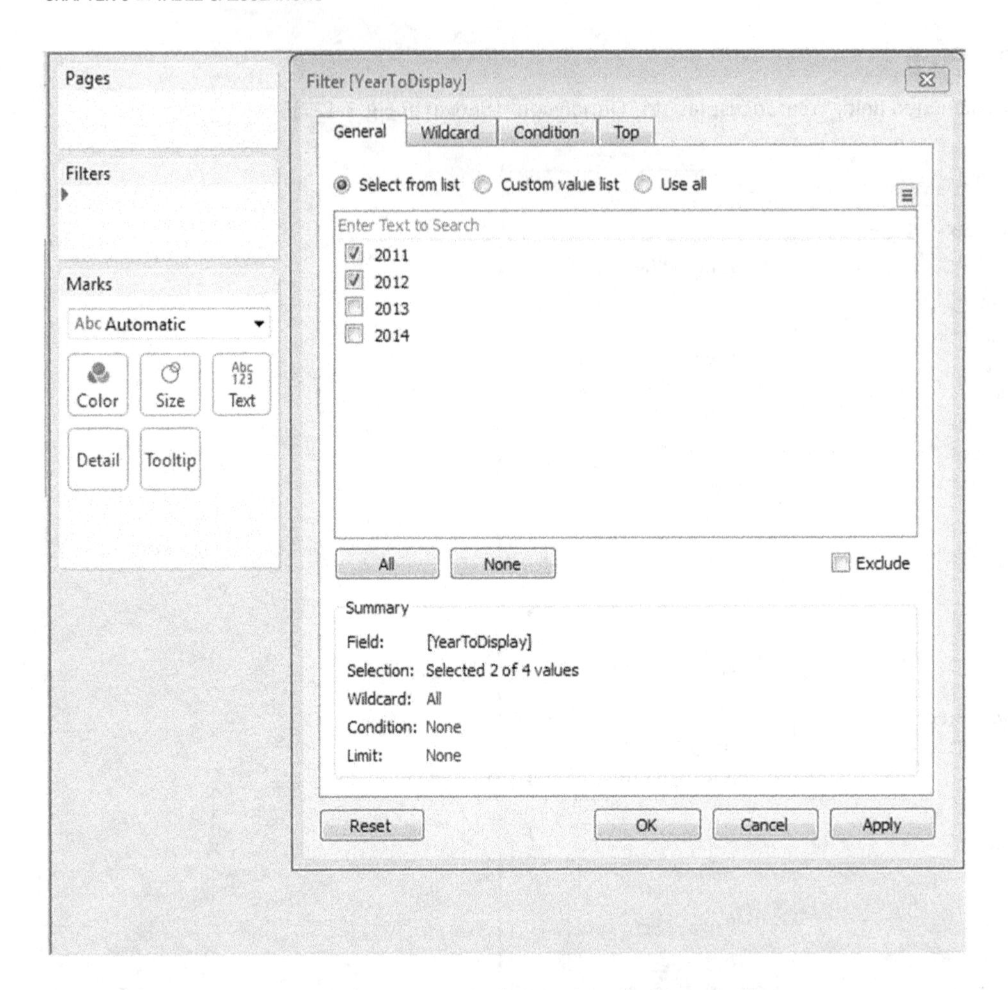

Figure 5-128. *Calculated field "YearToDisplay" placed on the filters shelf*

5.8.2.1.4 Step 4

Place the calculated field, "YearToDisplay" on "Color" on the marks card (Shown in Fig. 5-129).

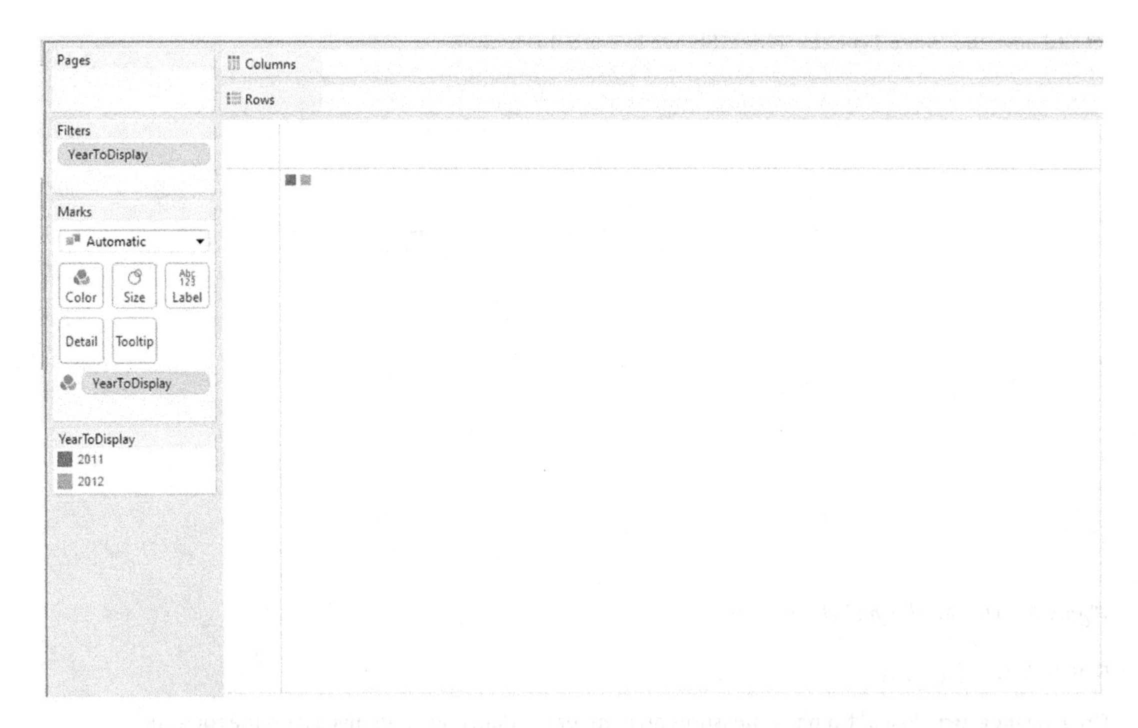

Figure 5-129. *Calculated field "YearToDisplay" placed on "Colors" on the marks card*

5.8.2.1.5 Step 5

Drag the **dimension** "Order Date" from the dimensions area under the data pane to the columns shelf. Change the date hierarchy to "Month". Retain it as "Discrete" (Shown in Fig. 5-130).

Figure 5-130. *Dimension "Order Date" placed on columns shelf*

Change the "Mark Type" to "Line" (Shown in Fig. 5-131).

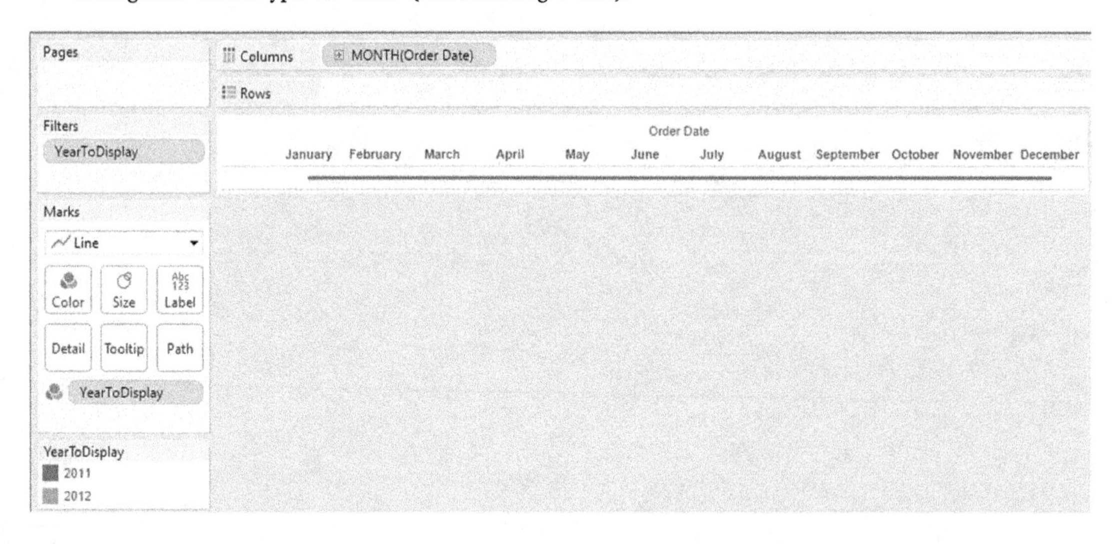

Figure 5-131. *"Mark Type" changed to "Line"*

5.8.2.1.6 Step 6

Drag the **measure** "Sales" from the measures area under the data pane and place it on the rows shelf (Shown in Fig. 5-132).

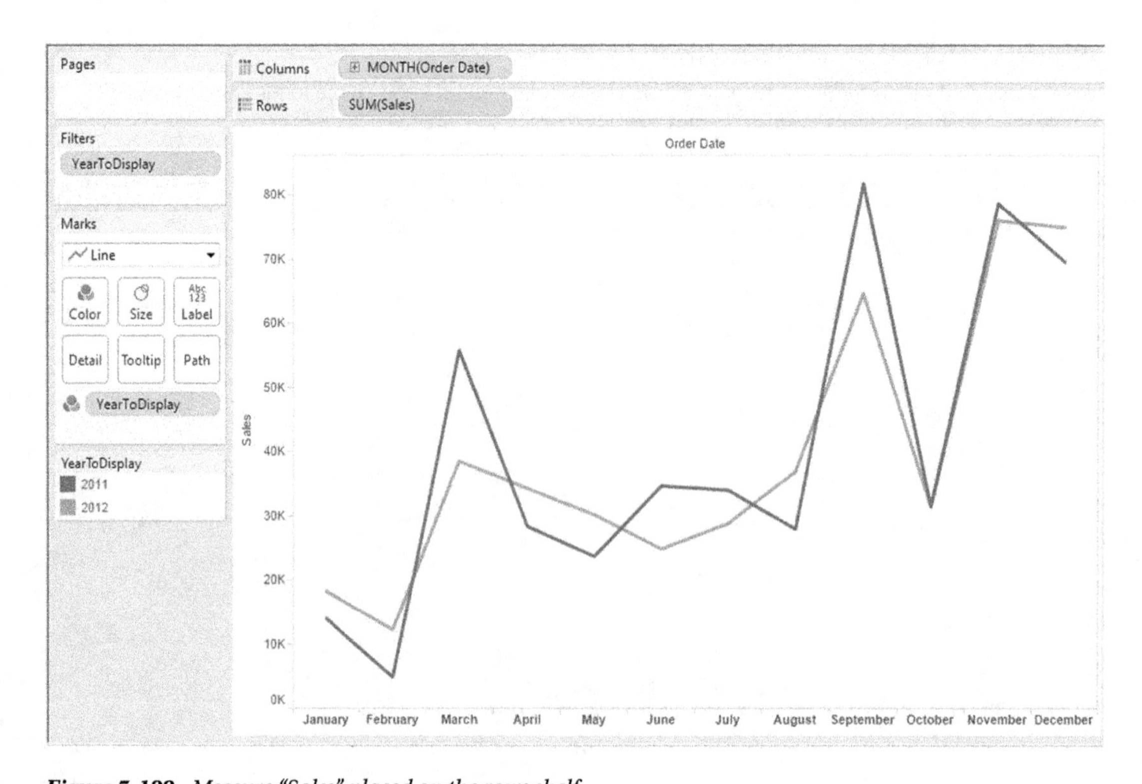

Figure 5-132. *Measure "Sales" placed on the rows shelf*

5.8.2.1.7 Step 7

Drag the **measure** "Sales" from the measures area under the data pane to "Label" on the marks card (Shown in Fig. 5-133).

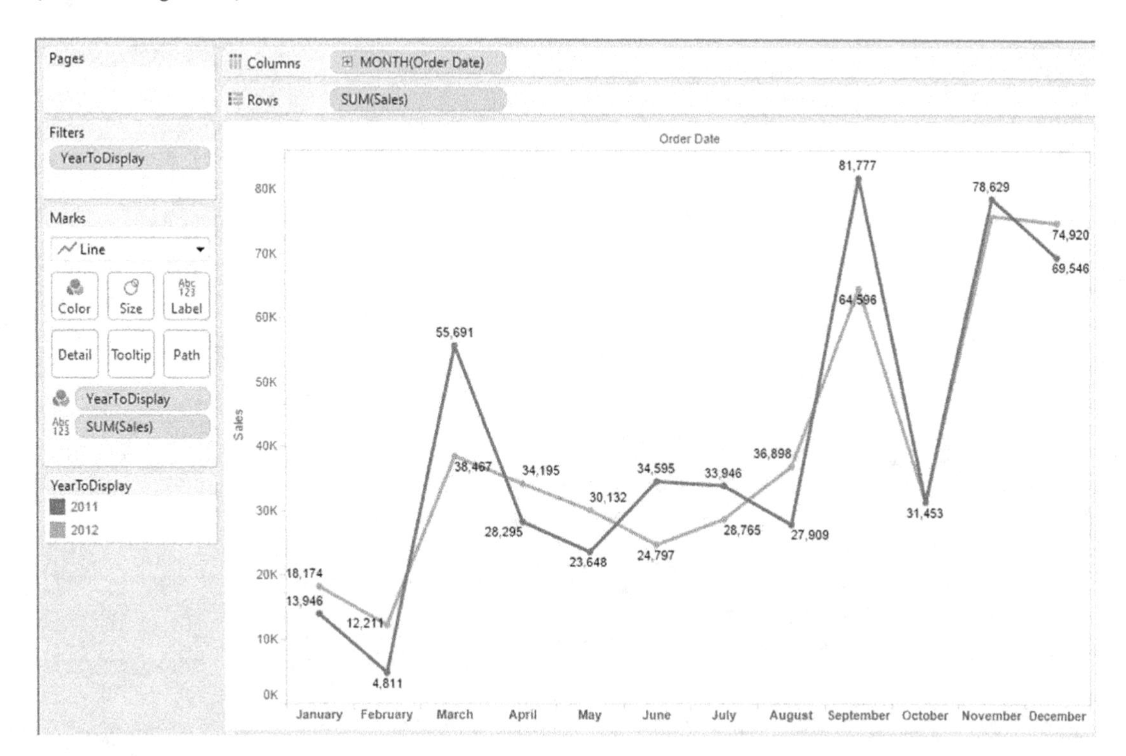

Figure 5-133. *Measure "Sales" placed on "Label" on the marks card*

5.8.2.1.8 Step 8

Add a "Quick Table Calculation - Year over Year Growth" to the **measure** "Sales" which is placed on the **"Label"** on the marks card (Shown in Fig. 5-134).

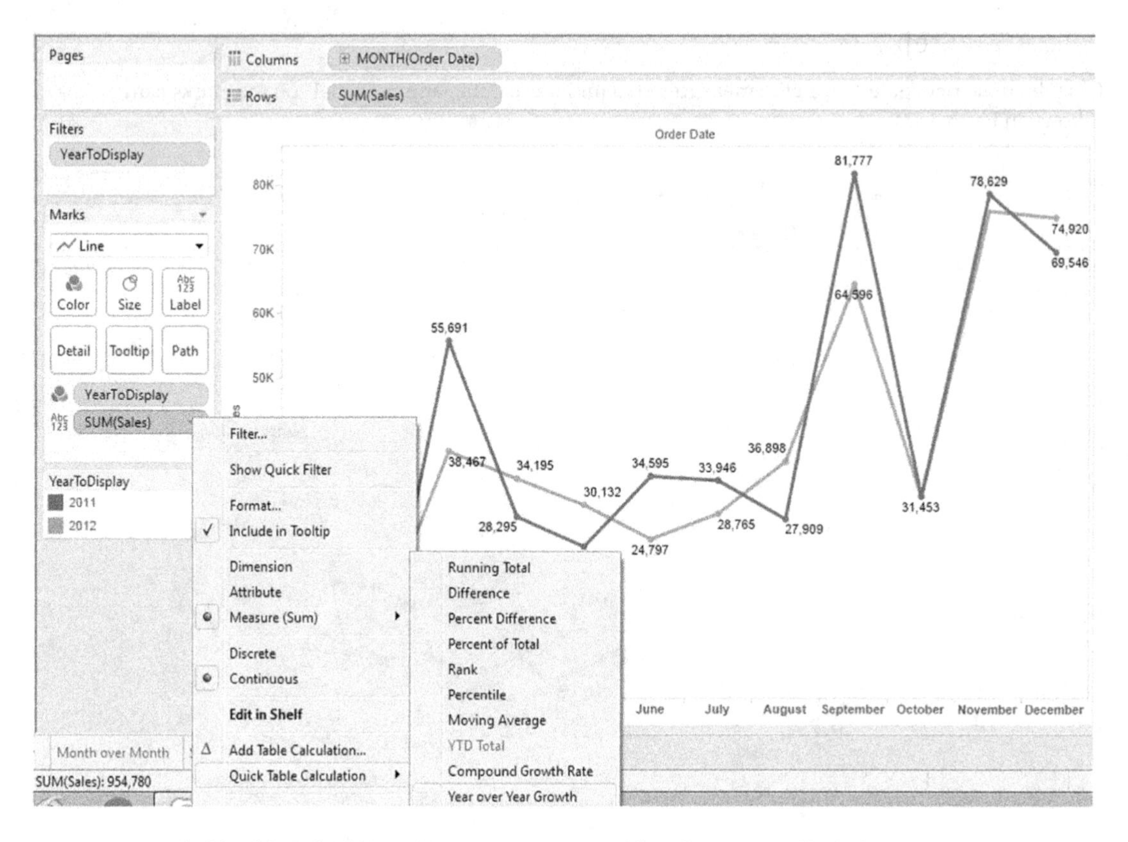

Figure 5-134. *"Add Table Calculation - Year over Year Growth" to the measure "Sales"*

Look at the "Table Calculation" dialog box (Shown in Fig. 5-135).

Table Calculation [% Difference in Sales]

Calculation Type: Percent Difference From

Calculation Definition

Calculate the difference along: Order Date

At the level: Month of Order Date

Display the value as a percent difference from: **Previous**

Compute compounded rate

OK Cancel Apply

Figure 5-135. *"Table Calculation [% Difference in Sales]" dialog box*

Note that the “Percent Difference” is calculated at the level, “Month of Order Date” and displays the value as a percent difference from “Previous”

The final output: (Shown in Fig. 5-136).

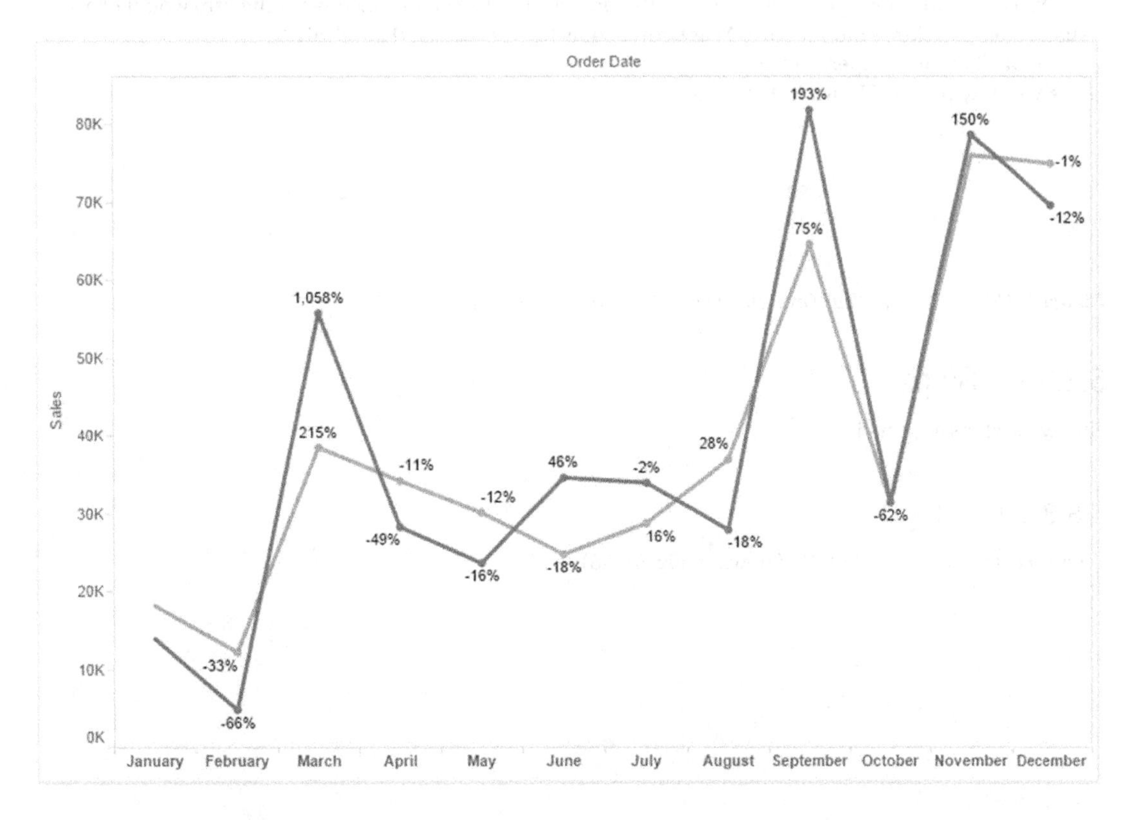

Figure 5-136. *Year over Year Growth – Demo 2 – final output*

Verify the output as follows:

Year	Month	Sales	Percent difference
2011	Jan	13,946	((4811- 13946) / 13946) * 100 = -66%
	Feb	4,811	
	Mar	55,691	((55691-4811)/4811) *100 = 1058%
	Apr	28,295	
	May	23,648	((28295 – 55691) / 55691) * 100 = -49%
	June	34,595	
			((23648 – 28295) / 28295) *100 = -16%

And so on...

Verify for the rest of the values.

5.8.3 Demo 3

Objective:

Data is provided for 4 years (2011, 2012, 2013 and 2014). The senior executive at the firm would like a visualization that shows the Year over Year Growth only for the years, 2011 and 2012.

Input: "Sample - Superstore.xls"

Expected output: Shown in Fig. 5-137.

2011	2012	YearOverYearGrowth
484,247	470,533	-2.83%

Figure 5-137. *Year over Year Growth - Demo 3 - expected output*

5.8.3.1 Steps

Follow the steps as provided.

5.8.3.1.1 Step 1

Create a calculated field, "2011" (Shown in Fig. 5-138).

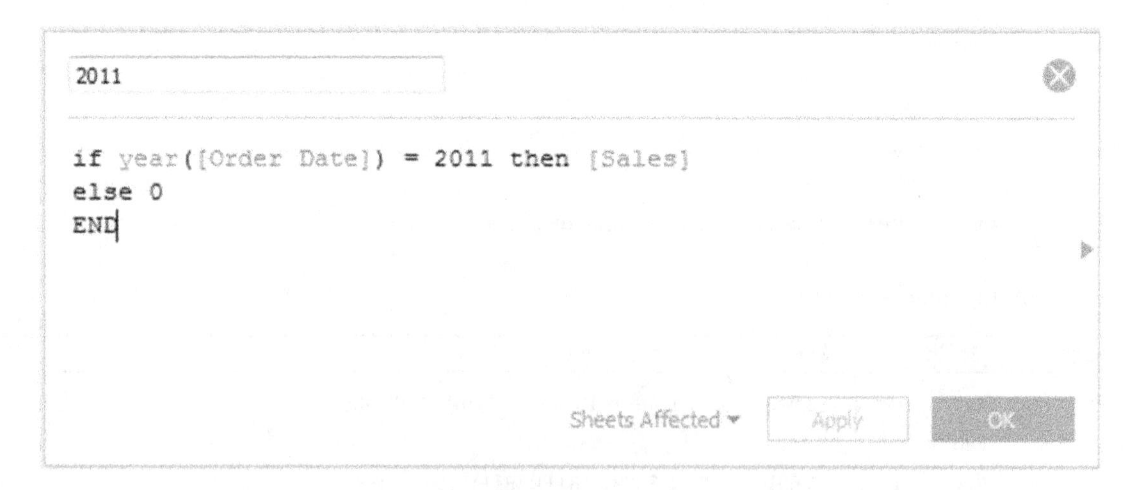

Figure 5-138. *Calculated field "2011" being created*

Create a calculated field, "2012" (Shown in Fig. 5-139).

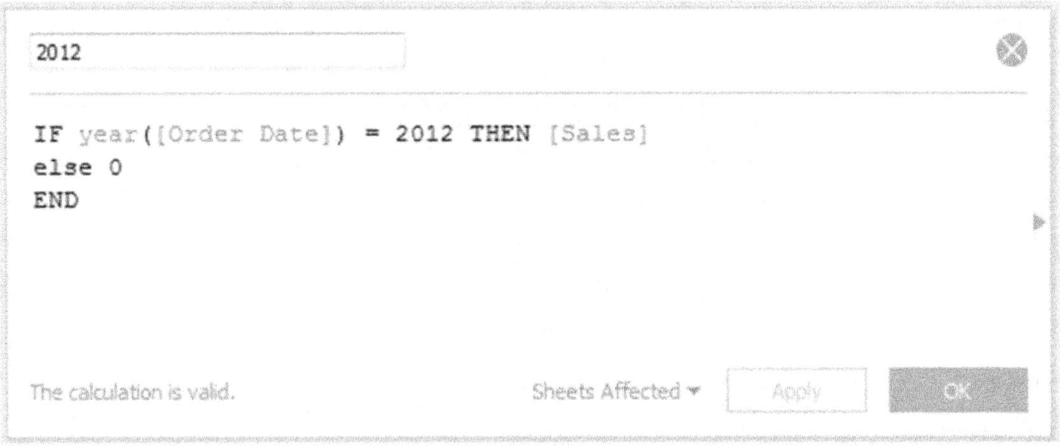

```
2012

IF year([Order Date]) = 2012 THEN [Sales]
else 0
END

The calculation is valid.        Sheets Affected ▾   Apply   OK
```

***Figure 5-139.** Calculated field "2012" being created*

Create a calculated field, "YearOverYearGrowth" (Shown in Fig. 5-140).

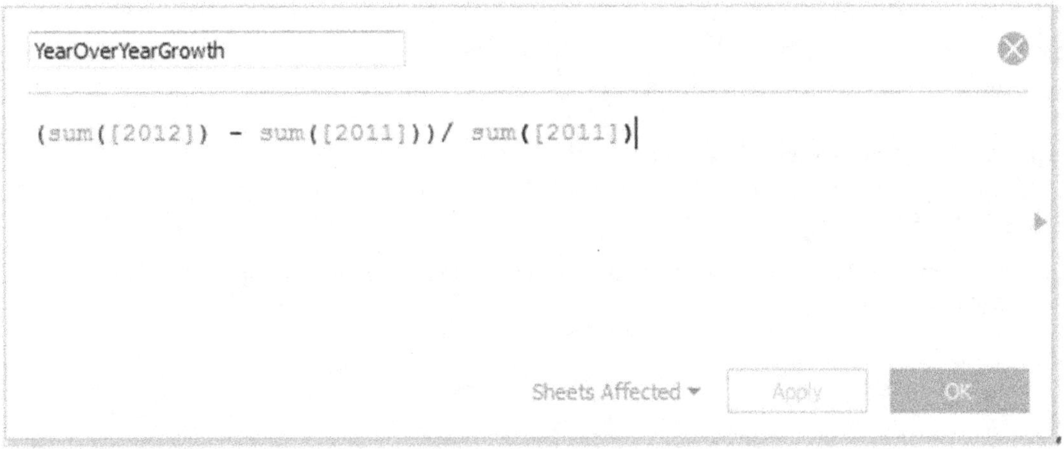

```
YearOverYearGrowth

(sum([2012]) - sum([2011]))/ sum([2011])

Sheets Affected ▾   Apply   OK
```

***Figure 5-140.** Calculated field "YearOverYearGrowth" being created*

5.8.3.1.2 Step 2

Drag the **dimension** "Measure Names" from the dimensions area under the data pane to the columns shelf (Shown in Fig. 5-141).

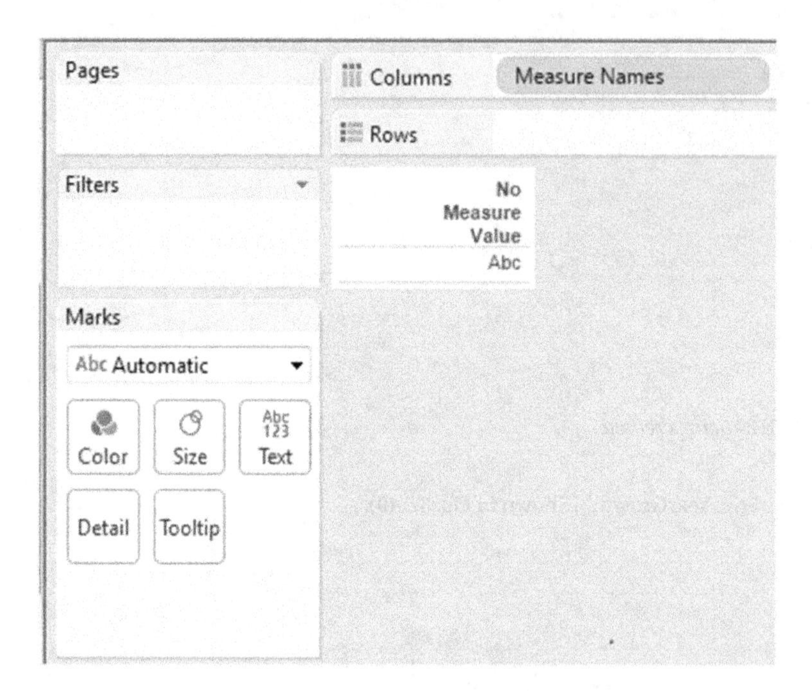

Figure 5-141. *"Measure Names" placed on the columns shelf*

5.8.3.1.3 Step 3

Drag the **dimension** "Measure Names" from the dimensions area under the data pane to the filters shelf (Shown in Fig. 5-142).

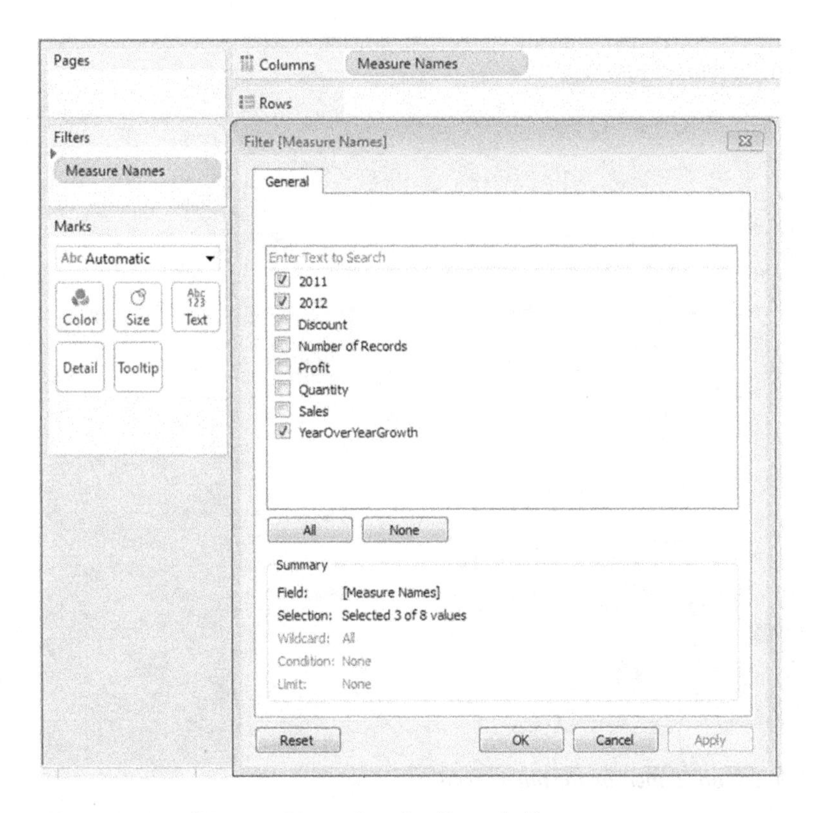

Figure 5-142. *"Measure Names" on the filters shelf*

Select the measures, "2011", "2012" and "YearOverYearGrowth".

5.8.3.1.4 Step 4

Drag the **measure** "Measure Values" from the measures area under the data pane and place it on **"Label"** on the marks card (Shown in Fig. 5-143).

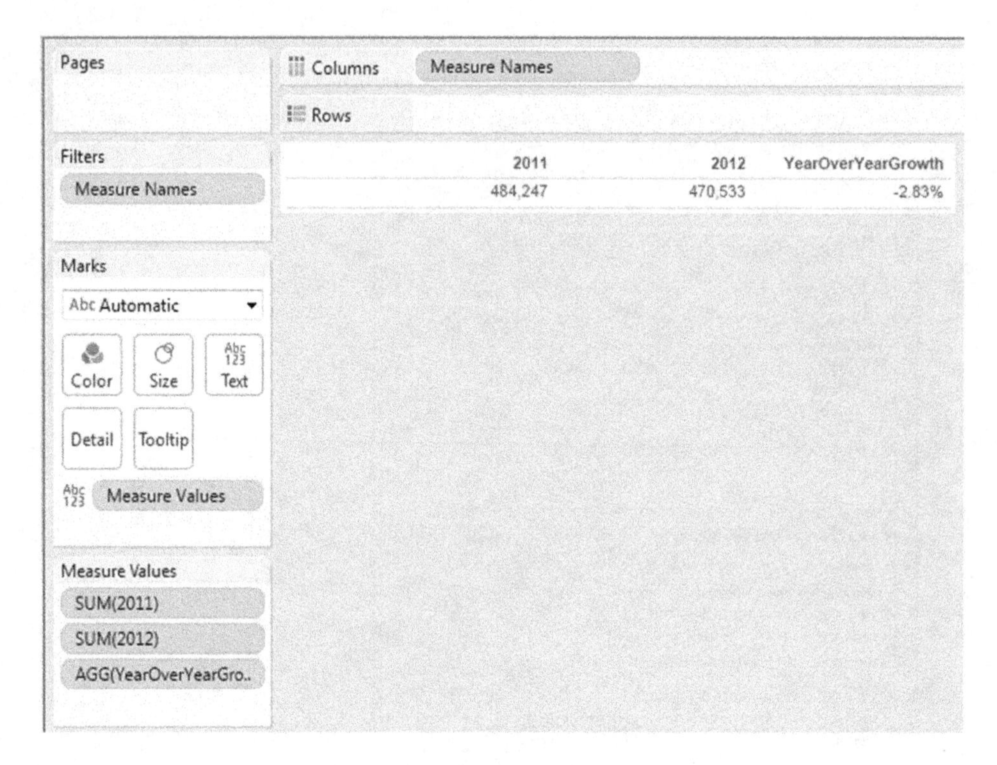

Figure 5-143. *"Measure values" placed on "Label" on the marks card*

The final output: Shown in Fig. 5-144.

	2011	2012	YearOverYearGrowth
	484,247	470,533	-2.83%

Figure 5-144. *Year over Year Growth – Demo 3 – final output*

5.9 Points to remember

- Table calculations aid in deriving additional insights from data. For example: (a) it helps to compare growth or differences across time periods (Year over year growth). (b) It helps to compute running total (running total) of inventory. The product list grows as products are added each day.
- There are ways in which table calculations can be customized such as by using its context menu or the calculation editor. To bring up the context menu, click on any field.
- Table calculations are generally applied to values in the entire table. For example to compute the running total or running average, a single method of calculation needs to be applied to the entire table.

5.10 Next Steps

This chapter familiarized us with table calculations. The next chapter will introduce us to string, numeric, date and logical functions.

CHAPTER 6

Customizing Data

"You can achieve simplicity in the design of effective charts, graphs and tables by remembering three fundamental principles: restrain, reduce, emphasize."

— Garr Reynolds, internationally acclaimed communications consultant and the author of best-selling books including the award-winning Presentation Zen, Presentation Zen Design

Chapter 5 introduced us to the various table calculations in Tableau. This chapter will help to explore and understand the following functions in Tableau:

- Number functions
- String functions
- Logical functions
- Date functions
- Aggregate functions
- Table calculation functions

6.1 Number functions

Tableau provides quite a few "Number" functions. Refer to Table 6-1 for Numeric functions supported by Tableau.

S. Acharya and S. Chellappan, *Pro Tableau*, DOI 10.1007/978-1-4842-2352-9_6

Table 6-1. *Numeric functions supported by Tableau*

Function Name	Description	Examples
ABS(number)	Absolute value of a given number is returned	ABS(-5) = 5
CEILING(number)	Rounds a number to the nearest integer of equal or greater value	CEILING(5.2345) = 6
DIV(integer1, integer2)	Integer part of a division operation is returned. Here integer1 is divided by integer2	DIV(13,2) = 6
FLOOR	Rounds a number to the nearest integer of equal or lesser value	FLOOR(5.3143) = 5
MIN(number, number)	Returns the minimum of the two arguments. The two arguments must be of same type. Returns Null if either argument is Null	MIN(6,5)= 5
MAX(number, number)	Returns the maximum of the two arguments. The two arguments must be of same type. Returns Null if either argument is Null	MAX(6,5)= 6
PI()	Returns a numeric constant value	3.14159
POWER(number, power)	Raises the number to the specified power.	POWER(6,2) = 36
ROUND(number,[decimals])	Rounds the number to a specified number of digits	ROUND(4.1567) = 4 ROUND(4.6567) = 5
SQRT(number)	Returns the square root of a number	SQRT(25) = 5
SQUARE(number)	Returns the square of a number	SQUARE(5) = 25

To learn more about number functions, refer to the link below.

`https://onlinehelp.tableau.com/current/pro/desktop/en-us/functions_functions_number.html`

Let us discuss a few number functions.

6.1.1 CEILING(number) and FLOOR(number)

Refer to Table 6-1 for description of the CEILING () and FLOOR() functions.

Let us learn to work with CEILING(number), FLOOR(number) functions.

Consider the below "Trainer Feedback" data set (Shown in Fig. 6-1).

	A	B
1	TrainerName	Feedback
2	John	3.14
3	James	4.78
4	Jack	3.35
5	Joshi	4.56
6	Joseph	4.23

Figure 6-1. "Trainer Feedback" data set

6.1.1.1 Steps to demonstrate the use of CEILING() and FLOOR() functions

Perform the following steps.

6.1.1.1.1 Step 1

Read in data from "Trainer Feedback" data set as shown in Fig. 6-2.

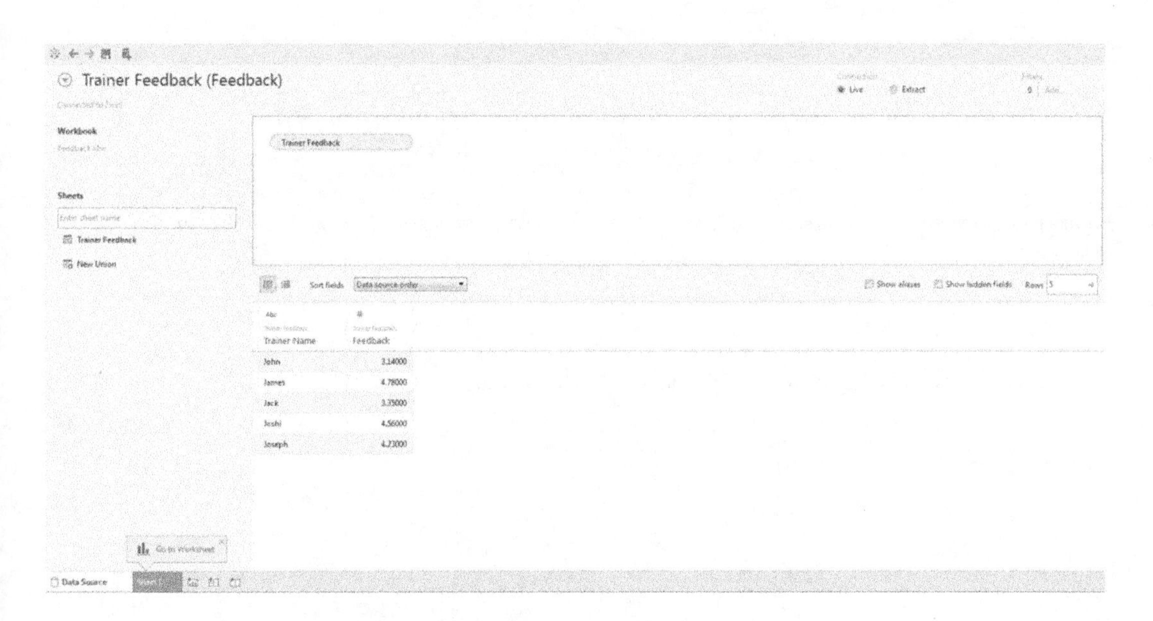

Figure 6-2. Data source page showing the "Trainer Feedback" data set

6.1.1.1.2 Step 2

Drag the dimension "Trainer Name" from the dimensions area under the data pane to the rows shelf. Drag the measure "Feedback" from the measures area under the data pane and place it on "Text" on the marks card (Shown in Fig. 6-3).

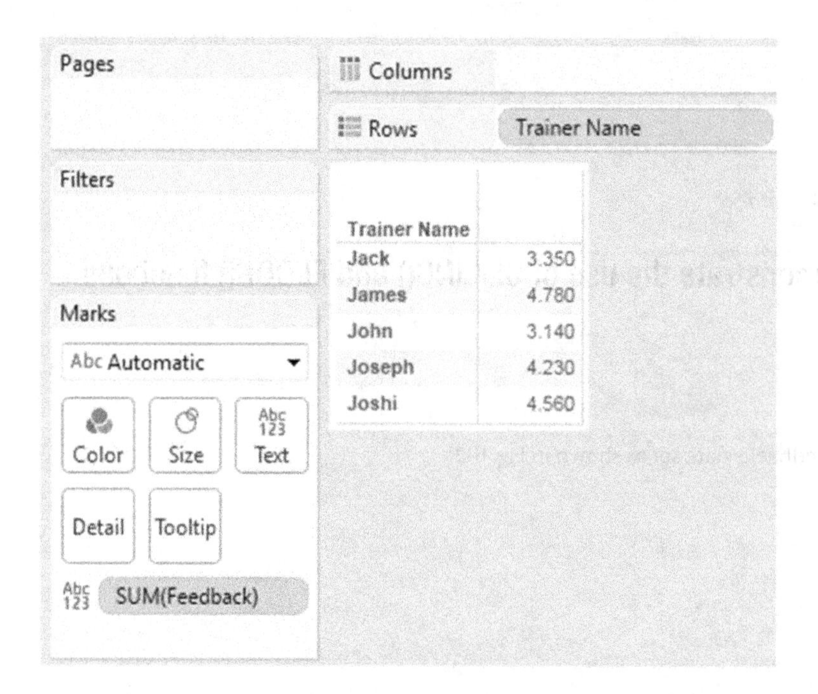

Figure 6-3. *Dimension, "Trainer Name" placed on the rows shelf, measure "Feedback" placed on "Text" on the marks card*

6.1.1.1.3 Step 3

Create a calculated field "Tableau CEILING" as shown in Fig. 6-4.

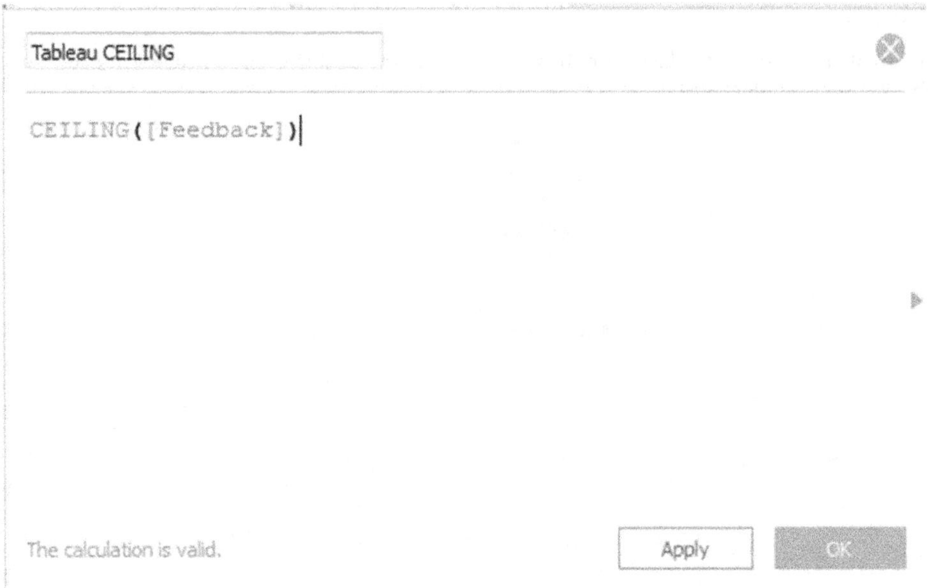

Tableau CEILING

CEILING([Feedback])

The calculation is valid. Apply OK

Figure 6-4. *Calculated field "Tableau CEILING" being created*

6.1.1.1.4 Step 4

Create a calculated field "Tableau FLOOR" as shown in Fig. 6-5.

Tableau FLOOR

FLOOR([Feedback])

The calculation is valid. Sheets Affected ▾ Apply OK

Figure 6-5. *Calculated field "Tableau FLOOR" being created*

6.1.1.1.5 Step 5

Double click on the calculated fields to place it on the view. Observe the CEILING and FLOOR value for the measure "Feedback" (Shown in Fig. 6-6).

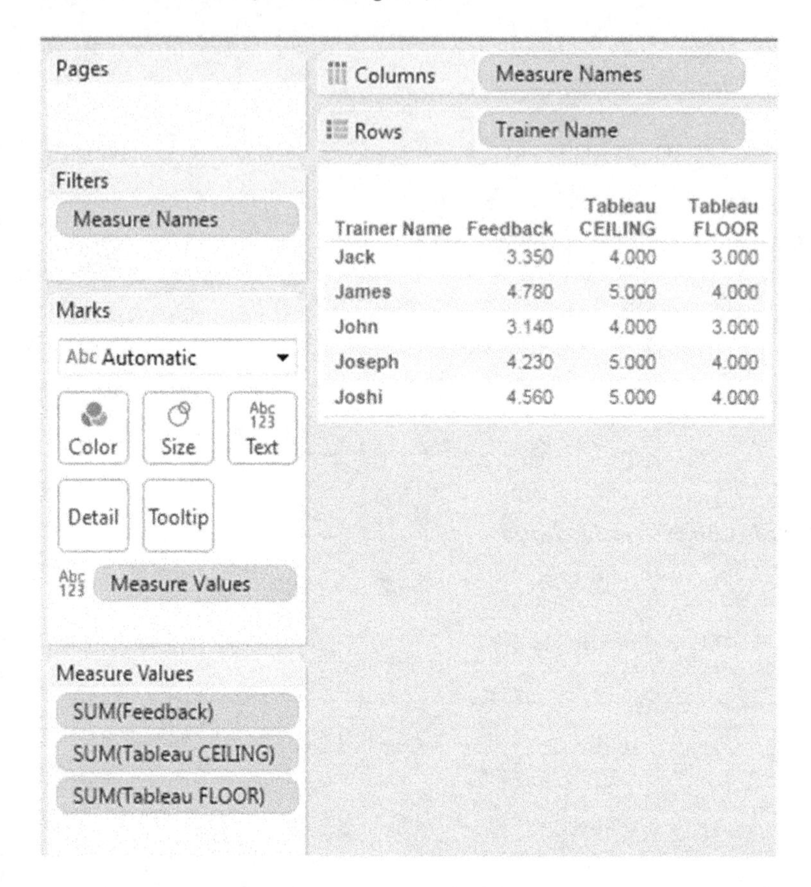

Trainer Name	Feedback	Tableau CEILING	Tableau FLOOR
Jack	3.350	4.000	3.000
James	4.780	5.000	4.000
John	3.140	4.000	3.000
Joseph	4.230	5.000	4.000
Joshi	4.560	5.000	4.000

Figure 6-6. *View using the calculated fields "Tableau CEILING" and "Tableau FLOOR"*

6.1.2 MAX(number, number), MIN(number, number)

Refer to Table 6-1 for a description for the functions.

Consider the "Student" data set (Shown in Fig. 6-7).

	A	B	C
1	Stud Name	Mark 1	Mark 2
2	Smith	23	14
3	Jack	18	24
4	John	20	21
5	Scott	22	24
6	James	17	24

Figure 6-7. *"Student" data set*

6.1.2.1 Steps to demonstrate MAX() and MIN() functions

Perform the following steps:

6.1.2.1.1 Step 1

Read data from "Student" data set into Tableau (Shown in Fig. 6-8).

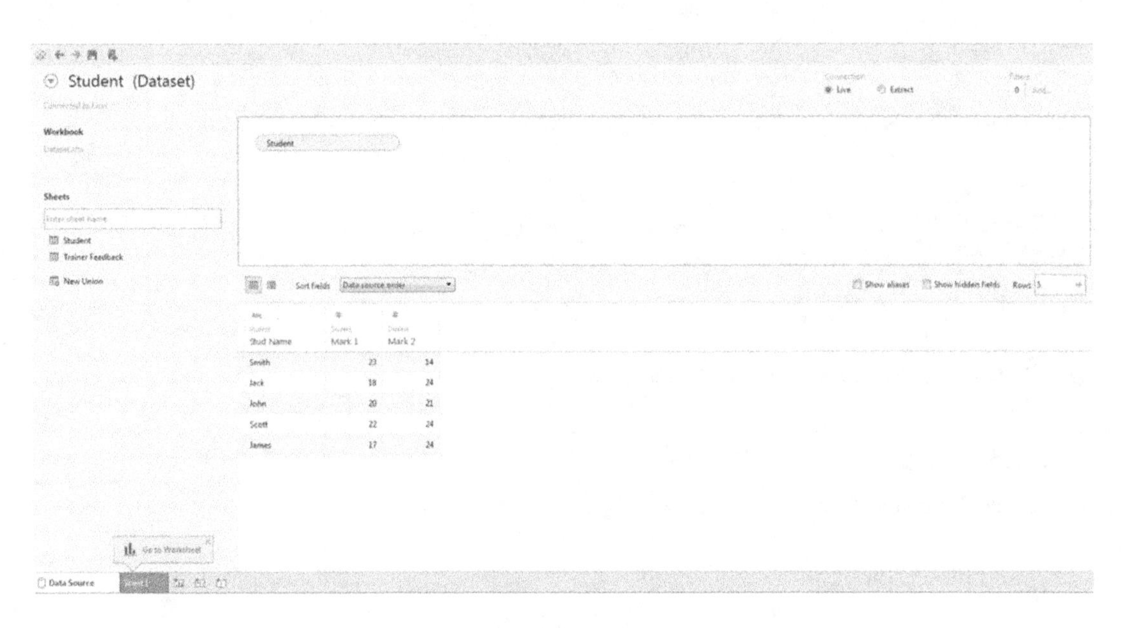

Figure 6-8. *Data source page showing the "Student" data set*

6.1.2.1.2 Step 2

Create a view as shown in Fig. 6-9.

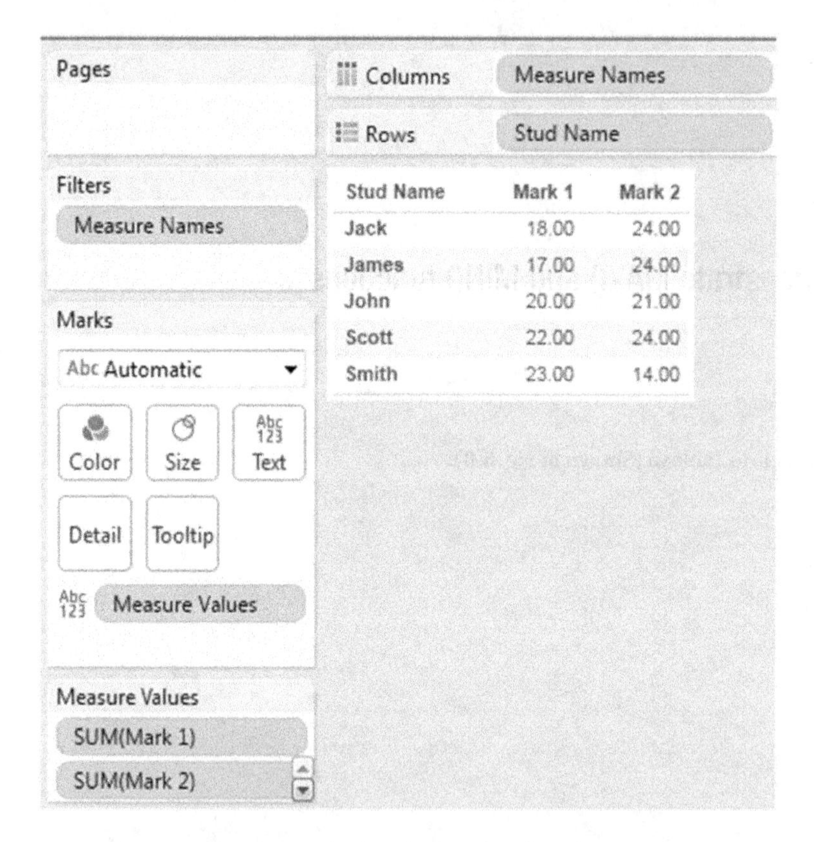

Figure 6-9. *View displaying the details of students' performance*

6.1.2.1.3 Step 3

Create a calculated field "Max Mark" (Shown in Fig. 6-10).

Max Mark

Student (Dataset)

MAX([Mark 1],[Mark 2])

The calculation is valid.

Apply

OK

Figure 6-10. *Calculated field "Max Mark" being created*

6.1.2.1.4 Step 4

Create a calculated field "Min Mark" (Shown in Fig. 6-11).

Min Mark

Min([Mark 1],[Mark 2])

The calculation is valid.

Apply

OK

Figure 6-11. *Calculated field "Min Mark" being created*

6.1.2.1.5 Step 5

Double click on the calculated fields to display "Max Mark" and "Min Mark" (Shown in Fig. 6-12).

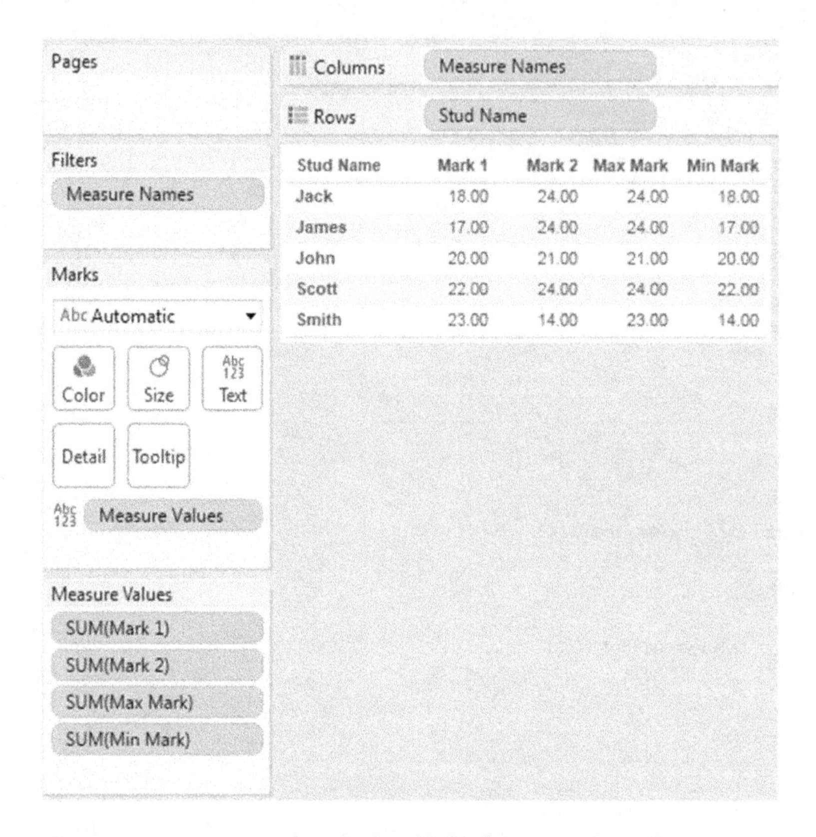

Stud Name	Mark 1	Mark 2	Max Mark	Min Mark
Jack	18.00	24.00	24.00	18.00
James	17.00	24.00	24.00	17.00
John	20.00	21.00	21.00	20.00
Scott	22.00	24.00	24.00	22.00
Smith	23.00	14.00	23.00	14.00

Figure 6-12. *View using the calculated fields "Max Mark" and "Min Mark"*

6.1.3 ABS(number)

Refer to Table 6-1 for description of the function.

Consider the "Items" data set (Shown in Fig. 6-13).

	A	B
1	Item Name	Profit
2	Books	-7800
3	Tables	9000
4	Chairs	-3500
5	Papers	-2400
6	Pens	7800

Figure 6-13. *"Items" data set*

6.1.3.1 Steps to demonstrate the use of ABS() function

Perform the following steps.

6.1.3.1.1 Step 1

Read data from "Items" data set into Tableau (Shown in Fig. 6-14).

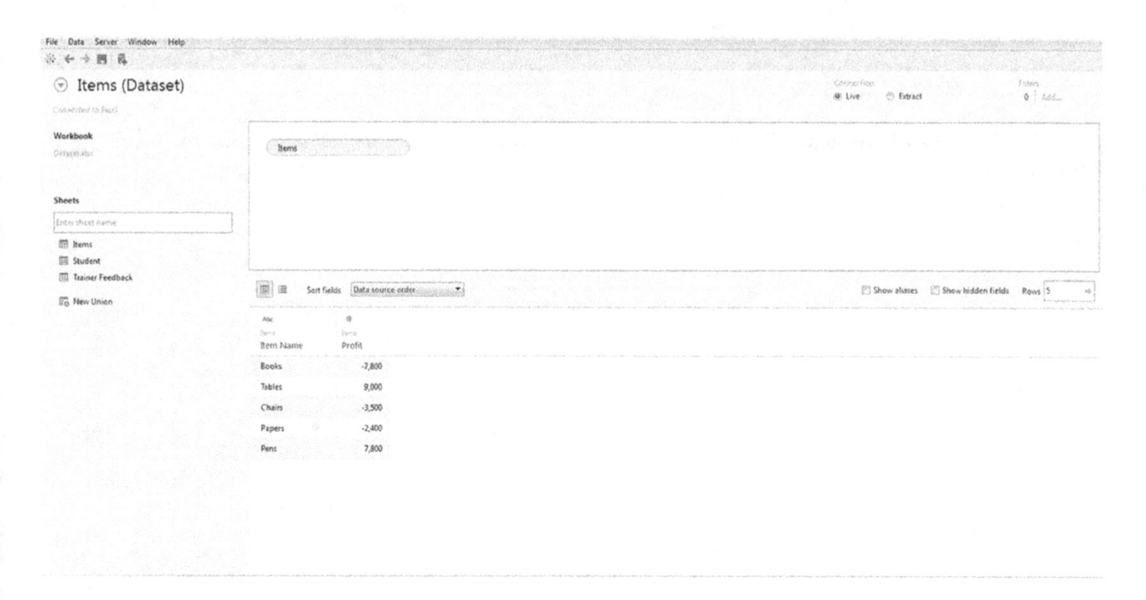

Figure 6-14. *Data source page showing the "Items" data set*

6.1.3.1.2 Step 2

Create a view as shown in Fig. 6-15.

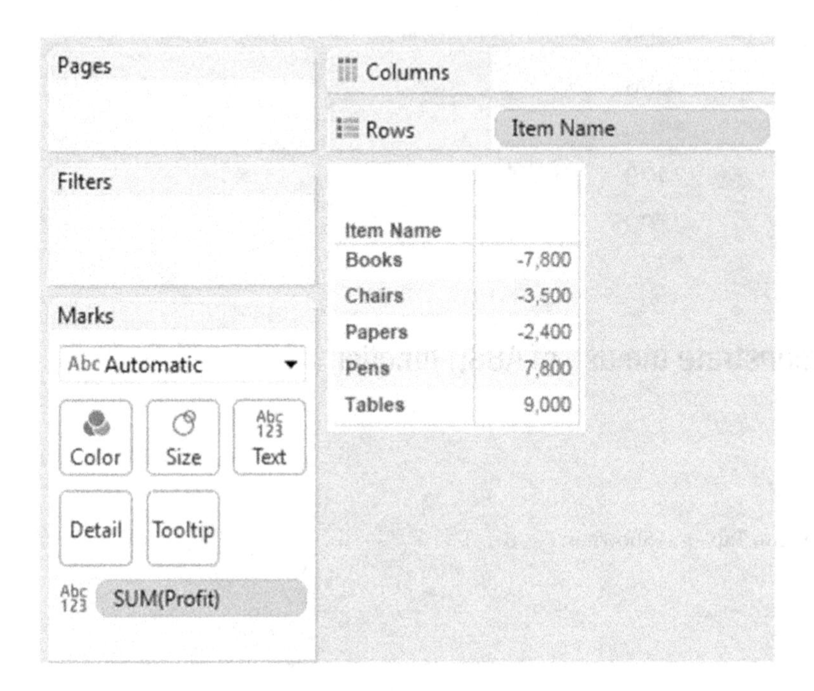

Figure 6-15. *View displaying the dimension "Item Name" placed on the rows shelf and measure, "Profit" placed on "Text" on the marks card*

6.1.3.1.3 Step 3

Create a calculated field "Tableau ABS()" as shown in Fig. 6-16.

Tableau ABS()

Items (Dataset)

```
ABS([Profit])
```

The calculation is valid.

Apply

OK

Figure 6-16. *Calculated field "Tableau ABS()" being created*

6.1.3.1.4 Step 4

Double click on the calculated field "Tableau ABS()" to place it on the view (Shown in Fig. 6-17).

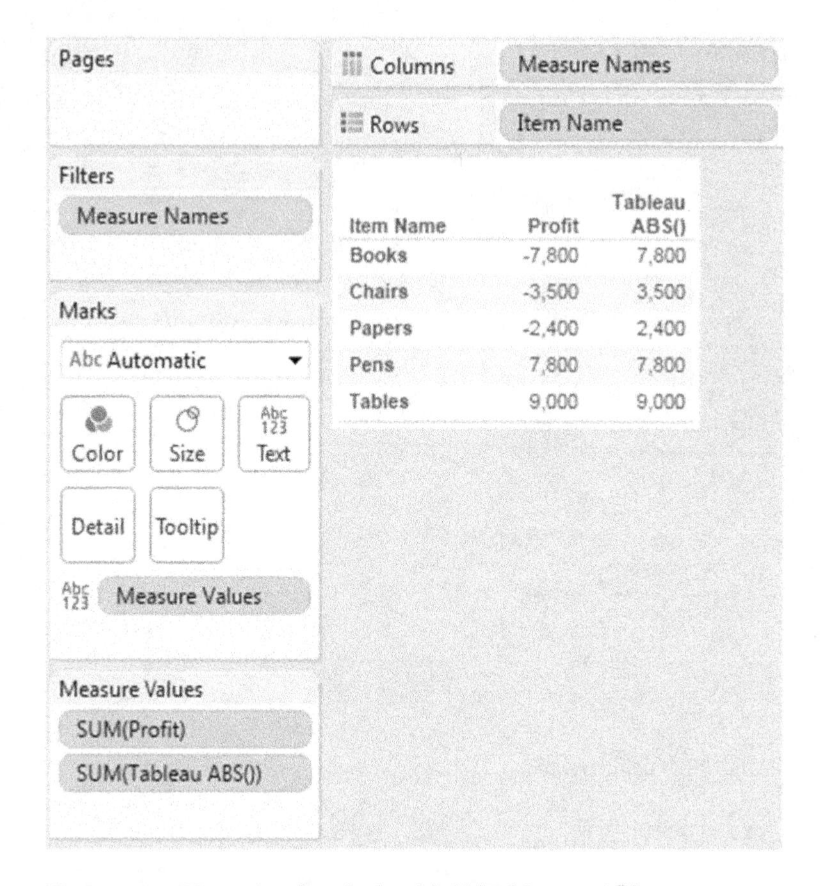

Figure 6-17. *View using the calculated field "Tableau ABS()"*

6.2 String functions

Tableau supports various string functions to enable working with strings.

Refer to Table 6-2 for string functions.

Table 6-2. String functions in Tableau

Function syntax	Example	
Ascii(character)	ASCII("T")	Returns 84
Char(integer)	CHAR(84)	Returns T
Len(string)	LEN("Tableau")	Returns 7
Max(a, b)	MAX("Tableau", "TABLEAU")	Returns Tableau
Min(a, b)	MIN("Tableau", "TABLEAU")	Returns TABLEAU
Replace(string, substring, replacement)	REPLACE("Visualisation", "sation","zation")	Returns Visualization
Startswith()	STARTSWITH("TABLEAU","T")	Returns TRUE
Upper(string)	UPPER("tableau")	Returns TABLEAU
Lower(string)	LOWER("TABLEAU")	Returns tableau
Left(string, num_characters)	LEFT("TABLEAU",3)	Returns TAB
Right(string, num_characters)	RIGHT("COMPASS",4)	Returns PASS
Trim()	TRIM(" Visualization ")	Returns Visualization
Rtrim()	RTRIM("Visualization ")	Returns "Visualization"
Ltrim()	LTRIM(" Visualization")	Returns "Visualization"

Let us discuss a few string functions in Tableau.

6.2.1 Concatenation

Objective: To concatenate the dimension "Customer_ID" with the dimension "Customer Name".
Formula: "Customer ID" + " : " + str([Customer ID]) + " , " + [Customer Name]

6.2.1.1 Steps to demonstrate concatenation

Perform the following steps:

6.2.1.1.1 Step 1

Read data from "Sample - Superstore" data set.

6.2.1.1.2 Step 2

Create the calculated field "Customer ID + Customer Name" (Shown in Fig. 6-18).

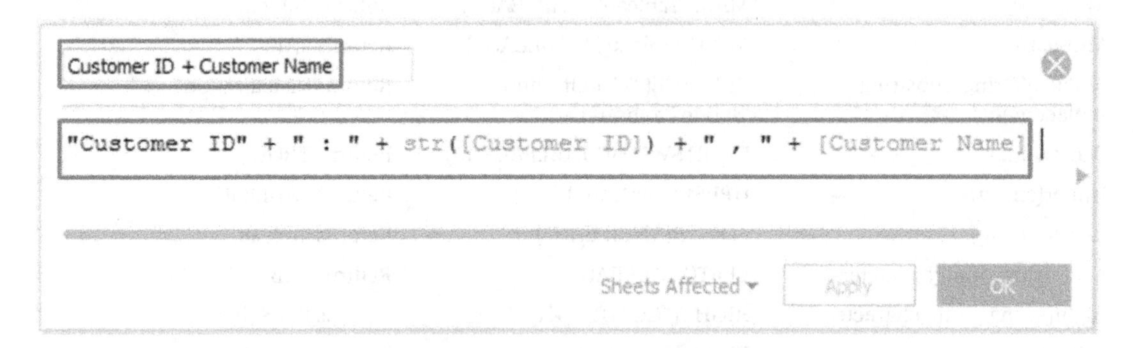

Figure 6-18. *Calculated field "Customer ID + Customer Name" being created*

6.2.1.1.3 Step 3

Drag the calculated field "Customer ID + Customer Name" to the rows shelf (Shown in Fig. 6-19).

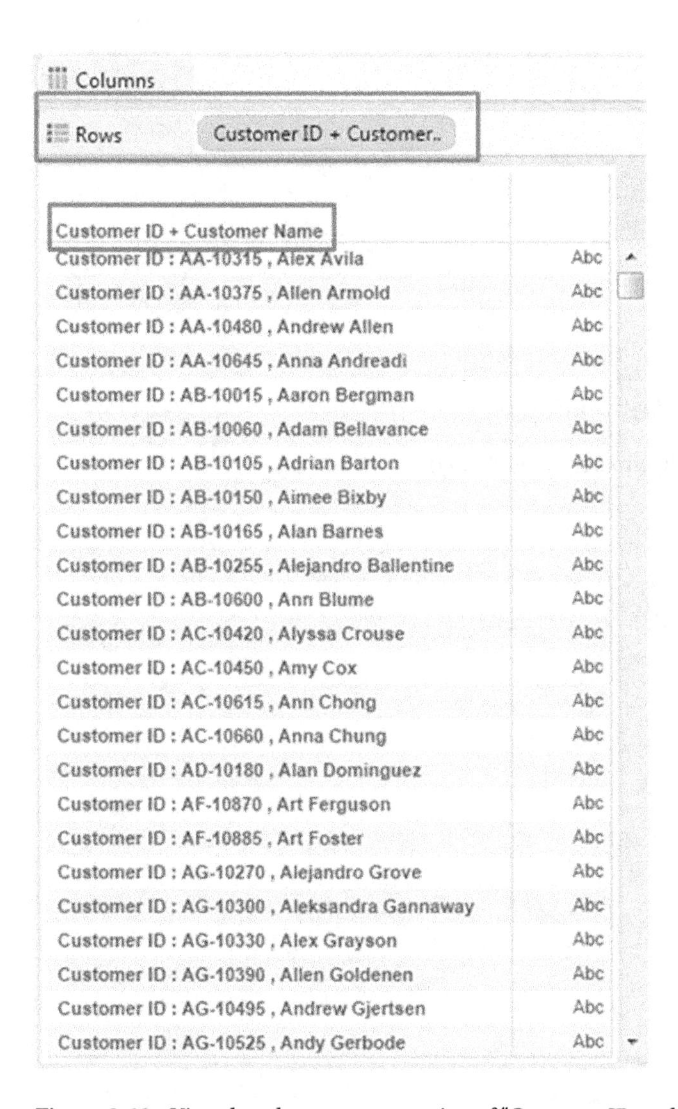

Figure 6-19. *View that shows concatenation of "Customer ID and Customer Name"*

6.2.2 Left() and Find() functions

Objective: To extract the first name from the "Customer Name" dimension.

Functions used: Left() and Find()

Syntax of Left()

Left(String, Number of characters to extract)

Syntax of Find()

Find(String, Substring, Start position)

Create a calculated field by the name, "FirstName" (Shown in Fig. 6-20).

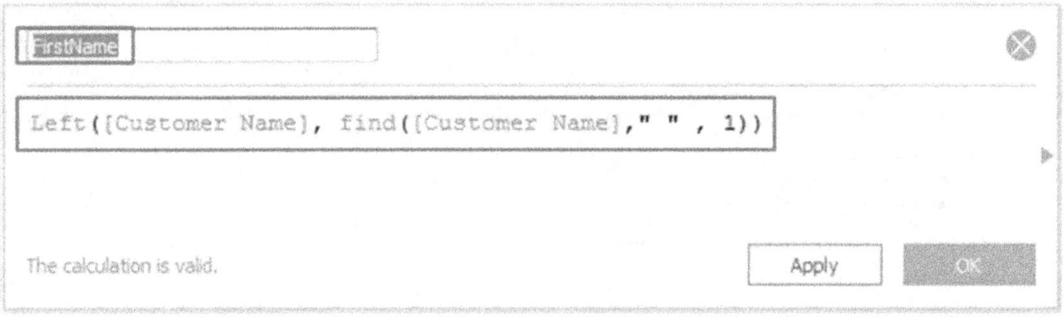

FirstName

```
Left([Customer Name], find([Customer Name]," " , 1))
```

The calculation is valid. Apply OK

Figure 6-20. Calculated field "FirstName" being created

Formula: Left([Customer Name], find([Customer Name],“ ” , 1))
Refer to Fig. 6-21 for output.

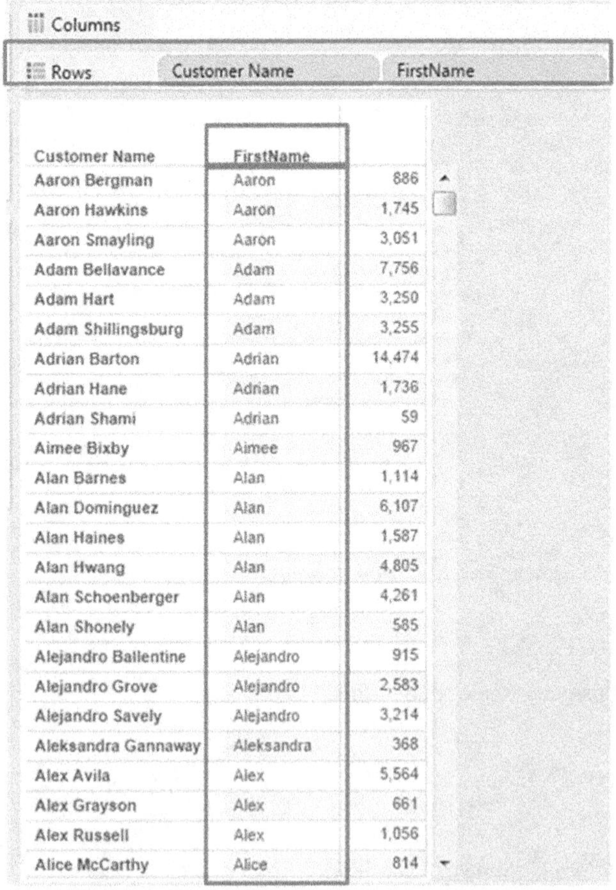

Columns
Rows: Customer Name | FirstName

Customer Name	FirstName	
Aaron Bergman	Aaron	886
Aaron Hawkins	Aaron	1,745
Aaron Smayling	Aaron	3,051
Adam Bellavance	Adam	7,756
Adam Hart	Adam	3,250
Adam Shillingsburg	Adam	3,255
Adrian Barton	Adrian	14,474
Adrian Hane	Adrian	1,736
Adrian Shami	Adrian	59
Aimee Bixby	Aimee	967
Alan Barnes	Alan	1,114
Alan Dominguez	Alan	6,107
Alan Haines	Alan	1,587
Alan Hwang	Alan	4,805
Alan Schoenberger	Alan	4,261
Alan Shonely	Alan	585
Alejandro Ballentine	Alejandro	915
Alejandro Grove	Alejandro	2,583
Alejandro Savely	Alejandro	3,214
Aleksandra Gannaway	Aleksandra	368
Alex Avila	Alex	5,564
Alex Grayson	Alex	661
Alex Russell	Alex	1,056
Alice McCarthy	Alice	814

Figure 6-21. View that shows "FirstName" from the dimension "Customer Name"

6.2.3 Contains() function

Returns true if the given string contains the specified substring.

6.2.3.1 Problem statement

Given below is a list of product names. We are looking for those product names that contain the word "Wall Clock" in it. Display a list of only those product names that contains the word "Wall Clock".

Example:

Product Name	Sales
12-1/2 Diameter Round Wall Clock	$551
"6""Cubicle Wall Clock, Black"	$125

Input: Refer to Fig. 6-22.

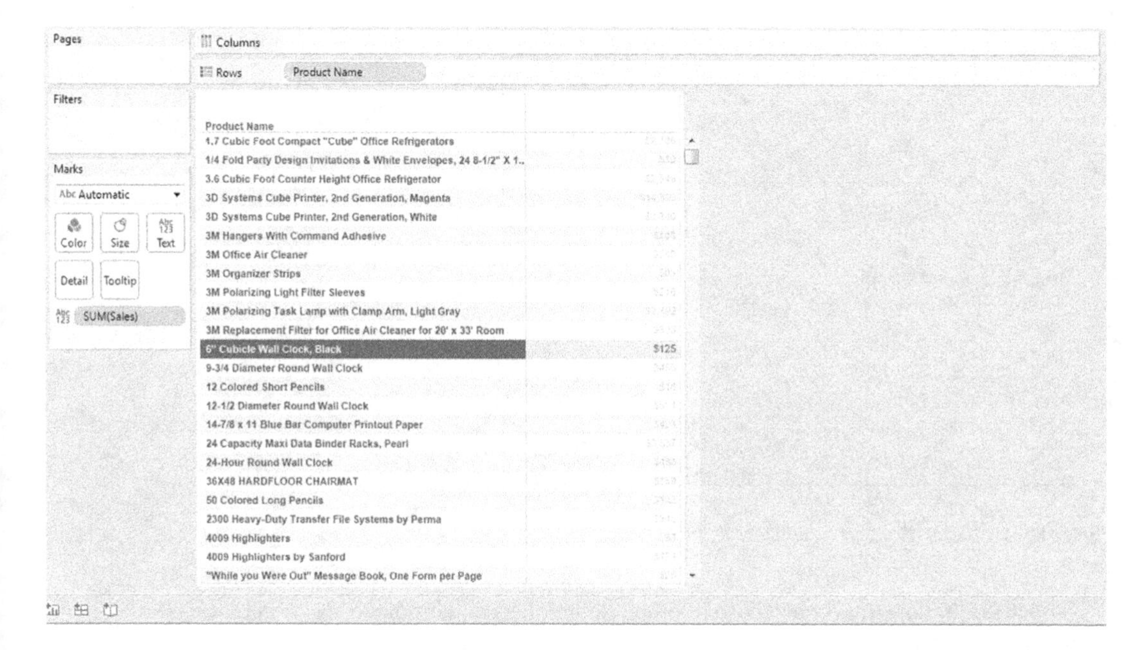

Figure 6-22. Dimension "Product Name" placed on the rows shelf

Create a calculated field "Product-Wall Clock" (Shown in Fig. 6-23).

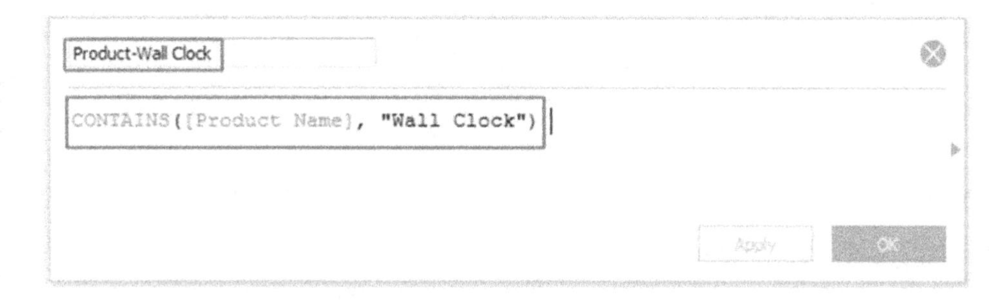

Figure 6-23. Calculated field "Product-Wall Clock" being created

Output:

To display the list of only those products that contain the string "Wall Clock".

Drag the dimension "Product-Wall clock" into the filters shelf and select only "True" (Shown in Fig. 6-24).

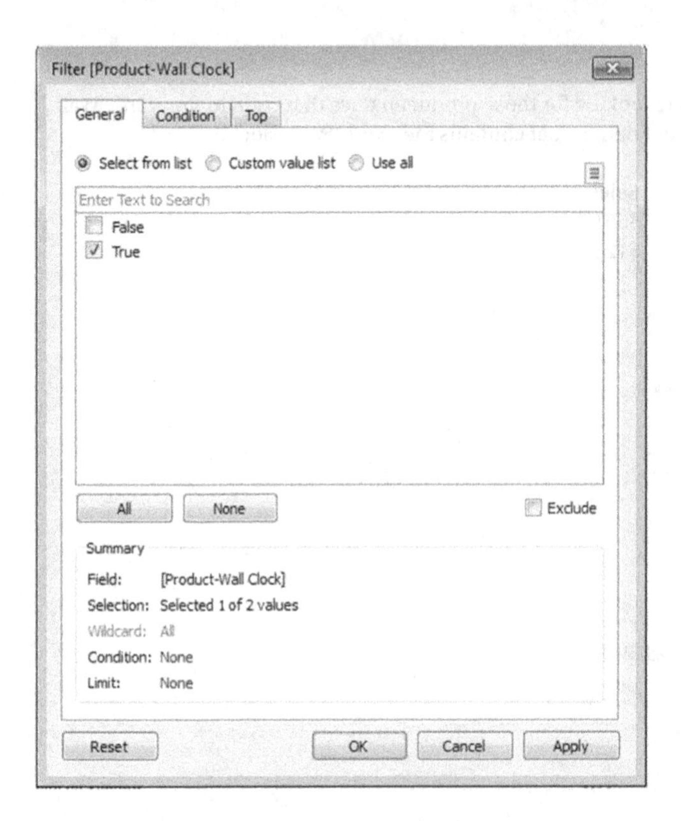

Figure 6-24. *"Filter[Product-Wall clock]" dialog box*

Notice the list now shows only those "Product Name" that contains the string "Wall Clock". (Shown in Fig. 6-25).

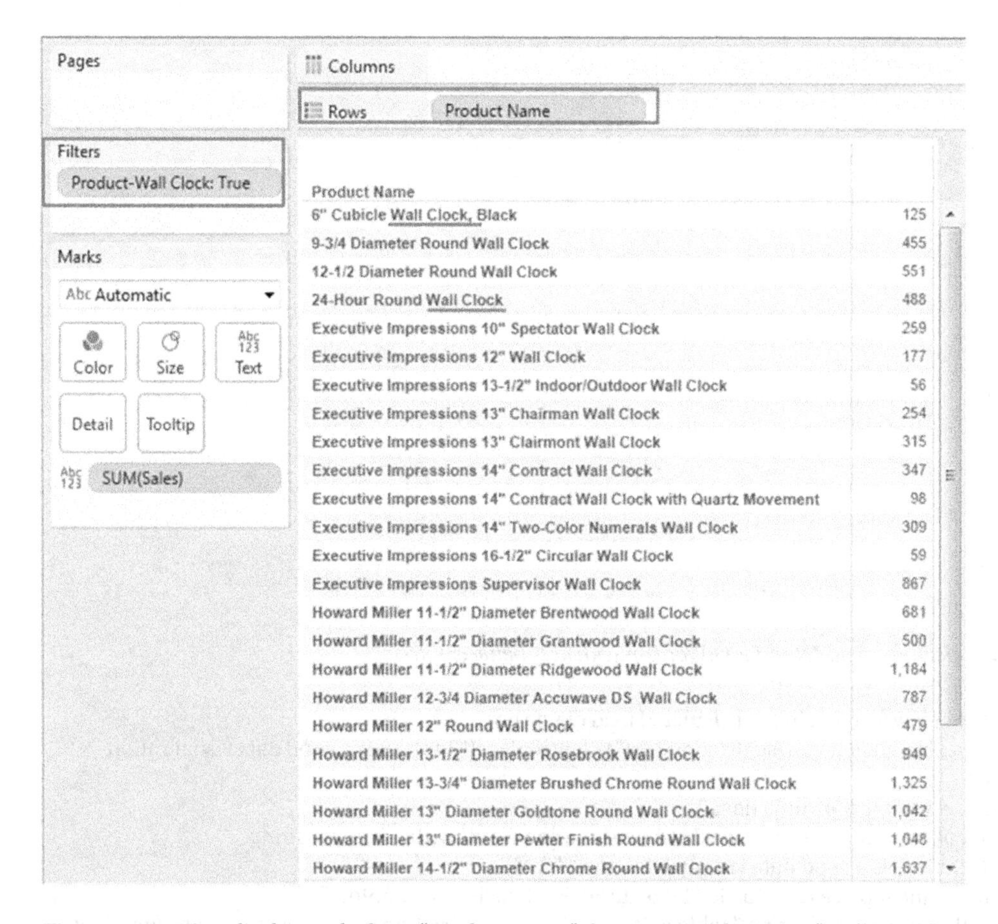

Figure 6-25. *View displays only those "Product Name" that contains the string "Wall Clock"*

6.2.4 Len() function

Returns the length of the string.

Objective: To count the number of words in a sentence.
Hint: Use len() function.
For example, count the number of words in the sentence, "Tableau is a good data visualization tool." Output expected is 7.

6.2.4.1 Steps to demonstrate the use of Len() function

Perform the following steps.

6.2.4.1.1 Step 1

Create a calculated field "NumberOfWordsinSentence" (Shown in Fig. 6-26).

NumberOfWordsinSentence Sample - Superstore

len("Tableau is a good Data Visualization Tool") - len(replac

The calculation is valid. Apply OK

Figure 6-26. *Calculated field "NumberOfWordsinSentence" being created*

Type in the below formula in the calculated field dialog box:

len("Tableau is a good data visualization tool") - len(replace("Tableau is a good data visualization tool"," ","")) + 1

Let us look at what the formula does:

We use a replace function to replace the space between words with an empty string.

Replace("Tableau is a good data visualization tool", " ", "")

The output of the replace function is "Tableauisagooddatavisualizationtool"

We count the length of the string "Tableau is a good data visualization tool" using the below function:

len("Tableau is a good data visualization tool")

The output is 41. This means that there are 41 characters in the string, "Tableau is a good data visualization tool"

Next, we determine the length of the string "Tableauisagooddatavisualizationtool"

The output of the above is 35.

Let us take a look at the formula again:

len("Tableau is a good data visualization tool") - len(replace("Tableau is a good data visualization tool", " ","")) + 1

Substituting the values returned by the functions:

41 - len(replace("Tableau is a good data visualization tool", " ","")) + 1

41 – len("Tableauisagooddatavisualizationtool") + 1

41 – 35 + 1

Why are we adding one at the end? That is because the last word does not have a space after it.

41 – 35 + 1 returns 7.

This is the number of words in the sentence "Tableau is a good data visualization tool".

6.2.4.1.2 Step 2

Because the calculated field "NumberOfWordsinSentence" will return a numeric value, by default it is placed under measures in the measure area under the Data Pane. Convert it to a "Dimension" as it is not required to run aggregation on it (Shown in Fig. 6-27).

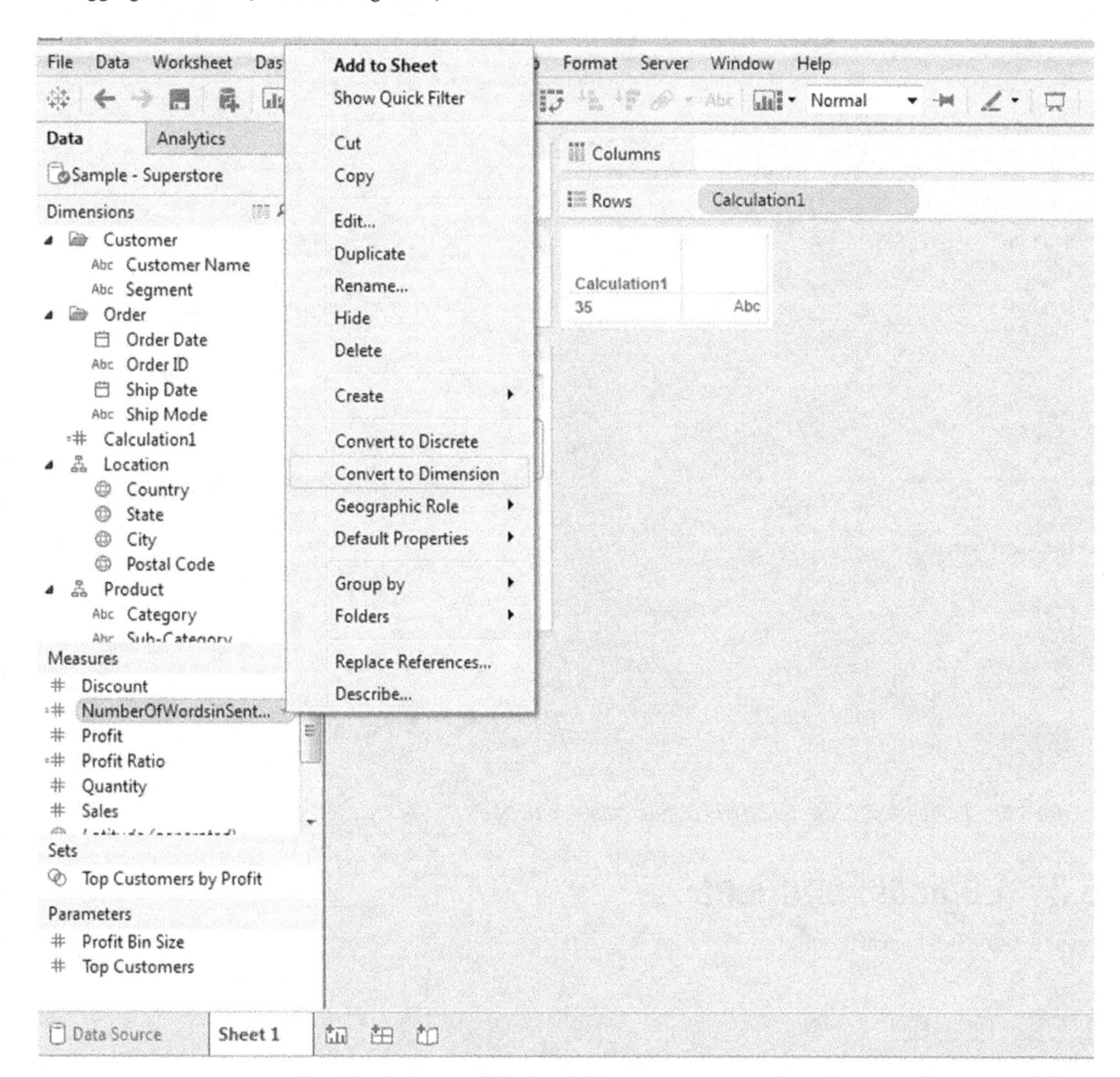

Figure 6-27. *Converting the measure "NumberOfWordsinSentence" to a dimension*

6.2.4.1.3 Step 3

Then drag the dimension "NumberOfWordsinSentence" to the rows shelf (Shown in Fig. 6-28).

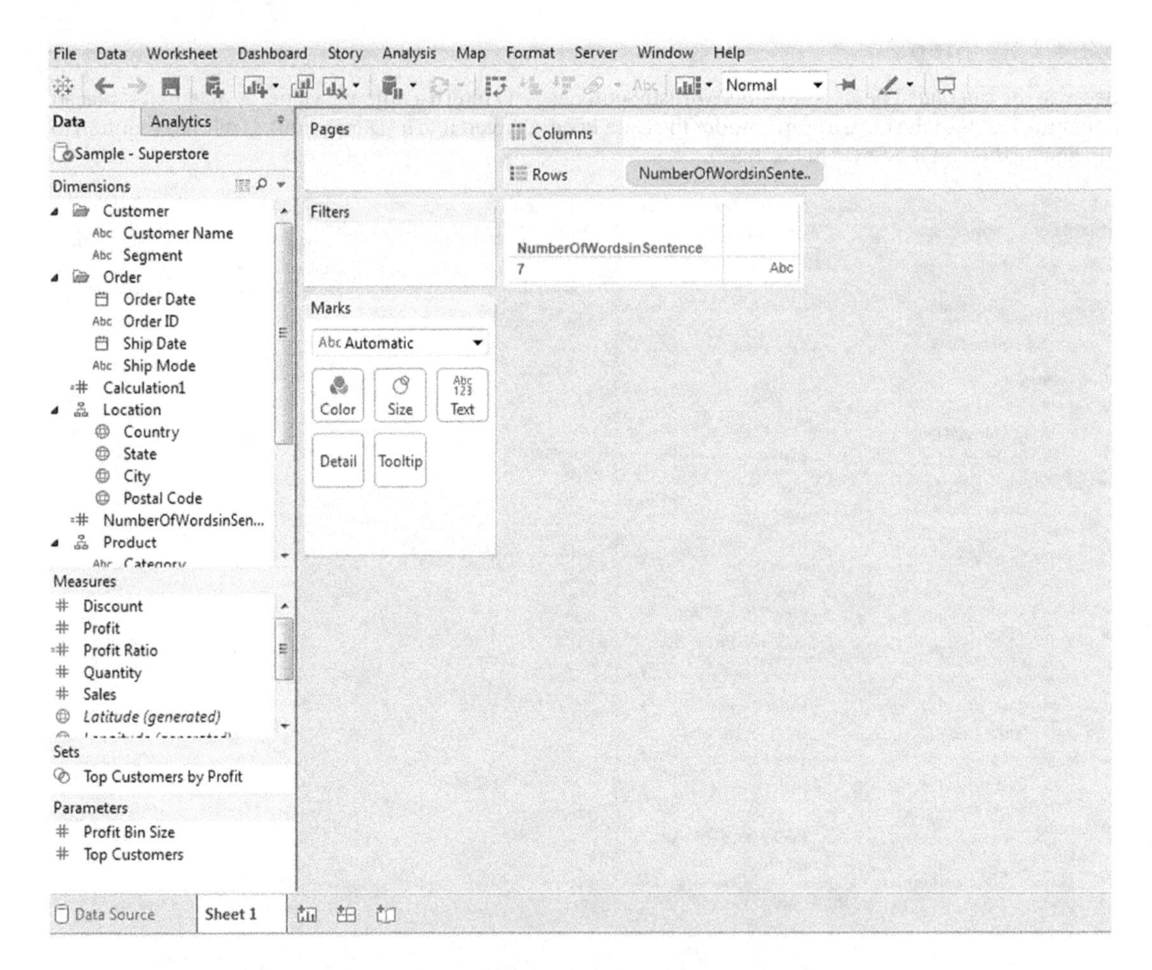

Figure 6-28. *Dimension, "NumberOfWordsinSentence" placed on the rows shelf*

6.3 Logical Functions

Let us explore few logical functions.

6.3.1 CASE

This function evaluates the expression and compares the sequence of values, value1, value2, etc. If there is a match, CASE returns the corresponding return value. Otherwise, it returns the default value. If there is no default value, it returns Null.

6.3.1.1 Steps to demonstrate CASE

Perform the following steps.

6.3.1.1.1 Step 1

Read the data from "Sample-Superstore" data set (Shown in Fig. 6-29).

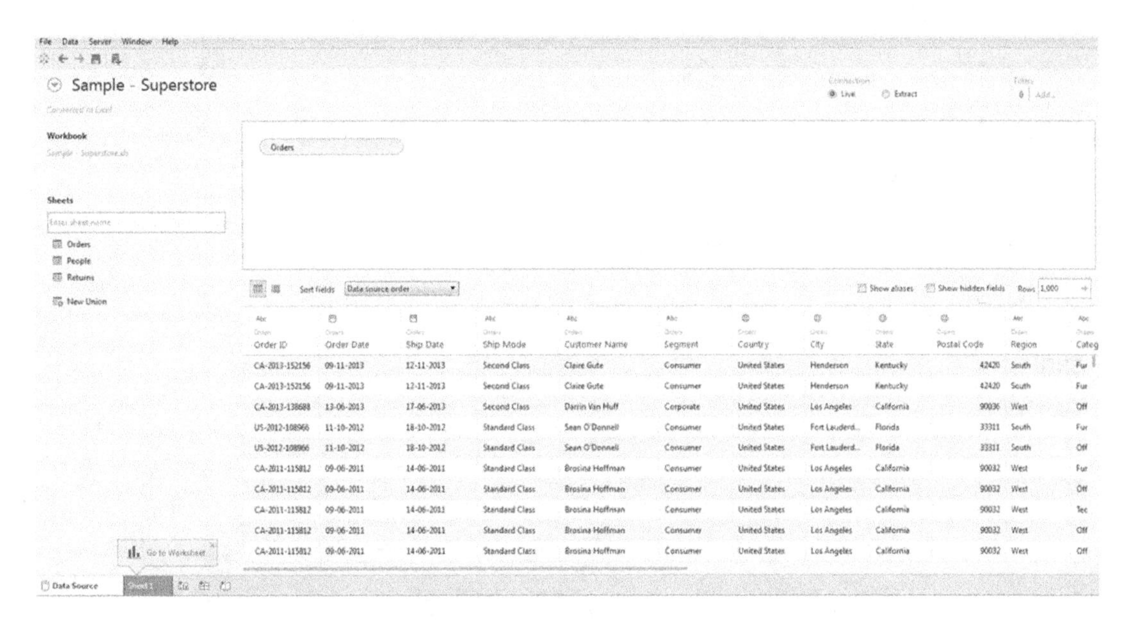

Figure 6-29. *Data source page showing "Orders" data set from "Sample – Superstore.xls"*

6.3.1.1.2 Step 2

Create a view as shown in Fig. 6-30.

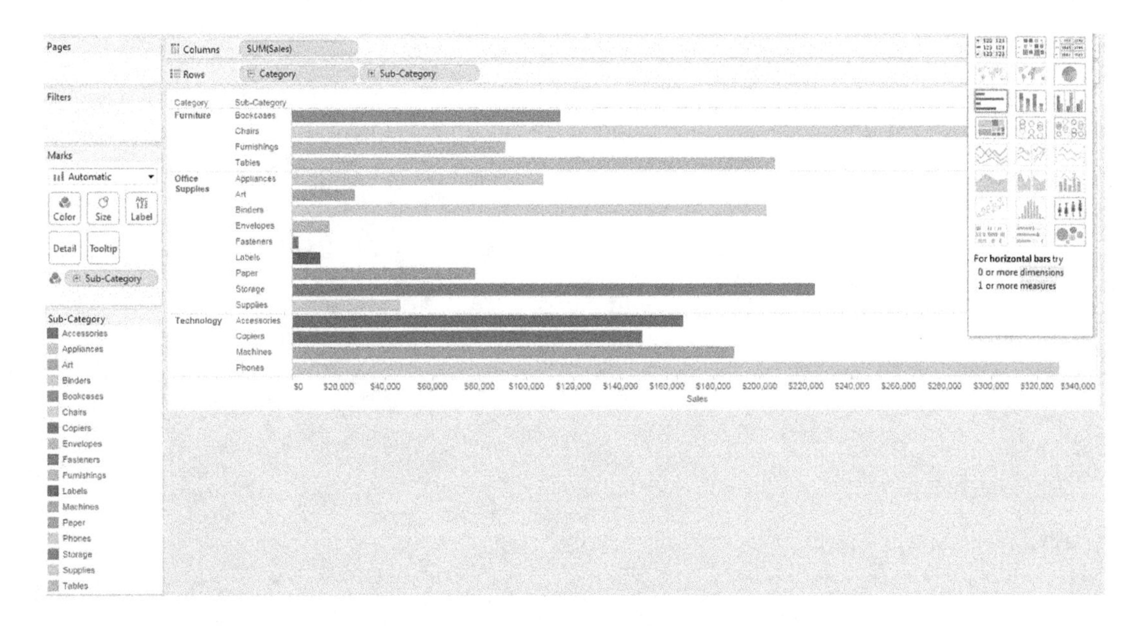

Figure 6-30. *View displaying "Sales" by "Category" and "Sub-Category"*

6.3.1.1.3 Step 3

Create a calculated field “Shipping Expense” (Shown in Fig. 6-31).

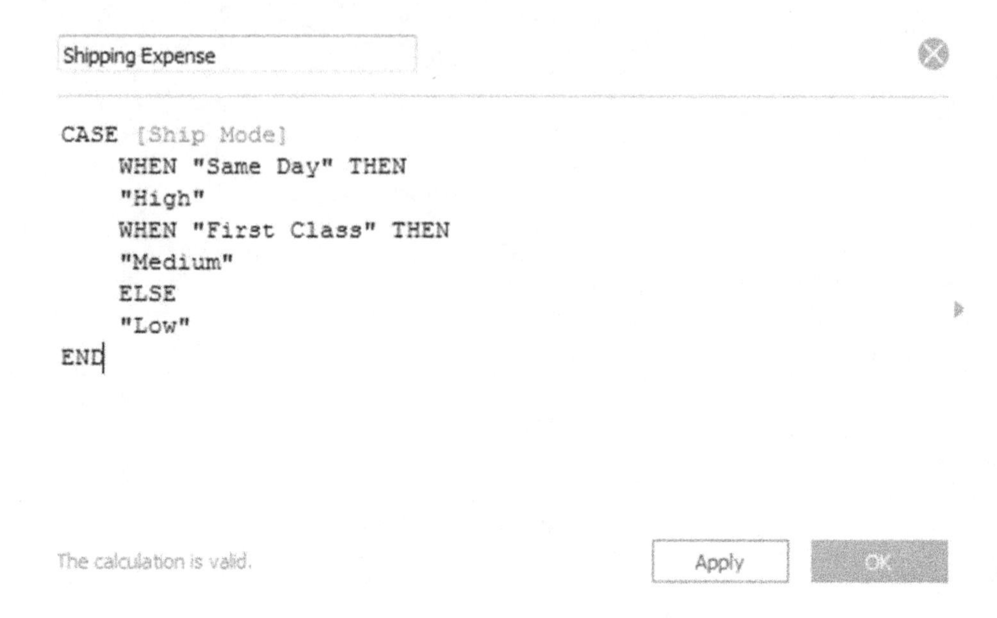

Figure 6-31. *Calculated field “Shipping Expense” being created*

6.3.1.1.4 Step 4

Drag the calculated field “Shipping Expense” to the rows shelf (Shown in Fig. 6-32).

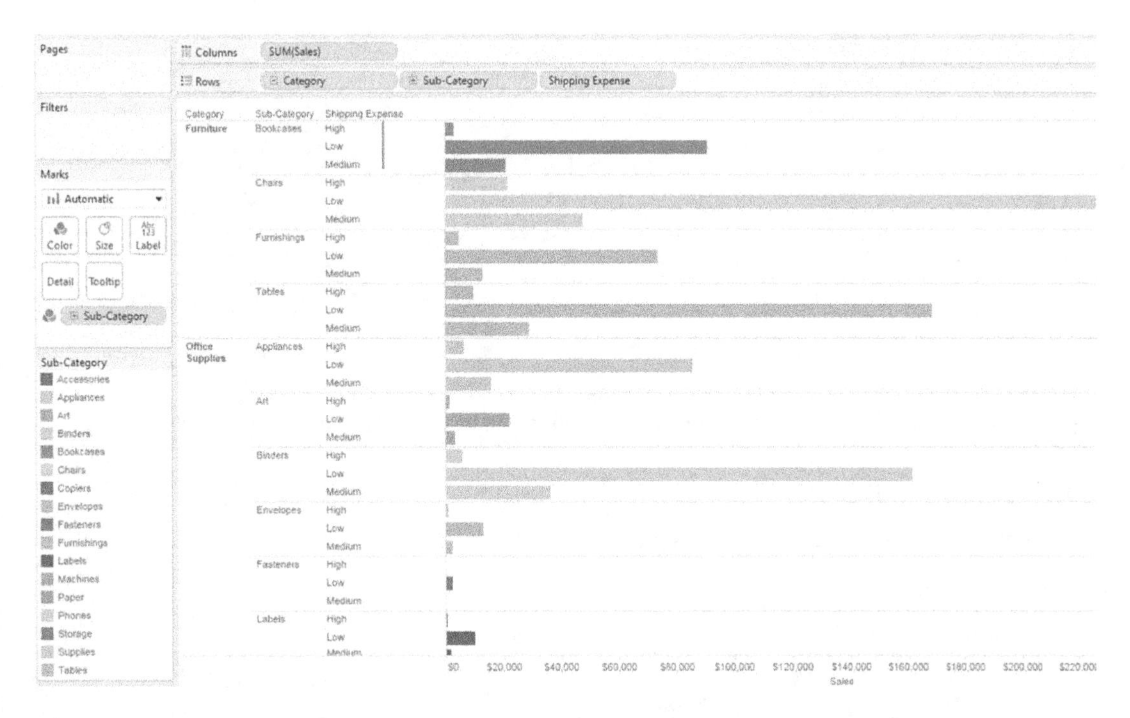

Figure 6-32. *View shows "Shipping Expense" by "Category" and "Sub-Category"*

CASE statement is useful when you need to test a single value. However, it is not suitable for comparison.

6.3.2 IIF() function

Formula: IIF (test, then, else, [unknown])

If the test evaluates to TRUE, then IIF returns the "then" value. If the test evaluates to FALSE, then IIF returns the "else" value.

6.3.2.1 Steps to demonstrate IIF() function

Perform the following steps.

6.3.2.1.1 Step 1

Read the data from "Sample-Superstore" data set.

6.3.2.1.2 Step 2

Construct the view as shown in Fig. 6-33.

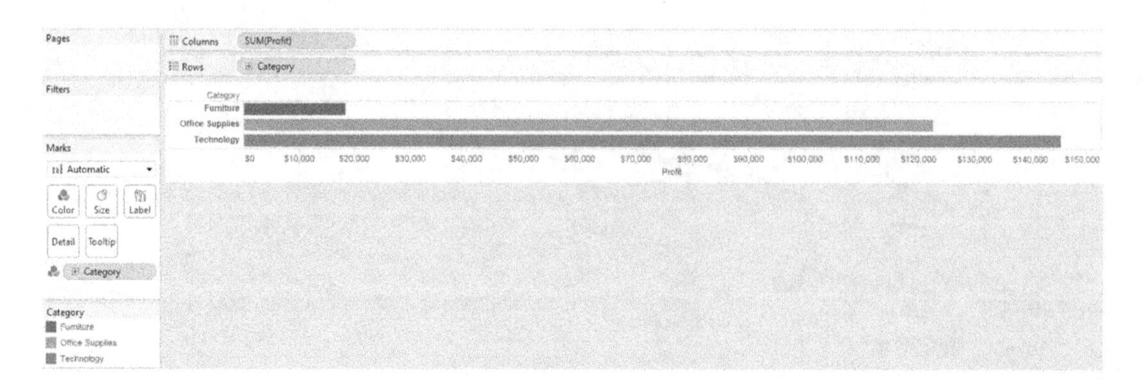

Figure 6-33. *Measure "Profit" placed on the rows shelf and Dimension placed on the columns shelf*

6.3.2.1.3 Step 3

Create a calculated field "Profit Category" (Shown in Fig. 6-34).

Figure 6-34. *Calculated field "Profit Category" being created*

6.3.2.1.4 Step 4

Drag the calculated field "Profit Category" to the columns shelf (Shown in Fig. 6-35).

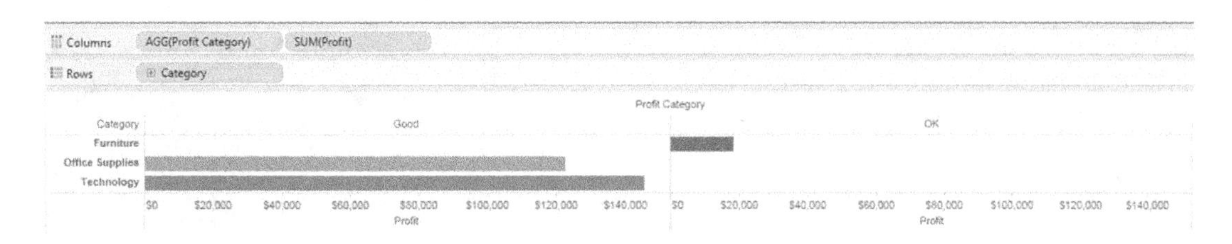

Figure 6-35. *Calculated field "Profit Category" placed on the columns shelf*

6.3.3 IF ELSE

Formula: IF test THEN value ELSE value END

This function evaluates a test condition and returns the THEN value for the condition that evaluates to "True". If no condition evaluates to True, the ELSE value is returned.

6.3.3.1 Steps to demonstrate IF THEN ELSE END

Perform the following steps.

6.3.3.1.1 Step 1

Consider the "Population" data set (Shown in Fig. 6-36).

	A	B
1	State	Population
2	Michigan	98,95,622
3	New Jersey	88,99,339
4	New York	1,96,51,127
5	Ohio	1,15,70,808
6	Illinois	1,28,82,135

Figure 6-36. *"Population" data set*

6.3.3.1.2 Step 2

Read the data from "Population" data set (Shown in Fig. 6-37).

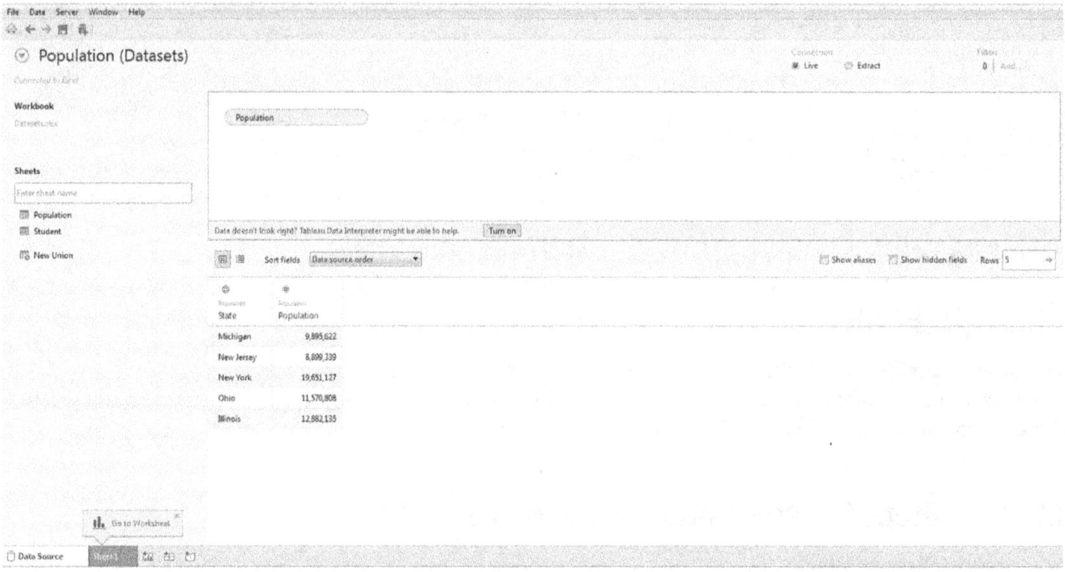

Figure 6-37. Data source page showing the "Population" data set

6.3.3.1.3 Step 3

Create a calculated field "Population Category" (Shown in Fig. 6-38).

Figure 6-38. Calculated field "Population Category" being created

6.3.3.1.4 Step 4

Drag the calculated field "Population Category" to the rows shelf (Shown in Fig. 6-39).

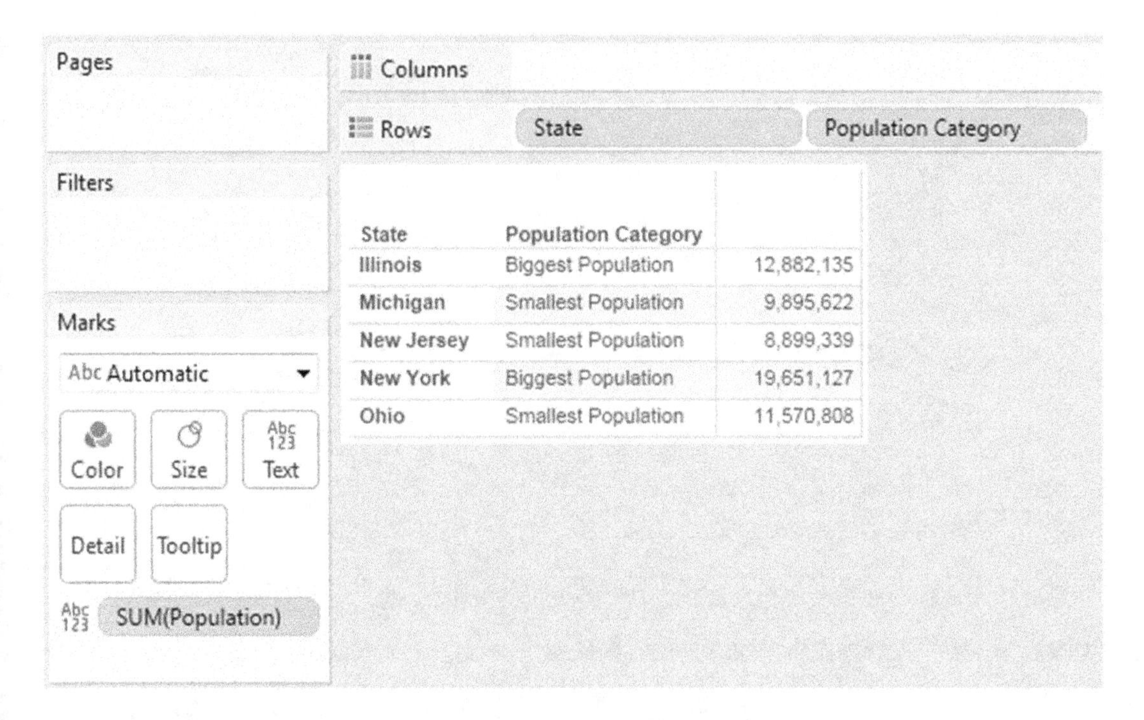

***Figure 6-39.** Calculated field "Population Category" placed on the rows shelf*

6.3.4 IF ELSEIF

Formula: IF test THEN value1 ELSEIF test2 THEN value2 ELSE value3 END.

You can use this version of IF function, when you need to perform logical tests recursively.

6.3.4.1 Steps to demonstrate IF ELSEIF

Perform the following steps.

Consider the "Student" data set (Shown in Fig. 6-40).

	A	B	C
1	Stud Name	Mark 1	Mark 2
2	Smith	23	14
3	Jack	18	24
4	John	20	21
5	Scott	22	24
6	James	17	24

***Figure 6-40.** "Student" data set*

6.3.4.1.1 Step 1

Read data from "Student" data set (Shown in Fig. 6-41).

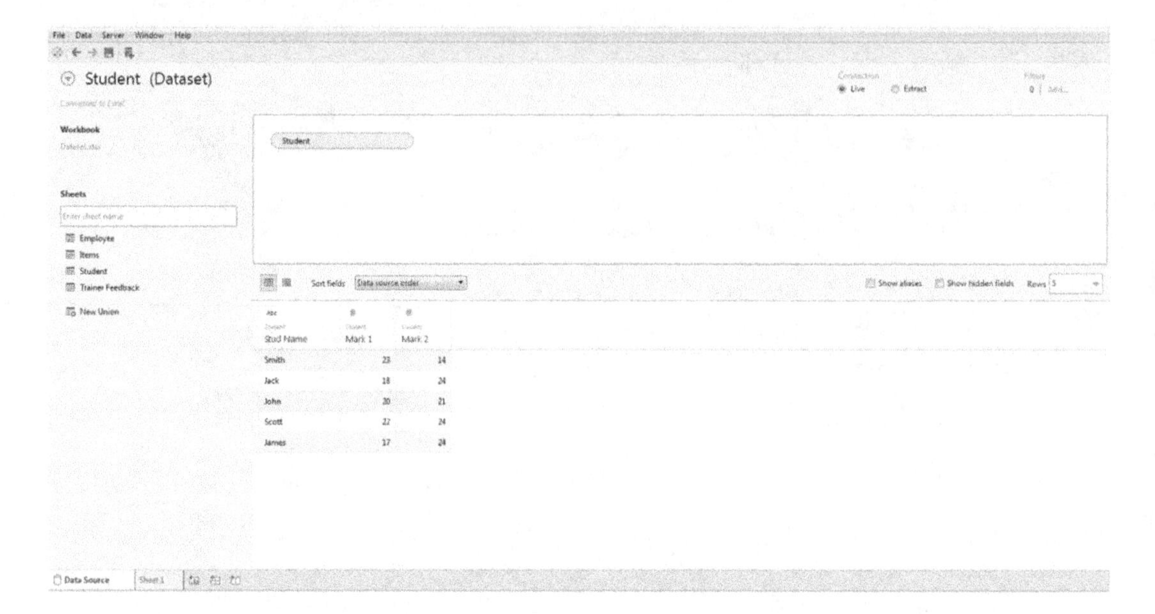

Figure 6-41. *Data source page showing "Student" data set*

6.3.4.1.2 Step 2

Create a view as shown in Fig. 6-42.

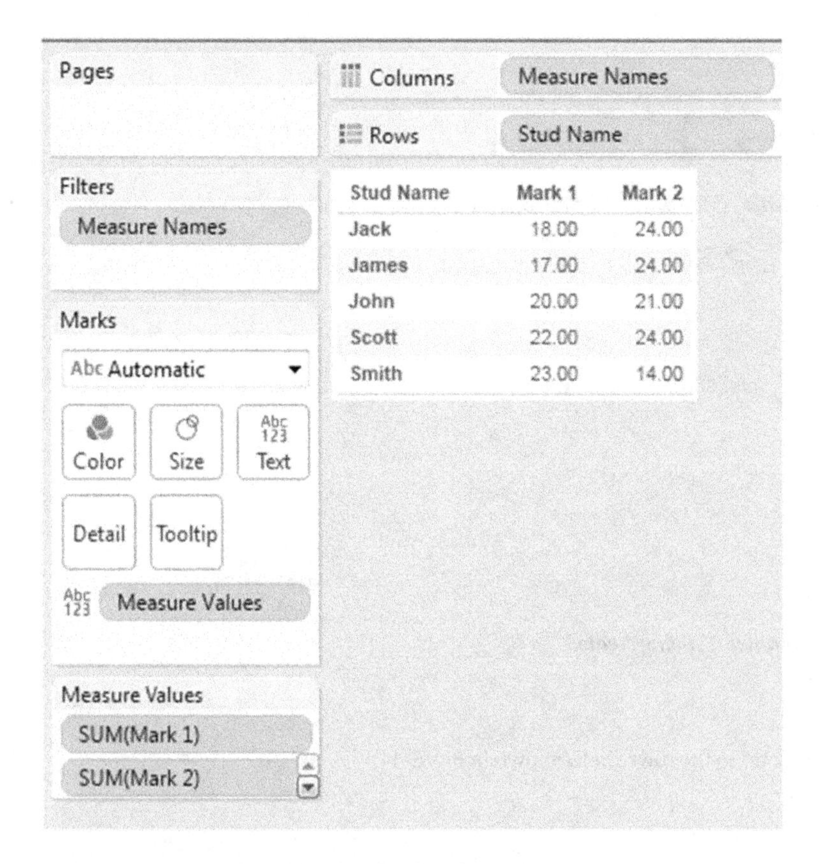

Figure 6-42. *View shows the details of students*

6.3.4.1.3 Step 3

Create a calculated field "Comments" as shown in Fig. 6-43.

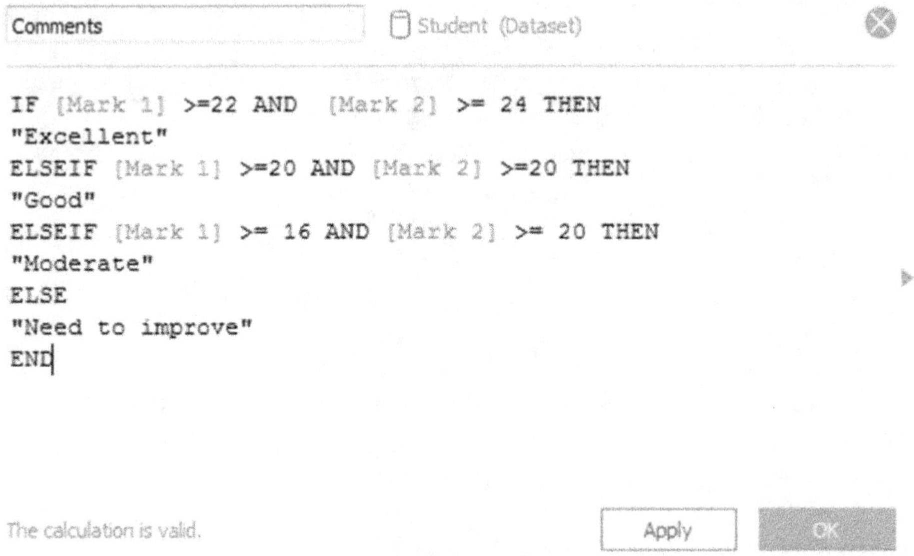

Comments   Student (Dataset)

```
IF [Mark 1] >=22 AND  [Mark 2] >= 24 THEN
"Excellent"
ELSEIF [Mark 1] >=20 AND [Mark 2] >=20 THEN
"Good"
ELSEIF [Mark 1] >= 16 AND [Mark 2] >= 20 THEN
"Moderate"
ELSE
"Need to improve"
END
```

The calculation is valid.   Apply   OK

Figure 6-43. *Calculated field "Comments" being created*

6.3.4.1.4 Step 4

Drag the calculated field "Comments" to the rows shelf (Shown in Fig. 6-44).

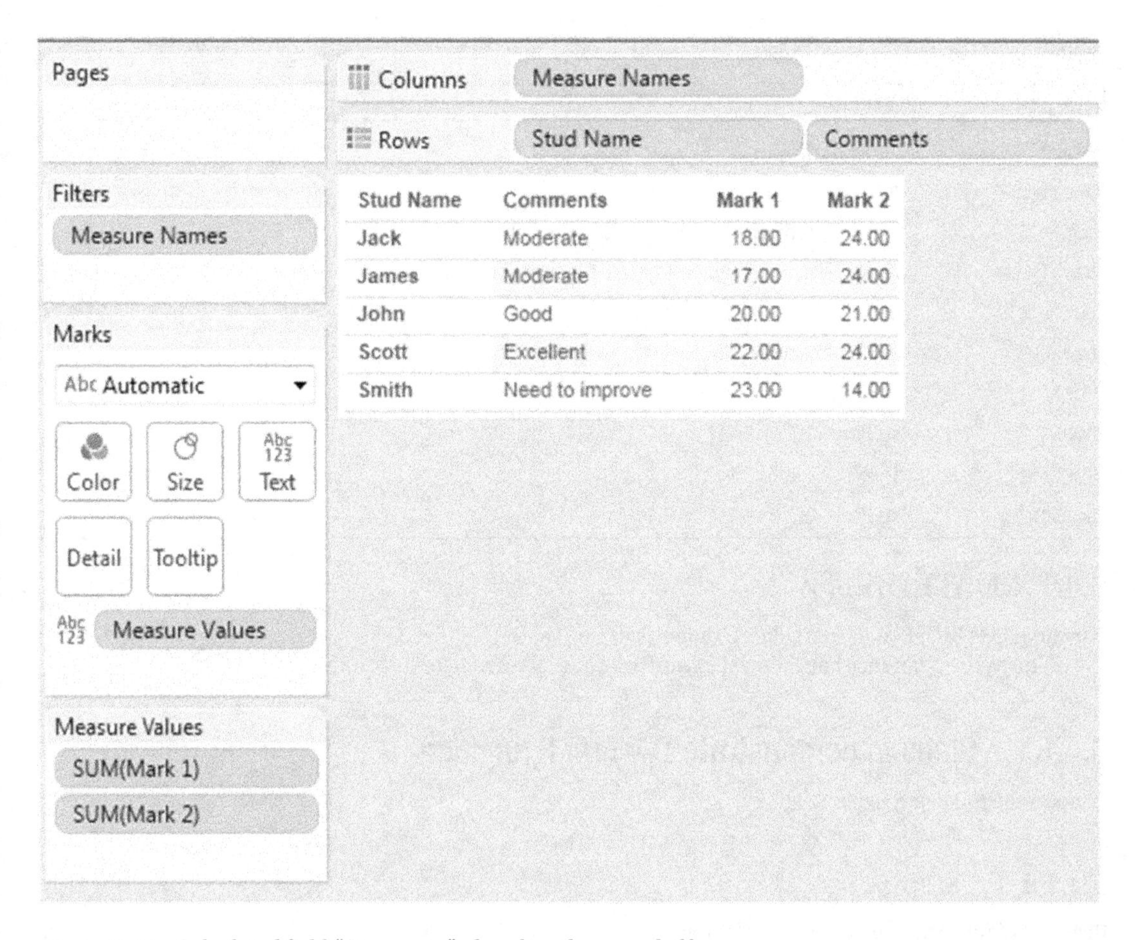

Figure 6-44. *Calculated field "Comments" placed on the rows shelf*

6.4 Date functions

Tableau provides a variety of date functions. Many date functions, use date_part, which is a constant string argument.

Refer to Table 6-3 for date_part and its value.

Table 6-3. *date_part and its value*

date_part	Values
`'year'`	Four-digit year
`'quarter'`	1-4
`'month'`	1-12 or "January", "February", and so on
`'dayofyear'`	Day of the year; Jan 1 is 1, Feb 1 is 2, and so on
`'day'`	1-31
`'weekday'`	1-7 or "Sunday", "Monday", and so on
`'week'`	1-52
`'hour'`	0-23
`'minute'`	0-59
`'second'`	0-59

6.4.1 DATEDIFF()

Formula: DATEDIFF (date_part, date1, date2, [start_of_week])

Returns the difference between `date1` and `date2` , expressed in units of `date_part`.

6.4.1.1 Steps to demonstrate DATEDIFF function

Perform the following steps.

6.4.1.1.1 Step 1

Read data from “Sample-Superstore.xls” data set.

6.4.1.1.2 Step 2

Create a view as shown in Fig. 6-45.

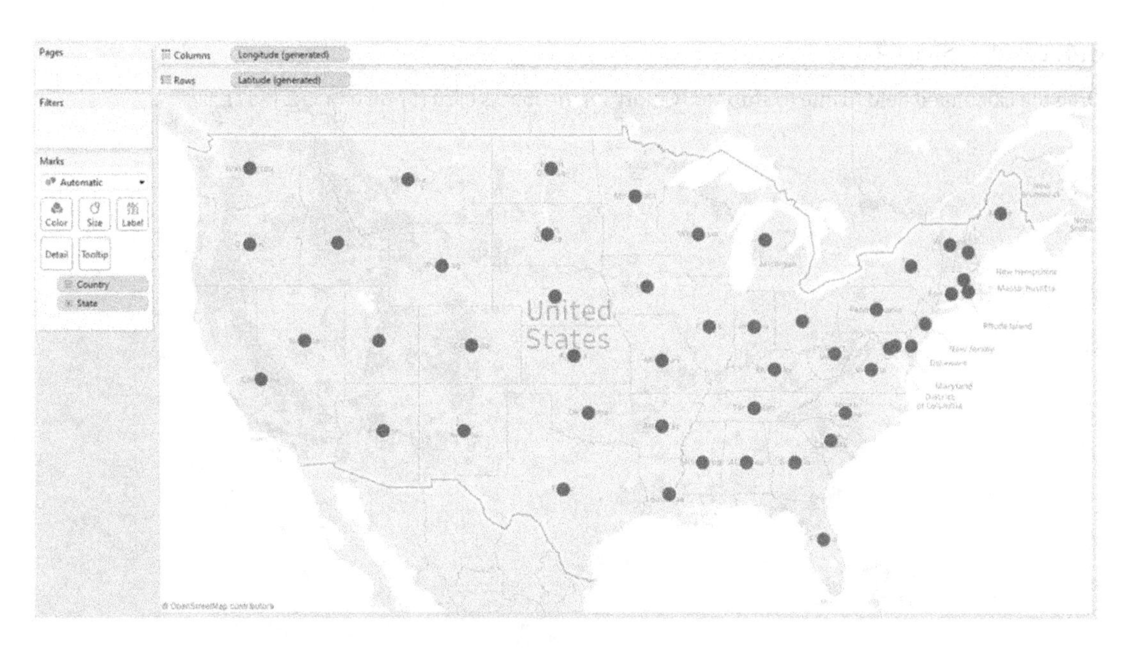

Figure 6-45. *View shows the States of United States*

6.4.1.1.3 Step 3

Create a calculated field "Time to ship" as shown in Fig. 6-46.

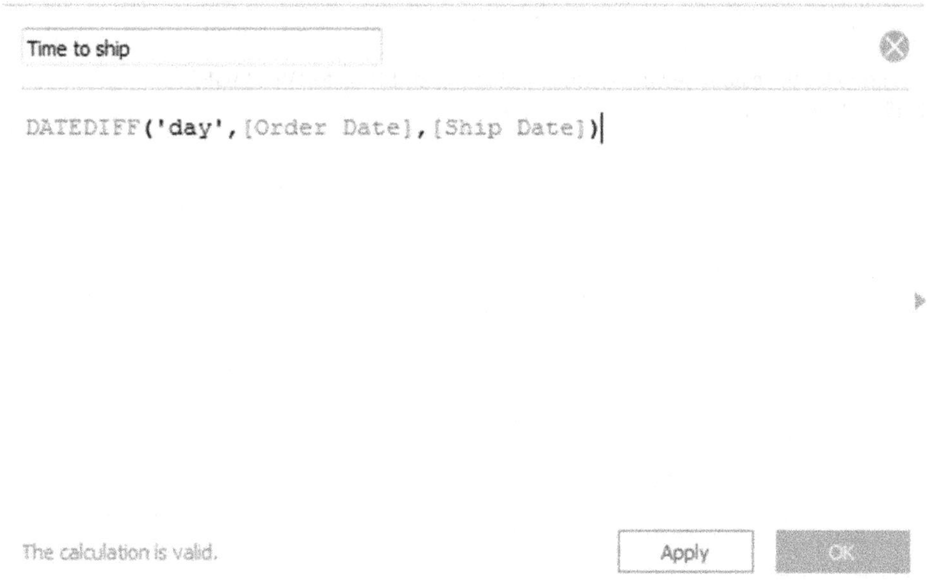

Time to ship

```
DATEDIFF('day',[Order Date],[Ship Date])
```

The calculation is valid.

Apply OK

Figure 6-46. *Calculated field "Time to ship"*

6.4.1.1.4 Step 4

Drag the calculated field "Time to ship" to "Color" on the marks card (Shown in Fig. 6-47).

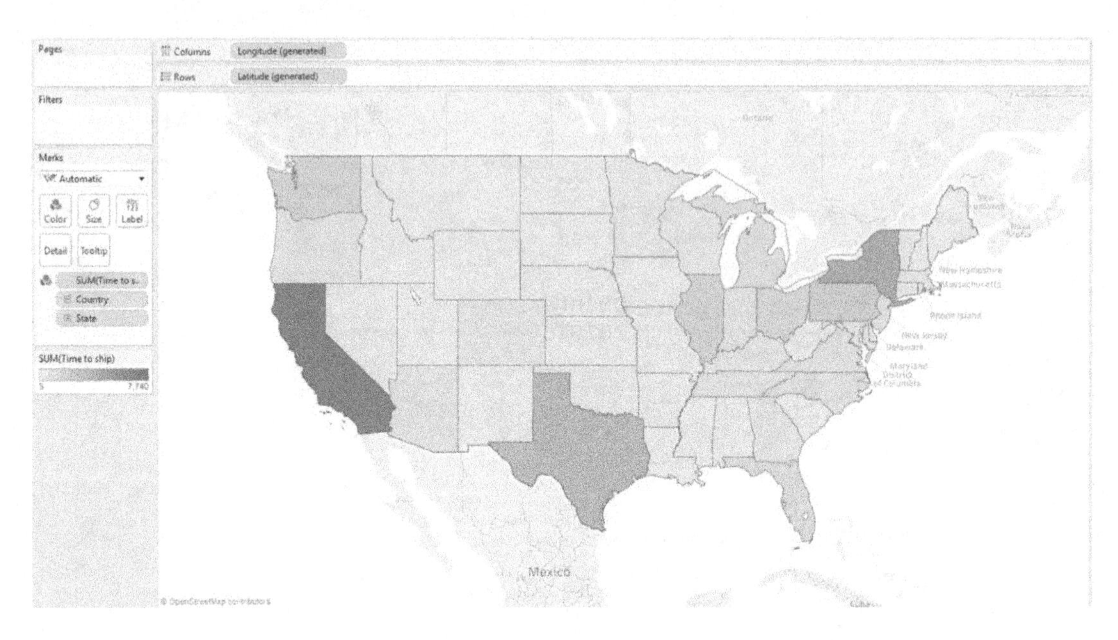

Figure 6-47. *Calculated field "Time to ship" placed on "Color" on the marks card*

6.4.1.1.5 Step 5

Convert the aggregation for the calculated field "Time to ship" from "SUM" to "AVERAGE" (Shown in Fig. 6-48 and Fig. 6-49).

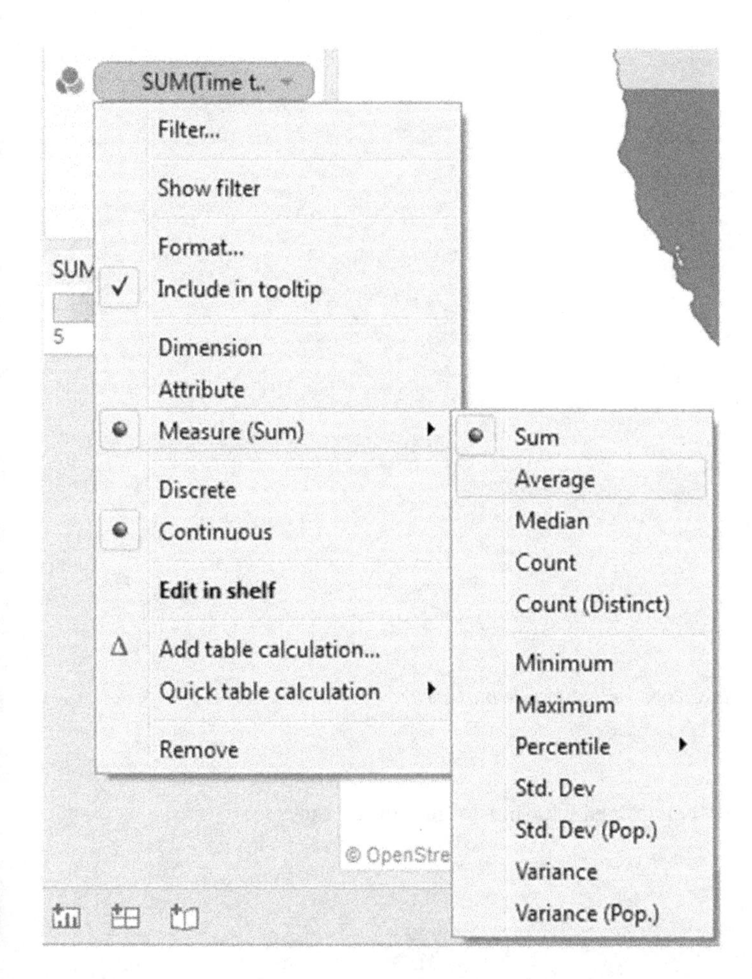

Figure 6-48. *Converting the aggregation for measure, "Time to ship" from "SUM" to "AVERAGE"*

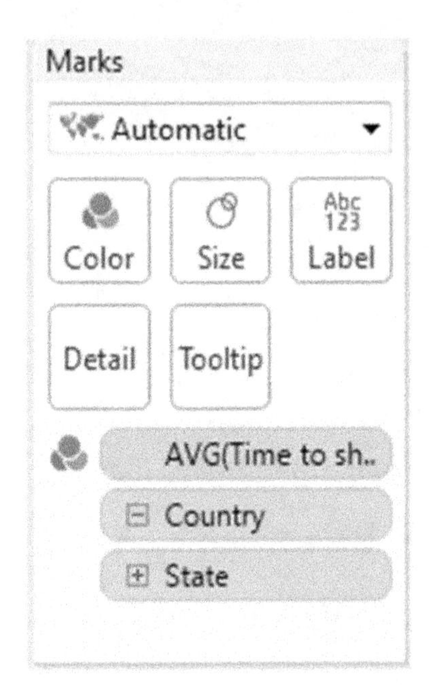

Figure 6-49. *Measure, "Time to ship" placed on "Color" on the marks card*

6.4.1.1.6 Step 6

Right click on "AVG(Time to ship)", select "Edit Colors[Avg. Time to ship] to edit the color (Shown in Fig. 6-50).

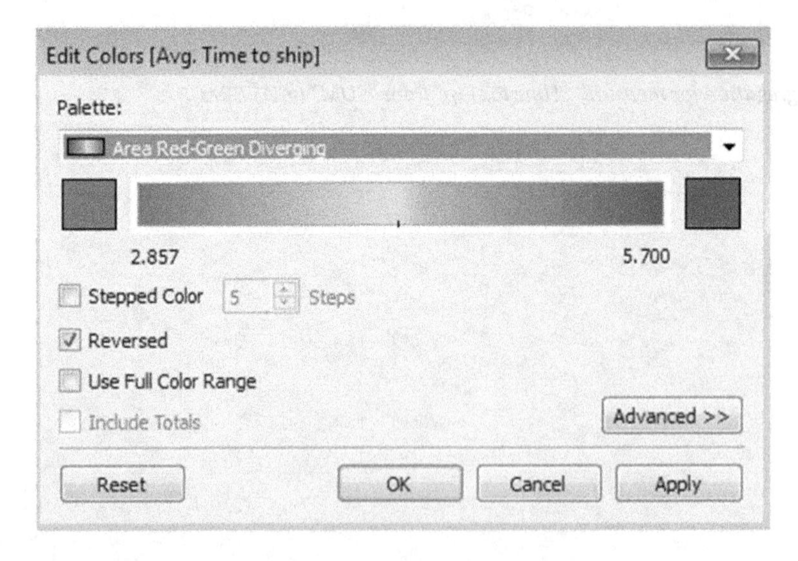

Figure 6-50. *"Edit Colors[Avg. Time to ship]" dialog box*

6.4.1.1.7 Step 7

Place the calculated field "Time to ship" to "Label" on the marks card (Shown in Fig. 6-51).

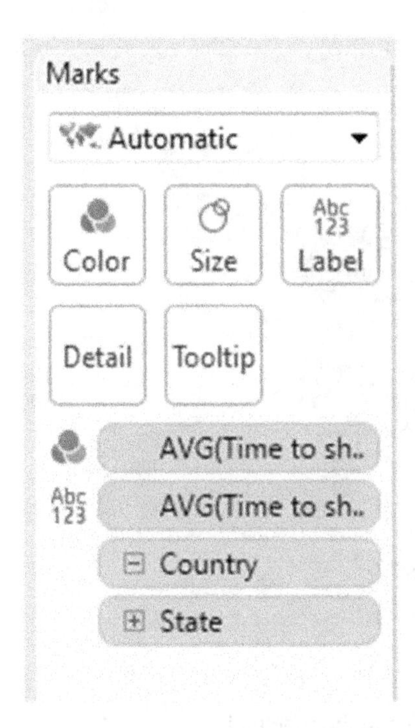

Figure 6-51. *Calculated field "Time to ship" to "Label" on the marks card*

6.4.1.1.8 Step 8

You can see the calculated field "Time to ship" in the visualization (Shown in Fig. 6-52).

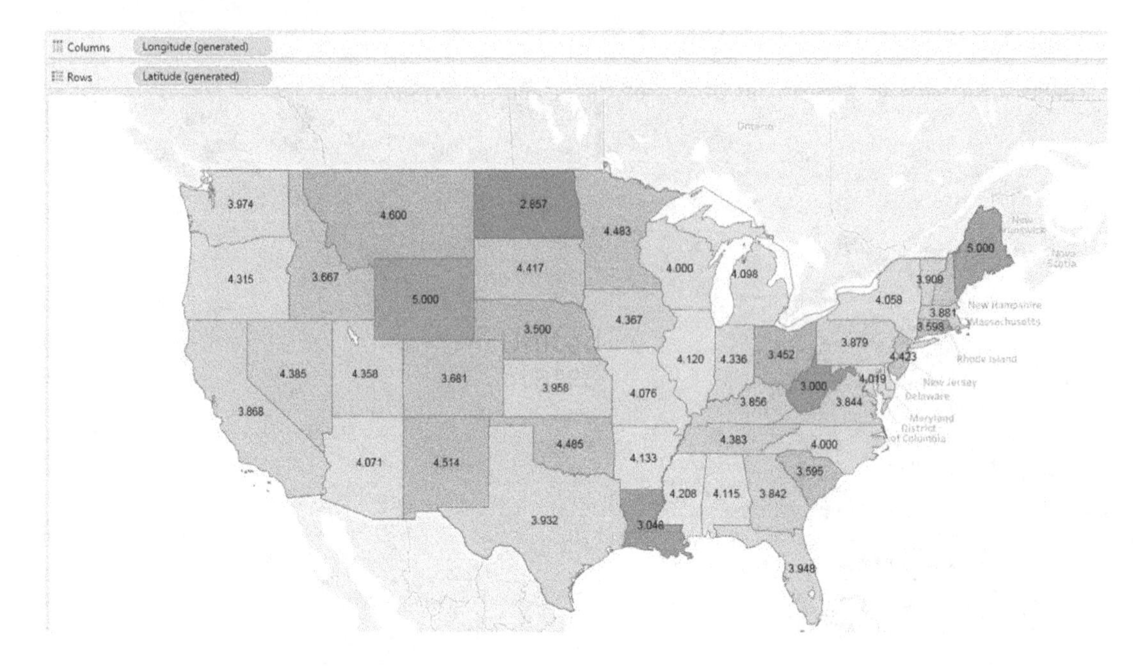

Figure 6-52. Visualization that shows the calculated field "Time to ship"

6.4.2 DATEADD() function

Formula: DATEADD (date_part, interval, date)

Returns the specified date with the specified number interval added to the specified date_part of that date.

6.4.2.1 Steps to demonstrate DATEADD() function

Perform the following steps.

6.4.2.1.1 Step 1

Read data from "Sample-Superstore" data set.

6.4.2.1.2 Step 2

Drag the dimension "Order Date" from the dimensions area under the data pane to the rows shelf. Set the hierarchy as "YEAR". Right click on "Order Date", select "Exact Date" as shown in Fig. 6-53. Refer to Fig. 6-54 for output.

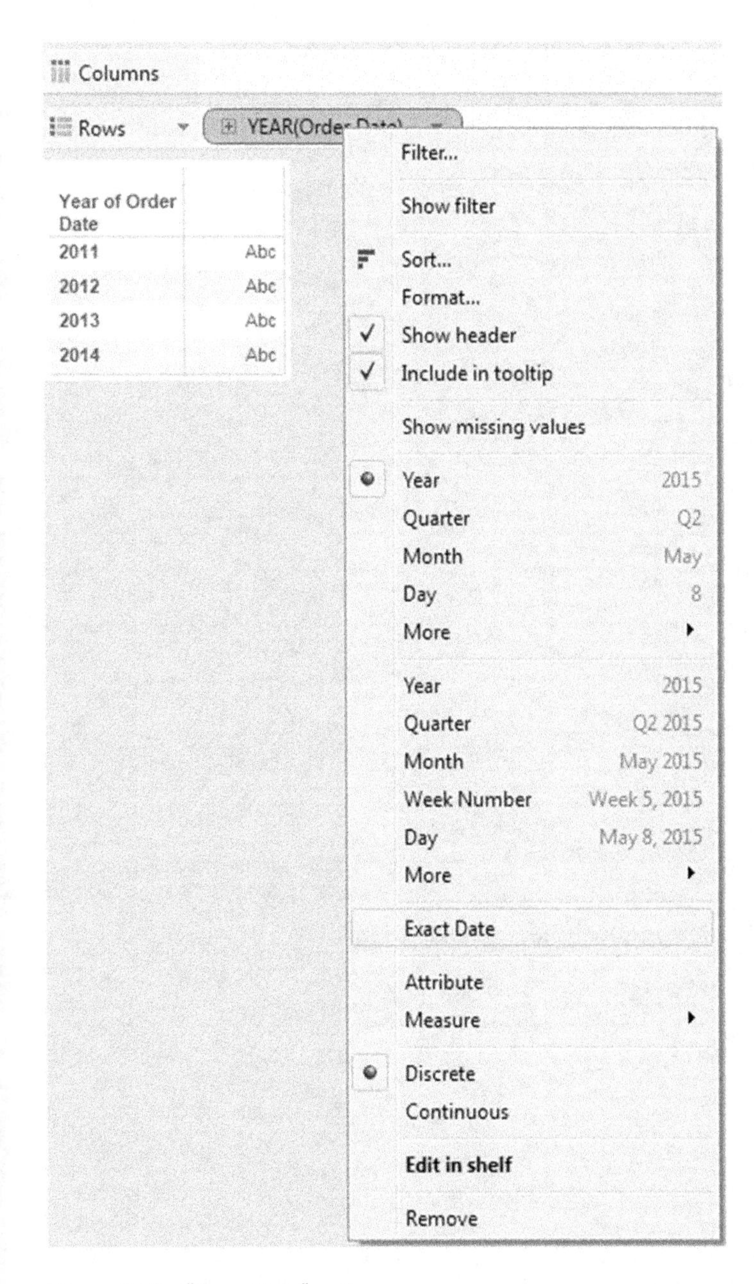

***Figure 6-53.** "Exact Date" option*

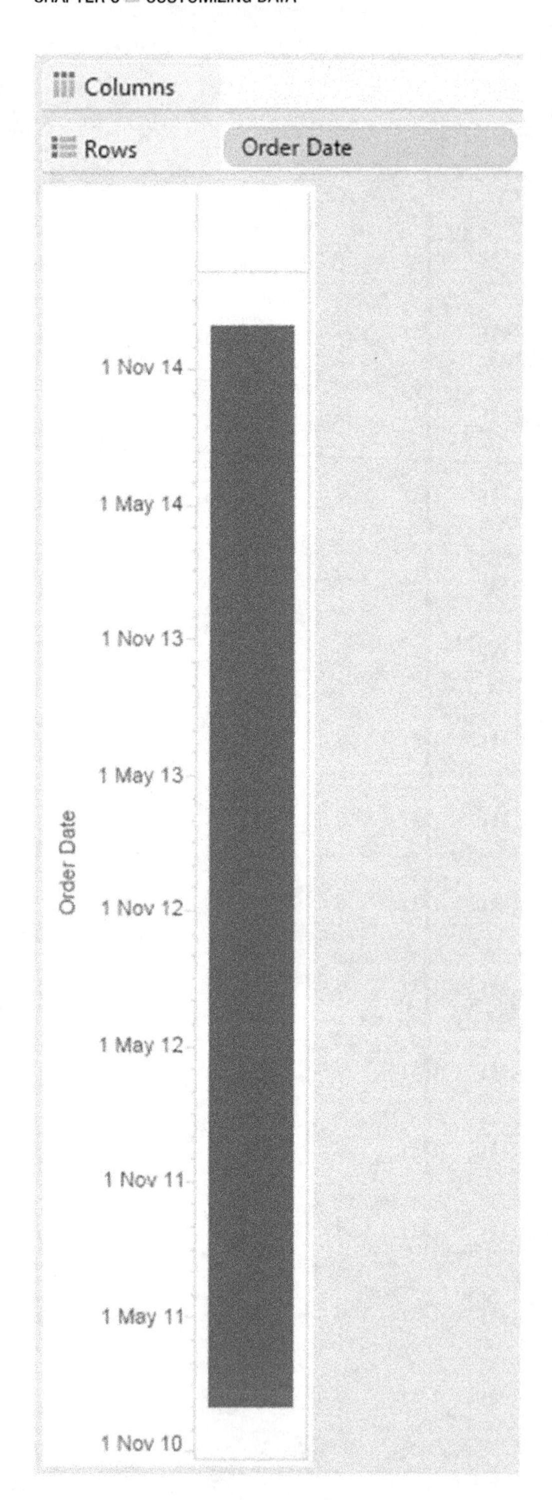

Figure 6-54. *Dimension "Order Date" is set to "Exact Order" date*

6.4.2.1.3 Step 3

Right click on "Order Date" select "Discrete" (Shown in Fig. 6-55). The output is shown in Fig. 6-56.

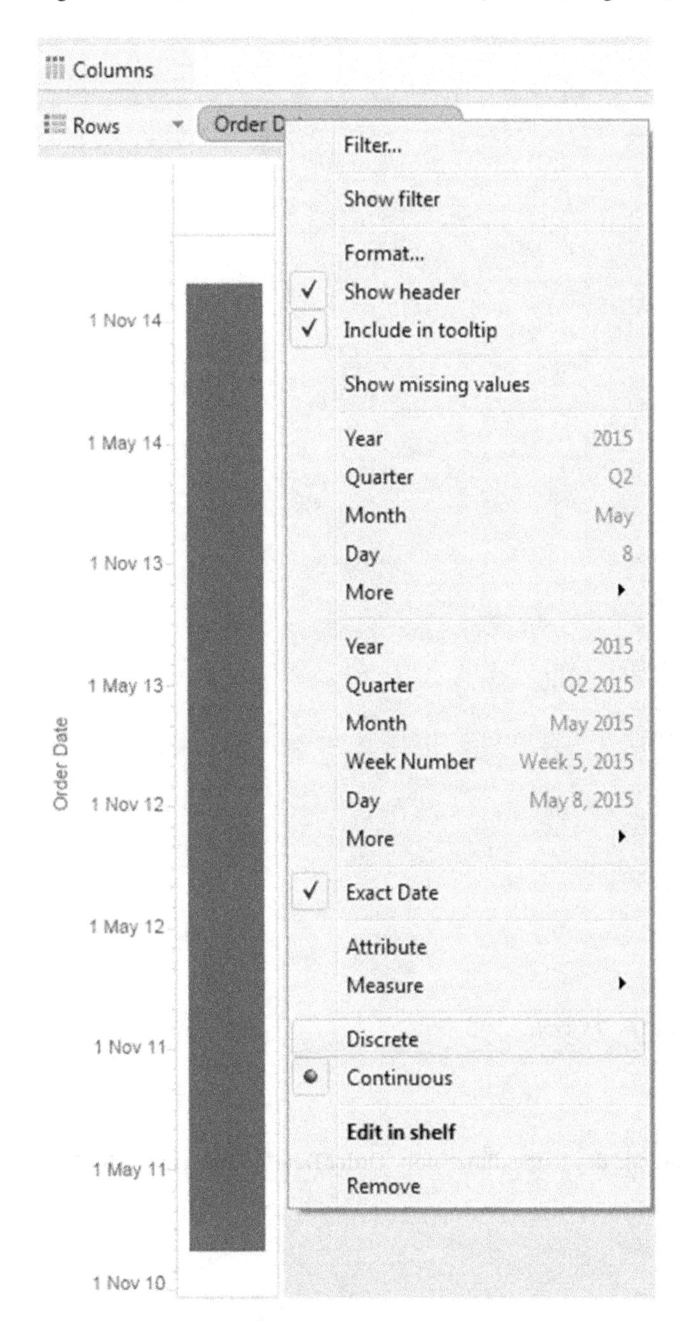

Figure 6-55. *Converting the dimension "Order Date" from "Continuous" to "Discrete"*

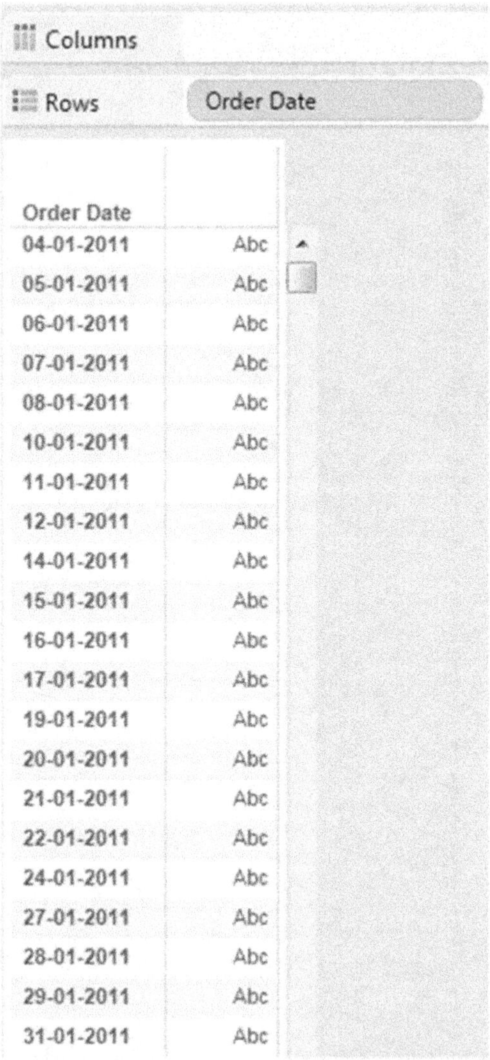

Figure 6-56. *View after setting the "Order Date" to "Discrete"*

6.4.2.1.4 Step 4

Create the calculated field "DATEADD" to add one day to the dimension "Order Date" (Shown in Fig. 6-57).

DATEADD

```
DATEADD('day',1,[Order Date])
```

The calculation is valid.

Apply OK

Figure 6-57. *Calculated field "DATEADD" being created*

6.4.2.1.5 Step 5

Drag the calculated field "DATEADD" to the rows shelf (Shown in Fig. 6-58).

Columns

Rows | Order Date | DATEADD

Order Date	DATEADD	
04-01-2011	05-01-2011 00:00:00	Abc
05-01-2011	06-01-2011 00:00:00	Abc
06-01-2011	07-01-2011 00:00:00	Abc
07-01-2011	08-01-2011 00:00:00	Abc
08-01-2011	09-01-2011 00:00:00	Abc
10-01-2011	11-01-2011 00:00:00	Abc
11-01-2011	12-01-2011 00:00:00	Abc
12-01-2011	13-01-2011 00:00:00	Abc
14-01-2011	15-01-2011 00:00:00	Abc
15-01-2011	16-01-2011 00:00:00	Abc
16-01-2011	17-01-2011 00:00:00	Abc
17-01-2011	18-01-2011 00:00:00	Abc
19-01-2011	20-01-2011 00:00:00	Abc
20-01-2011	21-01-2011 00:00:00	Abc
21-01-2011	22-01-2011 00:00:00	Abc
22-01-2011	23-01-2011 00:00:00	Abc
24-01-2011	25-01-2011 00:00:00	Abc
27-01-2011	28-01-2011 00:00:00	Abc
28-01-2011	29-01-2011 00:00:00	Abc
29-01-2011	30-01-2011 00:00:00	Abc
31-01-2011	01-02-2011 00:00:00	Abc
01-02-2011	02-02-2011 00:00:00	Abc
02-02-2011	03-02-2011 00:00:00	Abc
03-02-2011	04-02-2011 00:00:00	Abc
04-02-2011	05-02-2011 00:00:00	Abc
05-02-2011	06-02-2011 00:00:00	Abc
07-02-2011	08-02-2011 00:00:00	Abc
08-02-2011	09-02-2011 00:00:00	Abc
09-02-2011	10-02-2011 00:00:00	Abc
12-02-2011	13-02-2011 00:00:00	Abc
13-02-2011	14-02-2011 00:00:00	Abc

Figure 6-58. *Calculated field "DATEADD" placed on the rows shelf*

6.4.2.1.6 Step 6

Right click on the calculated field "DATEADD", select "Format..." to format the date format (Shown in Fig. 6-59).

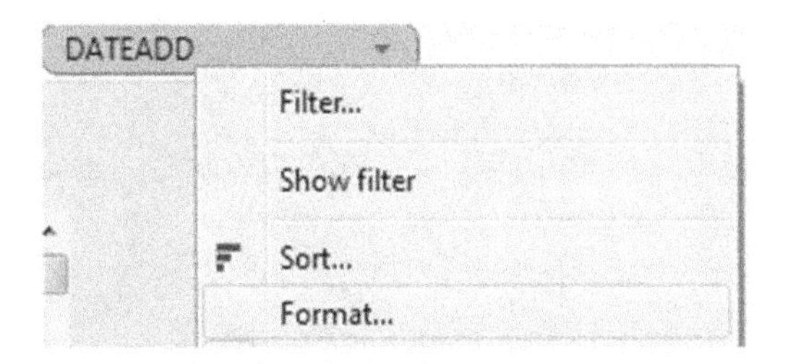

***Figure 6-59.** "Format..." option*

6.4.2.1.7 Step 7

"Format Order Date" window opens. Select the format as "Standard Long Date" (Shown in Fig. 6-60).

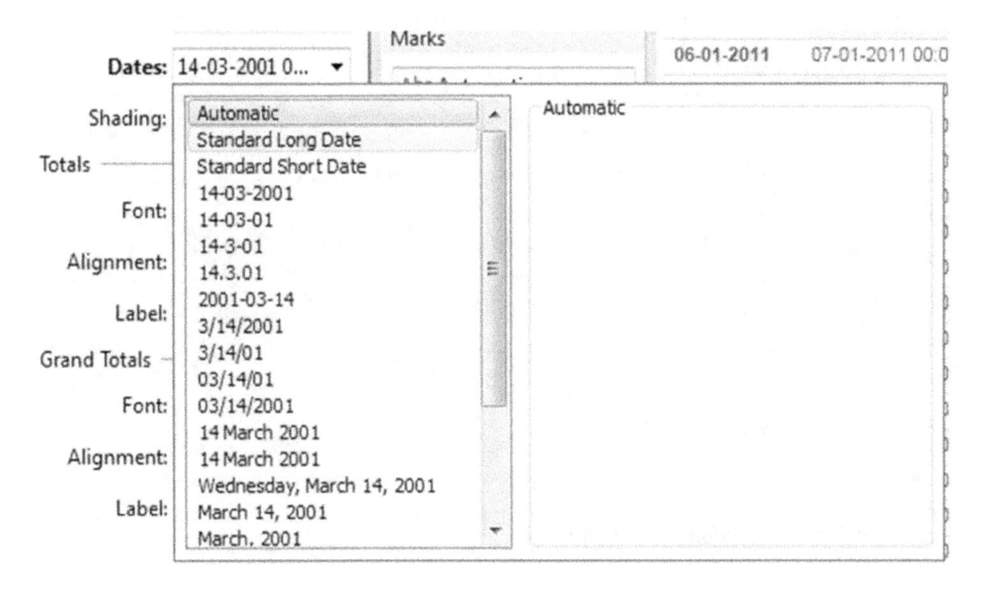

***Figure 6-60.** "Format Order Date" Window*

6.4.2.1.8 Step 8

The final output is shown in Fig. 6-61.

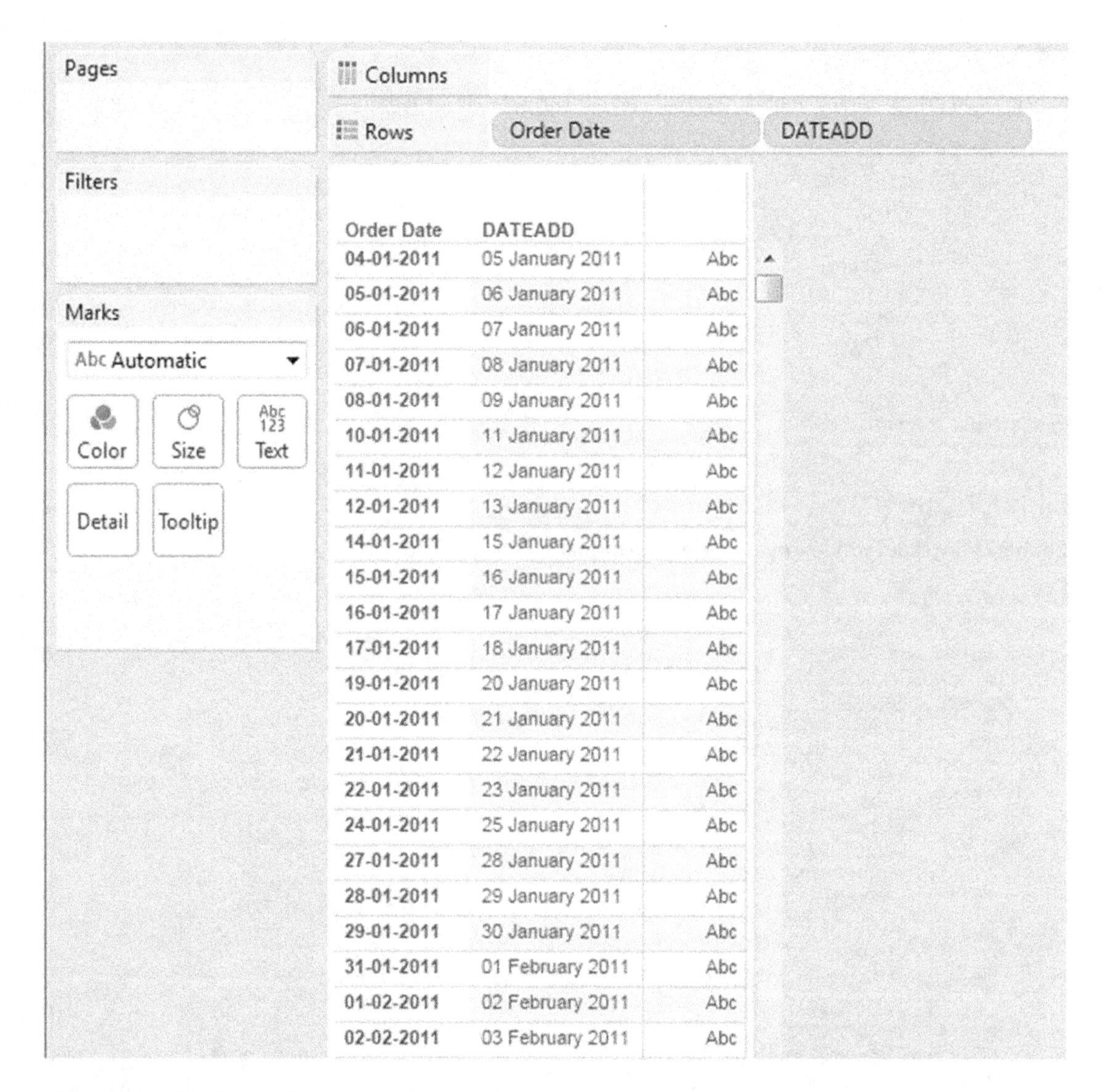

Figure 6-61. *"DATEADD" demonstration - final output*

6.4.3 DATENAME

DATENAME function returns date_part of date as a string.

Formula: DATENAME (date_part, date, [start_of_week])

Returns `date_part` of `date` as a string.

6.4.3.1 Steps to demonstrate DATENAME function

Perform the following steps.

6.4.3.1.1 Step 1

Read data from “Sample-Superstore” data set.

6.4.3.1.2 Step 2

Create a calculated field “DATENAME” as shown in Fig. 6-62.

DATENAME

```
DATENAME('day',[Order Date])
```

The calculation is valid.

Apply

OK

Figure 6-62. Calculated field “DATENAME” being created

6.4.3.1.3 Step 3

Drag the calculated field “DATENAME” to the rows shelf (Shown in Fig. 6-63).

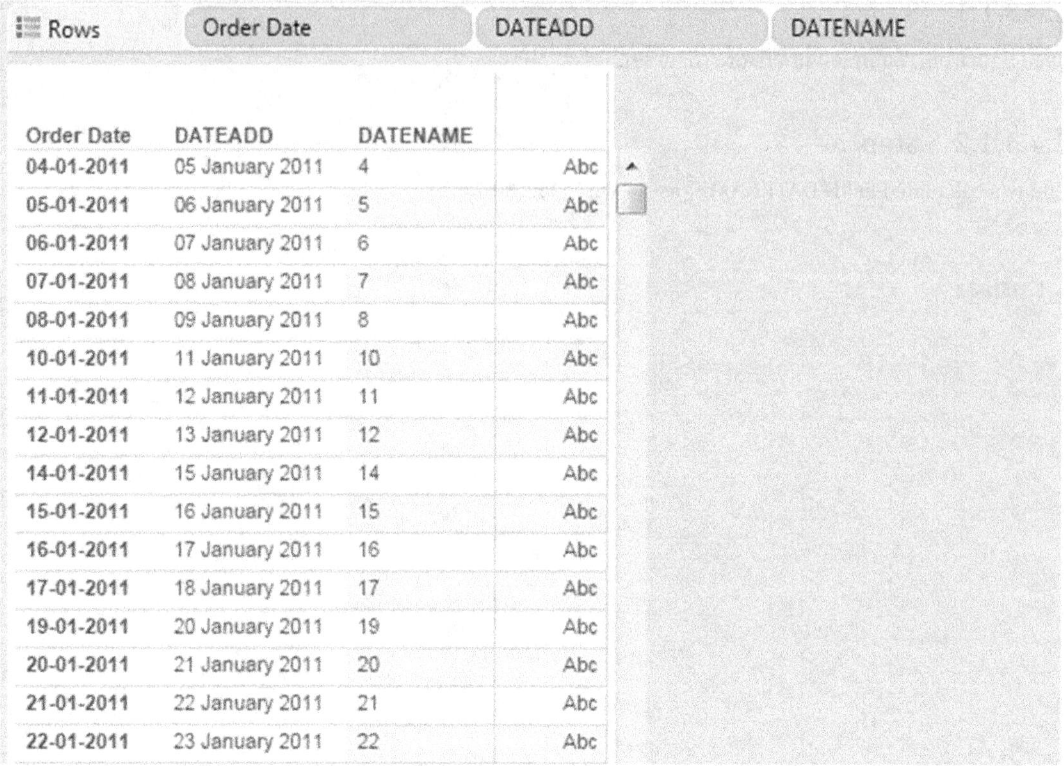

Rows | Order Date | DATEADD | DATENAME

Order Date	DATEADD	DATENAME	
04-01-2011	05 January 2011	4	Abc
05-01-2011	06 January 2011	5	Abc
06-01-2011	07 January 2011	6	Abc
07-01-2011	08 January 2011	7	Abc
08-01-2011	09 January 2011	8	Abc
10-01-2011	11 January 2011	10	Abc
11-01-2011	12 January 2011	11	Abc
12-01-2011	13 January 2011	12	Abc
14-01-2011	15 January 2011	14	Abc
15-01-2011	16 January 2011	15	Abc
16-01-2011	17 January 2011	16	Abc
17-01-2011	18 January 2011	17	Abc
19-01-2011	20 January 2011	19	Abc
20-01-2011	21 January 2011	20	Abc
21-01-2011	22 January 2011	21	Abc
22-01-2011	23 January 2011	22	Abc

Figure 6-63. *Calculated field "DATENAME" placed on the rows shelf*

To learn more about date refer to the link below.

`http://onlinehelp.tableau.com/current/pro/desktop/en-us/functions_functions_date.html`

6.5 Aggregate functions

Tableau has support for various aggregate functions.

Let us discuss ATTR function.

6.5.1 ATTR(expression)

The ATTR() function returns the value of the expression if it has a single value for all rows. Otherwise returns an asterisk. Null values are ignored.

6.5.1.1 Steps to demonstrate the use of ATTR() function

Perform the following steps.

6.5.1.1.1 Step 1

Specify the Sum([Profit])/Sum[Sales]) calculation on the columns shelf (Shown in Fig. 6-64).

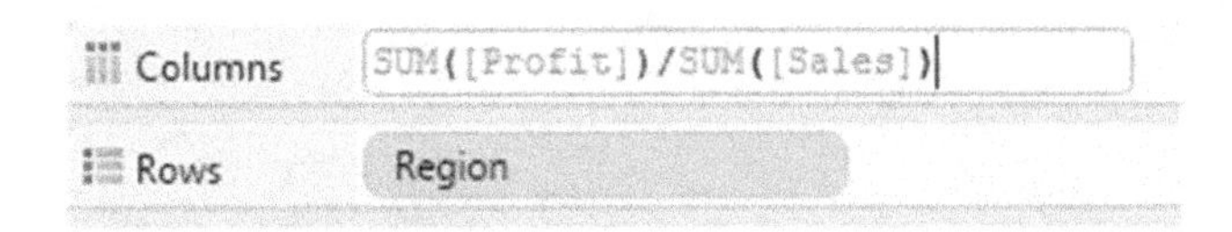

Figure 6-64. *Calculation placed on the columns shelf*

6.5.1.1.2 Step 2

The view shows that aggregation is part of calculation (Shown in Fig. 6-65).

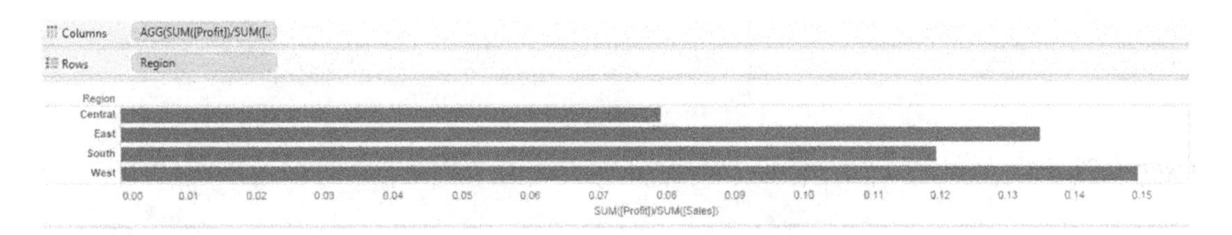

Figure 6-65. *Aggregation is part of calculation*

6.5.1.1.3 Step 3

Create a calculated field "Region Sales" (Shown in Fig. 6-66). This calculation is to find "AVERAGE" sales for "South" and "West" Region and "MEDIAN" sales for other regions.

Region Sales

```
IF [Region] = "SOUTH" AND [Region]="WEST" THEN
AVG(Sales)
ELSE
MEDIAN(Sales)
END
```

The calculation contains errors ▾ Apply OK

Figure 6-66. Calculated field "Region Sales" being created. Observe that the calculation contains errors

---

We cannot mix aggregate and non-aggregate functions in an expression. In such a situation, we can use ATTR function (Shown in Fig. 6-67).

Region Sales

```
IF ATTR([Region]) = "SOUTH" AND ATTR([Region])="WEST" THEN
AVG(Sales)
ELSE
MEDIAN(Sales)
END
```

The calculation is valid. Apply OK

Figure 6-67. ATTR() function being used with calculated field "Region Sales"

6.5.1.1.4 Step 4

The final output is shown in Fig. 6-68.

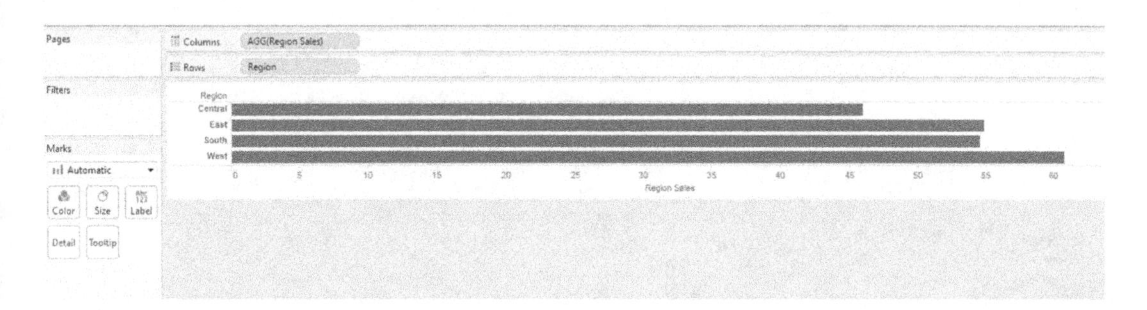

Figure 6-68. *ATTR() function - final output*

Attribute checks whether there is only one value for a given field for all rows in the result set. If there is only one value for the data selected, ATTR() function returns that value. If there is more than one value for that subset of data, it returns an asterisk.

Tableau computes attributes using the following formula:

IF MIN([dimension]) = MAX([dimension]) THEN MIN([dimension]) ELSE "*" END

6.6 Table calculation functions

Table calculations are computations that are applied to the entire table. Basically, table calculations are applied to values that come back from the database at some aggregation level.

Let us explore few table calculation functions.

6.6.1 First(), Index()

First()

This function returns the number of rows from the current row to the first row in the partition.

Index()

This function returns the index of the current row in the partition, without any sorting concerning the value. The index value starts at 1.

6.6.1.1 Steps to demonstrate the table calculations

Perform the following steps.

6.6.1.1.1 Step 1

Construct a view as shown in Fig. 6-69.

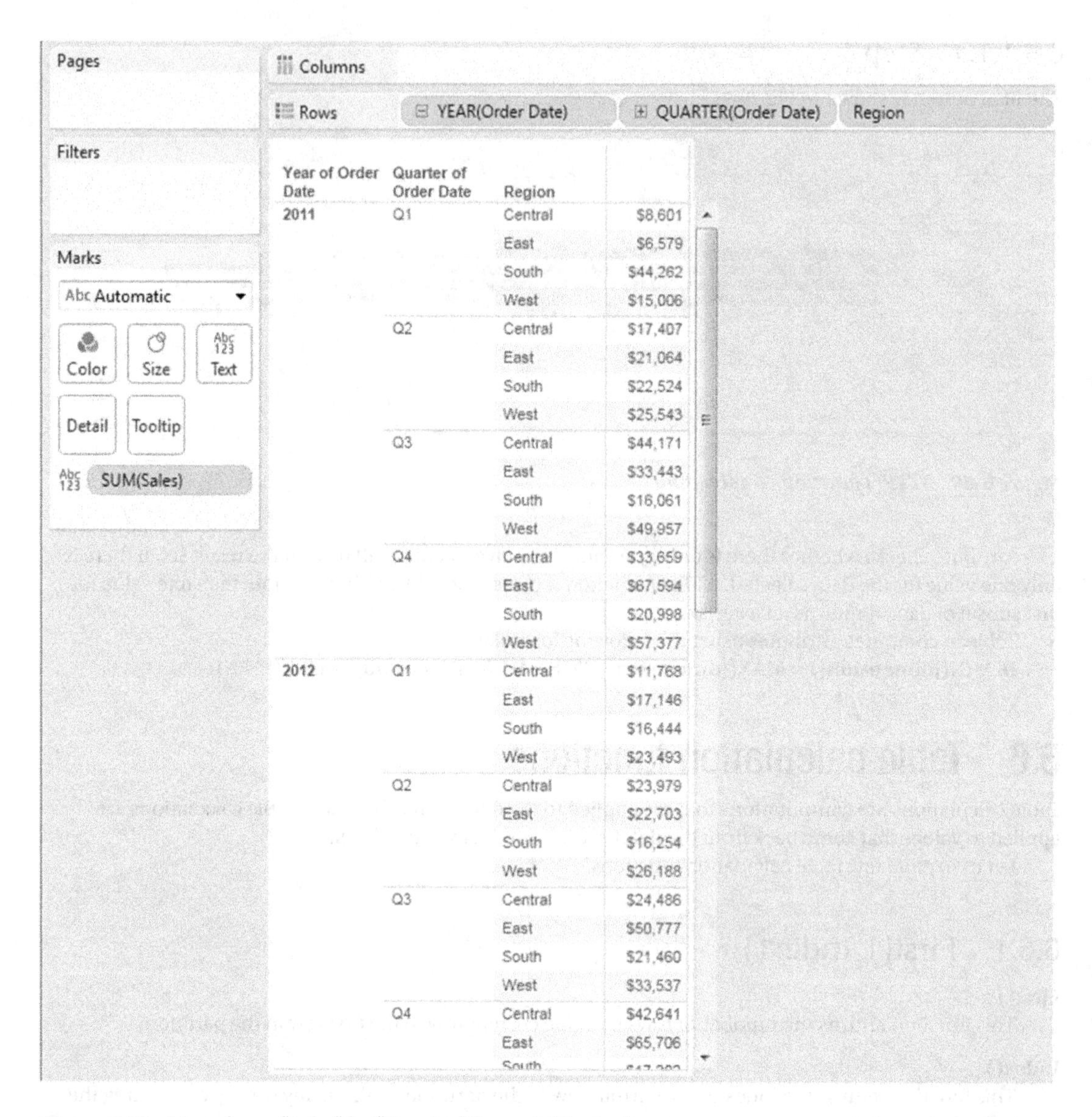

Year of Order Date	Quarter of Order Date	Region	
2011	Q1	Central	$8,601
		East	$6,579
		South	$44,262
		West	$15,006
	Q2	Central	$17,407
		East	$21,064
		South	$22,524
		West	$25,543
	Q3	Central	$44,171
		East	$33,443
		South	$16,061
		West	$49,957
	Q4	Central	$33,659
		East	$67,594
		South	$20,998
		West	$57,377
2012	Q1	Central	$11,768
		East	$17,146
		South	$16,444
		West	$23,493
	Q2	Central	$23,979
		East	$22,703
		South	$16,254
		West	$26,188
	Q3	Central	$24,486
		East	$50,777
		South	$21,460
		West	$33,537
	Q4	Central	$42,641
		East	$65,706

Figure 6-69. *View shows "Sales" by "Quarter" for each "Region"*

6.6.1.1.2 Step 2

Create a calculated field "First" as shown in Fig. 6-70.

Figure 6-70. *Calculated field "First" being created*

6.6.1.1.3 Step 3

Drag the calculated field "First" to the rows shelf (Shown in Fig. 6-71).

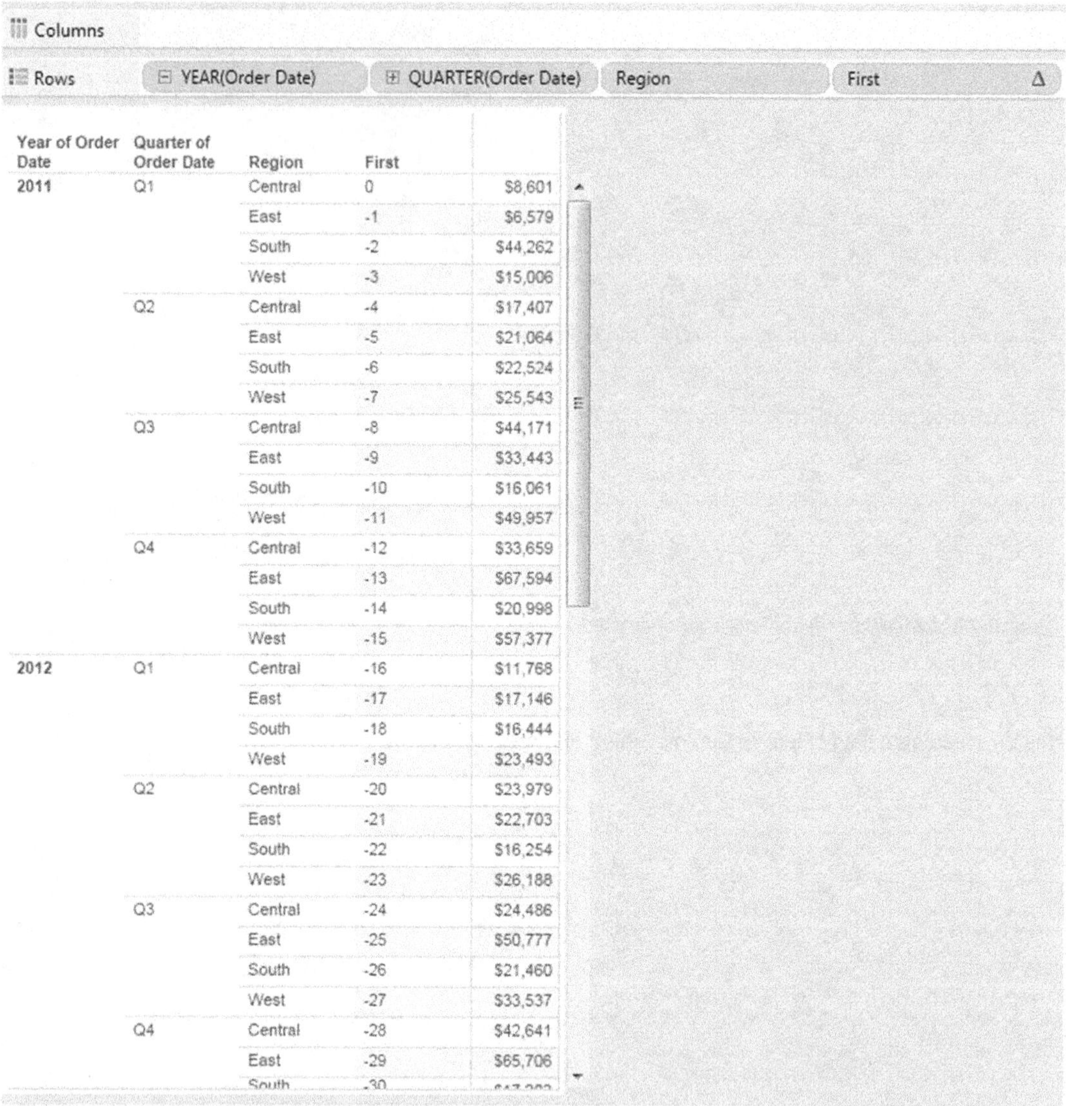

Columns

Rows: YEAR(Order Date) | QUARTER(Order Date) | Region | First Δ

Year of Order Date	Quarter of Order Date	Region	First	
2011	Q1	Central	0	$8,601
		East	-1	$6,579
		South	-2	$44,262
		West	-3	$15,006
	Q2	Central	-4	$17,407
		East	-5	$21,064
		South	-6	$22,524
		West	-7	$25,543
	Q3	Central	-8	$44,171
		East	-9	$33,443
		South	-10	$16,061
		West	-11	$49,957
	Q4	Central	-12	$33,659
		East	-13	$67,594
		South	-14	$20,998
		West	-15	$57,377
2012	Q1	Central	-16	$11,768
		East	-17	$17,146
		South	-18	$16,444
		West	-19	$23,493
	Q2	Central	-20	$23,979
		East	-21	$22,703
		South	-22	$16,254
		West	-23	$26,188
	Q3	Central	-24	$24,486
		East	-25	$50,777
		South	-26	$21,460
		West	-27	$33,537
	Q4	Central	-28	$42,641
		East	-29	$65,706
		South	-30	[illegible]

Figure 6-71. *Calculated field "First" placed on the rows shelf*

6.6.1.1.4 Step 4

Create a calculated field "Index" as shown in Fig. 6-72.

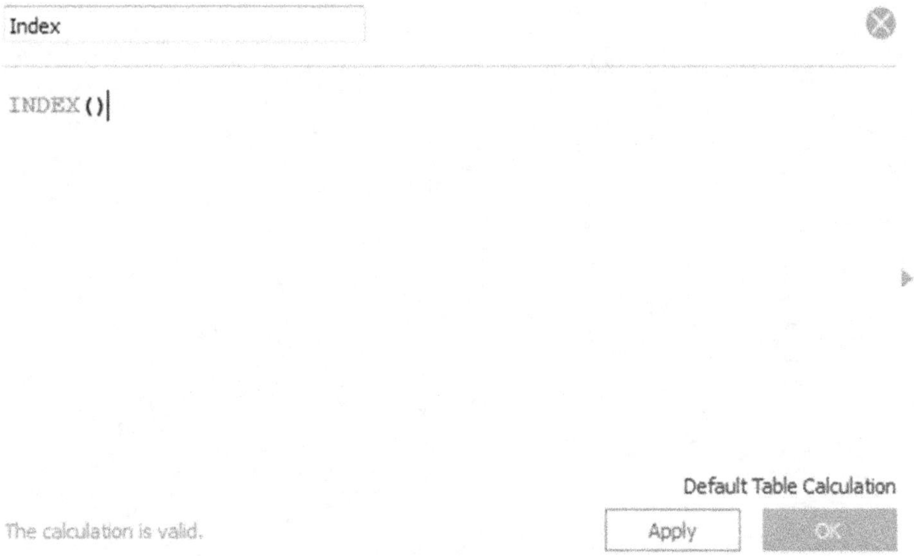

Index

```
INDEX()
```

Default Table Calculation

The calculation is valid.

Apply OK

Figure 6-72. *Calculated field "Index" being created*

6.6.1.1.5 Step 5

Drag calculated field "Index" to the rows shelf (Shown in Fig. 6-73).

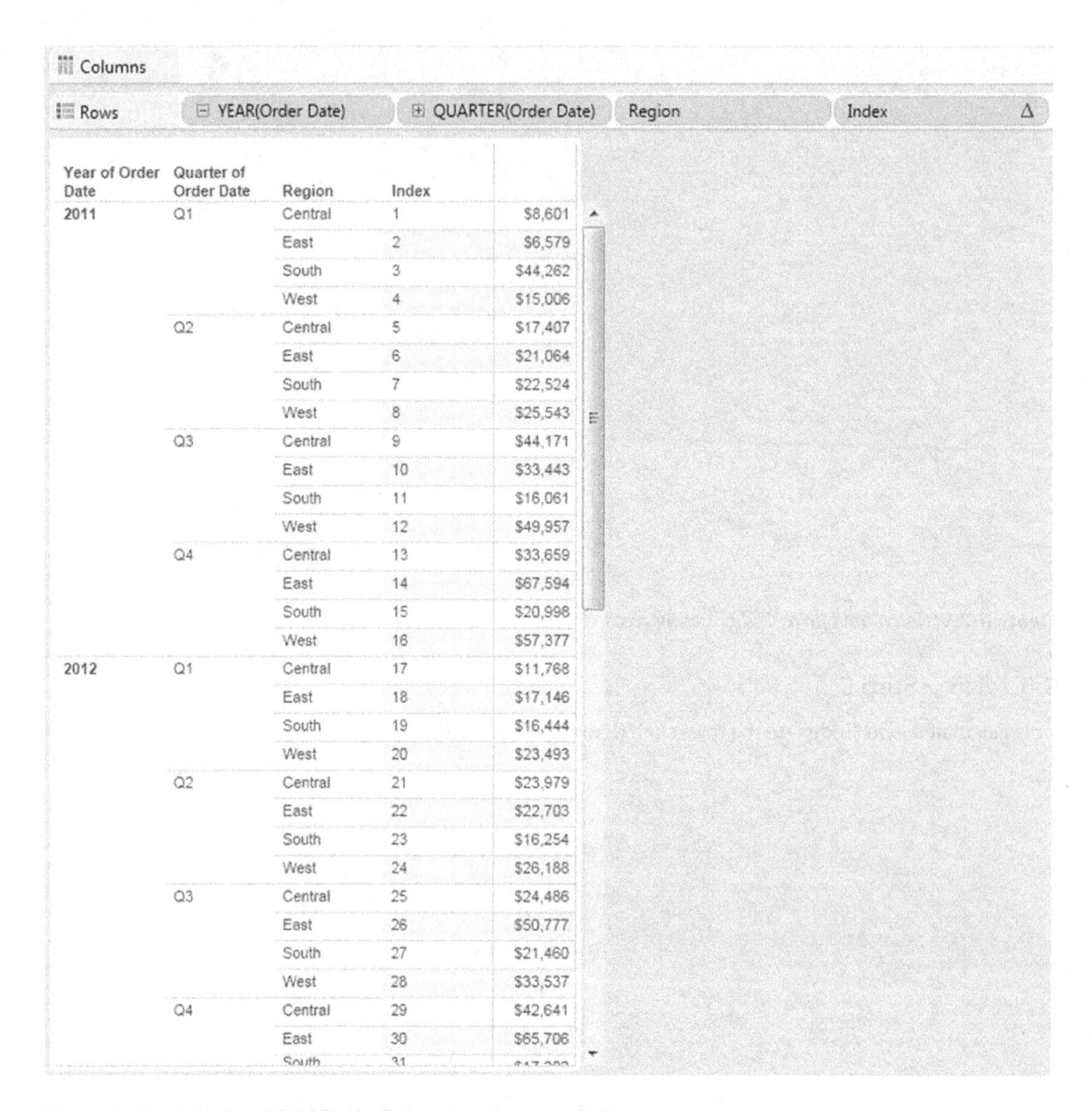

Year of Order Date	Quarter of Order Date	Region	Index	
2011	Q1	Central	1	$8,601
		East	2	$6,579
		South	3	$44,262
		West	4	$15,006
	Q2	Central	5	$17,407
		East	6	$21,064
		South	7	$22,524
		West	8	$25,543
	Q3	Central	9	$44,171
		East	10	$33,443
		South	11	$16,061
		West	12	$49,957
	Q4	Central	13	$33,659
		East	14	$67,594
		South	15	$20,998
		West	16	$57,377
2012	Q1	Central	17	$11,768
		East	18	$17,146
		South	19	$16,444
		West	20	$23,493
	Q2	Central	21	$23,979
		East	22	$22,703
		South	23	$16,254
		West	24	$26,188
	Q3	Central	25	$24,486
		East	26	$50,777
		South	27	$21,460
		West	28	$33,537
	Q4	Central	29	$42,641
		East	30	$65,706

Figure 6-73. *Calculated field "Index" placed on the rows shelf*

6.7 Points to remember

- Tableau supports a variety of functions such as number, string, logical and date functions, etc.
- Tableau also has support for user functions, type conversions, etc.
- Tableau computes attribute using the following formula:

IF MIN([dimension]) = MAX([dimension]) THEN MIN([dimension]) ELSE "*" END.

6.8 Next steps

In the next chapter, we will learn statistics. We will introduce the following:

- Basics of statistics
- Five magic number summary
- Box plot
- Statistics tools in Tableau
- Forecasting

CHAPTER 7

Statistics

Numbers have an important story to tell. They rely on you to give them a clear and convincing voice.

—Stephen Few, founder - Perceptual Edge

Chapter 6 introduced us to various functions in Tableau, such as string, number, date and logical functions. This chapter will help us to understand the basics of statistics and provide us with insights to perform advanced analysis in Tableau. We will explore the following:

- Why use statistics?
- What is statistics?
- Descriptive statistics
- Inferential statistics
- A few terms in statistics
- Why do we use inferential statistics?
- Why do we use descriptive statistics?
- Five magic number summary
- Spread of data
- Box plot
- Statistic tools in Tableau
- Forecasting

7.1 Why use statistics?

There are several branches in mathematics, including arithmetic, algebra, and geometry. Two subjects closely associated with mathematics "Statistics" and "Probability" come in very handy when one wishes to study, understand and analyze data.

Let us begin by analyzing few data-related questions that in order to answer may or may not need the concepts associated with statistics. Look at the following set of questions:

1. How much are your monthly expenditures?
2. Are data scientists paid more than IT programmers?

S. Acharya and S. Chellappan, *Pro Tableau*, DOI 10.1007/978-1-4842-2352-9_7

3. How fast can your dog run?
4. Do leopards run faster than tigers?
5. How much rain did Bangalore receive in July 2015?
6. Does it rain more in North India than in South India?
7. What is the probability of rain over the coming weekend?
8. What is the probability of flights taking off on schedule over the weekend?

Few of these questions are easier to answer. One can rely on recall or dig deep into information archives to find the answer. However questions 2, 4, and 6 have variability; there can / may be more data points and the questions demand the application of statistics to answer them.

The last two questions will require the concepts of probability to predict the possibility using a large set of data points garnered over a period of time.

Let us consider few business-related practical scenarios and answer questions that require statistics.

- Your city has recently witnessed the outbreak of an epidemic caused by a virus. You are a research scholar studying medicine. You have been working on a medicine that could possibly counter the effects of the virus. You have collected data from patients to whom your medicine was administered. Based on your study, analysis and evaluation, you are required to determine if your medicine can be labeled as effective in curing the effects of the virus.
- You work at the packaging unit of a water distillery in the capacity of a quality controller. Your unit packages 10,000 bottles of distilled water every day. The bottles are 1 litre each. Last Wednesday, the consignment delivered at a resort was rejected owing to the bottles being of smaller size than the usual 1 litre size. It is your responsibility to find out if there are more bottles of smaller size that were shipped. If yes, then how many and to which destinations.
- You are on the marketing team of an online retailer. You have been asked to spearhead digital campaigns that can advertise the products appropriately, offer suitable deals and promise 500 new customers every week. You have worked up a few campaigns and are thinking of the probability of success of these campaigns.
- You are the service in-charge of a motorcycle repair shop. You have seen several customers complain about the quality of service. This has you worried and you want a way to determine how many customers are likely to switch over to your competitor in the next three months.

Probability can come to your rescue by providing "Numbers" that will help you make decisions for remedial actions in business.

Let us look at the actions that can be taken in the above mentioned business scenarios.

- The drug researcher may need to focus on plan B if his current drug is ineffective in curing the patients.
- The quality controller will need to schedule recall of the dispatched consignments and check with the bottle suppliers on the size of the bottles.
- The service in-charge may want to talk to dissatisfied customers immediately to prevent churn.

Statistics and probability have a very significant role to play in analyzing business data and supporting decision-making.

7.2 What is statistics?

For a layman, "statistics" implies numerical information expressed in quantitative terms. This information may relate to objects, processes, business activities, scientific phenomena or sports.

In simpler terms, statistics can be defined as the science of collecting large number of facts (or real-world observations) and analyzing them with the purpose of summarizing the collection and drawing inferences.

Probability can be defined as a measure or estimate of the degree of confidence one may have in the occurrence of an event, measured on a scale of impossibility to certainty. It may be defined as the proportion of favorable outcomes to the total number of possibilities.

We study statistics under two broad areas.

- Summarization or aggregation or descriptive statistics and
- Probability statistics or inferential statistics.

7.3 Descriptive statistics

Descriptive statistics allows one to show, describe or present data in a meaningful way such that patterns might emerge from the data. It can only be used to describe the group that is being studied. It cannot be generalized to a larger group. In other words, it is just a way to describe our data. For example:

- Measures of central tendency, such as mean, median and mode.
- Measures of spread, such as range, IQR (inter-quartile range), variance and standard deviation.

7.4 Inferential statistics

Inferential statistics helps one to make predictions or inferences about a population based on the observation or analysis of a sample. However, it is imperative that the sample is representative.

Inferential statistics can be used for two purposes: to aid scientific understanding by estimating the probability that a statement is true or not, and to aid in making sound decisions by estimating which alternative among a range of possibilities is most desirable.

It is important to note that if the raw data sets are of poor quality, probabilistic and statistical calculations will not be very useful. Hence, decisions based on such erroneous foundations will be flawed.

There are *n* number of tools available in statistics and probability, such as t-test, Z-test, F-test, histogram, rank and percentile calculation, sampling, curve fitting, correlation, covariance, regression, random number generation, ANOVA, etc.

7.5 Few terms in statistics

Population: Any set of people or objects with something in common could be a population. If we are studying the feedback of students of a VI grade class and if the class size was 50, we need to study the feedback given by all 50 students of the VI grade.

Sample: A sample is a subset of a population. For example, we are required to study the feedback of students in a class. The class size is 50; we divide the class into high performers, average performers and poor performers. Then study the feedback of at least five students in each group (which means 15 students in all).

When we measure something in a population, it is called a parameter. When we measure something in a sample, it is called a statistic. A population is to a parameter as a sample is to a statistic.

For example, out of the 350 randomly selected people in the city of Pune, India, 250 people had the last name "Sen". An example of descriptive statistics is the following statement:

"80% of these people have the last name "Sen"".

For example, on the last 3 Sundays, Thomas sold 5, 4, and 3 new cars, respectively. An example of descriptive statistics is the following statement:

"Thomas averaged 4 new car sold for the last 3 Sundays."

These are both descriptive statements because they can actually be verified from the information provided.

For example, out of the 350 randomly selected people in the city of Pune, India, 250 people had the last name "Sen". An example of inferential statistics is the following statement:

"80% of all people living in India have the last name "Sen"".

We have no information about all people living in India, just about the 250 living in Pune. We have taken that information and generalized it to talk about all people living in India. The easiest way to tell that this statement is not descriptive is by trying to verify it based upon the information provided.

For example, on the last 3 Sundays, Thomas sold 5, 4, and 3 new cars respectively. The following statements are examples of inferential statistics:

"Thomas never sells more than 5 cars on a Sunday."

Although this statement is true for the last 3 Sundays, we do not know that this is true for all Sundays.

"Thomas is selling fewer cars lately because people have caught on to his dirty tricks."

There is nothing in the information given that tells us that this statement is true.

The major use of inferential statistics is to use information from a sample to infer something about a population.

7.6 Why do we use inferential statistics?

We use inferential statistics to generalize from our samples and descriptive statistics to describe our samples.

7.7 Why do we use descriptive statistics?

It is about describing the data in such a way that it becomes meaningful to whomever it is presented, without sharing the actual set of data. If we do not share the actual set of data, then what is it that we share? It could be the summary information about the data such as the mean, median and mode of the set of data.

Let us understand this with a scenario.

John is a transport manager. He is looking at introducing an office conveyance to ferry employees to the office. He is keeping a record of the miles that an employee commutes to work. Here is the data that he collected.

10 7 5 9 6 3 3 8 9 10 10

7.7.1 What is the measure of central tendency here?

7.7.1.1 Mean or arithmetic mean

Arithmetic mean is computed by summing up all the numbers and then dividing it by the total numbers in the data set.

10+7+5+9+6+3+3+8+9+10+10 / 11

= 80/11

= 7.27

7.7.1.2 Median

Median is just the middle value.

Arrange the numbers in ascending order: 3 3 5 6 7 8 9 9 10 10 10

The median value is at position 6. The value at position 6 is the number 8.

7.7.1.3 Mode

Mode is the given by the number which appears the most often in the data set

Given the data set: 3 3 5 6 7 8 9 9 10 10 10

The number 10 appears three times in the data set, which is more often than any other number in the data set.

Therefore, the mode of the data set is 10.

Mode has applications in printing. For example, it is important to print more of the most popular books, because printing different books in equal numbers would cause a shortage of some books and an oversupply of others.

Likewise, mode has applications in manufacturing. For example, it is important to manufacture more of the most popular shoes, because manufacturing different shoes in equal numbers would cause a shortage of some shoes and an oversupply of others.

7.7.1.4 Few practice examples

1. Given below are the number of baskets scored by the members of the national basketball team (Shown in Table 7-1).

Table 7-1. *National Basketball Team*

Name of the player	Number of baskets scored
Adrian	3
Jeff	6
Kris	6
Edwin	2
David	1
Thomas	7

Arithmetic Mean: 4.166
Median: 4.5
Mode: 6

2. A class test was conducted for 6th graders. The class had nine students. Here are their scores on 100:

Given below are the set of nine scores: 45 55 66 54 75 82 91 79 98
Median: 75
Arithmetic mean: 71.66
Mode: NA

3. In his mathematics class, Soham took six quizzes. His scores were 90, 92, 91, 89, 84 and 82. What was his average score on the quizzes?

Answer: (90 + 92 + 91 + 89 + 84 + 82) / 6
= 528/6
= 88
His average score was 88.

4. Jeet has taken six quizzes and his average score so far is 85. If he gets 100, i.e. a perfect score, on the remaining four quizzes, what will be his new average?

Step 1

Six quizzes with an average score of 85.

6 * 85 = 510

Step 2

Four quizzes with a perfect score of 100

4 * 100 = 400

Step 3

New average = total score / total no. of quizzes

New Average = 910 / 10
New Average = 91

5. On the first three quizzes of his Algebra class, Rohan got an average score of 84. What does he need on the next quiz to have an overall average of 87?

Step 1

Three quizzes with an average score of 84.

3 * 84 = 252

Step 2

Let us say he scores x in his next quiz.

That is 252 + x is his total score after the next quiz.

Step 3

To get an average score of 87, he needs

252 + x = 4 * 87
252 + x = 348
Therefore x = 348 – 252
= 96

6. On the first two quizzes of his geometry class, Wendy got an average score of 75. What does she need on the next quiz to have an overall average of 82?

Step 1

Two quizzes with an average score of 75.

2 * 75 = 150

Step 2

Let us say she scores x in her next quiz.

That is 150 + x is her total score after the next quiz.

Step 3

To get an average score of 82, she needs

150 + x = 3 * 82

150 + x = 246

Therefore x = 246 - 150

= 96

7.8 Five magic number summary

The five number summary includes the following:

- Minimum
- Maximum
- Median
- First quartile
- Third quartile

7.8.1 Mean

Refer to Table 7-2.

Table 7-2. *Age of the class*

Age	19	20	21
Frequency	1	3	1

Mean age of the class: 20

Mean is Σx / n, where *x* represents data values and *n* represents the total number of data values. Histogram for the age of the class is shown in Fig. 7-1.

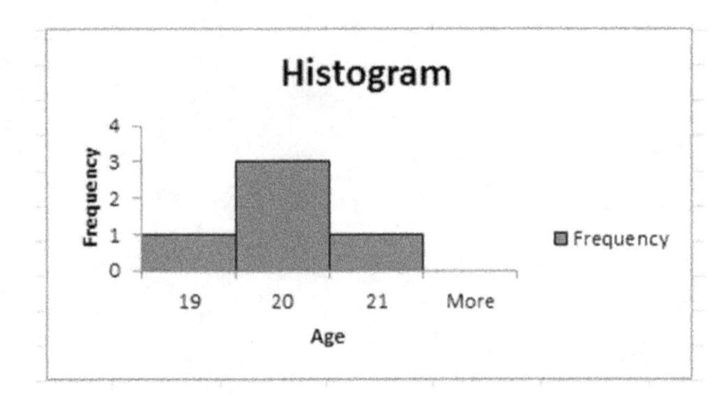

Figure 7-1. Age of the class

7.8.2 Median

Data values: 19 20 20 20 21

The median is always in the middle. It is the middle value, i.e. 20.

Another example:

Data values: 19 20 20 20 21 21 100 102

Median: (20+21)/2 = 20.5

Points to note:

- If you have an odd number of values, the median is the one in the middle. If you have n numbers, the middle number is at position (n + 1) / 2.
- If you have an even number of values, get the median by adding the two middle ones together and dividing by 2. You can find the midpoint by calculating (n + 1) / 2. The two middle numbers are on either side of this point.

7.8.2.1 Center of data - median

Refer to Table 7-3.

Table 7-3. Center of data - median data set

Values	1	2	3	4	5	6	7	8
Frequency	4	6	4	4	3	2	1	1

There are 25 numbers, and if you line them all up, the median is half way along, i.e., 13 numbers along. The median is 3. The data is skewed to the right, which pulls the mean higher. Therefore, the mean is higher than the median (Shown in Fig. 7-2).

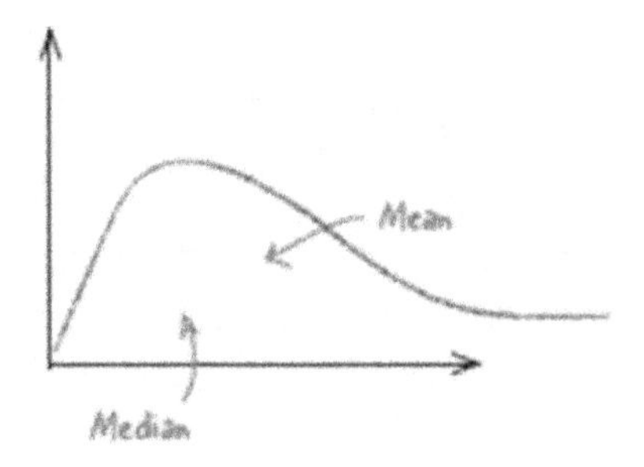

Figure 7-2. *Mean is higher than the median*

Refer to Table 7-4.

Table 7-4. *Center of data - median data set*

Values	1	4	6	8	9	10	11	12
Frequency	1	1	2	3	4	4	5	5

The median here is 10. The data is skewed to the left, so the mean is pulled to the left. Therefore, the mean is lower than the median (Shown in Fig. 7-3).

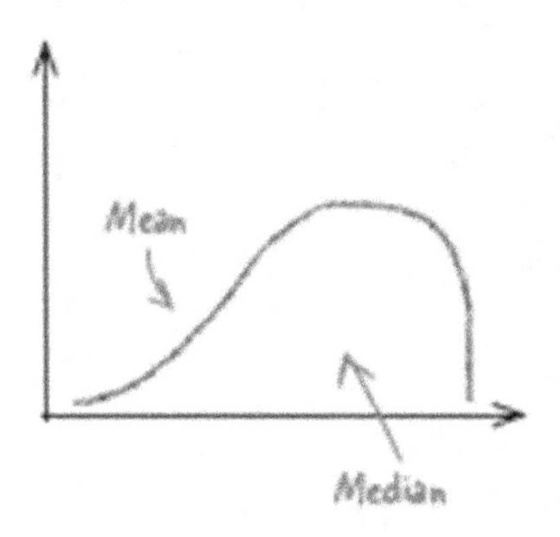

Figure 7-3. *The mean is lower than the median*

7.8.3 Mode

The mode has to be in the data set. It is the only average that works with categorical data. Refer to Table 7-5.

Table 7-5. *Mode data set*

Values	1	2	3	31	32	33
Frequency	3	4	2	2	4	3

Refer to Fig. 7-4.

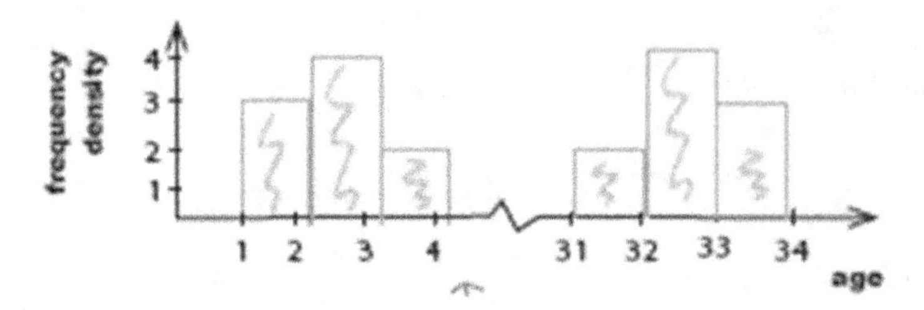

Figure 7-4. *Frequency density*

7.8.4 When to use which average?

Refer to Table 7-6.

Table 7-6. *When to use which average?*

Average	When the data is fairly symmetrical
Mean	When the data is skewed because of outliers
Median	When the data shows two or more clusters
Mode	While working with categorical data

7.9 Spread of data

Spread of data includes the following:

- Range
- Interquartile range
- Variance
- Standard deviation

7.9.1 Range

The range is a way of measuring how spread out a set of values is. It is given by an upper bound - lower bound where the upper bound is the highest value, and the lower bound the lowest.

7 8 9 9 10 10 11 12 13

Range = upper bound - lower bound

= 13 – 7

= 6

Thus, the range of this set of data is 6.

The primary problem with the range is that it only describes the width of your data. Because the range is calculated using the most extreme values of the data, it is impossible to tell what that data actually looks like and whether it contains outliers.

7.9.2 Interquartile range

Interquartile range = Upper quartile – Lower quartile

Quartiles are values that split your data into quarters. The lowest quartile is called the lower quartile, and the highest quartile is called the upper quartile. The middle quartile is the median.

Finding the position of the lower quartile

Calculate n ÷ 4. (If not an integer, round it off)

Finding the position of the upper quartile

Calculate 3n ÷ 4. (If not an integer, round it off)

Example:

Data values: 3 3 6 7 7 10 10 10 11 13 30

Median = 10

Lower quartile = 6

Upper quartile = 11

Interquartile range = upper quartile – lower quartile

Interquartile range = 11 - 6 = 5

7.9.3 Variance and standard deviation

Variance is defined as the "average of the squared differences from the mean".

Steps to compute variance:

- Determine the mean of the numbers.
- For each number, subtract the mean and square the difference (squared differences).
- Determine the average of those squared differences.

Refer to Table 7-7.

Table 7-7. *Data set*

600
470
170
430
300

1. Determine the mean of the numbers

   Mean: (600 + 470 + 170 + 430 + 300) / 5 = 394

2. For each number, subtract the mean and square the difference (squared differences)

   Mean: (600 + 470 + 170 + 430 + 300) / 5 = 394. Refer to Table 7-8.

Table 7-8. *Square differences*

600 - 394	206
470 - 394	76
170 - 394	-224
430 - 394	36
300 - 394	-94

3. Determine the average of those squared differences

   = (42436 + 5776 + 50176 + 1296 + 8836) / 5

   = (108520)/5 = 21704

   Variance is 21704.

7.9.4 Standard deviation

Standard deviation is just the square root of variance.

The square root of the variance, 21704 is 147.

Standard deviation is 147.

Standard deviation is a measure of disbursement in statistics. "Disbursement" just means how much your data is spread out. Specifically, it shows you how much your data is spread out around the mean or average.

In the example stated the data was for a population (the five students whose scores we considered were the only students in which we were interested).

However, if the data is a sample (a selection taken from a bigger population), then there is a change in the calculation!

When you have "N" data values that are:

- The population: divide by *N* when calculating variance (like we did)
- A sample: divide by N-1 when calculating variance

Going by the above, if the data is considered for a population, the standard deviation is 147.

However if the data is considered for a sample, the standard deviation is 164.

A quartile is one of three values (lower quartile, median and upper quartile) which divided data into four equal groups.

A percentile is one of 99 values that divided data into 100 equal groups.

The lower quartile corresponds to the 25th percentile. The median corresponds to the 50th percentile. The upper quartile corresponds to the 75th percentile.

7.9.5 Assignment 1

Objective: A gardener buys 10 packets of seeds from two different companies. Each pack contains 20 seeds and he records the number of plants that grow from each pack. Refer to Table 7-9.

Table 7-9. *Seeds data set*

Company	Packet No.	No of plants from the packet
Company A	1	20
	2	5
	3	20
	4	20
	5	20
	6	6
	7	20
	8	20
	9	20
	10	8
Company B	1	17
	2	18
	3	15
	4	16
	5	18
	6	18
	7	17
	8	15
	9	17
	10	18

- Find the mean, median and mode for each company's seeds.
- Which company does the "mode" suggest is the best?
- Which company does the "mean" suggest is the best?
- Find the range for each company's seeds.

7.9.6 Assignment 2

Objective: The scores of five students (Roll No ranging from 1 to 5) in three subjects, "Introduction to BI", "Statistics", and "Introduction to Analytics" are given below. Compute the following for EACH subject:

- Sum
- Mean
- Median
- Count (The number of students who took the test in the subject)
- StdDeviation
- Variance
- Min
- Max

Data Set is as follows: Refer to Table 7-10.

Table 7-10. *StatsData.xlsx*

	A	B	C
1	**RollNo**	**SubjectName**	**Score**
2	1	Introduction to BI	78
3	2	Introduction to BI	82
4	3	Introduction to BI	81
5	4	Introduction to BI	67
6	5	Introduction to BI	83
7	1	Statistics	69
8	2	Statistics	74
9	3	Statistics	78
10	4	Statistics	82
11	5	Statistics	85
12	1	Introduction to Analytics	82
13	2	Introduction to Analytics	87
14	3	Introduction to Analytics	90
15	4	Introduction to Analytics	85
16	5	Introduction to Analytics	83

7.9.6.1 Solution

7.9.6.1.1 Step 1

Read in the data from "StatsData.xlsx" into Tableau (Shown in Fig. 7-5).

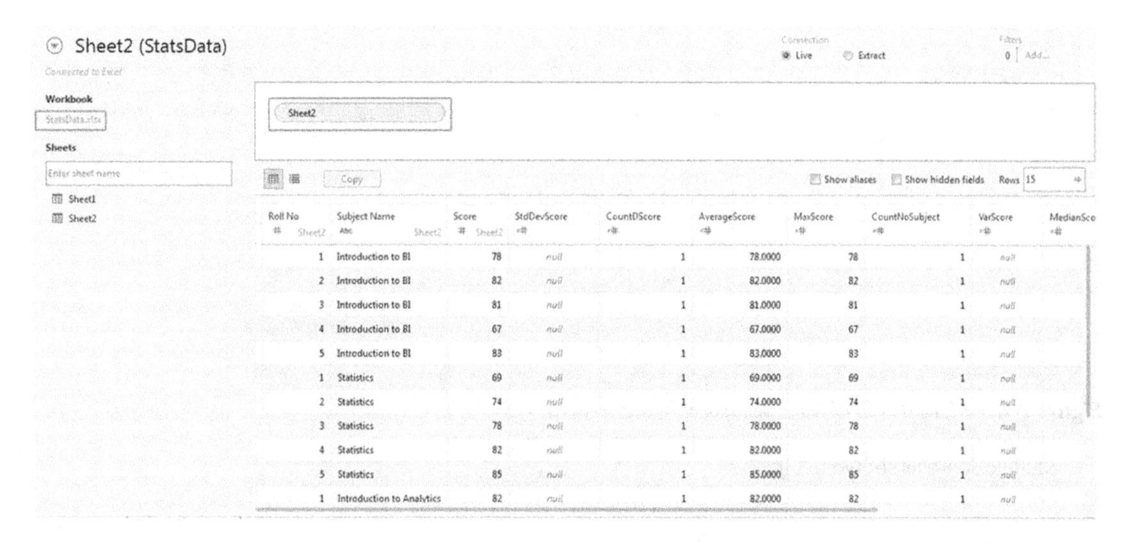

Roll No	Subject Name	Score	StdDevScore	CountDScore	AverageScore	MaxScore	CountNoSubject	VarScore	MedianSco
1	Introduction to BI	78	null	1	78.0000	78	1	null	
2	Introduction to BI	82	null	1	82.0000	82	1	null	
3	Introduction to BI	81	null	1	81.0000	81	1	null	
4	Introduction to BI	67	null	1	67.0000	67	1	null	
5	Introduction to BI	83	null	1	83.0000	83	1	null	
1	Statistics	69	null	1	69.0000	69	1	null	
2	Statistics	74	null	1	74.0000	74	1	null	
3	Statistics	78	null	1	78.0000	78	1	null	
4	Statistics	82	null	1	82.0000	82	1	null	
5	Statistics	85	null	1	85.0000	85	1	null	
1	Introduction to Analytics	82	null	1	82.0000	82	1	null	

***Figure 7-5.** Data from "StatsData.xlsx" read into Tableau*

7.9.6.1.2 Step 2

Create the calculated fields below:

SumScore is shown in Fig. 7-6.

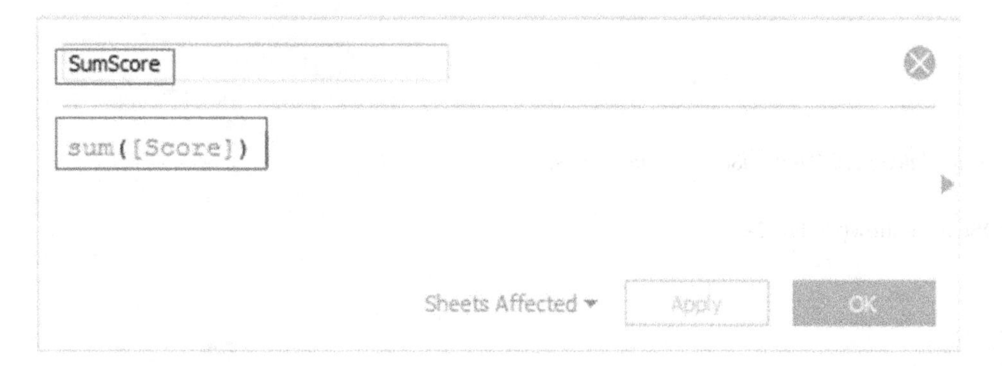

***Figure 7-6.** Calculated field "SumScore" being created*

AverageScore as shown in Fig. 7-7.

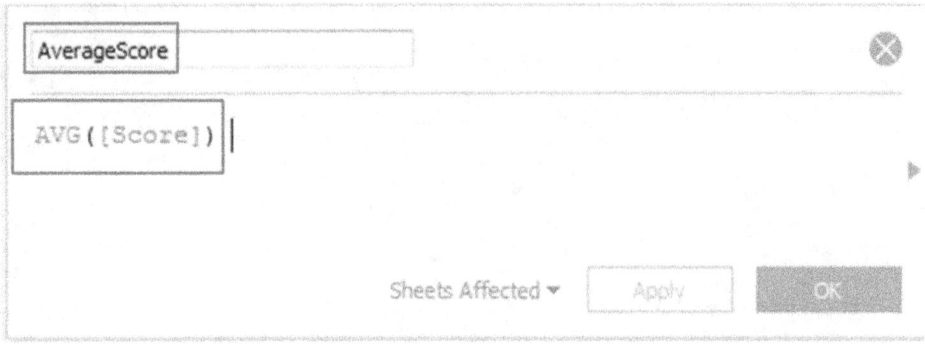

AverageScore

AVG([Score])

Sheets Affected ▾ Apply OK

Figure 7-7. *Calculated field "AverageScore" being created*

MedianScore as shown in Fig. 7-8.

MedianScore

median([Score])

Sheets Affected ▾ Apply OK

Figure 7-8. *Calculated field "MedianScore" being created*

CountDScore is shown in Fig. 7-9.

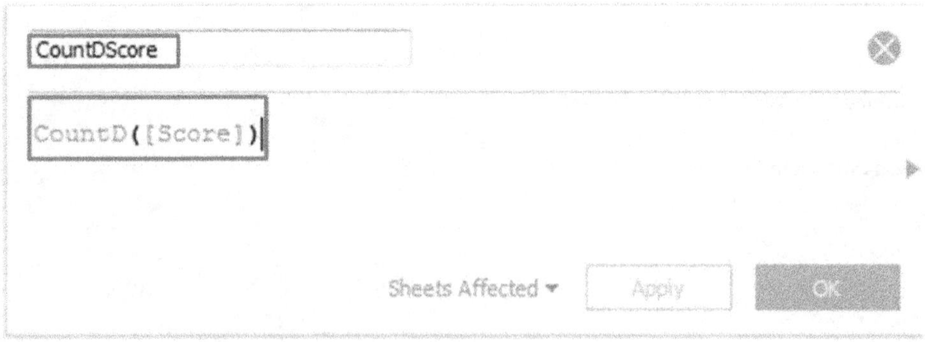

CountDScore

CountD([Score])

Sheets Affected ▾ Apply OK

Figure 7-9. *Calculated field "CountDScore" being created*

StdDevScore as shown in Fig. 7-10.

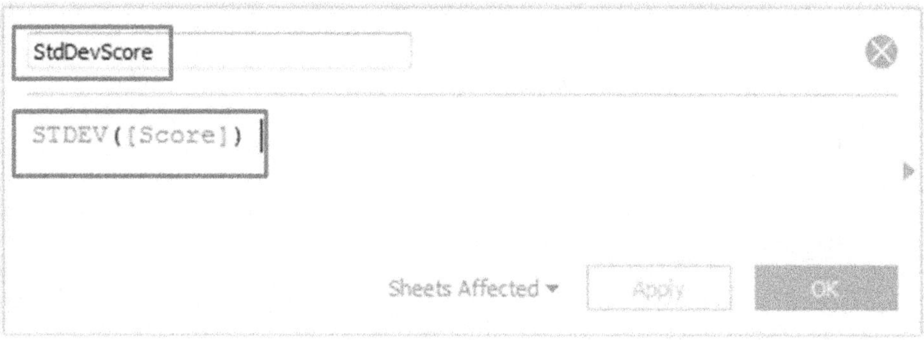

StdDevScore

STDEV([Score])

Sheets Affected ▾ Apply OK

Figure 7-10. *Calculated field "StdDevScore" being created*

VarScore as shown in Fig. 7-11.

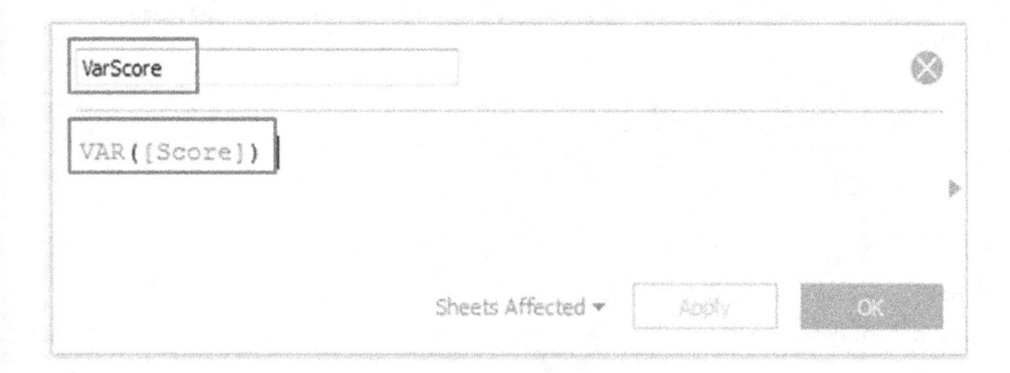

Figure 7-11. *Calculated field "VarScore" being created*

MinScore as shown in Fig. 7-12.

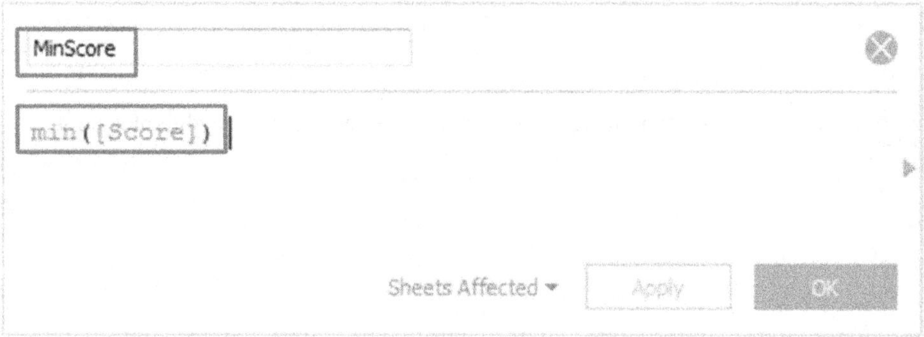

MinScore

min([Score])

Sheets Affected ▾ Apply OK

Figure 7-12. *Calculated field "MinScore" being created*

MaxScore as shown in Fig. 7-13.

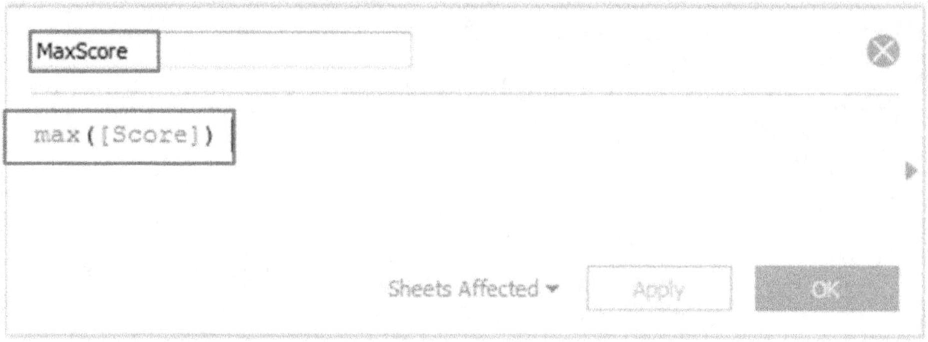

MaxScore

max([Score])

Sheets Affected ▾ Apply OK

Figure 7-13. Calculated field "MaxScore" being created

7.9.6.1.3 Step 3

Drag the dimension "Subject Name" from the dimensions area under the data pane to the rows shelf (Shown in Fig. 7-14).

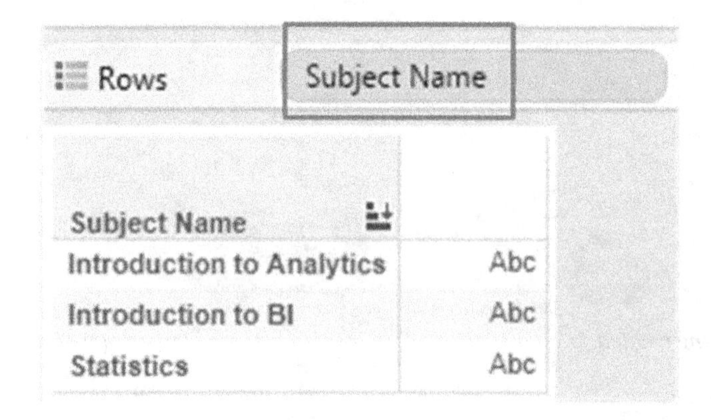

Figure 7-14. Dimension "Subject Name" placed on the rows shelf

7.9.6.1.4 Step 4

Drag the dimension "Measure Names" from the dimensions area under the data pane to the columns shelf (Shown in Fig. 7-15).

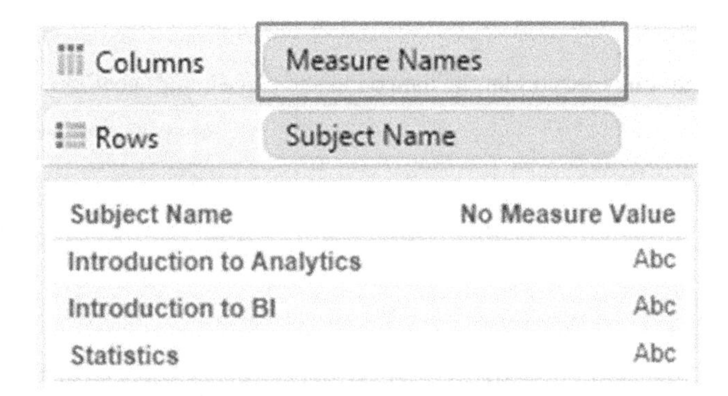

Figure 7-15. *Dimension "Measure Names" placed on the columns shelf*

7.9.6.1.5 Step 5

Drag the dimension "Measure Names" from the dimensions area under the data pane to the filters shelf (Shown in Fig. 7-16).

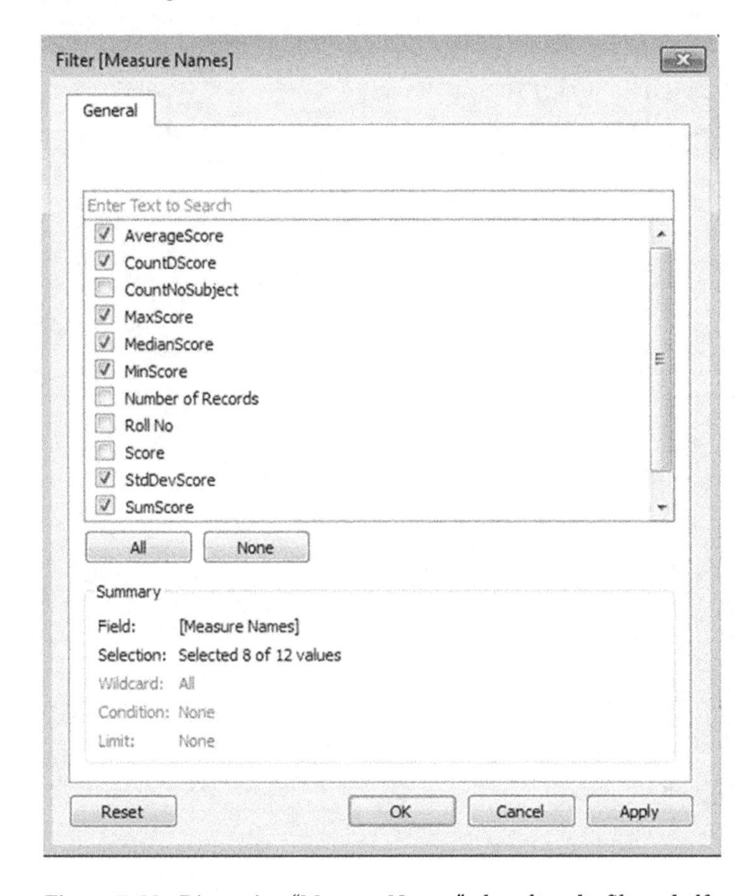

Figure 7-16. *Dimension "Measure Names" placed on the filters shelf*

Select the values to be displayed on the view / worksheet.

7.9.6.1.6 Step 6

Drag the measure "Measure Values" from the measures area under the data pane and place it on "Label" on the marks card (Shown in Fig. 7-17).

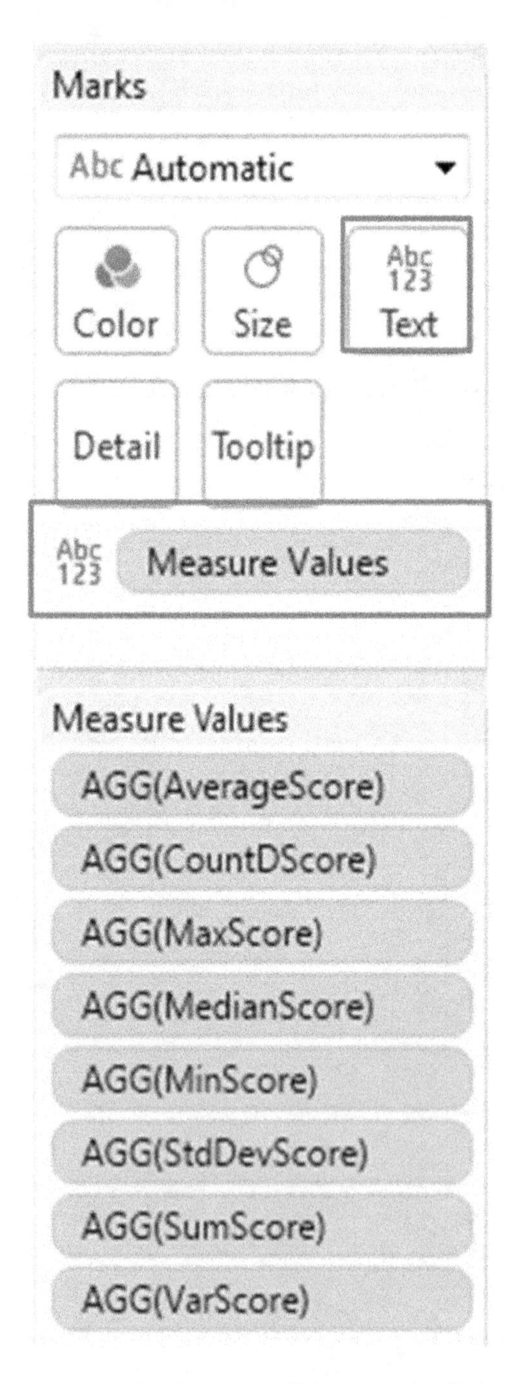

Figure 7-17. *The measure "Measure Values" placed on the "Label" on the marks card*

The final output: Shown in Fig. 7-18.

Subject Name	AverageScore	CountDScore	MaxScore	MedianScore	MinScore	StdDevScore	SumScore	VarScore
Introduction to Analytics	85.4	5.0	90.0	85.0	82.0	3.2	427.0	10.3
Introduction to BI	78.2	5.0	83.0	81.0	67.0	6.5	391.0	42.7
Statistics	77.6	5.0	85.0	78.0	69.0	6.3	388.0	40.3

Figure 7-18. *Final output*

7.10 Box plot

It is also called a box and whiskers plot. It was invented by John Tukey, father of exploratory data analysis, in 1977.

Purpose of box plot: To efficiently display the five magic numbers or statistical measures. Refer to Fig. 7-19.

- Minimum or low value
- Lower quartile or 25th percentile
- Median or 50th percentile
- Upper quartile or 75th percentile
- Maximum or high value

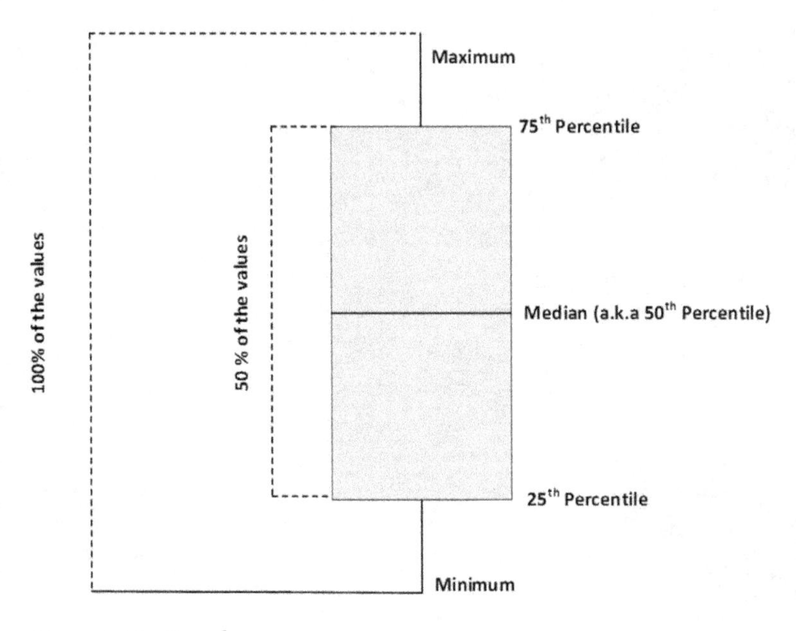

Figure 7-19. *Box plot*

Box plot

- Can be drawn either vertically or horizontally.
- Often used in conjunction with histogram.

Advantages of a box plot:

- Provides a fair idea about the data's symmetry and skewness (Shown in Figs. 7-20, 7-21 and 7-22).
- It shows outliers.
- Allows for an easy comparison of data sets

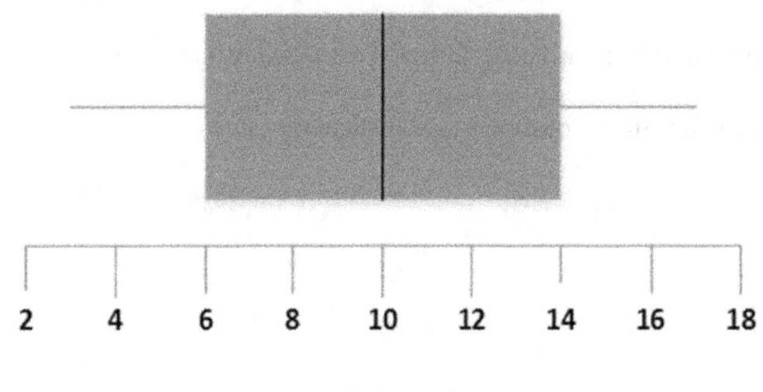

Figure 7-20. *Symmetric*

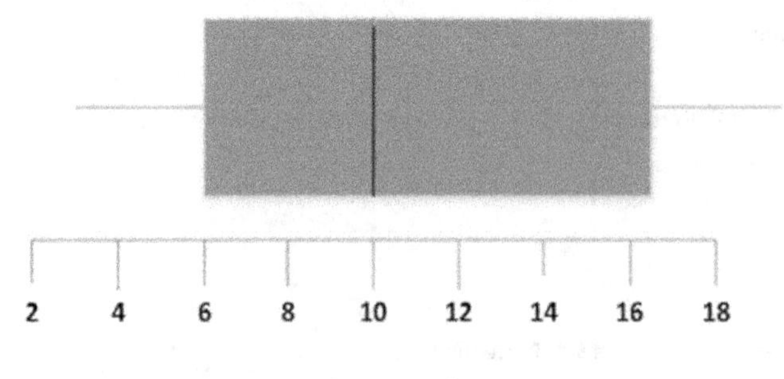

Figure 7-21. *Skewed right*

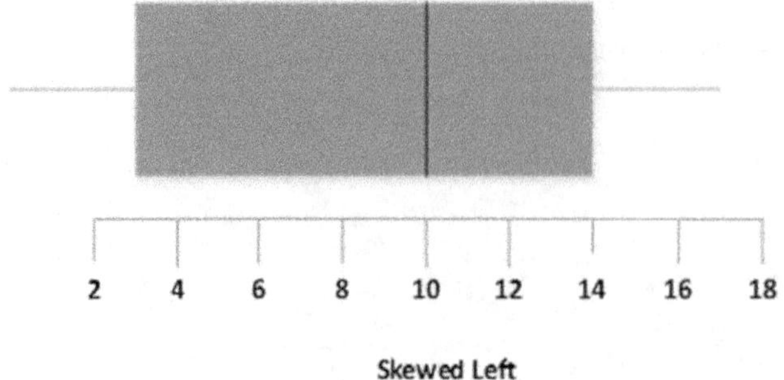

Figure 7-22. *Skewed left*

7.10.1 Plotting box and whiskers plot in Tableau

Seventeen students of sixth grade were given a quiz on "data visualization" and were given a score on a scale of 0 to 5. The following is the dataset: Refer to Fig. 7-23.

A	B
RollNo	Scores
1	4.3
2	5.1
3	3.9
4	4.5
5	4.4
6	4.9
7	5
8	4.7
9	4.1
10	4.6
11	4.4
12	4.3
13	4.8
14	4.4
15	4.2
16	4.5
17	4.4

Figure 7-23. *BoxPlot.xlsx*

Let us plot a box and whisker plot in Tableau.
The above dataset is read into an excel sheet, "BoxPlot.xlsx".

7.10.1.1 Steps to create a box and whiskers plot

Follow the following steps.

7.10.1.1.1 Step 1

Bring in the data from the excel sheet, "BoxPlot.xlsx" into Tableau (Shown in Fig. 7-24).

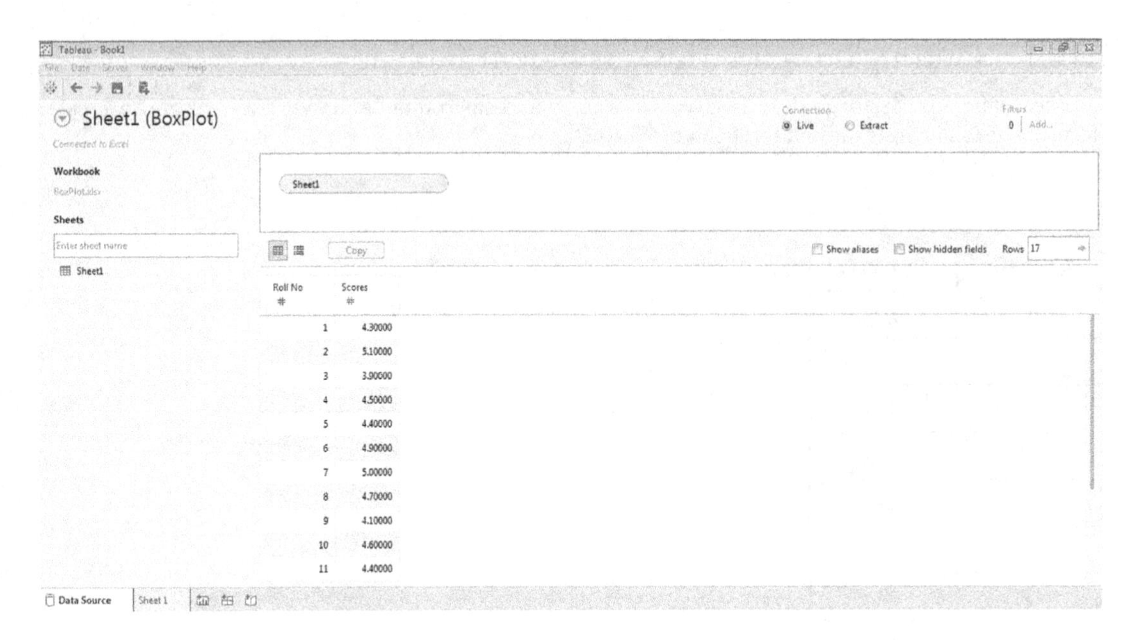

Figure 7-24. Read in data from "BoxPlot.xlsx"

7.10.1.1.2 Step 2

Select the dimension "RollNo" and the measure "Scores". Select "box-and-whisker plot" from "Show Me" (Shown in Fig. 7-25).

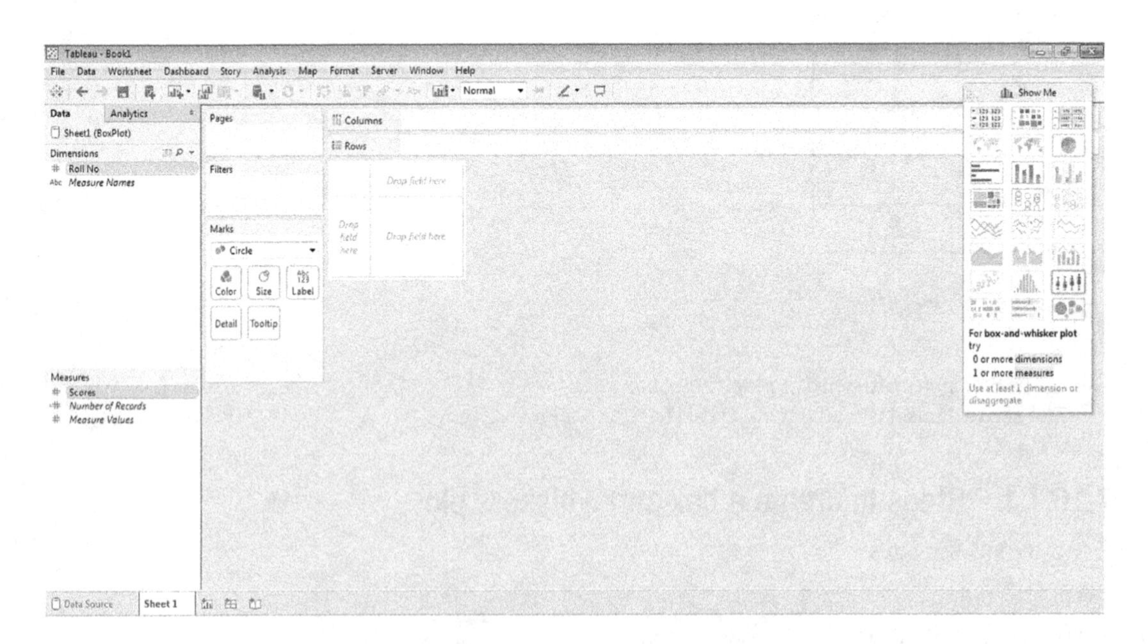

Figure 7-25. Selection of dimension "Roll No", measure "Scores" and "box-and-whisker" plot

7.10.1.1.3 Step 3

A box plot is plotted as shown in Fig. 7-26.

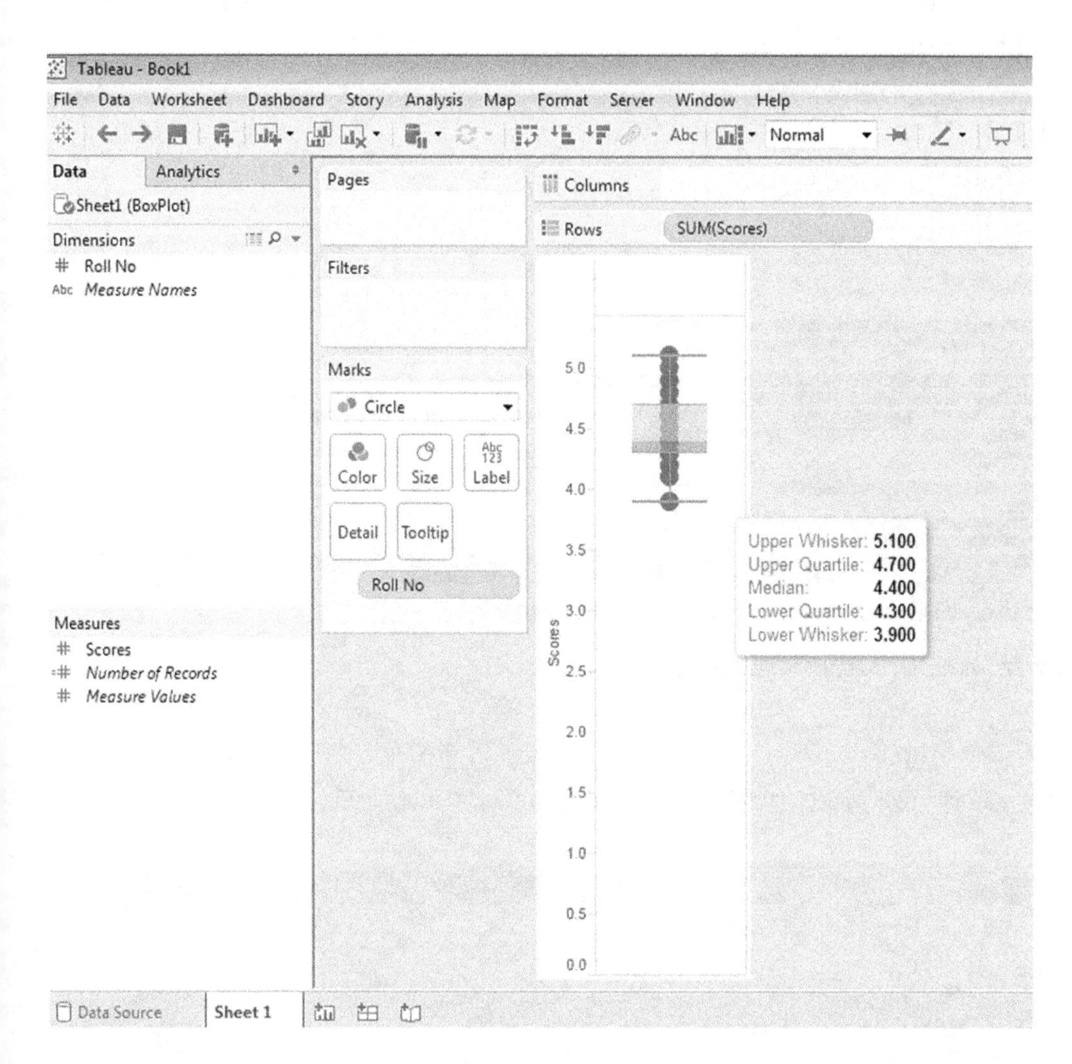

Figure 7-26. *Box plot*

7.11 Statistics tools in Tableau

Let us discuss how to use built-in statistics tools in Tableau.

7.11.1 Reference lines

Reference lines help you to mark a specific value or region on an axis. For example, you are analyzing sales for products. In this scenario, you can add reference line to indicate how each product is performing against the average sales.

7.11.1.1 Adding a reference line: average line

7.11.1.1.1 Step 1

Read in data from Sample - Superstore.xls into Tableau.

7.11.1.1.2 Step 2

Construct a view as shown in Fig. 7-27.

Figure 7-27. *View that displays "Sales" by "Sub-Category"*

7.11.1.1.3 Step 3

Select the sales axis, right click on it and select "Add reference line" (Shown in Fig. 7-28).

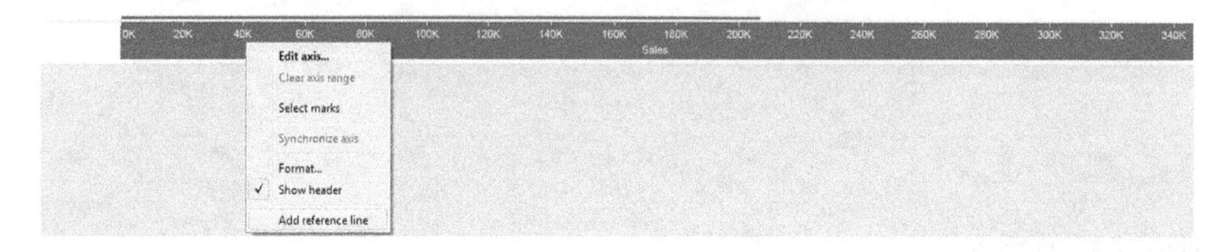

Figure 7-28. *"Add reference line" option*

7.11.1.1.4 Step 4

"Add reference line, band or box" dialog box appears. Select reference type as "Line", scope as "Entire Table" (adds a reference line to the entire table across all panes) and line options as shown below to display "Average" sales value for all products (Shown in Fig. 7-29).

Add reference line, band, or box

Line | Band | Distribution | Box Plot

Scope

Entire Table | Per Pane | Per Cell

Line

Value: SUM(Sales) | Average

Total
Sum
Constant
Minimum
Maximum
Average
Median

Label: Computation

Line only | 95

Formatting

Line:

Fill Above: None

Fill Below: None

Show recalculated line for highlighted or selected data points

OK

Figure 7-29. *Selection of "Average reference line"*

7.11.1.1.5 Step 5

You can see the average sales value as a reference line in the view (Shown in Fig. 7-30).

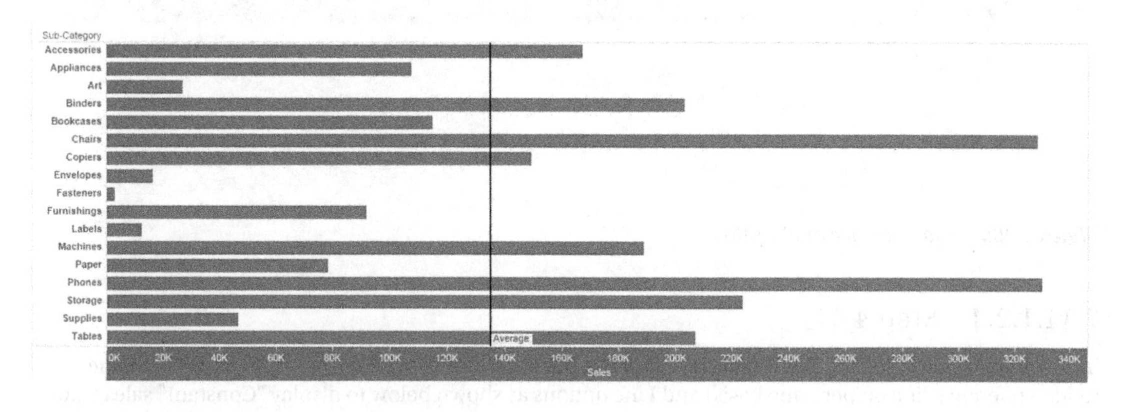

Figure 7-30. *View that shows "Average reference line" for "Sales"*

7.11.1.2 Adding a reference line: constant line

7.11.1.2.1 Step 1

Read in data from Sample - Superstore.xlsx into Tableau.

7.11.1.2.2 Step 2

Construct a view as shown in Fig. 7-31.

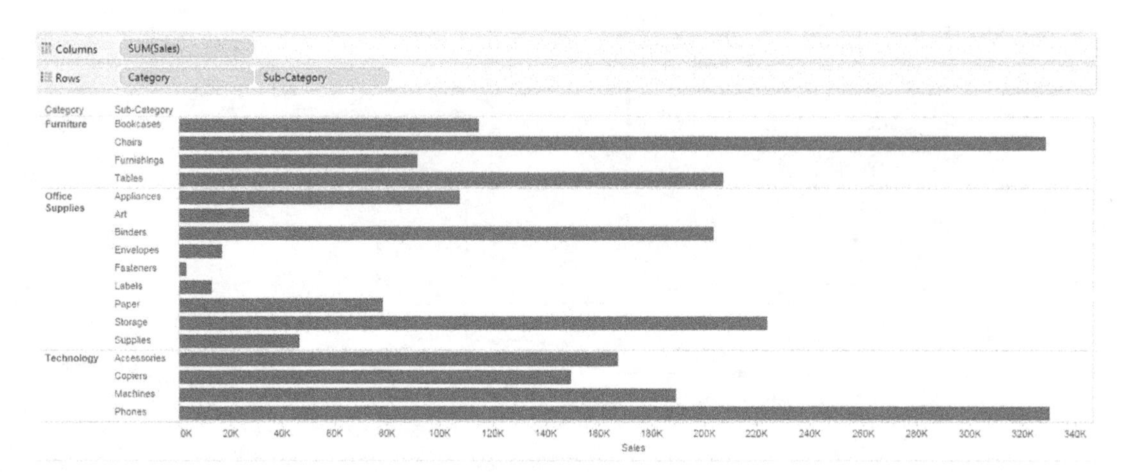

Figure 7-31. *View that displays "Sales" by "Sub-Category" for each "Category"*

7.11.1.2.3 Step 3

Select Sales axis, Right click on it and select "Add reference line" (Shown in Fig. 7-32).

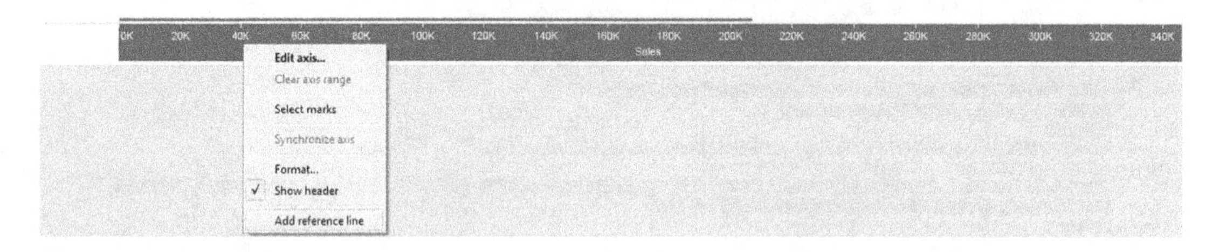

Figure 7-32. *"Add reference line" option*

7.11.1.2.4 Step 4

"Add reference line, band or box" dialog box appears. Select reference type as "Line", scope as "Per Pane" (adds a reference line on per pane basis) and Line options as shown below to display "Constant" sales value for all products (Shown in Fig. 7-33).

Add reference line, band, or box

Line | Band | Distribution | Box Plot

Scope

Entire Table | Per Pane | Per Cell

Line

Value: 3,024.28 | Constant

Label: Value

Line only | 95

Formatting

Line:

Fill Above: None

Fill Below: None

Show recalculated line for highlighted or selected data points

OK

Figure 7-33. *Selection of "Constant" line*

7.11.1.2.5 Step 5

You can see the "Constant" sales value as a reference line in the view (Shown in Fig. 7-34).

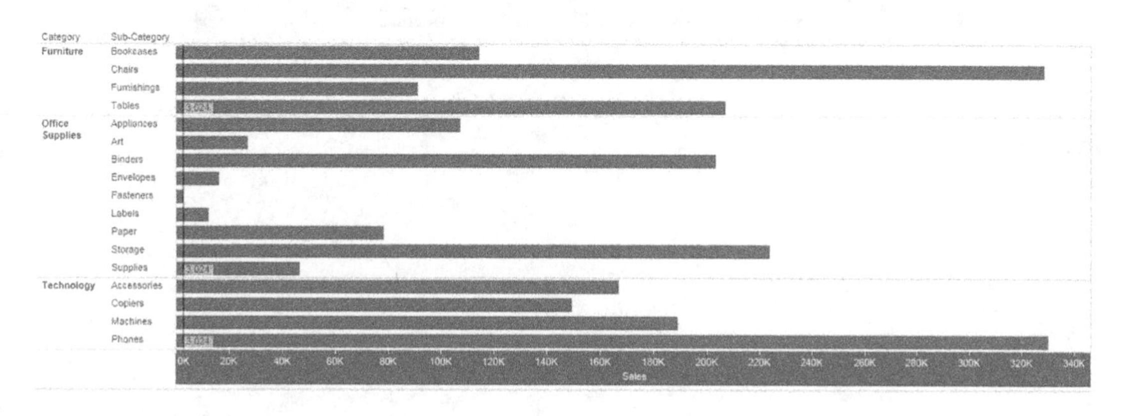

Figure 7-34. *View that shows "Constant" line for "Sales"*

7.11.1.3 Types of reference line, band and box

Line: Adds a line at a constant or computed value on the axis (Refer to Fig. 7-35).

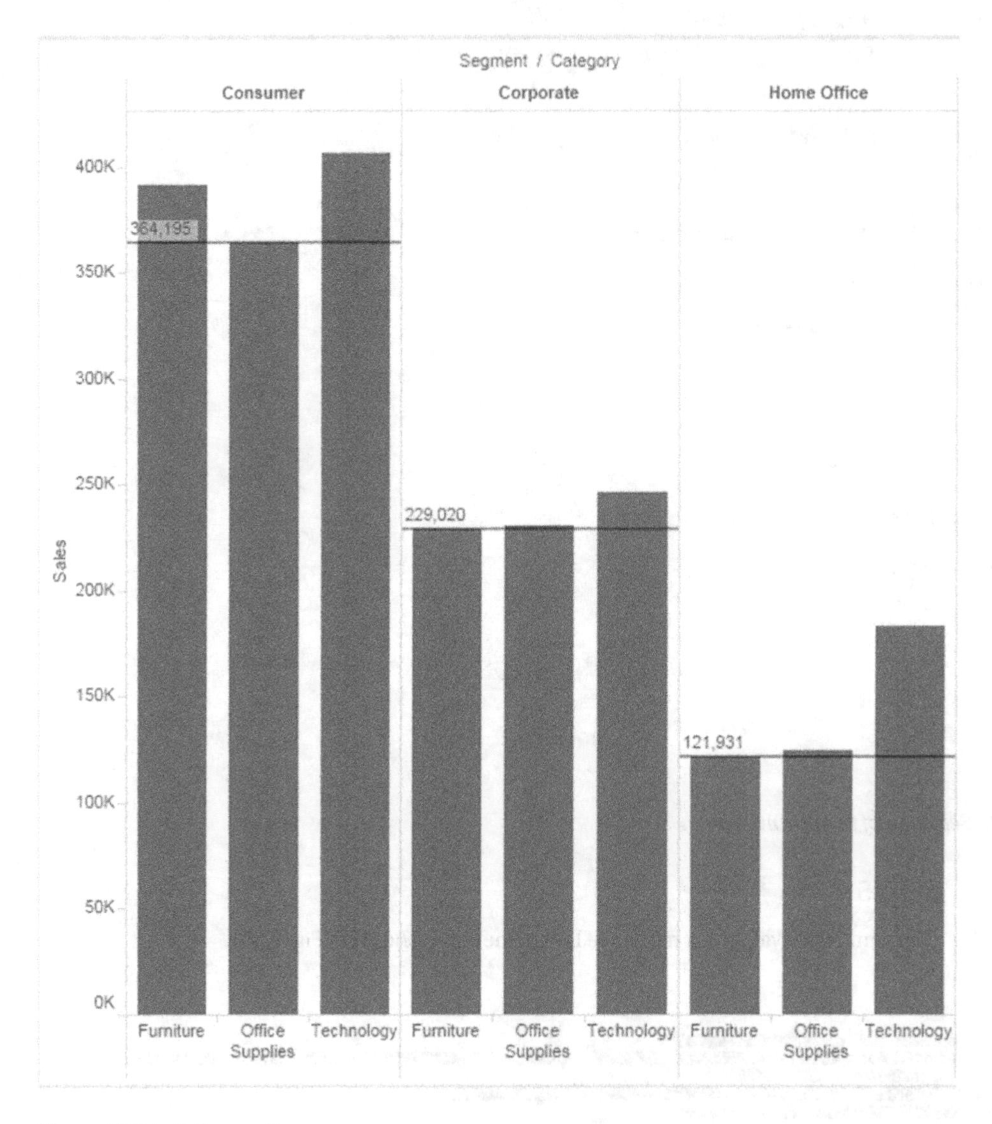

Figure 7-35. *Reference line*

Band: Band shades an area behind the marks in the view between two constants or computed values on the axis. Refer to Fig. 7-36 and Fig. 7-37.

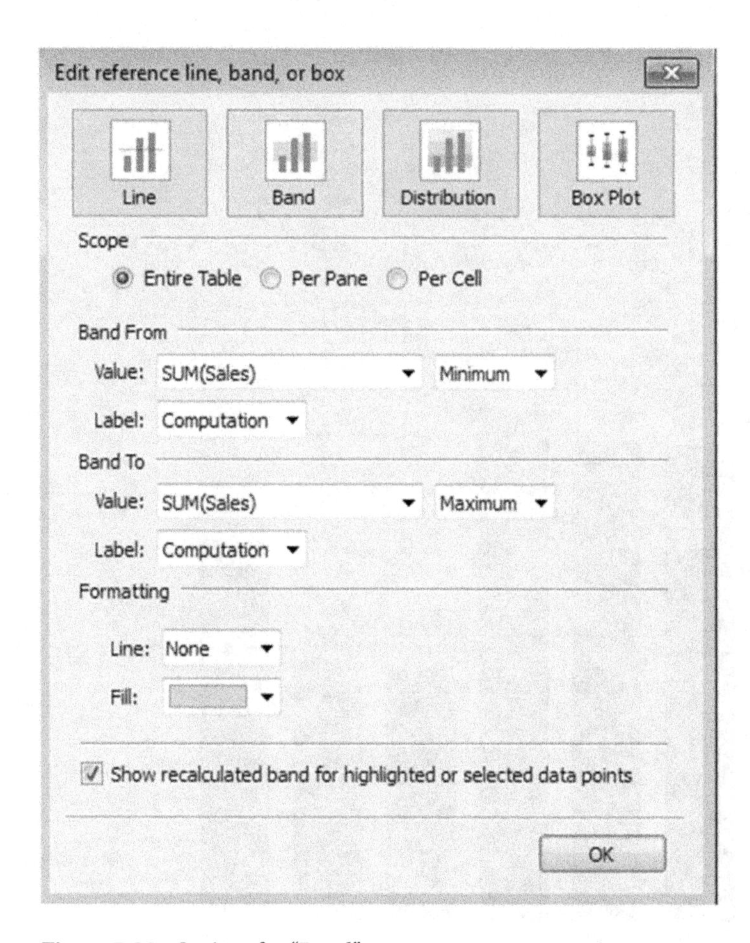

Figure 7-36. *Options for "Band"*

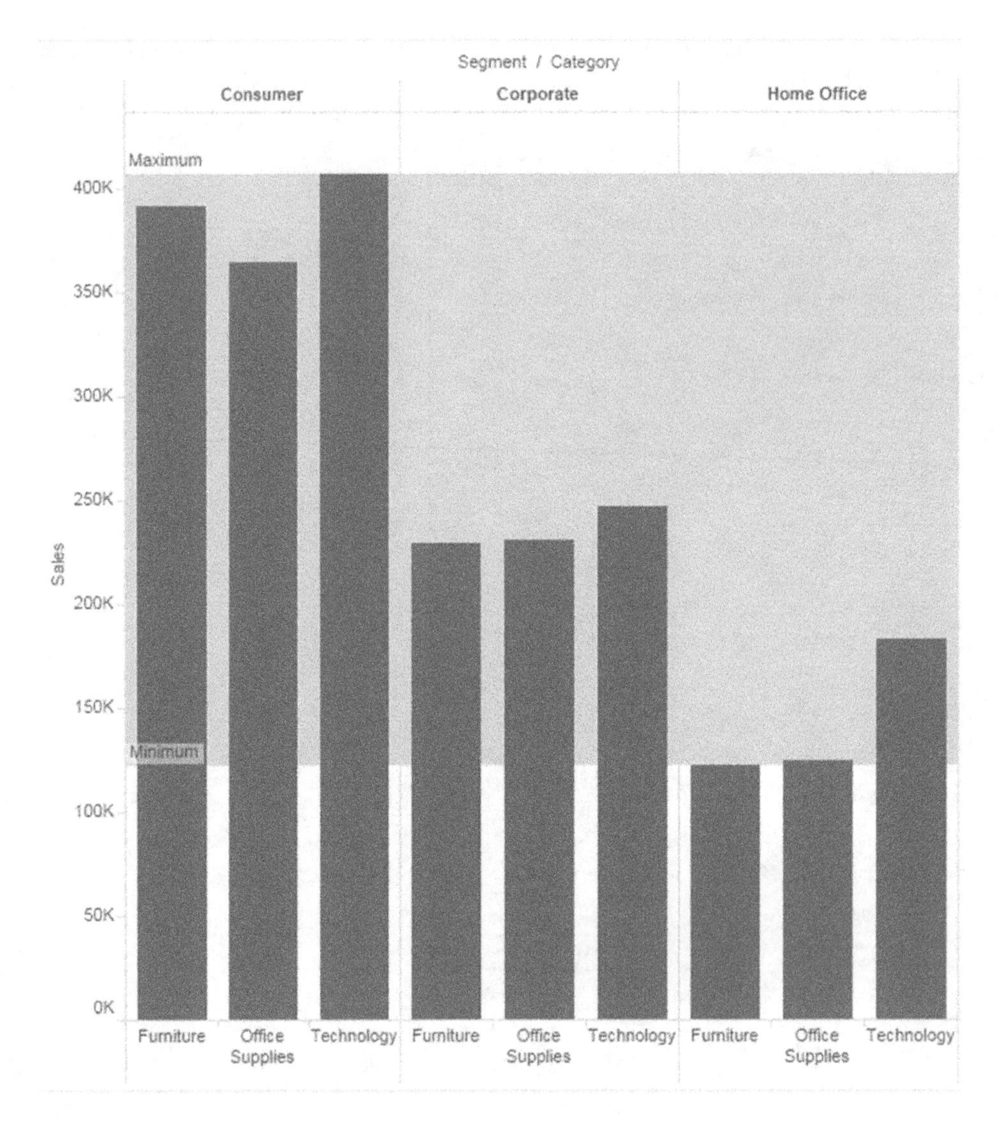

Figure 7-37. *"Band" type*

Distribution: Distribution adds a gradient of shading to indicate the distribution of values along the axis. Distribution can be defined by percentages, percentiles, quantiles or standard deviation. Refer to Figs. 7-38 and 7-39.

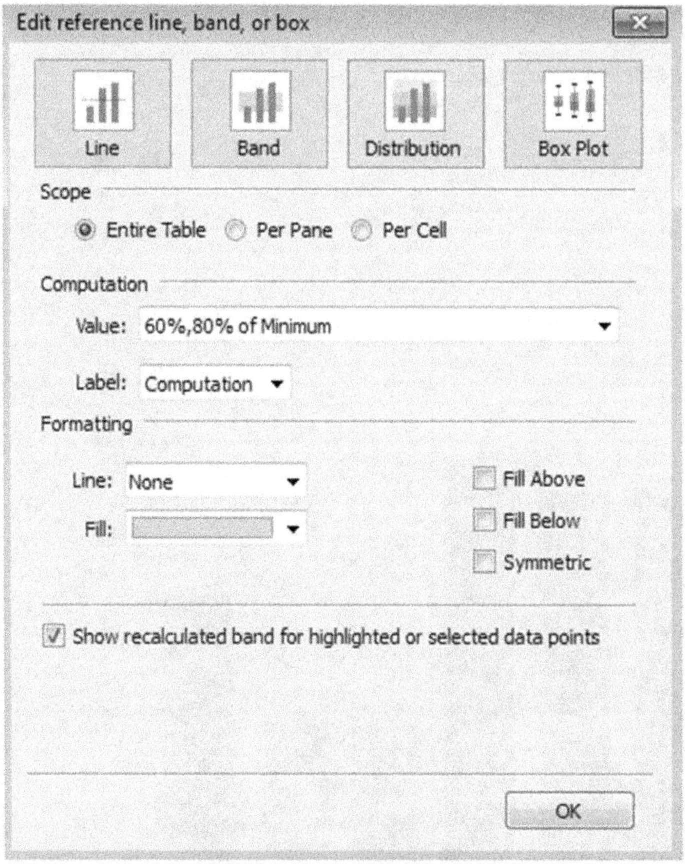

Figure 7-38. *Options for "Distribution" type*

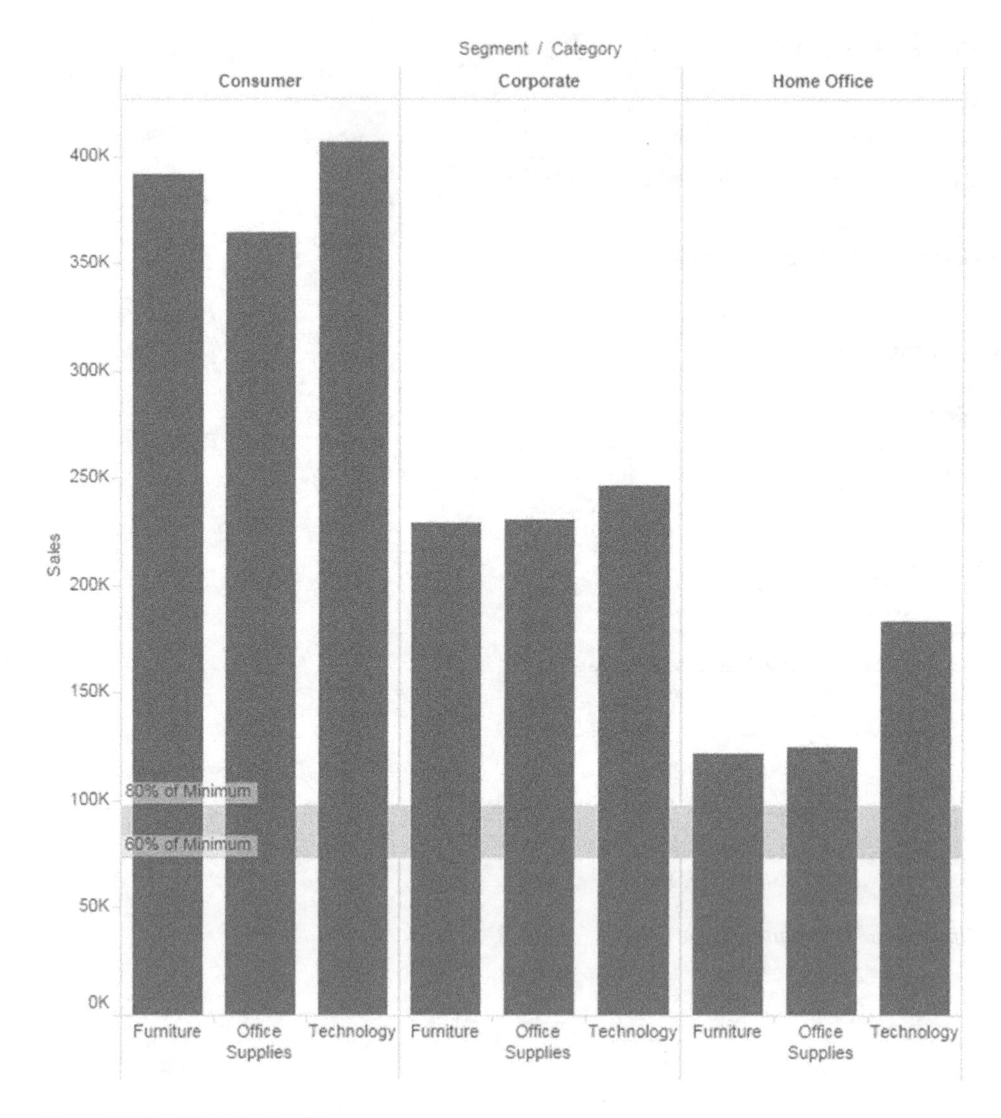

Figure 7-39. *"Distribution" type*

7.12 Trend lines

Trend lines allow you to incrementally construct interactive models of behavior. With the help of trend lines, you can answer questions like whether profit is predicted by sales.

7.12.1 Answering questions with trend lines

Let us start with a question:

Objective: What is causing high discount ratio at superstore?

7.12.1.1 Steps to plot trend lines to answer a question

Take the following steps to answer the question.

7.12.1.1.1 Step 1

Construct the view as shown in Fig. 7-40.

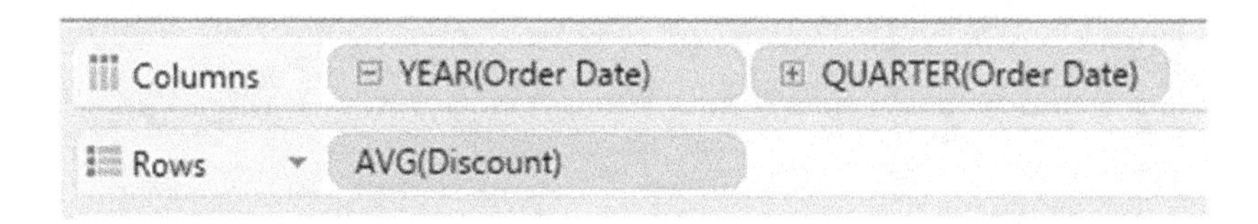

Figure 7-40. *Dimension "Order Date" placed on the columns shelf; hierarchy is set to "Quarter" and measure "AVG(Discount)" placed on the rows shelf*

7.12.1.1.2 Step 2

First, we will begin the test with the "Ship Mode" variable. The assumption is that discount rates may be high for few "Ship Mode". Drag the dimension "Ship Mode" from the dimensions area under the data pane to the rows shelf (Shown in Fig. 7-41).

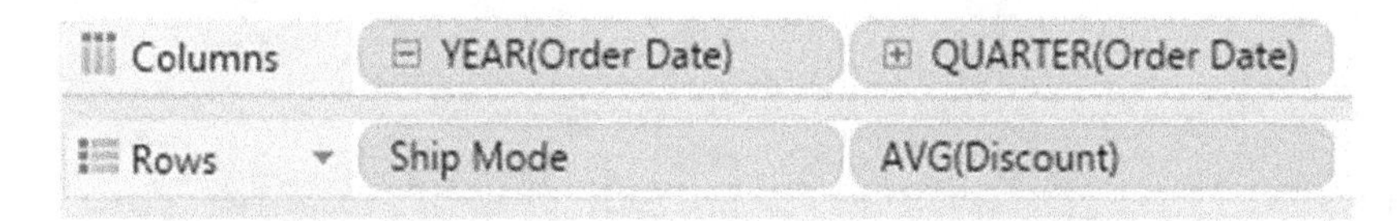

Figure 7-41. *Dimension "Ship Mode" placed on the rows shelf*

7.12.1.1.3 Step 3

Go to the analytics pane, double click on "Trend Line" (Shown in Fig. 7-42) to display trend lines (Shown in Fig. 7-43).

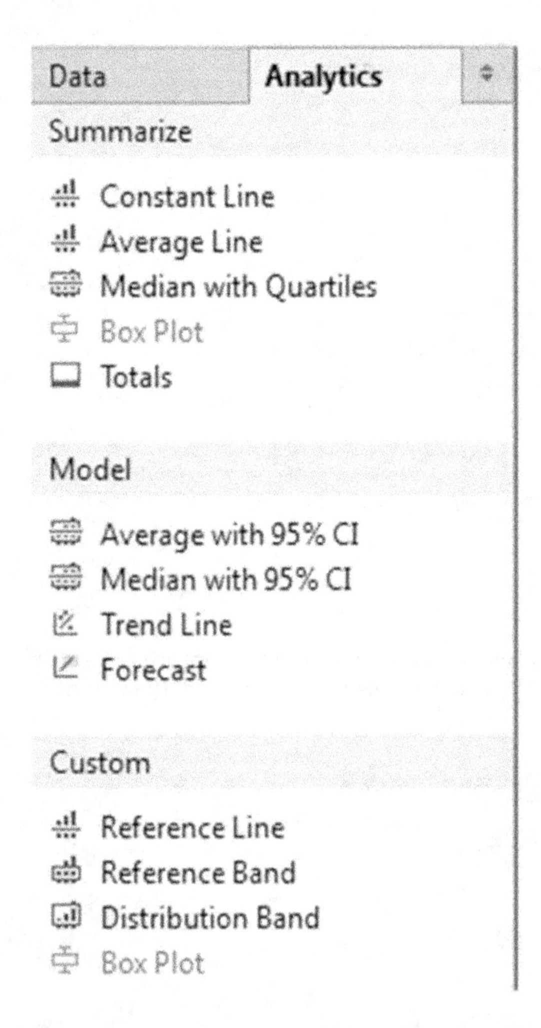

Figure 7-42. *Analytics pane showing "Trend Line"*

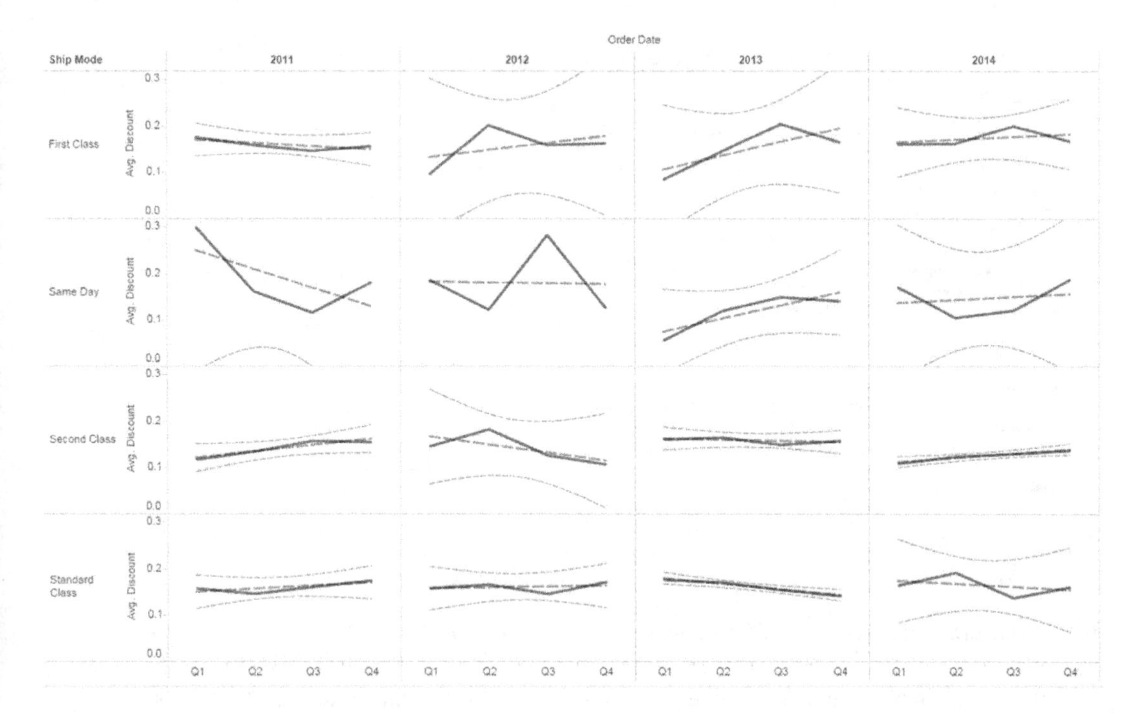

Figure 7-43. *View that shows trend lines for "Ship Mode"*

7.12.1.1.4 Step 4

Right click any trend line.

7.12.1.1.5 Step 5

Select "Describe trend model..." (Shown in Fig. 7-44) to describe trend model (Shown in Fig. 7-45).

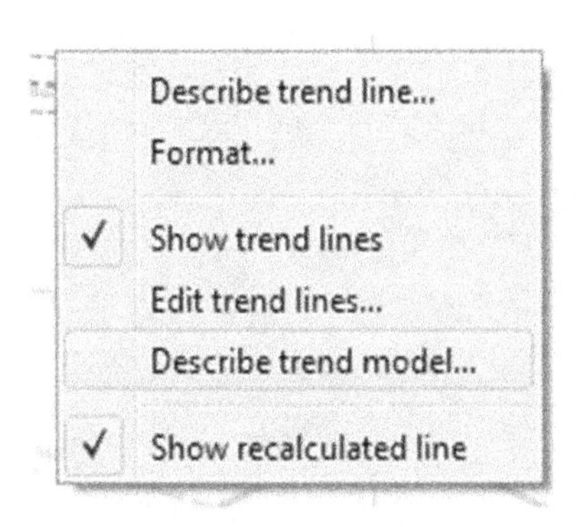

Figure 7-44. *Describe trend model*

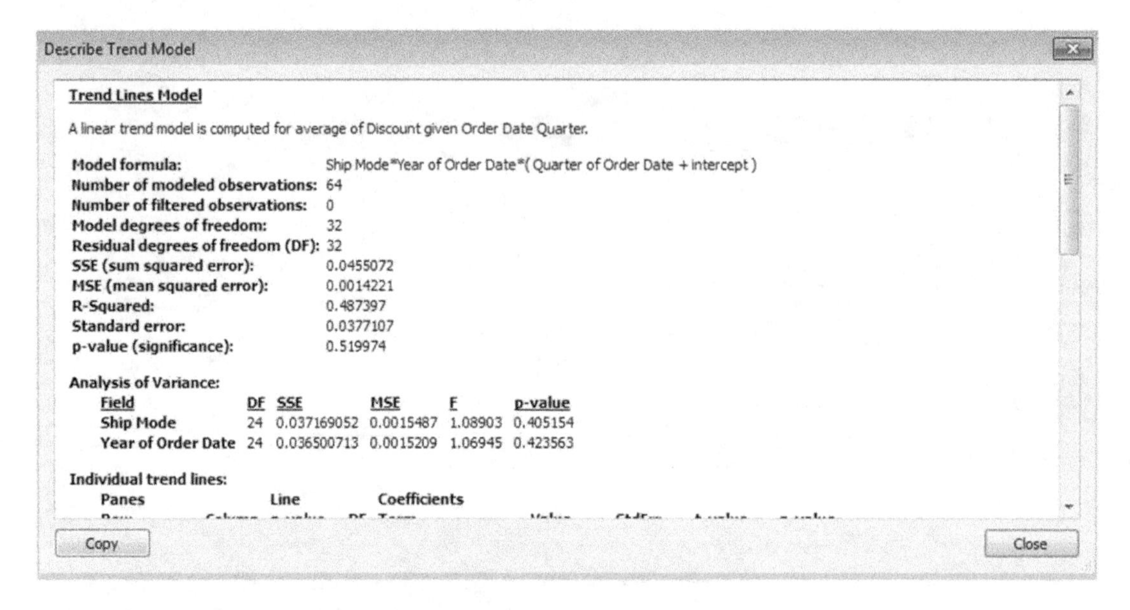

Figure 7-45. *"Describe Trend Model" showing p-value for "Ship Mode"*

The description of the trend line model shows that the p-value for ship mode is 0.40, which is not significant. Therefore, we cannot use "Ship Mode" to predict discount rates.

7.12.1.1.6 Step 6

Next, let us consider the dimension "Manufacturer". Some manufacturers occasionally offer substantial discounts. Remove "Ship Mode" from the rows shelf and drag "Manufacturer" to the rows shelf (Shown in Fig. 7-46).

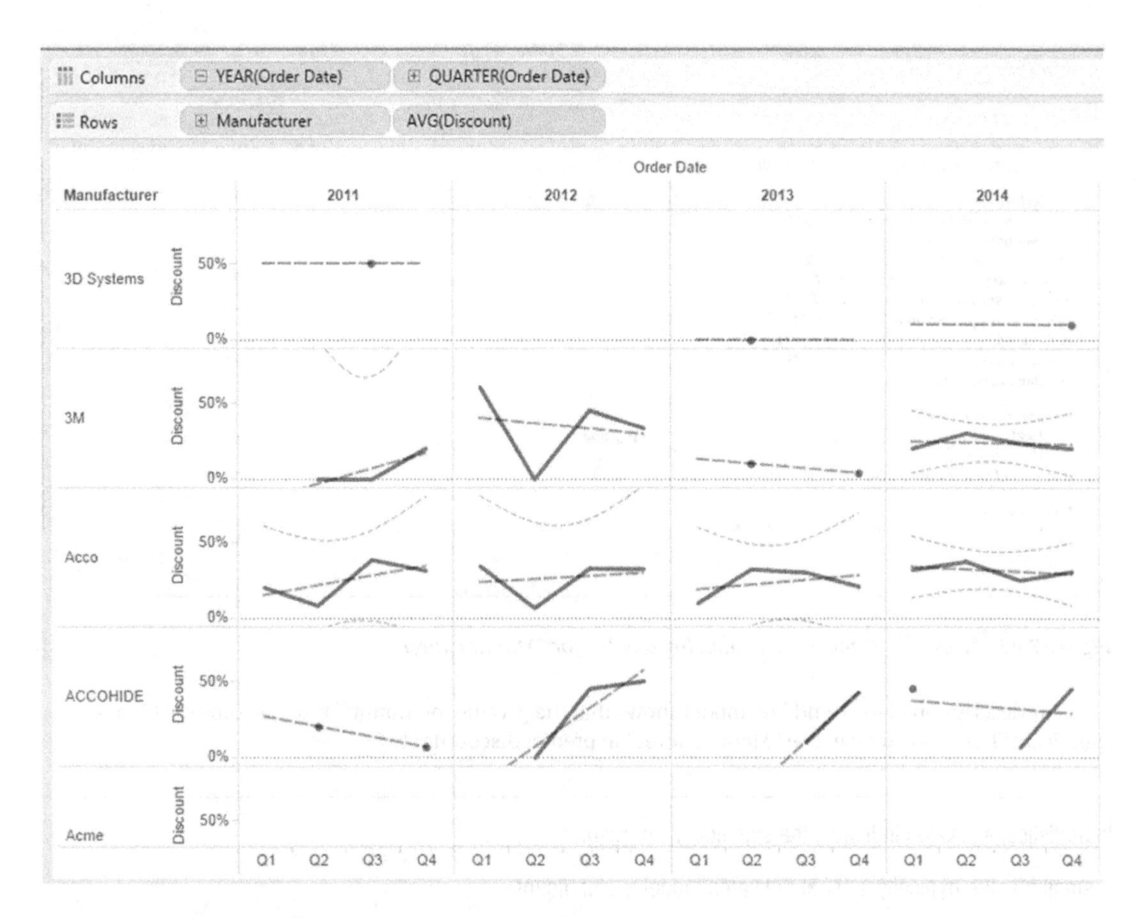

Figure 7-46. *Dimension "Manufacturer" placed on the rows shelf*

7.12.1.1.7 Step 7

Open the described trend model. Refer to Fig. 7-47.

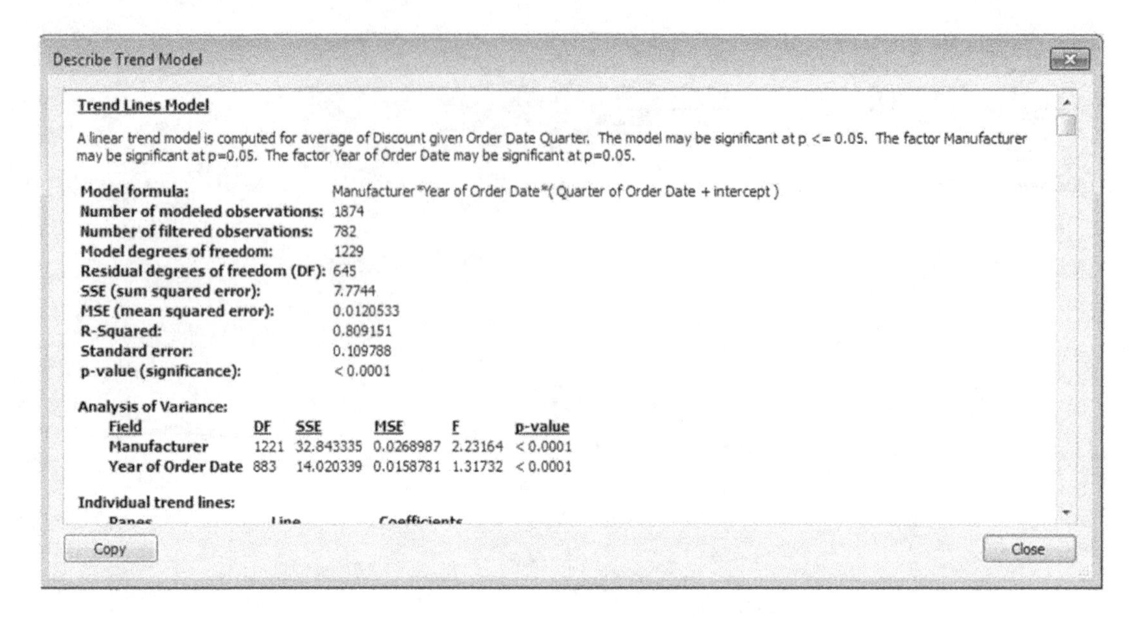

Figure 7-47. *"Describe Trend Model" showing p-value for "Manufacturer"*

The description of the trend line model shows that the p-value for Manufacturer is 0.0001, which is significant. Therefore, we can use "Manufacturer" to predict discount rates.

In statistics, a p-value indicates the significance of results.

A small p-value (typically ≤ 0.05) indicates model is significant.

A large p-value (> 0.05) indicates model is not significant.

P-values very close to the cut-off (0.05) are considered to be marginal.

7.13 Forecasting

Forecasting helps you to estimate future values of the measures along with the historical values. Tableau displays estimated values in a lighter shade of the color.

The techniques used in forecasting are exponential smoothing, which looks at the trends in the past to help predict future results. To create a forecast, you need at least one dimension and one measure.

7.13.1 Demo 1

Objective: To predict sales over time.

7.13.1.1 Steps to create a forecast

Follow the following steps.

7.13.1.1.1 Step 1

Connect to Sample-Superstore.xls data source.

7.13.1.1.2 Step 2

Drag the dimension "Order Date" from the dimensions area under the data pane to the columns shelf. Drag the measure "Sales" from the measures area under the data pane to the rows shelf (Shown in Fig. 7-48).

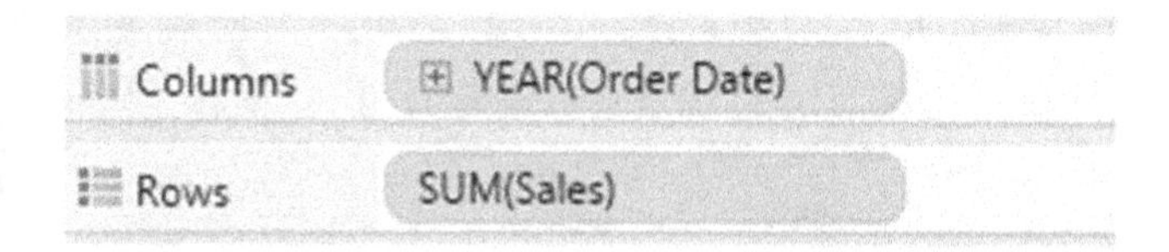

Figure 7-48. *The dimension "Order Date" placed on the columns shelf, measure "Sales" placed on the rows shelf*

7.13.1.1.3 Step 3

Set the "Order Date" hierarchy to continuous month (Shown in Fig. 7-49).

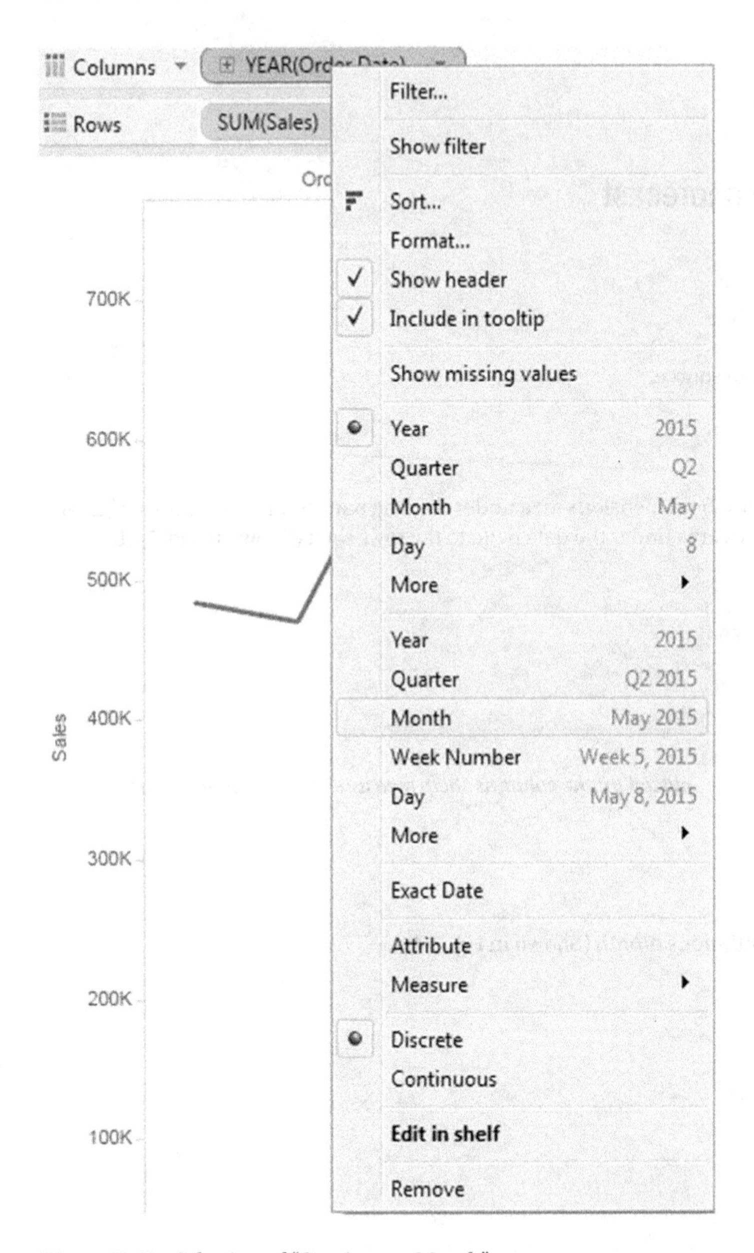

Figure 7-49. *Selection of "Continuous Month"*

7.13.1.1.4 Step 4

Go to the analysis menu select "Forecast" ➤ Select "Show Forecast" (Shown in Fig. 7-50).

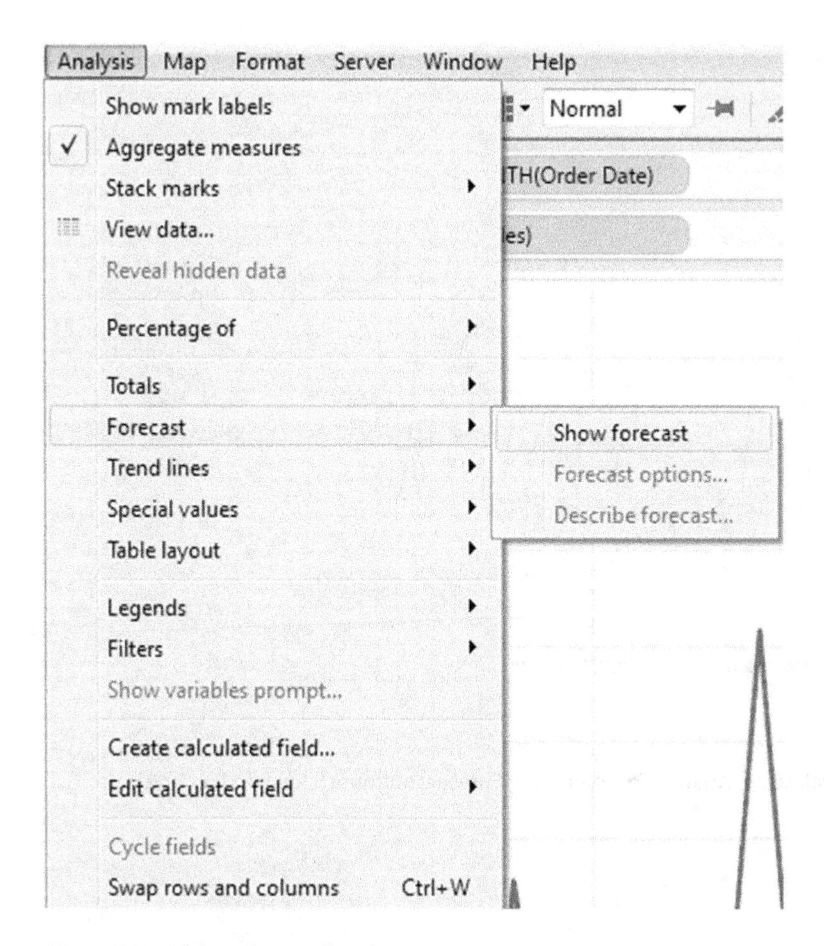

Figure 7-50. *"Show forecast" option*

7.13.1.1.5 Step 5

You can see the estimated values in lighter shade. You can also see the "Forecast indicator" below the marks card. The prediction interval level is 95% for the forecast (Shown in Fig. 7-51).

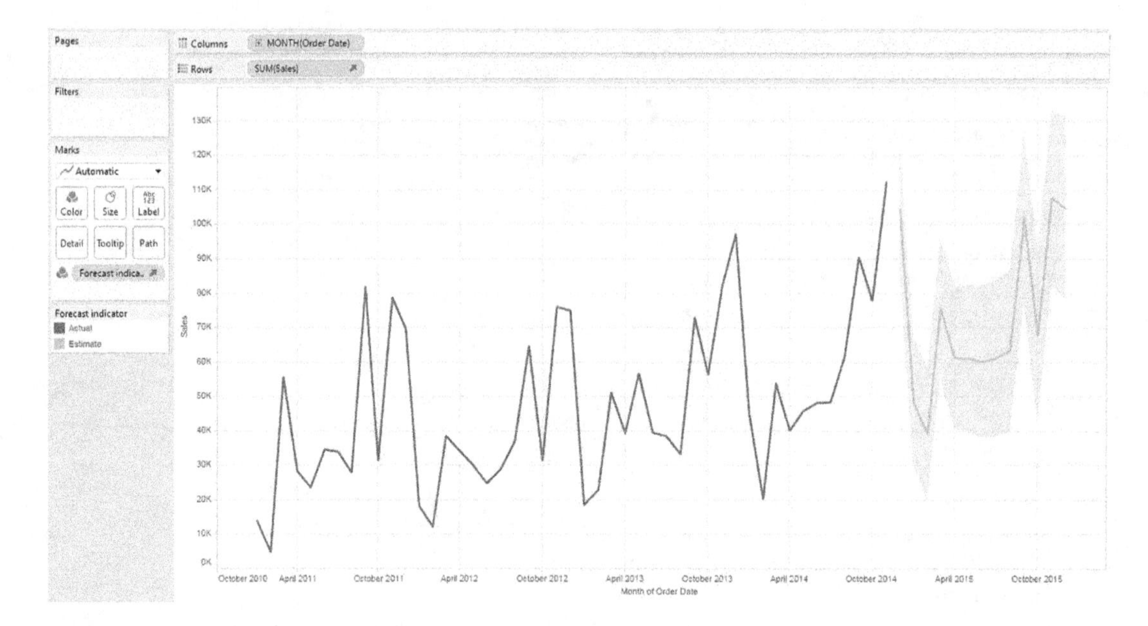

Figure 7-51. *View that shows estimated value in lighter shade*

7.13.1.1.6 Step 6

To configure prediction interval, select Analysis ➤ Forecast ➤ Forecast options (Shown in Fig. 7-52).

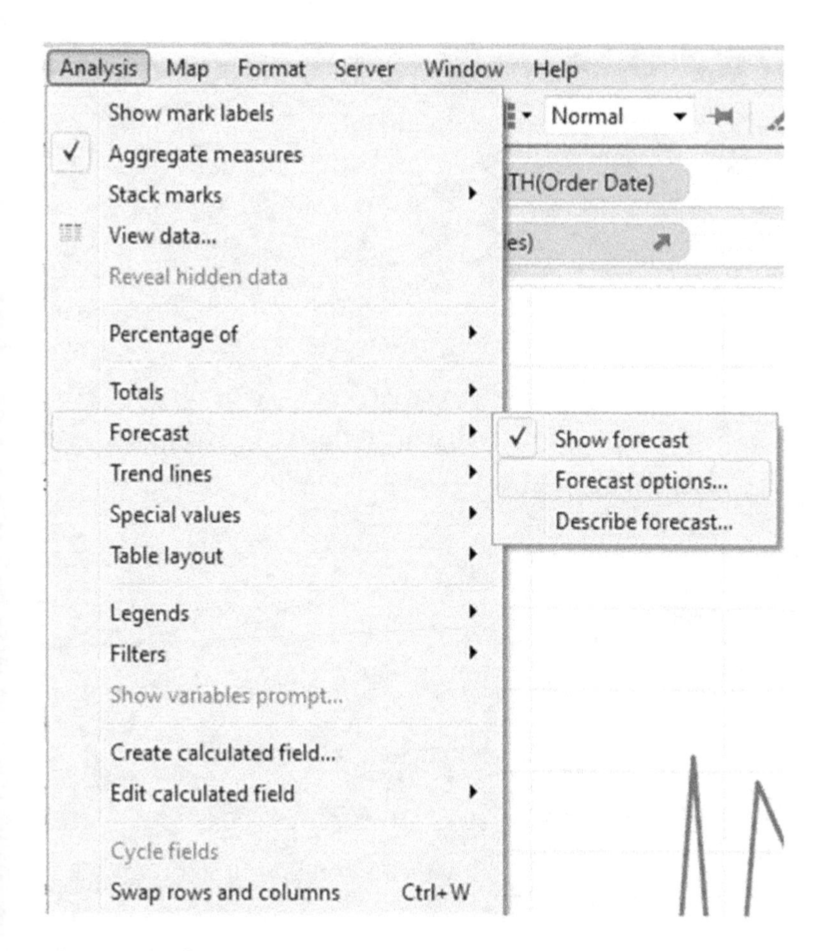

Figure 7-52. *"Forecast options..."*

7.13.1.1.7 Step 7

The "Forecast Options" dialog box appears. Select "99%" as prediction interval. That is, the model has determined that there is a 99% likelihood that the value of "Sales" will be within the shaded area for the forecast period. The other values are 90%, 95%, 99% (Shown in Fig. 7-53).

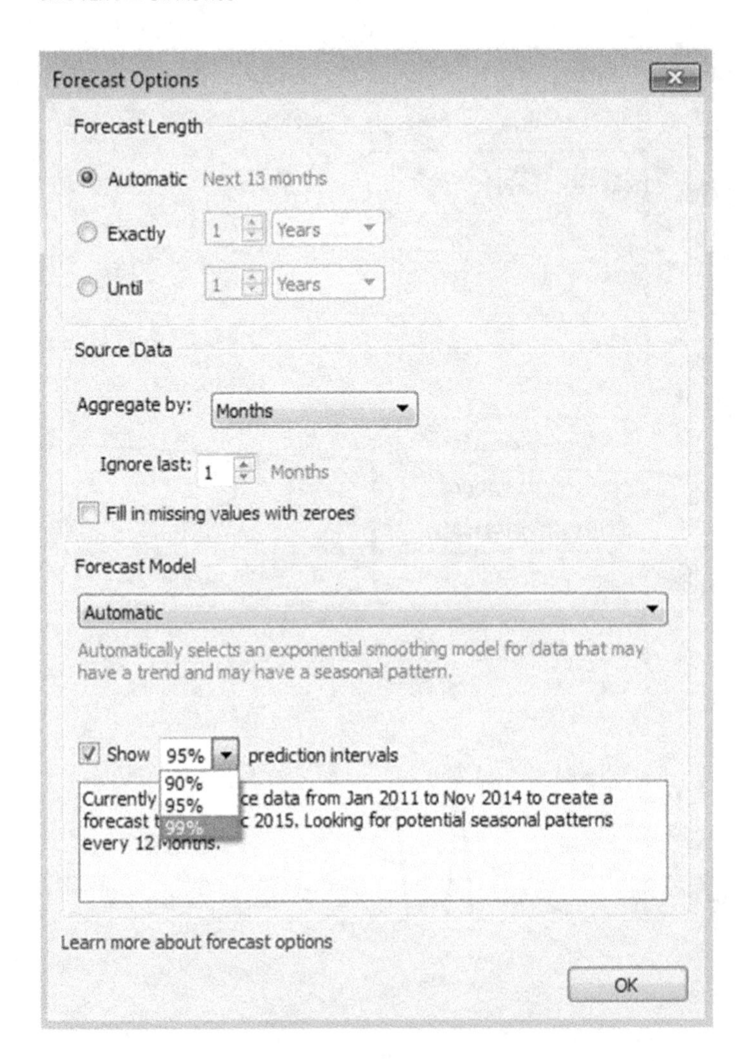

Figure 7-53. *Forecast options - "prediction intervals"*

7.13.1.1.8 Step 8

Prediction intervals are displayed based on the mark you select for the forecast. Refer to Table 7-11.

Table 7-11. *Prediction intervals mark type and display*

Forecast Mark Type	Prediction intervals displayed using
Line	Bands
Shape, Square, Circle, Bar, or Pie	Whiskers

7.13.1.1.9 Step 9

Change the mark type to circle. Tableau displays forecast data in lighter shades "Circles" and prediction intervals are displayed as "Whiskers" (Shown in Fig. 7-54).

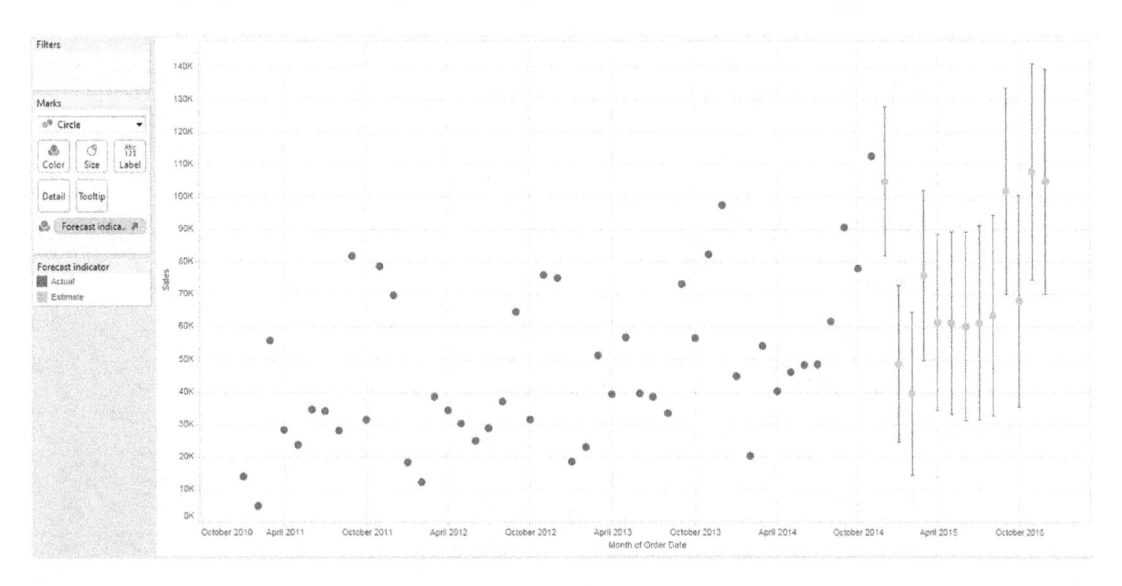

Figure 7-54. *View that shows prediction intervals as "Whiskers"*

7.13.1.1.10 Step 10

You can enhance forecast by adding additional information such as "Precision". Right click on measure select Forecast result ➤ Precision (Shown in Fig. 7-55, 7-56).

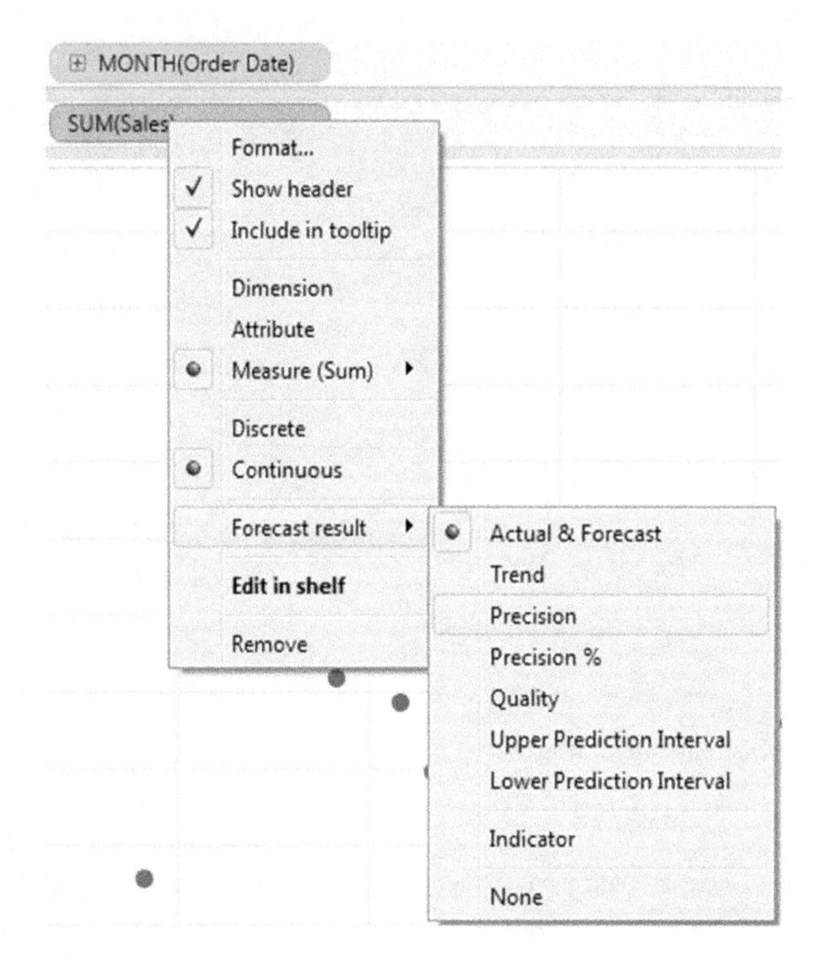

Figure 7-55. *Selection of forecast result as "Precision"*

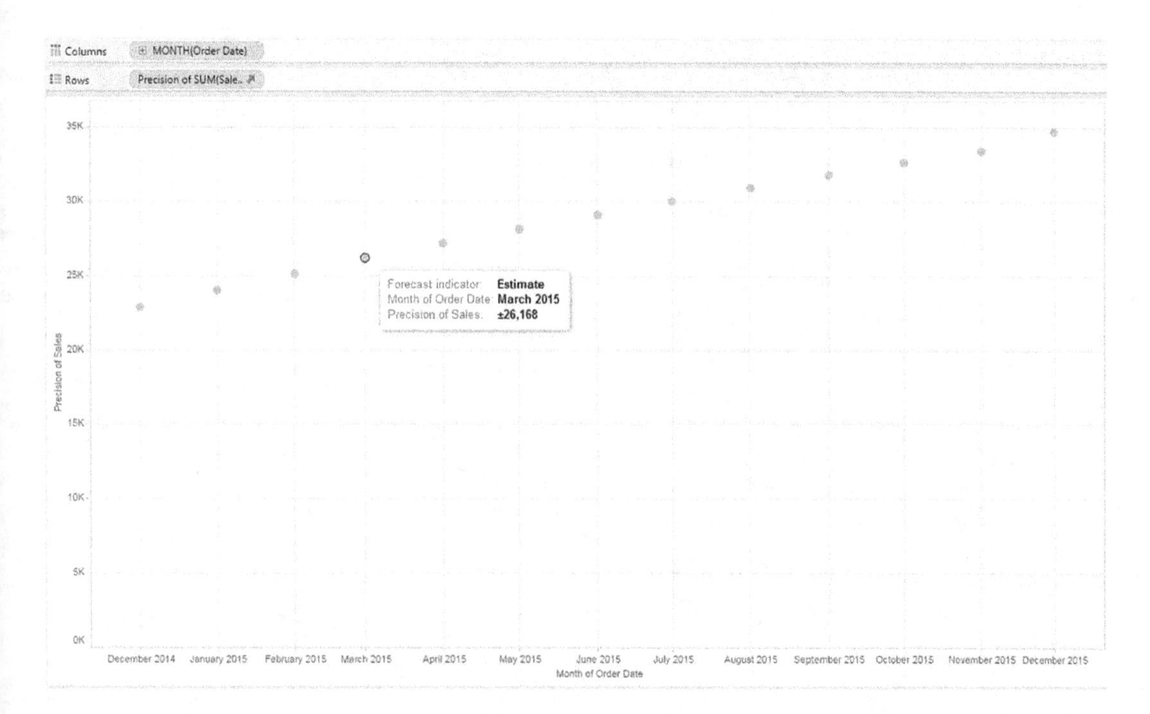

Figure 7-56. *View that shows "Precision of Sales"*

7.13.1.1.11 Step 11

To see the details of forecast summary and statistical model, go to Analysis ➤ Forecast ➤ Describe forecast (Shown in Fig. 7-57, 7-58).

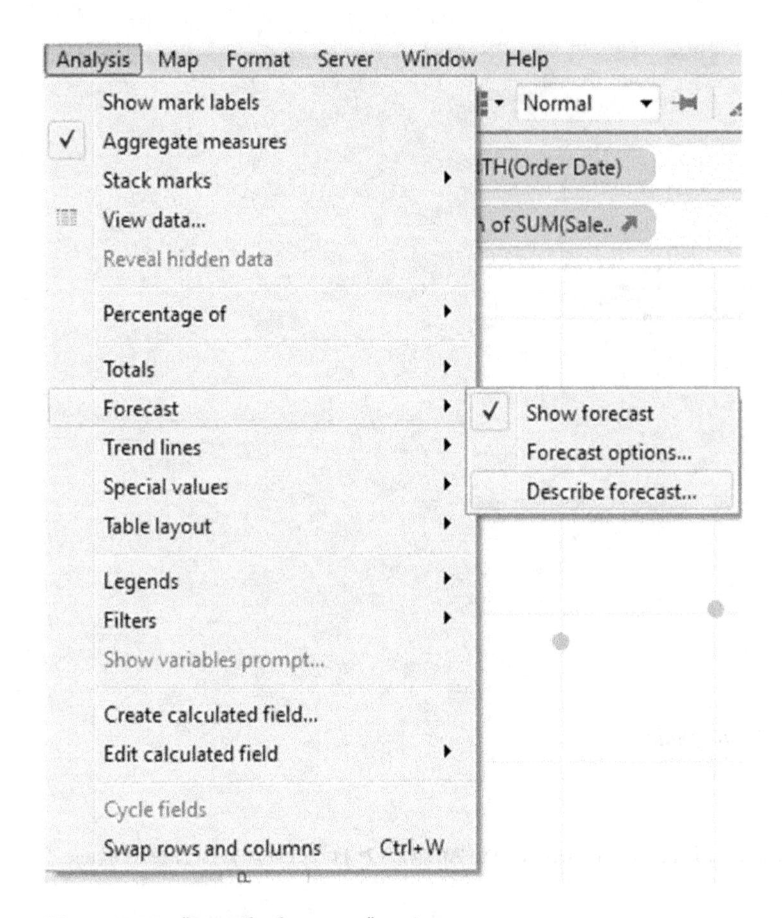

Figure 7-57. *"Describe forecast..." option*

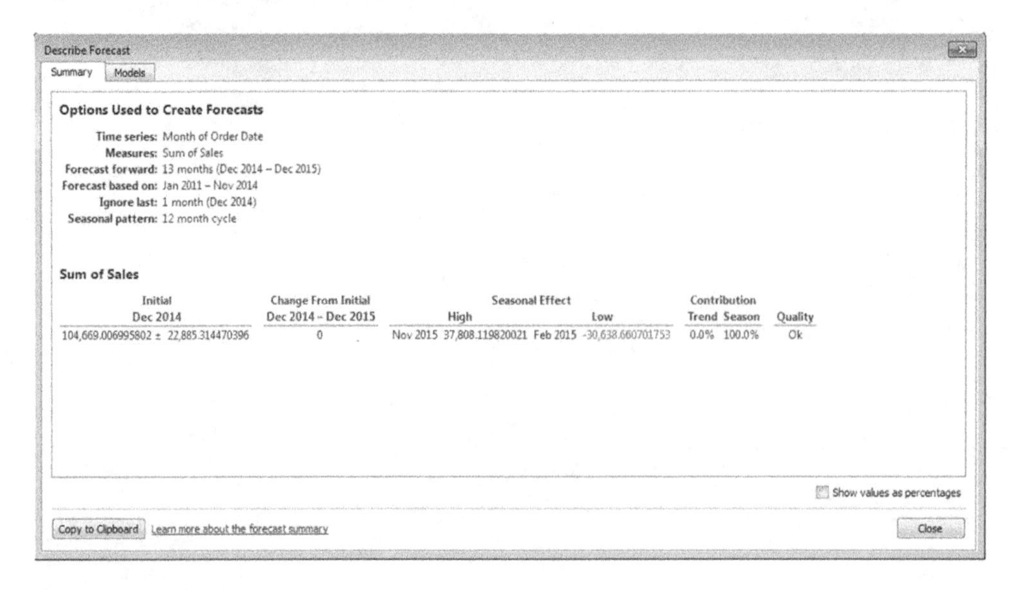

Figure 7-58. *Describe forecast*

7.14 Points to remember

- Statistics can be defined as the science of collecting large number of facts (or real-world observations) and analyzing them with the purpose of summarizing the collection and drawing inferences.
- Statistics allow one to show, describe or present data in a meaningful way such that patterns might emerge from the data.
- The five number summary includes minimum, maximum, median, first quartile and third quartile.
- Reference lines help one to mark a specific value or region on an axis.
- Trend lines allow one to construct incrementally interactive models of behavior.
- Forecasting helps one to estimate future values of the measures along with the historical values.

7.15 Next steps

In the next chapter, we will learn about the following chart forms in Tableau:

- Bar chart
- Pie chart
- Line graph
- Scatter plot
- Histogram
- Heat map
- Treemap
- Highlight table
- Gantt view

CHAPTER 8

Chart Forms

"Signals always point to something. In this sense, a signal is not a thing but a relationship. Data becomes useful knowledge of something that matters when it builds a bridge between a question and an answer. This connection is the signal."

—Stephen Few, founder - Perceptual Edge, in his book, Signal: Understanding What Matters in a World of Noise

Chapter 7 introduced us to the basics of descriptive statistics. We learned about the varied forms in which measures can be aggregated, such as SUM, AVERAGE, MEDIAN, MIN, MAX, COUNT, STANDARD DEVIATION, VARIANCE, etc. It also introduced us to the analytics pane in Tableau. We learned about the five magical numbers introduced by way of a box-and-whiskers plot. We were able to use constant lines, trend lines, forecasts, etc. to perform analysis.

In this chapter, we will hone our knowledge of chart forms, such as pie charts, Gantt charts, line graphs, stacked bar charts, histograms, and word clouds.

8.1 Pie chart

In the year 1901, William Playfair, a Scottish engineer and political economist, invented the pie chart.

8.1.1 What is a pie chart?

A pie chart is a circular chart divided into slices. The slices illustrate proportion based on percentages. The slices need to be mutually exclusive; they cannot overlap. The data should not only sum up to a meaningful whole but the values should be categorized such that they are not counted several times.

8.1.2 When to use a pie chart?

A pie chart should be used in the following circumstances.

- To show the relationship of a part to the whole.
- To display data that could otherwise be represented in a small table.
- When data is available in six or fewer categories.
- When your data set is very small and does not need to show progression over time.

S. Acharya and S. Chellappan, *Pro Tableau*, DOI 10.1007/978-1-4842-2352-9_8

- When data is either nominal or ordinal. Data in a nominal category is one in which it can be classified based on descriptive or qualitative information, such as country residing in, workplace location, etc. Data under the ordinal category is one in which it can be ranked. For example, participants can be asked to provide ratings such as excellent, very good, good, poor, very poor, etc. to measure the effectiveness of a learning program.

8.1.3 How to read a pie chart?

Consider two features while reading a pie chart:

- The angle that a slice covers (compared to the full circle).
- The area of a slice.

8.1.4 Pros

Few benefits of pie charts are stated below:

- Simple to read if there are only a few categories represented on the pie chart.
- Simple to understand even by an uninformed audience. Pie charts are visually simpler than other types of graphs, and it can be easily understood due to widespread use in business and the media.
- Does not require a lot of explanation when used in a report or presentation.

8.1.5 Cons

- The main purpose of the pie chart is to be able to perform relative comparison. The purpose gets defeated as it is not easy to decipher the angles created by the slice.
- Is ill-suited if there is too much data (read categories) to be represented on the pie chart.
- Do not easily reveal exact values.

Pie charts fail to reveal key assumptions, causes, effects, or patterns.

8.1.6 Five tips for using pie charts

If you still decide that a pie chart is the right data visualization for what you are trying to achieve, look at the following five tips before throwing that pie into your presentation:

- A pie chart with more than six individual segments is going to look too cluttered and lose its visual impact.
- Order your segments from largest to smallest. Do not make people work to see the scale.

- Do not try to make pie charts more visually appealing by adding effects like shadows or 3D perspectives. These actually make it more difficult to understand the data.
- Make sure everything adds up to 100 percent.
- Put a name or number onto each segment of your pie chart. A surprisingly large portion of the population is color blind, so the chart is meaningless to them without labels.

8.1.7 A critique's view

Edward Tufte, in his book, *The Visual Display of Quantitative Information*, states,

"A table is nearly always better than a dumb pie chart; the only worse design than a pie chart is several of them, for then the viewer is asked to compare quantities located in spatial disarray both within and between charts... Given their low density and failure to order numbers along a visual dimension, pie charts should never be used."

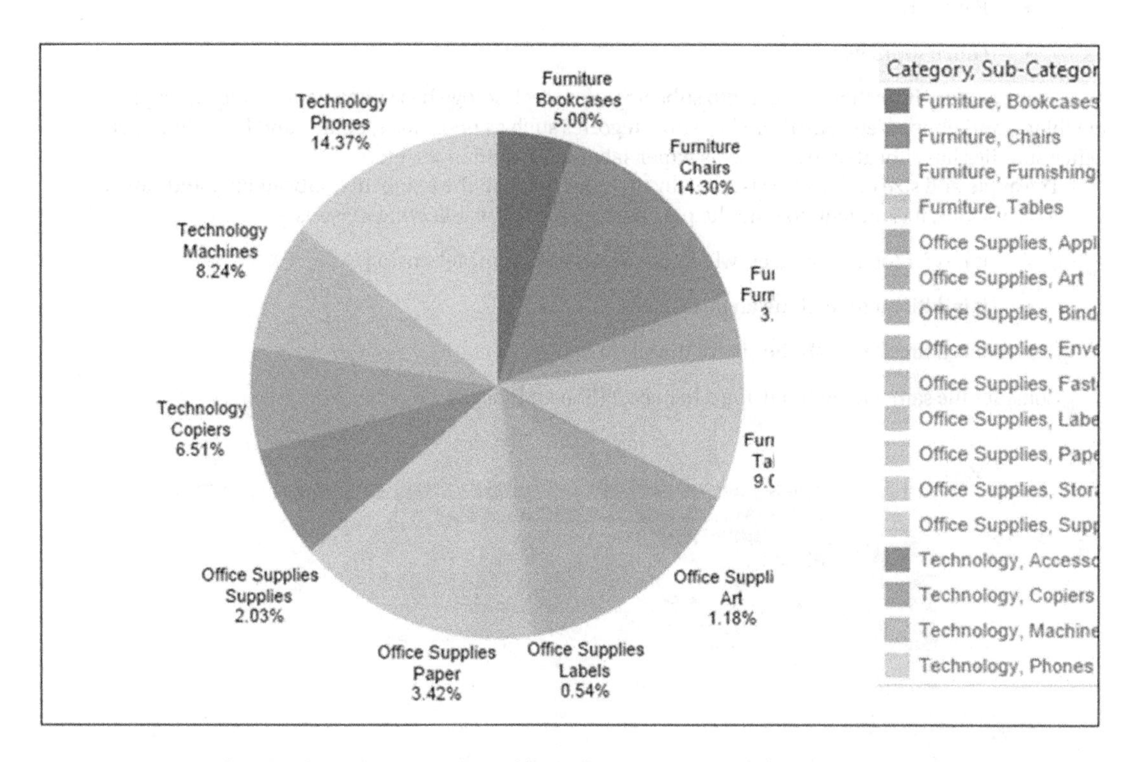

Figure 8-1. *A pie chart to show the percentage of "Sales" made in the each "Sub-Category" and "Category"*

8.1.8 An alternative for a pie chart

The alternative for pie chart is a simple bar graph. Why? Because it is easier to compare only one dimension, i.e. length. We are better at comparing lengths (bar chart) than comparing angles (pie chart).

8.1.9 What can further add to the woes?

- Pie charts used as a display of quantitative information
- Glossy color
- 3D effect
- Radius effects

Below are a few examples where pie charts should have been avoided.

8.1.9.1 Example 1

Refer to Fig. 8-1.

Data is available for three categories:

- Technology
- Furniture
- Office supplies

Each category is further divided into subcategories. Technology has subcategories such as copiers, machines, and phones, etc. Furniture has subcategories such as bookcases, chairs, and furnishings, etc. Office supplies has subcategories, such as paper, labels and appliances, etc.

The angle and size of the slices is given by the sales made in the respective subcategory and category. As is obvious, it is difficult to read the pie chart, owing to the following reasons:

- It has far too many slices, which dissuades easy comprehension.
- It is difficult to read the angle of slices.
- It is difficult to read the size of the slices.

Consider the same depiction using a bar chart (Shown in Fig. 8-2):

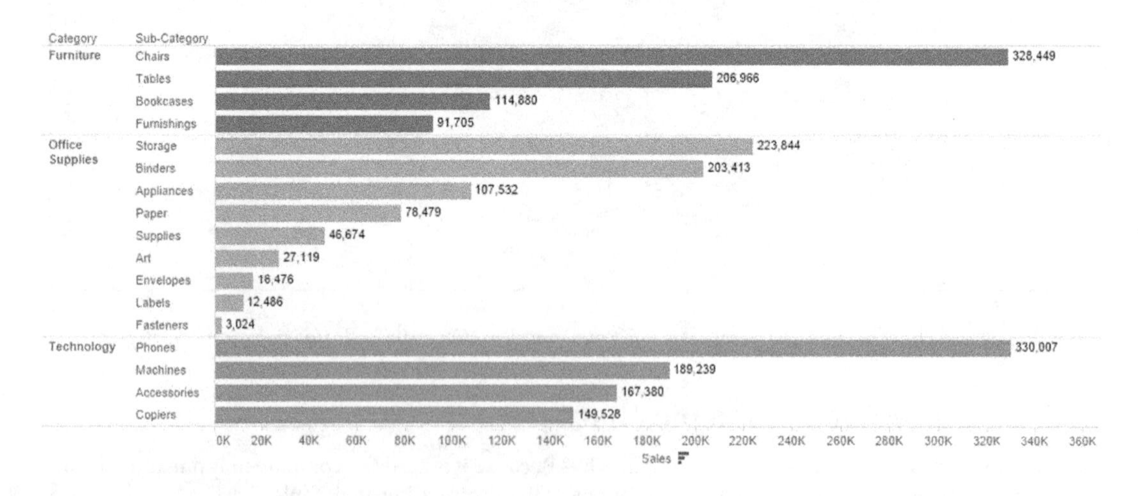

Figure 8-2. *Bar chart showing the same data as in Fig. 8.1*

This is so much easier to read. There is absolutely no confusion regarding which "Sub-Category" in the respective "Category" has had the maximum "Sales" ("Chairs" in "Furniture", "Storage" in "Office Supplies" and "Phones" in "Technology".

8.1.9.2 Example 2

We have the sales data of several cities of US, such as Colorado Springs, Philadelphia, San Diego, San Francisco, etc. We would like to determine the top three cities by sales. The angle and size of the slices is given by the sales made in the respective cities. Refer to Fig. 8-3.

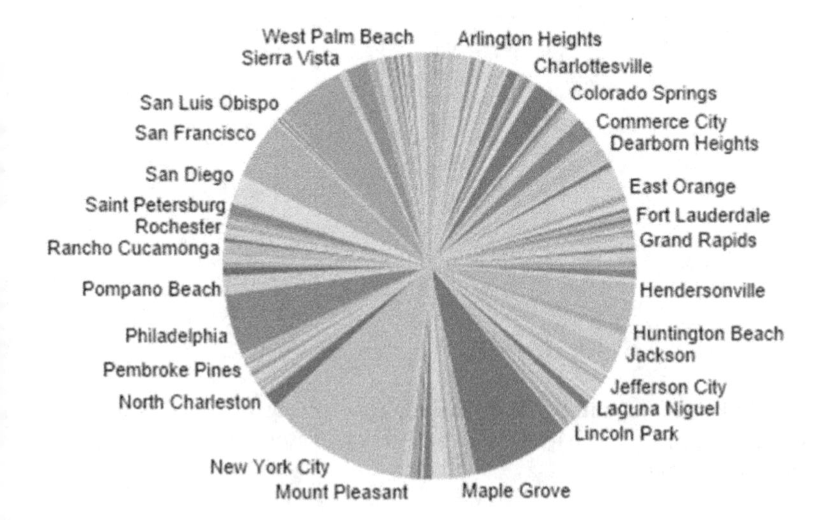

Figure 8-3. *A pie chart to show the "Sales" made in several US cities*

As is obvious, it is difficult to read the pie chart, owing to the following reasons:

- It has far too many slices, which prevents easy comprehension.
- It is difficult to read the angle of slices.
- It is difficult to read the size of the slices.

Consider the same plot using a bar chart (Shown in Fig. 8-4):

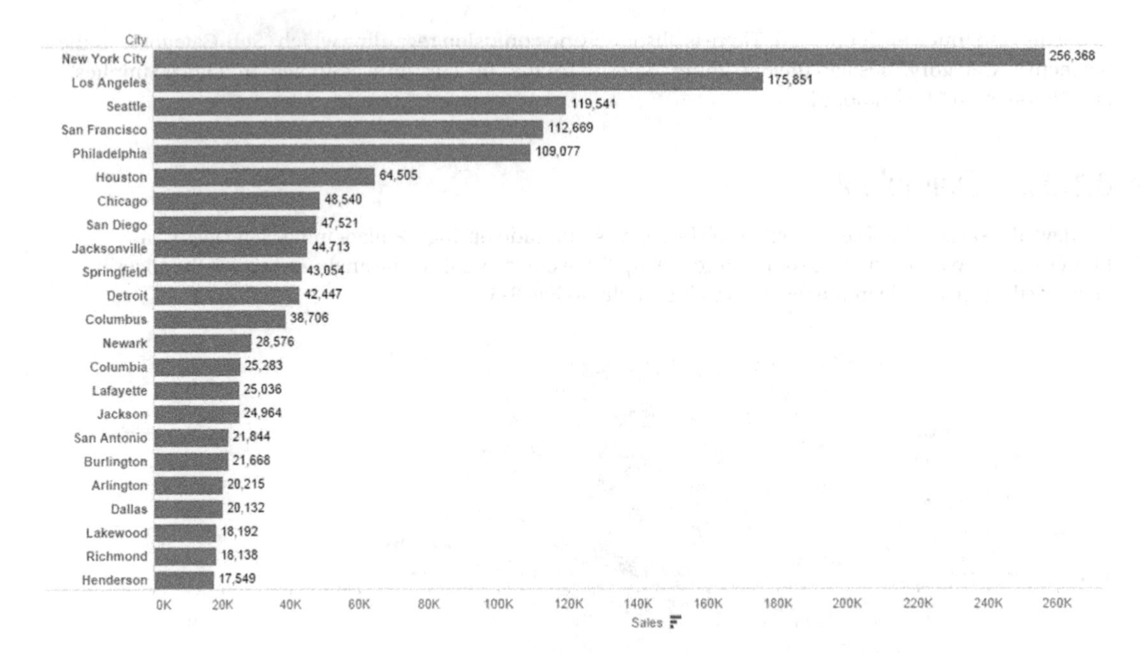

Figure 8-4. *Bar chart showing the same data as in Fig. 8-3*

One can clearly decipher from Fig. 8-4, the top three cities by sales (New York City, Los Angeles and Seattle). We did not have to use different colors for the bars, because one differentiating factor is more than enough. The length of the bars serves the purpose.

8.1.9.3 Demo 1

Objective: To demonstrate graphically the percentage of "Sales" by "Customer Segment".

Input: "Sample - Superstore.xls"

8.1.9.3.1 Steps to create a pie chart

8.1.9.3.2 Step 1

Select "Pie" in the marks card. Notice that “Angle” is added as another feature in the marks card (Shown in Fig. 8-5).

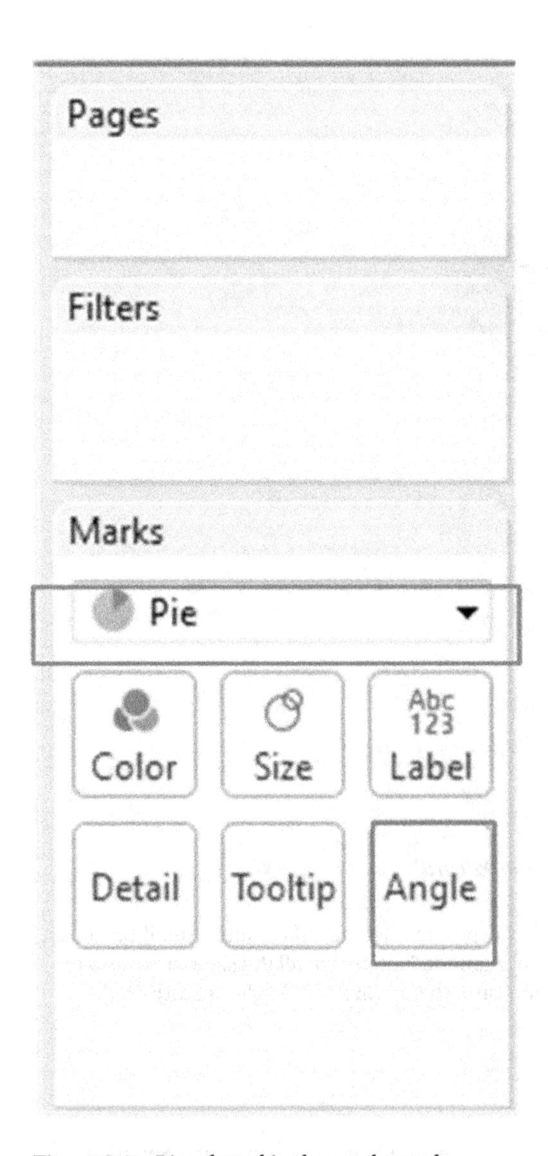

Figure 8-5. *Pie selected in the marks card*

8.1.9.3.3 Step 2

Drag the dimension "Customer Segment" from the dimensions area under the data pane and place it on "Color" on the marks card (Shown in Fig. 8-6).

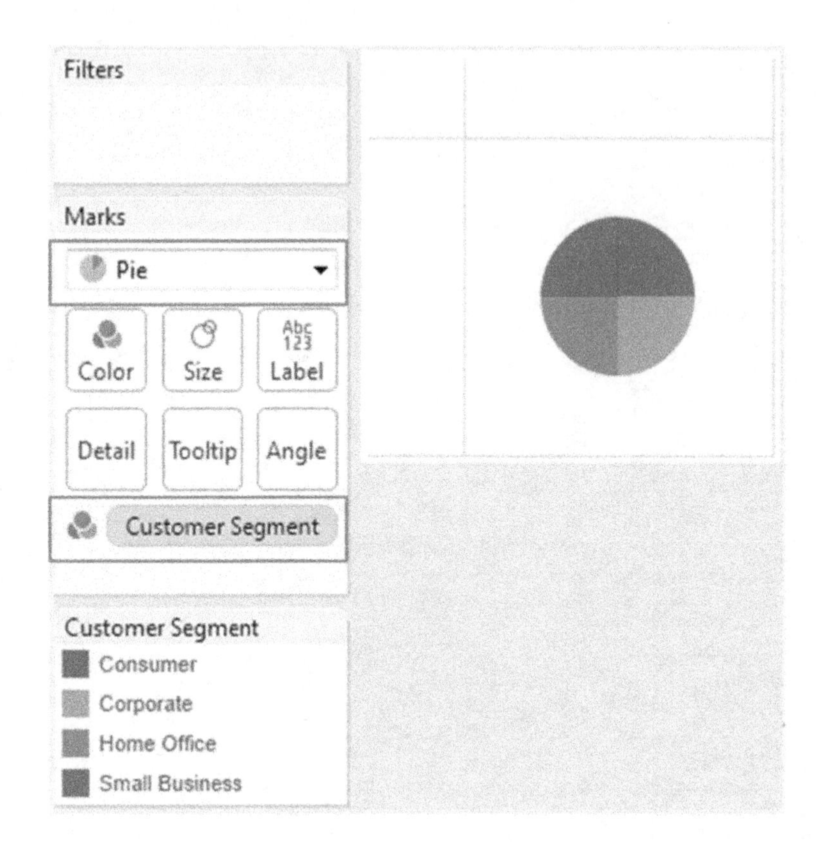

Figure 8-6. *"Customer Segment" placed on "Color" on the marks card*

We have data for four segments, namely, "Consumer", "Corporate", "Home Office" and "Small Business". The pie chart, therefore, is split into four slices, one for each segment. Notice that all the slices are the same size. The reason is that up to now we have not added any measure to the "Size" or "Angle" feature on the marks card.

8.1.9.3.4 Step 3

Let us have the size of the slices determined by the amount of "Sales" made in each "Customer Segment". Drag the measure "Sales" from the measures area under the data pane and place it on "Size" on the marks card. (Shown in Fig. 8-7).

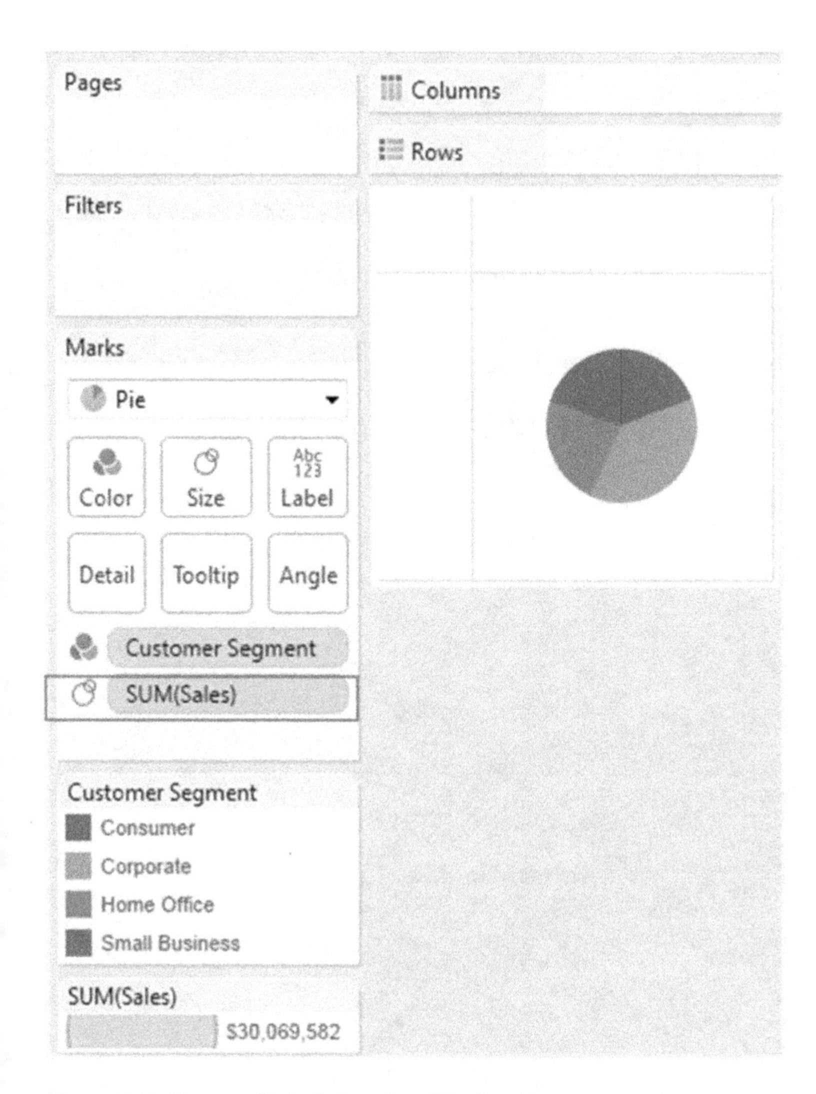

Figure 8-7. *Measure "Sales" placed on "Size" on the marks card*

Notice the change in the size of the slices. The size is as per the "Sales" made in each "Customer Segment".

8.1.9.3.5 Step 4

Drag the measure "Profit" from the measures area under the data pane and place it on "Angle" on the marks card (Shown in Fig. 8-8). The angle of the slices is now as per the "Profit" made in each "Customer Segment".

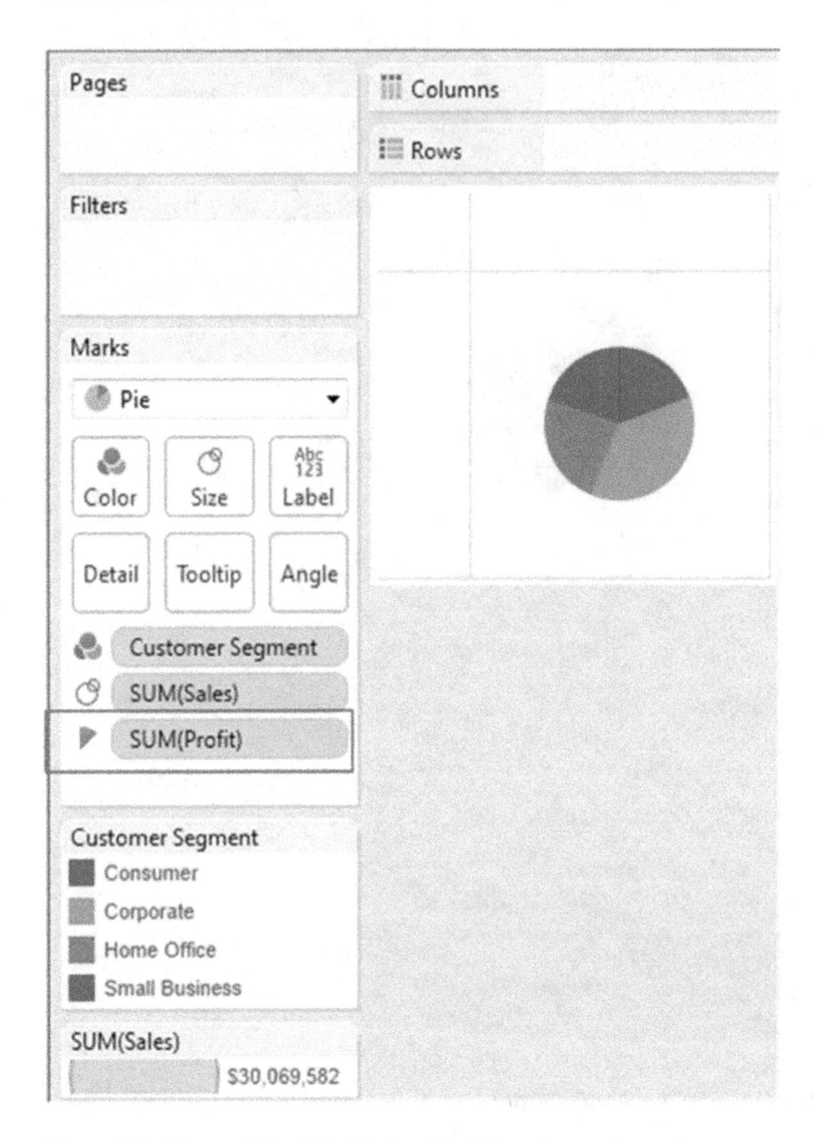

Figure 8-8. *Measure "Profit" placed on "Angle" on the marks card*

8.1.9.3.6 Step 5

Drag the measure "Sales" from the measures area under the data pane and place it on "Label" on the marks card (Shown in Fig. 8-9).

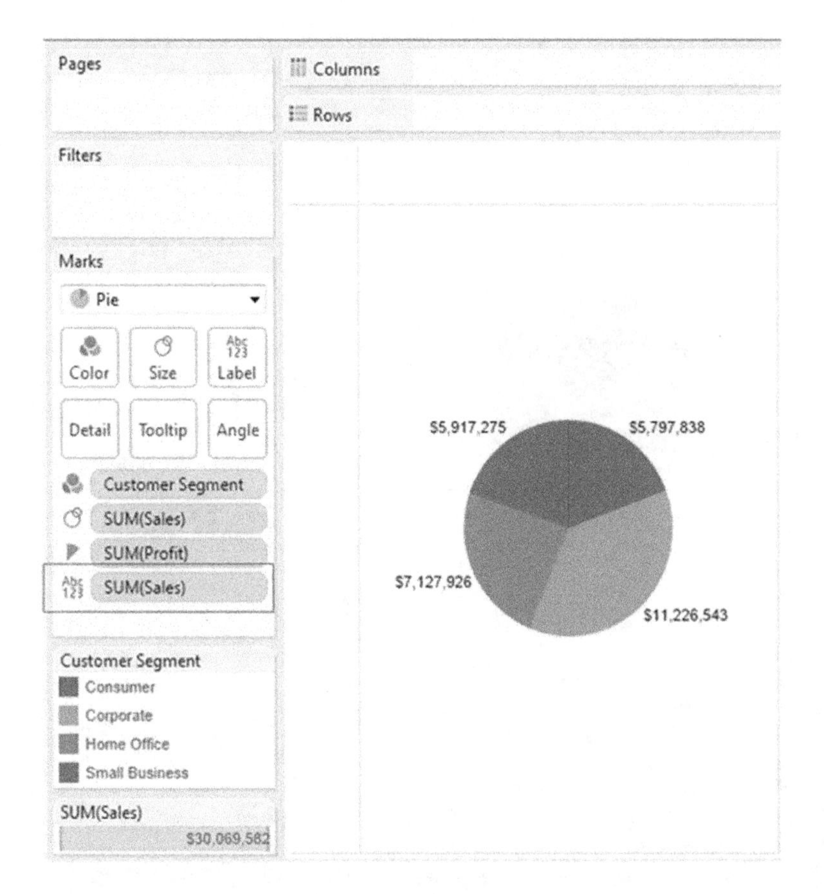

Figure 8-9. *Measure "Sales" placed on "Label" on the marks card*

8.1.9.3.7 Step 6

Let us transform the sales figures into percentages. Right click on measure "Sales" on the "Label" to bring up a drop down menu. Select "Quick Table Calculation" (Shown in Fig. 8-10).

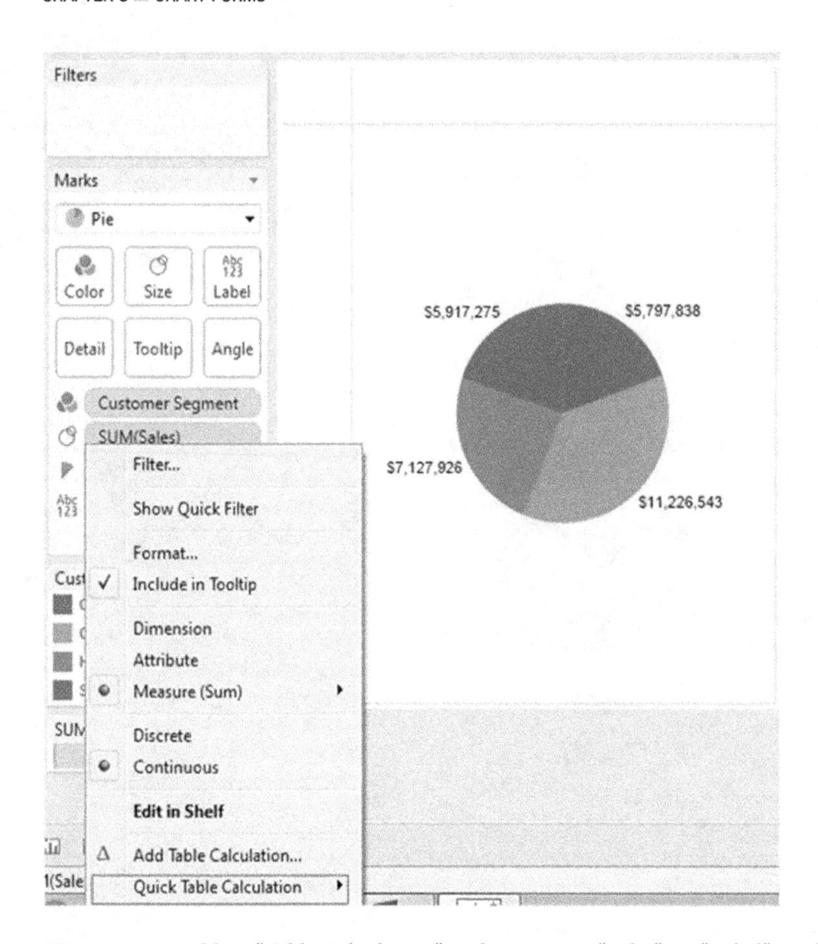

Figure 8-10. *Adding "Table Calculation" to the measure "Sales" on "Label" on the marks card*

Selecting "Quick Table Calculation" brings up another set of options. Select "Percent of Total" (Shown in Fig. 8-11).

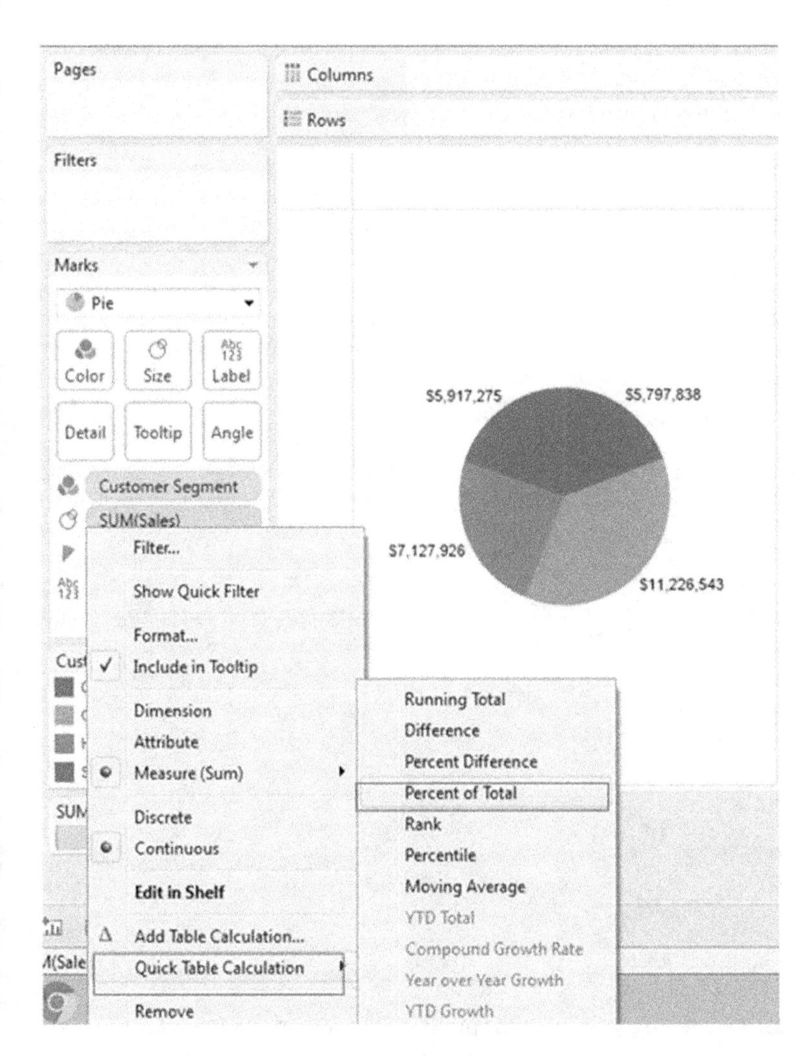

***Figure 8-11.** "Percent of Total" option on "Quick Table Calculation"*

The displayed labels on the pie chart in the worksheet / view changes to reflect the percentage of total (Shown in Fig. 8-12).

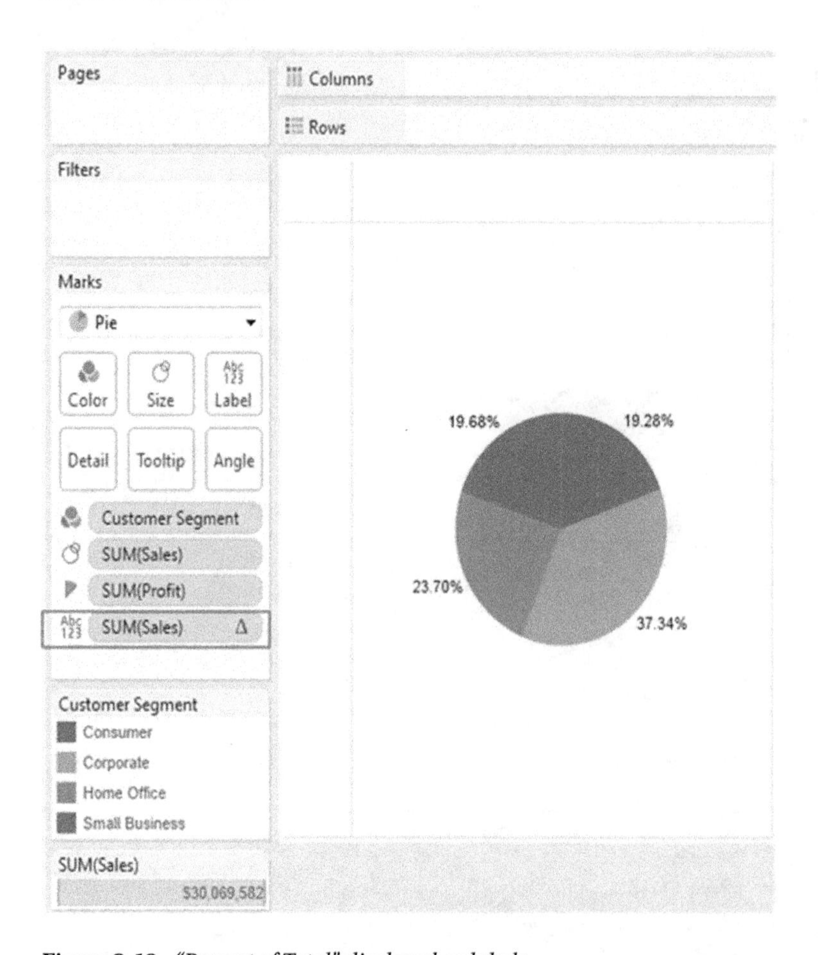

Figure 8-12. *"Percent of Total" displayed as labels*

If you happen to add up the percentages of each slice in the pie, it should add up to a 100%. Let us validate.

(19.68 + 19.28 + 37.34 + 23.70) = 100

8.1.9.4 Reference

A paper called "No Humble Pie: The Origins and Usage of a Statistical Chart," by Ian Spence, professor of psychology at the University of Toronto, published in the *Journal of Educational and Behavioral Statistics* in 2005.

8.2 Treemaps

Treemaps are used to show hierarchical data. Hierarchical data is also referred to as tree-structured data. It is depicted as a set of nested rectangles. The rectangles are sized and ordered by a measure. The largest sized rectangle is positioned in the top left corner and the smallest rectangle in the bottom right corner. If the rectangles are nested, the same ordering is maintained with the smaller rectangles within the larger rectangles.

This idea was invented by Professor Ben Shneiderman at the University of Maryland Human Computer Interaction Lab in the early 1990s.

The original motivation of treemaps was to display the content of the hard disk. SequoiaView is a disk browsing tool based on the principal of treemaps. The area denotes the file size and the file type decides the color.

SmartMoney was the first widely available treemap implementation for non-hierarchical data.

8.2.1 Pros

- Treemaps can display hundreds to thousands of items in a meaningful organized display. Such a display facilitates easy spotting of patterns and exceptions.
- By design they make efficient use of space.
- Useful to show "part to whole" relationship.

8.2.2 References

`http://www.cs.umd.edu/hcil/treemap-history/index.shtml`

`http://www.cs.umd.edu/hcil/treemap/`

8.2.2.1 Demo 1

Picture this...

We are a retail store with several categories of items. Each category is further subdivided into several sub categories. If we were to represent the entire data at once, we should use treemaps. Each category constitutes the branch of a tree. Each category is represented by a rectangle. Each category rectangle then is tiled by several small rectangles representing sub-branches. The sub-branches are constituted of sub-categories.

8.2.2.2 Example

The data in the data set is available in three categories.

- Furniture
- Office supplies
- Technology

Each Category further drills down into few subcategories as shown in Fig. 8-13.

Category	Sub-Category	
Furniture	Bookcases	114,880
	Chairs	328,449
	Furnishings	91,705
	Tables	206,966
Office Supplies	Appliances	107,532
	Art	27,119
	Binders	203,413
	Envelopes	16,476
	Fasteners	3,024
	Labels	12,486
	Paper	78,479
	Storage	223,844
	Supplies	46,674
Technology	Accessories	167,380
	Copiers	149,528
	Machines	189,239
	Phones	330,007

Figure 8-13. *Categories and sub-categories of data*

The above data set has data in 17 subcategories, which rolls up into three categories. However, tomorrow there can be several more subcategories of data. We need a chart form that allows one to view all the data at once. Treemaps clearly fulfil our need! (Refer to Fig. 8-14).

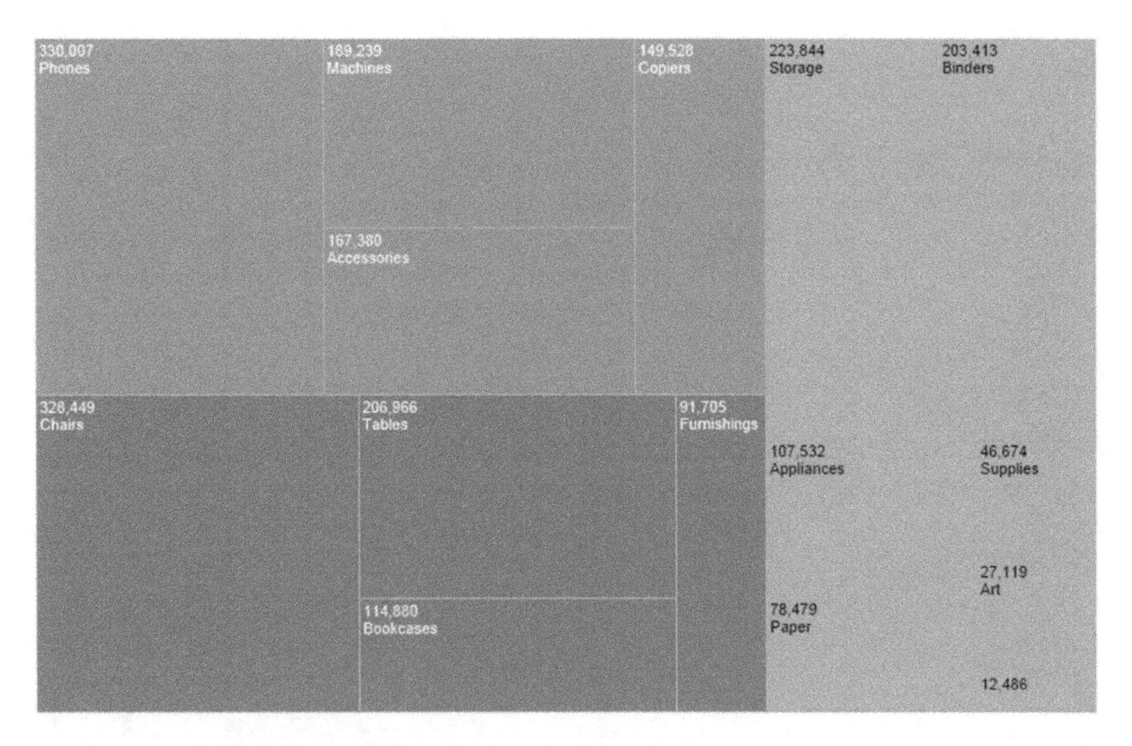

Figure 8-14. *Treemap depicting categories and sub-categories of data*

How to arrive at the above visualization?

To plot a treemap, the following is required:

- One or more dimensions
- One or two measures

Quite often a quantitative measure is used to provide size and another (same or different) measure is used to provide color to the treemap. If a measure is used to provide color, stick to using a single color if the measure begins at zero. However if the measure can have a negative value, such as the measure "Profit", one can play with two colors and also use the "Stepped Color" option.

The above treemap is colored by a dimension (Category).

8.2.2.2.1 Steps to create a treemap

8.2.2.2.2 Step 1

Drag the dimension "Category" from the dimensions area under the data pane and drop it on "Color" on the marks card (Shown in Fig. 8-15).

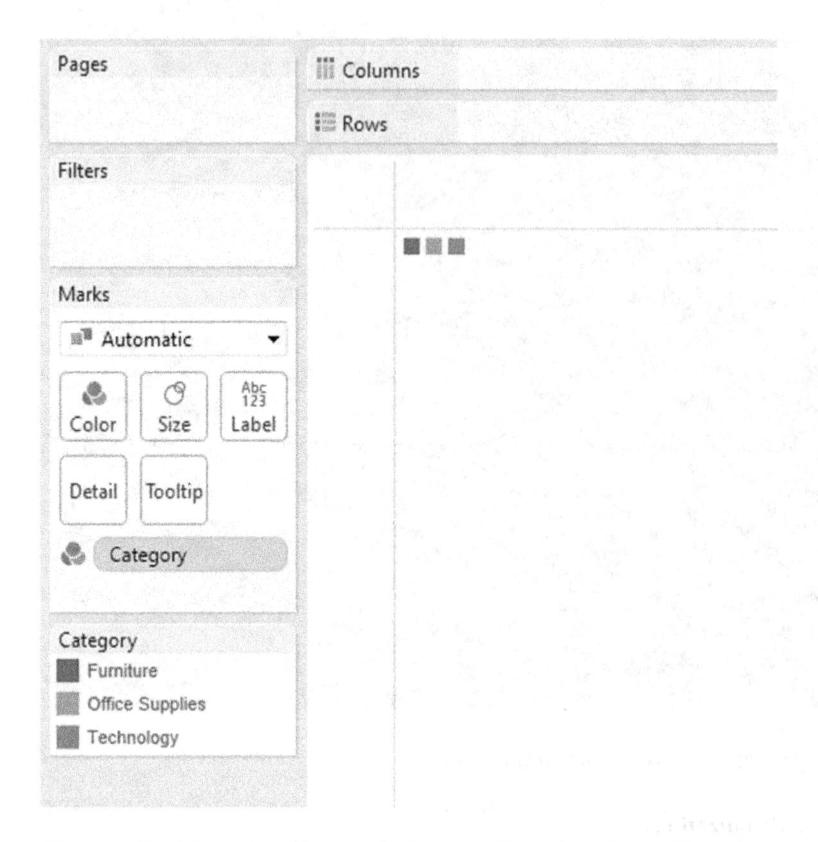

Figure 8-15. *Dimension "Category" placed on "Color" on the marks card*

Data is available in three categories as evident by three squares in the worksheet / view.

8.2.2.2.3 Step 2

Drag the measure "Sales" from the measures area under the data pane and drop it on "Size" on the marks card. The size of the rectangle is based on the amount of sales in each "Category" (Shown in Fig. 8-16).

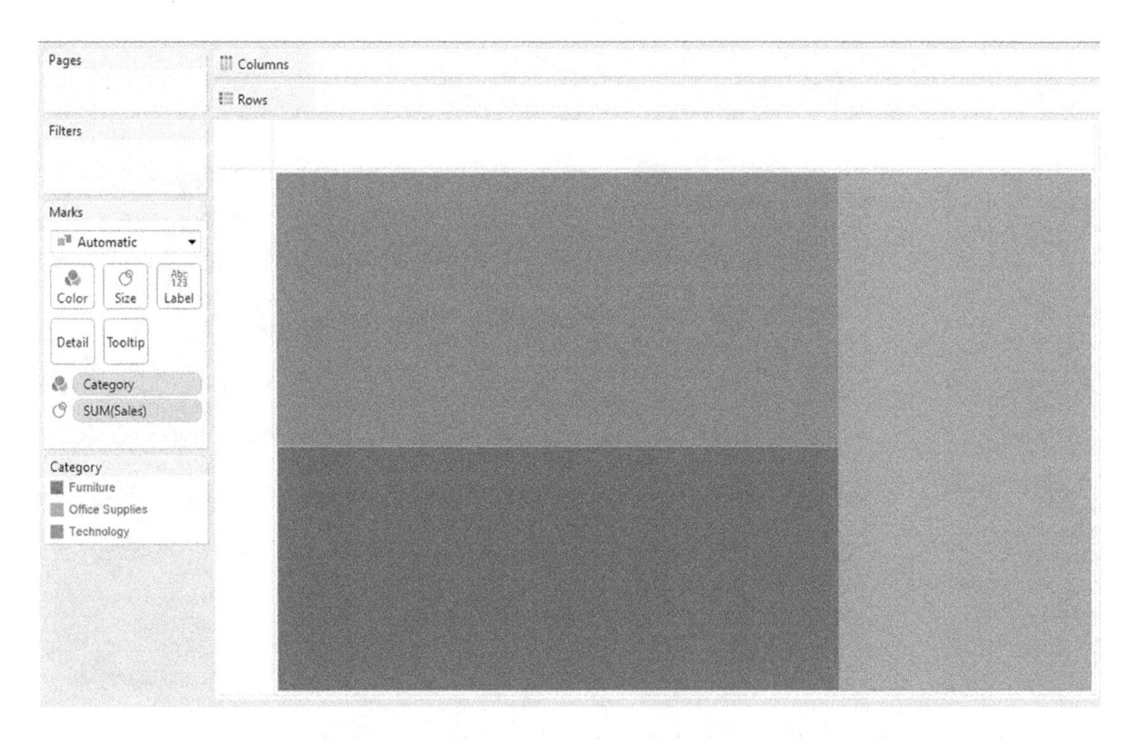

Figure 8-16. *Measure "Sales" placed on "Size" on the marks card*

8.2.2.2.4 Step 3

Drag the measure "Sales" from the measures area under the data pane and drop it on "Label" on the marks card (Shown in Fig. 8-17).

__Figure 8-17.__ Measure "Sales" placed on "Label" on the marks card

8.2.2.2.5 Step 4

Drag the dimension "Sub-Category" from the dimensions area under the data pane and drop it on "Label" on the marks card (Shown in Fig. 8-18).

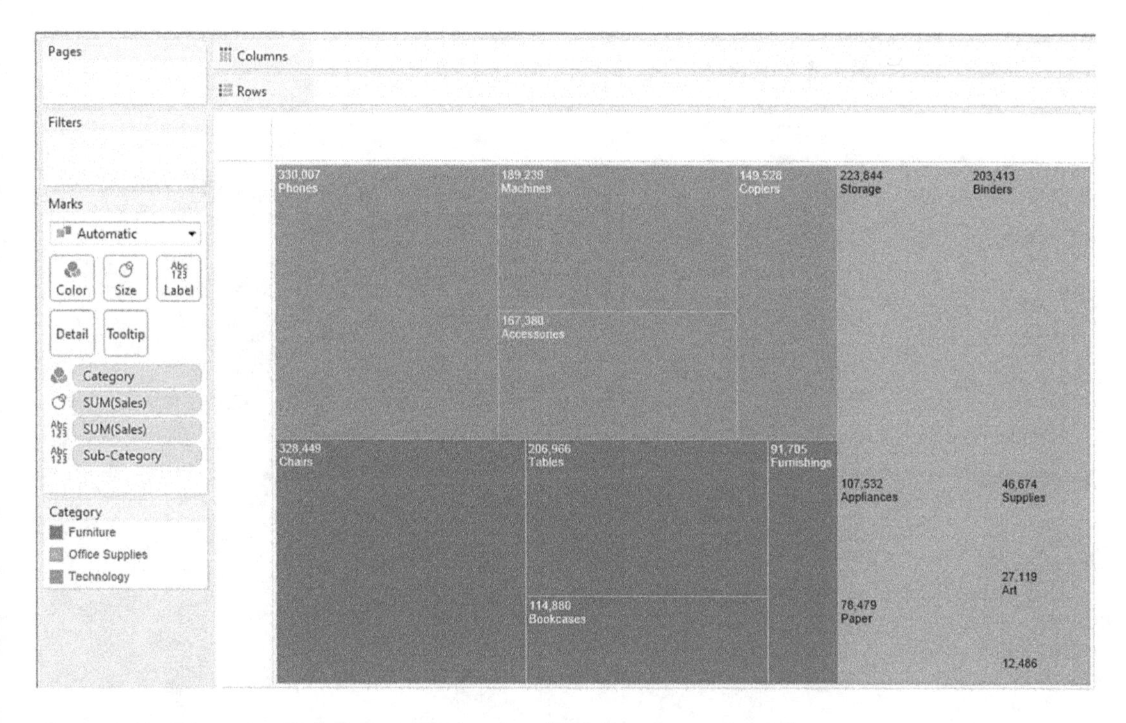

Figure 8-18. *Dimension "Sub-Category" placed on "Label" on the marks card*

8.2.2.2.6 Step 5

We can further add more details to the above visualization by adding the dimension "Product Name" from the dimensions area under the data pane to "Label" on the marks card (Shown in Fig. 8-19).

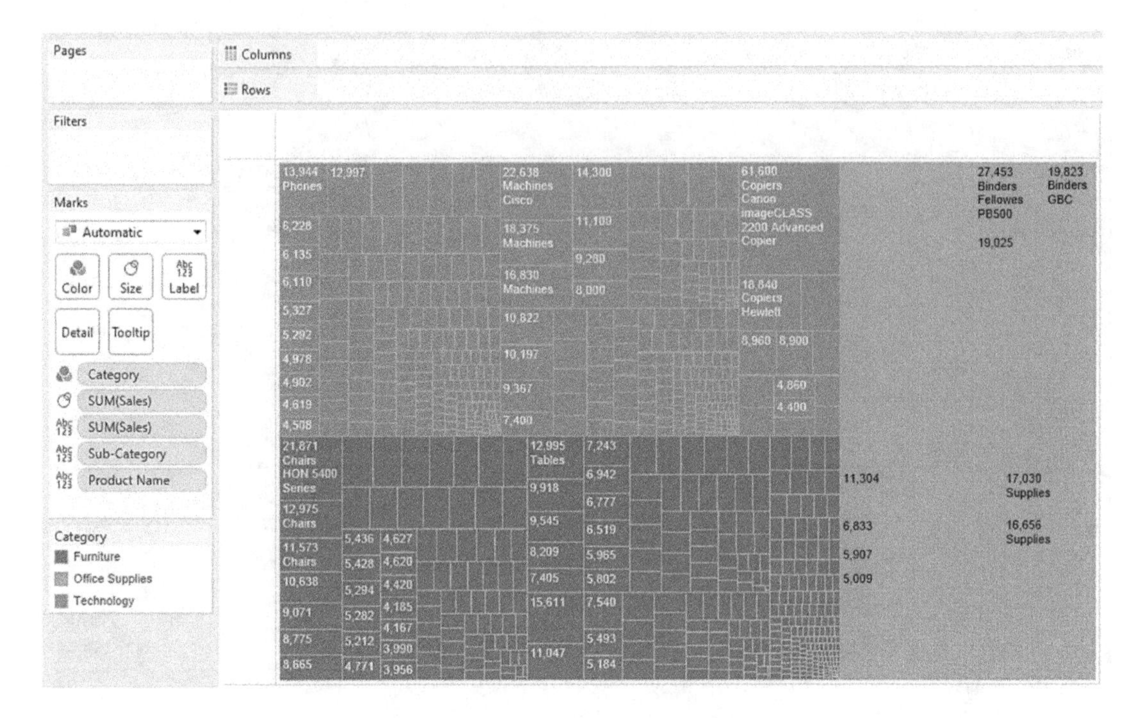

Figure 8-19. *Dimension "Product Name" placed on "Label" on the marks card*

8.3 Heat Map

A heat map is a **two-dimensional** representation of data. Heat maps use color to display values.

8.3.1 Why use heat maps?

- They help one to comprehend complex data sets quiet easily.
- They differ from fractal and treemaps, which are similar to heat maps in the use of color; however, they are better with hierarchies.

The term "heat map" was originally coined by the software designer Cormac Kinney in 1991. It was used the first time to describe a 2D display depicting real time financial market information.

Heat maps are often confused with choropleth maps. However, choropleth maps are constrained by geographical boundaries to show proportion of a variable of interest. Heat maps, however, do not have any such constraints.

8.3.2 How to create a heat map?

Data set used: The data lists each day of the year with a ranking for how many babies were born in the United States on each date from 1973 to 1999.

Data source: NYTimes.com, Amitabh Chandra, Harvard University
Fields in the dataset:

- Rank
- Month
- Day

A subset of the data from the data set (Shown in Fig. 8-20). The data set has 366 records.

	A	B	C
1	Rank	Month	Day
2	364	1	1
3	362	1	2
4	356	1	3
5	350	1	4
6	338	1	5
7	301	1	6
8	324	1	7
9	347	1	8
10	351	1	9
11	349	1	10
12	341	1	11
13	306	1	12
14	316	1	13
15	260	1	14
16	304	1	15
17	322	1	16
18	337	1	17
19	317	1	18
20	302	1	19

Figure 8-20. *A subset of the data from the data set*

8.3.2.1 Steps to create a heat map

8.3.2.1.1 Step 1

Connect the Excel sheet "Most Common Birthdays.xls" to Tableau (Shown in Fig. 8-21).

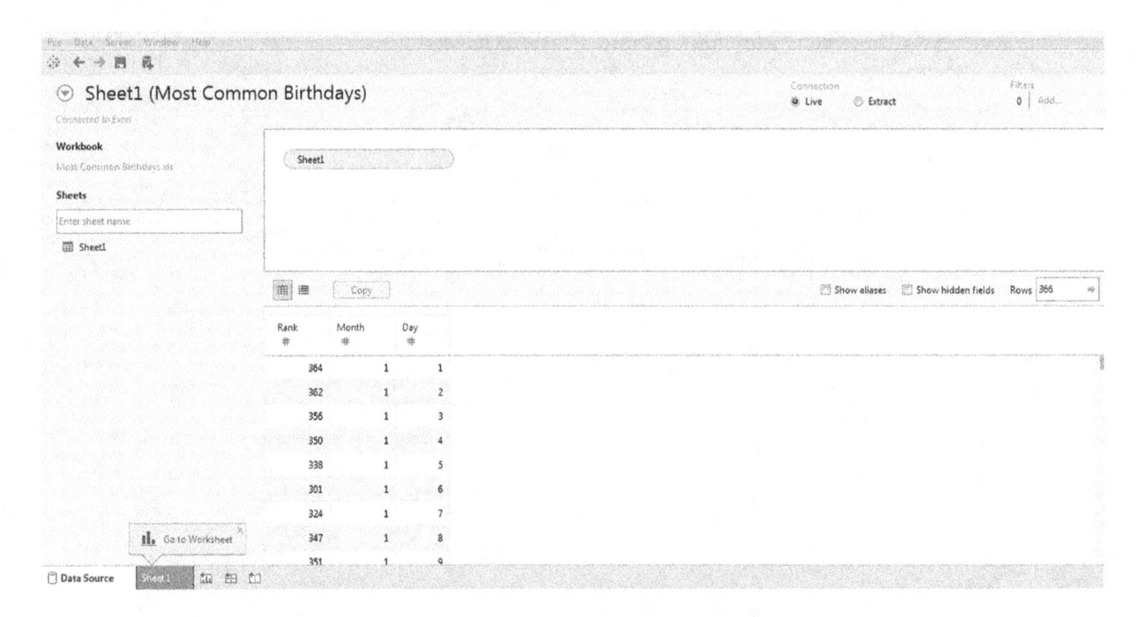

Figure 8-21. *Data from "Most Common Birthdays.xls" read into Tableau*

Go to the worksheet or view (Shown in Fig. 8-22).

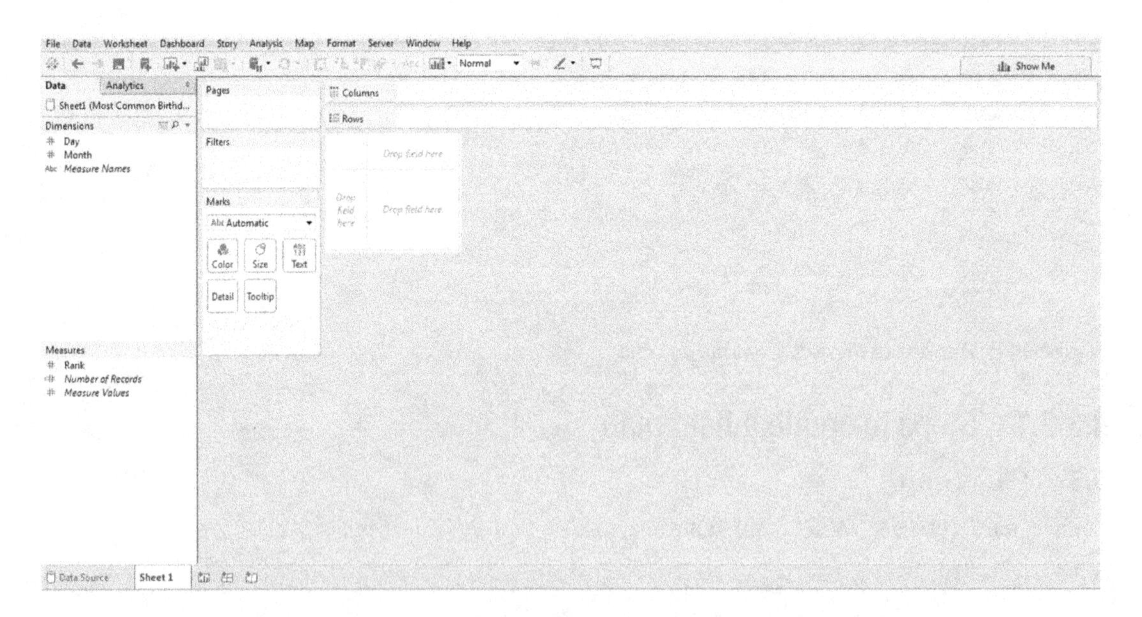

Figure 8-22. *Tableau Worksheet / View after connecting to "Most common Birthdays.xls"*

8.3.2.1.2 Step 2

Create a calculated field "MonthNames" which stores the month's name corresponding to the months. For example, "Jan" for 1, "Feb" for 2, etc. (Shown in Fig. 8-23).

Figure 8-23. *Calculated field "MonthNames" being created*

The formula used is as follows:

```
If [Month] = 1 then "Jan"
ELSEIF [Month] = 2 then "Feb"
ELSEIF [Month] = 3 then "Mar"
ELSEIF [Month] = 4 then "Apr"
ELSEIF [Month] = 5 then "May"
ELSEIF [Month] = 6 then "Jun"
ELSEIF [Month] = 7 then "Jul"
ELSEIF [Month] = 8 then "Aug"
ELSEIF [Month] = 9 then "Sep"
ELSEIF [Month] = 10 then "Oct"
ELSEIF [Month] = 11 then "Nov"
ELSEIF [Month] = 12 then "Dec"
END
```

8.3.2.1.3 Step 3

Convert the dimension "Day" to "Discrete" (Shown in Fig. 8-24).

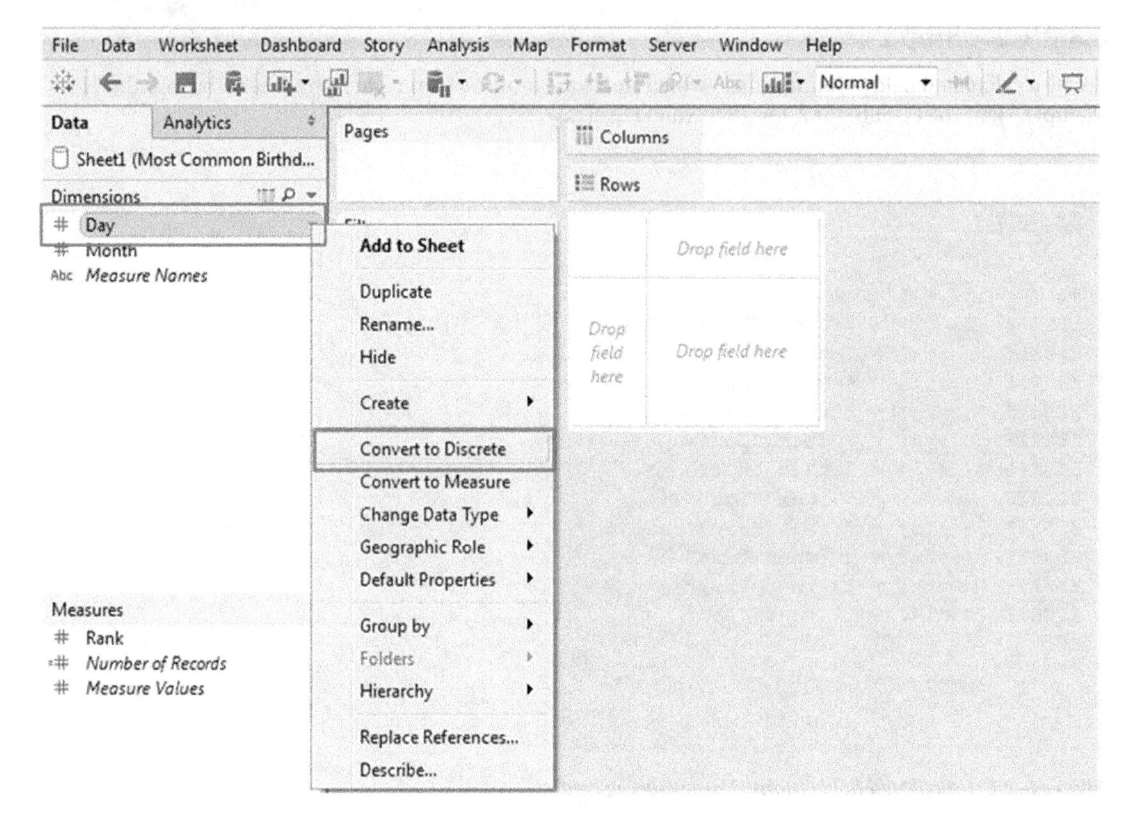

Figure 8-24. *Dimension "Day" being converted to "Discrete"*

8.3.2.1.4 Step 4

Drag and drop the dimension "MonthNames" from the dimensions area of the data pane on the columns shelf (Shown in Fig. 8-25).

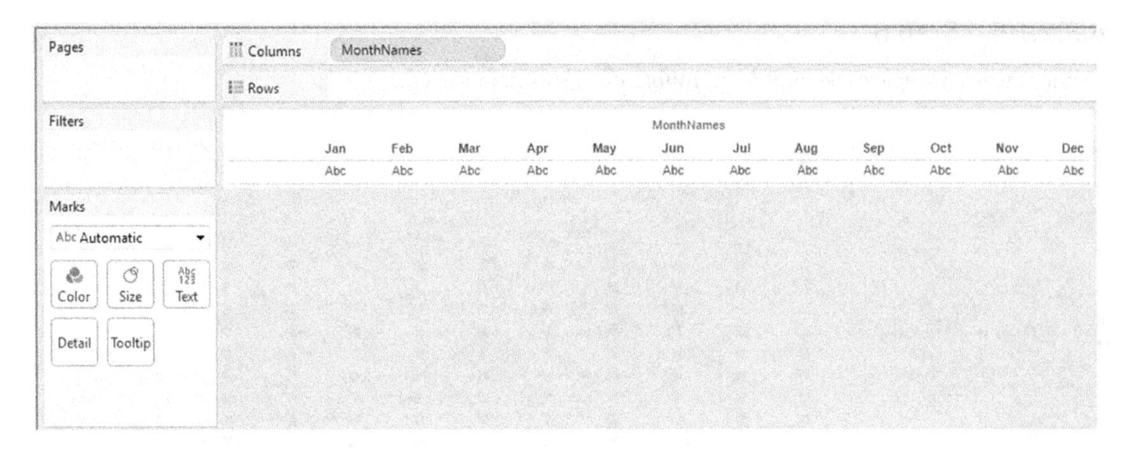

***Figure 8-25.** Calculated field "MonthNames" placed on the columns shelf*

8.3.2.1.5 Step 5

Drag and drop the dimension "Day" from the dimensions area of the data pane on the rows shelf (Shown in Fig. 8-26).

Columns: MonthNames

Rows: Day

MonthNames

Day	Jan	Feb	Mar	Apr	May	Jun	Jul	Aug	Sep	Oct	Nov	Dec
1	Abc	Abc	Abc	Abc	Abc	Abc	Abc	Abc	Abc	Abc	Abc	Abc
2	Abc	Abc	Abc	Abc	Abc	Abc	Abc	Abc	Abc	Abc	Abc	Abc
3	Abc	Abc	Abc	Abc	Abc	Abc	Abc	Abc	Abc	Abc	Abc	Abc
4	Abc	Abc	Abc	Abc	Abc	Abc	Abc	Abc	Abc	Abc	Abc	Abc
5	Abc	Abc	Abc	Abc	Abc	Abc	Abc	Abc	Abc	Abc	Abc	Abc
6	Abc	Abc	Abc	Abc	Abc	Abc	Abc	Abc	Abc	Abc	Abc	Abc
7	Abc	Abc	Abc	Abc	Abc	Abc	Abc	Abc	Abc	Abc	Abc	Abc
8	Abc	Abc	Abc	Abc	Abc	Abc	Abc	Abc	Abc	Abc	Abc	Abc
9	Abc	Abc	Abc	Abc	Abc	Abc	Abc	Abc	Abc	Abc	Abc	Abc
10	Abc	Abc	Abc	Abc	Abc	Abc	Abc	Abc	Abc	Abc	Abc	Abc
11	Abc	Abc	Abc	Abc	Abc	Abc	Abc	Abc	Abc	Abc	Abc	Abc
12	Abc	Abc	Abc	Abc	Abc	Abc	Abc	Abc	Abc	Abc	Abc	Abc
13	Abc	Abc	Abc	Abc	Abc	Abc	Abc	Abc	Abc	Abc	Abc	Abc
14	Abc	Abc	Abc	Abc	Abc	Abc	Abc	Abc	Abc	Abc	Abc	Abc
15	Abc	Abc	Abc	Abc	Abc	Abc	Abc	Abc	Abc	Abc	Abc	Abc
16	Abc	Abc	Abc	Abc	Abc	Abc	Abc	Abc	Abc	Abc	Abc	Abc
17	Abc	Abc	Abc	Abc	Abc	Abc	Abc	Abc	Abc	Abc	Abc	Abc
18	Abc	Abc	Abc	Abc	Abc	Abc	Abc	Abc	Abc	Abc	Abc	Abc
19	Abc	Abc	Abc	Abc	Abc	Abc	Abc	Abc	Abc	Abc	Abc	Abc
20	Abc	Abc	Abc	Abc	Abc	Abc	Abc	Abc	Abc	Abc	Abc	Abc
21	Abc	Abc	Abc	Abc	Abc	Abc	Abc	Abc	Abc	Abc	Abc	Abc
22	Abc	Abc	Abc	Abc	Abc	Abc	Abc	Abc	Abc	Abc	Abc	Abc
23	Abc	Abc	Abc	Abc	Abc	Abc	Abc	Abc	Abc	Abc	Abc	Abc
24	Abc	Abc	Abc	Abc	Abc	Abc	Abc	Abc	Abc	Abc	Abc	Abc

***Figure 8-26.** Dimension "Day" placed on the rows shelf*

8.3.2.1.6 Step 6

Change the marks type to "Square" from "Automatic" (Shown in Fig. 8-27).

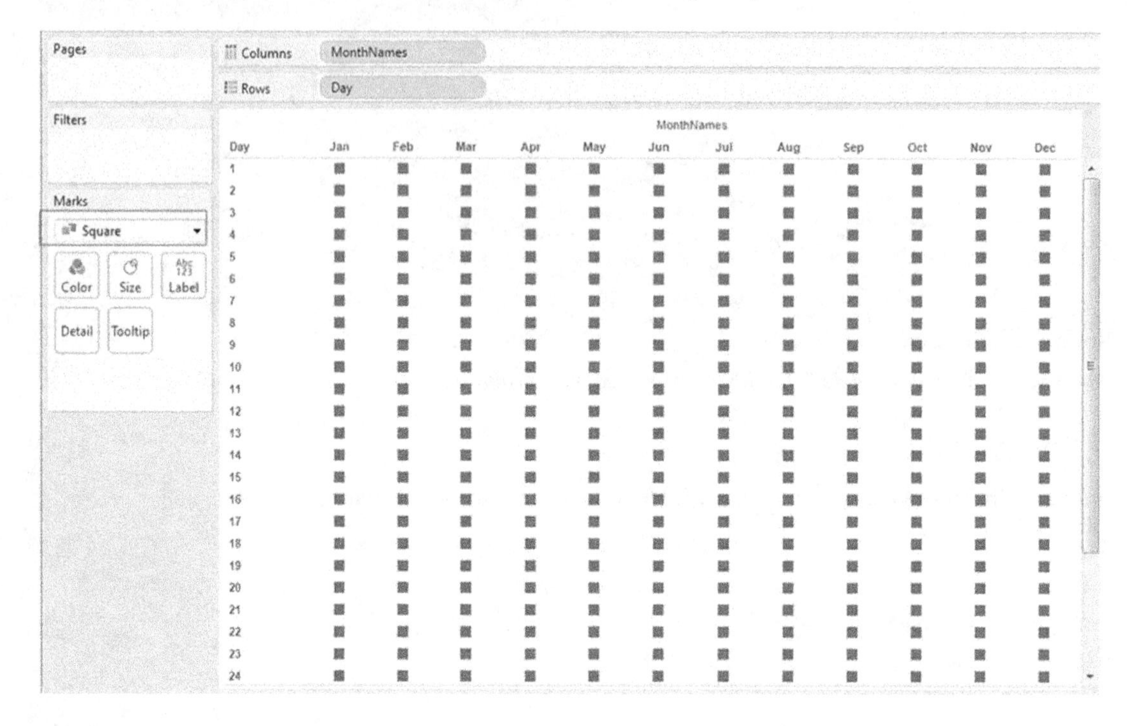

Figure 8-27. *Marks type changed to "Square"*

8.3.2.1.7 Step 7

Drag the measure "Rank" from measures area under the data pane and drop it on "Color" on the marks card (Shown in Fig. 8-28).

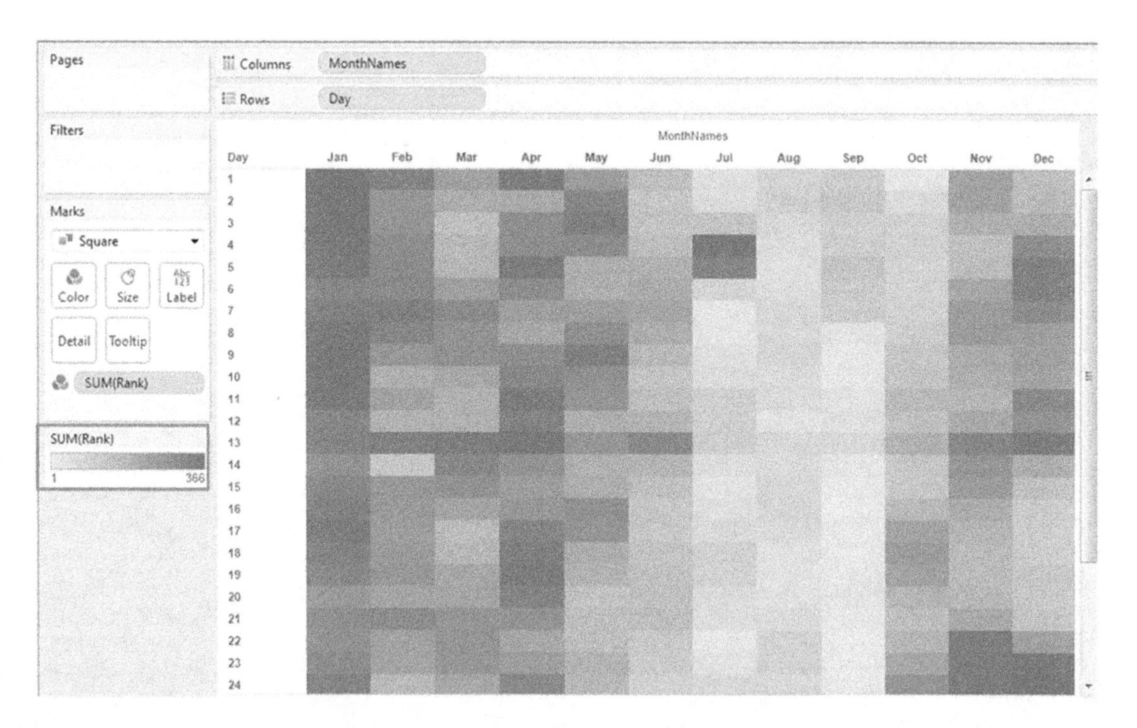

Figure 8-28. *Measure "Rank" placed on "Color" on the marks card*

8.3.2.1.8 Step 8

Double click on "Color" to bring up the edit colors dialog box. Change the color from "Green" to "Orange" (Shown in Fig. 8-29).

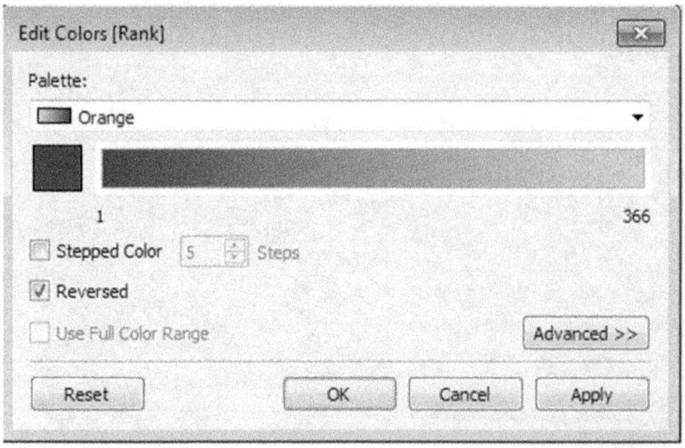

Edit Colors [Rank]

Palette:

Orange

1 366

Stepped Color 5 Steps

Reversed

Use Full Color Range

Advanced >>

Reset OK Cancel Apply

Figure 8-29. *"Edit Colors" dialog box for "Rank"*

Click on "Apply" and then ok. The output is as shown in Fig. 8-30.

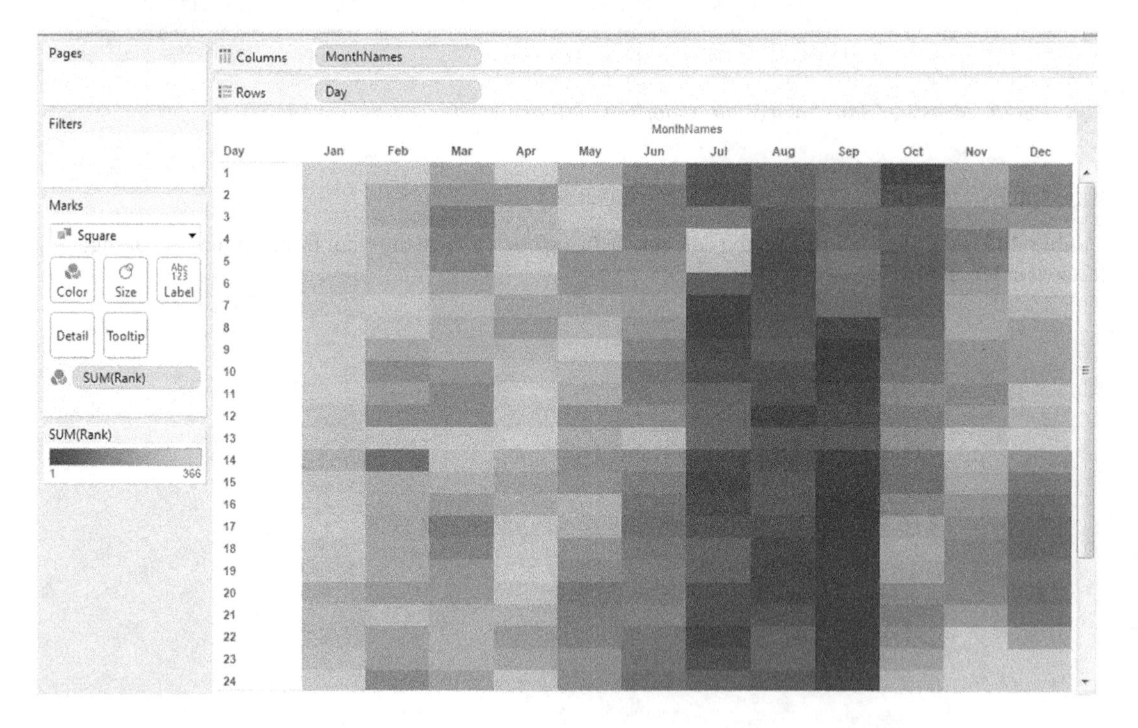

Figure 8-30. *Heat map - Demo 1 - final output*

8.3.2.1.9 Conclusion

The 16th day of September is ranked first, meaning the maximum number of babies was born on 16 September. (Shown in Fig. 8-31).

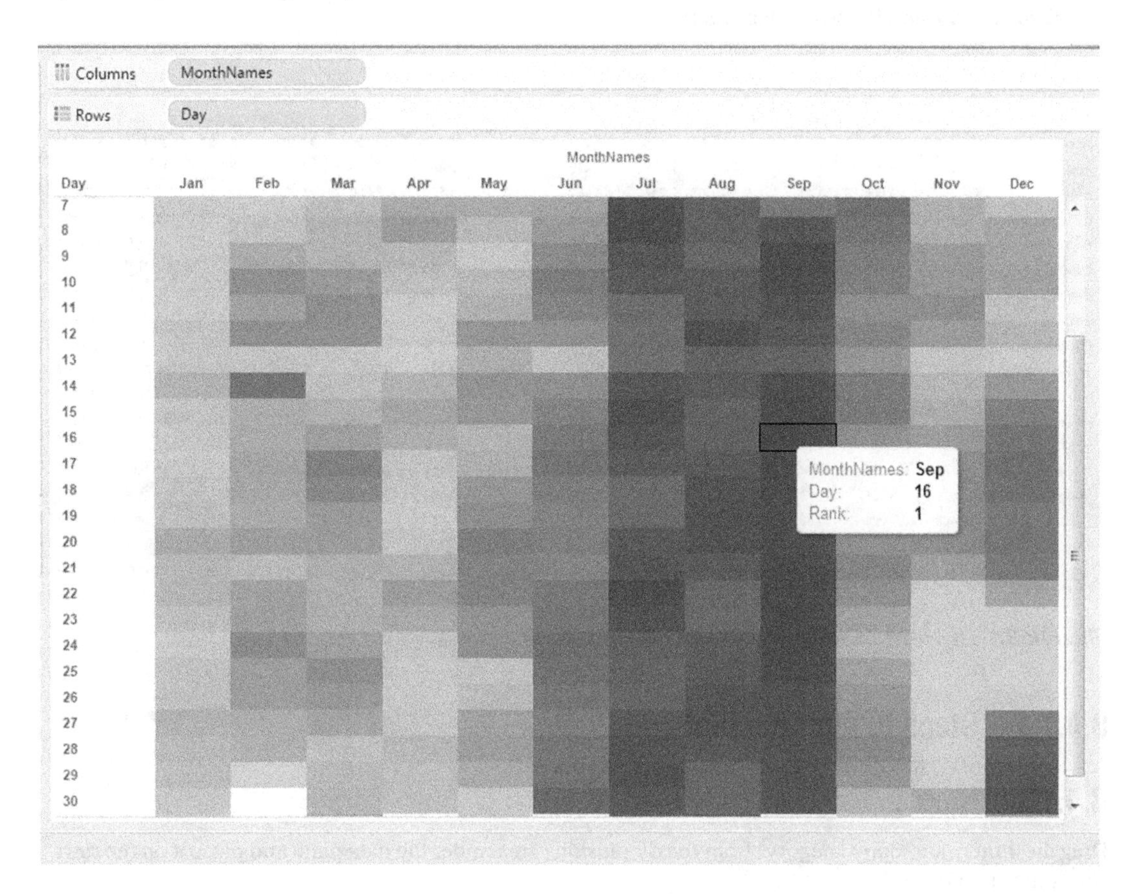

Figure 8-31. *Interpretation of heat map - Demo 1*

Likewise the 29th day of Feb had the least number of babies who were born on that day.

8.4 Highlight Table

Highlight table is simply a large text table wherein the data values are encoded by color. For example, the higher the profit values, the darker the color. They help to reduce the time to insight and improve the accuracy of the insights. They are suitable for

- Quickly identifying highs and lows or other points of interest in your data.
- A means of enhancing a crosstab.

Highlight tables are created using one or more dimensions and exactly one measure placed on "Color" on the marks card.

8.4.1 Demo 1

Objective: To study the measure "Sales" across the dimensions "Sub-Category" and "Order Date".

Input: "Sample - Superstore.xls"

Expected output: (Shown in Fig. 8-32).

	Order Date												
Sub-Category	January	February	March	April	May	June	July	August	September	October	November	December	Grand Tot..
Accessories	5,478	5,369	8,735	7,981	9,615	8,858	17,135	11,758	25,400	13,087	25,477	28,486	167,380
Appliances	3,176	4,933	6,700	6,075	7,526	7,479	3,384	12,862	10,828	9,155	18,306	17,107	107,532
Art	966	1,006	1,413	2,382	2,256	2,182	2,102	1,690	3,660	1,905	3,816	3,740	27,119
Binders	12,412	4,286	13,728	13,384	9,245	13,218	7,755	21,302	37,337	18,090	20,789	31,867	203,413
Bookcases	5,062	1,940	7,147	4,926	6,290	7,445	10,292	5,622	22,849	8,771	23,561	10,977	114,880
Chairs	11,285	7,768	20,832	18,855	25,703	21,145	23,585	17,770	52,147	21,905	47,314	60,141	328,449
Copiers	3,960		22,590	6,880	18,400	900	9,780	5,730	10,320	37,020	15,150	18,800	149,528
Envelopes	750	669	1,657	852	1,190	514	1,200	701	2,177	1,393	2,917	2,458	16,476
Fasteners	88	159	150	258	109	116	182	235	414	326	548	441	3,024
Furnishings	3,980	2,316	5,068	7,185	7,305	5,900	7,355	4,343	11,805	5,447	16,757	14,244	91,705
Labels	207	300	940	408	885	1,207	1,692	876	1,476	1,269	1,850	1,376	12,486
Machines	7,215	8,990	35,052	18,190	11,268	12,183	4,065	6,262	26,386	10,613	33,807	15,210	189,239
Paper	2,287	2,805	6,218	3,865	6,359	6,546	4,319	6,360	10,575	5,309	12,563	11,274	78,479
Phones	13,772	9,000	26,712	18,647	24,859	25,492	23,807	28,046	38,464	25,963	56,075	39,169	330,007
Storage	9,374	6,125	14,793	15,806	14,670	17,272	13,768	17,421	29,866	15,822	37,418	31,510	223,844
Supplies	4,403	289	10,607	6,246	1,154	1,267	8,816	859	6,442	816	1,372	4,402	46,674
Tables	10,952	4,218	16,913	9,913	9,288	15,360	10,344	17,752	19,626	20,223	31,401	40,975	206,966
Grand Total	95,366	60,173	199,253	141,852	156,122	147,083	149,581	159,589	309,770	197,115	349,120	332,177	2,297,201

Figure 8-32. *Highlight Table - Demo 1 - expected output*

8.4.1.1 Steps to create a highlight table

8.4.1.2 Step 1

Drag the dimension "Sub-Category" from the dimensions area under the data pane and place it on the rows shelf (Shown in Fig. 8-33).

Columns

Rows Sub-Category

Sub-Category	
Accessories	Abc
Appliances	Abc
Art	Abc
Binders	Abc
Bookcases	Abc
Chairs	Abc
Copiers	Abc
Envelopes	Abc
Fasteners	Abc
Furnishings	Abc
Labels	Abc
Machines	Abc
Paper	Abc
Phones	Abc
Storage	Abc
Supplies	Abc
Tables	Abc

Figure 8-33. *Dimension "Sub-Category" placed on the rows shelf*

Drag the dimension "Order Date" from the dimensions area under the data pane and place it on the columns shelf. Change the granularity to Month (Order Date) (Shown in Fig. 8-34).

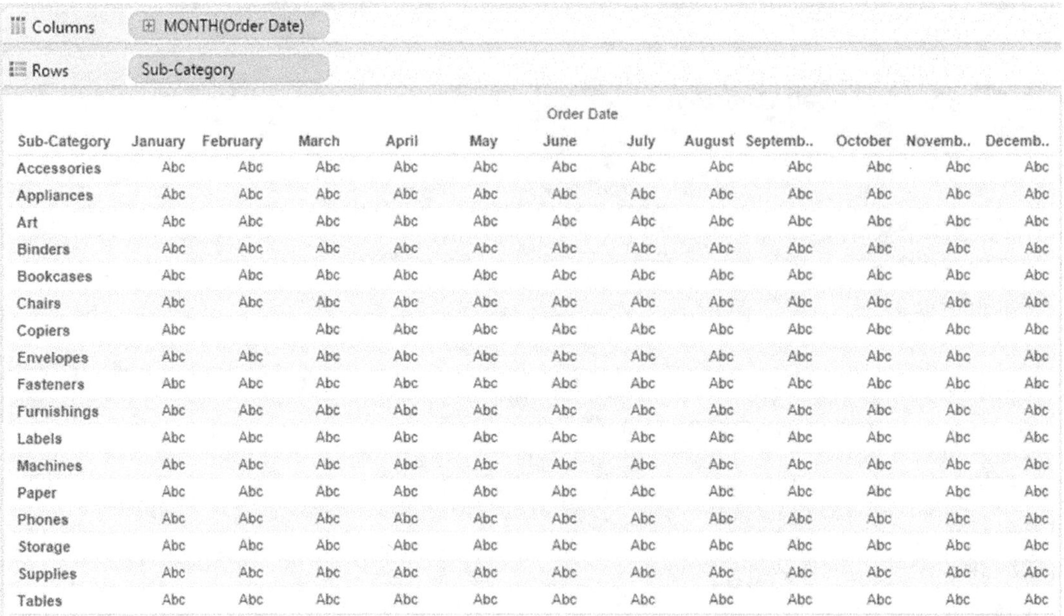

Columns: MONTH(Order Date)

Rows: Sub-Category

	Order Date											
Sub-Category	January	February	March	April	May	June	July	August	Septemb..	October	Novemb..	Decemb..
Accessories	Abc	Abc	Abc	Abc	Abc	Abc	Abc	Abc	Abc	Abc	Abc	Abc
Appliances	Abc	Abc	Abc	Abc	Abc	Abc	Abc	Abc	Abc	Abc	Abc	Abc
Art	Abc	Abc	Abc	Abc	Abc	Abc	Abc	Abc	Abc	Abc	Abc	Abc
Binders	Abc	Abc	Abc	Abc	Abc	Abc	Abc	Abc	Abc	Abc	Abc	Abc
Bookcases	Abc	Abc	Abc	Abc	Abc	Abc	Abc	Abc	Abc	Abc	Abc	Abc
Chairs	Abc	Abc	Abc	Abc	Abc	Abc	Abc	Abc	Abc	Abc	Abc	Abc
Copiers	Abc		Abc	Abc	Abc	Abc	Abc	Abc	Abc	Abc	Abc	Abc
Envelopes	Abc	Abc	Abc	Abc	Abc	Abc	Abc	Abc	Abc	Abc	Abc	Abc
Fasteners	Abc	Abc	Abc	Abc	Abc	Abc	Abc	Abc	Abc	Abc	Abc	Abc
Furnishings	Abc	Abc	Abc	Abc	Abc	Abc	Abc	Abc	Abc	Abc	Abc	Abc
Labels	Abc	Abc	Abc	Abc	Abc	Abc	Abc	Abc	Abc	Abc	Abc	Abc
Machines	Abc	Abc	Abc	Abc	Abc	Abc	Abc	Abc	Abc	Abc	Abc	Abc
Paper	Abc	Abc	Abc	Abc	Abc	Abc	Abc	Abc	Abc	Abc	Abc	Abc
Phones	Abc	Abc	Abc	Abc	Abc	Abc	Abc	Abc	Abc	Abc	Abc	Abc
Storage	Abc	Abc	Abc	Abc	Abc	Abc	Abc	Abc	Abc	Abc	Abc	Abc
Supplies	Abc	Abc	Abc	Abc	Abc	Abc	Abc	Abc	Abc	Abc	Abc	Abc
Tables	Abc	Abc	Abc	Abc	Abc	Abc	Abc	Abc	Abc	Abc	Abc	Abc

Figure 8-34. *Dimension "Order Date" placed on the columns shelf*

8.4.1.3 Step 2

Drag the measure "Sales" from the measures area under the data pane and place it on "Label" on the marks card (Shown in Fig. 8-35).

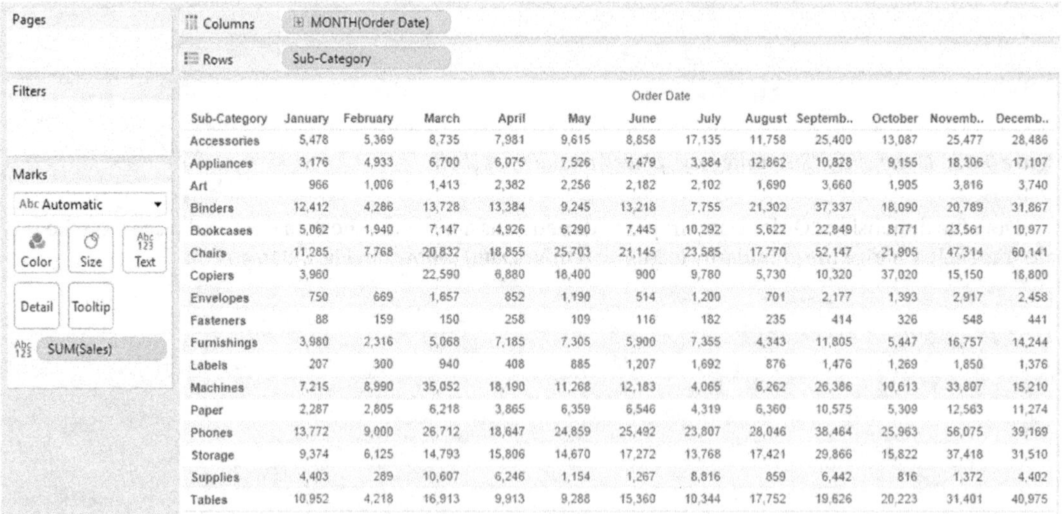

Pages

Filters

Marks

Abc Automatic

Color Size Text

Detail Tooltip

SUM(Sales)

Columns: MONTH(Order Date)

Rows: Sub-Category

	Order Date											
Sub-Category	January	February	March	April	May	June	July	August	Septemb..	October	Novemb..	Decemb..
Accessories	5,478	5,369	8,735	7,981	9,615	8,858	17,135	11,758	25,400	13,087	25,477	28,486
Appliances	3,176	4,933	6,700	6,075	7,526	7,479	3,384	12,862	10,828	9,155	18,306	17,107
Art	966	1,006	1,413	2,382	2,256	2,182	2,102	1,690	3,660	1,905	3,816	3,740
Binders	12,412	4,286	13,728	13,384	9,245	13,218	7,755	21,302	37,337	18,090	20,789	31,867
Bookcases	5,062	1,940	7,147	4,926	6,290	7,445	10,292	5,622	22,849	8,771	23,561	10,977
Chairs	11,285	7,768	20,832	18,855	25,703	21,145	23,585	17,770	52,147	21,905	47,314	60,141
Copiers	3,960		22,590	6,880	18,400	900	9,780	5,730	10,320	37,020	15,150	18,800
Envelopes	750	669	1,657	852	1,190	514	1,200	701	2,177	1,393	2,917	2,458
Fasteners	88	159	150	258	109	116	182	235	414	326	548	441
Furnishings	3,980	2,316	5,068	7,185	7,305	5,900	7,355	4,343	11,805	5,447	16,757	14,244
Labels	207	300	940	408	885	1,207	1,692	876	1,476	1,269	1,850	1,376
Machines	7,215	8,990	35,052	18,190	11,268	12,183	4,065	6,262	26,386	10,613	33,807	15,210
Paper	2,287	2,805	6,218	3,865	6,359	6,546	4,319	6,360	10,575	5,309	12,563	11,274
Phones	13,772	9,000	26,712	18,647	24,859	25,492	23,807	28,046	38,464	25,963	56,075	39,169
Storage	9,374	6,125	14,793	15,806	14,670	17,272	13,768	17,421	29,866	15,822	37,418	31,510
Supplies	4,403	289	10,607	6,246	1,154	1,267	8,816	859	6,442	816	1,372	4,402
Tables	10,952	4,218	16,913	9,913	9,288	15,360	10,344	17,752	19,626	20,223	31,401	40,975

Figure 8-35. *Measure "Sales" placed on "Label" on the marks card*

8.4.1.4 Step 3

Drag the measure "Sales" from the measures area under the data pane and place it on "Color" on the marks card (Shown in Fig. 8-36).

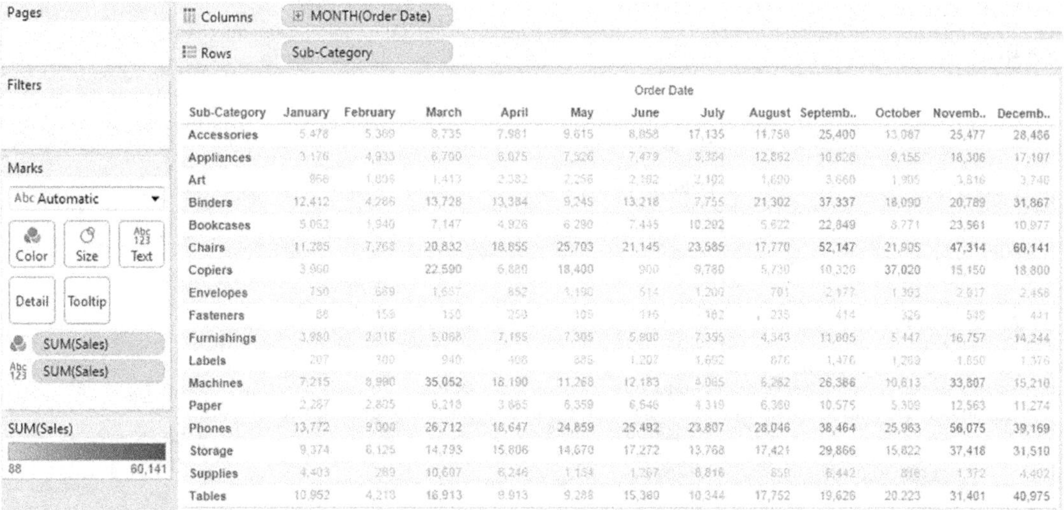

Pages | Columns: MONTH(Order Date) | Rows: Sub-Category
Filters
Marks: Automatic — Color, Size, Text, Detail, Tooltip — SUM(Sales), SUM(Sales)
SUM(Sales): 88 – 60,141

Sub-Category	January	February	March	April	May	June	July	August	Septemb..	October	Novemb..	Decemb..
						Order Date						
Accessories	5,478	5,369	8,735	7,981	9,615	8,858	17,135	11,758	25,400	13,087	25,477	28,486
Appliances	3,176	4,933	6,700	6,075	7,526	7,479	3,384	12,862	10,828	9,155	18,306	17,107
Art	966	1,006	1,413	2,382	2,256	2,182	2,102	1,690	3,660	1,905	3,816	3,740
Binders	12,412	4,286	13,728	13,384	9,245	13,218	7,755	21,302	37,337	18,090	20,789	31,867
Bookcases	5,062	1,940	7,147	4,926	6,290	7,445	10,292	5,622	22,849	8,771	23,561	10,977
Chairs	11,285	7,768	20,832	18,855	25,703	21,145	23,585	17,770	52,147	21,905	47,314	60,141
Copiers	3,960		22,590	6,880	18,400	900	9,780	5,730	10,320	37,020	15,150	18,800
Envelopes	750	669	1,657	852	1,190	514	1,200	701	2,177	1,393	2,917	2,458
Fasteners	88	159	150	258	109	116	182	235	414	326	548	441
Furnishings	3,980	2,316	5,068	7,185	7,305	5,900	7,355	4,343	11,805	5,447	16,757	14,244
Labels	207	300	940	408	885	1,207	1,692	876	1,476	1,269	1,850	1,376
Machines	7,215	8,990	35,052	18,190	11,268	12,183	4,065	6,262	26,386	10,613	33,807	15,210
Paper	2,287	2,805	6,218	3,865	6,359	6,546	4,319	6,360	10,575	5,309	12,563	11,274
Phones	13,772	9,000	26,712	18,647	24,859	25,492	23,807	28,046	38,464	25,963	56,075	39,169
Storage	9,374	6,125	14,793	15,806	14,670	17,272	13,768	17,421	29,866	15,822	37,418	31,510
Supplies	4,403	289	10,607	6,246	1,154	1,267	8,816	859	6,442	816	1,372	4,402
Tables	10,952	4,218	16,913	9,913	9,288	15,360	10,344	17,752	19,626	20,223	31,401	40,975

Figure 8-36. *Measure "Sales" placed on "Color" on the marks card*

8.4.1.5 Step 4

Change the mark type to "Square" (Shown in Fig. 8-37).

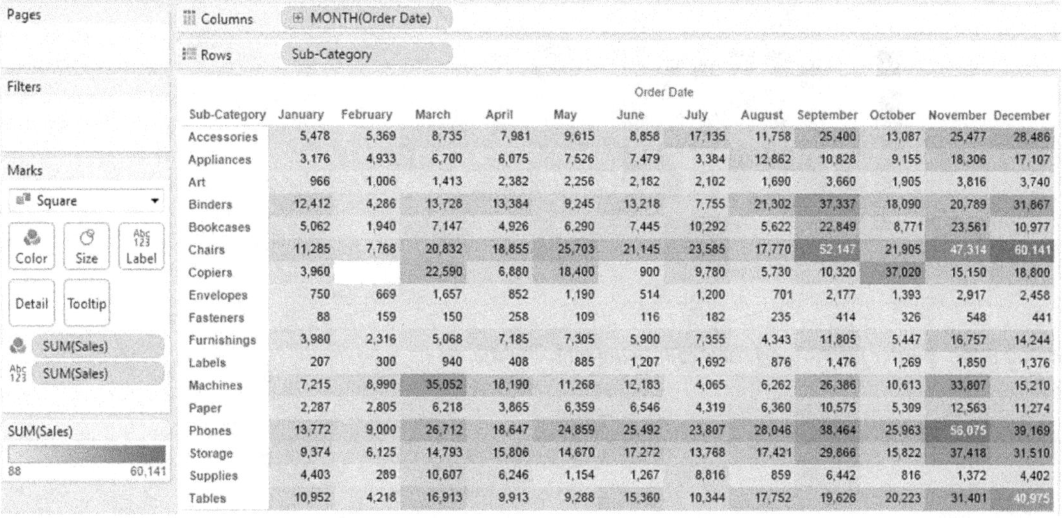

Pages | Columns: MONTH(Order Date) | Rows: Sub-Category
Filters
Marks: Square — Color, Size, Label, Detail, Tooltip — SUM(Sales), SUM(Sales)
SUM(Sales): 88 – 60,141

Sub-Category	January	February	March	April	May	June	July	August	September	October	November	December
						Order Date						
Accessories	5,478	5,369	8,735	7,981	9,615	8,858	17,135	11,758	25,400	13,087	25,477	28,486
Appliances	3,176	4,933	6,700	6,075	7,526	7,479	3,384	12,862	10,828	9,155	18,306	17,107
Art	966	1,006	1,413	2,382	2,256	2,182	2,102	1,690	3,660	1,905	3,816	3,740
Binders	12,412	4,286	13,728	13,384	9,245	13,218	7,755	21,302	37,337	18,090	20,789	31,867
Bookcases	5,062	1,940	7,147	4,926	6,290	7,445	10,292	5,622	22,849	8,771	23,561	10,977
Chairs	11,285	7,768	20,832	18,855	25,703	21,145	23,585	17,770	52,147	21,905	47,314	60,141
Copiers	3,960		22,590	6,880	18,400	900	9,780	5,730	10,320	37,020	15,150	18,800
Envelopes	750	669	1,657	852	1,190	514	1,200	701	2,177	1,393	2,917	2,458
Fasteners	88	159	150	258	109	116	182	235	414	326	548	441
Furnishings	3,980	2,316	5,068	7,185	7,305	5,900	7,355	4,343	11,805	5,447	16,757	14,244
Labels	207	300	940	408	885	1,207	1,692	876	1,476	1,269	1,850	1,376
Machines	7,215	8,990	35,052	18,190	11,268	12,183	4,065	6,262	26,386	10,613	33,807	15,210
Paper	2,287	2,805	6,218	3,865	6,359	6,546	4,319	6,360	10,575	5,309	12,563	11,274
Phones	13,772	9,000	26,712	18,647	24,859	25,492	23,807	28,046	38,464	25,963	56,075	39,169
Storage	9,374	6,125	14,793	15,806	14,670	17,272	13,768	17,421	29,866	15,822	37,418	31,510
Supplies	4,403	289	10,607	6,246	1,154	1,267	8,816	859	6,442	816	1,372	4,402
Tables	10,952	4,218	16,913	9,913	9,288	15,360	10,344	17,752	19,626	20,223	31,401	40,975

Figure 8-37. *Mark type set to "Square"*

Click on "Color" on the marks card and change the border to "White" (Shown in Fig. 8-38).

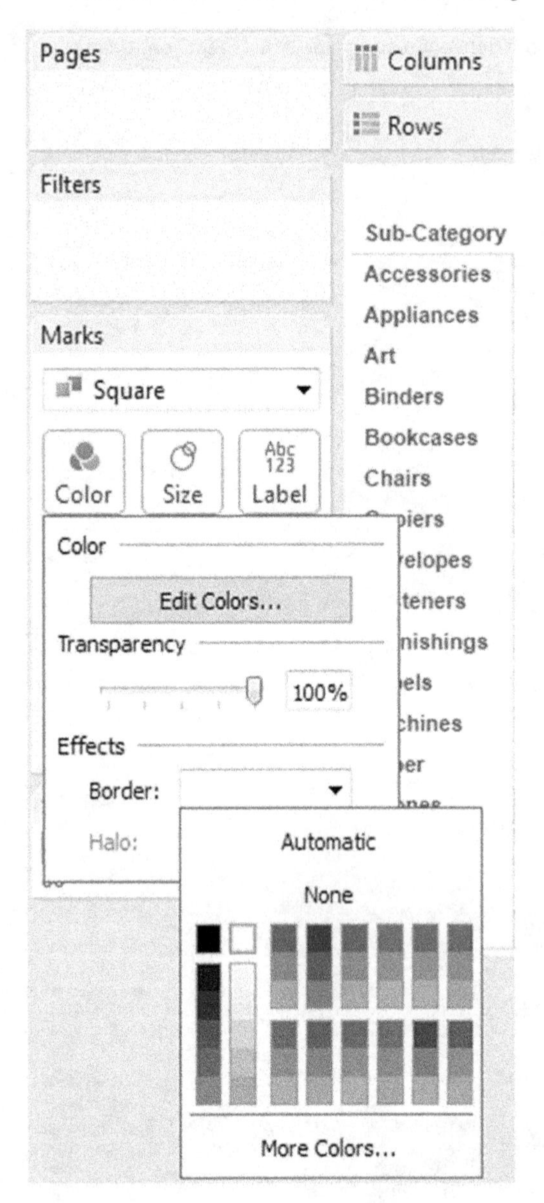

Figure 8-38. Border for the worksheet cells being set to "white"

The output after setting the border to "White" (Shown in Fig. 8-39).

	Order Date											
Sub-Category	January	February	March	April	May	June	July	August	September	October	November	December
Accessories	5,478	5,369	8,735	7,981	9,615	8,858	17,135	11,758	25,400	13,087	25,477	28,486
Appliances	3,176	4,933	6,700	6,075	7,526	7,479	3,384	12,862	10,828	9,155	18,306	17,107
Art	966	1,006	1,413	2,382	2,256	2,182	2,102	1,690	3,660	1,905	3,816	3,740
Binders	12,412	4,286	13,728	13,384	9,245	13,218	7,755	21,302	37,337	18,090	20,789	31,867
Bookcases	5,062	1,940	7,147	4,926	6,290	7,445	10,292	5,622	22,849	8,771	23,561	10,977
Chairs	11,285	7,768	20,832	18,855	25,703	21,145	23,585	17,770	52,147	21,905	47,314	60,141
Copiers	3,960		22,590	6,880	18,400	900	9,780	5,730	10,320	37,020	15,150	18,800
Envelopes	750	669	1,657	852	1,190	514	1,200	701	2,177	1,393	2,917	2,458
Fasteners	88	159	150	258	109	116	182	235	414	326	548	441
Furnishings	3,980	2,316	5,068	7,185	7,305	5,900	7,355	4,343	11,805	5,447	16,757	14,244
Labels	207	300	940	408	885	1,207	1,692	876	1,476	1,269	1,850	1,376
Machines	7,215	8,990	35,052	18,190	11,268	12,183	4,065	6,262	26,386	10,613	33,807	15,210
Paper	2,287	2,805	6,218	3,865	6,359	6,546	4,319	6,360	10,575	5,309	12,563	11,274
Phones	13,772	9,000	26,712	18,647	24,859	25,492	23,807	28,046	38,464	25,963	56,075	39,169
Storage	9,374	6,125	14,793	15,806	14,670	17,272	13,768	17,421	29,866	15,822	37,418	31,510
Supplies	4,403	289	10,607	6,246	1,154	1,267	8,816	859	6,442	816	1,372	4,402
Tables	10,952	4,218	16,913	9,913	9,288	15,360	10,344	17,752	19,626	20,223	31,401	40,975

Figure 8-39. *Output after worksheet cell borders have been set to "White"*

8.4.1.6 Step 5

Select "Analysis" on the menu bar. Select "Totals" and then select, "Show Row Grand Totals" and "Show Column Grand Totals" (Shown in Fig. 8-40).

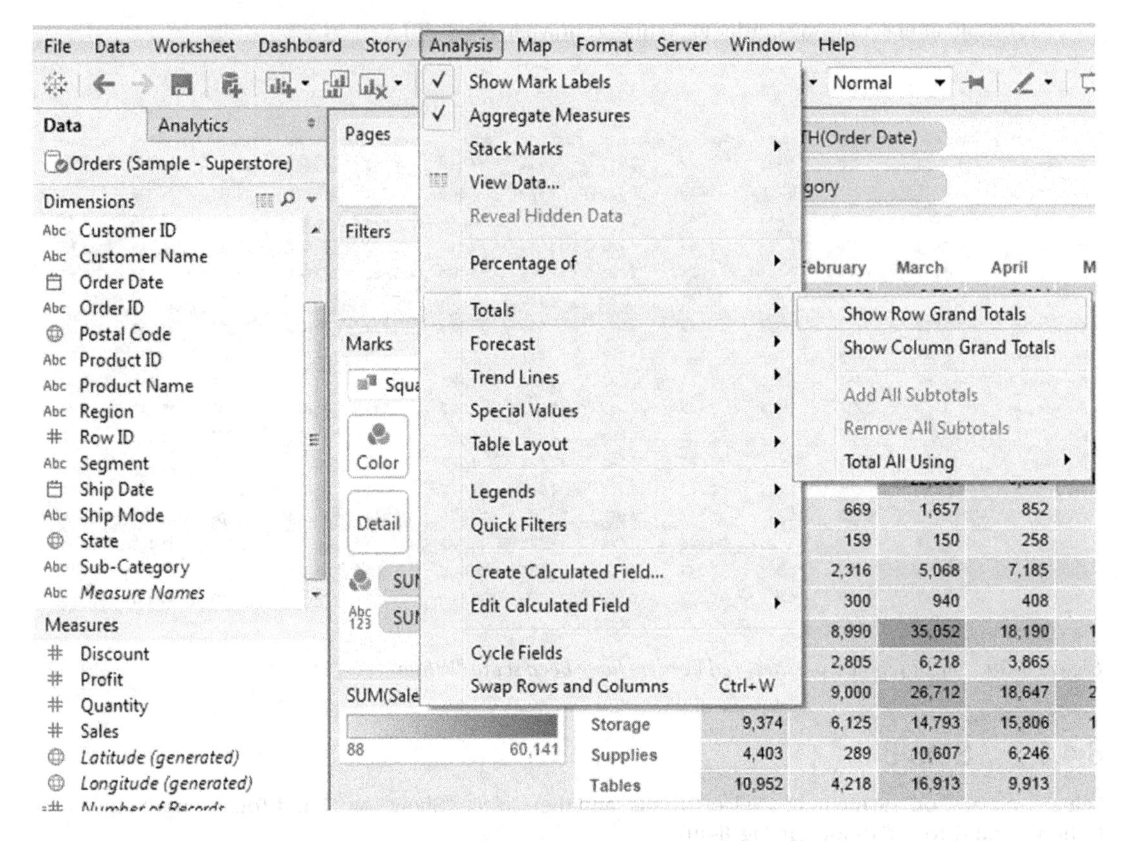

Figure 8-40. *Setting "Row Grand Totals and Column Grand Totals"*

The output after adding "Row Grand Totals" and "Column Grand Totals" (Shown in Fig. 8-41).

	Order Date												
Sub-Category	January	February	March	April	May	June	July	August	September	October	November	December	Grand Tot..
Accessories	5,478	5,369	8,735	7,981	9,615	8,858	17,135	11,758	25,400	13,087	25,477	28,486	167,380
Appliances	3,176	4,933	6,700	6,075	7,526	7,479	3,384	12,862	10,828	9,155	18,306	17,107	107,532
Art	966	1,006	1,413	2,382	2,256	2,182	2,102	1,690	3,660	1,905	3,816	3,740	27,119
Binders	12,412	4,286	13,728	13,384	9,245	13,218	7,755	21,302	37,337	18,090	20,789	31,867	203,413
Bookcases	5,062	1,940	7,147	4,926	6,290	7,445	10,292	5,622	22,849	8,771	23,561	10,977	114,880
Chairs	11,285	7,768	20,832	18,855	25,703	21,145	23,585	17,770	52,147	21,905	47,314	60,141	328,449
Copiers	3,960		22,590	6,880	18,400	900	9,780	5,730	10,320	37,020	15,150	18,800	149,528
Envelopes	750	669	1,657	852	1,190	514	1,200	701	2,177	1,393	2,917	2,458	16,476
Fasteners	88	159	150	258	109	116	182	235	414	326	548	441	3,024
Furnishings	3,980	2,316	5,068	7,185	7,305	5,900	7,355	4,343	11,805	5,447	16,757	14,244	91,705
Labels	207	300	940	408	885	1,207	1,692	876	1,476	1,269	1,850	1,376	12,486
Machines	7,215	8,990	35,052	18,190	11,268	12,183	4,065	6,262	26,386	10,613	33,807	15,210	189,239
Paper	2,287	2,805	6,218	3,865	6,359	6,546	4,319	6,360	10,575	5,309	12,563	11,274	78,479
Phones	13,772	9,000	26,712	18,647	24,859	25,492	23,807	28,046	38,464	25,963	56,075	39,169	330,007
Storage	9,374	6,125	14,793	15,806	14,670	17,272	13,768	17,421	29,866	15,822	37,418	31,510	223,844
Supplies	4,403	289	10,607	6,246	1,154	1,267	8,816	859	6,442	816	1,372	4,402	46,674
Tables	10,952	4,218	16,913	9,913	9,288	15,360	10,344	17,752	19,626	20,223	31,401	40,975	206,966
Grand Total	95,366	60,173	199,253	141,852	156,122	147,083	149,581	159,589	309,770	197,115	349,120	332,177	2,297,201

Figure 8-41. *Output after setting "Row Grand Totals and Column Grand Totals"*

Notice that the “Grand Total” appears in the darkest shade of green. This is because this is the highest value in any cell, but our requirement is not to have the “Grand Total” displayed in the darkest shade of green.

How to change that?

Click on Sum (Sales) placed on “Color” on the marks card. Select “Total Using (Automatic)” ➤ “Hide” (Shown in Fig. 8-42).

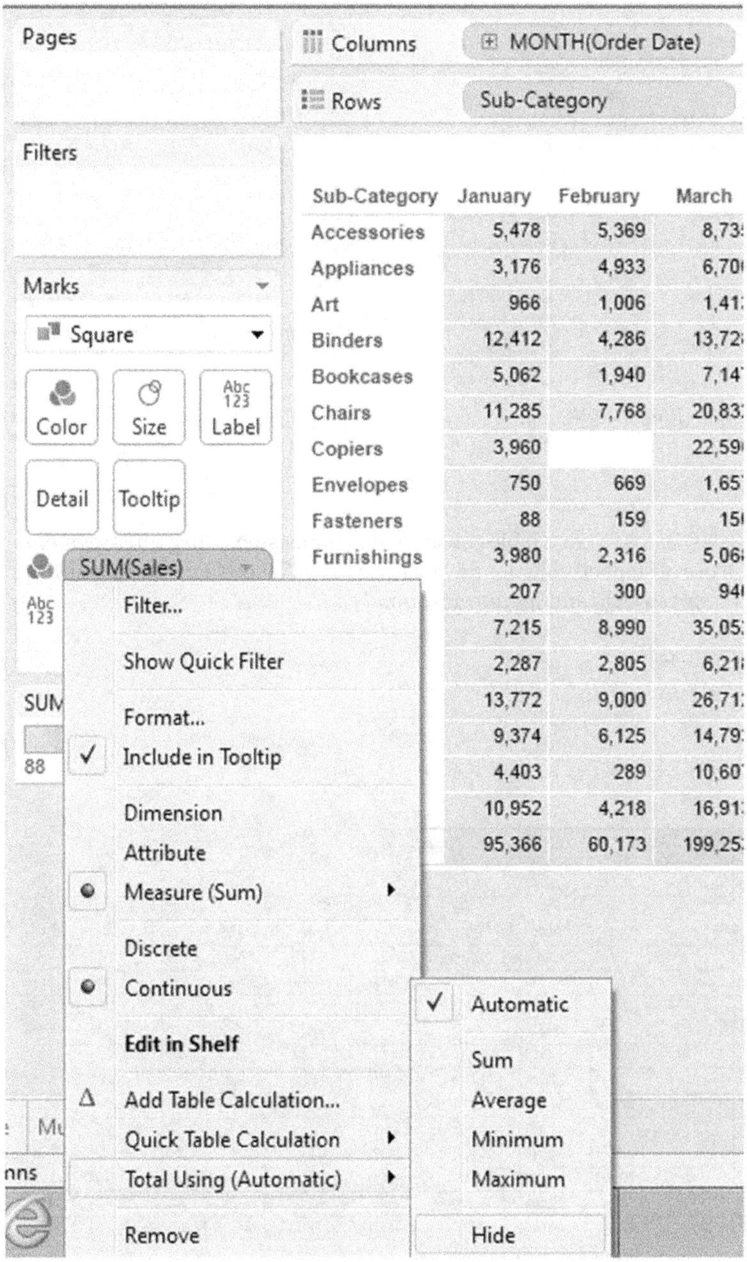

Figure 8-42. *Setting the “Total Using (Automatic) option for measure “Sales” placed on “Color” on the marks card*

Click on "Hide". The final output is as shown in Fig. 8-43.

	Order Date												
Sub-Category	January	February	March	April	May	June	July	August	September	October	November	December	Grand Tot..
Accessories	5,478	5,369	8,735	7,981	9,615	8,858	17,135	11,758	25,400	13,087	25,477	28,486	167,380
Appliances	3,176	4,933	6,700	6,075	7,526	7,479	3,384	12,862	10,828	9,155	18,306	17,107	107,532
Art	966	1,006	1,413	2,382	2,256	2,182	2,102	1,690	3,660	1,905	3,816	3,740	27,119
Binders	12,412	4,286	13,728	13,384	9,245	13,218	7,755	21,302	37,337	18,090	20,789	31,867	203,413
Bookcases	5,062	1,940	7,147	4,926	6,290	7,445	10,292	5,622	22,849	8,771	23,561	10,977	114,880
Chairs	11,285	7,768	20,832	18,855	25,703	21,145	23,585	17,770	52,147	21,905	47,314	60,141	328,449
Copiers	3,960		22,590	6,880	18,400	900	9,780	5,730	10,320	37,020	15,150	18,800	149,528
Envelopes	750	669	1,657	852	1,190	514	1,200	701	2,177	1,393	2,917	2,458	16,476
Fasteners	88	159	150	258	109	116	182	235	414	326	548	441	3,024
Furnishings	3,980	2,316	5,068	7,185	7,305	5,900	7,355	4,343	11,805	5,447	16,757	14,244	91,705
Labels	207	300	940	408	885	1,207	1,692	876	1,476	1,269	1,850	1,376	12,486
Machines	7,215	8,990	35,052	18,190	11,268	12,183	4,065	6,262	26,386	10,613	33,807	15,210	189,239
Paper	2,287	2,805	6,218	3,865	6,359	6,546	4,319	6,360	10,575	5,309	12,563	11,274	78,479
Phones	13,772	9,000	26,712	18,647	24,859	25,492	23,807	28,046	38,464	25,963	56,075	39,169	330,007
Storage	9,374	6,125	14,793	15,806	14,670	17,272	13,768	17,421	29,866	15,822	37,418	31,510	223,844
Supplies	4,403	289	10,607	6,246	1,154	1,267	8,816	859	6,442	816	1,372	4,402	46,674
Tables	10,952	4,218	16,913	9,913	9,288	15,360	10,344	17,752	19,626	20,223	31,401	40,975	206,966
Grand Total	95,366	60,173	199,253	141,852	156,122	147,083	149,581	159,589	309,770	197,115	349,120	332,177	2,297,201

Figure 8-43. *Highlight table - Demo 1 - final output*

8.4.2 Demo 2

Objective: To study the measure "Sales" across a cross table constituted of dimensions, "Region", "Segment", "Category" and "Sub-Category". The use of stepped color for the measure "Sales" allows one to see quickly the items that are "performing well", "not so well" and "performing poorly".

Input: "Sample - Superstore.xls"

Expected output: (Shown in Fig. 8-44).

		Region / Segment											
		Central			East			South			West		
Category	Sub-Category	Consumer	Corporate	Home Office	Consumer	Corporate	Home Office	Consumer	Corporate	Home Office	Consumer	Corporate	Home Office
Furniture	Bookcases	12,961	8,390	2,807	27,307	13,169	3,343	8,717	1,154	1,029	19,648	11,294	5,062
	Chairs	42,933	27,990	14,308	59,645	23,505	13,110	26,804	11,367	7,005	43,480	36,279	22,022
	Furnishings	9,500	3,144	2,610	13,177	9,983	5,911	10,871	4,505	1,931	16,072	7,369	6,631
	Tables	20,835	12,563	5,757	14,082	17,552	7,506	24,409	12,620	6,888	40,607	28,138	16,009
Office Supplies	Appliances	9,819	5,225	8,539	19,390	10,981	3,817	8,287	10,124	1,114	15,324	10,259	4,654
	Art	3,272	1,479	1,014	3,847	2,278	1,360	2,655	1,332	668	4,478	3,501	1,234
	Binders	39,694	8,880	8,350	31,623	12,048	9,827	18,618	11,828	6,585	28,227	18,805	8,929
	Envelopes	2,365	1,236	1,036	2,096	1,579	701	1,192	1,471	683	2,119	1,657	343
	Fasteners	422	232	123	424	286	110	276	81	146	558	184	180
	Labels	743	1,095	613	1,653	509	442	1,868	365	120	2,445	2,133	501
	Paper	7,564	5,952	3,976	8,661	5,975	5,536	7,603	3,345	3,202	12,496	8,611	5,557
	Storage	24,672	12,657	8,602	24,994	30,968	15,650	16,882	11,362	7,524	33,944	24,804	11,785
	Supplies	4,560	4,383	525	8,568	1,849	343	2,124	6,022	173	10,490	7,181	456
Technology	Accessories	19,062	6,195	8,699	22,861	11,748	10,424	11,652	10,175	5,450	33,530	20,072	7,512
	Copiers	4,520	29,680	3,060	30,400	4,220	18,600	4,440	4,860		30,460	8,070	11,220
	Machines	12,000	11,717	3,081	30,668	23,892	11,546	18,339	12,001	23,551	18,536	12,667	11,241
	Phones	37,109	17,181	18,114	51,512	29,866	19,237	30,846	19,274	8,184	50,466	24,833	23,386

Figure 8-44. *Highlight Table - Demo 2 - expected output*

8.4.2.1 Steps to create a highlight table

8.4.2.2 Step 1

Drag the dimensions “Region” and “Segment” from the dimensions area under the data pane and place it on the columns shelf (Shown in Fig. 8-45).

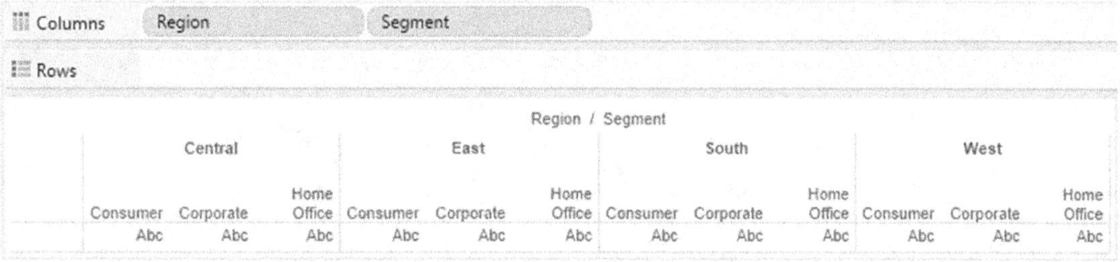

Columns: Region | Segment

Rows:

Region / Segment											
Central			East			South			West		
Consumer	Corporate	Home Office	Consumer	Corporate	Home Office	Consumer	Corporate	Home Office	Consumer	Corporate	Home Office
Abc	Abc	Abc	Abc	Abc	Abc	Abc	Abc	Abc	Abc	Abc	Abc

Figure 8-45. *Dimensions, “Region” and “Segment” placed on the columns shelf*

Drag the dimensions “Category” and “Sub-Category” from the dimensions area under the data pane and place it on the rows shelf (Shown in Fig. 8-46).

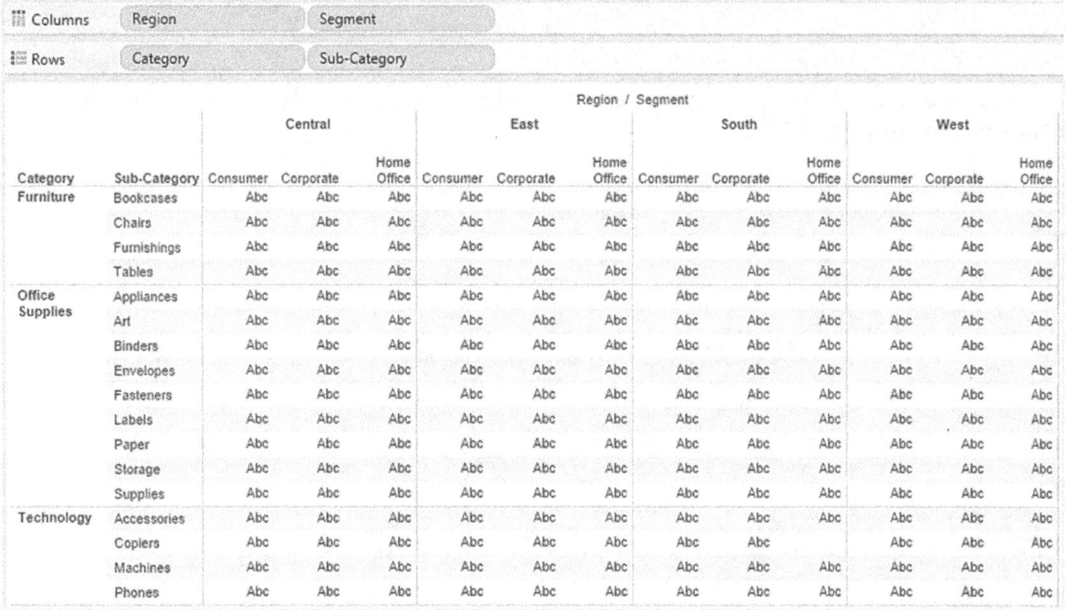

Columns: Region | Segment

Rows: Category | Sub-Category

		Region / Segment											
		Central			East			South			West		
Category	Sub-Category	Consumer	Corporate	Home Office	Consumer	Corporate	Home Office	Consumer	Corporate	Home Office	Consumer	Corporate	Home Office
Furniture	Bookcases	Abc	Abc	Abc	Abc	Abc	Abc	Abc	Abc	Abc	Abc	Abc	Abc
	Chairs	Abc	Abc	Abc	Abc	Abc	Abc	Abc	Abc	Abc	Abc	Abc	Abc
	Furnishings	Abc	Abc	Abc	Abc	Abc	Abc	Abc	Abc	Abc	Abc	Abc	Abc
	Tables	Abc	Abc	Abc	Abc	Abc	Abc	Abc	Abc	Abc	Abc	Abc	Abc
Office Supplies	Appliances	Abc	Abc	Abc	Abc	Abc	Abc	Abc	Abc	Abc	Abc	Abc	Abc
	Art	Abc	Abc	Abc	Abc	Abc	Abc	Abc	Abc	Abc	Abc	Abc	Abc
	Binders	Abc	Abc	Abc	Abc	Abc	Abc	Abc	Abc	Abc	Abc	Abc	Abc
	Envelopes	Abc	Abc	Abc	Abc	Abc	Abc	Abc	Abc	Abc	Abc	Abc	Abc
	Fasteners	Abc	Abc	Abc	Abc	Abc	Abc	Abc	Abc	Abc	Abc	Abc	Abc
	Labels	Abc	Abc	Abc	Abc	Abc	Abc	Abc	Abc	Abc	Abc	Abc	Abc
	Paper	Abc	Abc	Abc	Abc	Abc	Abc	Abc	Abc	Abc	Abc	Abc	Abc
	Storage	Abc	Abc	Abc	Abc	Abc	Abc	Abc	Abc	Abc	Abc	Abc	Abc
	Supplies	Abc	Abc	Abc	Abc	Abc	Abc	Abc	Abc	Abc	Abc	Abc	Abc
Technology	Accessories	Abc	Abc	Abc	Abc	Abc	Abc	Abc	Abc	Abc	Abc	Abc	Abc
	Copiers	Abc	Abc	Abc	Abc	Abc	Abc	Abc	Abc		Abc	Abc	Abc
	Machines	Abc	Abc	Abc	Abc	Abc	Abc	Abc	Abc	Abc	Abc	Abc	Abc
	Phones	Abc	Abc	Abc	Abc	Abc	Abc	Abc	Abc	Abc	Abc	Abc	Abc

Figure 8-46. *Dimensions “Category” and “Sub-Category” placed on the rows shelf*

8.4.2.3 Step 2

Drag the measure “Sales” from the measures area under the data pane and place it on “Label” on the marks card (Shown in Fig. 8-47).

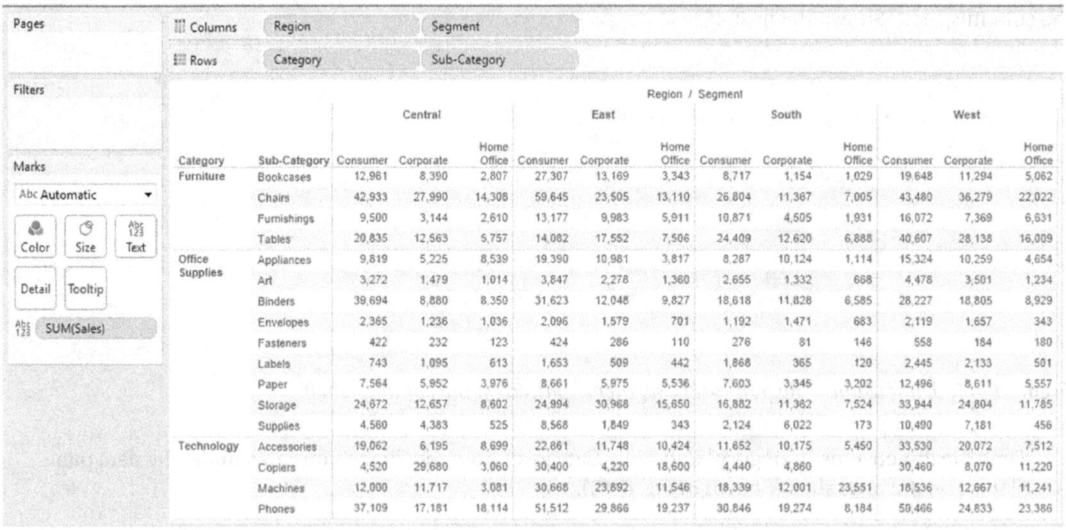

Pages | Filters | Marks: Abc Automatic | Color | Size | Text | Detail | Tooltip | SUM(Sales)

Columns: Region, Segment
Rows: Category, Sub-Category

Region / Segment

Category	Sub-Category	Central Consumer	Central Corporate	Central Home Office	East Consumer	East Corporate	East Home Office	South Consumer	South Corporate	South Home Office	West Consumer	West Corporate	West Home Office
Furniture	Bookcases	12,961	8,390	2,807	27,307	13,169	3,343	8,717	1,154	1,029	19,648	11,294	5,062
	Chairs	42,933	27,990	14,308	59,645	23,505	13,110	26,804	11,367	7,005	43,480	36,279	22,022
	Furnishings	9,500	3,144	2,610	13,177	9,983	5,911	10,871	4,505	1,931	16,072	7,369	6,631
	Tables	20,835	12,563	5,757	14,082	17,552	7,506	24,409	12,620	6,888	40,607	28,138	16,009
Office Supplies	Appliances	9,819	5,225	8,539	19,390	10,981	3,817	8,287	10,124	1,114	15,324	10,259	4,654
	Art	3,272	1,479	1,014	3,847	2,278	1,360	2,655	1,332	668	4,478	3,501	1,234
	Binders	39,694	8,880	8,350	31,623	12,048	9,827	18,618	11,828	6,585	28,227	18,805	8,929
	Envelopes	2,365	1,236	1,036	2,096	1,579	701	1,192	1,471	683	2,119	1,657	343
	Fasteners	422	232	123	424	286	110	276	81	146	558	184	180
	Labels	743	1,095	613	1,653	509	442	1,868	365	120	2,445	2,133	501
	Paper	7,564	5,952	3,976	8,661	5,975	5,536	7,603	3,345	3,202	12,496	8,611	5,557
	Storage	24,672	12,657	8,602	24,994	30,968	15,650	16,882	11,362	7,524	33,944	24,804	11,785
	Supplies	4,560	4,383	525	8,568	1,849	343	2,124	6,022	173	10,490	7,181	456
Technology	Accessories	19,062	6,195	8,699	22,861	11,748	10,424	11,652	10,175	5,450	33,530	20,072	7,512
	Copiers	4,520	29,680	3,060	30,400	4,220	18,600	4,440	4,860		30,460	8,070	11,220
	Machines	12,000	11,717	3,081	30,668	23,892	11,546	18,339	12,001	23,551	18,536	12,667	11,241
	Phones	37,109	17,181	18,114	51,512	29,866	19,237	30,846	19,274	8,184	50,466	24,833	23,386

Figure 8-47. *Measure “Sales” placed on “Label” on the marks card*

Drag the measure “Sales” from the measures area under the data pane and place it on “Color” on the marks card (Shown in Fig. 8-48).

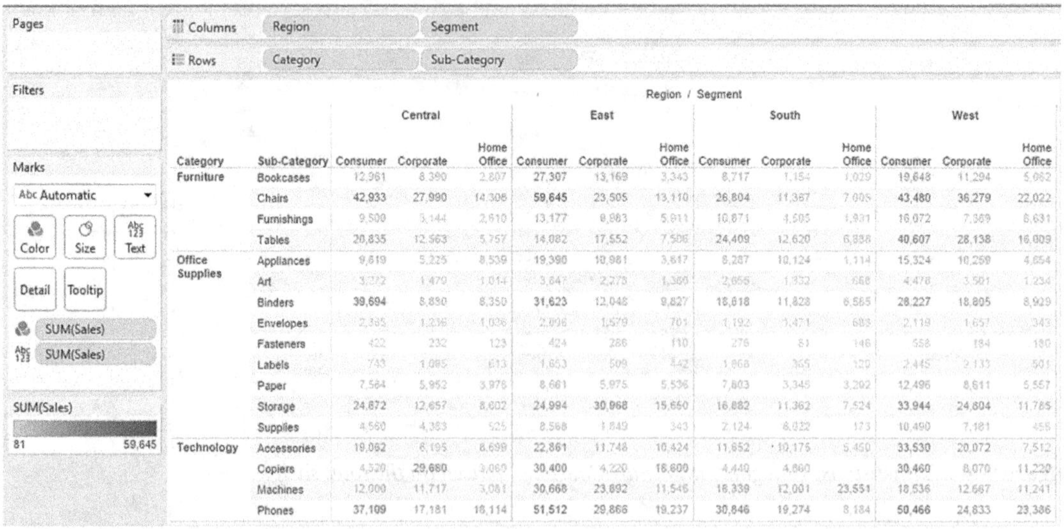

Pages | Filters | Marks: Abc Automatic | Color | Size | Text | Detail | Tooltip | SUM(Sales) | SUM(Sales)

SUM(Sales): 81 – 59,645

Columns: Region, Segment
Rows: Category, Sub-Category

Region / Segment

Category	Sub-Category	Central Consumer	Central Corporate	Central Home Office	East Consumer	East Corporate	East Home Office	South Consumer	South Corporate	South Home Office	West Consumer	West Corporate	West Home Office
Furniture	Bookcases	12,961	8,390	2,807	27,307	13,169	3,343	8,717	1,154	1,029	19,648	11,294	5,062
	Chairs	42,933	27,990	14,308	59,645	23,505	13,110	26,804	11,367	7,005	43,480	36,279	22,022
	Furnishings	9,500	3,144	2,610	13,177	9,983	5,911	10,871	4,505	1,931	16,072	7,369	6,631
	Tables	20,835	12,563	5,757	14,082	17,552	7,506	24,409	12,620	6,888	40,607	28,138	16,009
Office Supplies	Appliances	9,819	5,225	8,539	19,390	10,981	3,817	8,287	10,124	1,114	15,324	10,259	4,654
	Art	3,272	1,479	1,014	3,847	2,278	1,360	2,655	1,332	668	4,478	3,501	1,234
	Binders	39,694	8,880	8,350	31,623	12,048	9,827	18,618	11,828	6,585	28,227	18,805	8,929
	Envelopes	2,365	1,236	1,036	2,096	1,579	701	1,192	1,471	683	2,119	1,657	343
	Fasteners	422	232	123	424	286	110	276	81	146	558	184	180
	Labels	743	1,095	613	1,653	509	442	1,868	365	120	2,445	2,133	501
	Paper	7,564	5,952	3,976	8,661	5,975	5,536	7,603	3,345	3,202	12,496	8,611	5,557
	Storage	24,672	12,657	8,602	24,994	30,968	15,650	16,882	11,362	7,524	33,944	24,804	11,785
	Supplies	4,560	4,383	525	8,568	1,849	343	2,124	6,022	173	10,490	7,181	456
Technology	Accessories	19,062	6,195	8,699	22,861	11,748	10,424	11,652	10,175	5,450	33,530	20,072	7,512
	Copiers	4,520	29,680	3,060	30,400	4,220	18,600	4,440	4,860		30,460	8,070	11,220
	Machines	12,000	11,717	3,081	30,668	23,892	11,546	18,339	12,001	23,551	18,536	12,667	11,241
	Phones	37,109	17,181	18,114	51,512	29,866	19,237	30,846	19,274	8,184	50,466	24,833	23,386

Figure 8-48. *Measure “Sales” placed on “Color” on the marks card*

8.4.2.4 Step 3

Change the mark type to “Square” (Shown in Fig. 8-49).

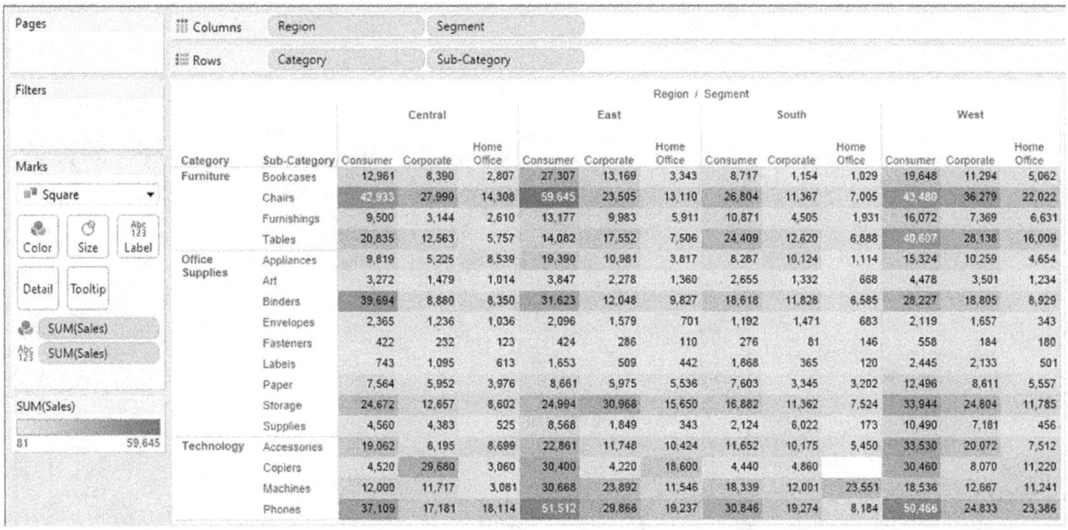

Pages

Columns: Region | Segment

Rows: Category | Sub-Category

Filters

Marks: Square

Color | Size | Label

Detail | Tooltip

SUM(Sales)

SUM(Sales)

SUM(Sales): 81 – 59,645

Region / Segment

Category	Sub-Category	Central Consumer	Central Corporate	Central Home Office	East Consumer	East Corporate	East Home Office	South Consumer	South Corporate	South Home Office	West Consumer	West Corporate	West Home Office
Furniture	Bookcases	12,961	8,390	2,807	27,307	13,169	3,343	8,717	1,154	1,029	19,648	11,294	5,062
	Chairs	42,933	27,990	14,308	59,645	23,505	13,110	26,804	11,367	7,005	43,480	36,279	22,022
	Furnishings	9,500	3,144	2,610	13,177	9,983	5,911	10,871	4,505	1,931	16,072	7,369	6,631
	Tables	20,835	12,563	5,757	14,082	17,552	7,506	24,409	12,620	6,888	40,607	28,138	16,009
Office Supplies	Appliances	9,819	5,225	8,539	19,390	10,981	3,817	8,287	10,124	1,114	15,324	10,259	4,654
	Art	3,272	1,479	1,014	3,847	2,278	1,360	2,655	1,332	668	4,478	3,501	1,234
	Binders	39,694	8,880	8,350	31,623	12,048	9,827	18,618	11,828	6,585	28,227	18,805	8,929
	Envelopes	2,365	1,236	1,036	2,096	1,579	701	1,192	1,471	683	2,119	1,657	343
	Fasteners	422	232	123	424	286	110	276	81	146	558	184	180
	Labels	743	1,095	613	1,653	509	442	1,868	365	120	2,445	2,133	501
	Paper	7,564	5,952	3,976	8,661	5,975	5,536	7,603	3,345	3,202	12,496	8,611	5,557
	Storage	24,672	12,657	8,602	24,994	30,968	15,650	16,882	11,362	7,524	33,944	24,804	11,785
	Supplies	4,560	4,383	525	8,568	1,849	343	2,124	6,022	173	10,490	7,181	456
Technology	Accessories	19,062	6,195	8,699	22,861	11,748	10,424	11,652	10,175	5,450	33,530	20,072	7,512
	Copiers	4,520	29,680	3,060	30,400	4,220	18,600	4,440	4,860		30,460	8,070	11,220
	Machines	12,000	11,717	3,081	30,668	23,892	11,546	18,339	12,001	23,551	18,536	12,667	11,241
	Phones	37,109	17,181	18,114	51,512	29,866	19,237	30,846	19,274	8,184	50,466	24,833	23,386

Figure 8-49. *Mark type set to “Square”*

Change the “Color” by clicking on “Color” on the marks card (Shown in Fig. 8-50).

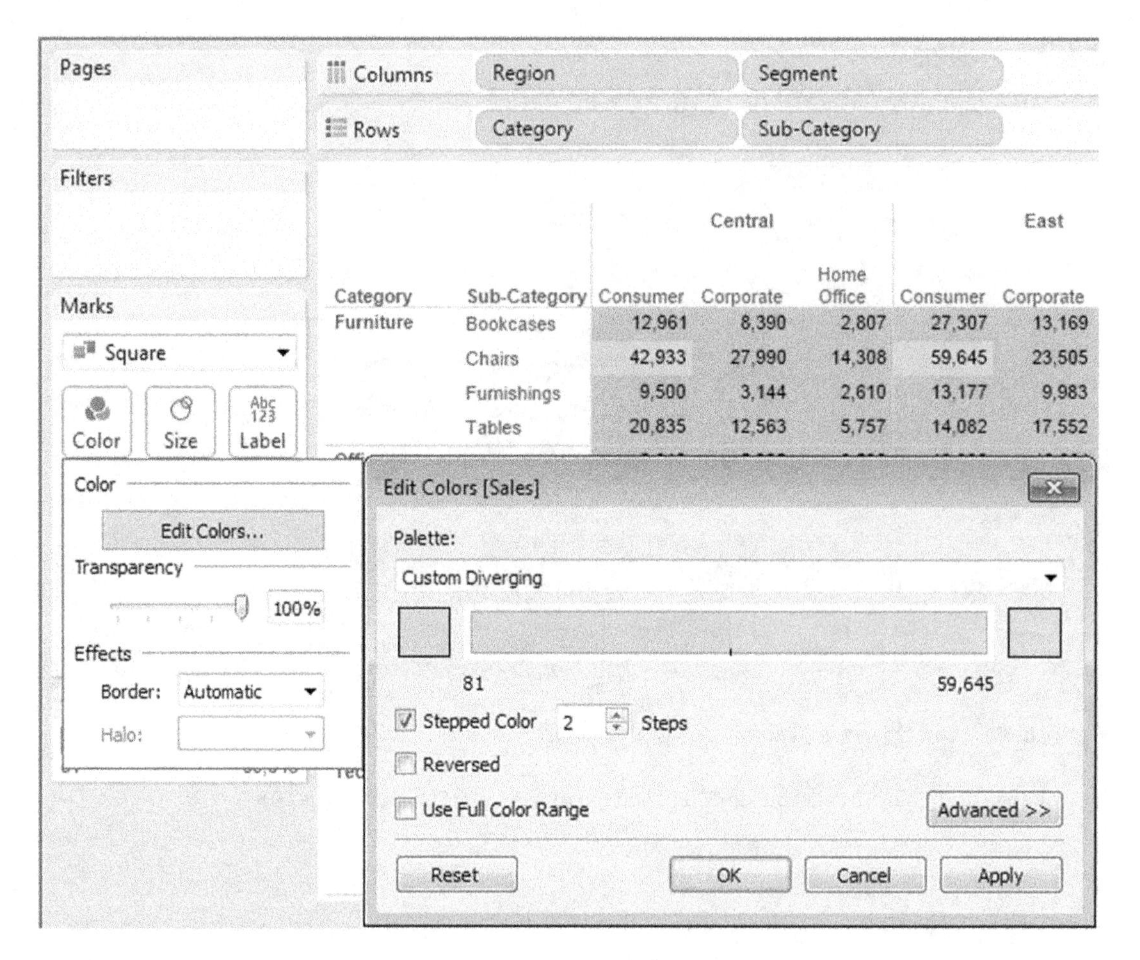

Figure 8-50. *"Edit Colors" dialog box*

Change the cell border to "Black" (Shown in Fig. 8-51).

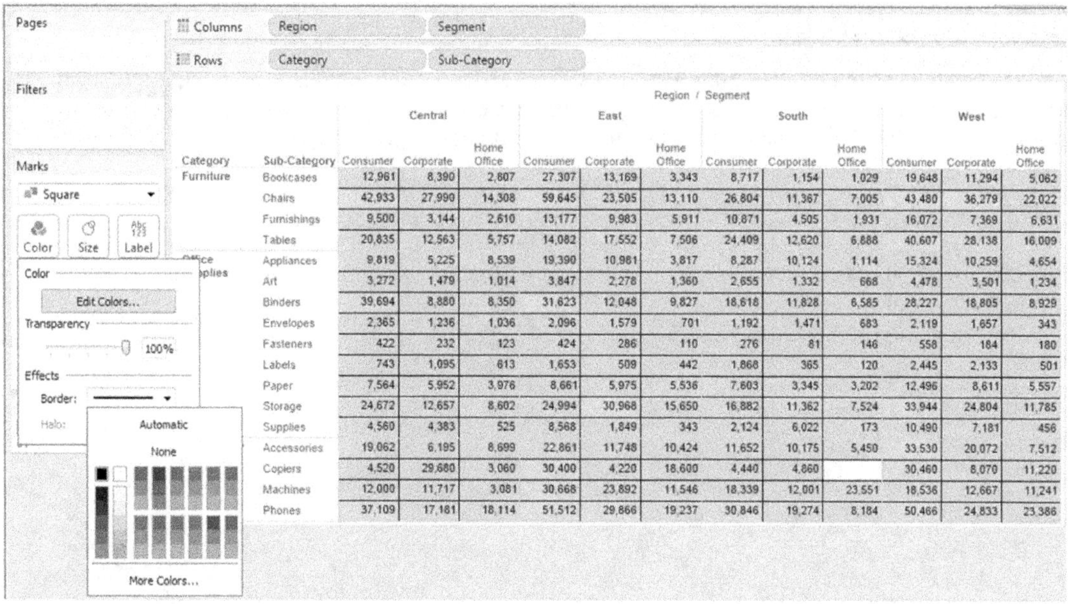

Pages | Columns: Region, Segment | Rows: Category, Sub-Category

Filters

Marks: Square — Color, Size, Label

Color: Edit Colors... | Transparency: 100% | Effects: Border: | Halo: | Automatic | None | More Colors...

Category	Sub-Category	Central Consumer	Central Corporate	Central Home Office	East Consumer	East Corporate	East Home Office	South Consumer	South Corporate	South Home Office	West Consumer	West Corporate	West Home Office
Furniture	Bookcases	12,961	8,390	2,807	27,307	13,169	3,343	8,717	1,154	1,029	19,648	11,294	5,062
	Chairs	42,933	27,990	14,308	59,645	23,505	13,110	26,804	11,367	7,005	43,480	36,279	22,022
	Furnishings	9,500	3,144	2,610	13,177	9,983	5,911	10,871	4,505	1,931	16,072	7,369	6,631
	Tables	20,835	12,563	5,757	14,082	17,552	7,506	24,409	12,620	6,888	40,607	28,138	16,009
Office Supplies	Appliances	9,819	5,225	8,539	19,390	10,981	3,817	8,287	10,124	1,114	15,324	10,259	4,654
	Art	3,272	1,479	1,014	3,847	2,278	1,360	2,655	1,332	668	4,478	3,501	1,234
	Binders	39,694	8,880	8,350	31,623	12,048	9,827	18,618	11,828	6,585	28,227	18,805	8,929
	Envelopes	2,365	1,236	1,036	2,096	1,579	701	1,192	1,471	683	2,119	1,657	343
	Fasteners	422	232	123	424	286	110	276	81	146	558	184	180
	Labels	743	1,095	613	1,653	509	442	1,868	365	120	2,445	2,133	501
	Paper	7,564	5,952	3,976	8,661	5,975	5,536	7,603	3,345	3,202	12,496	8,611	5,557
	Storage	24,672	12,657	8,602	24,994	30,968	15,650	16,882	11,362	7,524	33,944	24,804	11,785
	Supplies	4,560	4,383	525	8,568	1,849	343	2,124	6,022	173	10,490	7,181	456
	Accessories	19,062	6,195	8,699	22,861	11,748	10,424	11,652	10,175	5,450	33,530	20,072	7,512
	Copiers	4,520	29,680	3,060	30,400	4,220	18,600	4,440	4,860		30,460	8,070	11,220
	Machines	12,000	11,717	3,081	30,668	23,892	11,546	18,339	12,001	23,551	18,536	12,667	11,241
	Phones	37,109	17,181	18,114	51,512	29,866	19,237	30,846	19,274	8,184	50,466	24,833	23,386

***Figure 8-51.** Cell Border for worksheet cells set to "Black"*

Final output: (Shown in Fig. 8-52).

Category	Sub-Category	Central Consumer	Central Corporate	Central Home Office	East Consumer	East Corporate	East Home Office	South Consumer	South Corporate	South Home Office	West Consumer	West Corporate	West Home Office
Furniture	Bookcases	12,961	8,390	2,807	27,307	13,169	3,343	8,717	1,154	1,029	19,648	11,294	5,062
	Chairs	42,933	27,990	14,308	59,645	23,505	13,110	26,804	11,367	7,005	43,480	36,279	22,022
	Furnishings	9,500	3,144	2,610	13,177	9,983	5,911	10,871	4,505	1,931	16,072	7,369	6,631
	Tables	20,835	12,563	5,757	14,082	17,552	7,506	24,409	12,620	6,888	40,607	28,138	16,009
Office Supplies	Appliances	9,819	5,225	8,539	19,390	10,981	3,817	8,287	10,124	1,114	15,324	10,259	4,654
	Art	3,272	1,479	1,014	3,847	2,278	1,360	2,655	1,332	668	4,478	3,501	1,234
	Binders	39,694	8,880	8,350	31,623	12,048	9,827	18,618	11,828	6,585	28,227	18,805	8,929
	Envelopes	2,365	1,236	1,036	2,096	1,579	701	1,192	1,471	683	2,119	1,657	343
	Fasteners	422	232	123	424	286	110	276	81	146	558	184	180
	Labels	743	1,095	613	1,653	509	442	1,868	365	120	2,445	2,133	501
	Paper	7,564	5,952	3,976	8,661	5,975	5,536	7,603	3,345	3,202	12,496	8,611	5,557
	Storage	24,672	12,657	8,602	24,994	30,968	15,650	16,882	11,362	7,524	33,944	24,804	11,785
	Supplies	4,560	4,383	525	8,568	1,849	343	2,124	6,022	173	10,490	7,181	456
Technology	Accessories	19,062	6,195	8,699	22,861	11,748	10,424	11,652	10,175	5,450	33,530	20,072	7,512
	Copiers	4,520	29,680	3,060	30,400	4,220	18,600	4,440	4,860		30,460	8,070	11,220
	Machines	12,000	11,717	3,081	30,668	23,892	11,546	18,339	12,001	23,551	18,536	12,667	11,241
	Phones	37,109	17,181	18,114	51,512	29,866	19,237	30,846	19,274	8,184	50,466	24,833	23,386

***Figure 8-52.** Highlight Table - Demo 2 - final output*

8.5 Line Graph

A line graph or line chart displays information as a series of data points connected by straight-line segments. The data points are also referred to as "markers". A line chart is often used to visualize data trends over intervals of time. In other words, they are helpful to visualize data as it changes continuously over time.

8.5.1 Demo 1

Objective: To see the trend of "Sales" and "Profit" over time using a line graph and learn about forecasting.

Input: "Sample - Superstore.xls"

Expected output: (Shown in Fig. 8-53).

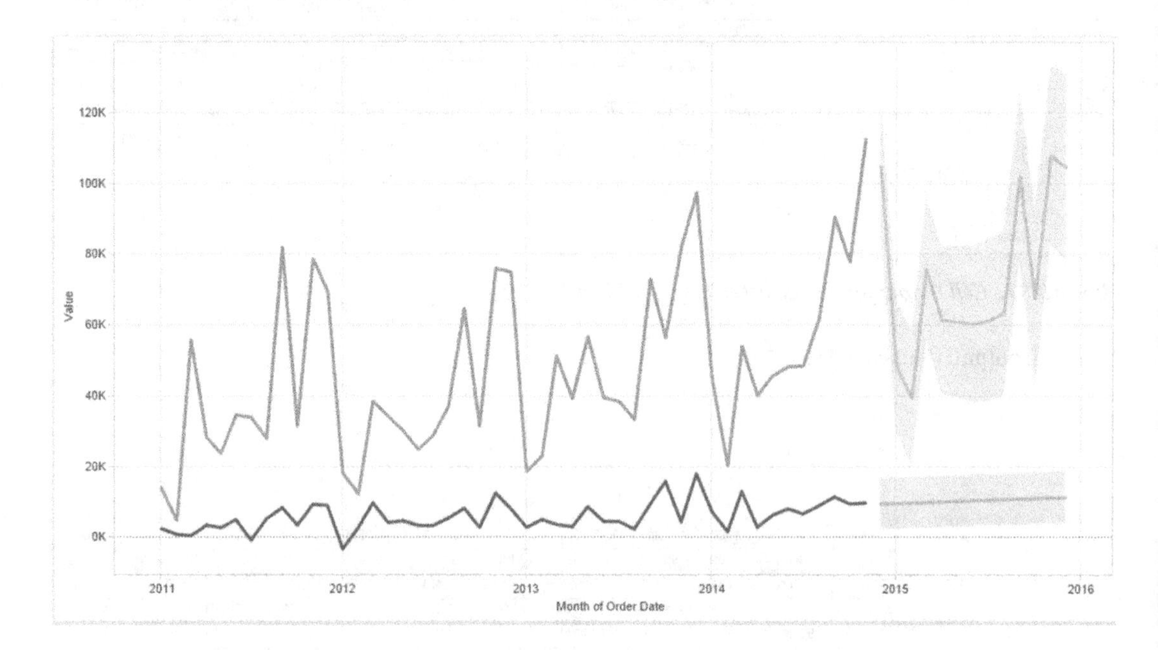

Figure 8-53. *Line Graph - Demo 1 - expected output*

8.5.1.1 Steps to create a line graph

8.5.1.2 Step 1

Drag the dimension "Order Date" from the dimensions area under the data pane and place it on the columns shelf. Let the date hierarchy remain at "Year". It is "Discrete" data as evident from the visual cue. (Date appears blue in color on the columns shelf) (Shown in Fig. 8-54).

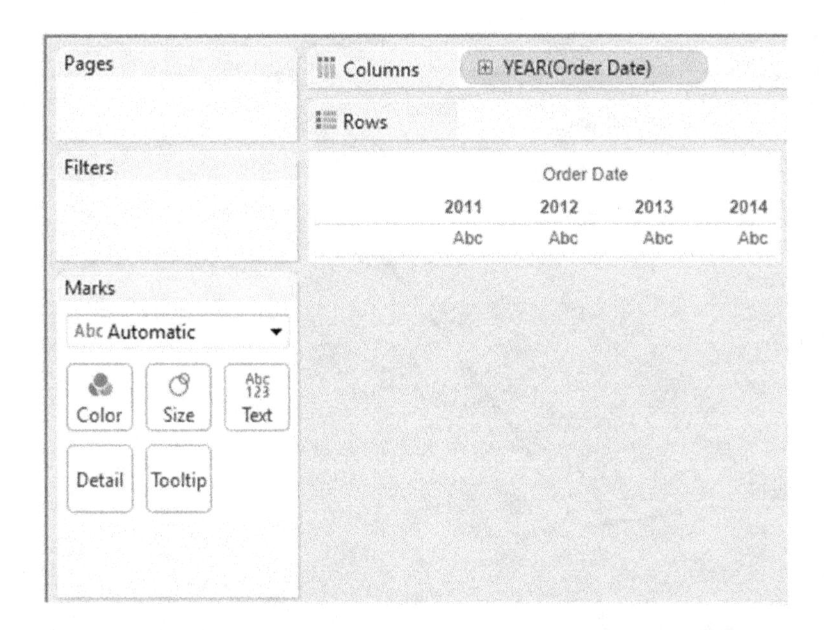

Figure 8-54. *Dimension "Order Date" placed on the columns shelf*

8.5.1.3 Step 2

Drag the measure "Sales" from the measures area under the data pane and place it on the rows shelf (Shown in Fig. 8-55).

Figure 8-55. *Measure "Sales" placed on the rows shelf*

Drag the measure "Profit" from the measures area under the data pane and place it on the rows shelf to the right of the measure "Sales" (Shown in Fig. 8-56).

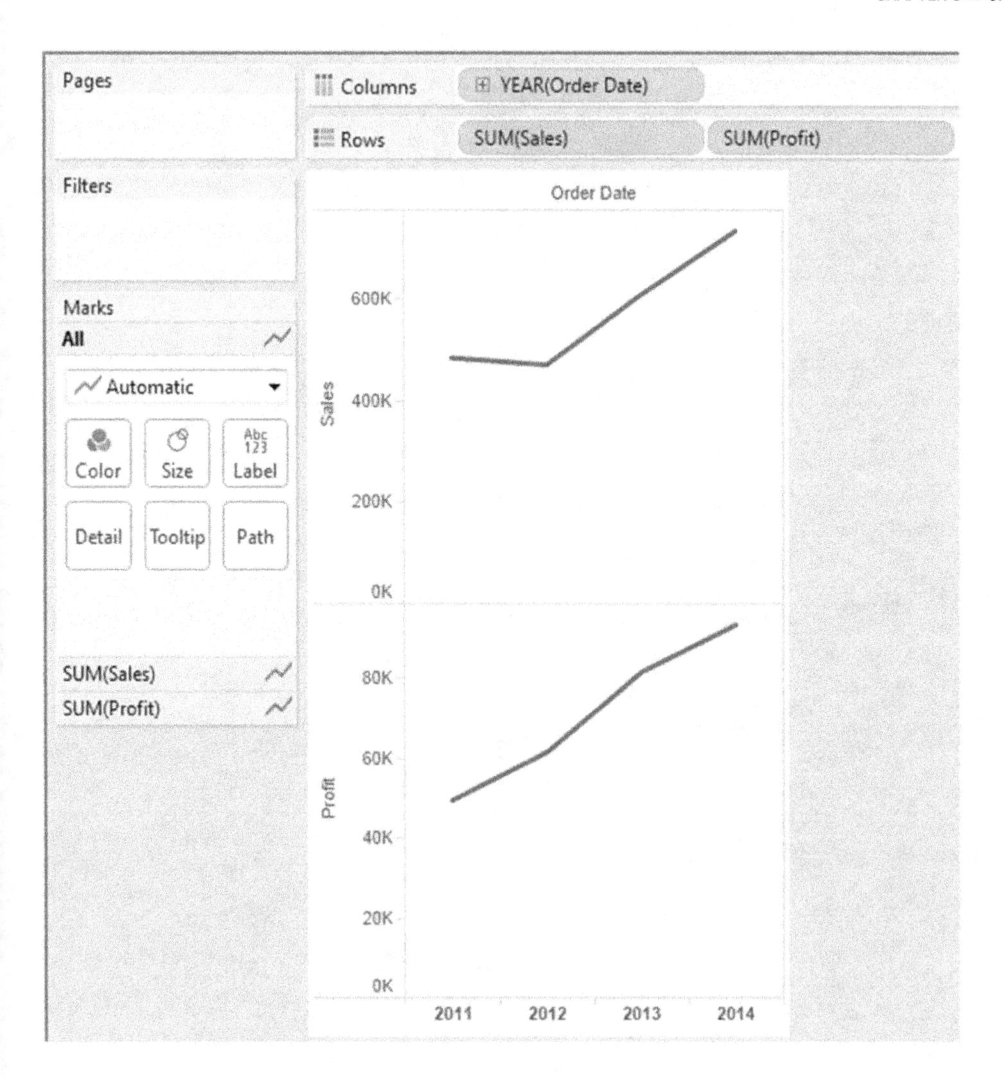

Figure 8-56. *Measure "Profit" placed on the rows shelf to the right of the measure "Sales"*

Pull the measure "Profit" from the rows shelf and place it on the same axis as "Sales" (Shown in Fig. 8-57).

Figure 8-57. *Measure "Profit" placed on the same axis as the measure "Sales"*

8.5.1.4 Step 3

Change the date from "Discrete" to "Continuous" and the grain from "Year" to "Month" (Shown in Fig. 8-58).

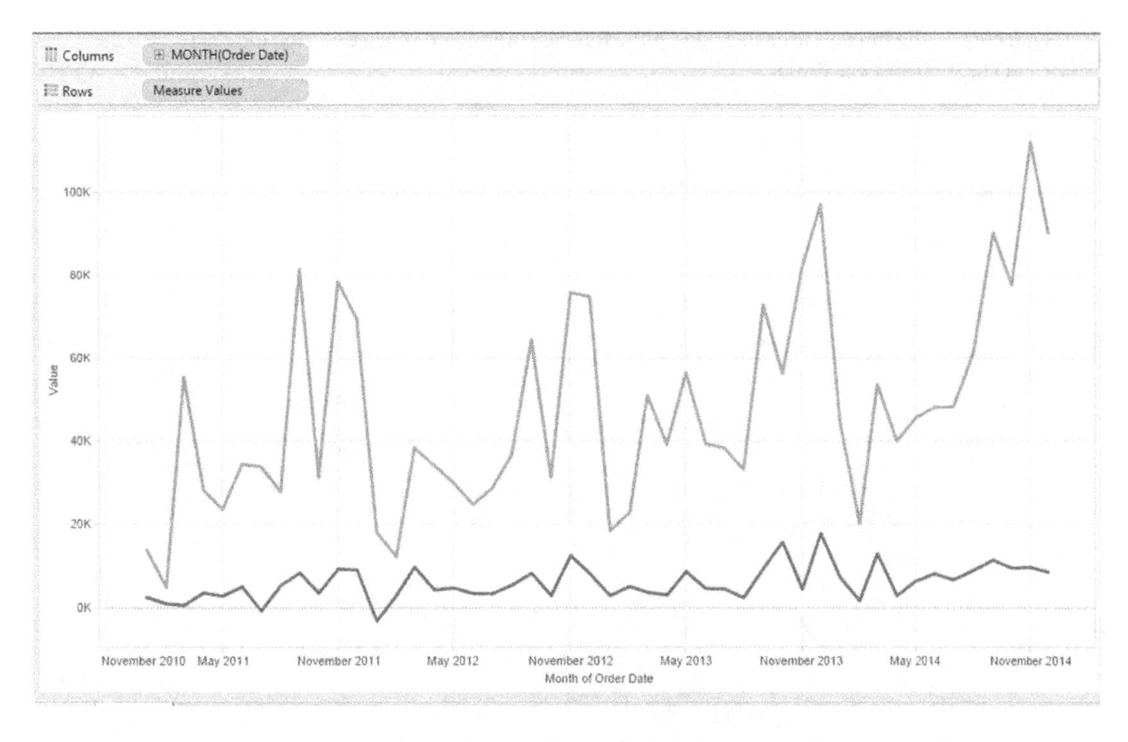

Figure 8-58. *Dimension "Order Date" converted to "Continous" and granularity changed to "Month"*

8.5.1.5 Step 4

Go to the analytics pane and drag "Forecast" to the view (Shown in Fig. 8-59).

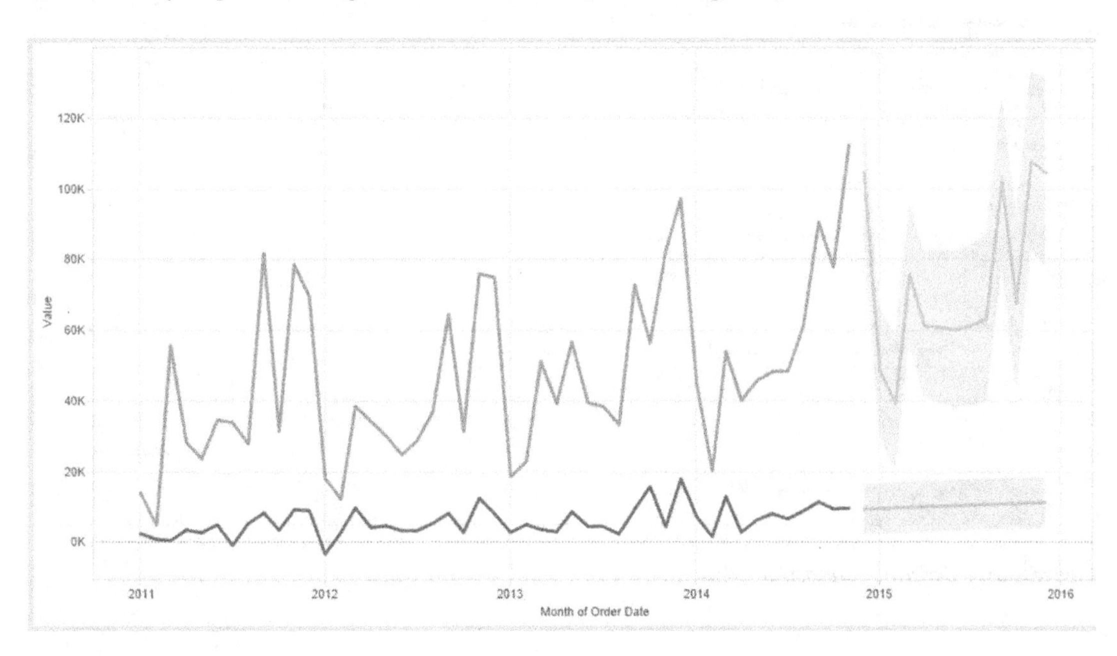

Figure 8-59. *Forecast added to the view/worksheet*

Right click in the forecast area to bring up the forecast menu and select describe forecast (Shown in Fig. 8-60 and Fig. 8-61).

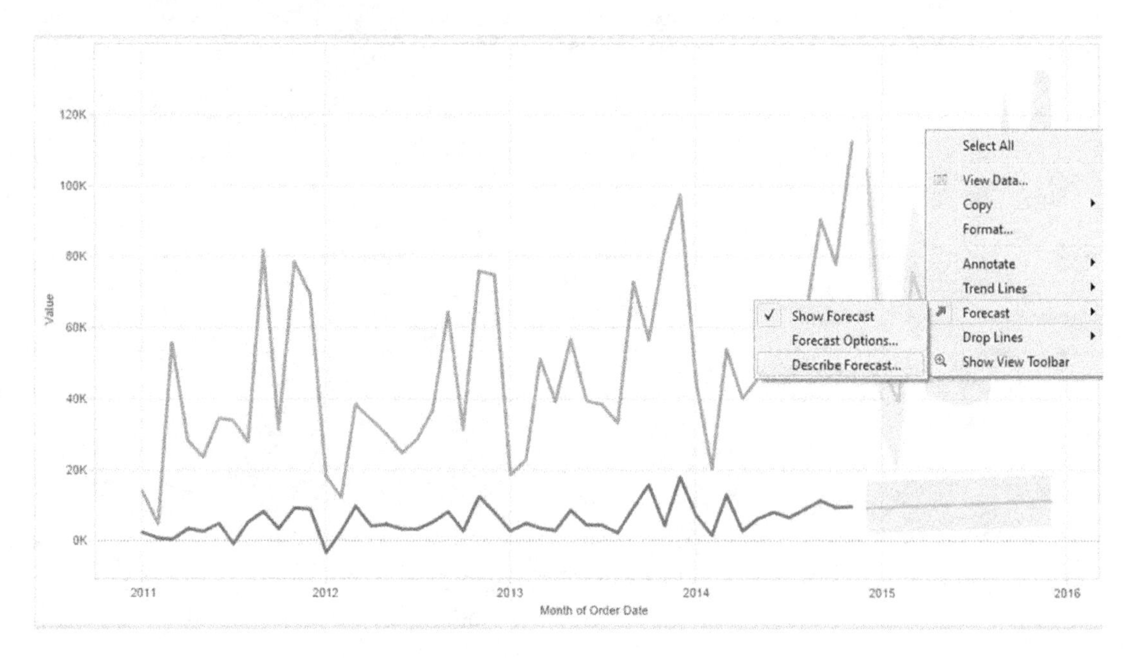

Figure 8-60. *Selecting "Describe Forecast" from the forecast menu*

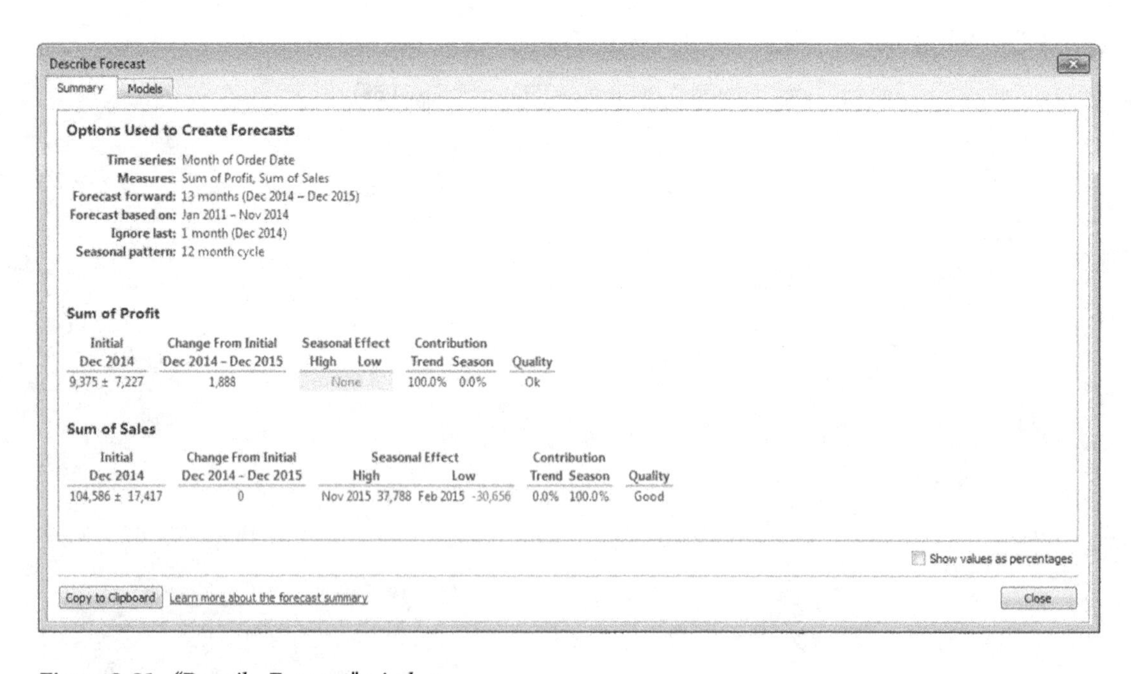

Figure 8-61. *"Describe Forecast" window*

8.5.2 Demo 2

Objective: To plot a line graph without using a date dimension.

Input: "Sample - Superstore.xls"

Expected output: (Shown in Fig. 8-62).

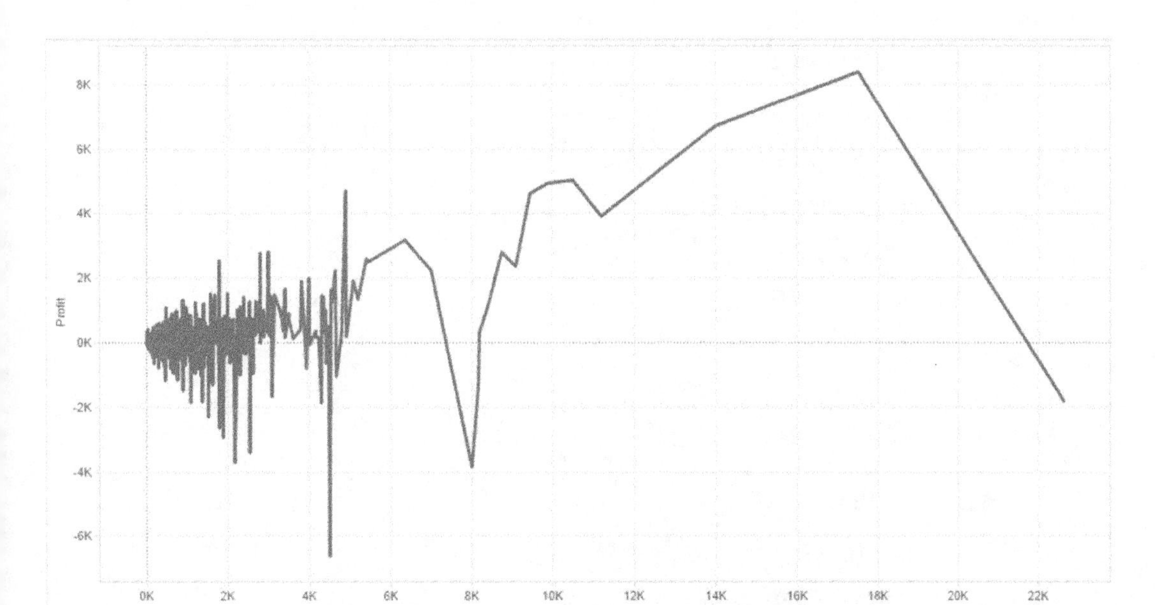

Figure 8-62. *Line Graph - Demo 2 - expected output*

8.5.2.1 Steps to create a line graph

8.5.2.2 Step 1

Drag the measure "Sales" from the measures area under the data pane to the columns shelf (Shown in Fig. 8-63).

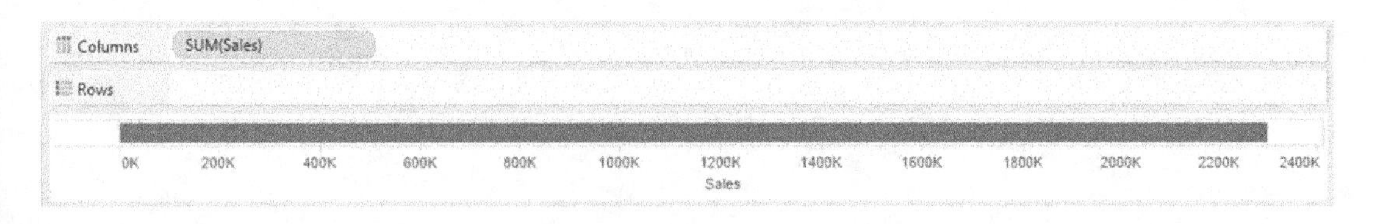

Figure 8-63. *Measure "Sales" placed on the columns shelf*

Convert the measure "Sales" on the columns shelf to "Dimension" (Shown in Fig. 8-64).

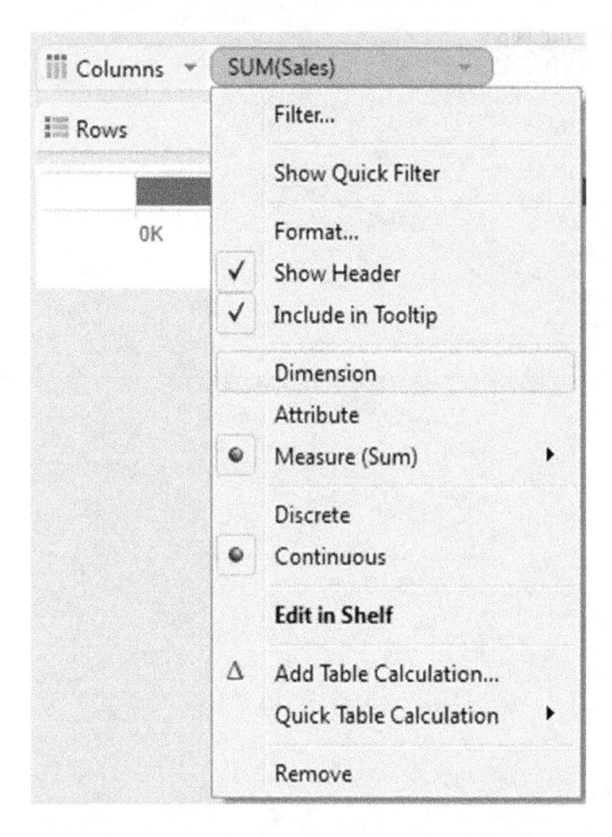

Figure 8-64. *Measure "Sales" on the columns shelf converted to "Dimension"*

8.5.2.3 Step 2

Drag the measure "Profit" from the measures area under the data pane and place it on the rows shelf (Shown in Fig. 8-65).

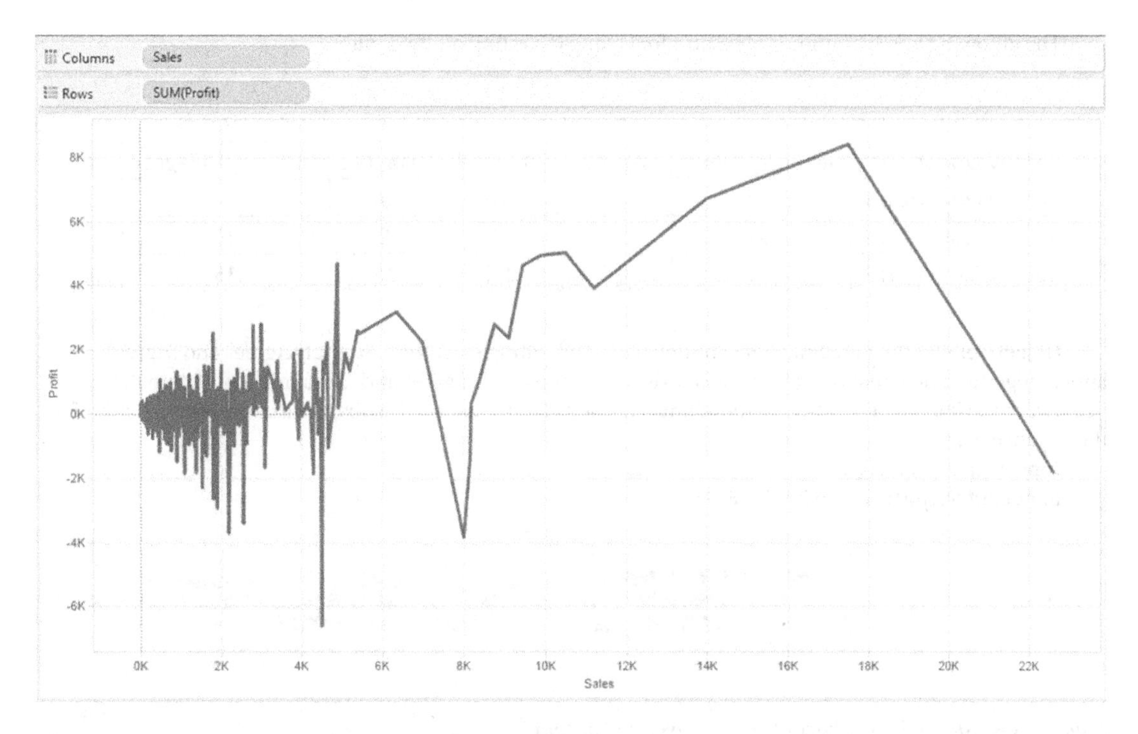

Figure 8-65. *Measure "Profit" placed on the rows shelf*

8.6 Stacked Bar Chart

A stacked bar chart or a stacked bar graph is used to break down and compare parts of a whole. Each bar represents the whole and the segments within a bar represent the categories or parts of the whole. The categories or parts are colored differently for ease of comprehension.

8.6.1 Demo 1

Imagine this ... A survey was conducted by the public health department. The survey questionnaire had three questions. There were 100 respondents to the survey. The responses were on a Likert scale, which had the following ratings:

- Strongly disagree
- Disagree
- Neutral
- Agree
- Strongly agree

The following are the number of respondents for each rating on the scale.

Survey Questions	Strongly Agree	Agree	Neutral	Disagree	Strongly Disagree
Survey Question 1	40	30	10	10	10
Survey Question 2	60	20	2	8	10
Survey Question 3	25	25	20	15	15

Objective: Plot the percentage respondents for each rating such that "Strongly Agree" and "Agree" appear together and constitute the positive axis and "Strongly Disagree" and "Disagree" constitute the negative axis. Half the percentage respondents for "Neutral" appear on the positive axis and the other half on the negative axis.

Input: StackedBar.xlsx

Expected output: Shown in Fig. 8-66.

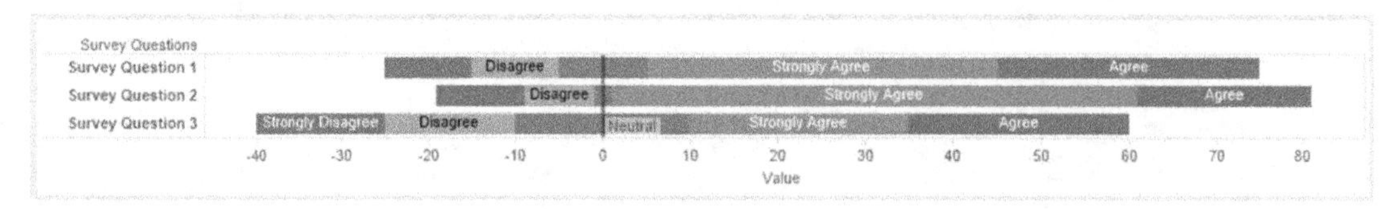

Figure 8-66. Stacked Bar Chart - Demo 1 - expected output

8.6.2 Steps to create a stacked bar chart

8.6.2.1 Step 1

Read in data from "Stacked Bar.xlsx" into Tableau (Shown in Fig. 8-67).

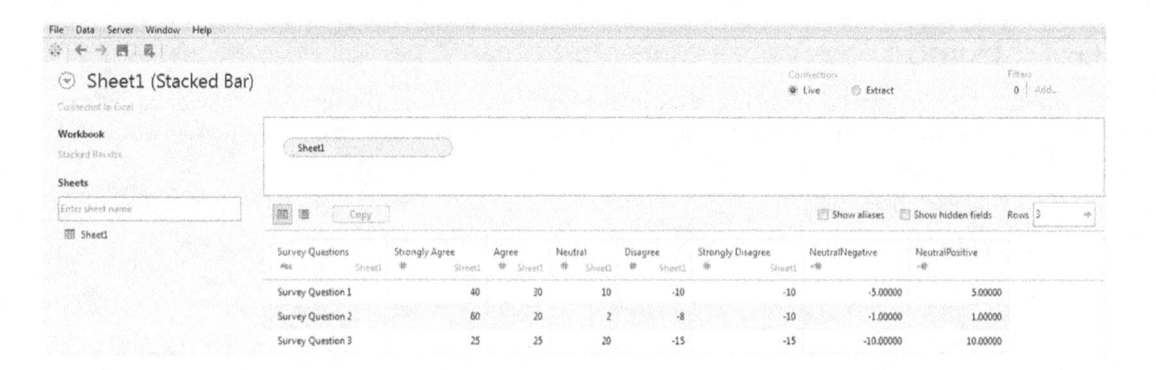

Figure 8-67. Data from "Stacked Bar.xlsx" read into Tableau

8.6.2.2 Step 2

Drag the dimension "Survey Questions" from the dimensions area under the data pane and place it on the rows shelf (Shown in Fig. 8-68).

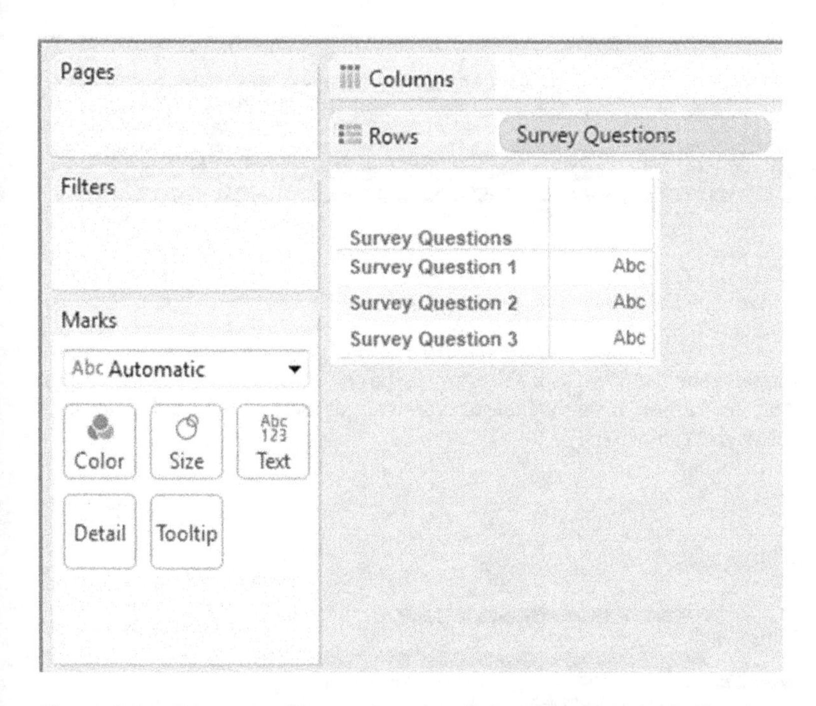

Figure 8-68. Dimension "Survey Questions" placed on the rows shelf

Drag the measure "Strongly Disagree" from the measures area under the data pane and place it on the columns shelf (Shown in Fig. 8-69).

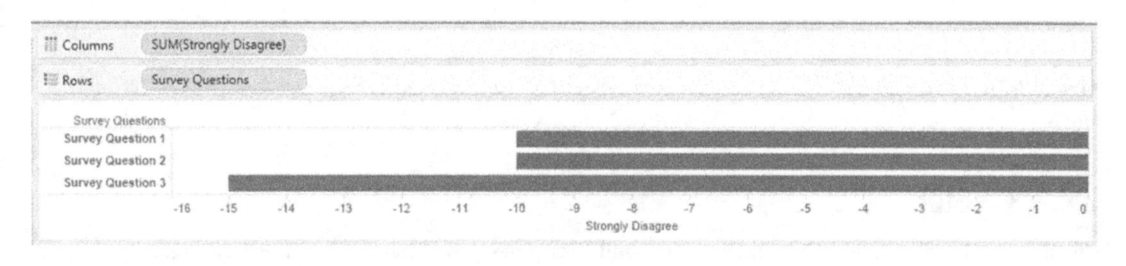

Figure 8-69. Measure "Strongly Disagree" placed on the columns shelf

Drag the measure "Disagree" from the measures area under the data pane and place it on the same axis as the measure "Strongly Disagree" (Shown in Fig. 8-70).

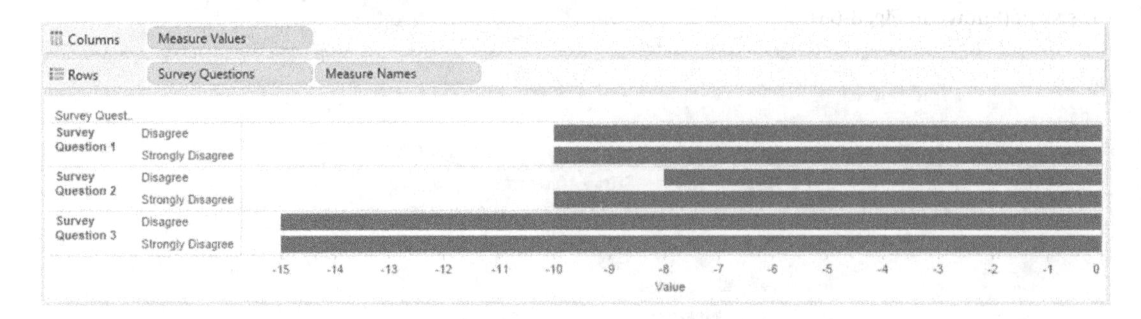

Figure 8-70. Measure "Disagree" placed on the same axis as measure "Strongly Disagree"

Note "Measure Names" on the rows shelf and "Measure Values" on the columns shelf.

Drag the measures, "Agree", "Strongly Agree" from the measures area under the data pane and place it on the same axis as the previous measures (Shown in Fig. 8-71).

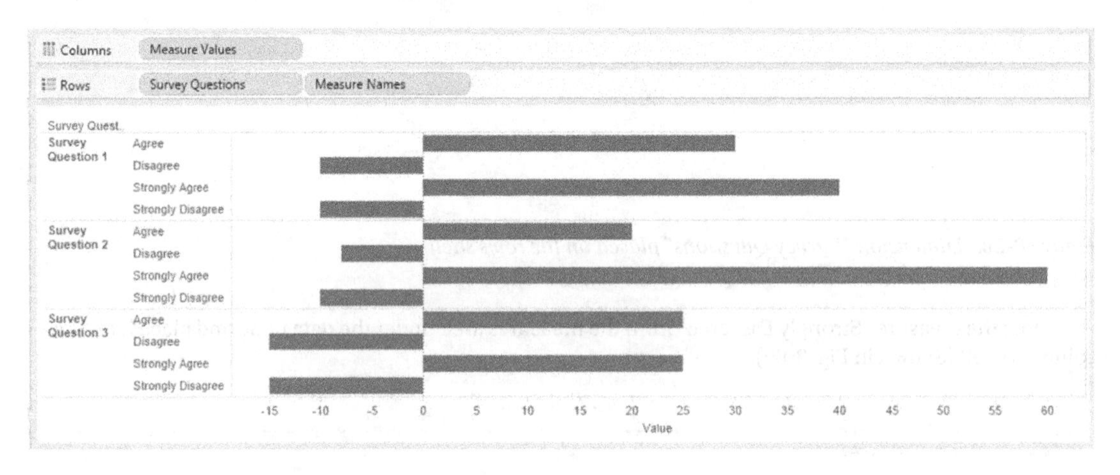

Figure 8-71. Measures "Agree" & "Strongly Agree" placed on same axis as measures, "Disagree" & "Strongly Disagree"

8.6.2.3 Step 3

Create two calculated fields, "NeutralNegative" and "NeutralPositive" (Shown in Fig. 8-72 & Fig. 8-73).

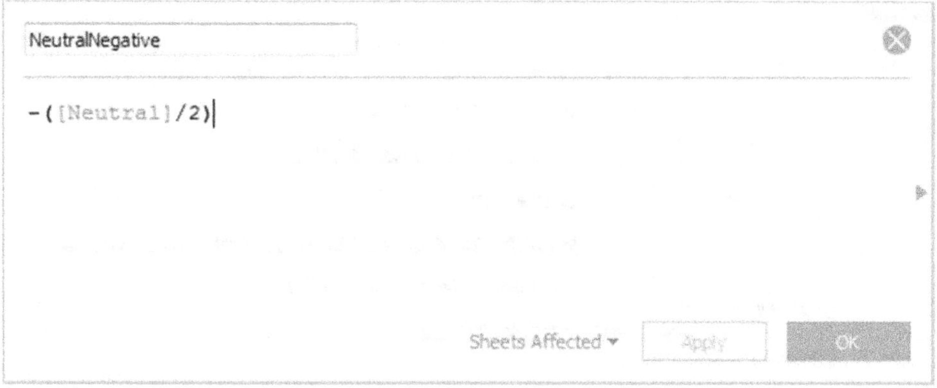

NeutralNegative

-([Neutral]/2)

Sheets Affected ▾ Apply OK

Figure 8-72. *Calculated field "NeutralNegative" being created*

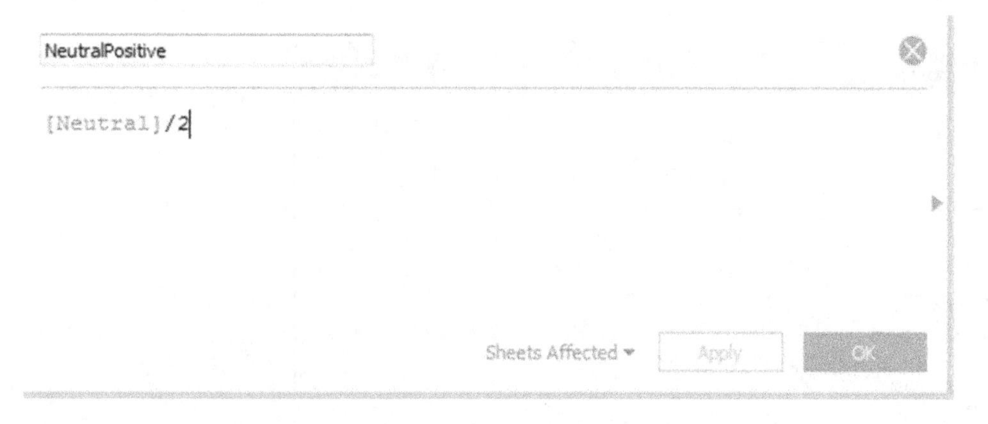

Figure 8-73. *Calculated field "NeutralPositive" being created*

8.6.2.4 Step 4

Manually sort the dimension "Measure Names" as shown in Fig. 8-74 and Fig. 8-75. The sort order should be as follows:

- Strongly disagree
- Disagree
- Strongly agree
- Agree

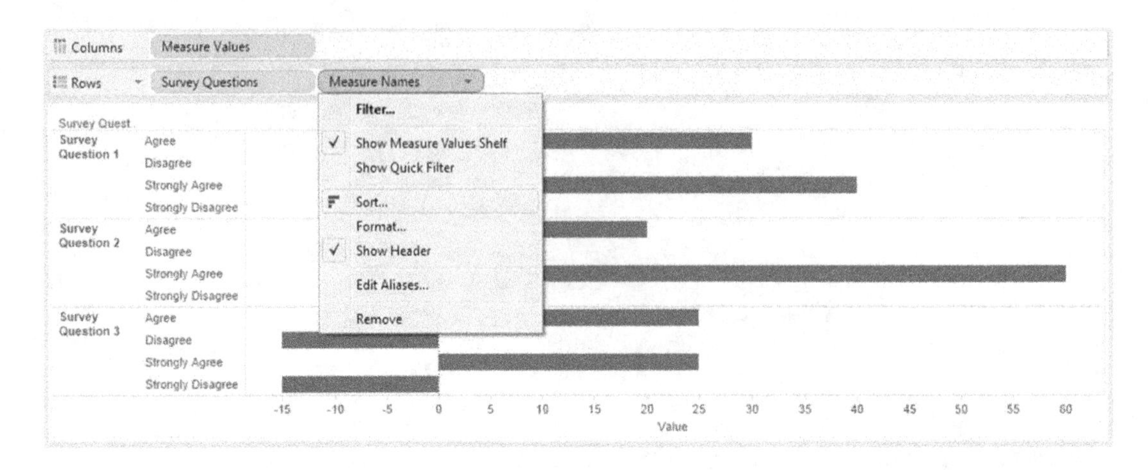

Figure 8-74. *Sorting Measure Names*

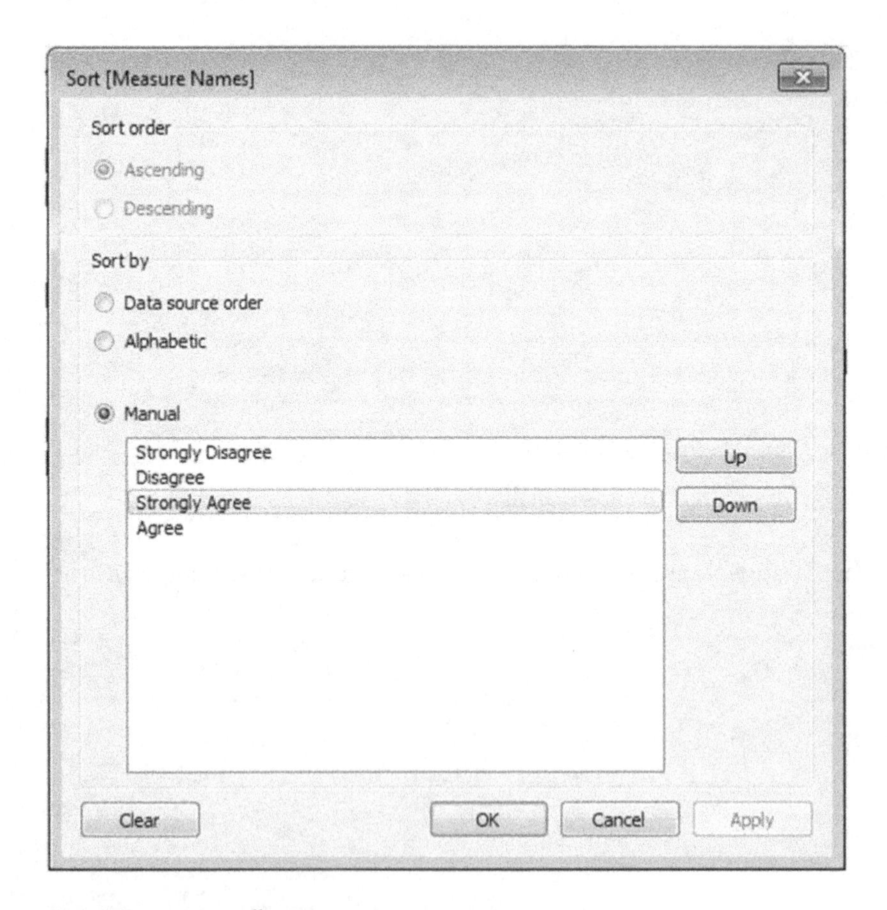

Figure 8-75. *Manually sort measure names*

The output after sorting shown in Fig. 8-76.

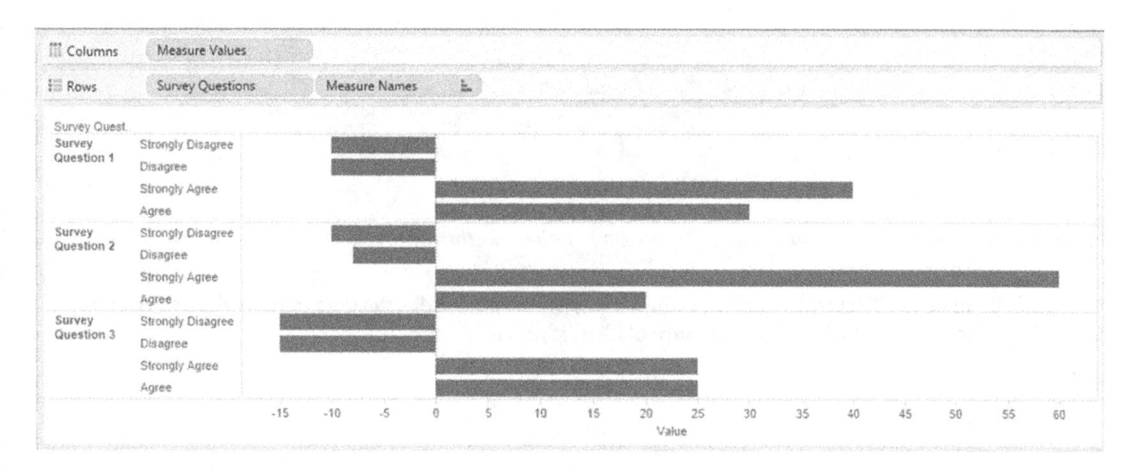

Figure 8-76. *Output after sorting on measure names*

8.6.2.5 Step 5

Drag "Measure Names" from the rows shelf and place it on "Color" on the marks card (Shown in Fig. 8-77).

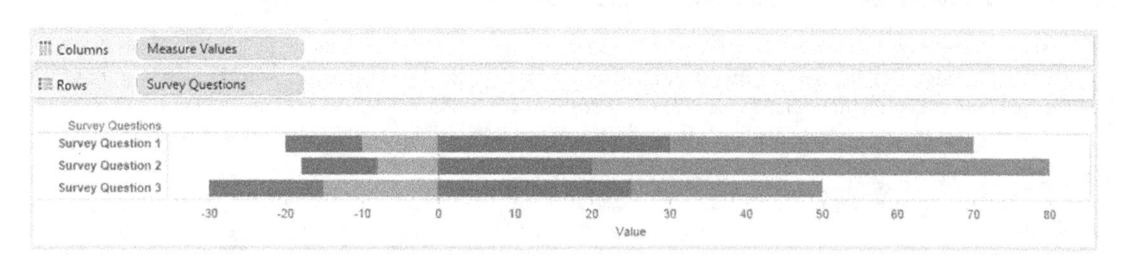

Figure 8-77. *Dimension "measure names" placed on "Color" on the marks card*

Drag the dimension "Measure Names" from the dimensions area under the data pane and place it on "Label" on the marks card (Shown in Fig. 8-78).

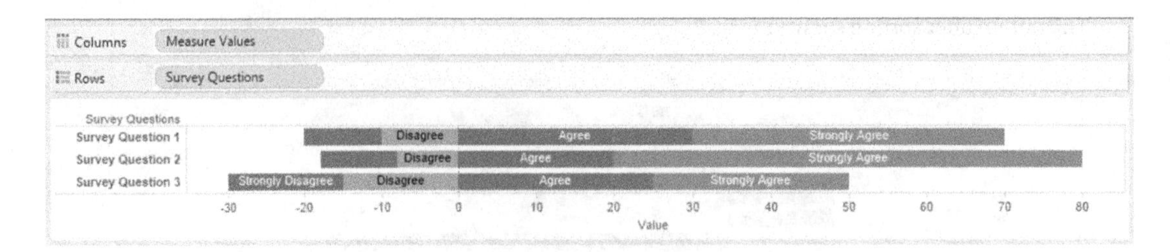

Figure 8-78. Dimension "Measure Names" placed on "Label" on the marks card

Drag the measure "NeutralNegative" from the measures area under the data pane and place it on the "X Axis" (the same axis on which all measures are placed) (Shown in Fig. 8-79).

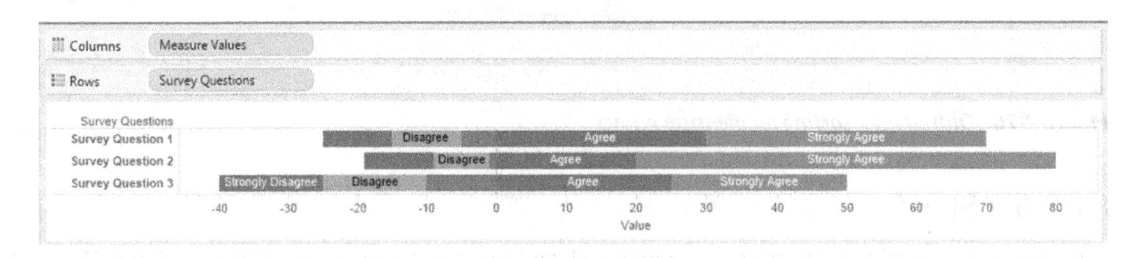

Figure 8-79. Calculated field "NeutralNegative" placed on "X Axis"

Drag the measure "NeutralPositive" from the measures area under the data pane and place it on the "X axis" (the same axis on which all measures are placed) (Shown in Fig. 8-80).

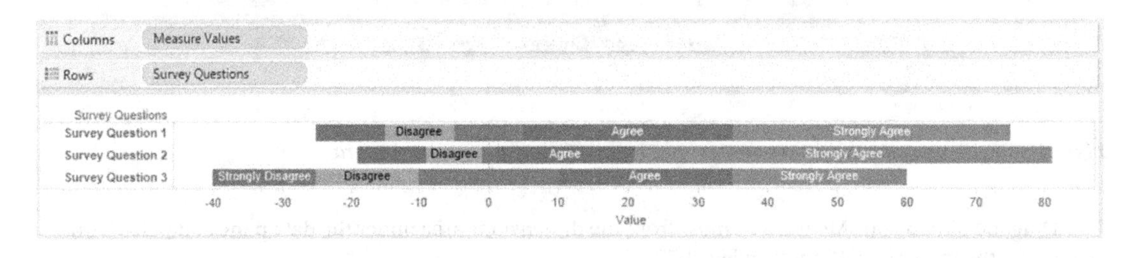

Figure 8-80. Calculated field "NeutralPositive" placed on "X Axis"

8.6.2.6 Step 6

Go to the "Analytics" Pane and drag "Constant Line" on the view (Shown in Fig. 8-81).

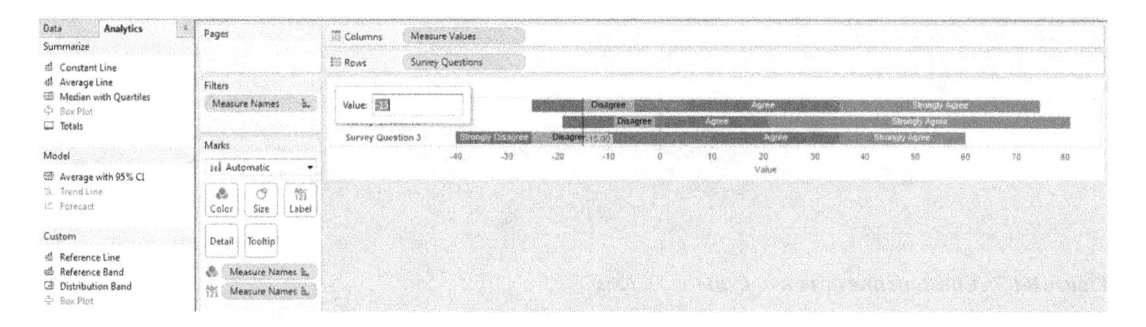

Figure 8-81. Constant line on the view

The above action draws a constant line with value at -15. Let us edit the constant line. Fill in the values as shown in the Figure 8-82.

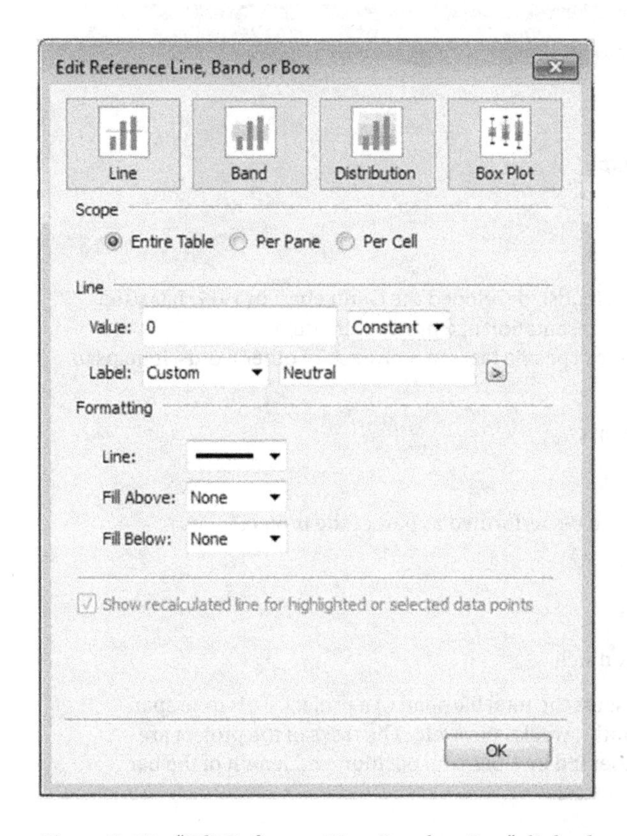

Figure 8-82. "Edit Reference Line, Band, or Box" dialog box

The above settings will draw a constant line at 0 on the "X Axis" (Shown in Fig. 8-83).

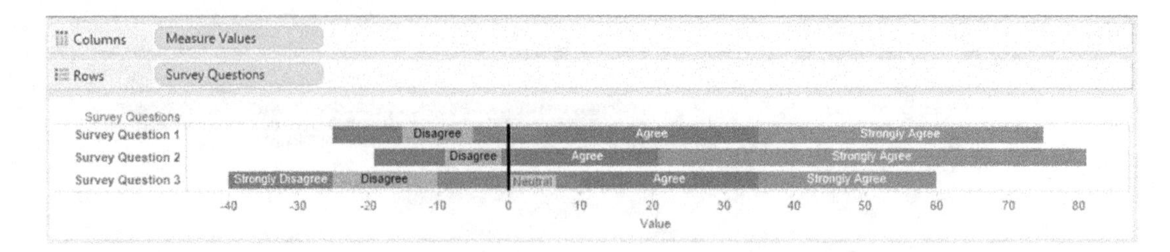

Figure 8-83. *Constant line drawn at 0 on the "X Axis"*

The measure "Neutral" has been split into two so that one half appears on the negative scale and the other half appears on the positive scale.

The final output is shown in Fig. 8-84.

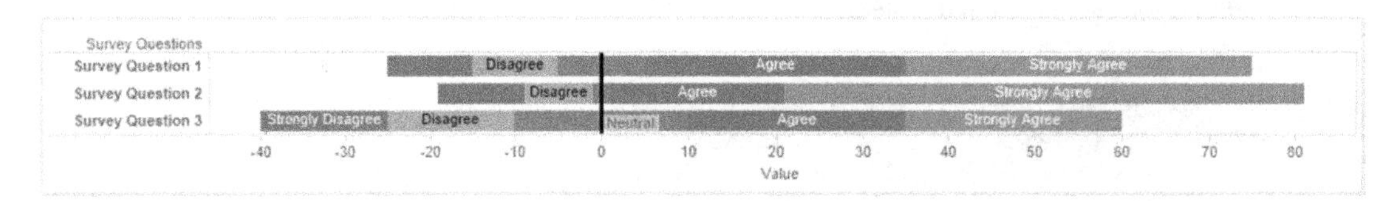

Figure 8-84. *Stacked bar chart - Demo 1 - final output*

8.7 Gantt chart

Henry L. Gantt, an American engineer and social scientist, developed the Gantt chart in 1917. It is used extensively in project management for graphical representation of a project schedule.

A Gantt chart helps to plan, coordinate and track specific tasks in a project. In other words, it helps to quickly visualize the following:

- The start and end date of the entire project.
- The duration of the entire project.
- What are the various tasks or activities to be performed as part of the project?
- When each task / activity begins and ends?
- The duration of each activity.
- The activities that overlap and by how much.

A Gantt chart has a horizontal axis that represents the total life span of a project. This time span is broken down into small increments such as months, weeks, days, etc. The tasks of the project are represented on the vertical axis. Each task is represented by a bar. The position and length of the bar indicates the start date, duration and end date of the task.

8.7.1 Shortcomings of Gantt charts

- They do not indicate task dependencies. They do not tell us how a task or tasks that are behind schedule will affect the other tasks.
- They might have been developed early in the planning stage. As the project undergoes change, there is a need to update constantly the Gantt chart. They should not be viewed as immutable.

- They do not take into consideration the cost factor.
- They can become really complex for large projects with several milestones with each milestone requiring several tasks to be completed.
- They can become difficult to view if the chart runs into more than one page.
- The size of the bar does not indicate the amount of work that is required to complete the task.

8.7.1.1 Demo 1

Objective:

You are a project manager with the responsibility of managing three projects ("Project 1", "Project 2" and "Project 3". You have in your team three technical architects. Let us call them as "Person 1", "Person 2" and "Person 3". You are working on allocating them to the three projects. You have worked out the below schedule for them. You would like a visualization that helps to depict clearly their schedule at a quick glance. You decide to accomplish this using "Gantt Chart".

	A	B	C	D
1	Resources	Project	Start Date	End Date
2	Person 1	Project 1	1-Apr-16	31-Jul-16
3	Person 2	Project 1	1-Apr-16	30-Jun-16
4	Person 3	Project 1	1-Jun-16	31-Jul-16
5	Person 1	Project 2	1-Aug-16	30-Sep-16
6	Person 2	Project 2	1-Jul-16	31-Aug-16
7	Person 3	Project 2	1-Aug-16	30-Sep-16
8	Person 2	Project 3	1-Sep-16	30-Sep-16
9	Person 3	Project 3	1-Apr-16	31-May-16

Input: "GanttChart.xlsx"
Expected output: Shown in Fig. 8-85.

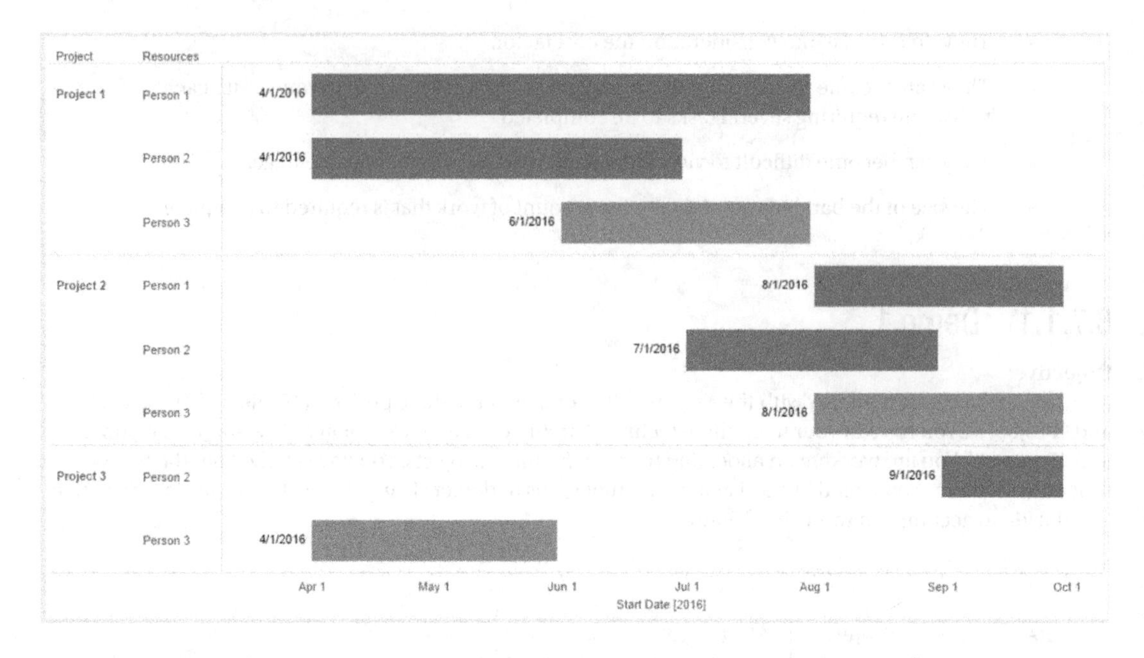

Figure 8-85. *Gantt chart - Demo 1 - expected output*

8.7.1.1.1 Steps to create a Gantt chart

8.7.1.1.2 Step 1

Read in data from "GanttChart.xlsx" into Tableau (Shown in Fig. 8-86).

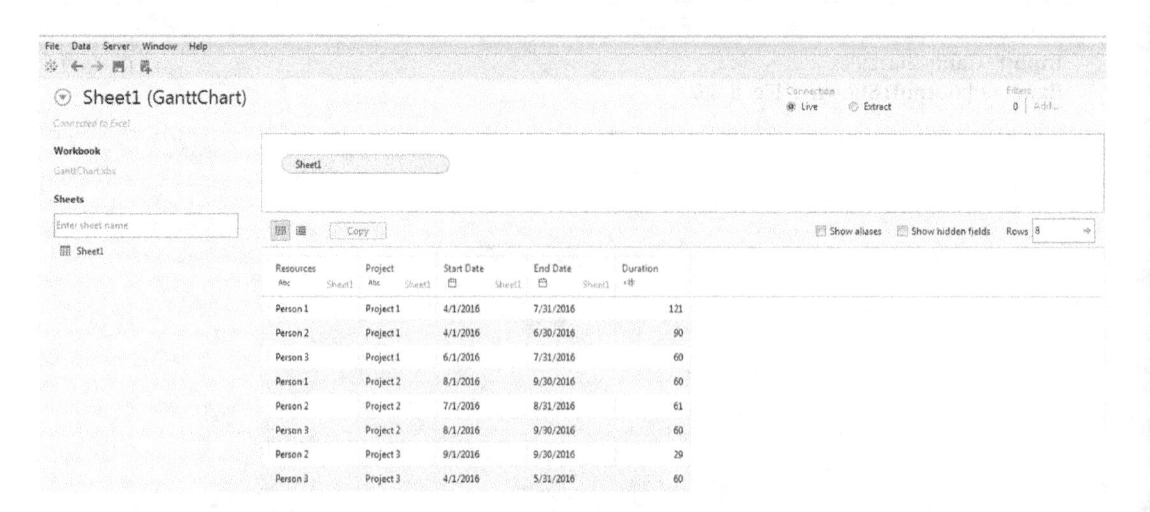

Resources	Project	Start Date	End Date	Duration
Person 1	Project 1	4/1/2016	7/31/2016	121
Person 2	Project 1	4/1/2016	6/30/2016	90
Person 3	Project 1	6/1/2016	7/31/2016	60
Person 1	Project 2	8/1/2016	9/30/2016	60
Person 2	Project 2	7/1/2016	8/31/2016	61
Person 3	Project 2	8/1/2016	9/30/2016	60
Person 2	Project 3	9/1/2016	9/30/2016	29
Person 3	Project 3	4/1/2016	5/31/2016	60

Figure 8-86. *Data from "GanttChart.xlsx" read into Tableau*

8.7.1.1.3 Step 2

Drag the dimension "Start Date" from the dimensions area under the data pane and place it on the columns shelf (Shown in Fig. 8-87).

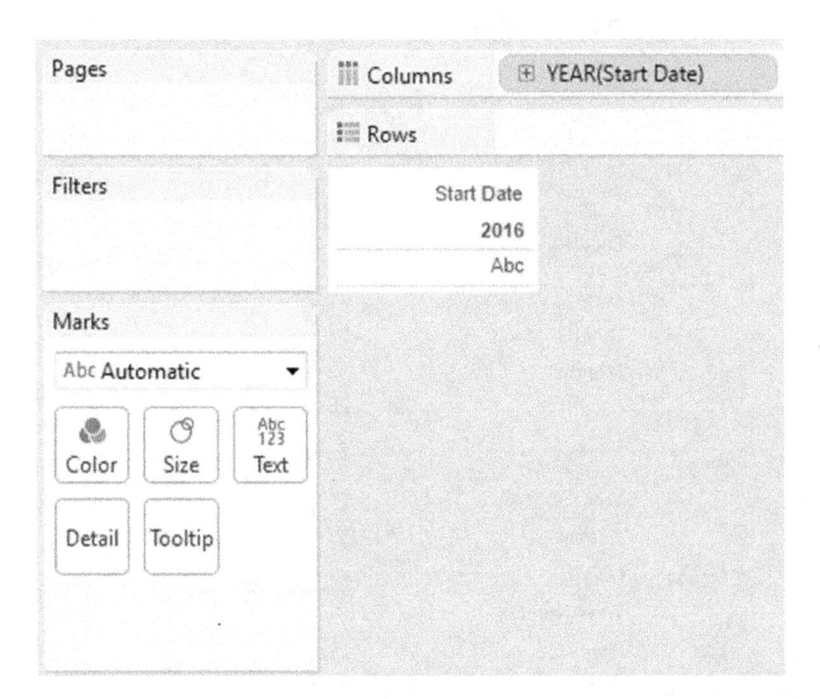

Figure 8-87. *Dimension "Start Date" placed on the columns shelf*

Click on the drop down next to YEAR(Start Date) on the columns shelf. Select Exact Date (Shown in Fig. 8-88).

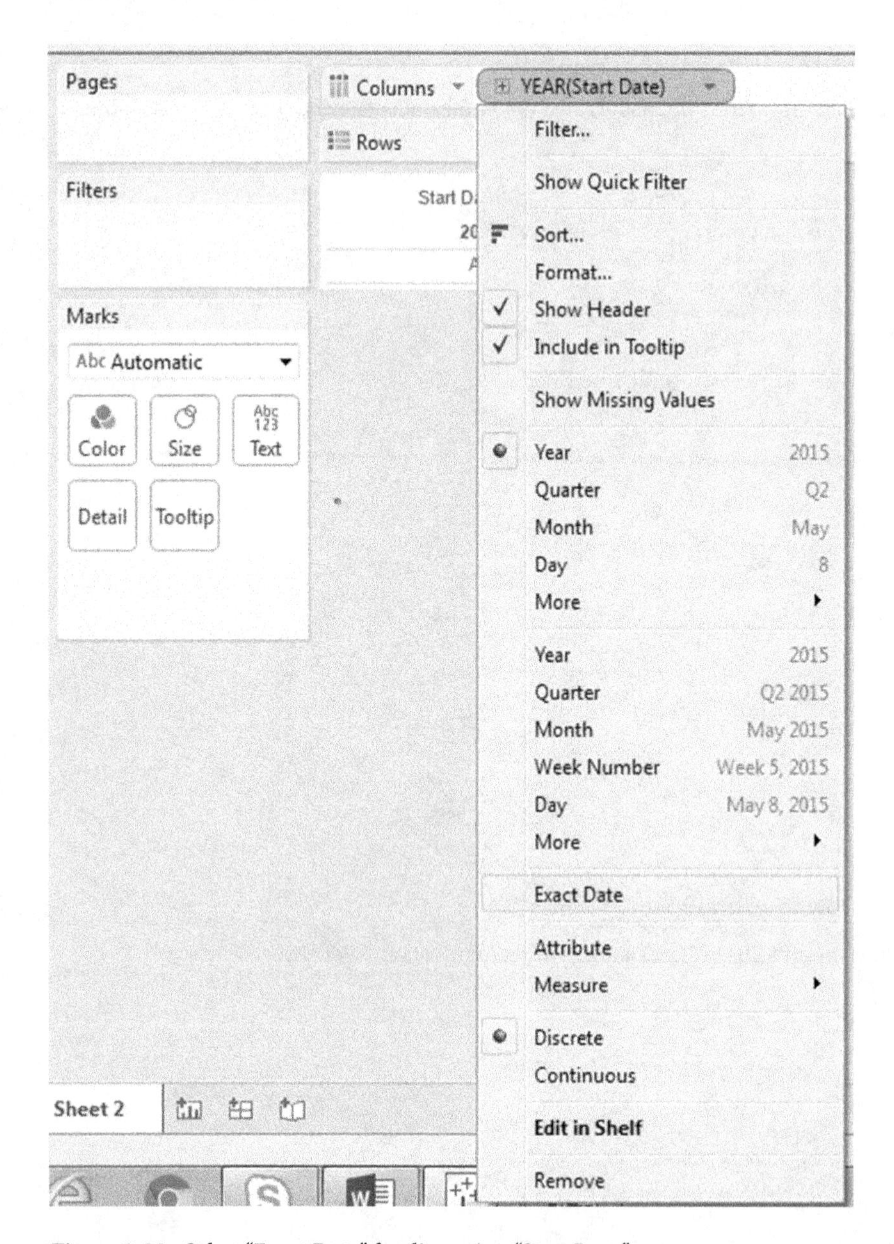

Figure 8-88. *Select "Exact Date" for dimension "Start Date"*

The output after selecting "Exact Date" (Shown in Fig. 8-89).

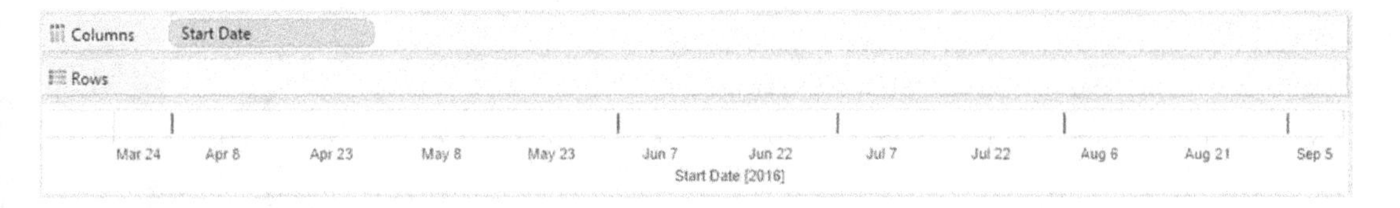

Figure 8-89. *Output after selecting "Exact Date"*

8.7.1.1.4 Step 3

Drag the dimensions "Project" and "Resources" from the dimensions area under the data pane and place it on the rows shelf (Shown in Fig. 8-90).

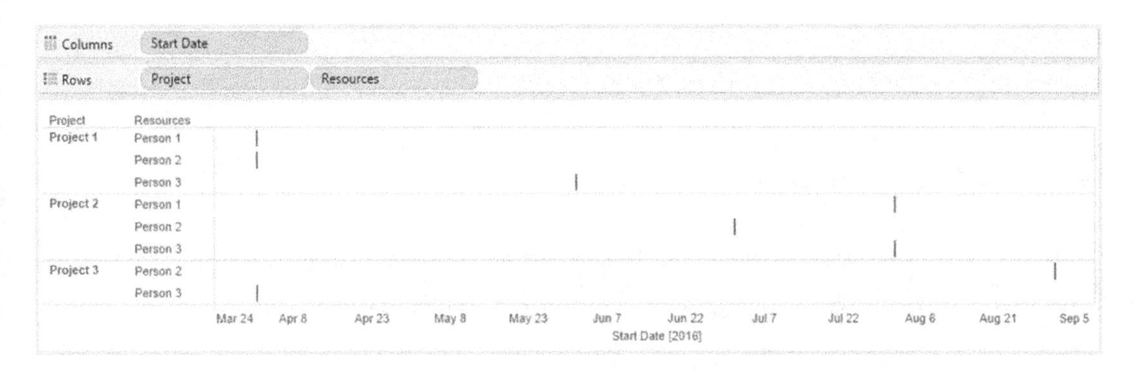

Figure 8-90. *Dimensions "Project" & "Resources" placed on the rows shelf*

8.7.1.1.5 Step 4

Change the mark type to "Gantt Bar" (Shown in Fig. 8-91).

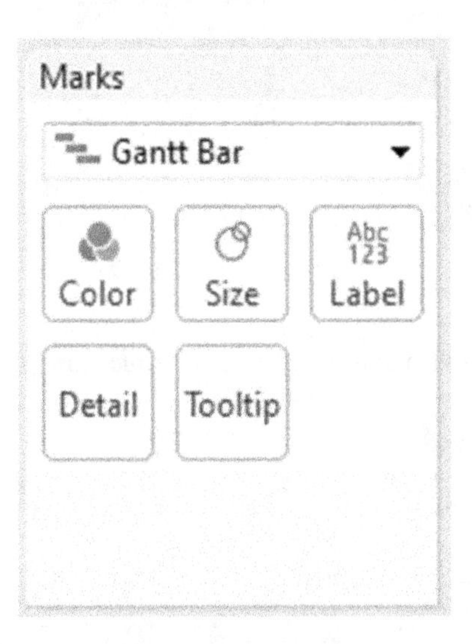

Figure 8-91. *"Mark Type" set to "Gantt Bar"*

8.7.1.1.6 Step 5

Create a calculated field "Duration" (Shown in Fig. 8-92).

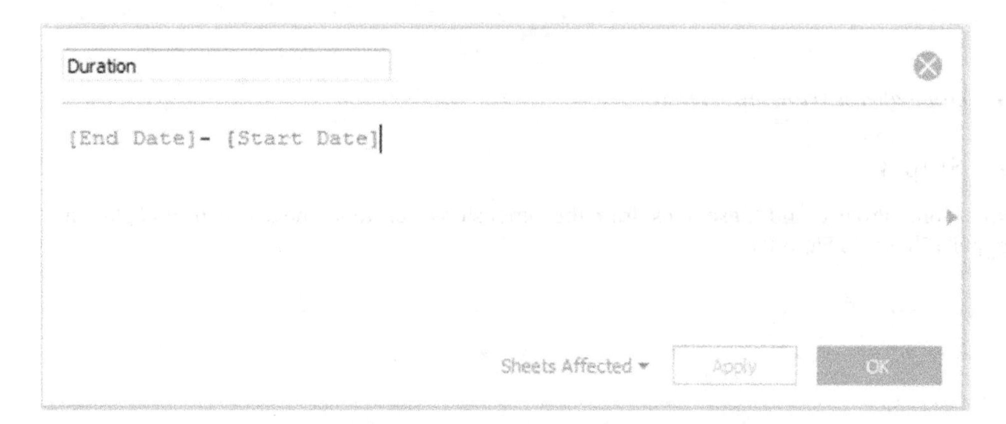

Figure 8-92. Calculated field "Duration" being created

Drag the calculated field "Duration" from the measures area under the data pane and place it on "Size" on the marks card (Shown in Fig. 8-93).

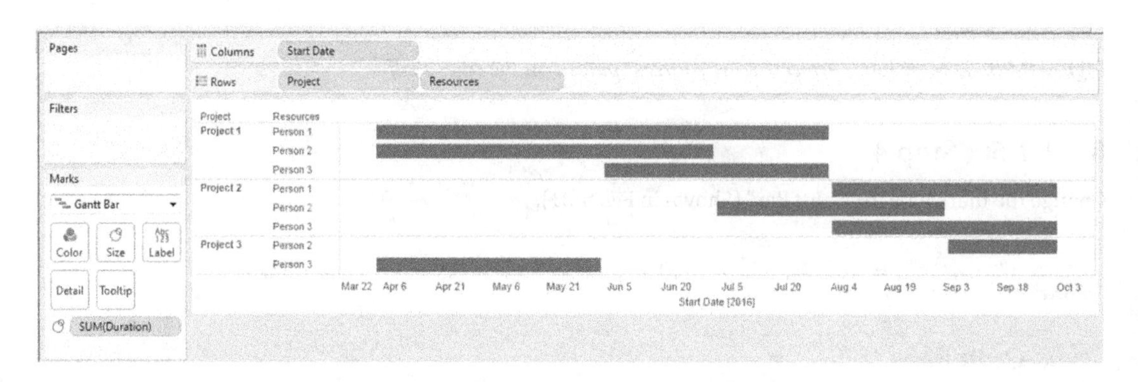

Figure 8-93. Calculated field "Duration" placed on "Size" on the marks card

8.7.1.1.7 Step 6

Drag the dimension "Resources" from the dimensions area under the data pane and place it on "Color" on the marks card (Shown in Fig. 8-94).

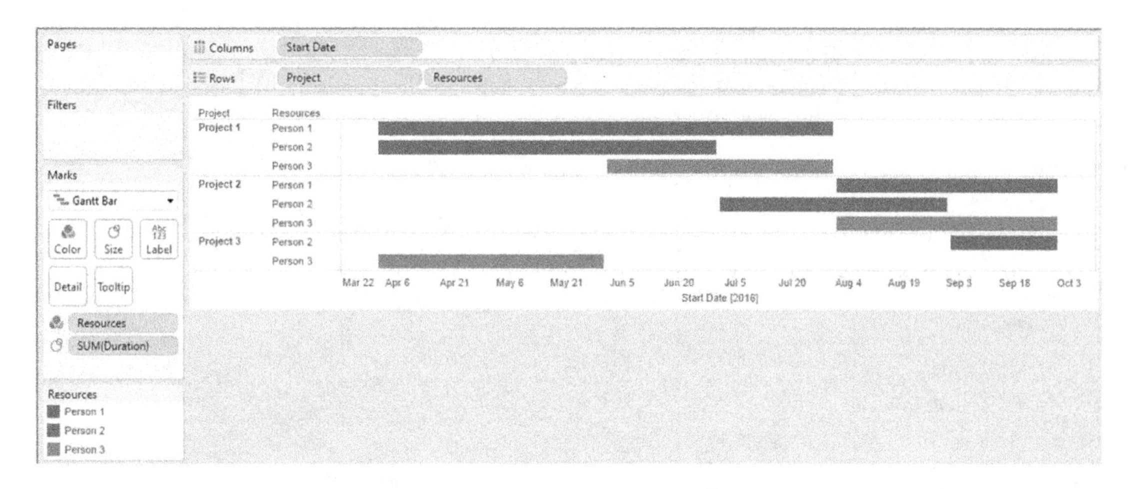

Figure 8-94. *Dimension "Resources" placed on "Color" on the marks card*

Drag the dimension "Start Date" from the dimensions area under the data pane and place it on "Label" on the marks card (Shown in Fig. 8-95).

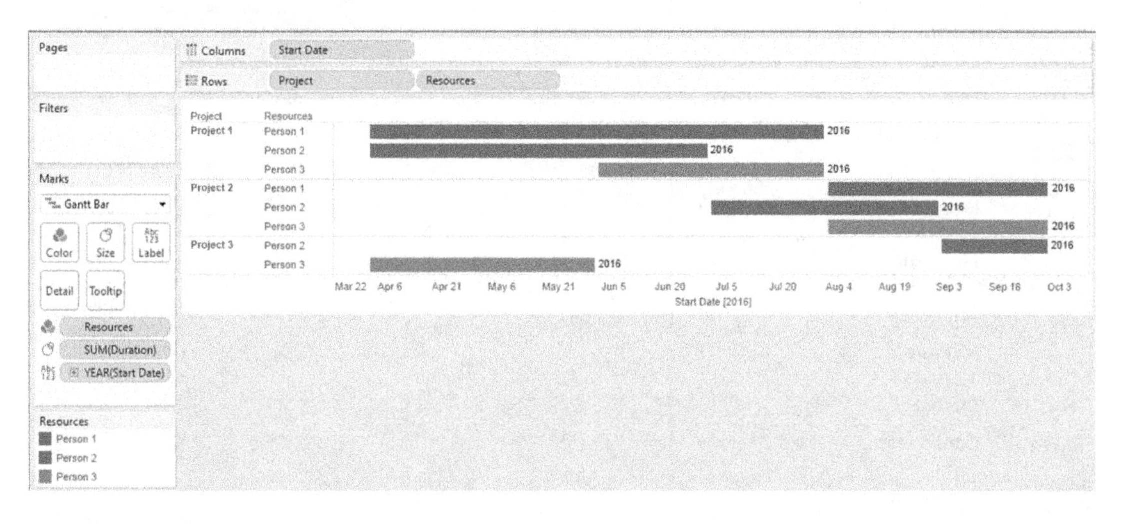

Figure 8-95. *Dimension "Start Date" placed on "Label" on the marks card*

Click on YEAR(Start Date) placed on "Label" on the marks card to bring up the menu. Select "Exact Date" (Shown in Fig. 8-96).

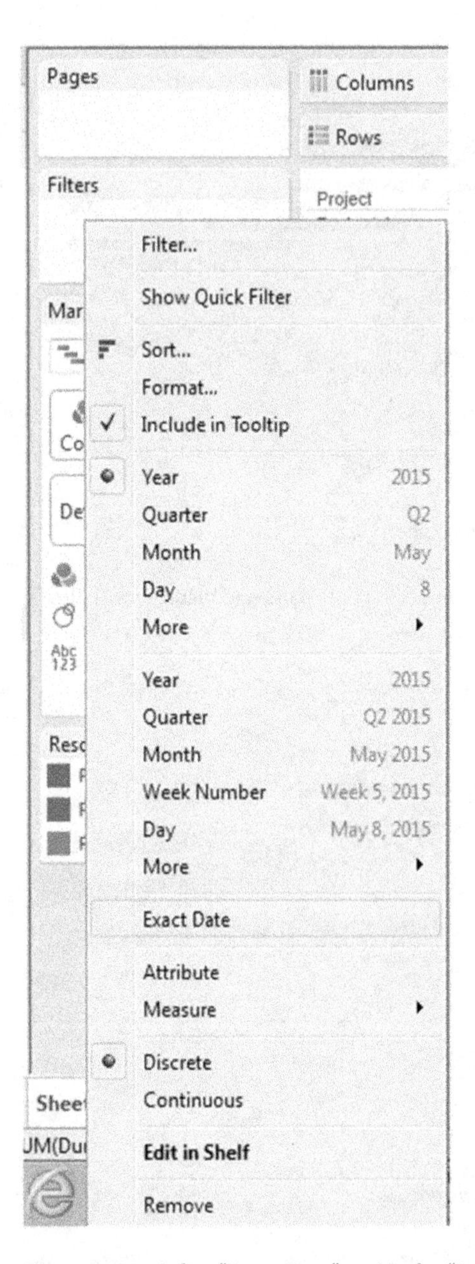

Figure 8-96. *Select "Exact Date" option for "Start Date"*

The output after setting the "Start Date" to "Exact Date" and placing it on "Label" on the marks card (Shown in Fig. 8-97).

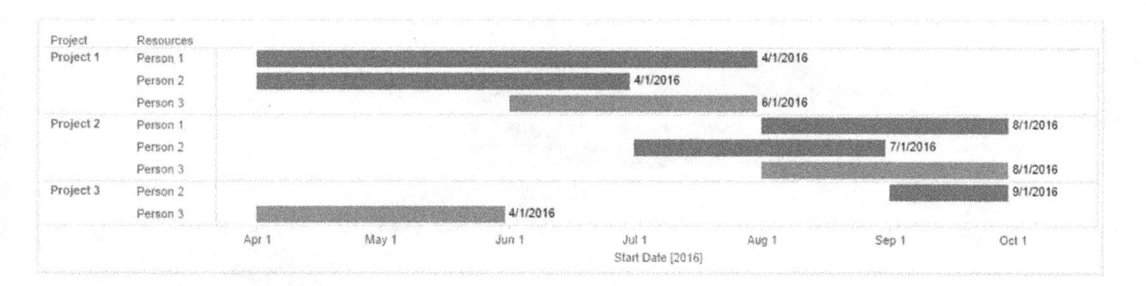

Figure 8-97. *Output after setting the "Start Date" to "Exact Date"*

Click on "Label" to change the alignment to "Left" (Shown in Fig. 8-98).

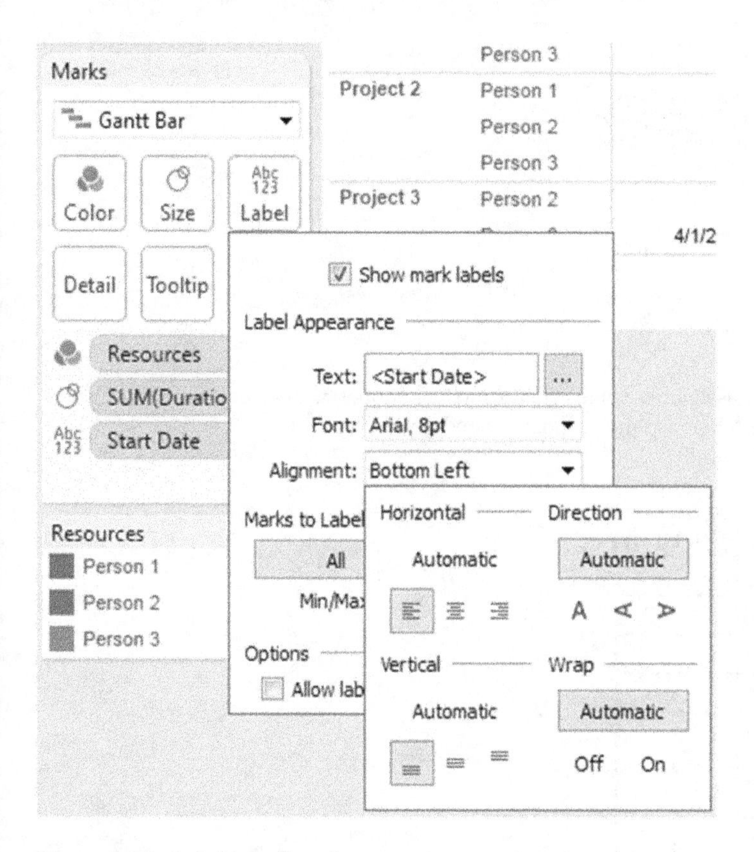

Figure 8-98. *Label left aligned*

Set the Fit to "Entire View".

The output is as shown in Fig. 8-99.

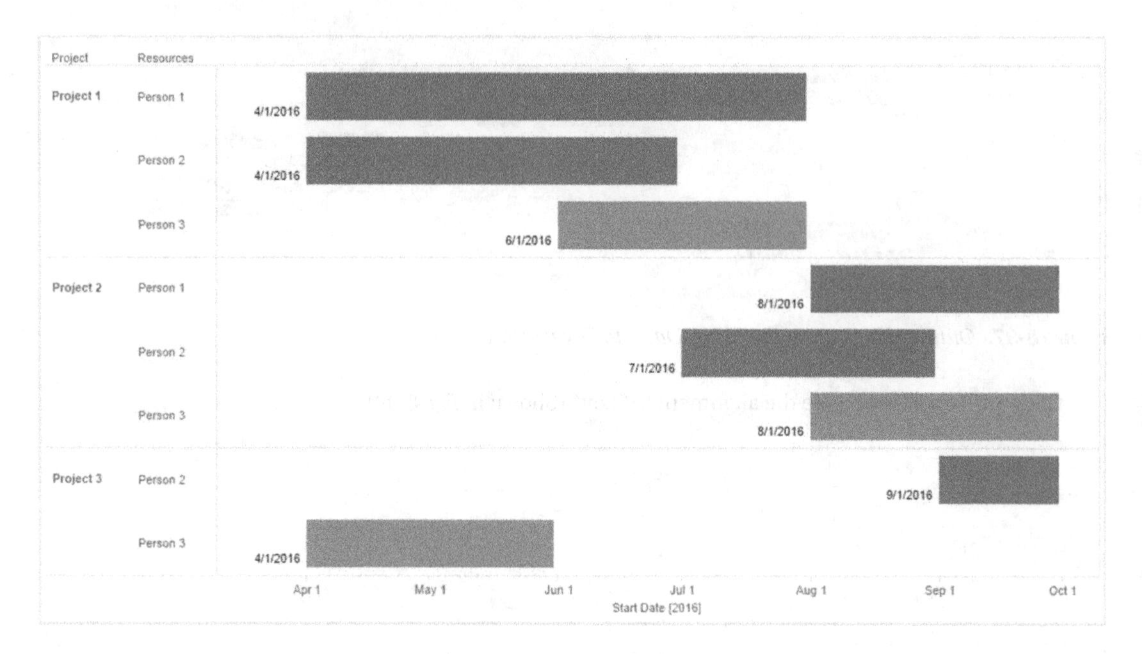

Figure 8-99. *Gantt chart - Demo 1 - final output*

8.7.2 Demo 2

Objective:

You are a builder working on the assignment of restructuring the banquet hall of a popular uptown restaurant. You have it all planned out. Below is the proposed schedule.

	A	B	C
1	**Tasks**	**Start Date**	**Duration**
2	Foundation	1-Jun-16	10
3	Walls	12-Jun-16	7
4	Roof	20-Jun-16	10
5	Windows, Doors	1-Jul-16	5
6	Plumbing	7-Jul-16	3
7	Electric	7-Jul-16	3
8	Painting	11-Jul-16	2
9	Flooring	13-Jul-16	2

In order to present the schedule to all the stakeholders, you decide to sketch a Gantt chart for an easy view and understanding.

Input: "GanttChartAssignment.xlsx".

Expected output: Shown in Fig. 8-100.

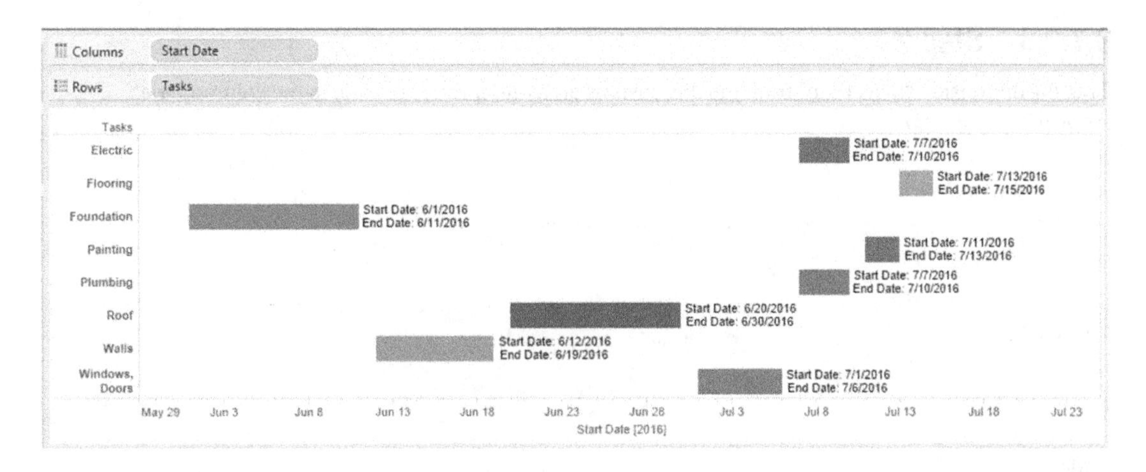

Figure 8-100. *Gantt Chart - Demo 2 - expected output*

8.7.2.1 Steps to create a Gantt chart

8.7.2.2 Step 1

Read in the data from "GanttChartAssignment.xlsx" into Tableau (Shown in Fig. 8-101).

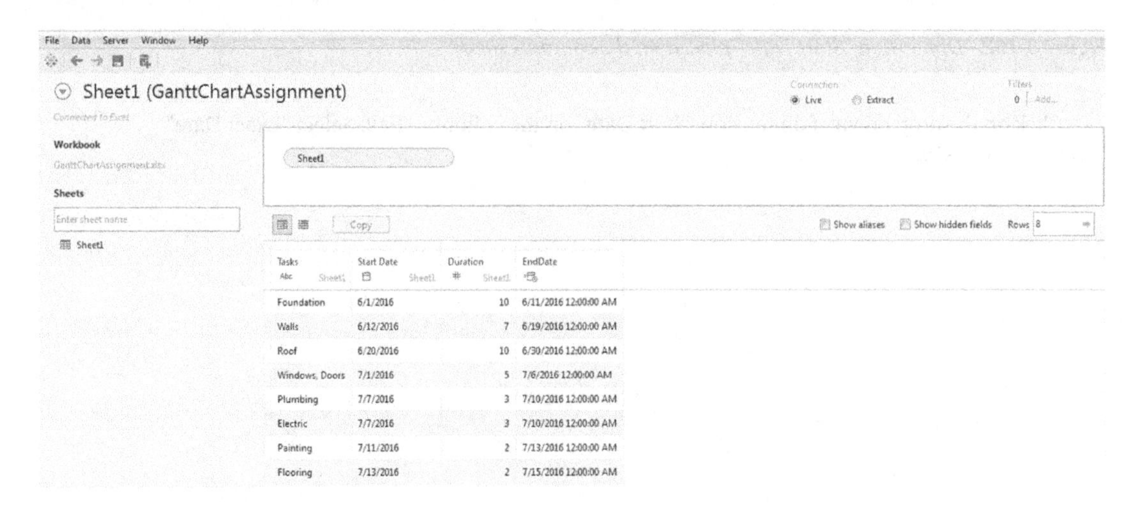

Tasks	Start Date	Duration	EndDate
Foundation	6/1/2016	10	6/11/2016 12:00:00 AM
Walls	6/12/2016	7	6/19/2016 12:00:00 AM
Roof	6/20/2016	10	6/30/2016 12:00:00 AM
Windows, Doors	7/1/2016	5	7/6/2016 12:00:00 AM
Plumbing	7/7/2016	3	7/10/2016 12:00:00 AM
Electric	7/7/2016	3	7/10/2016 12:00:00 AM
Painting	7/11/2016	2	7/13/2016 12:00:00 AM
Flooring	7/13/2016	2	7/15/2016 12:00:00 AM

Figure 8-101. *Data from "GanttChartAssignment.xlsx" read into Tableau*

8.7.2.3 Step 2

Drag the dimension "Start Date" from the dimensions area under the data pane to the columns shelf (Shown in Fig. 8-102).

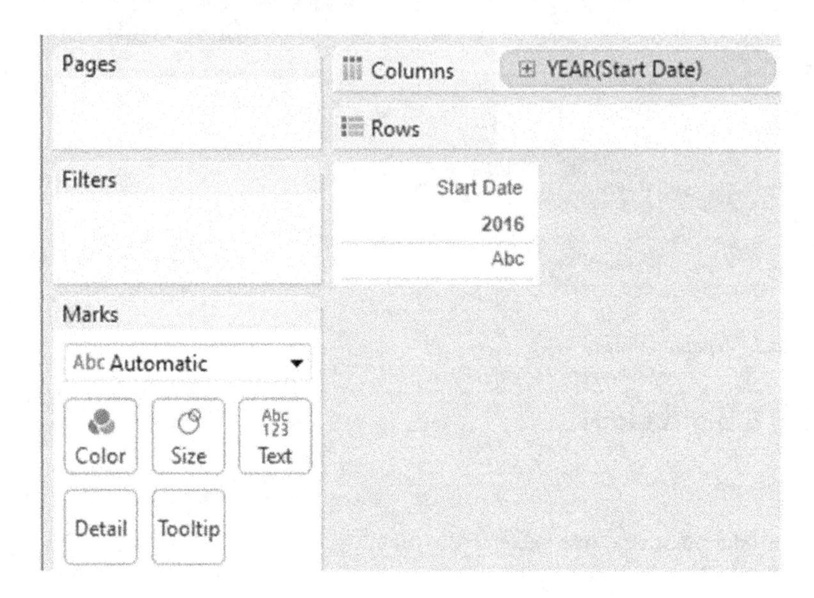

Figure 8-102. *Dimension "Start Date" placed on the columns shelf*

Click on the drop down of dimension "Start Date" on the columns shelf. Select "Exact Date" (Shown in Fig. 8-103).

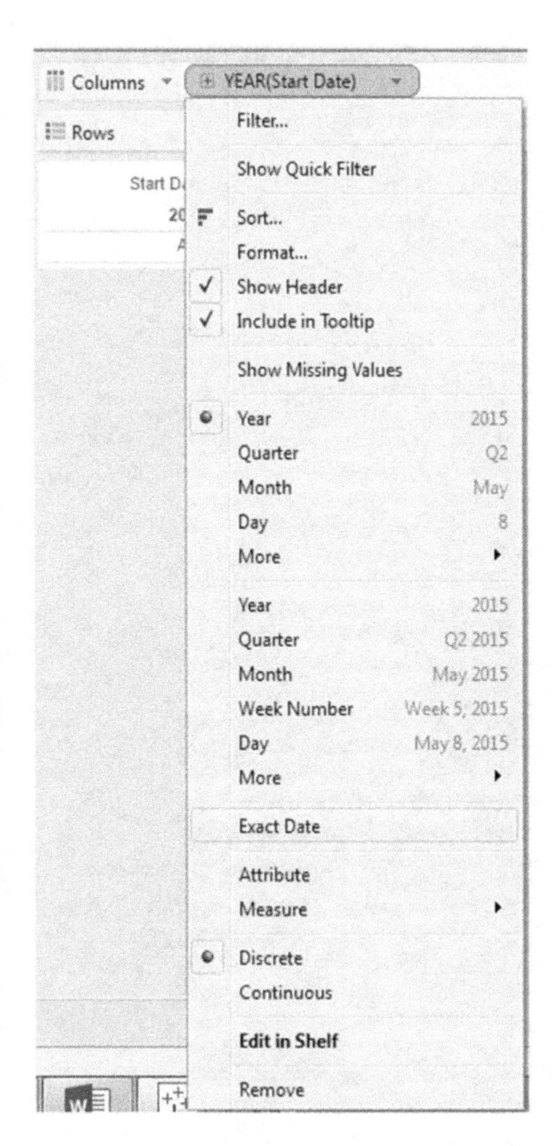

Figure 8-103. *"Exact Date" option selected for dimension "Start Date"*

The output after the dimension "Start Date" was set to "Exact Date" (Shown in Fig. 8-104).

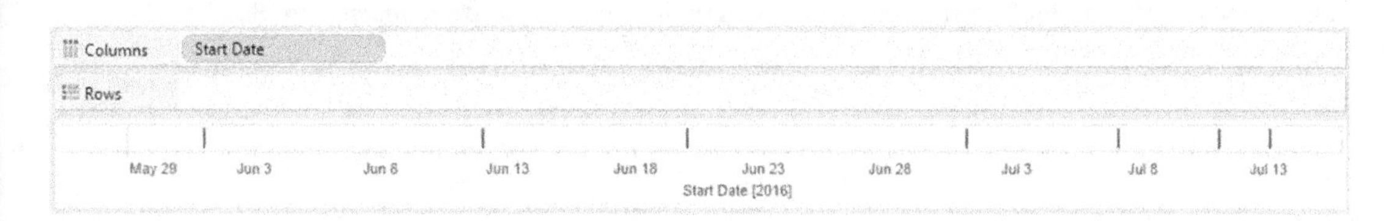

Figure 8-104. *Output after dimension "Start date" is set to "Exact Date"*

Drag the dimension “Tasks” from the dimensions area under the data pane and place it on the rows shelf (Shown in Fig. 8-105).

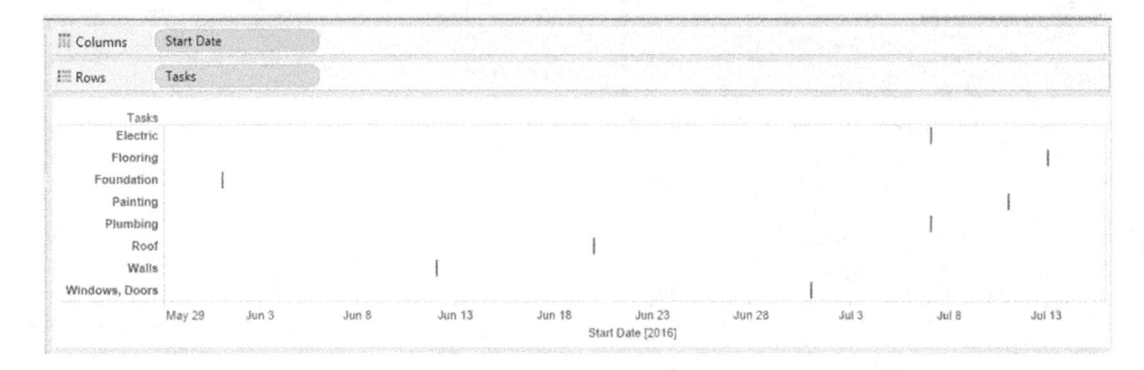

Figure 8-105. *Dimension “Tasks” placed on the rows shelf*

8.7.2.4 Step 3

Create a calculated field “EndDate” (Shown in Fig. 8-106).

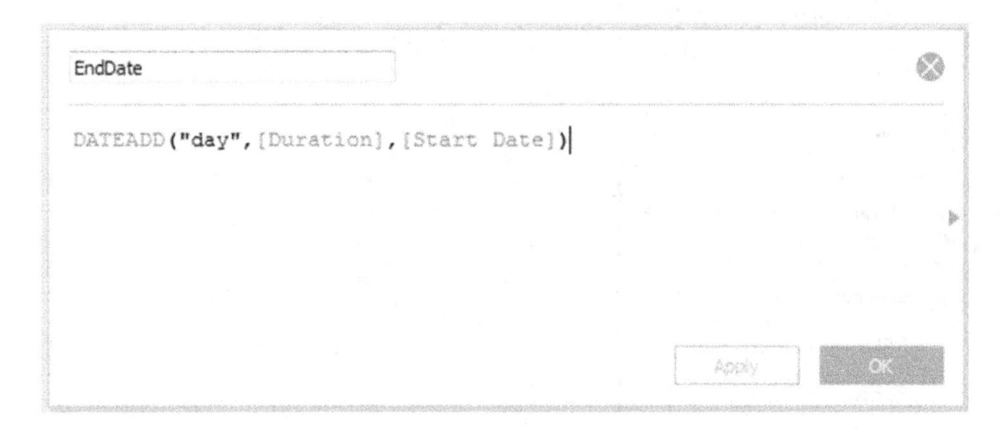

Figure 8-106. *Calculated field “EndDate” being created*

8.7.2.5 Step 4

Drag the measure “Duration” from the measures area under the data pane and place it on “Size” on the marks card (Shown in Fig. 8-107).

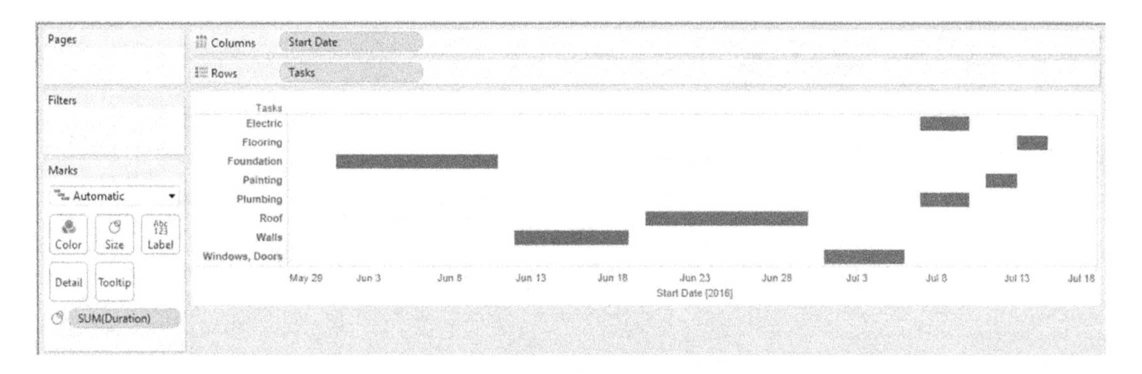

Figure 8-107. *Measure "Duration" placed on "Size" on the marks card*

Drag the dimension "Tasks" from the dimensions area under the data pane and place it on "Color" on the marks card (Shown in Fig. 8-108).

Figure 8-108. *Dimension "Tasks" placed on "Color" on the marks card*

Drag the dimension "Start Date" from the dimensions area under the data pane and place it on "Label" on the marks card (Shown in Fig. 8-109).

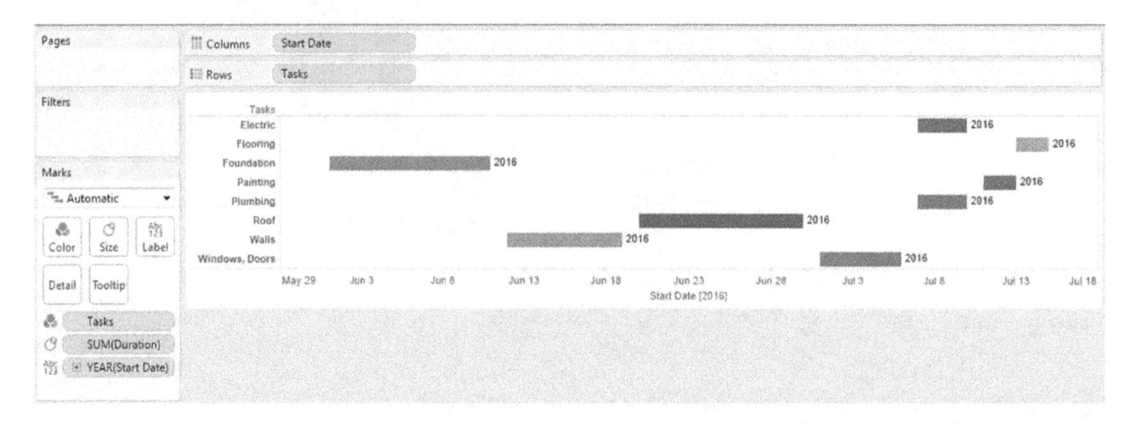

***Figure 8-109.** Dimension "Start Date" placed on "Label" on the marks card*

Click on the drop down of the dimension "Start Date" on "Label" and change it to "Exact Date" (Shown in Fig. 8-110).

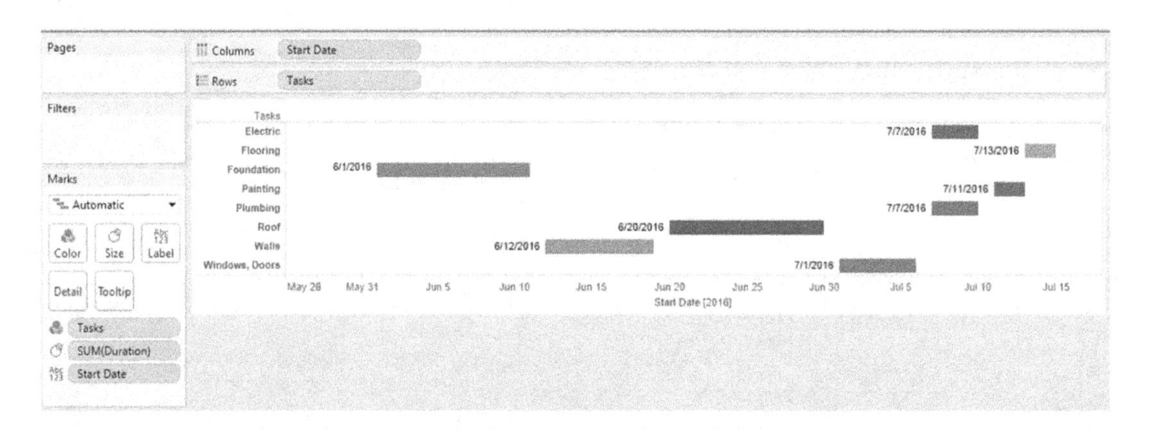

***Figure 8-110.** Dimension "Start Date" placed on "Label" and set to "Exact Date"*

Repeat it for the "End Date". Drag the calculated field "End Date" from the dimensions area under the data pane and place it on "Label" on the marks card (Shown in Fig. 8-111).

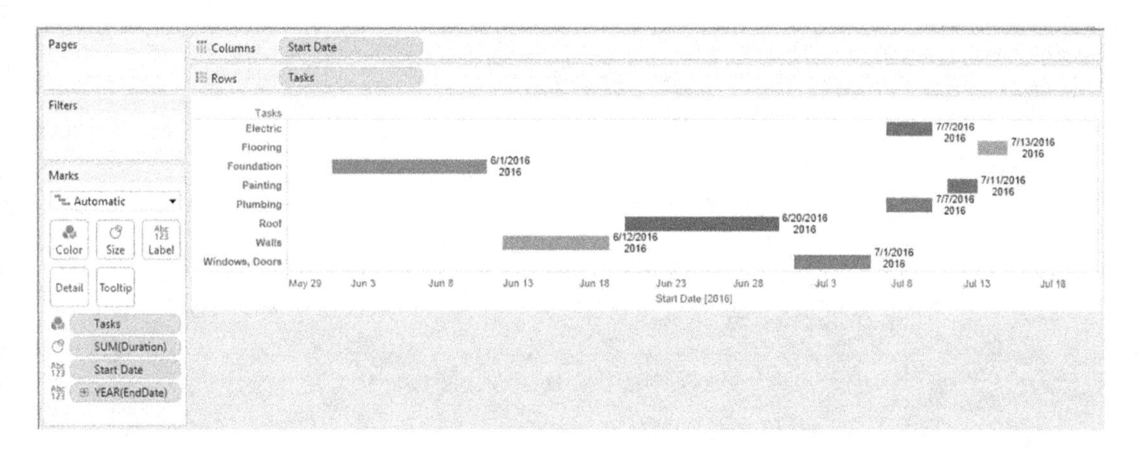

***Figure 8-111.** Calculated field "EndDate" placed on "Label" on the marks card*

Change the "EndDate" on "Label" to "Exact Date" (Shown in Fig. 8-112).

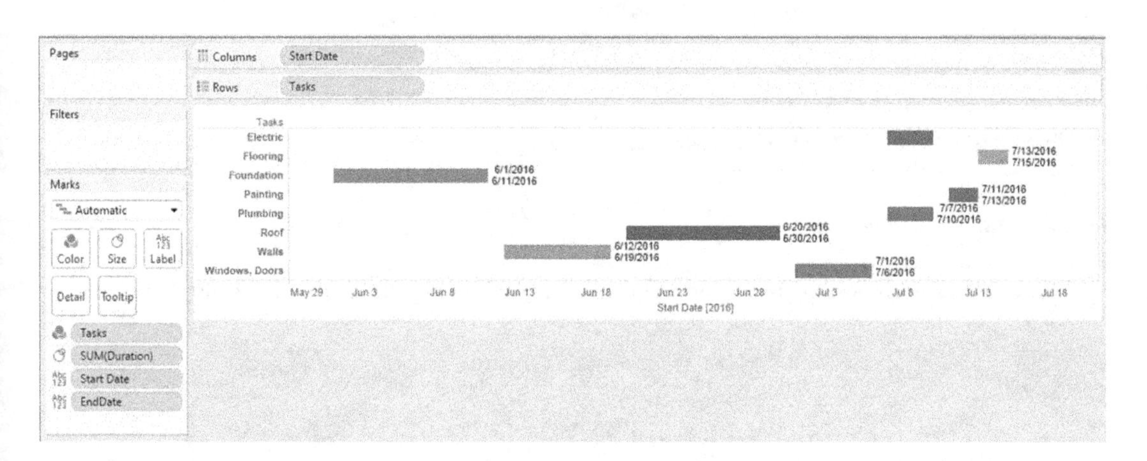

***Figure 8-112.** Calculated field "EndDate" set to "Exact Date"*

Click on "Label" on the marks card to bring up the label dialog box (Shown in Fig. 8-113).

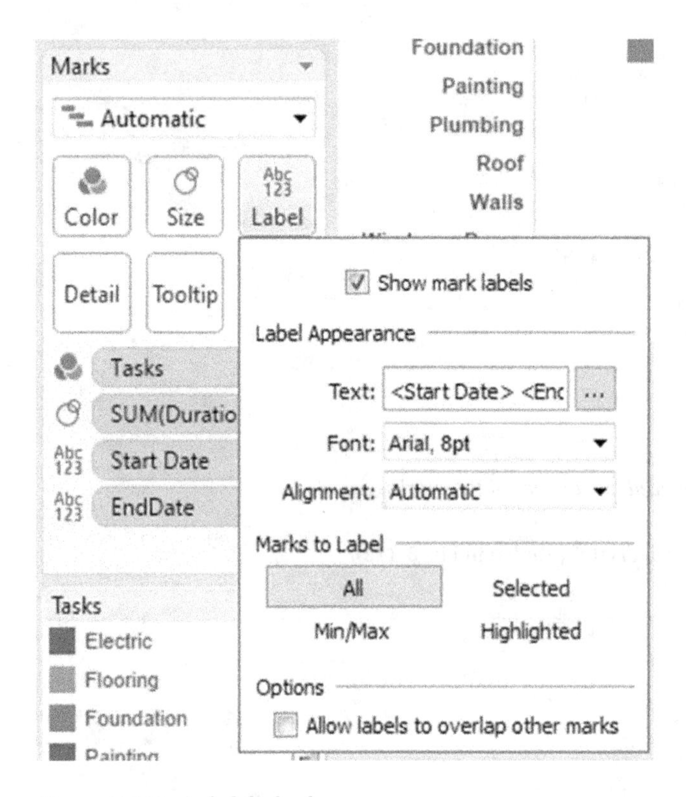

Figure 8-113. *Label dialog box*

Click on the ellipsis next to "Text" (Shown in Fig. 8-114) to bring up the "Edit Label" dialog box.

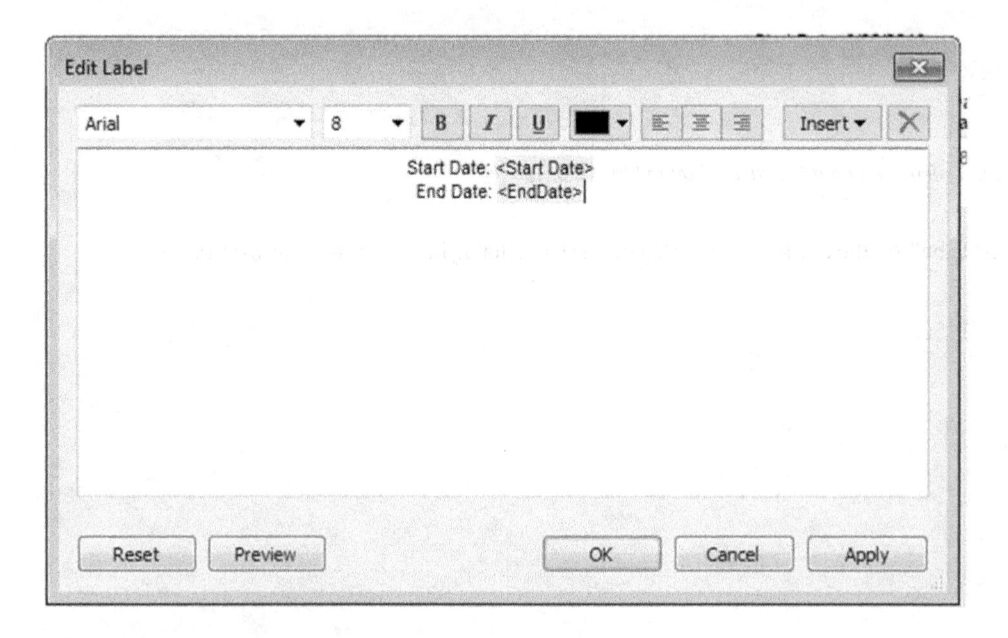

Figure 8-114. *"Edit Label" dialog box*

The final output is shown in Fig. 8-115.

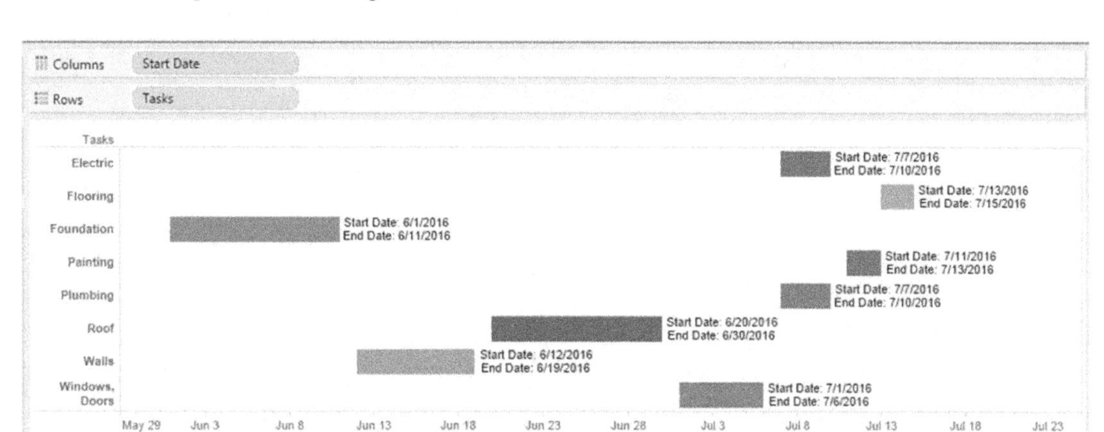

Figure 8-115. *Gantt Chart - Demo 2 - final output*

8.8 Scatter plot

8.8.1 Why use a scatter plot?

To visualize relationships between numerical variables.

8.8.2 What is a scatter plot?

A scatter plot displays many points scattered in the Cartesian plane.

8.8.3 Correlation coefficient

The strength of the correlation is determined by the correlation coefficient (R). It is sometimes referred to as the "Pearson Product Moment Correlation Coefficient". Correlation is expressed in a range from +1 to -1. +1 denotes the perfect positive correlation, whereas -1 denotes the perfect negative correlation. A value of zero indicates there is no correlation. Correlation does not imply causation. There may be an unknown factor that influences both variables similarly.

8.8.3.1 Positive correlation

A positive correlation is the correlation in the same direction. This implies that if the values of one variable increases, the values of the other variable also increases. Likewise if the values of one variable decreases, the values of the other variable also decreases. In other words, the two variables move in tandem (Shown in Fig. 8-116).

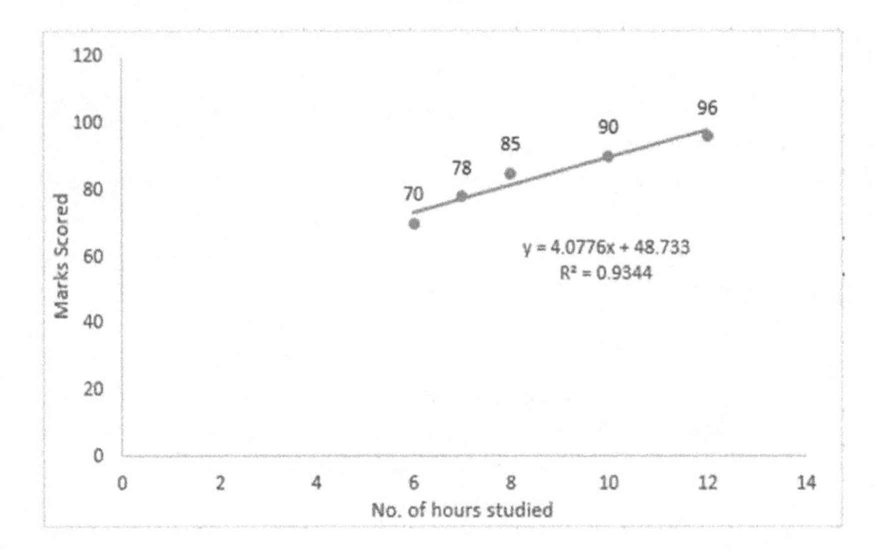

Figure 8-116. Positive correlation

8.8.3.2 Examples

As the temperature increases, the length of the iron bar also increases.
As the price of the fuel increases, the cost of air tickets also increases.
As the number of hours studied by a student increases, so do the marks scored.

8.8.3.3 Negative correlation

Negative correlation is the correlation in the opposite direction. This implies that if the values of one variable increases, the values of the other variable decreases and vice-versa (Shown in Fig. 8-117).

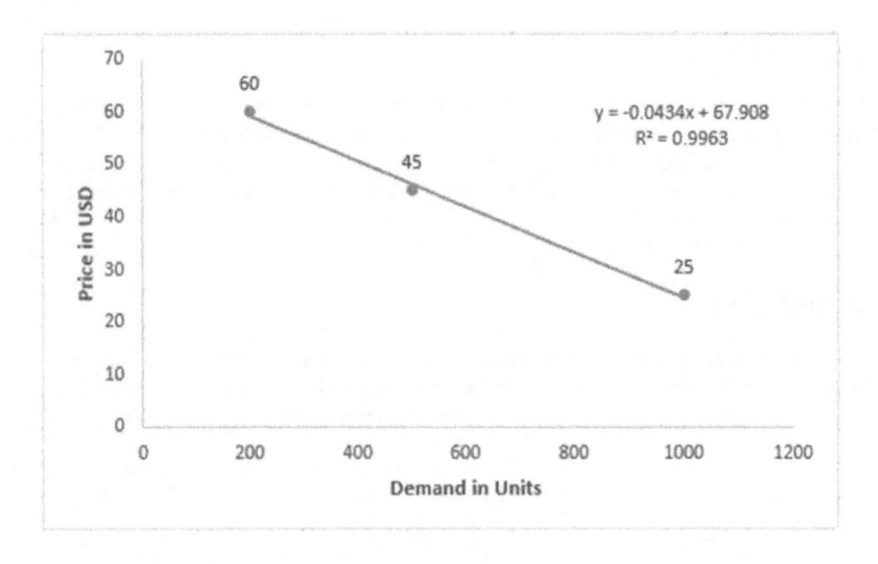

Figure 8-117. Negative correlation

Example: As the price of product rises, the demand for the product declines.

8.8.3.3 No correlation or zero correlation

If there is no relationship between the two variables such that the value of one variable changes and the value of the other variable remains constant, it is said to have no correlation or zero correlation (Shown in Fig. 8-118).

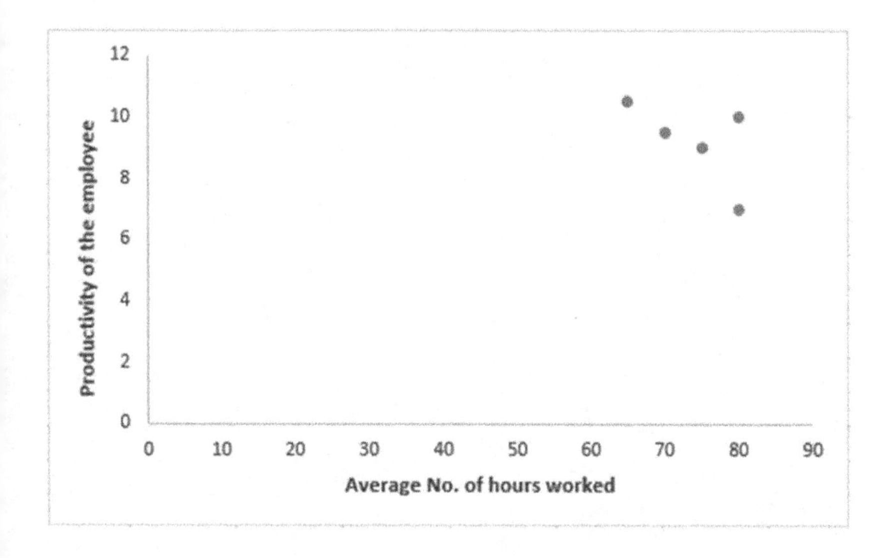

Figure 8-118. No correlation or zero correlation

8.8.4 How to plot scatter plots in Tableau?

8.8.4.1 Demo 1

Objective: Plot a simple scatter plot using two measures, "Sales" and "Profit" on the columns shelf and the rows shelf, respectively.

Input: "Sample - Superstore.xls"

Expected output: (Shown in Fig. 8-119).

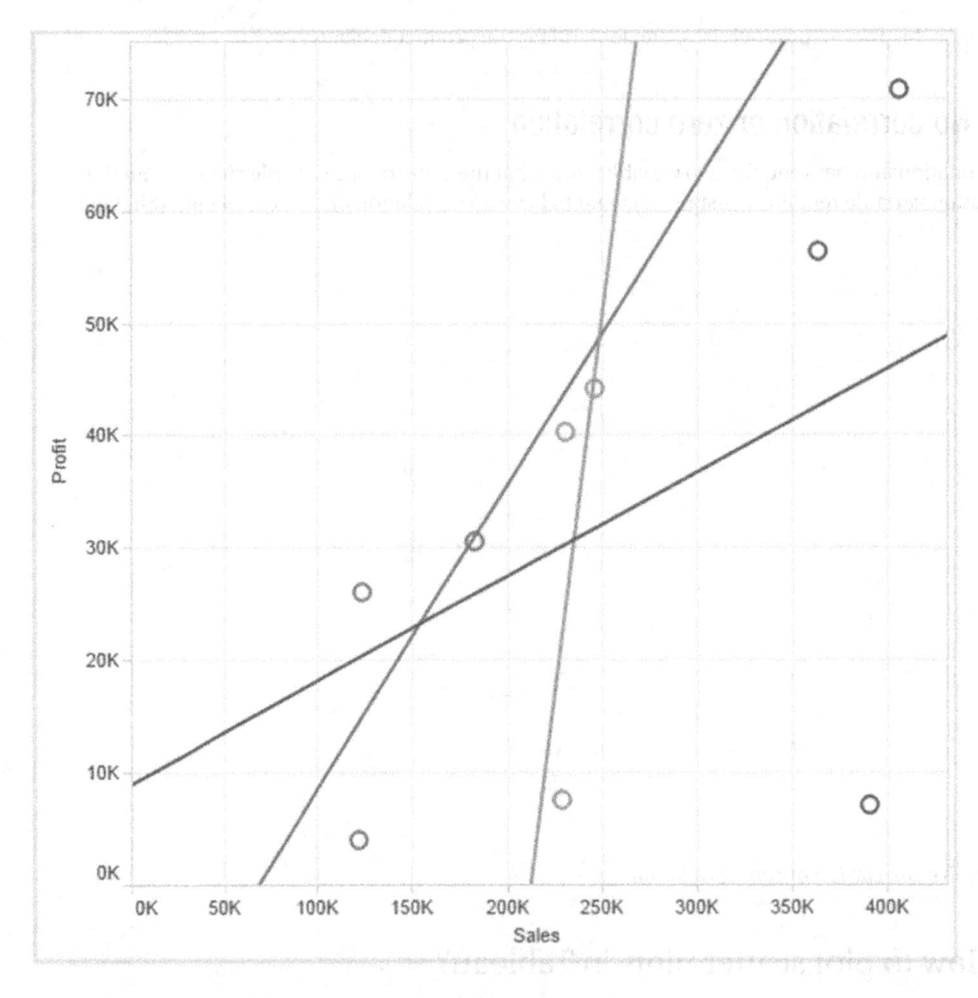

Figure 8-119. *Scatter Plot - Demo 1 - expected output*

8.8.4.2 Steps to create a scatter plot

8.8.4.2.1 Step 1

Drag the measures "Sales" and "Profit" from the measures area under the data pane and place it on the columns shelf and rows shelf, respectively (See Fig. 8-120).

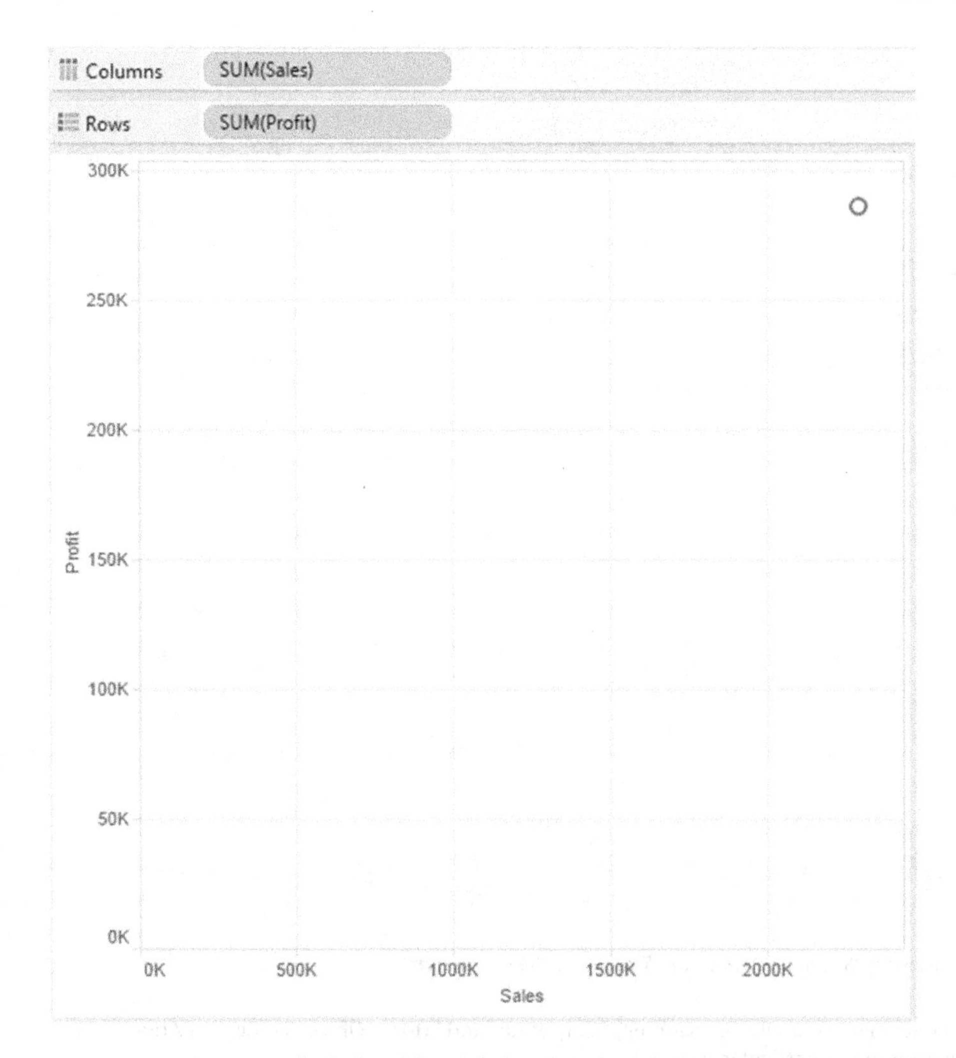

Figure 8-120. *Measures "Sales" and "Profit" placed on the columns shelf and rows shelf, respectively*

8.8.4.2.2 Step 2

Drag the dimension "Segment" from the dimensions area under the data pane and place it on "Color" on the marks card (Shown in Fig. 8-121).

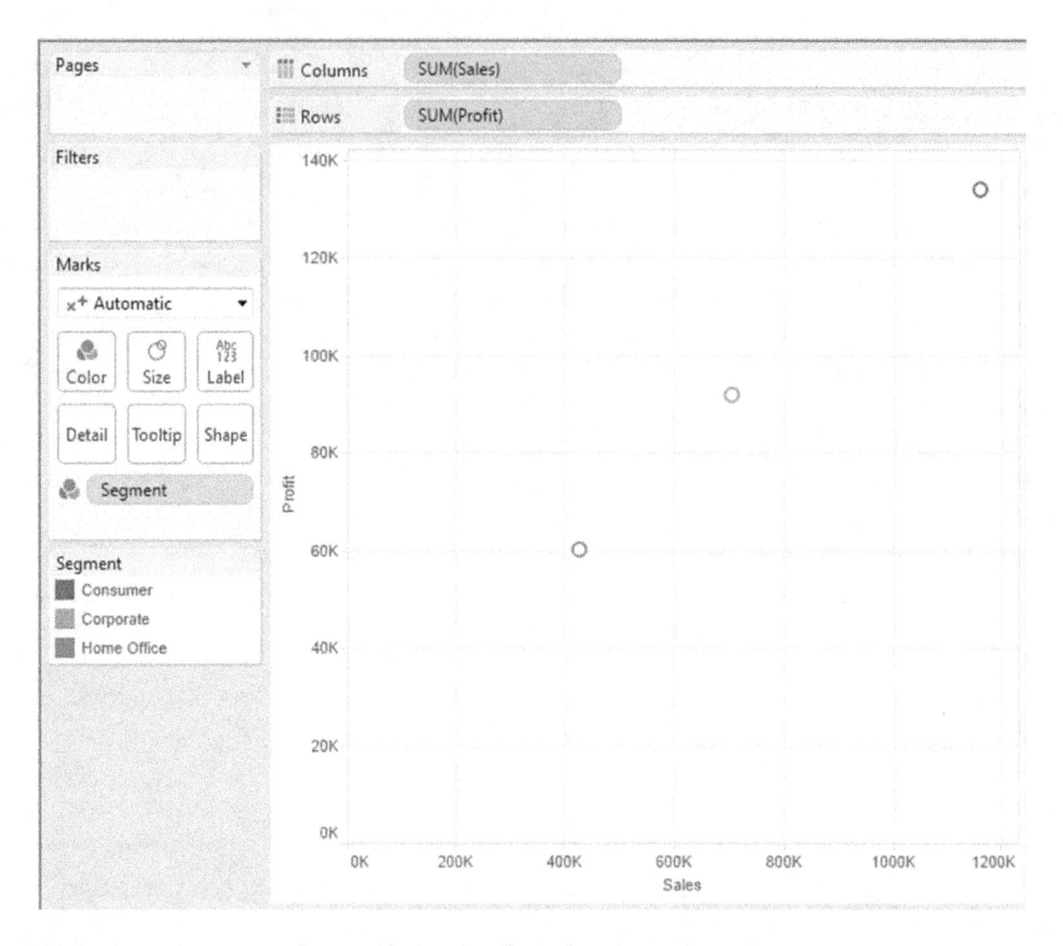

Figure 8-121. *Dimension "Segment" placed on "Color" on the marks card*

There are three segments namely, "Consumer", "Corporate" and "Home Office". Three dots / marks are displayed, each representing a particular segment..

Drag the dimension "Category" from the dimensions area under data pane to "Detail" on the marks card. Data is available for three categories in our data set, namely, "Furniture", "Office Supplies" and "Technology" (Shown in Fig. 8-122).

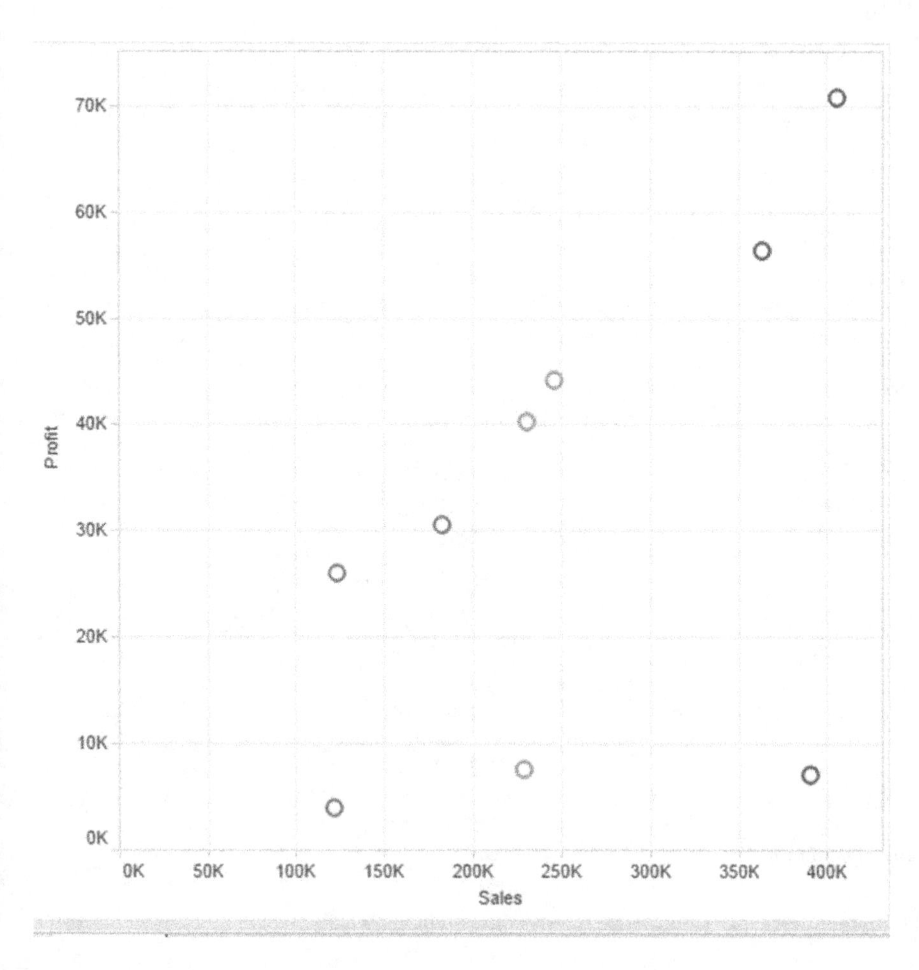

Figure 8-122. *Dimension "Category" placed on "Detail" on the marks card*

Notice: The number of marks is 9.
Three segments * three categories = 9 marks on the view.
Right click in the view area to show "Trend Lines" (Shown in Fig. 8-123).

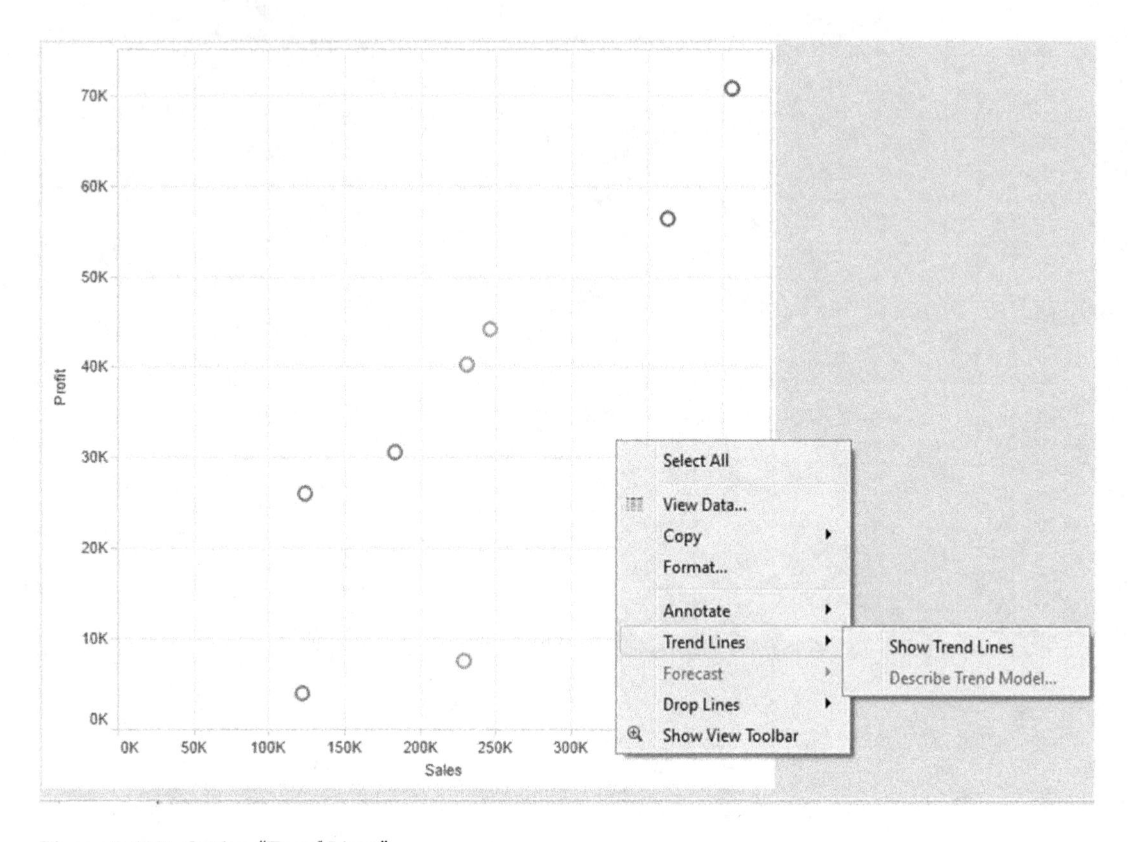

Figure 8-123. *Setting "Trend Lines"*

Edit the trend lines to remove the "Confidence Bands" (Shown in Fig. 8-124).

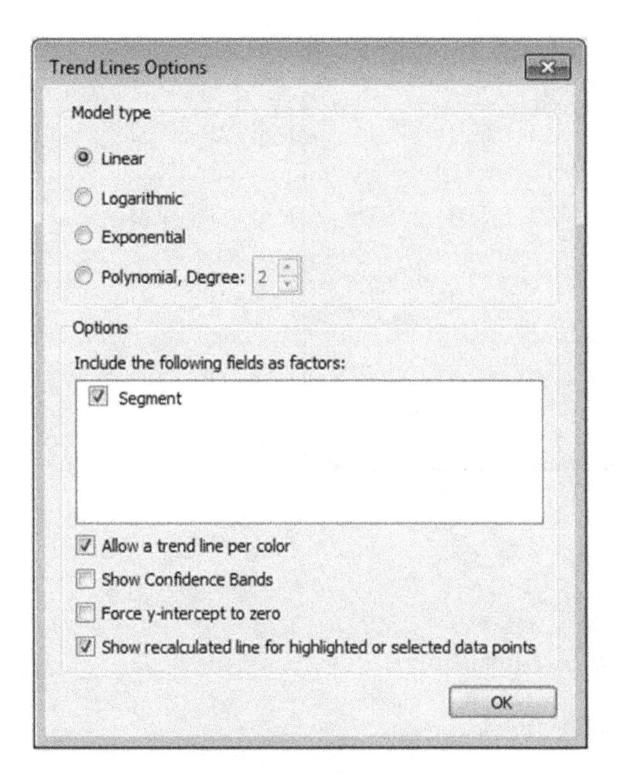

Figure 8-124. *Trend lines options*

Right click a trend line to describe the trend lines (Shown in Fig. 8-125).

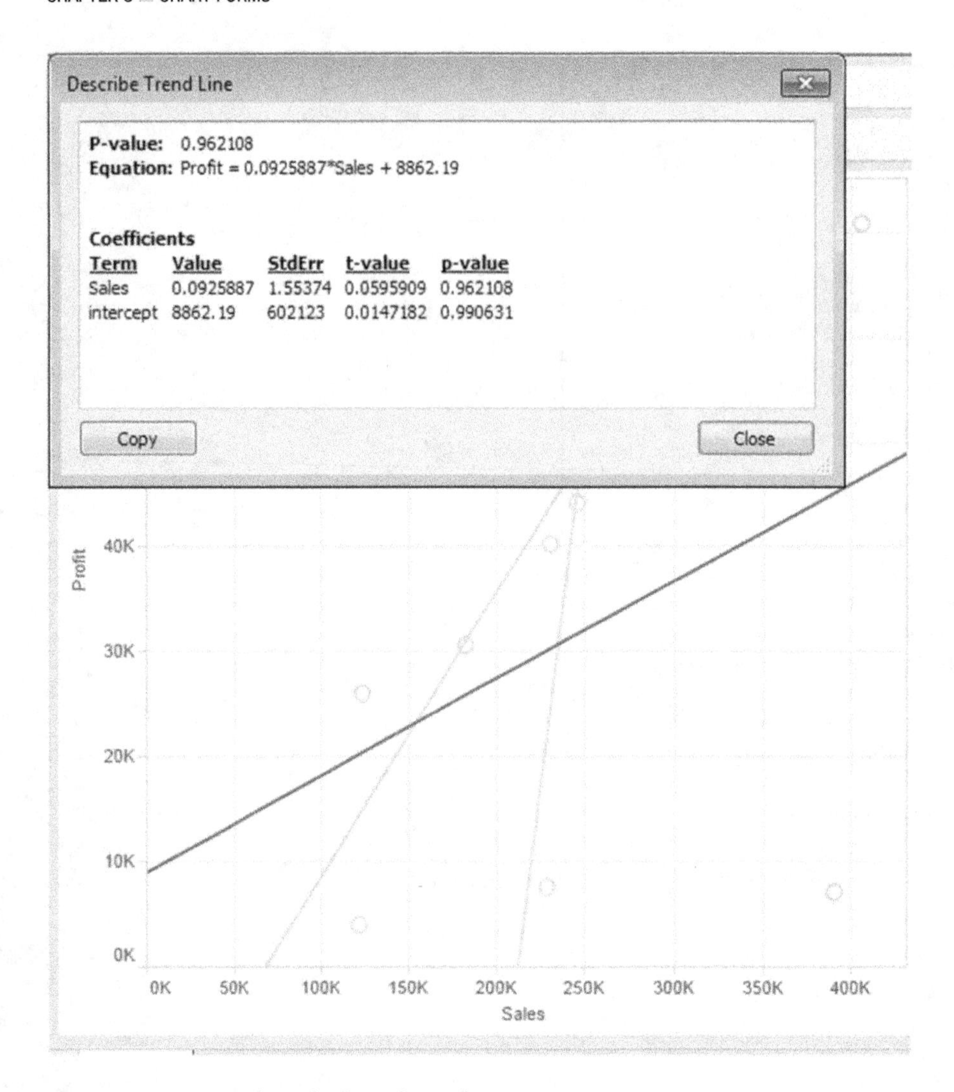

Figure 8-125. Describing the "Trend Lines"

8.8.4.3 Demo 2

Objective: Plot a matrix of scatter plots using dimensions on the marks card and on the columns shelf and rows shelf, respectively. .

Input: "Sample - Superstore.xls"

Expected output: Shown in Fig. 8-126.

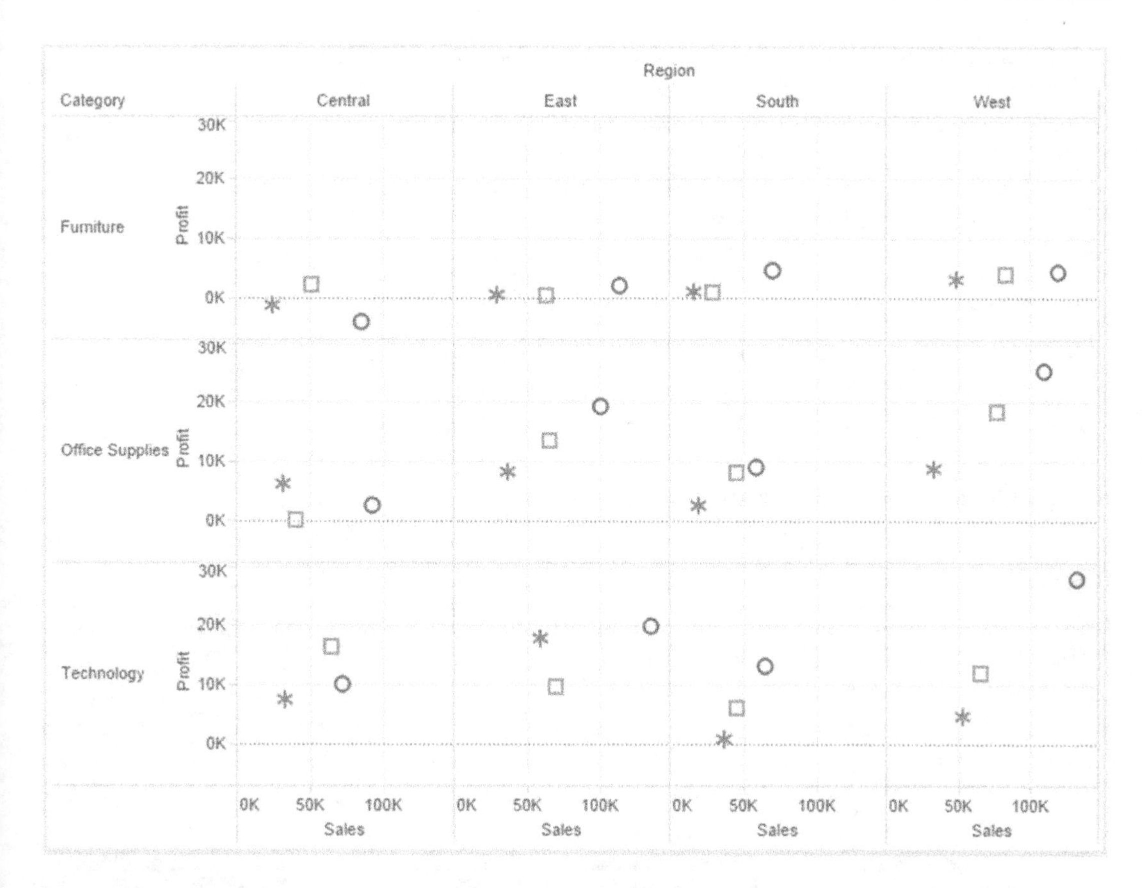

Figure 8-126. *Scatter plot - Demo 2 - expected output*

8.8.4.3.1 Steps to create a scatter plot

8.8.4.3.2 Step 1

Drag the dimension "Region" from the dimensions area under the data pane and place it on the columns shelf. Data is available for four regions, such as "Central", "East", "South" and "West" (Shown in Fig. 8-127).

Columns Region

Rows

Region			
Central	East	South	West
Abc	Abc	Abc	Abc

Figure 8-127. *Dimension "Region" placed on the columns shelf*

Drag the dimension "Category" from the dimensions area under the data pane and place it on the rows shelf (Shown in Fig. 8-128).

Columns Region

Rows Category

	Region			
Category	Central	East	South	West
Furniture	Abc	Abc	Abc	Abc
Office Supplies	Abc	Abc	Abc	Abc
Technology	Abc	Abc	Abc	Abc

Figure 8-128. *Dimension "Category" placed on the rows shelf*

8.8.4.3.3 Step 2

Drag the measure "Sales" from the measures area under the data pane and place it on the columns shelf, to the right of the dimension "Region" (Shown in Fig. 8-129).

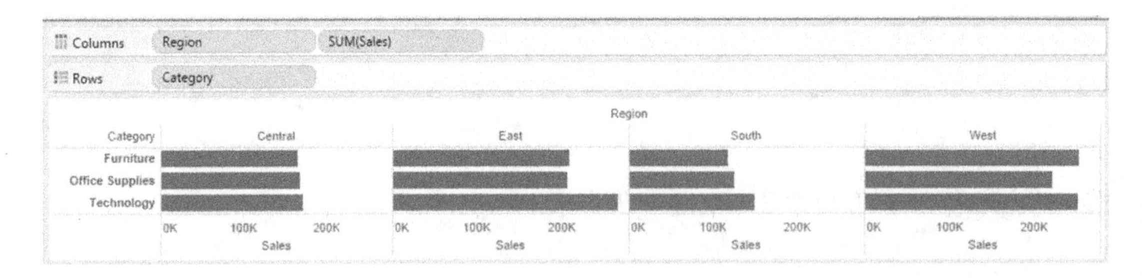

Figure 8-129. *Measure "Sales" placed on the columns shelf to the right of dimension "Region"*

Drag the measure "Profit" from the measures area under the data pane and place it on the rows shelf, to the right of the dimension "Category" (Shown in Fig. 8-130).

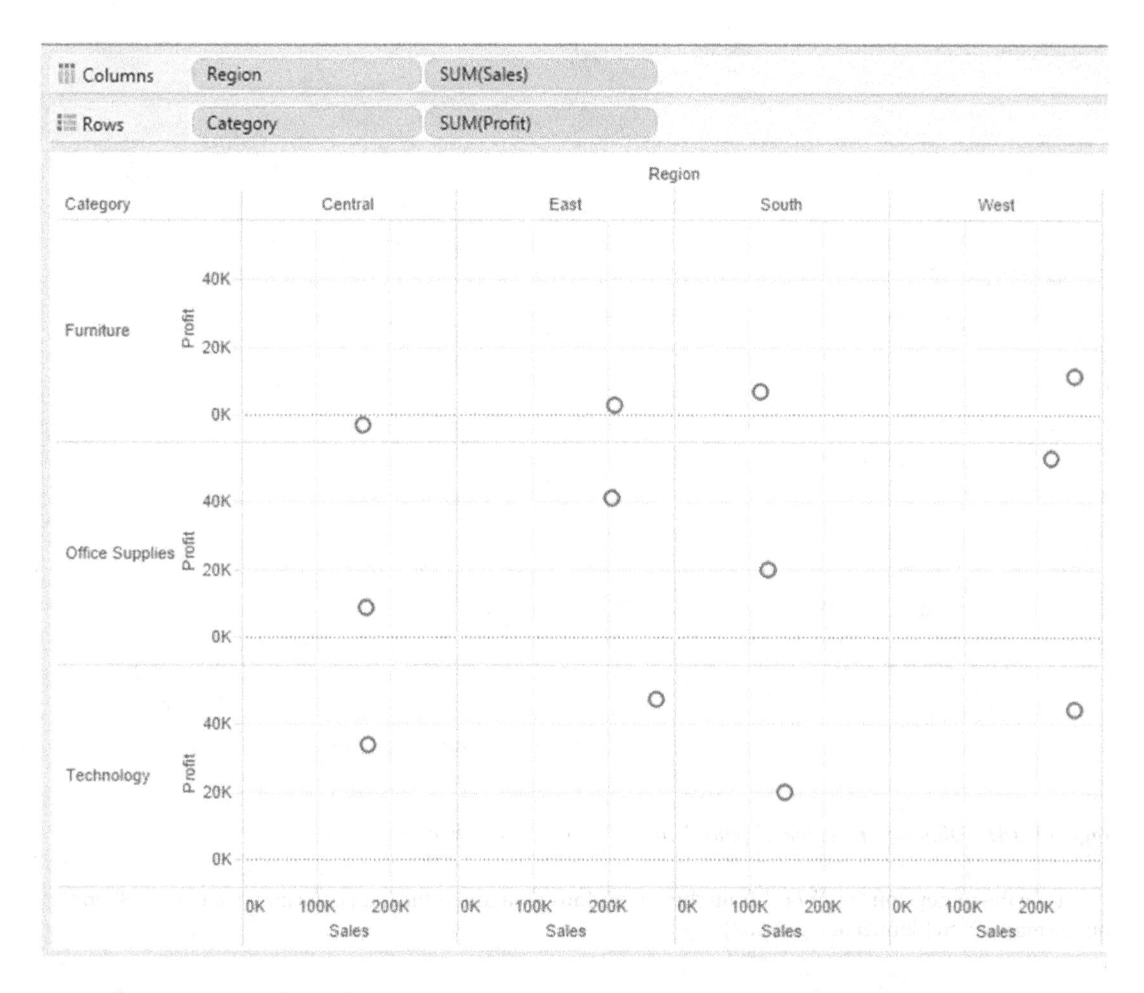

Figure 8-130. *Measure "Profit" placed on the rows shelf to the right of dimension "Category"*

8.8.4.3.4 Step 3

Drag the dimension "Segment" from the dimensions area under the data pane and place it on "Color" on the marks card (Shown in Fig. 8-131).

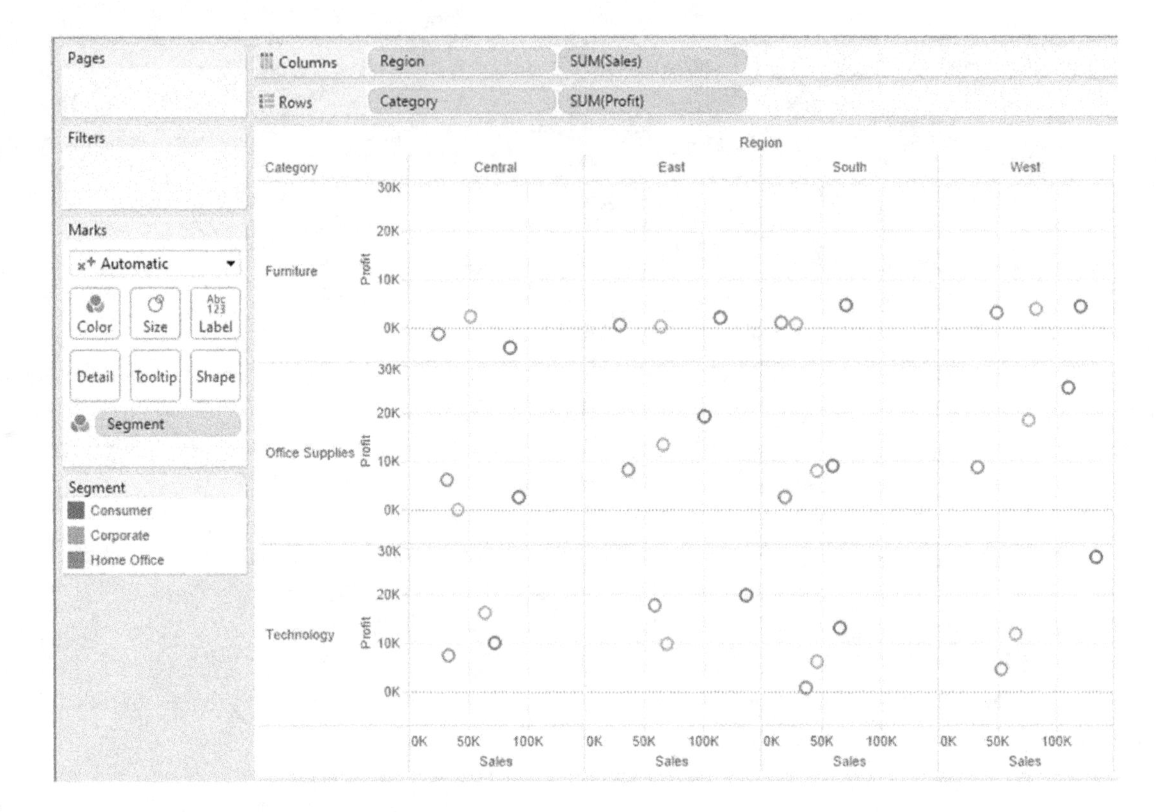

Figure 8-131. *Dimension "Segment" placed on "Color" on the marks card*

Drag the dimension "Segment" from the dimensions area under the data pane and place it on "Shape" on the marks card (Shown in Fig. 8-132).

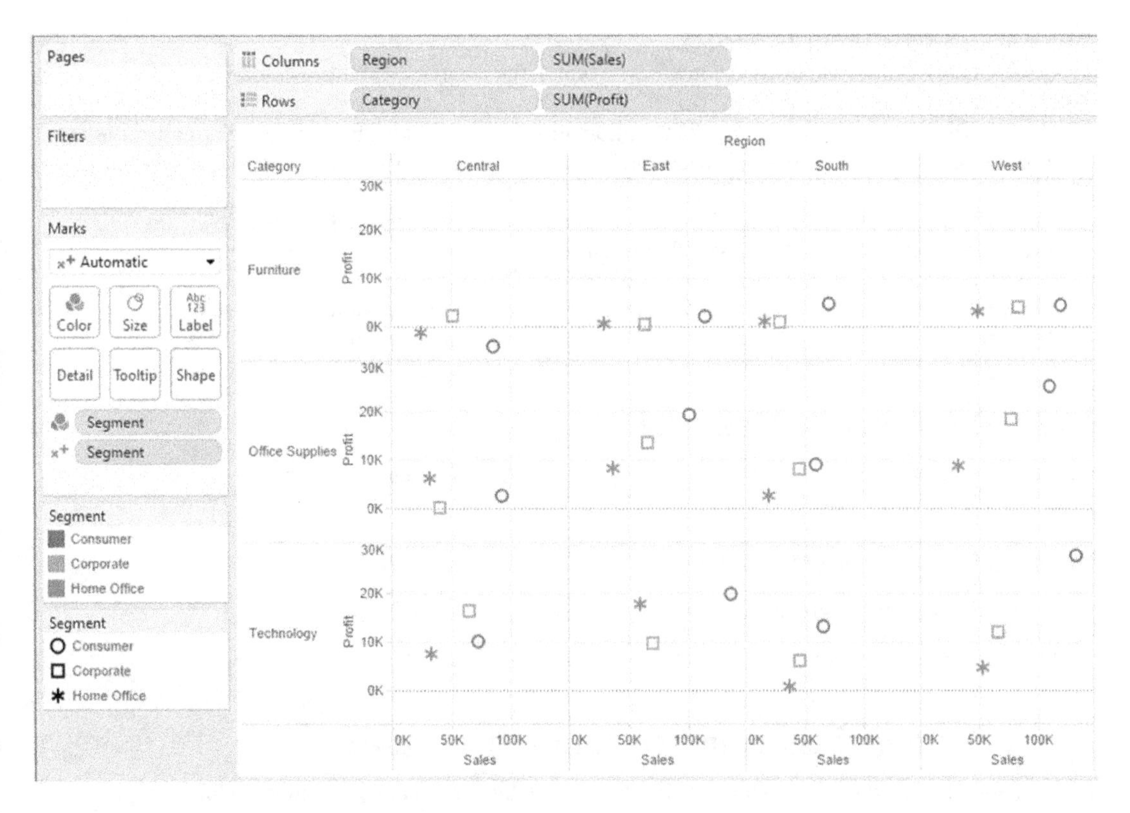

Figure 8-132. Dimension "Segment" placed on "Shape" on the marks card

8.9 Histogram

It is used to display the frequency distribution of continuous data.

Histograms were first introduced by Karl Pearson in the year 1891.

The word "histogram" can be spilt into two. The Greek word, "histos" meaning "anything set upright" and "gramma" meaning "drawing/writing".

8.9.1 What is required to plot a histogram?

The first step is to decide on the number of bins/baskets into which one would like to divide the range of values. The next step would be to count the number of values that fall into each bin/basket.

Bins or intervals can be equal or unequal.

Usually the bins are consecutive, equal-sized and non-overlapping.

If the bins are of equal size, a rectangle is erected over the bin with its height proportional to the frequency, i.e. the number of values that fall into each bin. If the bins are of unequal size, the rectangle erected over the bin has an area proportional to the frequency.

Terms generally used with histogram:

Symmetric: A symmetric distribution is one in which the two halves of the histogram are mirror images of each other.

Non-symmetric: In non-symmetric distribution, the two halves of the histogram are not mirror images.

Skewed distribution: A skewed distribution is one where one tail of the distribution is longer than the other is.

Skewed right: Skewed right is one in which the distribution has a tail on the right side.

Skewed left: Skewed left is one in which the distribution has a tail on the left side.

Unimodal: Unimodal is the distribution that has a single mode.

Bimodal: A bimodal is a continuous probability distribution with two different modes.

Multimodal: A multimodal is a continuous probability distribution with multiple modes.

8.9.2 Difference with bar charts

A histogram represents continuous data, whereas bar charts are for categorical data.

8.9.3 Pros of histogram

- Spotting outliers is easy
- Makes the distribution of scores easier to interpret

8.9.4 Plotting a histogram (customized bin size)

Dataset used for the plot: 2014 Olympic data. Data is available for 494 athletes under the following headings:

- Country
- Athlete
- Sex
- Age
- Sport
- Gold
- Silver
- Bronze
- Total

A subset of the data is shown in Fig. 8-133.

	A	B	C	D	E	F	G	H	I
1	Country	Athlete	Sex	Age	Sport	Gold	Silver	Bronze	Total
2	Australia	Torah Bright	Female	27	Snowboarding	0	1	0	1
3	Australia	David Morris	Male	29	Freestyle Skiing	0	1	0	1
4	Australia	Lydia Ierodiaconou-Lassila	Female	32	Freestyle Skiing	0	0	1	1
5	Austria	Anna Fenninger	Female	24	Alpine Skiing	1	1	0	2
6	Austria	Nicole Hosp	Female	30	Alpine Skiing	0	1	1	2
7	Austria	Dominik Landertinger	Male	26	Biathlon	0	1	1	2
8	Austria	Julia Dujmovits	Female	26	Snowboarding	1	0	0	1
9	Austria	Mario Matt	Male	34	Alpine Skiing	1	0	0	1
10	Austria	Matthias Mayer	Male	23	Alpine Skiing	1	0	0	1
11	Austria	Thomas Diethart	Male	21	Ski Jumping	0	1	0	1
12	Austria	Michael Hayböck	Male	22	Ski Jumping	0	1	0	1
13	Austria	Marcel Hirscher	Male	24	Alpine Skiing	0	1	0	1
14	Austria	Daniela Iraschko-Stolz	Female	30	Ski Jumping	0	1	0	1
15	Austria	Andreas Linger	Male	32	Luge	0	1	0	1
16	Austria	Wolfgang Linger	Male	31	Luge	0	1	0	1
17	Austria	Thomas Morgenstern	Male	27	Ski Jumping	0	1	0	1
18	Austria	Marlies Schild	Female	32	Alpine Skiing	0	1	0	1
19	Austria	Gregor Schlierenzauer	Male	24	Ski Jumping	0	1	0	1
20	Austria	Christoph Bieler	Male	36	Nordic Combined	0	0	1	1

Figure 8-133. *Subset of "Olympic Data Set"*

What is it that we wish to do?
We wish to determine the number of athletes in the following age groups:

- < 25
- >= 25 and <=35
- > 35

What is required?
Create customized bins and then compute the frequency, i.e. the count of the number of athletes in each category.

8.9.4.1 Steps to create a histogram

8.9.4.1.1 Step 1

Connect to the 2014 Olympics data set (Shown in Fig. 8-134).

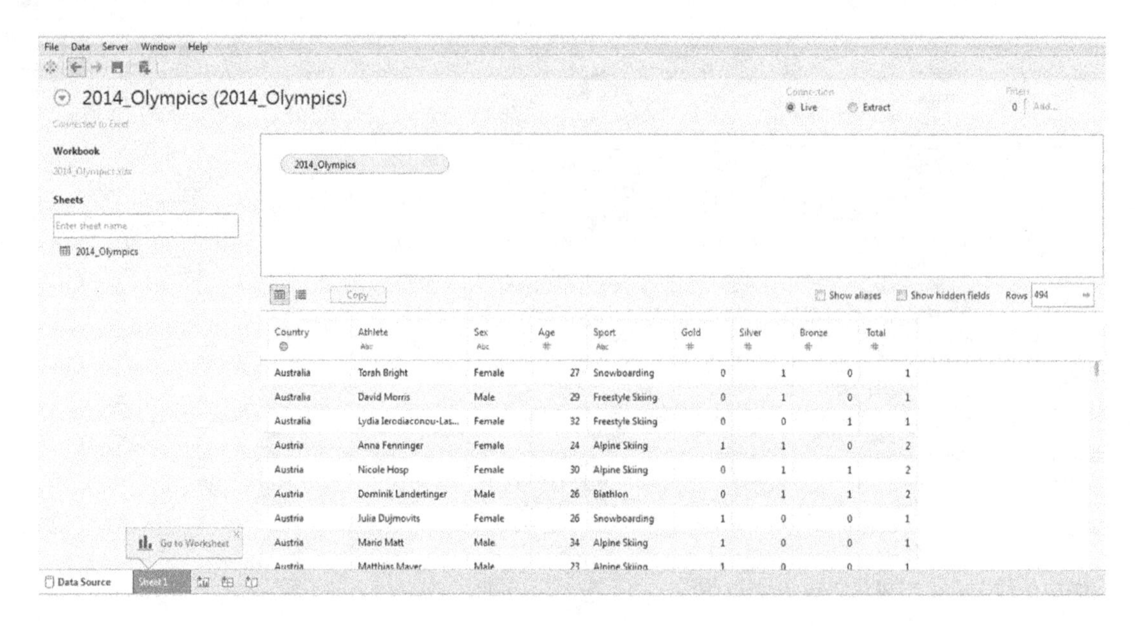

Figure 8-134. *Data from Olympic data set read into Tableau*

8.9.4.1.2 Step 2

Go to the worksheet / view (Shown in Fig. 8-135).

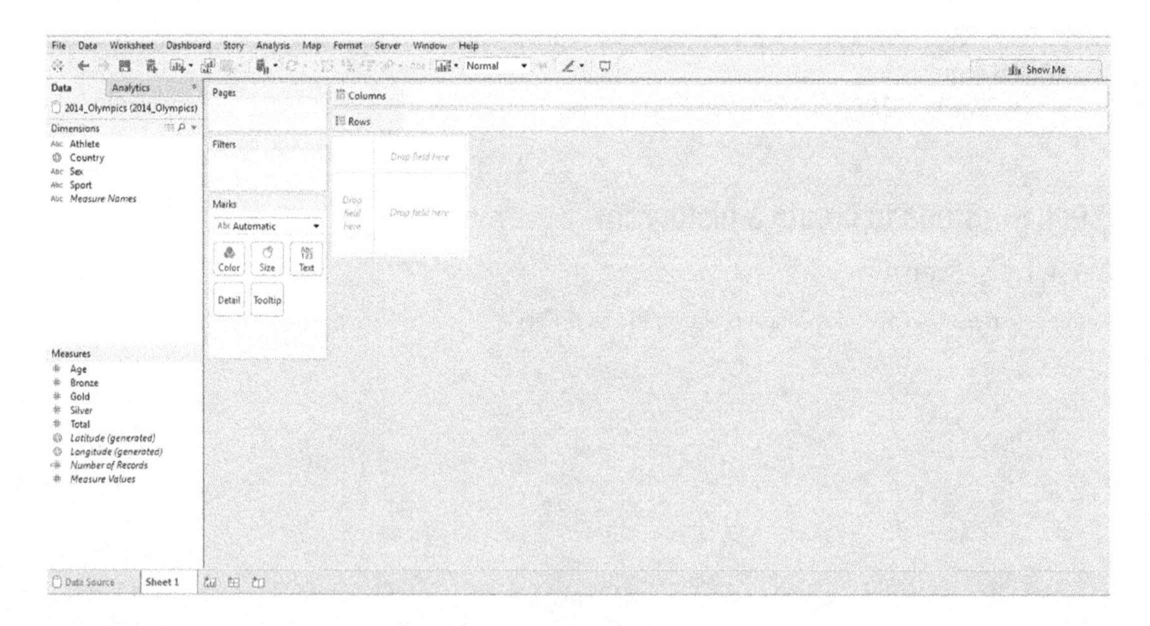

Figure 8-135. *Worksheet / View after reading in values from "Olympics" data set*

8.9.4.1.3 Step 3

Create a calculated field "Age Group" (Shown in Fig. 8-136).

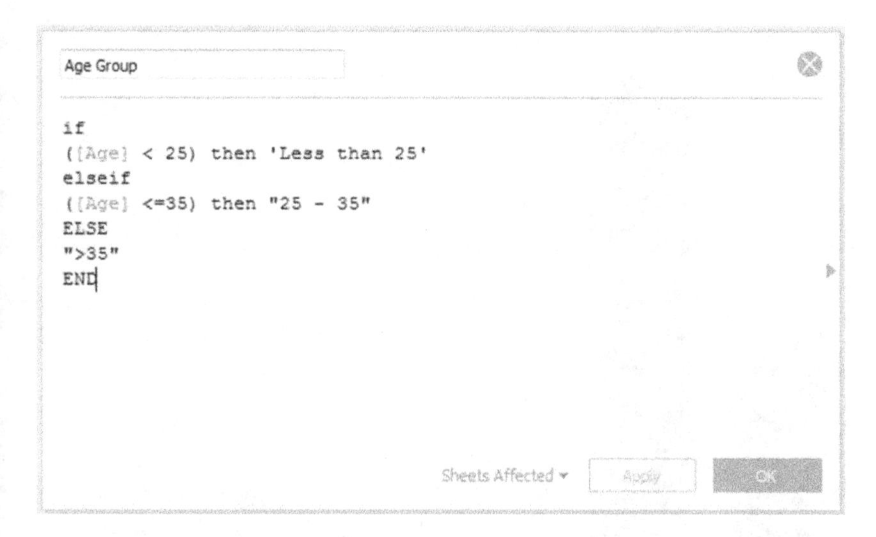
Age Group
if
([Age] < 25) then 'Less than 25'
elseif
([Age] <=35) then "25 - 35"
ELSE
">35"
END
Sheets Affected
Apply
OK

Figure 8-136. *Calculated field "Age Group" being created*

Note: The newly created calculated field "Age Group" appears under dimensions in the data pane.

8.9.4.1.4 Step 4

Drag and drop the dimension "Age Group" on the columns shelf. Drag and drop the measure "Number of Records" on the rows shelf (Shown in Fig. 8-137).

Columns shelf	Age group (Calculated field)
Rows shelf	Number of records. The aggregation applied to the measure is SUM.

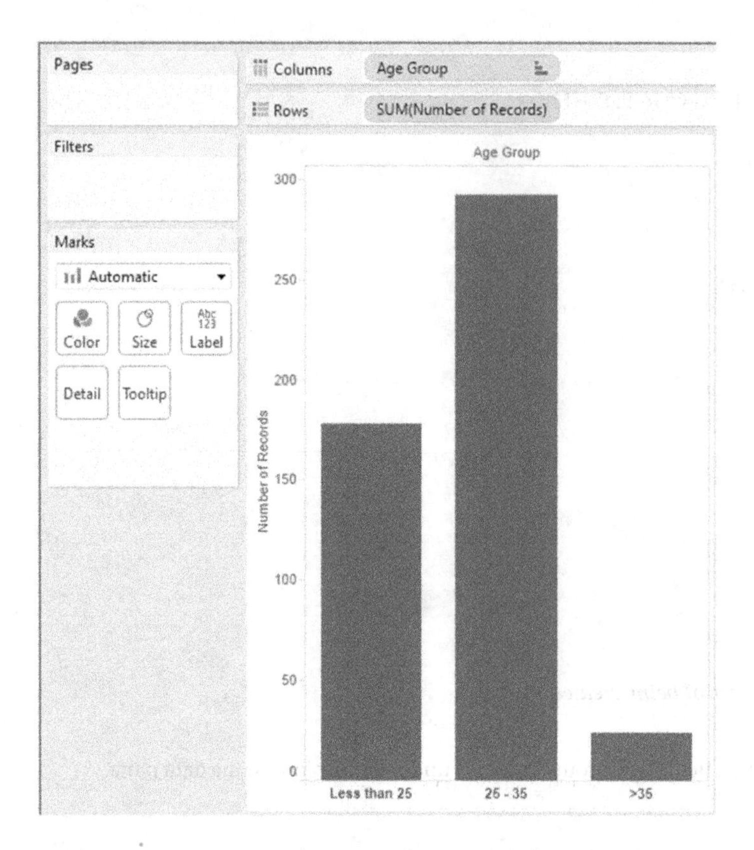

Figure 8-137. *Dimension "Age Group" & measure "Number of Records" placed on the columns and rows shelf*

8.9.4.1.5 Step 5

Drag the measure "Number of Records" to "Label" on the marks card (Shown in Fig. 8-138).

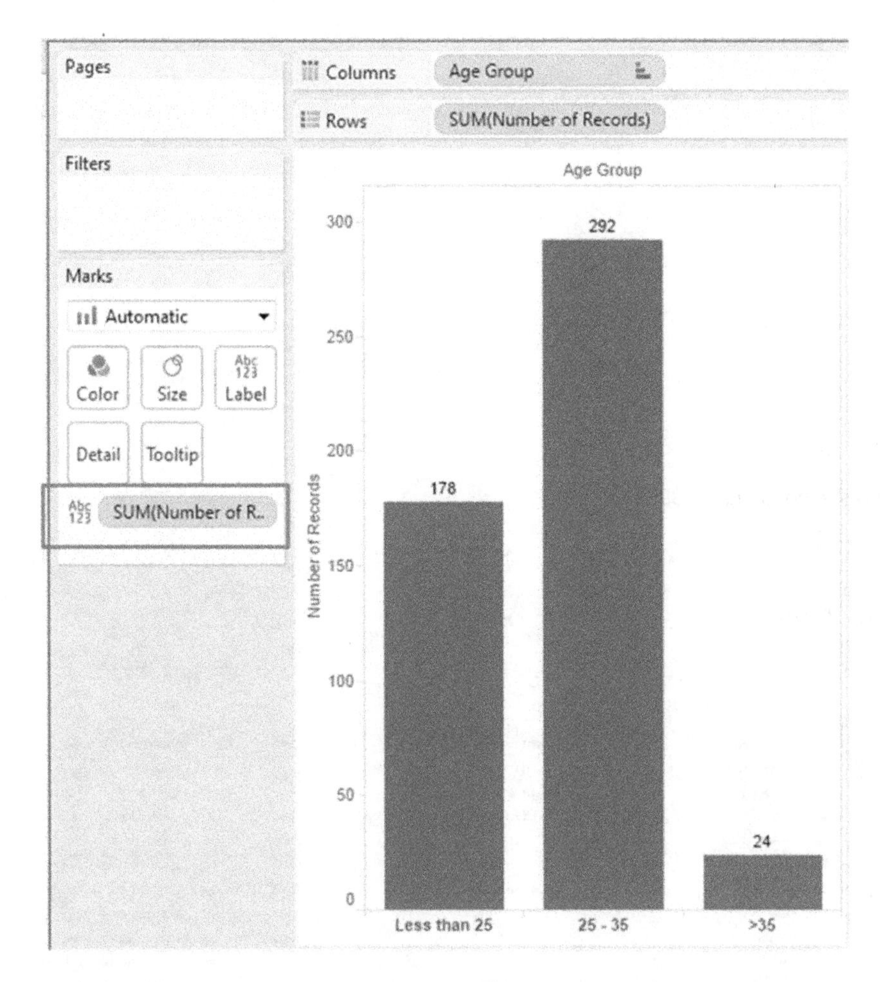

Figure 8-138. Measure "Number of Records" placed on "Label" on the marks card

8.9.4.2 Plot histogram (equal sized bin)

Dataset used for the plot: "TestResultsSample.tde". Data is available for 51,185 students under the following headings:

- Exams_ID
- Exams_Name
- SubjectArea
- Name
- School_Id
- State
- Date
- Exam

- Score
- Scores_Id
- Student
- DateOfBirth
- FirstName
- Id
- School
- LastName
- SubjectAreas_1
- SubjectAreas_Name

A subset of the data is shown in Fig. 8-139.

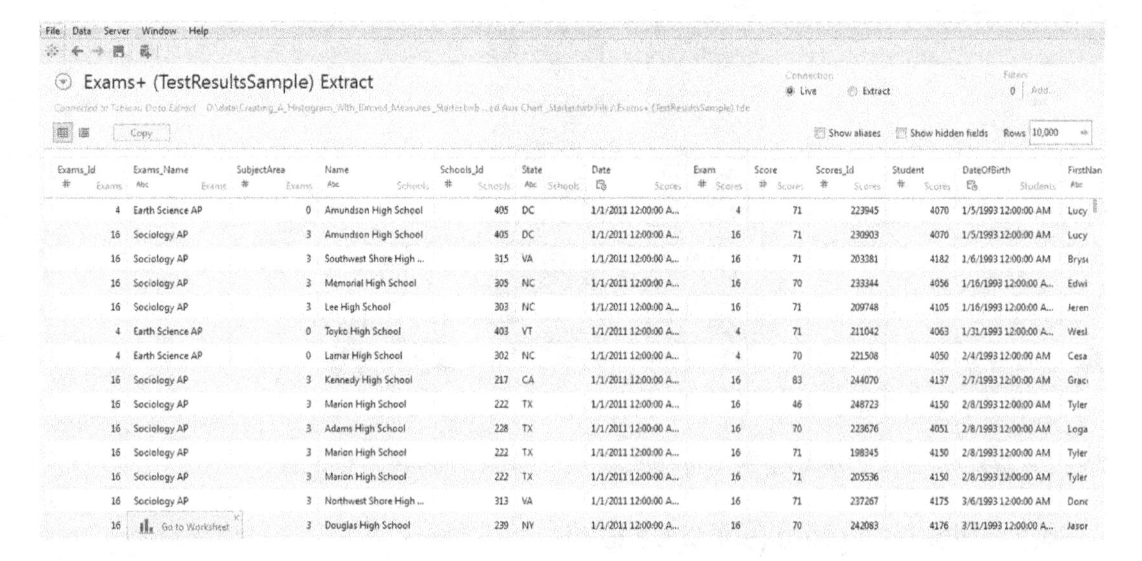

Exams_Id	Exams_Name	SubjectArea	Name	Schools_Id	State	Date	Exam	Score	Scores_Id	Student	DateOfBirth	FirstNan
4	Earth Science AP	0	Amundson High School	405	DC	1/1/2011 12:00:00 A...	4	71	223945	4070	1/5/1993 12:00:00 AM	Lucy
16	Sociology AP	3	Amundson High School	405	DC	1/1/2011 12:00:00 A...	16	71	230903	4070	1/5/1993 12:00:00 AM	Lucy
16	Sociology AP	3	Southwest Shore High ...	315	VA	1/1/2011 12:00:00 A...	16	71	203381	4182	1/6/1993 12:00:00 AM	Bryse
16	Sociology AP	3	Memorial High School	305	NC	1/1/2011 12:00:00 A...	16	70	233344	4056	1/16/1993 12:00:00 A...	Edwi
16	Sociology AP	3	Lee High School	303	NC	1/1/2011 12:00:00 A...	16	71	209748	4105	1/16/1993 12:00:00 A...	Jeren
4	Earth Science AP	0	Toyko High School	403	VT	1/1/2011 12:00:00 A...	4	71	211042	4063	1/31/1993 12:00:00 A...	Wesl
4	Earth Science AP	0	Lamar High School	302	NC	1/1/2011 12:00:00 A...	4	70	221508	4050	2/4/1993 12:00:00 AM	Cesa
16	Sociology AP	3	Kennedy High School	217	CA	1/1/2011 12:00:00 A...	16	83	244070	4137	2/7/1993 12:00:00 AM	Grac
16	Sociology AP	3	Marion High School	222	TX	1/1/2011 12:00:00 A...	16	46	248723	4150	2/8/1993 12:00:00 AM	Tyler
16	Sociology AP	3	Adams High School	228	TX	1/1/2011 12:00:00 A...	16	70	223676	4051	2/8/1993 12:00:00 AM	Loga
16	Sociology AP	3	Marion High School	222	TX	1/1/2011 12:00:00 A...	16	71	198345	4150	2/8/1993 12:00:00 AM	Tyler
16	Sociology AP	3	Marion High School	222	TX	1/1/2011 12:00:00 A...	16	71	205536	4150	2/8/1993 12:00:00 AM	Tyler
16	Sociology AP	3	Northwest Shore High ...	313	VA	1/1/2011 12:00:00 A...	16	71	237267	4175	3/6/1993 12:00:00 AM	Donc
16	Go to Worksheet	3	Douglas High School	239	NY	1/1/2011 12:00:00 A...	16	70	242083	4176	3/11/1993 12:00:00 A...	Jasor

Figure 8-139. *Subset of "TestResultsSample" data*

8.9.4.2.1 Step 1

Connect to the TestResultsSample data set (Shown in Fig. 8-140).

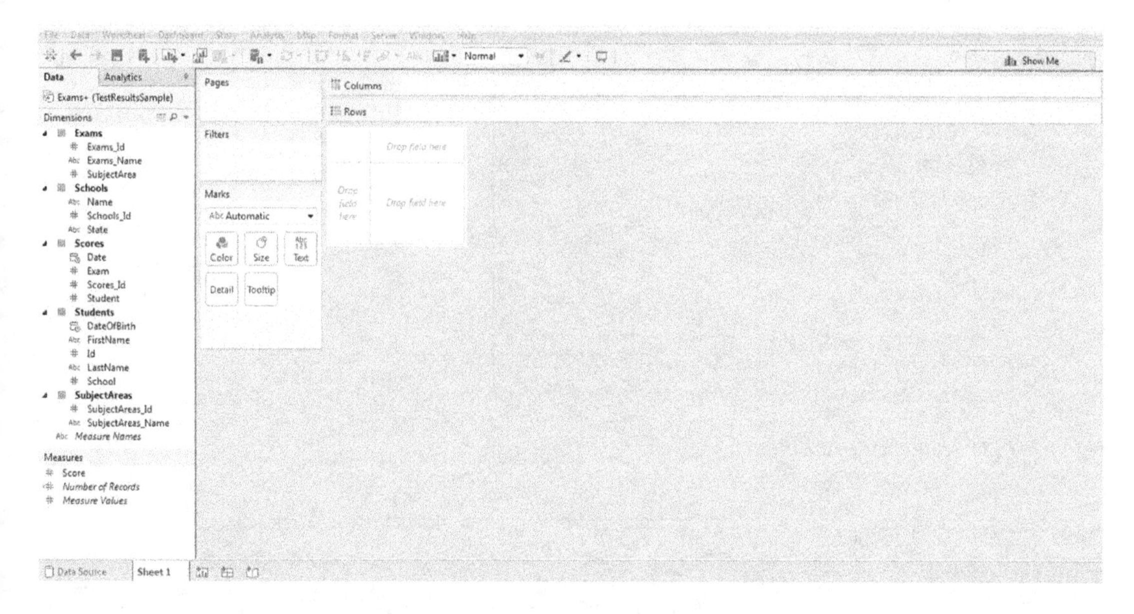

Figure 8-140. *Data from "TestResultsSample" data set read into Tableau*

8.9.4.2.2 Step 2

Define the bin. Right click on the measure "Score" and select "Create" and then "Bins" (Shown in Fig. 8-141).

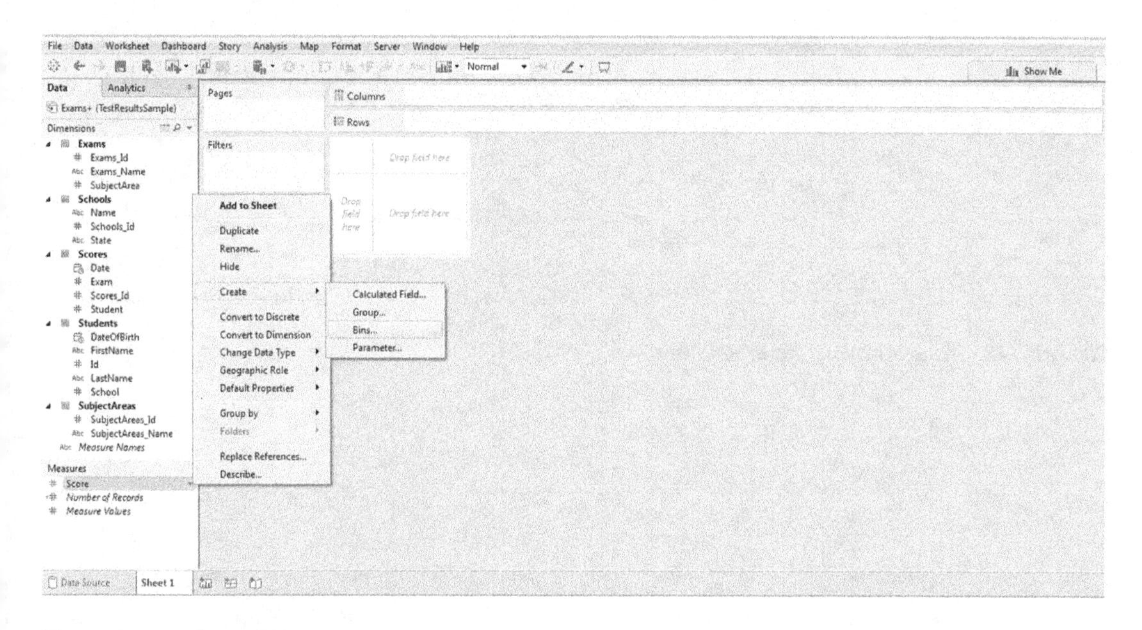

Figure 8-141. *Create Bins*

Fill in the values in the "Create Bins [Score]" dialog box as shown in Fig. 8-142. The size of the bins is specified as 10. Notice the Histogram icon next to the field, "BinSize" (Shown in Fig. 8-143).

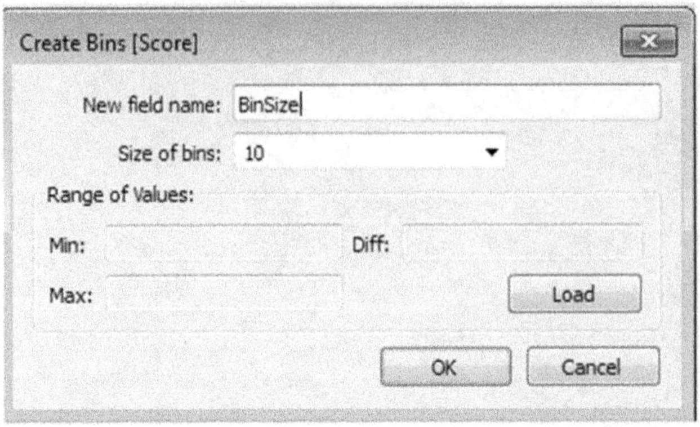

Figure 8-142. Set the size of the bins

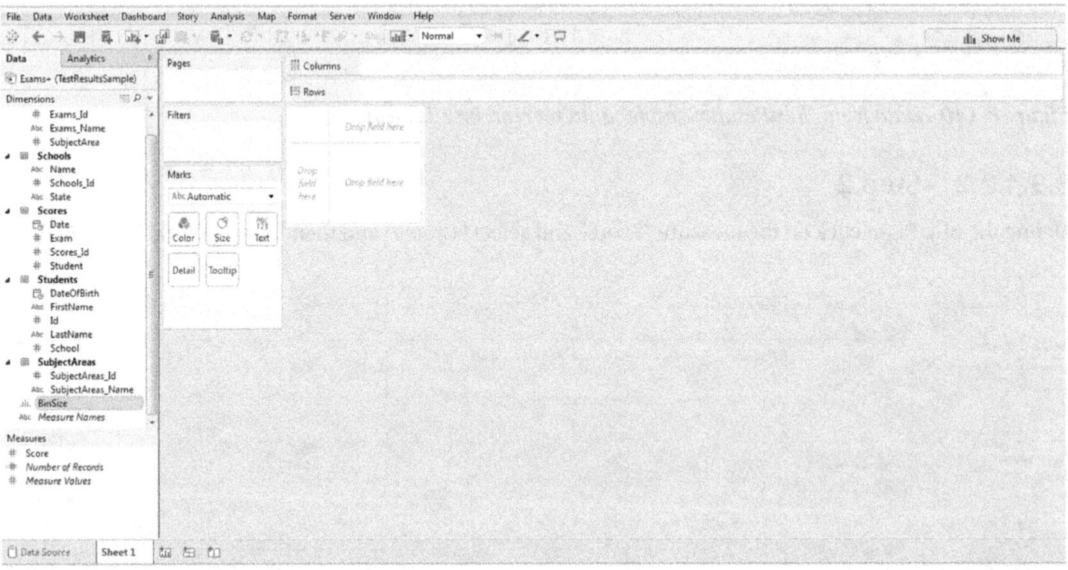

Figure 8-143. Histogram icon next to BinSize

8.9.4.2.3 Step 3

Drag and drop the dimension “BinSize” to the columns shelf. Drag and drop the measure “Score” to the rows shelf (Shown in Fig. 8-144).

Columns Shelf	BinSize (Calculated field)
Rows Shelf	Score. The aggregation applied to the measure is COUNT.

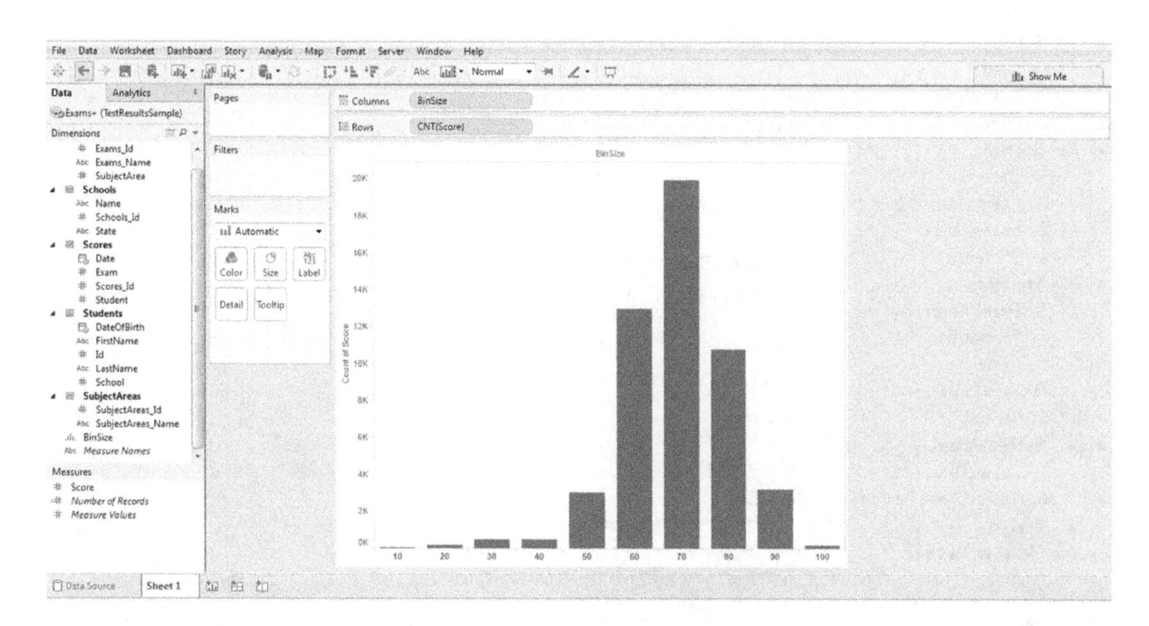

Figure 8-144. *Dimension “BinSize” placed on the columns shelf. Measure “Score” placed on the rows shelf*

Change the value of “BinSize” to 2. To do so, right click on the dimension “BinSize” and select “Edit”. It brings up the “Edit Bins [Score]” dialog box. Change the size of the bin to 2 (Shown in Fig. 8-145 & Figure 8-146).

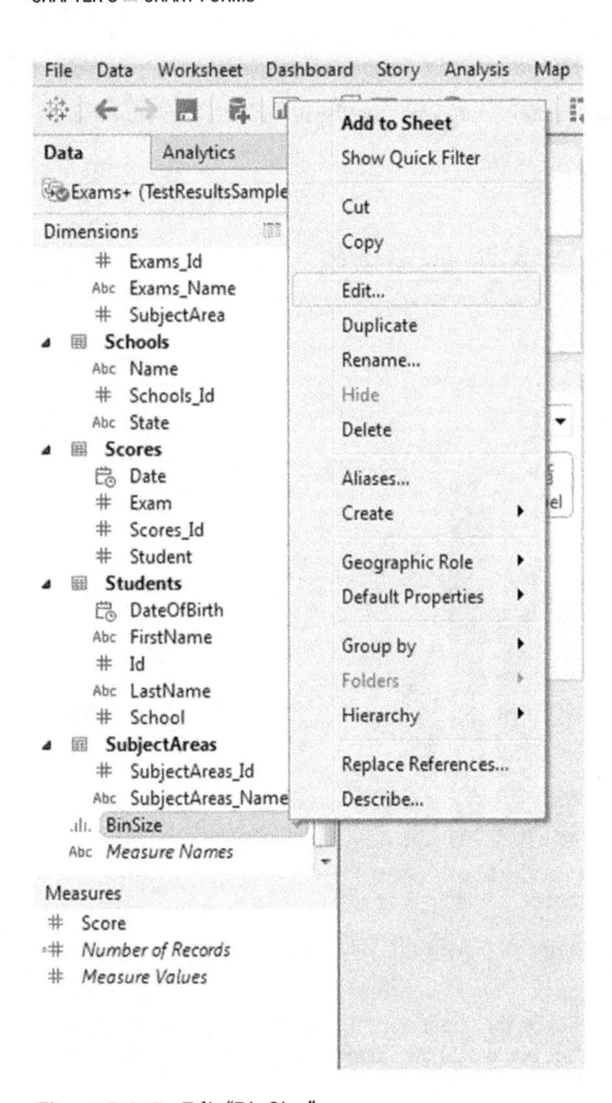

Figure 8-145. Edit "BinSize"

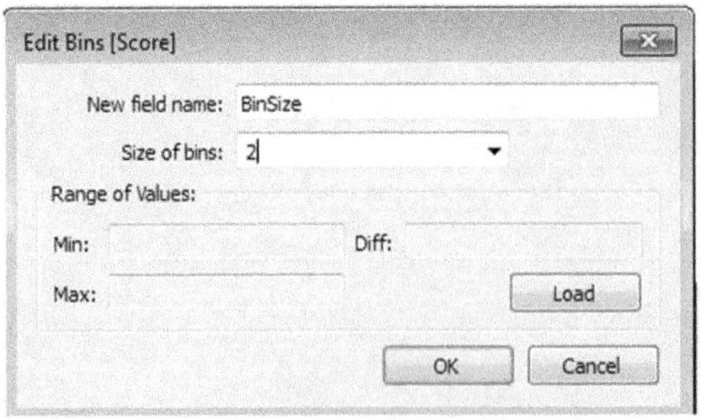

Edit Bins [Score]

New field name: BinSize

Size of bins: 2

Range of Values:

Min: Diff:

Max: Load

OK Cancel

Figure 8-146. "Edit Bins [Score]" dialog box"

The output after the size of the bins is changed to "2" is shown in Fig. 8-147.

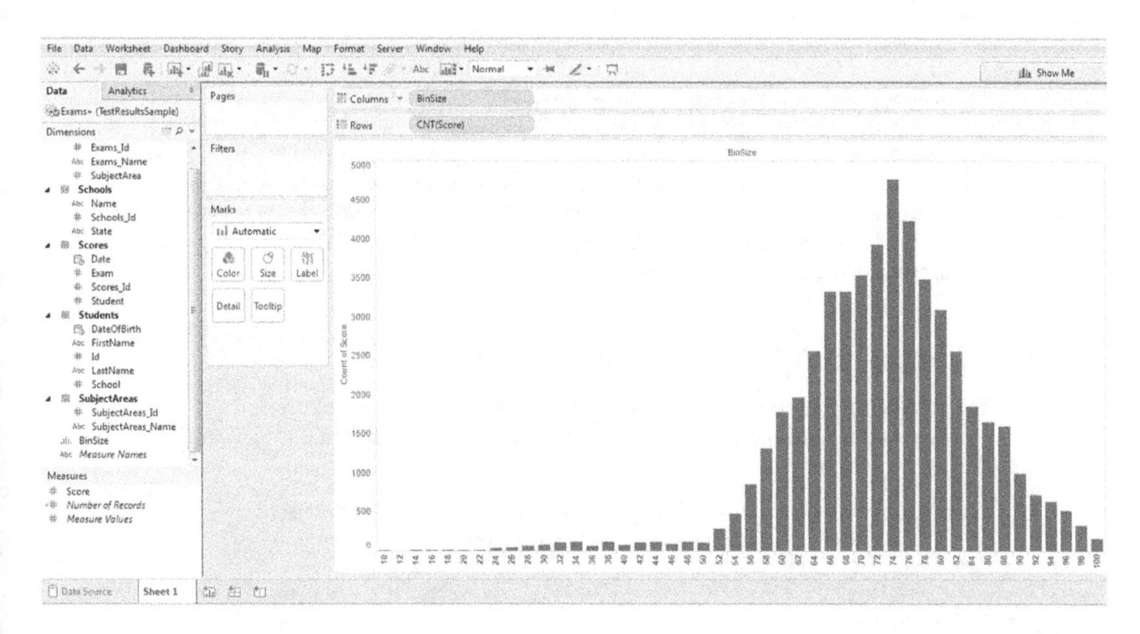

Figure 8-147. *Output after size of the bins is set to "2"*

8.10 Word Cloud

A "Word Cloud" is also called as a "Tag Cloud" or a "Weighted List" in visual design or "Text Cloud".

8.10.1 Why should you use a word cloud?

- You wish to improve the quality of customer experience (QCE).
- Picture this...

You have taken over as the Head of a retail firm. One of the objectives that you have set for yourself is to enhance the quality of customer experience. To this effect, you have conducted an online survey. To encourage more and more customers to give you feedback on the quality of service offered by your firm, you had come up with a crisp customer survey. One of the questions was on the pain points experienced by the customers. The customers had to choose from amongst words like "Inhospitable staff", "Poor Quality", "High Price", etc. You want to see quickly the biggest concern that your customers have. It can very easily be seen with the word cloud.

- You wish to retain and grow your employee base.

8.10.2 When should you use a word cloud?

When there is a need to highlight important textual data point.

8.10.3 For what should you use a word cloud?

A word cloud is a visual display of text data. A kind of simple text analysis. It is usually used to depict keywords or metadata on websites. The more a word appears (frequency) in a source of textual data such as underlying database or a speech or a blog post, the bigger and bolder it appears. This is achieved with a play of font and color. In other words, the font size of the word or /and the color of the word depicts the relative frequency of occurrence of the target word in the source.

8.10.4 Where should you not use a word cloud?

- If your data is not optimized for context. It is not simply about dumping any data into the word cloud generator. The data should have been optimized through careful sieving by data analysts.
- These should not be overused simply because, it is easy to create and use. There could be a different visualization that might work better.

8.10.4.1 Example

Let us create a word cloud of the below Shakespeare Sonnet.

"Love is too young to know what conscience is, Yet who knows not conscience is born of love? Then, gentle cheater, urge not my amiss, Lest guilty of my faults thy sweet self prove: For, thou betraying me, I do betray My nobler part to my gross body's treason; My soul doth tell my body that he may Triumph in love; flesh stays no farther reason, But rising at thy name doth point out thee, As his triumphant prize. Proud of this pride, He is contented thy poor drudge to be, To stand in thy affairs, fall by thy side. No want of conscience hold it that I call Her love, for whose dear love I rise and fall".

The "word cloud" for the sonnet above is shown in Fig. 8-148.

Figure 8-148. *Word cloud for the given Shakespeare sonnet*

To validate, the frequency of the words in descending order is as shown in Fig. 8-149.

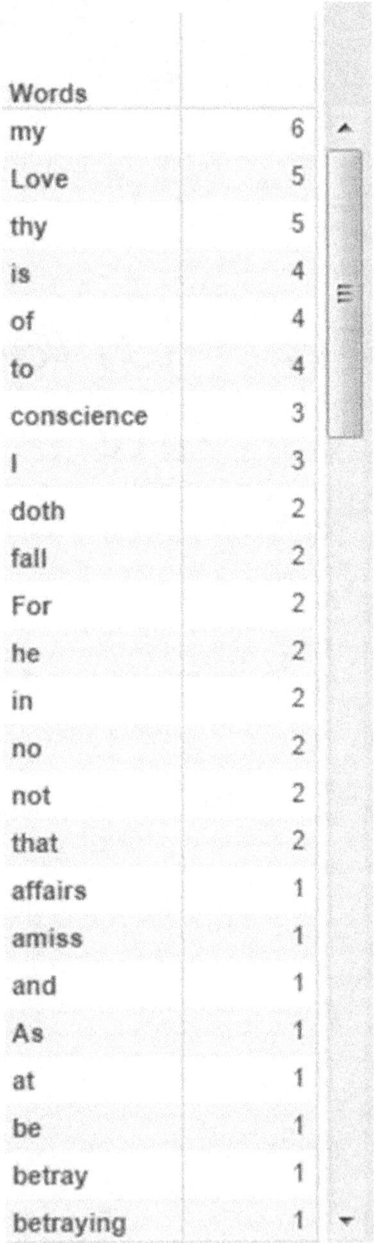

Words	
my	6
Love	5
thy	5
is	4
of	4
to	4
conscience	3
I	3
doth	2
fall	2
For	2
he	2
in	2
no	2
not	2
that	2
affairs	1
amiss	1
and	1
As	1
at	1
be	1
betray	1
betraying	1

Figure 8-149. *Frequency of the words from Shakespeare sonnet in descending order*

8.10.5 How to plot a word cloud?

8.10.5.1 Steps to create the word cloud

8.10.5.1.1 Step 1

The Shakespeare sonnet has been broken down into words and stored in an Excel sheet. There is only one column in the sheet, named, "Words". Connect to the Excel sheet (Shown in Fig. 8-150).

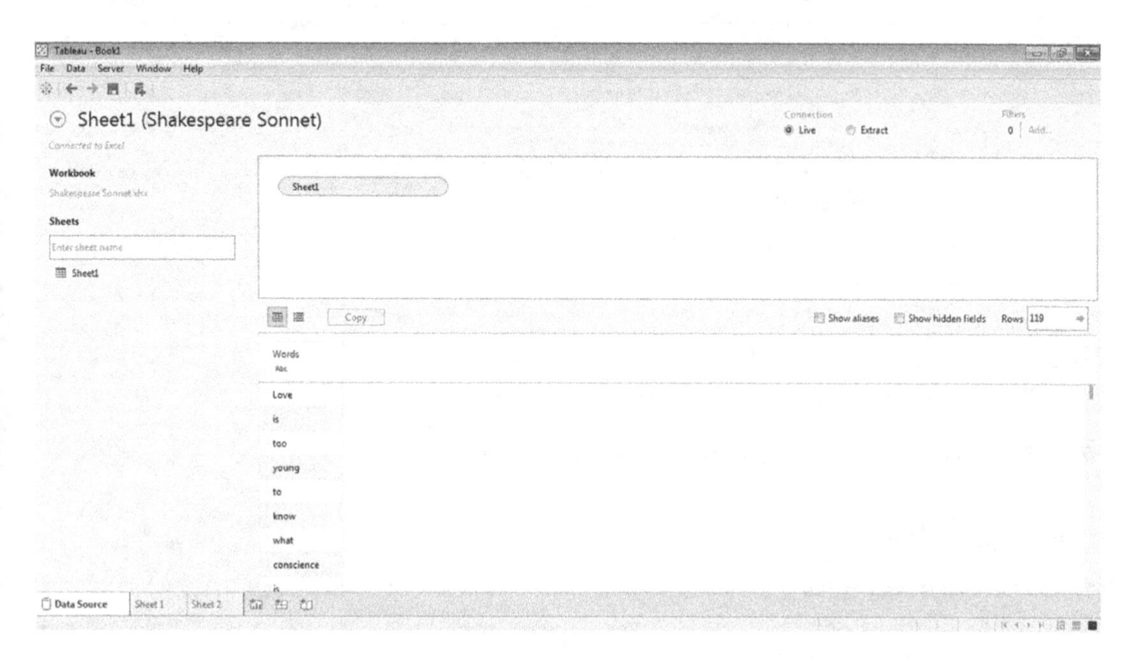

Figure 8-150. Excel sheet containing words from Shakespeare sonnet connected to Tableau

8.10.5.1.2 Step 2

Select "Text" in the marks card.

8.10.5.1.3 Step 3

Select the dimension "Words". Drag and drop it on "Label" on the marks card (Shown in Fig. 8-151).

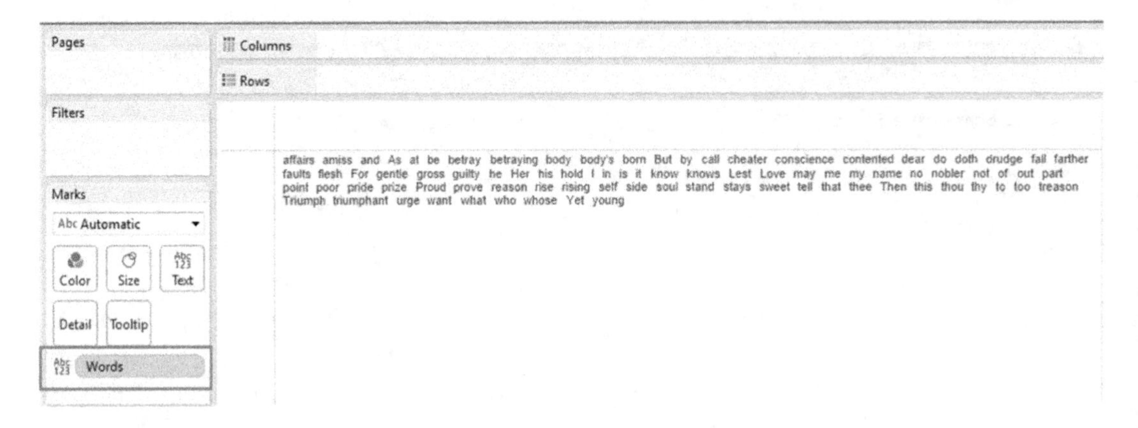

Figure 8-151. Dimension "Words" placed on "Label" on the marks card

8.10.5.1.4 Step 4

Change the default aggregation of the measure "Number of Records" to "Count" (Shown in Fig. 8-152).

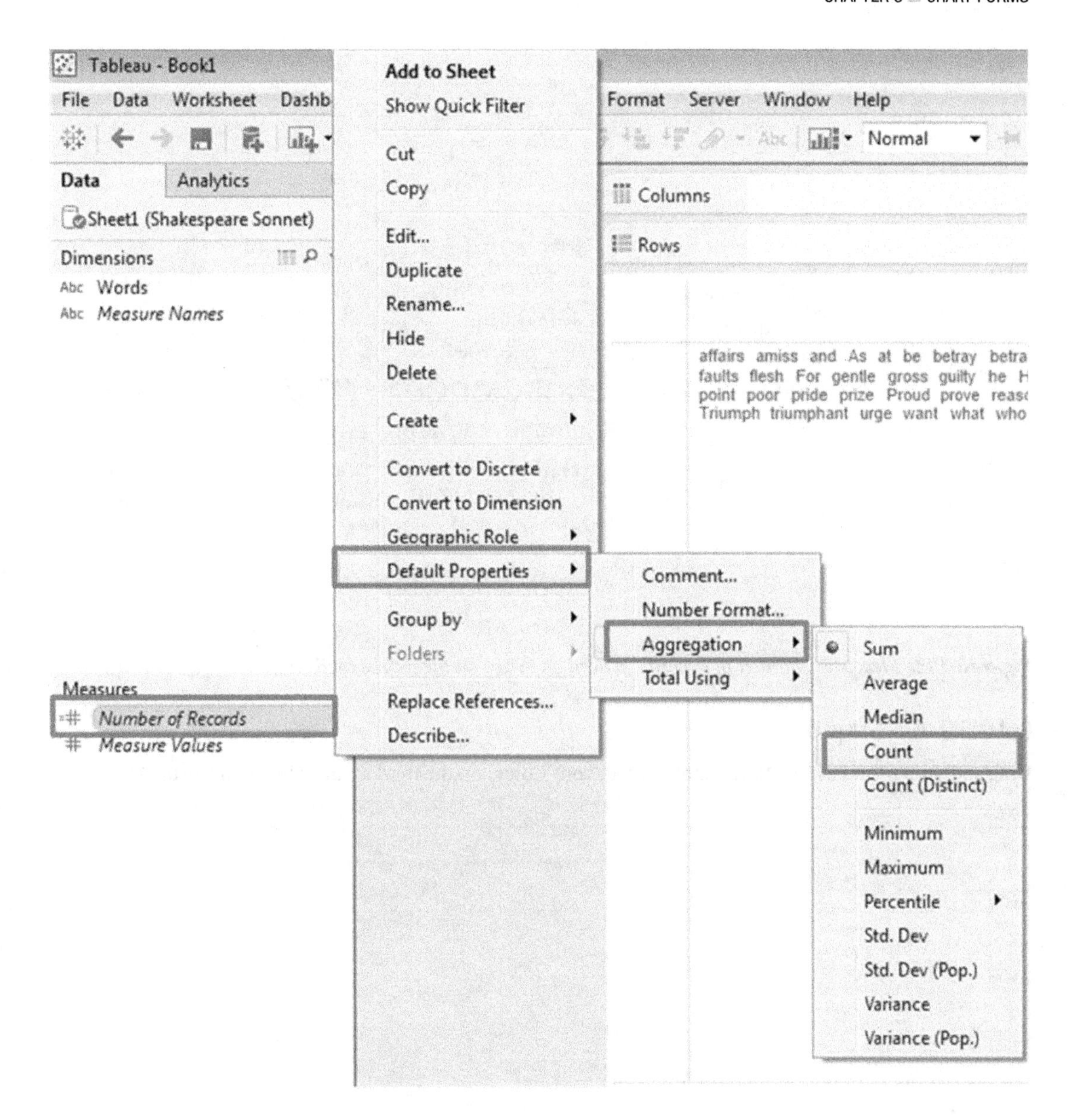

Figure 8-152. *Aggregation for measure "Number of Records" set to "Count"*

8.10.5.1.5 Step 5

Select the measure "Number of Records". Drag and drop it on "Size" in the marks card (Shown in Fig. 8-153).

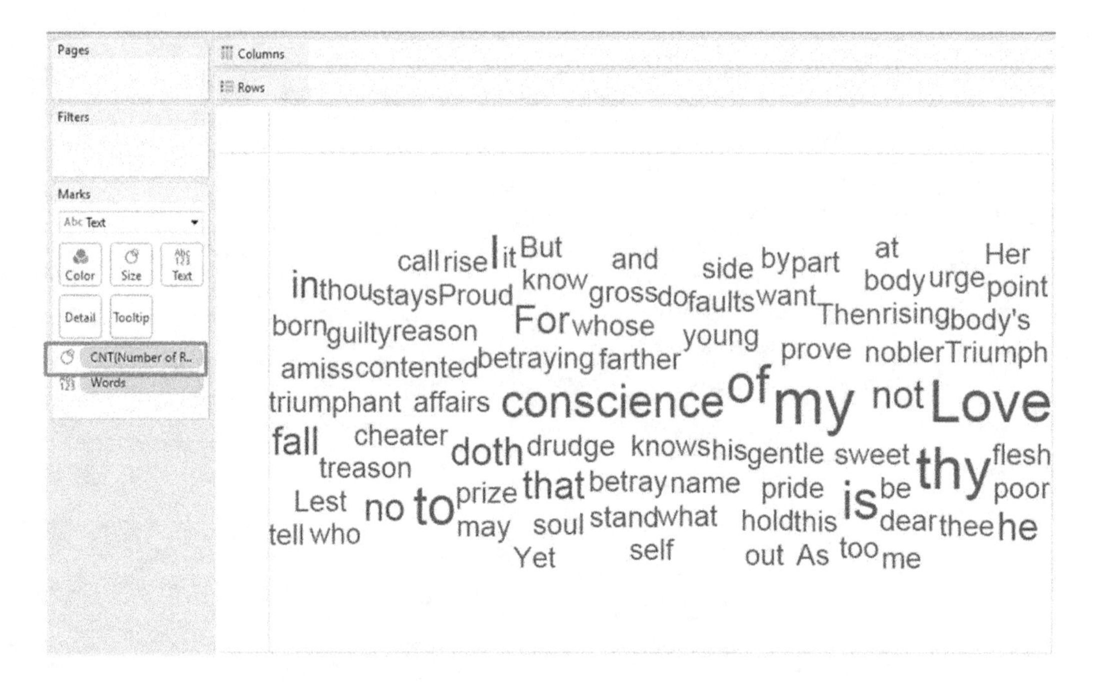

Figure 8-153. *Measure "Number of Records" placed on "Size" on the marks card*

8.10.5.1.6 Step 6

Drag and drop the measure "Number of Records" on "Color" on the marks card (Shown in Fig. 8-154).

Figure 8-154. *Measure "Number of Records" placed on "Color" on the marks card*

As is evident from Fig. 8-155, the word "my" appears in the darkest color and the font size for it is the largest. The frequency of **the word** "my" is 6. It is six times that the word "my" appears in the sonnet. This is followed by the word "Love", whose font size is the second largest as it occurs 5 times in the sonnet.

8.11 Points to remember

- Pie charts are simple to comprehend even by an uninformed audience.
- Hierarchical data is also referred to as tree-structured data. Tree maps can be used to represent hierarchical data.
- A heat map is a two-dimensional representation of data. Heat maps use color to display values.
- A highlight table is simply a large text table wherein the data values are encoded by color.
- A line graph or line chart displays information as a series of data points connected by straight-line segments.
- Scatter plots are used to visualize relationship between numeric variables.
- A Gantt chart is a graphical representation of a project schedule. They help one to plan, coordinate and track specific tasks in a project.
- A stacked bar chart or a stacked bar graph is used to break down and compare parts of a whole.
- Use a tag cloud to highlight important textual data point.

8.12 Next steps

The next chapter will take you deeper into advanced visualizations. You will learn about the following visualizations:

- Waterfall chart
- Bump chart
- Bullet chart

CHAPTER 9

Advanced Visualization

"In good information visualization, there are no rules, no guidelines, no templates, no standard technologies, no stylebooks... You must simply do whatever it takes."

—Edward Tufte, data scientist, pioneered the field of data visualization

In the previous chapter, we covered few chart forms such as bar chart, pie chart, line graph, scatter plot, histogram, etc. In this chapter, we will cover the following:

- Waterfall charts
- Bump charts
- Bullet graphs

9.1 Waterfall charts

A waterfall chart is a powerful tool for portraying how sequential processes contribute to the whole. In other words, it helps to show how one arrived at the net value by breaking down the cumulative effect of positive and negative contributions. Other names for waterfall chart are the flying brick chart, the Mario chart and particularly in finance, a bridge chart.

9.1.1 Where can waterfall charts be used?

Waterfall charts are useful in many different situations.

Here are just a few examples:

- Imagine that you work for the legal department of a leading corporation. You are in charge of all the deals and contracts of the firm. You start the financial year carrying forward the deals and contracts from the previous year. Add to this the deals and contracts won throughout the year. Then you drop off the contracts that were cancelled throughout the year. Finally, at the end of the year, you will have the total number of contracts that materialized.

S. Acharya and S. Chellappan, *Pro Tableau*, DOI 10.1007/978-1-4842-2352-9_9

- Waterfall charts help to visualize financial statements.
- Waterfall charts are used to navigate data about population, births and deaths.
- Imagine that you are a student who had undertaken the GRE graduate school admissions examination. You would like to assess your performance. You wish to see the number of questions that you answered correctly in the verbal and quantitative sections as well as the number of questions that you got wrong in both sections (verbal + quantitative). Then you wish to add up both scores to arrive finally at the final score.

Now let us go through some demos. The demos will provide step-by-step instructions to create a waterfall chart.

9.1.1.1 Demo 1

Objective: Let us plot a waterfall chart of the following scenario...

You work for the men's furnishing department of a leading retail store. You have received the latest inventory of men's t-shirts. The department has 225 units in stock. You place 100 t-shirts on display. Customers can also try on t-shirts to check the size, fabric and fitting. You realize that due to several people trying on t-shirts, a few units are damaged. You move to get the units repaired immediately. Out of the 55 units that were damaged, 18 units have been repaired. That leaves you with 188 saleable units (See Table 9-1 and Fig. 9-1).

Table 9-1. *Inventory of men's t-shirt stock*

Units in stock	225
Damaged	55
Refurbished	18
Saleable Units	188

Expected output:

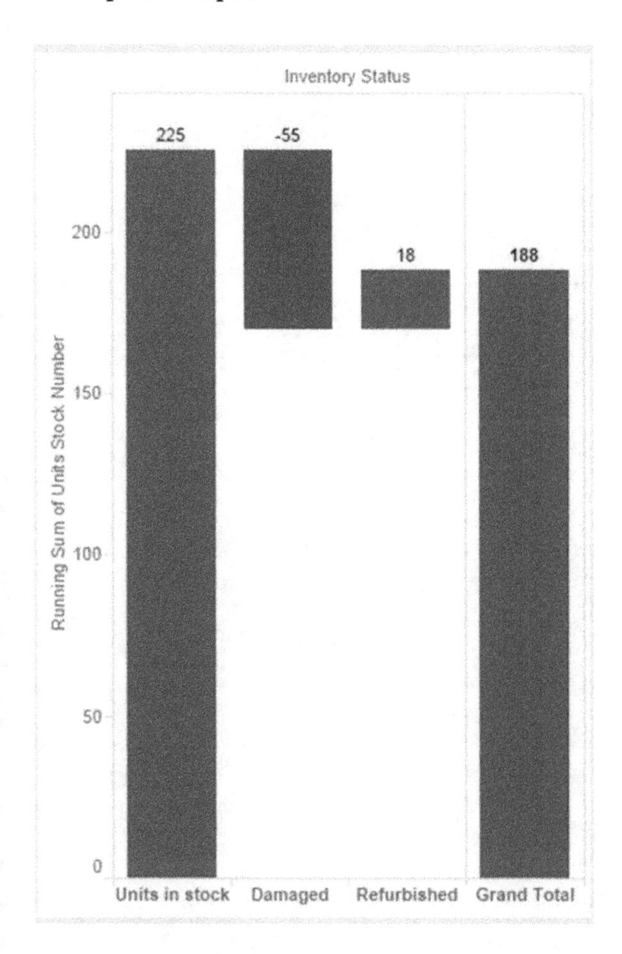

Figure 9-1. *Waterfall chart displaying the change in men's t-shirt inventory*

9.1.1.1.1 Steps to create a waterfall chart

The following steps should be taken. Data is available in the "WaterfallChart.xlsx" file (See Table 9-2).

Table 9-2. *Data as available in the Excel file*

	A	B
1	InventoryStatus	UnitsStockNumber
2	Units in stock	225
3	Damaged	-55
4	Refurbished	18

9.1.1.1.2 Step 1

Read the data from Excel into Tableau (Shown in Fig. 9-2).

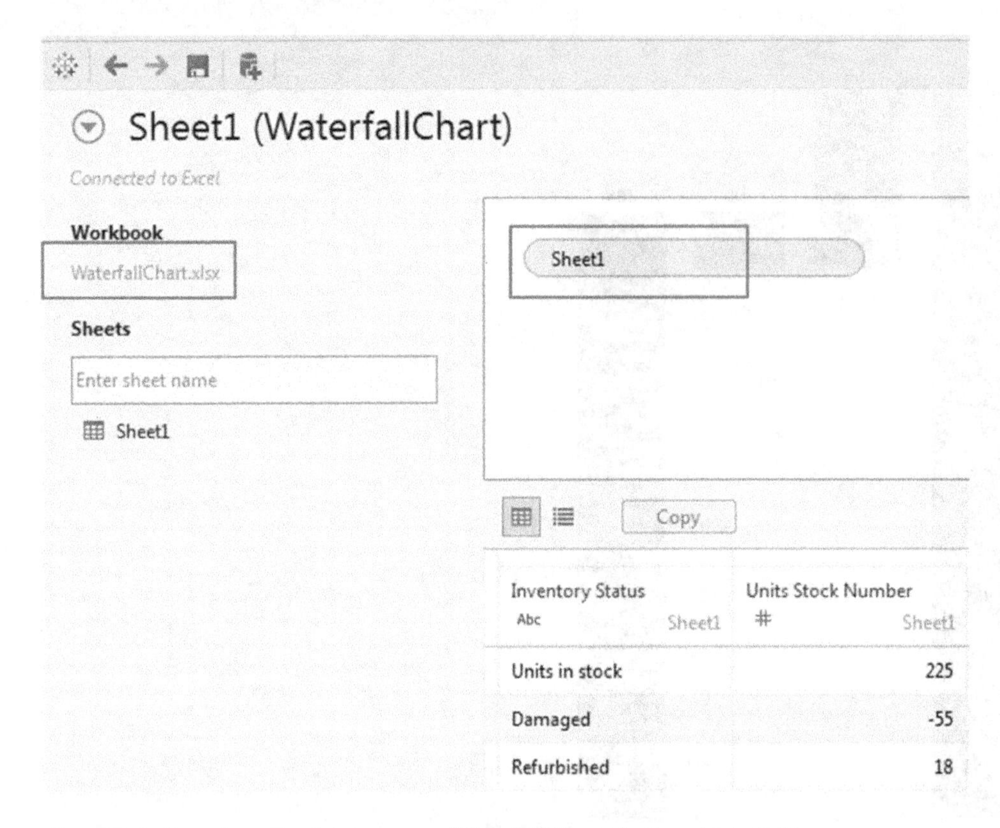

Figure 9-2. *Data from Excel data source, "WaterfallChart.xlsx" read into Tableau*

9.1.1.1.3 Step 2

Go to "Sheet1" (Shown in Fig. 9-3).

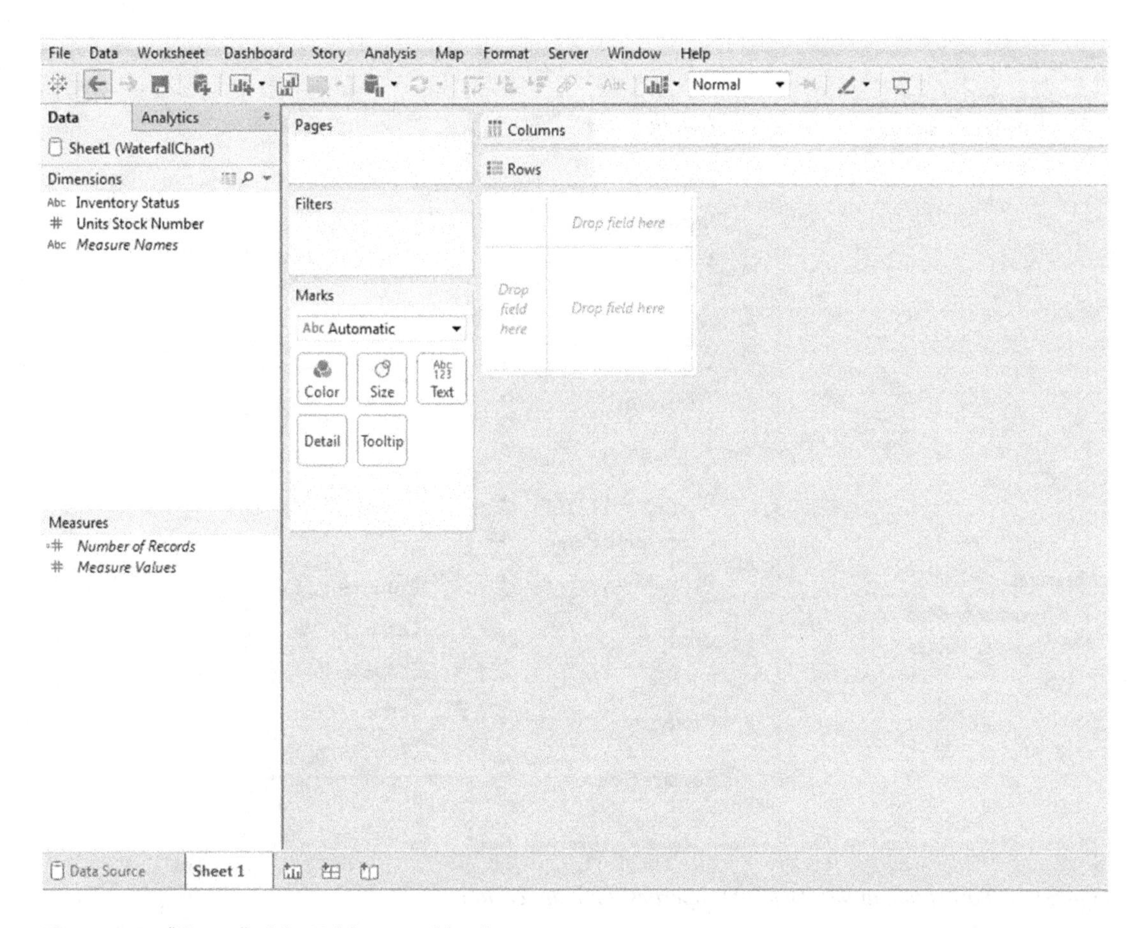

Figure 9-3. *"Sheet1" of the Tableau workbook*

9.1.1.1.4 Step 3

Sort the dimension "Inventory Status" (Shown in Fig. 9-4 and Fig. 9-5). The data members of "Inventory Status" should be sorted as follows:

- Units in stock
- Damaged
- Refurbished

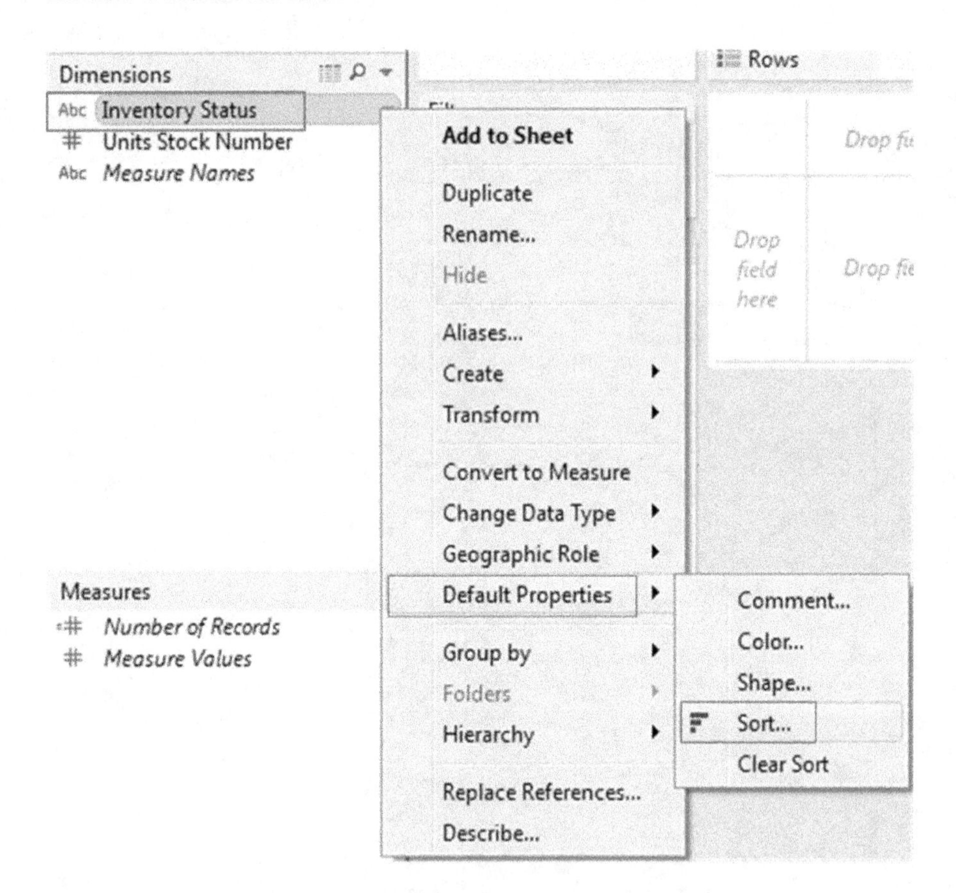

Figure 9-4. *Sort option under "Default Properties" for a dimension*

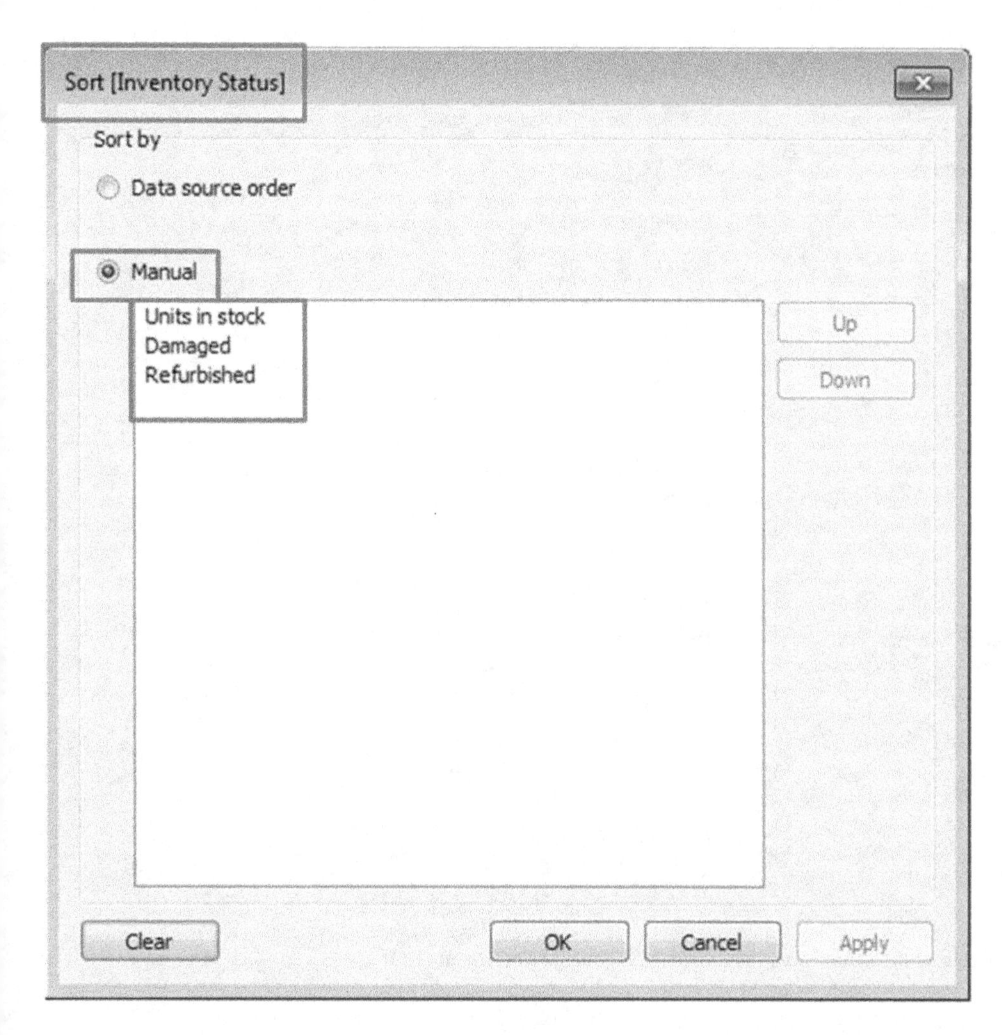

Figure 9-5. *Perform "Manual Sort"*

Convert the dimension "Units Stock Number" to measure (Shown in Fig. 9-6).

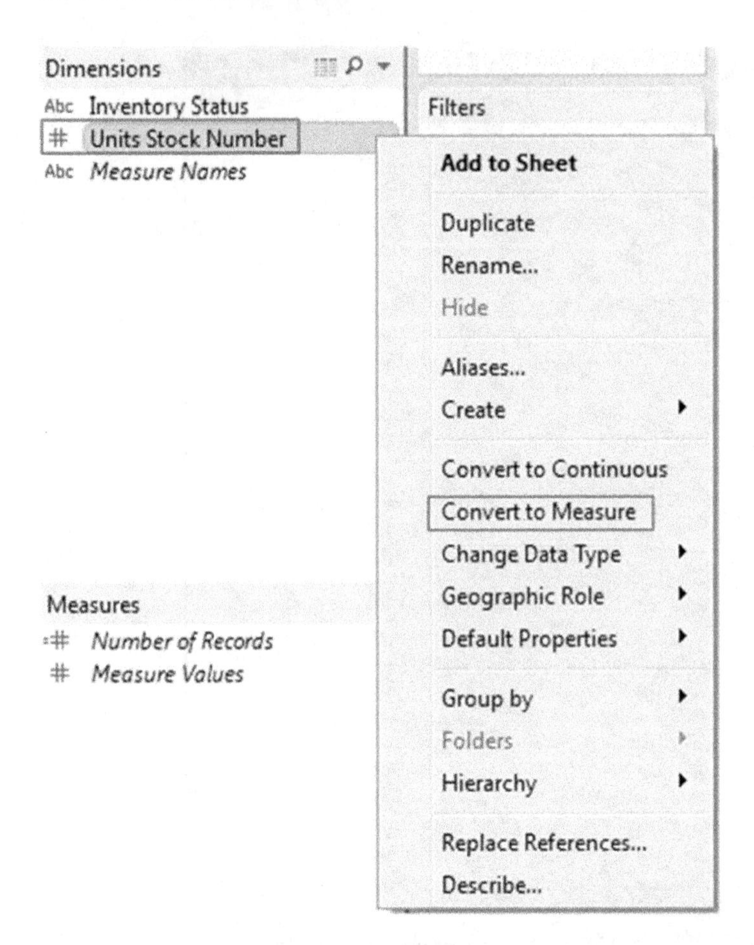

Figure 9-6. *Convert the dimension "Units Stock Number" to measure*

The worksheet after the conversion is as follows (Shown in Fig. 9-7):

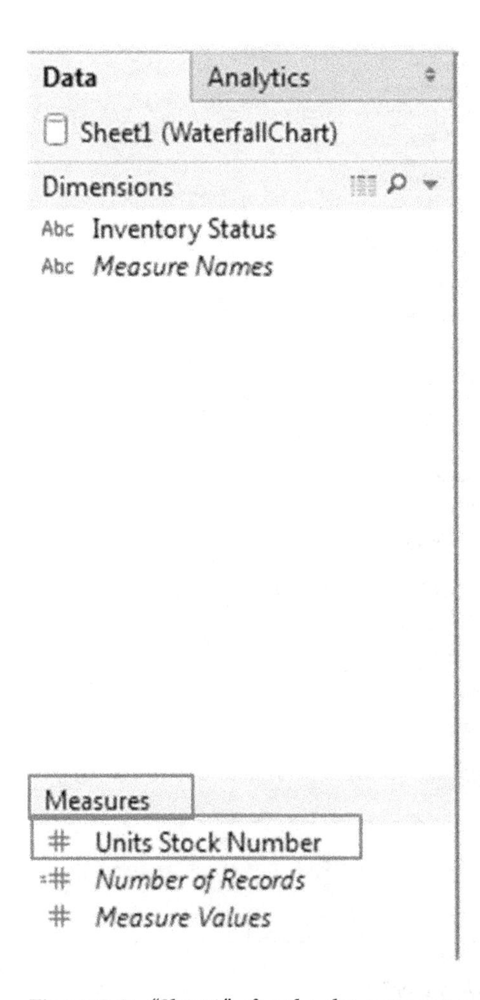

Figure 9-7. *"Sheet1" after the changes were made*

9.1.1.1.5 Step 4

Drag the dimension "Inventory Status" to the columns shelf. Drag the measure "Units Stock Number" to the rows shelf. Add a table calculation "Running Total" to the measure "Units Stock Number" (See Table 9-3, Fig. 9-8 and Fig. 9-9).

Table 9-3. *Activities to perform*

Columns shelf	Inventory status
Rows shelf	Units stock number
Table calculation Computed along	Running total Table across

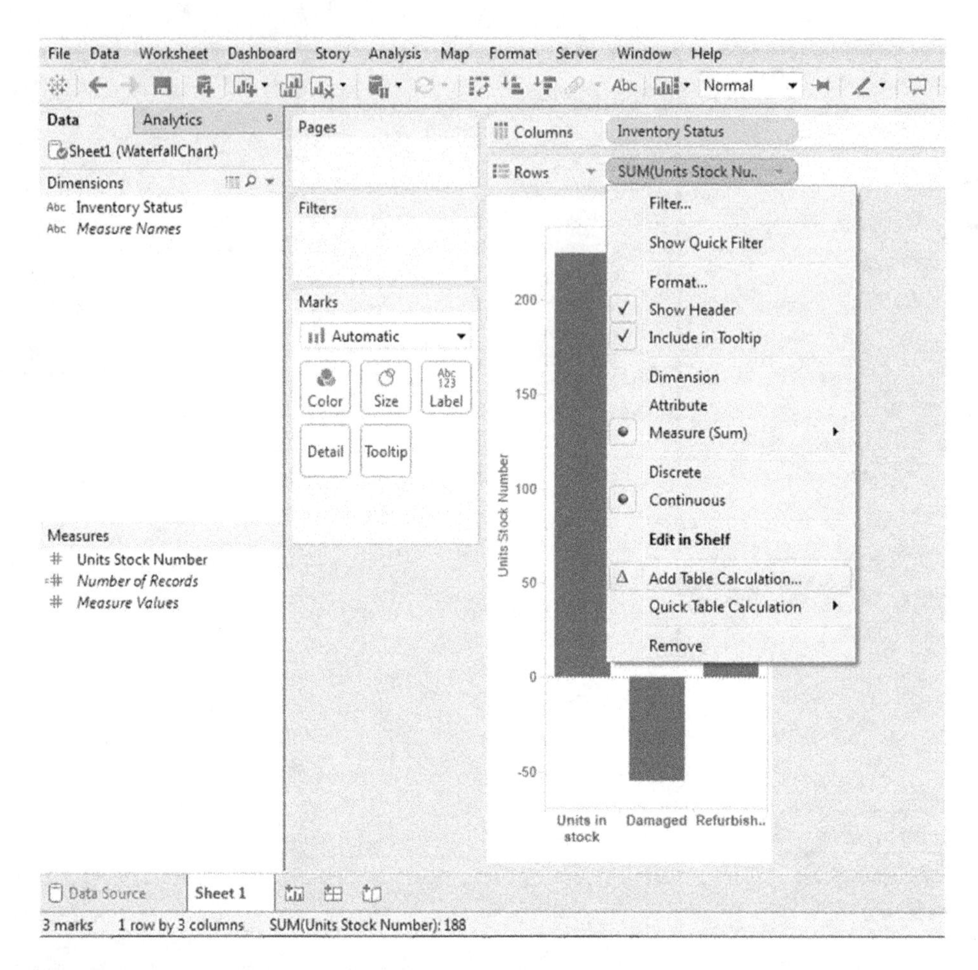

Figure 9-8. *Add a table calculation to the measure "Units Stock Number"*

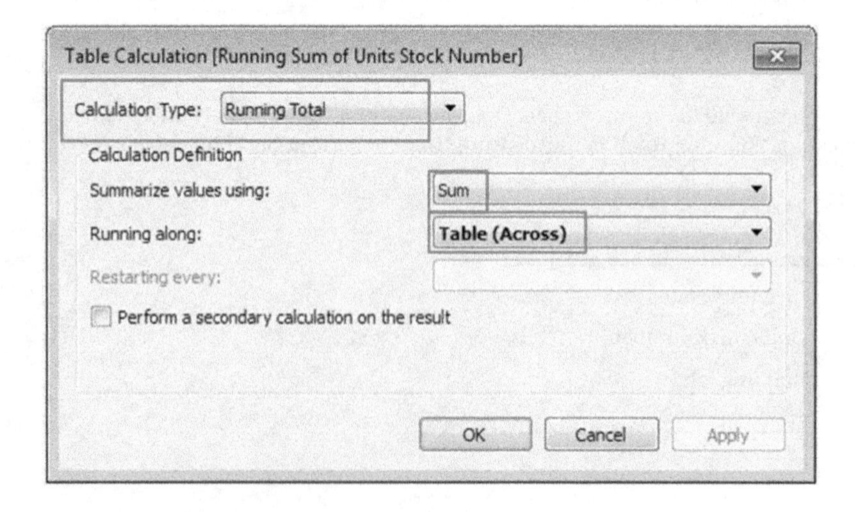

Figure 9-9. *Table calculation "Running Total" Running along "Table(Across)"*

The output is shown in Fig. 9-10.

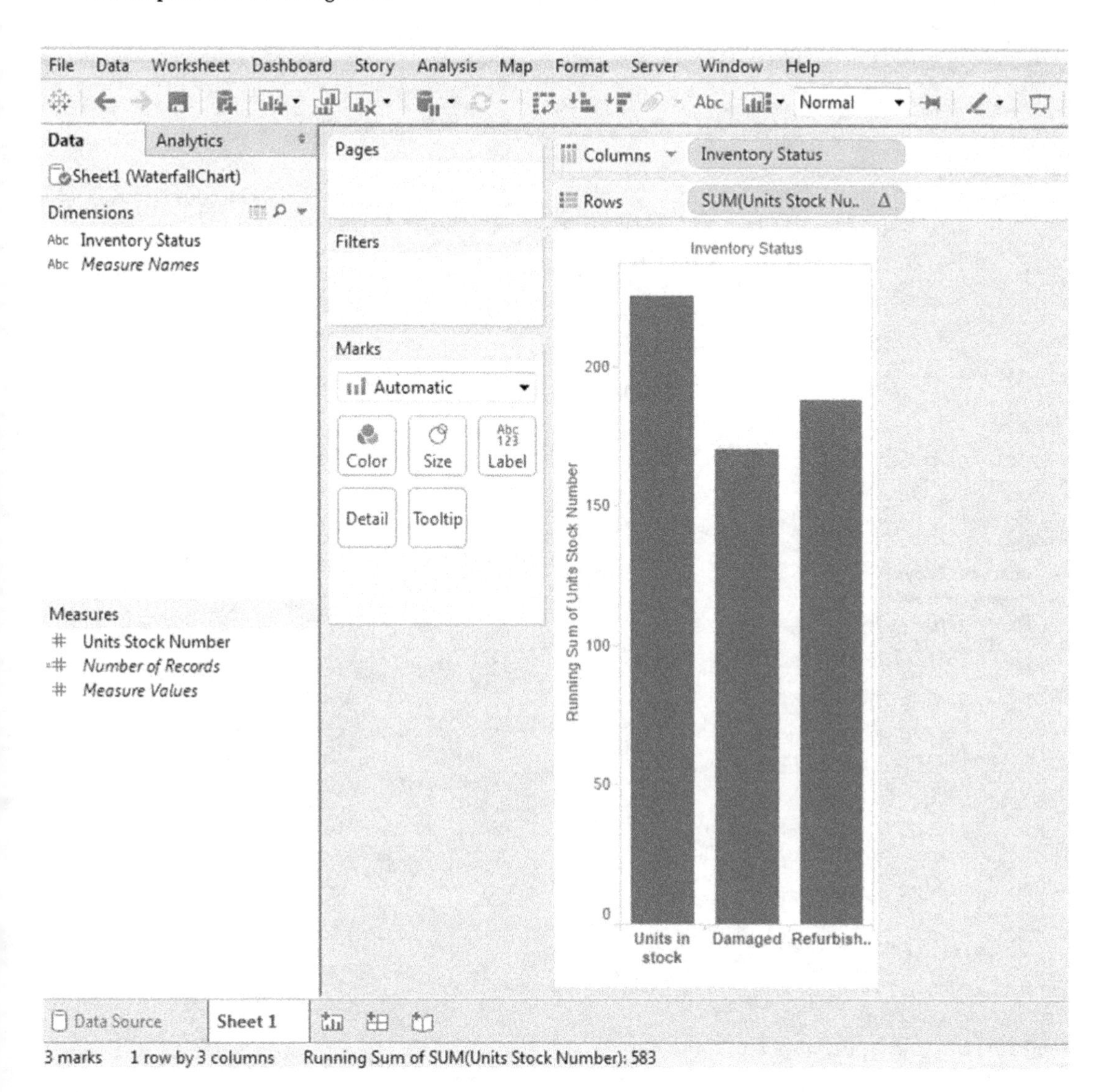

Figure 9-10. *Output after applying the table calculation "Running Total" to the measure "Units Stock Number"*

9.1.1.1.6 Step 5

Change the marks type from "Automatic" to "Gantt Bar" (Shown in Fig. 9-11).

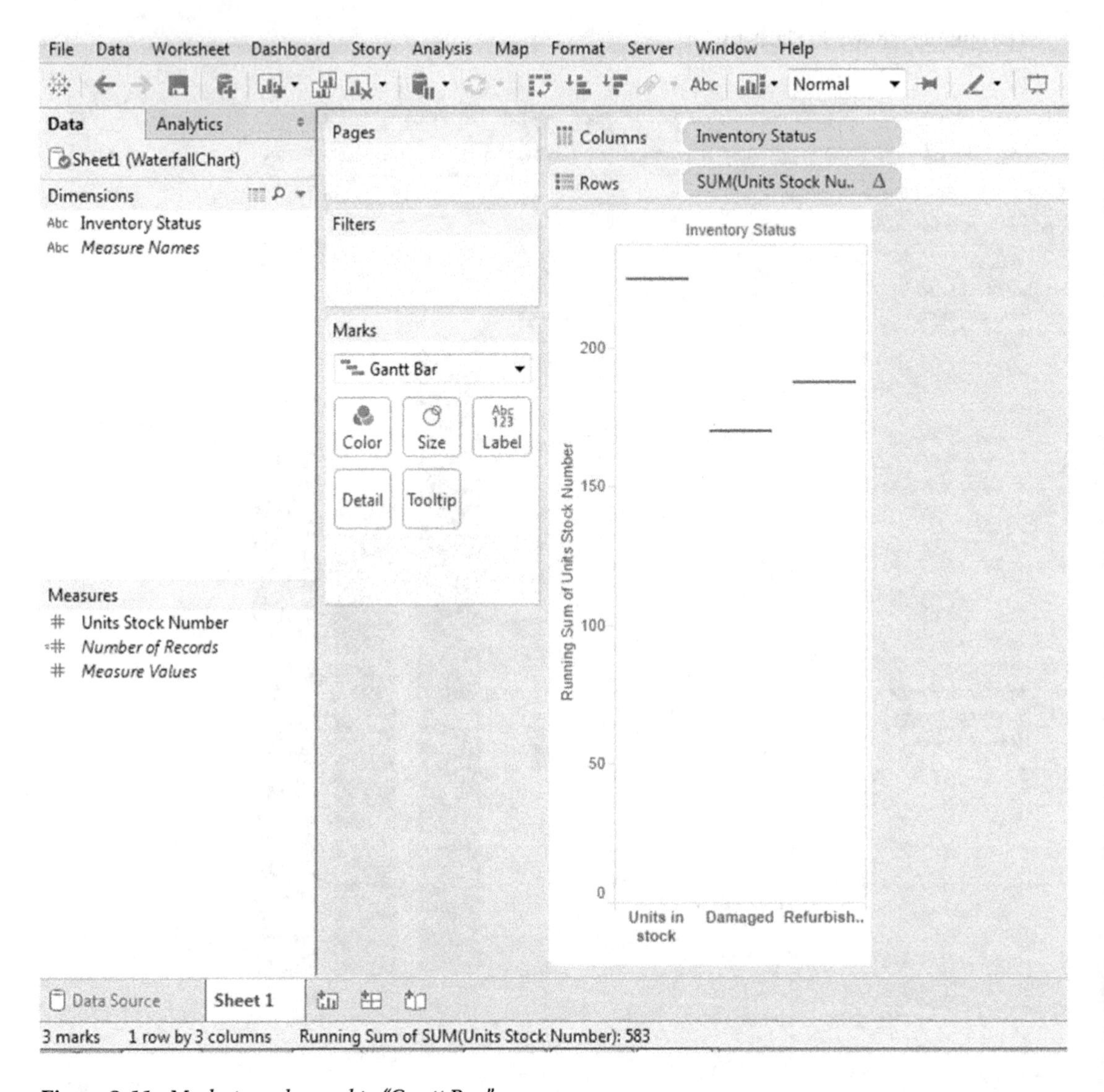

Figure 9-11. *Marks type changed to "Gantt Bar"*

9.1.1.1.7 Step 6

Create a calculated field "-UnitsStockNumber" (Shown in Fig. 9-12):

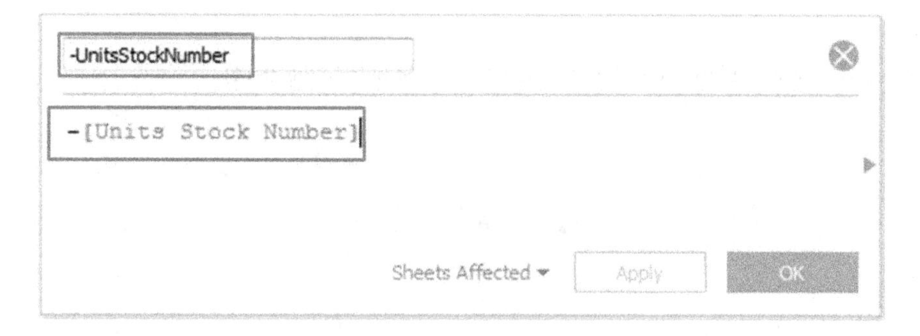

Figure 9-12. Calculated field "- Units Stock Number" created

9.1.1.1.8 Step 7

Drag the calculated field "- Units Stock Number" to "Size" on the marks card (Shown in Fig. 9-13).

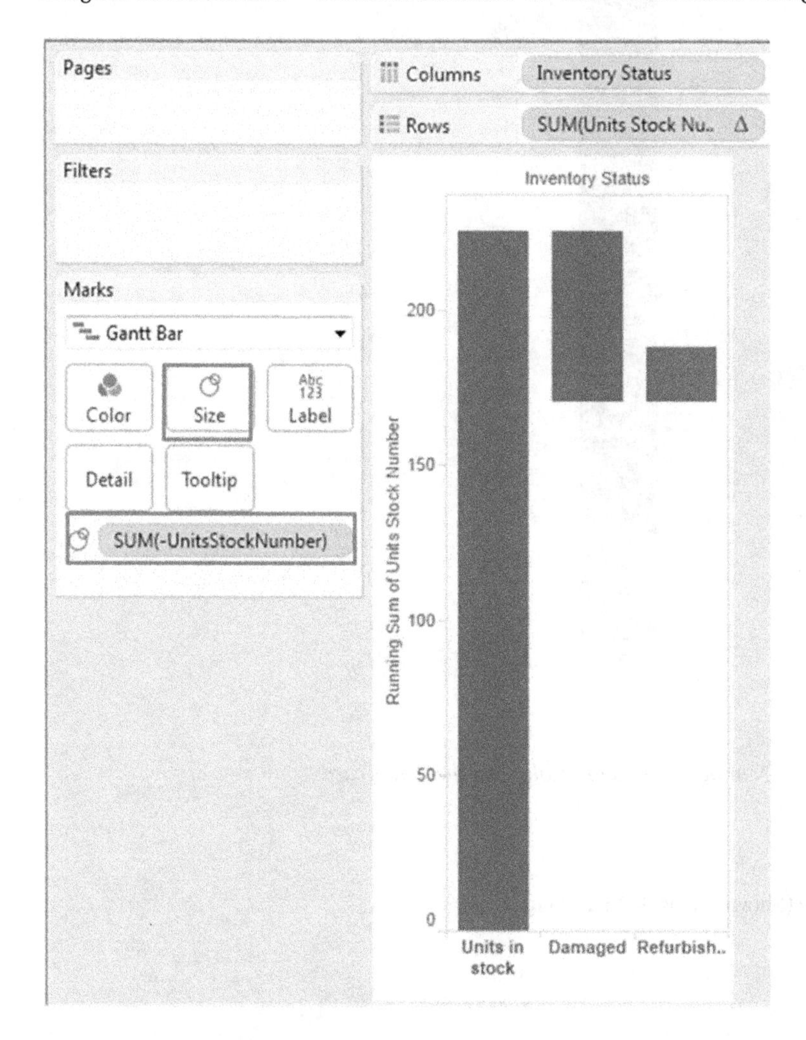

Figure 9-13. Calculated field "- Units Stock Number" placed on "Size" on the marks card

9.1.1.1.9 Step 8

Drag the measure "Units Stock Number" to "Color" on the marks card (Shown in Fig. 9-14).

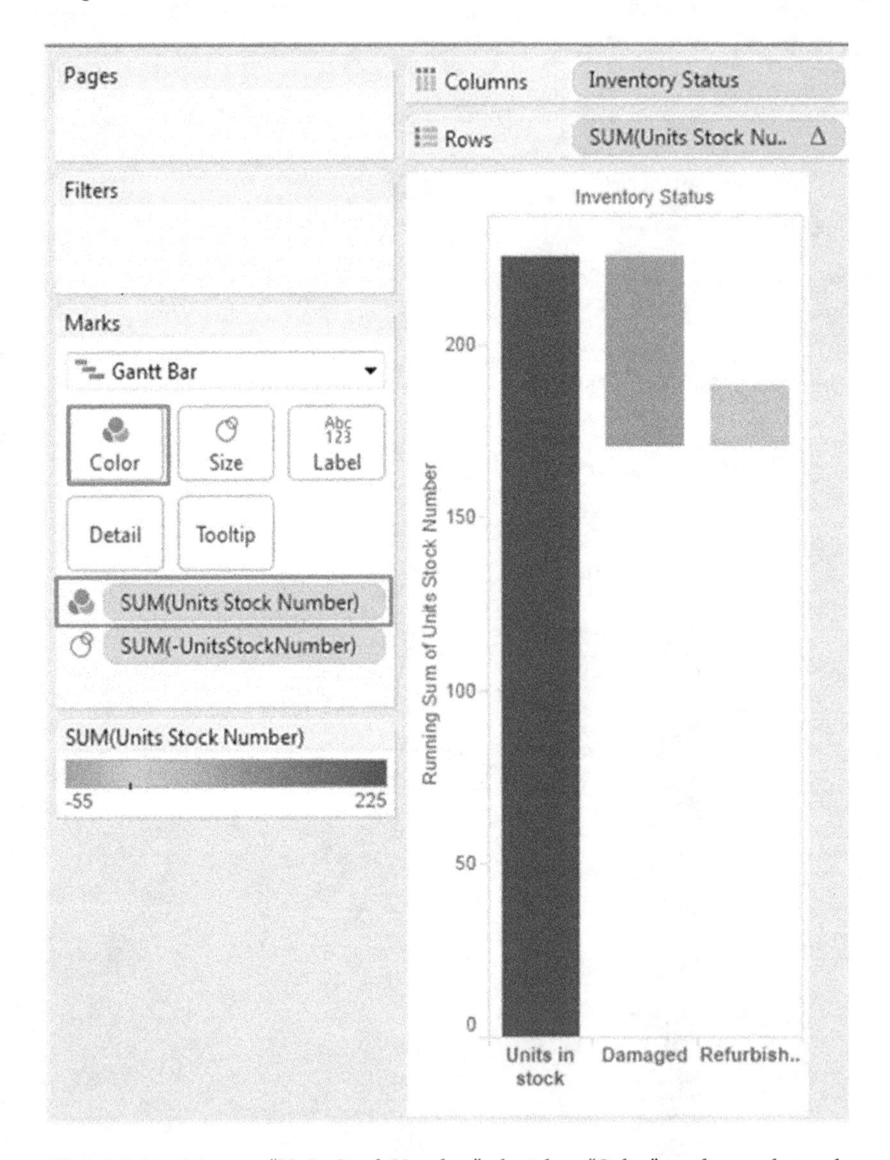

Figure 9-14. *Measure "Units Stock Number" placed on "Color" on the marks card*

9.1.1.1.10 Step 9

Change the "Stepped Color" to 2 (Shown in Fig. 9-15 and Fig. 9-16).

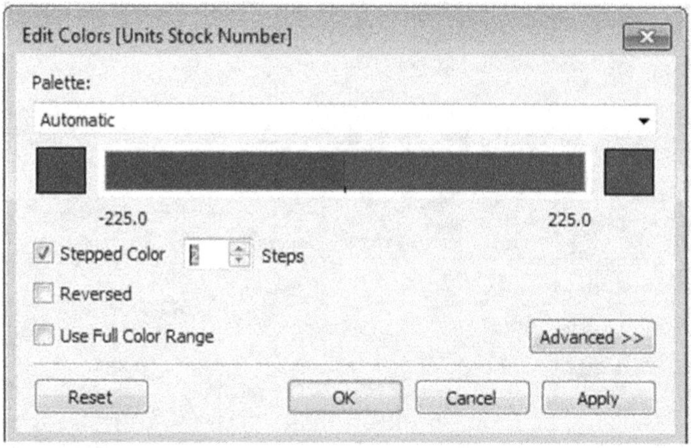

Edit Colors [Units Stock Number]

Palette:

Automatic

-225.0 225.0

Stepped Color 2 Steps

Reversed

Use Full Color Range

Advanced >>

Reset OK Cancel Apply

Figure 9-15. *Stepped Color set to 2*

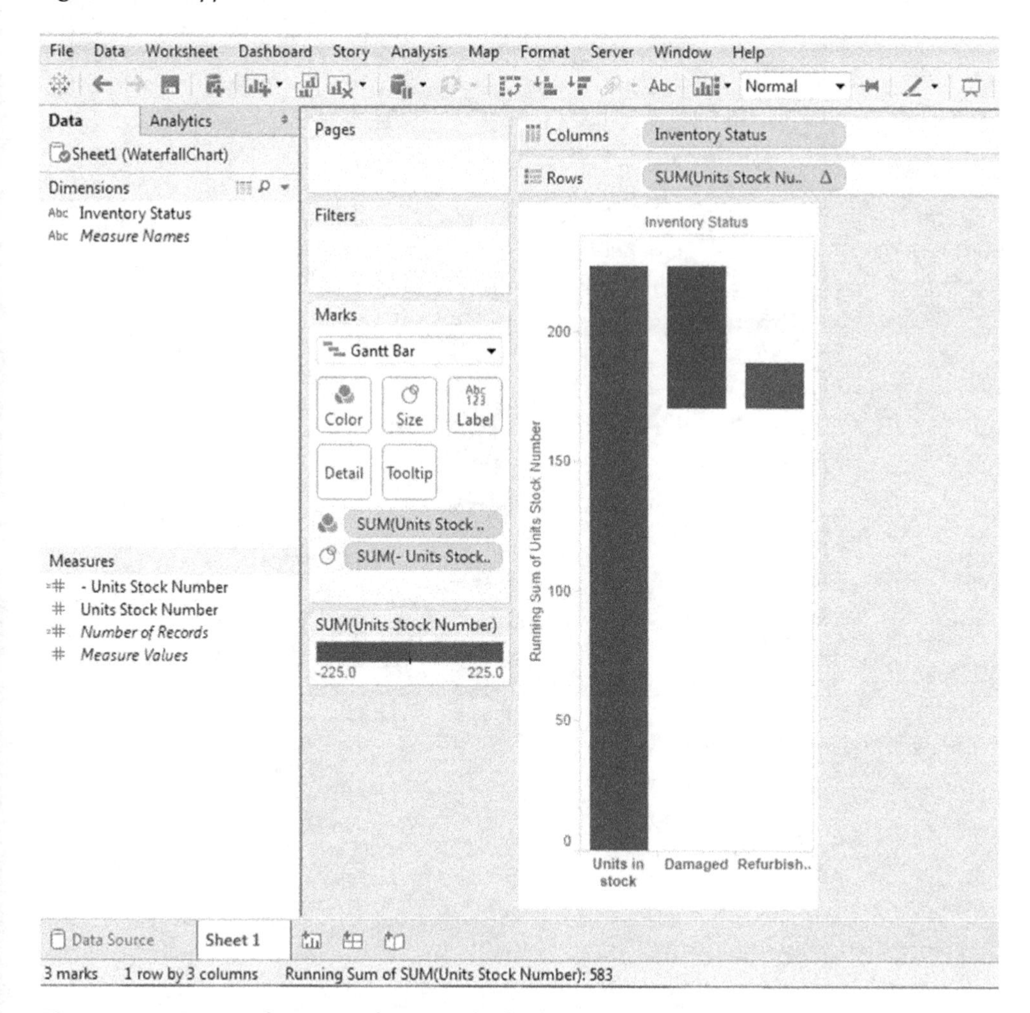

Figure 9-16. *Output after setting the stepped color to 2*

9.1.1.1.11 Step 10

Select Analysis ➤ Totals ➤ Show Row Grand Totals (Shown in Fig. 9-17).

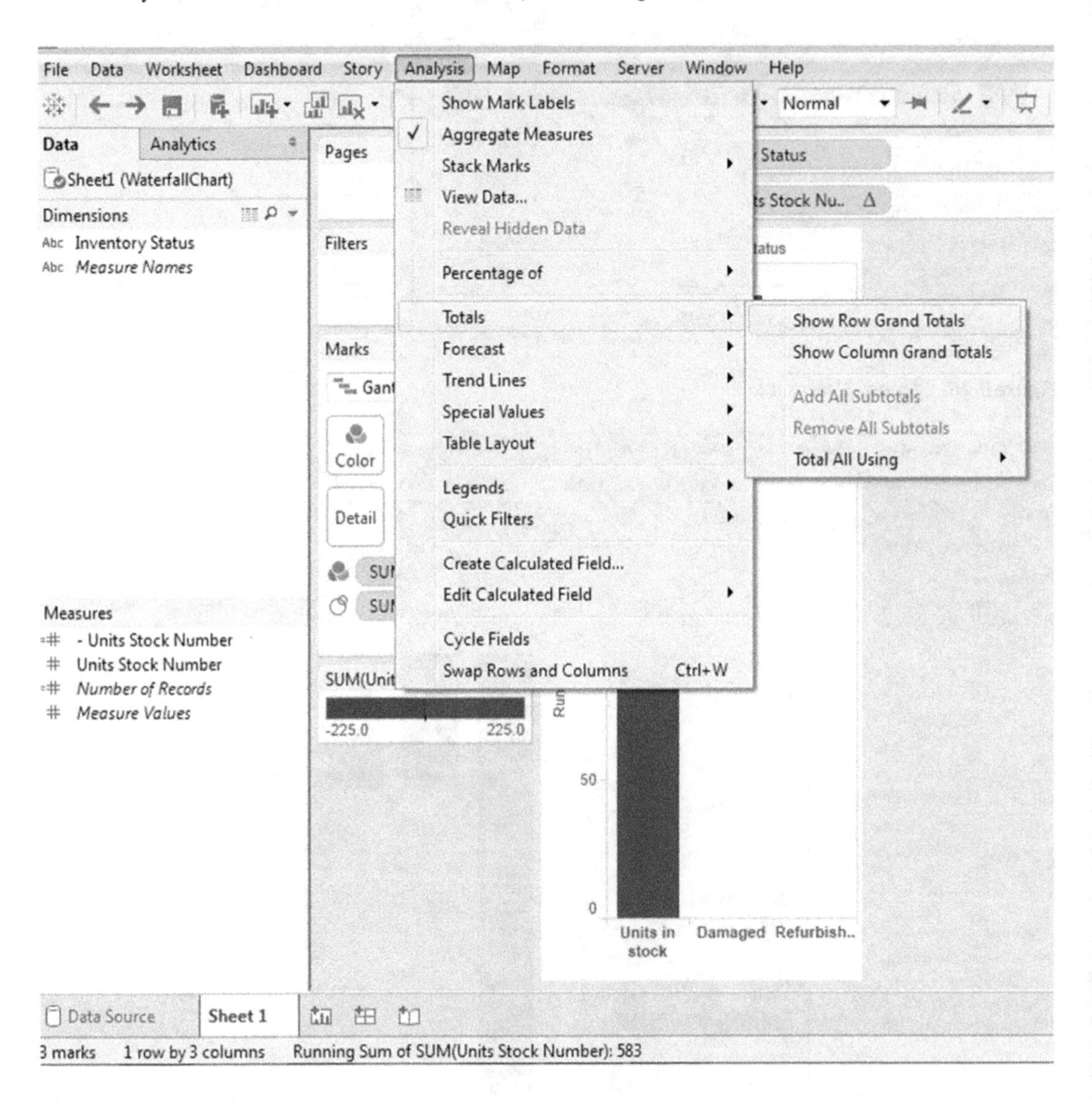

Figure 9-17. *Compute the "Row Grand Totals"*

The output (Shown in Fig. 9-18):

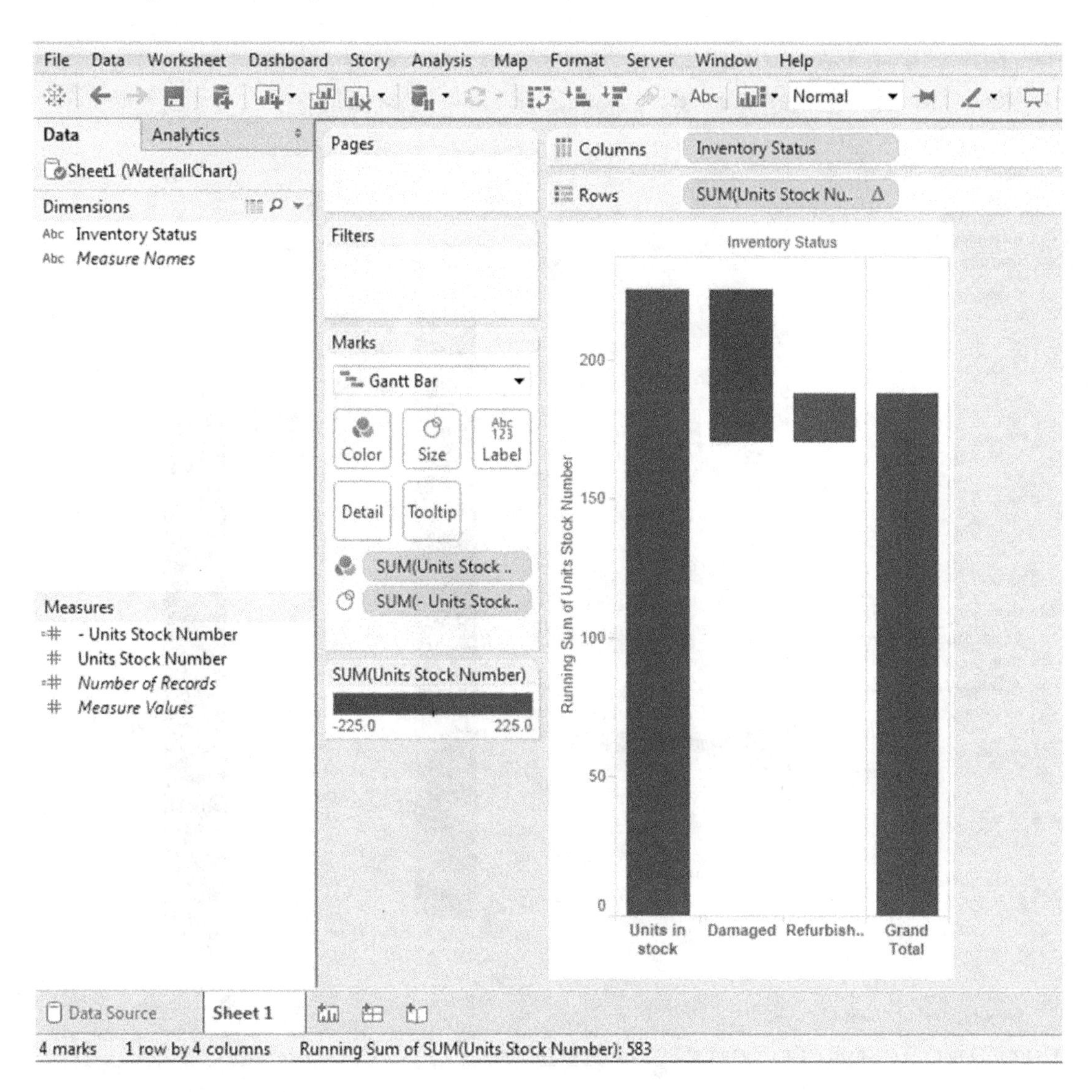

Figure 9-18. *Output after computing the "Row Grand Totals"*

Finally drag the measure "Units Stock Number" to "Label" on the marks card (Shown in Fig. 9-19).

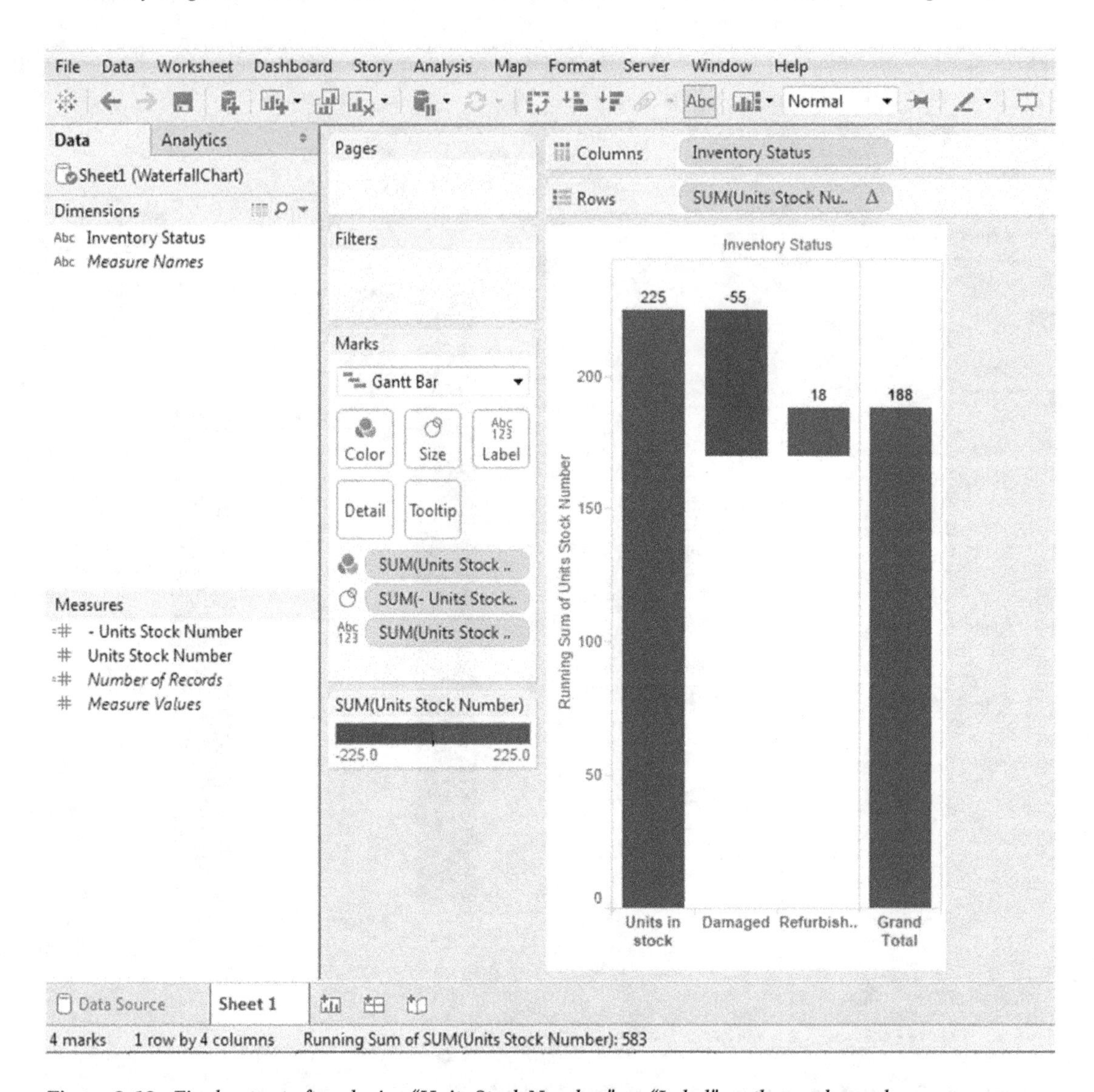

Figure 9-19. *Final output after placing "Units Stock Number" on "Label" on the marks card*

9.1.1.2 Demo 2

You are the project manager of "XYZ" account of a leading IT company. You are also in charge of staffing for your account. At the beginning of the financial year, you have 100 employees. Your HR department has helped you recruit 30 employees, and there have been eight transfers from other units of the company to your unit. Twelve employees have moved out of your unit to other units, and another ten have left the company. You would like to show your manager the headcount at the end of the financial year (See Table 9-4 and Fig. 9-20).

Table 9-4. *Status of employee strength of "XYZ" account*

Headcount at the beginning of the year	100
New hires	30
Transfer-ins	8
Transfer-outs	12
Exits	10
Headcount at the end of the year	116

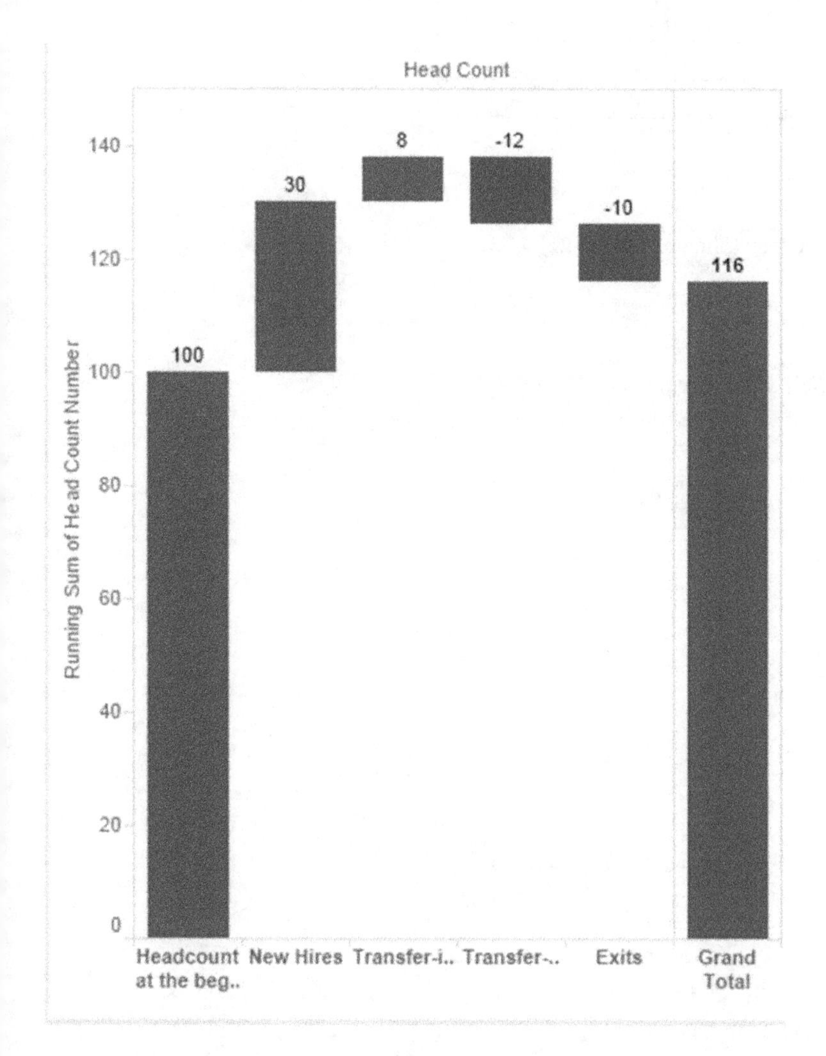

Figure 9-20. *Waterfall chart showing the changes in employee strength for "XYZ" project account*

9.1.1.2.1 Steps to create a waterfall chart

The below is the data in the "Waterfall Chart Assignment2.xlsx" (Shown in Table 9-5).

Table 9-5. *Data as available in the Excel sheet*

	A	B
1	HeadCount	HeadCountNumber
2	Headcount at the beginning of the year	100
3	New Hires	30
4	Transfer-ins	8
5	Transfer-outs	-12
6	Exits	-10

9.1.1.2.2 Step 1

Read the data from Excel into Tableau (Shown in Fig. 9-21).

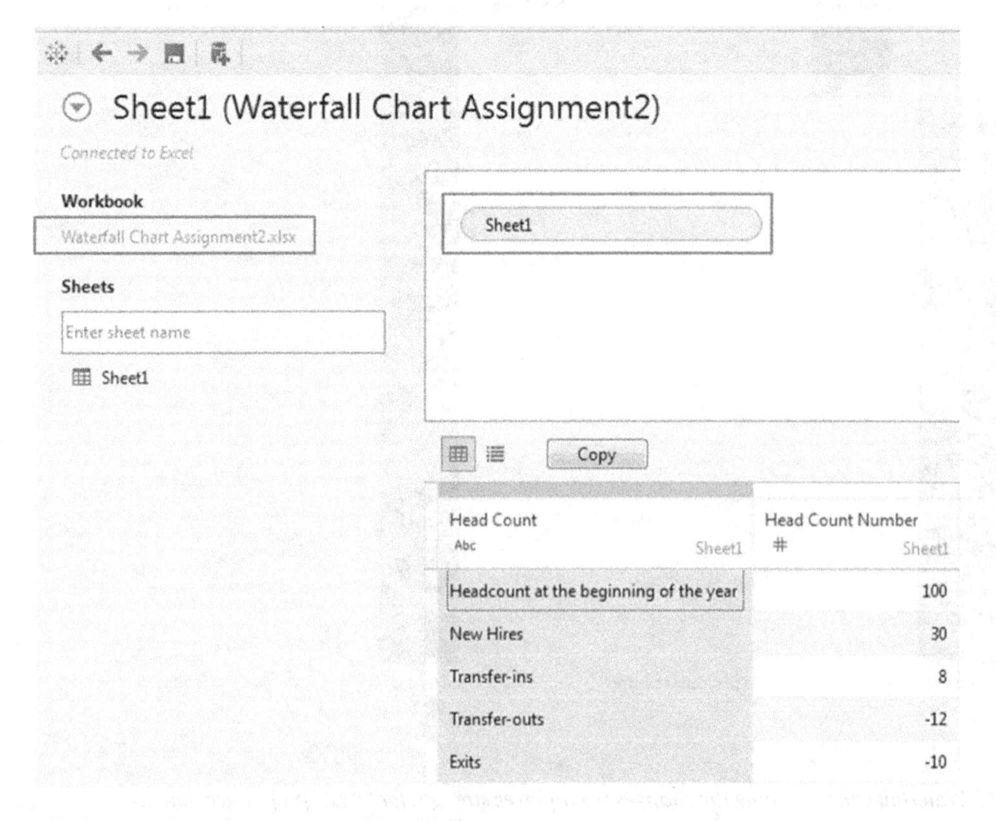

Figure 9-21. *Data from Excel read into Tableau*

9.1.1.2.3 Step 2

Go to "Sheet1" (Shown in Fig. 9-22).

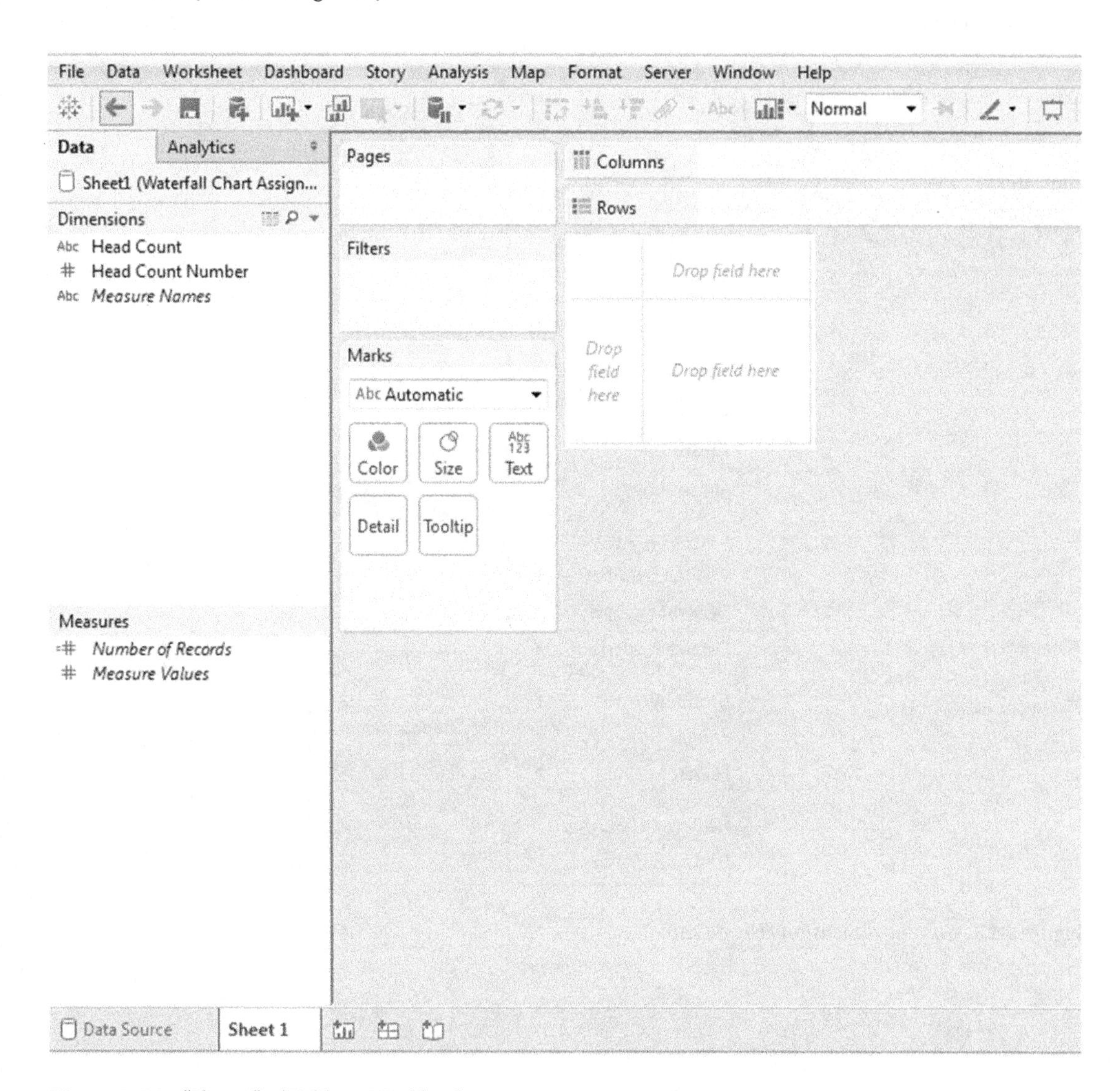

Figure 9-22. "Sheet1" of Tableau Workbook

9.1.1.2.4 Step 3

Sort the dimension "Head Count" (Shown in Fig. 9-23 and Fig. 9-24). The data members of "Head Count" should be sorted as follows:

- Headcount at the beginning of the year
- New hires

- Transfer-ins
- Transfer-outs
- Exits

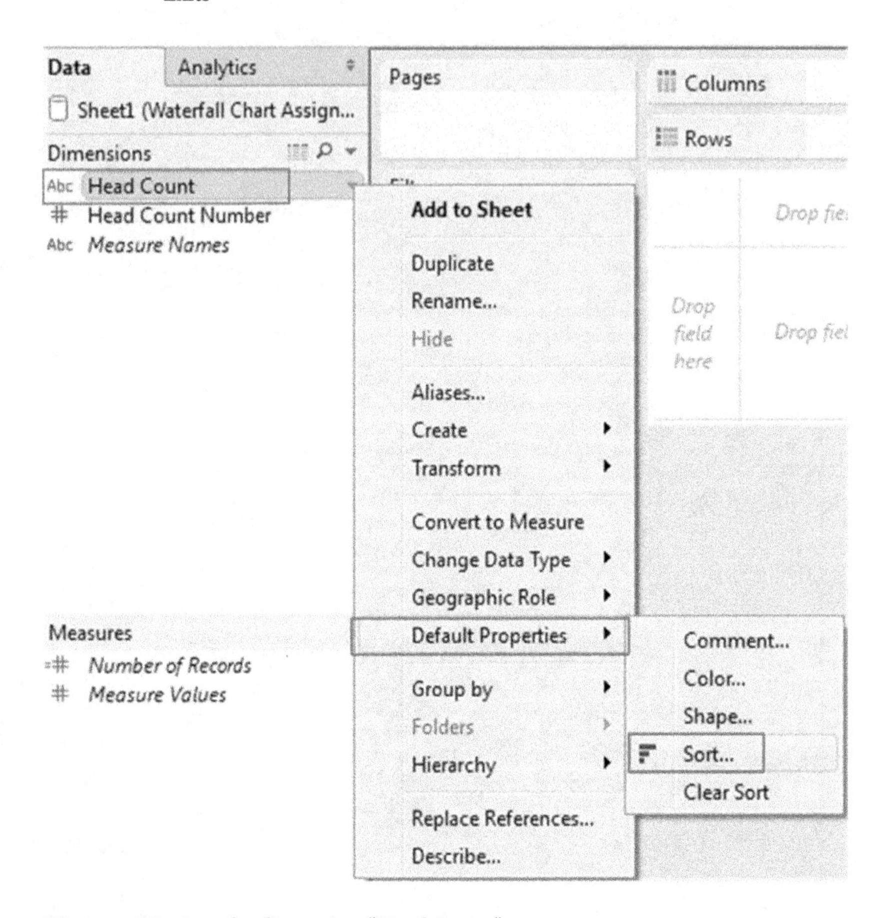

Figure 9-23. *Sort the dimension "Head Count"*

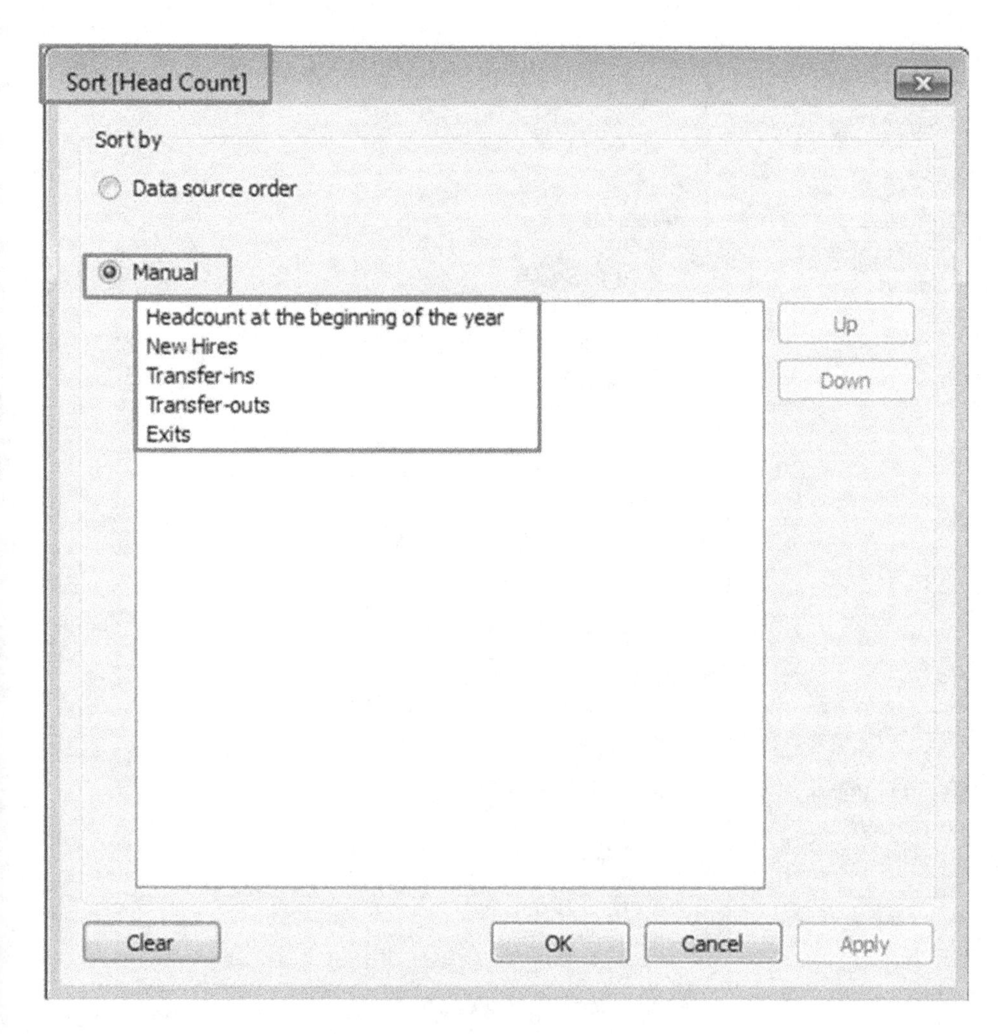

Figure 9-24. *Perform "Manual Sort" of the dimension "Head Count"*

Convert the dimension "Head Count Number" to measure (Shown in Fig. 9-25).

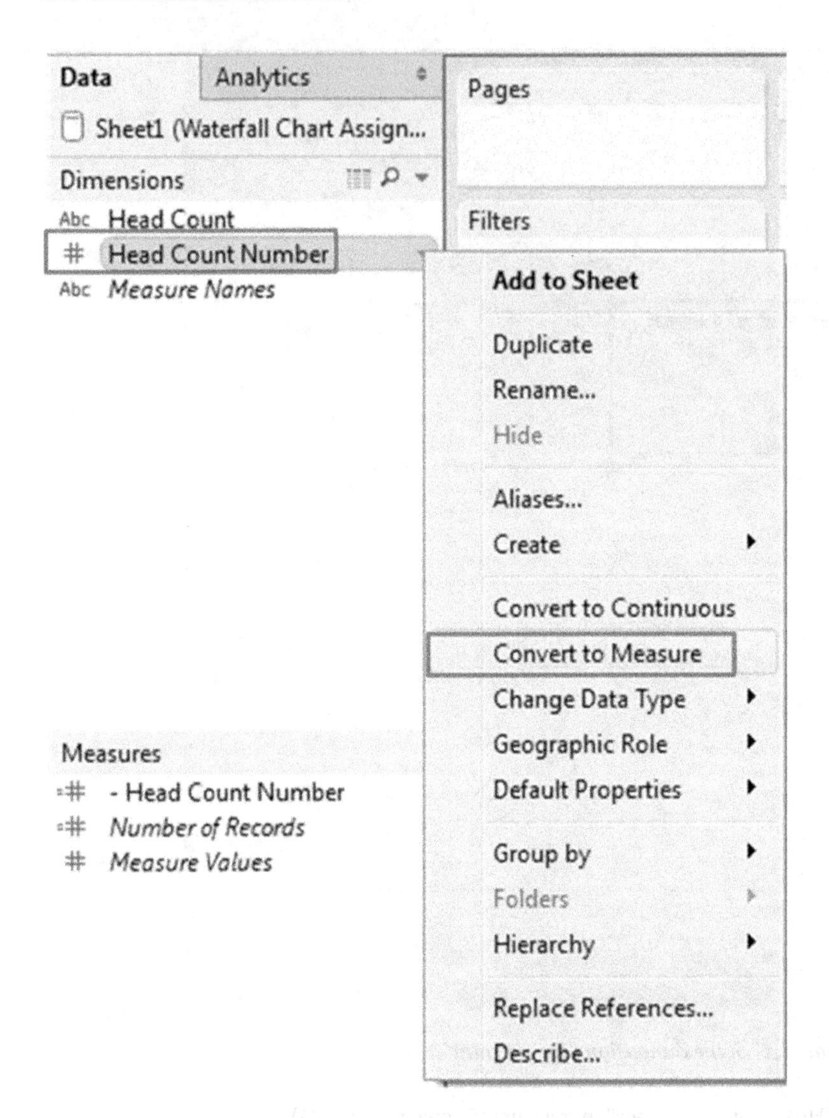

Figure 9-25. *Convert the dimension "Head Count Number" to measure*

The worksheet after the conversion (Shown in Fig. 9-26):

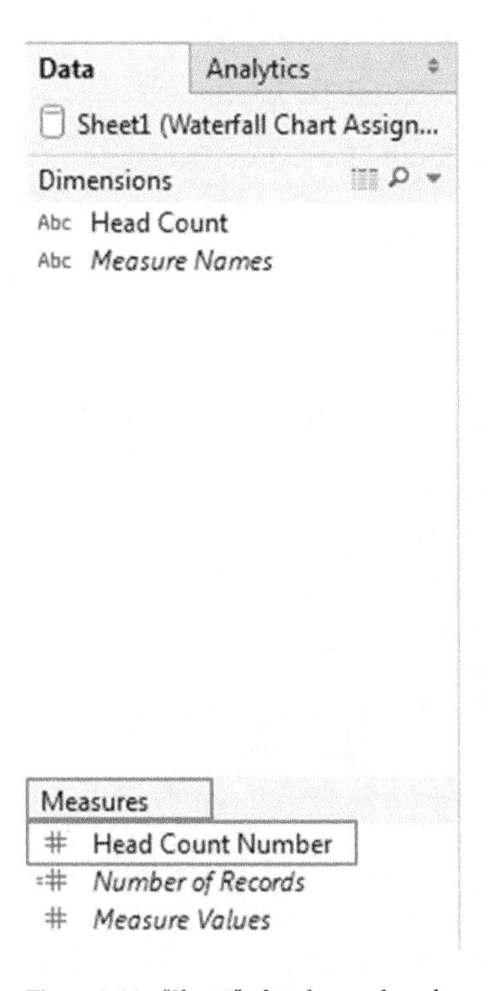

Figure 9-26. *"Sheet1" after changes have been made*

9.1.1.2.5 Step 4

Drag the dimension "Head Count" to the columns shelf. Drag the measure "Head Count Number" to the rows shelf. Add a table calculation "Running Total" to the measure "Head Count Number" (See Table 9-6, Fig. 9-27 and Fig. 9-28).

Table 9-6. *List of activities to perform*

Columns shelf	Head count
Rows shelf	Head count number
Table calculation Running along	Running total Table across

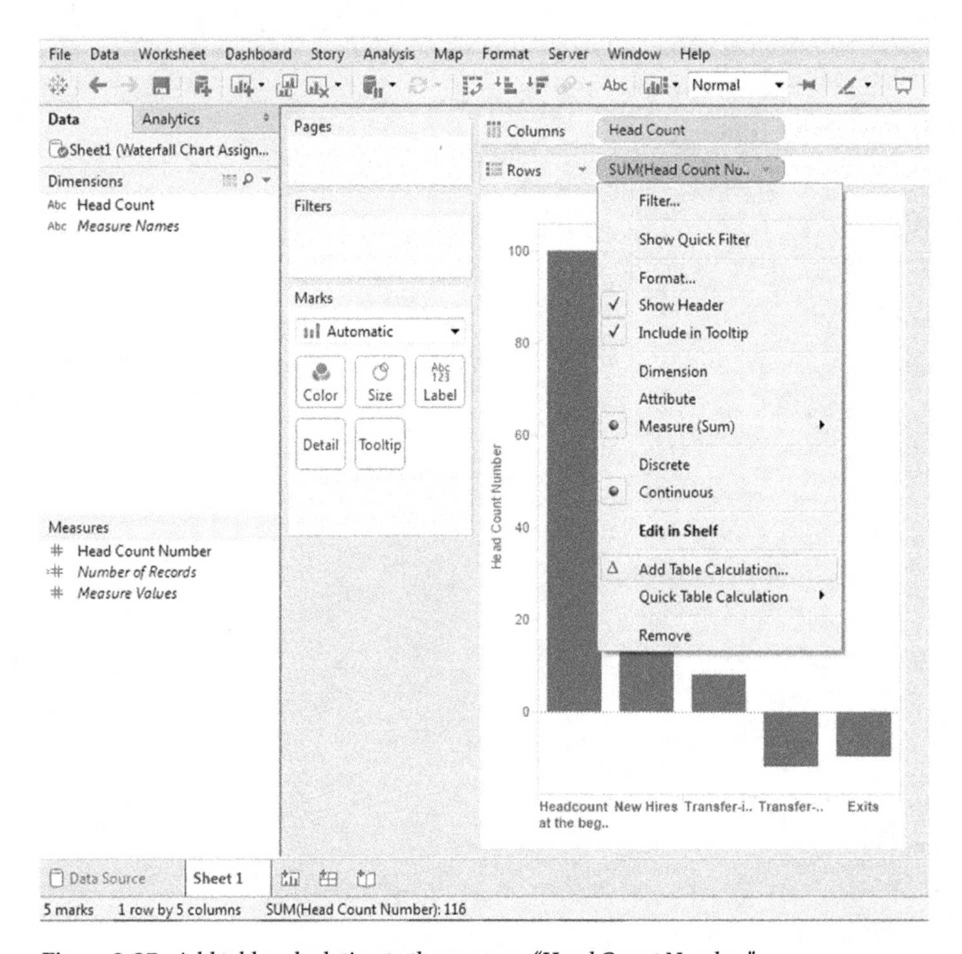

Figure 9-27. *Add table calculation to the measure "Head Count Number"*

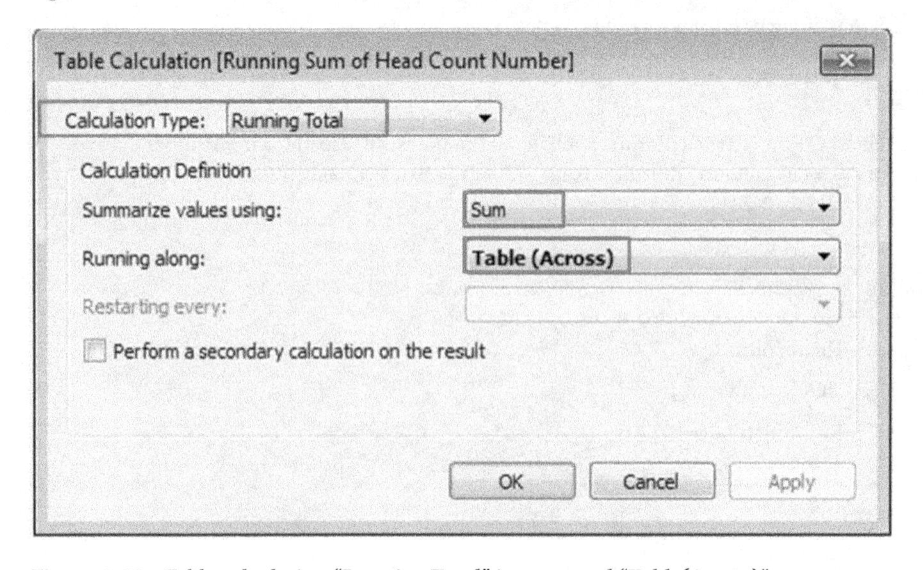

Figure 9-28. *Table calculation "Running Total" is computed "Table(Across)"*

The output is shown in Fig. 9-29.

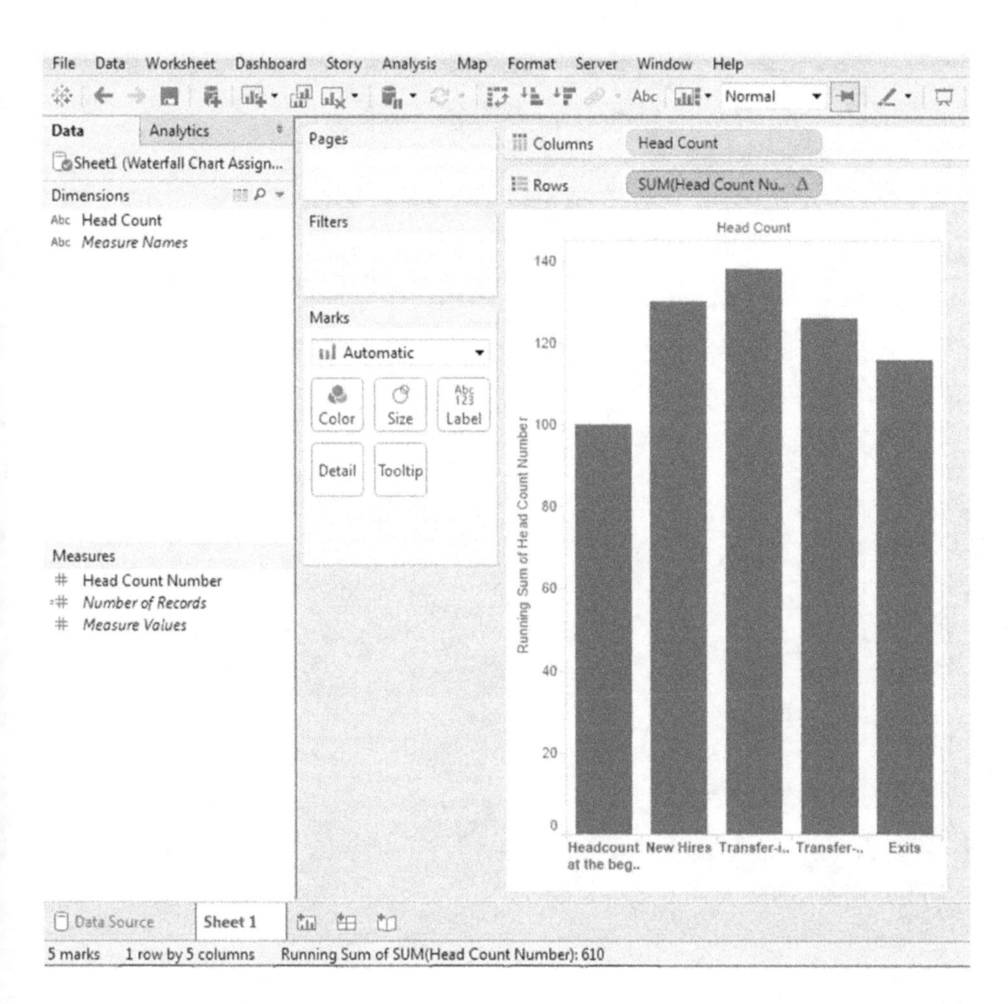

Figure 9-29. *Output after the table calculation has been applied to the measure "Head Count Number" on the rows shelf*

9.1.1.2.6 Step 5

Change the marks type from "Automatic" to "Gantt Bar" (Shown in Fig. 9-30).

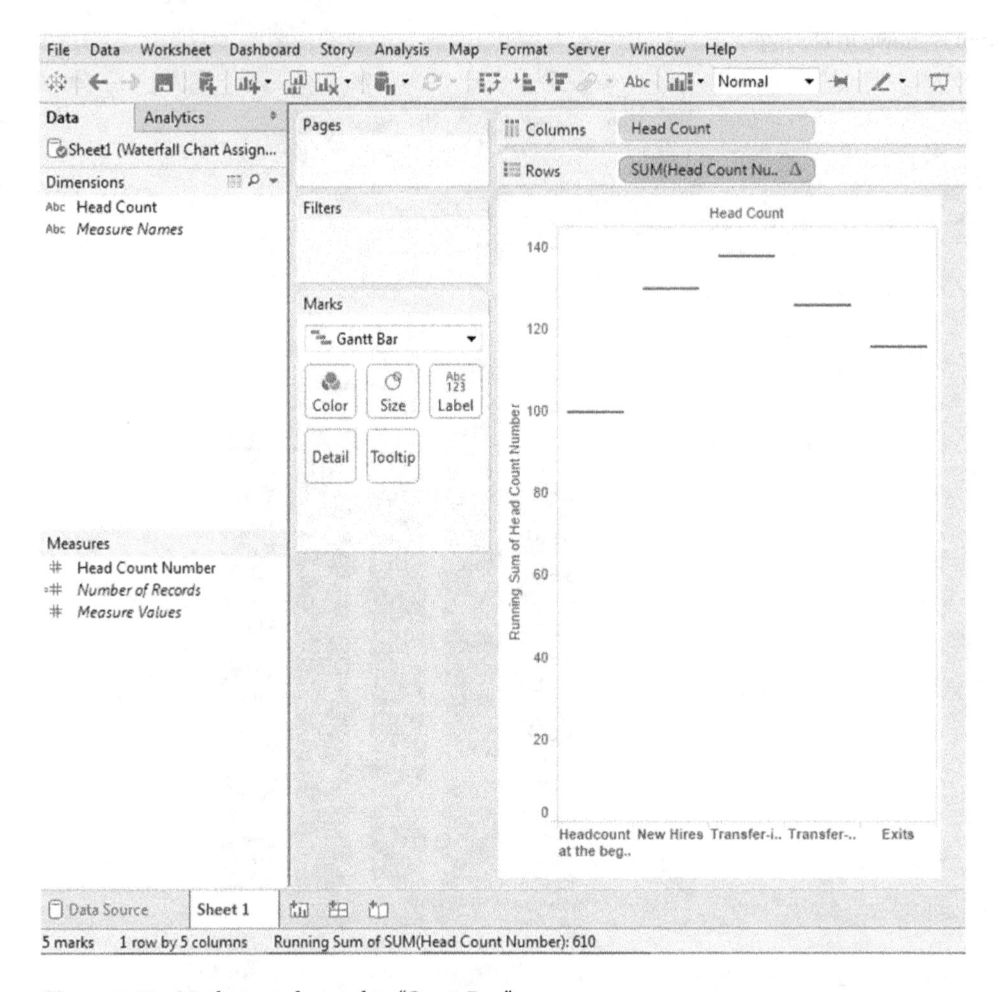

Figure 9-30. *Marks type changed to "Gantt Bar"*

9.1.1.2.7 Step 6

Create a calculated field "- Head Count Number" (Shown in Fig. 9-31):

- Head Count Number

```
-[Head Count Number]
```

Sheets Affected ▾ Apply OK

Figure 9-31. *Calculated field "- Head Count Number" created*

9.1.1.2.8 Step 7

Drag the calculated field "- Head Count Number" to "Size" on the marks card (Shown in Fig. 9-32).

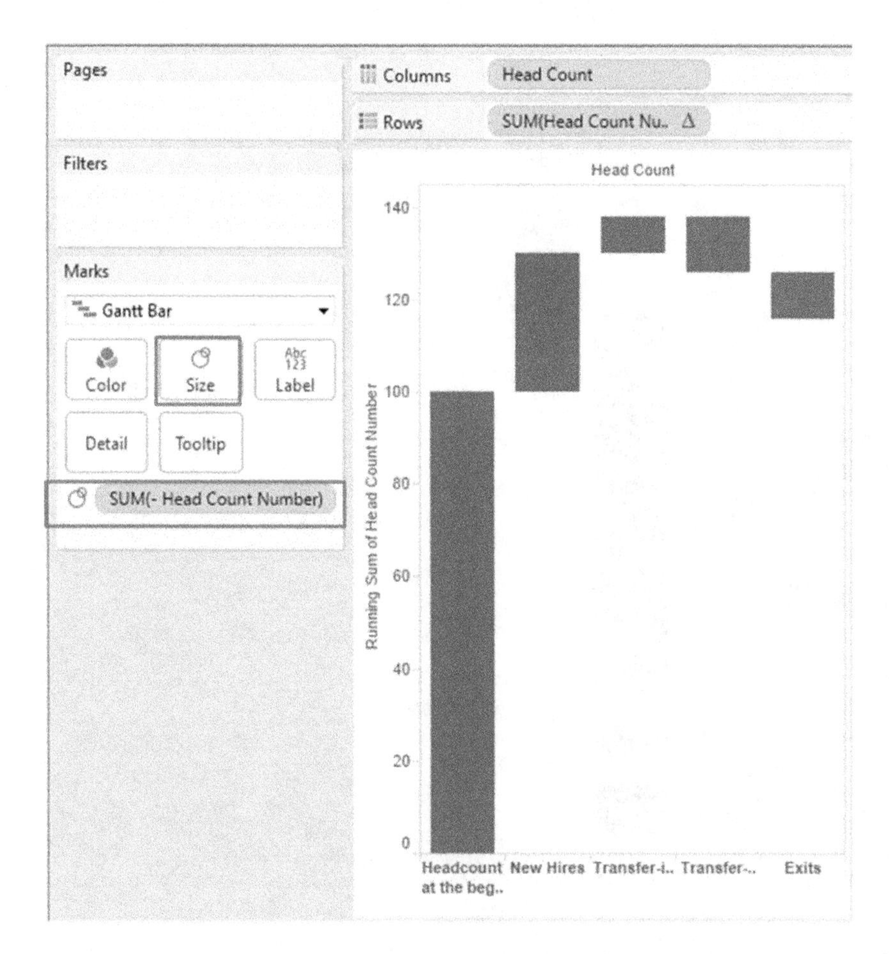

Figure 9-32. *Calculated field "- Head Count Number" placed on "Size" on the marks card*

9.1.1.2.9 Step 8

Drag the measure "Head Count Number" to "Color" on the marks card (Shown in Fig. 9-33).

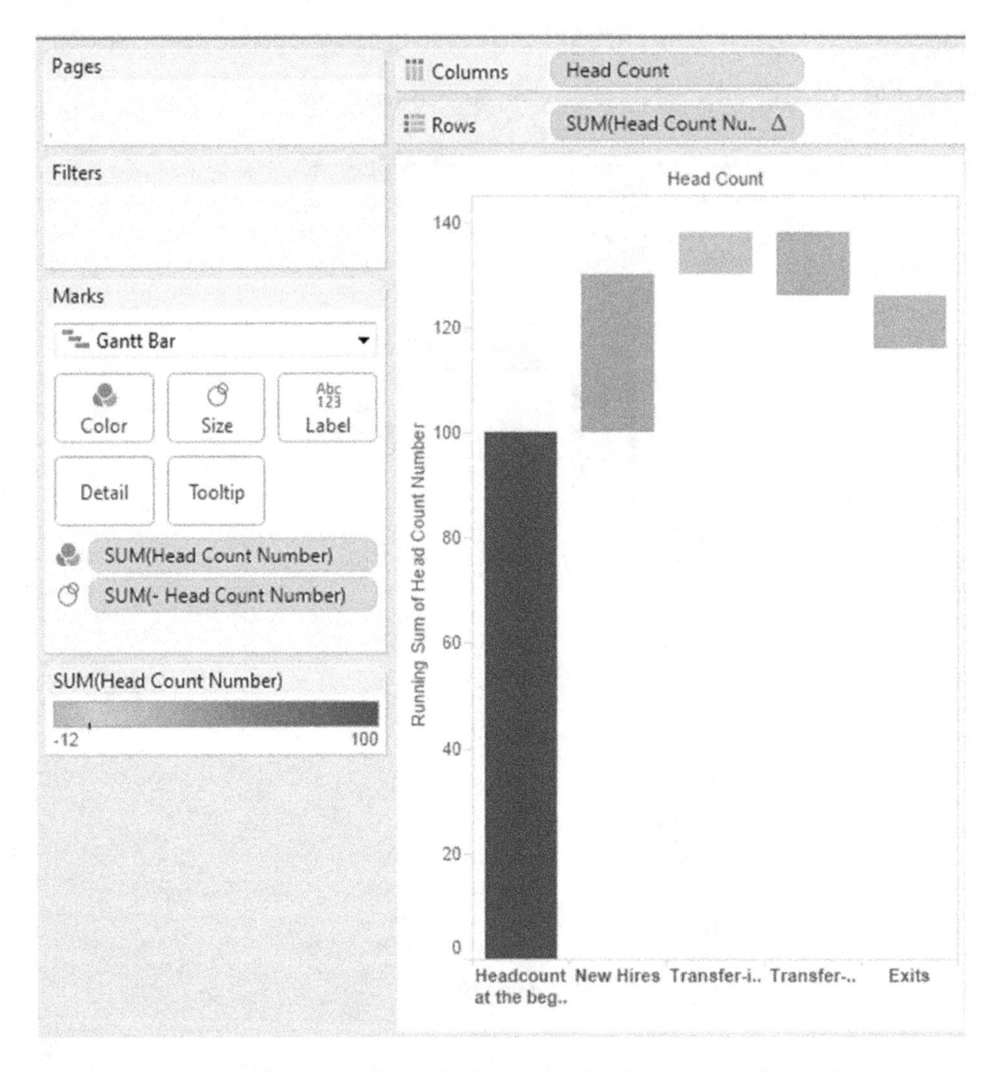

Figure 9-33. *Measure "Head Count Number" placed on "Color" on the marks card*

9.1.1.2.10 Step 9

Change the "Stepped Color" to 2 (Shown in Fig. 9-34 and Fig. 9-35).

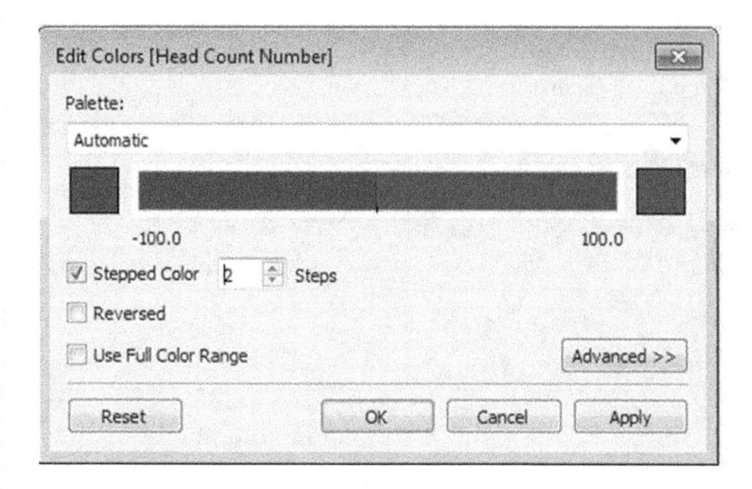

Figure 9-34. *Stepped Color set to 2*

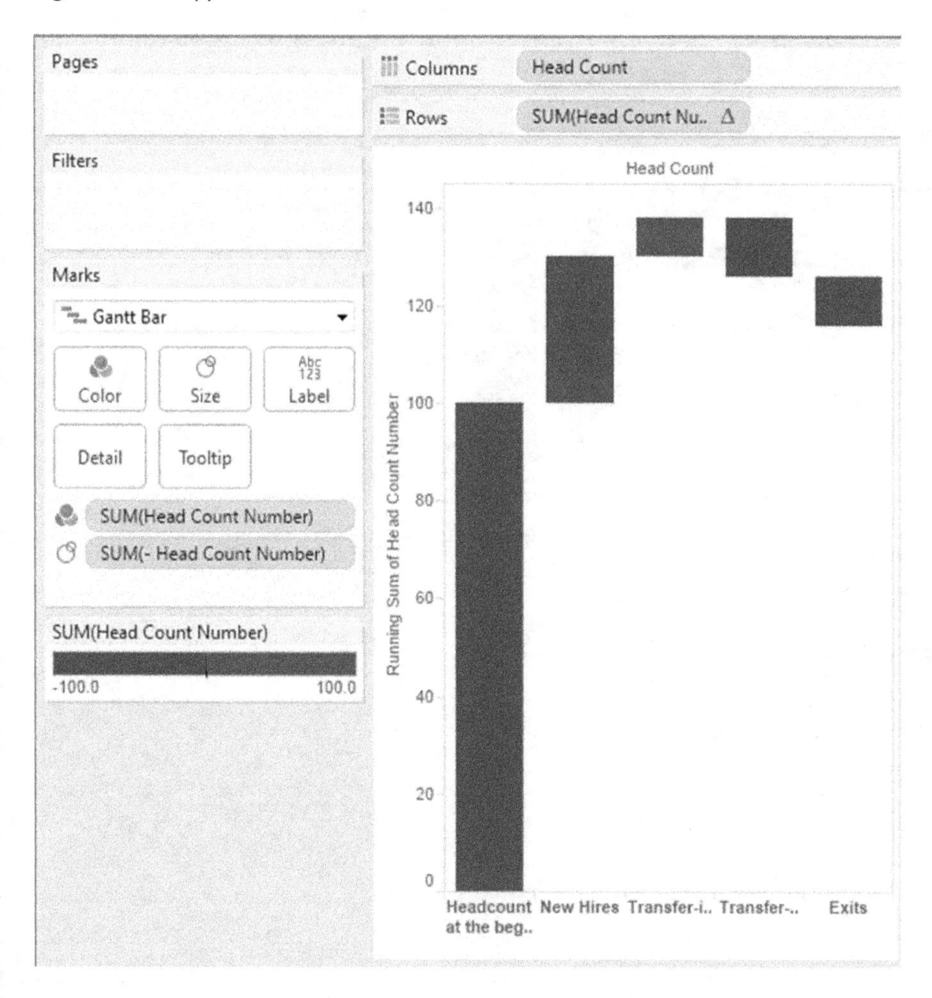

Figure 9-35. *Output after stepped color is set to 2*

9.1.1.2.11 Step 10

Select Analysis ➤ Totals ➤ Show Row Grand Totals (Shown in Fig. 9-36).

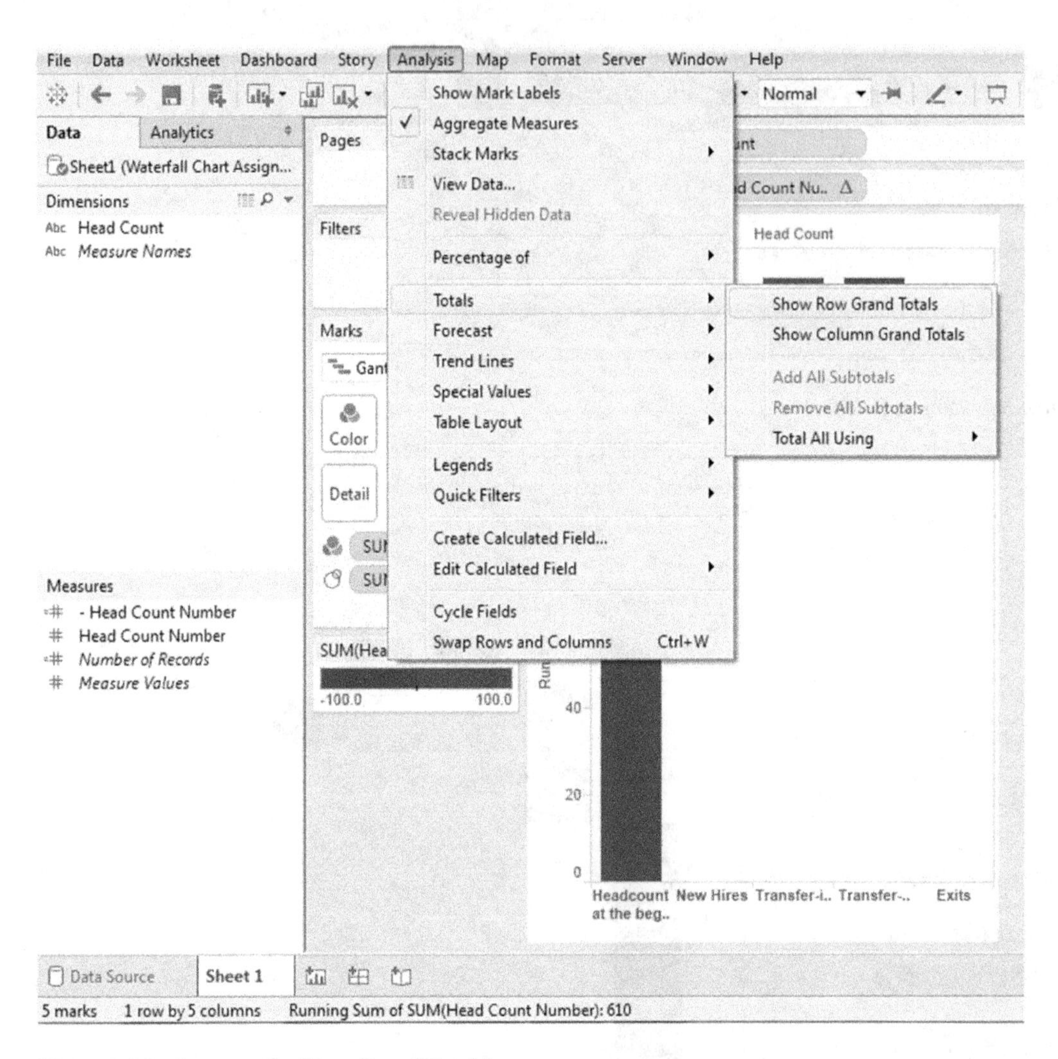

Figure 9-36. *Compute the "Row Grand Totals"*

The output (Shown in Fig. 9-37):

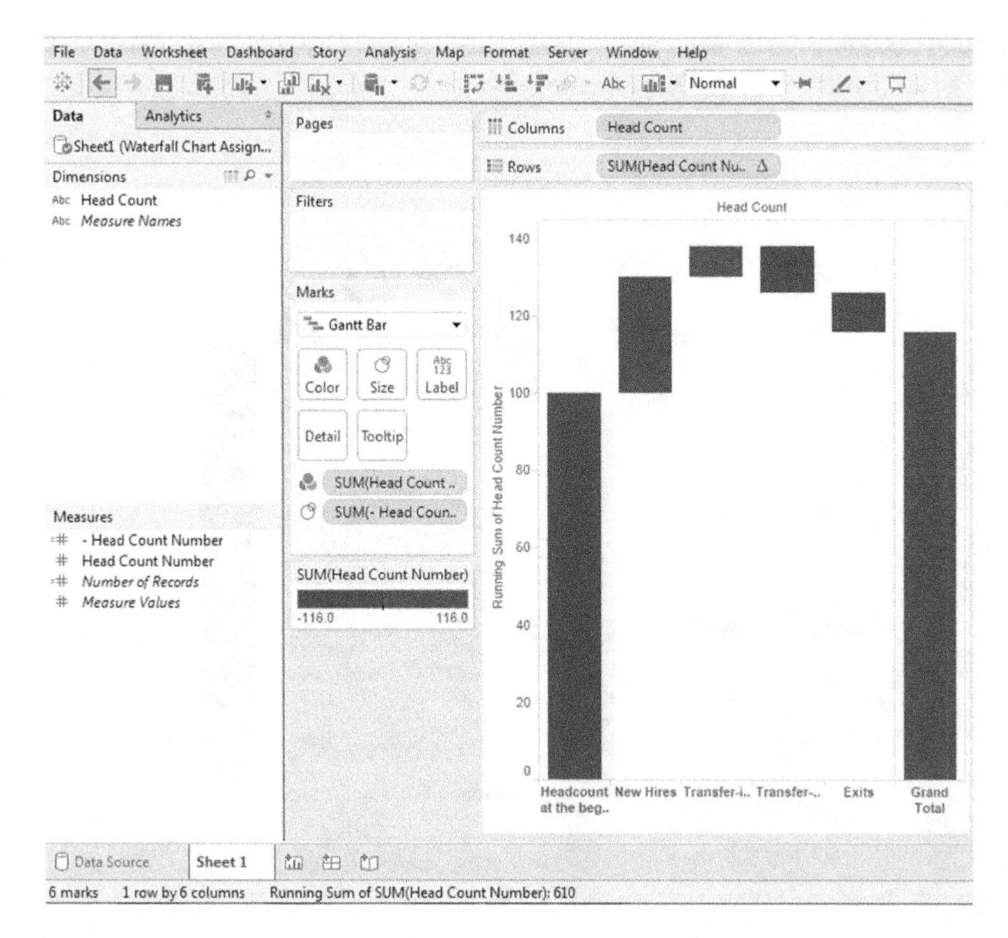

Figure 9-37. *Output after computing the "Row Grand Totals"*

Finally drag the measure "Head Count Number" to "Label" on the marks card (Shown in Fig. 9-38).

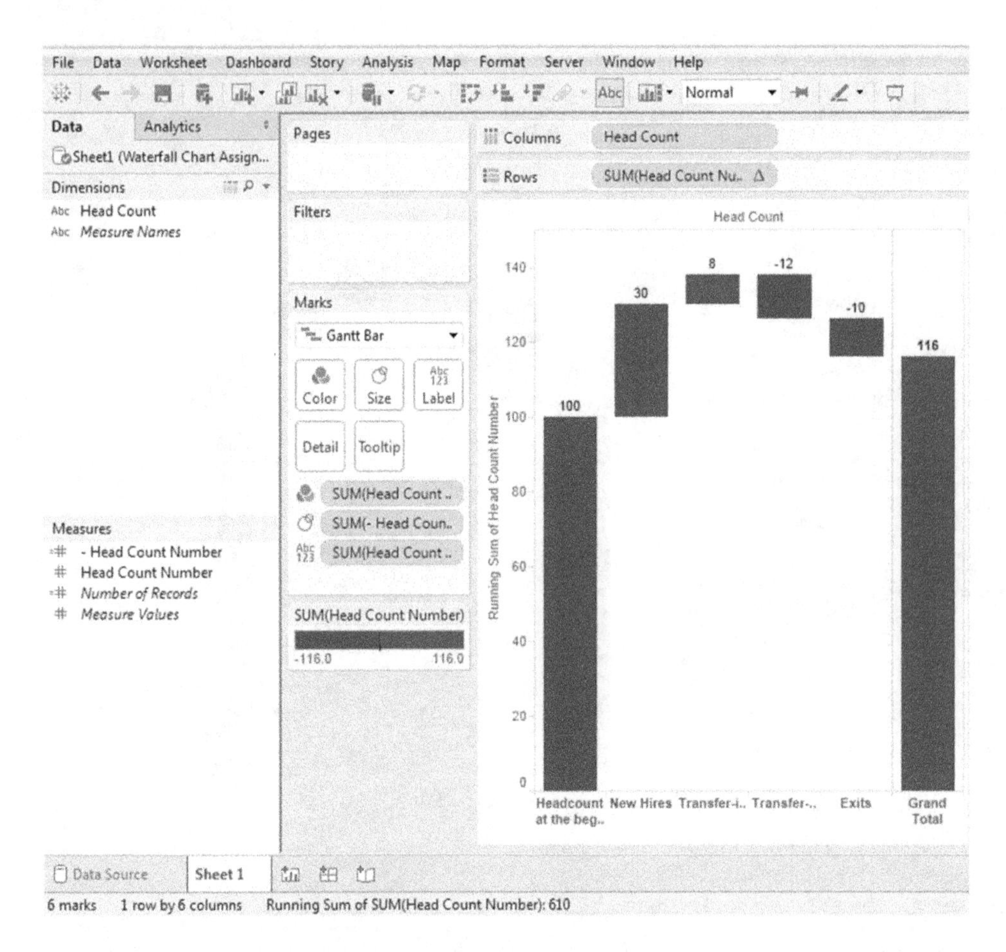

Figure 9-38. *Final output after placing the measure "Head Count Number" on "Label" on the marks card*

9.2 Bump charts

Bump charts got their name from "bumps race", a term used to refer to a boat race where each boat tries to "bump" the one in front and move up the chart. Bump charts have become quite common of late and are typically used to represent changes in the position of a given number of competing entities over a fixed duration.

9.2.1 Where to use a bump chart?

Use a bump chart when it is required to compare **two dimensions** against each other using **one measure** value. Bump charts help to show the changes in the rank of a value usually over a time dimension or it could be any other dimension that is relevant to the analysis.

To create a bump chart, use at least two dimensions with zero or more measures. Refer to Fig. 9-39 for a sample bump chart.

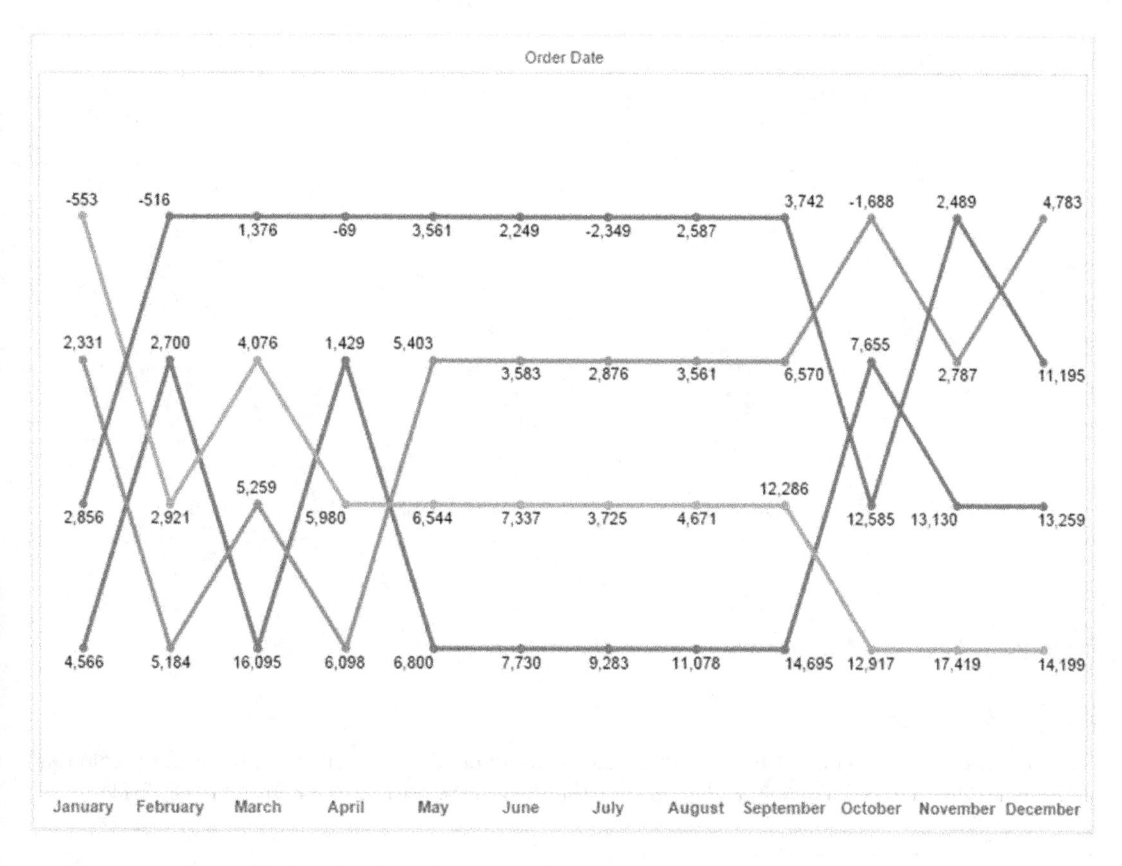

Figure 9-39. *A sample "Bump Chart"*

This chart is also called a "slope graph".

Now let us go through some demos. The demos will provide step-by-step instructions to create a bump chart.

9.2.1.1 Demo 1

Steps to create a bump chart.

9.2.1.1.1 Step 1

Read in the data from "Sample - Superstore.xls" into Tableau (Shown in Fig. 9-40).

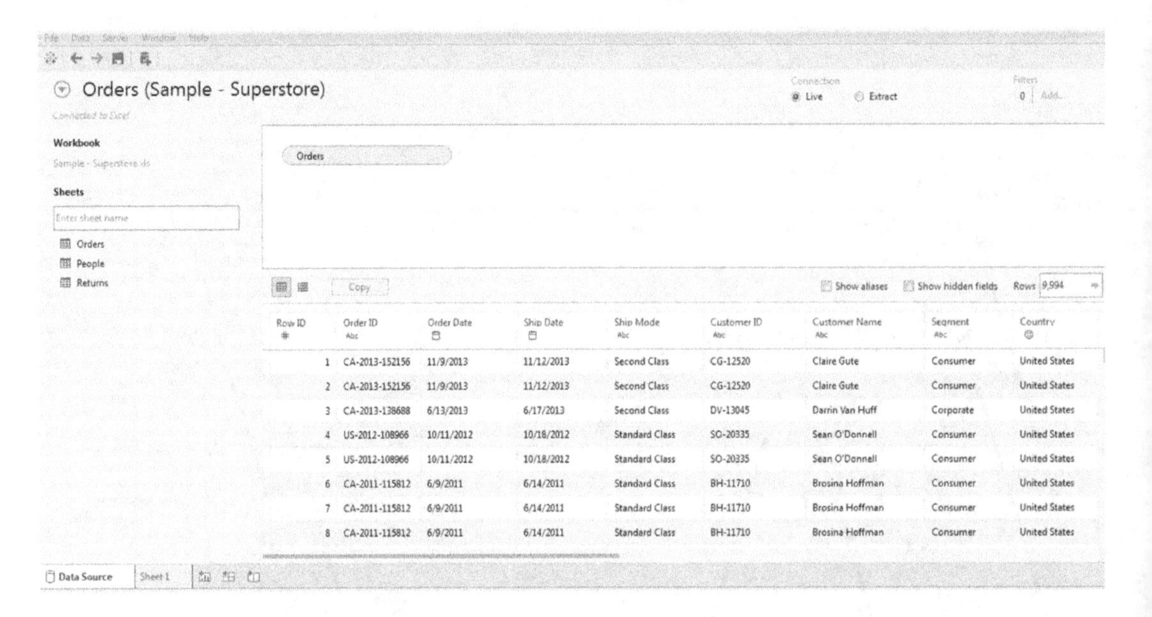

Figure 9-40. *Data from "Sample - Superstore.xls" read into Tableau*

9.2.1.1.2 Step 2

Drag the dimension "Order Date" from the dimensions area under the data pane and drop it on the columns shelf. The date should be "Discrete" and the granularity should be set to "Month" (Shown in Fig. 9-41).

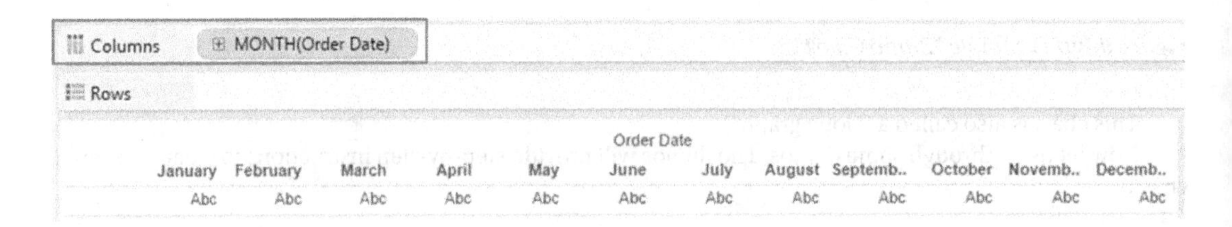

Figure 9-41. *Dimension, "Order Date" placed on the columns shelf*

Drag the dimension "Region" from the dimensions area under the data pane and drop it on "Color" on the marks card. Data is available for four regions: "Central", "East", "South" and "West" (Shown in Fig. 9-42).

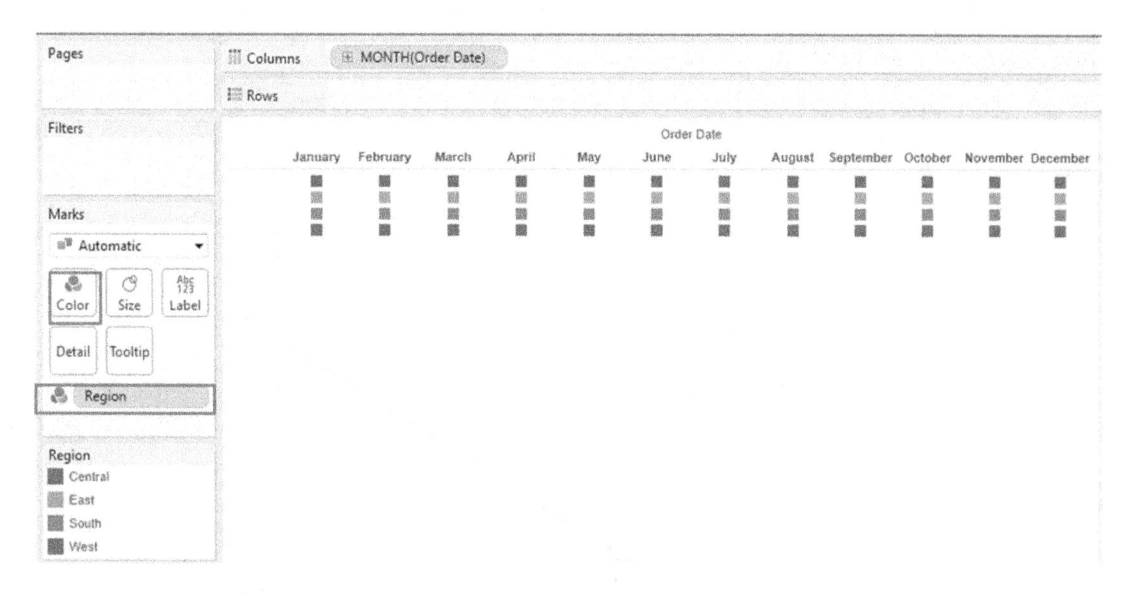

Figure 9-42. *Dimension, "Region" placed on "Color" on the marks card*

9.2.1.1.3 Step 3

Create a calculated field "RankRegion" (Shown in Fig. 9-43).

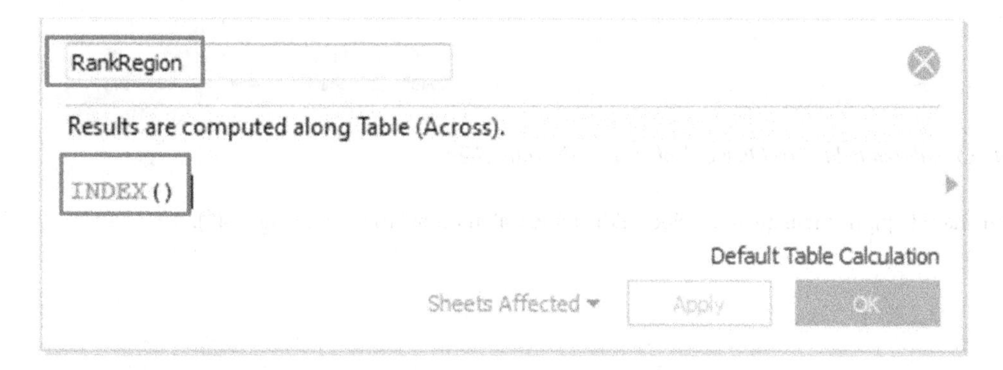

Figure 9-43. *Calculated field "RankRegion" is created*

Index function returns the index position of the current row in the partition. Example: For the first row in the partition, the index function will return 1.

Click on "Apply" and then OK.

Drag the newly created calculated field, "RankRegion" and drop it on the rows shelf (Shown in Fig. 9-44).

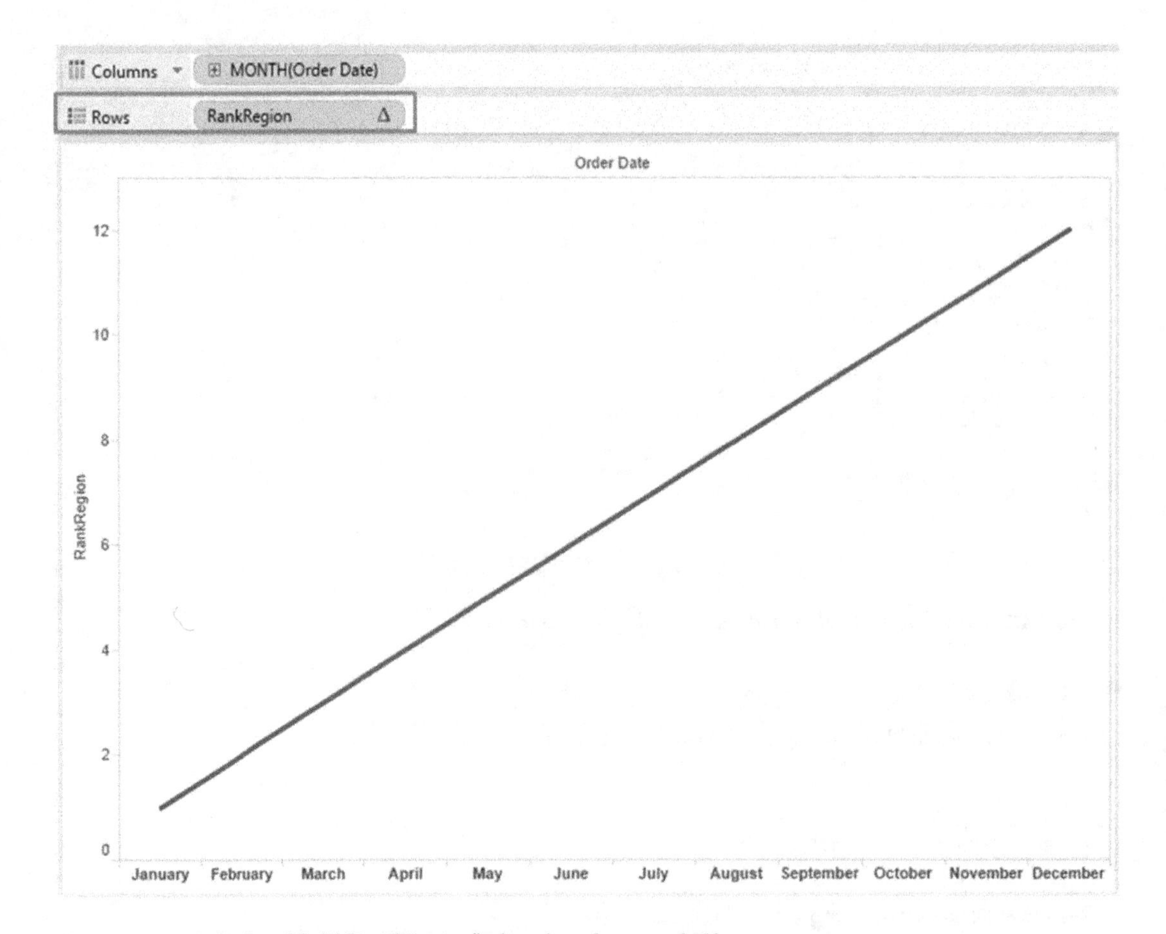

Figure 9-44. *Calculated field "RankRegion" placed on the rows shelf*

Click on "RankRegion" measure and select "Edit Table Calculation" (Shown in Fig. 9-45).

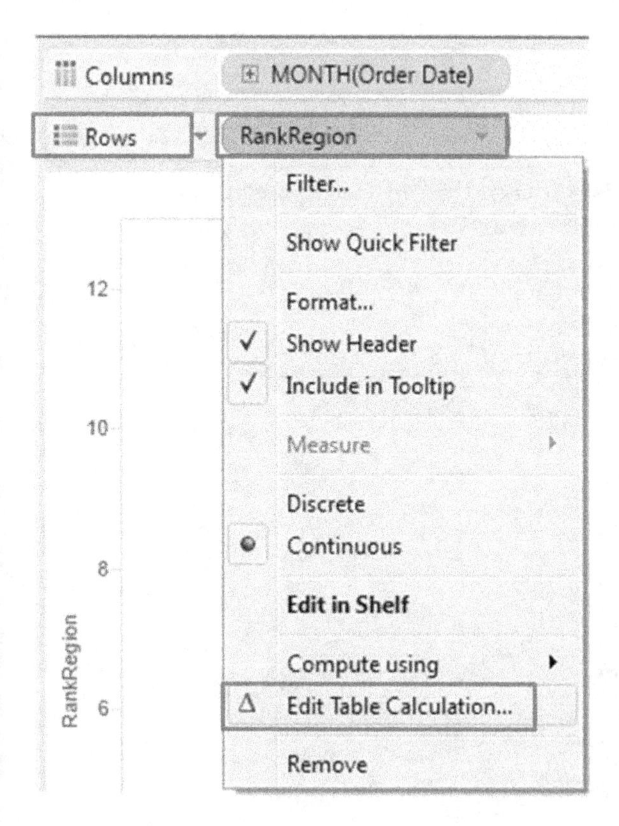

Figure 9-45. *Edit table calculation option for the calculated field "RankRegion"*

Table calculation [RankRegion] dialog box appears. Select the drop down next to "Compute using" and select "Advanced" (Shown in Fig. 9-46).

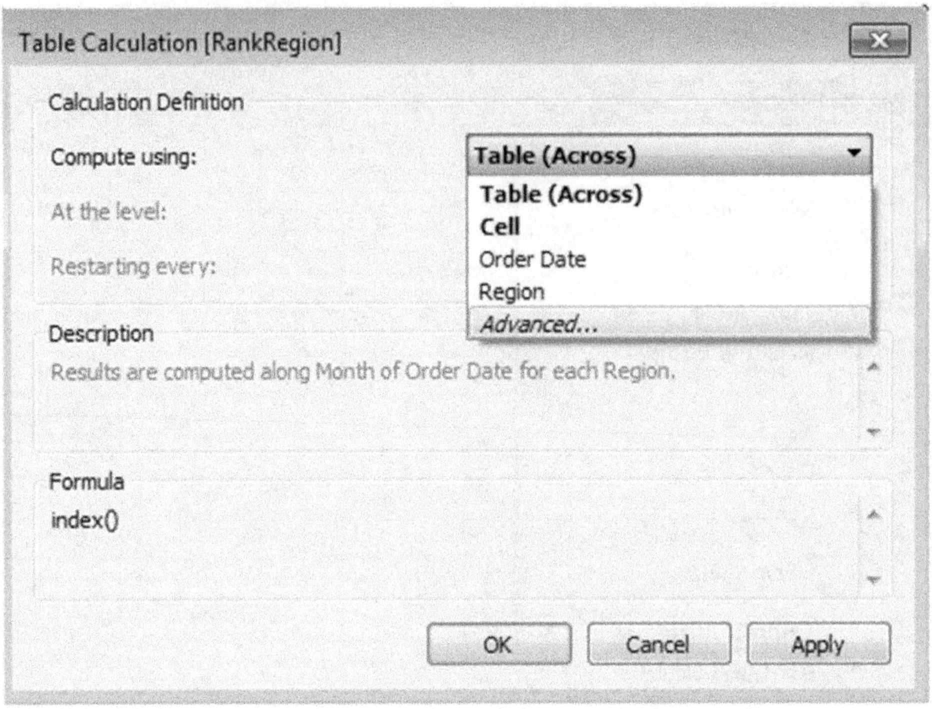

Figure 9-46. *Table calculation dialog box for the calculated field "RankRegion"*

By selecting the "Advanced" option, "Advanced dialog box" appears. In the "Advanced" dialog box, do the settings as follows and click on OK button (Shown in Fig. 9-47).

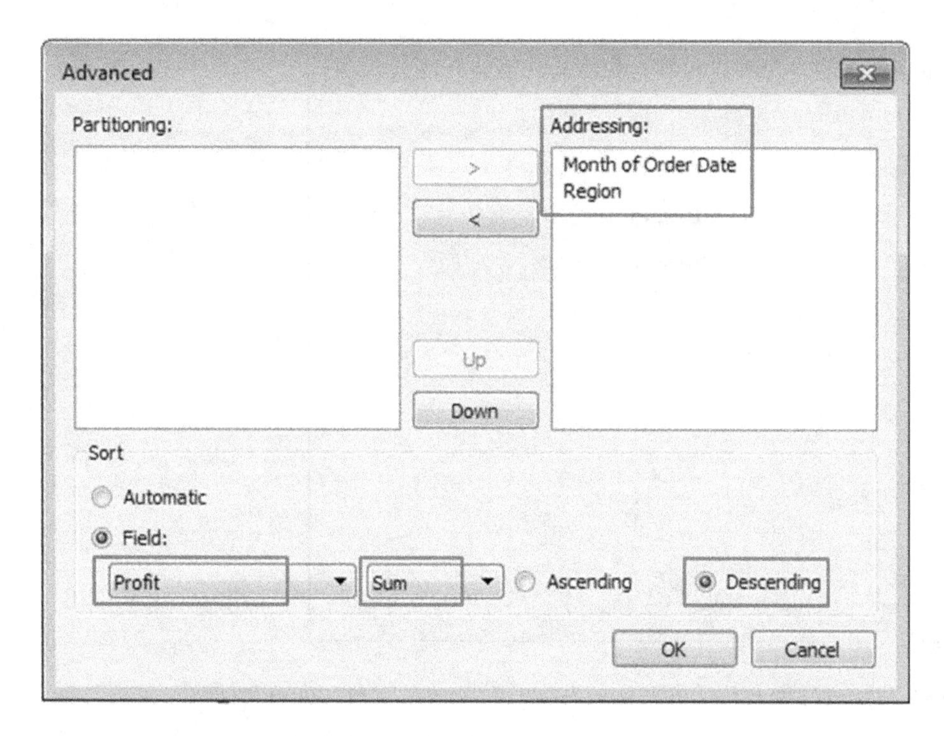

Figure 9-47. *Settings in the "Advanced" dialog box*

For the table calculation [RankRegion], do the settings for the remaining fields as shown and click on the OK button (Shown in Fig. 9-48).

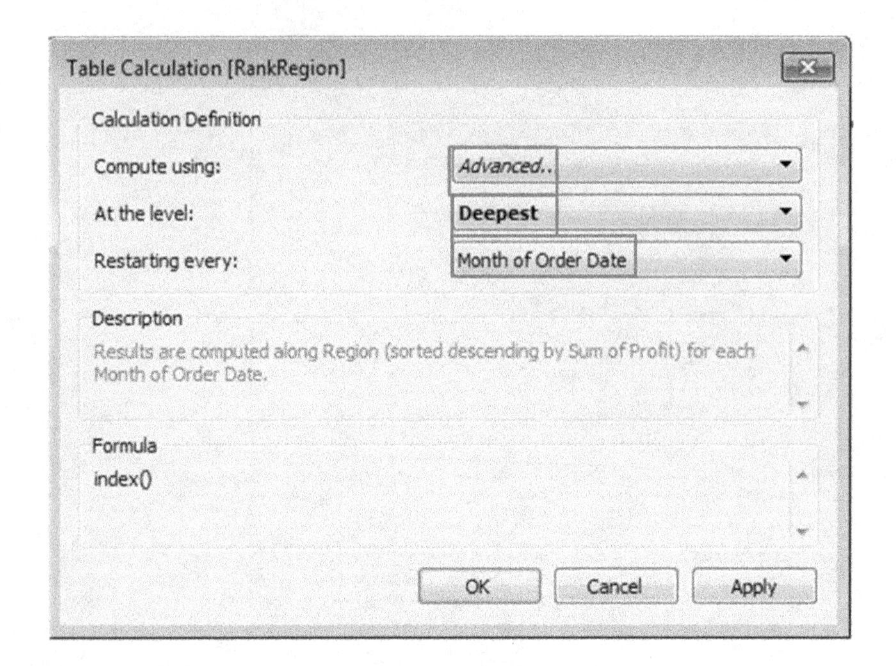

Figure 9-48. *Perform other settings in the "Table Calculation" dialog box for the calculated field "RankRegion"*

Drag the measure "Profit" available under the data pane and drop it on "Label" on the marks card. The final output (Shown in Fig. 9-49):

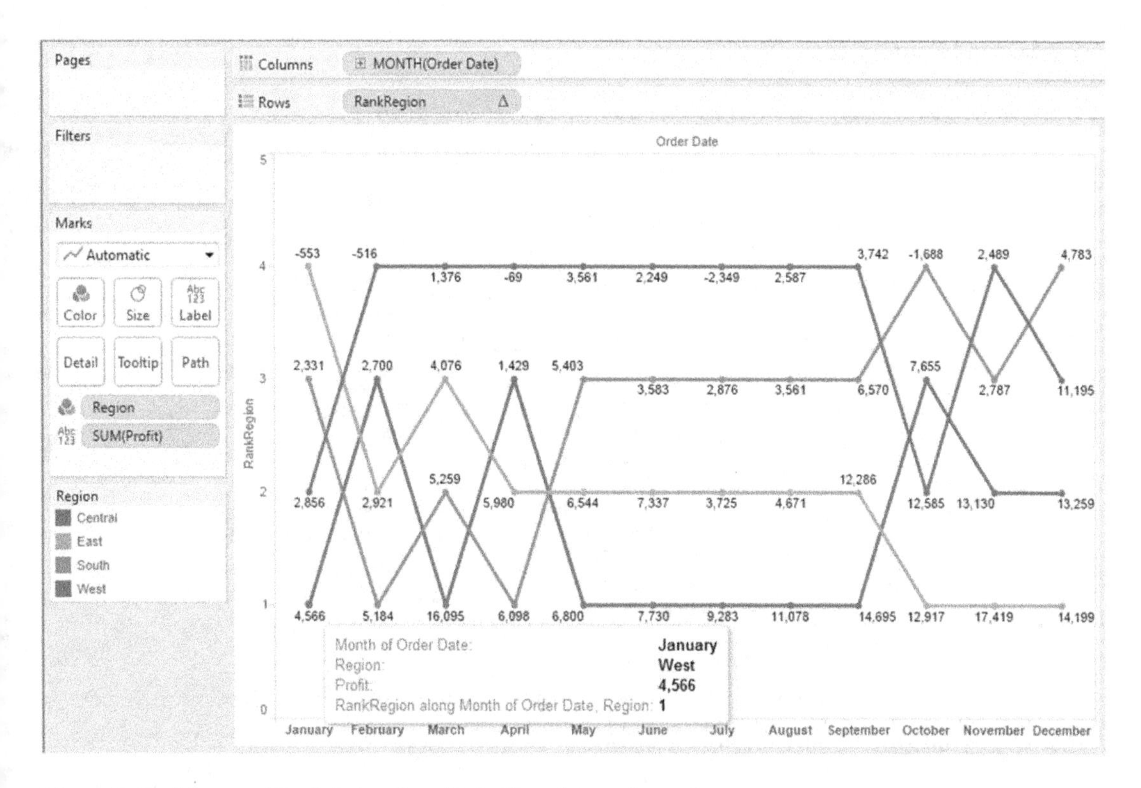

__Figure 9-49.__ Measure "Profit" placed on "Label" on the marks card

Interpretation of the bump chart:

The line graphs are arranged along the dimension "Region" (sorted descending by sum of "Profit") for each month of "Order Date".

The "West" region has the maximum "Profit" for the month of "January" followed by the "Central", "South" and "East".

9.2.1.2 Demo 2

Read the data from "Bump.xls" into Tableau and pull the following bump chart (Shown in Fig. 9-50).

Expected output:

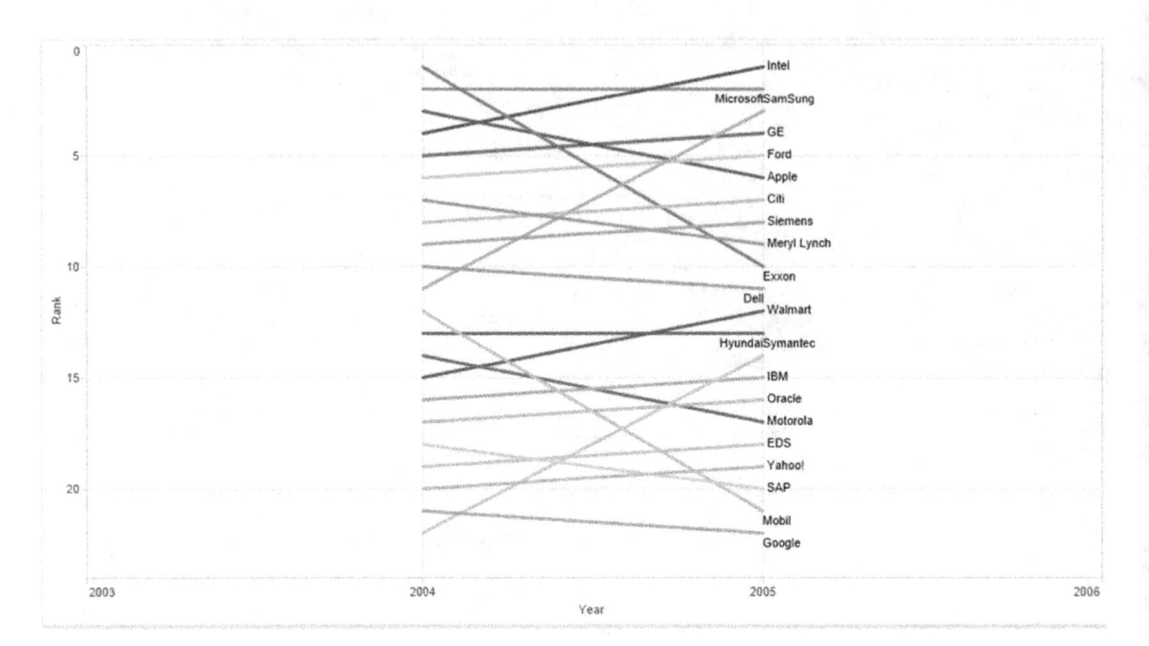

Figure 9-50. *Bump chart - Demo 2 - expected output*

9.2.1.2.1 Step 1

Read the data from "Bump.xls" into Tableau (Shown in Fig. 9-51).

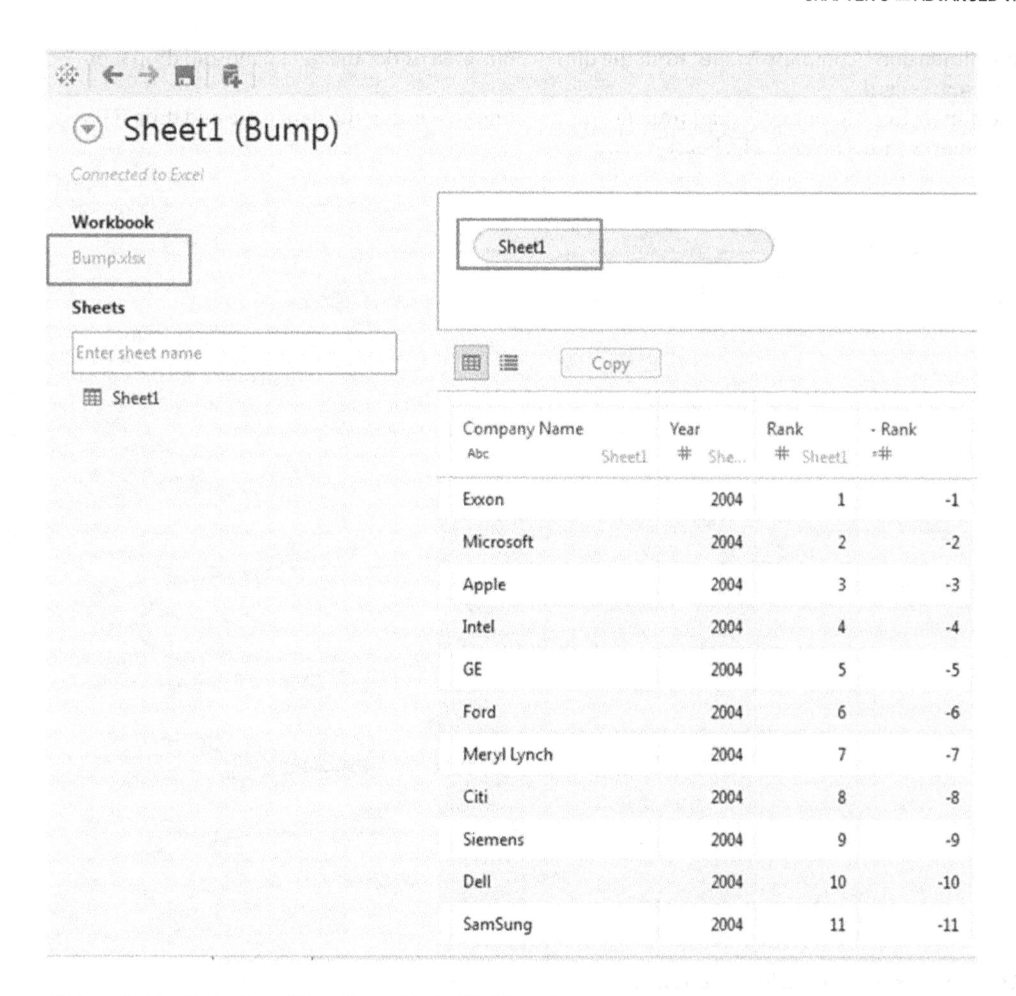

Figure 9-51. *Data from "Excel" read into "Tableau"*

9.2.1.2.2 Step 2

Drag the dimension "Year" from the dimensions area under the data pane and drop it on the columns shelf. Drag the measure "Rank" from the measures area under the data pane and drop it on the rows shelf (Refer to Table 9-7).

Table 9-7. *Activities to perform*

Columns shelf	Year
Rows shelf	Rank. Default aggregation : Sum

Drag the dimension "Company Name" from the dimensions area under the data pane and drop it on "Color" on the marks card.

Drag the dimension "Company Name" from the dimensions area under the data pane and drop it on "Label" on the marks card. (Refer to Fig. 9-52)

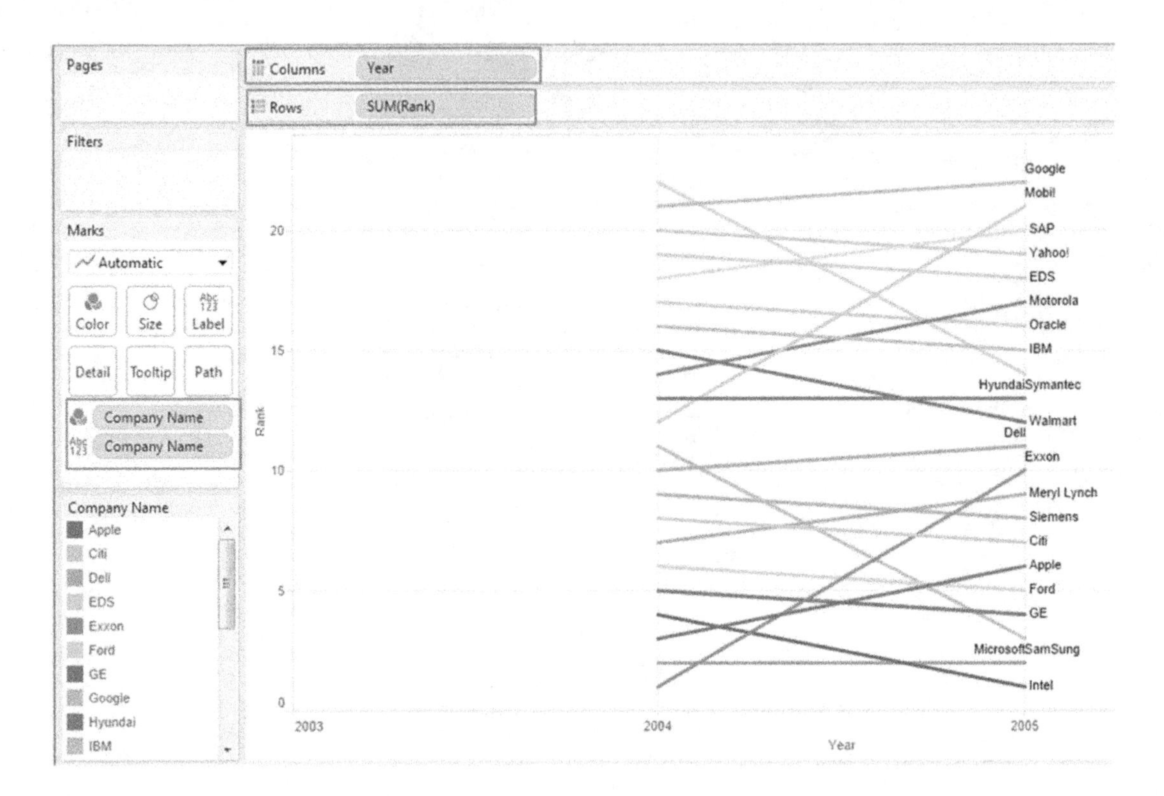

Figure 9-52. Activities stated in Table 9-7 executed

9.2.1.2.3 Step 3

Select the "Rank" axis. Right click and select "Edit Axis". Check the checkbox next to "Reversed" (Shown in Fig. 9-53).

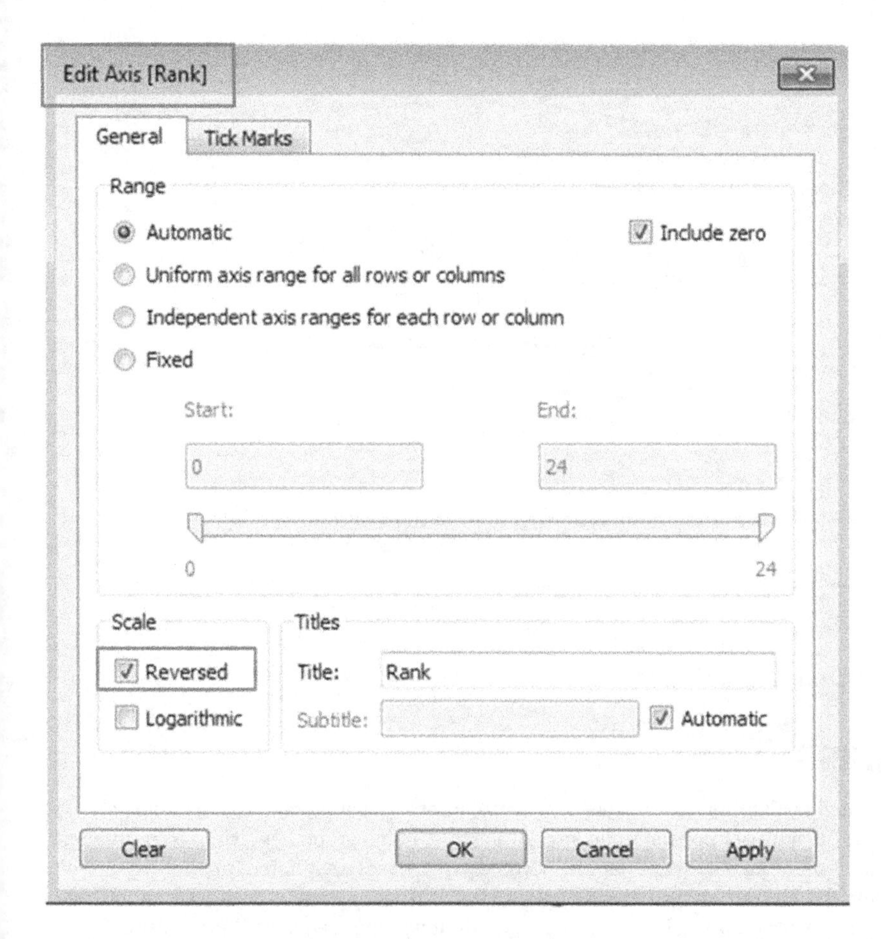

Figure 9-53. *Set the "Rank Axis" to "Reversed"*

The final output (Shown in Fig. 9-54):

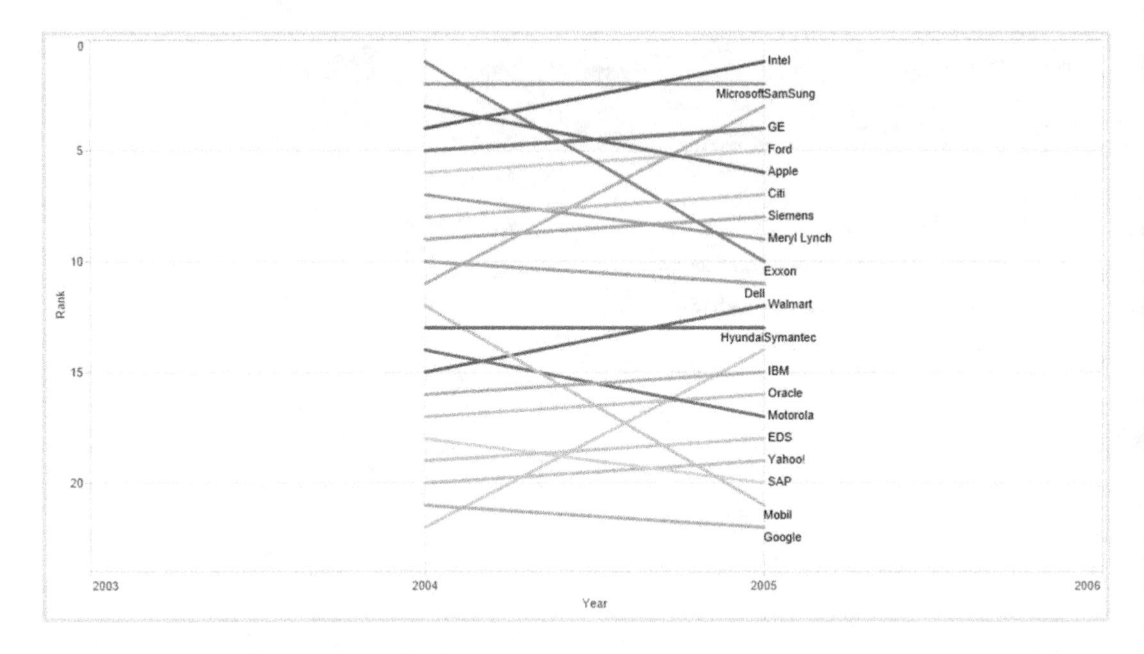

Figure 9-54. Bump chart - Demo 2 - final output

9.3 Bullet graph

The bullet graph was invented by Stephen Few in 2005 as a replacement for gauges and radial meters on dashboards. They convey quite a bit of information in a small space. A bullet graph has a single quantitative performance measure that is compared against a target. However, we can set up different qualitative measure ranges to measure the KPI performance, to decipher whether the measure is on-track or not. Often they show the primary quantitative measure in the context of qualitative ranges such as "Very satisfied", "satisfied", "neither satisfied nor dissatisfied", "dissatisfied", and "very dissatisfied".

What is wrong with a gauge chart or radial meters?

- They lack context
- They waste space
- They are far too simplistic (dangerously so!)

On the other hand, bullet graphs leverage our perceptual and cognitive predispositions. Humans are better at perceiving and comparing lengths or parallel positions rather than angles.

An important question here is if there is a need to display multiple measures, should one go for multiple bullet graphs. The answer is "NO" if all the measures share the same quantitative scale and qualitative ranges. In such a case, a single bar chart will suffice. This will help to keep all the measures together and help to compare them.

Refer to Fig. 9-55 for a simple bullet graph.

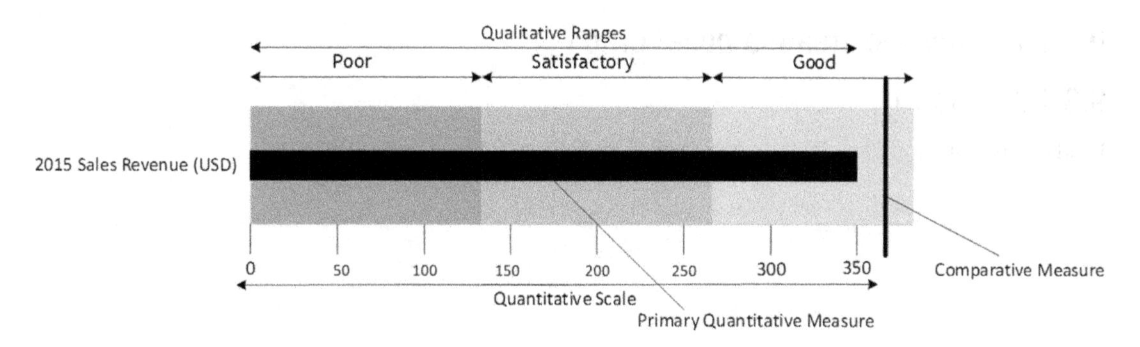

Figure 9-55. *A sample bullet graph*

9.3.1 Demo 1

Objective: To compare the "Sales" of 2014 against the "Sales" of 2013.

Input: "Sample - Superstore.xls"

Expected output: Shown in Fig. 9-56.

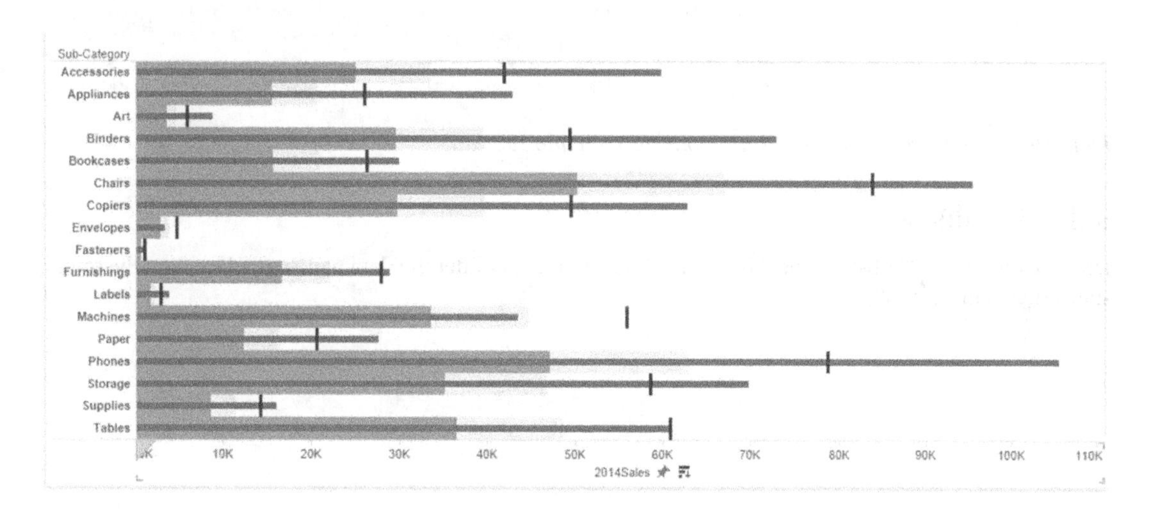

Figure 9-56. *Bullet graph - Demo 1 - expected output*

9.3.1.1 Steps to create a bullet graph

9.3.1.2 Step 1

Read in the data from "Sample - Superstore.xls" into Tableau (Shown in Fig. 9-57).

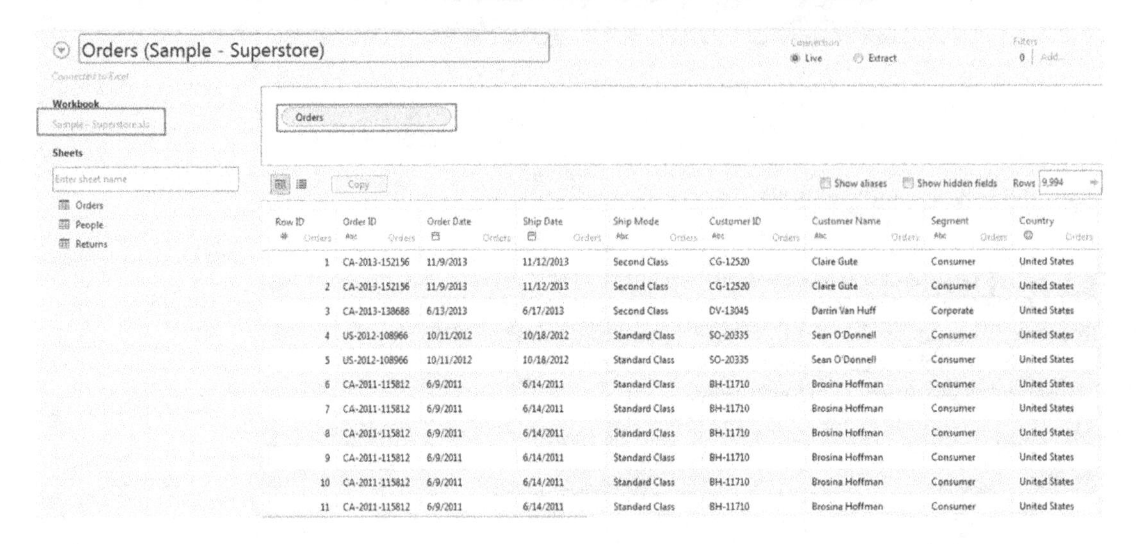

Figure 9-57. *Data from "Sample - Superstore.xls" read into Tableau*

9.3.1.3 Step 2

Drag the dimension "Sub-Category" from the dimensions area under the data pane and place it on the rows shelf (Shown in Fig. 9-58).

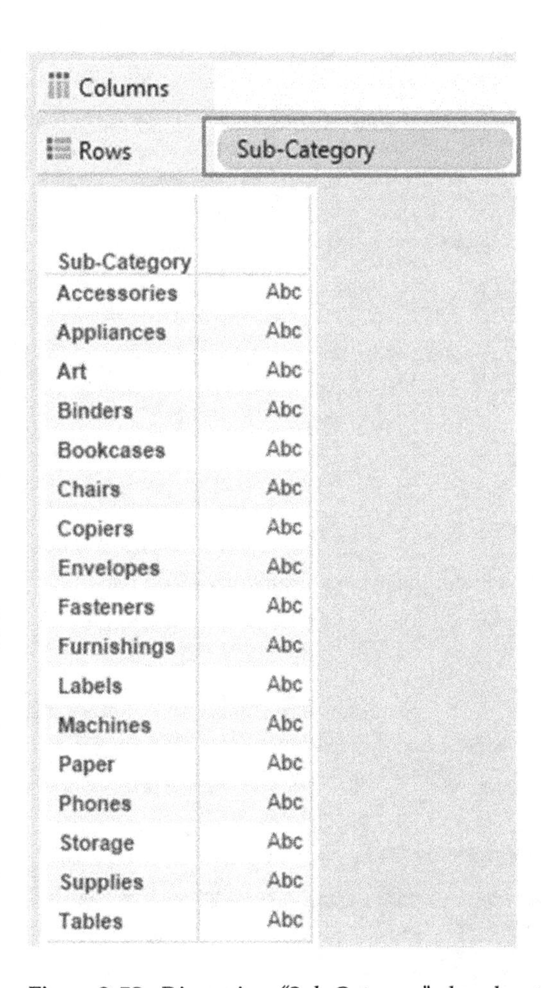

Figure 9-58. *Dimension, "Sub-Category" placed on the rows shelf*

9.3.1.4 Step 3

Create a calculated field "2013Sales" (Shown in Fig. 9-59).

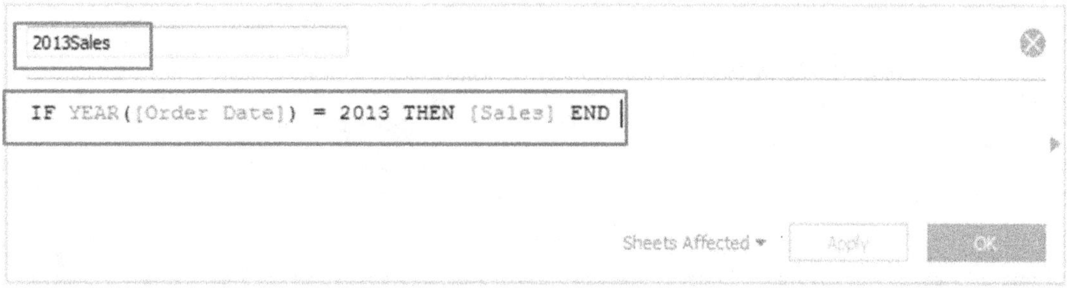

2013Sales

```
IF YEAR([Order Date]) = 2013 THEN [Sales] END
```

Sheets Affected ▾ Apply OK

Figure 9-59. *Calculated field "2013Sales" being created*

Create a calculated field "2014Sales" (Shown in Fig. 9-60).

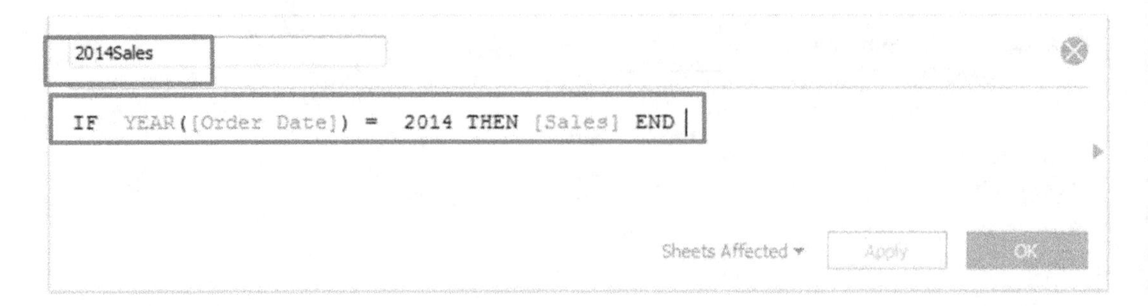

Figure 9-60. *Calculated field "2014Sales" being created*

9.3.1.5 Step 4

Drag the calculated field "2014Sales" from the measures area under the data pane and place it on the columns shelf (Shown in Fig. 9-61).

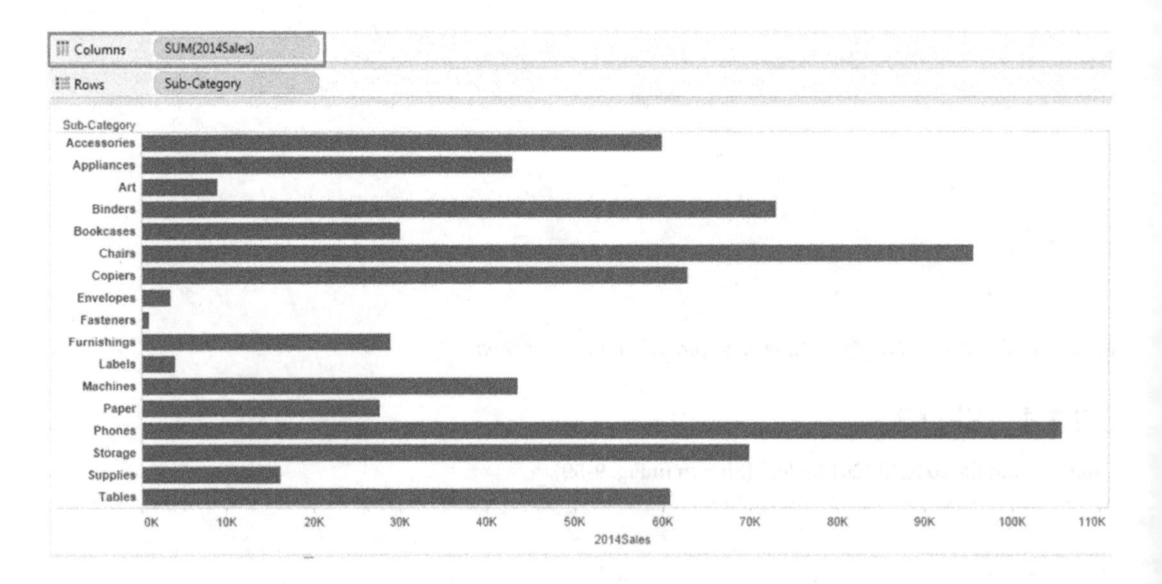

Figure 9-61. *Calculated field "2014Sales" placed on the columns shelf*

9.3.1.6 Step 5

Drag the calculated field "2013Sales" from the measures area under the data pane and place it on "Detail" on the marks card (Shown in Fig. 9-62).

Figure 9-62. *Calculated field "2013Sales" placed on "Detail" on the marks card*

9.3.1.7 Step 6

Go to the analytics pane. Drag the reference line to the view / worksheet. Perform the settings as shown in Fig. 9-63.

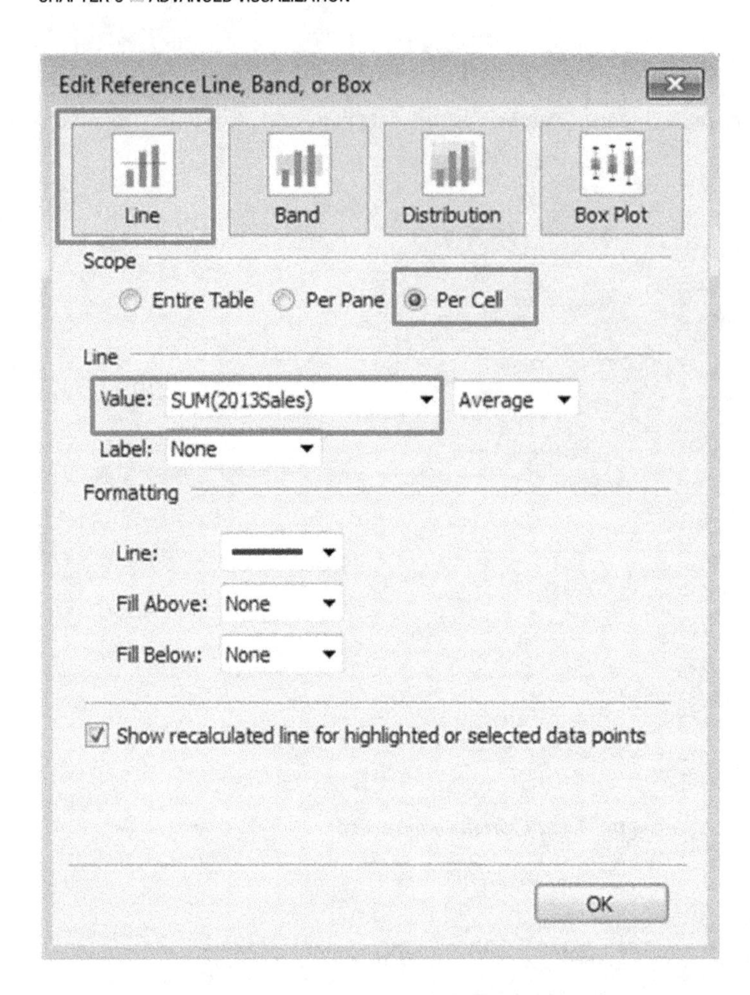

Figure 9-63. *Perform the settings in "Edit Reference Line, Band, or Box" dialog box*

Ensure that the value is set to "SUM(2013Sales)". This was essentially the reason for placing "2013Sales" on "Detail" on the marks card.

The output after adding the "Reference Line" "Per Cell" (Shown in Fig. 9-64).

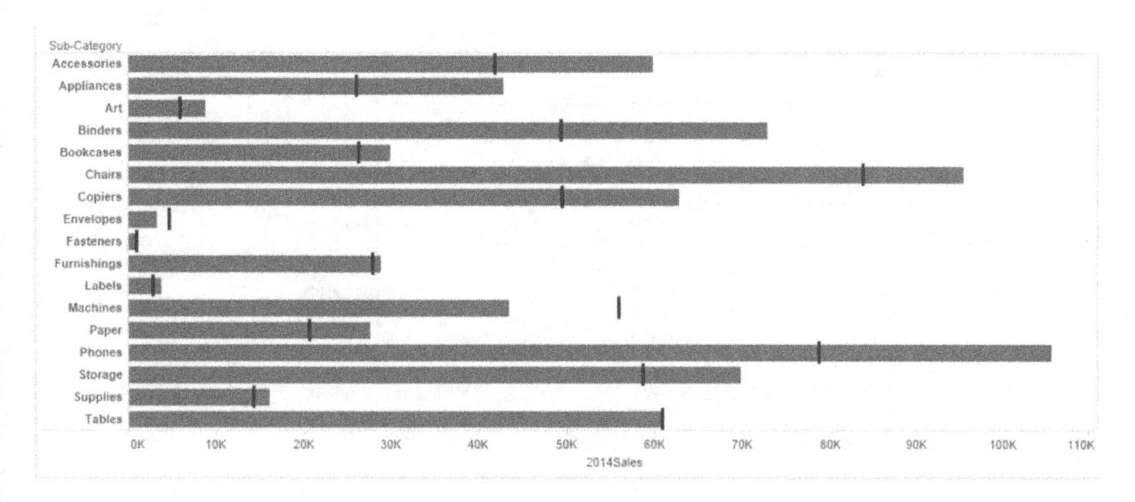

Figure 9-64. *Output after adding the reference line*

9.3.1.8 Step 7

Drag the "Distribution Band" from the analytics pane and place it on the worksheet / view. Perform the settings as shown in Fig. 9-65.

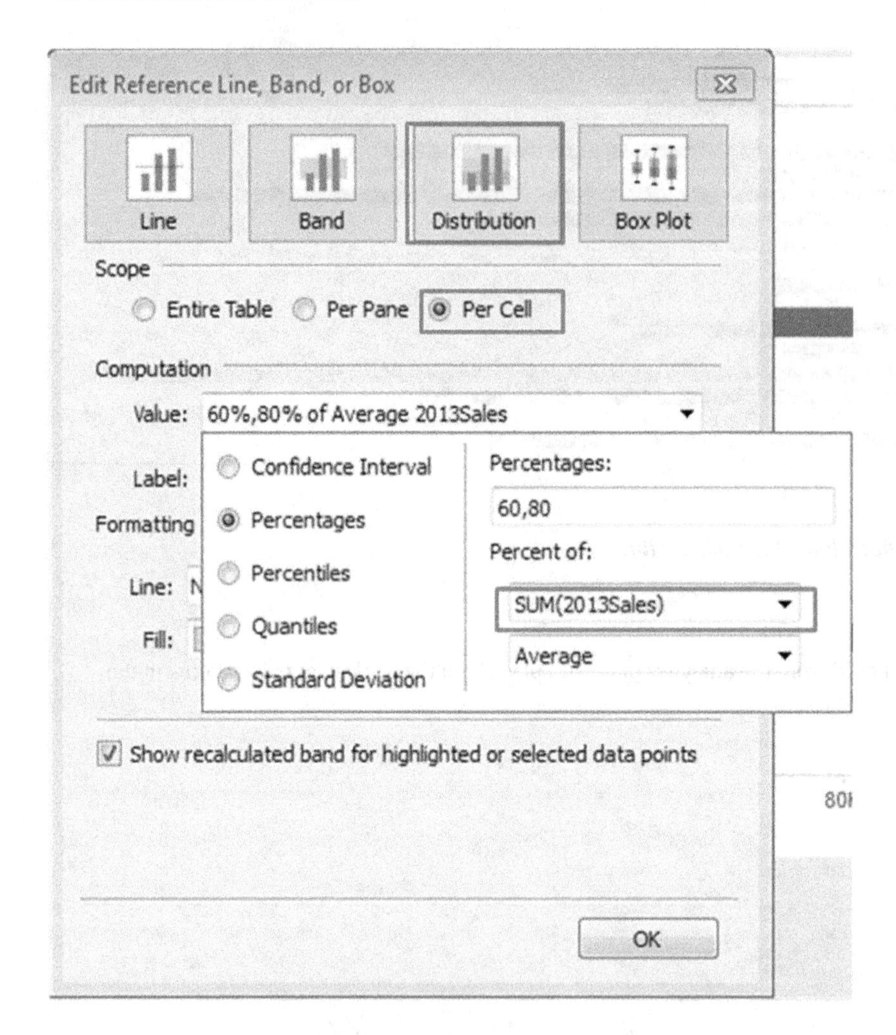

Figure 9-65. *Perform the settings in "Edit Reference Line, Band, or Box" dialog box*

Select the color "Gray" for "Fill" and check the "Fill Below" check box. Adjust the size of the bars appropriately by using "Size" on the marks card.

The output after using the "Distribution Band" (Shown in Fig. 9-66).

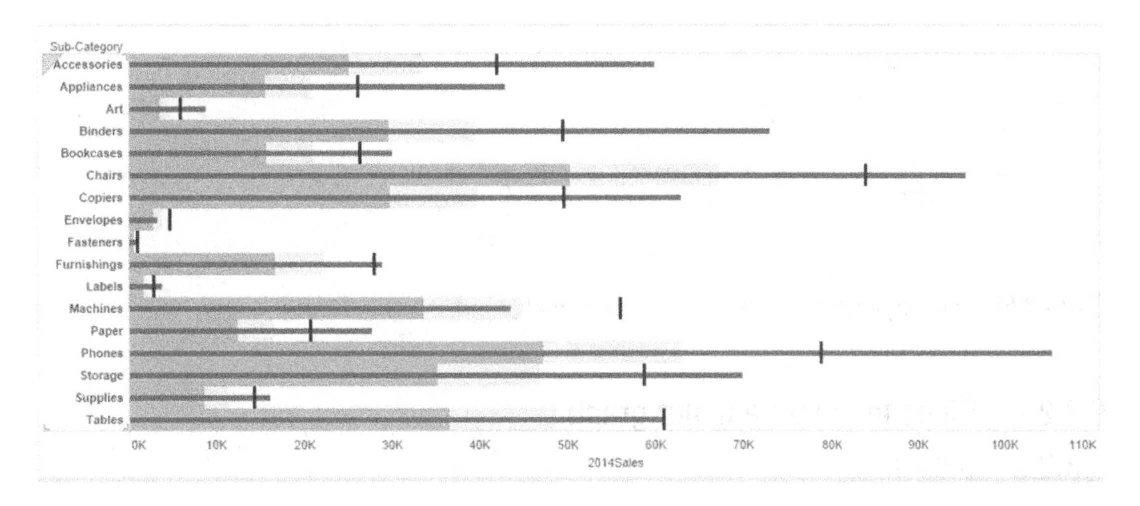

Figure 9-66. *Output after using the "Distribution Band"*

Conclusion

Let us consider the sub-category, "Machines". We see that 60% of average 2013Sales = 33,544; 80% of average 2013Sales = 44,726. 2014Sales is almost near the 80% mark at 43,545.

9.3.2 Demo 2

Objective: To visualize the staged progress towards the set goals.

Input: "Bar-in-BarChart.xls".

	A	B	C
1	Goals	Planned	Actuals
2	NoofTrainingsonNewTechnologies	40	15
3	NoofTrainingsonExistingTechnologies	80	55
4	NoofAssignments	12	6
5	NoofDemos	12	7
6	NoofTestPapers	6	3
7	NoofELearningArtifacts	20	15
8	NoofPapersPublished	2	1

Expected output: Shown in Fig. 9-67.

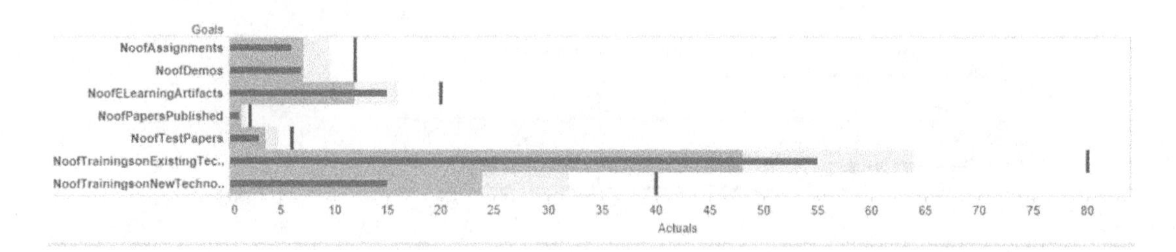

Figure 9-67. Bullet graphgraph - Demo 2 - expected output

9.3.2.1 Steps to create a bullet graph

9.3.2.2 Step 1

Read in the data from "Bar-inBarChart.xlsx" into Tableau (Shown in Fig. 9-68).

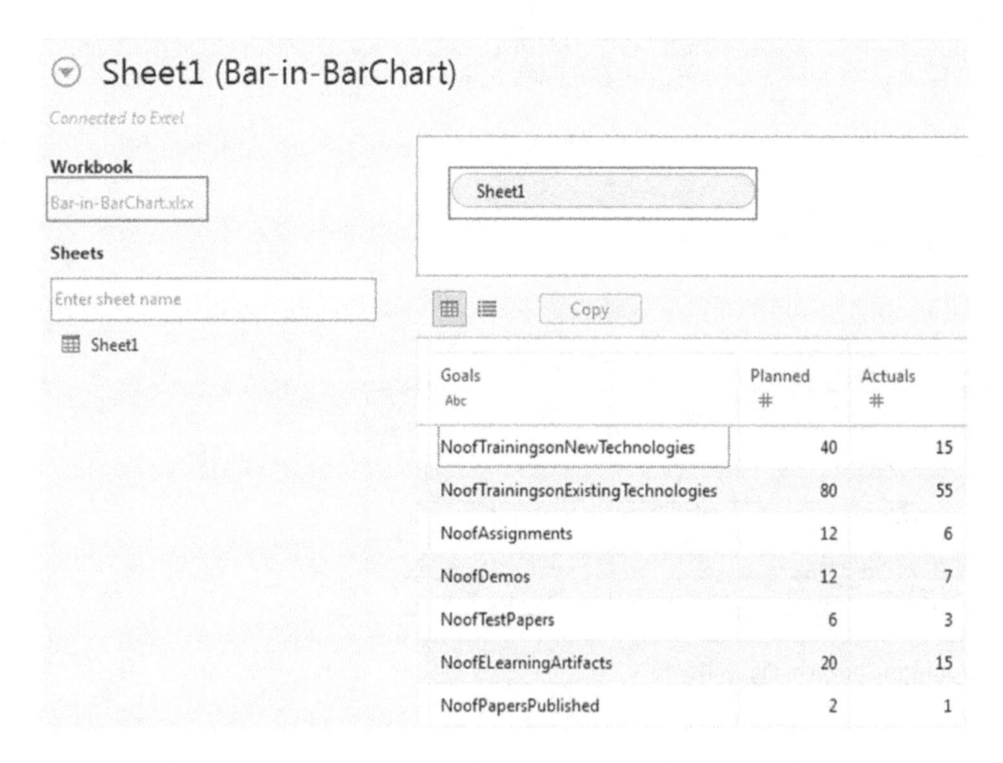

Goals (Abc)	Planned (#)	Actuals (#)
NoofTrainingsonNewTechnologies	40	15
NoofTrainingsonExistingTechnologies	80	55
NoofAssignments	12	6
NoofDemos	12	7
NoofTestPapers	6	3
NoofELearningArtifacts	20	15
NoofPapersPublished	2	1

Figure 9-68. Data from "Bar-inBarChart.xlsx" read into Tableau

9.3.2.3 Step 2

Drag the dimension "Goals" from the dimensions area under the data pane to the rows shelf (Shown in Fig. 9-69).

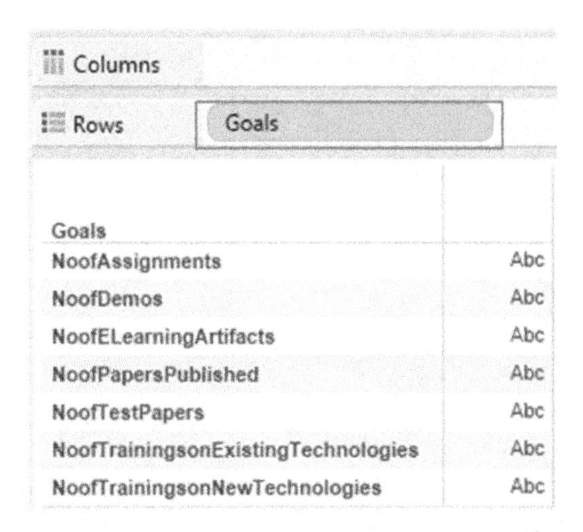

Figure 9-69. *Dimension, "Goals" placed on the rows shelf*

Drag the measure "Actuals" from the measures area under the data pane and place it on the columns shelf (Shown in Fig. 9-70).

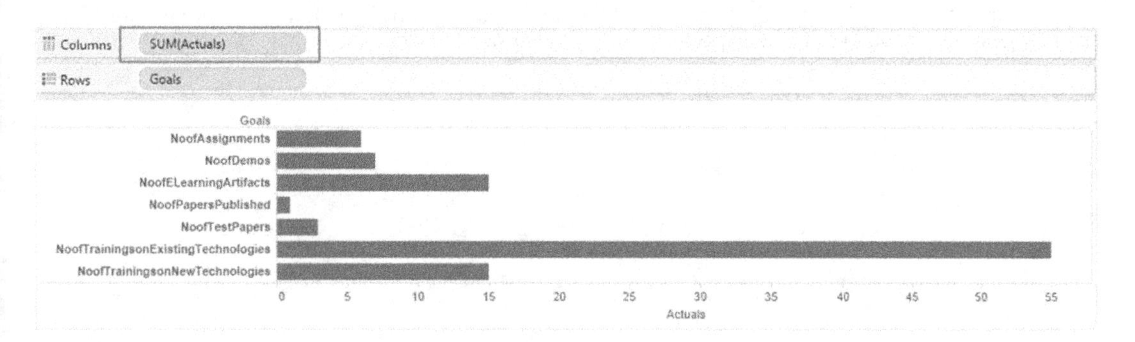

Figure 9-70. *Measure "Actuals" placed on the columns shelf*

Drag the measure "Planned" from the measures area under the data pane and place it on "Detail" on the marks card (Shown in Fig. 9-71).

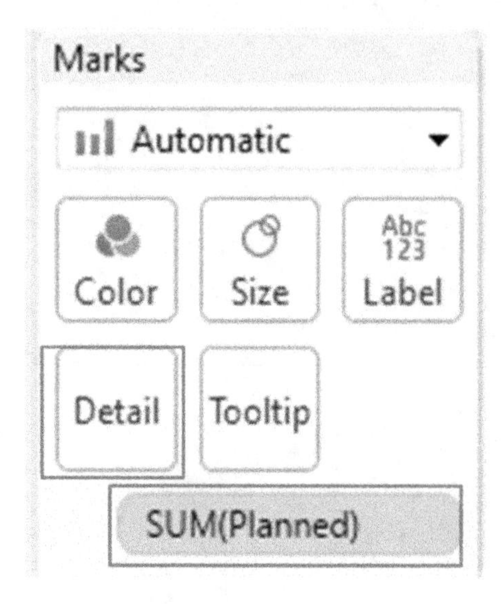

Figure 9-71. *Measure "Planned" placed on "Detail" on the marks card*

9.3.2.4 Step 3

Go to the analytics pane and drag the "Reference Line" to the view / worksheet. Edit the reference line in the "Edit Reference Line, Band or Box" dialog box (Shown in Fig. 9-72).

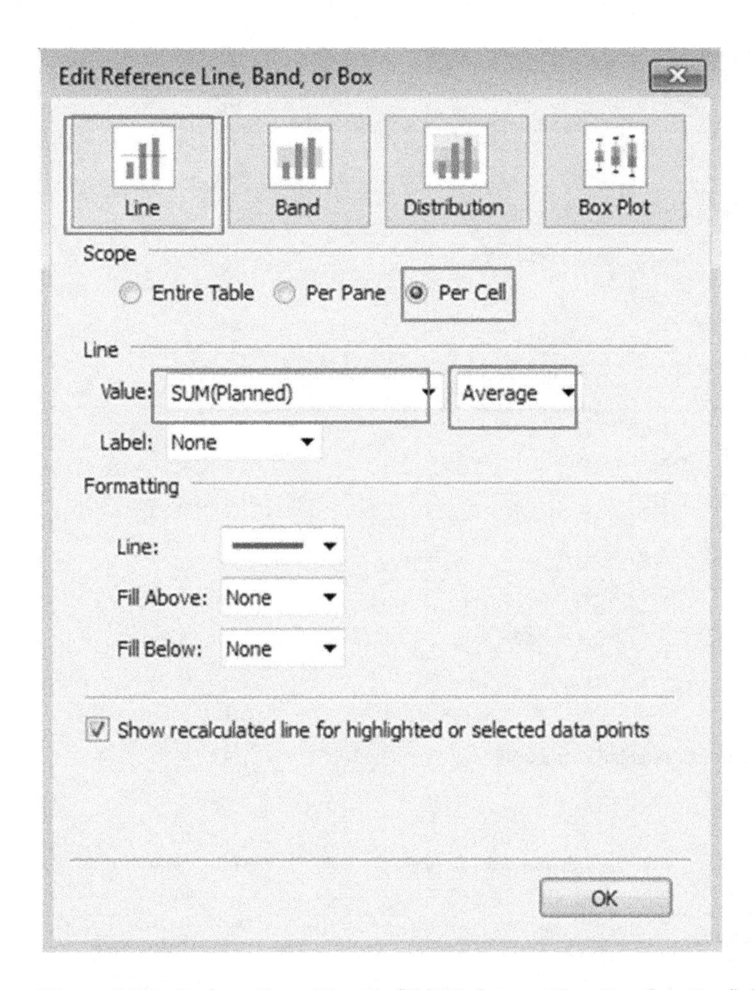

Figure 9-72. *Perform the settings in "Edit Reference Line, Band, or Box" dialog box*

9.3.2.5 Step 4

Drag the "Distribution Band" under the analytics pane and place it on the view / worksheet. Perform the settings as shown in Fig. 9-73.

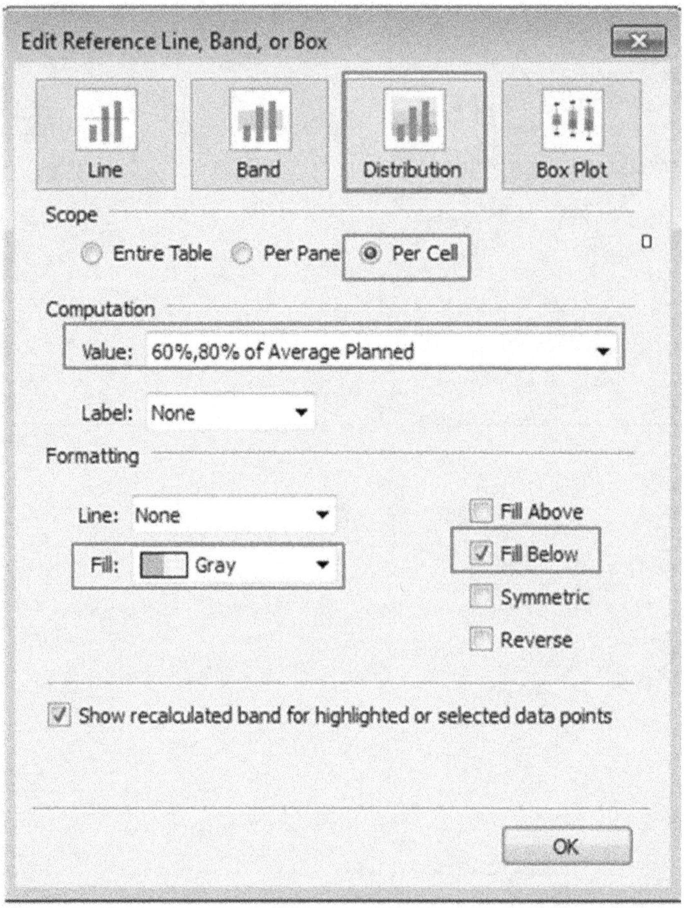

Edit Reference Line, Band, or Box

Line
Band
Distribution
Box Plot

Scope

Entire Table
Per Pane
Per Cell

Computation

Value: 60%,80% of Average Planned

Label: None

Formatting

Line: None
Fill: Gray

Fill Above
Fill Below
Symmetric
Reverse

Show recalculated band for highlighted or selected data points

OK

Figure 9-73. *Perform the settings in "Edit Reference Line, Band, or Box" dialog box*

The final output after setting the distribution band and reference line (Shown in Fig. 9-74).

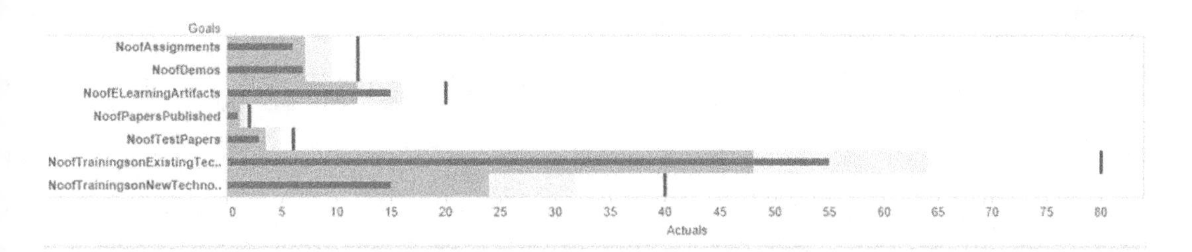

***Figure 9-74.** Bullet graph - Demo 2 - final output*

Conclusion

Let us look at one of the goals, "NoofTrainingsonNewTechnologies". We are far below the 60% mark.

The "Planned" for this goal was 40

The "Actuals" for this goal is 15

Accomplishing 60% of the target meant completing 24 days of training delivery, and accomplishing 80% of the target meant completing 32 days of training delivery.

9.4 Points to remember

In this chapter, we learned when to use each type of chart:

Waterfall charts are used to show how an initial value is affected by the series of positive or negative values. They are used in inventory management, performance analysis, finance, etc.

Bump charts are used to show the changes in ranks very effectively. By their very nature, bump charts are like spaghetti. Reduce the clutter in the bump chart to make it more legible.

You can test yourself by answering these questions:

1. This chart is also called a brick chart.
2. This chart represents changes in the position of a given number of competing entities over a fixed duration.
3. This chart is used as a replacement for gauges and radial meters on a dashboard.

Did you get them right?

1. Waterfall chart
2. Bump chart
3. Bullet graph

9.5 Next steps

In the next chapter, we move on to study how to create interactive dashboards and learn to tell stories.

CHAPTER 10

Dashboard and Stories

"You can achieve simplicity in the design of effective charts, graphs and tables by remembering three fundamental principles: restrain, reduce, emphasize."

—Garr Reynolds, internationally acclaimed communications consultant and the author of best-selling books, including the award-winning Presentation Zen and Presentation Zen Design

Chapter 9 introduced us to advanced visualization techniques, such as waterfall charts, bump charts and bullet graphs. This chapter will help us to understand how to create an interactive dashboard and weave a story. We will explore the following:

- Creating and organizing a dashboard
- Dashboard actions
- Creating a story

10.1 Why use a dashboard?

A dashboard helps one to show several worksheets in a single space. In addition to this, one can provide supporting information, compare and monitor varieties of data simultaneously.

10.2 What is a dashboard?

A dashboard is a visual display of the most important information needed to achieve one or more objectives that fits entirely on a single computer screen, so that it can be viewed, monitored and managed at a glance.

In other words, we can say that a dashboard is a user interface that organizes and presents information in a way that is easy to read. It consolidates and arranges numbers, metrics and sometimes performance scorecards on a single screen.

You can see the dashboard tab at the bottom of the workbook similar to the worksheet tab. You can add views from your workbook to the dashboard. You can add web pages, images and text areas to your dashboard as well. A dashboard allows you to format, edit, drill-down, and edit axes on your view. When you add a view to the dashboard, it automatically connects to the corresponding worksheet. When you modify the worksheet, the dashboard is updated automatically and vice-versa.

S. Acharya and S. Chellappan, *Pro Tableau*, DOI 10.1007/978-1-4842-2352-9_10

10.3 Creating a dashboard

One or more views/worksheets can be pulled into a dashboard. You can also add interactivity to the dashboard and can do much more.

10.3.1 Opening a dashboard sheet

You can create a dashboard in the same way you create a new worksheet. After creating a new dashboard sheet, you can add one or more views and objects to the dashboard.

10.3.1.1 Steps to create a dashboard

Let us explore the steps to create a dashboard.

10.3.1.1.1 Step 1

Select dashboard ➤ New dashboard (Shown in Fig. 10-1).

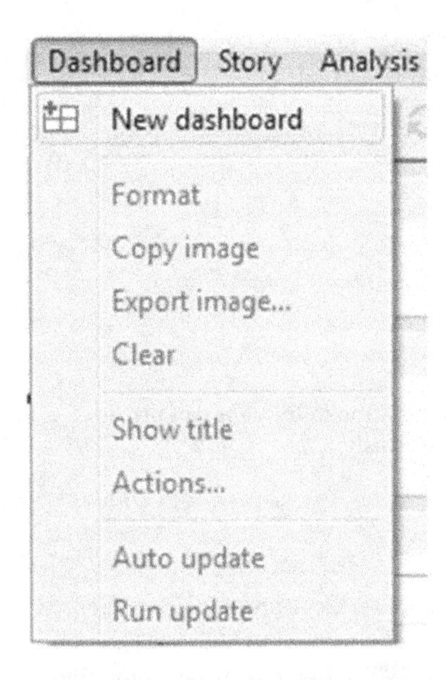

Figure 10-1. *"New dashboard" option*

10.3.1.1.2 Step 2

A new dashboard worksheet is created (Shown in Fig. 10-2).

Figure 10-2. *A dashboard worksheet*

10.3.1.1.3 Step 3

Alternatively, you can click on the "New dashboard" tab at the bottom of the workbook (Shown in Fig. 10-3).

Figure 10-3. *"New dashboard" tab*

10.3.2 Adding views to the Dashboard

You can pull one or more view/worksheet to the dashboard.

10.3.2.1 Steps to add views / worksheets to the dashboard

Perform the following steps to add a view to the dashboard.

10.3.2.1.1 Step 1

Click on "Dashboard 1" to open the dashboard (Shown in Fig. 10-4).

Figure 10-4. *"Dashboard 1" Sheet*

10.3.2.1.2 Step 2

You will see a dashboard window on the left side of the workbook (Shown in Fig. 10-5). All the worksheets currently present in the workbook are displayed. If you create a new worksheet, that worksheet is added to the dashboard window automatically. This feature helps you to view all the worksheets available in the workbook.

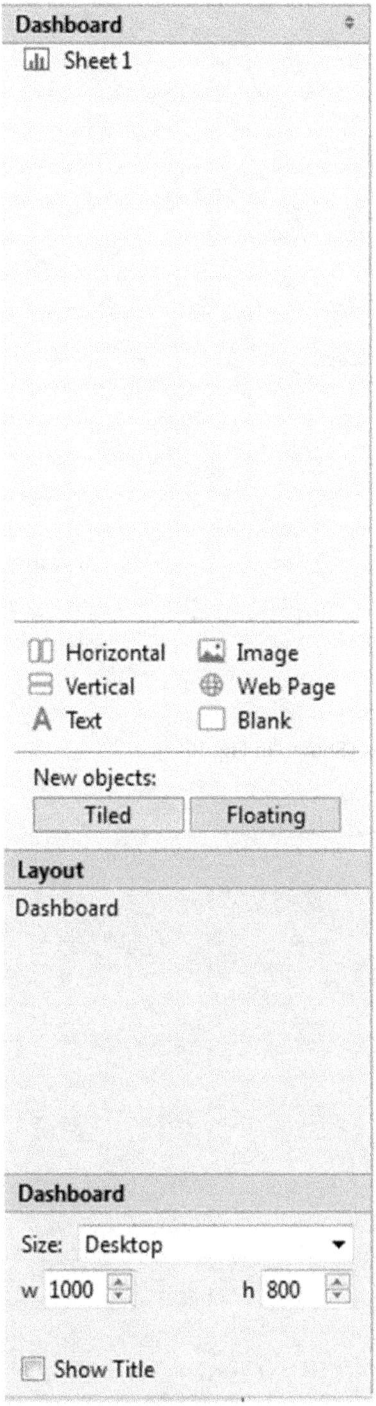

Figure 10-5. *Dashboard window*

10.3.2.1.3 Step 3

Create a view as shown in Fig. 10-6; rename the sheet as "Sales over year".

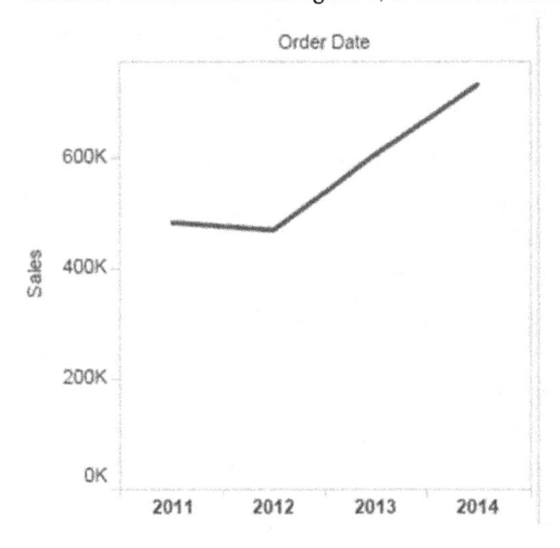

Figure 10-6. *"Sales over year"worksheet*

10.3.2.1.4 Step 4

You will notice the new sheet "Sales over year" in the dashboard window (Shown in Fig. 10-7).

Figure 10-7. *"Sales over year" worksheet in the dashboard window*

10.3.2.1.5 Step 5

Rename the dashboard as "Sales Details" (Shown in Fig. 10-8).

Figure 10-8. *"Sales Details" dashboard*

10.3.2.1.6 Step 6

To add a view to the dashboard, click the view and drag it to your dashboard sheet on the right (Shown in Fig. 10-9).

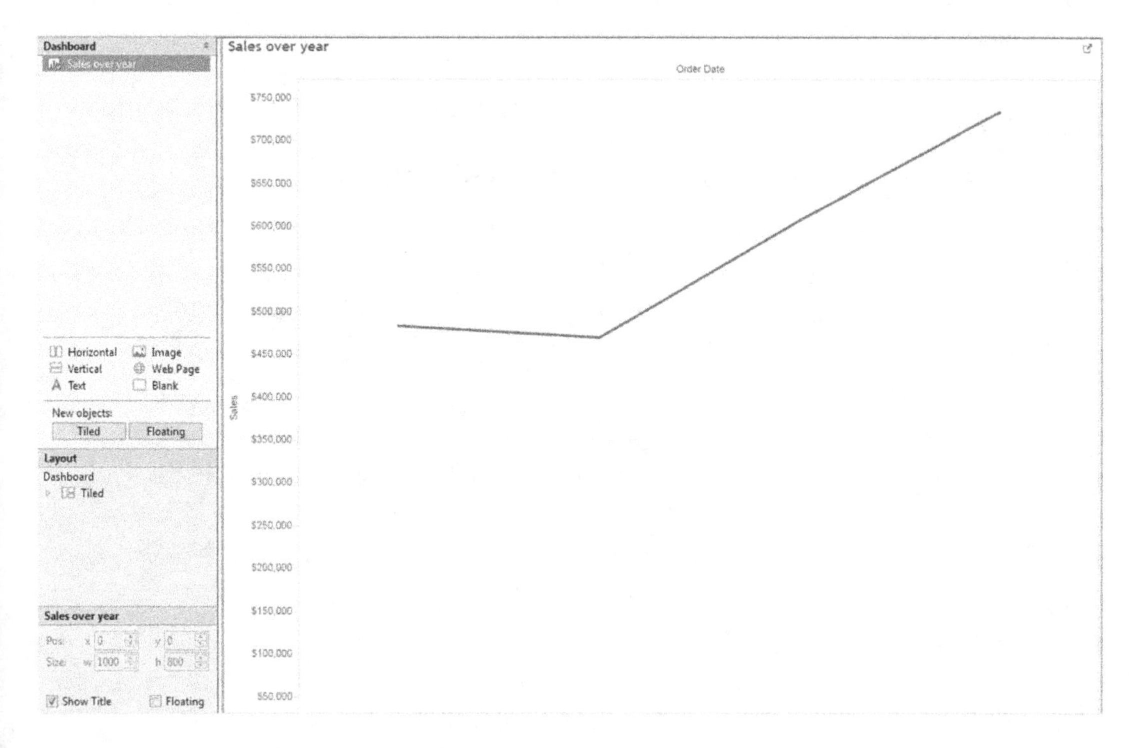

Figure 10-9. *Dashboard sheet containing the "Sales over year" view*

We have successfully added the view "Sales over year" to the dashboard. Let us explore how to add interactivity to the dashboard.

10.3.3 Adding interactivity to the dashboard

If you have a complex data source, it is difficult to review, monitor and manage the views. To achieve this, you can create an interactive dashboard to limit the data that is displayed on the dashboard. Let us create an interactive dashboard to show details of customers. We will work with dimensions “Category”, “Sub-Category” and “Region”.

To create an interactive dashboard take the following steps:

- Create an overview worksheet
- Create a detailed worksheet
- Create a dashboard

10.3.3.1 Create an overview worksheet

Create an overview worksheet.

10.3.3.1.1 Step 1

Create a view named “Sales by Sub-Category for each Region” as shown in Fig. 10-10.

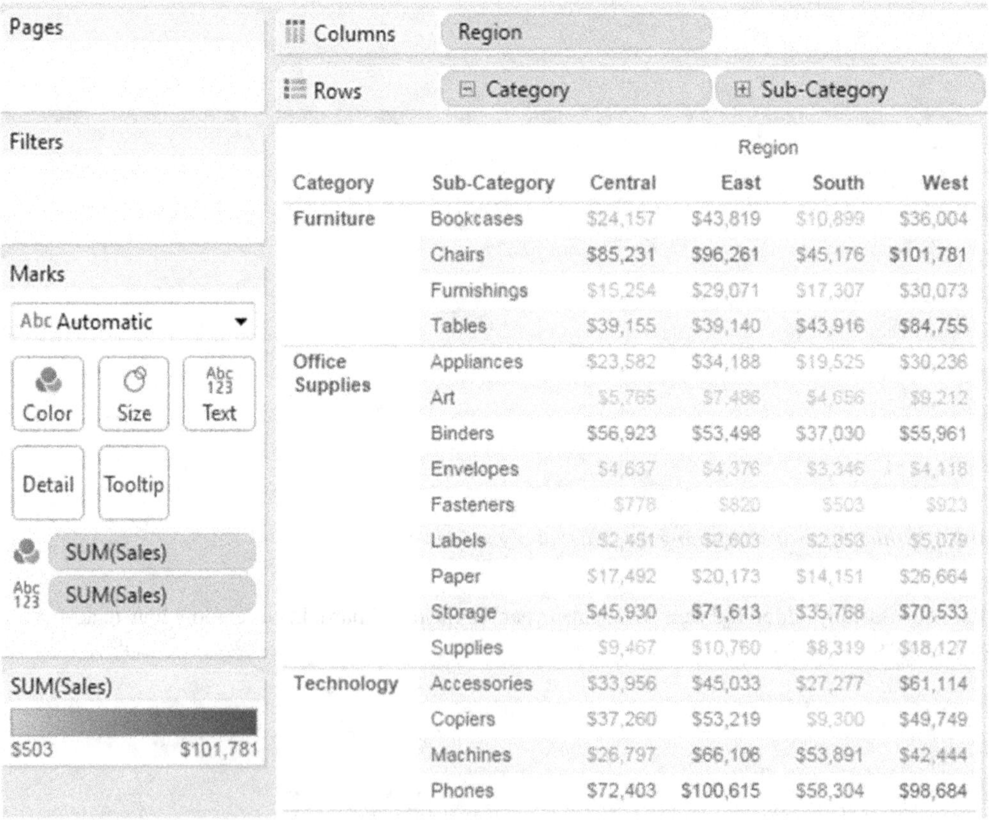

Pages

Columns: Region

Rows: Category, Sub-Category

Filters

Marks: Automatic — Color, Size, Text, Detail, Tooltip — SUM(Sales), SUM(Sales)

SUM(Sales): $503 – $101,781

Category	Sub-Category	Central	East	South	West
Furniture	Bookcases	$24,157	$43,819	$10,899	$36,004
	Chairs	$85,231	$96,261	$45,176	$101,781
	Furnishings	$15,254	$29,071	$17,307	$30,073
	Tables	$39,155	$39,140	$43,916	$84,755
Office Supplies	Appliances	$23,582	$34,188	$19,525	$30,236
	Art	$5,765	$7,486	$4,656	$9,212
	Binders	$56,923	$53,498	$37,030	$55,961
	Envelopes	$4,637	$4,376	$3,346	$4,118
	Fasteners	$778	$820	$503	$923
	Labels	$2,451	$2,603	$2,353	$5,079
	Paper	$17,492	$20,173	$14,151	$26,664
	Storage	$45,930	$71,613	$35,768	$70,533
	Supplies	$9,467	$10,760	$8,319	$18,127
Technology	Accessories	$33,956	$45,033	$27,277	$61,114
	Copiers	$37,260	$53,219	$9,300	$49,749
	Machines	$26,797	$66,106	$53,891	$42,444
	Phones	$72,403	$100,615	$58,304	$98,684

Figure 10-10. *“Sales” by “Category” and “Sub-Category” for each “Region”*

10.3.3.1.2 Step 2

Select heat map in "Show Me" window to convert the above view to a heat map (Shown in Fig. 10-11).

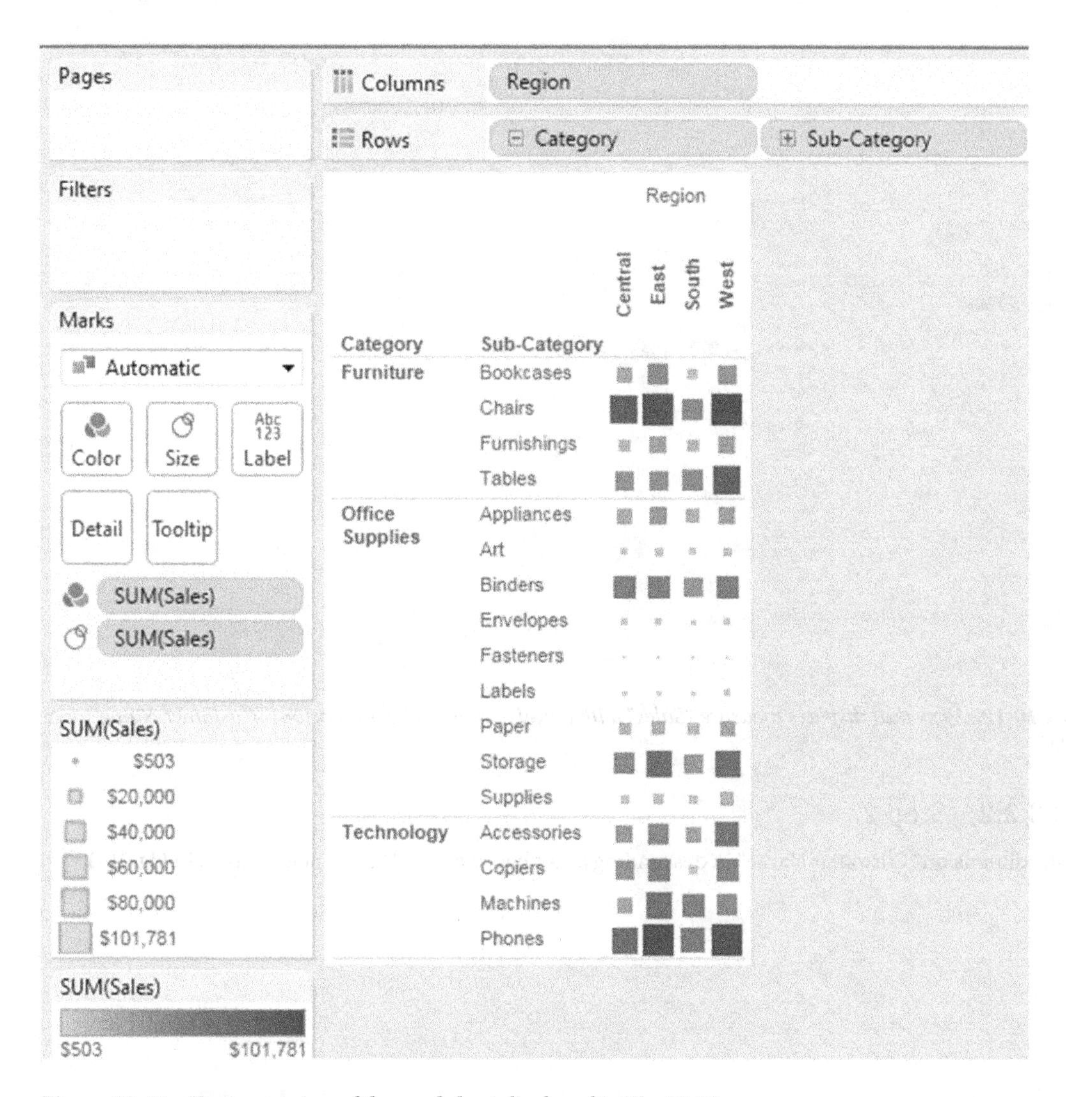

Figure 10-11. *Heat map view of the worksheet displayed in Fig. 10-10*

10.3.3.2 Create a worksheet with more details

Steps to create a detailed worksheet.

10.3.3.2.1 Step 1

Create a view named "Customer Details" as shown in Fig. 10-12.

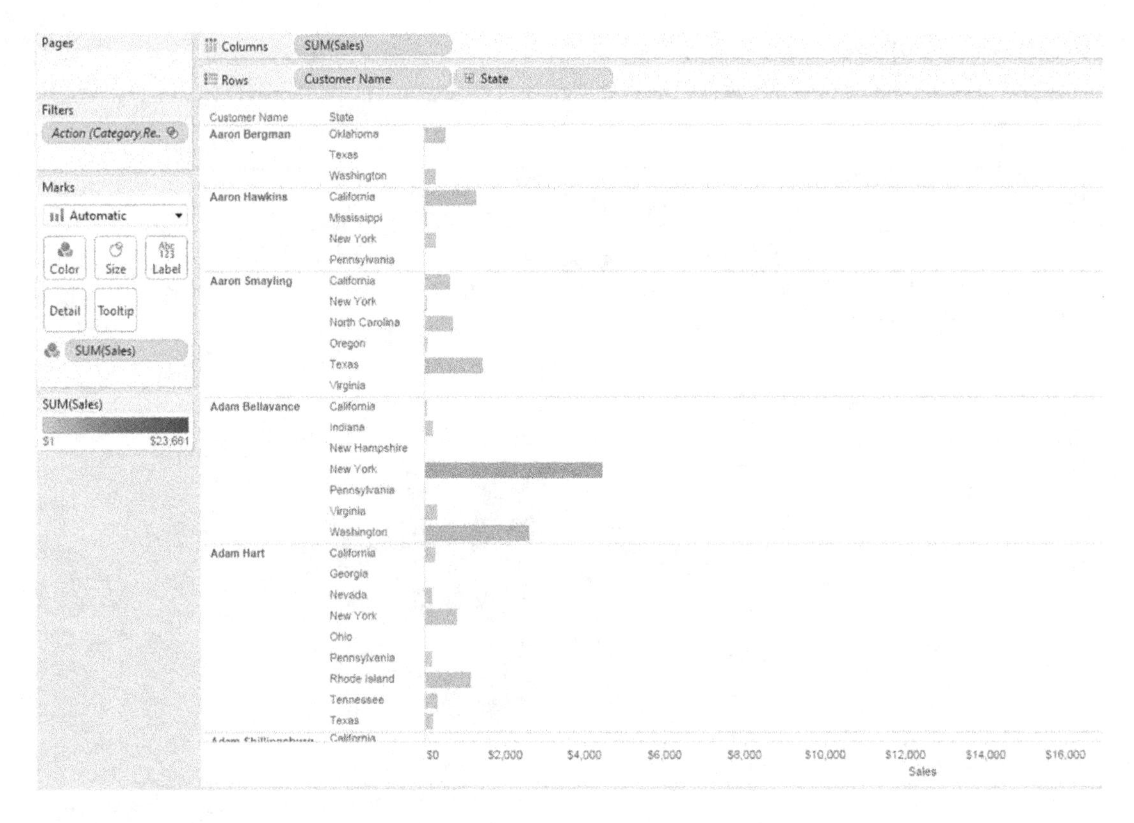

Figure 10-12. *View that displays measure "Sales" with details provided by dimensions "Customer Name" and "State"*

10.3.3.2.2 Step 2

Sort the dimension "Customer Name" in descending order based on the "Sales" field (Shown in Fig. 10-13).

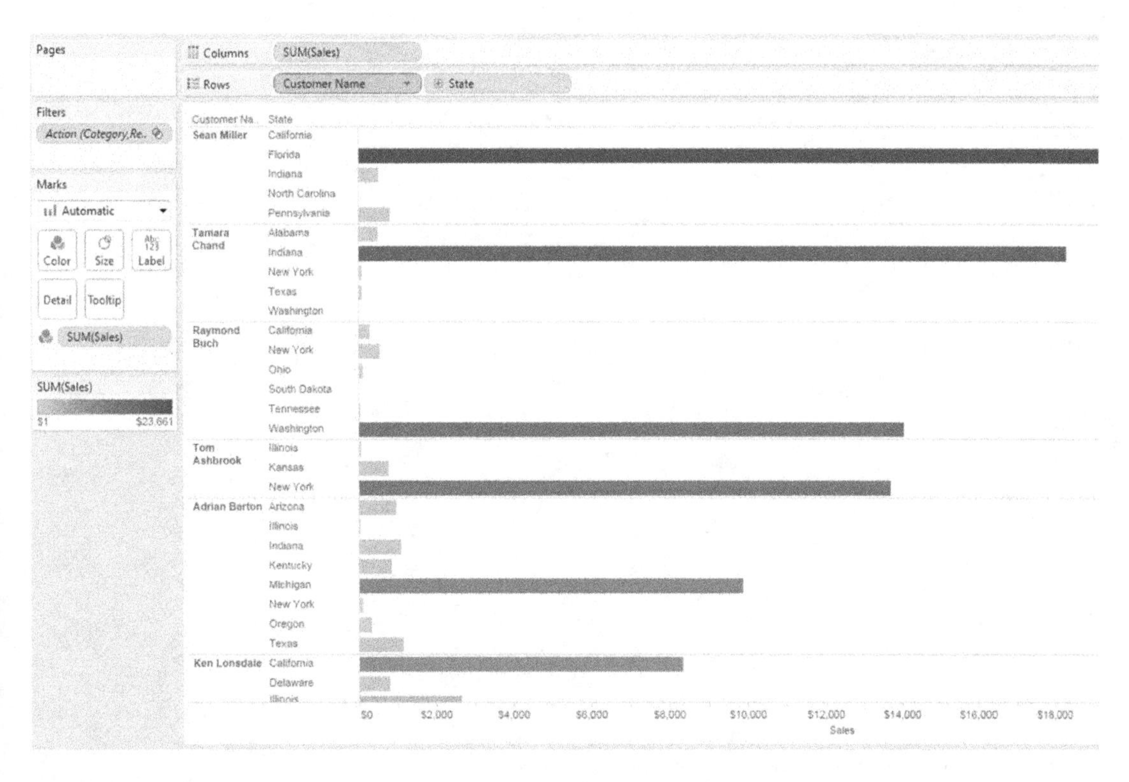

Figure 10-13. *"Customer Name" arranged as per "Sales" in descending order*

10.3.3.3 Create a dashboard

Follow the steps mentioned below to create a dashboard.

10.3.3.3.1 Step 1

Create a dashboard named "Customer Level Detail - Sales Info" (Shown in Fig. 10-14).

Figure 10-14. *"Customer Level Detail - Sales Info" dashboard*

10.3.3.3.2 Step 2

Click on the "Sales by Sub-Category for each Region" view and drag them to your dashboard sheet on the right (Shown in Fig. 10-15).

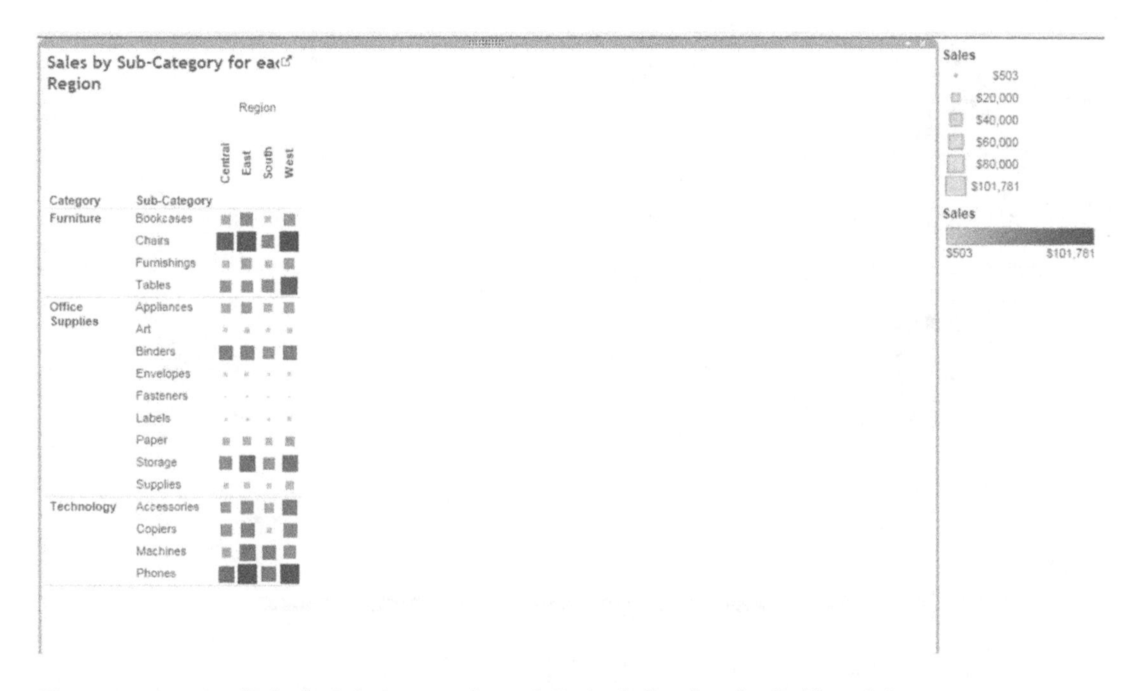

Figure 10-15. *View "Sales by Sub-Category for each Region" placed on the dashboard sheet*

10.3.3.3.3 Step 3

Drop the "Size legend" and "Color legend" to the bottom of the dashboard sheet (Shown in Fig. 10-16).

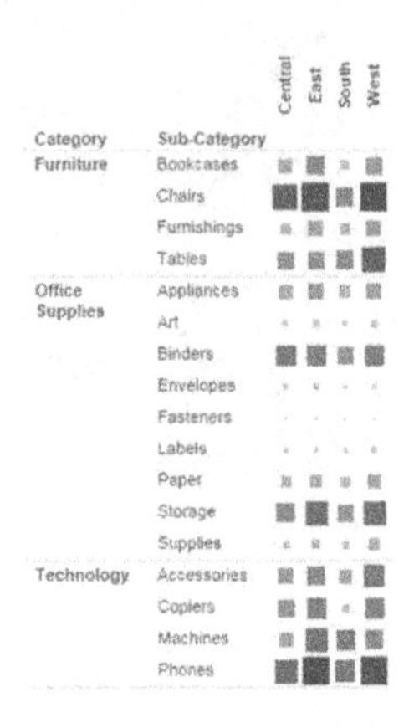

Figure 10-16. *"Size legend" and "Color legend" placed at the bottom of the dashboard sheet*

10.3.3.3.4 Step 4

Right click on dashboard sheet, select Fit ➤ Entire view (Shown in Fig. 10-17). Refer to Fig. 10-18 for the output.

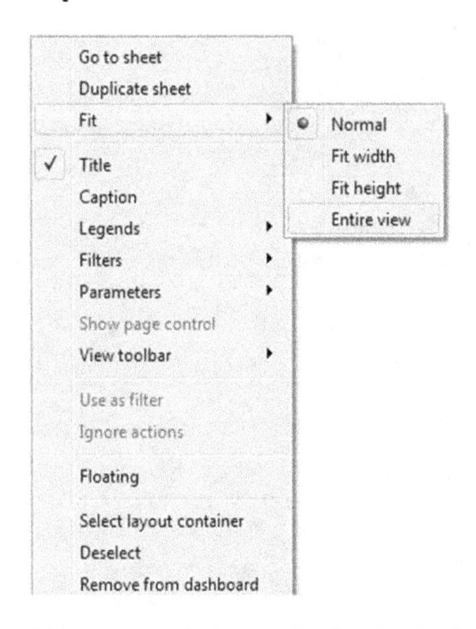

Figure 10-17. *Option to select "Entire view"*

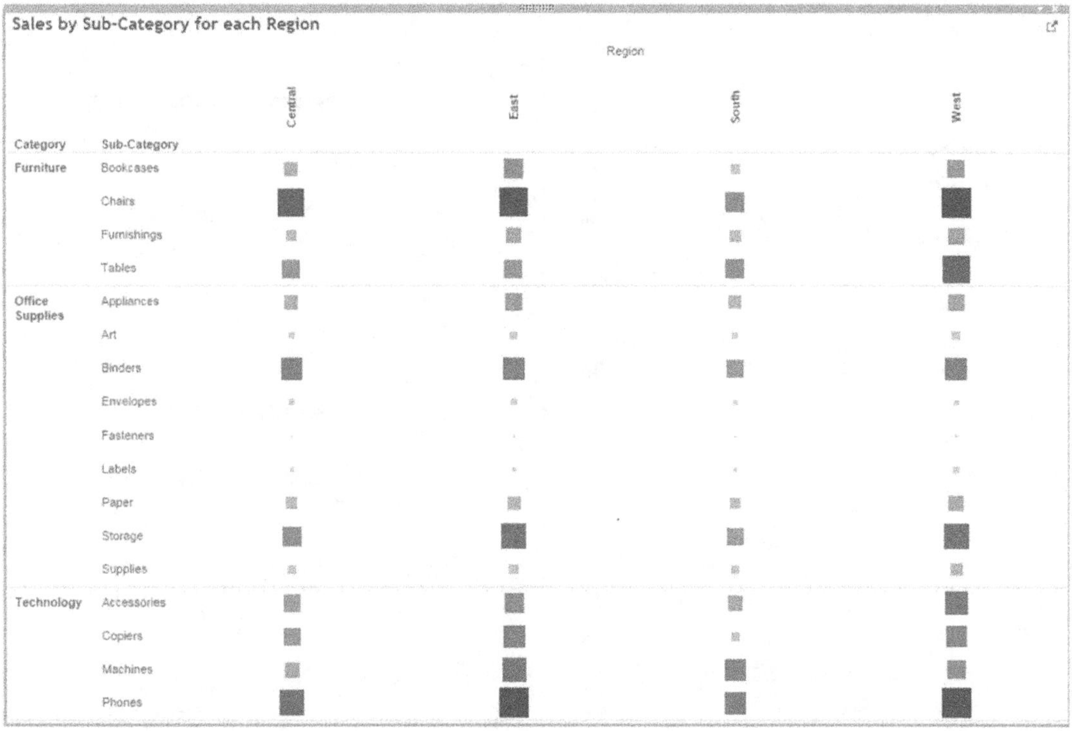

Figure 10-18. *Display of "Sales by Sub-Category for each Region" in "Entire view" mode*

10.3.3.3.5 Step 5

Click on the "Customer Details" view and drag them to your dashboard sheet on the right (Shown in Fig. 10-19).

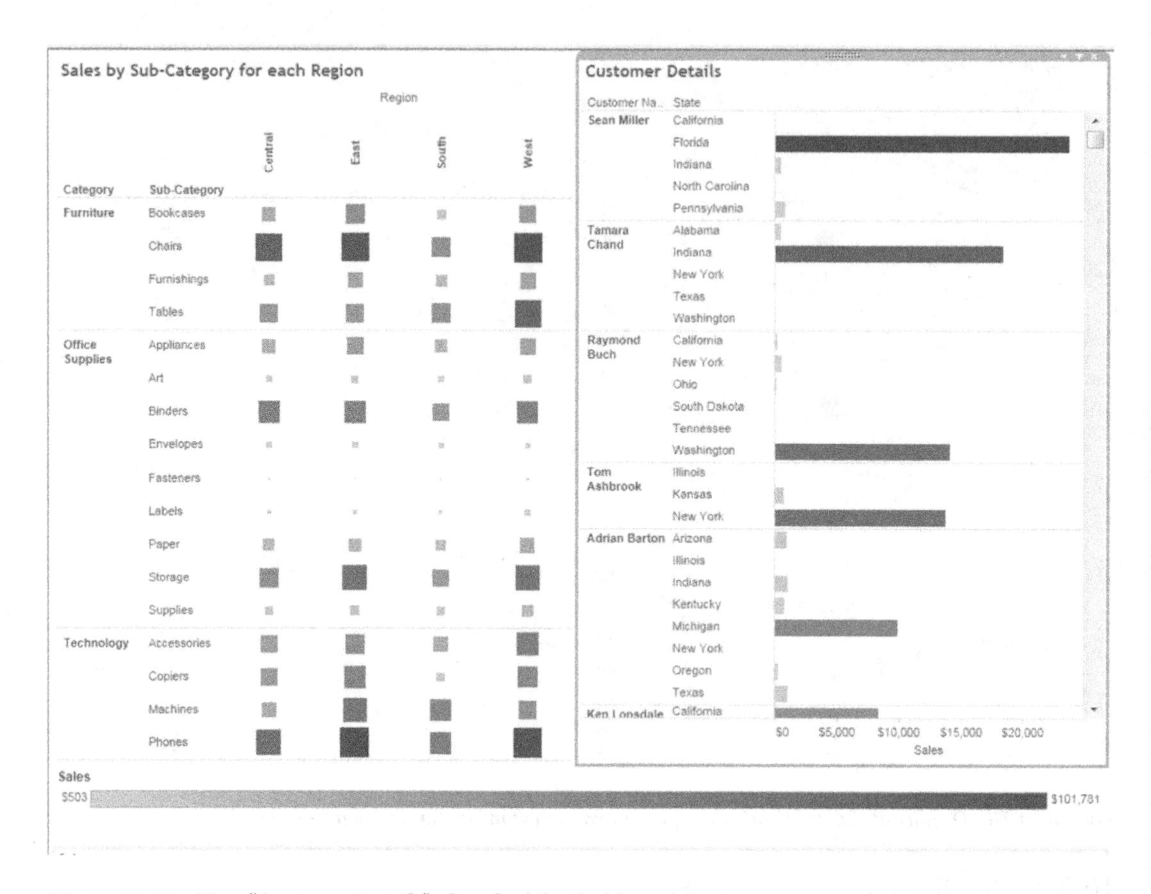

Figure 10-19. *View "Customer Details" placed on the dashboard sheet*

10.3.3.3.6 Step 6

Select "Sales by Sub-Category for each Region" view and select **"Use as filter"** option (Shown in Fig. 10-20).

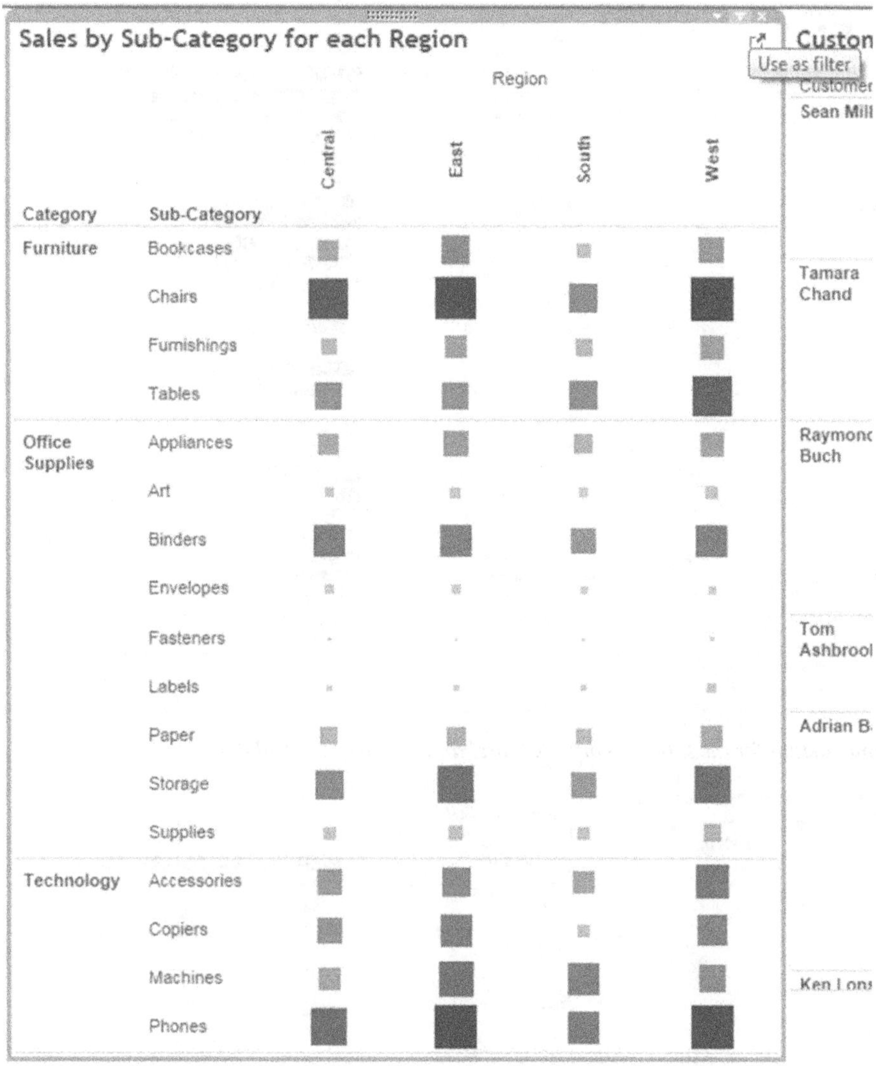

Figure 10-20. "Use as filter" option

10.3.3.3.7 Step 7

Based on the selection ("Sub-Category" and "Region"), relevant customer details will be displayed (Shown in Fig. 10-21 and 10-22).

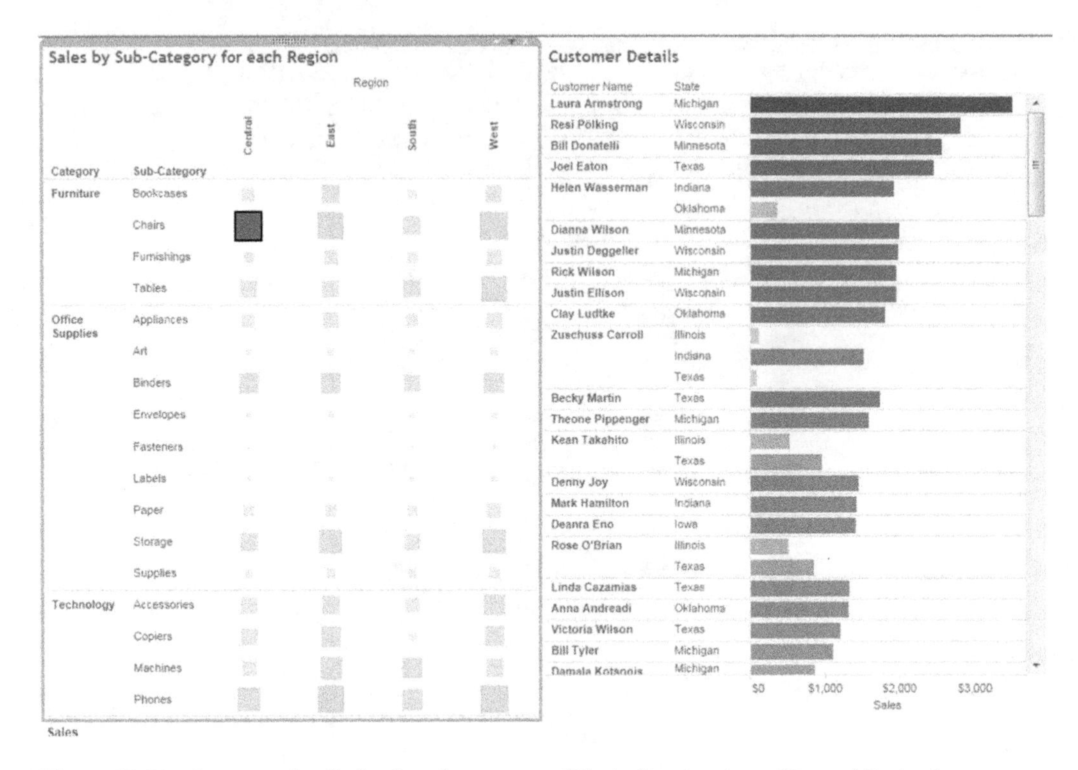

Figure 10-21. *Customer details for the sub-category - "Chairs" and region – "Central Region"*

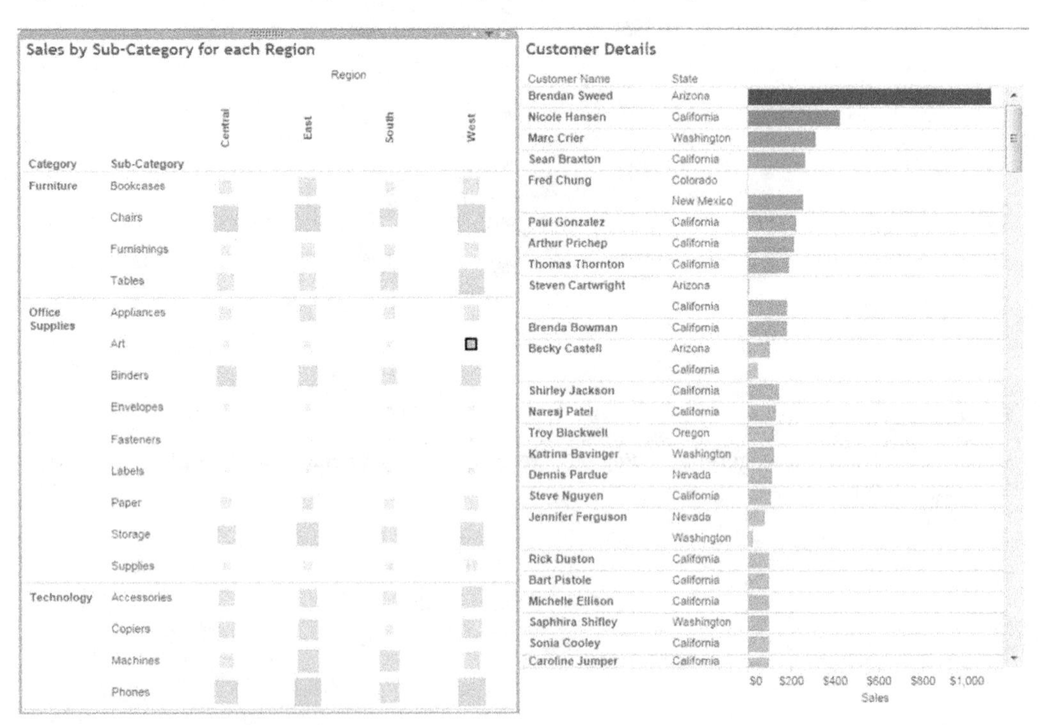

Figure 10-22. *Customer details for the sub-category - "Art" and region – "West Region"*

In the next section, we will discuss how to add supporting information for the view.

10.3.4 Adding an object to the dashboard

A dashboard object is an area in the dashboard. It helps to provide supporting information that is not present in the view. Dashboard objects are versatile, and you can use them to reflect the overall theme of your graphic composition.

For example, you can use the "Text" area to add detailed information about your view. You can see the dashboard objects in the dashboard pane (Shown in Fig. 10-23).

Horizontal	Image
Vertical	Web Page
Text	Blank

Figure 10-23. *Dashboard objects in the dashboard pane*

Let us discuss how to add dashboard objects to the dashboard sheet.

10.3.4.1 Steps to add dashboard objects

Follow the following steps to add dashboard objects to the dashboard sheet.

10.3.4.1.1 Step 1

Drag the dashboard object "Text" under the dashboard pane to the dashboard sheet.

10.3.4.1.2 Step 2

An "Edit Text" dialog box will show up. Type the text as shown in Fig. 10-24.

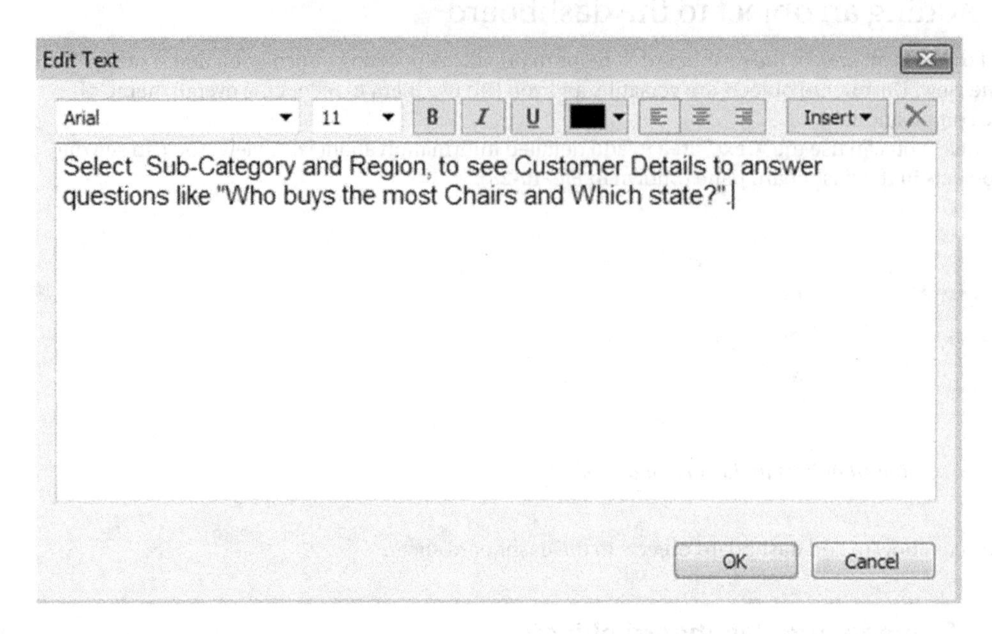

Figure 10-24. *"Edit Text" dialog box*

10.3.4.1.3 Step 3

You can see the "Text" object in the dashboard (Shown in Fig. 10-25).

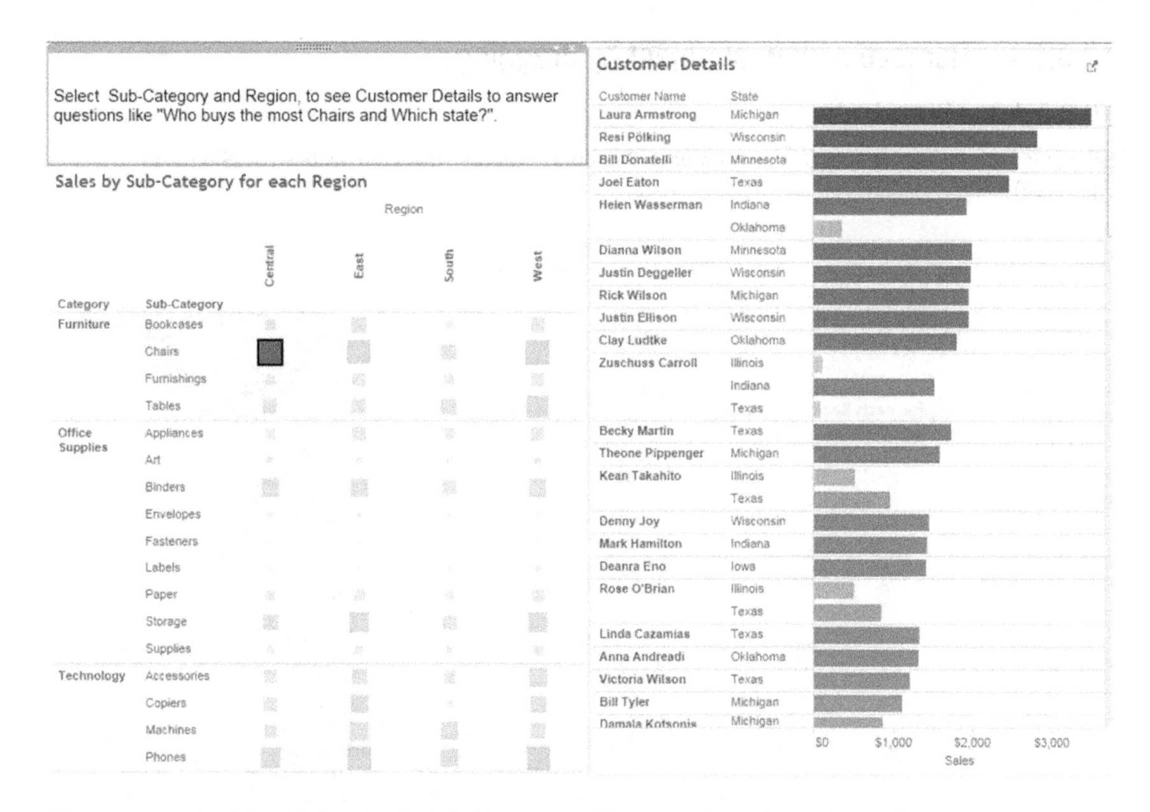

Figure 10-25. *Dashboard showing "Text" object to provide supporting information to the view*

There is a list of dashboard objects that can be added to a view. Each object is used for a different purpose.

- **Image:** An image object allows you to add a logo, branding elements and descriptive information.
- **Blank:** A blank object allows you to add a blank area in the dashboard. These objects can be used to control space. Blank containers are transparent. So, the background colors that you use will show through them.
- **Web Page:** A web page object allows you to add useful hyperlinks and dynamic content from the Internet to support your view.

In the next section, we will explain how to remove a worksheet from the dashboard.

10.3.5 Remove a view or an object from the dashboard

There are a number of ways to remove a view or an object from the dashboard.

10.3.5.1 Remove a view or object by dragging

10.3.5.1.1 Step 1

Select the view ("Sales by Sub-Category for each Region") to remove from the dashboard (Shown in Fig. 10-26).

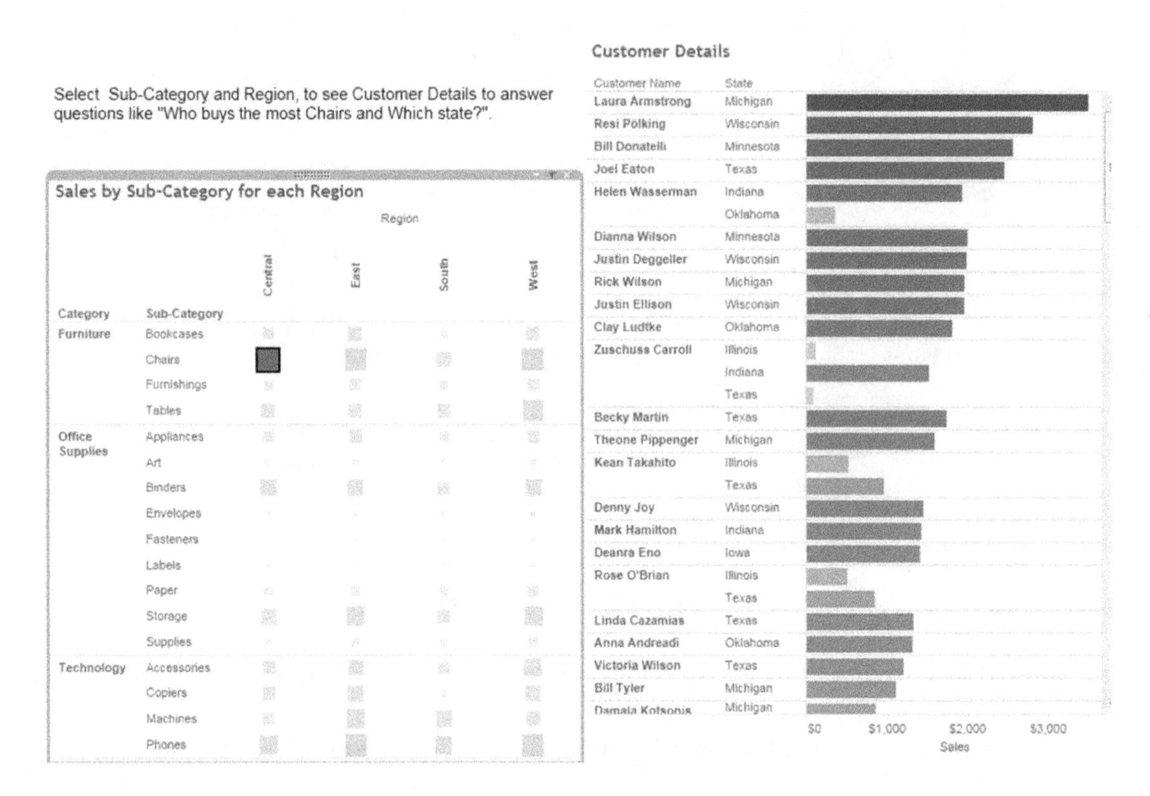

Figure 10-26. *Selected view "Sales by Sub-Category for each Region"*

10.3.5.1.2 Step 2

Click the move handle at the top of the view and drag it off to the dashboard pane.

10.3.5.2 Remove a view using the dashboard window

10.3.5.2.1 Step 1

Right click on the view that you want to remove, select "Remove from dashboard" (Shown in Figure 10-27).

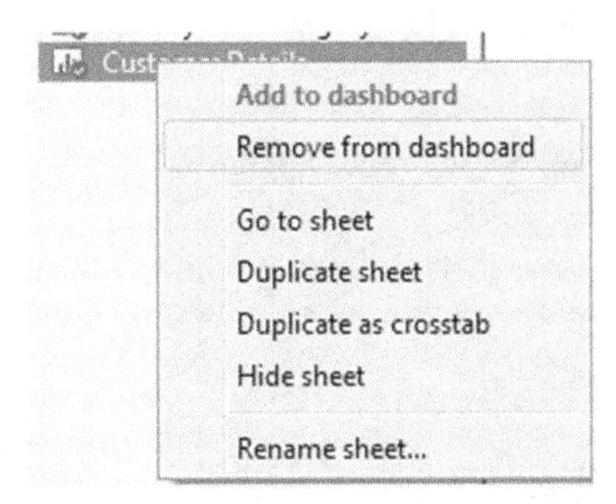

Figure 10-27. *"Remove from dashboard" option to remove view from the dashboard*

10.3.5.3 Remove a view or object using the dashboard view

10.3.5.3.1 Step 1

Select the view or object that you want to remove from the dashboard.

10.3.5.3.2 Step 2

Click on the dashboard view menu and select "Remove from dashboard" option as shown in Fig. 10-28.

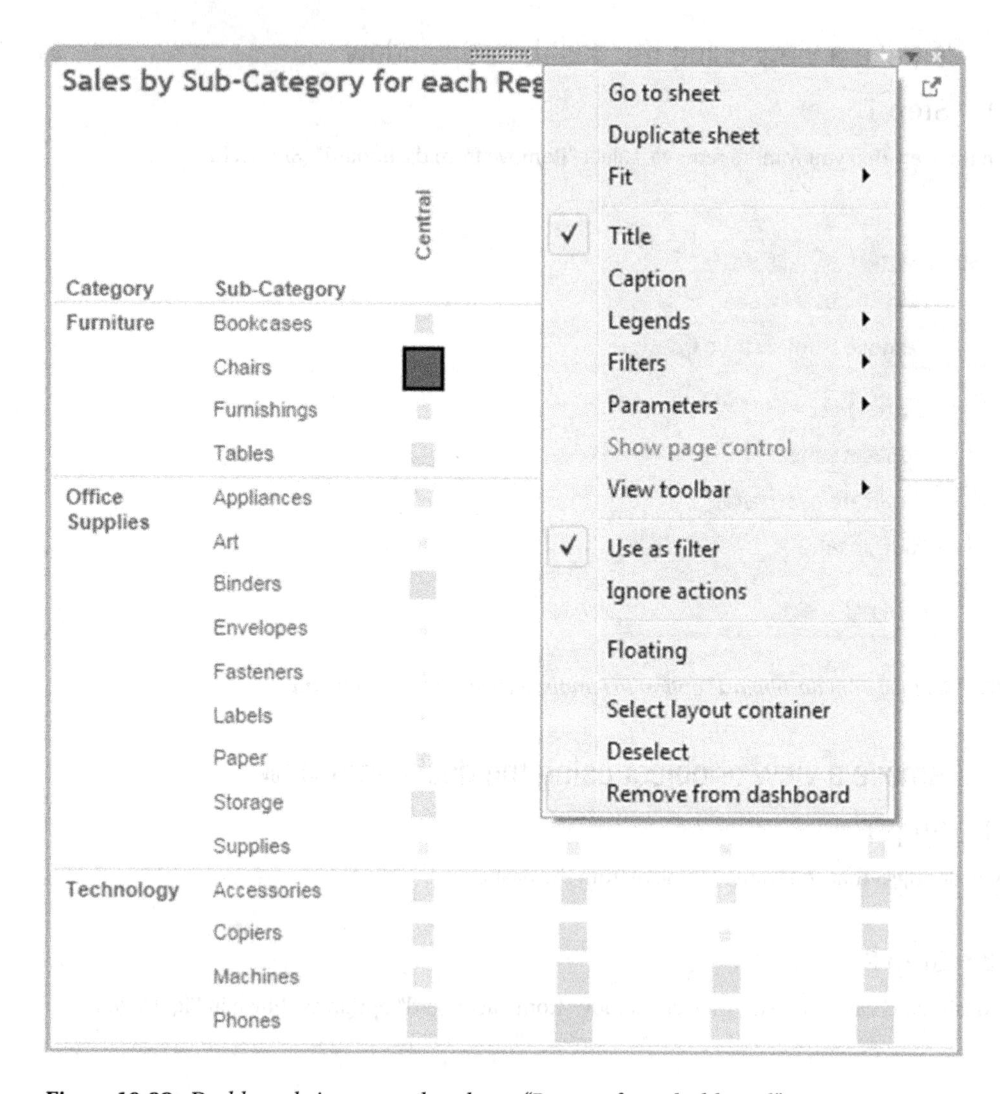

Figure 10-28. *Dashboard view menu that shows "Remove from dashboard" option*

10.3.6 Organizing a dashboard

You can organize your dashboard in several ways to tell a story or highlight/emphasize information.
Tableau provides two types of layouts:

- Tiled
- Floating

10.3.6.1 Tiled layout

By default, dashboard uses tiled layout. In tiled layout, all views and objects are arranged on a single layer as shown in Fig. 10-29.

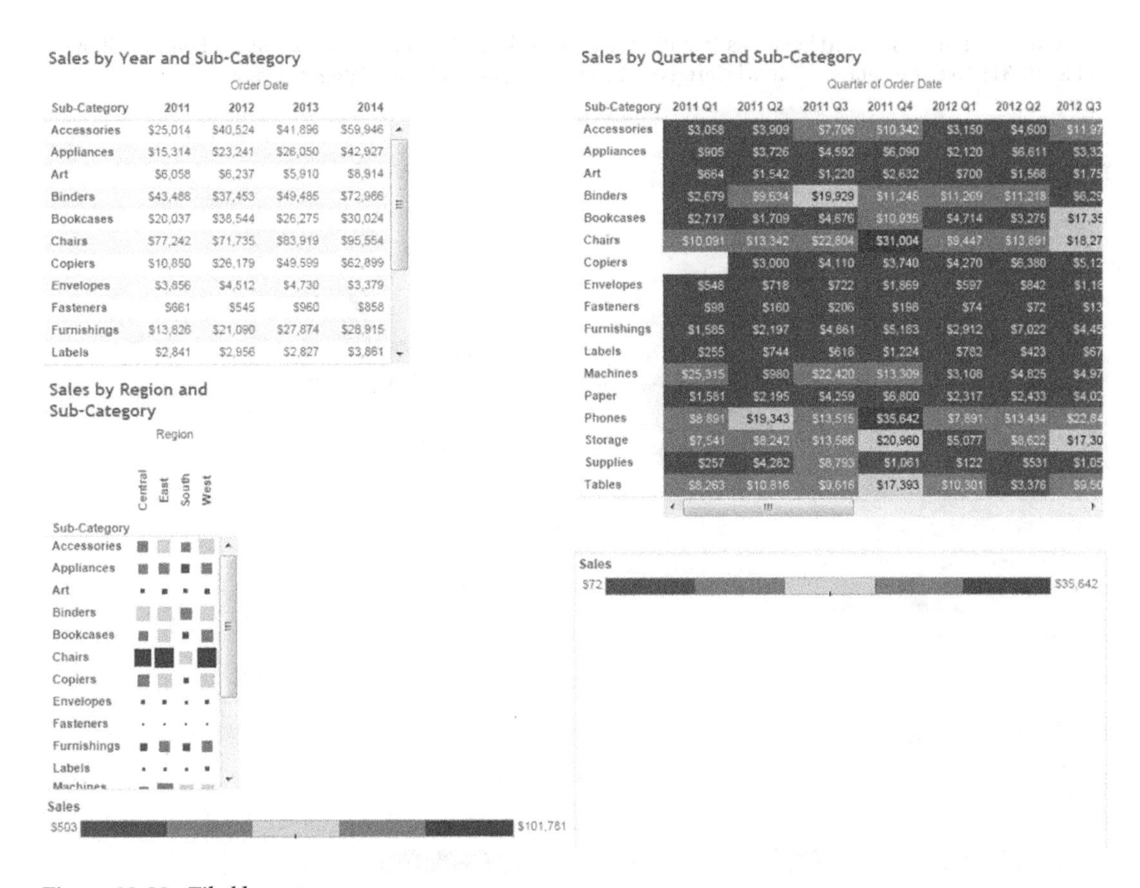

Figure 10-29. *Tiled layout*

10.3.6.2 Floating layout

In a floating layout, views or objects can be layered on other objects. To change default layout, click on "Floating" button in the middle of the dashboard pane as shown in Fig. 10-30.

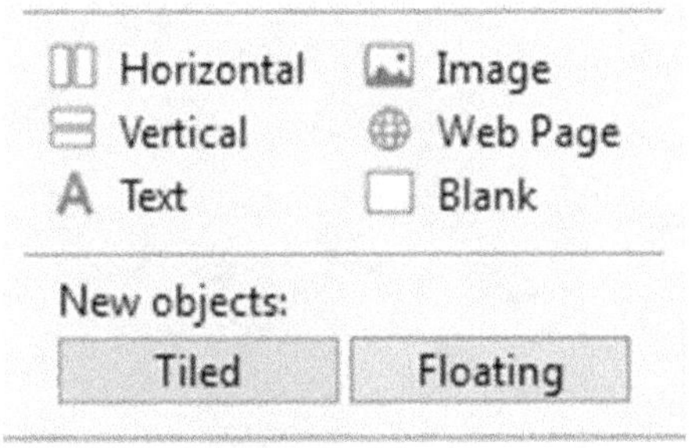

Horizontal
Image
Vertical
Web Page
Text
Blank
New objects:
Tiled
Floating

Figure 10-30. *"Floating" layout button*

When you set dashboard layout as "Floating", a new worksheet or an object is added as floating (Shown in Fig. 10-31). "Sales by State" view is layered on "Sales by Region and Sub-Category" view.

Figure 10-31. *Floating "Sales by State" view*

To switch between layouts, select the view. Select "Floating" option at the bottom of the dashboard pane as shown in Fig. 10-32.

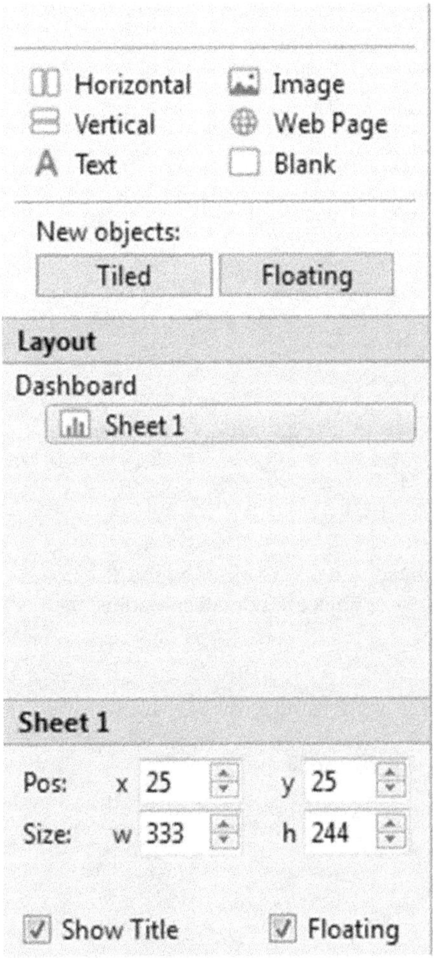

Horizontal
Vertical
Text
Image
Web Page
Blank
New objects:
Tiled
Floating
Layout
Dashboard
Sheet 1
Sheet 1
Pos: x 25 y 25
Size: w 333 h 244
Show Title
Floating

Figure 10-32. Floating option

The view layout is changed to "floating" layout as shown in Fig. 10-33.

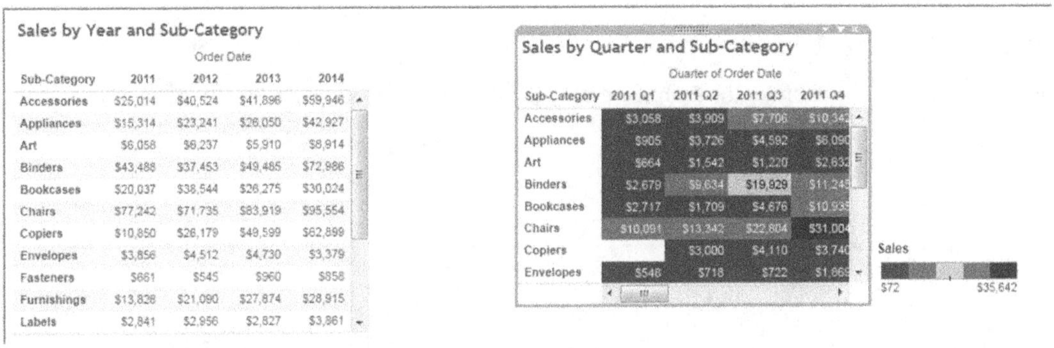

Sales by Year and Sub-Category

Sub-Category	2011	2012	2013	2014
Accessories	$25,014	$40,524	$41,896	$59,946
Appliances	$15,314	$23,241	$26,050	$42,927
Art	$6,058	$6,237	$5,910	$8,914
Binders	$43,488	$37,453	$49,485	$72,986
Bookcases	$20,037	$38,544	$26,275	$30,024
Chairs	$77,242	$71,735	$83,919	$95,554
Copiers	$10,850	$26,179	$49,599	$62,899
Envelopes	$3,856	$4,512	$4,730	$3,379
Fasteners	$661	$545	$960	$858
Furnishings	$13,826	$21,090	$27,874	$28,915
Labels	$2,841	$2,956	$2,827	$3,861

Sales by Quarter and Sub-Category

Sub-Category	2011 Q1	2011 Q2	2011 Q3	2011 Q4
Accessories	$3,058	$3,909	$7,706	$10,342
Appliances	$905	$3,726	$4,592	$6,090
Art	$664	$1,542	$1,220	$2,632
Binders	$2,679	$9,634	$19,929	$11,245
Bookcases	$2,717	$1,709	$4,676	$10,935
Chairs	$10,091	$13,342	$22,804	$31,004
Copiers		$3,000	$4,110	$3,740
Envelopes	$546	$718	$722	$1,869

Sales
$72 $35,642

Figure 10-33. View converted from tiled layout to floating layout

You can use "**Pos**" field present in the bottom of the dashboard pane (Shown in Fig. 10-34) to specify position for the floating layout. You can also use "**Size**" field to specify dimension for your floating view as shown in Fig. 10-34.

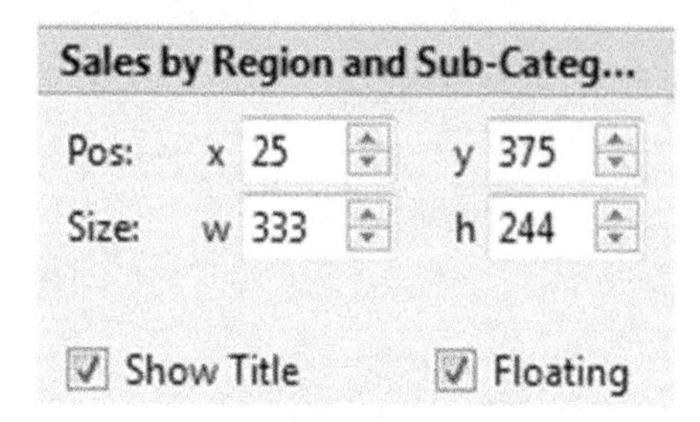

Figure 10-34. *"Pos" and "Size" field*

10.3.6.3 A layout container

A "Layout container" is a type of object. It helps to organize worksheets and other objects on the dashboard. Layout containers create an area in the dashboard sheet where views or objects automatically adjust their size and position based on other views or objects in the dashboard.

10.3.6.3.1 Adding layout containers

You can see the layout containers on the dashboard pane as shown in Fig. 10-35.

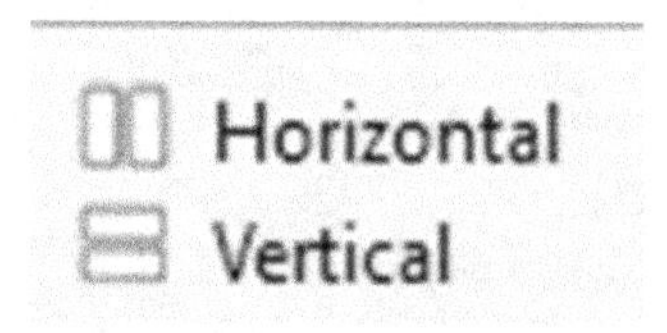

Figure 10-35. *Layout containers*

Steps to add "Layout Containers".

10.3.6.3.2 Step 1

Drag the layout container under the dashboard pane to the dashboard (Shown in Fig. 10-36, 10-37).

Figure 10-36. *Layout container*

10.3.6.3.3 Step 2

You can see the “Horizontal” layout container in the layout window under the dashboard pane (Shown in Fig. 10-37).

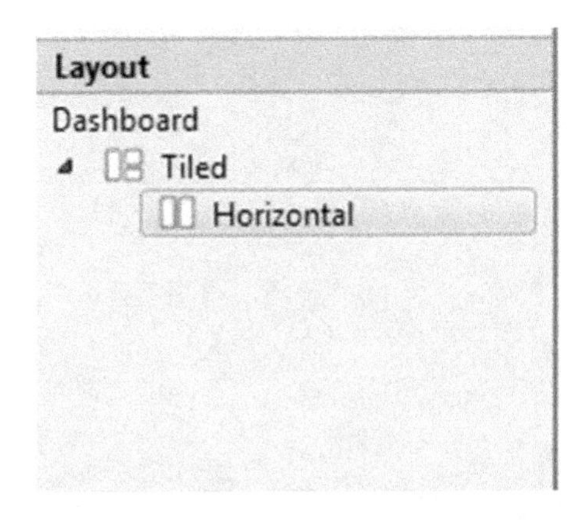

Figure 10-37. *Layout window showing “Horizontal Container”*

10.3.6.3.4 Step 3

Drag the "Running Total Shipping Costs" and "Shipping Cost" views under the dashboard pane to the horizontal containers (Shown in Fig. 10-38).

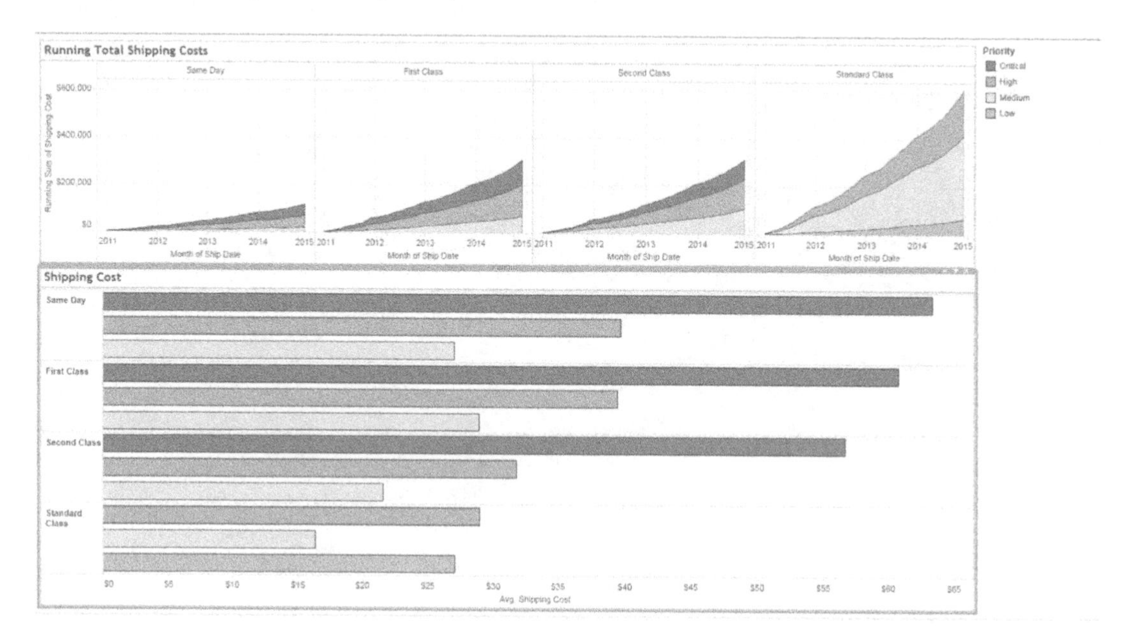

Figure 10-38. *Views, "Running Total shipping Costs" and "Shipping Cost" placed in horizontal layout containers*

10.3.6.3.5 Step 4

Click on the caret in the upper right corner of the filter to bring up the dashboard menu for "Running Total Shipping Costs" view (Shown in Fig. 10-39). From the menu, select Filters ➤ Ship Mode.

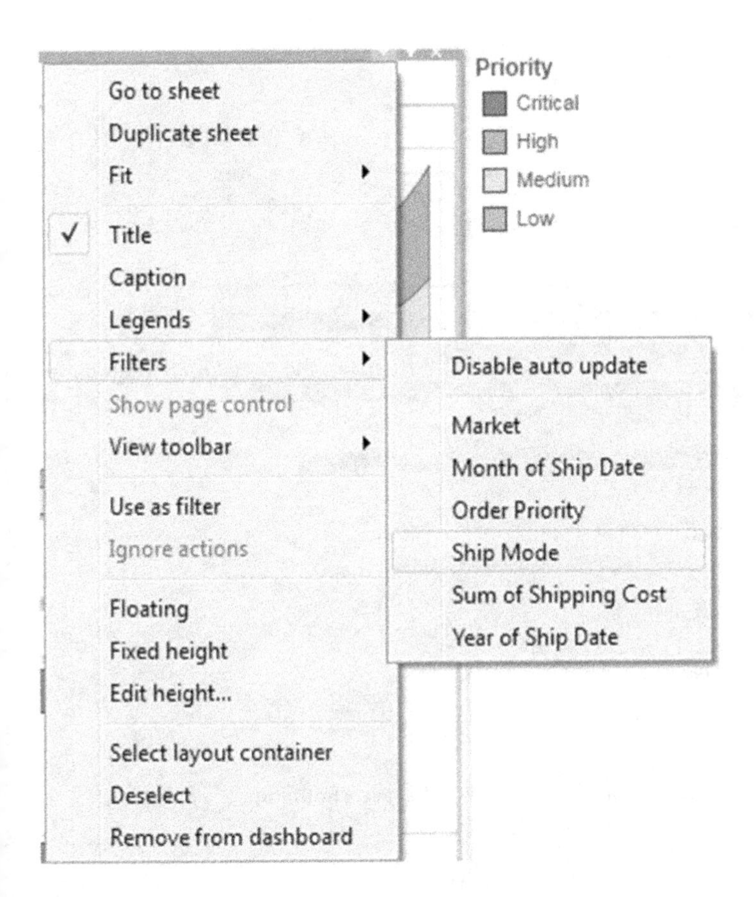

Figure 10-39. *Menu showing "Filters ➤ Ship Mode"*

10.3.6.3.6 Step 5

Observe the "Filters" on the right side of the layout (Shown in Fig. 10-40).

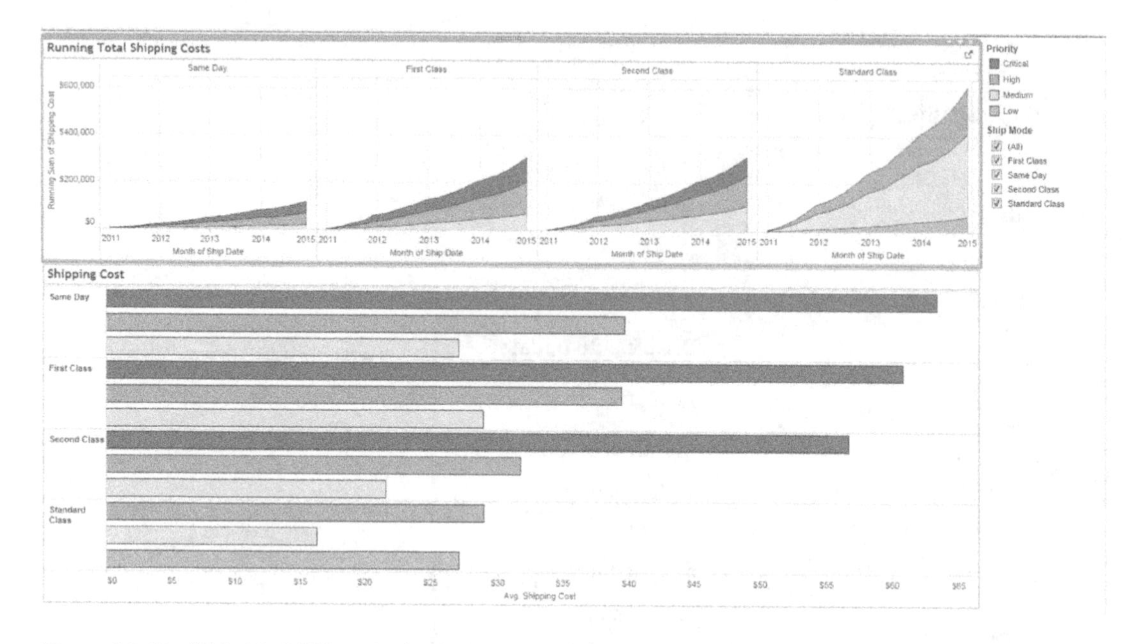

Figure 10-40. *"Ship Mode" filters in the view*

10.3.6.3.7 Step 6

When you deselect some "Ship Mode" value, "Shipping Cost "view collapses, "Running Total Shipping Costs" view fills that space automatically (Shown in Fig. 10-41).

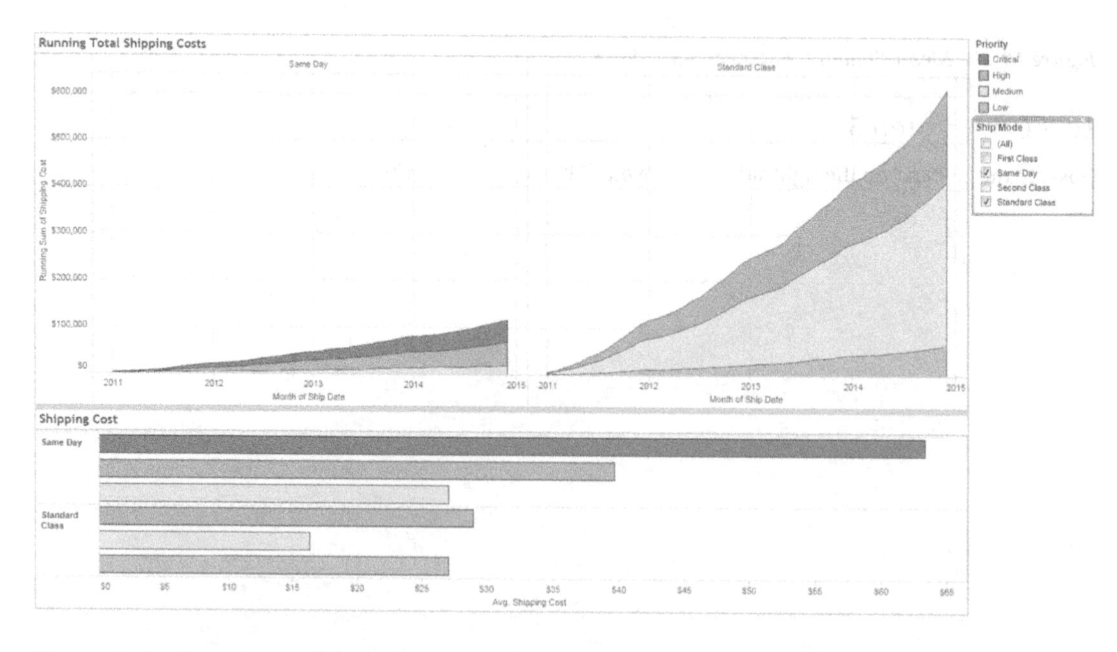

Figure 10-41. *"Running Total Shipping Costs" view automatically fills the "Shipping cost" view space when we deselect any or some "Ship Mode"*

A layout container helps you to control dashboards in an efficient way.

To add a layout container as floating, select "Floating" button as shown in Fig. 10-42.

Horizontal
Vertical
Text
Image
Web Page
Blank
New objects:
Tiled
Floating

Figure 10-42. *"Floating" button*

Next, drag the "Horizontal" container under the dashboard pane to dashboard sheet (Shown in Fig. 10-43).

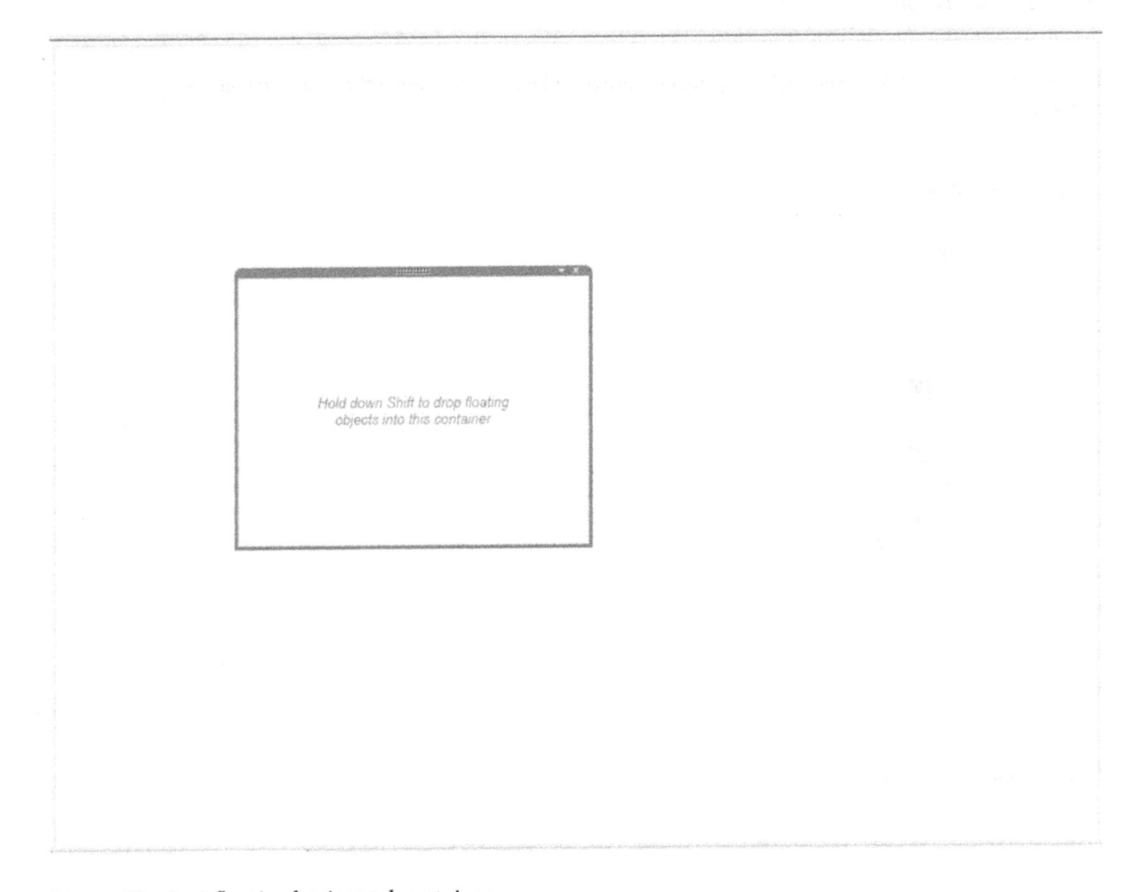

Figure 10-43. *A floating horizontal container*

You can move the floating layout anywhere in the dashboard.

To format the container, click on the header to get the drop down menu and select "Format container..." (Shown in Fig. 10-44).

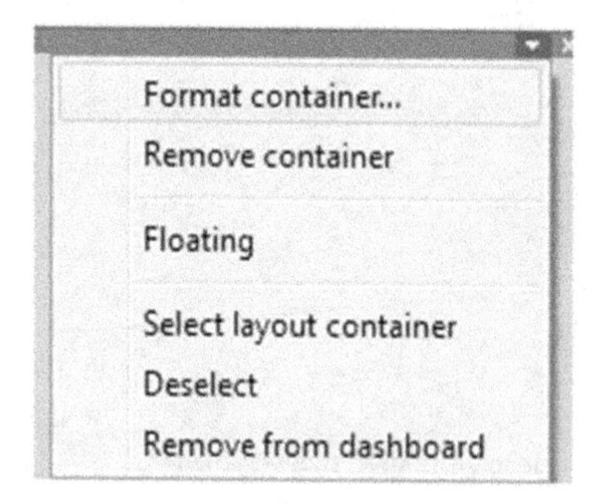

Figure 10-44. *"Format container..." option*

The Format container dialog box appears (Shown in Fig. 10-45). Select the shading color for your container.

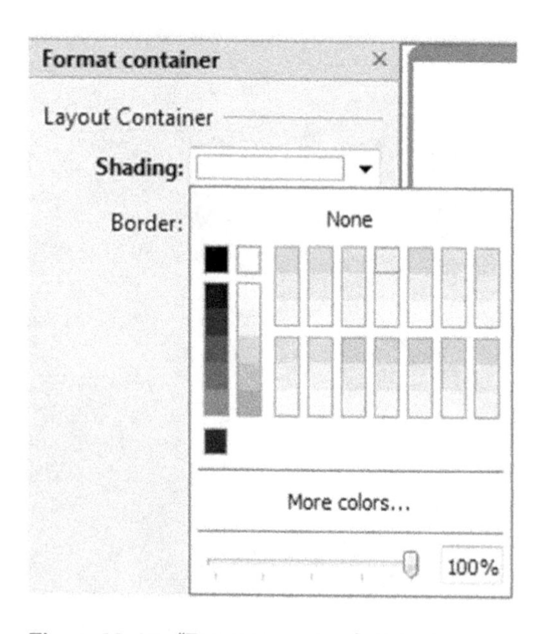

Figure 10-45. *"Format container" dialog box*

10.3.6.3.8 Set the dashboard size

You can specify the size of the dashboard using the dashboard window at the bottom of the dashboard pane (Shown in Fig. 10-46). The default size is "Desktop". You can select new size by using the drop down menu (Shown in Fig. 10-47).

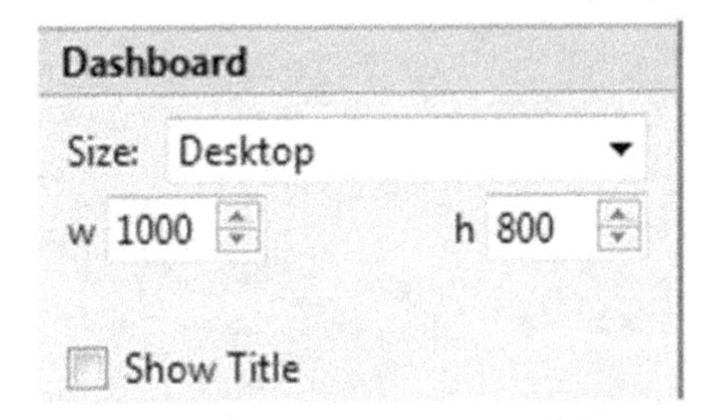

Figure 10-46. *Dashboard window showing "Desktop" as size option*

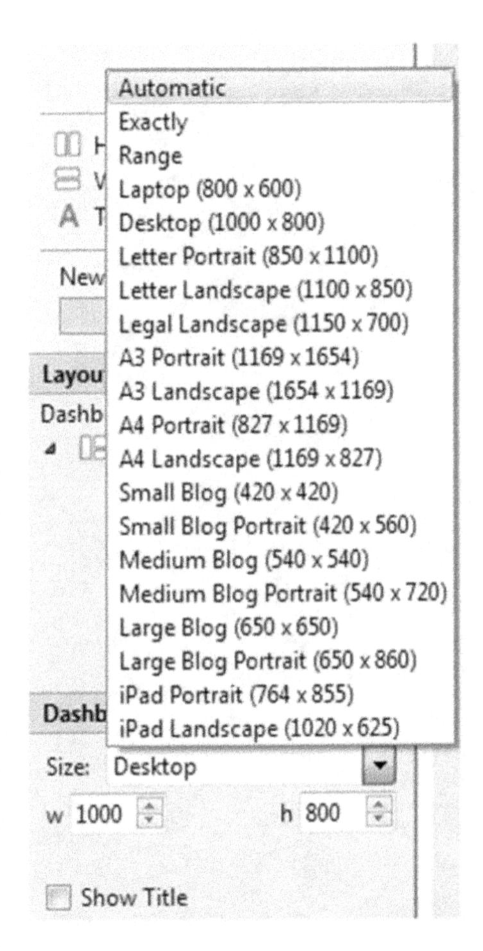

Figure 10-47. *Various options for "Dashboard Size"*

10.3.6.3.9 Rearrange dashboard views and objects

You can rearrange the view object, filter and legend by using move handle present at the top of the selected view, object, filter or legend.

10.3.6.3.10 Show or hide parts of the worksheet

Steps to show or hide parts of the worksheet.

10.3.6.3.11 Step 1

Select a view or object.

10.3.6.3.12 Step 2

Click on the drop down menu at the upper right corner of the view or object (Shown in Fig. 10-48). Select the items you want to show or hide.

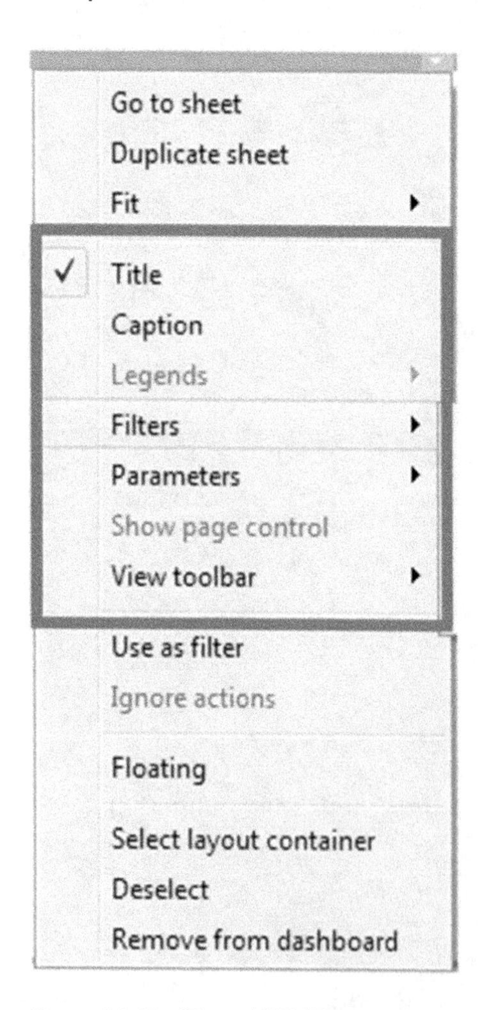

Figure 10-48. *"Show / Hide" items*

10.4 Dashboard actions

Tableau allows you to add interactivity to the dashboard using actions. With the help of actions, you can use data in one view to filter data in another view, to link external web pages, to highlight specific results.

There are three types of actions.

- **Filter Action:** Filter actions are defined by a source sheet(s) that passes one or more dimensional values as filters to target sheets upon an action.
- **Highlight Action:** Highlight actions allow you to call attention to marks of interest by coloring specific marks and dimming all others.
- **URL Action:** URL actions allow you to generate dynamically a URL based on an action and open it within a web object in the dashboard or in a new browser window or a tab.

10.4.1 Filter action

Let us discuss how to create filter action.

Begin with creating a control view to filter data in another view.

10.4.1.1 Steps to perform filter action

Follow the following steps to add filter action.

10.4.1.1.1 Step 1

Drag the dimension "Sub-Category" from the dimensions area under the data pane to the rows shelf (Shown in Fig. 10-49).

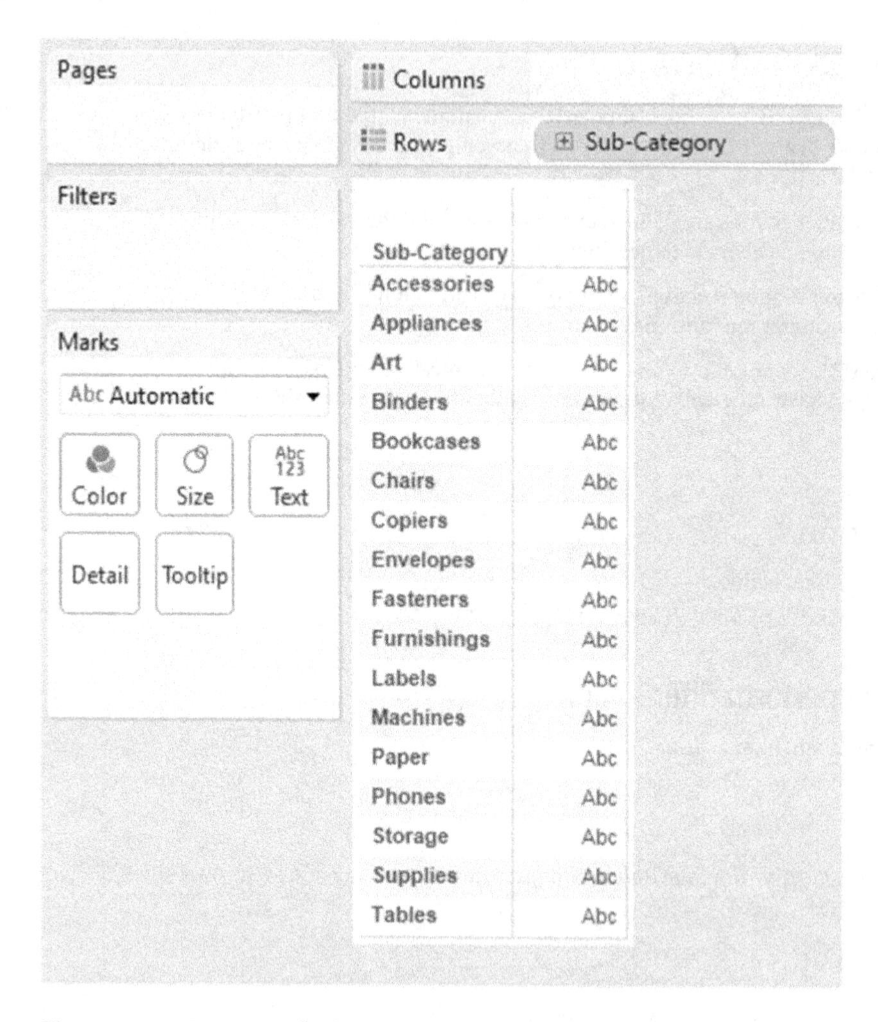

Figure 10-49. *Dimension "Sub-Category" placed on the rows shelf*

10.4.1.1.2 Step 2

Drag the dimension "Sub-Category" from the dimensions area under the data pane to "Text" on the marks card (Shown in Fig. 10-50).

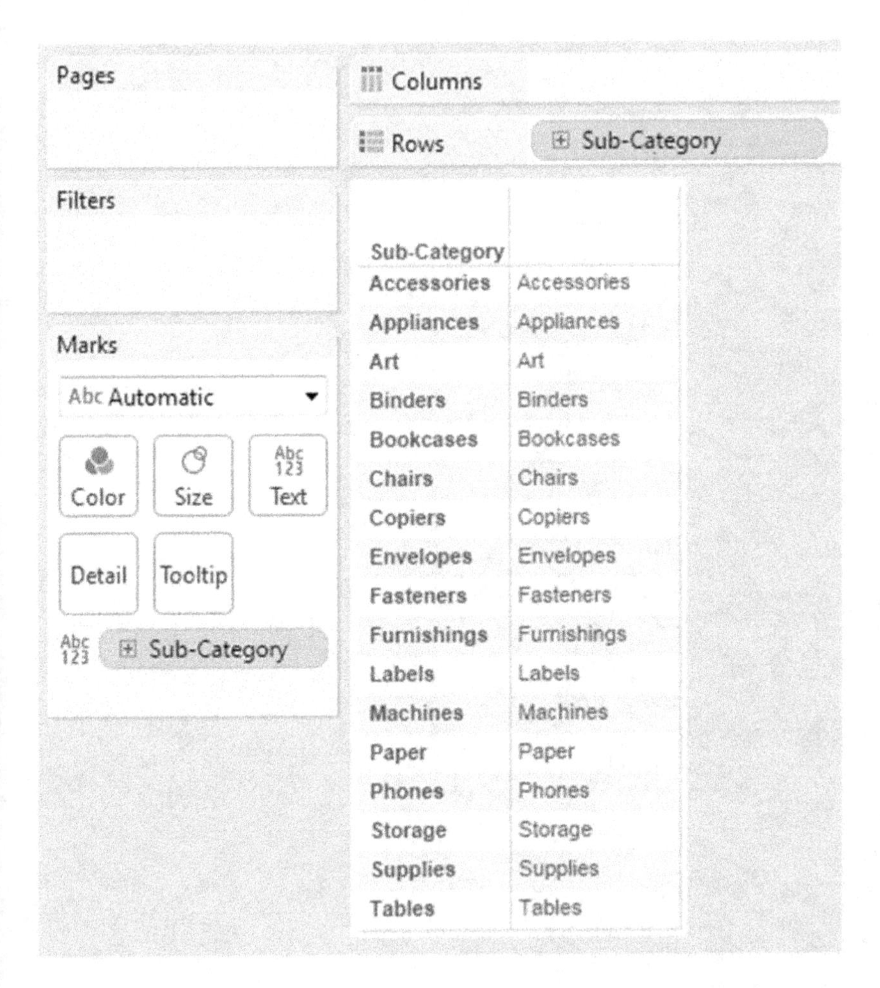

Figure 10-50. *Dimension "Sub-Category" placed on "Text" on the marks card*

10.4.1.1.3 Step 3

Right click on "Sub-Category" and unselect "Show header" option (Shown in Fig. 10-51) to remove the header.

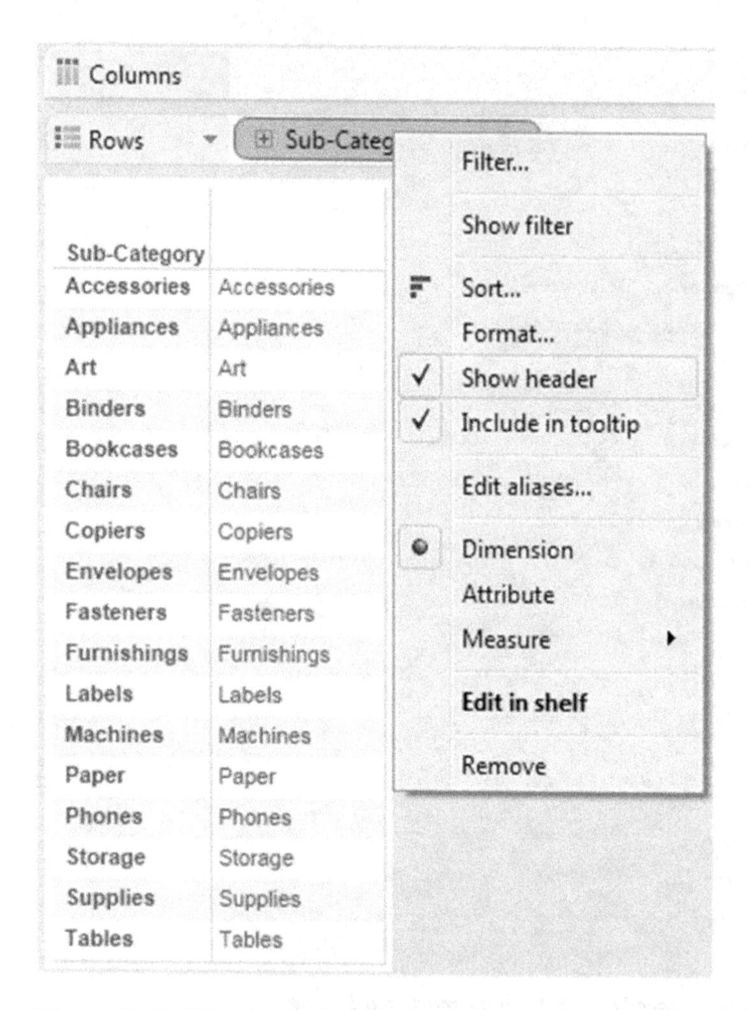

Figure 10-51. *"Show header" option*

10.4.1.1.4 Step 4

You can see the "Sub-Category" field without the header (Shown in Fig. 10-52).

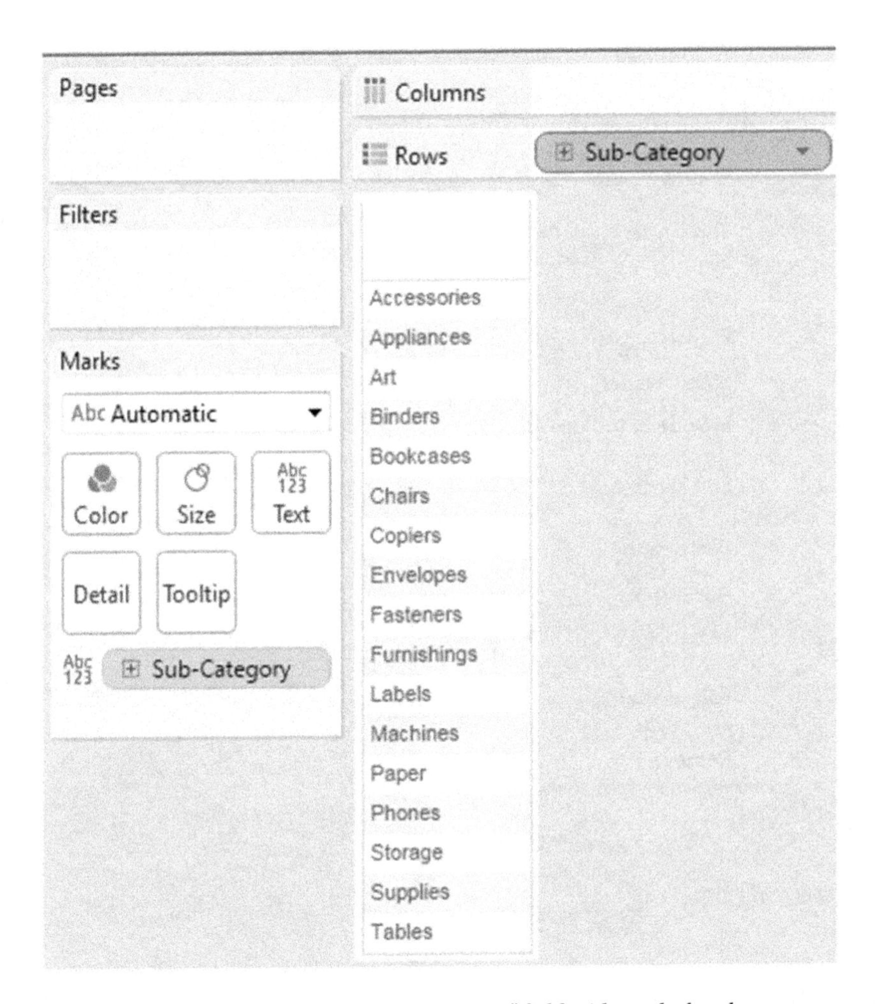

Figure 10-52. *View that displays "Sub-Category" field without the header*

10.4.1.1.5 Step 5

Right click on "Sub-Category", select "Format..." option to format "Sub-Category" (Shown in Fig. 10-53).

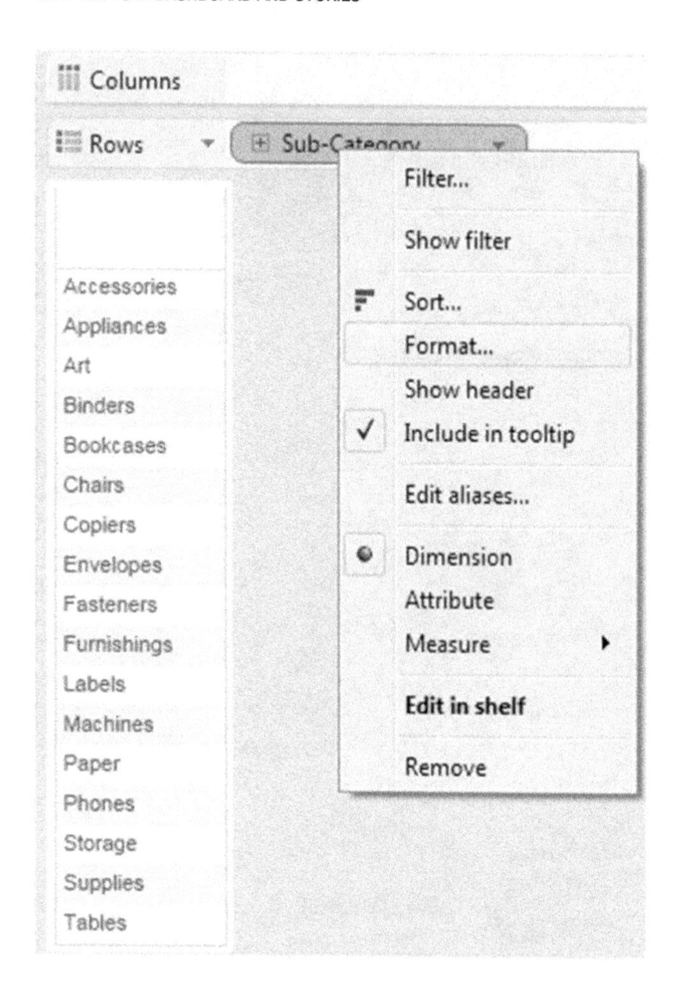

Figure 10-53. *"Format..." option*

10.4.1.1.6 Step 6

"Format Sub-Category" dialog box opens (Shown in Fig. 10-54). Select pane tab and select Font ➤ Underline.

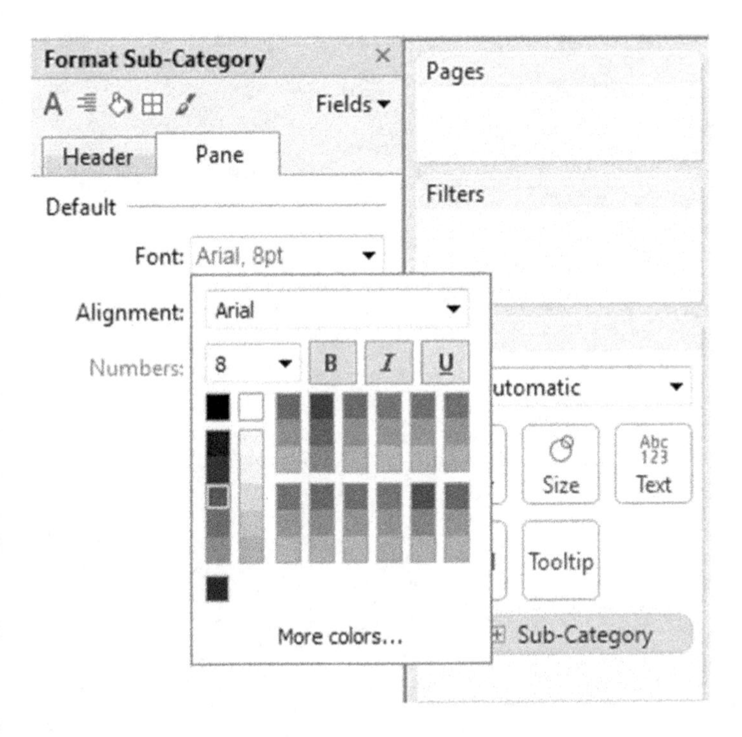

Figure 10-54. *"Format Sub-Category" dialog box*

10.4.1.1.7 Step 7

Right click in the blank area select "Title" to add title to the view (Shown in Fig. 10-55).

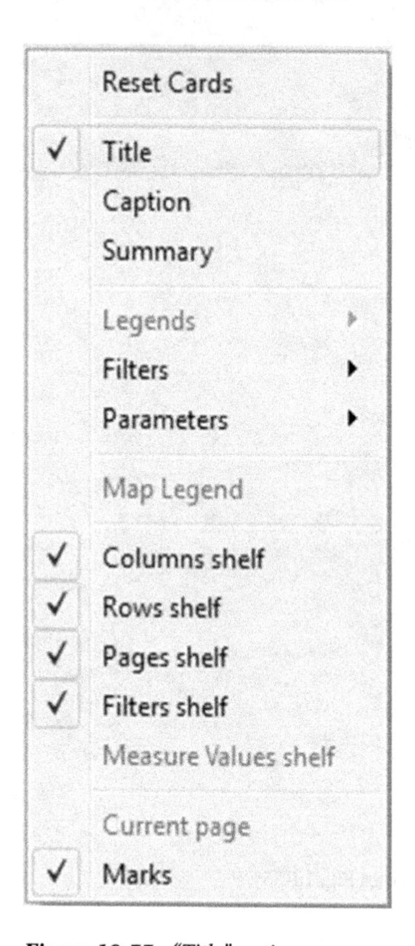

Figure 10-55. *"Title" option*

10.4.1.1.8 Step 8

"Edit Title" dialog box opens, type in the required title for your view (Shown in Fig. 10-56).

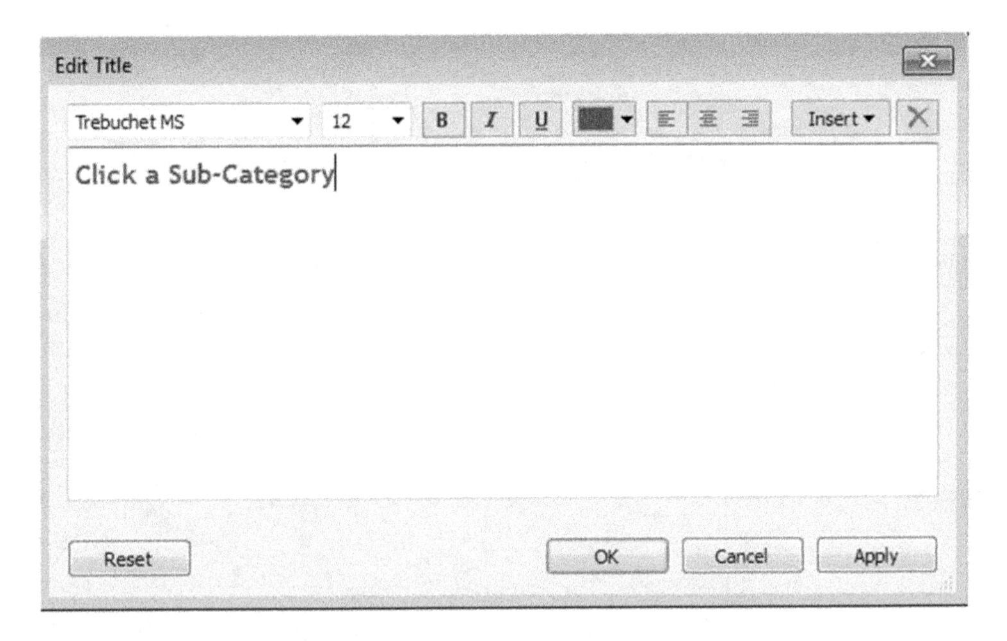

Figure 10-56. *"Edit Title"dialog box*

10.4.1.1.9 Step 9

Drag the dimension "Category" from the dimensions area under the data pane to the rows shelf (Shown in Fig. 10-57).

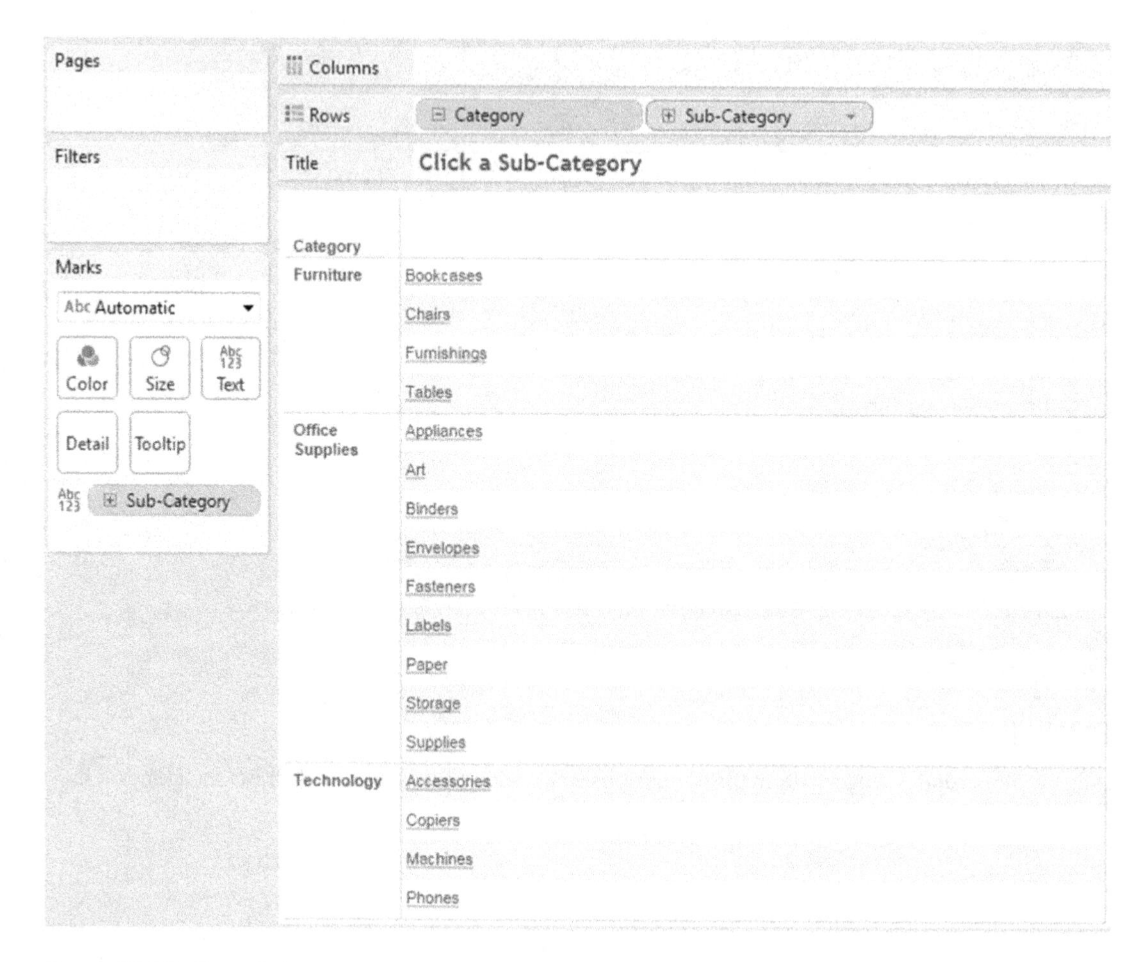

Figure 10-57. *Control view*

Let us create a “Sales vs Profit” and a “Sales Map” view.

10.4.1.1.10 Step 10

Create a “Sales vs Profit” view (Shown in Fig. 10-58). Apply filter to region - west.

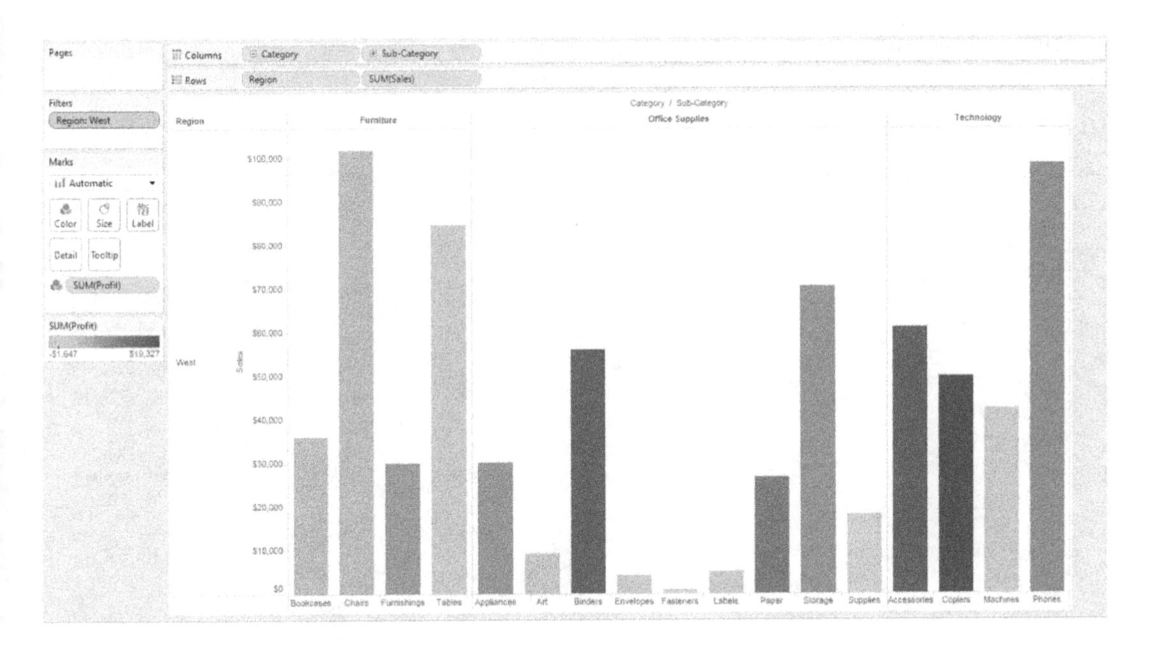

Figure 10-58. *Sales vs profit view*

10.4.1.1.11 Step 11

Create sales map view (Shown in Fig. 10-59). Apply filter to the “Region -West”.

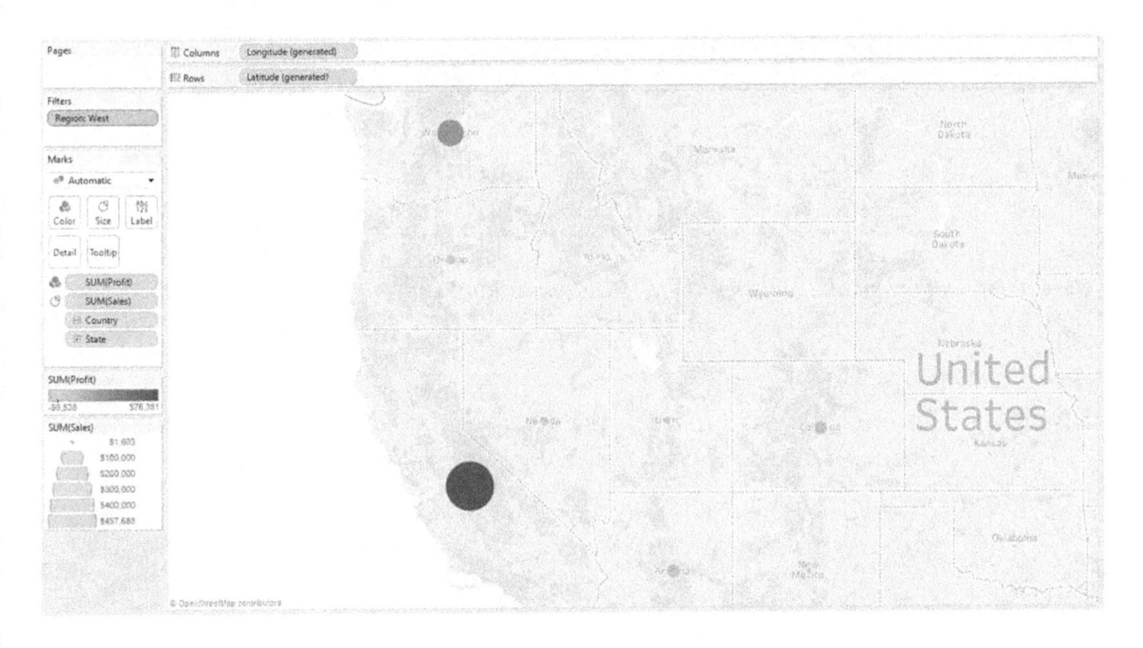

Figure 10-59. *Sales map view*

Create a dashboard.

10.4.1.1.12 Step 12

Create a dashboard as shown in Fig. 10-60.

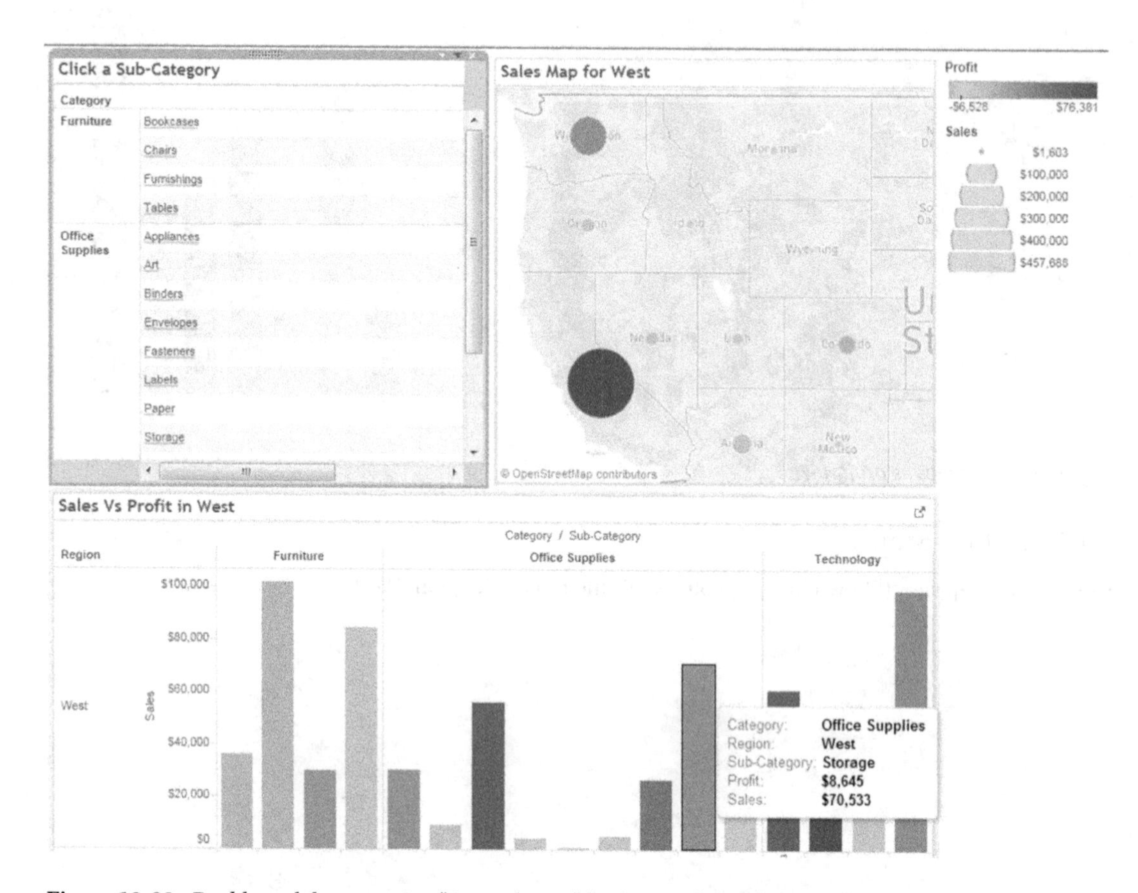

Figure 10-60. *Dashboard that contains "Control view", "Sales Map" and "Sales vs Profit" views*

10.4.1.1.13 Step 13

Select Control view, click on caret to bring up the menu, select **"Use as Filter"** (Shown in Fig. 10-61).

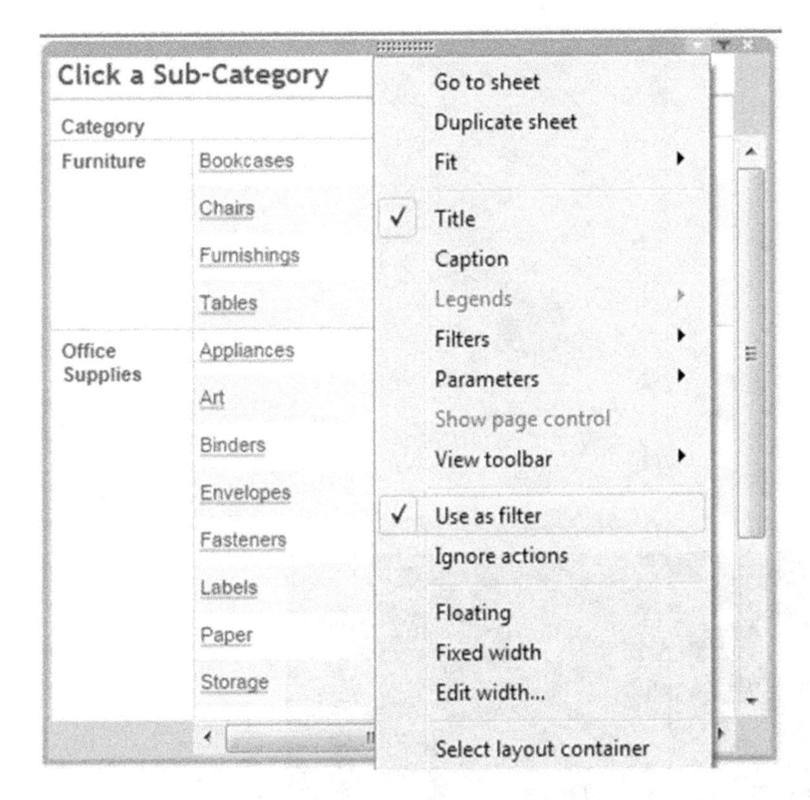

Figure 10-61. *Menu showing "Use as filter" selected*

10.4.1.1.14 Step 14

When you select a specific "Sub-Category" value, other views in the dashboard are updated automatically (Shown in Fig. 10-62).

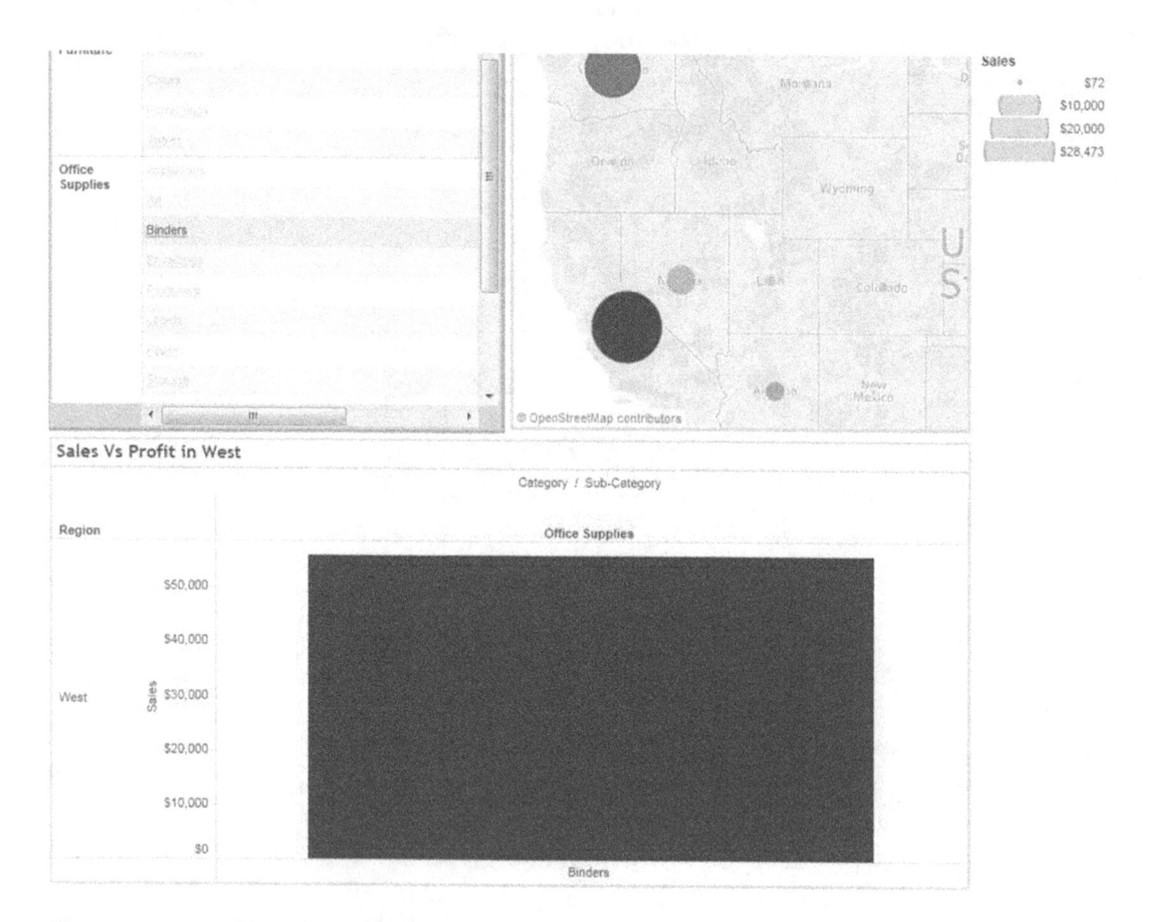

Figure 10-62. *Dashboard with "Filter" action*

10.4.2 Highlight Action

Steps to create highlight actions.

10.4.2.1 Steps to create highlight actions

10.4.2.1.1 Step 1

Consider the dashboard shown in Fig. 10-63.

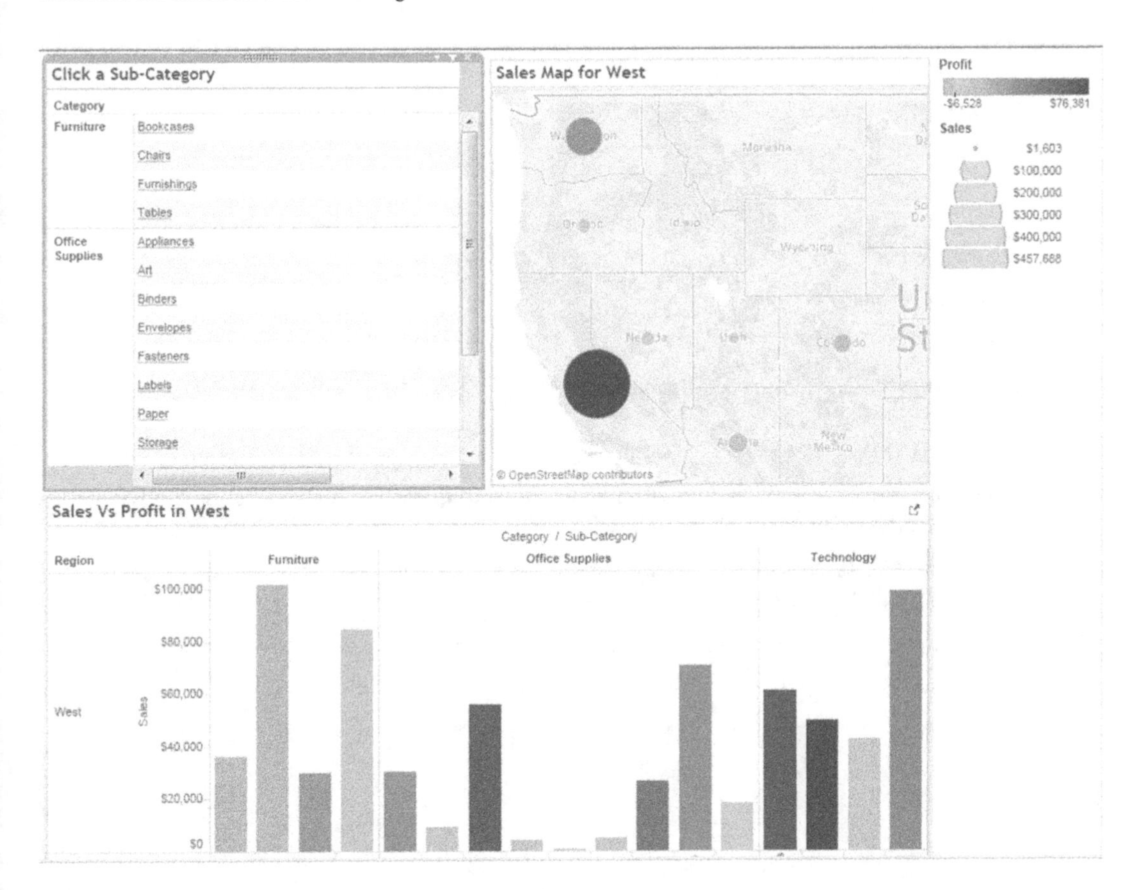

Figure 10-63. Dashboard that shows "Control view", "Sales Map" and "Sales vs Profit" views

10.4.2.1.2 Step 2

Go to the dashboard Menu and select "Actions ..." (Shown in Fig. 10-64).

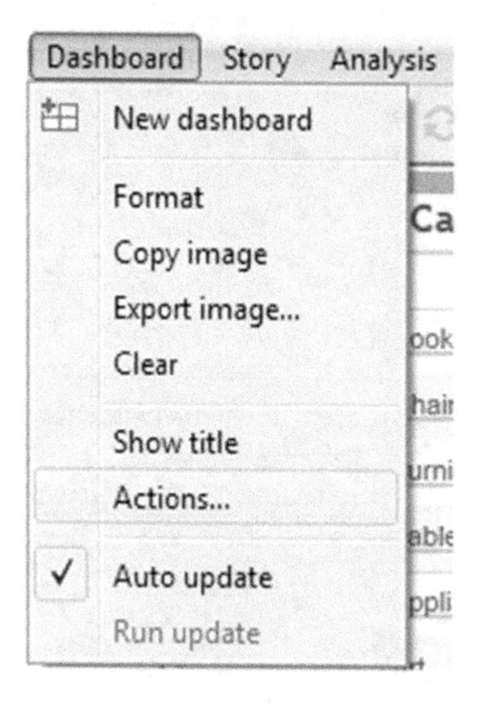

Figure 10-64. *Dashboard menu showing "Actions..." item*

10.4.2.1.3 Step 3

When you click on Actions, "Action [Dashboard with filter Action]" dialog box opens. Click on "Add Action", select "Highlight..." action (Shown in Fig. 10-65).

Actions [Dashboard with filter Action]

Connect sheets to external web resources using URL actions, or to other sheets in the same workbook using Filter actions and Highlight actions.

Name	Run On	Source	Fields

Add Action > Filter... Highlight... URL...

Edit... Remove

Show action[s] ...ok

OK Cancel

Figure 10-65. *"Actions [Dashboard with filter Action]" dialog box showing different actions*

10.4.2.1.4 Step 4

When you click on highlight action, the "Edit highlight action" dialog box opens. Select "Control View" as "Source Sheets" and "Sales Vs Profit" as "Target Sheets" and "Run action on" as "Select" (Shown in Fig. 10-66).

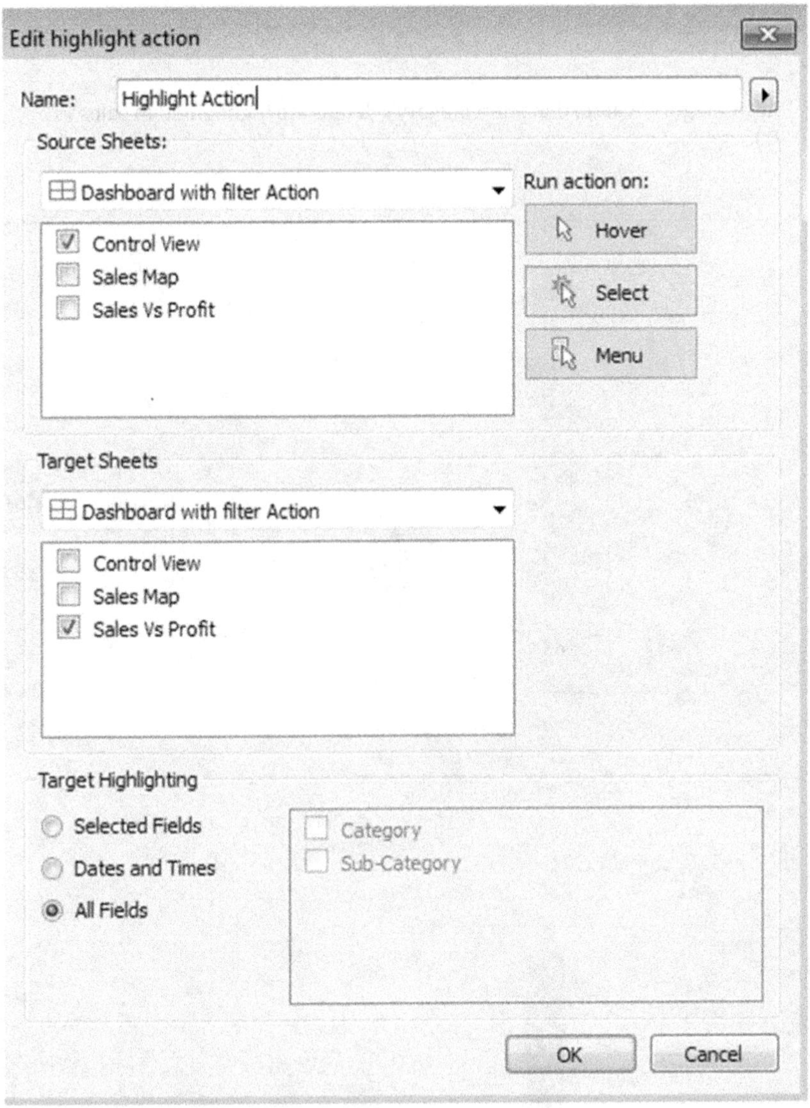

Figure 10-66. *"Edit highlight action" dialog box*

Filter actions can be set to occur on any one of the three possible actions:

Hover: The user moves the mouse cursor over a mark.

Select: The user clicks on a mark or lassos multiple marks by clicking and dragging a rectangle around them.

Menu: The user selects the menu option for the action on the tooltip.

10.4.2.1.5 Step 5

When user selects a specific "Sub-Category" value, that sub-category's details are highlighted in sales vs profit view (Shown in Fig. 10-67).

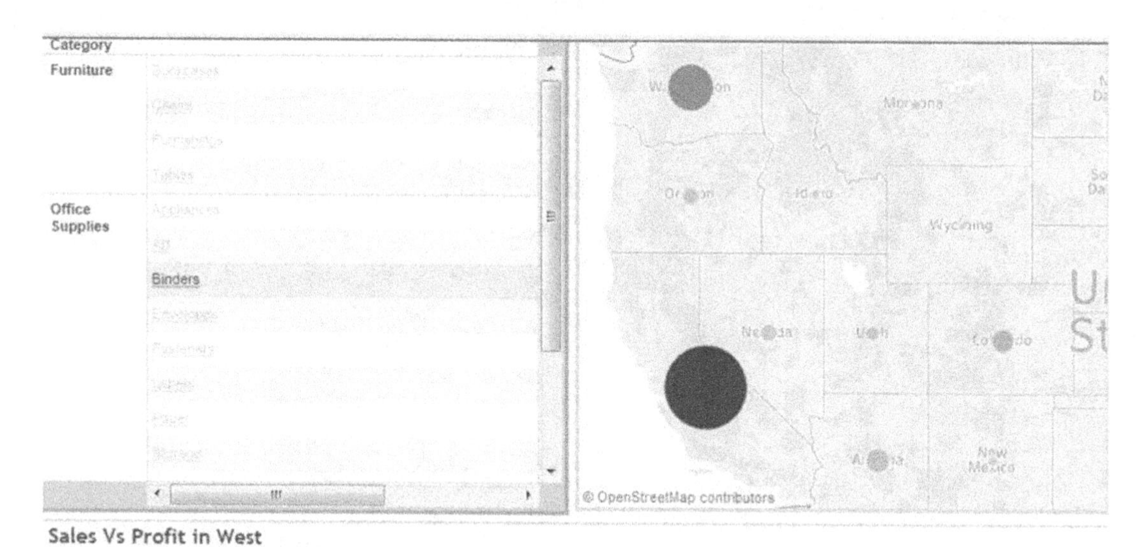

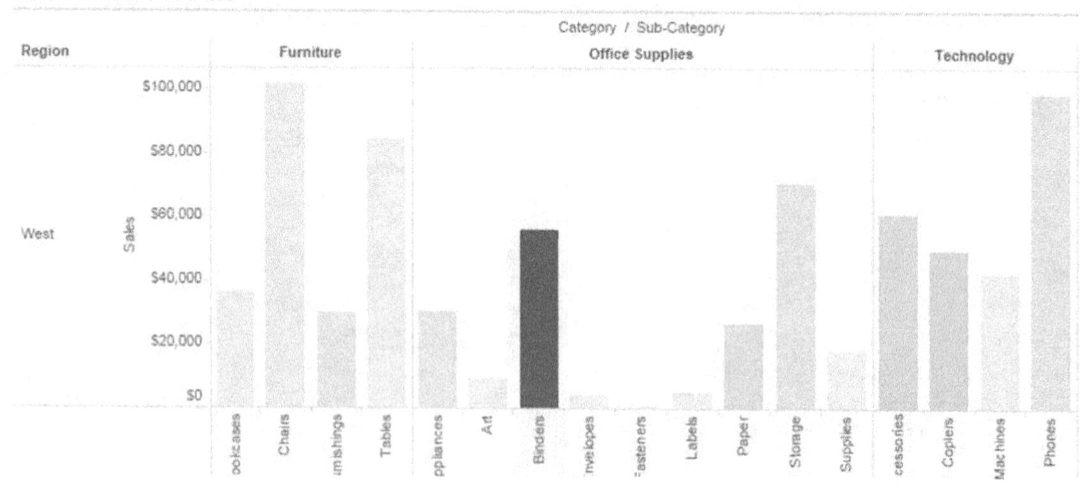

Figure 10-67. Highlighted sub-category "Binders" in "Sales vs Profit"

10.4.2.2 URL Action

Steps to add URL action to the dashboard to specify an external link.

10.4.2.2.1 Step 1

Select URL action as shown in Fig. 10-68.

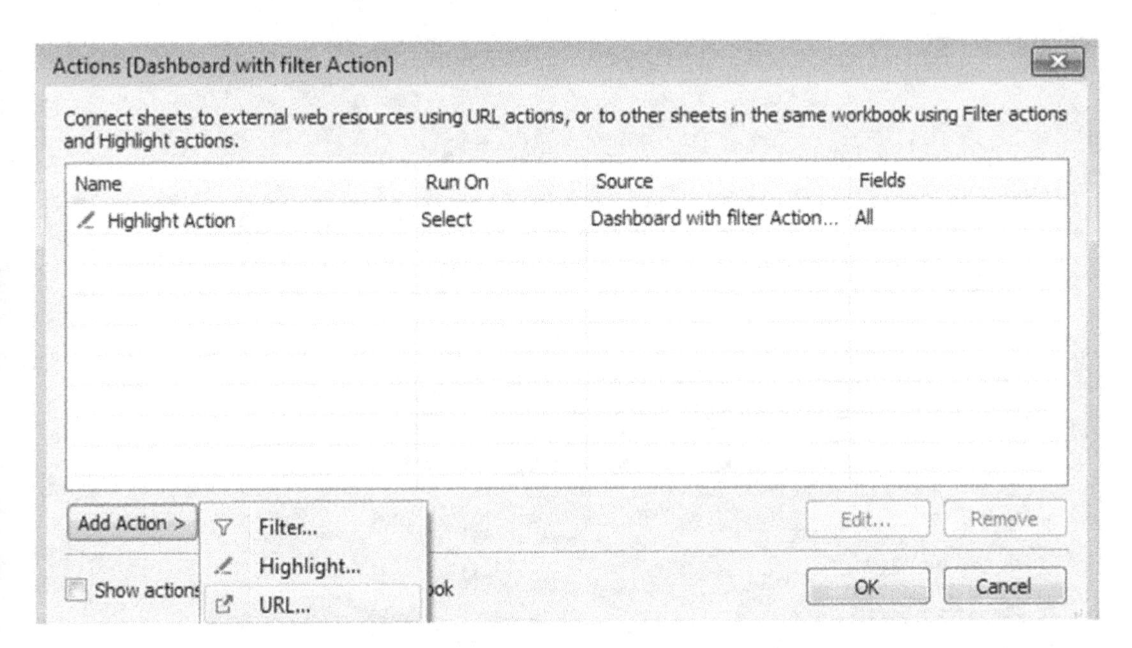

Figure 10-68. *Actions [dashboard with filter action] - URL action*

10.4.2.2.2 Step 2

When you click on URL action, “Edit URL Action” dialog box opens as shown in Fig. 10-69. From the “Source Sheets”, select control view, type the required URL to link to a particular Web page.

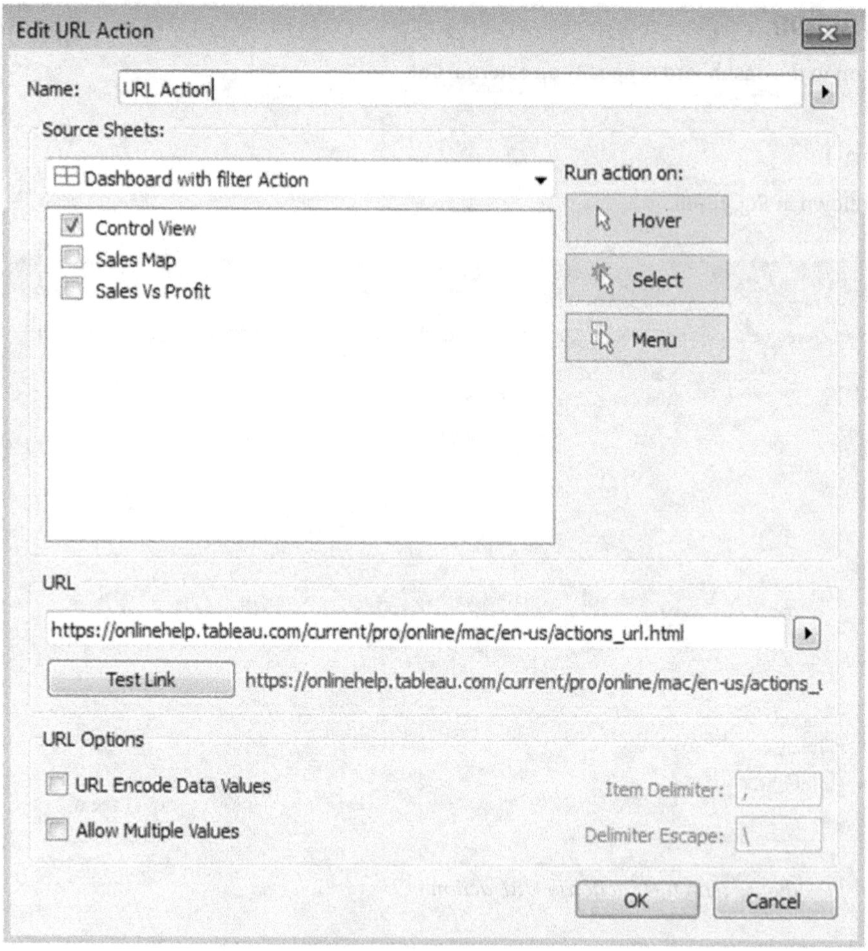

Figure 10-69. "Edit URL Action" dialog box

10.4.2.2.3 Step 3

When you select a specific "Sub-Category" value, you can see the URL Action link as shown in Fig. 10-70, which is linked to the "URL Action" (that leads to Tableau documentation).

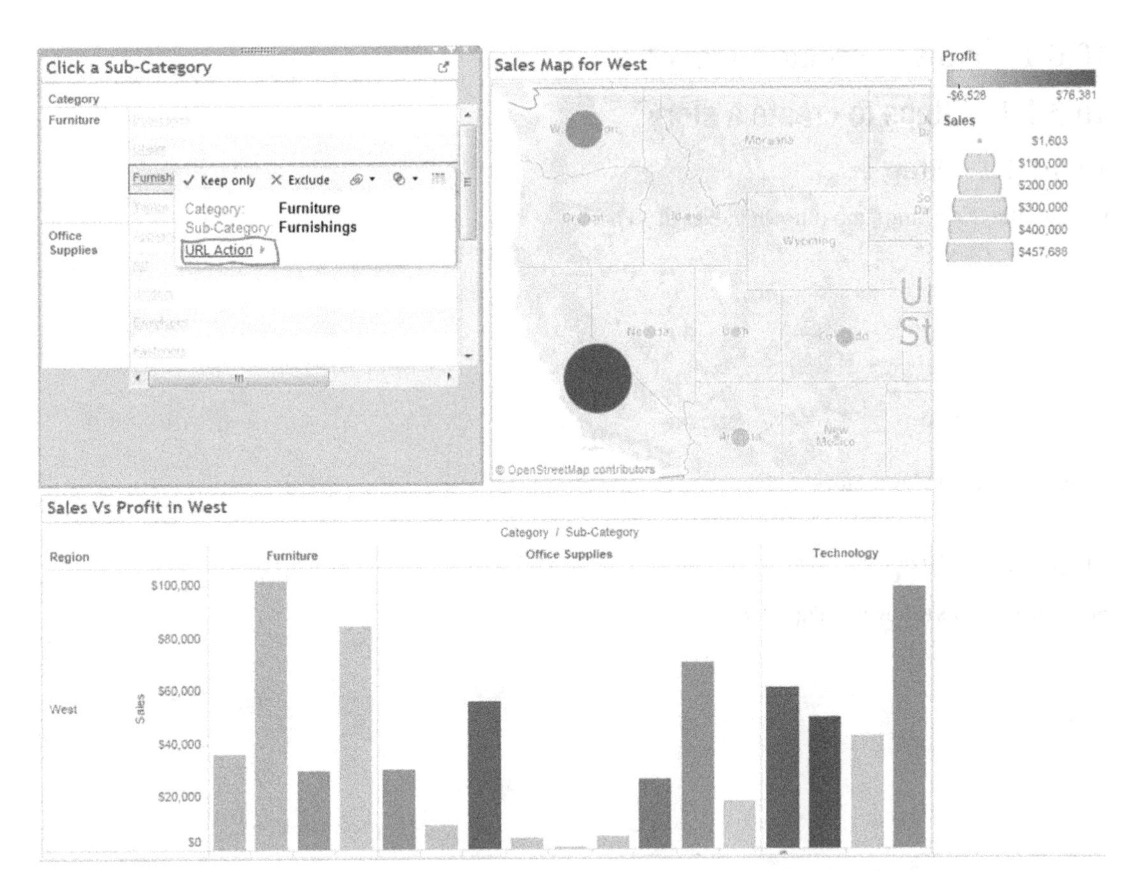

Figure 10-70. *"URL Action" for "Sub- Category"*

10.5 Creating a story

A story is a sheet that can contain a sequence of worksheets or dashboards to convey certain information. You can create stories to tell how facts are connected, to provide context and to tell compelling cases.

10.6 What is a story?

A story is a sheet or a collection of worksheets arranged in sequence. Each individual sheet in a story is known as story point.

In Tableau, stories are not just a collection of static sheets. You can make your story points remain connected to the underlying data to reflect data changes.

You can use stories in different ways.

- Use stories for collaborative analysis: You can assemble the sequence sheet to perform what-if analysis.
- Use stories as presentation tools: You can use stories to present history of views or dashboards to audience.

10.6.1 How to create a story?

10.6.1.1 Steps to create a story

10.6.1.1.1 Step 1

Click on the "New Story" tab (Shown in Fig. 10-71).

Figure 10-71. *The "New Story" tab*

10.6.1.1.2 Step 2

Story sheet opens (Shown in Fig. 10-72).

Story Title

Add a caption

Drag a sheet here

Figure 10-72. *"Story Sheet" on display*

10.6.1.1.3 Step 3

You can choose "Size" for your story from the lower-left corner of the dashboards and worksheets pane as shown in Fig. 10-73.

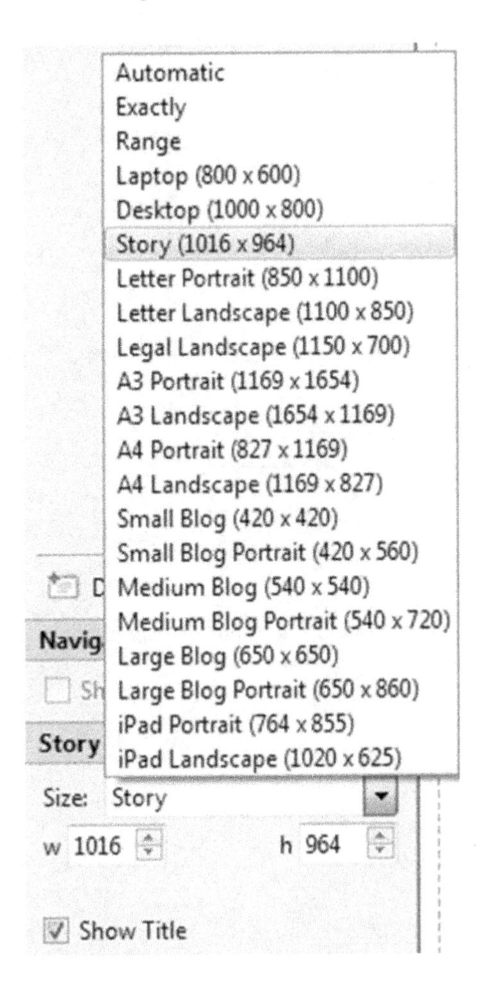

Figure 10-73. *"Story Size" option*

10.6.1.1.4 Step 4

Click on the story title to edit the title of the story as shown in Fig. 10-74.

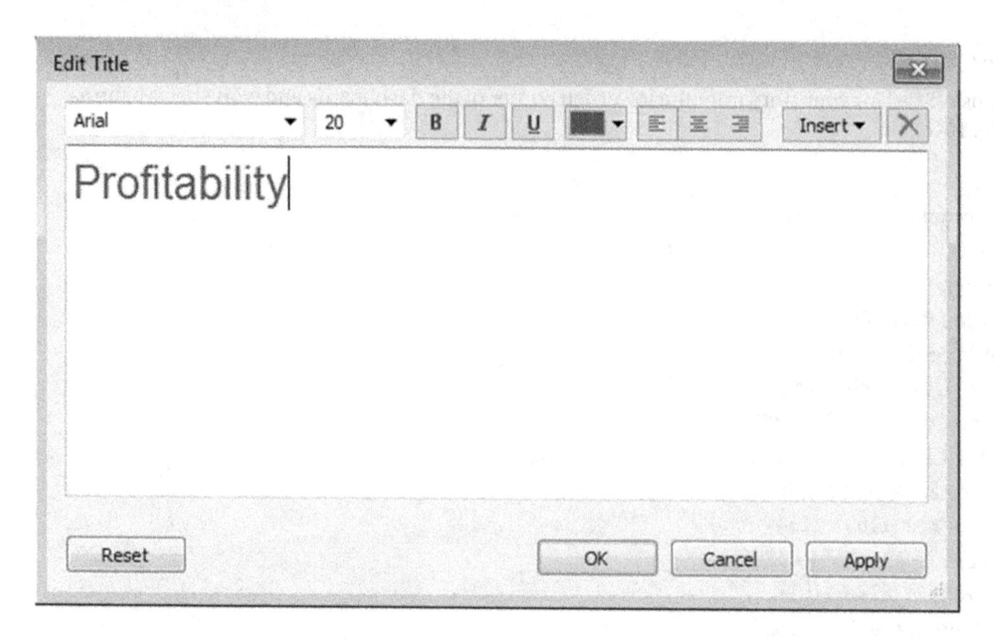

Figure 10-74. *Story "Edit Title" dialog box*

10.6.1.1.5 Step 5

Drag the "Sales vs Profit by Sales" view under dashboards and worksheets pane to the center of the view (Shown in Fig. 10-75).

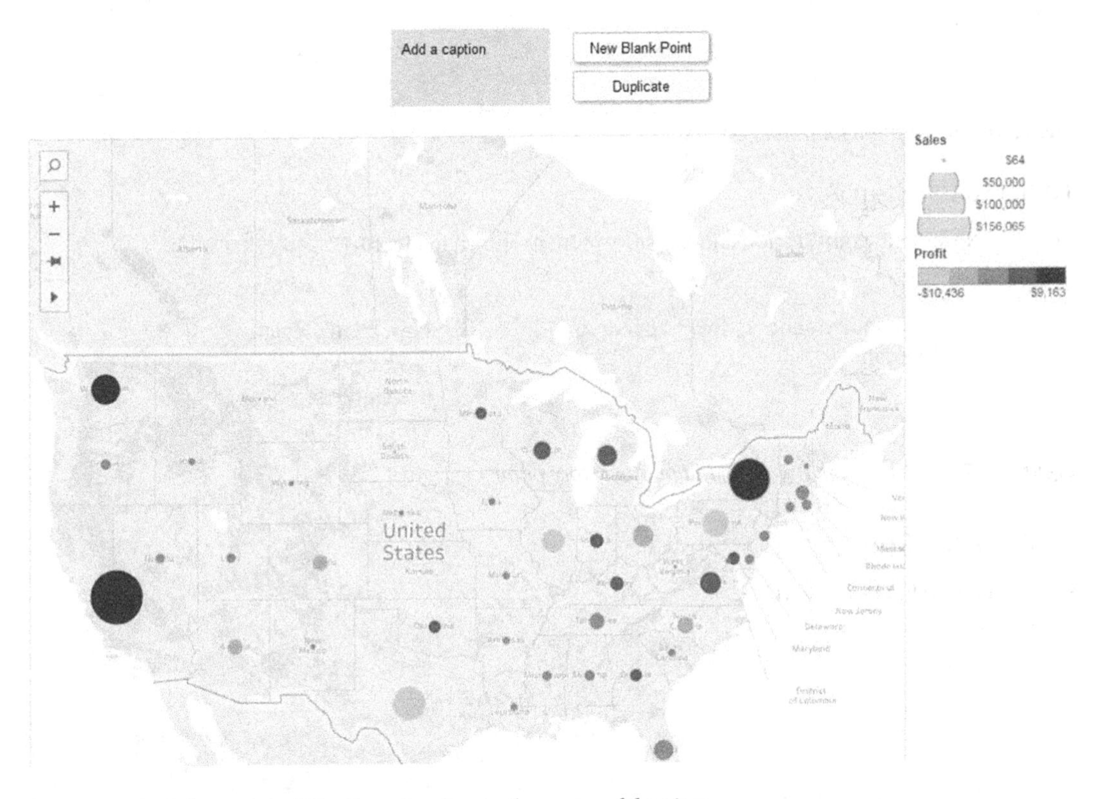

Figure 10-75. *Sales and profit by the sales view in the center of the view*

10.6.1.1.6 Step 6

Click on "Add a caption" to summarize the story point as shown in Fig. 10-76.

The overall profit looks good. Is this happening across all Category?

New Blank Point

Duplicate

Figure 10-76. *Caption for a story point*

10.6.1.1.7 Step 7

Click on "New Blank Point" to add another story point as shown in Fig. 10-77.

The overall profit looks good. Is this happening across all Category?

New Blank Point

Duplicate

Figure 10-77. *"New Blank Point" to add another story point*

10.6.1.1.8 Step 8

Sheet opens as shown in Fig. 10-78.

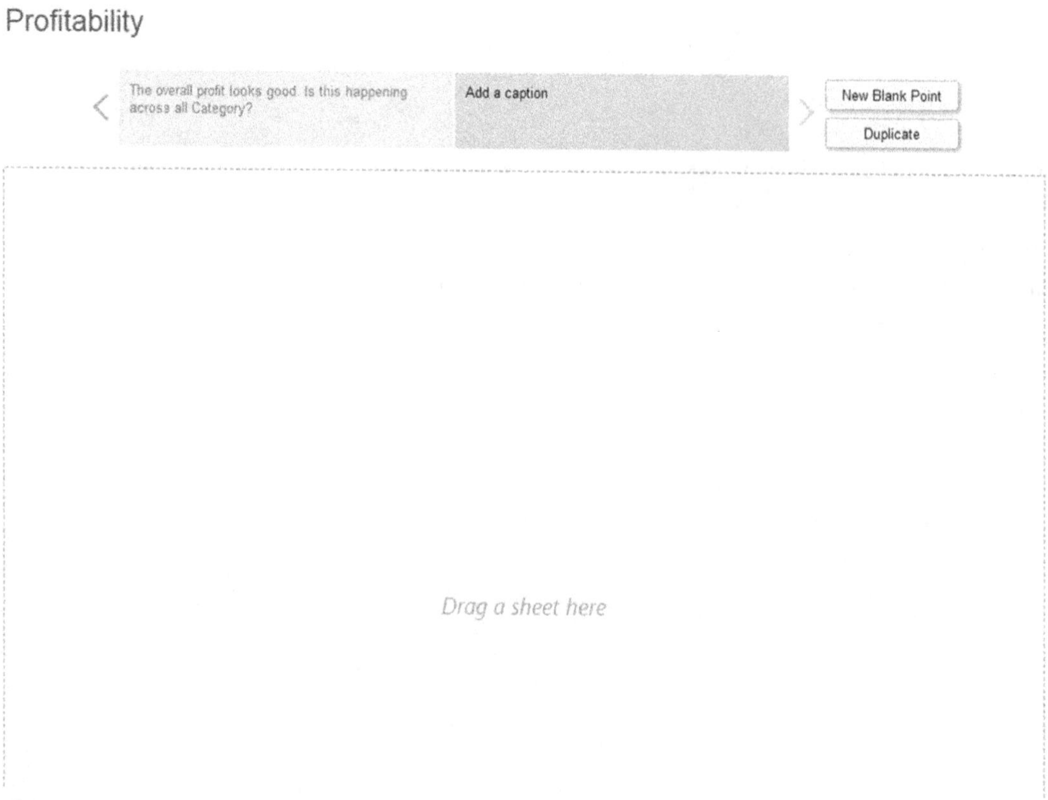

Profitability

The overall profit looks good. Is this happening across all Category?

Add a caption

New Blank Point

Duplicate

Drag a sheet here

Figure 10-78. *"New story sheet" with caption*

10.6.1.1.9 Step 9

Add a sheet and caption as shown in Fig. 10-79.

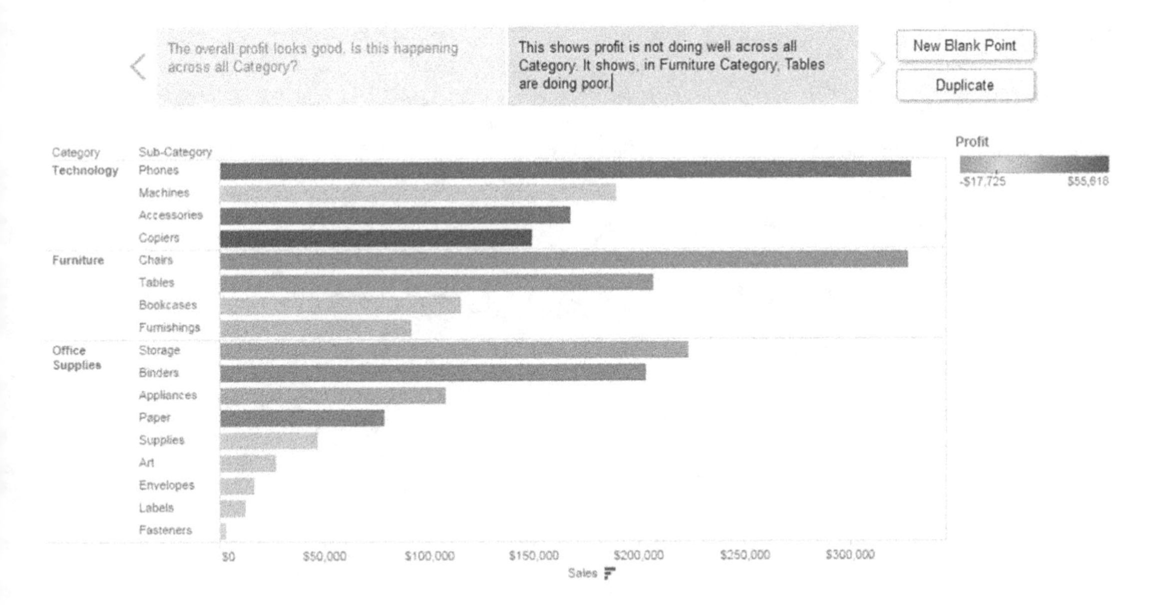

Figure 10-79. *"Sales by Category" story point*

10.6.1.1.10 Step 10

You can continue to add story points until your story is complete.

The sheets remain connected to the original sheet. If you make any changes to the original sheet, your changes will be updated in the story points automatically. But changes you make in a story point do not automatically update the original sheet."

10.6.2 Description

10.6.2.1 Steps to add description to the "Story Point"

Follow the following steps.

10.6.2.1.1 Step 1

To add description to the story point, double click on the description present in the lower left corner under the dashboards and worksheets (Shown in Fig. 10-80).

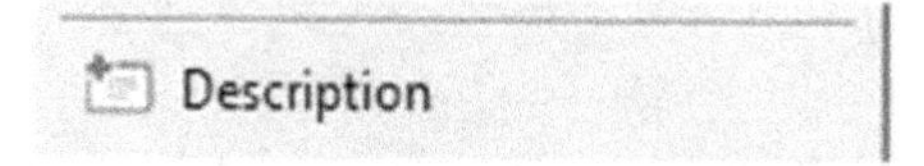

Figure 10-80. "Description" option

10.6.2.1.2 Step 2

"Edit Description" dialog box opens as shown in Fig. 10-81.

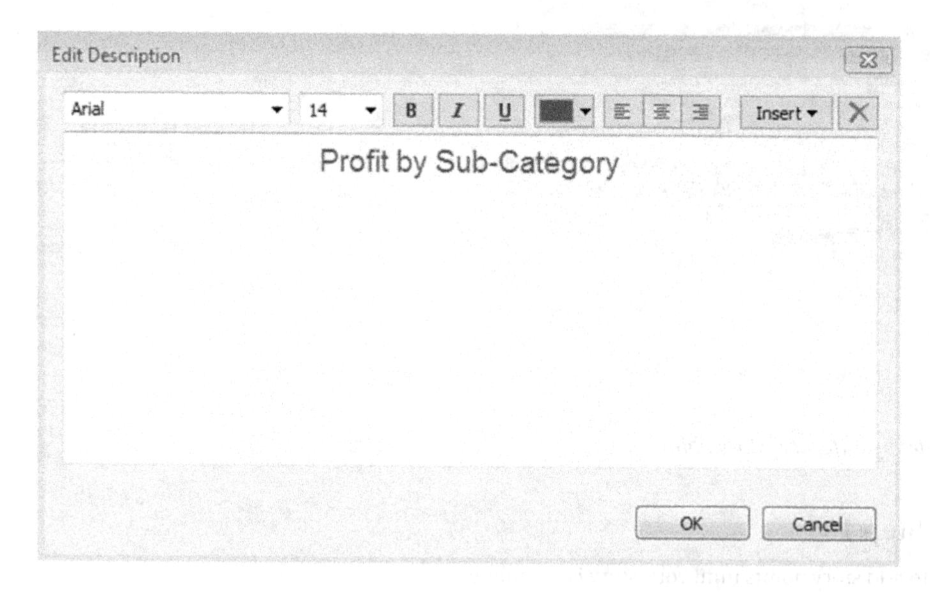

Figure 10-81. "Edit Description" Window

10.6.2.1.3 Step 3

You can drag and drop description anywhere in the sheet as shown in Fig. 10-82.

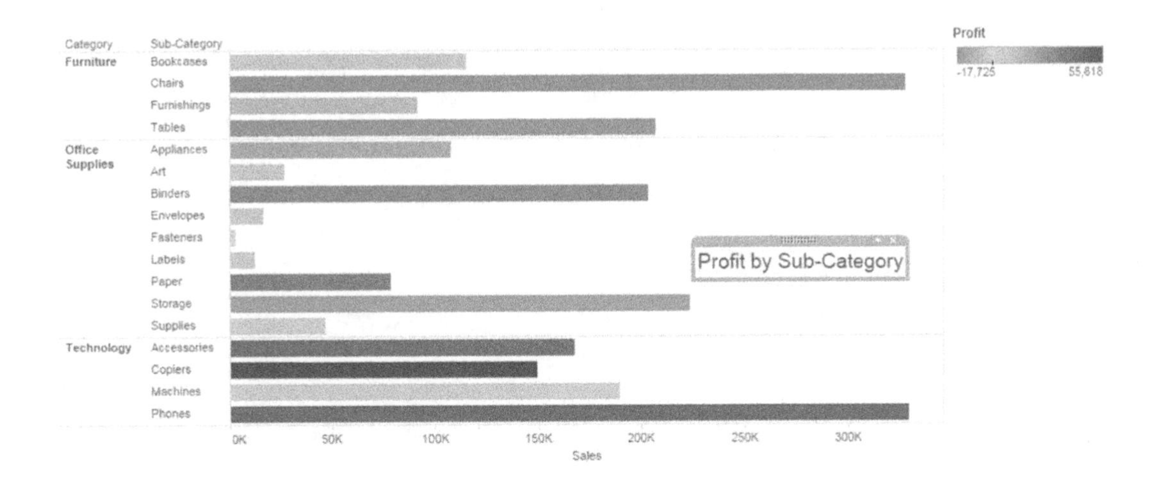

Figure 10-82. *Description "Profit by Category"*

To present a story use presentation mode button on the tool bar.

10.7 Points to remember

- A dashboard is a visual display of the most important information needed to achieve one or more objectives that fits entirely on a single computer screen so it can be monitored at a glance.
- You can add interactivity to the dashboard using actions.
- A story is a sheet that can contain a sequence of worksheets or dashboards to convey certain information. You can create stories to tell how facts are connected, to provide context and to tell compelling cases.

10.8 Next steps

In the next chapter, we will learn about integration of R with Tableau and be introduced to the following:

- Functions such as (SCRIPT_INT(), SCRIPT_REAL(), SCRIPT_BOOL(), SCRIPT_STR())
- Data mining
- Affinity analysis
- K-means clustering

CHAPTER 11

Integration of Tableau with R

The last chapter introduced us to dashboards in Tableau. We learned to create interactive dashboards and tell stories using data. This chapter will introduce us to the integration of Tableau with R.

R is an open-source statistical analysis tool. Tableau Desktop can connect to R through calculated fields and leverage the benefits provided by R functions, libraries, and packages. These calculations dynamically invoke the R engine and pass values to R via the Rserve package, and return the computed results back to Tableau. This integration of R with Tableau harnesses the statistical analytical abilities of R with the drag and drop visualization power of Tableau.

11.1 Steps to bring about this integration

- Start R and Rserve package.

  ```
  > install.packages("Rserve")
  > library(Rserve)
  > Rserve()
  ```

- Start Tableau.
- Connect to Rserve.

  Help ➤ Settings and Performance ➤ Manage R Connection...

  Refer to Fig. 11-1 and Fig. 11-2.

S. Acharya and S. Chellappan, *Pro Tableau*, DOI 10.1007/978-1-4842-2352-9_11

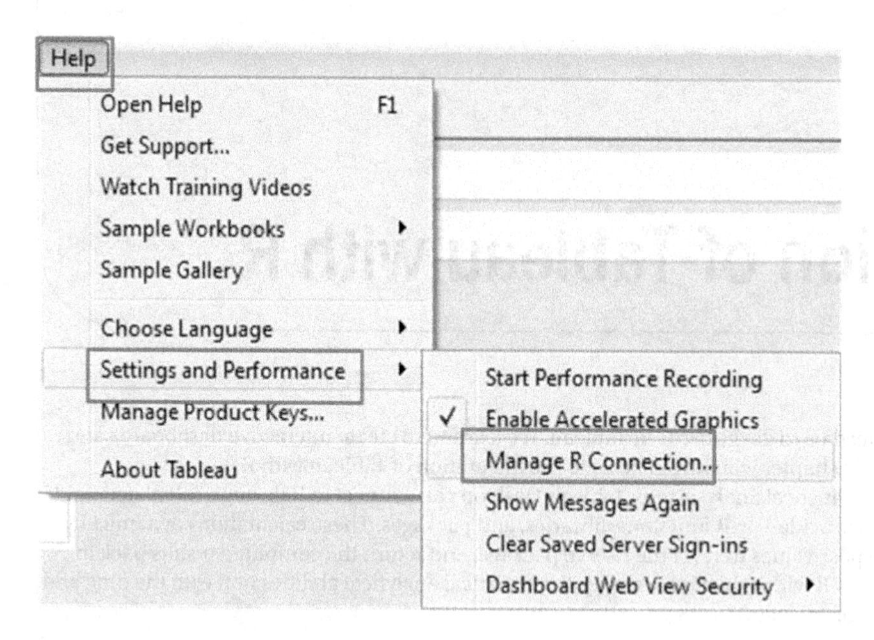

Figure 11-1. *Managing R connection*

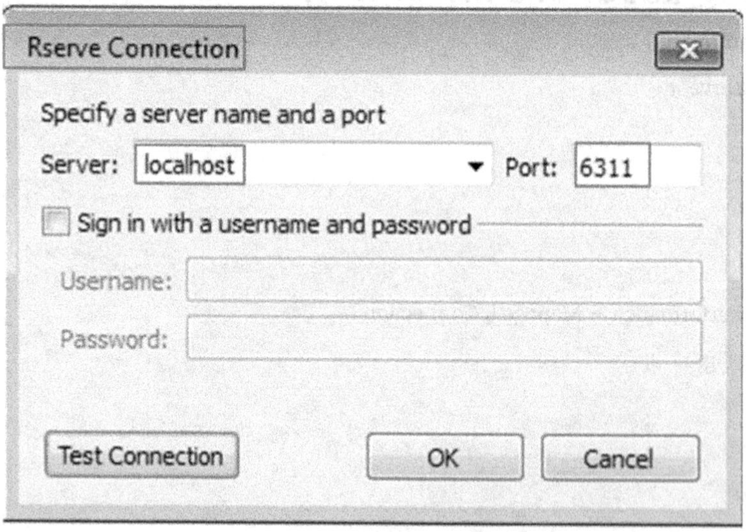

Rserve Connection
Specify a server name and a port
Server: localhost Port: 6311
Sign in with a username and password
Username:
Password:
Test Connection OK Cancel

Figure 11-2. *Setting up Rserve connection*

- Read in data into Tableau.
- Create calculated fields to invoke R functionality.
- Create charts and dashboards using dimensions, measures and calculated fields.

There are four new built-in functions that are used to call specific R models and functions:

- SCRIPT_REAL - returns a real number
- SCRIPT_STR - returns a string
- SCRIPT_INT - returns an integer
- SCRIPT_BOOL - returns a boolean value

The arguments to these functions include R language scripts and R function calls.

11.1.1 SCRIPT_STR function

SCRIPT_STR() returns a **string** result from a given R expression. The R expression is passed directly to a running Rserve instance. Use .arg# in the R expression to reference parameters.

Syntax: SCRIPT_STR(string, expression, ...)

11.1.1.1 Demo 1

Objective: To extract the first name from "Customer Name" and display it on the worksheet/view.

Input: "Sample - Superstore.xls"

Expected output: Refer to Fig. 11-3.

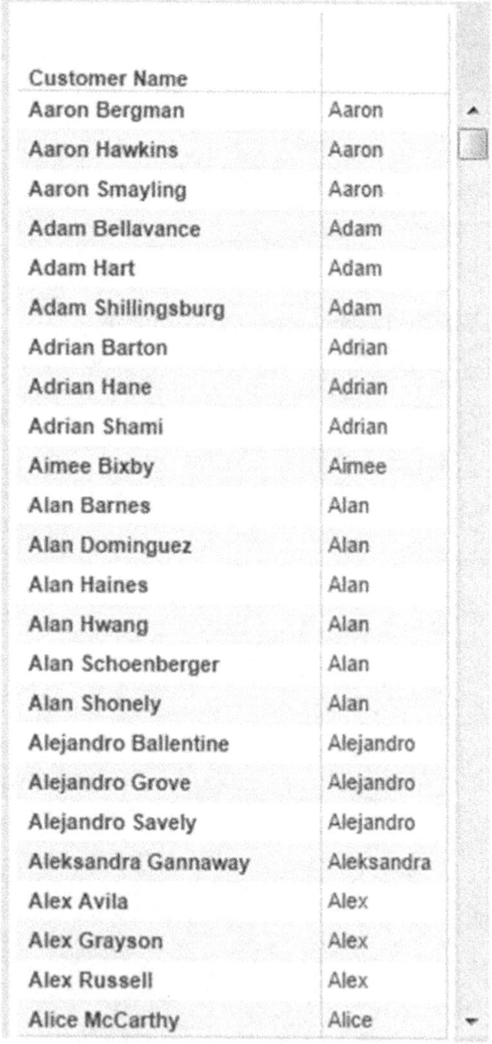

Customer Name	
Aaron Bergman	Aaron
Aaron Hawkins	Aaron
Aaron Smayling	Aaron
Adam Bellavance	Adam
Adam Hart	Adam
Adam Shillingsburg	Adam
Adrian Barton	Adrian
Adrian Hane	Adrian
Adrian Shami	Adrian
Aimee Bixby	Aimee
Alan Barnes	Alan
Alan Dominguez	Alan
Alan Haines	Alan
Alan Hwang	Alan
Alan Schoenberger	Alan
Alan Shonely	Alan
Alejandro Ballentine	Alejandro
Alejandro Grove	Alejandro
Alejandro Savely	Alejandro
Aleksandra Gannaway	Aleksandra
Alex Avila	Alex
Alex Grayson	Alex
Alex Russell	Alex
Alice McCarthy	Alice

Figure 11-3. *SCRIPT_STR function - Demo 1 - expected output*

11.1.1.1.1 Steps to extract first name from "Customer Name" using SCRIPT_STR()

Perform the below steps to retrieve the first name from the dimension "Customer Name".

11.1.1.1.2 Step 1

Start the Rserve services in R by performing the following:

- Library(Rserve)
- Rserve()

11.1.1.1.3 Step 2

Create a calculated field "CustFirstName" (Shown in Fig. 11-4).

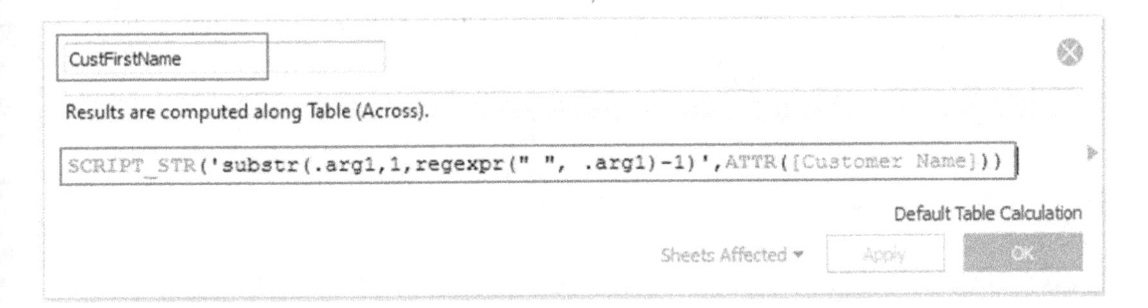

Figure 11-4. *Calculated field "CustFirstName" is created that invokes SCRIPT_STR function*

The SCRIPT_STR function calls the "substr" function in R with "Customer Name" as argument 1 and extracts characters from "Customer Name", starting at character position 1 until the first occurrence of space.

---

substr() function is used to retrieve or replace a substring from a character string. The syntax of the function is as follows:

substr(x, start, stop) where

x – a character string,

start – the index position at which the extraction of characters should begin and

stop – number of characters to return

The ATTR() function in Tableau returns the value of the given expression if all rows in the group has ONLY a single value otherwise it returns an asterisk (*). Null values are ignored.

11.1.1.1.4 Step 3

Drag the dimension "Customer Name" from the dimensions area under the data pane and place it on the rows shelf (Shown in Fig. 11-5).

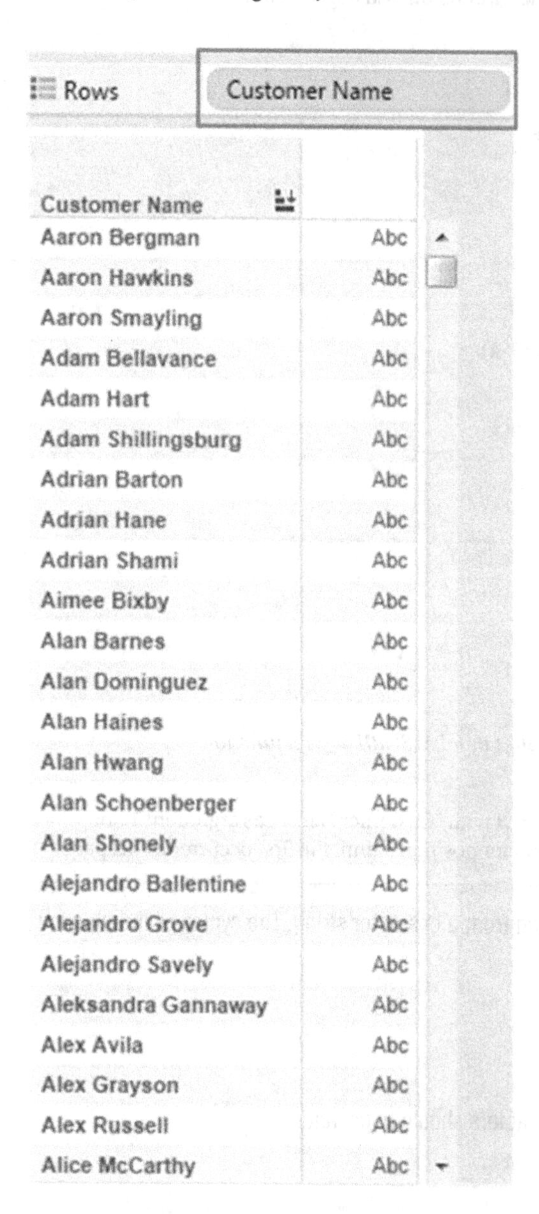

Figure 11-5. *Dimension "Customer Name" placed on the rows shelf*

11.1.1.1.5 Step 4

Drag the calculated field "CustFirstName" from the measures area under the data pane and place it on "Label" on the marks card (Shown in Fig. 11-6).

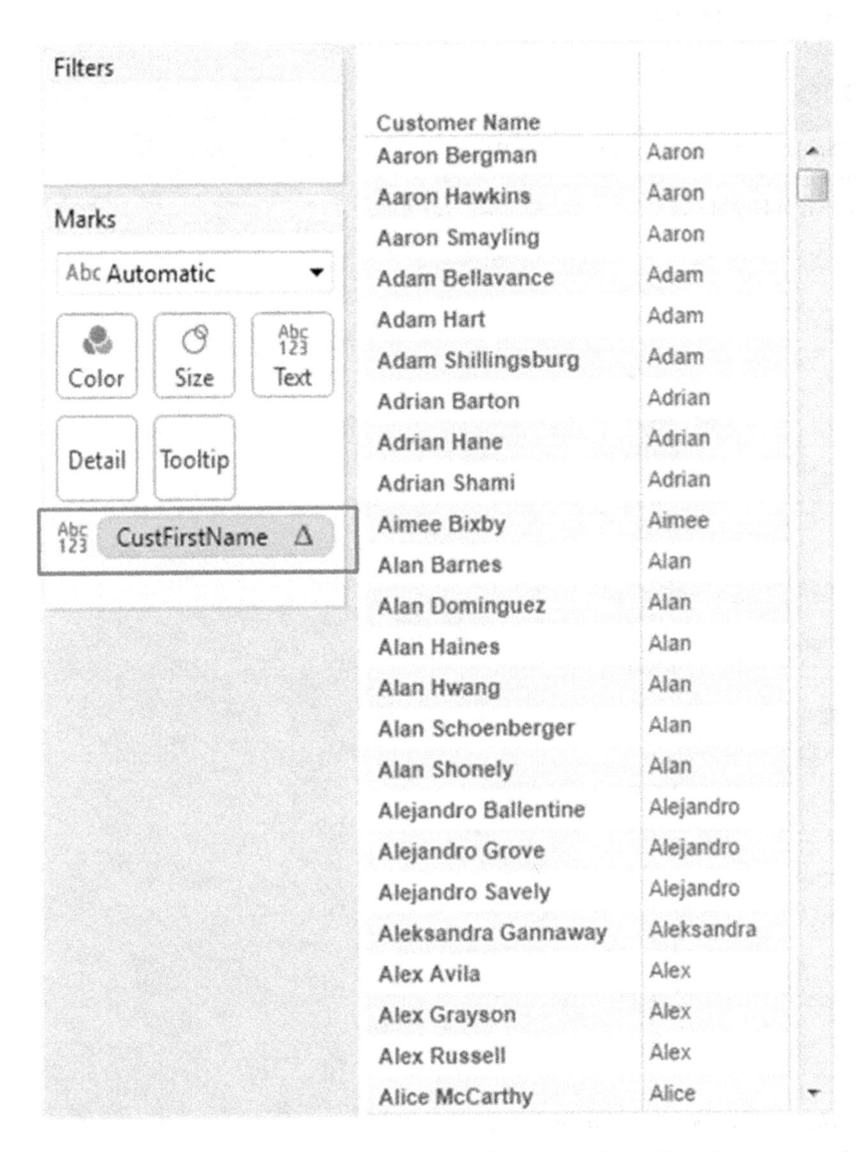

Figure 11-6. *Calculated field, "CustFirstName" placed on "Label" on the marks card*

11.1.2 SCRIPT_BOOL function

SCRIPT_BOOL() returns a **Boolean** result from a given R expression. The R expression is passed directly to a running Rserve instance. Use .arg# in the R expression to reference parameters.

Syntax: SCRIPT_BOOL(string, expression, ...)

11.1.2.1 Demo 2

Objective: To display the "Sales" for all cities in Washington State.

Input: "Sample - Superstore.xls"

Expected output: Shown in Fig. 11-7.

State	City	
Washington	Auburn	4
	Bellevue	104
	Bellingham	3,790
	Covington	414
	Des Moines	3,454
	Edmonds	2,524
	Everett	4
	Kent	1,351
	Longview	119
	Marysville	102
	Olympia	1,020
	Pasco	2,201
	Redmond	55
	Renton	1,243
	Seattle	119,541
	Spokane	2,028
	Vancouver	687

Figure 11-7. *SCRIPT_BOOL function - Demo 2 - expected output*

11.1.2.1.1 Steps to display "Sales" for all cities in "Washington" state using SCRIPT_BOOL()

Perform the below steps to retrieve a list of all cities in Washington state.

11.1.2.1.2 Step 1

Read in the data from "Sample - Superstore.xls" into Tableau (Shown in Fig. 11-8).

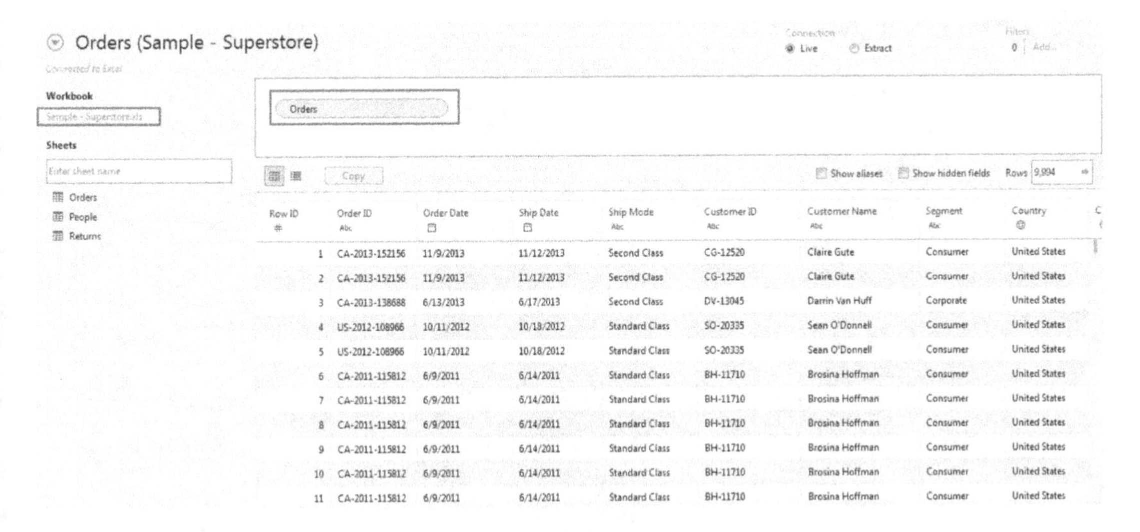

Row ID	Order ID	Order Date	Ship Date	Ship Mode	Customer ID	Customer Name	Segment	Country
1	CA-2013-152156	11/9/2013	11/12/2013	Second Class	CG-12520	Claire Gute	Consumer	United States
2	CA-2013-152156	11/9/2013	11/12/2013	Second Class	CG-12520	Claire Gute	Consumer	United States
3	CA-2013-138688	6/13/2013	6/17/2013	Second Class	DV-13045	Darrin Van Huff	Corporate	United States
4	US-2012-108966	10/11/2012	10/18/2012	Standard Class	SO-20335	Sean O'Donnell	Consumer	United States
5	US-2012-108966	10/11/2012	10/18/2012	Standard Class	SO-20335	Sean O'Donnell	Consumer	United States
6	CA-2011-115812	6/9/2011	6/14/2011	Standard Class	BH-11710	Brosina Hoffman	Consumer	United States
7	CA-2011-115812	6/9/2011	6/14/2011	Standard Class	BH-11710	Brosina Hoffman	Consumer	United States
8	CA-2011-115812	6/9/2011	6/14/2011	Standard Class	BH-11710	Brosina Hoffman	Consumer	United States
9	CA-2011-115812	6/9/2011	6/14/2011	Standard Class	BH-11710	Brosina Hoffman	Consumer	United States
10	CA-2011-115812	6/9/2011	6/14/2011	Standard Class	BH-11710	Brosina Hoffman	Consumer	United States
11	CA-2011-115812	6/9/2011	6/14/2011	Standard Class	BH-11710	Brosina Hoffman	Consumer	United States

***Figure 11-8.** Data from "Sample - Superstore.xls" read into Tableau*

11.1.2.1.3 Step 2

Drag the dimensions, "State" and "City" from the dimensions area under the data pane and place it on the rows shelf (Shown in Fig. 11-9).

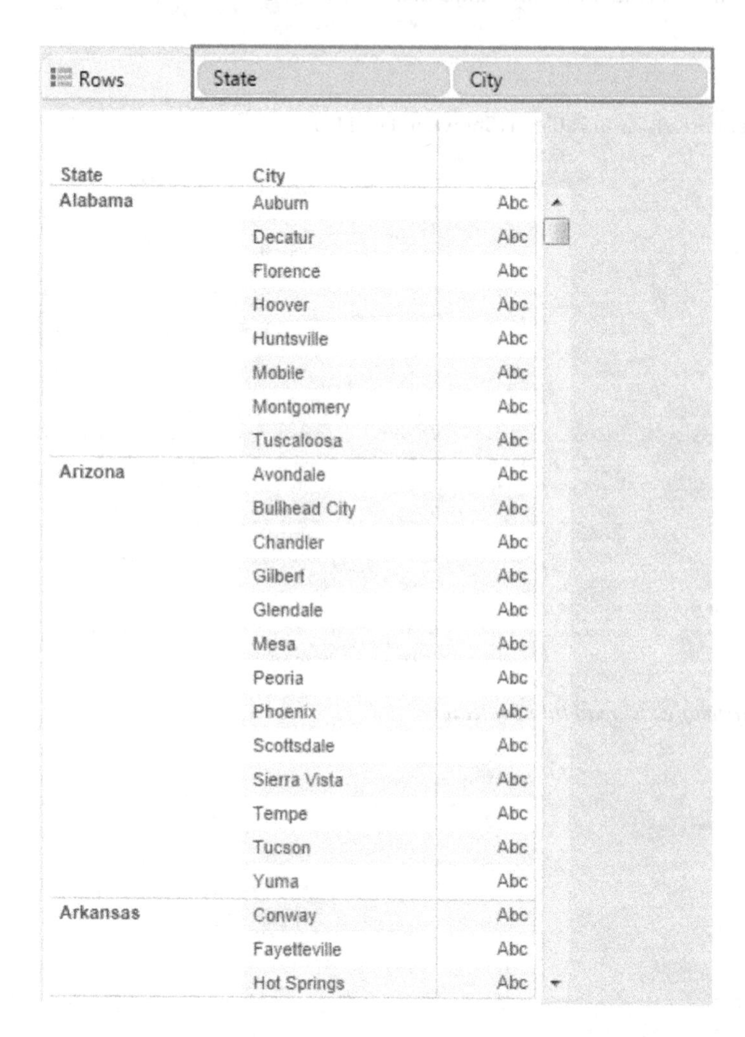

Figure 11-9. Dimensions "State" and "City" placed on the rows shelf

Drag the measure "Sales" from the measures area under the data pane and place it on "Label" on the marks card (Shown in Fig. 11-10).

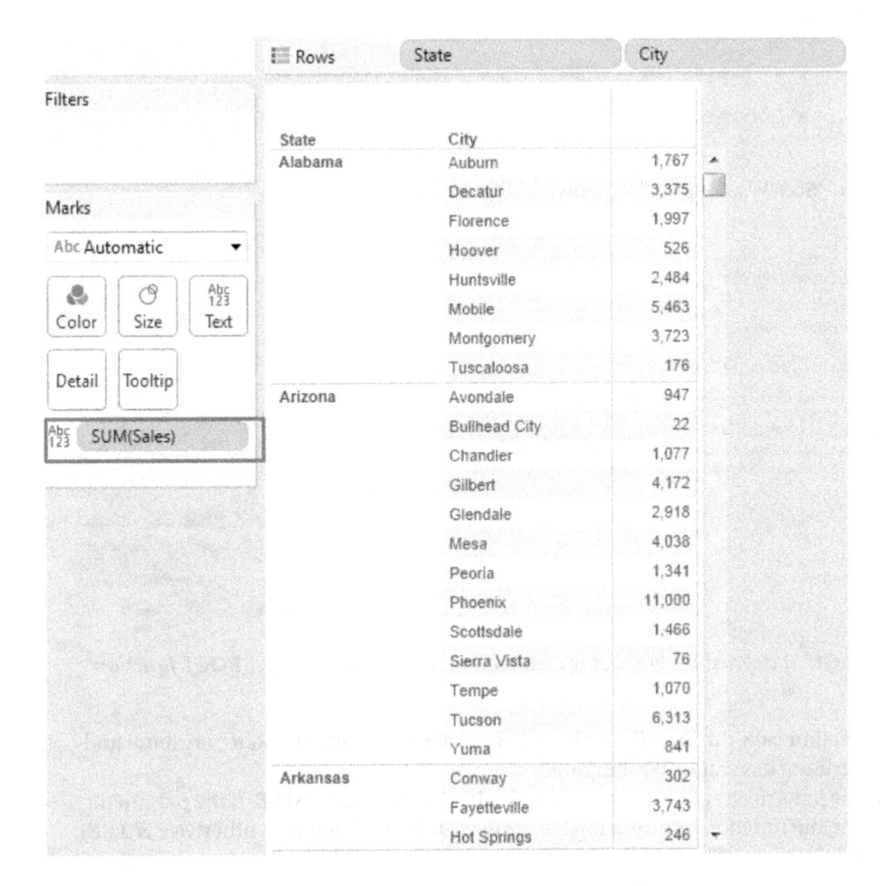

***Figure 11-10.** Measure "Sales" placed on "Label" on the marks card*

11.1.2.1.4 Step 3

Start the Rserve services in R as

```
> library(Rserve)
> Rserve()
Starting Rserve...
 "C:\Users\seema_acharya\Documents\R\win-library\3.2\Rserve\libs\i386\Rserve.exe"
```

Create a calculated field "StateWashington" as shown in Fig. 11-11.

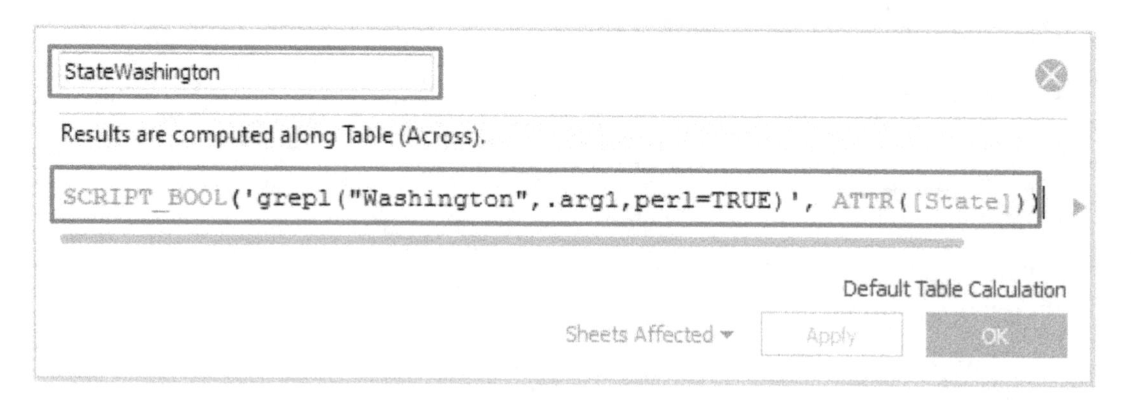

Figure 11-11. *Calculated field "StateWashington" being created that invokes the SCRIPT_BOOL function*

The SCRIPT_BOOL function calls the "grepl" function in R, which evaluates the "State" attribute and returns "TRUE" if "State" attribute has value "Washington".

If the string contains the pattern, grepl function returns TRUE else it returns FALSE. If the parameter happens to be a string vector, the function returns a logical vector (TRUE if it is a match otherwise FALSE). The syntax of the function:

grepl(pattern, x, ignore.case = FALSE, perl = FALSE, fixed = FALSE, useBytes = FALSE)

pattern: regular expression, or string for fixed=TRUE

x: string, the character vector

ignore.case: case sensitive or not

perl: logical. Should perl-compatible regexps be used?

fixed: logical. If TRUE, pattern is a string to be matched as is. Overrides all conflicting arguments

useBytes: logical. If TRUE the matching is done byte-by-byte rather than character-by-character

11.1.2.1.5 Step 4

Drag the calculated field "StateWashington" to the filters shelf and set it to "True" (Shown in Fig. 11-12).

The final output is shown in Fig. 11-13.

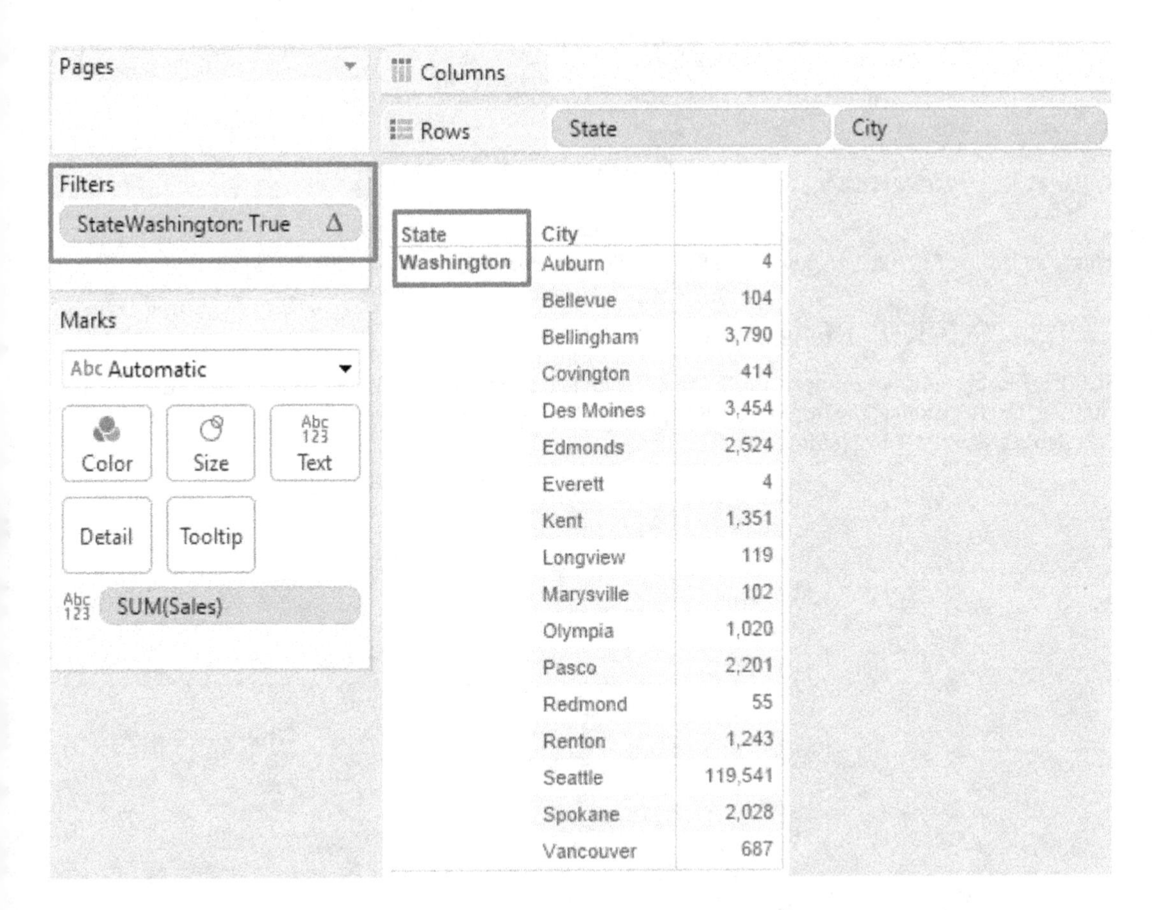

Figure 11-12. *Calculated field "StateWashington" placed on the filters shelf and set to "True"*

State	City	
Washington	Auburn	4
	Bellevue	104
	Bellingham	3,790
	Covington	414
	Des Moines	3,454
	Edmonds	2,524
	Everett	4
	Kent	1,351
	Longview	119
	Marysville	102
	Olympia	1,020
	Pasco	2,201
	Redmond	55
	Renton	1,243
	Seattle	119,541
	Spokane	2,028
	Vancouver	687

Figure 11-13. *SCRIPT_BOOL function - Demo 2 - final output*

11.1.3 SCRIPT_REAL function

SCRIPT_REAL() returns a **numeric** result from a given R expression. The R expression is passed directly to a running Rserve instance. Use .arg# in the R expression to reference parameters.

Syntax: SCRIPT_REAL(string, expression, ...)

11.1.3.1 Demo 3

Objective: To compute the correlation coefficient between two datasets in Excel, R and Tableau.

Input: "Correlation.xlsx"

Expected output: Shown in Fig. 11-14.

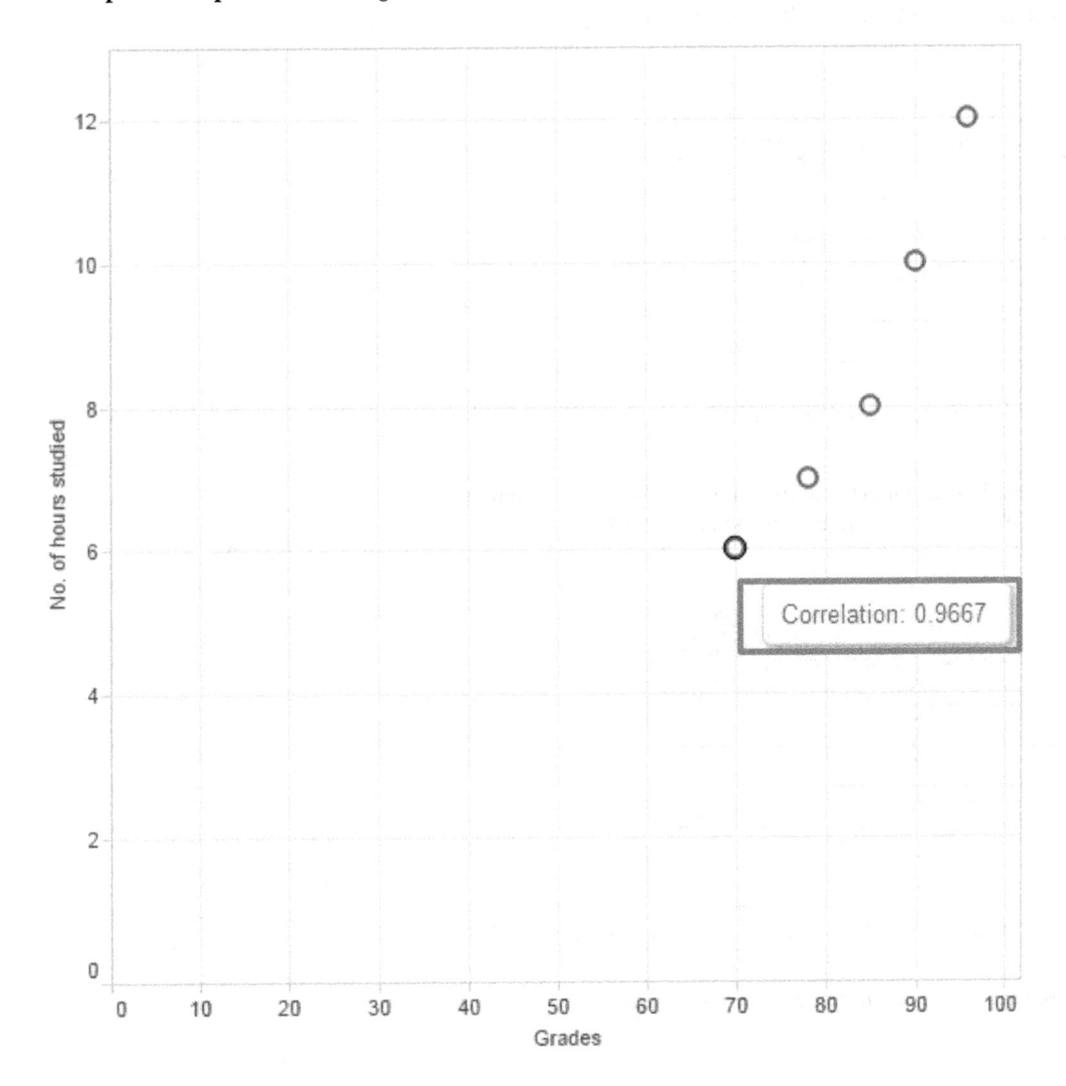

***Figure 11-14.** SCRIPT_REAL function - Demo 3 - expected output*

11.1.3.1.1 Steps to determine correlation coefficient between the two datasets in Excel, R and Tableau.

Perform the steps below to determine the correlation coefficient between two datasets in Excel, R and Tableau.

Let us consider the below data set (Shown in Fig. 11-15).

	A	B
1	No. of hours studied	Grades
2	6	70
3	7	78
4	8	85
5	10	90
6	12	96

Figure 11-15. *Data set used in Demo 3*

In Excel the formula to compute the correlation coefficient is CORREL(array 1, array 2)(Shown in Fig. 11-16).

	A	B	C
1	No. of hours studied	Grades	
2	6	70	
3	7	78	
4	8	85	
5	10	90	
6	12	96	
7			
8	Correlation Coefficient	=CORREL(A2:A6,B2:B6)	

Figure 11-16. *Formula to compute correlation coefficient*

The output in Excel (Shown in Fig. 11-17):

	A	B
1	No. of hours studied	Grades
2	6	70
3	7	78
4	8	85
5	10	90
6	12	96
7		
8	Correlation Coefficient	0.966668

Figure 11-17. *Correlation coefficient computed in Excel*

To determine the correlation coefficient in R

```
> NoofHoursStudied = c(6,7,8,10,12)
> Grade = c(70,78,85,90,96)
> cor(NoofHoursStudied,Grade,method="pearson")
[1] 0.9666679
```

To determine the correlation coefficient in Tableau:

Correlation is a statistical measure that indicates the extent to which two or more variables fluctuate together. A positive correlation is one when the values of variables increase or decrease in parallel; a negative correlation is when the value of one variable increases as the value of the other variable decreases.

Correlation does not specify causation.

11.1.3.1.2 Steps to compute the correlation coefficient in Tableau

11.1.3.1.3 Step 1

Read the data from "Correlation.xlsx" into Tableau (Shown in Fig. 11-18).

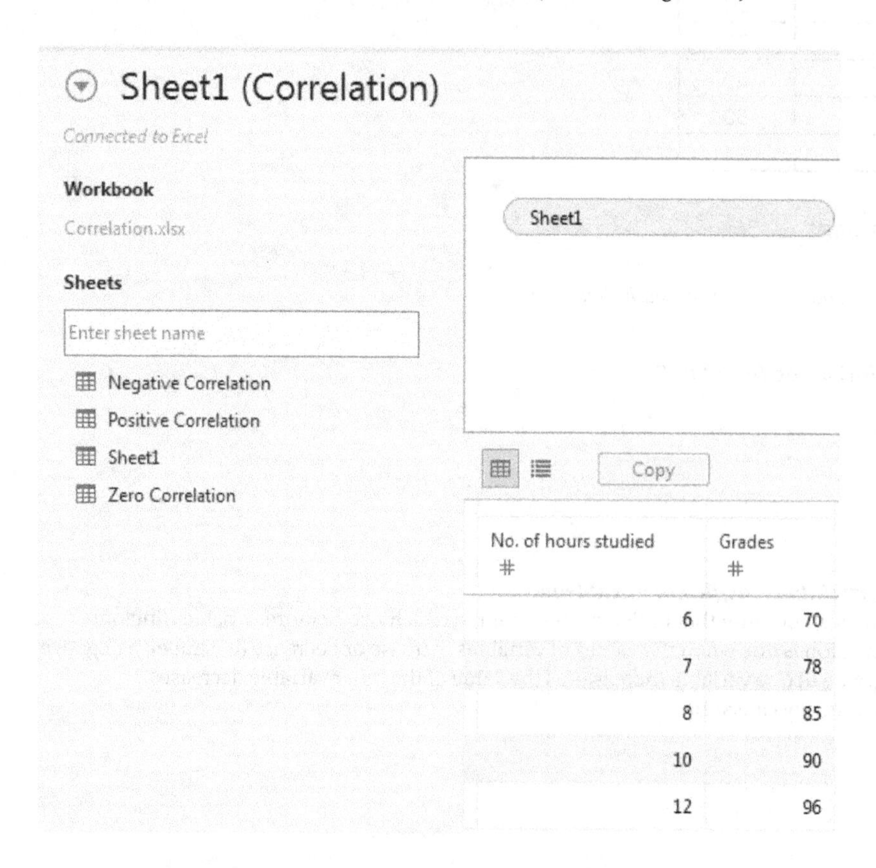

Figure 11-18. *Data from "Correlation.xlsx" read into Tableau*

11.1.3.1.4 Step 2

Drag the measures "Grades" and "No. of hours studied" from the measures area under the data pane to the columns shelf and the rows shelf, respectively (Shown in Fig. 11-19).

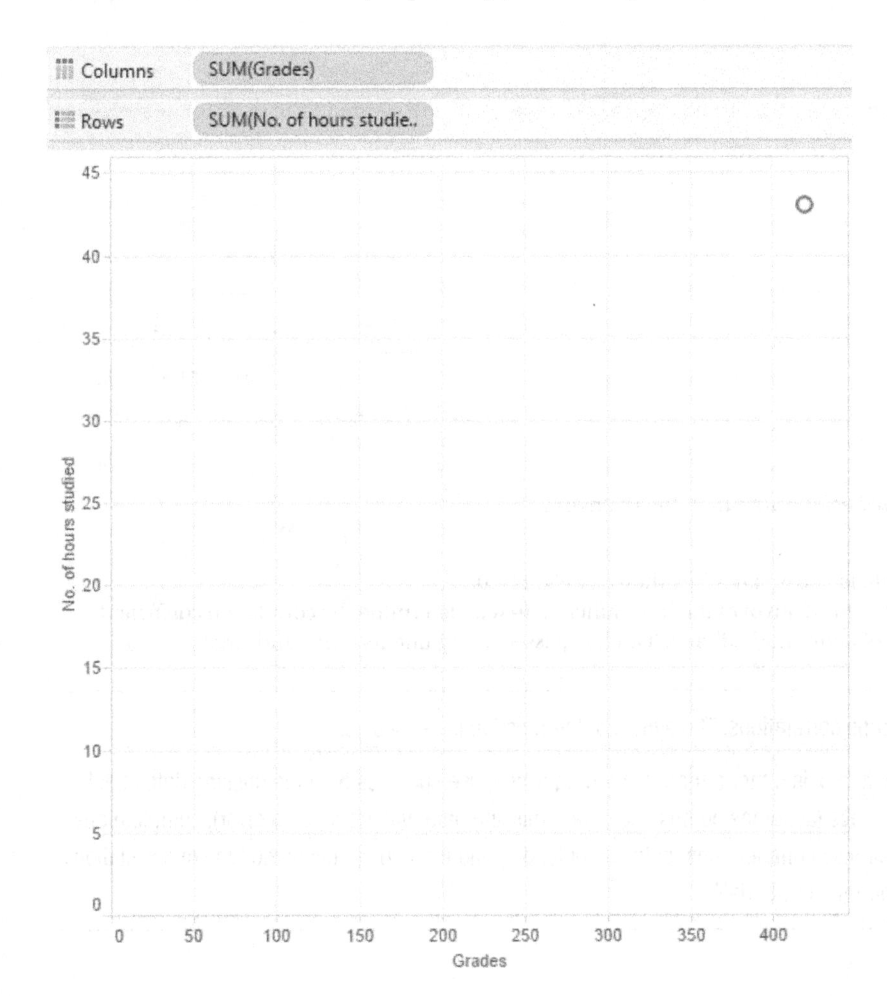

***Figure 11-19.** Scatter plot between measures "Grade" and "No. of hours studied"*

11.1.3.1.5 Step 3

Start the Rserve by performing the following

```
> library(Rserve)
> Rserve()
Starting Rserve...
 "C:\Users\seema_acharya\Documents\R\win-library\3.2\Rserve\libs\i386\Rserve.exe"
```

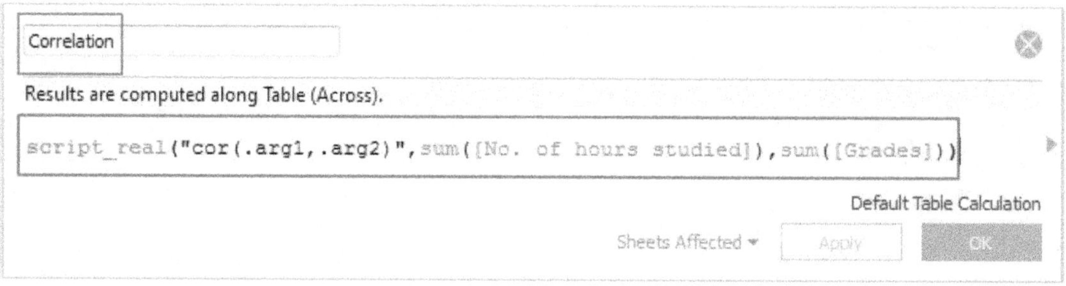

```
Correlation

Results are computed along Table (Across).

script_real("cor(.arg1,.arg2)",sum([No. of hours studied]),sum([Grades]))

Default Table Calculation
Sheets Affected    Apply    OK
```

Figure 11-20. "Calculated field "Correlation" being created

Create a calculated field "Correlation" as shown in Fig. 11-20.

The function SCRIPT_REAL invokes the "cor" function in R to determine the correlation coefficient between measures "No. of hours studied" and "Grades" passed as arguments (.arg1 and .arg2).

---

cor() function in R returns correlations. The syntax of the function is as follows:

cor(x, use=, method=), where x is either a matrix or a data frame, use specifies how the missing data should be handled (options are all.obs (assumes no missing data – missing data will produce an error), complete.obs (listwise deletion), and pairwise.complete.obs (pairwise deletion)) and method specifies the type of correlation (options are pearson, spearman or kendall)

Drag the calculated field "Correlation" to "Detail" on the marks card (Shown in Fig. 11-21).

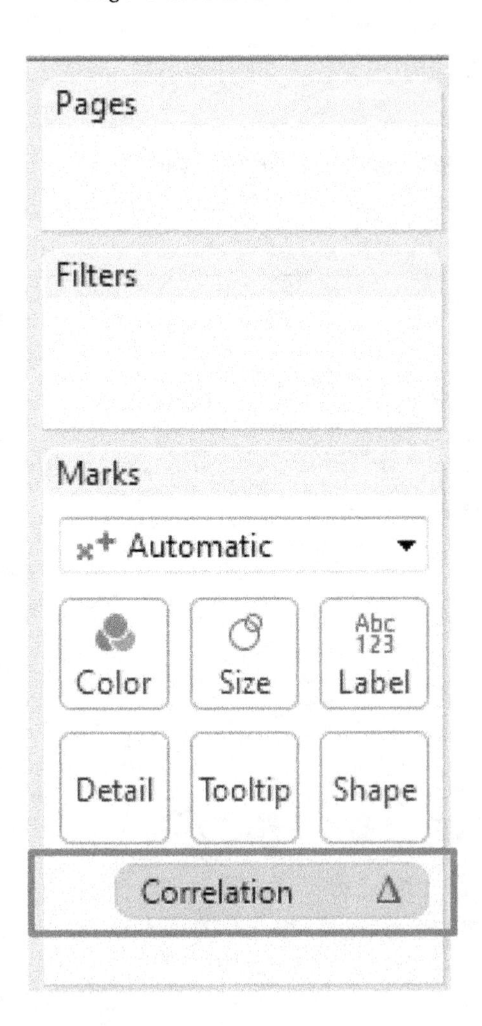

Figure 11-21. *Calculated field "Correlation" placed on "Detail" on the marks card*

11.1.3.1.6 Step 4

Disaggregate the measures as follows:

Analysis ➤ Aggregate Measures; Aggregate Measures (uncheck it).

Tableau displays a separate mark for every row data value in the data source if disaggregating the data. Exercise caution when disaggregating data as it can lead to significant performance degradation particularly if the data source is huge.

Set the tooltip to display correlation as shown in Fig. 11-22.

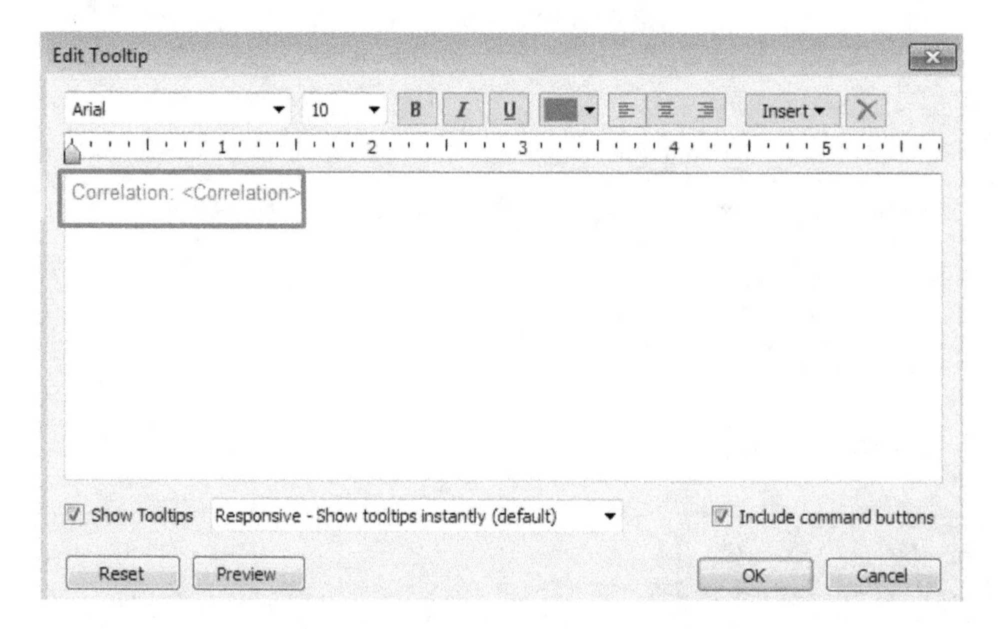

Figure 11-22. *Editing the tooltip to display "Correlation"*

The final output (Shown in Fig. 11-23):

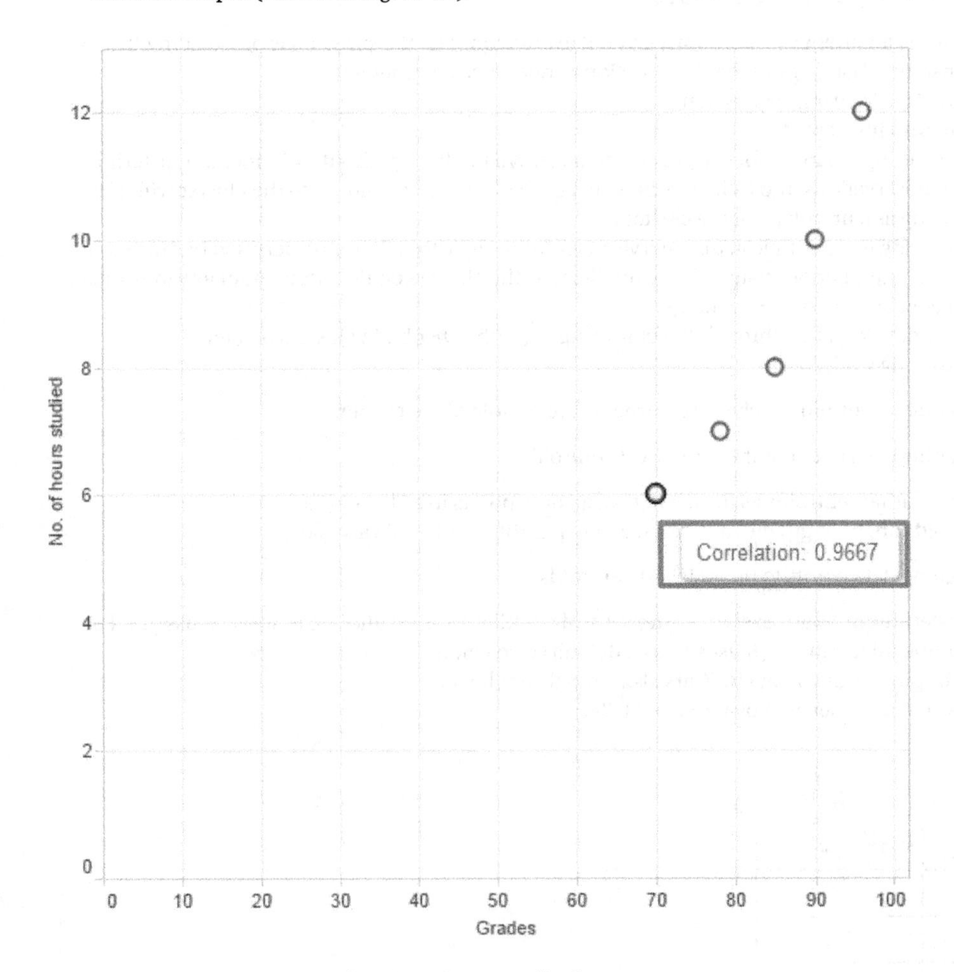

Figure 11-23. *SCRIPT_REAL function - Demo 3 - final output*

11.1.4 SCRIPT_INT function

SCRIPT_INT() returns an **integer** result from a given R expression. The R expression is passed directly to a running Rserve instance. Use .arg# in the R expression to reference parameters.

Syntax: SCRIPT_INT(string, expression, ...)

What is k-means clustering?

A k-means clustering means to form k groups/clusters. Wikipedia explains it as "k-means clustering aims to partition n observations into k clusters in which each observation belongs to the cluster with the nearest mean, serving as a prototype of the cluster".

K-means clustering is the simplest, unsupervised learning algorithm. It is unsupervised because one has to specify only the number of clusters. K-means "learns" the clusters on its own without any information about to which cluster an observation belongs.

Begin with raw data ➤ pass it through the clustering algorithm ➤ obtain clusters of data

K-means clustering works by:

1. Selecting K centroids. A cluster centroid is the middle of the cluster.
2. Assigning each data point to its closest centroid.
3. Recalculating centroids as the average of all data points in a cluster (i.e., centroids are p-length mean vectors, where p is the number of variables)
4. Assigning data points to their closest centroids.

K-means clustering continues executing steps 3 and 4 until the observations cannot be reassigned or the maximum number of iterations (R uses 10 as a default) is reached.

Objective: To split the given data in "Cars.xlsx" into three clusters.

Input: "Cars.xlsx". Data set as shown in Fig. 11-24:

	A	B
1	Petrol	Kilometers
2	1.1	60
3	6.5	20
4	4.2	40
5	1.5	25
6	7.6	15
7	2	55
8	3.9	39

Figure 11-24. *Data set used in the demonstration on k-means clustering*

Expected output: Shown in Fig. 11-25.

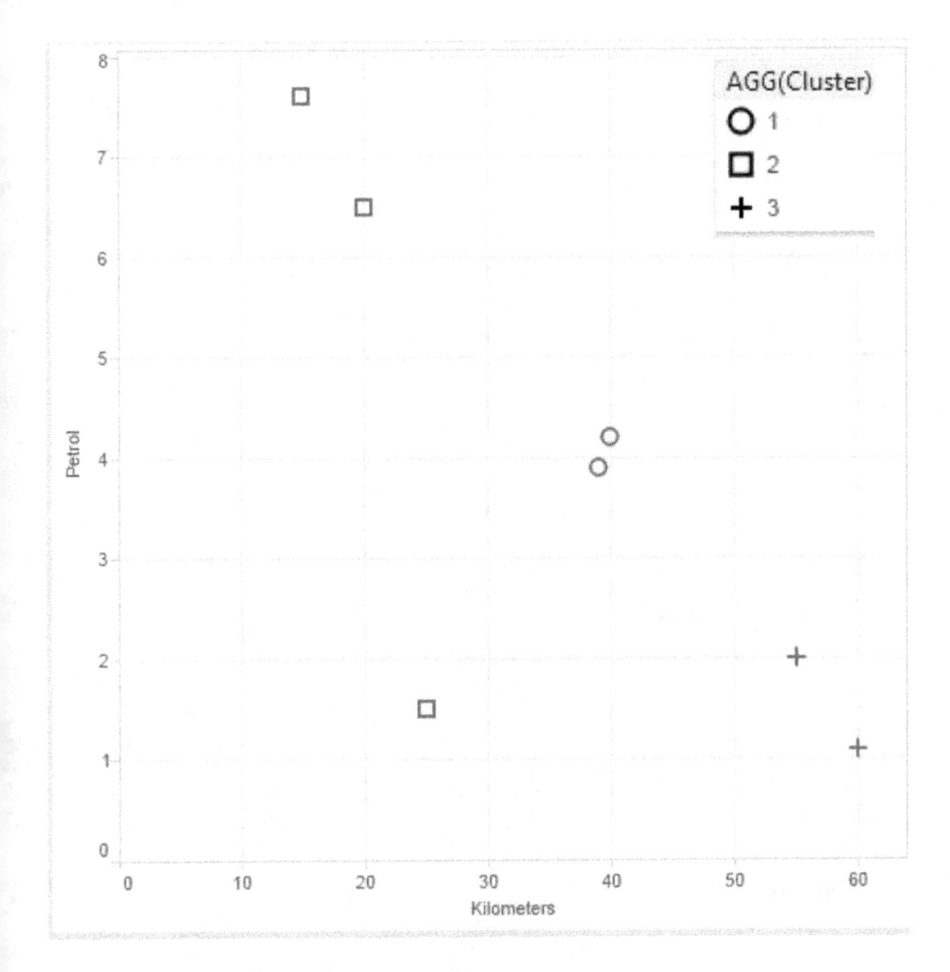

Figure 11-25. *SCRIPT_INT function - K-means clustering - expected output*

11.1.4.1 Steps to create clusters

11.1.4.1.1 Step 1

Read in data from "Cars.xlsx" into Tableau (Shown in Fig. 11-26).

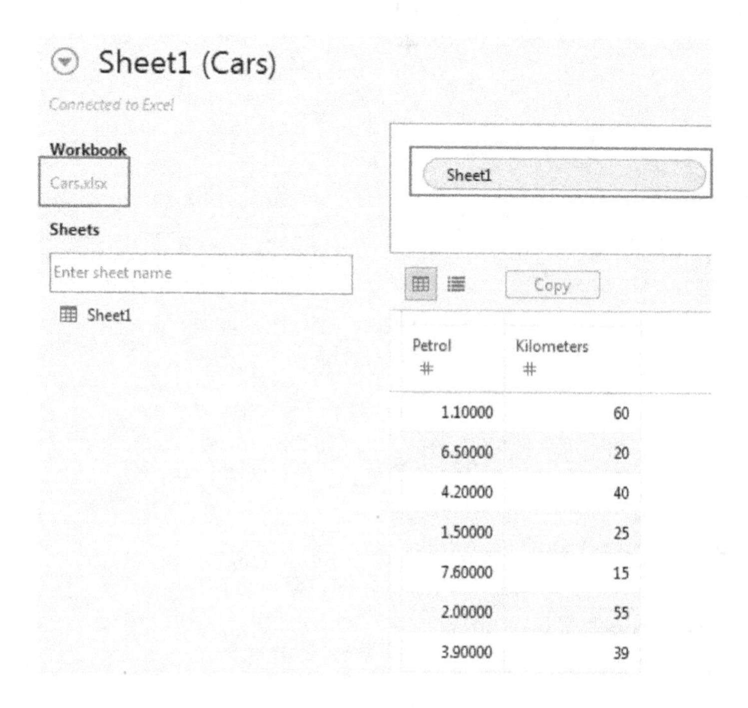

Figure 11-26. *Data from "Cars.xlsx" read into Tableau*

11.1.4.1.2 Step 2

Create a calculated field "Cluster" as shown in Fig. 11-27.

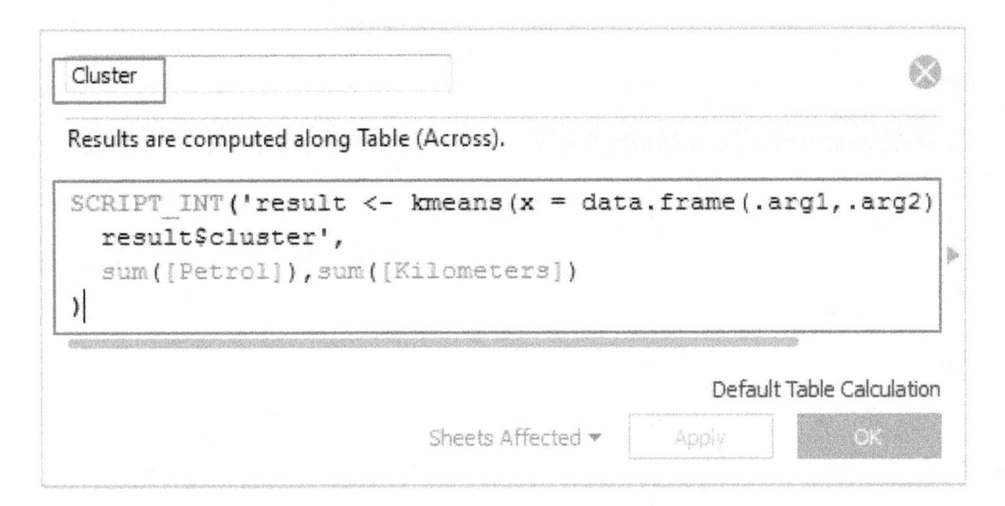

Figure 11-27. Calculated field "Cluster" being created.

The SCRIPT_INT function calls the kmeans function in R, passing it the dimensions "Petrol" and "Kilometers" as arguments. The kmeans function in R returns with clusters of data.

Syntax of kmeans() function in R:

Kmeans(x, centers, iter.max=10) where

x is a data frame or matrix. It is mandatory that all values be numeric.

centers is the K of K Means. For example, centers = 5 results in 5 clusters being created.

iter.max is the number of times the algorithm repeats the cluster assignment and moving of centroids.

11.1.4.1.3 Step 3

Go to the "Analysis" menu option and uncheck the "Aggregate Measures". It is important to disaggregate the measure (Shown in Fig. 11-28).

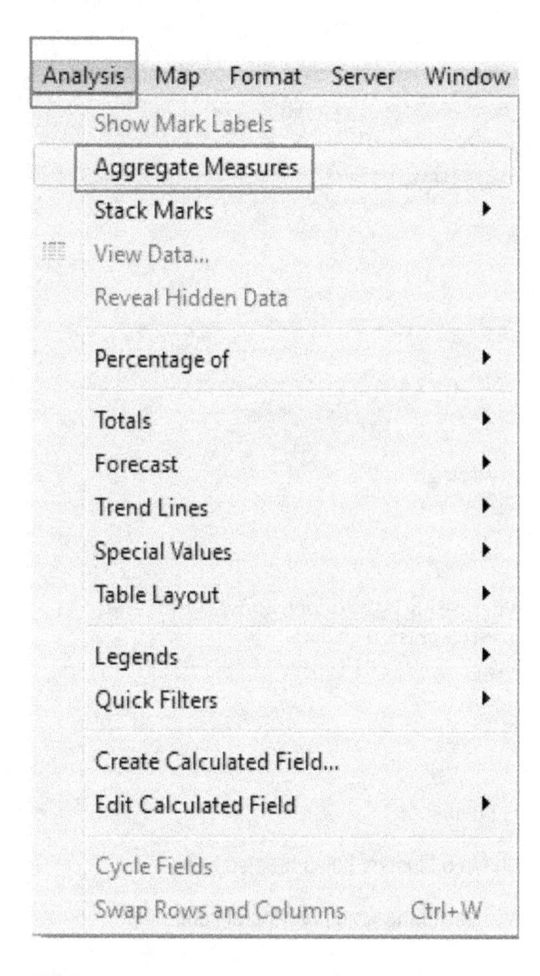

Figure 11-28. *Disaggregating the measure*

11.1.4.1.4 Step 4

Drag the measure "Kilometers" from the measures area under the data pane and place it on the columns shelf (shown in Fig. 11-29).

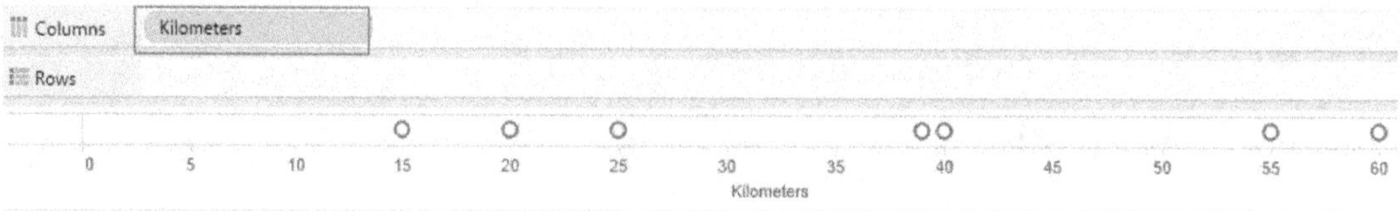

Figure 11-29. *Measure, "Kilometers" placed on the Columns Shelf*

11.1.4.1.5 Step 5

Drag the measure "Petrol" from the measures area under the data pane and place it on the rows shelf (Shown in Fig. 11-30).

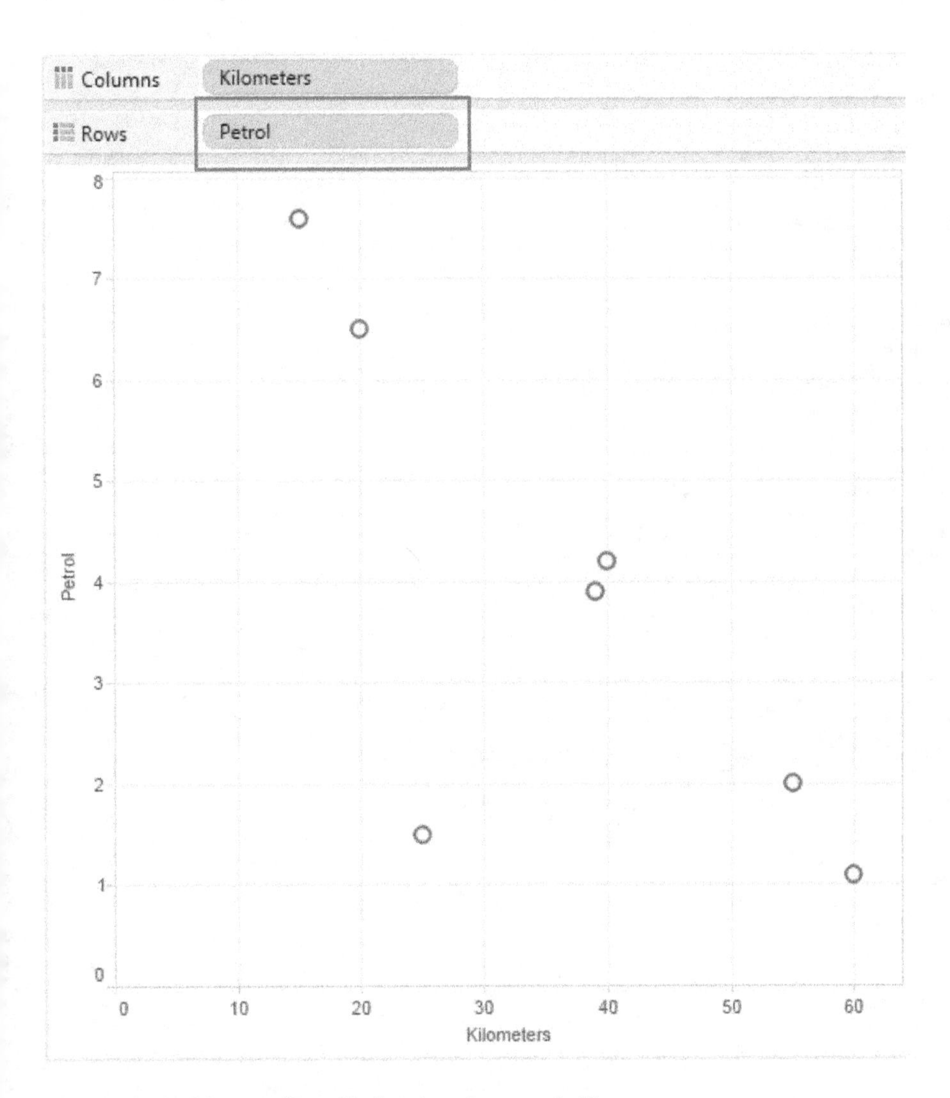

Figure 11-30. *Measure "Petrol" placed on the rows shelf*

11.1.4.1.6 Step 6

Drag the calculated field "Cluster" to "Shape" on the marks card (Shown in Fig. 11-31).

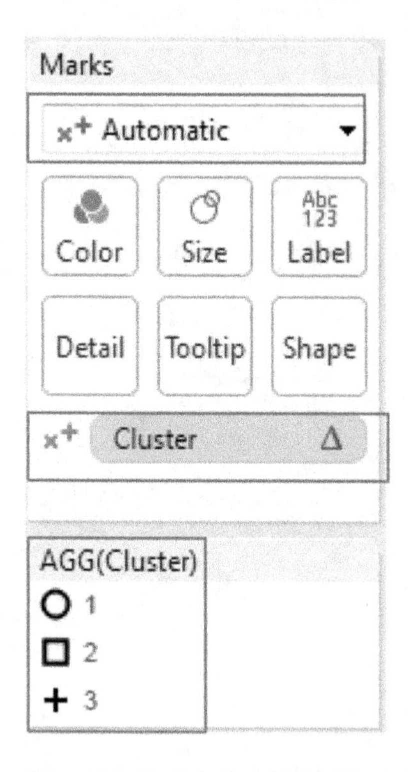

Figure 11-31. *Calculated field "Cluster" placed on "Shape" on the marks card*

The final output after placing the calculated field "Cluster" on "Shape" on the marks card (Shown in Fig. 11-32).

Conclusion: Data has been spilt into three clusters.

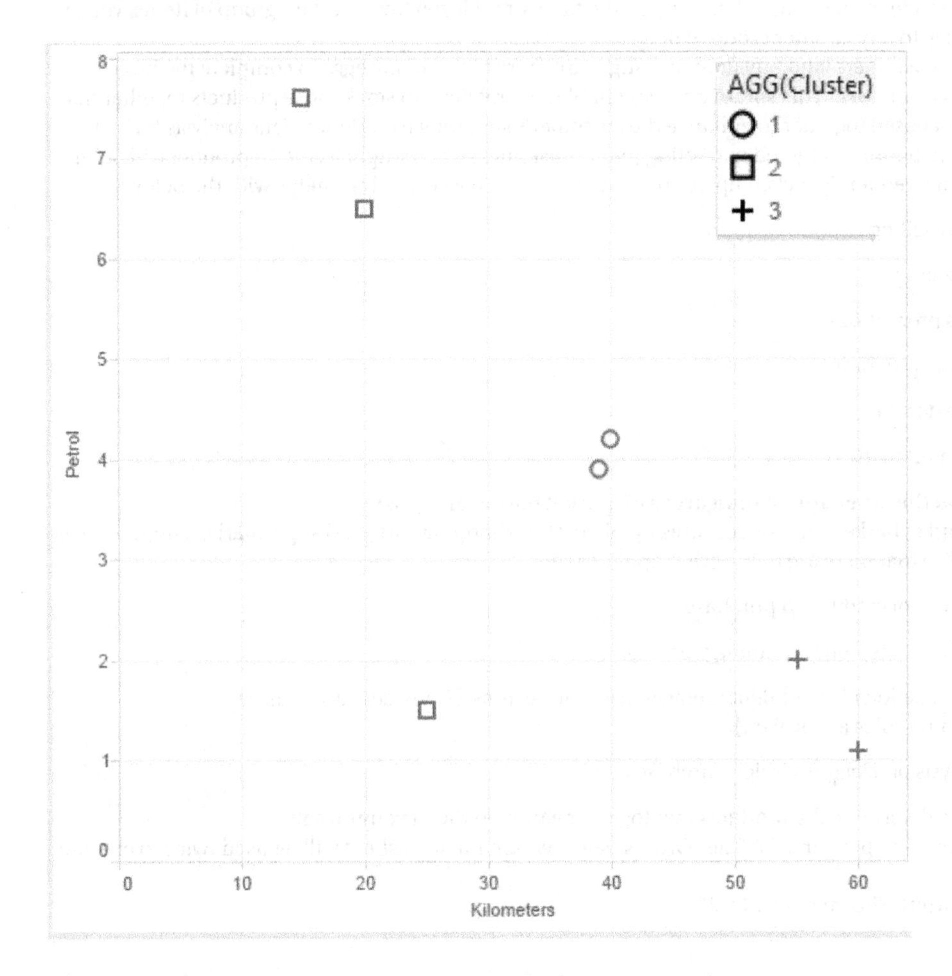

Figure 11-32. *SCRIPT_INT function - K-means clustering - expected output*

11.1.5 Market basket analysis

Market basket analysis helps to spot the combination of products that are frequently bought together by customers. It is a modeling technique based upon the theory that if you buy a certain group of items, you are more (or less) likely to buy another group of items.

For example, customers who buy flour and sugar are more likely to buy eggs to complete the basic ingredients for baking a cake. This sort of analysis enables the retailers to stock those products together that are frequently purchased together to enhance the customer's shopping experience. This analysis further helps the retailer to perform targeted marketing (email customers who bought a certain product with offers on another product frequently picked up by customers). Market basket analysis helps with the below:

- Cross-selling
- Up-selling
- Sales promotions
- Loyalty programs
- Store design
- Discount plans

Let us look at the other application areas of market basket analysis:

Although market basket analysis conjures up pictures of shopping carts and supermarket shoppers, it is important to realize that there are many other applications:

- Analysis of credit card purchases.
- Analysis of telephone calling patterns.
- Identification of fraudulent medical insurance claims. (Consider cases where common rules are broken).
- Analysis of telecom service purchases.

Objective: To determine the products that together garnered the maximum sales.

Input: "Sample - Superstore.xls". The "Orders" sheet within the worksheet will be used twice to execute a self-join.

Expected output: Shown in Fig. 11-33.

	Sub-Category														
Sub-Categor..	Accesso..	Applianc..	Art	Binders	Bookcas..	Chairs	Copiers	Envelop..	Fasteners	Furnishi..	Labels	Machines	Paper	Phones	Storage
Accessories		514	944	1,767	249	703	57	316	270	1,106	411	128	1,587	1,014	955
Appliances	514		589	1,068	130	403	36	181	165	624	210	84	937	620	572
Art	944	589		1,760	258	736	79	282	270	1,083	404	134	1,598	1,013	973
Binders	1,767	1,068	1,760		473	1,383	152	625	506	2,073	754	282	3,049	1,918	1,842
Bookcases	249	130	258	473		207	26	89	66	293	139	30	428	300	270
Chairs	703	403	736	1,383	207		64	242	211	896	315	120	1,226	809	760
Copiers	57	36	79	152	26	64		29	19	93	39	15	146	104	81
Envelopes	316	181	282	625	89	242	29		78	380	137	41	566	325	346
Fasteners	270	165	270	506	66	211	19	78		324	125	51	454	315	290
Furnishings	1,106	624	1,083	2,073	293	896	93	380	324		532	176	1,908	1,328	1,230
Labels	411	210	404	754	139	315	39	137	125	532		66	734	470	458
Machines	128	84	134	282	30	120	15	41	51	176	66		255	159	159
Paper	1,587	937	1,598	3,049	428	1,226	146	566	454	1,908	734	255		1,771	1,695
Phones	1,014	620	1,013	1,918	300	809	104	325	315	1,328	470	159	1,771		1,105
Storage	955	572	973	1,842	270	760	81	346	290	1,230	458	159	1,695	1,105	
Supplies	262	151	248	441	69	169	16	77	74	245	81	39	388	232	214
Tables	374	204	349	722	97	282	26	129	106	419	160	49	613	410	409

Figure 11-33. Market basket analysis - expected output

11.1.5.1 Steps to perform market basket analysis

Perform the below steps to determine the products that together garnered the maximum sales.

11.1.5.1.1 Step 1

Read in the data from the "Orders" sheet of "Sample - Superstore.xls" into Tableau (Shown in Fig. 11-34).

***Figure 11-34.** Data from "Orders" sheet of "Sample - Superstore.xls" read into Tableau*

11.1.5.1.2 Step 2

Drag the "Orders" sheet a second time into the "Drop sheets here". The "Inner Join" dialog window shows up (Shown in Fig. 11-35).

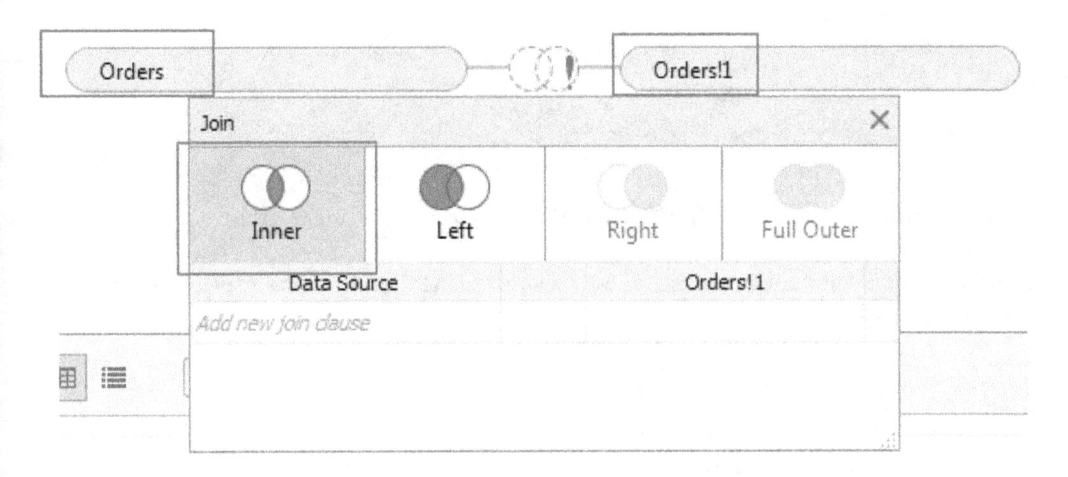

***Figure 11-35.** "Inner Join" dialog box*

Fill in the condition for the "Inner Join" (Shown in Fig. 11-36).
Move to "Sheet1".

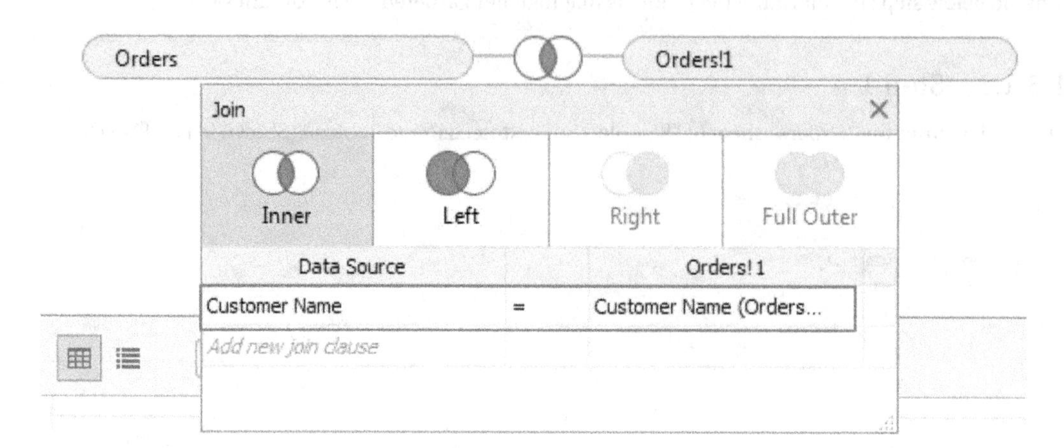

Figure 11-36. *Setting the condition for "Inner Join"*

11.1.5.1.3 Step 3

Drag the dimension "Sub-Category" from "Orders$" sheet and place it on the columns shelf (Shown in Fig. 11-37).

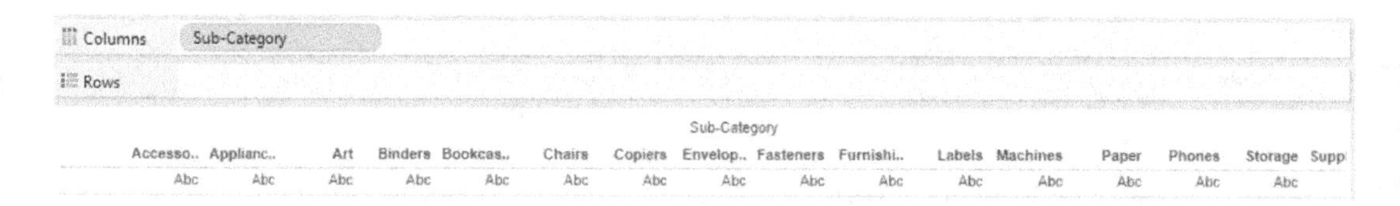

Figure 11-37. *Dimension "Sub-Category" from "Orders$" sheet placed on the columns shelf*

Drag the dimension "Sub-Category" from "Sub-Category from "Orders$1" sheet and place it on the rows shelf (Shown in Fig. 11-38).

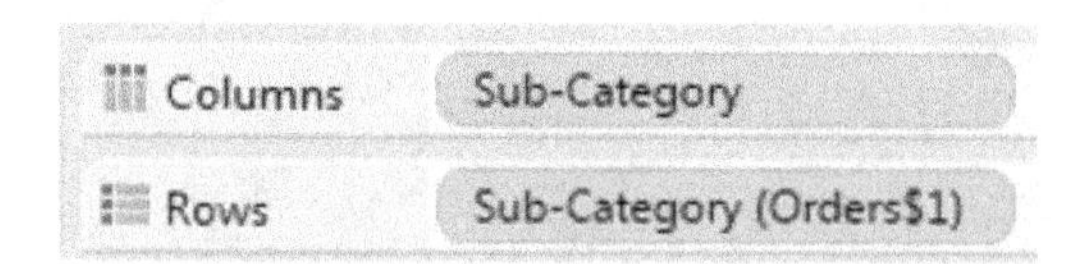

Figure 11-38. *Dimension "Sub-Category" from "Orders$1" sheet placed on the rows shelf.*

11.1.5.1.4 Step 4

Drag the dimension "Customer Name" from "Orders$" to "Label" on the marks card (Shown in Fig. 11-39).

Change the aggregation of the dimension "Customer Name" on "Label" on the marks card to "Count" (Shown in Fig. 11-40)

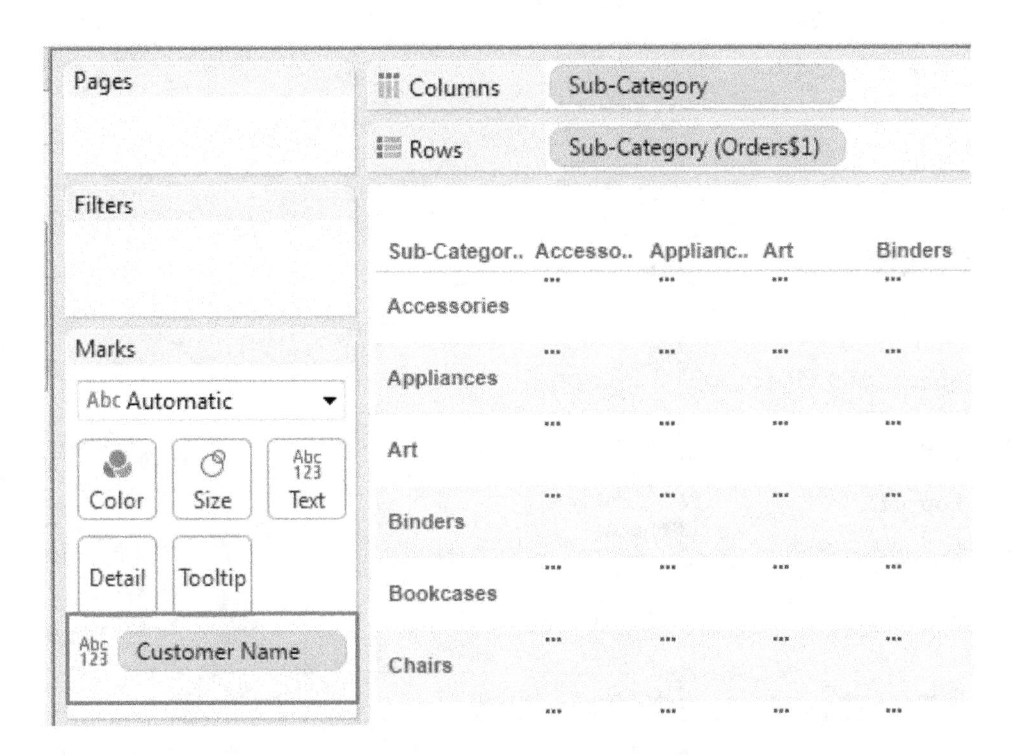

Figure 11-39. Dimension "Customer Name" from "Orders$" sheet placed on "Label" on the marks card

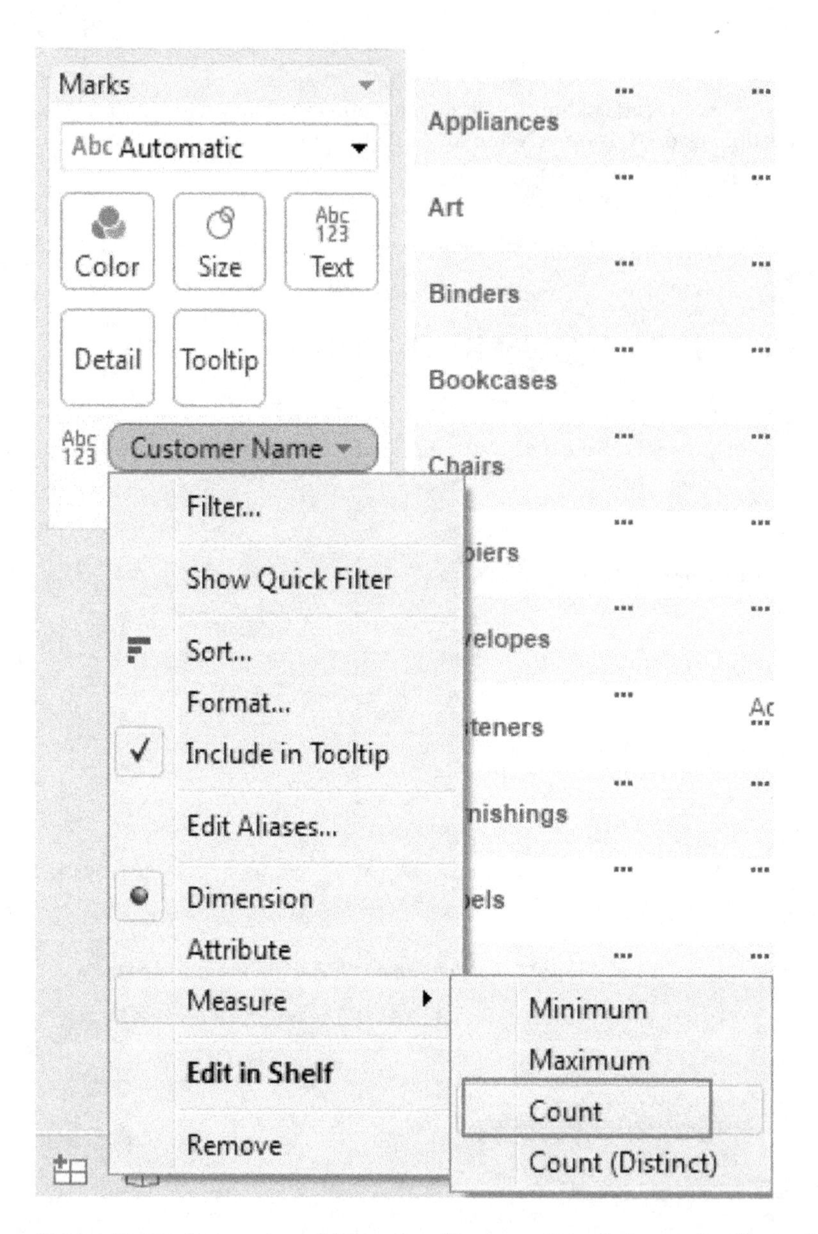

Figure 11-40. *Aggregation of dimension, "Customer Name" changed to "Count"*

The output after setting the aggregation to "Count" (Shown in Fig. 11-41).

	Sub-Category													
Sub-Categor..	Accesso..	Applianc..	Art	Binders	Bookcas..	Chairs	Copiers	Envelop..	Fasteners	Furnishi..	Labels	Machines	Paper	Phones
Accessories	1,623	514	944	1,767	249	703	57	316	270	1,106	411	128	1,587	1,014
Appliances	514	744	589	1,068	130	403	36	181	165	624	210	84	937	620
Art	944	589	1,658	1,760	258	736	79	282	270	1,083	404	134	1,598	1,013
Binders	1,767	1,068	1,760	4,937	473	1,383	152	625	506	2,073	754	282	3,049	1,918
Bookcases	249	130	258	473	302	207	26	89	66	293	139	30	428	300
Chairs	703	403	736	1,383	207	1,231	64	242	211	896	315	120	1,226	809
Copiers	57	36	79	152	26	64	76	29	19	93	39	15	146	104
Envelopes	316	181	282	625	89	242	29	364	78	380	137	41	566	325
Fasteners	270	165	270	506	66	211	19	78	277	324	125	51	454	315
Furnishings	1,106	624	1,083	2,073	293	896	93	380	324	2,307	532	176	1,908	1,328
Labels	411	210	404	754	139	315	39	137	125	532	560	66	734	470
Machines	128	84	134	282	30	120	15	41	51	176	66	155	255	159
Paper	1,587	937	1,598	3,049	428	1,226	146	566	454	1,908	734	255	4,158	1,771
Phones	1,014	620	1,013	1,918	300	809	104	325	315	1,328	470	159	1,771	2,065
Storage	955	572	973	1,842	270	760	81	346	290	1,230	458	159	1,695	1,105
Supplies	262	151	248	441	69	169	16	77	74	245	81	39	388	232
Tables	374	204	349	722	97	282	26	129	106	419	160	49	613	410

Figure 11-41. *Output after changing aggregation of "Customer Name" to "Count".*

11.1.5.1.5 Step 5

Remove the duplicates. For removing the duplicates, create a calculated field "Duplicates" (Shown in Fig. 11-42).

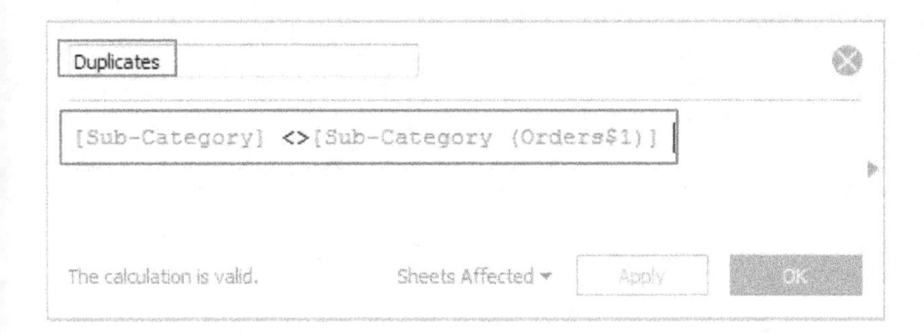

Figure 11-42. *Calculated field "Duplicates" being created*

11.1.5.1.6 Step 6

Drag the calculated field "Duplicates" to the filters shelf and set it to "True"(Shown in Fig. 11-43).

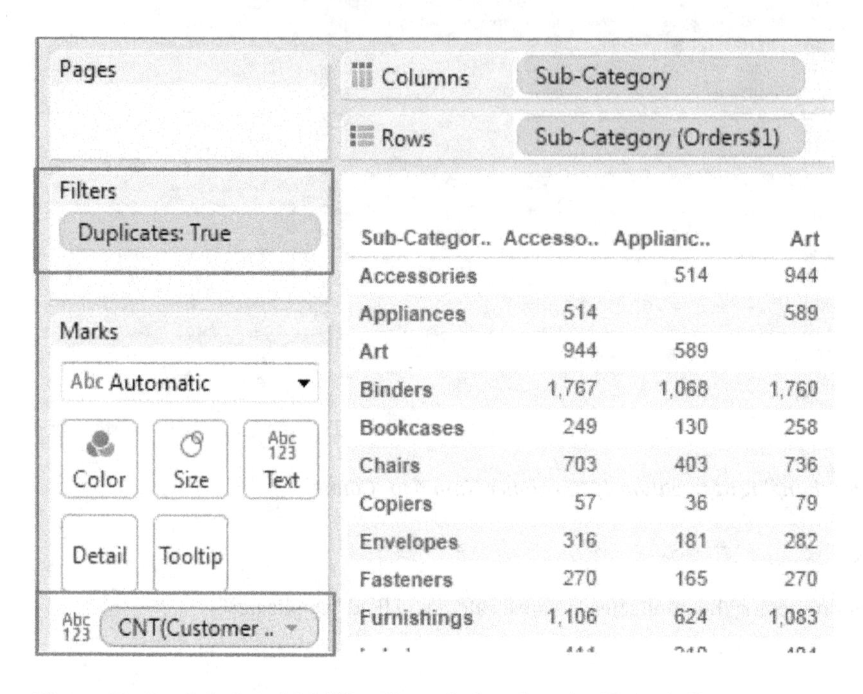

Figure 11-43. Calculated field "Duplicates" placed on the filters shelf

The final output after removing the duplicates (where the sub-category from the Orders$ sheet equals the sub-category from Orders$1 sheet) (Shown in Fig. 11-44).

	Sub-Category														
Sub-Categor..	Accesso..	Applianc..	Art	Binders	Bookcas..	Chairs	Copiers	Envelop..	Fasteners	Furnishi..	Labels	Machines	Paper	Phones	Storage
Accessories		514	944	1,767	249	703	57	316	270	1,106	411	128	1,587	1,014	955
Appliances	514		589	1,068	130	403	36	181	165	624	210	84	937	620	572
Art	944	589		1,760	258	736	79	282	270	1,083	404	134	1,598	1,013	973
Binders	1,767	1,068	1,760		473	1,383	152	625	506	2,073	754	282	3,049	1,918	1,842
Bookcases	249	130	258	473		207	26	89	66	293	139	30	428	300	270
Chairs	703	403	736	1,383	207		64	242	211	896	315	120	1,226	809	760
Copiers	57	36	79	152	26	64		29	19	93	39	15	146	104	81
Envelopes	316	181	282	625	89	242	29		78	380	137	41	566	325	346
Fasteners	270	165	270	506	66	211	19	78		324	125	51	454	315	290
Furnishings	1,106	624	1,083	2,073	293	896	93	380	324		532	176	1,908	1,328	1,230
Labels	411	210	404	754	139	315	39	137	125	532		66	734	470	458
Machines	128	84	134	282	30	120	15	41	51	176	66		255	159	159
Paper	1,587	937	1,598	3,049	428	1,226	146	566	454	1,908	734	255		1,771	1,695
Phones	1,014	620	1,013	1,918	300	809	104	325	315	1,328	470	159	1,771		1,105
Storage	955	572	973	1,842	270	760	81	346	290	1,230	458	159	1,695	1,105	
Supplies	262	151	248	441	69	169	16	77	74	245	81	39	388	232	214
Tables	374	204	349	722	97	282	26	129	106	419	160	49	613	410	409

Figure 11-44. *Market basket analysis - final output*

Conclusion:

The sub-categories "Binders" and "Paper" garnered the maximum sales.

11.2 Points to Remember

- Tableau Online and Tableau Public do not support R.
- Data cannot be exported directly from Tableau into R.
- Datasets cannot be imported from R into Tableau.
- Visualizations created in R cannot be imported into Tableau; however, one can use the image files of R visualizations or URL to R visualizations in the Tableau Dashboard.

Index

A

ABS(number) function, 442–443, 445–446
Actions, dashboard
  filter action
    Control view, 773–774, 776
    dimension "Sub-Category", 765–767
    Edit Title dialog box, 772–773
    format option, 769–770
    format sub-category dialog box, 770–771
    Sales map view, 775, 776
    Sales *vs.* Profit" view, 774–776
    Show header option, 767–768
    sub-category field
      without header, 768–769
    Sub-Category value, 778
    Title option, 771–772
    Use as filter, 777
  highlight action
    Actions [dashboard with filter action]
      dialog box, 780
    Actions item, 779–780
    Control view, Sales Map and Sales *vs.*
      Profit views, 779
    edit highlight action dialog box, 781
    Sub-Category value, 782
  URL action
    Edit URL Action dialog box, 783–784
    filter action, 783
    Sub- Category, 784–785
Aggregate functions, 484
ATTR(expression) function, 484–487

B

Bar chart, 550
Bookmark file
  copy sheet, dashboard, 32
  creation, 31
  paste sheets, 33
  worksheets, 34
Bullet graph
  creating, 714–720
  expected output, 713
  staged progress towards set goals, 721
    actuals, 723
    distribution band, 725
    expected output, 722
    final output, 726
    planned, 724
    training delivery, 727
Bump charts, 698
  activities to perform, 709–710
  creating, 699–700
    order date, 700
    RankRegion, 701–705
  expected output, 708
  index function, 701
  interpretation, 707
  uses, 698–699

C

Cache memory, 50
Calculation filters, 153–155
Cascading filters, 150–153
CASE function, 456–459
Cassandra No SQL database
  Cassandra CQL shell, 81
  data source page, 85
  DataStax Community Edition, 80
  ODBC Driver option, 82
  Other Databases (ODBC), 84–85
  Simba Cassandra OBDC DSN, 82
  Simba Cassandra ODBC
      driver DSN setup, 83
  System DSN, 82
  test results window, 84
CEILING(number) function, 434–438
Chart forms
  descriptive statistics, 547
  knowledge, 547

S. Acharya and S. Chellappan, *Pro Tableau*, DOI 10.1007/978-1-4842-2352-9

Cloud
aggregation, 661
conscience, 656
customer experience , 655
default aggregation, 660
Excel sheet, 659
frequency, 658
Number of Records, 662
visual display, 656
Cloudera Hadoop, 50
Computed sets, 225–230
Computed sorting, 159–164
Concatenation, string functions, 447–449
Constant sets
create set option, 217
Filter [In / Out of Random Sub-Category], 222
IN/OUT (Random Sub-Category), 220–221
Random Sub-Category, 218–219, 225
show members in set option, 224
total sales of "IN/OUT" members, 223
view, sales and sub-category, 216–217
Contains() function, 451–453
Context filters, 146–150
Continuous dates
convert, discrete date, 188
create, 190–192
definition, 178
marks card, 189
maximum sales, 179
measure, sales, 189
Order Date, 187
Custom SQL
edit custom SQL query, 106
edit custom SQL statement, 105
grade table, 103
MySQL data source, 103
New Custom SQL, 104
student database, 103–104
student table, 103
view data window, 105

D, E

Dashboard
actions (*see* Actions, dashboard)
adding an object
blank object, 749
edit text dialog box, 748
image object, 749
text object, 748–749
web page object, 749
adding views
dashboard 1, 732
dashboard window, 732–733
Sales Details dashboard, 735
Sales over year view, 735
Sales over year worksheet, 734
and stories (*see* Story creation, Dashboard)
creation
new dashboard option, 730
new dashboard tab, 731–732
new dashboard worksheet, 730–731
description, 729
interactivity (*see* Interactive dashboard)
layout container (*see* Layout container, dashboard)
organization
floating layout, 753–756
tiled layout, 752–753
remove a view/object
dashboard view, 751–752
dashboard window, 751
sales by sub-category, region, 750
Data blending
CoffeeChain Query, 107
custom option, 109
edit relationships option, 107
market, 111–112
market, region mapping, 110
orders, 107
Region, 112–113
relationships dialog box, 108, 110–111
"Sales" by "Market" view, 113
Sample-Coffee chain Access, 106, 107
Sample-Superstore data source, 107
Sample-Superstore Excel data source, 106
Data extracts
aggregate data, visible dimensions, 117–118
data grid, California details, 118, 119
data source, 114
extract data dialog box, 115
extract option, 114
filter condition, 115–116
filter[State] dialog box, 116–117
Save Extract As, 119
Data layer, 50
Data source filters, 155–158
Data visualization
enterprises, 6
expectations, 7
graphicl depiction, 4
history, 7
market leading data visualization tools, 1
negative % Change, 3
online survey, 2
power of data visualization, 6
power of visualization, 5
purposes, 1
sample - superstore data subset, 5
sample -visitors dataset, 3

Tableau, 7-8
tools, 1, 6
usage, 2
DATEADD() function, 474-482
DATEDIFF() function, 468-470, 472-474
Date functions, 467-468
DATEADD(), 474-482
DATEDIFF(), 468-470, 472-474
DATENAME, 482-484
DATENAME function, 482-484
Desktop architecture
data connectors, 50
data layer, 50
in-memory data, 50
n-tier client server architecture, 49
Tableau Desktop Architecture, 50
Discrete dates
continuous bucket, 183
definition, 178
discrete bucket, 183
drop down menu, 182, 184
marks card, 185
maximum sales, 179-180
measure sales, 185
Order Date, 181
sales data, descending order, 186-187
Sort[Month of Order Date] dialog box, 186
Distribution Band, 719-721, 725
Dual axis
bar, 295, 310
Color, 311
data, 302
expected output, 292, 301
final output, 300
granularity, Order Date to Month, 293-294
Line, 310
lollipop chart
bar, 318
color, 315
expected output, 312
final output, 319
output, 317
reading data, 312
sales, 316-317
segment, 314
State, 313
Order Date, 292
profit, 296
Profit Axis, Sales Axis, 299
Profit to Line, 300
sales, 304-306
sales, rows shelf, 294
sample - Superstore.xls, 291
second measure, sales, 307-308
sorting, Sub-Category, 304-305
sub-category, 303
synchronize, secondary to primary axis, 297
table calculation, 309

F

Filtering
calculation, 153-155
cascading, 150-153
conditions, 130-132
context, 146-150
dashboard, 121
data source, 155-158
description, 121
dimensions, 122
general tab
category and color, 123-124
Edit filter option, 127
furniture and technology, 124-126
Order Date, 123
options, 122
quantitative, 134-139
quick, 139-145
Top tab, 133-134
Wildcard tab, 127-129
Floating layout, dashboard
floating layout button, 753
Floating option, 754-755
Pos and Size field, 756
sales by state" view, 754
view layout, 755
FLOOR(number) function, 434-438

G

Gantt Bar, 675-676, 691-692
Gantt chart
alignment, 619
American engineer, 610
assignment, 620
calculated field, 616, 624
color, 625
columns shelf, 613, 622
data pane, 617
demo 2, 629
dimensions area, 622, 624, 625
drop down, 613, 626
duration, 616
Edit Label, 628
ellipsis, 628
EndDate, 627
Exact Date, 614, 618, 623
Gantt Bar, 615
GanttChartAssignment.xlsx, 621
GanttChart.xlsx, 611-612

Gantt chart (*cont.*)
horizontal axis, 610
label dialog box, 627
resources, 616
rows shelf, 615
size, 611
task dependencies, 610
technical architects, 611
Groups
creation
category and states, 193
data pane, 198
data source, 193
dimension member, 196
group members CA & California, 197–198
pop-up menu, group members icon, 195–196
select CA and California, 194
State(group), 198
sub-category dimension, 199–200, 202
definition, 192
editing and existing, 203, 205–208
removing, member, 208–209
renaming, 210–212

H

Heat map
creation, 568
fractal and treemaps, 568
geographical boundaries, 568
steps
Birthdays.xls, 570
calculated field, 571
color, 575
data pane, 572, 574
demo 1, 576–577
dimension, 573
discrete, 571
Edit Colors, 576
Excel sheet, 569
marks type, 574
MonthNames, 571
two-dimensional representation, 568
worksheet/view, 570
Hierarchies, 212–215
Highlight table
accuracy, 577
color, 577
demo 2
expected output, 586
sales, 586
steps
analysis, 583
cell borders, 583
color, 585, 588–589
columns shelf, 580
darkest shade, 585
data pane, 581, 588
demo, 586, 591
dimensions, 587
Edit Colors, 590
grand totals, 584
label, 580
marks card, 579, 588
mark type, 581, 589
Order Date, 578
rows shelf, 579
sales, 588
square, 581
sub-category, 578, 587
worksheet, 591
worksheet cells, 582
sub-category, 578
text table, 577
Histogram
bar charts, 644
frequency distribution, 643
plot, 643–645
steps
age group, 647
BinSize, 652, 654
calculated field, 647
columns shelf, 647
dialog box, 651, 653
number of records, 648
Olympic data set, 646
output, 655
score, 651
subset, 650
TestResultsSample, 650
TestResultsSample.tde, 649
worksheet/view, 646

I

IF ELSE function, 461–463
IF ELSEIF function, 463, 465–467
IIF()function, 459–461
Independent axis
blended axes
blended measures, expected output, 253
Color, marks card, 259
expected output, 261
Filters Shelf, 263–264
Fit, Entire View, 270
Label, marks card, 260
marks type, bar, 255
measure names and values, 256
measure names, color, 267

measure names, Label, 268, 269
measure values, rows shelf, 265
Order Date, columns shelf, 254, 264
Sales and Profit, 256–258
Sample - Superstore.xls, 262
blended measures
final output, 283
calculated field performance, 287–288
combined axis chart
activities, 271
Bronze, 274
Bronze and Silver and Gold measure, 275, 276
changed sequence, measure values, 277
choosing colors, 281
country, rows shelf, 273
data set, 270
editing colors, 281
expected output, 272
filter data, 274
measure names, Color, 278
measures, 270
measure values, Label, 278
measure values, marks card, 276
number format, 282
re-sequencing, 277
sort, country, 279–280
special values, measure bronze, 275
stacked marks, 271
Marks Type, Line, 287
measure, 251–252
measure names and values, 285
number of records, 286
Sales and Profit, 252
slope graph, expected output, 291
Slope Graph.xls, 284
units placed on detail, 289
units placed on label, 290
visualization, 252
Index function, 701
Interactive dashboard
creation
customer details, 743–744, 746
customer level detail - sales info dashboard, 739–740
Entire view, 742–743
filter option, 744–745
sales by sub-category, each region, 740–741
Size legend and Color legend, 741–742
Sub-Category and Region, 745–746
detailed worksheet
Customer Details, 737–738
customer name, descending order, 738–739
overview worksheet
heat map view, 737
Sales by Sub-Category, each Region, 736

J, K

Joins
fields, data pane, 93–95
inner join, 97–98
left join, 98–99
right join, 99–100
types, 95–97

L

Layout container, dashboard, 756–757
dashboard size, 763
filters, ship mode", 758–759
Floating button, 761
Format container option, 762
horizontal, 757
horizontal container, 761
rearrange dashboard views and objects, 764
running total shipping costs and shipping cost, 758
Ship Mode filters, 759–760
Ship Mode value, Shipping Cost view, 760
show/hide parts, worlsheet, 764
Left() and Find()functions, 449–450
Len()function, 453–456
Level of Detail (LOD)
category, 361
Customer Name, 363
demo 1, 364
demo 3, 384–385, 387
expressions, 363
region, 360
sales and profit, 362, 365
state, 365
steps
aggregation, 392
amount, 397
Column Grand Totals, 399, 400
CTRL key, 395
Customer_Average_Amount, 388, 393
Customer Name, 396
data pane, 377
demo 2, 374–375
dimensions, 374, 389
Exclude_Region_Sales, 369–370
Exclude_State_Sales, 371
Filter dialog box, 390
final output, 373
Fixed_Region_Sales, 378–379
Fixed_State_Sales, 381
LOD.xls, 387
measure values, 391, 395
region, 367, 376
rows shelf, 367

Level of Detail (LOD) (*cont.*)
sales, 372, 382
sales by state, 370
Sample - Superstore.xls, 366, 375
Show Column Grand Totals, 398
state, 368, 377
Total All Using, 399
view/worksheet, 380
view/worksheet, 360, 364, 366
Line graph
demo, 592
steps
columns shelf, 592, 599–600
data pane, 593
demo , 599
forecast, 597–598
line graph, 599
Order Date, 592–593, 597
profit, 595, 600
rows shelf, 594, 601
sales, 594
view/worksheet, 597
straight-line segments, 592
Logical functions, 456
CASE, 456–459
IF ELSE, 461–463
IF ELSEIF, 463, 465–467
IIF()function, 459–461

M

Manual sorting, 164
drag and drop, 165–166
tool bar, 164–165
Market basket analysis, 826–833
MAX(number, number) function, 438–442
Measure names and values, 389
built-in Tableau fields, 237
container, 238
data set, 238
data source, 238
dimension
Filter Shelf, 242
Label, marks card, 240
Profit and Sales, 243
Quick Filter, 242–243
Region, columns shelf, 241
rows shelf, 239
Sample - Superstore.xls, 239
requirements, 237
worksheet/view
Order Date, columns shelf, 245
Profit and Segment, 250
Quick Filter, 244, 248
rows shelf, 246
Sample - Superstore.xls, 245
Segment and Color, marks card, 247
single and dual axis, 250
single value list, 249
Metadata grid, 92
MIN(number, number) function, 438–442
MongoDB NoSQL Database
client, 87
connectivity test results, 90
Data source page, 91
MongoDB installation directory path, 86
ODBC Driver, 87
Other Databases (ODBC), 90–91
Server, 86
Simba MongoDB ODBC Driver DSN Setup, 89
Simba MongoDB ODBC DSN, 88, 91
System DSN, 88
Moving average
asset's momentum, 341
demo 1, 342
final output, 350
steps
add table calculation, 345
data pane, 348
default position, 347
dialog box, 346
Order Date, 343
primary axis, 349
sales, 345
secondary axis, 348
special values, 347
tasks, 343
visual cue, 344
technical analysis, 341
trading strategy, 341
types, 341
MS SQL Server
connection details, 79
connection window, 74
data source page, 76, 80
employee (Test) table, 77
installer, 77
MySQL command line client, 77, 78
MySQL command prompt, 77–78
MySQL Driver, 78
Server Connection, 75
Tableau Desktop, 78
Test database, 76

N, O

Nested sorting
clear sort and sub-category, 171–172
combined field, 169–170
region and sub-category, 166–168, 172–173

show header option, 176
sort field dialog box, 174
sub-category and sales, 175
View, sub-category and region, 177
Number functions, 433–434
ABS(number), 442–443, 445–446
CEILING(number)
and FLOOR(number), 434–438
MAX(number, number) and
MIN(number, number), 438–442

P

Parameters, 231
create, 233
dimension "Sub-Category", 231
Show parameter control, 234
TOP N, 233, 235–236
TOP N Sub-Category, 236
Percentiles
class, 400
Demo 1, 401
population, 400
steps
CGPA, 403–405
Percentile.xlsx, 402
Roll No, 402
table calculation, 404
PieChart.tbm file, 30
Pie chart
alternative, 549
angle and size, 548, 550–551
categories, 548
circumstances, 547
colors, 552
data visualization, 548
3D effect, 550
features, 548
glossy color, 550
percentages, 547
Phones, 551
political economist, 547
quantitative information, 549–550
radius effects, 550
reasons, 551
steps
Angle, 552
Customer Segment, 553–555
data pane, 556
label, 557
marks card, 553
percentages, 557, 560
profit, 555
reference, 560
sales, 557
slices, 554
table calculation, 558
worksheet/view, 559
sub-category, 549
uninformed audience, 548
visual dimension, 549
Probability, 496–497
Profitability, Percent of Total
demo 1, 333–334
steps
apply button, 339
category, 337
control key, 340
data pane, 335
dialog box, 338
dimensions area, 339
Order Date, 335
Profit/Loss, 339
rows shelf, 336
Sample - Superstore.xls, 334
segments, 336
Sum (Sales), 337

Q

Quantitative filters, 134–139
Quick filters, 139–145

R

R
description, 795
Manage R Connection, 795–796
market basket analysis, 826–833
models and functions, 797
Rserve connection, 796
SCRIPT_BOOL(), 802–808
SCRIPT_INT(), 818–825
SCRIPT_REAL(), 808–817
SCRIPT_STR(), 797, 799–801
RAM, 50
Rank
calculations, 351
demo 1, 351
steps
Add Table Calculation, 355
category, 358
Demo 1, 360
dialog box, 355
drop down button, 355
Edit Table Calculation, 359
Label, 356
sales, 353–354
Sub-Category, 352
Sum (Sales), 357, 359
worksheet /view, 356
type, 351

RDataFile, 72
RDataSet, 70
R Interface, 69
Running Total of Sales
  demo 1, 324
  steps
    calculation, 327
    data pane, 329
    Demo , 333
    dialog box, 328
    dimensions, 326
    label, 327, 331
    measure names, 330
    measure values, 330
    Order Date, 326
    running sum, 332
    sample - Superstore.xls, 324
    subtotals, 332
    Table Calculation dialog box, 328
    tasks, 325
    verification, 331

S

Sample - Superstore, 52
Scatter plot
  Cartesian plane, 629
  coefficient, 629
  negative correlation, 630
  numerical variables, 629
  positive correlation, 629–630
Sets
  computed, 225–230
  constant (*see* Constant sets)
  definition, 215
  vs. groups, 231
Slope graph, 699
Sorting
  computed, 159–164
  description, 159
  manual, 164–166
  nested (*see* Nested sorting)
Stacked bar chart
  calculated fields, 604
  categories, 601
  columns shelf, 603
  constant line, 609
  dialog box, 609
  likert scale, 601
  measure names, 604, 606, 607
  negative axis, 602
  NeutralNegative measurement, 608
  NeutralPositive, 605
  rating, 602
  sorting, 607
  sort order, 605
  Stacked Bar.xlsx, 602
  Strongly Disagree, 604
  Survey Questions, 603
  X Axis, 610
Statistical File, 71
Statistics
  AverageScore, 510
  box plot, 515–516
  business-related practical scenarios, 496
  CountDScore, 510
  definition, 497
  descriptive, 497–498
  final output, 515
  forecasting
    Continuous Month, 536
    describe, 544
    options, 539, 544
    Order Date, 535
    Precision of Sales, 542, 543
    prediction intervals, 540
    Sample-Superstore data source, 535
    Show forecast, 537
    shows estimated value,
      lighter shade, 538
    Whiskers, 541
  inferential, 497–498
  magic number
    age of the class, 502
    average, 504
    mean, 501
    median, 502–503
    mode, 503
  mathematics, 495
  MaxScore, 512
  mean/arithmetic mean, 498
  Measure Names, columns shelf, 513
  Measure Names, filters shelf, 513
  Measure Values, Label, 514
  median, 499
  MedianScore, 510
  MinScore, 511
  mode, 499
  National Basketball Team, 499
  parameter, 498
  plotting box and whiskers plot, 517–519
  population, 497
  quizzes, 500
  reference lines
    average line, 520–521
    band, 525–526
    constant line, 522–523
    distribution, 526–528

sample, 497
seeds data set, 507
skewed left, 516
skewed right, 516
spread of data
data set, 505
interquartile range, 505
range, 505
square differences, 506
standard deviation, 506
variance and standard deviation, 505
StatsData.xlsx, 508-509
StdDevScore, 511
Subject Name, 512
SumScore, 509
symmetric, 516
trend lines
analytics pane, 530
AVG(Discount), 529
describe trend model, 531-532
manufacturer, 533-534
ship mode, 529, 531
VarScore, 511
Story creation, Dashboard
add description, 791-793
Caption, 789-790
New Blank Point, 790
New story sheet, caption, 790
New Story tab, 786
Sales by Category story point, 791
Sales *vs.* Profit by Sales view, 788-789
Story Edit Title dialog box, 787-788
Story Sheet on display, 786
Story Size option, 787
String functions, 446-447
concatenation, 447-449
Contains() function, 451-453
Left() and Find()functions, 449-450
Len()function, 453-456
Student_Grade, 101

T

Tableau
categories, 9
data connection page, 10
data source
file-open option, 24-25
open dialog box, 26
Sample Superstore, 27
Gartner's Magic Quadrant, 9
glance, 13
products line, 12
self-service analytics, 8
Tableau bookmark, 28-30
Tableau Online, 11
Tableau Public, 11
Tableau Reader, 12
Tableau Server, 11
tds, 19
tdsx, 20-24
twb, 13, 15-17
twbx, 17-19
versions, 10
visualization and/or dashboards tasks, 10
Tableau data extract
creating packaged workbooks, 44
data security, 44
data source connection page, 40
data source page, 35
extract or add another filter, 37
filter condition, 35, 38, 41
filters, 42-43
full extracts, 44
incremental extracts, 44
large text file/Excel file, 42
live data connection, 34
optimizing, 42
other files option, 40
performance, 42
portability, 42
publishing to Tableau Public, 44
save extract as dialog box, 36
.tde file, 41
use extract implies, 39
value(s), 36
Tableau Data Source (tds) file, 19
Tableau environment
close button, 51
connecting to RData files, 67-72
connect page, 53
connect to MS Access, 64-67
connect to server, 73
dashboard and worksheet pane, 62
data source icons, 60
data source page
canvas, 58
connect section, 56
grid, 58
left pane, 58
metadata grid, 58
open Sample - Superstore, 56
orders sheet, 58
processing request window, 57
types of connections, 59
viewing, 57

Tableau environment (*cont.*)
- discover page, 54–55
- fields, data pane, 61
- MS SQL Server 2014
  - Management Studio, 73
- pin option, 53–54
- Tableau start page, 52
- text File, 62–64
- to open, 51
- visual cues and icons, 60
- workbooks and sheets, 60
- workspace, 59

Tableau Packaged Data
Source (tdsx) file, 20–24
Tableau Packaged Workbook (twbx), 17–19
Tableau Workbook (twb), 13, 15–17
Table calculation function, 487
- First(), Index(), 487–492

Table calculations
- annual profit, 322
- dimensions, 321–322
- partitioning and addressing, 323
- triangular mark, 323

TedEx video, 4
Teradata, 50
Tiled layout, dashboard, 752–753
Trainer Feedback data set, 435
Treemaps
- categories, 561–562
- category, 561
- data, 563
- data set, 561
- hierarchical data, 560
- motivation, 561
- references, 561
- steps
  - category, 563
  - data pane, 564–565
  - dimension, 564
  - Product Name, 568
  - sales, 565
  - Sub-Category, 566
  - visualization, 567
- sub-branches, 561
- sub-categories, 561, 562
- visualization, 563

U

Union operation
- inner join, 100
- new union option, 101
- result, 102
- Union window, 101

V

Vertica, 50

W, X

Waterfall charts, 665
- activities to perform, 673
- employee strength of project account, 682–683, 685
  - activities to perform, 689
  - data in Excel, 684
  - Gantt Bar, 691–692
  - Head Count, 685–687
  - Head Count Number, 687–693, 697–698
  - Row Grand Totals, 696–697
  - Running Total, 690
  - Stepped Color, 694–695
- inventory of men's t-shirts, 666–667
  - data in Excel file, 667–668
  - data members, 669
  - Default Properties, 670
  - expected output, 667
  - Gantt Bar, 675–676
  - Manual Sort, 671
  - Row Grand Totals, 680–681
  - Running Total, calculation, 674–675
  - Stepped Color, 678–679
  - Units Stock Number, 671–674, 677–678, 682
- uses, 665–666

Y

Year over year (YOY)
- company's financial performance, 406
- Demo 1, 406
- steps
  - Add Table Calculation, 424
  - calculated field, 426
  - Colors, 414
  - CTRL, 414, 416
  - Demo 1, 417, 426
  - Dimension, 419
  - filters shelf, 419
  - Label, 423, 430
  - Mark Type, 409, 422
  - Measure Names, 428
  - Measure Values, 429

1 null, 412
Order Date, 408, 421
output, 425
Percent Difference, 425
Quick Table
  Calculation, 410, 423
rows shelf, 422
Sales, 416, 422
Sample - Superstore.xls, 407, 408
special values, 413
stepped color, 415
table calculation, 411, 414
YearOverYearGrowth, 427
YearToDisplay, 418, 420

Z

Zero correlation
  columns shelf, 631
  demo , dimensions, 638
  steps
    category, 640
    columns shelf, 632, 639
    dimensions area, 633, 639, 642
    lines options, 637
    marks card, 635
    sales measure, 640
    segment and shape, 641, 643, 634
  variables, 631

be obtained

18

V00010B/209/P

9 781484 223512